HEATING, VENTILATING and AIR CONDITIONING

George E. Clifford
Professor of Mechanical Engineering
University of Maine at Orono
Orono, Maine

Reston Publishing Company, Inc.
A Prentice-Hall Company
Reston, Virginia 22090

To my wife, Sally.

Library of Congress Cataloging in Publication Data

Clifford, George E.
Heating, ventilating and air conditioning.

1. Air conditioning. 2. Heating. 3. Ventilation.
I. Title.
TH7687.C54 1984 697 84-4795
ISBN 0-8359-2812-8

Freon is a registered trademark of E.I. DuPont, Freon Products Division
Genetron is a registered trademark of Allied Chemical Corporation,
General Chemical Division

10 9 8 7 6

Printed in the United States of America

Table of Contents

Chapter 1—Fundamental Concepts

Chapter 2—Review of Thermodynamic Principles

Chapter 3—Refrigeration for Comfort Cooling

Chapter 4—Psychrometrics

Chapter 5—Applied Psychrometrics

Chapter 6—Comfort in Air Conditioning

Chapter 7—Heat Transfer in Building Sections

Chapter 8—Heating and Ventilating Loads

Chapter 9—The Cooling Load

Chapter 10—Heat Transfer Devices

Chapter 11—Fluid Flow Fundamentals and Piping Systems

Chapter 12—Pumps, and Piping System Design

Chapter 13—Principles of Air Distribution

Chapter 14—Air Flow in Duct Systems and Fans

Chapter 15—Fundamental of Automatic Control and Control Components

Chapter 16—Automatic Control Systems

Chapter 1

Fundamental Concepts

1-1 *INTRODUCTION*

Air conditioning has made it possible for man to exist with comfort under adverse climate conditions. Application of systems for heating, ventilating, and refrigeration to control living and working conditions has assisted in giving us a better standard of living and has measurably increased production in industry. Application of controlled ventilation and refrigeration in the care and processing of foods has greatly improved food quality and appearance and has made it possible to expand the distribution of perishable foods, as well as increased greatly the permissible storage time for foods and other products.

In past years extensive use of air conditioning has made new products and processes possible. In many industrial plants air conditioning has become essential to the manufacture of products from both a quality and quantity standpoint. More recently, the exploration of space with its many problems, both known and unknown, has opened up a new and interesting field of application of air conditioning.

Energy demand to produce comfort conditions for people and process development has been ever increasing over the years. The availability of inexpensive raw energy has become a thing of the past. All of us must become aware of the limitations of our diminishing energy supplies, and as engineers and technologists, we must design better and more efficient systems which use energy. Furthermore, existing systems must be reevaluated and, when required, should be redesigned to bring them up to acceptable operating levels of efficiency.

Since heating, ventilating, and air-conditioning systems are component parts of a building designed to overcome building heat losses and gains, the *entire building system* should be analyzed, including the basic structure itself. It is desirable, then, for the designer of air-conditioning systems to carefully study the *complete* building system in order to design the best possible mechanical systems.

1-2 DEFINITIONS OF AIR CONDITIONING

Air conditioning as an inclusive term means the control of *temperature, moisture content, circulation,* and *purity* of the air within a space to produce desired effects on the occupants of that space or on products and materials manufactured or stored there.

Complete air conditioning involves the control of all four factors and implies the establishment of year-round conditions for the comfort of people and for stability of products and processes. In addition to the term *complete air conditioning* are found such terms as *summer air conditioning* or *summer cooling* and *winter air conditioning* or *heating*.

Summer cooling implies the maintenance of a space temperature *below* that of the surroundings and may include *dehumidification* of the air.

Winter heating implies the maintenance of a space temperature *above* that of the surroundings and normally should include *humidification* of the air.

The term *air conditioning* shall be used as an inclusive term when discussing year-round control of environmental space conditions, regardless of the time or season of the year. More and more application of complete year-round air conditioning to industrial and commercial buildings is being employed, as well as in private homes, because the only addition to satisfactorily operating heating systems is a means for chilling water, or air, and making use of existing piping or air ducts to convey the cooled medium to the various spaces to be cooled.

1-3 DIMENSION AND UNITS

A physical quantity is something that can be measured. A dimension is a property or quality of the quantity or entity that characterizes or describes that quantity. Thus, distance has the characteristic property, or dimension, of length *(L)*. Area is a function of length and breadth or, dimensionally, $A = L^2$. Also, volume is a function of length × breadth × height, or dimensionally, $V = L^3$.

The dimension of length may be expressed in various *units,* such as inches, feet, yards, miles, centimeters, meters, and kilometers. Any physical quantity is exactly specified by a dimension and by a multiple or fraction of a defined unit, such as a length of 14 ft.

A dimensional system that will completely describe an event can be constructed from a relatively small number of fundamental dimensions. The usual fundamental dimensions are time *(T)*, length *(L)*, force *(F)*, and mass *(M)*.

For many years engineering calculations in the United States have been made using the unit *second (s)* for time, the unit *feet (ft)* for length, the unit *pound force (lbf)* for force, and the unit *pound mass (lbm)* for mass. This system of units is called the English Engineering System (EES).

Another system of units called the Engineering Dynamical System makes use of second for time, foot for length, pound mass for mass, and poundal for force. The English Gravitational System makes use of second for time, foot for length, slug for mass, and pound force for force. The cgs system uses the centimeter (cm) for length, second for time, gram for mass, and dyne for force.

As world technology grew, it became apparent that there was a need for international standardization of units. The official name of the international standardization of units is *Le Système International d' Unités,* abbreviated SI. Do not call it the "SI system" because the *S* in SI means "system."

SI is an absolute system. The basic unit of length is the meter (m), the unit of mass is the kilogram (kg), and the unit of time is the second (sec). The unit of force is derived and is called newton (N) to distinguish it from kilogram which, as indicated, is the unit of mass.

The SI base units, with their symbols, are shown in Table 1–1. These are dimensionally independent. Lowercase letters are used for the symbols unless they are derived from a proper name, then a capital is used for the first letter of the symbol. Note that the unit of mass uses

Table 1-1

SI Base Units

Quantity	Name	Symbol
Length	meter	m
Mass	kilogram	kg
Time	second	s
Electric current	ampere	A
Thermodynamic temperature	kelvin	K
Amount of matter	mole	mol
Luminous intensity	candela	cd

Table 1-2

Derived SI Units

Quantity	Unit	SI Symbol	Formula
Acceleration	meter per second squared		m/s^2
Area	square meter		m^2
Density	kilogram per cubic meter		kg/m^3
Energy	joule	J	N m
Force	newton	N	$(kg\ m)/s^2$
Frequency	hertz	Hz	1/s
Power	watt	W	J/s^1
Pressure	pascal	Pa	N/m^2
Specific heat capacity			J(kg K)
Quantity of heat	joule	J	N m
Speed	revolutions per second		1/s
Velocity	meter per second		m/s
Volume	cubic meter		m^3
Work	joule	J	N m

the prefix kilo; this is the only base unit having a prefix. Table 1–1 shows that the SI unit of temperature is the kelvin. The Celsius temperature scale (once called Centigrade) is not part of SI, but a difference of one degree on the Celsius scale equals one on the Kelvin scale.

A second class of SI units comprises the derived units, many of which have special names. Table 1–2 is a list of derived units of primary concern in our work. Note that each unit is spelled with an initial lowercase letter, except when it occurs at the beginning of a sentence. The first letter is then capitalized. If the unit is named for an individual, the first letter of the symbol is capitalized; otherwise the symbol is lowercase.

1-4 FORCE, MASS, AND WEIGHT

In Section 1-3 different systems of units were discussed briefly. In this text we will be concerned primarily with the English Engineering System of units and the SI units. A brief discussion of what is meant by force, mass, and weight will help us to understand the relationships and conversions from one unit system to the other.

A *force* may be defined as a push or pull tending to change the motion or direction of a body on which it acts. A *mass* is a quantity of matter. A given mass has the same value any place in the universe. A body has the tendency to resist a change in motion, and we say that it has *inertia*. Force and mass are related by Newton's second law of motion. *Weight* is a force exerted by a mass due to the pull of gravity, which varies with geographical location.

Newton's second law of motion states that if an unbalanced force acts on a body, the body will accelerate in the direction of the force, and the acceleration is directly proportional to the unbalanced force and inversely proportional to the body mass. Stated mathematically, this would be

$$a \propto \frac{F}{m} \quad \text{or} \quad a = g_c\left(\frac{F}{m}\right) \quad \text{and} \quad F = \frac{m}{g_c}a \qquad (1\text{-}1)$$

where F is the unbalanced force, m is the body mass, a is the acceleration, and g_c is the proportionality constant.

The numerical value of g_c depends on the units used for the terms in Eq. (1-1). A newton (N) is a unit of force which will acclerate 1 kg of mass at the rate of 1 m/s^2 (meter per second squared). Using Eq. (1-1) and solving for the units of g_c, we have

$$g_c = \frac{ma}{F} = \frac{(1\ \text{kg})(\text{m/s}^2)}{\text{N}} = 1\ \frac{\text{kg m}}{\text{N s}^2} \tag{1-2}$$

Equation (1-2) gives the value and units of g_c for SI.

The English Engineering System (EES) uses the unit of pound mass (lbm), which is equal to 0.453 592 37 kg. The pound force (lbf) is defined as the weight of one pound mass when subjected to standard acceleration of gravity of 32.174 048 56 ft/s^2. Substitution into Eq. (1-1) gives us

$$g_c = \frac{ma}{F} = \frac{(1\ \text{lbm})(32.174\ 048\ 56\ \text{ft/s}^2)}{\text{lbf}}$$

or

$$g_c = 32.2\ \text{lbm ft/1bf s}^2 \quad \text{(approximate)} \tag{1-3}$$

Equation (1-3) gives the value and units of g_c for EES.

If the only force acting on a body is the force of gravity (its weight, w), the resulting acceleration will be the local acceleration of gravity, g. From Eq. (1-1) we would have

$$w = \frac{1}{g_c} mg \quad \text{or} \quad \frac{w}{g} = \frac{m}{g_c} \tag{1-4}$$

Combining Eqs. (1-1) and (1-4) produces the following important relationship:

$$F = \frac{w}{g} a \tag{1-5}$$

Equation (1-4) is important because the ratio of g/g_c appears in many equations for the purpose of obtaining proper engineering units. If the average acceleration of gravity, g, on the surface of the earth is assumed to be 32.2 ft/s^2, the ratio becomes

$$\frac{g}{g_c} = \frac{32.2\ \text{ft/s}^2}{32.2\ \ \text{lbm ft/lbf s}^2}$$

$$= 1\ \text{lbf/lbm} \quad \text{(approximately)}$$

and, if standard gravity is 9.806 m/s^2 or about 9.80 m/s^2 using SI units,

$$g/g_c = \frac{9.80\ \text{m/s}^2}{1\ \text{kgm/M s}^2} = 9.80\ \text{N/kg}$$

1-5 MATTER

The molecule is the smallest division of matter that has all the chemical and physical properties of a large quantity of matter. Molecules, in turn, are composed of smaller particles known as electrons, protons, and neutrons. The study of atoms and subatomic particles is beyond the scope of this book, and the discussion will be limited for the most part to the study of molecules and their behavior.

When chemical reactions occur, the atoms are regrouped to form molecules of different substances without basic changes in the atoms themselves. In the combustion process the rearrangement of atoms into different molecules is accompanied by the release of energy, part of which may be converted into work in a suitable engine. This energy could also have been used to heat a building. When energy is released by chemical reactions, there is a very slight decrease in mass, but the change is too small to be measured. Accordingly, chemical reactions may be said to take place in accordance with the *law of conservation of matter,* which states that matter is indestructible, that the mass of material is the same before and after the chemical reactions occur, and that the mass of each chemical element remains unchanged during the chemical reactions. Thus, if fuel oil and air are supplied to the combustion chamber of a furnace, the mass of flue gas leaving the furnace in a given time period equals the mass of fuel and air supplied; also, the mass of oxygen, hydrogen, carbon, nitrogen, and other elements entering the combustion chamber is equal to the mass of those same elements leaving, although an entirely new set of chemical compounds may result from the combustion process.

Matter ordinarily exists as a solid, liquid, or gas. These are called *phases* of matter. In the solid phase or condition the molecules are held in fixed positions by powerful forces. However, the atoms may vibrate about mean positions within the molecules. Solids, therefore, have definite shape and volume, and relatively great external forces are required to deform them. In the liquid phase the molecules have a motion of translation with frequent collisions from which they rebound as perfectly elastic bodies. A liquid will occupy a definite volume at a given temperature but conforms to the shape of the confining vessel. In the gaseous phase the molecules are far apart in comparison to their size, move in straight lines between collisions, and will fill a confining vessel regardless of the amount of gas present.

Many substances, such as water, may exist in the solid, liquid, and gaseous phase, depending upon the pressure and temperature. Thus, ice may melt to form water, and water may evaporate to form steam.

When a substance is in the gaseous phase, but at a temperature and pressure not far removed from the temperature and pressure at which it can be liquified, it is usually referred to as a *vapor*. A *perfect gas* is at a temperature far removed from the pressure and temperature at which it can be liquified.

1-6 PROPERTIES OF MATTER

Work is obtained from an engine by causing a working fluid such as steam or gas to undergo heating, cooling, expansion, and compression processes or changes in a suitable mechanism. Heat is added to a fluid, such as water, in a water boiler, and the heated water is forced to circulate through a system of piping to heating devices where some of its heat is given up to heat a building. A refrigerant absorbs heat as it boils at low temperature. The resulting vapor is compressed to a high pressure and temperature and is then condensed back to liquid by rejection of heat.

It is important that the *state* or condition of a fluid be known while it is being subjected to changing conditions. Among the easily measured characteristics or properties that define the state or condition of a fluid at any instant are volume per unit mass, pressure, and temperature.

In general, if a proper combination of two of these properties is known, the state or molecular condition of the fluid is defined, and the values of the other properties are fixed. Also, if a given mass of fluid initially occupying some known volume at a known pressure and temperature undergoes some change or process after which the final volume, pressure, and temperature can be determined, then the *change* in volume, pressure, and temperature may be computed by subtracting the final values of these properties from the initial and final states.

Referring to Fig. 1-1, let the coordinates of point 1 be pressure p_1 and volume V_1 representing to some scale the pressure and volume of a given mass of gas to the left of the gas-tight frictionless piston in the cylinder. If the piston moves to the right to a final position, point 2, the final pressure and volume would be p_2 and V_2. During the expansion process of the gas within the cylinder, the relationship between the pressure and volume at each successive position of the piston might be represented by some curve such as *a* or *b*, which is called the *path* of the process. The increase in volume $(V_2 - V_1)$ during the process depends

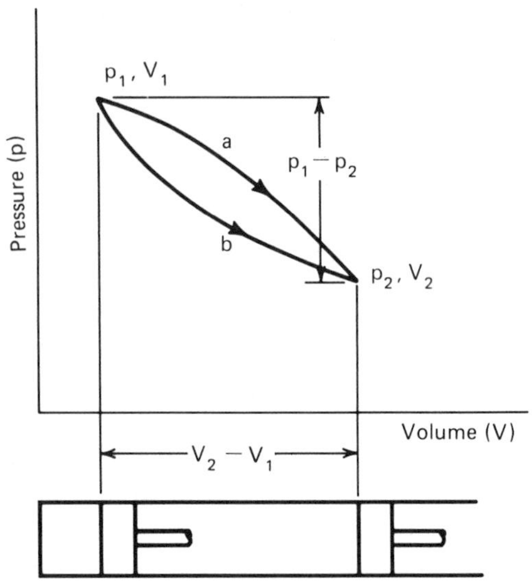

Figure 1–1

Change in pressure and volume during expansion of a gas

only on the initial and final positions of the piston and is independent of how the pressure varied with the volume, that is, the path. Also, the change in pressure for the process is dependent only on the difference between the initial and final pressures and is independent of the path. Such properties of matter as pressure and volume are known as *point functions* since the change in their values during any process depends upon the initial and final values of the properties and are independent of the process path.

1-7 DENSITY

Density is a fluid property and is defined as the mass per unit of volume. It is calculated by

$$\rho \text{ (rho)} = m/V \qquad (1\text{-}6)$$

where m is the total mass and V is the total volume occupied by the mass.

Definition: Mass per unit of volume

Symbol: ρ (rho)

Units: EES: lbm/ft^3; SI: kg/m^3

Conversion: To convert lbm/ft^3 to kg/m^3 multiply by 16.02

Illustrative Problem 1-1

Determine the density of water in SI units, if the density at 100° F is 61.996 lbm/ft^3.

Solution:

$$61.996 \times 16.02 = 993.18 \text{ kg/m}^3 \qquad \textit{(ans)}$$

1-8 SPECIFIC WEIGHT

Specific weight is defined as weight (force) per unit of volume. It is calculated by

$$\gamma \text{ (gamma)} = w/V \qquad (1\text{-}7)$$

where w is the total weight and V is the total volume.

Definition: Weight (force) per unit volume

Symbol: γ (gamma)

Units: EES: lbf/ft^3; SI: N/m^3

Conversion: To convert lbf/ft^3 to N/m^3 multiply by 157

Relationship to density. By definition, specific weight is defined by Eq. (1-7). By Eq. (1-4),

$$w/g = m/g_c \quad \text{or} \quad w = m(g/g_c)$$

if the latter expression is substituted into Eq. (1-7), we have

$$\gamma = \frac{m\ (g/g_c)}{V}$$

But, $\rho = m/V$, therefore

$$\gamma = \rho \left(\frac{g}{g_c}\right) \tag{1-7}$$

It should be remembered that the ratio of g/g_c has units of lbf/lbm; therefore, if the local acceleration of gravity, g, is equal to 32.2 ft/s^2 and g_c is constant at 32.174 (approximately), then the numerical value of specific weight (γ) and density (ρ) are equal. But it must be remembered that the *units are different*.

1-9 SPECIFIC VOLUME

Specific volume is defined as the volume per unit of mass. It may be calculated by

$$v = V/m \tag{1-8}$$

In many references specific volume is defined as the reciprocal of density, or $v = 1/\rho$

Definition: Volume per unit of mass

Symbol: v

Units: EES: ft^3/lbm; SI: m^3/kg

Conversion: To convert ft^3/lbm to m^3/kg multiply by 1/16.02.

Relationship to specific weight. If $v = 1/\rho$, we may substitute into Eq. (1-7a) and obtain

$$\gamma = \frac{1}{v}\left(\frac{g}{g_c}\right) \quad \text{or} \quad v = \frac{1}{\gamma}\left(\frac{g}{g_c}\right) \tag{1-8a}$$

Repeating once again, it must be remembered that the ratio of g/g_c has units of lbf/lbm, and if the ratio is unity, $v = 1/\gamma$, numerically.

Illustrative Problem 1-2

Determine the specific volume, density, and specific weight of a gas when 5 lbm of gas occupies 75 ft^3 at a location where the local acceleration of gravity, g, is 30 ft/s^2, using (a) EES units and (b) SI units.

Solution: (a) EES units

Specific volume $(v) = \frac{V}{m} = \frac{75 \text{ ft}^3}{5 \text{ lbm}} = 15 \text{ ft}^3/\text{lbm}$

Density $(\rho) = \frac{1}{v} = \frac{1}{15} = 0.066 \text{ lbm/ft}^3$

Specific weight $(\gamma) = \rho\left(\frac{g}{g_c}\right) = 0.0667\left(\frac{30}{32.174}\right) = 0.0662 \text{ lbf/ft}^3$(b) SI units

$\rho = (0.0667)(16.02) = 1.0684 \text{ Kg/m}^3$

$v = 1/\rho = 1/1.0684 = 0.936 \text{ m}^3/\text{kg}$

$\gamma = (0.0622)(157) = 9.765 \text{ N/m}^3$

1-10 PRESSURE

Definition: Force per unit area

Symbol: p

Units: EES: lbf/ft^2, or lbf/in.^{2}; (psf) or psi. SI: newton/m^2 (N/m^2), 1 N/m^2 is called 1 Pa (pascal)

Conversion: To convert lbf/in.^{2} to N/m^2 multiply by 6895. To convert lbf/ft^2 to N/m^2 multiply by 47.88

Pressure is the force exerted per unit of area. It may be defined as a measure of the intensity of a force at a given point on the contact surface. Whenever a force is evenly distributed over a given area, the pressure at all points on the contact surface is the same and can be calculated by dividing the total force exerted by the total area over which the force is applied, or

$$p = F/A \qquad (1\text{-}9)$$

where

p = pressure expressed in units of F per unit of A

F = the total force in any units of force

A = the total area in any units of area

The most frequently used units of pressure are $lbf/in.^2$ (psi). However, since the units of force and area may be any appropriate unit, the unit of pressure may be lbf/ft^2 (psf), $tons/ft^2$, and in SI units $newton/m^2$ (N/m^2) or pascal (Pa).

We will find in our work that it is often convenient in many types of problems to express pressure in units of "length." For example, a column of fluid of given length h above a given surface exerts a pressure on that surface of γh, where γ is the specific weight of the fluid. Therefore, $p = \gamma h$, or $h = p/\gamma$ expressed in feet or meters of fluid.

Atmosphere pressure is measured by a barometer and may be expressed as either inches of mercury or pounds per square inch. Standard atmospheric pressure is exerted by the atmosphere (air) which extends outward from the earth's surface. Air has weight and, because of its weight, exerts a pressure on the surface of the earth. The pressure of the atmosphere does not remain constant but will usually vary from hour to hour dependent on the air temperature, water vapor content, and elevation above sea level.

Barometers are instruments used to measure atmospheric pressure. Figure 1-2 illustrates the basic principle of a mercury manometer.

Gauge pressure is the difference between the pressure of a fluid in a container, such as a tank or pipe, and that of the atmosphere. Gauge pressures may be either positive or negative. Negative gauge pressure is usually called *vacuum*.

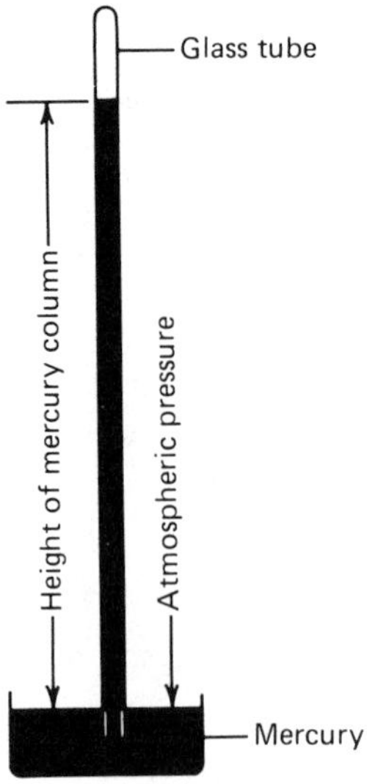

Figure 1–2

Barometer

Absolute pressure is the algebraic sum of the gauge pressure and the atmospheric pressure.

Absolute pressure = Gauge pressure + Atmospheric pressure, or

Absolute pressure = Atmospheric pressure − vacuum.

Figure 1-3 illustrates the relationship between gauge pressure and absolute pressure. In usual practice gauge pressure has units of *pounds per square inch gauge* (psig) and absolute pressure in *pounds per square inch absolute* (psia). A useful conversion factor for converting gauge pressure in inches of mercury to pounds per square inch is

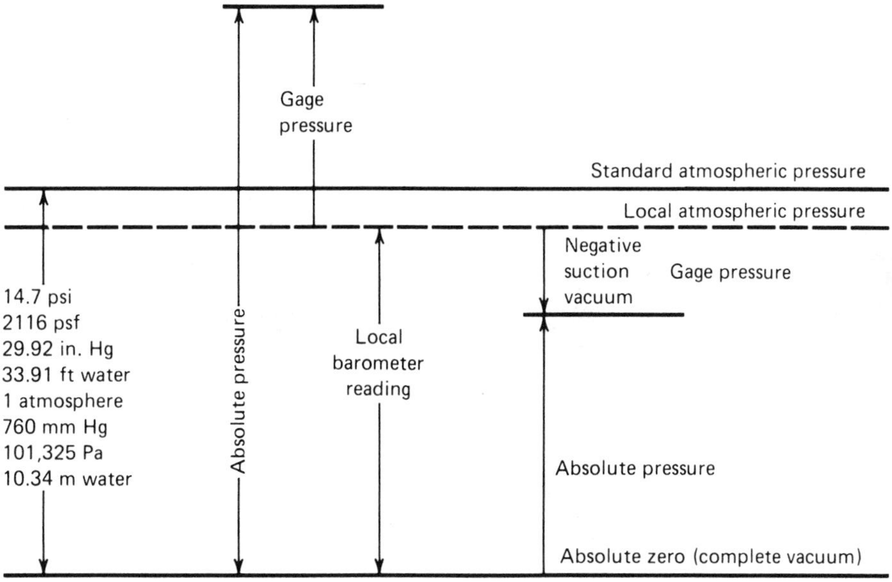

Figure 1–3

Units and scales for absolute and gage pressures

0.491 psi/inch of mercury (in. Hg)

Pressure gauges are instruments used to measure the fluid pressure (either gas or liquid) inside a closed container (tank or pipe). Pressure-measuring devices are available in many different types. The simplest type is called a *manometer*. Figure 1-4 shows various manometer arrangments which are in common use. The manometer usually consists of a hollow glass or clear plastic tube bent into a *U* shape and partially filled with a liquid. A calibrated scale between the legs of the manometer indicates the difference in elevation of the liquid in two manometer legs. One leg is attached to the point in the container where the pressure is to be determined. The other leg is open to the atmosphere or, in the case of a *differential manometer*, it is attached to the other pressure source. The pressure indicated on a manometer is always a differential pressure; that is, between the pressure source and the atmosphere or between any two pressure sources. The displacement of the fluid in the manometer is the pressure in units of *feet or inches of manometer fluid*. The manometer fluid is usually a liquid having properties quite different from the properties of the fluid where the pressure is to be determined. Water is frequently used as a manometer fluid (sometimes colored with dye) when the pressure to be measured is small. Mercury is a liquid used when greater pressures are to be indicated.

Illustrative Problem 1-3

A mercury manometer (vacuum gauge) reads 18.60 in. Hg when the local barometer reads 30.06 in. Hg. Determine the absolute pressure in (a) in. Hg, (b) psia, and (c) SI units.

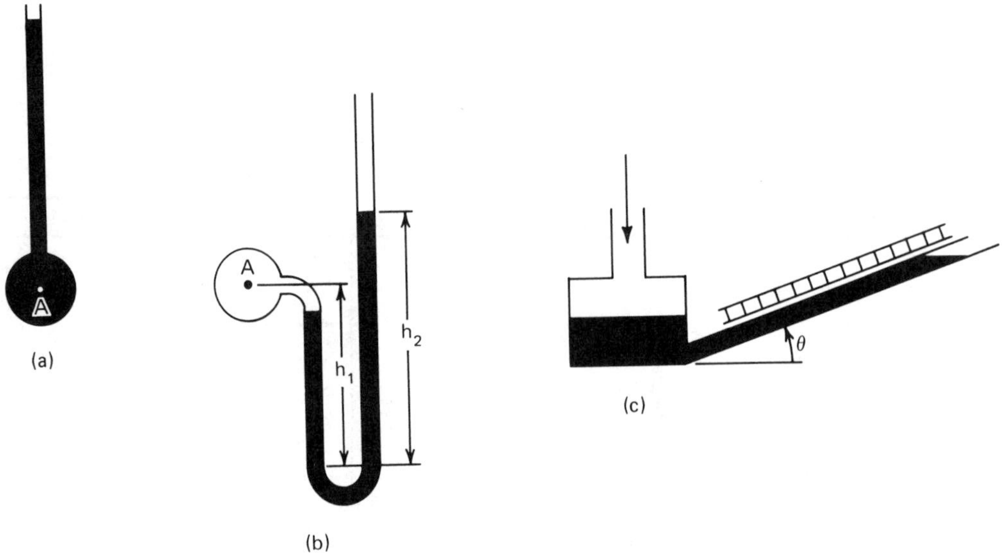

Figure 1–4

Manometers. (a) Piezometer (b) U-tube manometer (c) Inclined manometer

Solution:

(a) $30.06 - 18.60 = 11.46$ in. Hg absolute

(b) $(30.06 - 18.60)(0.491) = 5.63$ psia

(c) $5.63 \text{ psia} \times 6895 = 38{,}817.5 \text{ N/m}^2$
$= 38.817 \text{ kN/m}^2$
$= 38.817 \text{ kPa}$

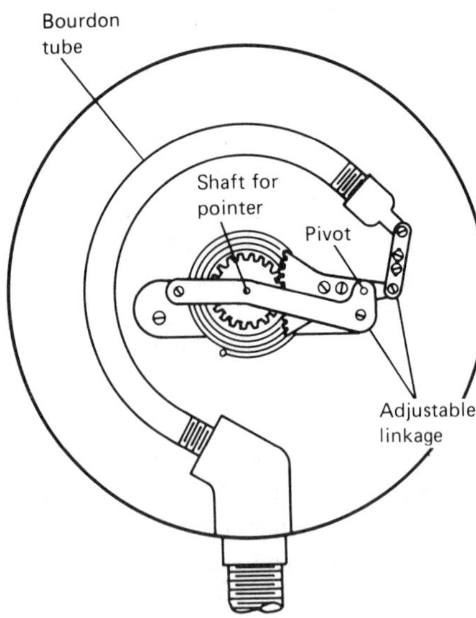

Figure 1–5

Bourdon pressure gage

The *Bourdon tube gauge* is another commonly used pressure-measuring device (see Fig. 1-5). The Bourdon gauge is used for measurement of pressures of considerable magnitude. The tube is elliptical in cross section and bent into the arc of a circle, rigidly attached to the case at one end and free to move at the other end. When a pressure is applied to the inside of the Bourdon tube, the elliptical cross section tends to become circular, and the free end of the tube moves outward. By means of a suitable linkage, pivoted gear sector, and pinion, a needle is moved around a graduated scale on the face of the gauge indicating the pressure. If the pressure within the Bourdon tube is less than atmospheric pressure on the outside of the tube, the elliptical cross section tends to be flattened, the free end of the tube moves nearer the center, and the pointer is moved in the opposite direction. Thus, the gauge is operated by the difference between atmospheric pressure on the outside of the Bourdon tube and the pressure within the tube.

Pressure reading obtained by either a manometer or a Bourdon gauge are called *gauge* pressures and may be expressed in pounds per square inch gauge (psig).

Illustrative Problem 1-4

A Bourdon pressure gauge attached to a tank indicates a pressure of 65.0 psig. If the barometric pressure is 29.98 in. Hg, what is the absolute pressure of the fluid within the tank in (a) psia and (b) kN/m^2?

Solution:

$$\text{(a) Atmospheric pressure} = (29.98)(0.491) = 14.72$$
$$= \underline{65.00 \text{ psig}}$$
$$\text{absolute pressure} = 79.72 \text{ psia}$$

$$\text{(b) N/m}^2 = 79.72 \times 6895 = 549{,}650 \text{ N/m}^2 = 594.65 \text{ kN/m}^2 = 594.65 \text{ kPa}$$

Illustrative Problem 1-5

A compound Bourdon gauge (indicates vacuum as well as pressures above atmospheric pressures) is attached to the suction pipe of a centrifugal water pump and indicates a vacuum of 3.0 in. Hg. If the barometric pressure is 29.66 in. Hg, what is the absolute pressure of the water in the suction pipe in (a) in. Hg absolute, (b) psia, and (c) kN/m^2?

Solution:

Atmospheric pressure = 29.66 in. Hg
Vacuum gauge reading = 3.00 in. Hg

Absolute pressure = atmospheric pressure minus vacuum

(a) Absolute pressure = 26.66 in. Hg absolute
(b) Absolute pressure = 26.66 × 0.491 = 13.09 psia
(c) Absolute pressure = 13.09 × 6895 = 90,252 N/m^2 = 90,252 kN/m^2

1-11 TEMPERATURE

Definition: A measure of level of heat intensity; "thermal pressure"

Symbol: EES: t_F (degrees Fahrenheit; °F)
T_R (degrees Rankine; °R)

Units: SI: t_c (degrees Celsius; °C)
T_K (degrees Kelvin; K)

Conversion: to convert Fahrenheit to Rankine
$T_R = t_f + 460$
to convert Celsius to Kelvin
$T_K = t_c + 273$
To convert Fahrenheit to Kelvin
$T_K = (t_f + 460)/1.8$
To convert Rankine to Kelvin
$T_K = T_R/1.8$
To convert Fahrenheit to Celsius
$t_c = (t_f - 32)/1.8$

Temperature is a property of matter. It is a common experience that bodies feel *hot* or *cold* to the sense of touch, depending upon their temperatures. In comparing the temperatures of two bodies, A and B, a third body called a *thermometer,* which has some observable property that changes with temperature, is brought in contact with body A until the thermometer and that body reach thermal equilibrium, that is, until there is no further change in the thermometer scale reading. Then the thermometer may be brought into contact with body B until thermal equilibrium is reached as evidenced by no further change in the scale reading. The *difference* in the two scale readings of the thermometer is then taken as an indication of the difference in temperature of bodies A and B.

Thermometers. The most commonly used instrument for measuring temperature is the mercury thermometer. The operation of most thermometers depends upon the fact that a liquid or gas will expand or contract as its temperature is increased or decreased, respectively. Because of their low freezing temperatures and relatively constant coefficients of thermal expansion, alcohol or mercury are the liquids most commonly used in thermometers. The mercury thermometer is the more accurate of the two because its coefficient of expansion is more constant through a greater temperature range. However, mercury thermometers have the disadvantage of being more expensive.

The constant-volume gas thermometer contains a fixed volume of gas, such as hydrogen or helium, under relatively low pressure which can be accurately measured. The pressure of the gas will be found to vary in a linear manner with temperature, and this type of thermometer is used as a basis for temperature calibration for other types of thermometers.

Temperature scales. The Fahrenheit temperature scale has been used in the United States for many years for ordinary temperature measurements. On this scale water freezes at 32°F (sometimes called ice point) and boils at 212°F (steam point) when the pressure above the liquid is at standard atmospheric pressure of 14.696 psia.

The *Celsius temperature scale* (formally Centigrade) has an ice point of 0°C and a steam point of 100°C.

The *Kelvin temperature scale* is the absolute Celsius scale. The kelvin (K with no degree sign) is defined as the SI unit of temperature and is 1/273.16 of the fraction of the thermodynamic temperature of the triple point of water. The International Practical Temperature Scale of 1968 assigns a value of 0.01°C to the triple point of water.

The *Rankine temperature scale* is the absolute Fahrenheit scale.

Relationships for temperature scales. The temperature relationship of the four temperature scales are shown in Fig. 1-6. The Celsius scale has 100 degrees between the ice and steam points, and the Fahrenheit scale has 180 degrees between the same two points. The relationship between the two scales is $\Delta t_f/\Delta t_c = 180/100$, where t_f is Fahrenheit temperature and t_c is Celsius. At the ice point $t_f = 32$°F and $t_c = 0$°C, therefore

$$\frac{\Delta t_f}{\Delta t_c} = \frac{t_f - 32}{t_c - 0} = 1.8$$

Solving first for t_f and then for t_c we would have

$$t_f = 1.8t_c + 32 \tag{1-10}$$

and

$$t_c = \frac{t_f - 32}{1.8} \tag{1-11}$$

The triple point of water on the Celsius scale is 0.01°C, and on the Kelvin scale 273.16 K;

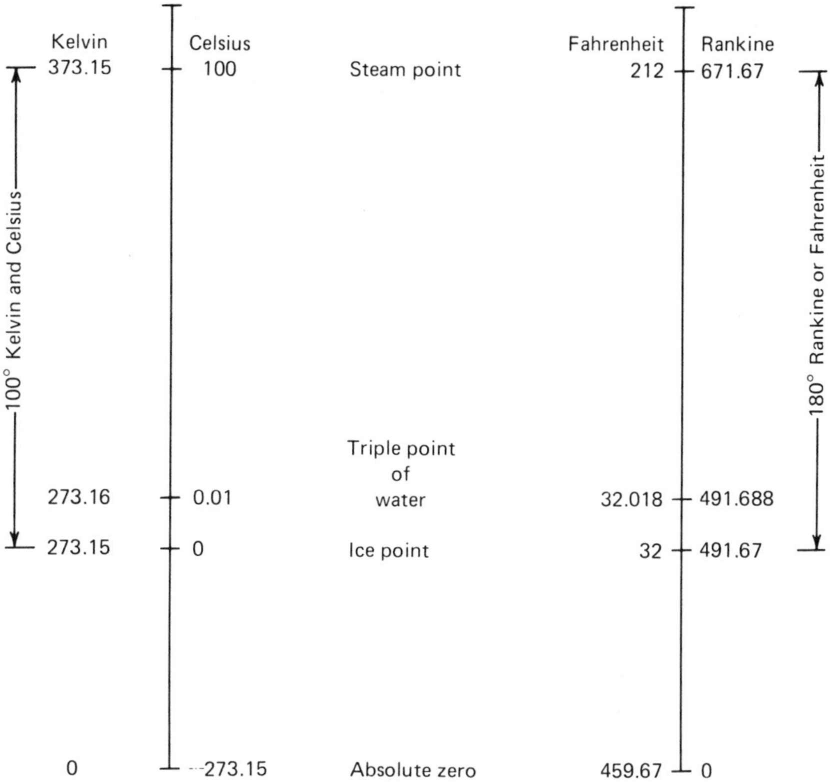

Figure 1–6

Temperature scales

therefore $T_K = t_c + (273.16 - 0.01)$, or $T_K = t_c + 273.15$. For our work we will use

$$T_K = t_c + 273 \qquad (1\text{-}12)$$

where T_K is the absolute temperature in Kelvin.

Since the Kelvin and Rankine are the absolute scales of the Celsius and Fahrenheit scales, respectively, with the same differences between ice and steam points, we may write $\Delta t_f/\Delta t_c = \Delta T_R/\Delta T_K = 180/100 = 1.8$, where T_R is the difference temperature in degrees Rankine. Since both are to absolute zero

$$\frac{\Delta T_R}{\Delta T_K} = \frac{T_R - 0}{T_K - 0} = 1.8$$

or

$$T_R = 1.8T_K \qquad (1\text{-}13)$$

Substituting Eq. (1-12) into (1-13) we have $T_R = 1.8T_K = 1.8(t_c + 273)$ or

$$T_R = 1.8t_c + 491.4 \qquad (1\text{-}14)$$

Substituting Eq. (1-10) into (1-14) we have

$$T_R = \frac{1.8(t_f - 32)}{1.8} + 491.4$$

or

$$T_R = t_f + 459.4$$

For our work we will use

$$T_R = t_f + 460 \qquad (1\text{-}15)$$

1-12 SPECIFIC GRAVITY

Definition: Fluid density/reference fluid density

Symbol: SG

Dimensions: Dimensionless ratio

Units: None

Reference fluids: Water for solids and liquids; air for gases

Since the density of water changes with temperature and, at very high pressures, for a precise definition of specific gravity, the temperatures and pressures of the fluid and reference fluid should be stated. In practice, two temperatures are stated, for example, 60/60°F where the upper temperature refers to the fluid and the lower to water. If no temperatures are stated, it must be assumed that reference is made to water at its maximum density. The maximum density of water at atmospheric pressure is at 3.98°C and has a value of 999.97 kg/m^3 or 62.4 lbm/ft^3.

For gases it is common practice to use the ratio of the molecular weight of the gas to that of air (28.96), thus eliminating the necessity of stating the pressure and temperature for ideal gases.

Hydrometer scales have been established by industry which have arbitrary graduations. In the petroleum and chemical industries the Baumé (°B) and the American Petroleum Institute (°API) are used. Conversions of units are as follows.

Baumé Scale

$$\text{Heavier than water } SG_{60/60} = \frac{145}{145 - °B} \qquad (1\text{-}16)$$

$$\text{Lighter than water } SG_{60/60} = \frac{140}{140° + °B} \qquad (1\text{-}17)$$

American Petroleum Institute Scale

$$SG_{60/60} = \frac{141.5}{131.5 + °API} \qquad (1\text{-}18)$$

Since the specific gravity of a fluid is greatly influenced by temperature, the following equation may be used to determine the specific gravity of a liquid at some temperature, t, other than 60°F.

$$(SG)_t = SG_{60/60} - 0.00035(t - 60) \qquad (1\text{-}19)$$

1-13 ENERGY

Definition: Capacity to produce an effect. Capacity to do work

Symbol: Depends on type of energy discussed

Units: EES: Foot-pounds force (ft lbf)
British thermal unit (Btu)

SI: joule (J)

Conversion: To convert Btu to joule multiply by 1.054. To convert ft lbf to joule multiply by 1.356.

A body is said to possess energy when it is capable of doing work. More generally, energy is the capacity for producing an effect. Energy may be classified as *stored energy* or *energy in transition*. Chemical energy is stored in high explosives and fuels such as coal, gas, and oil. Energy is stored in an automobile battery.

When a switch in an electric circuit is closed, electrical energy flows from a power plant to an electric motor or other electrical device. Energy flows from an automobile engine to the wheels through a gear train to drive the vehicle.

The heating engineer is concerned with the flow of heat energy in systems to provide thermal comfort. The refrigeration engineer is concerned with the removal of heat from areas where it is undesirable and the discharge of this energy to the atmosphere or to cooling water.

The complete or partial transformation of energy from one of the many forms in which it may exist into other forms of energy takes place in accordance with the *law of conservation of energy*. This law states that *energy can be neither created nor destroyed;* therefore, when energy is transformed either completely or partially from one form to another, the total amount of energy remains the same. In all types of engineering problems except those dealing with nuclear reactions, the laws of conservation of energy and conservation of matter apply.

1-14 MECHANICAL POTENTIAL ENERGY

Definition: Stored energy due to position relative to some selected datum

Symbol: pe

Units: EES: ft lbf; SI: J

Conversion: To convert ft lbf to joule multiply by 1.356

Stored energy associated with the position of bodies is called *mechanical potential energy*. In order to establish the value of potential energy, it is necessary to define a reference datum, usually the surface of the earth. A body on or near the earth's surface is attracted to the earth by the gravitational force, which is called the weight *(w)* of the body. If the body is lifted from the earth's surface to an elevation of Z ft above the earth, the body has stored (potential) energy equivalent to the work required to lift the body, or

$$\text{PE} = wZ \text{ (ft lbf)} \qquad (1\text{-}20)$$

It should be noted that potential energy is a function of gravitational force or weight. If mass is known, instead of weight, we may use Eq. (1-4), $w/g = m/g_c$, and potential energy becomes

$$\text{PE} = \left(\frac{g}{g_c}\right) mZ \text{ (ft lbf)} \qquad (1\text{-}21)$$

Since potential energy depends upon the location of an arbitrary datum of reference, the absolute value of PE is relative. In most problems we are interested in the *difference* in values of *PE* between two conditions of the system; therefore the absolute value is of little importance in most cases.

1-15 MECHANICAL KINETIC ENERGY

Definition: Energy possessed by a body due to its velocity

Symbol: KE

Units: EES: ft lbf; SI: J

Conversion: To convert ft lbf to joule multiply by 1.356.

Kinetic energy is the energy a body possesses as a result of its motion or velocity. The amount of kinetic energy a body possesses is a function of its mass and its velocity squared, or

$$\text{KE} = \frac{1}{2g_c} mV^2 \text{ (ft lbf)} \qquad (1\text{-}22)$$

Remembering again from Eq. (1-4) that $w/g = m/g_c$, Eq (1-22) may be written as

$$\text{KE} = \frac{wV^2}{2g} \text{ (ft lbf)} \qquad (1\text{-}23)$$

As with potential energy, we are normally concerned with a *change* in kinetic energy, if and when the velocity changes.

1-16 INTERNAL ENERGY

Definition: Molecular energy due to molecular motion and molecular attractive forces

Symbol: EES: u (specific); U (total)

Units: EES: u (Btu/lbm); U (Btu)
SI: J/kg (specific); J (total)

Conversion: To convert Btu/lbm to J/kg multiply by 2324. To convert Btu to J multiply by 1054

Matter was discussed as being composed of molecules. Molecules, like tangible bodies, have mass. In the solid state they are held together in relatively fixed position by powerful forces. However, the atoms of which they are composed vibrate. In the liquid and gaseous states the molecules have motion of translation and rotation and spin. Consequently, because of their mass and motion, the molecules have kinetic energy stored within them. Since the activity of the molecules is some function of temperature, any change in temperature is accompanied by a change in the kinetic energy stored in the molecules.

Also, molecules are attracted to each other by forces that are very large in the solid phase, but much less in the liquid phase, and tend to vanish in the gas phase where the molecules are far apart in comparison to their size. In the melting of a solid or the evaporation of a liquid, it is necessary that the powerful molecular attractive forces be overcome. The energy required to bring about this change of phase is stored in the molecules as *molecular potential energy* and will be released when the substance returns to its initial state.

1-17 ENTHALPY

Definition: A property combining the properties of internal energy, pressure, and specific volume in a defined way

Symbol: h (specific); H (total)

Units: EES: Btu/lbm (specific); Btu (total)
SI: J/kg (specific); J (total)

Conversion: To convert Btu/lbm to J/kg multiply by 2324
To convert Btu to J multiply by 1054

Pressure, temperature, specific volume, and internal energy are properties which define the state of a substance. In general, if a proper combination of two of the properties is known for a given state, the others are fixed and can be computed. Frequently, in equations in thermodynamics the internal-energy term and the product of pressure and specific volume appear in the same equation. Therefore, it has been found convenient to group these three properties into a new, single term called *enthalpy, h,* which is defined as

$$h = u + \frac{Pv}{J} \text{ (Btu/lbm)} \qquad (1\text{-}24)$$

where

u = internal energy (Btu/lbm)
P = pressure (lbf/ft^2)
v = specific volume (ft^3/lbm)
J = conversion unit (778 ft lbf/Btu)

1-18 THE SYSTEM

In discussions of thermodynamics and in other areas of study it has been found convenient to define what is called *the system*. The system means a particular region in space that is surrounded by real or imaginary boundaries that can be readily specified. Only that which takes place within the defined boundaries of a system and what flows into and out of the system across the boundaries is of consequence. The region outside the boundaries is called the *surroundings*. Systems may be divided into two general types: (a) the *closed* or constant-mass system and (b) the *open* system.

A closed tank containing a gas or liquid could be classified as a fixed-volume closed system. The gas within a cylinder where a piston moves up and down could be classified as a variable-volume closed system. The open system consists of a region surrounded by specified boundaries so arranged that a fluid may enter and leave the system across the system boundaries. In the general case of the open system, the mass of fluid within the system and the fluid properties at inlet and outlet could change with time.

The *steady-flow system* will be of most interest to us. A system may be called steady flow if such factors as flow velocity, pressure, temperature, and flow rate are constant with respect to time.

1-19 WORK, POWER, AND HEAT

Work may be defined as the product of the displacement of a body multiplied by the component of the force in the direction of the displacement. In equation form it would be

$$\text{wk} = F \times L \tag{1-25}$$

where F is the force component in the direction of motion in units of pound force (lbf) and L is the distance the body moves in units of feet when using the EES units. In SI units work is expressed in joules.

Another way of defining work is to say that it is energy that is being transferred across the boundaries of a system because of a force acting through a distance.

Definition: Energy in transition across the boundaries of a system caused by a force acting through a distance

Symbol: wk

Units: EES: ft lbf; SI: J

Conversion: To convert ft lbf to joule multiply by 1.356

A "sign convention" has been established in thermodynamics concerning work. Positive and negative work quantities must be distinguished when equations involving work are used. *Positive* work is work *leaving* a system; *negative* work is work *entering* a system from the surroundings.

Power is the *rate* of doing work. The horsepower (hp) is the common unit of power in the EES of units. One horsepower is defined as work being done at the rate of 550 ft lbf/sec or 33,000 ft lbf/min. One horsepower-hour (hp hr) is defined as the quantity of work done in 1 hr if the work is done continuously at an average rate of 33,000 ft lbf/min for a period of 1 h. Therefore, 1 hp hr equals 1,980,000 ft lbf/h.

The SI unit of power is the watt (W), which is J/sec.

POWER

Definition: The time rate of doing work

Symbol: Power (hp) or Power (W)

Units: EES: ft lbf/sec or ft lbf/min
SI: Watt (W)

Conversion: To convert ft lbf/sec to watt multiply by 1.356
To convert ft lbf/min to watt multiply by 0.026
To convert horsepower (hp) to watt multiply by 746

Heat is a form of energy and as such may be converted into other forms of energy. Also, other forms of energy can be converted into heat. Heat is often defined as energy that is being transferred across the boundaries of a system because of a temperature difference. If the surroundings of a system are at higher temperature than the system, internal energy stored initially in the molecules of the surroundings will be transferred across the system boundaries and increase the stored internal energy within the system. Transfer of energy will occur until equilibrium conditions are established between the system and surroundings.

The similarities in the definitions of heat and work should be noted. Both are forms of energy in the process of *being transferred* across the system boundaries. In the case of heat the driving potential is the *temperature difference*. In the case of work the driving potential is a *force*. Since both heat and work are forms of energy in transition, the quantity of energy transferred as heat or as work may be expressed in the same units.

The unit of heat in EES units is the British thermal unit (Btu), which has been defined as $1/180$ the quantity of heat required to change the temperature of 1 lbm of pure water from 32°F to 212°F, a temperature increase of 180 degrees.

Since energy may be transferred across system boundaries as either heat or work, the units in which heat and work are expressed must be consistent in any equation in which both quantities appear. Heat is normally expressed in Btu, and work is normally expressed in ft lbf. The relationship between Btu and ft lbf is

$$1 \text{ Btu} = 778.26 \text{ ft lbf} = 778 \text{ ft lbf}$$

SI units do not distinguish between work and heat. They are both classified as energy and the name is joule (J), which has the formula N m, or expressed in SI base units, it is $m^2 \text{ kg s}^{-2}$

HEAT

Definition: Energy in transition caused by a temperature difference

Symbol: q

Units: EES: Btu; SI: J

Conversion: To convert Btu to joule multiply by 1055

1-20 SPECIFIC HEAT

For small temperature change, experiments show that the temperature rise of a substance due to heat addition or removal is given by

$$q = mc(t_2 - t_1) \qquad \textit{(1-26)}$$

where

q = heat transferred (Btu)
m = mass of substance (lbm)
$t_2 - t_1$ = temperature change (°F)
c = experimental factor, Btu/lbm°F, called *specific heat*

Basically we may define the specific heat of a substance as the amount of heat required to change the temperature of 1 lbm of the substance 1°F. The specific heat of a substance varies somewhat with temperature; however, over the normal range of temperatures encountered in air-conditioning, a mean or average specific heat may be used with reasonable accuracy. The specific heat of a substance varies with the *state* of the substance. For instance, the specific heat of ice is 0.5 Btu/lbm°F, water 1.0 Btu/lbm°F, and steam 0.45 Btu/lbm°F.

Specific Heat:

Definition: Quantity of heat energy required to raise the temperature of a unit mass of a substance one degree of temperature

Symbol: c

Units EES: Btu/lbm °F
SI: joule/kilogram-kelvin (J/kg K)

Conversion: To convert Btu/lbm°F to J/kg K multiply by 4187

Illustrative Problem 1-6

Fifty pounds of oil having an average specific heat of 0.45 Btu/lb°F are heated from 40°F to 120 °F. How much heat was added to the oil?

<u>*Solution:*</u>

By Eq. (1-26)

$$q = mc\,(t_2 - t_1)$$
$$= (50)(0.45)(120 - 40)$$
$$q = 1800 \text{ Btu (added)}$$
$$= 1800 \times 1055 = 1{,}899{,}000 \text{ J}$$

Illustrative Problem 1-7

If water weighs 8.34 lb/gal, how much heat must be removed from the water to cool 100 gal from 190°F to 150°F?

<u>*Solution:*</u>

By Eq. (1-26),

$$q = mc\,(t_2 - t_1)$$
$$= (100)(8.34)(1)(150 - 190)$$
$$q = -33{,}360 \text{ Btu} \quad \text{(removed)}$$

NOTE: The minus sign in the answer of Problem 1-7 indicates that the heat is *leaving* the system. When heat is *added* to the system the sign is positive.

1-21 SPECIFIC HEAT OF GASES

If heat is added to a compressible fluid such as gas, the volume of the gas may be either constant or changing during the process of heat addition. If the volume is constant, all the heat added is used to increase the gas temperature. The heat required to change the temperature of 1 lbm of a gas 1°F at constant volume is called the constant-volume specific heat, c_v. Figure 1-7a illustrates a closed constant-volume system such as a sealed tank.

A gas may expand at constant pressure during the heating process, as would be the case in a vertical cylinder closed by a gas-tight movable piston as illustrated in Fig. 1-7b. In this case work would be done by the gas as it expands, lifting the piston weight. The work so done would be stored as the mechanical potential energy of the weight in its final position shown by the dotted outline. Sufficient heat energy must be added to increase the gas temperature and also to perform the work of expansion. The heat required to raise the temperature of 1 lbm of gas 1°F during a constant-pressure process is called the constant-pressure specific heat, c_p. It should be obvious that c_p is greater than c_v for a particular gas because no work is done during a constant-volume heat addition process.

Many of the processes involved in air con-

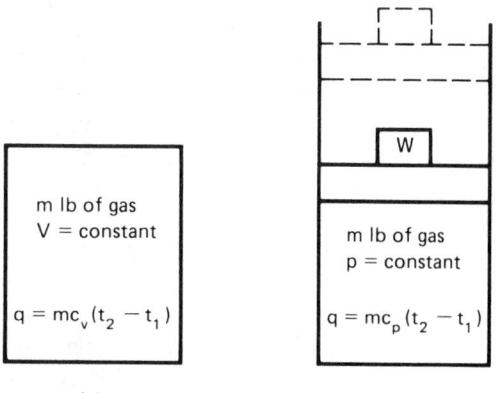

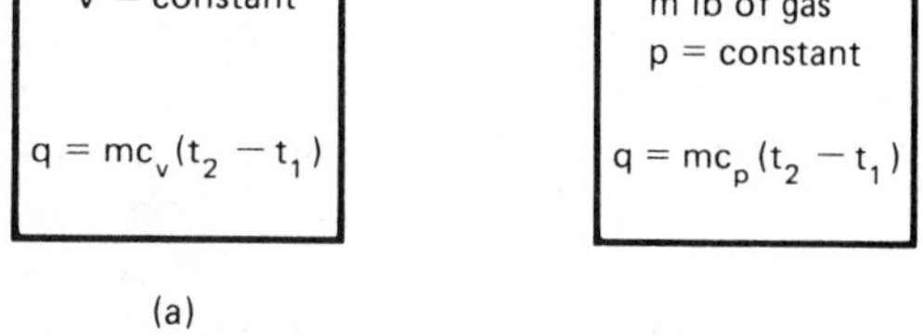

Figure 1–7

Constant volume and constant pressure heating of gas

ditioning may be considered constant-pressure processes.

1-22 EFFECT OF HEAT ADDITION OR REMOVAL ON THE STATES OF MATTER

Matter can exist in three different phases, or states: solid, liquid, and gas. Many materials, under proper conditions of pressure and temperature, can exist in any or all of these three physical states of matter. The amount of energy possessed by the molecules of a substance (internal energy) determine not only the temperature of the substance but also which of the three physical states the substance will assume at any particular time. In other words the addition or removal of heat can bring about a change in the physical state of a substance, as well as a change in its temperature as noted in Sections 1-20 and 1-21.

A change in the physical state that is familiar to all of us is the melting of ice to form water and the boiling of water to form steam. Each change of state or phase requires a relatively large amount of heat exchange.

When heat, either absorbed or rejected by a material, causes or accompanies a change in temperature of the material, the heat quantity is called *sensible heat* (q_s). The name sensible heat comes from the fact that the change of temperature may be detected by the sense of touch, and the temperature can be measured with a thermometer.

When heat, either absorbed or rejected by a material, brings about or accompanies a change in the physical state of the material, the heat quantity is sometimes called *latent heat*, q_L. Latent heat addition or removal is accompanied by *no change in temperature* of the substance. When a liquid changes to a solid, we say that a *fusion* process has taken place; when a liquid changes to a vapor, there is *vaporization;* and, when a solid changes directly to a vapor, a *sublimation* process is said to have taken place. In each of these processes energy must be exchanged between the substance and its surroundings to affect the change in phase. The temperature at which these changes will occur is dependent on the pressure exerted on the substance, as well as the substance itself.

Figure 1-8 is a sketch of the pressure-temperature diagram for water. The fusion line represents the solid-liquid mixture; the vaporization line represents the liquid-vapor mixture; and the sublimation line the solid-vapor mixture.

The *triple point* is the condition where it is possible to maintain an equilibrium mixture of all three phases. The *critical point* is the state where the vapor phase has the same properties as the liquid phase. It is not possible to distinguish between the liquid and vapor phases at pressures and temperatures above this point.

The three equilibrium lines in Fig. 1-8 (fusion, vaporization, and sublimation) may be said to designate *saturation* regions. We may say that the vaporization line represents the saturation region between liquid and vapor. The vapor present in such a mixture is said to be *saturated vapor,* and the liquid present in the mixture is said to be *saturated liquid.*

Consider the constant-pressure line 1234 of Fig. 1-8, located at 14.696 psia. At point 1 the material is solid (ice). As heat is added at constant pressure to the solid, the temperature will increase (process 1-2). At point 2 any addition of heat will cause a change in phase (melting). The temperature will remain constant until the phase change has been completed. Further addition of heat will raise the temperature of the liquid (process 2-3). At point 3 any addition of heat will cause some of the liquid to vaporize. The vaporization process will continue without an increase in temperature until all the liquid has been vaporized. Any further addition of heat will produce *superheated va-*

por. Processes 1-2, 2-3, and 3-4 are all sensible heating processes. The change of phase processes occurring at 2 and 3 are latent heat processes.

The quantity of heat absorbed by a given weight of material at the fusion temperature corresponding to the pressure in melting to the liquid phase, or conversely the heat rejected by a given weight of liquid at the fusion temperature in solidifying, can be determined by

$$q_L = mh_{fu} \tag{1-27}$$

where q_L = quantity of heat (Btu)

m = mass of liquid or solid (lbm)

h_{fu} = latent heat of fusion (Btu/lbm)

Illustrative Problem 1-8

The latent heat of fusion of ice at atmospheric pressure is approximately 144 Btu/lbm. How much heat is required to melt 1 ton of ice at 32°F to water at 32°F in the period of one day.

Solution:

By Eq. (1-27),

$$q = mh_{fu}$$

$$\text{Btu/day} = (2000 \text{ lbm/day})(144 \text{ Btu/lbm})$$

$$= 288{,}000 \text{ Btu/day} \qquad \textit{(ans)}$$

NOTE: In refrigeration work this is the basis for the term *1 ton of refrigeration,* which is defined as the removal of heat at the rate of 288,000 Btu/day, 12,000 Btu/hr, or 200 Btu/min.

The temperature at which a liquid will change into the vapor phase is called the *saturation temperature,* sometimes called boiling point or boiling temperature. A liquid whose temperature is at the saturation temperature corresponding to the existing pressure is called saturated liquid, as previously noted. The sat-

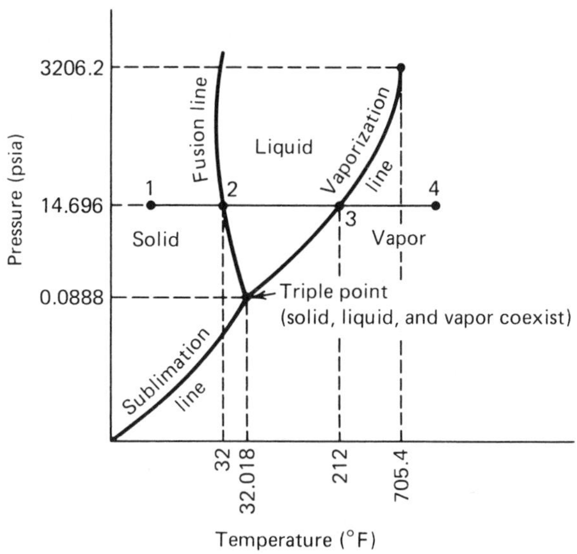

Figure 1–8

Pressure-temperature diagram for water (no scale)

uration temperature is different for each liquid and varies with pressure. Water boils at 212°F when the pressure above it is 14.696 psia, but it will boil at 180°F if the pressure above it is 7.515 psia. If the pressure above the liquid is maintained at 100 psia, the water will not boil until the temperature is 327.86°F. If the pressure above the refrigerant-12 is maintained at 51.667 psia, the liquid will boil at 40°F.

The quantity of heat absorbed by a given weight of liquid at its saturation temperature to cause it to change to a gas at the same temperature is called the *latent heat of evaporation* and may be calculated by

$$q_L = mh_{fg} \tag{1-28}$$

where q_L = quantity of heat (Btu)

m = mass of liquid (lbm)

h_{fg} = latent heat of evaporation (Btu/lbm)

Illustrative Problem 1-9

The specific volume of water at 30 psia and 250.34°F (saturation) is 0.017004 ft³/lbm. How much heat must be added to evaporate the water at the rate of 5 gal/hr? The latent heat of evaporation is 945.4 Btu/lbm.

Solution:

By Eq. (1-28),

$$q_L = mh_{fg}$$

$$m = \frac{5 \text{ gal/hr}}{(7.48 \text{ gal/ft}^3)(0.017004 \text{ ft}^3\text{/lbm})}$$

$$m = 39.31 \text{ lbm/hr}$$

$$q_L = 39.31(945.4)$$

$$= 37{,}160 \text{ Btu/hr} \qquad \textit{(ans)}$$

Once a liquid has been completely vaporized, the temperature of the vapor may be increased by the further addition of heat. The head added to the saturated vapor is called *superheat*. When the temperature of the vapor is above the saturation temperature corresponding to the pressure, the vapor is said to be *superheated* vapor.

BIBLIOGRAPHY

1. Jordan, Richard C., and Priester, Gayle B., *Refrigeration and Air Conditioning,* 2nd ed., Prentice-Hall, Inc., Englewood Cliffs, NJ, 1956.

2. Granet, Irving, *Thermodynamics and Heat Power,* Reston Publishing Company, Inc., Reston, VA, 1974.

CHAPTER 1

Review

1-1. Convert 20°C, 40°C, and 60°C to equivalent degrees Fahrenheit.

1-2. Convert 0°F, 10°F, and 50°F to equivalent degrees Celsius.

1-3. A pressure gauge indicates 25 psi when the barometer is at a pressure equivalent to 14.50 psia. Compute the absolute pressure in psia and inches of mercury when the specific weight of mercury is 13.0 g/cm^3.

1-4. A vacuum gauge reads 8 in. Hg when the atmospheric pressure is 29.00 in. Hg. If the specific weight of mercury is 13.6 g/cm^3, compute the absolute pressure in psia.

1-5. A skin diver descends to a depth of 80 ft in fresh water. What is the pressure on his body? The specific weight of fresh water can be taken as 62.4 lbf/ft^3.

1-6. Determine the density and specific volume of the contents of a 10-ft^3 tank if the contents weigh 250 lb. The local acceleration of gravity is 31.1 ft/sec^2.

1-7. A *U* tube mercury manometer, open at one end to atmospheric pressure (14.7 psia), is connected to a pressure source. If the difference in mercury levels in the tube is 7.6 in., determine the unknown pressure in psia.

1-8. Convert 500°R, 500 K, and 650°R to degrees Celsius.

Chapter 2

Review of Thermodynamic Principles

2-1 *INTRODUCTION*

Thermodynamics is that branch of science that deals with energy and its transformations. In our work in air conditioning we deal constantly with energy and its exchanges. Two other areas of engineering science are also very much involved. They are heat transfer and fluid mechanics, which will be discussed in later chapters.

Components found in all air-conditioning systems are heat exchanges of various types, pumps, fans, and piping and duct systems, through which fluids flow and in which heat is transferred or work is done. The transfer and conversion of energy in such devices takes place in accordance with the *law of conservation of energy*.

2-2 THERMODYNAMIC LAWS

There are two major premises upon which the science of thermodynamics is based. Because no major exceptions to these rules have ever been experienced, either accidentally or through controlled experiments, and because all attempts to disprove them have failed, they are termed thermodynamic laws and have been used as a foundation for further developments.

The *first law of thermodynamics* states that heat and mechanical energy are interconvertible and neither can be created nor destroyed. This is a limited form of the more general law of conservation of energy, which states that all forms of energy are interconvertible and can neither be created nor destroyed. Although these statements are all-inclusive, no inference can be drawn that the conversion of one energy form to another is necessarily complete. Only a portion of a given definite amount of heat energy can be converted to mechanical or to electrical energy, whereas all energy forms can be converted completely to heat energy. This ability for complete conversion into heat accompanied by a general tendency for all energy to be dissipated eventually in this manner has sometimes led to the description of heat as a "low-grade energy." For example, the power input to an electric motor operating a refrigeration compressor is all converted, eventually, into heat and is absorbed as heat through the condenser or lost to the surroundings by convection, conduction, and radiation. All electric-motor losses, such as resistance, windage, hysteresis, and friction, are transformed into heat. All electric power converted into mechanical energy by the motor is supplied to the compressor for compression of the refrigerant. This mechanical energy is eventually extracted at the condenser as heat, and all frictional losses are dissipated as heat. Even the energy transformed into sound in the operation of the plant is so dissipated.

The *second law of thermodynamics,* according to Clausius, states that it is impossible for a self-acting machine unaided by any external agency to transfer heat from one body to another at higher temperatures. Other statements of this same premise are that heat will not of itself flow from one body to another body maintained at a higher temperature and that no machine, actual or ideal, can both completely and continuously transform heat into mechanical energy. All these statements imply the same principle, but Clausius' is the most closely allied to air conditioning. The second law enables direct limitations to be placed upon the theoretical maximum operating efficiency that can be attained by a thermodynamic system operating under any specified set of conditions.

2-3 ENERGY EQUATIONS

It is frequently necessary to evaluate the energy interchanges occurring to or from a working medium during a process. The development of the quantitative relationship necessary to accomplish such an evaluation is based upon five basic equations, known as the *specific heat* equation (Eq. 1-26), the *nonflow* and the *steady-flow work* equations for an expanding substance, the *nonflow* or *simple energy* equation, and the *steady-flow* or *general energy* equation.

When there is no transfer of the working substance during a process, it is termed *nonflow;* when there is a continuous and steady flow of the working medium it is termed *steady flow*. Processes involving intermittant flow with rapid cycling, such as the compression of a refrigerant vapor or air, are usually treated as steady flow, although if desired, the individual parts of the cycle may be treated as nonflow. The nonflow work equation for an expanding substance may be derived from the energy required to move a piston of known area through a known distance against a known force and is equal to

$$ {}_1wk_2 = \int pdv \quad \text{(nonflow work)} \qquad (2\text{-}1) $$

In the compression process represented in Fig. 2-1 this is equivalent to the area under curve 1-2 (area 01230). If, during compression, however, steady-flow conditions exist, this area represents only a portion of the work required for compression. The working medium

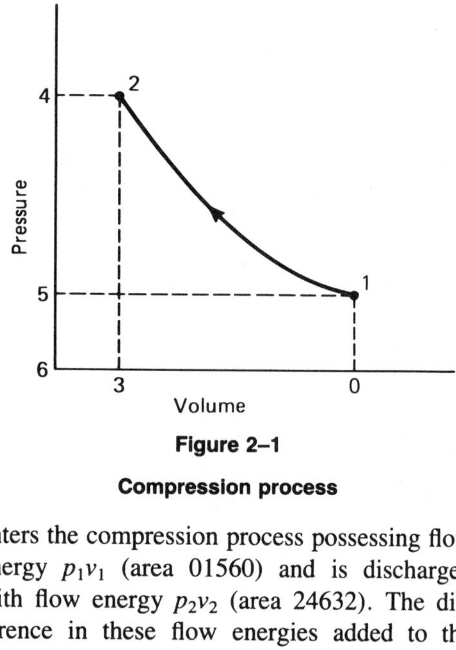

Figure 2–1

Compression process

enters the compression process possessing flow energy p_1v_1 (area 01560) and is discharged with flow energy p_2v_2 (area 24632). The difference in these flow energies added to the expression of Eq. (2-1) results in the steady-flow work equation equivalent to the process

$$ {}_1wk_2 = \int vdp \quad \text{(steady-flow work)} \qquad (2\text{-}2) $$

The quantity pv represents energy associated with a unit mass of fluid by virtue of the displacement of its boundary from one position to another at constant pressure. One element of mass in transport exerts pressure on the element immediately ahead and in turn has exerted upon it pressure by the immediately following elements. The distance of displacement of the element times the pressure introduces *flow work* or displacement energy.

Figure 2-2 represents an elementary-type steady-flow system existing between the limiting surfaces 1 and 2. Fluid in the inlet pipe is at pressure p_1 (psfa) and has a specific volume v_1 (ft^3/lbm). To push the fluid into the system across the system boundary requires work from further upstream; otherwise the fluid would not be flowing. If the cross-hatched volume of the inlet pipe contains 1 lbm of fluid, and if A_1 is the internal area of the inlet pipe in square feet, we may say that $A_1\ell_1 = v_1$ (ft^3). The force pushing on this volume of fluid is P_1A_1 pounds. The work done in pushing this fluid across the system boundary at section 1 is

$$\text{wk} = \text{force} \times \text{distance} = (P_1A_1)(\ell_1)$$

but $A_1\ell_1 = v_1$, therefore

$$\text{wk} = p_1v_1 \text{ (ft lbf/lbm of fluid)} \qquad (2\text{-}3)$$

Thus, when a fluid is flowing past a given section into a system, it is flowing because of work being done on it to push it past the section, and the amount of work is equal to p_1v_1 (ft lbf/lbm) of fluid. This form of energy is called *flow energy* or flow work.

The same analysis could be done at the exit pipe section. In that case the exit flow energy or work would be p_2v_2 (ft lbf/lbm).

The net flow work for a given system would be

$$\text{Net flow work} = p_2v_2 - p_1v_1 \qquad (2\text{-}4)$$

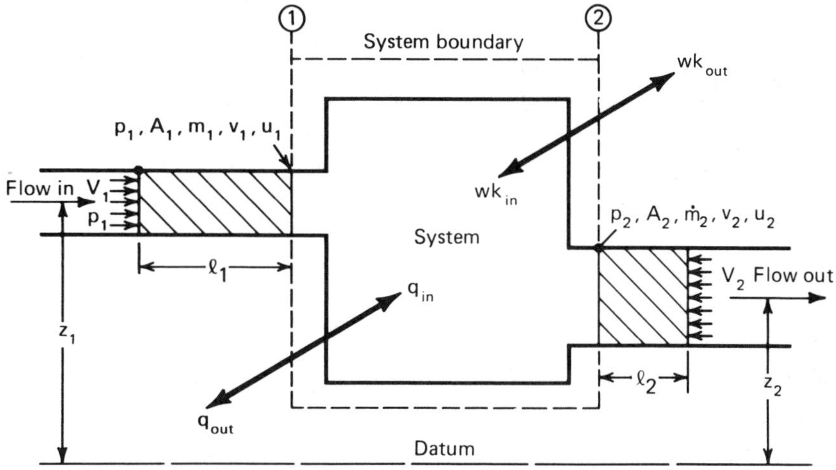

Figure 2–2

Flow work–steady flow system

The steady-flow or general energy equation is a mathematical statement of the energy conservation law including all flow process terms normally encountered in heat engineering. It is merely an energy balance equating all energy entering a system to that leaving a system. The only restrictions on its application are as follows:

1. The mass of fluid in the system must be constant. If a 1 lbm of fluid enters, 1 lbm must leave at the same time.

2. Pressure, temperature, specific volume, and flow velocity are all constant with respect to time at the entrance section. Values of the same quantities at the exit section must also be constant with respect to time.
3. Energy transferred across the system boundaries as heat and work must be at a constant rate. Any process which meets these limitations may be called a steady-flow system. Steady-flow processes are the most frequently encountered processes in air conditioning.

According to requirement (item 1) for a steady-flow system, the mass flow of fluid entering the system of Fig. 2-2 would be

$$\dot{m}_1 = \rho_1 A_1 V_1$$

where

$\dot{m}_1$ = mass flow rate (lbm/sec or kg/s)

ρ_1 = mass density (lbm/ft^3 or kg/m^3)

A_1 = flow area (ft^2 or m^2)

V_1 = flow velocity (ft/sec or m/s)

Also, the mass flow of fluid leaving the system would be

$$\dot{m}_2 = \rho_2 A_2 V_2$$

Since mass entering and leaving must be equal for a steady-flow system, we have

$$\dot{m}_1 = \dot{m}_2 = \rho_1 A_1 V_1 = \rho_2 A_2 V_2 \qquad (2\text{-}5)$$

Equation (2-5) is commonly referred to as the *equation of continuity,* or simply continuity equation. When the fluid flowing through the system is considered *incompressible* (usually liquids are considered incompressible at ordinary pressures), $\rho_1 = \rho_2$ and Eq. (2-5) becomes

$$\dot{Q} = A_1 V_1 = A_2 V_2 \qquad (2\text{-}6)$$

where $\dot{Q}$ = volume flow rate (ft^3/sec or m^3/s).

In applying the first law of thermodynamics to the steady-flow system of Fig. 2-2, it is important to include *all* energy terms identified in Chapter 1, such as potential energy, kinetic energy, internal energy, and flow work. The fluid entering the device can do *mechanical work* in the amount $\pm\, {}_1wk_2$ or mechanical work can be added to the fluid in the amount $\pm\, {}_1wk_2$. The subscript 1 and 2 simply signify that work occurs between the fluid inlet 1 and outlet 2 points and is read "work one to two." In the case of a turbine or engine work energy leaves the system through a driven shaft in the amount $+\, {}_1wk_2$. In the case of a compressor or pump work is supplied to the system in the amount $-\, {}_1wk_2$. In the engine example some of the work ${}_1wk_2$ generated in the device may be dissipated as friction in the bearings; not all of it will appear as delivered energy from the shaft. Such a condition also exists in an opposite sense for a compressor or pump.

Heat in the amount of $+\, {}_1q_2$ can be added to certain devices, such as a boiler, or perhaps may leave, in amount $-\, {}_1q_2$, by radiation or convection to cooling coils, colder surroundings, or the like.

The steady-flow energy equation for Fig. 2-2 is written by equating the energy leaving the system to that entering, or

$$\frac{Z_1}{J}\left(\frac{g}{g_c}\right) + \frac{V_1^2}{2g_cJ} + u_1 + \frac{p_1v_1}{J} \pm \qquad (2\text{-}7)$$
$${}_1q_2 \pm \frac{{}_1wk_2}{J} = \frac{Z_2}{J}\left(\frac{g}{g_c}\right) + \frac{V_2^2}{2g_cJ} + u_2 + \frac{p_2v_2}{J}$$

It should be noted that each term in Eq. (2-7) has units of Btu/lbm, which are heat units.

Since $J = 778$ ft lbf/Btu we may multiply Eq. (2-7) by J and obtain mechanical energy units of ft lbf/lbm or

$$Z_1\left(\frac{g}{g_c}\right) + \frac{V_1^2}{2g_c} + 778u_1 + p_1v_1 \pm 778\ {}_1q_2 \pm {}_1wk_2 = Z_2\left(\frac{g}{g_c}\right) + \frac{V_2^2}{2g_c} + 778u_2 + p_2v_2 \quad (2\text{-}7a)$$

where

p = static pressure (lbf/ft^2 or N/m^2)
v = specific volume (ft^3/lbm or m^3/kg)
V = average flow velocity (ft/sec or m/s)
g = local acceleration of gravity (ft/sec^2 or m/s^2)
g_c = constant (32.2 lbm ft/lbf sec^2 or 1.0 kg m/N s^2
Z = elevation (ft or m)
wk = work (ft lbf/lbm J/kg)

Recalling that enthalpy h is equal to $u + pv$, Eq. (2-7) may be written as

$$\frac{Z_1}{J}\left(\frac{g}{g_c}\right) + \frac{V_1^2}{2g_cJ} + h_1 \pm {}_1q_2 \pm \frac{{}_1wk_2}{J} = \frac{Z_2}{J}\left(\frac{g}{g_c}\right) + \frac{V_2^2}{2g_cJ} + h_2 \quad (2\text{-}8)$$

where h = enthalpy (Btu/lbm or J/kg).

Equation (2-8) is usually more convenient than Eq. (2-7) for steady-flow processes involving steam or a refrigerant vapor, whereas Eq. (2-7a) may be more convenient for steady-flow processes involving compressors, fans, pumps, and turbines.

The nonflow or simple energy equation is a simplified mathematical statement of the general energy equation eliminating all terms not involved in a nonflow process as well as those which are usually negligible. Its expression is

$${}_1q_2 = u_2 - u_1 + \frac{{}_1wk_2}{J} \quad (2\text{-}9)$$

Figure 2-3 illustrates two constant-mass nonflow systems. Figure 2-3a represents a cylinder containing a frictionless, movable piston. As the piston moves, the gas volume changes. In Fig. 2-3b the tank, considered to be rigid, represents a constant-mass and constant-volume system. In Eq. (2-9) subscripts 1 and 2 identify the limits of the process, that is, process 1 to 2.

${}_1q_2$ represents that heat energy transferred to or from the system during the process. A *positive* sign indicates heat addition, and a *negative* sign indicates heat removal.

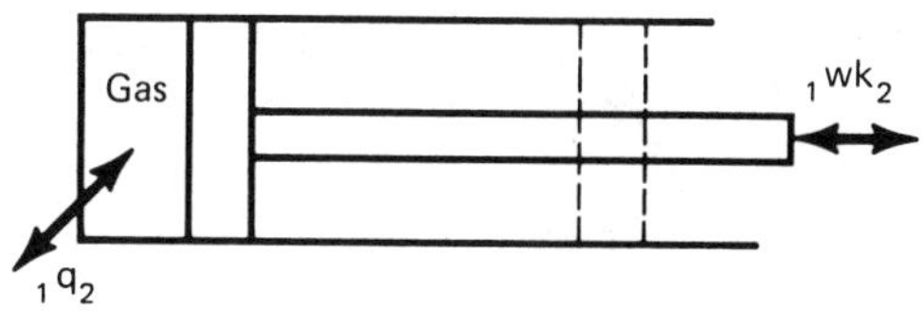

(a)

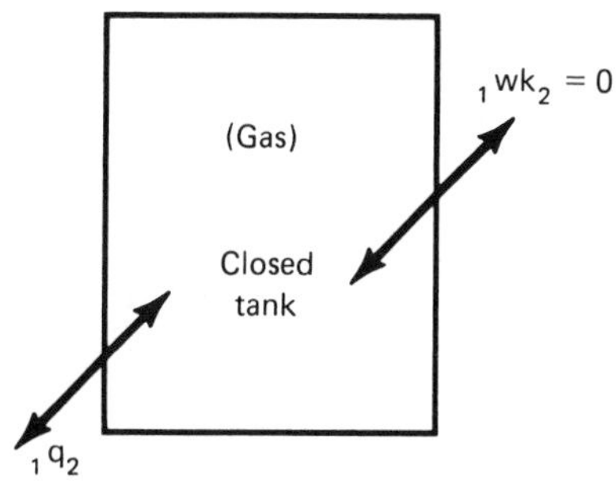

(b)

Figure 2–3

Nonflow systems–fixed mass

${}_1wk_2$ represents the work transferred from or to the system during the process. A *positive* sign indicates work done *by* the system as the gas expands against the piston. A *negative* sign indicates work done *on* the system by the piston compressing the gas.

$u_2 - u_1$ represents the change in internal energy stored in the gas within the system.

NOTE: As with all equations, it must be remembered that units must be consistent in equations. Equation (2-9) is normally written in such a way that all terms are expressed in heat units. Therefore, if work is expressed in ft lbf units, the term, ${}_1wk_2$ must be divided by the conversion unit from ft lbf to Btu (778) to make it consistent with the units of q and u terms.

Illustrative Problem 2-1

A nonflow system performs work by expansion of a gas behind a piston (Fig. 2-3a) in the amount of 75,000 ft lbf. At the same time heat is added to the gas in the amount of 100 Btu. Determine the change in internal energy of the gas within the system.

Solution:

By Eq. (2-9)

$${}_1q_2 = u_2 - u_1 + \frac{{}_1wk_2}{J}$$

$$100 = u_2 - u_1 + \frac{75{,}000}{778}$$

$$u_2 - u_1 = +3.6 \text{ Btu} \quad \text{(increase)}$$

Illustrative Problem 2-2

In a closed system work is done on the system in the amount of 80,000 ft lbf to compress the gas in a cylinder (Fig. 2-3a). At the same time heat in the amount of 50 Btu leaves the system. Determine the change in internal energy of the gas within the system.

Solution:

By Eq. (2-9)

$${}_1q_2 = u_2 - u_1 + \frac{{}_1wk_2}{J}$$

$$-50 = u_2 - u_1 + \left(\frac{-80{,}000}{778}\right)$$

$$u_2 - u_1 = +\ 52.8 \text{ Btu} \quad \text{(increase)}$$

Illustrative Problem 2-3

The burning gas in a single-cylinder engine expands against the piston and does 250,000 ft lbf of work, and heat in the amount of 30 Btu is transferred to the cylinder-cooling water. Determine the change in internal energy for the system.

Solution:

By Eq. (2-9)

$${}_1q_2 = u_2 - u_1 + \frac{{}_1wk_2}{J}$$

$$-30 = u_2 - u_1 + \frac{250{,}000}{778}$$

$$u_2 - u_1 = -351.3 \text{ Btu} \quad \text{(decrease)}$$

2-4 APPLICATIONS OF THE STEADY-FLOW ENERGY EQUATIONS

Let us now look briefly at some applications of the steady-flow energy equation to several system components frequently encountered in air-conditioning work.

Heat Exchanger. A heat exchanger could be a boiler, fin tube radiation unit, heat-transfer coil, and so forth. Normally, we can assume that for a heat exchanger (a) the difference in elevation between entrance and exit is relatively small $(Z_1 - Z_2)$, (b) the flow velocity into and out of the unit is approximately the same $(V_1 - V_2)$, and (c) no work is done. Therefore, applying Eq. (2-8) and eliminating unnecessary terms we have

$$q = h_2 - h_1 \qquad \textit{(2-10)}$$

where h_1 and h_2 are the fluid enthalpy values at entrance and exit, respectively.

Pumps. A pump is a device for transferring liquid from one place to ar ther against an external pressure. Normally, ıt may be assumed that (a) 1 lb of liquid enters the pump as 1 lb leaves, (b) the heat transfer from the liquid to the surroundings or from the surroundings to the liquid is negligible, (c) the liquid is incompressible, and (d) the increase in temperature of the liqiud as the liquid passes through the pump is negligible. From Eq. (2-8) we have

$$\frac{Z_1}{J}\left(\frac{g}{g_c}\right) + \frac{V_1^2}{2g_cJ} + \frac{P_1v_1}{J} = \frac{Z_2}{J}\left(\frac{g}{g_c}\right) + \frac{V^2}{2g_cJ} + \frac{p_2v_2}{J} + \frac{{}_1\text{wk}_2}{J} \qquad \textit{(2-11)}$$

The work term ${}_1\text{wk}_2/J$ appears on the right side of the equation; however, it should be remembered that for a pump work is *negative* because work is being added to the system fluid from the surroundings. Therefore, Eq. (2-11) for a pump may be arranged to establish work as a positive input to the system fluid, and at the same time we will multiply the equation through by J to obtain ft lbf/lbm units; thus

$${}_1\text{wk}_2 = \frac{g}{g_c}(Z_2 - Z_1) + \frac{V_2^2 - V_1^2}{2g_c} + p_2v_2 - p_1v_1 \qquad \textit{(2-11a)}$$

Since liquid may be considered incompressible, $v_2 = v_1$ or we may say $\rho_2 = \rho_1$. Also, multiplying the terms on the right side by g_c/g we have

$${}_1\text{wk}_2 = Z_2 - Z_1 + \frac{V_2^2 - V_1^2}{2g} + \frac{p_2 - p_1}{\rho(g/g_c)} \qquad \textit{(2-11b)}$$

but $\rho(g/g_c)$ is equal to the specific weight γ, therefore

$${}_1\text{wk}_2 = Z_2 - Z_1 + \frac{V_2^2 - V_1^2}{2g} + \frac{P_2 - P_1}{\gamma} \qquad \textit{(2-12)}$$

At this point we may note that each term of Eq. (2-12) has units of feet, or better *feet of fluid flowing.* The first term on the right is called *static head,* the second term is called *velocity head,* and the third term is called *pressure head.* We will discuss pumps in detail in a later chapter.

Gas Compressors and Fans. Compressors and fans are used to compress or move a gas at pressures that vary from a fraction of 1 psi to more than 10,000 psi. Since the operating conditions vary over such a wide range, and since gas is compressible, an analysis of their performance should always start with the complete Eq. (2-7). Terms may be eliminated if they do not apply to the specific problem. We will be primarily concerned with fans where pressure increases in the gas passing through the fan are relatively small. Under such conditions the gas may be considered incompressible and a variation of Eq. (2-12) may be used.

Illustrative Problem 2-4

A fan delivers 10,250 cfm of air, having a density of 0.0750 lbm/ft^3 at an outlet velocity of 1800 fpm and at a static pressure of 2.50 in. of water. Air enters the fan at 1400 fpm at a negative static pressure of -0.25 in. of water. Calculate the work per pound of air, disregarding friction and turbulence, for delivery under the conditions indicated. Also, calculate the work per minute or power requirement.

Solution:

For the small pressure rise involved, [2.50 − (−0.25)] or 2.75 in. of water, there would be no significant change in air density; therefore, the density remains at 0.0750 lbm/ft^3.

$$\text{Specific volume } (v) = \frac{1}{\rho} = \frac{1}{0.0750} =$$

13.33 ft^3/lbm

$$\text{Mass flow of air } (\dot{m}) = \frac{10{,}250}{13.33} =$$

768.9 lbm/min

Since the term $(g/g_c)(Z_2 - Z_1)$ of Eq. (2-11a) is relatively negligible for air, we may write

$$_1\text{wk}_2 = \frac{V_2^2 - V_1^2}{2g_c} + (p_2 - p_1)v$$

The quantity $p_2 - p_1$ must have units of lbf/ft^2; therefore we must convert 2.75 in. of water to lbf/ft^2. Recall that 1 ft^3 of water at 60°F weighing 62.32 lbf would exert a pressure of 62.32 lbf/ft^2 if its height were 12 in., so a 1-in. column exerts a pressure one-twelfth as great. Therefore

$$p_2 - p_1 = \frac{2.75}{12}(62.32) = 14.28 \text{ lbf/ft}^2$$

Then, for 1 lb of air,

$$_1\text{wk}_2 = \left[\left(\frac{1800}{60}\right)^2 - \left(\frac{1400}{60}\right)^2\right]\frac{1}{2(32.2)} +$$

$$(14.28)(13.33)$$

$$_1\text{wk}_2 = 5.52 + 190.35 = 195.87 \text{ ft lbf}$$

$$\text{work per minute} = (768.9)(195.87) =$$

$$150{,}604 \text{ ft lbf/min}$$

$$\text{horsepower} = \frac{150{,}604}{33{,}000} = 4.56$$

$$\text{kilowatts} = (4.56)(0.746) = 3.40$$

This result is the so-called *fan air horsepower* or *fan air kilowatts* and as such represents the power directly needed to move and pressurize the air. The motor shaft necessarily delivers more power to offset turbulence in the airstream passages, overcome bearing friction, and contribute to the slight increase in the internal energy of the air.

Nozzles. Nozzles are flow channels that are so proportioned as to permit expansion of a fluid from a higher pressure to a lower pressure with a conversion of as much as possible of the energy stored in the fluid into the kinetic energy of the high-velocity jet at exit. For nozzle flow it is normally assumed that (a) the change in potential energy is small, (b) work is zero, and (c) no heat is transferred. Therefore Eq. (2-8) becomes

$$\frac{V_1^2}{2g_cJ} + h_1 = \frac{V_2^2}{2g_cJ} + h_2$$

or

$$\frac{V_2^2 - V_1^2}{2g_c J} = h_1 - h_2 \qquad (2\text{-}13)$$

Equation (2-13) shows that for a nozzle the change in kinetic energy of the fluid is equal to the change in fluid enthalpy.

Orifice. An orifice is a throttling device placed in a fluid flow stream. It may be a partially open valve, or a circular thin plate with a concentric hole in its center, inserted into a pipe to meter the fluid flowing. The orifice produces a static pressure drop in the fluid as the flow is restricted. Normally, it is assumed that (a) no heat is transferred, (b) no work is done, and (c) kinetic and potential energy changes are negligible. Equation (2-8) then becomes

$$h_1 = h_2 \qquad (2\text{-}14)$$

2-5 THERMODYNAMIC CYCLE AND THERMAL EFFICIENCY

A thermodynamic cycle consists of a series of events or processes in a system carried out in such a manner that the system and the surroundings return to their initial state at the end of each cycle. The cycle may take place once or repeat over and over again.

Another statement of the first law of thermodynamics could be, *the work done in a cyclic process is equal to the heat that disappears.*

Thermal efficiency may be defined as the ratio of the energy converted into work in a given period of time to the energy supplied in the same period of time, both expressed in consistent units, or

Thermal efficiency (η_t)

$$= \frac{\text{Net work done by a system}}{\text{Heat supplied to system}} \cdot 100 \qquad (2\text{-}15)$$

In Section 2-2 it was noted that the second law of thermodynamics places direct limitations upon the theoretical maximum operating efficiency of a thermodynamic cycle. This maximum thermal efficiency may be stated as

$$\eta_t\,(\max) = 1 - (T_L/T_H) \qquad (2\text{-}16)$$

where T_L is the lowest absolute temperature and T_H is the highest absolute temperature for the cycle.

Reversibility. Analysis based on the second law, in addition to temperature considerations, leads to the concept of *reversibility*. Reversiblity implies, for a thermodynamic process, that merely changing the direction of a driving force by a small amount will cause a process to reverse itself and return a system essentially to its initial condition. Reversibility is an idealized condition that can never be realized because of natural phenomena which inhibit its realization, the most obvious being *friction* and *fluid turbulence*.

Entropy. The concept of a reversible process in thermodynamics brings about the development of a fluid properly called *entropy*. On a unit mass basis the specific entropy is s. The defining equation for entropy is given as

$$\Delta s = (q/T) \quad \text{reversible process} \qquad (2\text{-}17)$$

where Δs is the change in entropy; that, ($s_2 - s_1$), q is the heat added or removed.

In the EES system of units entropy has units of Btu/lbm °R and the SI units are J/kg K.

2-6 THE CARNOT CYCLE

The Carnot cycle is an ideal, thermodynamically reversible power cycle, first investigated by Sadi Carnot in 1824 as a measure of the maximum possible conversion of heat energy into mechanical energy. In its reversed form it is used as a measure of the maximum possible performance of refrigeration equipment. Although it cannot be applied in an actual machine because of the impossibility of obtaining a completely reversible engine, it is nevertheless extremely valuable as a criterion of inherent limitations.

The direct Carnot cycle is shown graphically in Fig. 2-4 on *p-v* and *T-s* coordinates. The cycle consists of an isothermal (constant-temperature) heat addition process *BC*, an adiabatic (frictionless and no heat loss) expansion process *CD*, an isothermal heat rejection process *DA*, and an adiabatic compression process *AB* to form a closed cycle. Areas on the *T-s* diagram represent heat quantities because from Eq. (2-17), $T \times \Delta s = q$.

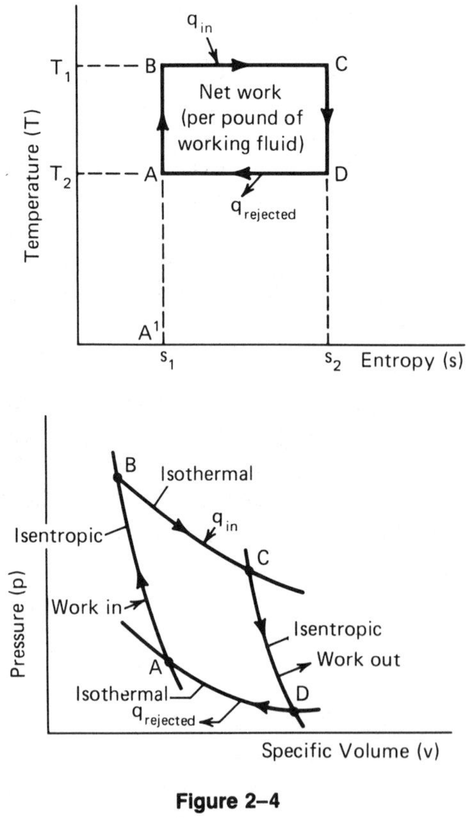

Figure 2–4

Direct Carnot cycle

From Fig. 2-4 we may see that $T_1(s_2 - s_1)$ represents the total area $A'BCD'$ and indicates the total heat supplied at constant temperature T_1. The quantity $T_2(s_2 - s_1)$ represents the area $A'ADD'$ and indicates the total heat rejected at constant temperature T_2. The difference between the two areas is area *ABCD* which indicates the net cycle work. Equation (2-15) gives a statement of thermal efficiency. Therefore, for the Carnot cycle

$$\eta_{\text{carnot}} = \frac{T_1(s_2 - s_1) - T_2(s_2 - s_1)}{T_1(s_2 - s_1)} = \frac{T_1 - T_2}{T_1} \qquad (2\text{-}18)$$

You may note that Eq. (2-18) is a different form of Eq. (2-16). The temperatures indicated in Eqs. (2-16), (2-17), and (2-18) are always in degrees absolute.

2-7 THERMODYNAMIC RELATIONSHIPS FOR GASES

In our study of air conditioning we are involved with the behavior of liquids, gases, and vapors as various processes take place. We will now review briefly these processes and the changes in fluid properties which occur.

An *ideal* or perfect gas may be defined as one which obeys the *equation of state* and whose internal energy is a function of temperature only. The equation of state relates the properties of pressure, specific volume, and temperature as

$$pv = RT \qquad (2\text{-}19)$$

where

Fluid Property	EES Units	SI Units
p = pressure	lbf/sq ft	N/m^2
v = specific volume	ft^3/lbm	m^3/kg
R = gas constant	ft lbf/lbm °R	J/kg K
T = absolute temperature	°R	K

There are no ideal (perfect) gases, but many common gases such as dry air, oxygen, nitrogen, hydrogen, and carbon dioxide follow closely the equation of state. As gases approach the condition of pressure and temperature required for liquification (saturation), they no longer obey Eq. (2-19), and they are called *vapors*.

If Eq. (2-19) is multiplied by mass m, the quantity of gas in pound mass, we obtain

$$pmv = mRT$$

Since the product mv is equal to the total volume V, we have

$$pV = mRT \tag{2-20}$$

For a fixed mass or closed system the relationship between pressure, volume, and tempera-

Table 2-1

Values of Gas Constant *R*, Molecular Weight *M*, and Specific Heats of Various Gases

Gas	R	M	Specific Heat of Vapor (Btu/lb F) Constant Pressure C_p	Constant Volume C_v
Acetylene	59.34	26.02	0.36	0.29
Air	53.34	28.97	0.24	0.17
Ammonia	90.50	17.03	0.52	0.40
Carbon Dioxide	35.12	44.00	0.21	0.16
Carbon Monoxide	55.14	28.00	0.24	0.17
Refrigerant-12	12.77	120.92	0.15	0.13
Refrigerant-21	15.00	102.92	0.18	0.16
Ethane	51.38	30.05	0.41	0.35
Hydrogen	765.86	2.02	3.42	2.44
Isobutane	26.58	58.08	0.55	0.50
Methane	96.31	16.03	0.52	0.39
Methyl chloride	30.59	50.48	0.24	0.20
Nitrogen	54.99	28.02	0.25	0.18
Oxygen	48.25	32.00	0.22	0.16
Propane	35.04	44.06	0.58	0.51
Steam	85.78	18.01	0.46	0.36
Sulphur dioxide	24.10	64.07	0.15	0.12
Refrigerant-II	11.24	137.37	0.14	0.12
Water vapor	85.78	18.01	0.46	0.36

Used by permission (Courtesy of Industrial Press).

ture at the beginning and end of any processes involving a perfect gas may be shown to be

$$\frac{p_1V_1}{T_1} = \frac{p_2V_2}{T_2} = constant \tag{2-21}$$

or for 1 lbm we may write

$$\frac{p_1v_1}{T_1} = \frac{p_2v_2}{T_2} = constant \tag{2-21a}$$

Since the internal energy of a perfect gas is dependent on absolute temperature, the change in internal energy in a nonflow constant-volume process may be given by

$$u_2 - u_1 = c_v(T_2 - T_1) \quad \text{(Btu/lbm)} \tag{2-22}$$

Many processes encountered in air conditioning involve working fluids that alternate between the vapor and liquid phase and therefore cannot be treated thermodynamically as ideal gases. Some fluids, such as air, may be treated as a perfect gas. Table 2-2 summarizes thermodynamic relationships for your reference. You should refer to any good textbook in thermodynamics for further information. Table 2-2 has been developed using Eq. (2-19) through (2-24), the nonflow energy equation (2-9), and the polytropic pressure-volume relationship, Eq. (2-25).

$$pv^n = \text{constant} \tag{2-25}$$

Values for the exponent n for the various processes are

Isentropic* (s = constant)	frictionless adiabatic	$n = k = Cp/C_v$
Isothermal (t = constant)	constant temperature	$n = 1$
Isobaric (p = constant)	constant pressure	$n = 0$
Isometric (v = constant)	constant volume	$n = \infty$
Polytropic	any	$n = 0$ *to* ∞

*If a process takes place without heat transfer and is frictionless, it is reversible and follows a path of constant entropy (s) and hence is called an *isentropic* process. Without heat transfer a process is called *adiabatic*.

The relationship between the constant-pressure specific heat c_p and the constant-volume specific heat c_v for a perfect gas may be shown to be

$$c_p = c_v + \frac{R}{J} \tag{2-23}$$

The enthalpy change of a perfect gas during all processes (nonflow and steady flow) may be shown to be

$$h_2 - h_1 = c_p(T_2 - T_1) \quad \text{(Btu/lbm)} \tag{2-24}$$

2-8 THERMODYNAMIC RELATIONSHIPS FOR REAL GASES AND VAPORS

Most fluids used in air-conditioning systems must be treated as imperfect gases or as vapors. Refrigerants and steam are the two most common fluids. In most systems these fluids alternate between the gaseous phase and the liquid phase and sometimes may coexist in varying proportions in the two phases at the same time. Although this greatly complicates the thermodynamic relationships, the actual solution of problems is simplified by the use of

Table 2-2

Thermodynamic Relationships for Gases (Constant Specific Heat)

Process	Isometric or Constant Volume	Isobaric or Constant Pressure	Isothermal or Constant Temperature	Isentropic of Constant Entropy (Reversible Adiabatic)	Polytropic	Free Expansion (Irreversible Adiabatic)
Pressure-Volume Relationship pv^n = Const.	$n = \infty$ $V_1 = V_2$	$n = 0$ $p_1 = p_2$	$n = 1$ $p_1V_1 = p_2V_2$	$n = k$ $p_1V_1^k = p_2V_2^k$ $\frac{T_2}{T_1} = \left(\frac{V_1}{V_2}\right)^{k-1} = \left(\frac{P_2}{P_1}\right)^{(k-1)/k}$	$p_1V_1^n = p_2V_2^n$ $\frac{T_2}{T_1} = \left(\frac{V_1}{V_2}\right)^{n-1} = \left(\frac{P_2}{P_1}\right)^{(n-1)/n}$	
Heat Added to Gas	$mC_v(T_2 - T_1)$	$mC_p(T_2 - T_1)$	$\frac{mRT}{J} \ln\left(\frac{V_2}{V_1}\right)$	0	$m\left(\frac{n-k}{n-1}\right) C_v(T_2 - T_1)$	0
Work done by gas (nonflow process)	0	$p(V_2 - V_1)$	$pV \ln\left(\frac{V_2}{V_1}\right)$	$\frac{p_1V_1 - P_2V_2}{k-1}$	$\frac{P_1V_1 - P_2V_2}{n-1}$	0
Work done by gas (steady-flow process)	$V(p_2 - p_1)$	0	$pV \ln\left(\frac{P_2}{P_1}\right)$	$\frac{k}{k-1}(p_1V_1 - p_2V_2)$	$\frac{n}{n-1}(p_1V_1 - p_2V_2)$	0
Gain in entropy of gas	$mC_v\ln\left(\frac{T_2}{T_1}\right)$	$mC_p\ln\left(\frac{T_2}{T_1}\right)$	$\frac{mR}{J} \ln\left(\frac{V_2}{V_1}\right)$	0	$m\left(\frac{n-k}{n-1}\right)C_v\ln\left(\frac{T_2}{T_1}\right)$	$\frac{mR}{J} \ln\left(\frac{V_2}{V_1}\right)$
Gain in internal energy of gas	$mC_v(T_2 - T_1)$	$mC_v(T_2 - T_1)$	0	$-mC_v(T_2 - T_1)$	$mC_v(T_2 - T_1)$	0

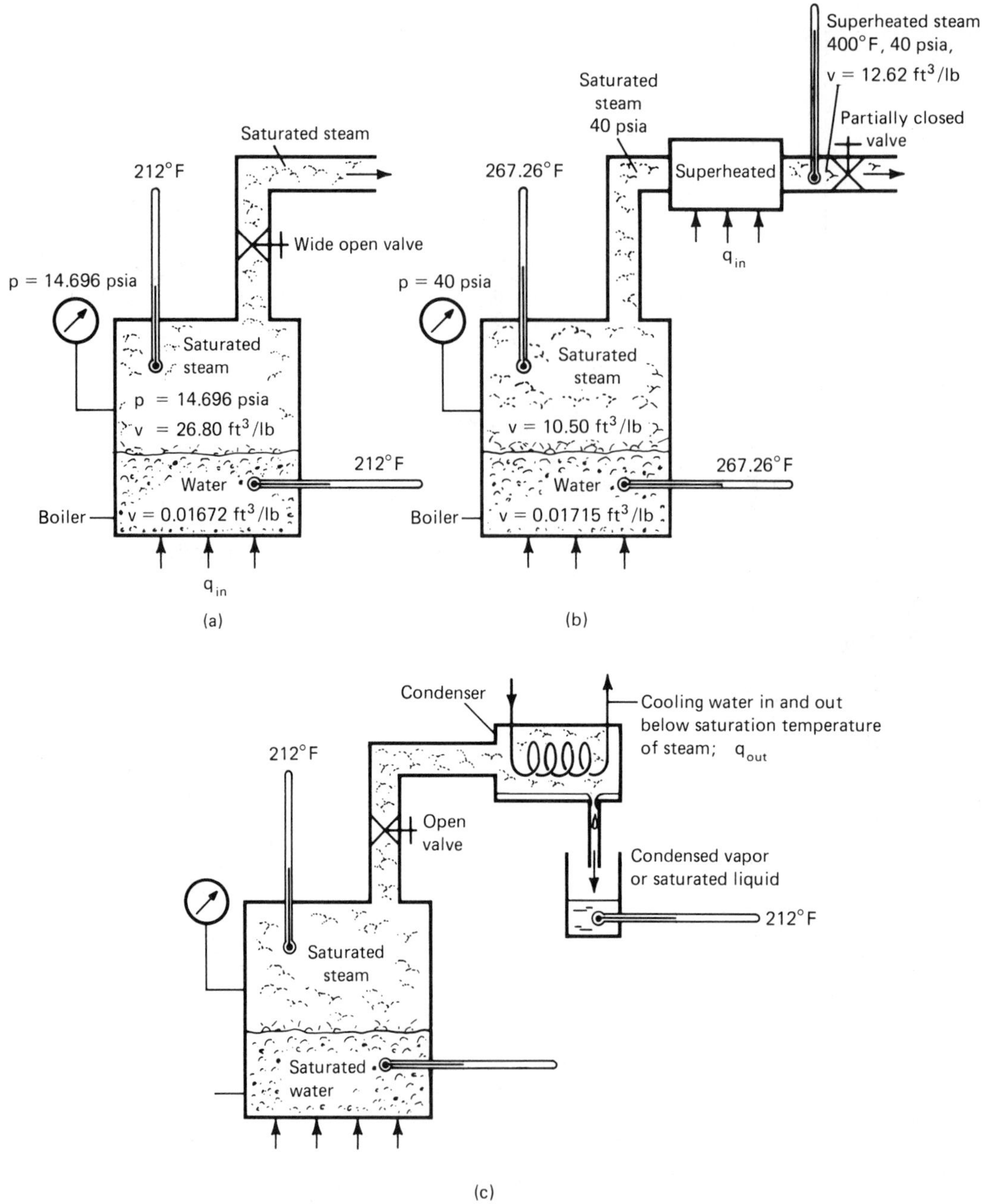

Figure 2–5

Saturation, superheating, and condensation

tabular data of the thermodynamic properties of these fluids, since the process equations are not feasible.

Before discussing tabular data on real fluids we should review briefly what is meant by *saturation* and *superheat*. If the temperature of a liquid is at a point when any small amount of heat addition will cause it to boil, the liquid is said to be "saturated" and the corresponding temperature is called the *saturation temperature*. The pressure on the liquid is called the *saturation pressure*.

Figure 2-5 illustrates the processes of saturation, superheating, and condensation using the familiar fluid, water. Figure 2-5a shows the water at the saturation temperature of 212°F (211.99°F exactly) when the pressure above the liquid is 14.696 psia. At this condition, if heat is added (q_{in}), even a small amount, saturated steam (vapor) will be produced above the liquid. The liquid is said "to boil." Please note that this process of evaporation (boiling) takes place at constant pressure and constant temperature. As long as any liquid remains in the container, the temperature and pressure will remain at 212°F and 14.696 psia, respectively.

In Figure 2-5b the outlet valve is partially closed, restricting the escape of steam. Under these conditions pressure above the liquid will increase to a pressure above standard atmospheric pressure, to a pressure of say 40.0 psia. At this pressure the water will not start to boil until the temperature is 267.26°F. Again we have a condition of saturated liquid and saturated vapor. Therefore, we can say that the saturation temperature of water for 40.0 psia is 267.26°F.

If the saturated steam is heated further in a heat-exchange device, at constant pressure, the temperature of the steam is increased above the saturation temperature, and the steam is said to be "superheated." The heating device is called a "superheater."

In Figure 2-5c heat is added to saturated water at 14.696 psia and 212°F to produce saturated steam at the same pressure and temperature. The saturated steam enters a heat-exchange device where heat is extracted from the steam (q_{out}) to produce saturated liquid. The heat-exchange device is called a *condenser*, and the process carried out at constant pressure is called *condensation*. The heat that is removed from the steam in the condenser is exactly equal to the amount of heat added to the water in the boiler ($q_{in} = q_{out}$) neglecting radiation losses.

Certain things may be noted at this point. First, for each pressure there is a corresponding saturation temperature; second, as the water changes to steam, there is a large change in specific volume, approximately 1600 times at atmospheric pressure; third, as saturated steam is superheated at constant pressure, there is an increase in specific volume.

As mentioned previously, properties of real fluids (water, refrigerants, air, etc.) such as pressure, temperature, specific volume, enthalpy, internal energy, and entropy have been tabulated in so-called gas tables. It is sometimes convenient to show fluid properties in graphical form, called *phase diagrams*. Figure 2-6 shows four such phase diagrams for water in its three phases of solid, liquid, and vapor. Of the several different combinations of properties for the coordinates, the ones shown are the most common. Figure 2-6a is the *pressure-volume* diagram, Fig 2-6b is the *temperature-entropy* diagram, Fig. 2-6c is the *pressure-enthalpy* diagram, and Fig 2-6d is the *enthalpy-entropy* diagram. Figure 2-6a, b, and c has boundary curves separating the phase areas and are identically numbered. The *saturated-liquid line* (3-4) and *saturated-vapor line* (4-6), together with the triple-point line (2-3-5) bound the area in which the liquid and vapor phases coexist in varying proportions. In this area the weight proportionality of the homogeneous

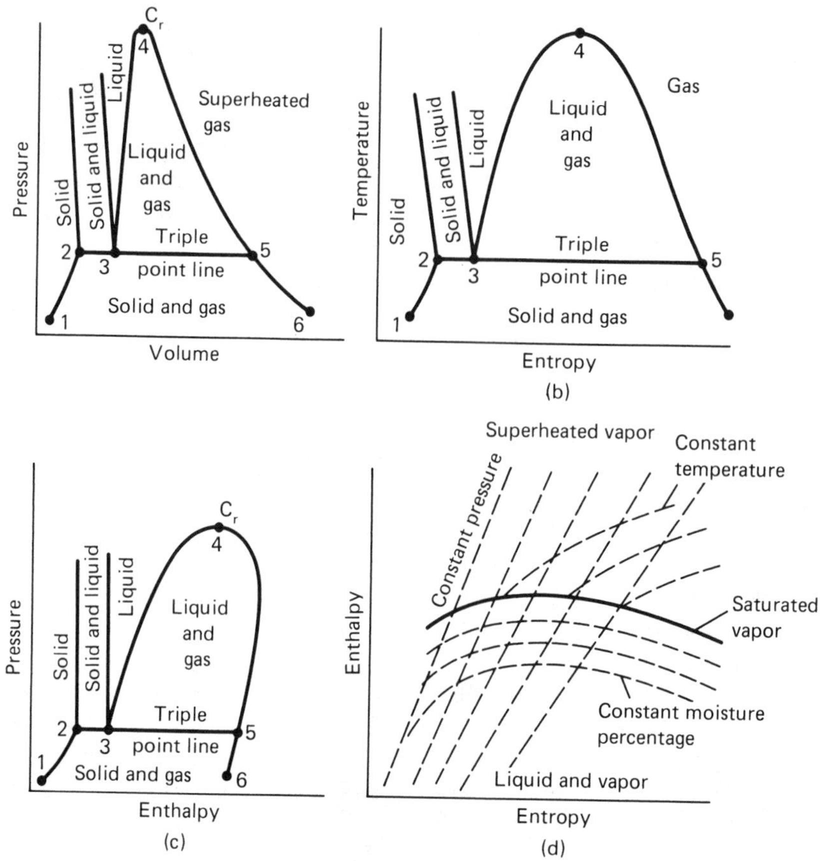

Figure 2–6

Phase diagrams for water (not to scale)

mixture of the two phases is called *quality;* a quality of 90 percent would indicate a mixture of 10 percent saturated liquid and 90 percent saturated vapor by weight.

In the area to the left of the saturated-liquid line and above the triple-point temperature the substance is called a *subcooled liquid.* In the area to the right of the saturated-vapor line, both above and below the triple-point temperature, the substance is a superheated vapor or gas.

The critical point (point 4) at the junction of the saturated-liquid and saturated-vapor lines defines the *critical temperature* above which liquification of the substance cannot occur, except under extreme pressures. Above the *critical pressure* the latent heat of vaporization becomes zero, the boundary line between liquid and vapor phases disappears, and there is no recognizable phenomenon such as vaporization (boiling) and condensation. There is no observable evidence of a phase change when the water passes from liquid to vapor.

The *saturated-solid line* (1-2), the lower portion of the saturated-vapor line (5-6), and the triple-point isotherm (2-3-5) bound the area in which the solid and vapor phases coexist in varying proportions. In the area to the left of the saturated-solid line and below the triple-point (2-3) (at triple-point temperature) is a unique series of state points at which the substance may exist in all phases (solid, liquid, and gas) in equilibrium. Below the triple-point temperature the heat required to change the substance from a solid directly to a gas is termed *latent heat of sublimination*. At the triple-point temperature the heat required to change from a solid to a liquid (along 2-3) is termed *latent heat of fusion*. Above this temperature the heat required to change from a liquid to a vapor is termed *latent heat of vaporization*.

Figure 2-6d is the enthalpy-entropy diagram for steam. The coordinates and steam property lines plotted on the diagram make it especially useful in steam power work, where isentropic expansion and compression processes, throttling processes, and so forth, may be illustrated graphically. Its usefullness for our work in air conditioning is rather limited.

Figure 2-6a and b is useful in showing processes in a graphical way when discussing steam. Figure 2-6c is used most often in studying the processes involved in compression refrigeration using refrigerants.

2-9 TABULATED FLUID PROPERTIES

Tables A-1 and A-2 (see Appendix A) list the saturation properties of pressure, temperature, internal energy, specific volume, enthalpy, and entropy for steam. Table A-3 lists the properties of pressure, temperature, specific volume, internal energy, enthalpy, and entropy for superheated steam.

Tables B-3 and B-5 (see Appendix B) list the saturation properties for refrigerant-12 and refrigerant-22, respectively. Subscripts used in the tables are:

f for saturated liquid
g for saturated vapor
fg for change in specific property from liquid to vapor

It may be seen, when studying the tables, that the following is true:

$$v_g = v_f + v_{fg} \qquad u_g = u_f + u_{fg}$$

$$h_g = h_f + h_{fg} \qquad s_g = s_f + s_{fg}$$

As is the case with several working fluids, such as water, refrigerants, and so forth, it is possible for the fluid to coexist in the liquid and vapor phase during some processes. The two-phase fluid properties may be calculated from saturation properties if the quality, x, is known by one of the following equations where x is the percent quality expressed as a decimal.

$$h = h_f + xh_{fg} \qquad (2\text{-}26)$$

$$v = xv_g + (1 - x)v_f \qquad (2\text{-}27)$$

$$s = s_f + xs_{fg} \qquad (2\text{-}28)$$

2-10 THERMODYNAMIC PROCESSES FOR REAL FLUIDS

The thermodynamic processes for real fluids will be discussed in this section. The processes will be applied to specific air-conditioning equipment, components, and systems to which they apply. The various processes are constant pressure (isobaric), constant volume (isometric), constant temperature (isothermal), throttling, and adiabatic. All processes represented will utilize steam as the working fluid; however, any fluid which normally operates in a system in both the liquid and vapor phase, such as refrigerants, will follow the same types of processes.

CONSTANT-PRESSURE PROCESS: Figure 2-7 illustrates a constant-pressure process on the p-v and T-s coordinates. The constant-pressure process is one of the more important processes in air-conditioning work. Although no equipment actually works at constant pres-

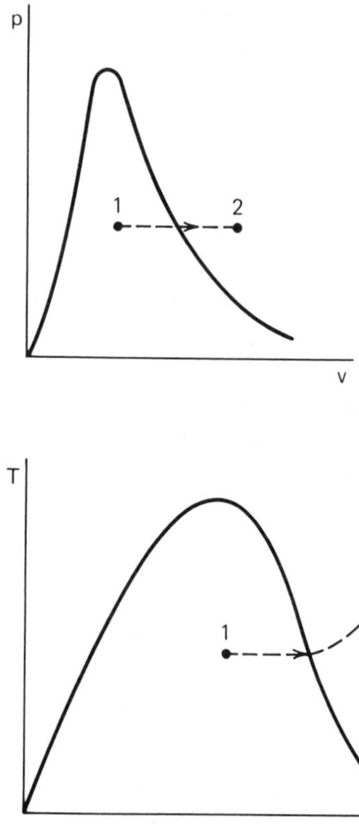

Figure 2–7

The constant pressure process (no scale)

sure because of flow pressure losses, a close approximation to the actual process may be obtained by the idealized process. The constant-pressure process is typical of the process of heating water in a boiler, condensing steam in a steam coil, evaporating a refrigerant in an evaporator, or condensing refrigerant vapor in the condenser. The equation for the constant-pressure process applies to both nonflow and steady-flow conditions and is

Energy Equation $${}_1q_2 = u_2 - u_1 + \frac{{}_1\text{wk}_2}{J}$$

but $$\frac{{}_1\text{wk}_2}{J} = \frac{p\Delta v}{J} = p\,\frac{(v_2 - v_1)}{J}$$

therefore $${}_1q_2 = u_2 - u_1 + \frac{p_2v_2 - p_1v_1}{J}$$

since $h = u + (pv/J)$ and $p_1 = p_2$, we have $_1q_2 = h_2 - h_1$ (Both nonflow and steady-flow) ***(2-10) Repeated***

Illustrative Problem 2-5

Water under a pressure of 16 psia and at a temperature of 180°F enters a steam boiler where heat is added to produce saturated steam. How much heat must be added to each pound of water?

Solution:

By Eq. (2-10) $_1q_2 = h_2 - h_1$

Data: From Table A-2 at 16 psia, find

$$h_2 = h_g = 1152.1 \text{ Btu/lbm}$$

From Table A-1 at 180°F, find

$$h_1 = h_f = 147.99 \text{ Btu/lbm}$$

Therefore,

$$_1q_2 = 1152.1 - 147.99$$

$$_1q_2 = 1004.1 \text{ Btu/lbm} \quad \text{(added)}$$

Illustrative Problem 2-6

Saturated steam is supplied to a heat-transfer coil at a gauge pressure of 3.30 psig. The steam condenses and the condensate drains from the coil at a temperature of 190°F. How much heat in Btu/lbm was given up to the surroundings by condensing steam?

Solution:

By Eq. (2-10)

$$_1q_2 = h_2 - h_1$$

Data: From Table A-2 at $p = 14.7 + 3.30$

$= 18.0$ psia, find $h_1 = h_g$

$= 1154.4$ Btu/lbm

From Table A-1 at $t = 190°F$, find

$h_2 = h_f = 158.03$ Btu/lbm.

Therefore,

$$q_2 = 158.03 - 1154.4$$

$$_1q_2 = -996.37 \text{ Btu/lbm} \quad \text{(removed)}$$

NOTE: The negative sign simply indicates that the direction of heat flow is from the system to the surroundings.

CONSTANT-VOLUME PROCESS: Figure 2-8 illustrates a *constant-volume process* on the *p-v* and *T-s* coordinates. It is a nonflow process only and may be considered to be that process which takes place when a fluid is heated or cooled in a closed tank. The process illustrated in Fig. 2-8 is one where wet steam at condition 1 is heated at constant volume to superheated steam at condition 2. The energy equation for a nonflow process is

$$_1q_2 = u_2 - u_1 + \frac{p\Delta v}{J} \qquad \text{(2-9) Repeated}$$

But since $\Delta v = 0$

$$_1q_2 = u_2 - u_1 \qquad \text{(2-29)}$$

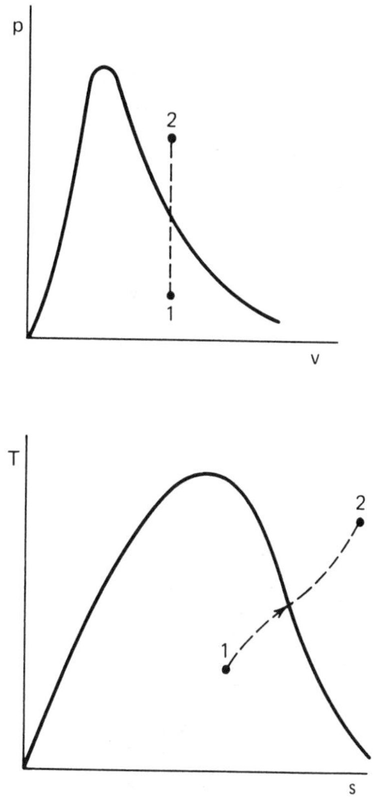

Figure 2–8

Constant volume process (no scale)

Illustrative Problem 2-7

A closed tank has a volume of 10 ft^3. It contains 1 lbm of saturated water at 170°F. If the water in the tank is heated until the tank is filled with saturated steam, determine the final pressure of the steam and the heat added.

Solution:

Since the tank volume is 10 ft^3, the specific volume of the saturated steam will be V/m = 10 ft^3/lbm. Interpolation in Table A-2 at 10 ft^3/lbm gives approximately 42.0 psia, the final pressure. By Eq. (2-29),

$$_1q_2 = u_2 - u_1$$

From Table A-1 at 170°F, find
$u_1 = u_f = 137.95$ Btu/lb
From Table A-2 at 42.0 psia,
find $u_2 = u_g = 1093.0$ Btu/lbm

$$_1q_2 = 1093.0 - 137.95$$

$$_1q_2 = 955.1 \text{ Btu/lbm} \quad \text{(added)}$$

CONSTANT-TEMPERATURE PROCESS: Figure 2-9 illustrates a constant-temperature process on the *p-v* and *T-s* coordinates. The constant-temperature process is typical of the condensation of steam in a steam-heating coil. The process shown indicates slightly superheated steam at condition 1 and saturated liquid (water) at condition 2. Please observe that in the wet region (liquid and vapor coexisting) the constant-temperature process is also a constant-pressure process. The energy equation is

$$_1q_2 = u_2 - u_1 + \frac{_1\text{wk}_2}{J} \qquad \text{(2-9) Repeated}$$

If the process takes place at constant pressure, as well as constant temperature (condensation), then

$$_1q_2 = h_2 - h_1 \qquad \text{(2-10) Repeated}$$

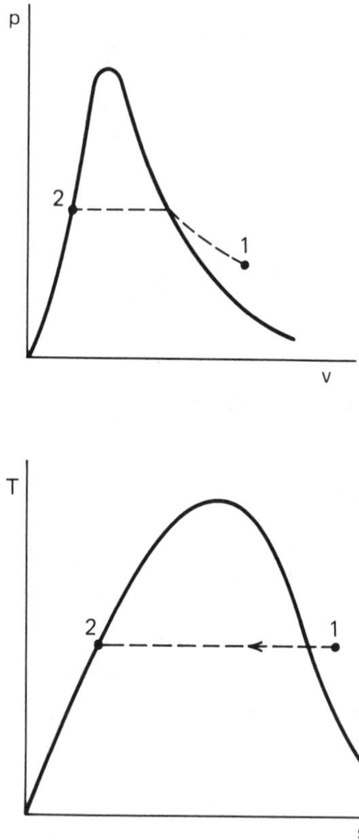

Figure 2–9

Constant temperature process (no scale)

Illustrative Problem 2-8

Steam, initially wet, containing 5 percent liquid and at a pressure of 15 psia, condenses in a steam-heating coil, and leaves as saturated liquid. How much heat is given up by the steam if the steam enters the coil at the rate of 10 lb/min, with condensate leaving at the same rate?

Solution:

By Eq. (2-10),

$$_1q_2 = h_2 - h_1 \quad \text{(Btu/lb of steam)}$$

For $\dot{m}$ (lb/min) of steam we have

$$_1\dot{q}_1 = \dot{m}(h_2 - h_1)$$

Since the steam is initially wet (quality = 1.0 − 0.05 = 0.95), the enthalpy h_1 must be calculated by use of Eq. (2-26)

$$h_1 = h_f + xh_{fg_1}$$

From Table A-2 at 15 psia, find $h_{f_1} =$ 181.19 and $h_{fg_1} = 969.7$

Then

$$h_1 = 181.19 + (0.95)(969.7)$$

$$h_1 = 1102.4 \text{ Btu/lbm}$$

Saturated water at the final condition has an enthalpy $h_2 = h_f = 181.19$ Btu/lbm. Therefore,

$$_1q_2 = h_2 - h_1 = 181.19 - 1102.4$$

$$_1q_2 = -921.2 \text{ Btu/lbm}$$

and

$$_1\dot{q}_2 = \dot{m}\,(-921.2) = 10\,(-921.2)$$

$$_1\dot{q}_2 = -9212 \text{ Btu/min} \quad \text{(removed)}$$

THROTTLING PROCESS: Figure 2-10 illustrates a throttling process on the *T-s* and *h-s* coordinates. The throttling process is typical of the process that takes place when a fluid passes through a partially open valve, which causes a local disturbance of the flow. It is a necessary component process involved in the vapor compression refrigeration cycle (expansion valve). In steam work the process may be used to determine the wetness, or quality of the steam. The process line 3-4 is the line showing the typical expansion process of liquid refrigerant through the expansion valve if the *T-s* diagram were for a refrigerant. Saturated liquid at point 3 enters the expansion valve and expands to the lov'er pressure point 4 in a constant-enthalpy process. Some of the liquid flashed to vapor, as indicated by the location of point 4 in the wet region.

The process line 1-2 indicates slightly wet steam at point 1 expanding to the lower pressure in a constant-enthalpy process. If the initial steam is not too wet, the expansion process

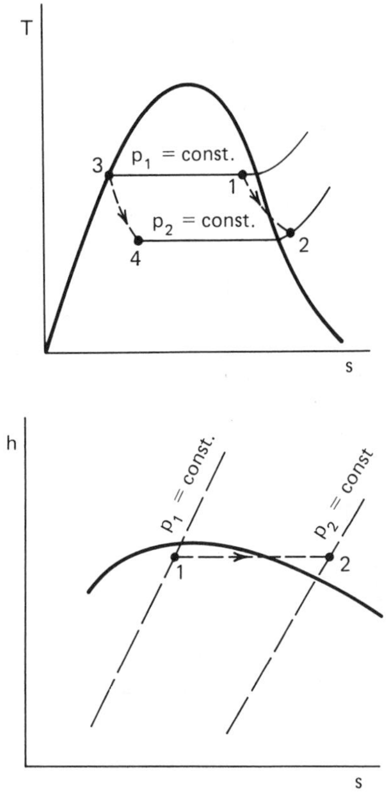

Figure 2–10

Throttling process (no scale)

may produce superheated steam at condition 2. This is the principle of operation of the throttling calorimeter, a device used to determine the quality of the steam. The throttling process takes place at constant enthalpy, or

$$h_1 = h_2 \qquad \textbf{\textit{(2-14) Repeated}}$$

Illustrative Problem 2-9

Steam is flowing in a steam pipe where the pressure is 160 psia. A throttling calorimeter attached to the pipe permits a sample of steam to expand to a pressure of 14.696 psia and the resulting steam temperature of 230°F. Determine the percent of moisture by weight flowing in the steam pipe.

Solution:

Referring to Table A-1 at a temperature of 230°F we find that the saturation pressure is 20.78 psia, which is greater than the actual pressure of 14.696 psia. Therefore, the steam at the final condition is superheated. From Table A-3 at $p_2 = 14.696$ psia and $t_2 = 230°F$, find $h_2 = 1159.2$ Btu/lbm.

From Table A-2 at $p_1 = 160$ psia, find $h_{f_1} = 336.16$, $h_{fg_1} = 859.8$. By Eq. (2-14),

$$h_1 = h_2$$

or

$$h_2 = 1159.2 = h_1 = h_{f_1} + x_1 h_{fg_1}$$

$$1159.2 = 336.16 + x_1(859.8)$$

$$x_1 = 0.957 \quad \text{or} \quad 95.7 \text{ percent quality}$$

$$\text{percent moisture} = 100 - 95.7$$
$$= 4.3 \text{ percent}$$

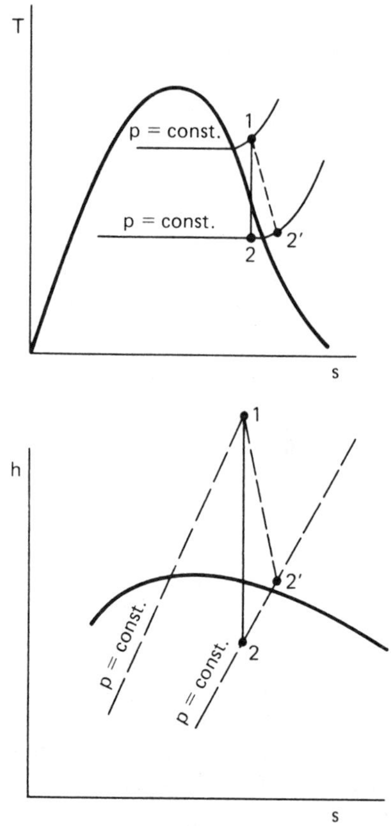

Figure 2–11

Isentropic and adiabatic processes (no scale)

ADIABATIC PROCESS: The adiabatic process is one of the more important processes involving the compression or expansion of gases where the primary purpose is to determine the work required to compress the gas or the work obtained by the expansion of a gas. The name "adiabatic" means no heat transfer ($q = 0$). If the process is also frictionless (an ideal), the process is isentropic ($\Delta s = 0$) and it becomes a reversible process.

Figure 2-11 illustrates an isentropic process (1-2). It also shows an irreversible adiabatic process (1-2′) where some of the available energy at point 1 leaves as lost work. The coordinates shown are the *T-s* and *h-s* coordinates for steam, and the two processes are typical of the expansion of steam through a turbine. For process 1-2 steam enters the turbine at condition 1 and expands isentropically to a final condition 2. If the process were considered nonflow, the energy equation is

$${}_1q_2 = u_2 - u_1 + \frac{{}_1\text{wk}_2}{J}$$

But for an adiabatic process, ${}_1q_2 = 0$; therefore

$$\frac{{}_1\text{wk}_2}{J} = u_1 - u_2 \qquad \textit{(2-30)}$$

If the process were considered to be steady flow, energy equation (2-8) reduces to

$$\frac{{}_1\mathrm{wk}_2}{J} = h_1 - h_2 \qquad (2\text{-}31)$$

For process 1-2′ steam enters at condition 1 but expands without friction to condition 2′. The energy equations for the nonflow and steady-flow processes are

$$\frac{{}_1\mathrm{wk}_2}{J} = u_1 - u'_2 \qquad \text{(nonflow)} \qquad (2\text{-}32)$$

$$\frac{{}_1\mathrm{wk}_2}{J} = h_1 - h'_2 \qquad \text{(steady flow)} \qquad (2\text{-}33)$$

Compression of a gas, such as refrigerant vapor, in a piston compressor or a centrifugal compressor will be discussed in the following chapter.

CHAPTER 2

Review

2-1. A gas initially at 150°F and having C_p = 0.23 Btu/lbm°R and C_v = 0.170 Btu/lbm°R is placed within a cylinder containing a frictionless piston. If 900 Btu are added to 15 lbm of the gas in a nonflow constant-pressure process, determine the final gas temperature and the work done on or by the gas.

2-2. Air enters a gas turbine at 150 psia and 540°F and leaves at 15 psia and 140°F. If the average value of the constant-pressure specific heat is 0.24 Btu/lbm°R, determine the work output of the turbine per lbm of working fluid.

2-3. A gas flows in a pipe. At an upstream section (1), the gas properties are p_1 = 100 psia, t_1 = 950°F, and v = 4.0 ft^3/lbm. At a downstream section (2), the gas properties are p_2 = 76 psia, t_2 = 580°F, and v = 3.86 ft^3/lbm. The assumed specific heat at constant volume is 0.32 Btu/lbm°R. If no work is done and if velocities are small, determine the magnitude and direction of the heat transfer. Assume pipe is horizontal.

2-4. A steam boiler is required to produce 5000 lbm/hr of superheated steam at 200 psia and 820°F when supplied with feed water at 200 psia and 100°F. How much heat was added to the water to convert it to steam at these conditions.

2.5. Water flows through a heat exchanger and heat is transferred from the water to heat air. The entering water is at 300°F and 100 psia. The leaving water is at 200°F and 80 psia. Determine the heat transferred from the water.

2-6. Dry saturated steam at 200 psia flows adiabatically in a constant-diameter pipe. At the pipe outlet the pressure is 100 psia and the temperature is 400°F. (a) How much energy was transferred; (b) was this into or out of the system; (c) was this heat or work?

2-7. A sample of steam at 100 psia passes through a throttling calorimeter. The pressure and temperature in the calorimeter are 14.696 psia and 240°F, respectively. What are the quality and moisture content of the 100 psia steam?

2-8. A refrigerant gas is compressed from 30 to 180 psia, and the specific volume decreased from 9 to 2 ft^3/lb. Determine the polytropic exponent of compression.

2-9. A refrigerant gas has a specific volume of 10 ft^3/lbm and is compressed from 10 psia at 66 psia. If n = 1.25, determine the specific volume at the end of compression.

2-10. One pound of saturated liquid ammonia is expanded through a throttling valve from a condenser pressure of 180.6 psia to an evaporator pressure of 38.51 psia. Determine: (a) the refrigerant quality at discharge, (b) the change in specific volume, and (c) the final quality if the refrigerant is subcooled 10°F before expansion.

BIBLIOGRAPHY

1. Jordan, R. C., and Priester, G. B., *Refrigeration and Air Conditioning,* 2nd ed., Prentice-Hall, Inc., Englewood Cliffs, NJ, 1956.

2. Granet, Irving, *Thermodynamics and Heat Power,* Reston Publishing Company, Inc., Reston, VA, 1974.

3. Burghardt, M. D., *Engineering Thermodynamics with Applications,* Harper and Row, Publishers, New York, 1978.

4. Van Wylen, G. J., and Sontag, R. E., *Fundamentals of Classical Thermodynamics,* John Wiley and Sons, Inc., 1973.

Chapter 3

Refrigeration for Comfort Cooling

3-1 INTRODUCTION

In Chapter 2 consideration was given to a heat cycle which produces useful work from a heat source. By simply rearranging the sequence of events in such a cycle, it is possible (in principle) to remove heat from a region of lower temperature and to deliver it to a region of higher temperature by the input of mechanical work. This removal of heat by the use of mechanical work has been called *refrigerating effect*. In more general terms refrigeration can be defined as the art of maintaining a body or space at a temperature below that of its surroundings or, alternatively, as the removal of heat from a place in which it is undesirable to a place in which it is not.

In the artificial production of low temperatures ranging from normal ambient air temperature down to absolute zero, one or more refrigeration cycles or processes may be involved. For cooling in the temperature range down to $-200°F$, *vapor compression* refrigeration using *reciprocating, rotary,* or *centrifugal* compressors may be used. For some applications within this band, air-cycle refrigeration, steam-jet refrigeration, or absorption refrigeration may be applied. Such systems are commonly used for the "production of cold." For the production of extremely low temperatures down to absolute zero, irreversible or reversible adiabatic expansion of a gas, vaporization of a liquified gas, or magnetic cooling may be used. A study of these specialized and limited processes is beyond the scope of this text.

Although refrigeration is primarily an application of thermodynamics, many other phases of engineering are involved in the design, manufacture, application, and operation of refrigeration systems. Some knowledge of the chemistry of refrigerants is desirable in determining their reactions with metals and other materials with which they may come in contact. The thermodynamic properties of refrigerants must be known before the cycle analyses can be made. A study of evaporators and condensers, those portions of the system used for the absorption and rejection of heat, respectively, involves the fields of heat transmission and fluid mechanics. In addition, steady-state and, in some cases, periodic and transient heat transfer are involved in the determination of cooling-load requirements. The calculations of cooling loads also requires a knowledge of psychrometrics. Fluid mechanics forms the basis for the sizing of the lines through which the gaseous and liquid refrigerants must flow. The design of reciprocating, rotary, or centrifugal compressors or of an air-cycle expansion turbine involves a variety of machine-design problems. The physical capacity of a compressor or expander will be determined from thermodynamic factors, but the physical design must also involve structural considerations.

3-2 STANDARD RATING OF REFRIGERATING MACHINES

The standard unit used in the rating of refrigerating machines, condensing units, and other parts of a refrigeration system is the *ton of refrigeration,* defined as the removal of heat at the rate of 12,000 Btu/h or 200 Btu/min. The *ton-day of refrigation,* infrequently used, is defined as the heat removed by 1 ton of refrigeration operating for one day, or 288,000 Btu. These somewhat ambiguous terms have their origin in the concept of the amount of heat absorbed by 1 ton of ice when melting from the solid to the liquid phase at 32°F. This derivation assumes a latent heat of ice of 144 Btu/lb, whereas it is actually slightly less than this. However, no error is introduced in the calculations because the units defined in terms of heat removal are no longer associated with the melting of a definite weight of ice.

3-3 ELEMENTARY VAPOR COMPRESSION REFRIGERATION CYCLE WITH RECIPROCATING COMPRESSOR

At atmospheric pressure liquid ammonia evaporates at −28°F (saturation temperature corresponding to 14.7 psia), and under these conditions 1 lb of liquid ammonia in changing to vapor absorbs 589.3 Btu (latent heat of evaporation). Thus, the simplest form of a vapor refrigeration system consists of an open vessel containing a liquid refrigerant such as ammonia. The ammonia evaporates at temperatures below those surrounding the container and in so doing absorbs the heat from the surroundings. However, such an uncontrolled refrigeration system is uneconomical since the refrigerant is not recovered and the vaporizing temperature is limited to that corresponding to the atmospheric pressure. Also, such a system would be a hazard to life because the ammonia vapor is extremely hazardous.

With the proper auxiliary equipment, however, the refrigerant can be recovered and reused in a cyclic process; moreover, the temperature of evaporation of the refrigerant can be controlled by controlling the pressure. Thus, if liquid ammonia is maintained at a pressure of 30.42 psia, the saturation or evaporating temperature is 0°F, and the latent heat of vaporization is 568.9 Btu/lb; if the absolute pressure is 59.74 psia, the evaporating temperature is 30°F, and the latent heat of evaporation is 544.8 Btu/lb. The refrigerant change from liquid to vapor or from vapor to liquid may be controlled by controlling the pressure of the refrigerant. If the ambient temperature surrounding the refrigerant and its container is above the saturation temperature corresponding to the refrigerant pressure, then evaporation and, consequently, absorption of heat takes place. If the refrigerant is already in the vapor state and if the temperature surrounding the refrigerant and its container is below the saturation temperature corresponding to the refrigerant pressure, condensation occurs. Other standard refrigerants serve as well as ammonia, and the pressure, volume, temperature, enthalpy, and entropy may be determined from standard refrigerant tables.

Figure 3-1 represents diagramatically a complete compression refrigeration system. The system is closed. That is, the refrigerant flows through the system components in sequence in steady-flow processes, alternating between the liquid and gaseous phases. The *receiver,* in which the high-pressure liquid is stored, maintains a constant supply of refrigerant for the system. On leaving the receiver, the liquid refrigerant flows through the *expansion valve,* which is essentially a needle valve. The *compressor* maintains a difference in pressure between the vaporator and condenser. Without the expansion valve this difference in pressure could not, of course, be maintained. The expansion valve separates the high-pressure part

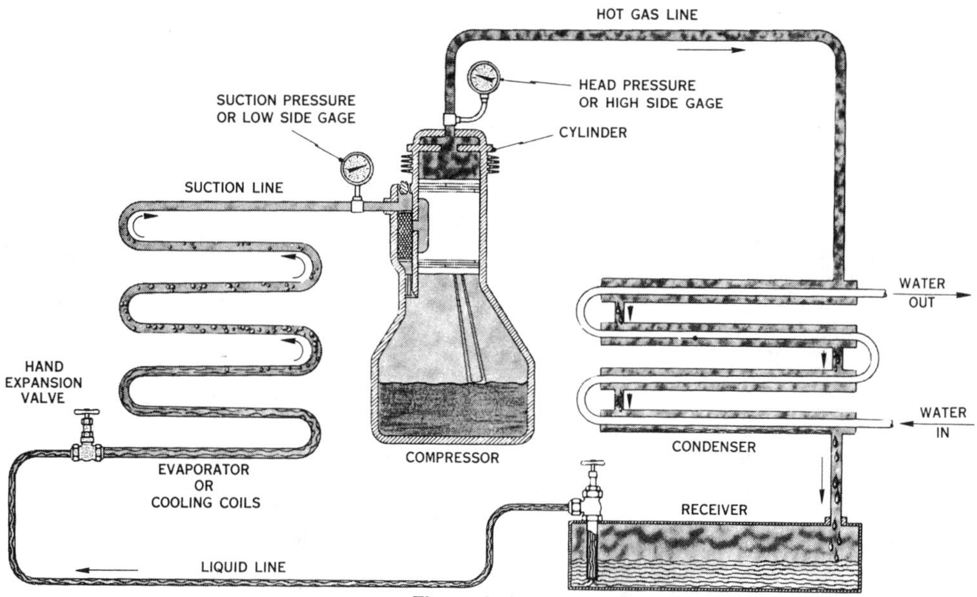

Figure 3–1

Mechanical vapor-compression refrigeration system (Courtesy of the Trane Company)

of the system from the low-pressure part. It acts as a pressure-reducing valve because the pressure of the liquid flowing through it is lowered. Only a small trickle of refrigerant fluid flows through the valve into the evaporator. In fact, the valve is always so adjusted that only just as much liquid can pass through it as can be vaporized in the evaporator.

The liquid that flows through the evaporator is entirely vaporized by the heat flowing through the walls of the evaporator. This heat is absorbed from the air or other fluid being cooled. After leaving the evaporator, the vaporized refrigerant flows to the compressor where its pressure and temperature are raised to such a point that it can be condensed by relatively warm water or air. After being compressed, the vapor flows to the *condenser*. Here the walls of the condenser tubes are cooled by water or air and, as a result, the vapor is liquified. Latent heat and any superheat of the vapor is transferred from the vapor to the water or air through the walls of the condenser tubes. From the condenser the liquid refrigerant flows back to the receiver and the cycle is repeated.

In Fig. 3-1, then, we may identify four essential components of the refrigeration cycle: expansion valve, evaporator, compressor, and condenser. The receiver is essential for storage of liquid refrigerant, but no thermodynamic changes occur in the receiver. These components are represented in Fig. 3-2 in block form, which will be used to illustrate the vapor compression cycle in problems to follow.

3-4 COMMON REFRIGERANTS

Any substance that absorbs heat through expansion or vaporization may be termed a *refrigerant*. In the broadest sense of the word, the term *refrigerant* is also applied to such secondary cooling mediums as brine solutions or cold water. As commonly interpreted, however, refrigerants include only those working mediums which pass through the cycle of evaporation, recovery, compression, and condensation.

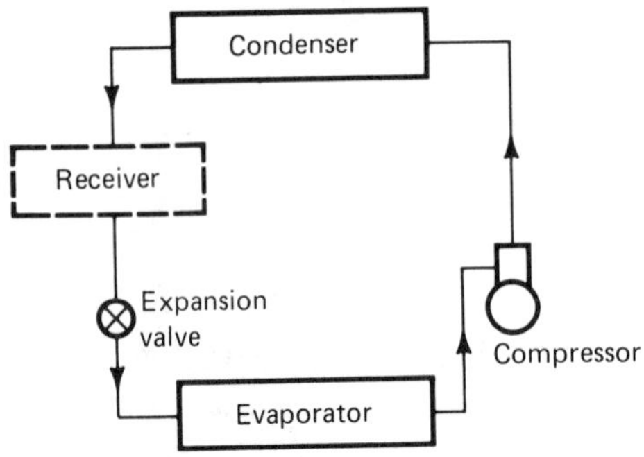

Figure 3–2

Vapor compression cycle system components

Thus, circulating cold mediums are not primary refrigerants, nor are cooling mediums such as ice and solid carbon dioxide.

Desirable refrigerants are those which possess chemical, physical, and thermodynamic properties that permit their efficient application and service in practical designs of refrigerating equipment. In addition, if the volume of the charge is large, there should be no danger to health and property in case of its escape. A great variety of substances, among which are butane, carbon tetrachloride, ethane, hexane, methane, pentane, propane, and chloroform, have been applied to refrigeration systems but found to be of little practical use. These and similar materials are either highly explosive and flammable or possess other combinations of undesirable properties.

Many of the refrigerants used in the United States during past years are listed in Table 3-1, along with their chemical formulas and a few of their more important thermal and physical properties. Those with widest application at present are ammonia, the Freon® group, methyl chloride, and water vapor. The fields of application for which the refrigerants listed in Table 3-1 have been commonly selected are summarized in the following paragraphs.

†*Air*—Air was one of the earliest refrigerants and was widely used even as late as World War I whenever a completely nontoxic medium was needed. Although air is free of cost and completely safe, its low coefficient of performance makes it unable to compete with the present-day nontoxic refrigerants. Only when operating efficiency is secondary, as in aircraft refrigerants, does air find any modern application as a refrigerant.

Ammonia—Ammonia is one of the oldest and widely used of all refrigerants. It is highly toxic and flammable. It has a boiling point of −28.0°F and a liquid specific gravity of 0.684 at atmospheric pressure. Its greatest application has been in large industrial and commercial reciprocating compression systems where high toxicity is secondary. For industrial application it has been found valuable because of its low volumetric displacement, low cost, low weight of liquid circulated per ton of refrigeration, and high efficiency. However, in recent years there has been some decrease in the volume of new installations using ammonia. It is also widely used as the refrigerant in absorption systems.

Carbon Dioxide—Carbon dioxide, a colorless and odorless gas, is heavier than air. It has a boiling point of −109.3°F and a liquid specific gravity of 1.56 at atmospheric pressure. It is nontoxic and nonflammable but has extremely high operating pressures. Because of the high horsepower requirements per ton of refrigeration and the high operating pressures, it has received only limited usage. In former years it was selected for marine refrigeration, for theater air-conditioning systems, and for hotel and institutional refrigeration systems instead of ammonia because it is nontoxic. The Freon group has largely supplanted it for these applications. At the present time its use is limited primarily to the manufacture of dry ice (solid carbon dioxide). The cycle efficiency can be improved by the use of two-stage compression, and when applied in this manner carbonic refrigeration has received some usage. It has also been used for very low temperature work by adapting it to a binary or "cascade" cycle in which the efficiency is improved by using carbon dioxide in the low-temperature stage and ammonia or some other refrigerant in the high-temperature stage.

The Freon and Genetron® Refrigerants—In 1928

Table 3-1*

Physical and Thermal Properties of Common Refrigerants

Refrigerant	Chemical Formula	Boiling Point, F	Freezing Point, F	Critical Point: Temper-ature, F	Critical Point: Pressure, lb per sq in. abs.	Specific Gravity of Liquid at Atmos. Press.	Specific Heat of Liquid Avg. 5 F to 86 F
Ammonia	NH_3	−28.0	−107.9	271.4	1657.0	0.684	1.12
Azeotropic mixture Freon-12 and unsymmetrical difluoroethane (Carrene-7)	78.3% F-12 26.2% G-100	−28.0	−254.0	221.1	631.0	.....	.....
Bromotrifluoromethane (Kulene-131)	CF_3Br	−73.6	−226.0	153.5	587.0	.....	0.19
Carbon dioxide	CO_2	−109.3	−69.9	87.8	1069.9	1.560	0.77
Dichlorodifluoromethane (Freon-12: Genetron-12)	CCl_2F_2	−21.6	−252.0	233.6	596.9	1.480	0.23
Ethylene	C_2H_4	−155.0	−272.0	48.8	731.8	.....	.....
Isobutane	C_4H_{10}	10.3	−229.0	272.7	537.0	0.549	0.62
Methyl chloride	CH_3Cl	−10.8	−144.0	289.4	968.7	1.002	0.38
Methylene chloride (Carrene-1)	CH_2Cl_2	105.2	−142.0	480.0	670.0	1.291	0.33
Monochlorodifluoromethane (Freon-22: Genetron-141)	$CHClF_2$	−41.4	−256.0	204.8	716.0	1.411	0.30
Sulfur dioxide	SO_2	14.0	−103.9	314.8	1141.5	1.357	0.34
Trichloromonofluoromethane (Freon-11, Genetron-11)	CCl_3F	74.7	−168.0	388.4	635.0	1.468	0.21
Trichlorotrifluoroethane (Freon-113)	$CCl_2F\text{-}CClF_2$	117.6	−31.0	417.4	495.0	1.559	0.21
Water	H_2O	212.0	32.0	706.1	3226.0	1.000	1.00

**From* Refrigeration and Air Conditioning *by R. C. Jordan and G. B. Priester, 2nd ed., Prentice-Hall, Englewood Cliffs, N.J., 1956.*

Charles Kettering and Dr. Thomas Migley, Jr., instigated research to find a nontoxic, nonflammable refrigerant. These efforts culminated in the development of a series of fluorinated compounds using ethane and methane as bases. These fluorinated hydrocarbons, commonly known by the trade names Freon and Genetron, are probably the most important group of refrigerants in use today. The commonest of these are: trichloromonofluoromethane (Freon-11, Genetron-11), CCl_3F; dichlorodifluoromethane (Freone-12, Genetron-12), CCl_2F_2; monochlorotrifluoromethane (Freon-13), $CClF_3$; tetrafluoromethane (Freon-14), CF_4; $CHCl_2F$; monochlorodifluoromethane (Freone-22, Genetron-141), $CHClF_2$; trichlorotrifluoroethane (Freon-113), $CCl^2F\text{-}CC_3F_2$; and dichlorotetrafluoroethane (Freon-114), $C_2Cl_2F_4$. They are almost universally referred to by their trade names. Freon-11 or Genetron-11 has a boiling point of 74.7°F; Freon-12 or Genetraon-12, −21.6°F; Freon-13, −114.5°F; Freon-14, −198.2°F; Freon-22 or Genetron-141, −41.4°F; Freon-113, 117.6°F; and Freon-114, 38.4°F. The entire group is clear and water-white in color and has a somewhat ethereal odor similar to that of carbon tetrachloride. They are all nonflammable and for all practical purposes nontoxic. Freon-11, or Genetron-11, is widely used for centrifugal compression refrigeration. In this field it has almost completely supplanted dichloroethylene and methylene chloride. Freon-12, or Genetron-12, the most widely used of the group, is generally applied to reciprocating compression refrigeration. It has received its widest case in air-conditioning applications where nontoxic, nonflammable refrigerants are required. Freon-114 has been applied by several manufacturers to rotary compressors in domestic refrigerators and has also been used experimentally for absorption refrigeration. Freon-22 or Genetron-141 has been developed for reciprocating compressor applications below −20 °F, and Freon-13 and Freon-14 are intended for extremely low-temperature usage. Freon-22, however, may receive wide application in higher-temperature installations in the future if increased volume of production will permit a lowering of cost. Its favorable characteristics may allow it to compete successfully with ammonia and possible with Freon-12 for general refrigeration purposes.

Methyl Chloride—Methyl chloride, CH_3Cl, is a colorless liquid with a faint, sweet, nonirritating odor. It was introduced about 1920 in the United States for refrigeration purposes and is now widely used. It replaced ammonia and carbon dioxide for many new installations in the 1920s and early 1930s and was widely used during World War II as a substitute for Freon, then unavailable. Its use, however, appears to be definitely on the decline. It has a boiling point of −10.6°F and a liquid specific gravity of 1.002 at atmospheric pressure. It is to a certain degree both flammable and toxic. Methyl chloride has been used in domestic units with both reciprocating and rotary compressors and in commercial units with reciprocating compressors up to approximately 10-ton capacity.

Sulfur Dioxide—Sulfur dioxide, SO_2, a colorless gas or liquid, is extremely toxic and has a pungent irritating odor. It is nonexplosive and nonflammable and has a boiling point of 13.8°F and a liquid specific gravity of 1.36. It was used extensively in domestic systems built in the 1930s. It has been applied to both reciprocating and rotary compressors. With such applications the volume of refrigerant charge was small, and there was little danger of fatal concentrations resulting through refrigerant leakage. Sulfur dioxide has also been used to a considerable extent in small-tonnage commercial machines. However, the volume of new units using sulfur dioxide as a refrigerant is small.

Water Vapor—Water vapor, H_2O, is the cheapest and probably the safest of all refrigerants. However, because of its high freezing temperature of 32°F, it is limited in application to high-temperature refrigeration. Its application has been to steam-jet refrigeration and to centrifugal compression refrigeration. It is, of course, nontoxic, nonflammable, and nonexplosive. Because of its high-temperature limitations and its complete safety, it has been used principally for comfort air-conditioning applications and to some extent for water cooling.

Hydrocarbon Refrigerants—Many of the hydrocarbons are used as refrigerants in industrial installations where they are frequently available at low cost. These include butane (C_4H_{10}), isobutane (C_4H_{10}), propane (C_3H_8), propylene (C_3H_6), ethane (C_2H_6), and ethylene (C_2H_4). However, they are all highly flammable and explosive, and therefore their use has been limited principally to the chemical and refining industries where similar hazards already exist. They all possess satisfactory thermodynamic properties.

Several of the hydrocarbons, including propane, ethane, and ethylene, exhibit promise for use as refrigerants at −100°F or lower. Ethylene has a saturation temperature of −176.8°F at a pressure of 6.75 psia and makes an excellent refrigerant when used in a cascade system with an auxiliary cycle.

Isobutane was used by one manufacturer of domestic refrigerators in a rotary compressor until 1933, but it has no such application at the present time. Propane has received limited use as both a motive fuel and the refrigerant in a refrigerated truck unit. Such a cycle is termed *transitory,* the propane first passing through the evaporator and then to the engine.

Halogenated Hydrocarbons and Other Refrigerants—Chemical compounds formed from methane (CH_4) and ethane (C_2H_6) by the substitution of chlorine, fluorine, or bromine for part of their hydrogen content are termed *halogenated hydrocarbons*. Many refrigerants, including methyl chloride and the Freon and Genetron groups, are included in this class. Some of the others are described below.

Dichloroethylene (Dielene), $C_2H_2Cl_2$, is a colorless liquid with a boiling point of 118.0°F and a liquid specific gravity of 1.27. It has a strong, nonirritating odor similar to that of chloroform and is to a limited extent both toxic and explosive. Its principle application has been in centrifugal compression systems in which it received limited usage in the 1920s.

Ethyl chloride, C_2H_5Cl, is a colorless liquid with a boiling point of 54.5°F. It is to a certain degree both toxic and flammable and is similar in many respects to methyl chloride but with lower operating pressures. It is not used in refrigerating equipment at the present time but has in the past been selected for both rotary and reciprocating compressors.

Methylene chloride, CH_2Cl_2, is a clear, water-white liquid with a sweet, nonirritating odor similar to that of chloroform. It has a boiling point of 103.6°F and a liquid specific gravity of 1.291. It is nonflammable and nonexplosive and is toxic only at comparatively high concentrations. Methylene chloride is used by several manufacturers in domestic rotary compressors and has found some use commercially for absorption refrigeration because of its low toxicity and absence of fire hazard. It has been used successfully in centrifugal compressors but has been supplanted in recent years by Freon-11 (Genetron-11).

Bromotrifluoromethane (Kulene-131), CF_3Br, is a nontoxic, nonflammable, noncorrosive, halogenated hydrocarbon particularly suitable for low-temperature refrigeration. It is nicely adapted to use in a two-stage cascade system in the lower stage with an evaporatory temperature of about −75°F. Under such conditions the evaporator pressure is approximately atmospheric and the compressor displacement desirably low.

Azeotropes - Mixtures of refrigerants which do not separate into their components with pressure or temperature changes and have fixed thermodynamic properties unlike those of their components are classed as azeotropes. Several are known, but only one is in common use. This is Carrene-7, consisting of 73.8 percent Freon- or Genetron-12 (CCl_2F_2) and 26.2 percent ethylidene fluoride or Genetron-100 (CH_2CHF_2). This particular azeotrope mixture of refrigerants is important since it has a theoretical refrigeration capacity and horsepower requirement approximately 18 percent greater than that of Freon-12, with the same compressor displacement and with the same evaporator and condenser temperatures. A Freon-12 system designed for 60-Hz current may be shifted to 50-Hz current and Carrene-7 to result in approximately the same refrigerating capacity and evaporator and condenser conditions.

---

†Extracted from *Refrigeration and Air Conditioning* by R. C. Jordan and G. B. Priester, Prentice-Hall, Inc., Englewood Cliffs, NJ, 2*nd* Ed. 1956, p. 81-87 with permission.

Freon-12 and Freon-22 are good refrigerants for air-conditioning systems. They operate at moderate pressures and are easily handled. Freon-22 was originally developed for low-temperature applications since it has a boiling point of −41.4°F at atmospheric pressure. With Freon-22 the cfm per ton of capacity is less than with Freon-12, so design engineers have recently applied it extensively to refrigeration systems for comfort cooling to achieve greater compactness.

Different refrigerants can be used to vary the capacity of a compressor. For example, a line of reciprocating compressors may have capacity ratings of 5, 10, 15, and 20 tons with Freon-12. If Freon-22 is used, the capacity of the same compressors will be 8, 15, 23, and 30 tons. *It is not recommended that the refrigerant in a system be changed without approval of the compressor manufacturer.*

3-5 COMPARISON OF REFRIGERANTS

In the choice of a refrigerant it should be remembered that as yet no one substance has proved the ideal working medium under all operating conditions. The characteristics of some refrigerants make them desirable for use with reciprocating compressors. Other refrigerants are best adapted to centrifugal or rotary compressors. In some applications toxicity is of negligible importance, whereas in others, such as comfort cooling, a nontoxic and nonflammable refrigerant is essential. The requirements of a refrigerant to be used for low-temperature work are different from those for high-temperature applications. Therefore, in selecting the correct refrigerant, it is necessary to determine those properties which are most desirable and to choose the one most closely approaching the ideal for the particular application. Other properties of refrigerants which influence the selection are:

1. The heat of vaporization of the refrigerant should be high. The higher the h_{fg} the greater the refrigerating effect per pound of refrigerant circulated.
2. The specific heat of the liquid should be low. The lower the specific heat of the liquid, the less heat it will pick up for a given change in temperature during either throttling or in flowing through the piping, and consequently the greater the refrigerating effect per pound of refrigerant.
3. The specific volume of the refrigerant should be low to minimize the work required per pound of refrigerant circulated.
4. The critical temperature of the refrigerant should be higher than the condensing pressure to prevent excessive power consumption.
5. It is desirable that both condenser and evaporator pressures be positive, yet not too high above atmospheric pressure.

In addition to the foregoing it is necessary to consider items of toxicity, corrosiveness, dielectric strength for hermetically sealed systems, viscosity, thermal conductivity, explosiveness, effects on food products, stability, inertness, and cost. No one refrigerant has been found to meet all of these requirements.

The Carnot-cycle coefficient of performance (discussed in a later section) is 5.74 for an ideal refrigerant operating between 86°F condenser temperature and 5°F evaporator temperature. The coefficient of performance for common refrigerants operating between these same temperatures on the vapor refrigerant cycle are arranged in decreasing order in Table 3-2. The last column of this table also shows the theoretical horsepower required per standard ton of refrigeration. If the volumetric and thermal efficiences remain constant, the horsepower required per standard ton is inversely proportional to the coefficient of performance.

Table 3-2*

Coefficient of Performance and Power Requirement for Various Refrigerants

Refrigerant	Coefficient of Performance (Standard Cycle, 86°F Condenser, 5°F Evaporator)	Efficiency (% Carnot Cycle)	HP per ton (Standard Cycle, 86°F Condenser 5°F Evaporator)
Carnot	5.74	100.0	0.82
Freon-11, Genetron-11	5.09	88.8	0.93
Methyl chloride	4.90	85.3	0.96
Sulfur dioxide	4.87	84.9	0.97
Freon-113	4.79	83.5	0.96
Ammonia	4.76	83.0	0.99
Freon-12, Genetron-12	4.70	82.0	1.00
Freon-22, Genetron-141	4.66	81.3	1.01
Carrene-7	4.61	80.4	1.02
Kulene-131	4.25	74.1	1.03
Carbon dioxide	2.56	44.6	1.84

**Extracted from* Refrigeration and Air Conditioning *by R. C. Jordan and G. B. Priester, Prentice-Hall, Inc., Englewood Cliffs, New Jersey. 2nd Ed., 1956, pg. 92 with permission.*

Practically all refrigerants in common use have approximately the same coefficient of performance and horsepower requirements. The one exception shown in Table 3-2 is carbon dioxide, with a power requirement of 1.84 hp per ton and a coefficient of performance of 2.56. These poor properties may be attributed directly to carbon dioxide's very low critical point.

3-6 TABLES OF REFRIGERANT PROPERTIES

The properties of various refrigerants can be tabulated in exactly the same way as the properties of steam (see Tables A-1, A-2, and A-3, Appendix A). The properties of saturated and superheated ammonia are tabulated in Tables B-1 and B-2, respectively (see Appendix B). The properties of saturated and superheated Freon-12 are tabulated in Tables B-3 and B-4, respectively, and the properties of Freon-22 are tabulated in Tables B-5 and B-6, respectively.

In the first column of Tables B-1, B-3, and B-5 are listed the saturation temperatures corresponding to the various pressures shown in the second column. The volume, density, enthalpy, and entropy are given in the remaining columns. Superheat tables (B-2, B-4, and B-6) list values of volume, enthalpy, and entropy for various combinations of pressures and temperatures. Note that pressures are given in both pounds per square inch absolute (psia) and in pounds per square inch gauge (psig), or in some cases, inches of mercury gauge when so indicated. These tables of refrigerant properties are used in exactly the same way as those listing steam properties.

Before proceeding with illustrative problems, we must consider the accuracy level at which we should be working when using tabular data for refrigerant properties. We may note that the fluid properties listed in the tables indicate an accuracy level of at least three decimal places. Whenever tabular data is available, it is recommended that the *full table value* be inserted into equations where indicated and used to calculate the results. At the end of such calculations, the answer may be rounded-off, if this seems advisable.

Illustrative Problem 3-1

One pound of liquid Freon-12 at 20°F is to be heated at constant pressure to saturated liquid at 70°F and is then vaporized to produce saturated vapor. How much heat must be added to the refrigerant?

Solution:

Since this is a constant-pressure process, the heat addition will be equal to the change in enthalpy. From Table B-3:

At 20°F find $p_1 = 35.736$ psia (saturation pressure)

$h_{f_1} = 12.863$ Btu/lb

At 70°F find $p_2 = 84.888$ psia (saturation pressure)

$h_{g_2} = 84.359$ Btu/lb

At condition 1 the liquid is subcooled because the saturation pressure is lower than the constant pressure for the process, which is 84.888 psia.

The heat addition would be

$$q_{in} = h_{g_2} - h_{f_1} = 84.359 - 12.863$$

$$q_{in} = 71.496 \text{ Btu/lb} \quad \text{(say 71.5 Btu/lb)}$$

Illustrative Problem 3-2

The gauge on a receiver tank containing Freon-22 indicates a pressure of 190 psig. If the liquid in the tank is saturated, what is its temperature?

Solution:

From Table B-5 the gauge pressure of 190.18 psig may be found in column 3. The saturation temperature corresponding to this pressure is 98°F, found in column 1. Therefore, the saturation temperature at 190 psig is slightly lower than 98°F, about 97.9°F by interpolation.

Illustrative Problem 3-3

Freon-12 leaves a compressor and enters a condenser at 185.30 psig and at a temperature of 190°F. What is the enthalpy of the refrigerant?

Solution:

Referring to Table B-3, we may find a gauge pressure of 185.94 psig in column 3, and the corresponding saturation temperature is 132°F. Since the problem states a temperature of 190°F, the fluid must be superheated.

Refer to Table B-4. We may note that the pressure is in psia. Therefore, the pressure for our problem is 185.3 + 14.7 = 200 psia. Locating a temperature of 190°F, we may read an enthalpy of 100.78 Btu/lb.

3-7 PHASE DIAGRAMS

A good working knowledge of the vapor compression refrigeration cycle requires an intensive study not only of the individual processes that make up the cycle but also of the relationships that exist between the several processes and of the effects that changes in any one process in the cycle have on all the other processes in the cycle. This is greatly simplified by the use of charts and diagrams upon which the complete cycle may be shown graphically. Graphical representation of the refrigeration cycle permits the desired simultaneous consideration of all the various changes in the condition of the refrigerant which occur during the cycle and the effect that these changes have on the cycle without the necessity of holding in mind all the different numerical values involved in cyclic problems.

The diagrams frequently used in the analysis of the vapor compression cycle are the temperature-entropy (*T-s*) diagram and the pressure-enthalpy *(p-h)* diagram. Of the two, the pressure-enthalpy diagram seems to be the most useful. In Chapter 2 phase diagrams were discussed where the concern was water in its various phases. Figure 3-3 shows sketches of the *T-s* and *p-h* diagrams for a refrigerant such as Freon-12 showing only the areas of concern to us for refrigeration cycle analysis.

The *p-h* chart shows graphically the table values of refrigerant properties. Thus, it is easy to visualize the changes that take place as the refrigerant flows from one part of the cycle to another. Figures B-1, B-2, and B-3 (see Appendix B) are *p-h* diagrams for ammonia, Freon-12 and Freon-22, respectively. A careful study of Figs. 3-3, B-1, B-2, and B-3 should be made to determine the lines representing constant properties of the refrigerants, the lines separating the various phase areas.

Horizontal lines on the *p-h* diagram are lines of constant pressure, and vertical lines are lines of constant enthalpy. (Note the "scale change" on the enthalpy scale.) The curved line marked "saturated liquid" is a plot of the pressure and enthalpy of the saturated liquid. The curved line marked "saturated vapor" is a plot of the pressure and enthalpy of saturated vapor. Marked on the saturated-liquid and saturated-vapor curves are the values of the saturation temperature corresponding to the pressure. Lines of constant temperature are vertical in the subcooled liquid region, horizontal in the liquid-vapor region, and curve downward in the superheat region. In the liquid-vapor mixture region lines of constant pressure and of constant temperature are horizontal because the liquid is changing phase (boiling). As long as the pressure is constant during boiling, the temperature must also be constant.

The *p-h* diagram is of value in showing graphically the processes of a refrigeration cycle. However, it must be realized that, because of the scales used, it is difficult to read accurate thermodynamic properties from the diagram. It can give approximate values, which are sufficiently precise for many problems, but table values should be used whenever possible.

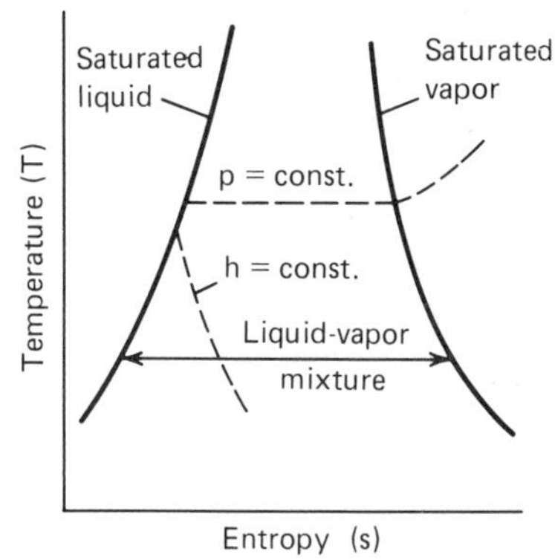

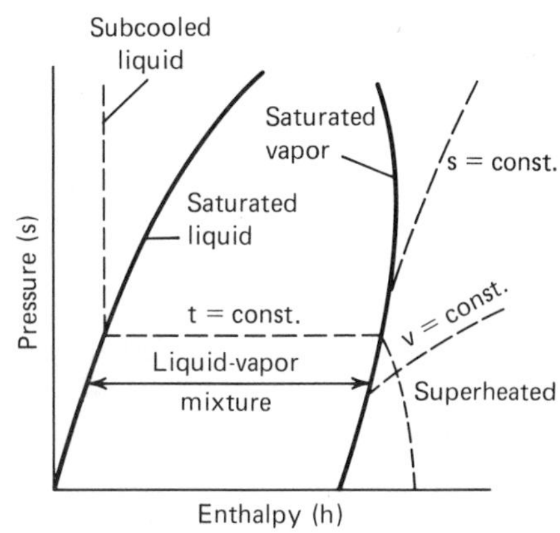

Figure 3–3

Sketch of T-s and p-h diagrams for a refrigerant

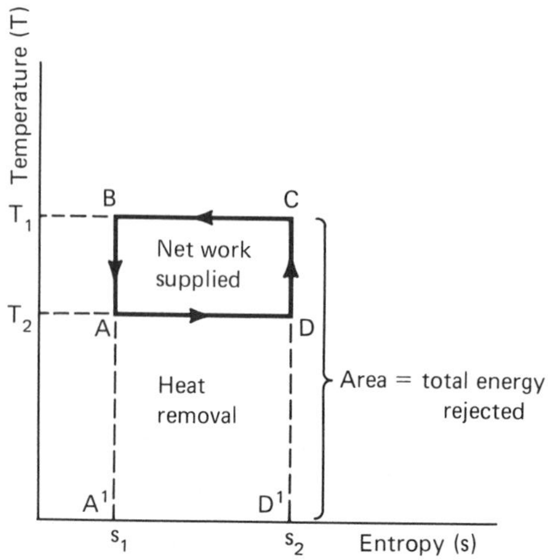

Figure 3–4

Reversed Carnot cycle

3-8 REVERSED CARNOT CYCLE AND COEFFICIENT OF PERFORMANCE

In Chapter 2 the direct Carnot cycle was briefly discussed. The Carnot cycle is an ideal, thermodynamically reversible cycle and represents a measure of the maximum possible conversion of heat energy into mechanical energy. The maximum thermal efficiency of the direct Carnot cycle was stated as

$$\eta = \frac{T_1 - T_2}{T_1} \qquad \textit{(2-18) Repeated}$$

where T_1 was the highest absolute temperature of the cycle and T_2 was the lowest absolute temperature of the cycle.

Since the direct Carnot cycle is, by definition, a reversible cycle, it is possible to use the cycle in a reversed sense to obtain a measure of the maximum performance to be obtained from a refrigerating machine. In this case shaft work must be applied to the cycle which represents an adiabatic compression process from T_2 to T_1, process *DC* of Fig. 3-4. Heat is discharged from the cycle isothermally at T_1 with a decrease in entropy s_2 to s_1, process *CB*. The working medium is then expanded adiabatically from T_1 to T_2, process *BA*, Heat absorption takes place isothermally at T_2, process *AD*, with an increase in entropy s_1 to s_2. Thus, when operating in one direction or the other, the Carnot cycle may be used for three purposes:

a. for converting heat energy into mechanical energy when used as a heat engine

b. for using mechanical energy to absorb heat at some undesirable location and to reject it at some unobjectionable one (a refrigeration machine)

c. for using mechanical energy to absorb heat at some ineffective location and to discharge it at a desirable one (a heat pump)

The second and third applications are based upon the reversed Carnot cycle and differ only in the result desired.

The term *coefficient of performance* (COP) has been devised to measure the effectiveness of refrigerating machines and is usually defined as the ratio of the refrigeration produced in Btu to the net work supplied in Btu. The term can be applied to an actual refrigeration machine or to a theoretical cycle such as the reversed Carnot cycle. Reference to Fig. 3-4 shows that the refrigeration produced in the reversed Carnot cycle is $T_2\,(s_2 - s_1)$ Btu/lbm, and the net work supplied is $[T_1(s_2 - s_1) - T_2\,(s_2 - s_1)]$ Btu/lbm. Thus, the coefficient of performance is

$$\text{COP} = \frac{\text{heat absorbed}}{\text{heat equivalent of the work supplied}}$$

$$= \frac{T_2(s_2 - s_1)}{T_1(s_2 - s_1) - T_2(s_2 - s_1)}$$

$$\text{COP} = \frac{T_2}{T_1 - T_2} \qquad (3\text{-}1)$$

Since the term *coefficient of performance* is analogous to the term *efficiency* applied to heat engines, a broader concept of COP is sometimes used in defining it as the ratio of the "desired effect" in Btu/lbm to the net energy supplied in Btu/lbm. For a Carnot-cycle refrigeration machine this is still

$$\text{COP} = \frac{T_2}{T_1 - T_2}$$

For a Carnot-cycle heat pump in which the desired effect is the heat rejected,

$$\text{COP} = \frac{\text{heat rejected}}{\text{net work supplied}}$$

$$= \frac{T_1(s_2 - s_1)}{T_1(s_2 - s_1) - T_2(s_2 - s_1)}$$

$$\text{COP} = \frac{T_1}{T_1 - T_2} \qquad (3\text{-}2)$$

The Carnot-cycle COP for both a refrigerating machine and a heat pump increases as the spread between source temperature *(T_1)* and sink temperature *(T_2)* decreases, whereas the reverse is true for the COP or efficiency of a heat engine. If the cycle condition under which T_2 is equal to absolute zero is included, the COP for either a heat pump or a refrigerating machine is always greater than 1.0 and that for a heat engine is always less than 1.0.

Illustrative Problem 3-4

A Carnot-cycle machine operates between the temperature limits of $t_1 = 86°F$ and $t_2 = 5°F$. Determine the COP when it is operated as (a) a refrigerating machine, (b) a heat pump, and (c) a heat engine.

Solution:

$$T_1 = t_1 + 460 = 86 + 460 = 546°R$$

$$T_2 = t_2 + 460 = 5 + 460 = 465°R$$

(a) As a refrigerating machine, by Eq. (3-1),

$$\text{COP} = \frac{T_2}{T_1 - T_2} = \frac{465}{546 - 465} = 5.74$$

(b) As a heat pump, by Eq. (3-2),

$$\text{COP} = \frac{T_1}{T_1 - T_2} = \frac{546}{546 - 465} = 6.74$$

(c) As a heat engine, by Eq. (2-18),

$$\text{COP} = \frac{T_1 - T_2}{T_1} = \frac{546 - 465}{546} = 0.148$$

In this problem the COP for the refrigerating machine is 38.3 times greater than that for the heat engine, whereas that for the heat pump is 1.18 times that for the refrigerating machine. This result is to be expected because the heat engine is converting heat energy into mechanical energy, whereas both the refrigerating machine and the heat pump are merely increasing the energy level of heat absorbed from a low-temperature source. The COP for the heat pump is greater than that for the refrigerating machine because all the mechanical shaft work required to operate the pump is dissipated as heat and added to the "desired effect," or heat discarded at the high temperature.

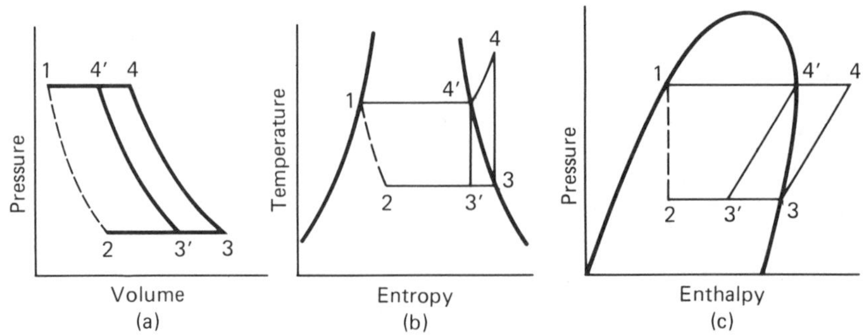

Figure 3–5

Theoretical vapor-compression refrigeration cycle

3-9 THEORETICAL VAPOR COMPRESSION

The theoretical vapor compression refrigeration cycle as actually applied and approached in practice (Rankine form of cycle) is shown in Fig. 3-5a, b, and c on *p-V, T-s,* and *p-h* coordinates. Here 1-2 is the throttling or expansion process, 2-3 is evaporation, 3-4 is compression, and 4-1 is condensation. Two alternate compression paths, 3-4 and 3′-4′ are shown, the former termed *dry compression* with the charge entering the compressor initially dry and saturated, the latter *wet compression* with the charge dry and saturated (theoretically) at the end of compression. Despite the theoretically slightly lower COP of dry compression, it is generally utilized in the United States because, among other reasons, there is less danger of damage to the compressor through the entrance of slugs of liquid refrigerant that are not completely vaporized during compression.

No attempt is made to recover the work of expansion in process 1-2 since this positive work, represented in Fig. 3-5a by the area between process 1-2 and the pressure-coodinate axis, is generally small, and the equipment necessary for its recovery is not economically justifiable. Whereas heat engines are usually large devices centrally located for the production of power in manufacturing plants or central stations, refrigeration machinery is usually limited in application and therefore comparatively low in power input. The recovery of any small amounts of positive work by means of an expansion device or engine is therefore not attempted, and expansion occurs as an irreversible adiabatic process.

Both evaporation (process 2-3′ or 2-3) and condensation (process 4′-1) take place within the saturated vapor region and therefore occur under constant-pressure and constant-temperature conditions. In the case of dry compression, however, the gas leaving the compressor is superheated, and the heat of superheat must be removed in the condenser before condensation of the refrigerant can occur. This process 4-4′ is one of constant pressure (at condenser pressure) but not of constant temperature.

The compression process (3-4 or 3′-4′) in the theoretical cycle is assumed to be isentropic because this process is more nearly approached in practice. Since the charge entering the compressor has been cooled to a temperature greatly below that of the cylinder walls and ports, it is virtually impossible to approach isothermal compression practically. Theoretically, it would be desirable if isothermal compression could be assumed since the work for isentropic compression is greater than that for isothermal compression. However, from a practical standpoint isothermal compression would have no physical meaning except under conditions of extremely high suction superheat. With normal superheat, isothermal compression would result in the discharge of subcooled liquid refrigerant. This change of state would introduce a considerable reduction in volume and therefore a compression exponent, n, that is less than 1. It is also evident that if a cooling medium were available that would allow discharge at the suction temperature, there would be no need for the compressor because the cooling medium could be used to replace the refrigerating system.

A comparison of the theoretical vapor compression cycle with the reversed Carnot cycle is shown in Fig. 3-6 on T-s coordinates. This graph indicates the degree of deviation from the reversed Carnot cycle necessary for a

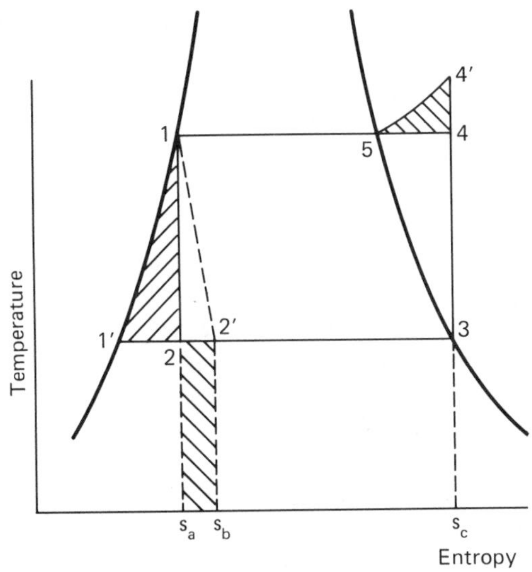

Figure 3–6

Comparison of reversed Carnot and theoretical vapor-compression cycles

theoretical cycle approachable in practice. In making such a comparison, T-s coordinates are of advantage since the areas shown represent actual heat quantities. In this figure 1-2-3-4-1 represents the Carnot cycle, and 1-2′-3-4′-5-1 represents the vapor compression cycle.

The vapor compression cycle represents three area deviations when compared with the Carnot cycle. The superheat "horn" 4-4′-5-4 represents an increase in work input to the cycle as well as an increase in heat rejected to the condenser as a result of contant-pressure rather than constant-temperature desuperheating between 4′ and 5. The area 1-1′-2-1 represents additional work input to the cycle resulting from failure to recover the expansion work between states 1 and 2. The area 2-s_a-s_b-2′-2 represents a loss in refrigerating effect as a result of the actual throttling process entropy increase. Thus, the superheat "horn" adds to the required work of the cycle and irreversible expansion both increases the required work by failing to recover the expansion work and reduces the refrigerating effect. Area 1-1′-2-1 may be proved equal to area 2-s_a-s_b-2′-2.

3-10 DEPARTURES FROM THEORETICAL VAPOR COMPRESSION CYCLE

The actual vapor compression cycle as applied in practice differs in several ways from the theoretical cycle discussed thus far in this chapter. These departures from the theoretical cycle discussed may be summarized as follows:

1. The refrigerant vapor leaving the evaporator is normally superheated a few degrees.
2. Liquid refrigerant entering the expansion valve is usually subcooled a few degrees.

3. Compression of the refrigerant in the compressor, assumed to be isentropic in the ideal cycle, may actually prove to be neither isentropic nor polytropic.

4. Both the compressor suction and discharge valves are actuated by pressure difference, and this process requires the actual suction pressure inside the compressor to be slightly below that of the evaporator and the discharge pressure to be above that of the condenser.

5. Heat exchanges, either positive or negative, between parts of the system and the surrounding air (usually considered negligible).

6. Pressure drops in long suction- and liquid-line piping and any vertical differences in head caused by locating the evaporator and condenser at different elevations.

Figure 3-7 shows the practical vapor compression cycle on the *p-h* and *T-s* coordinates. On these diagrams process 10-11-1-2 represents the passage of the refrigerant through the condenser with 10-11 indicating removal of superheat, 11-1 the removal of latent heat of vaporization, and 1-2 the removal of heat of the liquid or subcooling. Process 3-4-5 represents passage of the refrigerant through the evaporator, with 3-4 indicating gain of latent heat of evaporation and 4-5 the gain of superheat before entrance to the compressor. Both of these processes approach very closely the constant-pressure conditions assumed in theory.

Suction and Exhaust Pressures. Four constant-pressure lines are shown in Fig. 3-7, with p_e representing evaporator pressure, p_c condenser pressure, p_s suction pressure inside the compressor cylinder after passage through the suction valves, and p_d discharge pressure inside the compressor cylinder before passage through the exhaust valves. Path 5-6-7-8-9-10 represents the passage of the gas from entrance to discharge of the compressor. Path 5-6 represents the throttling action that occurs during passage through the suction valves, and path 9-10 represents the throttling during passage through the exhaust valves. Both of these actions are accompanied by an entropy increase and a slight drop in temperature.

Compression. Isentropic compression, assumed theoretically, presupposes that there is no transfer of heat between the refrigerant and the cylinder walls during compression. Actually, the cylinder walls assume a temperature at some point between those of the cold suction gases and the hot exhaust gases, and there is a transfer of heat from the walls to the gases during the first part of compression and a reversal of heat flow during the last part of compression. Moreover, after the cold refrigerant gases pass through the suction valves, the gases undergo, prior to compression, a rise in temperature upon contact with the cylinder walls, and these same gases experience a similar temperature drop after compression and prior to exhaust. These last two heat transfers occur essentially at constant pressure and are indicated by paths 6-7 and 8-9, respectively.

Compression of the refrigerant occurs along path 7-8, which is actually neither isentropic nor polytropic. If it is assumed that the quantity of heat absorbed by the gases during the first part of compression is equal to the quantity of heat rejected by these same gases during the last part of compression, then the form of compression curve shown in Fig. 3-7 (*T-s* diagram) will be approached. Here the quantity of heat absorbed by the gases is equal to the area under curve 6-7-*M*, and this quantity must be equal to the quantity of heat rejected, as represented by the area under curve *M*-8-9. Since the temperature at which rejection of heat from the gases takes place is higher than that at which absorption of heat occurs, and since the

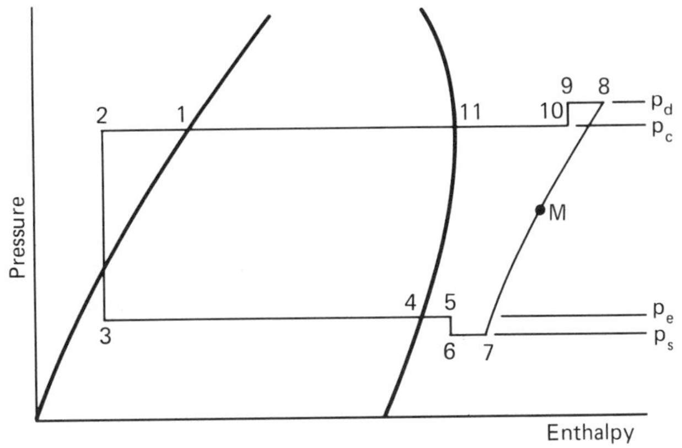

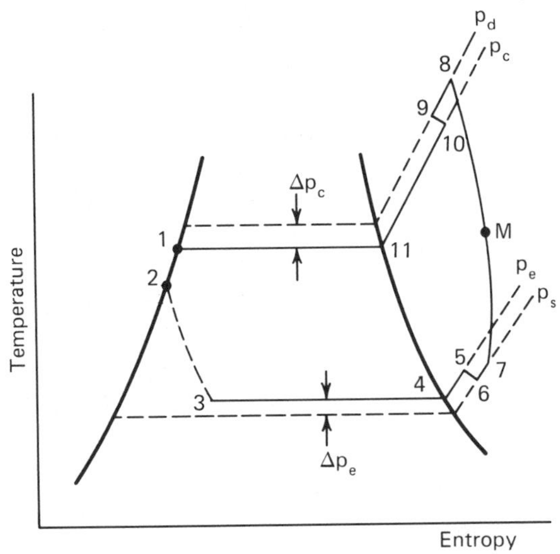

Figure 3–7

P-h and T-s diagrams of actual cycle (no scale)

quantities of heat absorbed and rejected are equal, there must be a net entropy increase.

If the cylinder walls are water jacketed and the temperature of the water is below the mean cylinder-wall temperature, there will be an absorption of some of the heat resulting from compression, a decrease of the cylinder-wall mean temperature, and a decrease in the temperature of the discharged gases. These changes will result in a shifting of curve 6-7-*M*-8-9 such that a greater amount of heat will be rejected to the cylinder walls than is absorbed by them, and there will be a smaller increase or even a decrease in entropy. The magnitude of these changes will of course depend upon the temperature of the cooling water and the effectiveness of the jacketing. These shifts will result in a reduction of power requirements and an increase in volumetric efficiency. The greatest advantage will be found when refrigerants resulting in high discharge temperatures are used. Thus, ammonia compressors are usually supplied with water jackets, whereas Freon-12 and Freon-22 compressors are usually only air cooled.

In most mathematical analyses of actual cycles many of the differences between theory and practice are disregarded as negligible. Thus, any pressure differences across the compressor valves may be neglected because their effect on calculation is small except at very low evaporator temperatures and pressures. Compression is usually assumed to be isentropic or at least polytropic. The effects of heat transfer between parts of the system and the ambient air are difficult to analyze and frequently small, so they are usually disregarded. The consideration of liquid subcooling and vapor superheating introduce no difficulties in calculations and their effects on cycle performance are significant, therefore they are included. The effects of actual compressor performance are extremely important and are included.

In the following sections we will discuss in detail the effects of subcooling, superheating, and compressor performance.

3-11 MATHEMATICAL ANALYSIS OF THE IDEAL VAPOR COMPRESSION CYCLE

Figure 3-8 represents a theoretical vapor compression refrigeration cycle on the *p-h* phase diagram using Freon-22 as the refrigerant. It is assumed in this cycle that

a. saturated liquid leaves the condenser and enters the expansion valve (point 1); saturated vapor leaves the evaporator and enters the compressor (point 3)

b. expansion of the fluid through the expansion valve is a throttling process from condenser pressure to evaporator pressure

c. evaporation and condensation processes take place at constant pressure

d. compression of the vapor is isentropic

e. the entire cycle operates under steady-flow conditions

For the discussion of the various processes of the cycle we will assume that the condensing temperature will be 100°F and the evaporation temperature will be 20°F. Referring to Table B-5 (Appendix B), we may find that the saturation pressure corresponding to the condensing temperature of 100°F is 210.60 psia, or 195.91 psig. The saturation pressure corresponding to the evaporator temperature of 20°F is 57.727 psia, or 43.031 psig.

Expansion Process (1-2). Process 1-2 is a throttling type of adiabatic expansion which is an irreversible steady-flow process. The refrigerant passes through a series of state points between 1 and 2 in such a way that there is no uniform distribution of any of the fluid properties. No true path could be drawn for the process, and the line 1-2 merely represents a process that starts at point 1 and ends at point 2. This type of process for steady flow is a constant-enthalpy process, or

$$h_1 = h_2 \tag{3-3}$$

Referring to the data of Table B-5, saturated liquid at 100°F has an enthalpy h_f = 39.267 Btu/lb; therefore, $h_1 = 39.267 = h_2$.

As one may see in Fig. 3-8, point 2 is located in the liquid-vapor mixture area of the *p-h* diagram. This indicates that during the expansion of saturated liquid to a lower pressure, *flash* vapor is produced. A small amount of energy in the saturated liquid provides the necessary latent heat of evaporation to evaporate a small portion of the liquid. The percentage of flash vapor formed is called the *quality,* the same term used when discussing wet steam which contains both liquid and vapor. The

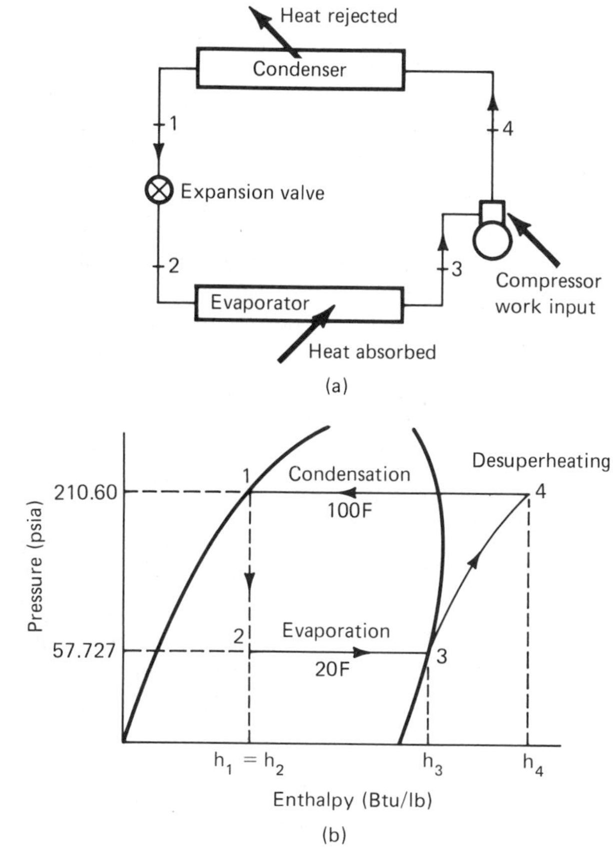

Figure 3–8 Pressure-enthalpy diagram (b) and mechanical equipment arrangement (a) for an ideal vapor compression refrigeration cycle using Freon-22 at an evaporator temperature of 20F and condensing temperature of 100F

quality of the liquid-vapor mixture may be easily calculated by

$$h_1 = h_2 = h_{f2} + xh_{fg2}$$

where h_1 = enthalpy of liquid entering expansion valve (Btu/lb)

h_{f2} = enthalpy of saturated liquid at evaporator pressure (Btu/lb)

h_{fg2} = latent heat of vaporization at evaporator pressure (Btu/lb)

x = quality of mixture, expressed as a decimal (pounds of vapor per pound of liquid vapor mixture)

Flash gas formed during the expansion process reduces the refrigerating effect.

For Fig. 3-8 the quality would be found as follows: From Table B-5 at evaporator pressure of 57.727 psia, find h_{f_2} = 15.837 Btu/lb and h_{fg_2} = 90.545 Btu/lb. Using Eq. (3-4) we have

$$39.267 = 15.837 + x(90.545)$$
$$x = 0.259 \quad \text{or} \quad 25.9 \text{ percent}$$

This indicates that for each pound of fluid that leaves the expansion valve and flows to the evaporator, 0.259 lb will be vapor and 0.741 lb will be liquid at the evaporator pressure and temperature. Thus, 25.9 percent of the liquid vaporizes as it passes through the expansion valve.

Vaporization Process (2-3). The quantity of heat that each pound of refrigerant absorbs while flowing through the evaporator is called the *refrigerating effect*. Each pound of liquid refrigerant flowing through the evaporator is able to absorb only the heat required to vaporize it if no superheating takes place.

The heat added to each pound of refrigerant in the evaporator is heat added at constant pressure, which is calculated from

$$q_{\text{evap}} = (h_3 - h_2) \quad \text{(Btu/lb)} \qquad (3\text{-}5)$$

or $q_{\text{evap}} = h_3 - h_1$, since $h_1 = h_2$. For Fig. 3-8 the vapor leaving the evaporator is saturated vapor at the evaporator pressure and temperature. From Table B-5 at 20°F, find h_g = 106.383 Btu/lb which is equal to h_3. Therefore,

$$\text{Refrigerating Effect} = q_{\text{evap}} = 106.383 - 39.267$$
$$q_{\text{evap}} = 67.116 \text{ Btu/lb}$$

Compression Process (3-4). In the ideal cycle it is assumed that saturated vapor leaves the evaporator and enters the compressor without change in the compressor suction line. Process 3-4 takes place within the compressor as the pressure and temperature of the vapor is increased by compression from the evaporator pressure to the condenser pressure. For this ideal compression process it is assumed that compression takes place isentropically. An isentropic process is a special type of adiabatic process which takes place without friction or heat transfer (see Section 2-10). An isentropic process is a process of no entropy change. Therefore, process 3-4 on Fig. 3-8 follows a line of constant entropy from the evaporator pressure to the condenser pressure.

At point 4 the refrigerant is a superheated vapor. The properties of the superheated vapor may be determined by interpolation of the superheat tables (Table B-6 for Freon-22), or, as usual, may be read approximately from the *p-h* diagram (Fig. B-3).

The isentropic work of compression may be calculated from

$$\text{iwk}_c = (h_4 - h_3) \quad \text{(Btu/lb)} \qquad (3\text{-}6)$$

where iwk_c is the ideal work in Btu/lb of refrigerant compressed when steady-flow conditions exist. Although the compression is intermittent in a piston-type compressor, if the rotational speed of the drive shaft is reasonably high, steady-flow conditions may be assumed. The work done on the gas by the piston during compression is often called the ''heat of compression'' and is absorbed by the gas. As a result of absorbing the heat of compression, the hot vapor discharged from the compressor is in a superheated condition if the vapor is saturated at the start of compression. Before the compressed, superheated vapor can be condensed, the superheat must be removed and the temperature of the vapor reduced to the saturation temperature at the condenser pressure.

Referring once again to Fig. 3-8 and point 3, we have determined that h_3 = 106.383 Btu/

lb. Also from Table B-5 at $t_3 = 20°F$ we may find the entropy $s_3 = 0.22379$ Btu/lb °R. Using Fig. B-3, locate point 4 by following a line of constant entropy ($s_4 = s_3$) upward and to the right until it intersects the condenser pressure line of 210.60 psia. At point 4 find $h_4 = 121$ Btu/lb, $t_4 = 140°F$. By Eq. (3-6),

$$\text{iwk}_c = (h_4 - h_3) = 121.0 - 106.383$$

$$\text{iwk}_c = 14.617 \text{ Btu/lb}$$

Condensation Process (4-1). As noted in the preceding sections, the vapor leaving the compressor (point 4) is superheated. Temperature t_4 is 140°F which is 40 degrees above the saturation temperature of 100°F at the condenser pressure. The enthalpy of saturated vapor at the condenser pressure from Table B-5 is 112.105 Btu/lb. Therefore, 121.0 − 112.105 = 8.895 Btu/lb of heat must be removed from the refrigerant before condensation starts. Usually, this process of desuperheating takes place in the upper portion of the condenser and, to some extent, in the hot-gas line from compressor discharge to the condenser inlet. The process of desuperheating and condensation takes place at constant pressure in the ideal cycle. Therefore, if the removal of superheat and condensation takes place in the condenser, the rejected heat to the condenser-cooling medium may be calculated from

$$q_{\text{cond}} = (h_4 - h_1) \quad \text{(Btu/lb)} \tag{3-7}$$

where q_{cond} is the heat rejected in Btu/lb of refrigerant circulated.

From our previous discussion we have found $h_1 = 39.267$ Btu/lb and $h_4 = 121.0$ Btu/lb. Therefore, applying Eq. (3-7)

$$q_{\text{cond}} = (h_4 - h_1) = 121.0 - 39.267$$

$$q_{\text{cond}} = 81.733 \text{ Btu/lb}$$

After the condensing process, if the refrigerant at point 1 has the same properties it had at the start of the cycle, the total heat rejected to the condenser-cooling medium must be equal to the total heat absorbed by the refrigerant during the other processes (law of conservation of energy), or

$$q_{\text{cond}} = q_{\text{evap}} + iwk_c \tag{3-8}$$

From the results of our previous calculations and Eq. (3-8), we have

$$81.733 = 67.116 + 14.617$$

$$81.733 = 81.733$$

Mass Rate of Refrigerant Flow. The closed, ideal vapor compression refrigeration cycle we are discussing is a steady-flow system where the refrigerant flows continuously through each system component. The mass rate of flow is frequently calculated on the basis of a unit of refrigeration capacity (1 ton). Therefore, the mass rate of refrigerant flow in pounds per minute per ton would be

$$\dot{m}_r = \frac{\text{heat absorbed in evaporator per ton min}}{\text{refrigerating effect in Btu/lb}}$$

or

$$\dot{m}_r = \frac{200 \text{ Btu/ton min}}{(h_3 - h_2) \text{ Btu/lb}}$$

or

$$\dot{m}_r = \frac{200}{h_3 - h_2} = \frac{200}{h_3 - h_1} \text{ lb/ton min} \tag{3-9}$$

The total mass rate of flow in a system may be readily determined by multiplying the results of Eq. (3-9) by the system capacity in tons.

Referring to the data of Fig. 3-8 and our calculations,

$$\dot{m}_r = \frac{200}{h_3 - h_1} = \frac{200}{106.383 - 39.267}$$

$$\dot{m}_r = 2.979 \text{ lb/ton min}$$

If the total system capacity was 10 tons, the total mass rate of refrigerant flow would be (10)(2.979) = 29.79 lb/min.

If Eq. (3-7) is multiplied by Eq. (3-9), the heat removed by the condenser may be calculated on the basis of Btu/ton min. Thus,

$$q_{\text{cond}} = \left(\frac{200}{h_3 - h_1}\right)(h_4 - h_1) \qquad \textit{(3-7a)}$$

For our illustration

$$q_{\text{cond}} = 2.979(81.733) = 243.482 \text{ Btu/ton min}$$

or for a 10-ton capacity,

$$q_{\text{cond}} = (243.482)(10) = 2434.82 \text{ Btu/min.}$$

Theoretical horsepower. The theoretical horsepower to drive the refrigeration compressor per ton of refrigeration capacity may be calculated by multiplying Eq. (3-6) by Eq. (3-9). Thus,

$$\text{iwk}_c = \dot{m}_r\,(h_4 - h_3)$$

$$= \left(\frac{200}{h_3 - h_1}\right)(h_4 - h_3) \text{ (Btu/ton min)} \qquad \textit{(3-10)}$$

To obtain horsepower we may multiply by 778 ft lb/Btu and divide by 33,000 ft lb/hp min. Thus

$$\text{ihp}_c = \left(\frac{200}{h_3 - h_1}\right)(h_4 - h_3)\left(\frac{778}{33{,}000}\right)$$

or

$$\text{ihp}_c = 4.717\left(\frac{h_4 - h_3}{h_3 - h_1}\right) \text{ hp/ton} \qquad \textit{(3-11)}$$

This ihp_c, calculated by Eq. (3-11), is the theoretical horsepower per ton of refrigeration required to compress the vapor. It assumes 100 percent compressor efficiency and does not take into account the power required to overcome friction and other power losses.

For our illustration the theoretical horsepower per ton of refrigeration using Eq. (3-11) would be

$$\text{ihp}_c = 4.717\left(\frac{h_4 - h_3}{h_3 - h_1}\right)$$

$$= 4.717\left(\frac{121.0 - 106.383}{106.383 - 39.267}\right)$$

$$\text{ihp}_c = 1.027 \text{ hp/ton}$$

The horsepower requirements may also be determined for polytropic compression by means of the polytropic steady-flow work equation. First, expressing the polytropic work as

$$\text{Work of compression} = \left(\frac{200}{h_3 - h_1}\right)\left(\frac{n}{n - 1}\right)(p_4v_4 - p_3v_3) \quad \text{(ft lb/ton min)} \qquad \textit{(3-10a)}$$

or

$$\text{Work of compression} = \left(\frac{200}{h_3 - h_1}\right)\left(\frac{n}{n - 1}\right)\left(\frac{p_3v_3}{J}\right)\left[\left(\frac{p_4}{p_3}\right)^{\frac{n-1}{n}} - 1\right] \quad \text{(Btu/ton min)} \qquad \textit{(3-10b)}$$

Then the theoretical power requirement is

$$\text{ihp}_c = \frac{\left(\frac{200}{h_3 - h_1}\right)\left(\frac{n}{n - 1}\right)p_3v_3\left[\left(\frac{p_4}{p_3}\right)^{\frac{n-1}{n}} - 1\right](144)}{33{,}000}$$

Table 3-3

Refrigerant	$k = c_p/c_{v7}$ (t, °F)
Ammonia	1.31 (70°F)
Dichlorodifluoromethane (Freon-12)	1.13 (50°F)
Monochlorodifluoromethane (Freon-22)	1.18 (118°F)

$$\text{ihp}_c = \frac{0.873 n p_3 v_3 \left[\left(\frac{p_4}{p_3} \right)^{\frac{n-1}{n}} - 1 \right]}{(n - 1)(h_3 - h_1)} \quad \text{(hp/ton)} \tag{3-11a}$$

If the compression process is isentropic, $n = k = c_p/c_v$. Values for k for some of the common refrigerants are given in Table 3-3.

Actual horsepower per ton of refrigeration can be approximated by using the correct value of n in Eq. (3-11a) by using the actual cylinder suction and discharge pressures and by dividing the resulting horsepower requirements by the mechanical efficiency. The mechanical efficiency is the ratio of the indicated horsepower of the compressor to the horsepower required to drive the compressor. All three of these factors cannot be calculated. They may be approximated but are best determined by actual test. Indicator cards may be used to determine the indicated horsepower on large, low-rpm compressors; on small compressors the volume of the indicator and its connecting piping is sufficiently large to add materially to the compressor clearance volume and thus reduce the volume of gas pumped and the work required. The actual capacity and power requirements of small condensing units are usually determined by their manufacturers through the aid of calorimeter tests.

If the compressor cylinders are provided with cooling water jackets, an appreciable amount of heat may be rejected to the cooling water during compression. If the suction and discharge conditions are known, this heat quantity can be determined from the general energy equation as the difference between the initial and final enthalpies. Thus, the heat rejected to the compressor coolant is

$$q_j = \frac{n}{(n - 1)J} (p_4 v_4 - p_3 v_3) - (h_4 - h_3) \quad \text{(Btu/lb)} \tag{3-12}$$

or

$$q_J = \frac{n}{(n - 1)J} p_3 v_3 \left[\left(\frac{p_4}{p_3} \right)^{\frac{n-1}{n}} - 1 \right] - (h_4 - h_3)^{(Btu/lb)} \tag{3-12a}$$

Coefficient of performance. The coefficient of performance of a refrigeration cycle was defined in Section 3-8 as

$$\text{COP} = \frac{\text{heat absorbed}}{\text{heat equivalent of work input}}$$

or

$$\text{COP} = \frac{\text{refrigeration effect}}{\text{heat equivalent of work input}}$$

Therefore, at this point we may say

$$\text{COP} = \frac{h_3 - h_1}{h_4 - h_3} \tag{3-13}$$

Referring to Eqs. (3-11) and (3-12), we may also see that

$$\text{COP} = \frac{4.717}{\text{ihp}_c} \tag{3-14}$$

Using data from Fig. 3-8 and Eq. (3-13),

$$\text{COP} = \frac{h_3 - h_1}{h_4 - h_3} = \frac{106.383 - 39.267}{121.0 - 106.383}$$

$$\text{COP} = 4.59$$

Theoretical Piston Displacement of Compressor. The refrigerant vapor formed at one certain temperature inside the evaporator has a definite specific volume. Inasmuch as the cylinders of a reciprocating compressor fill only during the downstroke of the piston (single-acting), the cylinder volume must be at least large enough to accommodate the mass of refrigerant vaporized in the interval between downstrokes. If the cylinder volume is less than this, the vapor that remains behind in the evaporator will increase the evaporator pressure and, hence, increase the boiling point of the liquid. The evaporator temperature would rise.

The volume swept by the piston during any certain interval of time is known as the *piston displacement*. The piston displacement must be at least equal to the volume of refrigerant vaporized. It is apparent, therefore, that the size of the cylinder (bore and stroke) and the piston speed must be based upon the volume of vapor to be removed from the evaporator. The theoretical volume of vapor produced in the evaporator is

$$\dot{Q}_t = \dot{m}_r \text{x} (v_g)_3 \quad \text{(cfm/ton min)} \qquad (3\text{-}15)$$

where $\dot{Q}$ is the ft^3/ton min of vapor produced, $\dot{m}_r$ is the mass of refrigerant circulated in lb/ton min, and $(v_g)_3$ is the specific volume of the refrigerant vapor entering the compressor in ft^3/lb.

The displacement of a single-acting reciprocating compressor is calculated from the dimensions of the cylinder, the number of cylinders, and the speed. In equation form this would be

$$\text{PD}_t = \frac{\pi \times d^2 \times s \times n \times \text{rpm}}{4 \times 1728} = \frac{d^2 \times s \times n \times \text{rpm}}{2200} \qquad (3\text{-}16)$$

where PD_t = displacement in ft^3/min
d = cylinder diameter (bore) in inches
s = length of piston stroke in inches
n = number of cylinders
rpm = revolutions per minute of the crank shaft

For steady operation $\dot{Q}_t$ must equal PD_t.

Again referring to Fig. 3-8 and data, the specific volume of the vapor leaving the evaporator and entering the compressor is the specific volume of saturated vapor at 20°F. From Table B-5 $(v_g)_3$ = 0.93631 ft^3/lb. By Eq. (3-15),

$$\dot{Q}_t = \dot{m}_r \times (v_g)_3 = 2.979 \times 0.93631$$

$$\dot{Q}_t = 2.789 \text{ ft}^3\text{/ton min}$$

This must also be the theoretical piston displacement (PD_t) of the compressor.

Illustrative Problem 3-5

An ideal vapor compression refrigeration system using Freon-12 has the condenser operating at 90°F and the evaporator operating at 38°F. The liquid leaving the condenser and entering the expansion valve is saturated liquid at condenser pressure, and the vapor leaves the evaporator and enters the compressor as saturated vapor at evaporator pressure. For a system capacity of 15 tons determine

a. the refrigerating effect
b. compressor work
c. refrigerant flow
d. theoretical piston displacement
e. theoretical compressor horsepower
f. coefficient of performance
g. heat rejected to condenser-cooling medium
h. theoretical bore and stroke of a four-cylinder, single-acting compressor operating at 1750 rpm with the bore and stroke equal

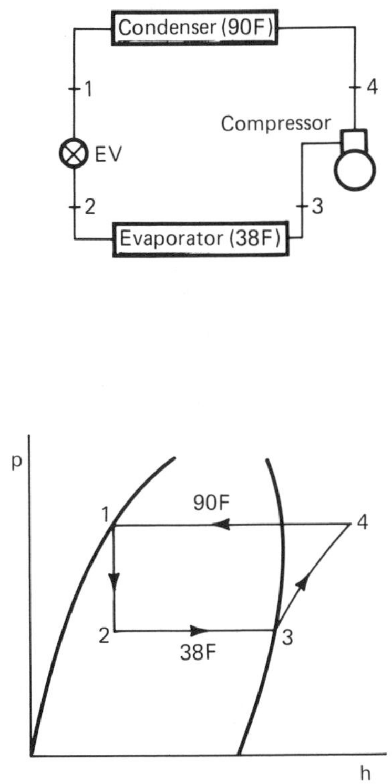

Figure 3–9
Diagrams for Illustrative Problem 3-5

Solution:

Figure 3-9 shows the sketch of the system and *p-h* diagram. From our previous work in this section you will note that most of the calculations depend upon the enthalpy of the refrigerant at the various points in the system. Therefore, the first step for our solution will be to determine all required fluid properties.

Step 1: Fluid properties.
From Table B-3 at 90°F find condenser pressure = 114.49 psia.

$$h_1 = h_{f1} = 28.713 \text{ Btu/lb}$$

From Table B-3 at 38°F find evaporator pressure = 49.870 psia.

$$h_3 = h_g = 81.234 \text{ Btu/lb}$$

$$(v_g)_3 = v_g = 0.80023 \text{ ft}^3\text{/lb}$$

$$s_3 = s_g = 0.16598 \text{ Btu/lb °R}$$

From Fig. B-2 at $s_4 = s_3 = 0.16598$ and $p_4 = 114.49$ psia, find $h_4 = 88.0$ Btu/lb (approximately), and $t_4 = 98$°F (approximately).

Step 2: Determine refrigeration effect.

$$h_1 = h_2 = 28.713 \text{ Btu/lb}$$

By Eq. (3-5),

$$q_{evap} = (h_3 - h_1) = 81.234 - 28.713$$

$$q_{evap} = 52.521 \text{ Btu/lb} \quad \text{(refrigerating effect)}$$

The *total* refrigerating effect is, of course, 15

tons or 15 × 200 = 3000 Btu/min or 200 Btu/ton min.

Step 3: Determine refrigerant flow rate. By Eq. (3-9),

$$\dot{m}_r = \frac{200}{h_3 - h_1}$$

where $h_3 - h_1$ is the refrigerating effect of Step 2 equal to 52.521. Therefore,

$$\dot{m}_r = \frac{200}{52.521} = 3.808 \text{ lb/ton min}$$

Step 4: Determine compressor work and theoretical horsepower. By Eq. (3-6),

$$\text{iwk}_c = h_4 - h_3 = 88.8 - 81.234$$

$$\text{iwk}_c = 6.766 \text{ Btu/lb}$$

By Eq. (3-10),

$$\text{iwk}_c = \left(\frac{200}{h_3 - h_1}\right)(h_4 - h_3)$$

$$= 3.808(6.766)$$

$$\text{iwk}_c = 25.765 \text{ Btu/ton min}$$

By Eq. (3-11),

$$\text{ihp}_c = 4.717\left(\frac{h_4 - h_3}{h_3 - h_1}\right)$$

From before, $(h_4 - h_3) = 6.766$ Btu/lb, and $(h_3 - h_1) = 52.521$ Btu/lb.

$$\text{ihp}_c = 4.717\left(\frac{6.766}{52.521}\right)$$

$$= 0.607 \text{ hp/ton}$$

Step 5: Determine theoretical piston displacement. By Eq. (3-15),

$$\dot{Q}_t = \dot{m}_r \times (v_g)_3 = 3.808 \times 0.80023$$

$$\dot{Q}_t = 3.047 \text{ ft}^3\text{/ton min}$$

Step 6: Determine coefficient of performance. By Eq. (3-13),

$$\text{COP} = \frac{h_3 - h_1}{h_4 - h_3} = \frac{52.521}{6.766}$$

$$COP = 7.762$$

Step 7: Heat rejected to condenser-cooling medium. By Eq. (3-7),

$$q_{\text{cond}} = h_4 - h_1 = 88.0 - 28.713$$

$$q_{\text{cond}} = 59.287 \text{ Btu/lb}$$

On a per ton-min basis,

$$\dot{q}_{\text{cond}} = 3.808 \times 59.287$$

$$= 225.765 \text{ Btu/ton min}$$

We should also be able to see Eq. (3-8) as a check for our calculations.

$$q_{\text{cond}} = q_{\text{evap}} + iwk_c$$

$$225.765 = 200 + 25.765$$

$$225.765 = 225.765 \quad \text{(check of system energy balance)}$$

Step 8: Determine theoretical bore and stroke of the compressor. From Step 5, $Q_t = 3.047$ ft^3/ton min. The total piston displacement would be (3.047 ft^3/ton min)(15 tons) = 45.705 ft^3/min. For a four-cylinder compressor, each cylinder would have a theoretical displacement of 45.705/4 = 11.426 ft^3/min. Since Q_t = PD_t we have, by Eq. (3-16),

$$PD_t = \frac{d^2 \times s \times n \times \text{rpm}}{2200}$$

With $s = d$

$$45.705 = \frac{d^2 \times d \times 4 \times 1750}{2200}$$

$$d^3 = 14.364$$

$$d = 2.43 \text{ in.}$$

Therefore, both the bore and stroke are equal to 2.43 in.

Step 9: Determine totals.
Total refrigerating effect = 15 tons

Total refrigerant flow = (3.808 lb/ton min)(15 tons) = 57.12 lb/min
Total compressor work = (25.765 Btu/ton min)(15 tons) = 386.475 Btu/min
Total compressor hp = (0.607 hp/ton)(15 tons) = 9.105 hp
Total piston displacement = 45.705 ft^3/min.
Total heat rejected = (225.765 Btu/ton min)(15 tons) = 3386.475 Btu/min.

3-12 COMPRESSOR VOLUMETRIC EFFICIENCY

In the preceding discussion it was assumed that at each stroke of the piston the cylinder would fill completely with vapor at exactly the same pressure and temperature at which it left the evaporator. A theoretically perfect compressor would have neither clearance nor losses of any type and would pump on each stroke a quantity of refrigerant vapor equal to the piston displacement (Eq. 3-16). No actual compressor is able to do this since it is impossible to construct a compressor without clearance or one that will have no wire-drawing (throttling) through the suction and discharge valves, no superheating of the suction gases upon contact with the hot cylinder walls, or no leakage of gases past the piston or valves. All these factors affect the volume of gas pumped or the capacity of the compressor. Some of them affect the horsepower requirements per ton of refrigeration developed.

The ratio of the actual volume (or weight) of gas drawn into the compressor (at evaporator pressure and temperature) on each stroke to the piston displacement is termed *volumetric efficiency* (η_v), or

$$\eta_v = \frac{\text{actual volume indicated}}{\text{theoretical displacement}} \times 100 = \frac{\text{actual weight}}{\text{theoretical weight}} \times 100 \qquad (3\text{-}17)$$

If the effect of clearance volume alone is considered, the resulting expression may be termed *clearance volumetric efficiency* (η_{vc}). The expression used for grouping into one constant all the factors affecting volumetric efficiency may be termed *total volumetric efficiency* (η_{vt}). The clearance volumetric efficiency may be calculated with reasonable accuracy; the total volumetric efficiency is best obtained by actual laboratory tests, although a fair approximation to it may be calculated if sufficient data is available.

Clearance Volumetric Efficiency (η_{vc}). The volume of the space between the end of the cylinder and the piston when the latter is in top-dead-center position is termed *clearance volume*. Figure 3-10 represents the ideal *p-v* diagram for a piston compressor. The clearance volume is represented by volume 3-4. The clearance, c, is the ratio, expressed in percent, of the clearance volume to the piston dis-

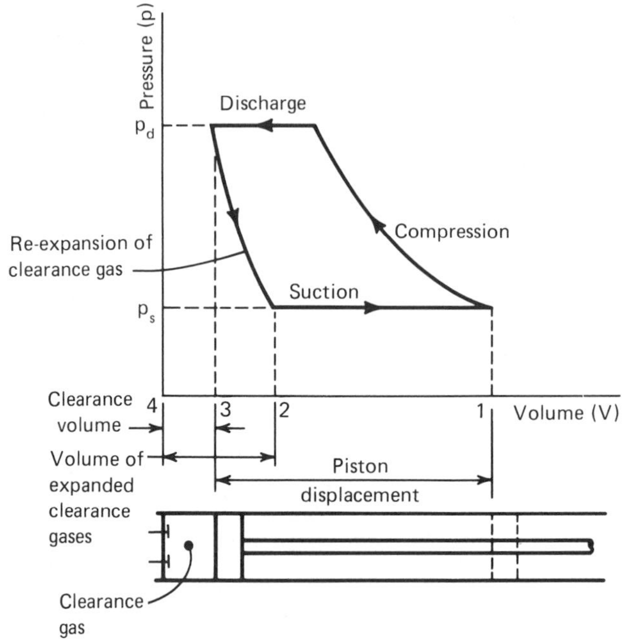

Figure 3–10

Ideal p-V diagram for piston compressor

placement, volume 1-3. The clearance usually ranges from 3 to 5 percent.

Upon expansion of the clearance gases from the discharge pressure, p_d, to the suction pressure, p_s, the volume of these gases will change from 4-3 to 4-2, and the portion of the displacement effective in drawing in a new charge of gas from the evaporator will be reduced by volume 3-2. When considering clearance volume reexpansion then, the ratio of this actual volume (weight) of new suction gases to the piston displacement volume (weight) is defined as the clearance volumetric efficiency. It may be shown that clearance volumetric efficiency may be calculated from

$$\eta_{vc} = 1 + C - C\left(\frac{p_d}{p_s}\right)^{1/n} \qquad (3\text{-}18)$$

where C is the percent clearance, expressed as a decimal, and n is the compression exponent.

The clearance volumetric efficiency has a marked effect upon the compressor piston displacement required per ton of refrigeration developed. The effect becomes even more marked as the compression ratio or spread between condenser and evaporator pressures increases. The compression ratio is defined by

Compression Ratio (CR)

$$= \frac{\text{head pressure, psia}}{\text{suction pressure, psi}} = \frac{p_d}{p_s}$$

Illustrative Problem 3-6

In Problem 3-5 the theoretical piston displacement for the compressor was calculated to be 45.705 ft^3/min, and the resulting bore and stroke for the compressor was 2.43 in.

Assume that the clearance for this compressor is 5 percent, determine the clearance volumetric efficiency and the actual bore and stroke required. Assume a compression exponent $n = 1.13$.

<u>Solution:</u>

By Eq. (3-18),

$$\eta_{vc} = 1 + C - C\left(\frac{p_d}{p_s}\right)^{1/n}$$

$$= 1 + 0.05 - 0.05\left(\frac{114.49}{49.870}\right)^{1/1.13}$$

$$\eta_{vc} = 0.946 \quad \text{or} \quad 94.6 \text{ percent}$$

$$\text{Actual piston displacement required} = \frac{PD_t}{\eta_{vc}}$$

$$\text{PD} = \frac{45.705}{0.946} = 48.31 \text{ ft}^3\text{/min}$$

By Eq. (3-16),

$$\text{PD} = \frac{d^2 \times s \times n \times \text{rpm}}{2200}$$

$$48.31 = \frac{d^2 \times d \times 4 \times 1750}{2200}$$

$$d^3 = 15.18$$

$$d = 2.48 \text{ in.} \quad \text{(bore and stroke)}$$

It is usually considered that clearance has no effect upon the horsepower requirements per ton of refrigeration because the alternate compression and expansion of a fixed quantity of vapor theoretically requires zero power. The process is analogous to the alternate compression and expansion of a spring with the work required during compression recovered during the following expansion (approximately).

In an actual system it is probable that clearance has some effect upon the power requirements since the compression and expansion of the clearance gases are not likely to follow the same path; that is, the work required during compression is not numerically identical to the work recovered during expansion. The difference, however, is usually small and is disregarded in calculating compressor power requirements; and reasonably accurate results are obtained by Eq. (3-18) to determine compressor displacement requirements.

<u>*Total Volumetric Efficiency*</u> *(η_{vt}).* An *approximation* to the total volumetric efficiency can be calculated if the pressure drop through the suction valves and the temperature of the gases at the end of the suction stroke are known and if it is assumed that there is no leakage past the pistons during compression. This approximation may be computed with a modification of Eq. (3-18); that is, multiplying it by the ratio of the cylinder and evaporator suction pressures and by the ratio of evaporator suction and cylinder absolute temperatures. Thus,

$$\eta_{vt} = \left[1 + C - C\left(\frac{p_d}{p_s}\right)^{1/n}\right]\left(\frac{p_{cy}}{p_s}\right)\left(\frac{T_s}{T_{cy}}\right) \tag{3-19}$$

$$\eta_{vt} = \left[1 + C - C\left(\frac{p_d}{p_s}\right)^{1/n}\right]\left(\frac{v_s}{v_{cy}}\right) \tag{3-19a}$$

where subscript cy refers to the condition of the vapor in the cylinder at the end of the piston suction stroke, and subscript s refers to the condition of the vapor in the suction pipe just adjacent to the compressor cylinder. The temperature rise of the suction gases passing through the suction line into the compressor is difficult to evaluate. If tests are to be made to determine this temperature rise, it is usually more practical to run complete tests and determine the total volumetric efficiency.

The average total volumetric efficiency of reciprocating compressors using Freon-12 or Freon-22 and running at 1750 rpm is shown in Table 3-4 for various compression ratios. These efficiencies will not fit all compressors exactly. However, the capacity of a compressor computed by means of the total volumetric efficiency obtained from this table will generally be close enough for most practical applications.

The *actual piston displacement* (PD) per ton of refrigeration per minute can be determined by dividing the theoretical piston displacement (PD_t) by the volumetric efficiency. Thus,

$$PD = \frac{PD_t}{\eta_{vc}} \quad (2\text{-}20)$$

or

$$PD = \frac{PD_t}{\eta_{vt}} \quad (2\text{-}20a)$$

depending upon whether η_{vc} or η_{vt} is known or calculated.

Table 3-4

Approximate Efficiencies of Modern High-Speed Freon-12 and Freon-22 Compressors*

Compression Ratio	Volumetric Efficiency (n_{vt} (Percent)	Performance Factor (hp/ton)
2.0	88.50	0.72
2.2	87.00	0.81
2.4	85.60	0.89
2.6	84.20	0.96
2.8	82.90	1.05
3.0	81.60	1.13
3.2	80.03	1.20
3.4	79.10	1.27
3.6	78.00	1.34
3.8	76.70	1.41
4.0	75.60	1.47
4.2	74.50	1.54
4.4	73.50	1.60
4.6	72.50	1.67
4.8	71.40	1.73
5.0	70.40	1.78
5.2	68.90	1.84
5.4	68.40	1.90
5.6	67.40	1.96
5.8	66.40	2.02

*Extracted by permission, *Trane Reciprocating Refrigeration Manual, The Trane Company, LaCrosse, Wisconsin.*

Illustrative Problem 3-7

The bore and stroke of a four-cylinder Freon-22 compressor are 3.66 in. and 2.75 in., respectively. The compressor operates at 1750 rpm. The evaporator temperature is 38°F and the condensing temperature is 105°F. Liquid reaches the expansion valve at the same temperature at which condensed. Determine the theoretical and actual refrigerating capacities of the compressor.

Solution:

Step 1: Determine the theoretical weight of refrigerant pumped by compressor. By Eq. (3-16),

$$PD_t = \frac{d^2 \times s \times n \times \text{rpm}}{2200}$$

$$= \frac{(3.66)^2(2.75)(4)(1750)}{2200}$$

$$PD_t = 117.2 \text{ cfm}$$

This may be converted to lb/min by multiplying by the refrigerant density at 38°F. From Table B-5 at 38°F find $1/v_g = 1.4697 \text{ lb/ft}^3$.

Theoretical weight of refrigerant pumped would be

$$\dot{m}_r = PD_t \times 1/v_g = 117.2 \times 1.4697$$

$$= 172.25 \text{ lb/min}$$

Step 2: Determine total volumetric efficiency and actual amount of refrigerant pumped.

To make use of Table 3-4, the compression ratio must be calculated. From Table B-5 at 105°F the condensing pressure is 225.45 psia, and at 38°F the evaporator pressure is 80.336 psia. Then,

$$\text{CR} = \frac{p_d}{p_s} = \frac{225.45}{80.336} = 2.80$$

From Table 3-4 at CR = 2.80 find $\eta_{vt} = 82.9$ percent. Therefore, the actual weight of refrigerant pumped would be

$$\text{Actual } \dot{m}_r = 172.25 \times 0.829$$

$$= 142.80 \text{ lb/min}$$

Step 3: Determine refrigeration capacity.

Each pound of refrigerant circulated will produce a certain refrigerating effect. The enthalpy of saturated vapor leaving the evaporator at 38°F is 107.974 Btu/lb. The enthalpy of saturated liquid at the condenser temperature of 105 °F is 40.847 Btu/lb.

The refrigerating effect is (107.974 − 40.847) = 67.127 Btu/lb.

Theoretical capacity (tons)

$$= \frac{(172.25 \text{ lb/min})(67.127 \text{ Btu/lb})}{200 \text{ Btu/ton min}}$$

$$= 57.8 \text{ tons}$$

Actual capacity (tons)

$$= \frac{(143.83 \text{ lb/min})(67.127 \text{ Btu/lb})}{200 \text{ Btu/ton min}}$$

$$= 48.3 \text{ tons}$$

3-13 SUBCOOLING AND SUPERHEATING OF THE REFRIGERANT

Subcooling is a term used to describe the cooling of the liquid refrigerant, at constant pressure, to a temperature below the saturation temperature corresponding to the condenser pressure.

Superheating is a term used to describe the sensible heating of the refrigerant vapor, at constant pressure, in the evaporator to a temperature above the saturation temperature corresponding to the evaporator pressure.

Figure 3-11 represents the *p-h* diagram from a refrigeration cycle using Freon-12 comparing the theoretical cycle *(abcd)* with a cycle using liquid subcooling and vapor superheating (1234) operating between the same condensing and evaporating pressures. Subcooling of the liquid is represented by the process a-1, and superheating of the vapor is represented by the process c-3.

We will assume for purposes of discussion that the condenser pressure p_c is 117.16 psig (131.86 psia) and the evaporator pressure p_e is 36.971 psig (51.667 psia). From Table B-3 we may find that the saturation temperature at the condenser pressure is 100°F and the saturation temperature at the evaporator pressure is 40°F. If, after the vapor in the condenser is completely condensed, the resulting liquid is still further cooled to 76°F without change in pressure, the liquid refrigerant will have been subcooled 24°. When a liquid refrigerant can be subcooled either by cold water circulating in a heat exchanger or by other means external to the refrigeration cycle, the refrigeration effect of the system will be increased. This is because the enthalpy of the subcooled liquid is less than the enthalpy of the saturated liquid.

Illustrative Problem 3-8

Determine the increase in refrigerating effect caused by subcooling the liquid as indicated in Fig. 3-11.

Solution:

From Table B-3

$$h_a = h_f \text{ at } 100°\text{F} = 31.100 \text{ Btu/lb}$$

$$h_1 = h_f \text{ at } 76°\text{F} = 25.435 \text{ Btu/lb}$$

$$h_b = h_a$$

$$h_2 = h_1$$

$$h_c = h_g \text{ at } 40°\text{F} = 81.436 \text{ Btu/lb}$$

Refrigerating effect *without* subcooling

$$= h_c - h_a$$

$$= 81.436 - 31.100$$

$$= 50.336 \text{ Btu/lb}$$

Refrigerating effect *with* subcooling

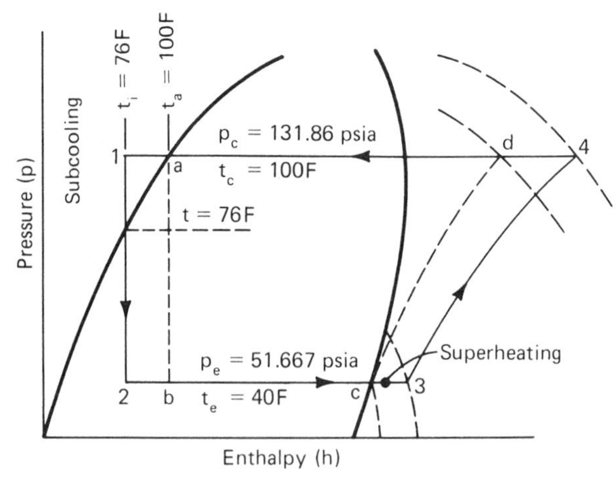

Figure 3–11

Schematic p-h diagram showing subcooling of liquid refrigerant and superheating of refrigerant vapor (Freon-12) (Not to scale.)

$$= h_c - h_1$$
$$= 81.436 - 25.435$$
$$= 56.001 \text{ Btu/lb}$$

The increase in refrigerating effect is 56.001 − 50.336 = 5.665 Btu/lb, or 5.665/50.336 = 11.25 percent.

From the results of the preceding problem it is evident that worthwhile gains in refrigerating capacity can be obtained through subcooling of the liquid refrigerant. As a generalization, it can be said that a 1 percent increase in system capacity can be anticipated for each 2° of liquid subcooling that is obtained from outside the refrigeration cycle. This fact has been responsible for design changes in all types of condensers to obtain significant amounts of liquid subcooling.

Two additional advantages derived from subcooling of the liquid are (1) less flash gas is produced during the expansion process and (2) the system has greater latitude in handling the liquid-line piping and location of the evaporator. Elevation changes in the liquid line and pressure drop can be more readily coped with if the liquid is subcooled. Other piping limitations such as velocity must be observed, however.

Further evidence of this is found in considerations of the pressure as related to Problem 3-8. Referring to Fig. 3-11 and Table B-3, we find the condensing pressure at 100°F is 117.16 psig, while the saturation pressure corresponding to 76°F is 78.391 psig. The difference between the two pressures is 38.769 psig, which the system designer can use in overcoming liquid-line rises (0.5 psig per foot rise), or line pressure drop, before flashing will occur. The designer, of course, should always allow a comfortable margin of from 5 to 10 psig for unseen contingencies.

Subcooling from a condenser can also create some problems for the designer, as will be discussed later under the subject of condensers.

Superheating. After liquid refrigerant has been admitted to an evaporator, it will usually be completely vaporized before it reaches the evaporator outlet connection. Inasmuch as the liquid is vaporized at low temperature, the vapor is still cold after the liquid has completely evaporated. As the cold vapor flows through the balance of the evaporator, it continues to absorb heat and becomes superheated.

The vapor absorbs sensible heat in the evaporator as it becomes superheated. Therefore, the refrigerating effect of each pound of refrigerant is increased. In other words each pound of refrigerant absorbs not only the heat required to vaporize it but also an additional amount of sensible heat, which superheats it. Figure 3-11 illustrates the superheating process (c-3). The refrigerating effect now becomes $h_3 - h_a$ in comparison to the theoretical cycle of $h_c - h_a$. Although the refrigerating effect has been increased over that of the theoretical cycle, the density of the vapor leaving the evaporator and entering the compressor is decreased. Superheating of any vapor always decreases its density. This means that the weight of vapor filling the compressor cylinders is decreased by superheating.

On the one hand the capacity of the refrigeration system is increased when superheating of the vapor takes place because the refrigerating effect of each pound of vapor removed has been increased by the sensible heat added during superheating. On the other hand the refrigerating capacity of a system is decreased because of the decrease in density during superheating. The effect of these two opposing tendencies must be computed in order to determine whether or not the refrigerating capacity of a system is increased by superheating the vapor in the evaporator. The various refrigerants are not alike in this respect. Each one must be checked individually. For superheats amounting to as much as 40 degrees, Freon-12 shows a negligibly small change in refrigerating capacity when superheating takes place inside the evaporator. The gain due to additional sensible heat absorbed by the vapor is almost equal to the loss due to its decrease in density.

Two additional reasons why superheating is frequently applied in practice are (1) superheating assures complete evaporation of all the liquid refrigerant before it enters the compressor and (2) in some types of control systems, variations in superheat temperature serves as a means of modulating the size of the opening of the expansion valve (automatic control).

The preceding discussion of superheating has dealt with the superheating that takes place within the evaporator. However, superheating of the vapor often takes place after it leaves the evaporator. The vapor may absorb sensible heat while it is flowing through the suction line that connects the evaporator to the compressor. This gain in sensible heat usually has no effect on the useful refrigeration because this heat has not been removed from the material being cooled. Thus, if vapor is superheated in an evaporator that is used to cool air, the sensible heat gained by the vapor is removed from the air being cooled. Useful cooling is done by the vapor. However, if the vapor is superheated in the suction line external to the airstream being cooled, no useful cooling is being done by the vapor while superheating. As a rule, it is only the heat absorbed by the vapor while it is in the evaporator that has any useful refrigerating value and which may properly be credited to the refrigerating effect.

When superheating of the vapor takes place in the suction line, there is no increase in refrigerating capacity to offset the loss due to the increase in the volume of the vapor. As a result, superheating of any refrigerant in the suction line decreases the capacity of the compressor in pounds pumped. For fluorinated hydrocarbons there is a theoretical loss in refrigerating capacity of 1 percent for each 4.3 degrees (approximately) of superheat. Thus, if the vapor entering a compressor has been superheated 21 degrees in the suction line, the refrigerating capacity of the system will be decreased by almost 5 percent. For this reason superheating of the vapor in the suction line of the system should be prevented. The suction line between the evaporator and the compressor is frequently insulated.

Subcooling by liquid suction-line heat exchanger. It is possible to subcool liquid refrigerant by means of a suction-line heat exchanger using the cold vapor leaving the evaporator as illustrated in Fig. 3-12b. In some cases the actual gain in refrigerating capacity may be small since the refrigerating capacity gained by liquid subcooling is partially offset by the loss in capacity due to the decrease in the density of the superheated vapor entering the compressor.

The maximum exchange of heat that can theoretically take place between fluorinated hydrocarbon liquid and vapor is limited in a refrigeration system because of the difference in their specific heats. For Freon-22 the specific heat of the liquid is 0.336 and of the vapor 0.163. Because of the lower specific heat of the vapor, its temperature rise is greater than the temperature drop of the liquid for each unit of heat exchanged. Thus, the greatest temperature range through which the liquid can theoretically be cooled in a counterflow cooler (with equal weights of gas and liquid) is 48.5 percent (0.163/0.336) of the difference between the initial and final temperature of the vapor. In practice, the temperature range through which the liquid can be cooled is even less than the 48.5 percent. This is because more of the heat that flows from the surroundings through the shell of the heat exchanger is absorbed by the cool refrigerant vapor than by the liquid refrigerant.

Actually, the gain in capacity with a liquid suction line heat exchanger is dependent upon the refrigerant used. For a system using Freon-12 there will be some gain in capacity. The gain will be negligible for Freon-22, however,

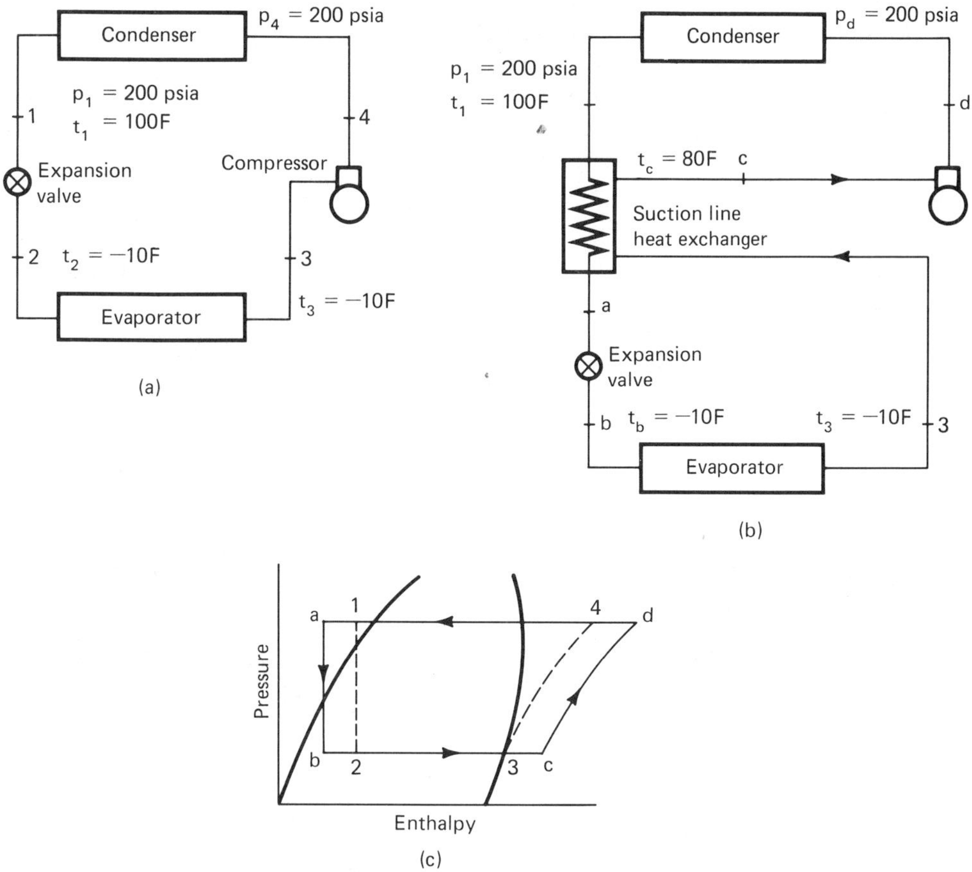

Figure 3–12

(a) Conventional refrigeration system (b) Subcooling by liquid-suction line heat exchanger (c) P-h diagram

because of a decrease in compressor capacity due to decreased gas density from superheat.

Subcooling by means of a liquid-suction heat exchanger does provide some benefit because of the increased capacity of the expansion valve. Each pound of liquid flowing through the expansion valve is able to absorb more heat in the evaporator (less flash gas produced). Therefore, the refrigerating capacity of an expansion valve, through which a given weight of liquid flows, increases as the temperature of the entering liquid is reduced.

Illustrative Problem 3-9

For a conventional refrigeration cycle (Fig. 3-12a) that uses Freon-12 as the refrigerant, the temperature in the evaporator is −10°F. The fluid leaves the evaporator as saturated vapor at −10°F and enters the isentropic compressor. The pressure in the condenser is 200 psia. The liquid leaves the condenser and enters the expansion valve at a temperature of 100°F.

It is proposed to modify the cycle by adding

a heat exchanger, as shown in Fig. 3-12b, in which the cold vapor leaving the evaporator cools the liquid before it enters the expansion valve.

Compare the coefficient of performance of these two cycles.

Solution:

The *p-h* diagram for the two cycles is shown in Fig. 3-12c. The conventional cycle is 12341. With the heat exchanger the cycle is *ab3cda*.

Conventional Cycle Calculations

From Table B-3

$$h_1 = h_f = 31.100 \text{ Btu/lb} \quad \text{at } 100°\text{F}$$

$$h_2 = h_1 = 31.100 \text{ Btu/lb}$$

$$h_3 = h_g = 76.196 \text{ Btu/lb} \quad \text{at } -10°\text{F}$$

$$s_3 = 0.16989 \text{ Btu/(lb°R)}.$$

From Fig. B-2:

$$h_4 = h \text{ at } p_4 = 200 \text{ psia} \quad \text{and}$$

$$s_4 = s_3 = 0.16989 \text{ Btu/(lb °R)}$$

$$h_4 = 94.5 \text{ Btu/lb}$$

By Eq. 3-13

$$\text{COP} = \frac{h_3 - h_1}{h_4 - h_3} = \frac{76.196 - 31.100}{94.5 - 76.196}$$

$$\text{COP} = 2.46$$

Cycle with Suction-Line Heat Exchanger Calculations

From before

$$h_1 = 31.100$$

$$h_3 = 76.196$$

From Table B-3

$$p_3 = p_c = 19.189 \text{ psia}$$

(saturation pressure at -10 °F)

$$t_c = 80°\text{F} \quad \text{(given)}$$

From Fig. B-2, at $p_c = 19.189$ psia and $t_c = 80$ °F find

$$h_c = 89.0 \text{ Btu/lb}$$

At $p_d = 200$ psia and $s_c = s_d$ find

$$h_d = 112 \text{ Btu/lb}$$

For the heat exchanger:

$$h_1 - h_a = h_c - h_3$$

$$31.100 - h_a = 89.0 - 76.196$$

$$h_a = 18.3 \text{ Btu/lb}$$

$$\text{COP} = \frac{h_3 - h_a}{h_d - h_c} = \frac{76.196 - 18.3}{112 - 89.0}$$

$$\text{COP} = 2.51$$

This represents a slight improvement in performance by use of the suction-line heat exchanger to subcool the liquid entering the expansion valve.

Before proceeding further with our discussion of refrigeration cycles, we will now put together a series of four problems that will illustrate the use of relationships presented thus far.

Illustrative Problem 3-10

A food-storage locker requires a refrigeration system of 12 tons capacity at an evaporator temperature of 20°F and a condenser temperature of 86°F. The refrigerant, ammonia, is subcooled 9°F before entering the expansion valve, and the vapor is superheated 9°F before leaving the evaporator coil. Compression of

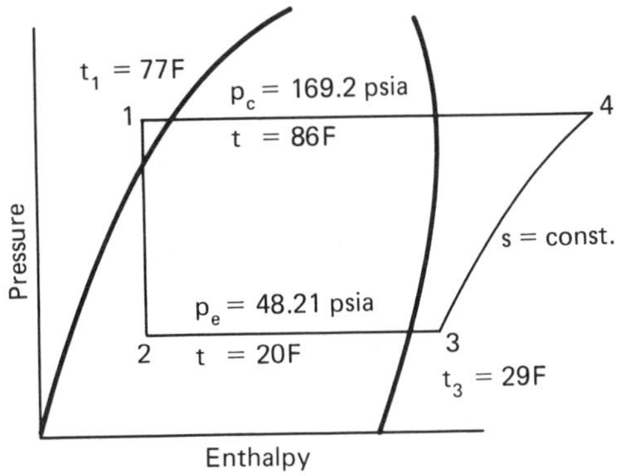

Figure 3–13

P-h diagram for Illustrative Problem 3-10 (Ammonia) (Not drawn to scale.)

the refrigerant is adiabatic, and the compressor valve throttling and compressor clearance are to be disregarded. A two-cylinder vertical single-acting compressor with stroke equal to 1.5 times the bore is to be used operating at 900 rpm. Determine the following: (a) refrigerating effect, (b) weight of refrigerant to be circulated per minute, (c) theoretical piston displacement per minute, (d) theoretical horsepower, (e) coefficient of performance, (f) heat removed through condenser, and (g) theoretical bore and stroke of the compressor.

Solution:

Step 1: Construct a sketch of the *p-h* diagram, process lines and identify points in cycle.

Step 2: Identify all fluid properties required for solution of problem, referring to Tables B-1, B-2, and Fig. B-1.

Step 3: Calculate individual quantities required.

(a) Refrigerating effect $= h_3 - h_1 = 623.3 - 128.5 = 494.8$ Btu/lb.

(b) Weight of refrigerant circulated

By Eq. (3-9) $$\dot{m}_r = \frac{200}{h_3 - h_1} = \frac{200}{494.8} = 0.404 \text{ lb/ton min}$$

Total $$\dot{m}_r = 0.404(12) = 4.86 \text{ lb/min}$$

(c) Theoretical piston displacement

By Eq. (3-15) $$\dot{Q}_t = \dot{m}_r \times (v_g)_3 = 0.404(6.1) = 2.46 \text{ ft}^3\text{/ton min}$$

$$(\dot{Q}_t)_{\text{total}} = 2.46(12) = 29.6 \text{ ft}^3\text{/min}$$

(d) Theoretical horsepower

By Eq. (3-11) $$\text{ihp}_c = 4.717\left(\frac{h_4 - h_3}{h_3 - h_1}\right) = 4.717\left(\frac{701.9 - 623.3}{623.3 - 128.5}\right) = 0.75 \text{ hp/ton}$$

$$(\text{ihp}_c)_{\text{total}} = 0.75(12) = 9.0 \text{ hp}$$

Point 1	Point 2	Point 3	Point 4
$t_1 = 86 - 9 = 77°\text{F}$	$t_2 = 20°\text{F}$	$t_3 = 20 + 9 = 29°\text{F}$	$t_4 = ?$
$p_1 = 169.2$ psia	$p_2 = 48.21$ psia	$p_3 = 48.21$ psia	$p_4 = 169.2$ psia
$h_1 = 128.5$	$h_2 = 128.5$	$h_3 = 623.3$	$h_4 = 701.9$
		$v_{g_3} = 6.1$	

Alternately, by Eq. (3-11a) with $n = 1.31$ (Table 3-3)

$$\text{ihp} = \frac{0.873\ np_3v[(p_4/p_3)^{(n-1)/n} - 1)}{(n - 1)(h_3 - h_1)}$$

$$= \frac{0.873(1.31)(48.21)(6.1)}{(1.31 - 1)(623.3 - 128.5)}$$

$$\times \frac{[(169.2/48.21)^{(1.31-1)/1.31} - 1]}{(1.31 - 1)(623.3 - 128.5)}$$

$$= 0.75 \text{ hp/ton}$$

In this case the alternate calculation for theoretical horsepower checks out much more closely with the calculations obtained by use of refrigerant tables and charts than can normally be expected. In the alternate calculation the value $n = k = 1.31$ has been used, although this value is not truly applicable when the perfect gas equations are used in the region close to the saturated vapor line.

(e) Coefficient of performance

By Eq. (3-1) $$\text{COP} = \frac{h_3 - h_1}{h_4 - h_3}$$

$$= \frac{623.3 - 128.5}{701.9 - 623.3}$$

$$= 6.28$$

(f) Heat removed through condenser

By Eq. (3-7a) $$q_{cond} = 200\left(\frac{h_4 - h_1}{h_3 - h_1}\right)$$

$$= 200\left(\frac{701.9 - 128.5}{623.3 - 128.5}\right)$$

$$= 232 \text{ Btu/ton min}$$

$$232 \times 12 = 2784 \text{ Btu/min}$$

(g) Theoretical cylinder dimensions.
Theoretical piston displacement per cylinder = 29.6/2 = 14.8 ft³/min.

By Eq. (3-16) $$\text{PD}_t = \frac{d^2 \times s \times n \times \text{rpm}}{2200}$$

$$= \frac{d^2(1.5d)(1)(900)}{2200}$$

$$14.8 = \frac{d^2(1.5d)(1)(900)}{2200}$$

$$d = 2.89 \text{ in.} \quad \text{(bore)}$$

$$s = 1.5(2.89) = 4.34 \text{ in.} \quad \text{(stroke)}$$

Illustrative Problem 3-11

If the compressor in Problem 3-10 has 2 percent clearance, determine (a) the clearance volumetric efficiency, (b) corrected piston displacement, and (c) required bore and stroke of the compressor.

<u>*Solution:*</u>

(a) By Eq. (3-18)

$$\eta_{vc} = 1 + C - C\left(\frac{p_d}{p_s}\right)^{1/n}$$

$$= 1 + 0.02 - 0.02\left(\frac{169.2}{48.21}\right)^{1/1.31}$$

$$= 0.968$$

(b) Actual piston displacement $= \dfrac{\text{PD}_t}{\eta_{vc}}$

$$= \frac{29.6}{0.968} = 30.6 \text{ ft}^3\text{/min}$$

(c) Cylinder dimensions.

By Eq. (3-16)

$$30.6 = \frac{d^2(1.5d)(2)(900)}{2200}$$

$$d = 2.92 \text{ in.} \quad \text{(bore)}$$

$$s = 1.5(2.92) = 4.38 \text{ in.} \quad \text{(stroke)}$$

Illustrative Problem 3-12

If the compressor in Problem 3-11 has a suction valve pressure drop of 4 psi and a discharge-valve pressure drop of 2 psi, determine (a) the piston displacement, (b) horsepower, (c) coefficient of performance, (d) heat removed through condenser, and (e) bore and stroke of compressor.

Solution:

(a) By Eq. (3-19)

$$\eta_{vt} = \left[1 + c - c\left(\frac{p_d}{p_s}\right)^{1/n}\right]\left(\frac{p_{cy}}{p_s}\right)\left(\frac{T_s}{T_{cy}}\right)$$

with $p_s = 48.21$, $p_{cy} = 48.21 - 4 = 44.21$ psia. Referring to Fig. B-1, $T_s = T_{cy}$ (approximately). Therefore,

$$\eta_{vt} = 0.968\left(\frac{44.21}{48.21}\right)$$
$$= 0.888$$

$$\text{Piston displacement} = \frac{29.6}{0.88} = 33.3 \text{ ft}^3/\text{min}$$

Alternate solution:

By Eq. (3-18), using cylinder pressure ($p_s = 44.21$ psia, $p_d = 169.2 + 2 = 171.2$ psia)

$$\eta_{vc} = 1 + 0.02 - 0.02\left(\frac{171.2}{44.21}\right)^{1/1.31}$$
$$= 0.964$$

By Eq. (3-9) and Eq. (3-20), using v_g at cylinder suction pressure,

$$\text{PD} = \frac{4.86(6.65)}{0.964} = 33.5 \text{ ft}^3/\text{min}$$

(b) Horsepower
By Eq. (3-11)

$$\text{ihp}_c = 4.717\left(\frac{h_4' - h_3'}{h_3' - h_1}\right)$$

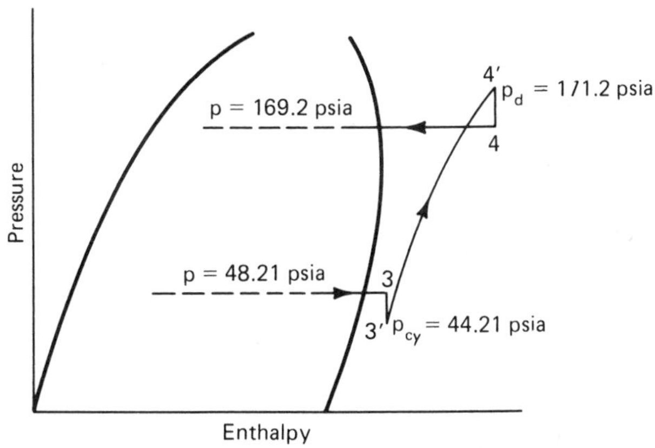

FIGURE 3–14

P-h diagram for Illustrative Problem 3-12 (Not drawn to scale.)

Referring to Fig. B-1 at p_d = 171.2 psia, $h_4{}'$ = 709 Btu/lb, and $h_3{}'$ = 623.3 Btu/lb = h_3

$$\text{ihp}_c = 4.717\left(\frac{709 - 623.3}{623.3 - 128.5}\right)$$
$$= 0.82 \text{ hp/ton}$$
$$(0.82)(12) = 9.84 \text{ hp}$$

(c) Coefficient of performance
By Eq. (3-14),

$$\text{COP} = \frac{4.717}{i\text{HP}_c} = \frac{4.717}{0.82} = 5.76$$

(d) Heat removed through condenser
By Eq. (3-7a)

$$\dot{q}_{\text{cond}} = 200\left(\frac{h_4 - h_1}{h_3 - h_1}\right)$$

With $h_4 = h_4{}'$ = 709 Btu/lb, and h_3 = 623.3 Btu/lb

$$\dot{q}_{\text{cond}} = 200\left(\frac{709 - 128.5}{623.3 - 128.5}\right)$$
$$= 235 \text{ Btu/ton min}$$
$$(235)(12) = 2820 \text{ Btu/min}$$

(e) Bore and stroke of compressor

By Eq. (3-16) $$33.5 = \frac{d^2(1.5d)(2)(900)}{2200}$$

$$d = 3.01 \text{ in.} \quad \text{(bore)}$$
$$(1.5)(3.01) = 4.51 \text{ in.} \quad \text{(stroke)}$$

Illustrative Problem 3-13

If the compressor of Problem 3-10 is equipped with a water jacket so that the compression follows a polytropic path n = 1.20 and the mechanical efficiency is 80 percent, determine (a) theoretical horsepower, (b) actual horsepower, (c) heat rejected to compressor cooling water, and (d) heat rejected to condenser coolant.

Solution:

(a) Theoretical horsepower
By Eq. (3-11a)

$$\text{ihp}_c = \frac{0.873 n p_3{}' v_3{}'\left[\left(\frac{p_4'}{p_3'}\right)^{(n-1)/n} - 1\right]}{(n - 1)(h_3 - h_1)}$$

$$= \frac{(0.873)(1.2)(44.21)(6.65)}{(1.2 - 1)(623.3 - 128.5)}$$

$$\times \frac{\left[\left(\frac{171.2}{44.21}\right)^{(1.2-1)/1.2} - 1\right]}{(1.2 - 1)(623.3 - 128.5)}$$

$$= 0.79 \text{ hp/ton}$$
$$(0.79)(12) = 9.48 \text{ hp}$$

(b) Actual horsepower

$$\text{Actual horsepower} = \frac{\text{ihp}}{\text{mechanical efficiency}} = \frac{9.48}{0.80} = 11.86 \text{ hp}$$

(c) Heat rejected to compressor cooling water.
By Eq. (3-12a)

$$\dot{q}_{\text{rej}} = \frac{n}{(n - 1)J}(p_3'v_3')\left[\left(\frac{p_4'}{p_3'}\right)^{(n-1)/n} - 1\right] - (h_4' - h_3')$$

$$= \frac{(1.2)(144)(44.21)(6.65)}{(0.2)(778)}$$

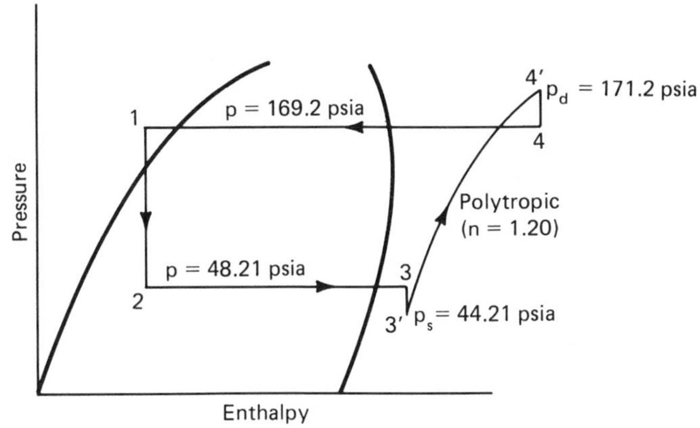

Figure 3–15

P-h diagram for Illustrative Problem 3-13 (Not drawn to scale.)

$$\times \frac{\left[\left(\frac{171.2}{44.21}\right)^{(1.2\ -\ 1)/1.2} - 1\right]}{(0.2)(778)}$$

$$- (676 - 623.3)$$

$$\dot{q}_{rej} = 30.3 \text{ Btu/lb}$$

$$30.3(4.86) = 2654 \text{ Btu/min}$$

For a final problem in this series, we will be using cycle temperatures and pressures that could have been obtained by actual measurements taken during the system operation. Certain assumptions will also be necessary.

Illustrative Problem 3-14

An ammonia vapor compression refrigeration system operates with a capacity of 150 tons. The condenser pressure is 200 psia and the evaporator pressure is 34 psia. Liquid ammonia leaves the condenser and enters the expansion valve at 80°F. Ammonia vapor leaves the evaporator at 20°F and is superheated to a temperature of 30°F in the suction line. Pressure drop in the compressor suction valves is 4 psi and in the discharge valves 10 psi. The vapor is superheated 10 degrees in the cylinder on the intake stroke after passing the suction valves. Compression will be assumed polytropic with $n = 1.20$. The discharge vapor is cooled 100°F in the compressor discharge line before entering the condenser. Ignore pressure drop in condenser, evaporator, and refrigerant piping. Ignore heat transfer in the liquid line to expansion valve and in the line between the expansion valve and the evaporator. The compressor is a four-cylinder, single-acting unit operating at 900 rpm, has a clearance of 5 percent, and has a stroke-to-bore ratio of 1.25. Determine the following items: (a) volumetric efficiency of compressor, (b) required bore and stroke for the compressor, (c) brake horsepower input to compressor for an assumed mechanical efficiency of 75 percent, (d) heat rejected to compressor jacket water in Btu/min, and (e) quantity of condenser-cooling water required in gallons per minute if the water temperature rise is 12°F.

<u>*Solution:*</u>

Construct a sketch of the mechanical equip-

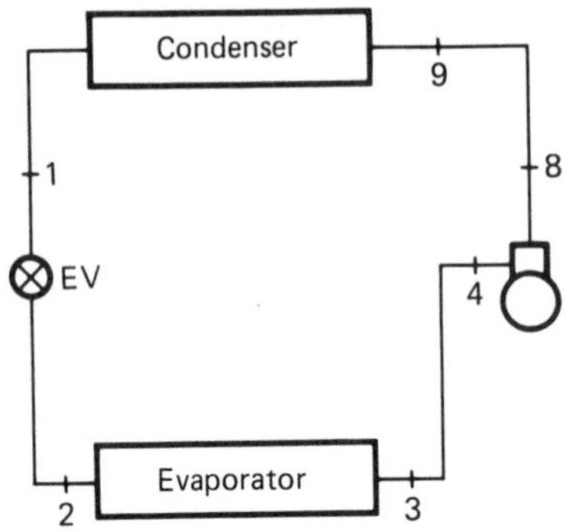

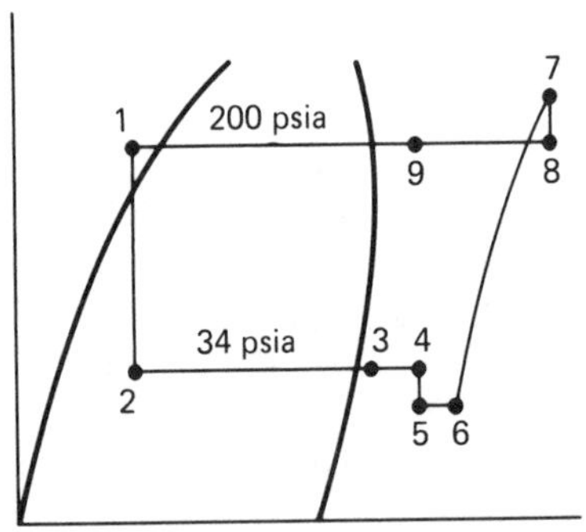

Figure 3–16

Diagrams for Illustrative Problem 3-14 (Ammonia). (Not drawn to scale.)

ment arrangement and *p-h* diagram for the system.

The next step is to identify the properties of all state points in the system. States (1), (2), (3), (4), (5), and (6) may be located from given data and using Tables B-1, B-2, and Fig. B-1. Thus,

$p_1 = 200$ psia $\quad t_1 = 80°F$

$h_1 = h_2 = 132.0$ Btu/lb

$p_2 = p_3 = p_4 = 34$ psia $\quad t_3 = 20°F$

$h_3 = 622$ Btu/lb

$t_4 = 30°F \quad h_4 = 628$ Btu/lb

$v_4 = 8.8$ ft³/lb

$p_5 = p_4 - \Delta p_s = 34 - 4 = 30$ psia

$h_5 = h_4 = 628$ Btu/lb

$t_5 = 28°F \quad p_6 = p_5 = 30$ psia

$t_6 = t_5 + 10 = 38°F$

$h_6 = 634 \quad v_6 = 10.2$ ft³/lb

State (7) may be located by the following method:

$$p_7 = p_8 + 10 = 200 + 10 = 210 \text{ psia}$$

$$p_6 v_6^n = p_7 v_7^n$$

$$v_7 = \frac{v_6}{\left(\dfrac{p_7}{p_6}\right)^{1/n}} = \frac{10.2}{\left(\dfrac{210}{30}\right)^{1/1.2}} = 2.02 \text{ ft}^3/\text{lb}$$

Thus,

$p_7 = 210$ psia $\quad v_7 = 2.02$ ft³/lb

$h_7 = 734$ Btu/lb $\quad t_7 = 250°F$

$p_8 = 200$ psia $\quad h_8 = h_7 = 734$ Btu/lb

$t_8 = 248°F$

$p_9 = p_8 = 200$ psia

$t_9 = t_8 - 100 = 248 - 100 = 148F$

$h_9 = 672$ Btu/lb

(a) Volumetric efficiency using v_4 instead of v_3
By Eq. (3-19a)

$$\eta_{vt} = \left[1 + C - C\left(\frac{p}{p}\right)^{1/n}\right]\frac{v_4}{v_6}$$

$$= \left[1 + 0.05 - 0.05\left(\frac{210}{30}\right)^{1/1.2}\right]\left(\frac{8.8}{10.2}\right)$$

$$\eta_{vt} = 0.688$$

(b) Compressor bore and stroke
By Eq. (3-9)

$$\dot{m}_r = \frac{200}{h_3 - h_1} = \frac{200}{622 - 132.0}$$

$$= 0.408 \text{ lb/ton min}$$

$$(150)(0.408) = 61.2 \text{ lb/min}$$

By Eq. (3-15) $Q_t = \dot{m}_r \times (v_4) = (61.2)(8.8) = 538.6$ cfm

$$\text{Actual } Q_t = \frac{Q_t}{\eta_{vt}} = \frac{538.6}{0.688} = 782.8 \text{ cfm}$$

$$\text{By Eq. (3-16)} \quad \text{PD} = \frac{d^2 \times s \times n \times \text{rpm}}{2200}$$

$$782.8 = \frac{(d^2)(1.25d)(4)(900)}{2200}$$

$$d = 7.26 \text{ in.} \quad \text{(bore)}$$

$$(1.25)(7.26) = 9.08 \text{ in.} \quad \text{(stroke)}$$

By Eq. (3-10b)

$$\text{Polytropic work} = \left(\frac{200}{h_3 - h_1}\right)\left(\frac{n}{n-1}\right)\left(\frac{p_6 v_6}{778}\right)\left[\left(\frac{p_7}{p_6}\right)^{\frac{n-1}{n}} - 1\right]$$

$$= \frac{(0.408)(1.2)(144)(30)(10.2)}{(0.20)(778)}\left[\left(\frac{210}{30}\right)^{(1.2-1)/1.2} - 1\right]$$

$$= 53.1 \text{ Btu/ton min}$$

$$(150)(53.1) = 7965 \text{ Btu/min}$$

(c) By Eq. (3-11a), or

$$\text{ihp} = \frac{7965 \times 778}{33000} = 187.8 \text{ hp}$$

$$\text{BHP} = \frac{\text{ihp}}{\eta_m} = \frac{187.8}{0.75} = 250.4 \text{ hp}$$

(d) Heat to compressor coolant jacket
By Eq. (3-12a) or

$$\dot{q}_{comp} = \text{polytropic work} - (h_8 - h_4)$$

From part (b), polytropic work = 53.1 Btu/ton min, or 7965 Btu/min. Therefore,

$$\dot{q}_{comp} = 7965 - (61.2 \text{ lb/min})(734 - 628)$$

$$= 1478 \text{ Btu/min}$$

(e) Quantity of condenser-cooling water required.

$$\dot{q}_{cond} = \dot{m}_r(\text{h}_9 - \text{h}_1) = 61.2(672 - 132.0)$$

$$= 33{,}048 \text{ Btu/min}$$

$$\dot{q}_{water} = \dot{q}_{cond}$$

$$\dot{m}_w c_p(\Delta t) = 33{,}048$$

$$\dot{m}_w = \frac{33{,}048}{1(12)} = 2754 \text{ lb/min}$$

$$\text{gpm} = \frac{2754}{8.345 \text{ lb/gal}} = 330 \text{ gpm}$$

3-14 VARIATION IN REFRIGERATION CAPACITY WITH SUCTION PRESSURE

The capacity of a system is not a constant quantity but depends upon the temperature of the liquid ahead of the expansion valve and the temperature at which the liquid vaporizes in the evaporator. The change in capacity due to the first of these two factors was discussed in Section 3-13 under subcooling. The variation of system capacity because of changes in suction pressure will be discussed in this section.

The higher the temperature of the liquid vaporizing in the evaporator, the higher will be the pressure. Also, by referring to the tables of properties, you may note the rapid increase in the density of the vapor with an increase in evaporation temperature and pressure. This means that 1 lb of vapor at high pressure occupies a smaller volume than at a low pressure. Therefore, for each cubic foot of piston displacement, a larger weight of vapor will flow into the compressor cylinders when the evaporated pressure is high than when it is low. It is important to remember that the refrigerating capacity of all reciprocating compressor systems increases with an increase in suction pressure, if other factors remain constant.

Another factor should be noted. Referring to Table 3-4, you may see that as the compression ratio of a compressor increases there is a decrease in volumetric efficiency. Therefore, for a constant discharge pressure, the volumetric efficiency will increase as the suction pressure increases.

Illustrative Problem 3-15

An eight-cylinder 2.75 by 2.00 in. refrigerant-12 compressor is directly driven at 1750 rpm. The condenser pressure (head pressure) is 110.0 psig and the temperature of the liquid at the expansion valve is 86°F. Determine the theoretical and actual refrigerating capacity of the system if the evaporator temperature is (a) 20°F and (b) 46°F. Assume saturated vapor leaves evaporator and enters compressor.

Solution:

Step 1: Draw the *p-h* diagram for the problem and number points of interest.

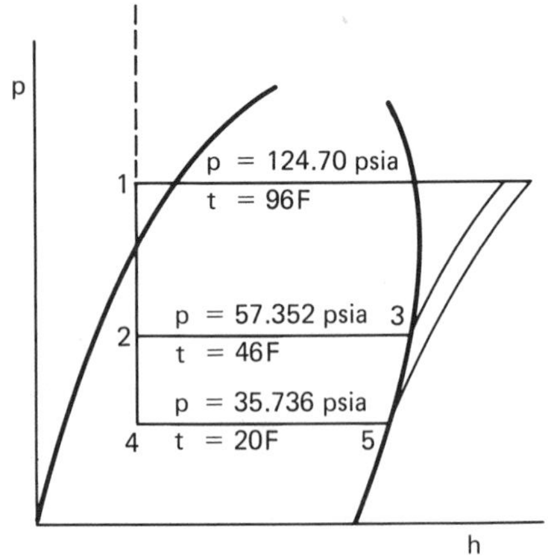

Figure 3–17

P-h diagram for Illustrative Problem 3-13 (Not drawn to scale.)

Step 2: Determine enthalpy values for numbered points on Fig. 3-17. From Table B-3:

$$h_1 = h_f \text{ at } 86°\text{F} = 27.769 \text{ Btu/lb}$$

$$h_1 = h_2 = h_4 = 27.769 \text{ Btu/lb}$$

$$h_3 = h_g \text{ at } 46°\text{F} = 82.037 \text{ Btu/lb}$$

$$h_5 = h_g \text{ at } 20°\text{F} = 79.385 \text{ Btu/lb}$$

Step 3: Determine theoretical and actual system capacity for 20°F evaporator temperature. By Eq. 3-16,

$$\text{PD}_t = \frac{d^2 \times s \times n \times \text{rpm}}{2200}$$

$$= \frac{(2.75)^2(2.00)(8)(1750)}{2200}$$

$$\text{PD}_t = 96.25 \text{ cfm}$$

By Eq. 3-5, refrigerating effect $= h_5 - h_1 = 79.385 - 27.769 = 51.616$ Btu/lb. By Eq. 3-9

$$\dot{m}_r = \frac{200}{h_5 - h_1} = \frac{200}{51.616}$$

$$= 3.87 \text{ lb/ton min}$$

From Table B-3 v_{g_5} at 20°F = 1.0988, or ρ_g = 1/1.0988 = 0.9100 lb/ft^3
Volume to be compressed = m_r/ρ_g = 3.87/0.911 = 4.25 ft^3/ton min
Theoretical system capacity = 96.25/4.25 = 22.65 tons *(ans)*

$$\text{CR} = 124.70/35.736 = 3.489$$

From Table 3-4 η_{vt} = 78.6 percent
Actual system capacity = 22.65 × 0.786 = 17.8 tons *(ans)*

Step 4: Determine theoretical and actual system capacity for 46°F evaporator temperature.

By Eq. 3-5

$$\text{Refrigerating effect} = h_3 - h_1$$
$$= 82.037 - 27.769$$
$$= 54.268 \text{ Btu/lb}$$

By Eq. 3-9

$$\dot{m}_r = \frac{200}{h_3 - h_1} = \frac{200}{54.268} = 3.68 \text{ lb/ton min}$$

From Table B-3 v_{g_3} at 46°F = 0.69982, or ρ_g = 1/0.69982 = 1.4289 lb/ft^3

Volume to be compressed = $\dot{m}_r/\rho_g$ = 3.68/1.4289 = 2.575 ft^3/ton min
Theoretical system capacity = 96.25/2.575 = 37.38 tons *(ans)*

$$\text{CR} = 124.70/57.352 = 2.17$$

From Table 3-4 η_{vt} = 87.2 percent

Actual system capacity = 37.38 × 0.872 = 32.6 tons *(ans)*

Notice in the previous problem that, in spite of the fact that the weight of refrigerant to be circulated is very nearly the same (3.87 versus 3.68) in both cases, the volume per ton is quite different. The difference in refrigerating capacity of the system at the two suction temperatures is due almost entirely to the difference in the specific volume of the vapor. At the higher suction temperature and pressure, the vapor is more dense. Even though the piston displacement remains the same, a greater *weight* of vapor fills the cylinder at the higher suction conditions. Also note that the volumetric efficiency is higher at lower compression ratios.

In order to maintain a given suction pressure, the evaporator must be so selected that it can vaporize as great a weight of refrigerant as the compressor can handle. If the load on the evaporator increases and the refrigerant vaporizes faster than the compressor can remove it, the suction pressure will rise because the excess vapor will remain in the evaporator. However, due to the rising suction pressure, the vapor becomes more dense and a point is soon reached where, even though the volume of vapor pumped by the compressor remains the same, the *weight removed* is equal to the *weight evaporated*. When this point is reached, equilibrium is established and the suction pressure will remain constant.

On the other hand, if the compressor is removing a larger weight of refrigerant than is being vaporized in the evaporator, the suction pressure (and temperature) will fall. As the pressure falls, the vapor expands and a smaller weight fills the cylinder at each succeeding stroke. Again a point will soon be reached where the weight removed is equal to the weight vaporized, and the suction pressure will then remain constant.

The suction pressure of any given compressor is not a constant quantity. It is continually changing as the load on the system changes. In an air-conditioning system the load on the compressor will increase with an increase in the temperature of the air entering the evaporator coil. An increase in load may also be due to an increase in the quantity of air flowing over the evaporator. The suction pressure will rise with an increase in load and fall with a decrease. The compressor must be so selected that, when operating at full load, it will maintain a suction pressure corresponding to the highest evaporator temperature needed. If the compressor should be overloaded, the pressure will rise above this point. However, if the load should be lighter than that for which the compressor was selected, the suction pressure will fall.

3-15 POWER REQUIREMENTS OF COMPRESSORS

In Section 3-11 the ideal work and ideal horsepower requirements for a compressor were discussed. Equation (3-6) states the isentropic work of compression in units of Btu/lb of refrigerant compressed. Equation (3-10) states the isentropic work in units of Btu/ton min. Equation (3-11) gives the ideal horsepower required to compress the refrigerant vapor per ton of refrigeration. Recall that isentropic compression means no heat transfer between the vapor being compressed and cylinder walls and that the compression process is frictionless.

In actual compressors the power required is greater than that computed theoretically because of the heat exchange that actually does take place between the vapor and the cylinder walls and also because of the inevitable friction losses in the compressor. The power that must actually be applied to the shaft of a compressor is called the *brake horsepower (BHP)*. The required brake horsepower for a compressor can be found from that computed theoretically by applying a factor called the *overall efficiency*. The overall efficiency (η_{oa}) is defined as follows:

$$\eta_{oa} = \frac{\text{Theoretical Horsepower } (\text{ihp}_c)}{\text{Brake Horsepower (BHP)}} \qquad (3\text{-}21)$$

The overall efficiency must be determined by performance tests on actual compressors. The overall efficiency takes into account not only the exchange of heat between the cylinder walls and the vapor but also the friction of the moving parts of the compressor.

Figure 3-18 gives the approximate average overall efficiencies that may be expected from modern compressors using refrigerant-12 or refrigerant-22. While the efficiency curves represent only the compressor of one manufacturer at 1750 rpm, they may be used for estimating purposes at other speeds and for other manufacturer's equipment. If more accurate performance data are required, it is recommended that the compressor manufacturer be consulted.

The power required to offset the friction losses of a compressor can best be determined by test. It is also possible to find by test both the *mechanical efficiency* and the *compression*

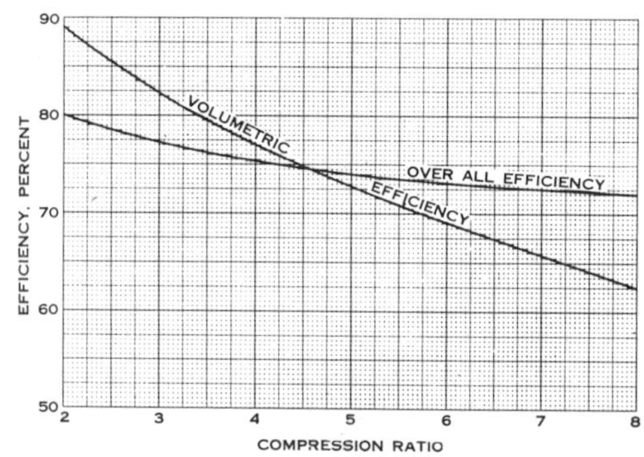

Figure 3–18

Average efficiencies for reciprocating compressors using Freon-12 and Freon-22

efficiency. The mechanical efficiency is an index of the friction losses, and the compression efficiency is an index of the departure of the actual compression process from the isentropic compression process. Multiplying the mechanical efficiency by the compression efficiency gives the overall efficiency. The overall efficiency is the only one that need actually be known in order to compute the brake horsepower required for a compressor. However, the performance of compressors is sometimes presented in terms of their mechanical and compression efficiencies and in such a case, it is necessary to use them in finding the overall efficiency. They are useful for analyzing the performance of compressors from test data.

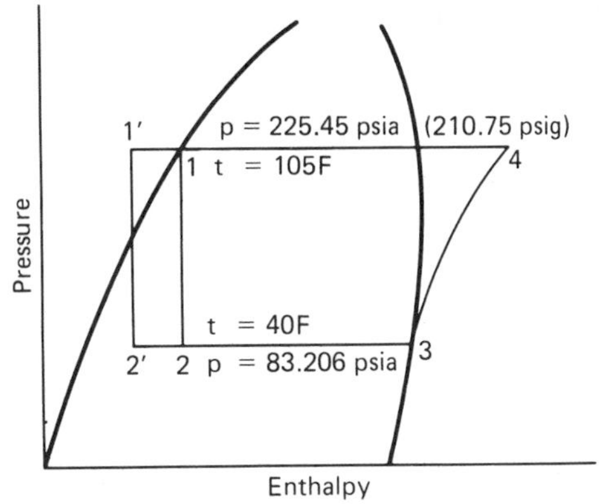

Figure 3–19

Sketch of p-h diagram for Illustrative Problems 3-16 and 3-17

Illustrative Problem 3-16

A 30-ton Freon-22 compressor maintains an evaporator temperature of 40°F while discharging against a head pressure of 210.75 psig. The liquid refrigerant enters the expansion valve as saturated liquid at condenser pressure. Saturated vapor at the evaporator temperature enters the compressor. Determine the following:

a. The total theoretical horsepower required

b. The total brake horsepower required

<u>Solution:</u>

Step 1: Make a sketch of the *p-h* diagram for Freon-22 and find pressure, temperature, and enthalpy values at cycle points.

From Table B-5 at the condenser pressure of 225.45 psia (210.75 psig) find saturation temperature of 105°F. $h_1 = h_f = 40.847$ Btu/lb. At the evaporator temperature of 40°F, find saturation pressure of 83.206 psia (68.51 psig), $h_3 = h_g = 108.142$ Btu/lb, and $s_3 = S_g = 0.21986$ Btu/lb°R. From Fig. B-3 with $s_3 = s_4 = 0.21986$, and $p_4 = 225.45$ psia, find $h_4 = 118.5$ Btu/lb.

Step 2: Determine refrigeration effect and mass flow of refrigerant. By Eq. (3-5),

$$\text{Refrigerating effect} = h_3 - h_1 = 108.142 - 40.847 = 67.295 \text{ Btu/lb}$$

By Eq. (3-9)

$$\dot{m}_r = \frac{200}{h_3 - h_1} = \frac{200}{67.295} = 2.97 \text{ lb/ton min}$$

$$\text{Total } \dot{m}_r = (2.97 \text{ lb/ton min})(30 \text{ tons}) = 89.1 \text{ lb/min}$$

Step 3: Determine theoretical horsepower required. By Eq. 3-6,

$$\text{iwk}_c = h_4 - h_3 = 118.5 - 108.142 = 10.358 \text{ Btu/}lb$$

$$\text{Total iwk}_c = \dot{m}_r(h_4 - h_3)$$
$$= 89.1(10.358)$$
$$= 922.9 \text{ Btu/min}$$

$$\text{Total ihp}_c = \frac{(922.9 \text{ Btu/min})(60 \text{ min/hr})}{2545 \text{ Btu/hp hr}}$$
$$= \frac{922.9}{42.4}$$
$$= 21.7 \text{ hp}$$

or By Eq. 3-11

$$\text{ihp}_c = 4.717\left(\frac{h_4 - h_3}{h_3 - h_1}\right)$$
$$= 4.717\left(\frac{10.358}{67.295}\right)$$
$$= 0.726 \text{ hp/ton}$$

$$\text{Total ihp}_c = (0.726 \text{ hp/ton})(30 \text{ tons})$$
$$= 21.7 \text{ hp}$$

Step 4: Determine total brake horsepower.

$$\text{CR} = \frac{p_d}{p_s} = \frac{225.45}{83.206}$$
$$= 2.71$$

From Fig 3-18 η_{oa} = 78 percent.
By Eq. (3-21),

$$\text{BHP} = \frac{ihp_c}{\eta_{oa}}$$
$$= \frac{21.7}{0.78}$$
$$\text{BHP} = 27.8 \text{ hp}$$

The compressor operating under the conditions of Problem 3-16 would have a 30-hp motor, the next larger commercial size obtainable. Compressors for air-conditioning duty are frequently equipped with motors selected on the basis of 1 hp/ton of refrigeration. Such compressors have been rated on the assumption that the evaporator temperature will be 40°F, a temperature commonly used in air-conditioning work and also specified in Groups IV and VIII of ASRAE Standard 23-R.

When the liquid refrigerant is subcooled after leaving the condenser, the horsepower required per ton of refrigeration is reduced. This is due to the fact that a smaller weight of refrigerant need be circulated and compressed in order to obtain 1 ton of refrigerating capacity. The following problem illustrates this.

Illustrative Problem 3-17

Use all the data of Problem 3-16, except that the liquid is subcooled to 80°F after leaving the condenser and before entering the expansion valve. Find the required brake horsepower.

Solution:

Step 1: Refer to Fig. 3-19 and locate pcint 1′ at 80°F and at the condenser pressure of 225.45 psia. From Table B-5 at 80°F find h_1 = 33.109 Btu/lb.

The refrigerating effect is now $h_3 - h_1' = 108.142 - 33.109 = 75.033$ Btu/lb. By Eq. (3-9),

$$\dot{m}_r = \frac{200}{h_3 - h_1'} = \frac{200}{75.033}$$
$$= 2.665 \text{ lb/ton min}$$

$$\text{Total } \dot{m}_r = 2.665 \times 30 = 79.95 \text{ lb/min}$$

Step 2: Determine theoretical horsepower. By Eq. (3-11),

$$\text{ihp}_c = 4.717\left(\frac{h_4 - h_3}{h_3 - h_1'}\right)$$

$$= 4.717\left(\frac{9.858}{75.033}\right)$$

$$\text{ihp}_c = 0.619 \text{ hp/ton}$$

$$\text{Total ihp}_c = (0.619)(30) = 18.57 \text{ hp}$$

Step 3: Determine brake horsepower. From problem 3-14, CR = 2.71, η_{oa} = 78 percent.

$$\text{BHP} = \frac{18.57}{0.78} = 23.8 \text{ hp}$$

From the preceding two problems, it is apparent that subcooling reduces the required horsepower as long as the system capacity remains the same before and after subcooling. The refrigerating capacity can remain the same only if the compressor speed is reduced when subcooling. However, if, as is usually the case, the compressor speed remains unchanged, subcooling will increase the refrigerating capacity of the system. Under these conditions, there will be no change in the total horsepower required. As long as the weight of refrigerant pumped by the compressor remains the same, the horsepower required will not change. Subcooling does not affect the weight of vapor compressed by a given compressor; it only increases the heat-absorbing capacity of each pound of refrigerant compressed. As a result, although subcooling provides an increase in refrigerating capacity of a given system, it does not affect the total horsepower requirements as long as the speed of the compressor remains unchanged.

If the total horsepower required remains constant while the system tonnage increases, it is evident that subcooling decreases the horsepower required per ton of capacity. The preceding problem illustrates this point.

In finding the brake horsepower required by a compressor, the actual tonnage should be used, not the theoretical tonnage. The energy required to compress the vapor depends only upon the weight of vapor actually in the cylinder, not the quantity that the cylinder can theoretically hold.

Illustrative Problem 3-18

A six-cylinder compressor using Freon-12 is driven at 1750 rpm. The bore and stroke of the compressor cylinders are 2.75 in. and 2.0 in., respectively. The condensing temperature is 90°F and the evaporator temperature is 40°F. The liquid is subcooled 10 degrees before reaching the expansion valve. Saturated vapor leaves the evaporator and enters the compressor. Find the following:

a. the actual refrigerating capacity of the system

b. the brake horsepower required

Solution:

Step 1: Draw the *p-h* diagram for the cycle and determine enthalpy values. From Table B-3, at condensing temperature of 90°F, find condensing pressure of 114.49 psia. At evaporator temperature of 40°F find evaporator pressure of 51.667 psia. Also, from Table B-1 find

$$h_1 = h_f \text{ at } 80° \text{ F} = 26.365 \text{ Btu/lb}$$

$$h_2 = h_1 = 26.365 \text{ Btu/lb}$$

$$h_3 = h_g \text{ at } 40°\text{F} = 81.436 \text{ Btu/lb}$$

$$v_3 = v_g = 0.77357 \text{ ft}^3\text{/lb}$$

$$h_4 = h \text{ at } s_3 = s_4 \text{ and } p_4 = 114.49 \text{ psia}$$

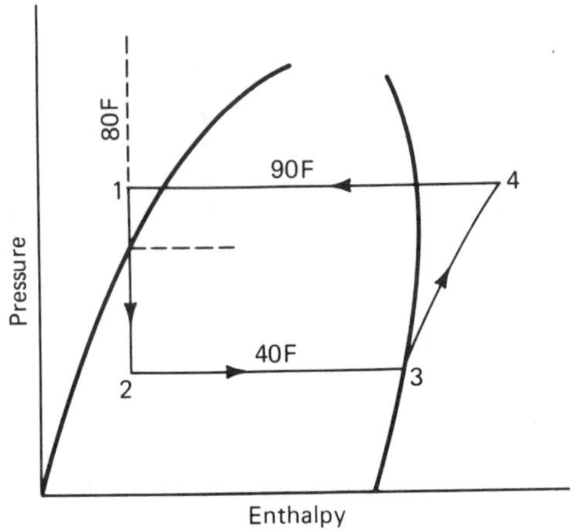

Figure 3–20

Sketch of p-h diagram for Illustrative Problem 3-18

From Fig. B-2 find h_4 = 88 Btu/lb.

Step 2: Determine refrigerating effect, mass, and volume rate of flow.

Refrigerating effect = $h_3 = h_1$ = 81.436 − 26.365 = 55.071 Btu/lb

$$\text{By Eq. (3-9)}\quad \dot{m}_r = \frac{200}{h_3 - h_1} = \frac{200}{55.071} = 3.63 \text{ lb/ton min}$$

$$\text{Volume flow rate} = (\dot{m}_r)(v_3) = (3.63)(0.77357) = 2.808 \text{ ft}^3\text{/ton min}$$

Step 3: Determine theoretical capacity and actual capacity. By Eq. (3-16),

$$PD_t = \frac{d^2 \times s \times n \times \text{rpm}}{2200} = \frac{(2.75)^2(2.0)(6)(1750)}{2200}$$

$$PD_t = 72.19 \text{ cfm piston displacement}$$

$$\text{Theoretical capacity} = \frac{72.19 \text{ cfm}}{2.808 \text{ ft}^3\text{/ton min}} = 25.7 \text{ tons} \quad \textit{(ans)}$$

$$CR = \frac{p_d}{p_s} = \frac{114.49}{51.667} = 2.22$$

From Fig. 3-18 at CR = 2.22 find η_{vt} = 87.5 percent, and η_{oa} = 79 percent.

$$\text{Actual capacity} = (25.7)(0.875) = 22.48 \text{ tons} \quad \textit{(ans)}$$

Step 4: Determine actual brake horsepower. By Eq. (3-11),

$$\text{ihp}_c = 4.717\left(\frac{h_4 - h_3}{h_3 - h_1}\right) = 4.717\left(\frac{88 - 81.436}{55.071}\right)$$

$$\text{ihp}_c = 0.562 \text{ hp/ton}$$

$$BHP = \frac{\text{ihp}_c}{\eta_{oa}} = \frac{0.562}{0.79} = 0.711 \text{ hp/ton}$$

$$\text{Total BHP} = 0.711 \times 22.48 = 15.98 \text{ hp} \quad \textit{(ans)}$$

The total horsepower required for any one compressor may either increase or decrease as the evaporator temperature changes. When using any one refrigerant, the total horsepower required by a compressor is determined primarily by only two factors:

a. the weight of refrigerant vapor compressed per minute

b. the compression ratio

The first of these two factors is self-evident. The second factor is defined in the section on *volumetric efficiency* where the method of computing it is also given. The horsepower required by any compressor increases as the difference in pressure between condenser and evaporator increases. However, as the pressure difference increases the compression ratio also increases. Hence, the larger the compression ratio of any compressor, the greater will be its power requirements.

As the suction pressure rises, the weight of vapor compressed increases, but the compression ratio becomes smaller if the condensing pressure is held constant. Any saving in power due to the smaller compression ratio is offset by the increased weight of vapor that is compressed. Whether the required total power will increase or decrease depends entirely upon which of these two factors has the greater effect.

For refrigerant-12 the total power required increases with rising suction pressures until a certain suction pressure is reached. Above this pressure, the saving in power due to the smaller compression ratio is larger than the additional power required for the greater weight of refrigerant compressed. If the suction pressure increases beyond this point, the power required will fall instead of continuing to rise.

The evaporator temperature (pressure) at which the total power required by a fluorinated hydrocarbon compressor reaches its peak value is not the same under all conditions; it varies with the condensing temperature. For ordinary condensing temperatures of about 100°F, the maximum point will be reached at an evaporator temperature of about 35°F. However, the evaporator temperature for which the horsepower is a maximum increases as the head pressure increases.

The power required by a fluorinated hydrocarbon compressor changes little over a wide range of evaporator temperatures. The power required by most fluorinated hydrocarbon compressors will vary less than 5 percent from the maximum within a range of roughly 15 degrees above the particular evaporator temperature at which the power required is a maximum. The variation is somewhat more than 5 percent as the evaporator temperature decreases.

The calculations for brake horsepower and capacity of a compressor have been illustrated in Problems 3-14 and 3-15. The following illustrative problem will use the same calculating methods and will illustrate the relationship between brake horsepower and capacity with variable suction temperature.

Illustrative Problem 3-19

An eight-cylinder refrigerant-12 compressor running at 1750 rpm is operating against a head pressure of 115.3 psig. The bore and stroke are 3.66 in. and 2.75 in., respectively. The temperature of the liquid refrigerant before the expansion valve is 86°F, and saturated vapor leaves the evaporator and enters the compressor.

Find the actual refrigerating capacity and power required when the evaporator temperature is 16°F, 24°F, 32°F, 40°F, and 48°F.

<u>*Solution:*</u>

Step 1: Determine the actual refrigerating capacity and power required for an evaporator temperature of 16°F. Make a sketch of the *p-h* diagram. From Table B-3 at 115.3 psig (130.0 psia) find condensing temperature of 99°F. At evaporator temperature of 16°F find evaporator pressure of 33.060 psia.

$$h_1 = h_f \text{ at } 86\ ^\circ\text{F} = 27.769 \text{ Btu/lb}$$

$$h_2 = h_1 = 27.769 \text{ Btu/lb}$$

$$h_3 = h_g \text{ at } 16^\circ\text{F} = 78.966 \text{ Btu/lb}$$

$$v_3 = v_g = 1.1828 \text{ ft}^3\text{/lb}$$

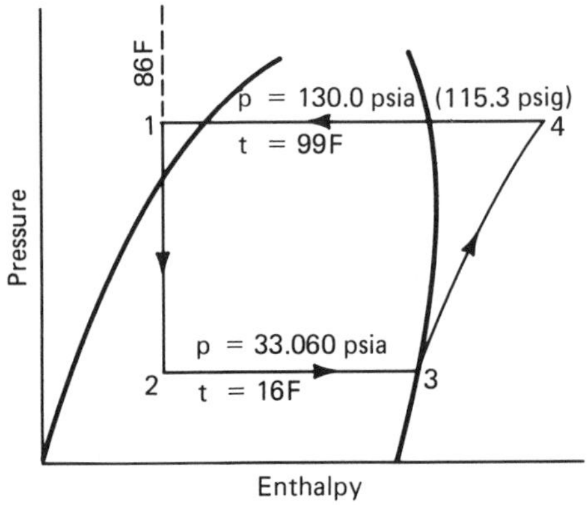

Figure 3–21

Sketch of p-h diagram for Illustrative Problem 3-19

$h_4 = h$ at 130 psia and $s_4 = s_3$

$h_4 = 90$ Btu/lb

By Eq. (3-16),

$$\text{PD}_t = \frac{d^2 \times s \times n \times \text{rpm}}{2200}$$

$$= \frac{(3.66)^2(2.75)(8)(1750)}{2200}$$

$$\text{PD}_t = 234.4 \text{ cfm}$$

Refrigerating effect $= h_3 - h_1 = 78.966 - 27.769 = 51.197$ Btu/lb.

$$\dot{m}_r = \frac{200}{h_3 - h_1} = \frac{200}{51.197}$$

$$= 3.91 \text{ lb/ton min.}$$

$$\text{ft}^3\text{/ton min} = (3.91)(1.1828)$$

$$= 4.62 \text{ ft}^3\text{/ton min}$$

$$\text{Theoretical capacity} = \frac{234.4 \text{ cfm}}{4.62 \text{ ft}^3\text{/ton min.}}$$

$$= 50.7 \text{ tons}$$

$$\text{CR} = \frac{130.0}{33.06} = 3.93$$

From Fig. 3-18 find $\eta_{vt} = 77$ percent and $\eta_{oa} = 76$ percent.

Actual capacity $= (50.7)(0.77)$

$= 39.04$ tons **(ans)**

$$\text{Theoretical Horsepower } (\text{ihp}_c) = 4.717 \frac{h_4 - h_3}{h_3 - h_1}$$

$$= 4.717 \frac{90 - 78.966}{51.197}$$

$$\text{ihp}_c = 1.017 \text{ hp/ton}$$

$$\text{Total ihp}_c = (1.017)(39.04) = 39.7 \text{ hp}$$

$$\text{Actual BHP} = \frac{39.7}{0.76} = 52.2 \text{ hp}$$ **(ans)**

The calculated results for evaporator temperature of 24°F, 32°F, 40°F, and 48°F are represented in the table on the following page, and shown in Fig. 3-22.

3-16 THE EVAPORATOR

Any device in which a refrigerant is boiled for the purpose of extracting heat from the surrounding medium is called an *evaporator*. Thus, cooling coils, chillers, unit coolers, or the ice cube maker in a domestic refrigerator can all be classified as evaporators.

Evaporators fall into two general types: flooded evaporators and dry or direct expansion evaporators. The flooded evaporator is designed to carry a constant level of liquid refrigerant within the evaporator; this level being maintained by a float valve or other suitable control. The dry expansion evaporator is designed to handle only that amount of refrigerant actually demanded by the load. Liquid refrigerant is fed to the dry expansion evaporator through an expansion valve in just the right

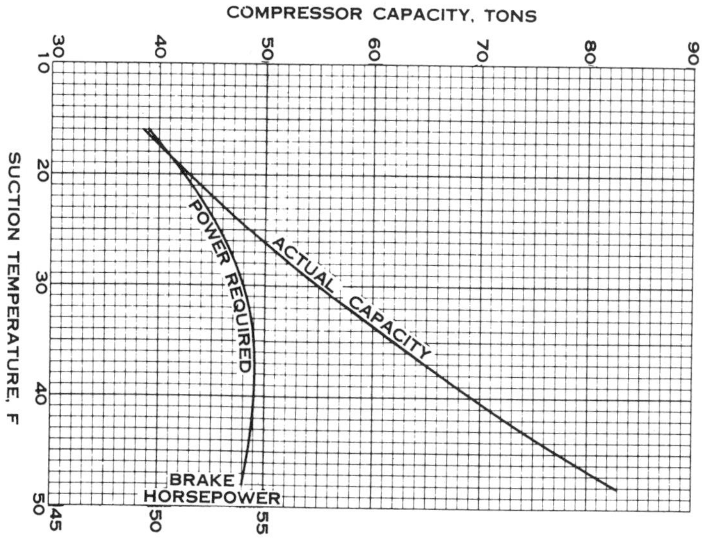

Figure 3–22

Variation of capacity and brake horsepower with suction temperature (See Illustrative Problem 3-19.)

Tabular Results for Illustrative Problem 3-19

Evaporator Temp., F	Evaporator Pres., psia	Vapor Density lb/ cu ft.	Enthalpy Sat. Vapor Btu/lb.	Refrig. Effect Btu/lb.	Refrig. Flow lb/min.	Theo. Capacity Tons
16	33.060	0.84544	78.966	51.197	3.91	50.7
24	38.574	0.97843	79.800	52.031	3.84	60.0
32	44.760	1.1271	80.624	52.855	3.79	70.3
40	51.667	1.2927	81.436	53.667	3.72	81.8
48	59.347	1.4768	82.236	54.467	3.68	95.0

Evaporator Temp., F	Evaporator Pres., psia	Compres-sion Ratio	Volume Effi-ciency Percent	Actual Capacity Tons	Theo. Power Reqd. HP	Overall Effic. Percent	Actual Power Reqd. BHP
16	33.060	3.93	77.0	39.04	39.7	76.0	52.2
24	38.574	3.37	80.0	48.1	40.3	76.5	52.7
32	44.760	2.90	82.8	58.2	42.1	77.5	54.4
40	51.667	2.52	85.3	68.9	42.7	78.5	54.4
48	59.347	2.19	87.5	83.0	43.1	79.5	54.1

amount so all liquid will be converted to gas before the refrigerant reaches the suction end of the evaporator.

The essential differences between the flooded- and dry-type evaporators can best be illustrated by comparison of water chillers of each type; the latter type is shown in Fig. 3-23. In the flooded chiller the water is passed through the tubes while a high level of liquid refrigerant is carried between the shell and the tubes. This level is maintained by a float valve that admits liquid to replace that boiled away. On the other hand the dry expansion chiller (Fig. 3-23) carries the water between the shell and the tubes. The passage is baffled to increase the velocity and enhance heat transfer. The refrigerant is passed through the tubes of the chiller. An expansion valve meters the refrigerant so only that actually required is admitted to the dry expansion chiller. Immediately upon entering the tubes, the liquid starts boiling and continues to boil as it passes through the tube circuits. As the refrigerant approaches the suction end of each circuit, all liquid will have been converted to gas.

The Direct Expansion Coil. To the air-conditioning specialist, the most familiar form of evaporator is the direct expansion cooling coil (sometimes called DX coils), shown in Fig. 3-24. The direct expansion cooling coils are used in duct work as part of a central system or in various types of air-conditioning units. Regardless of the type of evaporator or the application, the same conditions will affect the operation of the entire refrigeration system. We will not attempt to cover the subject of evaporator selection. Here our discussion will be limited to the evaporator only as it affects the balance of the refrigeration system.

The Capacity of the Direct Expansion Evaporator. The capacity of the direct expansion coil depends upon the quantity of air being passed through the coil, the dry- and wet-bulb temperatures of this air, and the temperature of the refrigerant. If the quantity of air handled is increased, or if the temperature difference between the air and the refrigerant is increased, the coil will have greater capacity. Because of this increased capacity, the coil will be able to evaporate a greater quantity of refrigerant in a given time. Conversely, if the quantity of air or the temperature difference between the air and refrigerant is reduced, the rate of refrigerant evaporation will decrease. If the compres-

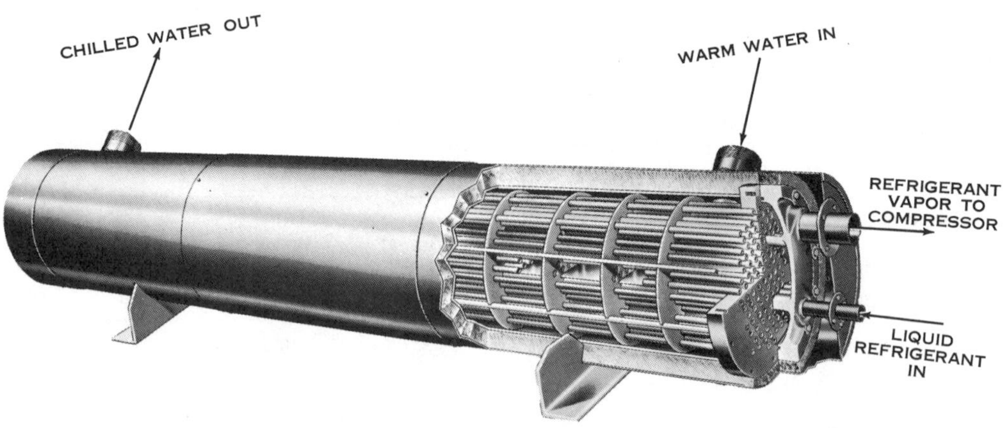

Figure 3–23

A direct expansion shell-and-tube evaporator (Courtesy of The Trane Company.)

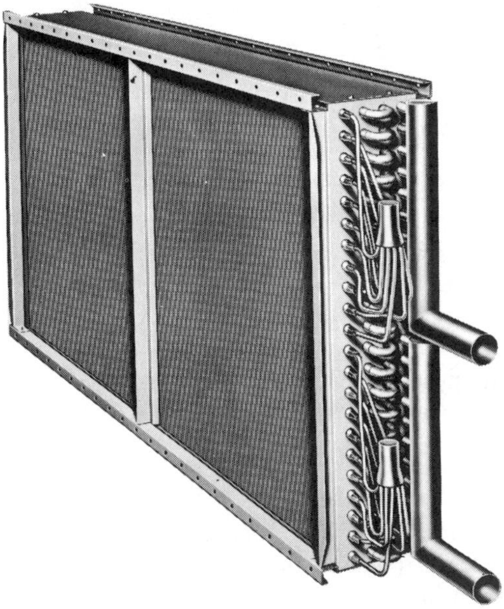

Figure 3–24

Direct expansion coil (Courtesy of The Trane Company.)

sor is running at constant speed, any increase in the rate of evaporation will, because of the additional gas created, raise the suction pressure. Any decrease in the rate of evaporation will lower the suction pressure.

Any direct expansion coil capacity table shows that as the suction pressure increases, the capacity of the coil is lowered because of the decrease in temperature difference between the refrigerant and the air. Conversely, on a decrease in suction temperature the capacity of the coil is increased. From any compressor capacity table, it is found that as the suction pressure or temperature is increased, the capacity of the compressor rises due to the increased density of the refrigerant gas, making it possible to pump a greater weight of refrigerant with each piston stroke. Conversely, as the suction pressure decreases the capacity of the machine decreases.

Here is a system in which identical conditions affect the two principle parts very differently. When the capacity of the direct expansion coil is increasing, the capacity of the compressor is decreasing. When the capacity of the direct expansion coil is decreasing, the capacity of the compressor is increasing. Obviously, the compressor and coil will balance at a point where each has exactly the same capacity. For every condition encountered by the coil, the compressor will respond by raising or lowering the suction pressure until a balance is restored.

Determining the Point of Balance. To illustrate clearly how any change of conditions at the evaporator will cause a definite change in operating conditions of the compressor, Fig. 3-25 has been constructed. Here the capacities of a Trane Model FC-215 condensing unit operating at 95°F condensing temperature and 1750 rpm and at 105°F condensing temperature and 1750 rpm have been plotted against suction temperature and capacity. The direction of these curves (*A* and *B*) indicates the increase in capacity when the suction temperature rises. On the same figure have been plotted the curves of a 30 by 48 in., five-row Type F direct expansion cooling coil handling 4000 cfm of air entering at 80°F dry bulb and 67°F wet bulb; a 30 by 48 in., six-row Type F direct expansion cooling coil handling 4000 cfm at the same entering air conditions; and a 30 by 48 in., six-row coil handling 5000 cfm at the same entering air conditions. Notice that the direction of curves *X*, *Y*, and *Z* indicates the increase in capacity of the coils as the suction of refrigerant temperature drops. Neglecting pressure drop, the point at which the coil curve and compressor curve intersect is the point at which the system will balance.

Effect of Changing Condensing Temperature. Curve *Y* represents the capacity of the six-row coil handling 4000 cfm at the stated entering conditions. This curve intersects curve *A*, the capacity curve of the compressor at 105°F condensing temperature, at point 2. At this balance point the suction temperature is

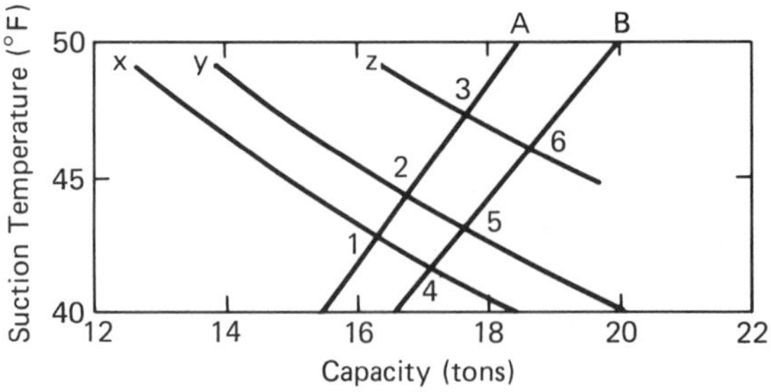

Figure 3–25

Evaporator—Compressor Balance (Courtesy of The Trane Company.)

44.4°F and the system capacity is 16.8 tons. If, however, the compressor were operating at 95°F condensing temperature instead of 105°F, the system would balance where the coil capacity, curve *Y*, intersects compressor capacity, curve *B*, at point 5. Under these conditions the suction temperature would be 43.2°F and the system capacity would be 17.6 tons. A change in condensing temperature changes the conditions throughout the system.

Effect of Increased Air Flow. To show the effect of increased air flow through the cooling coil, curve *Z* has been plotted for the same six-row coil handling 5000 cfm instead of 4000 cfm. The entering air conditions remain the same. Here the coil and compressor strike new balance points at points 3 and 6 depending upon the condensing temperature. Notice that with increased air flow the capacity of the coil has increased and the compressor must operate at higher suction temperature in order to balance the load.

Effect of Reduced Coil Surface Area. To show the effect of reduced coil surface area, or reduced evaporator capacity, curve *X* has been plotted for a coil of the same face area, 30 by 48 in., but with five rows of tubes intead of six. The air handled is 4000 cfm at the same entering conditions as for curves *Y* and *Z*. Here the capacity of the coil is reduced because of less surface area. The coil and compressor will balance at points 1 or 4 depending upon the condensing temperature. Notice that because of reduced coil capacity, less refrigerant is evaporated and the compressor must operate at lower suction temperature.

Charts similar to Fig. 3-25 are very easily constructed. The points for each curve may be computed from manufacturers published data. The scale of the chart may be modified to suit whatever cross-section paper is available. To avoid errors in the selection of equipment and systems that balance at some unexpected and possibly undesirable point, every system layout should be checked in this manner.

Serpentining. In addition to the proper sizing of the evaporator, proper "serpentining" or circuiting is of utmost importance. By dividing the flow of refrigerant through the coil into one or more circuits, it is possible to maintain the correct refrigerant velocities and keep the pressure drop, due to friction, within the desired limits. The velocity of the refrigerant must be great enough to "scrub" the walls of the tubes and break up the film of liquid refrigerant and oil that might adhere to them. The refrigerant velocity must also be sufficient to insure a continual movement of the oil that is always present in the system. The velocity must not be so great that it causes excessive pressure drop, which lowers the system capacity.

Pressure drop is, of course, proportional to refrigerant velocity. The velocity at pressure drops below the minimum is too low to insure proper oil movement and satisfactory heat transfer. The pressure drops above the maximum limit are generally too high for economical operation. Whenever there is a question concerning proper selection of evaporators, equipment manufacturers should be consulted.

3-17 THE COMPRESSOR

It was stated earlier that the compressor is a refrigerant gas pump. As such, it takes gaseous refrigerant at a low pressure from the evaporator and raises it to a higher pressure. The gaseous refrigerant must be delivered to the condenser at a pressure at which the condensing process can take place at a reasonable temperature.

Types of Compressors. There are three main types of compressors used in refrigeration work. These are *rotary, centrifugal,* and *reciprocating*. Rotary and reciprocating compressors are positive-displacement machines.

Rotary compressors, normally small-capacity machines, are usually designed in one or two different ways; a single vane located in the body of the compressor and sealed against the rotor which is offset on the shaft and a multivane rotary with vanes located in the rotor and sealing against an offset cylinder. These vanes may or may not be spring loaded. Rotary compressors are used mostly in household refrigerators and food freezers. Since they are not used in air conditioning, they will not be discussed further.

Centrifugal compressors rotate at high speed and add centrifugal force to the refrigerant to compress it. The rotor speeds range from 3500 rpm to 80,000 rpm. The centrifugal compressor is normally used with refrigerants having high specific volumes and which require low compression ratios, although multistage units may be used to obtain higher discharge pressures. The number of stages is limited only by the discharge temperature of the gas as it leaves the rotor. Centrifugal compressors are manufactured in sizes from 100 tons to over 4500 tons. They are used in many varied applications from water chilling for air conditioning to low-temperature freezing applications below −100°F.

The Reciprocating Compressor. Most of the following discussion will pertain to reciprocating compressors. Such compressors have automotive-type pistons that operate in cylinders and automatic suction and discharge valves. Reciprocating compressors are classified into two types: "open" compressors and hermetic compressors.

Open-type compressors are so designated because one end of the crankshaft extends outside the crankcase, see Fig. 3-26. The compressor, therefore, is adaptable to a variety of drives. Where the rotating shaft passes through the crankcase, a mechanical seal is used to prevent outward leakage of refrigerant and oil and inward leakage of air. Most open-type compressors are driven by electric motors but may also be driven by internal combustion engines. There are two types of drives commonly used: belt drive and direct drive. On belt-drive units speed variation is obtained by varying the sizes of the pulleys. For direct-drive units the compressor must be designed to run at motor speed.

Hermetic compressors are commonly classified as serviceable hermetic and welded hermetic or nonserviceable hermetic. The motor and compressor of the serviceable hermetic compressor are enclosed in a common housing. The cylinder heads, end bell, and crankcase cover plates may be removed for servicing the internal mechanism. The compressor section is basically the same as the open-type compressor. However, the compressor and

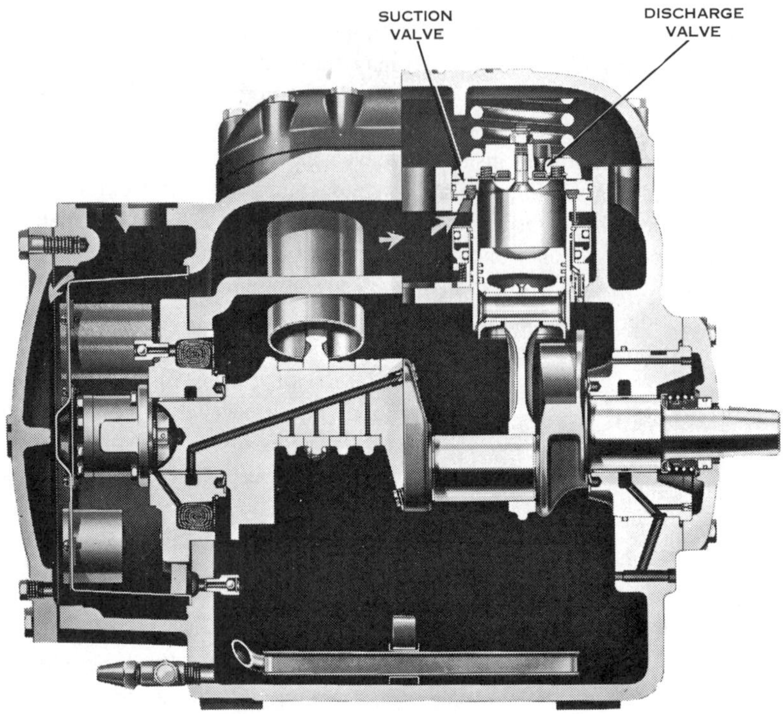

Figure 3–26

Open type compressor (Courtesy of The Trane Company.)

motor are connected by a common shaft inside the compressor housing. Suction gas is drawn through the motor section and is used to cool the motor, see Fig. 3-27.

The welded hermetic compressor and motor are sealed within a welded steel shell. The shell cannot normally be removed for service in the field. Usually, the motor and compressor are mounted in a vertical plane. The only basic difference between the compressor and the serviceable hermetic compressor is the means of access to the compressor and motor assembly.

Compressor Valves. There are two common types of valves used in reciprocating compressors: the nonflexing ring plate type and the flexing disc type. The suction valve is a ring plate that fits around the outside of the cylinder, a small distance below the top of the cylinder, and is held closed by small springs. The suction valve opens on the downstroke of the piston when the cylinder pressure becomes less than the suction pressure. The pressure in the crankcase of the compressor is the same as the suction pressure on the left-hand side of the compressor in Fig. 3-26. On the upstroke of the piston the increase in pressure causes the suction valve to close. As the piston continues upward, the cylinder pressure opens the discharge valve. The high-pressure vapor then passes into the compressor heads through the center hold as well as around the discharge valve cage. To avoid a metallic noise as the valve opens, and to increase valve life, a plas-

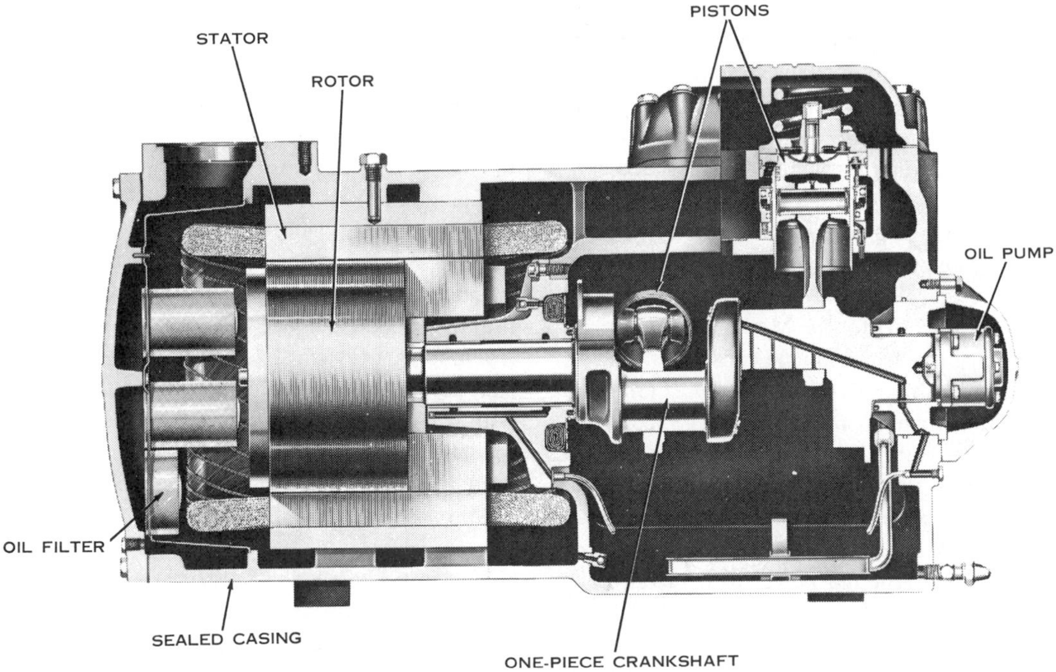

Figure 3–27 Hermetic compressor (Courtesy of The Trane Company.)

tic cushion is installed in the valve cage. The valve opens against this cushion rather than against metal.

Flexing disc-type valves operate similarly with the disc covering the discharge parts. The disc is fastened in the center, and the cylinder pressure causes the disc to ''flex'' around the center. This type of valve is generally less satisfactory because the flexing action causes high stresses in the valve material that can cause early failure.

Should a slug of liquid enter a cylinder, the head might be blown off when the piston is at the top of its stroke because of the small clearance space. To prevent this from happening, the compressors shown in Figs. 3-26 and 3-27 are equipped with safety heads. The head of the cylinder is not fastened in position but is held firmly in place by a strong spring. If a slug of liquid enters the cylinder, the entire head will lift and pass the liquid into the discharge pipe. The pressure produced during ordinary operation, with only vapor in the cylinder, is insufficient to lift the safety head.

3-18 COMPRESSOR PERFORMANCE FACTORS AND RATING TABLES

All three types of reciprocating compressors, the open type, the serviceable, and welded hermetic, function in the same manner. In applying any of these compressors, there are four fundamental factors that influence their performance and must be considered. They are: (a) rotative speed, (b) suction pressure, (c) discharge pressure, and (d) type of refrigerant.

If it were necessary to calculate the weight of refrigerant to be circulated, the compression ratio, and the displacement corrected for volumetric efficiency every time a compressor is to be selected, the job would be laborious and time consuming. For this reason most manufacturers publish rating tables prepared on the basis of capacity in tons of refrigeration under various operating conditions. In other words the displacement of each machine has been taken, corrected for volumetric efficiency, and translated into terms of refrigerating capacity of the refrigerant handled under the stated conditions.

Table 3-5 is a reproduction of one such table of ratings published by The Trane Company in their Trane Reciprocating Compressor Bulletin DS-361, for the Model F-215 and FC-215 units. To determine the capacity of an FC-215 condensing unit operating, for example, at 40°F suction temperature and 105°F condensing temperature, find the proper condensing temperature column at the top of the table. Then find, across from the suction temperature, the capacity that will be found to be 15.4 tons and the required power to be 15.6 brake horsepower.

In determining motor size, the use of a service factor of greater than 1.15 is not recommended. In this case a 15-hp motor would be satisfactory because the brake horsepower at full load is less than 15 × 1.15 = 17.25. At low suction temperatures a smaller motor may be satisfactory for design conditions. For example, at zero degrees suction temperature and 105°F condensing temperature, the same compressor has a capacity of 5.6 tons requiring 10.9 hp. The use of the smaller 10-hp motor is not recommended in this case, unless some provision is made to prevent overload when the system is operating at higher suction temperatures during the cooling down period.

The capacity table applies to Model FC condensing units and Model F compressor without condenser. In the case of the former the water-cooled condenser contained in the machine is of such size as to maintain the stated condensing temperature when supplied with adequate condensing water. In the case of the latter it is assumed that the machine will be used in connection with a condenser of adequate capacity to maintain the desired condensing temperature.

The effect of suction pressure (temperature) upon compressor capacity has already been discussed. In Table 3-5 this may be noted. For a constant condensing temperature, say 105°F, note that the capacity varies from 18.6 to 5.6 tons as the suction temperature varies from 50°F to 0°F. Note also that at 40°F suction temperature the brake horsepower required per ton of capacity is 15.6/15.4 = 1.01, and at 0°F suction temperature this becomes 10.9/5.6 = 1.95.

Table 3-5*

Typical Rating Table Refrigerant-22 (Model F-215 and FC-215 Compressor Capacity)

SUCTION		CONDENSING TEMPERATURE AND PRESSURE											
		85°		95°		105°		115°		125°		135°	
TEMP.	GAGE PRES.	155.7 LBS. GAGE		181.8 LBS. GAGE		210.8 LBS. GAGE		242.7 LBS. GAGE		277.9 LBS. GAGE		316.6 LBS. GAGE	
		TONS	BHP	TONS	BHP	TONS	BHP	TONS	BHP	TONS	BHP	TONS	BHP
0	24.1	6.8	10.3	6.1	9.0	5.6	10.9	—	—	—	—	—	—
5	28.3	7.5	10.8	7.1	11.3	6.3	11.7	5.7	12.1	—	—	—	—
10	32.9	8.9	11.2	8.1	11.9	7.3	12.4	6.6	13.0	—	—	—	—
15	38.0	10.1	11.7	9.3	12.4	8.4	13.2	7.6	13.8	6.8	14.3	—	—
20	43.3	11.4	12.1	10.5	13.0	9.6	14.0	8.7	14.5	7.9	15.2	6.8	15.7
25	49.0	12.9	12.4	11.9	13.4	11.0	14.5	9.9	15.3	9.0	16.0	8.1	16.6
30	55.2	14.4	12.5	13.3	13.8	12.4	14.9	11.3	15.8	10.3	16.8	9.2	17.4
35	61.9	16.1	12.5	14.9	14.0	13.8	15.3	12.7	16.4	11.6	17.5	10.5	18.3
40	69.0	17.7	12.6	16.6	14.1	15.4	15.6	14.2	16.9	13.0	18.1	11.8	19.0
45	76.6	19.6	12.7	18.3	14.3	17.0	15.9	15.6	17.5	14.4	18.5	13.2	19.7
50	84.7	21.5	12.7	20.0	14.4	18.6	15.9	17.3	17.7	16.0	18.8	14.5	20.0

*Reproduced with permission of The Trane Company.

Compressor Performance with Different Refrigerants

	Refrigerant	Capacity	Percent Capacity
Compressor	Freon-12	24,000 Btu/hr (2 tons)	100
No change in size	Freon-500	28,300 Btu/hr (2.36 tons)	118
or	Freon-22	38,400 Btu/hr (3.2 tons)	160
Operating conditions	Freon-502	42,000 Btu/hr (3.5 tons)	175

The effect of speed or rpm, although not shown in the table, is obvious. The slower the compressor runs, the smaller its displacement per minute and the smaller the volume of gas handled. The faster the compressor runs, the greater the volume of gas handled. This variation is nearly a straight-line function.

The effect of high discharge or condensing pressure (temperature) is also more or less obvious. As the discharge pressure is increased in relation to the suction pressure, the compression ratio is increased, the density of the gas remaining in the clearance space increases, and the volumetric efficiency decreases (see Table 3-4). As the volumetric efficiency decreases, the capacity of the machine decreases. This may be observed in Table 3-5 at a constant suction temperature, say 40°F, the capacity is 17.7 tons at a condensing temperature of 85°F, and becomes 11.8 tons at a condensing temperature of 135°F.

The fourth factor that effects compressor performance is the refrigerant which is used in the system. The commonly used refrigerants are Freon-12, Freon-500, Freon-502, and Freon-22.

Although these refrigerants have the same basic and desirable characteristics, they do possess varying refrigerating capacities per pound of refrigerant. For example, identical systems or compressors can be made to operate at varying capacities by changing refrigerants and compressor horsepower input. The table above shows how capacity varies in a given compressor by changing the refrigerant only.

This characteristic of refrigerants makes it possible to not only reduce the size of compressors used in a system but gives the engineer greater flexibility in the selection of compressors that will more nearly meet the system requirements.

Use of different refrigerants in a given compressor should not be arbitrary. Manufacturers of the compressors should always be consulted before a change of refrigerant is made.

3-19 COMPRESSOR OPERATING LIMITS

In the application of compressors to a refrigeration system, the engineer must select the compressor speed, suction and condensing temperatures, and the refrigerant. In doing this there are certain limitations placed on his choice by the manufacturer. With respect to speed, most manufacturers recommend a maximum of 1750 rpm on their open reciprocating compressors. For the serviceable hermetic units again 1750 rpm is most common. On the welded hermetic compressors speeds of 1750 or 3450 rpm are normally used. When using the open-type compressor, there is a choice of speeds. However, the manufacturers' recommendations should not be exceeded.

Manufacturers also place limits on suction and condensing temperatures. In the case of hermetic compressors a maximum allowable suction temperature of 50°F and a maximum superheated or actual suction temperature of 65°F is used. This is done to assure adequate motor cooling and to prevent actual discharge temperatures in excess of 275°F. Discharge temperatures in excess of 275°F result in possible breakdown of the lubrication oil. Similarly, suction temperatures below the minimum in the manufacturer's catalog should not be used. As the suction temperature is reduced, at a given condensing temperature, the compression ratio increases. With increasing pressure ratios, the actual discharge temperature will also increase. Thus, when compressors are selected below the minimum suction temperatures, the resulting increase in compression ratio can result in excessive discharge temperatures and compressor damage.

When selecting compressors within the limits of the manufacturer's published ratings, one need not worry about its performance. There are applications, however, like industrial process cooling, where suction temperatures above 50°F are encountered. In this case the use of hermetic compressors should be checked with the manufacturer. It may be perfectly feasible that due to lower condensing temperatures this application would be acceptable.

The need for operating at lower suction temperatures than listed in catalog ratings comes up quite often. In this case also the application of hermetic compressors should be checked with the manufacturer.

So far only the limitations on hermetic compressors have been considered. The open compressor does not have the same limitations as hermetics. Obviously, its application is not limited by motor cooling or horsepower. Its primary limitation is discharge temperature. Again, a maximum of 275°F discharge temperature is the controlling factor, 250°F with water-cooled heads. Compressors using Freon-22 are limited to a compression ratio of 5:1. Above this compression ratio, water-cooled heads on the compressor are required.

Open-type compressors are not only adaptable to air-conditioning systems but, due to their greater flexibility, are used for higher-temperature cooling as well as the low-temperature cooling or freezing applications. They have almost infinite applications, but again, whenever a selection falls outside the ratings, check it out with the manufacturer.

After a compressor has been installed, it is important that it will operate between established maximum and minimum limitations. Almost all compressors are, therefore, sold with suction and discharge pressurestats. If either the high or low limits are exceeded during normal operation, the compressor will be cut off the line. In some instances discharge line thermostats are used to prevent excessive discharge temperatures.

It must also be recognized when using Freon-22 that these limitations are particularly important. Being a higher-pressure refrigerant, Freon-22 results in higher discharge temperatures. For this reason liquid-suction heat exchangers (Section 3-13) should not be used on Freon-22 applications. Their use would only increase the actual suction gas temperature to a point where high discharge temperatures would result.

When compression ratios exceed the maximum permitted for a given compressor, as would occur in low-temperature applications, two compressors in series may be used. In this type of installation one compressor will discharge into a second compressor. This is referred to as direct staging, and the low-stage compressor is called a "booster" compressor. In direct staging some means of desuperheating the discharge gas leaving the low-stage compressor is necessary to prevent excessive discharge temperatures on the top-stage compressor. This may be accomplished with a liquid-suction interchanger, an open flash-type intercooler, liquid injection into the interstage vapor, or by some type of closed interchanger.

Cascade staging employs two or more refrigerants of progressively lower boiling points such as Freon-22 on the low stage and Freon-12 on the high stage. The compressed refrigerant of the low stage is condensed in an exchanger (cascade condenser) which is cooled by another lower-pressure refrigerant in the next higher stage.

3-20 COMPRESSOR CAPACITY CONTROL

The effect of changing evaporator conditions upon compressor capacity has been pointed out in Section 3-13. The manner in which the compressor and evaporator always strike a balance point where the capacity of each is equal has been shown by curves. If the load remained constant, this subject could be dismissed without further comment. Unfortunately, this is seldom, if ever, the case. Loads on any system usually vary from full load to a small percentage of full load.

Ice on Coil. Figure 3-25 shows that the compressor will operate at a lower suction pressure or temperature when the evaporator load is reduced. If the load on an air-conditioning system falls low enough, the suction temperature may drop well below 32°F before the balance point is reached. In this case the final temperature of the air will be undesirably low, causing moisture condensing on the evaporator to freeze. The ice and frost forming on the coil will restrict the air flow and aggravate the condition by forcing the suction temperature even lower. Obviously, some means must be used to regulate the capacity of the machine.

On-Off Control. The most simple form of compressor control is found in the low-pressure switch of the dual-pressure control furnished as standard equipment in most reciprocating compressors. This is nothing more than a simple pressure switch arranged to open an electric circuit at a predetermined low pressure (pressurestats mentioned in the previous section). The pressure tube of this switch is connected to the compressor suction manifold and the electrical contacts are wired in series with the holding coil of the compressor motor starter. Thus, on falling load, when the pressure in the suction manifold drops below the desired low limit, the switch opens and operates the motor starter to stop the compressor. When the pressure in the suction manifold rises to a predetermined level, the switch closes, energizes the motor starter, and starts the compressor. When sufficient gas has been pumped to lower the suction pressure, the machine again stops.

This system is satisfactory where the load is moderately constant and very light loads are not encountered. On very light loads the machine will "short-cycle" (stop and start frequently) and impose undesirable stress on the electrical equipment. If too frequently started, the motor may overheat and the contacts in the starter may suffer. The starting current drawn from the line may cause voltage fluctuations that may be detrimental to other equipment on the same distribution system.

Multispeed Compressor Motors. As the capacity of the compressor is proportional to the compressor speed, multispeed compressor motors are sometimes used. These may be used with a manual speed-changing switch and constant speed motor starter so arranged that the starter stops and starts the machine at a single speed. This is satisfactory where loads are seasonal and do not fluctuate widely. In spring and fall, when loads are light, the machine can be operated at low speed. In summer, when the load is heavy, the machine may be operated at high speed. More frequently encountered, however, is the somewhat more costly system employing two low-pressure switches and a two-speed magnetic starter. At high suction pressure, indicating a heavy load, the machine is operated at high speed. As the load decreases and the suction pressure falls, one of the low-pressure switches functions to switch the speed from high to low. If the load continues to decrease and the suction pressure falls to a minimum, the second low-pressure switch functions to stop the compressor.

Hot-Gas Bypass. Because multispeed motors and their controls are usually too costly to use on the larger machines, other means of capacity reduction have been devised. One of the earlier methods used to reduce capacity, without reduction of compressor speed, took the form of the hot-gas bypass in which part of the gas discharged by the compressor is returned to the suction side of the machine.

Figure 3-28 illustrates the hot-gas bypass used by some manufacturers. When full-capacity operation is desired, the solenoid stop valve in the bypass line is closed and the gas from both banks of cylinders must pass through the discharge line to the condenser. When half-capacity operation is desired, the solenoid stop valve is opened and the gas discharged by one bank of cylinders is carried by the hot-gas bypass to the suction line. The check valve prevents the gas from the other bank of cylinders from backing into the bypass, thus forcing it to travel through the discharge line to the condenser.

This arrangement is now completely satisfactory for a number of reasons:

1. The capacity reduction step is a large one and is not compatible with the highly variable air-conditioning load.
2. It is inefficient because the compressor must pump the same volume of gas through the same pressure difference for all conditions of load.
3. If the hot gas is introduced into the suction line too close to the compressor, the suction temperature may be high enough to cause abnormally high discharge temperatures with early discharge valve failure.

Refrigeration specialty manufacturers have been able to offer some help in overcoming points 1 and 3 above, but inefficiency must

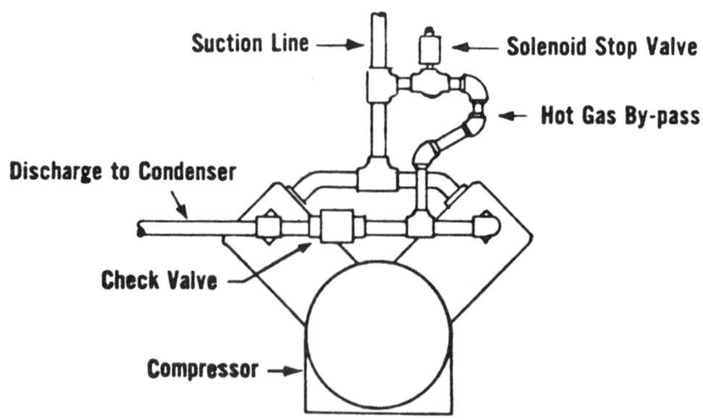

Figure 3–28

Hot gas bypass for reducing compressor capacity (Courtesy of The Trane Company.)

be tolerated whenever hot-gas bypass is employed.

The problem of discharge valve failure due to excessive discharge temperature can be solved by injecting liquid refrigerant into the suction line to desuperheat the hot gas that is bypassed. A thermal expansion valve is used for this purpose. The valve must be selected carefully in accordance with the manufacturer's instructions. The location and installation must be handled carefully. The critical aspect is that of introducing the hot gas to the suction line as far as practical from the compressor. This is done to insure complete mixing of desuperheating liquid, the bypass hot gas, and the suction gas. It may be necessary, where a close-coupled installation is encountered, to use a nozzle or other distribution device for proper mixing. Specific details can be obtained from the valve manufacturer. The remote bulb of the desuperheating valve is mounted on the suction line between the liquid injection connection and the compressor.

The importance of selecting a desuperheating valve with care cannot be overemphasized. If the valve is too small, the suction gas will not be fully desuperheated. With excessive superheat in the suction gas stream plus normal heat of compression, discharge gas temperatures may be high enough to cause early discharge valve failure. If the valve is oversized and is close to the compressor, liquid slugging can result. The least critical effect of liquid slugging on the compressor is excessively noisy operation. Liquid refrigerant in the compressor crankcase can cause excessive oil pumping and reduce oil viscosity, resulting in marginal lubrication. If the slugging is severe enough and occurs over a long period of time, the suction valves may fail.

The hot-gas bypass is most successful when introduced between the thermal expansion valve and the evaporator. The expansion valve in this case serves as a desuperheating control valve. The evaporator configuration contributes to good mixing of the refrigerant streams, and the separation of the evaporator and compressor helps to eliminate any chance of slugging.

An innovation by refrigeration specialty manufacturers is the adaptation of constant-pressure expansion valves or back-pressure regulators as hot-gas bypass regulators. Using these devices, the flow of hot gas is throttled to keep a reasonably constant evaporator pressure and temperature. At the same time a degree of system capacity modulation is also obtained. Figure 3-29 illustrates this arrangement. While variation in capacity can thus be achieved, the power to drive the compressor remains virtually constant so the hot-gas bypass is comparatively inefficient.

The hot-gas bypass method of capacity control is used quite widely with packaged air-conditioning apparatus up to 10 tons capacity. The compressors in these small-capacity packages are built without unloaders so the hot-gas bypass can be used for capacity modulation. In the larger packaged equipment using compressors with built-in unloaders, certain applications may make the use of hot-gas bypass desirable with the last step of compressor unloading.

Systems that will be operated in cool ambients may be loaded to less than the minimum unloaded step of the compressor. The hot-gas bypass thus can be used to provide continuous operation at variable load in this range without sacrificing too much efficiency. There are other cases, such as hospitals or industrial applications, that are associated with voltage-sensitive equipment where stopping and starting of a large compressor motor must be avoided.

Cylinder Cutoff Method. The cylinder cutoff method is less frequently encountered. In this method a solenoid stop valve is used to close the suction line to one bank of cylinders so no

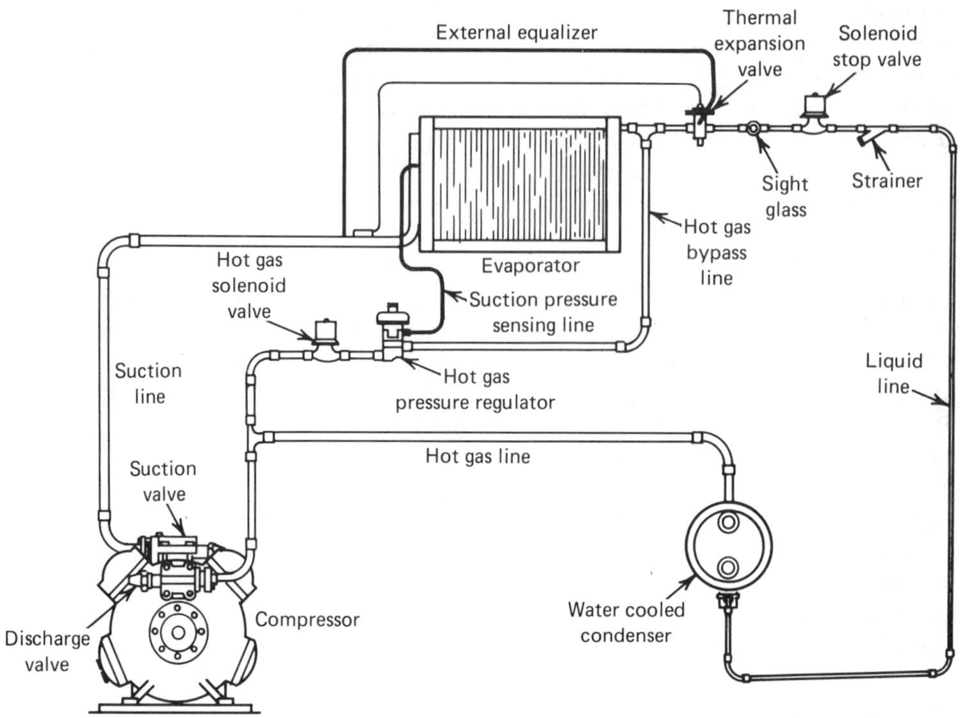

Figure 3–29

Typical schematic piping arrangements for hot gas bypass circuit using hot gas pressure regulator (Courtesy of The Trane Company.)

gas can be pumped from them. This is poor practice. The valves of the cylinders, to which the flow of gas is stopped, are permitted to operate. A relatively high vacuum is created in these cylinders and in the manifold behind the stop valve. This vacuum encourages oil pumping and may, in the case of a worn machine, cause serious trouble.

Cylinder Unloaders. Modern reciprocating compressors, with capacities of approximately 10 hp and larger, are generally provided with cylinder unloaders for capacity control. Cylinder unloading is usually accomplished by holding the suction valves open. This permits the piston to draw refrigerant gas from the suction manifold on the downstroke, but instead of compressing the gas on the upstroke, the gas is allowed to return uncompressed to the suction line through the open suction valve.

There are many advantages to cylinder unloading. One of the most important is that it prevents frequent starting and stopping, commonly called short cycling. It is estimated that 90 percent of the wear in a compressor occurs during the starting cycle, when lubrication is marginal. Also, frequent starting of a large motor may cause voltage fluctuation, with the result that lights dim in other parts of the building. This light flicker is annoying, and the voltage fluctuation can cause serious interference with sensitive equipment such as X-ray machines. For these reasons—and because motor and starter are under high stress when starting—minimizing frequent starting and stopping will result in longer compressor, motor, and starter life.

A typical hydraulic unloader mechanism is shown in Fig. 3-30. The top portion of this figure shows the cylinder with the suction valve in a raised position. When the compressor is unloaded, reservoir *R* is vented to the compressor crankcase, springs *V* push piston *S* upward, forcing the oil out of reservoir *R*, pushing the takeup ring *T* upward, forcing the lift pins *U* up, and lifting the suction valve *W* off its seat. The refrigerant now works in and out of the suction manifold as the piston moves up and down, and no compression takes place in that cylinder.

The top portion of Fig. 3-31 shows the same hydraulic unloader with the cylinder loaded. Oil pressure is supplied from the oil pump into reservoir *R* which depresses piston *S* against spring *V*. This allows takeup ring *T* to drop, lift pins *U* are forced down by their springs, and the suction valve *W* is allowed to seat. As the piston moves down, gas pressure from the suction manifold forces valve *W* upward and allows gas to enter the cylinder. As the piston moves upward, the springs behind suction valve *W* force it on its seat, closing it, and the gas is compressed and forced into the discharge manifold.

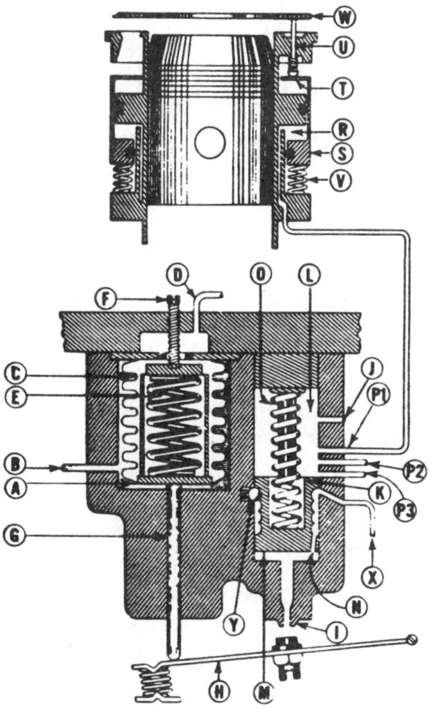

Figure 3–30

Schematic drawing of cylinder unloading mechanism (unloaded) (Courtesy of The Trane Company.)

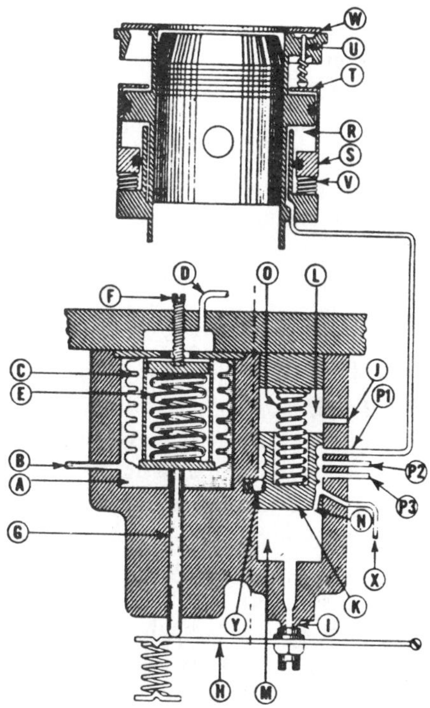

Figure 3–31

Schematic drawing of cylinder unloading mechanism (loaded) (Courtesy of The Trane Company.)

Oil flow from the oil pump to the hydraulic unloader on the cylinder is controlled in several ways. The lower portion of Fig. 3-30 and Fig. 3-31 shows a typical suction pressure actuator that will control three sets of cylinders. Suction pressure is allowed to enter chamber *A* on the outside of bellows *C* through part *B*. Atmospheric pressure is allowed into the inside of bellows *C* through port *D*.

The pressure of spring *E* is controlled by adjusting screw *F*. At design load, spring pressure plus atmospheric pressure balance against suction pressure to maintain bleed port *I* in the fully closed position (as shown in Fig. 3-31).

As suction pressure decreases, air pressure plus spring pressure is greater than suction pressure. Push rod *G* then depresses lever arm *H* which allows bleed port *I* to open. Oil pump pressure is applied into chamber *M* through connection *X*. As bleed port *I* is opened, the floating piston *K* is depressed by spring *O*. Oil can then bleed out of the hydraulic unloader through line *P*1 and into the compressor crankcase through port *J*.

The lower portion of Fig. 3-31 shows the suction pressure actuator in a fully loaded position. The suction pressure through port *B* is greater than the combined spring and air pressure inside the bellows, forcing the bellows and push rod *G* upward. Lever arm *H* closes bleed port *I*, and oil pump pressure is applied into chamber *M*, forcing the floating piston *K* upward. As the piston *K* moves upward, ball detent *Y* forces it to move in definite steps. Neglecting for the moment lines *P*2 and *P*3, the floating piston *K* first covers up the port to line *P*1 so that oil cannot bleed into the crankcase. Then it uncovers this port to oil pump pressure, which is applied through line *P*1 to the hydraulic cylinder unloader, causing it to load. Lines *P*2 and *P*3 function similarly and in sequence and are connected to other cylinder unloaders.

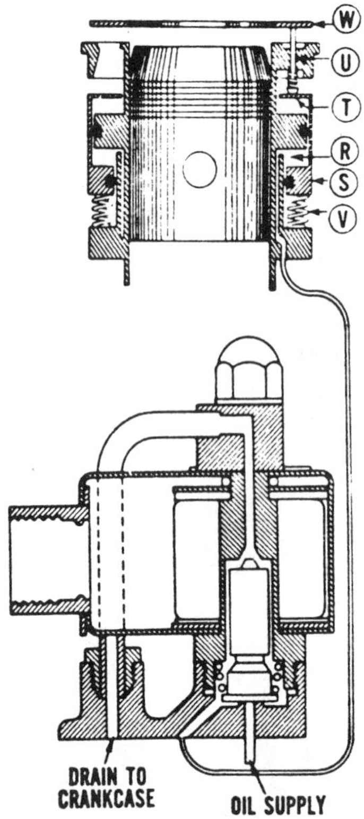

Figure 3–32

Controlling oil pressure to unloader with three-way solenoid valve (cylinder unloaded) (Courtesy of The Trane Company.)

Three-way Solenoid Valves. Figure 3-32 shows the method of directly controlling oil pressure to the hydraulic unloader by the use of three-way solenoid valves. In this case one three-way solenoid valve is required for each set of cylinders to be unloaded. When the valve is deenergized (Fig. 3-32), oil is bled from the hydraulic unloader through the common port of the three-way valve, out the normally open port, and into the crankcase. When the valve is energized (Fig. 3-33), the normally open port is closed. The normally closed port, which is connected to oil pump pressure, applies pressure through the common port to the cylinder unloader, causing the cylinder to load.

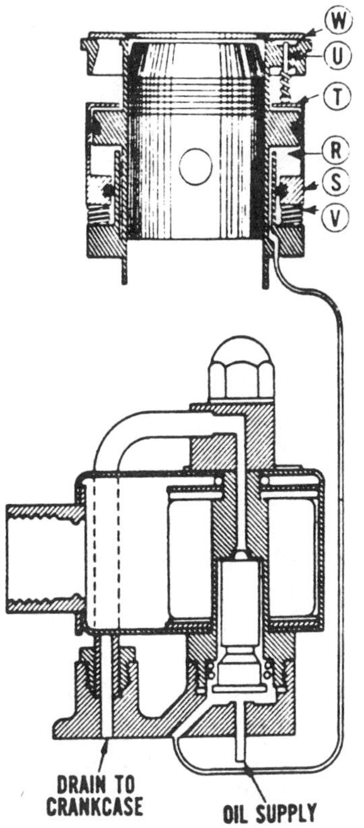

Figure 3–33

Controlling oil pressure to unloader with three-way solenoid valve (cylinder loaded) (Courtesy of The Trane Company.)

Since full oil pressure is not obtained until the compressor is up to speed; compressors using the hydraulic unloaders will start with all controlled cylinders unloaded. This feature makes it possible to use normal-starting torque motors instead of the more expensive high-starting torque motors. Several variations of this type of control are in use, but the basic principle of control is the same.

Multiple Compressors. There are some installations where complete shutdown due to failure of equipment would cause serious financial loss. In cases of this kind it is desirable to break up the refrigeration load into two or more parts, each handled by a single compressor. This provides standby service, and under partial-load conditions only the number of compressors are operated as required by the load on the system.

3-21 WATER-COOLED CONDENSERS

In the compression refrigeration cycle it is necessary to convert the hot gas, discharged from the compressor, to a liquid ready for use in the evaporator. The condenser must accomplish this by transferring enough heat from the hot gas to cause it to condense at the pressure existing in the condenser. The heat is transferred to some other medium, which is usually the water or air used to cool the condenser. We will first discuss the water-cooled condenser and then look at other methods of heat rejection.

Condensing water must be noncorrosive, clean, inexpensive, below a certain maximum temperature, and available in sufficient quantity. The use of corrosive or dirty water will result in high maintenance costs for condensers and piping. Dirty water, as from a river, can generally be economically filtered if it is noncorrosive; corrosive water can sometimes be economically treated to neutralize its corrosive properties if it is clean. An inexpensive source of water that must be filtered and chemically treated will probably not be economical to use without some means of conservation, such as an evaporative condenser or a cooling tower.

Water circulated in evaporative condensers and cooling towers must always be treated to reduce the formation of scale, algae, and chalky deposits. Overtreatment of water, however, can waste costly chemicals and result in just as much maintenance as undertreatment.

The four common types of water-cooled condensers are (1) double-pipe, (2) double-tube, (3) shell-and-coil, and (4) shell-and-tube.

Double-pipe condensers, seldom used at

present, consist of a pipe within a pipe or a tube within a tube. Units fabricated from 1¼- to 2-in. steel pipe in 10- to 20-ft lengths have been used in many ammonia systems. The tube sections were placed in a horizontal position to form a vertical bank, and either fittings or welded construction was used. The complete unit was usually mounted on a wall, occupying space that often was of comparatively little value. Water usually entered the bottom sections, flowing upward through the inner pipe. The compressed refrigerant vapor entered the top section flowing downward, thus providing counterflow.

A *double-tube condenser* is a relatively modern version of the double-pipe design used for systems of 7½ tons or less capacity. In most designs the water flows through the inside tube, which may be ¾- or ⅞-in. copper or steel. A 1- or 1¼-in. steel or copper tube surrounds the inside tube, and the refrigerant flows through the annular space between the two tubes. The double-tube arrangement is frequently formed into a coil 12- to 20-in. in diameter. A compact condensing unit may be formed by locating the hermetic compressor inside the condenser coil.

A *shell-and-coil* condenser is simply a continuous copper coil mounted inside a steel shell. Water flows through the coil, and refrigerant vapor from the compressor is discharged inside the shell to condense on the outside of the cold tubes. In many designs the shell also serves as a liquid receiver.

The shell-and-coil condenser has a low manufacturing cost, but this is offset by the disadvantage that this type condenser is difficult to service in the field. If a leak develops in the coil, the head of the shell must be removed and the entire coil must be pulled from the shell in order to find and repair the leak. A continuous coil is a nuisance to clean, whereas straight tubes can easily be cleaned with mechanical tube cleaners. Thus, with some types of cooling water, it may be difficult to maintain a high rate of heat transfer with the shell-and-coil condenser.

The *shell-and-tube* condenser, illustrated in Fig. 3-34, permits a large amount of condensing surface to be installed in a comparatively

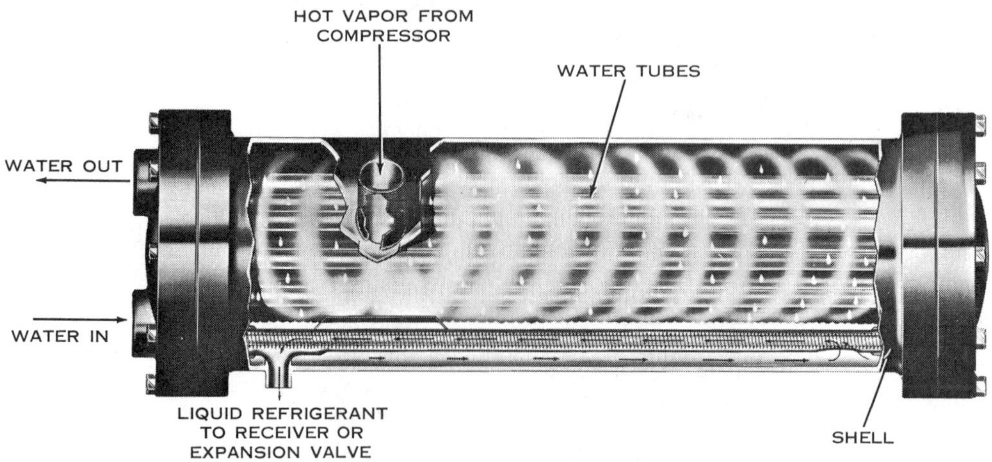

Figure 3–34

Shell-and-tube, water-cooled condenser with subcooling circuit (Courtesy of The Trane Company.)

small space. The condenser consists of a large number of ¾- or ⅝-in. tubes installed inside a steel shell. The water flows inside the tubes while the vapor flows outside, around the nest of tubes. The vapor condenses on the outside surface of the tubes and drips to the bottom of the condenser, which may be used as a receiver for the storage of liquid refrigerant. Shell-and-tube condensers are used for a large range of water-cooled refrigeration systems.

Condensing units for Freon-12 and Freon-22 (compressor and water-cooled condenser mounted on same frame) are generally offered by most manufacturers complete with the proper condenser for most applications. The condenser thus furnished will seldom, if ever, be too small as far as condenser capacity is concerned. On systems with a large amount of pipe, it may be too small to hold the entire charge of refrigerant when the system is "pumped down" in which case an auxiliary receiver should be used. More frequently the standard condenser is found to be too large if the system is operated at low temperatures. The fact that the receiver is then somewhat oversized does not impair the operation of the system. The main disadvantage lies in the fact that more refrigerant is required to charge the system than would otherwise be necessary.

The capacity of a water-cooled condenser is affected by the temperature of the water, the quantity of water circulated, and the temperature of the refrigerant gas. The capacity of the condenser will increase whenever the temperature difference between the refrigerant gas and the water is increased. This temperature difference may be increased by raising the condensing pressure, by lowering the temperature of the entering water, or by increasing the quantity of water in order to maintain a lower average water temperature.

At a constant suction pressure any increase in head or condensing pressure will require an increase in power input because of the additional work to be done in compressing the gas to the higher pressure. On the other hand the higher the condensing pressure or temperature the smaller the quantity of condenser water required. Thus, it is possible to save power at the expense of water or save water at the expense of power. The most economical conditions of operation can be determined only by a study of local power and water costs.

In order to obtain a high rate of heat transfer through the heat-transfer surface of a condenser, it is necessary for the water to pass through the tubes at a fairly high velocity. For this reason the tubes in shell-and-tube condensers are separated into several groups, called passes; the same water travels in series through each of the passes. A condenser having four groups of water tubes would be a four-pass unit because the water flows back and forth along its length four times. Four-pass condensers are common, although any reasonable number of passes may be used. The greater the number of tube passes, the greater will be pressure drop (friction) through the tubes, and consequently the greater will be the power required to pump the condenser water.

The allowable temperature rise in the condenser water is an important factor in circuiting the condenser. Knowing the allowable temperature rise in the condenser water determines the quantity of water required. Once this is known, the condenser is then selected that is circuited to give the best pressure drop for this required water flow.

As municipalities grew in population and air-conditioning systems employing mechanical refrigeration increased in number, it became increasingly more difficult to supply water for all needs. The sewage systems also became overtaxed. Many localities now prohibit the use of sanitary sewers for the disposal of waste water from condensing equipment because the large volumes of clean water upset the bacterial action at the sewage treatment plant. When confronted with these problems, the local authorities restricted more and more the use of "once-through" city water for condensing purposes. This forced the consideration of cooling towers and evaporative condensers where water is reused, and air-cooled condensers that completely eliminate water from the condensing process.

3-22 OTHER HEAT REJECTION METHODS

The first water conservation idea was to employ the evaporative effect of water when rejecting condenser heat to the air. This was attractive because of the high latent heat of vaporization of water, approximately 1000 Btu/lb evaporated.

When air is used to cool water, the rate of heat transfer depends upon (1) the difference between the air wet-bulb temperature and the water temperature, (2) the area of the water surface exposed to the air, and (3) the relative velocity of the air and water. Secondary factors also influence the performance of various types of systems. The size of the equipment required for the cooling of a given water quantity is affected by (1) the cooling temperature range of the water, (2) the difference between the entering air wet-bulb temperature and the leaving-water temperature, (3) design wet-bulb temperature itself, and (4) time of contact of air and water.

Spray ponds were one of the earliest devices used to obtain the evaporative effect. The removal of moisture-laden air from the spray pond area depended on natural air movement. Since winds are highly variable in most areas, spray ponds were not practical. They required large open spaces to catch the slightest air movement, and they had to be sufficiently separated from populated areas because of the spray drift nuisance.

The next step was to enclose the spray pond to prevent or reduce windage losses and the spray drift nuisance. To make this work required some sort of stack or tower effect that would expose the maximum water surface area to the air. The result was the *natural-draft cooling tower,* which overcame only a few of the spray pond problems.

When a fan that either pushed the air through the tower (forced-draft design) or pulled the air through (induced-draft design) was added to the cooling tower, the present concept of the cooling tower was reached.

Figure 3-35 illustrates a typical cooling tower installation. Water from the sump or tank of the tower flows by gravity to the circulating pump. The circulating pump forces the water through the condenser where it picks up heat from the condensing refrigerant. The warm water from the condenser is carried to the sprays of the cooling tower where it is broken into drops and falls over the slats or "packing" of the tower. The water, in its progress through the air and over the tower packing, is exposed to a blast of air induced by the cooling tower circulating fan. In this airstream part of the water evaporates, cooling the remaining water to a temperature satisfactory for condenser use. The cooled water collects in the sump and flows to the pump for reuse.

The advantages and limitations of the cooling tower should be well understood before selecting a tower for a refrigeration system. Water, cooled by spraying in air, cannot be cooled below the prevailing wet-bulb temperature of the air. It is obvious, therefore, that the condensing temperature at which the system will operate must be somewhat above the wet-bulb temperature of the air. High efficiency forced- or induced-draft cooling towers will cool the condenser water to a point within 5 to 8 degrees of the prevailing wet-bulb temperature. Natural-draft towers seldom approach closer than 10 or 12 degrees to the prevailing wet bulb. It should be noted that the "design" wet-bulb temperature is sometimes exceeded

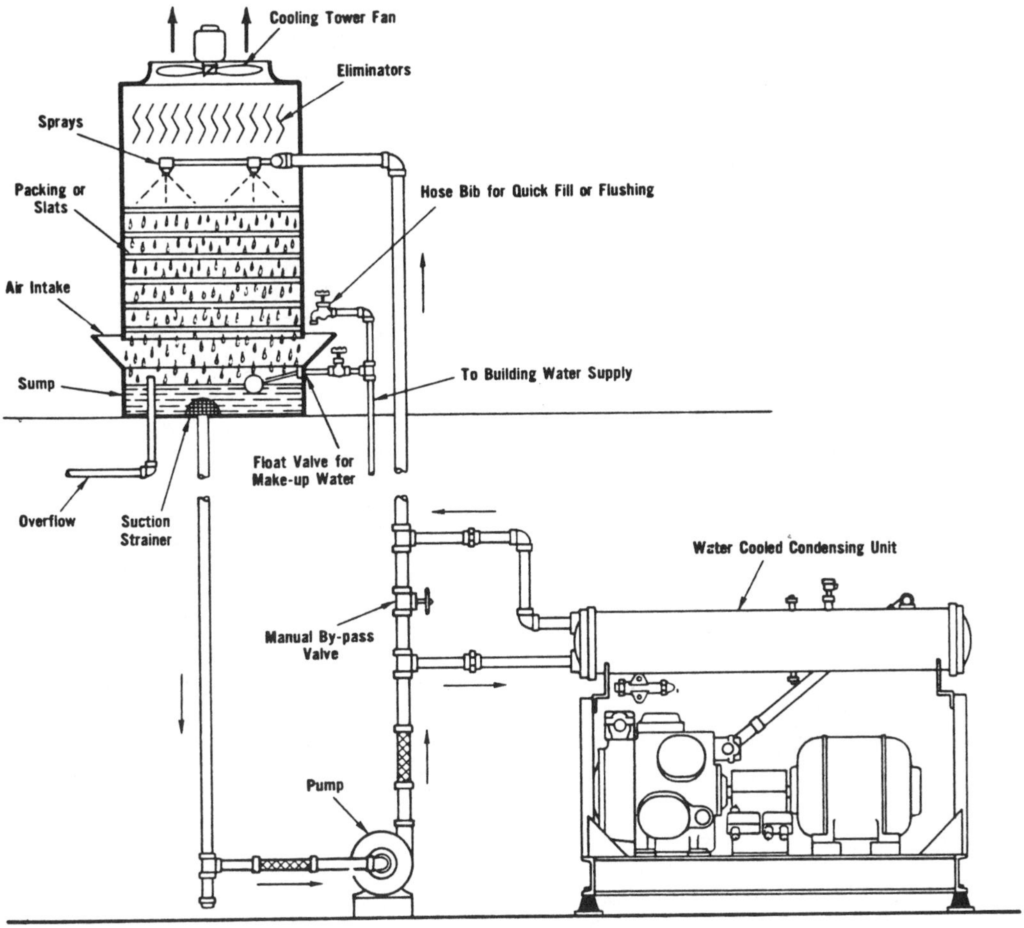

Figure 3–35

Typical cooling tower system (Courtesy of The Trane Company.)

by the actual wet-bulb temperature for appreciable lengths of time. Because of this, it is common practice to increase the usual "design" wet-bulb temperature by 5 percent before selecting a cooling tower or forecasting its operation.

Outdoor cooling towers are subject to other troubles. When the system must be operated during the winter, the quantity of air must be reduced to a point where the spray water will not freeze in the sump, on the packing, or at the tower discharge. Overflow and makeup water piping must be adequately protected. In severe climates it is sometimes necessary to provide an auxiliary tank within the heated part of the building in which reserve water is accumulated instead of in the tower sump. This auxiliary tank permits the spray to drain from the tower immediately.

Indoor cooling towers eliminate some of the trouble due to freezing but still must be carefully tended during severe weather. The quantity of air handled must be carefully regulated so freezing temperatures are not reached at periods of light load.

In summer algae or other organic slime may tend to grow in the cooling tower and, if permitted to develop, will seriously foul the condenser tubes. This can be eliminated by proper water treatment.

Evaporative Condensers. This type of condenser was developed extensively in the early 1930s to alleviate the problems arising from the use of numerous water-cooled condensers in small air-conditioning systems. As has been mentioned, in many communities the water supply and drainage facilities were becoming overburdened. In some localities the high cost of water was a serious objection to the extensive use of air conditioning, and the use of a cooling tower for small installations was not practical. Therefore, the evaporative condenser was designed to combine the functions of a condenser and a cooling tower.

A typical evaporative condenser installation is illustrated in Fig. 3-36. Notice that the conventional water-cooled condenser has been eliminated. Hot gas from the discharge of the compressor is carried directly to a condensing coil within the casing of the evaporative condenser where it is condensed to a liquid and drains to the liquid receiver. The condensing coil is usually of the extended surface or finned type. Heat is removed from the condensing gas by water evaporating directly on the coil. Evaporation is accelerated by the airstream induced by the unit fans. The water remaining after evaporation drains to the sump where it collects and flows to the spray pump for recirculation. Since 3 to 5 percent of the water circulated evaporates, makeup water is admitted to the sump tank through a float-operated valve (not shown). The quantity needed is from 5 to 10 percent of the water required for comparable water-cooled condensers. Partial evaporation increases the salt concentration of the water, and therefore treatment of the water and a continuous overflow (about 1/60 gpm/ton) is recommended. Water treatment should also be used to reduce scale formation on the warm condenser coil. Centrifugal fans are used, except for the very small units. Large multiple units have been developed and used for capacities of well over 100 tons.

Because of the very short piping and little resistance to spray water flow, the pump horsepower for an evaporative condenser is considerably lower than for a cooling tower system of equal capacity. The evaporative condenser, creating its own air movement, can be installed outdoors, or indoors if proper provisions are made to supply fresh air and carry away the humid waste air. When placed indoors, they may serve as exhaust units for interior spaces. They must be made weatherproof for outside locations. When they are so placed, provisions must be made for draining if subfreezing temperatures occur. In cool weather it may not be necessary to use water, and the unit then operates as an air-cooled condenser.

The effects of wet-bulb temperature of the air supply and the condensing temperature of the refrigerant are readily apparent from the capacity tables. As might be expected, any increase in condensing temperature and pressure will cause an increase in the evaporative capacity due to the increase in the temperature difference between the gas and the spray water. For any given condensing temperature the capacity of the evaporative condenser decreases as the wet-bulb temperature of the air rises. This decrease in capacity is due to a decrease in temperature difference between the gas and the spray water and a reduction of evaporation due to the greater initial moisture content in the air. Since the entering air wet-bulb temperature affects the refrigerant condensing temperature, the condenser and com-

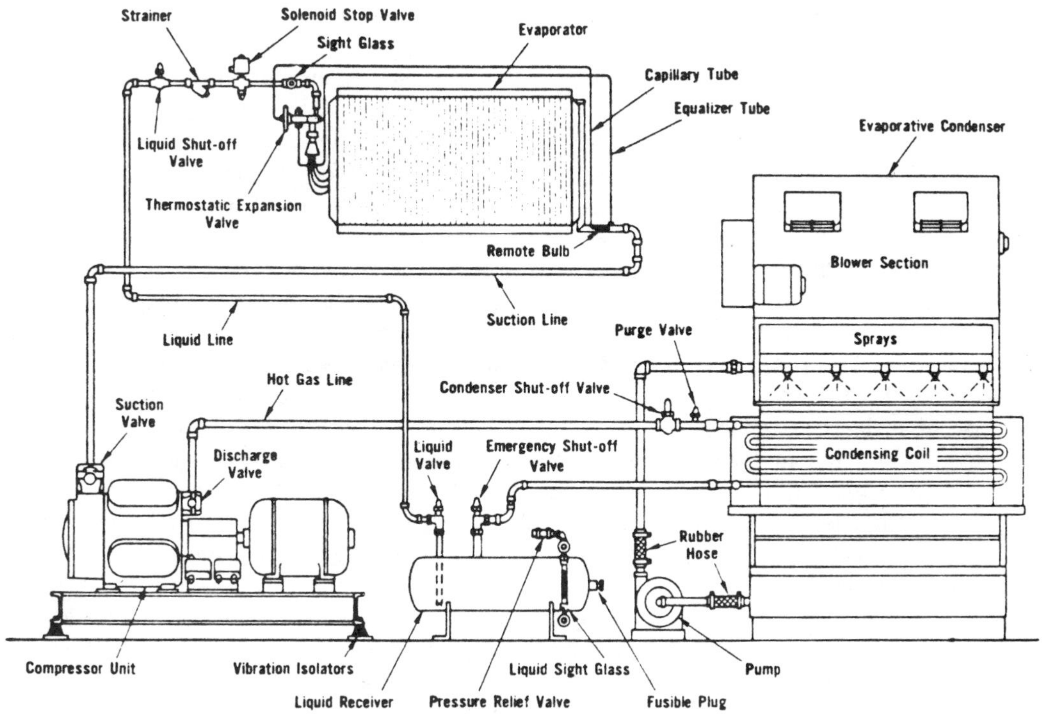

Figure 3–36

A typical evaporative condenser installation (Courtesy of The Trane Company.)

pressor used should be selected to handle the design load at a wet-bulb temperature 2 to 3 degrees above the design wet-bulb temperature.

In some instances evaporative condensers are furnished with subcooling coils. The benefits of using subcooling has already been discussed. The subcooler can be a coil or a loop of tubing immersed in the spray tank sump. It can also be additional surface on the air intake side of the condensing coil and immediately under it. Some manufacturers claim that their condensing surface is oversized and suggest that the system be charged so that the last pass of the coil remains full of liquid to obtain subcooling. Other manufacturers recommend the addition of a coil immersed in the evaporative condenser sump to provide subcooling. This arrangement will provide 7 to 10 degrees of subcooling.

Air Condensers. With increasing restrictions on the use of water in many regions, dry air is being used for condensing purposes. The primary reasons for this are (1) air is readily available and there is no disposal problem, (2) first cost is lower than other condensing means involving water conservation, and (3) maintenance costs are reduced. Air condensers are now built in sizes that match the largest reciprocating compressors so they may be found in almost any application.

There are offsetting disadvantages that must be evaluated, such as (1) large volumes of air are required which may cause noise problems; (2) operating costs are high because the power to drive the compressor at or near full load is high; (3) the air condenser increases in capacity when the system load falls off, creating operating problems at part load; and (4) start-up problems are encountered at low outdoor air temperatures.

Nothing can be done to overcome the first two disadvantages. However, the part-load problem has been attacked in various ways. The use of multispeed condenser fan motors reduces air flow at part load, thus reducing condenser capacity. This solution is costly and not wholly satisfactory. It is almost impossible to reduce the air quantity sufficiently to balance the reduced load.

Multilouvered dampers also have been used and are less expensive than variable-speed fan motors. When the dampers are coupled with a fan that does not overload its motor at throttled air flow, satisfactory part-load performance over a limited outdoor temperature range results.

The modern air condenser is available with either propeller- or centrifugal-type fans for moving the air through the condenser coil. Figure 3-37 illustrates a propeller-type unit, while Fig. 3-38 shows a unit with a centrifugal fan, the more costly of the two designs. When the air condenser can be installed out of doors, the propeller fan-type unit is generally used. If for some reason the condenser is to be installed indoors, with duct work to carry the air to and from the unit, the condenser with the centrifugal fan is more suitable. In addition, a Trane centrifugal fan air-cooled condenser may be used to provide auxiliary heating or ventilation in certain applications. Either type unit can be built with a liquid subcooling circuit. This is desirable as explained previously.

Air condensers are available in sizes as large as 125 tons in a single unit. Multiple units may be used for higher tonnages, but it is always desirable to use a single condenser, if possible, because the piping for multiple

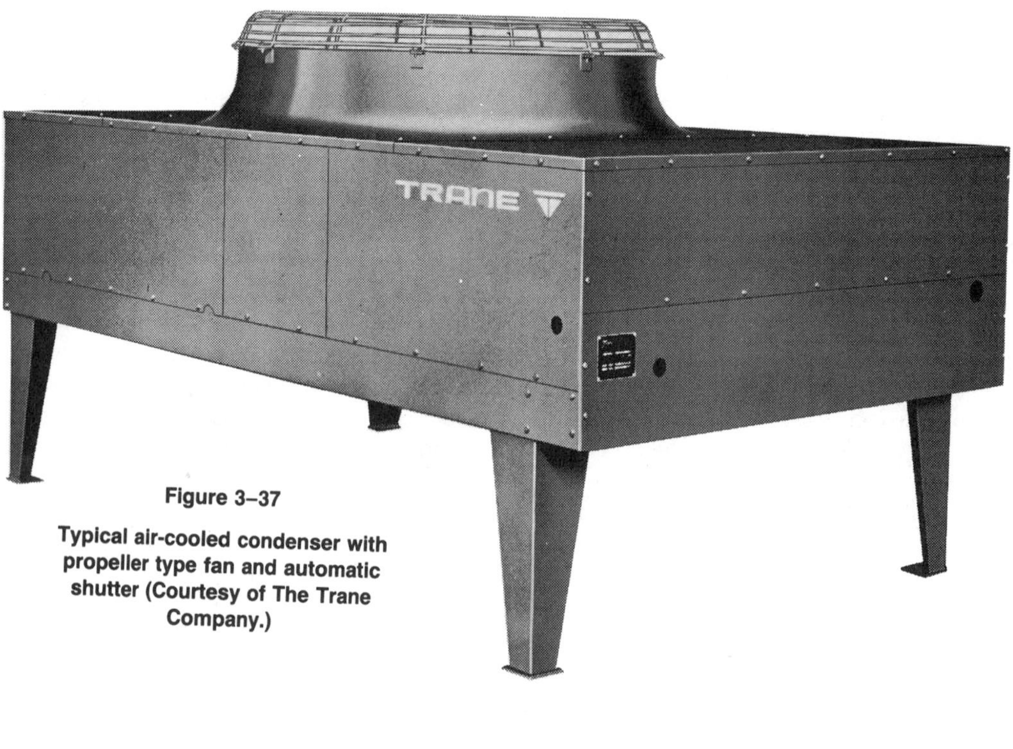

Figure 3–37

Typical air-cooled condenser with propeller type fan and automatic shutter (Courtesy of The Trane Company.)

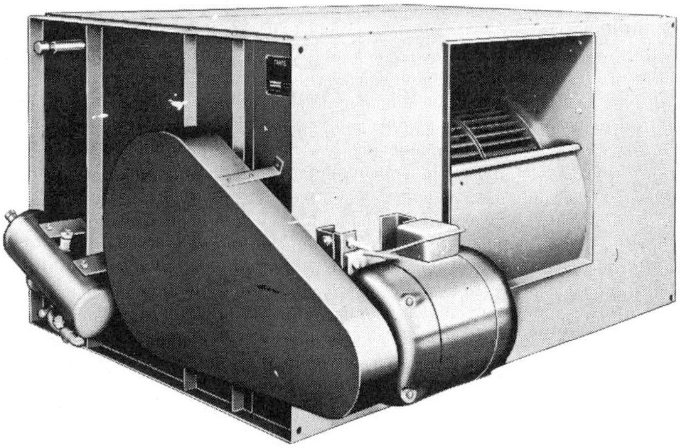

Figure 3–38

Typical air-cooled condenser with centrifugal type fan (Courtesy of The Trane Company.)

units is more involved and requires greater care in design and installation.

3-23 REFRIGERATION CONTROL

Automatic control of a refrigeration system requires complete control of (a) the flow of liquid refrigerant, (b) the on-off operation of the compressor, (c) the flow of condensing medium, and (d) safety controls for prevention of damage to equipment. In addition, special controls designed for specific applications are frequently required. The various types of devices used to accomplish these purposes are so numerous that it would be impossible to describe all their modifications.

As the refrigeration cycle is followed, it becomes obvious that some very effective means of metering the amount of liquid refrigerant entering the evaporator be provided. If too little liquid enters the evaporator, it is soon evaporated and too much of the evaporator surface becomes ineffective. If too much liquid enters the evaporator, some liquid will pass, unevaporated, completely through the evaporator and into the compressor suction line.

Should liquid in the suction line reach the compressor, severe damage may result. To all practical purposes, the liquid is incompressible, and should it enter the minute clearance space between the pistons and cylinder heads in too large a volume, it may cause broken pistons or badly damaged valves. Liquid in small quantities will cause an undesirable clattering noise, which can be annoying in itself. Unusual noise in a compressor should be investigated and corrective measures should be taken without delay. Liquid reaching the compressor through the suction line is commonly called *flooding back*. It should be avoided at all costs.

Methods of Feeding Liquid to the Evaporator. Several methods of accomplishing the proper feeding of liquid to the evaporator have been devised. The common method employs a thermostatic expansion valve. It operates in such a manner that the refrigerant starts evaporating it the instant it enters the evaporator and is completely evaporated before it can leave the evaporator. This is called *dry expansion,* and it is the principle on which direct expansion cooling coils operate. A second method, frequently applied on domestic refrigerators, replaces the thermostatic expansion valve with an automatic expansion valve set to maintain a constant pressure in the evaporator. This is also a dry expansion system. The third method, frequently used with ammonia on commercial or industrial refrigeration systems, employs a float and surge chamber and is designed to maintain a constant liquid level in the evaporator. This is called a *flooded* system. Inasmuch as practically all refrigeration systems using the Freon refrigerants use the dry expansion system and employ a thermostatic expansion valve, we will limit our discussion to this type.

3-24 THE THERMOSTATIC EXPANSION VALVE

Two typical thermostatic expansion valves are shown in Fig. 3-39 and 3-40; a schematic diagram of a thermostatic expansion valve is shown in Fig. 3-41. The valve consists of shell or body *A*, valve pin *B*, spring *C*, diaphram *D*, and remote bulb *E*. The remote bulb *E* and diaphram *D* are connected by means of a capillary tube and are charged with a volatile fluid, sometimes the same as the refrigerant used in the system. Obviously, if heat is applied to remote bulb *E*, the evaporating fluid in the remote bulb will cause an increase in pressure which is transmitted through the capillary tube to diaphram *D*. As the pressure is increased, the diaphram is forced downward moving pin *B* to open the valve port. If remote bulb *E* is then cooled, the condensing fluid will lower the pressure, the diaphram will recede and spring *C* will move pin *B* to close the valve port. Thus, the degree of heat available at remote bulb *E* determines the amount the valve port shall be opened and the amount of liquid refrigerant that will be passed to the evaporator. Most thermostatic expansion valves are furnished with an adjustment that changes the tension of the spring so that a greater or lesser degree of heat at the remote bulb will affect a corresponding movement of the valve pin.

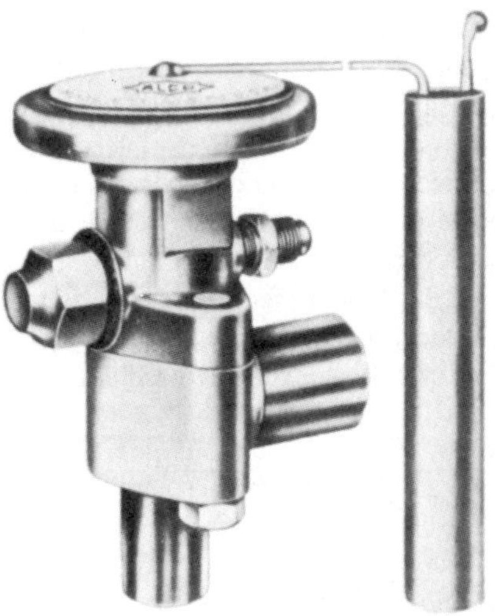

Figure 3–39

Typical thermostatic expansion valve (Courtesy of Alco Valve Company.)

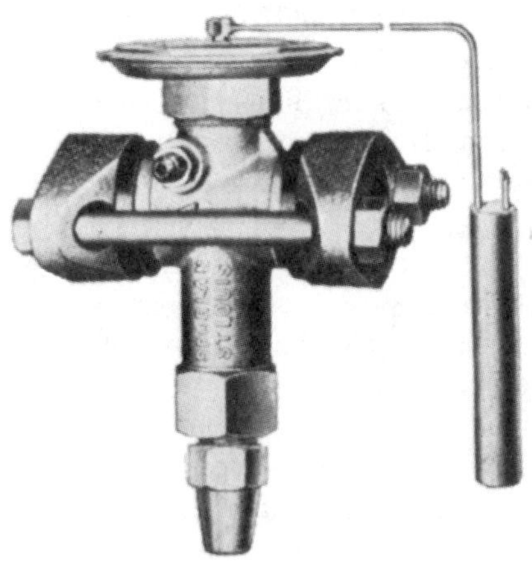

Figure 3–40

Typical thermostatic expansion valve (Courtesy of Sporlan Valve Company.)

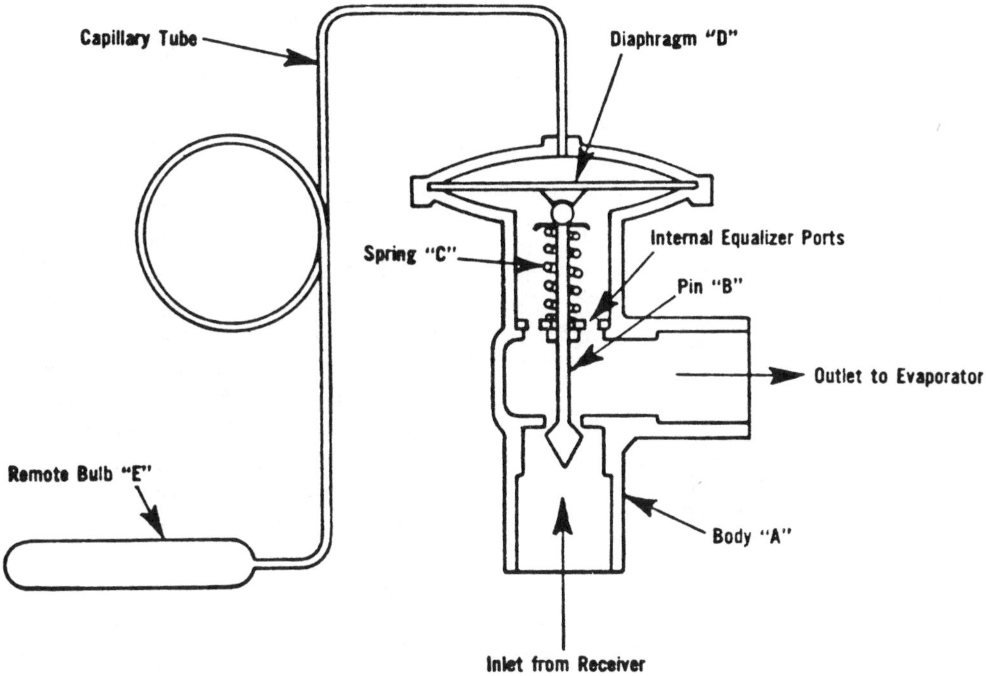

Figure 3–41

Schematic showing typical thermostatic valve having internal equilizer parts (Courtesy of The Trane Company.)

This adjustment is usually referred to as a superheat adjustment.

Figure 3-42 is a schematic showing the thermostatic expansion valve connected to an evaporator such as a direct expansion cooling coil. The valve is attached directly to the liquid line and will control the amount of liquid that can enter the coil. The remote bulb is attached to the suction pipe near the outlet of the coil. In operation, the valve will pass liquid refrigerant which is evaporated in the coil, the resulting gas leaving the coil through the suction line. If, for example, the valve is set for 10 degrees superheat, the gas passing the remote bulb must be 10 degrees warmer than the temperature of the evaporating refrigerant. This means that part of the coil immediately ahead of the suction header must be used to heat the completely evaporated refrigerant from a temperature corresponding to the suction pressure to a temperature 10 degrees higher. Under these conditions, the coil in Fig. 3-42 will contain a mixture of liquid and gas from the liquid header to point *X*. At point *X* the liquid will have been completely evaporated. From point *X* to the remote bulb, the coil surface will be used only to raise the temperature of the gas to the superheat setting of the valve. If the load on the coil decreases, the superheating section, between point *X* and the remote bulb, will absorb less heat so the temperature of the superheated gas will decrease. This decrease in gas temperature will cool the remote bulb, and the valve will function to decrease the flow of

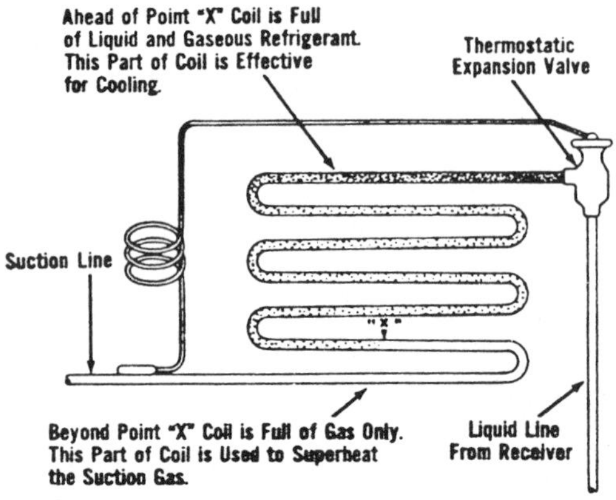
Ahead of Point "X" Coil is Full of Liquid and Gaseous Refrigerant. This Part of Coil is Effective for Cooling.
Thermostatic Expansion Valve
Suction Line
"X"
Beyond Point "X" Coil is Full of Gas Only. This Part of Coil is Used to Superheat the Suction Gas.
Liquid Line From Receiver

Figure 3–42

Effect of thermostatic expansion valve on evaporator operation (Courtesy of The Trane Company.)

liquid into the coil. If the load increases, the superheating section of the coil will absorb more heat, and the temperature of the superheated gas will rise. This increase in temperature will warm the remote bulb, and the valve will function to admit more liquid.

Superheat. As the load fluctuates, point *X* will move back and forth in such a manner that there is always enough of the evaporator between point *X* and the remote bulb to warm the gas up to the superheat setting of the valve, in this case 10 degrees above the evaporating temperature. If the superheat setting of the valve is changed manually, point *X* will move accordingly. Notice particularly that point *X* is determined by the *difference* in temperature (superheat) between the evaporating temperature and the gas temperature at the bulb, rather than by the evaporating temperature alone. This is important since, regardless of evaporating or suction temperature variations, the amount of superheat remains approximately constant to prevent liquid from entering the suction line.

The amount of heat that can be absorbed by evaporating liquid refrigerant is very much greater than the amount that can be absorbed by superheating the cold refrigerant gas. For that reason that part of the coil required to superheat the gas is almost worthless from a cooling-load standpoint. It is, therefore, desirable to hold the superheat reasonably low and use the greatest amount of the coil possible for useful heat absorption. There is always a lag, however, between the time the remote bulb senses a change in superheat and the time the valve can respond. Too low a superheat setting will cause liquid to spill into the suction line, and too high a superheat setting will cause a reduction in capacity. Thermostatic expansion valves are usually adjustable over a superheat range of 0 to 25 or 30 degrees but will be shipped by their manufacturers adjusted to the superheat desired. Direct expansion cooling coils are rated usually on the basis of 5 degrees superheat. If load variations are wide and subject to rapid changes, the valves should be set for superheats from 10 to 15 degrees or apply remote bulb in a bulb well for close control at low superheat setting.

Pressure Drop. In Fig. 3-41 the spring *C* is shown to be opposing the action of the diaphram of power element *D*. Notice that the valve outlet pressure is also exerted on this diaphram in the same direction as the spring pressure. Thus, using Freon-22 when the evaporating temperature is 40°F, the absolute pressure beneath the diaphram is 83.206 psia. As Freon-22 also charges the bulb and power element, when the bulb is subjected to the 40 degrees temperature, the pressure in the bulb and above the power element diaphram is also 83.206 psia. The gas pressures are balanced and only the spring pressure is holding the valve closed. If the spring pressure corresponds to 15.521 psi, the valve will open when the gas temperature at the remote bulb reaches 50 degrees, or 10 degrees superheat, as the pressure within the bulb and power element will then be 98.727 psia, or 83.206 plus 15.521.

Now, it is supposed this thermostatic expansion valve is used with an evaporator having a pressure drop of 6 psi and a distributor with a 15-psi drop. As the pressure at the suction header must be 83.205 psia for 40 degrees suction temperature, the pressure at the liquid header or inlet must be 21 psi higher, or 83.206 + 21.00 = 103.206 psi. This 103.206 psia pressure added to the 15.521-psi spring pressure, a total of 118.727 psia, is now imposed on the diaphram underside. To balance this pressure and open the valve, a temperature of approximately 61 degrees must exist at the bulb. It now can be seen that a superheat of 61 − 40 = 21 degrees will be required instead of the 10 degrees required when the 21-psi pressure drop did not exist. This additional superheat will result in a loss of coil capacity due to the large amount of coil surface used for superheating gas.

External Equalizer. To overcome the effect of pressure drop in larger evaporators, the internal equalizer ports shown in Fig. 3-41 are closed and an external equalizer is connected from the suction side of the coil to the space immediately below the diaphram of the valve, as shown in Fig. 3-43. The use of the external equalizer eliminates the effect of the pressure drop across the coil, and the superheat setting depends only on the adjustment of the spring tension.

All thermostatic expansion valves used in

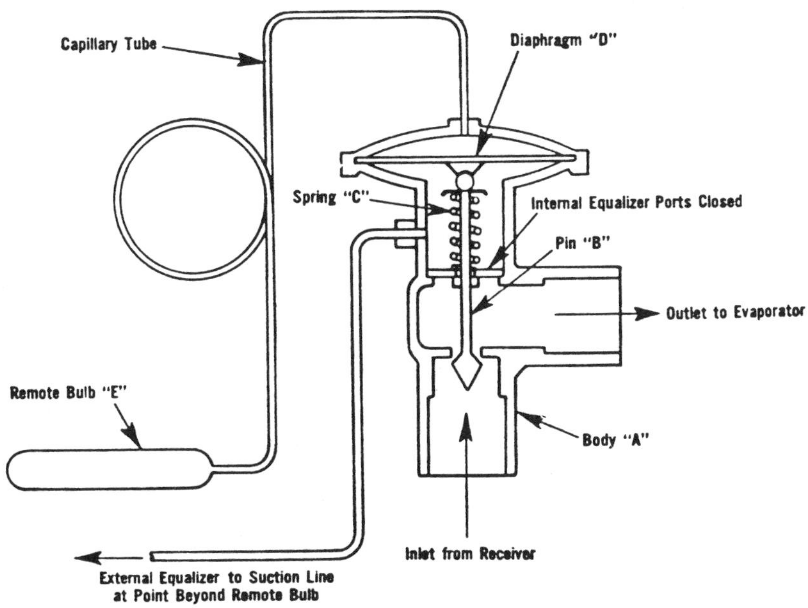

Figure 3–43

Schematic showing typical thermostatic expansion valve having external equilizer (Courtesy of The Trane Company.)

air-conditioning coils, having a pressure drop of 2 psi or greater, should have an external equalizer.

3-25 THERMOSTATIC EXPANSION VALVE SELECTION.

Selection of the correct expansion valve is not difficult. Tables 3-6 and 3-7 are reproductions of rating tables of the Alco Valve Company, covering in part the capacities of Alco thermostatic expansion valves. It should be noted that in both tables the valves are rated on the pressure difference between the valve inlet and the valve outlet.

The pressure at the expansion valve inlet will be the existing pressure in the receiver or condenser *less* the pressure drop in the liquid line. The liquid-line pressure drop is the total pressure drop from both friction and lift. The friction pressure drop can be estimated from refrigerant piping tables. The lift pressure drop can be calculated using the physical properties of the liquid refrigerants, or can be found from Table 3-8.

The pressure at the expansion valve outlet

Table 3-6

Alco T-Series Single-Outlet Thermo Valves*
(With external superheat adjustment)
Refrigerant-12

VALVE TYPE NUMBER	EVAPORATOR TEMPERATURE, °F +40					+20				
	PRESSURE DROP ACROSS VALVE, PSI									
	60	80	100	120	150	60	80	100	120	150
	CAPACITIES IN TONS OF REFRIGERATION									
TCL25F	.25	.29	.32	.35	.39	.24	.28	.31	.34	.38
TCL50F	.60	.69	.77	.85	.95	.58	.66	.74	.81	.91
TCL100F	1.3	1.5	1.7	1.8	2.0	1.2	1.4	1.6	1.8	2.0
TCL200F	2.0	2.3	2.6	2.8	3.2	1.9	2.2	2.5	2.7	3.0
TCL250F	2.5	2.9	3.2	3.5	3.9	2.4	2.8	3.1	3.4	3.8
TCL300F	3.5	4.0	4.5	4.9	5.5	3.4	3.9	4.3	4.7	5.3
TCL400F	4.3	5.0	5.5	6.1	6.8	4.1	4.8	5.3	5.8	6.5
TCL600F	6.0	6.9	7.7	8.5	9.5	5.8	6.6	7.4	8.1	9.1
TCL650F	6.5	7.5	8.4	9.2	10.3	6.2	7.2	8.0	8.8	9.9
TCL750F	7.5	8.6	9.7	10.6	11.8	7.2	8.3	9.3	10.2	11.4
TJL800F	8.5	9.8	11.0	12.0	13.4	8.2	9.4	10.5	11.5	12.9
TJL1100F	11.0	12.7	14.2	15.5	17.4	10.6	12.2	13.6	14.9	16.7
TER13F	13.0	15.0	16.8	18.4	20.5	12.5	14.4	16.1	17.6	19.7
TER15F	15.0	17.3	19.3	21.2	23.7	14.4	16.6	18.6	20.3	22.8
TER20F	20.0	23.1	25.8	28.3	31.6	19.2	22.1	24.8	27.1	30.4
TER25F	25.0	28.8	32.2	35.3	39.5	24.0	27.7	30.9	33.9	37.9
TIR35F	35.0	40.3	45.1	49.5	55.3	33.6	38.7	43.3	47.5	53.1
THR45F	45.0	51.9	58.0	63.6	71.1	43.2	49.8	55.7	61.1	68.3
THR55F	55.0	63.4	70.9	77.8	86.9	52.8	60.9	68.1	74.6	83.5

**Courtesy of Alco Valve Company.*

Table 3-6a

R-12 Expansion Valve Correction Factors

REFRIGERANT LIQUID TEMPERATURE °F	80	90	110	120	130	140
MULTIPLIER FACTOR	1.12	1.06	.94	.88	.81	.75

Table 3-7

Alco T-Series Single-Outlet Thermo Valves[a]*
(with external superheat adjustment)
Refrigerant-22

VALVE TYPE NUMBER	EVAPORATOR TEMPERATURE, °F									
	+40					+20				
	PRESSURE DROP ACROSS VALVE, PSI									
	75	100	125	150	175	100	125	150	175	200
	CAPACITIES IN TONS OF REFRIGERATION									
TCL50H	.43	.50	.56	.61	.66	.48	.54	.59	.64	.68
TCL100H	.87	1.0	1.1	1.2	1.3	.97	1.1	1.2	1.3	1.4
TCL200H	1.8	2.1	2.3	2.6	2.8	2.0	2.3	2.4	2.7	2.9
TCL300H	3.1	3.6	4.0	4.4	4.8	3.5	3.9	4.3	4.6	4.9
TCL400H	3.9	4.5	5.0	5.5	5.9	4.4	4.9	5.3	5.8	6.2
TCL500H	4.8	5.5	6.1	6.7	7.3	5.3	6.0	6.5	7.0	7.5
TCL700H	6.1	7.0	7.8	8.6	9.3	6.8	7.6	8.3	9.0	9.6
TCL900H	8.4	9.7	10.8	11.9	12.8	9.4	10.5	11.5	12.4	13.3
TCL1000H	9.1	10.5	11.7	12.9	13.9	10.2	11.4	12.5	13.5	14.4
TCL1200H	10.5	12.1	13.5	14.8	16.0	11.7	13.1	14.4	15.5	16.6
TJL1400H	11.9	13.8	15.4	16.9	18.2	13.4	14.9	16.4	17.7	18.9
TJL1800H	15.4	17.8	19.9	21.8	23.5	17.3	19.3	21.1	22.8	24.4
TER22H	19.0	22.0	24.6	26.9	29.1	21.3	23.8	26.1	28.2	30.2
TER26H	22.5	26.0	29.1	31.8	34.4	25.2	28.2	30.9	33.3	35.7
TER35H	30.3	35.0	39.1	42.9	46.3	33.9	37.9	41.6	44.9	48.0
TER45H	39.0	45.0	50.3	55.1	59.5	43.6	48.8	53.5	57.7	61.7
TIR55H	47.6	55.0	61.5	67.4	72.7	53.3	59.6	65.3	70.6	75.4
THR75H	64.9	75.0	83.8	91.9	99.2	72.7	81.3	89.1	96.2	102.9
THR100H	86.6	100.0	111.8	122.5	132.3	97.0	108.4	118.8	128.3	137.2

[a]Valve capacities in Table 3-6 and Table 3-7 are based on 100°F vapor-free refrigerant liquid entering the valve. To determine the valve capacities for other temperatures of vapor-free refrigerant liquid entering the valve, multiply the capacities listed above by the multiplier factors listed in Table 3-6a for Refrigerant-12 and Table 3-7a for Refrigerant-22.

Courtesy of Alco Valve Company.

Table 3-7A

R-22 Expansion Valve Correction Factors

REFRIGERANT LIQUID TEMPERATURE °F	80	90	110	120	130	140
MULTIPLIER FACTOR	1.13	1.06	.93	.87	.81	.75

Table 3-8

Factors Affecting Thermostatis Expansion Valve Selection
Pressure Drop of Liquid Refrigerant at Various Vertical Lifts

REFRIGERANT	LIQUID LINE RISE IN FEET							
	10′	15′	20′	30′	40′	50′	60′	70′
REFRIGERANT-12 PRESSURE DROP	5.5 LBS.	8.2 LBS.	11.0 LBS.	16.5 LBS.	22.0 LBS.	27.6 LBS.	33.2 LBS.	38.8 LBS.
REFRIGERANT-22 PRESSURE DROP	5.0 LBS.	7.5 LBS.	9.9 LBS.	14.9 LBS.	19.8 LBS.	24.8 LBS.	29.8 LBS.	34.7 LBS.

Alco Valve Company

will be the suction pressure at the compressor *plus* the pressure drop in the suction line and the pressure drop across the evaporator. The pressure drop in the suction line is considered to be due entirely to friction. The low density of the gas makes the difference in pressure due to lift so small that it may be neglected. The pressure drop due to suction-line friction can be estimated from refrigerant piping tables. The pressure drop through the evaporator must be figured from the manufacturer's data. A good average figure is 20 psi, which is close enough for general use if no other value is available.

Illustrative Problem 3-20

As an example of expansion valve selection, assume a system having an evaporator load of 132,000 Btu/hr (11.0 tons). The condensing temperature is 100°F, the suction temperature is 40°F, and the liquid will leave the condenser at the condensing temperature. There is no subcooling. The pressure drop through the evaporator and distributer is 20.0 psi, through the suction line 2.0 psi, and through the liquid line 2.0 psi. The expansion valve is 10 ft higher than the liquid level in the receiver. The refrigerant is Freon-22. Select the proper expansion valve.

Solution.:

Step 1: Determine the pressure difference across the valve.

Step 2: Consider formation of *flash gas* ahead of expansion valve.

If the refrigerant in the liquid line is to remain a liquid until it enters the thermostatic expansion valve, it must be at or below 97.55°F. This is the saturation temperature corresponding to the pressure at the valve inlet, 203.6 psia (see Table B-5). To accomplish this, the liquid refrigerant must be reduced in temperature after being condensed (subcooled). When no subcooling is done, the pressure will drop in the liquid line and the refrigerant will boil. Since there is no source of heat except the liquid refriger-

Pressure at valve inlet	**Psi**
Condensing pressure at 100°F (Table B-5)	210.60 psia
Less pressure drop in liquid line	2.00 psi
Less loss in pressure due to 10-ft elevation (Table 3-8)	5.00 psi
Resulting pressure at valve inlet =	203.60 psia

Pressure at valve outlet	**Psi**
Suction pressure at 40°F suction temperature (Table B-5)	83.206 psia
Plus pressure drop in suction line	2.00 psi
Plus pressure drop across coil and distributor	20.00 psia
Resulting pressure at valve outlet =	105.206 psia

ant, just enough liquid will flash into vapor to lower the temperature of the remaining liquid. This vapor is known as flash gas, and it is undesirable when it occurs in the liquid line ahead of the expansion valve.

Flash gas generated in the liquid line must pass through the expansion valve with the liquid. The gas displaces some of the liquid, thus reducing the expansion valve liquid capacity and hence the system capacity. Flash gas causes "wire drawing" of the valve seat as well as a hissing noise that is often objectionable. For these reasons the thermostatic expansion valve should not be oversized to compensate for flash gas. The proper approach is to eliminate the flash gas from the liquid line ahead of the expansion valve. Heat exchangers are commonly used for this purpose, as we have discussed. Thus, with the flash gas eliminated, the thermostatic expansion valve should be sized on the basis of evaporator temperature, system pressures, and pressure drop.

Step 3: Select the expansion valve.

From Table 3-7, an Alco TCL 1200 H has a capacity of 12.1 tons at 100 psi pressure difference across the valve and is, therefore, ample.

CHAPTER 3

Review

3-1. Calculate the number of pounds per minute of ammonia that must be circulated in a 20-ton vapor compression refrigeration system operating between a condenser temperature of 80°F and an evaporator temperature of −20°F. Assume saturated liquid leaves the condenser and enters the expansion valve, and saturated vapor leaves the evaporator.

3-2. A vapor compression refrigeration system uses Freon-12 as the refrigerant. The vapor leaves the evaporator as saturated vapor at −5°F and saturated liquid enters the expansion valve at 80°F. Determine the refrigerating effect per pound of Freon circulated.

3-3. A 10-ton vapor compression refrigeration system uses Freon-22 as the refrigerant. The evaporator temperature is 25°F and the condenser temperature is 100°F. Liquid enters the expansion valve at 90°F and vapor leaves the evaporator at 30°F and 60 psia. Calculate the required refrigerant flow rate in lb/min.

3-4. A food freezing system requires 20 tons of refrigeration at an evaporator temperature of −30°F and a condenser temperature of 72°F. The refrigerant, Freon-22, is subcooled 6 degrees before entering the expansion valve, and the vapor leaving the evaporator and entering the compressor is superheated 7 degrees. Compression of the vapor is isentropic, valve throttling and cylinder clearance are to be disregarded. A six-cylinder, single-acting compressor with bore and stroke equal is to be used operating at 1750 rpm. Determine the following: (a) refrigerating effect, (b) weight of refrigerant circulated per minute, (c) theoretical horsepower required, (d) coefficient of performance, (e) heat removed at the condenser, and (f) theoretical bore and stroke of the compressor cylinders.

3-5. Assume that the compressor of Problem 3-4 has a clearance of 4 percent, suction valve and discharge valve pressure drops of 5 and 10 psi, respectively, polytropic compression with $n = 1.15$, and the vapor is heated by the hot cylinder walls to a temperature of 0.0°F before the start of compression. The compressor has a mechanical efficiency of 80 percent. Calculate the following: (a) clearance volumetric efficiency; (b) total volumetric efficiency; (c) piston displacement per minute; (d) theoretical horsepower; (e) brake horsepower; (f) heat rejected to condenser; and (g) if the condenser is water cooled, how much water circulation in gpm would be required for a 10-degree water temperature rise?

3-6. An ammonia vapor compression refrigeration system has a capacity of 150 tons. Condenser temperature is 96°F, evaporation temperature in the brine cooler is -10°F. Ammonia vapor leaves the evaporator and enters the compressor at $+10$°F. Liquid ammonia enters the expansion valve at 86°F. Pressure drops through the suction and discharge valves are 4 and 8 psi, respectively. The polytropic compression exponent $n = 1.22$. Compressor has a clearance of 5 percent. Calculate the following: (a) the theoretical horsepower, (b) the heat transferred to the compressor cooling water jackets, (c) piston displacement in cfm, (d) heat transferred to condenser cooling water, and (e) coefficient of performance.

3-7. An ideal vapor compression refrigeration system using Freon-22 has the condenser operating at 90°F and the evaporator operating at 30°F. The liquid leaving the condenser and entering the expansion valve is saturated, and the vapor leaving the evaporator and entering the isentropic compressor is saturated. For a system capacity of 20 tons, calculate the following: (a) the refrigerant flow rate in lb/min; (b) the theoretical horsepower required; (c) the theoretical piston displacement, cfm; and (d) coefficient of performance.

3-8. A two-cylinder Freon-12 compressor has a bore of 2.5 in. and a stroke of 2.75 in. The compressor operates at 1750 rpm. The evaporator temperature is 38°F and the condensing temperature is 105°F. Saturated vapor at evaporator temperature enters the compressor, and saturated liquid at the condenser temperature enters the expansion valve. Determine the theoretical and the actual capacity of the compressor.

3-9. A theoretical vapor compression refrigeration system using ammonia has liquid leaving the condenser and entering the expansion valve at 240 psia and 104°F. Evaporator pressure is 20.0 psia, and vapor leaves the evaporator and enters the compressor at 0.0°F. The system produces 20 tons of refrigeration. Determine the following: (a) coefficient of performance; (b) piston displacement, assuming a volumetric efficiency of 100 percent; (c) the temperature of the vapor leaving the compressor; and (d) compressor power.

3-10. The temperature of the ammonia vapor in Problem 3-9 after compression is high. It is proposed to modify the cycle by using two compressors in series with a water-cooled intercooler between the two compressors. The low-pressure compressor will compress the vapor to 70 psia. The vapor then passes through the intercooler where it is cooled at constant pressure to a temperature of 40°F. The vapor is then compressed to the final condenser pressure of 240 psia. Determine the following: (a) the coefficient of performance, (b) the piston displacement of the two compressors for 100 percent volumetric efficiency, (c) the temperature of the vapor leaving the high-pressure compressor, and (d) total power required by the compressors. Compare the results with those of Problem 3-9.

3-11. A Freon-12 vapor compression refrigeration system operating at an evaporating temperature of 0.0°F and a condensing temperature of 80°F, uses a compressor having a 5 percent clearance. Under these conditions the plant produces 20 tons of refrigeration. Calculate the capacity of the plant if the condensing temperature were to be increased to 100°F. Assume that other conditions are the same and the compressor is operated at the same rpm.

3-12. An industrial plant has available a four-cylinder, 3-in. bore by 4-in. stroke, 800 rpm, single-acting compressor for use with Freon-12. Proposed operating conditions for the compressor are 100°F condensing temperature and 50 psia evaporator pressure. It is estimated that the refrigerant will enter the expansion valve as saturated liquid, that vapor will leave the evaporator at a temperature of 40°F, and that vapor will enter the compressor at 50°F. Assume a compressor volumetric efficiency from Table 3-4, frictionless flow. Calculate the refrigerating capacity in tons for a system equipped with this compressor.

3-13. A vapor compression refrigeration system is operating with 75°F condenser temperature, 10°F evaporator temperature, 5°F subcooling of the liquid refrigerant leaving the condenser, and 10°F superheating of the vapor leaving the evaporation. Compression is isentropic and the clearance is 2 percent. Neglect valve pressure drops and additional superheating of gases entering the compressor. Determine the required cylinder displacement per ton-minute when (a) the refrigerant is Freon-12, (b) the refrigerant is Freon-22, (c) the evaporator temperature is lowered to −40°F and the refrigerant is Freon-12, and (d) the evaporator temperature is 40°F and the refrigerant is Freon-22.

3-14. A food-freezing system requires 20 tons of refrigeration at an evaporator pressure of 20 psia and a condenser pressure of 140 psia. The refrigerant, Freon-22, is subcooled to 70°F before entering the expansion valve, the vapor is superheated to −20°F before leaving the evaporator. Compression is isentropic, and valve throttling and clearance are to be disregarded. A six-cylinder single-acting compressor with stroke equal to bore is to be used, operating at 1500 rpm. Determine the following: (a) the refrigerating effect, (b) weight of refrigerant to be circulated per minute, (c) theoretical piston displacement per minute, (d) theoretical horsepower, (e) coefficient of performance, (f) heat removed through condenser, and (g) theoretical bore and stroke of compressor.

3-15. You are in need of an ammonia compressor for a proposed installation in your plant. A surplus, used ammonia compressor in good condition is available from another plant if it has adequate capacity.

It is estimated that your proposed installation would operate under the following conditions: capacity 18.5 tons, condensing pressure = 195 psia, ammonia liquid leaves condenser at saturated conditions, ammonia liquid enters expansion valve at 75°F, evaporating temperature = 10°F, vapor leaves evaporator at saturated conditions, vapor is superheated 20°F in compressor suction line.

The data supplied to you about the surplus compressor are only the following: It is a four-cylinder, vertical, reciprocating, single-acting, single-stage, water-jacketed compressor, with maximum rpm of 600. Bore is 4 in. and stroke is 5 in.

Based on your past experience, you make the following supplemental assumptions to allow further calculations: clearance = 4 percent, pressure drop in suction valves = 4 psi, vapor is superheated 15°F in cylinder on intake stroke after passing the suction valves, polytropic compression with $n = 1.27$, pressure drop in discharge valves = 6 psi, compressor mechanical efficiency = 80 percent. (a) Determine whether the surplus compressor may be used and, if so, what rpm it should operate at. (b) Estimate the horsepower required to drive the compressor.

BIBLIOGRAPHY

1. *Trane Air Conditioning Manual,* The Trane Company, LaCrosse, Wisconsin, 1965.

2. *Trane Reciprocating Refrigeration Manual,* The Trane Company, LaCrosse, Wisconsin, 1966.

3. Jordan, R. C., and Priester, G. B., *Refrigeration and Air Conditioning,* 2nd ed., Prentice-Hall, Inc., Englewood Cliffs, NJ, 1956.

4. Threlkeld, J. L., *Thermal Environmental Engineering,* 2nd ed., Prentice-Hall, Inc., Englewood Cliffs, NJ, 1970.

Chapter 4

Psychrometrics

4-1 INTRODUCTION

Psychrometrics is the study of the state of the atmosphere with reference to moisture content. The earth's atmosphere, the air we breathe, and the air in most air-conditioned spaces, is mainly a mechanical mixture of dry air and water vapor. The definition of air conditioning includes the conditioning of air as to its *temperature* and *moisture* content. Therefore, it is of great significance for us to thoroughly understand and be able to determine the psychrometric properties of the air as it is being processed in heating and humidifying and cooling and dehumidifying equipment.

4-2 ATMOSPHERIC AIR

Atmospheric air is a complex mixture of several gases, including nitrogen, oxygen, carbon dioxide, water vapor, and traces of other gases. The air also usually contains some particulate matter, such as dust and pollen, and additional vapors in small concentrations.

Although the relative amount of water vapor in the atmosphere is very small (about 0.08 percent by weight at 0°F saturated and only 1.56 percent by weight at 70°F saturated), variations in this amount as well as changes in temperatures are very significant in air-conditioning work. Investigations by ASHRAE and other groups have shown that proper humidity has a direct bearing on the comfort as well as health of human beings.

Many industrial processes may require precise control of humidity for proper production. Poor register is often noticed in color printing when adjacent colors overlap or do not meet. A change in atmospheric humidity between the printing of different colors probably was the cause since paper shrinks and stretches as the humidity changes. Proper control of humidity has made it possible to greatly extend the safe storage time for food products and to maintain their quality. The textile industry requires careful control of humidity during the manufacturing processes to reduce the effect of static electricity on the fibers.

The study of psychrometry, including the properties of air-vapor mixtures and their measurement and control, is of great importance as a preliminary to the proper design and control of an air-conditioning system.

4-3 DEFINITIONS

Dry air (abbreviated da), as noted in Section 4-2, is a complex mixture of a number of different gases. Dry air contains no moisture vapor. Nitrogen accounts for about 78 percent of the total volume, and oxygen approximately 21 percent. In spite of the complexity, the proportions of the dry-air constituents remain practically constant, and for psychrometric purposes dry air may be treated as if it were a single gas with a molecular weight of 28.967.

The common designation, air, generally means moist air in air-conditioning work; however, because of its nearly constant constituents, dry air is used as the basis for defining the properties of air.

Moist air, as it usually occurs in nature, is

a mechanical mixture of dry air and water vapor. Atmospheric air is constantly in contact with liquid or solid water and picks up moisture from these sources through vaporization and sublimation. The quantity of water thus picked up depends, principally, upon the temperature and quantity of water available for evaporation. Conversely, when air containing large quantities of water vapor is cooled, it loses moisture through condensation. Thus, the amount of moisture vapor present in air varies from hour to hour, day to day, and with location on the earth's surface.

Humidity. In a broad sense the moisture content of air is referred to as its humidity, but since the moisture content may be expressed in terms of volumes, weights, moles, or pressures, the term *humidity* has been defined in various ways. For many years air-conditioning engineers have defined humidity as the number of pounds of water vapor carried by 1 lb of dry air and is called *humidity ratio*. Other names which have been used over the years have been *absolute humidity* or *specific humidity*. For our work we will use humidity ratio as defined above.

The amount of moisture contained in the air, when expressed in pounds per pound of dry air, is a relatively small number. Therefore, it is frequently expressed in grains per pound of dry air. There are 7000 grains to the pound.

Saturation. Saturation denotes the maximum amount of water vapor that can exist in 1 ft^3 of space at a given temperature and is essentially independent of the weight and pressure of the air which may simultaneously exist in the same space. Frequently, we will speak of "saturated air"; however, it must be remembered that the *air* is not saturated; it is the water vapor contained in the air which may be saturated at the air temperature.

4-4 PROPERTIES OF AIR-VAPOR MIXTURES

Atmospheric air has properties like those of its separate constituents. The properties of greatest interest for a given mixture in air conditioning include temperature, pressure, density or specific volume, ratio of water vapor to air quantity, and enthalpy. The pressure, temperature, and density or volume of an ideal gas are interdependent. From thermodynamics, this relationship was expressed by Eq. (2-19)

$$pV = mRT \quad \text{or} \quad p = \rho RT \quad \text{or} \quad pv = RT \tag{2-19}$$

Actually, there are no ideal gases, but many of the commonly encountered gases approach the ideal. The ideal-gas equation can be applied in solving most problems involving air and many gases. When greater accuracy is required, moist air tables should be used.

Vapors do not follow the ideal-gas law, and the ideal-gas equation cannot be applied to them. This is the main distinction between a vapor and a gas. For this reason tables and charts must be used to determine the properties of vapors such as steam and, in refrigeration work, refrigerants. However, under certain conditions some vapors come near to being ideal gases, and the ideal-gas equation can be applied with fair accuracy. For example, water vapor or steam that is highly superheated or that is at very low pressures can be treated as a gas with small error, and is well within the degree of accuracy that is realized in engineering calculations for air-conditioning work. The water vapor in the atmosphere is usually at a low partial pressure (0.01 to 0.05 psia).

4-5 VAPOR DENSITY (ρ_v)

Vapor density is the number of pounds or grains of water vapor contained in 1 ft^3 of space. The maximum amount depends on the air temperature.

4-6 DRY-BULB TEMPERATURE (t)

The dry-bulb (DB) temperature is the air temperature measured by an accurate thermometer or thermocouple. When measuring the dry-bulb temperature of the air, the thermometer should be shielded to reduce the effects of direct radiation.

4-7 WET-BULB TEMPERATURE (t_w)

Figure 4-1 is a schematic drawing of a device to measure the wet- and dry-bulb temperatures. The various instruments used to take these measurements are called *psychrometers*.

When unsaturated air is passed over a wetted thermometer bulb, water evaporates from the wetted surface and latent heat absorbed by the vaporizing water causes the temperature of the wetted surface and the enclosed thermometer bulb to fall. As soon as the wetted surface temperature drops below that of the surrounding atmosphere, heat begins to flow from the warmer air to the cooler surface, and the quantity of heat transferred in this manner increases with an increasing drop in temperature. On the other hand, as the temperature drops, the vapor pressure of the water becomes lower and, hence, the rate of evaporation decreases. Eventually, a temperature is reached where the rate at which heat is transferred from the air to the wetted surface by convection and conduction is equal to the rate at which the wetted surface loses heat in the form of latent heat of vaporization. Thus, no further drop in temperature can occur. This is known as the wet-bulb temperature (WB).

As moisture evaporates from the wetted bulb, the air surrounding the bulb becomes more humid. Therefore, in order to measure the wet-bulb temperature of the air in a given space, a continuous sample of the air must pass around the bulb. The purpose of the fan in Fig. 4-1 is to cause the air to be drawn across the wetted bulb. Conventional air velocities used are between 500 and 1000 fpm for normal size thermometer bulbs. Soft, fine-meshed cotton tubing is recommended for the wick; it should cover the bulb plus about an inch of the thermometer stem. The wick should be watched and replaced before it becomes dirty or crusty. Using distilled water is recommended to give greater accuracy for a longer period of time.

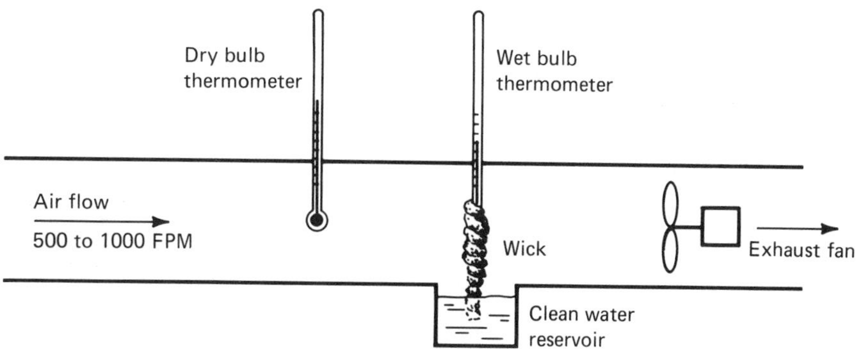

Figure 4–1

Steady-flow device for measuring wet and dry bulb temperature

Figure 4-2 shows a device called a *sling psychrometer*. It is a commonly used device especially for checking conditions on a job. The instrument is rotated by hand to obtain the air movement across the bulbs. The instrument is rotated until no further change is indicated on the wet bulb. The reading taken at that time is the air wet-bulb temperature.

Thermodynamic wet-bulb temperature (t^*), sometimes called *adiabatic saturation temperature* will be discussed in Section 4-14.

4-8 PARTIAL PRESSURE OF WATER VAPOR (p_v)

Low pressure air-water vapor mixtures follow closely the Gibbs-Dalton law of partial pressure which states that the total pressure of a mixture of gases is the sum of the partial pressures of each of its constituent gases and each constituent gas occupies the entire volume. Modern tables of air-water vapor mixture properties are based on corrected versions of the Dalton law. Most of the usual commercial problems encountered in air-conditioning work can be solved with sufficient accuracy by applying basic equations and using values from modern tables.

Moist air, as we have noted, consists of several gases, each having their own partial pressure. However, in air-conditioning work, we are primarily interested in treating dry air as a separate gas and water vapor as a separate gas. We may write the following equation relating the partial pressures

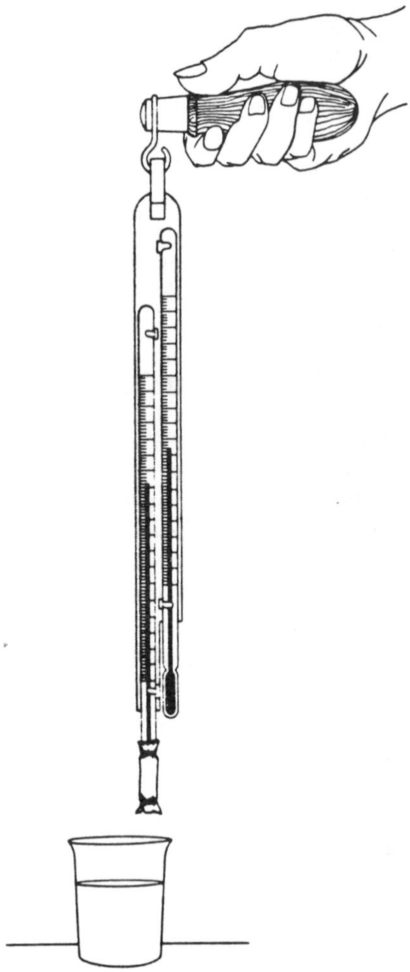

Figure 4–2

Sling psychrometer (Courtesy of The Trane Company)

$$p_b = p_a + p_v \tag{4-1}$$

where p_b is the total or barometric pressure, p_a is the partial pressure of the dry air, and p_v is the partial pressure of the water vapor in the mixture.

Several equations for calculating the partial pressure of water vapor in the air have been proposed and used. Dr. Carrier's equation,† first presented in 1911, has been frequently used with a high degree of accuracy, making use of the easily obtainable wet- and dry-bulb

†Carrier, W. H., "Rational Psychrometric Formulae." *Transaction ASME*, Vol. 33 (1911), p. 1005.

temperatures. The equation in its present form is

$$p_v = p_w - \frac{(p_b - p_w)(t - t_w)}{2831 - 1.43t_w} \quad (4\text{-}2)$$

where p_w = partial pressure of water vapor saturated at the wet-bulb temperature t_w

p_b = barometric pressure

t and t_w = dry- and wet-bulb temperatures, respectively, in °F

p_v, p_b, and p_w = the same units, either in. Hg or psia

At temperatures below 32°F this equation applies only for temperatures of and vapor pressures over supercooled water. For partial pressures of water vapor over ice the denominator becomes $3160 - 0.09t_w$; and p_w must be the partial pressure of water vapor over ice at t_w, the temperature of an iced wet bulb.

Illustrative Problem 4-1

Air has a dry-bulb temperature of 70°F and a wet-bulb temperature of 60°F. The barometric pressure is 29.90 in. Hg. Determine the partial pressure of the water vapor and of the dry air in the air.

Solution:

By Eq. (4-2),

$$p_v = p_w - \frac{(p_b - p_w)(t - t_w)}{2831 - 1.43t_w}$$

From Table A-1 at 60°F WB find $p_w = 0.2563$ psia, or from Table 4-1 at 60°F WB find $p_w = 0.5218$ in. Hg. (Recall that in Eq. (4-2), either in. Hg or psia may be used for the pressures, but we must be consistent.) Therefore, using in. Hg in Eq. (4-2) we have

$$p_v = 0.5218 - \frac{(29.90 - 0.5218)(70 - 60)}{2831 - 1.43(60)}$$

$$p_v = 0.4147 \text{ in. Hg} \quad \textit{(ans)}$$

Rearranging Eq. (4-1) we have

$$p_a = p_b - p_v = 29.90 - 0.4147 \quad \textit{(ans)}$$

$$p_a = 29.48 \text{ in. Hg}$$

4-9 DEW-POINT TEMPERATURE (t_{dp})

During the various seasons of the year, especially during the summer months, in localities where the water supply is cool, it is a common sight to see the outside surface of bare cold-water pipes covered with moisture. Another common sight is that of a glass of ice water with its outside surface covered with a film of moisture. The term commonly used to describe the appearance of moisture on a cold surface is *sweating,* as though the moisture came through the walls of the pipe or the glass.

What is actually happening is that the outside surface temperature of the pipe or the glass is at or below the saturation temperature corresponding to the partial pressure of water vapor in the surrounding air. This saturation temperature is called the *dew-point (DP)* temperature when condensation first starts to appear on the cold surface as the moist air is cooled at constant pressure.

In Problem 4-1 we calculated the partial pressure of the water vapor in the air to be 0.4147 in. Hg. Referring to Table 4-1, the saturation temperature in column 1 corresponding to 0.4147 in. Hg in column 2 is 53.3°F by interpolation. Therefore, this is the dew-point temperature. If any surface located in this air were at 53.3°F, moisture would start to condense on the surface.

In air-conditioning work the dew-point temperature is frequently called *air dew point*. However, this is a misnomer because the air does not condense nor does it have anything to do with the cooling and condensation of the water vapor. Actually, the same cooling and condensation of the water vapor would take place if there were no air present and the entire process were carried out in a closed vessel under vacuum. Since the term *air dew point* is in common use, we will also use it in our discussion.

Once the dew-point temperature has been reached, further cooling results in more condensation of water vapor on the cooling surface. The cooling surface will be covered with a film of moisture which, gathering in drops, will literally "rain" from the cooling surface.

At the dew-point temperature and below, the air is said to be saturated because the air is mixed with the maximum possible weight of moisture. If the mixture of air and water vapor is cooled, but not to as low a temperature as the dew point, there will be no condensation. However, as the air and water vapor are cooled, the volume of both will contract in the same proportion because both are cooled through the same temperature range. In other words, if a mixture consisting of 1 lb of air and 0.15 lb of water vapor is cooled, the smaller air volume will still contain 1 lb of air and 0.15 lb of water vapor as both gases will contract in the same proportion. Changes in the temperature of an air-water vapor mixture do not affect the amount of water vapor mixed with each pound of air, as long as the mixture is not cooled down to the dew-point temperature. Under these conditions the weight of moisture vapor per pound of dry air will remain the same regardless of temperature changes. An air-water vapor mixture at a dry-bulb temperature higher than its dew-point temperature is said to be unsaturated, and the moisture vapor in the mixture is superheated.

4-10 HUMIDITY RATIO *(W)*

Humidity ratio is defined as the pounds or grains of moisture vapor associated with 1 lb of dry air as defined in Section 4-3. By applying Dalton's law and the ideal-gas law, and neglecting intermolecular forces, an expression for calculating the humidity ratio can be derived as follows.

Step 1: The volume occupied by 1 lb of dry air at the partial pressure of the dry air is from

$$p_a V = m_a R_a T$$

or

$$V = \frac{m_a R_a T}{P_a} = \frac{1(53.3)(T)}{p_b - p_v} \qquad (a)$$

Step 2: The mass of water vapor, W, in 1 lb of air of V ft^3 is from

$$p_v V = W R_w T$$

or

$$W = \frac{p_v V}{R_w T} = \frac{p_v V}{(85.6)T} \qquad (b)$$

Step 3: Substitute Eq. (a) into Eq. (b) and we have

$$W = \frac{p_v\left(\dfrac{53.3T}{p_b - p_v}\right)}{85.6T}$$

which reduces to

$$W = 0.622\left(\frac{P_v}{p_b - p_v}\right) \text{ lb}_v\text{/lb da} \qquad (4\text{-}3)$$

If W is to be expressed in units of grains per pound of dry air, multiply Eq. (4-3) by 7000.

4-11 RELATIVE HUMIDITY (ϕ)

Relative humidity is the ratio of the actual partial pressure of the water vapor in the air to the partial pressure of water vapor saturated at the air dry-bulb temperature. In equation form it is

$$\phi \text{ (phi)} = \frac{p_v}{p_{vs}} \times 100 \qquad (4\text{-}4)$$

where p_{vs} is the partial pressure of saturated water vapor at the air dry-bulb temperature.

4-12 SATURATION RATIO (μ)

Saturation ratio is defined as the ratio of the actual humidity ratio of the air to the humidity ratio of air at saturation and is calculated from

$$\mu \text{ (mu)} = \frac{W}{W_s} \times 100 \qquad (4\text{-}5)$$

where W_s is calculated from Eq. (4-3) by using p_{vs} in place of p_v. Saturation ratio is sometimes called percent humidity, percent saturation, or degree of saturation.

Illustrative Problem 4-2

From the data of Problem 4-1 determine the relative humidity, humidity ratio, and saturation ratio of the air.

Solution:

From before $p_v = 0.4147$ in. Hg. Referring to Table 4-1 at the dry-bulb temperature of 70°F, find $p_{vs} = 0.7393$ in. Hg.

By Eq. (4-4),

$$\phi = \frac{p_v}{p_{vs}} = \frac{0.4147}{0.7393} = 0.56 \quad \text{or} \quad 56 \text{ percent}$$

By Eq. (4-3),

$$W = 0.622\left(\frac{p_v}{p_b - p_v}\right)$$

$$= 0.622\left(\frac{0.4147}{29.90 - 0.4147}\right)$$

$$W = 0.00875 \text{ lbv/lb da}$$

By Eq. (4-3),

$$W_s = 0.622\left(\frac{p_{vs}}{p_b - p_{vs}}\right)$$

$$= 0.622\left(\frac{0.7393}{29.90 - 0.7393}\right)$$

$$W_s = 0.01577 \text{ lbv/lb da}$$

By Eq. (4-5),

$$\mu = \frac{W}{W_s} = \frac{0.00875}{0.1577}$$

$$= 0.555 \quad \text{or} \quad 55.5 \text{ percent}$$

4-13 ENTHALPY OF AIR-VAPOR MIXTURES

When dealing with gases, vapors, and air-water vapor mixtures, the property *enthalpy* becomes an important property to evaluate the exchange of energy during a steady-flow process. The absolute enthalpy of a substance at any temperature t is, theoretically, the quantity of heat necessary to raise its temperature from absolute zero to temperature t. This quantity of heat includes not only the sensible heat but also the heat absorbed during phase changes. However, the term *enthalpy,* as ordinarily used, is not absolute enthalpy but is the difference between the absolute enthalpy of the substance at temperature t and the absolute enthalpy at an arbitrary datum temperature and in an arbitrary standard state.

In air-conditioning work we are very much concerned with both sensible heat and latent heat quantities of air-water vapor mixtures. Sensible heat manifests itself by a change in temperature of the substance. Latent heat is that heat required at a given pressure to produce changes in the physical state of a substance; that is, to change it from solid to liquid, from liquid to vapor, or the reverse of either, *without* changing its temperature. Therefore, the change of physical state may be either melting, freezing, vaporization (evaporation), or condensation. During melting or vaporization a relatively large amount of heat must be supplied to the substance, and during freezing and condensation a corresponding amount of heat must be removed.

The enthalpy of mixture of perfect gases is equal to the sum of the enthalpies of each constituent and is usually referenced to a unit mass of one constituent gas. For an air-water vapor mixture dry air is used as the reference because the amount of water vapor may vary during some processes. Therefore,

$$h_m = h_a + Wh_v \qquad (4\text{-}6)$$

Each term has units of energy per unit mass of dry air. If it is assumed that we have a perfect gas, the enthalpy is a function of temperature only $(h = c_p t)$. In most air-conditioning processes *enthalpy changes* alone are important. Thus, if zero Fahrenheit or Celsius is selected as the reference state where the enthalpy of dry air is zero, and if the specific heats c_{pa} and c_{pv} are assumed to be constant, simple relationships result.

$$h_a = c_{pa}t \qquad (4\text{-}6a)$$

$$h_v = h_g + c_{pv}t \qquad (4\text{-}6b)$$

where the enthalpy of saturated water vapor h_g at 0°F is 1061.2 Btu/lbm, and 2501.3 kJ/kg at 0°C.

Using Eqs. (4-6), (4-6a), and (4-6b) with $c_{pa} = 0.240$, $c_{pv} = 0.450$ Btu/lbm°F, respectively, and t is °F, we have

$$h_m = 0.240\,t + W(1061.2 + 0.45t) \quad \text{Btu/lb da} \qquad (4\text{-}7)$$

In SI units Eq. (4-7) becomes

$$h_m = 1.0t + W(2501.3 + 1.86t) \quad \text{kJ/kg da} \qquad (4\text{-}8)$$

where c_{pa} and c_{pv} are 1.0 and 1.86 kJ/kg°C, respectively, and t is in degrees Celsius. Since all of our work in psychrometrics is with moist air, the symbol for enthalpy of moist air will be h without subscript m.

4-14 THERMODYNAMIC WET-BULB TEMPERATURE (t*)

Figure 4-3 represents an idealized, fully insulated flow device where unsaturated moist air enters at dry-bulb temperature t_1, enthalpy h_1, and humidity ratio W_1. When this air is brought into contact with the water at a lower temperature, the air is both cooled and humidified. If the system is insulated so that no heat is transferred into or out of the system, the process is adiabatic; and if the water is at a constant temperature, the latent heat for evaporation can come only from the sensible heat given up by the air in cooling. The quantity of water present is assumed to be large (large surface area and quantity) compared to the amount evaporated into the air. We may assume that there is no temperature gradient in the body of water.

If the temperature reached by the air when it becomes saturated is identical with the temperature of the water, this temperature is called the *adiabatic saturation temperature*, or more commonly called the *thermodynamic wet-bulb temperature*.

Thus, in Fig. 4-3, the saturated air leaving the device will have properties t_2^*, h_2^*, and W_2^*. Liquid water must be supplied to the device having an enthalpy h_{f2}^* at t_2^* for the process to be steady flow. Assuming that steady-

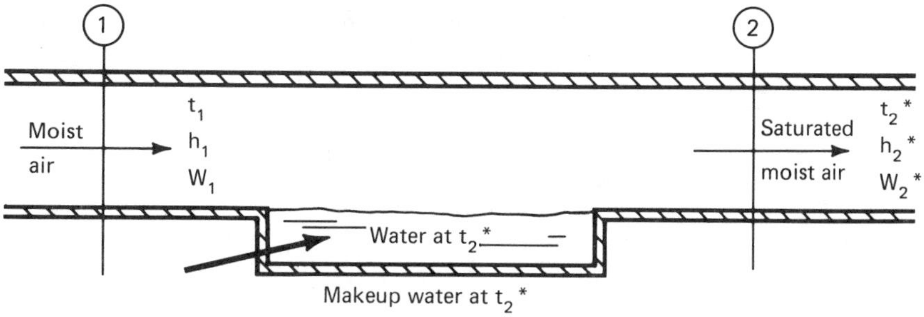

Figure 4–3

Adiabatic saturation of air

flow conditions do exist, the energy equation for the process is

$$h_1 + (W_2^* - W_1)h_{f2}^* = h_2^* \qquad (4\text{-}9)$$

The asterisk is used to denote properties at the thermodynamic wet-bulb temperature. The temperature corresponding to h_2^* for given values of h_1 and W_1 is the defined thermodynamic wet-bulb temperature.

The process discussed in this section is called the adiabatic saturation process (to be discussed in detail in the next chapter). The usefulness of the foregoing discussion lies in the fact that the temperature of the saturated air-water vapor mixture leaving the system is a function of the temperature, pressure, and relative humidity of the mixture entering and the exit pressure. Conversely, knowing the entering and exit pressures and temperatures, we may determine the relative humidity and humidity ratio of the entering mixture.

In principle, there is a difference between the wet-bulb temperature t_w and the temperature of adiabatic saturation t^*. The wet-bulb temperature is a function of both heat and mass diffusion rates, while the adiabatic saturation temperature is a function of a thermodynamic equilibrium process. However, in practice, it has been found that for air-water vapor mixtures at atmospheric pressures and temperatures, the wet-bulb and adiabatic saturation temperatures are essentially the same, numerically.

4-15 THE PSYCHROMETRIC TABLE†

The psychrometric properties of moist air at saturation conditions are shown in Table 4-1 and cover the range of temperature from −5°F to 200°F. This table may be used to solve almost any type of problem in psychrometry. Use of the table will be illustrated by the problems in the following section.

The table was calculated using vapor pressure data compiled by the U.S. Weather Bureau, the pressures below 32°F being over ice. Figures for heat content of saturated vapor, on which enthalpy data are based, are from Keenan and Keyes steam tables.

The table gives important properties of a mixture of dry air and saturated steam. Such a mixture is sometimes called *saturated air,* as previously noted. It could equally well be called "air at 100 percent relative humidity". Also, such a mixture (saturated) of dry air and saturated steam contains the greatest weight of water vapor which can be mixed with a given

†Table 4-1 and its description are taken from Strock and Koral, *Handbook of Air Conditioning Heating and Ventilating,* 2nd ed., The Industrial Press, with permission.

Table 4-1

Psychrometric Table*
(29.92 in. Hg)

Dry Bulb Temp., F	Saturated Water Vapor: Pressure, Inches of Mercury	Saturated Water Vapor: Weight, Grains per Cu. Ft. of Vapor	Saturated Water Vapor: Weight, Grains per Pound of Dry Air	Dry Air, Btu per Lb. Above oF	Air Saturated with Water Vapor: Volume, Cu. Ft. of 1 lb. Dry Air and Vapor to Saturate It	Air Saturated with Water Vapor: Weight, Grains, Dry Air in 1 Cu. Ft. of Mixture	Air Saturated with Water Vapor: Weight, Grains, Air and Vapor in 1 Cu. Ft. of Mixture	Heat Content per lb. Dry Air and Vapor to Saturate It: Heat of Vapor Above 32F, Btu	Heat Content per lb. Dry Air and Vapor to Saturate It: Total Heat, Btu	Wet Bulb Temp., F
(1)	(2)	(3)	(4)	(5)	(6)	(7)	(8)	(9)	(10)	(11)
−5	.0291	.370	4.239	−1.20	11.47	610.1	610.4	.641	−0.559	−5
−4	.0307	.389	4.472	−0.96	11.50	608.6	609.0	.677	−0.283	−4
−3	.0325	.412	4.734	−0.72	11.53	607.3	607.7	.717	−0.003	−3
−2	.0344	.434	5.001	−0.48	11.55	605.9	606.3	.759	+0.279	−2
−1	.0363	.457	5.289	−0.24	11.58	604.6	605.1	.802	0.562	−1
0	.0383	.481	5.58	0.00	11.60	603.2	603.7	.846	0.846	0
1	.0403	.505	5.87	0.24	11.63	601.8	602.3	.891	1.131	1
2	.0423	.529	6.16	0.48	11.66	600.5	601.1	.935	1.415	2
3	.0444	.554	6.47	0.72	11.68	599.2	599.7	.982	1.702	3
4	.0467	.581	6.81	0.96	11.71	597.8	598.4	1.03	1.994	4
5	.0491	.610	7.16	1.20	11.73	596.6	597.2	1.09	2.287	5
6	.0515	.638	7.51	1.44	11.76	595.2	595.9	1.14	2.581	6
7	.0542	.670	7.90	1.68	11.79	593.9	594.5	1.20	2.882	7
8	.0570	.703	8.31	1.92	11.81	592.5	593.2	1.26	3.184	8
9	.0600	.739	8.75	2.16	11.83	591.7	592.4	1.33	3.491	9
10	.0631	.775	9.19	2.40	11.86	590.4	591.1	1.40	3.799	10
11	.0665	.815	9.70	2.64	11.89	588.6	589.4	1.48	4.118	11
12	.0699	.855	10.19	2.88	11.92	587.3	588.2	1.55	4.433	12
13	.0735	.897	10.72	3.12	11.95	586.0	586.9	1.64	4.755	13
14	.0772	.941	11.26	3.36	11.97	584.7	585.7	1.72	5.078	14
15	.0810	.985	11.82	3.60	12.00	583.4	584.3	1.80	5.403	15
16	.0850	1.031	12.40	3.84	12.03	582.1	583.1	1.89	5.733	16
17	.0891	1.079	13.00	4.08	12.05	580.8	581.8	1.99	6.066	17
18	.0933	1.127	13.62	4.32	12.08	579.5	580.6	2.08	6.401	18
19	.0979	1.180	14.29	4.56	12.11	578.2	579.4	2.19	6.745	19
20	.103	1.239	15.04	4.80	12.14	576.8	578.0	2.30	7.099	20
21	.108	1.297	15.68	5.04	12.16	575.6	576.9	2.41	7.452	21
22	.113	1.354	16.51	5.28	12.19	574.2	575.6	2.53	7.806	22
23	.118	1.411	17.24	5.52	12.22	572.9	574.4	2.64	8.159	23
24	.124	1.480	18.12	5.76	12.25	571.7	573.1	2.78	8.535	24
25	.130	1.548	19.00	6.00	12.27	570.4	571.9	2.91	8.911	25
26	.136	1.616	19.88	6.24	12.30	569.1	570.7	3.05	9.289	26
27	.143	1.696	20.90	6.48	12.32	568.0	569.7	3.21	9.685	27
28	.150	1.775	21.94	6.72	12.36	566.5	568.3	3.37	10.09	28
29	.157	1.854	22.97	6.96	12.39	565.2	567.1	3.53	10.49	29

**Reproduced from* Handbook of Air Conditioning, Heating and Ventilating, *2nd ed., The Industrial Press, New York, 1965.*

Table 4-1 (Cont.)

Psychrometric Table
(29.92 in. Hg)

Dry Bulb Temp., F	Saturated Water Vapor			Dry Air, Btu per Lb. Above 0F	Air Saturated with Water Vapor					Wet Bulb Temp., F
	Pressure	Weight			Volume	Weight, Grains		Heat Content per lb. Dry Air and Vapor to Saturate It		
	Inches of Mercury	Grains per Cu. Ft. of Vapor	Grains per Pound of Dry Air		Cu. Ft. of 1 lb. Dry Air and Vapor to Saturate It	Dry Air in 1 Cu. Ft. of Mixture	Air and Vapor in 1 Cu. Ft. of Mixture	Heat of Vapor Above 32F, Btu	Total Heat, Btu	
(1)	(2)	(3)	(4)	(5)	(6)	(7)	(8)	(9)	(10)	(11)
30	.164	1.933	23.99	7.20	12.41	563.9	565.9	3.68	10.88	30
31	.172	2.023	25.17	7.44	12.44	562.6	564.6	3.87	11.31	31
32	.1803	2.117	26.39	7.68	12.47	561.5	563.6	4.06	11.74	32
33	.1878	2.201	27.50	7.92	12.49	560.4	562.6	4.23	12.15	33
34	.1955	2.289	28.63	8.16	12.52	559.1	561.3	4.40	12.56	34
35	.2035	2.375	29.81	8.40	12.55	557.7	560.1	4.59	12.99	35
36	.2118	2.467	31.04	8.64	12.58	556.5	559.0	4.78	13.42	36
37	.2203	2.562	32.29	8.88	12.60	555.4	557.9	4.97	13.85	37
38	.2292	2.660	33.61	9.12	12.64	554.0	556.6	5.18	14.30	38
39	.2383	2.760	34.95	9.36	12.66	552.8	555.6	5.39	14.75	39
40	.2478	2.864	36.36	9.60	12.69	551.4	554.3	5.61	15.21	40
41	.2576	2.971	37.81	9.84	12.73	550.1	553.1	5.83	15.67	41
42	.2677	3.082	39.30	10.08	12.75	548.9	552.1	6.07	16.15	42
43	.2782	3.196	40.86	10.32	12.78	547.6	550.8	6.31	16.63	43
44	.2891	3.314	42.48	10.56	12.82	546.2	549.5	6.56	17.12	44
45	.3004	3.437	44.15	10.80	12.84	545.0	548.4	6.82	17.62	45
46	.3120	3.564	45.87	11.04	12.87	543.8	547.3	7.09	18.13	46
47	.3240	3.694	47.66	11.28	12.90	542.5	546.2	7.37	18.65	47
48	.3364	3.828	49.51	11.52	12.93	541.3	545.1	7.66	19.18	48
49	.3493	3.967	51.43	11.76	12.97	539.9	543.9	7.96	19.72	49
50	.3626	4.110	53.41	12.00	13.00	538.7	542.8	8.27	20.27	50
51	.3764	4.257	55.47	12.24	13.03	537.3	541.6	8.59	20.83	51
52	.3906	4.409	57.59	12.48	13.06	535.9	540.4	8.92	21.40	52
53	.4052	4.565	59.77	12.72	13.09	534.7	539.2	9.26	21.98	53
54	.4203	4.727	62.03	12.96	13.12	533.4	538.1	9.62	22.58	54
55	.4359	4.893	64.37	13.20	13.16	532.1	537.0	9.98	23.18	55
56	.4520	5.064	66.78	13.44	13.19	530.8	535.8	10.36	23.80	56
57	.4686	5.240	69.27	13.68	13.22	529.5	534.8	10.75	24.43	57
58	.4858	5.422	71.85	13.92	13.25	528.2	533.6	11.16	25.08	58
59	.5035	5.609	74.52	14.16	13.29	526.8	532.4	11.58	25.74	59
60	.5218	5.801	77.27	14.40	13.32	525.5	531.3	12.01	26.41	60
61	.5407	5.999	80.12	14.64	13.36	524.1	530.1	12.46	27.10	61
62	.5601	6.203	83.05	14.88	13.39	522.9	529.1	12.92	27.80	62
63	.5802	6.414	86.09	15.12	13.42	521.5	528.0	13.40	28.52	63
64	.6009	6.631	89.23	15.36	13.46	520.2	526.8	13.89	29.25	64

Table 4-1 (Cont.)

Psychrometric Table (29.92 in. Hg)

Dry Bulb Temp., F	Saturated Water Vapor			Dry Air, Btu per Lb. Above oF	Air Saturated with Water Vapor					Wet Bulb Temp., F
	Pressure	Weight			Volume	Weight, Grains		Heat Content per lb. Dry Air and Vapor to Saturate It		
	Inches of Mercury	Grains per Cu. Ft. of Vapor	Grains per Pound of Dry Air		Cu. Ft. of 1 lb. Dry Air and Vapor to Saturate It	Dry Air in 1 Cu. Ft. of Mixture	Air and Vapor in 1 Cu. Ft. of Mixture	Heat of Vapor Above 32F, Btu	Total Heat, Btu	
(1)	(2)	(3)	(4)	(5)	(6)	(7)	(8)	(9)	(10)	(11)
65	.6222	6.853	92.46	15.60	13.49	518.9	525.7	14.40	30.00	65
66	.6442	7.082	95.80	15.84	13.53	517.5	524.6	14.93	30.77	66
67	.6669	7.318	99.25	16.08	13.56	516.1	523.4	15.47	31.55	67
68	.6903	7.560	102.82	16.32	13.60	514.7	522.3	16.03	32.35	68
69	.7144	7.810	106.50	16.56	13.64	513.3	520.9	16.61	33.17	69
70	.7392	8.065	110.28	16.80	13.67	511.9	520.0	17.21	34.01	70
71	.7648	8.329	114.21	17.04	13.71	510.5	518.9	17.83	34.87	71
72	.7912	8.601	118.25	17.28	13.75	509.1	517.7	18.47	35.75	72
73	.8183	8.879	122.42	17.52	13.79	507.7	516.6	19.12	36.64	73
74	.8462	9.165	126.71	17.76	13.83	506.3	515.5	19.81	37.57	74
75	.8750	9.460	131.16	18.00	13.87	504.9	514.3	20.51	38.51	75
76	.9046	9.762	135.73	18.24	13.91	503.4	513.2	21.23	39.47	76
77	.9352	10.07	140.47	18.48	13.94	502.0	512.1	21.98	40.46	77
78	.9666	10.39	145.35	18.72	13.99	500.5	510.9	22.75	41.47	78
79	.9989	10.72	150.37	18.96	14.03	499.0	509.8	23.55	42.51	79
80	1.0321	11.06	155.52	19.20	14.07	497.7	508.7	24.37	43.57	80
81	1.0664	11.40	160.88	19.44	14.11	496.1	507.5	25.21	44.65	81
82	1.1016	11.76	166.41	19.68	14.15	494.7	506.4	26.09	45.77	82
83	1.1378	12.12	172.11	19.92	14.20	493.1	505.2	26.99	46.91	83
84	1.1750	12.50	177.95	20.16	14.24	491.6	504.1	27.92	48.08	84
85	1.2133	12.88	184.00	20.40	14.29	490.0	502.9	28.88	49.28	85
86	1.2527	13.28	190.23	20.64	14.33	488.5	501.8	29.87	50.51	86
87	1.2931	13.68	196.63	20.88	14.37	487.0	500.7	30.89	51.77	87
88	1.3347	14.09	203.24	21.12	14.42	485.4	499.5	31.94	53.06	88
89	1.3775	14.52	210.08	21.36	14.47	483.4	498.3	33.03	54.39	89
90	1.4215	14.96	217.13	21.60	14.52	482.2	497.2	34.15	55.75	90
91	1.4667	15.40	224.40	21.84	14.57	480.5	495.9	35.31	57.15	91
92	1.5131	15.86	231.89	22.08	14.62	478.8	494.7	36.50	58.58	92
93	1.5608	16.34	239.60	22.32	14.67	477.3	493.6	37.73	60.05	93
94	1.6097	16.82	247.52	22.56	14.72	475.6	492.5	38.99	61.55	94
95	1.6600	17.31	255.75	22.80	14.77	473.9	491.2	40.30	63.10	95
96	1.7117	17.82	264.16	23.04	14.82	472.2	490.1	41.64	64.68	96
97	1.7647	18.34	272.87	23.28	14.88	470.5	488.8	43.03	66.31	97
98	1.8192	18.87	281.83	23.52	14.93	468.8	487.6	44.47	67.99	98
99	1.8751	19.42	291.06	23.76	14.99	467.0	486.4	45.94	69.70	99

Table 4-1 (Cont.)

Psychrometric Table
(29.92 in. Hg)

Dry Bulb Temp., F	Saturated Water Vapor: Pressure: Inches of Mercury	Saturated Water Vapor: Weight: Grains per Cu. Ft. of Vapor	Saturated Water Vapor: Weight: Grains per Pound of Dry Air	Dry Air, Btu per Lb. Above oF	Air Saturated with Water Vapor: Volume: Cu. Ft. of 1 lb. Dry Air and Vapor to Saturate It	Air Saturated with Water Vapor: Weight, Grains: Dry Air in 1 Cu. Ft. of Mixture	Air Saturated with Water Vapor: Weight, Grains: Air and Vapor in 1 Cu. Ft. of Mixture	Air Saturated with Water Vapor: Heat Content per lb. Dry Air and Vapor to Saturate It: Heat of Vapor Above 32F, Btu	Air Saturated with Water Vapor: Heat Content per lb. Dry Air and Vapor to Saturate It: Total Heat, Btu	Wet Bulb Temp., F
(1)	(2)	(3)	(4)	(5)	(6)	(7)	(8)	(9)	(10)	(11)
100	1.9325	19.98	300.60	24.00	15.05	465.2	485.2	47.46	71.46	100
101	1.9915	20.55	310.44	24.24	15.10	463.4	484.0	49.03	73.27	101
102	2.0519	21.14	320.54	24.48	15.16	461.7	482.8	50.65	75.13	102
103	2.1138	21.75	330.95	24.72	15.22	459.9	481.6	52.31	77.03	103
104	2.1775	22.36	341.70	24.96	15.28	458.0	480.4	54.03	78.99	104
105	2.2429	22.99	352.80	25.20	15.34	456.1	479.1	55.81	81.01	105
106	2.3099	23.63	364.21	25.44	15.41	454.2	477.9	57.64	83.08	106
107	2.3786	24.30	375.97	25.68	15.47	452.4	476.7	59.52	85.20	107
108	2.4491	24.97	388.12	25.92	15.54	450.4	475.4	61.47	87.39	108
109	2.5214	25.67	400.61	26.16	15.61	448.5	474.2	63.47	89.63	109
110	2.5955	26.38	413.50	26.40	15.68	446.5	472.9	65.54	91.94	110
111	2.6715	27.10	426.82	26.64	15.75	444.4	471.5	67.68	94.32	111
112	2.7494	27.84	440.69	26.88	15.82	442.4	470.3	69.90	96.78	112
113	2.8293	28.61	454.69	27.12	15.89	440.4	469.0	72.15	99.27	113
114	2.9111	29.39	469.23	27.36	15.97	438.4	467.8	74.48	101.84	114
115	2.9948	30.19	484.21	27.60	16.04	436.4	466.6	76.89	104.49	115
116	3.0806	31.00	499.71	27.84	16.12	434.3	465.3	79.38	107.22	116
117	3.1687	31.83	515.69	28.08	16.20	432.1	463.9	81.95	110.03	117
118	3.2589	32.68	532.19	28.32	16.29	430.0	462.5	84.60	112.92	118
119	3.3512	33.54	549.13	28.56	16.37	427.6	461.1	87.33	115.89	119
120	3.4458	34.44	566.67	28.80	16.46	425.4	459.8	90.16	118.96	120
121	3.5427	35.35	584.70	29.04	16.55	423.2	458.5	93.06	122.10	121
122	3.6420	36.28	603.38	29.28	16.63	420.9	457.2	96.06	125.34	122
123	3.7436	37.23	622.62	29.52	16.73	418.6	455.8	99.17	128.69	123
124	3.8475	38.20	642.48	29.76	16.82	416.2	450.4	102.37	132.13	124
125	3.9539	39.19	662.90	30.00	16.92	413.9	453.0	105.67	135.67	125
126	4.0629	40.21	684.10	30.24	17.01	411.4	451.6	109.08	139.32	126
127	4.1745	41.24	705.91	30.48	17.12	409.0	450.2	112.60	143.08	127
128	4.2887	42.30	728.47	30.72	17.22	406.5	448.8	116.24	146.96	128
129	4.4055	43.38	751.72	30.96	17.33	404.0	447.4	120.00	150.96	129
130	4.5251	44.49	775.80	31.20	17.44	401.4	445.9	123.90	155.10	130
131	4.6474	45.62	800.57	31.44	17.56	398.9	444.5	127.90	159.34	131
132	4.7725	46.77	826.26	31.68	17.67	396.2	443.0	132.04	163.72	132
133	4.9005	47.95	852.73	31.92	17.79	393.6	441.6	136.34	168.26	133
134	5.0314	49.15	880.12	32.16	17.91	390.9	440.1	140.75	172.91	134

Table 4-1 (Cont.)

Psychrometric Table
(29.92 in. Hg)

Dry Bulb Temp., F	Saturated Water Vapor			Dry Air, Btu per Lb. Above oF	Air Saturated with Water Vapor					Wet Bulb Temp., F
	Pressure	Weight			Volume	Weight, Grains		Heat Content per lb. Dry Air and Vapor to Saturate It		
	Inches of Mercury	Grains per Cu. Ft. of Vapor	Grains per Pound of Dry Air		Cu. Ft. of 1 lb. Dry Air and Vapor to Saturate It	Dry Air in 1 Cu. Ft. of Mixture	Air and Vapor in 1 Cu. Ft. of Mixture	Heat of Vapor Above 32F, Btu	Total Heat, Btu	
(1)	(2)	(3)	(4)	(5)	(6)	(7)	(8)	(9)	(10)	(11)
135	5.1653	50.38	908.42	32.40	18.04	388.2	438.6	145.34	177.74	135
136	5.3022	51.63	937.68	32.64	18.16	385.4	437.0	150.07	182.71	136
137	5.4421	52.91	967.89	32.88	18.30	382.7	435.6	154.97	187.85	137
138	5.5852	54.21	999.24	33.12	18.43	379.8	434.0	160.05	193.17	138
139	5.7316	55.55	1031.68	33.36	18.57	376.9	432.4	165.15	198.51	139
140	5.8812	56.91	1065.1	33.60	18.71	374.1	431.0	170.72	204.32	140
141	6.0341	58.29	1099.3	33.84	18.87	371.2	429.5	176.27	210.11	141
142	6.1903	59.71	1135.7	34.08	19.02	368.0	427.7	182.17	216.25	142
143	6.3500	61.16	1172.9	34.32	19.18	365.0	426.2	188.20	222.52	143
144	6.5132	62.63	1211.5	34.56	19.34	362.0	424.6	194.45	229.01	144
145	6.680	64.13	1251.4	34.80	19.51	358.7	422.9	200.96	235.76	145
146	6.850	65.67	1292.7	35.04	19.68	355.7	421.4	207.66	242.70	146
147	7.024	67.23	1335.6	35.28	19.86	352.4	419.6	214.63	249.91	147
148	7.202	68.82	1380.1	35.52	20.05	349.1	417.9	221.86	257.28	148
149	7.384	70.45	1426.4	35.76	20.25	345.7	416.2	229.39	265.15	149
150	7.569	72.11	1474.3	36.00	20.44	342.5	414.6	237.17	273.17	150
151	7.759	73.80	1524.9	36.24	20.66	338.8	412.6	245.41	281.65	151
152	7.952	75.53	1575.9	36.48	20.87	335.4	410.9	253.69	290.17	152
153	8.150	77.29	1629.8	36.72	21.10	331.9	409.2	262.47	299.19	153
154	8.351	79.08	1685.6	36.96	21.32	328.3	407.4	271.55	308.51	154
155	8.557	80.91	1743.8	37.20	21.56	324.8	405.7	281.03	318.23	155
156	8.767	82.76	1804.6	37.44	21.80	321.1	403.9	290.93	328.37	156
157	8.981	84.65	1867.3	37.68	22.06	317.3	402.0	301.14	338.82	157
158	9.200	86.59	1933.0	37.92	22.32	313.6	400.2	311.88	349.80	158
159	9.424	88.56	2001.8	38.16	22.60	309.7	398.3	323.08	361.24	159
160	9.652	90.57	2073.2	38.40	22.89	305.8	396.4	334.73	373.13	160
161	9.885	92.62	2148.0	38.64	23.19	301.8	394.4	346.92	385.56	161
162	10.122	94.70	2225.8	38.88	23.50	297.9	392.6	359.62	398.50	162
163	10.364	96.82	2307.2	39.12	23.83	293.7	390.6	372.92	412.04	163
164	10.611	98.97	2392.4	39.36	24.17	289.6	388.6	386.81	426.17	164
165	10.863	101.17	2481.6	39.60	24.53	285.4	386.5	401.04	440.64	165
166	11.120	103.41	2575.0	39.84	24.90	281.1	384.5	416.64	456.48	166
167	11.382	105.69	2673.0	40.08	25.29	276.8	382.5	432.65	472.73	167
168	11.649	108.03	2775.7	40.32	25.69	272.5	380.5	449.41	489.73	168
169	11.921	110.39	2883.4	40.56	26.12	267.9	378.4	467.02	507.58	169

Table 4-1 (Cont.)

Psychrometric Table
(29.92 in. Hg)

Dry Bulb Temp., F	Saturated Water Vapor: Pressure: Inches of Mercury	Saturated Water Vapor: Weight: Grains per Cu. Ft. of Vapor	Saturated Water Vapor: Weight: Grains per Pound of Dry Air	Dry Air, Btu per Lb. Above oF	Air Saturated with Water Vapor: Volume: Cu. Ft. of 1 lb. Dry Air and Vapor to Saturate It	Air Saturated with Water Vapor: Weight, Grains: Dry Air in 1 Cu. Ft. of Mixture	Air Saturated with Water Vapor: Weight, Grains: Air and Vapor in 1 Cu. Ft. of Mixture	Heat Content per lb. Dry Air and Vapor to Saturate It: Heat of Vapor Above 32F, Btu	Heat Content per lb. Dry Air and Vapor to Saturate It: Total Heat, Btu	Wet Bulb Temp., F
(1)	(2)	(3)	(4)	(5)	(6)	(7)	(8)	(9)	(10)	(11)
170	12.199	112.79	2996.9	40.80	26.57	263.5	376.2	485.59	526.39	170
171	12.483	115.25	3116.6	41.04	27.04	258.9	374.1	505.18	546.22	171
172	12.772	117.74	3242.5	41.28	27.54	254.2	371.9	525.74	567.02	172
173	13.066	120.28	3375.0	41.52	28.06	249.5	369.7	547.43	588.95	173
174	13.366	122.94	3515.1	41.76	28.61	244.7	367.6	570.36	612.12	174
175	13.671	125.49	3662.8	42.00	29.19	239.8	365.3	594.50	636.50	175
176	13.983	128.18	3819.7	42.24	29.79	234.9	363.2	620.20	662.44	176
177	14.301	130.89	3986.1	42.48	30.46	229.8	360.7	647.46	689.94	177
178	14.625	133.66	4162.7	42.72	31.14	224.8	358.5	676.36	719.08	178
179	14.955	136.48	4350.5	42.96	31.88	219.6	356.1	707.09	750.05	179
180	15.291	139.36	4550.4	43.20	32.65	214.4	353.7	739.81	783.01	180
181	15.633	142.28	4763.7	43.44	33.48	209.1	351.3	774.77	818.21	181
182	15.982	145.26	4991.8	43.68	34.36	203.7	349.0	812.16	855.84	182
183	16.337	148.27	5236.1	43.92	35.32	198.2	346.5	852.23	896.15	183
184	16.699	151.35	5498.5	44.16	36.32	192.7	344.1	895.22	939.38	184
185	17.068	154.49	5781.2	44.40	37.42	187.1	341.6	941.59	985.99	185
186	17.443	157.66	6085.9	44.64	38.60	181.3	339.0	991.58	1036.22	186
187	17.825	160.88	6415.6	44.88	39.88	175.5	336.4	1045.68	1090.56	187
188	18.214	164.17	6773.5	45.12	41.26	169.7	333.8	1104.37	1149.49	188
189	18.611	167.50	7164.1	45.36	42.77	163.7	331.2	1168.46	1213.82	189
190	19.014	170.90	7589.5	45.60	44.41	157.6	328.5	1237.93	1283.53	190
191	19.425	174.35	8057.1	45.84	46.21	151.5	325.8	1314.92	1360.76	191
192	19.843	177.85	8571.9	46.08	48.20	145.2	323.1	1399.93	1446.01	192
193	20.269	181.44	9142.5	46.32	50.39	138.9	320.4	1493.07	1539.39	193
194	20.703	185.14	9777.7	46.56	52.85	132.5	317.5	1597.61	1644.17	194
195	21.144	188.73	10487.9	46.80	55.57	125.9	314.7	1714.04	1764.84	195
196	21.593	192.47	11287.8	47.04	58.64	119.4	311.8	1845.92	1892.96	196
197	22.050	196.30	12195.9	47.28	62.13	112.7	309.0	1994.41	2041.69	197
198	22.515	200.17	13234.8	47.52	66.12	105.9	306.0	2165.38	2212.90	198
199	22.987	204.08	14431.8	47.76	70.72	98.9	303.1	2361.62	2409.38	199
200	23.467	208.09	15828.9	48.00	76.07	92.0	300.1	2591.88	2638.88	200

weight of dry air at any particular temperature and total pressure.

As already mentioned, in the process of conditioning air, water vapor is likely to be added or removed. Thus, the only component of the mixture that is carried through the whole process, in all cases, is the dry air. This makes the dry air the most convenient base to refer air properties to and, for this reason, dry air is the point of reference in the psychrometric table and chart in this chapter and practically all presently used psychrometric tables and charts.

COLUMNS (1) AND (11) of the psychrometric table are dry- or wet-bulb temperatures. These are the same because the table is based on saturated conditions.

COLUMN (2) is the partial pressure of the saturated water vapor in inches of mercury.

COLUMN (3) is the weight of water vapor in grains per ft^3. This value is nearly independent of the total pressure of the air-vapor mixture. We have defined this property as vapor density, ρ_v.

COLUMN (4) is the weight of water vapor (in grains), which is in the space occupied by 1 lb of dry air. We have defined this property as humidity ratio, W.

COLUMN (5) is the heat required to raise 1 lb of dry air from 0°F to the particular temperature. (For example, the value for 100°F is 0.24(100) = 24.00 Btu/lb da.

COLUMN (6) is the volume occupied by 1 lb of dry air when it is in the same space (mixed with) saturated vapor at the particular temperature. For example, the table of properties of dry air (Table A-4) shows the volume of 1 lb of dry air at 29.92 in. Hg and 70°F as 13.353 ft^3/lb. If the same space contains water vapor as well as air, the total pressure is 29.92 in. Hg, and the pressure on the air will be 29.92 less the vapor pressure. At lower pressure the pound of air occupies a larger volume. Referring to column 6 for 70°F saturated air, the volume of 1 lb of dry air is shown as 13.67 ft^3/lb.

COLUMN (7) shows the weight (grains) of dry air in 1 ft^3 of the mixture of air and saturated vapor at the particular temperature. The values in column 7 are 7000 times the reciprocal of those in column 6.

COLUMN (8) is the weight (grains) of dry air plus vapor in 1 ft^3. The values in column 8 are the total of those in columns 3 and 7.

COLUMN (9) is the heat of the liquid water above 32°F plus the latent heat of evaporation for the weight of vapor in column 4.

COLUMN (10) is the enthalpy, or total heat, of 1 lb of air plus that of the saturated vapor occupying the same space. The values of column 10 are the sum of those in columns 5 and 9.

This psychrometric table makes it possible to solve any problems for air and water vapor mixtures in the range of temperatures and the pressures for which it is made. From tables such as this, psychrometric charts are constructed. These charts make the solution of most psychrometric problems much simpler than it would be with the table alone. A discussion of the psychrometric chart and its use in solving psychrometric problems will follow in this chapter and in Chapter 5.

The calculation of a few air-water vapor problems, by means of the psychrometric table will help toward a more complete understanding of the subject. However, it is not recommended that one solve air-conditioning problems in this way. The time spent in learning to use the psychrometric chart (Chapter 5) will be generously rewarded by easier solutions and usually better and more accurate ones.

4-16 ILLUSTRATIVE PROBLEMS IN PSYCHROMETRICS

In the solutions to the following problems total heat at saturation, which corresponds to a particular wet-bulb temperature, is assumed to match the same wet-bulb temperature at lower than 100 percent relative humidity. This is not quite true but, up to about 90°F wet bulb and 120°F dry bulb, the deviation is less than 2 percent of the heat of the vapor. The correction (deviation) would add much work to the problem solutions.

Illustrative Problem 4-3

What is the relative humidity of moist air having a dry-bulb temperature of 80°F and a wet-bulb temperature of 72°F.

Solution:

At 72°F wet bulb, the total heat (column 10) is 35.75 Btu/lb. The heat content of dry air at 80°F (column 5) is 19.20 Btu/lb. The difference between the two is 16.55 Btu/lb, which is the heat of the vapor. Referring to column 9, it will be noted that 16.03 Btu/lb corresponds to 0.6903 in. Hg. By interpolation, 16.55 Btu/lb corresponds to 0.7119 in. Hg. (p_v). [Note: p_v may also be found by using Eq. (4-2)]. At 80°F saturated, the vapor pressure is 1.0321 in. Hg (p_{vs}). Since the relative humidity is the ratio of the actual vapor pressure p_v to the saturated vapor pressure p_{vs} at the same temperature, the relative humidity ϕ = 0.7119/1.0321 = 0.69 or 69 percent.

Illustrative Problem 4-4

Moist air exists at 80°F dry bulb and 60°F dew point. Determine the relative humidity.

Solution:

By definition, the 60°F dew-point temperature is the saturation temperature corresponding to the actual partial pressure of the water vapor in the air. Therefore, from Table 4-1 at 60°F find p_v = 0.5218 in. Hg. At 80°F dry bulb the value of p_{vs} = 1.0321 in. Hg. The relative humidity ϕ = 0.5218/1.0321 = 0.505 or 50.5 percent.

Illustrative Problem 4-5

What is the relative humidity of moist air if the wet-bulb temperature is 62°F and the dew-point temperature is 54°F?

Solution:

From Table 4-1 the total heat at 62°F is 27.80 Btu/lb. The heat of the vapor at 54°F is 9.62 Btu/lb. The difference (27.80 − 9.62 = 18.18 Btu/lb) is the heat content of the dry air. The dry-bulb temperature is the heat content of the dry air divided by the dry-air specific heat (Btu/lb °F). 18.18/0.24 = 75.8°F. (This could also be found by interpolation of column 5 and column 1). The vapor pressure at 54°F saturated is p_v = 0.4203 in. Hg; the vapor pressure at 75.8°F is p_{vs} = 0.8986 in. Hg (by interpolation). Therefore, the relative humidity ϕ = 0.4203/0.8986 = 0.468 or 46.8 percent.

Illustrative Problem 4-6

What is the dry-bulb temperature of air having a relative humidity of 76 percent and a dew-point temperature of 66°F?

Solution:

From Table 4-1 the vapor pressure at 66°F dew point is p_v = 0.6442 in. Hg. Relative humidity $\phi = p_v/p_{vs}$; therefore, $p_{vs} = p_v/\phi$ = 0.6442/0.76 = 0.8476 in. Hg. Table 4-1 with p_{vs} = 0.8476 in. Hg find dry bulb, t = 74°F (approximately).

Illustrative Problem 4-7

What is the dew-point temperature of moist air at 80°F dry bulb and 72°F wet bulb?

Solution:

The total heat at 72°F saturated is 35.75 Btu/lb. The heat content of dry air at 80°F is

19.20 Btu/lb. The heat of the vapor is 35.75 − 19.20 = 16.55 Btu/lb. By interpolation of column 9 and column 1, 16.55 corresponds to 68.9°F dew point.

Illustrative Problem 4-8

What is the dew-point temperature of air having a dry-bulb temperature of 80°F and a relative humidity of 50 percent.

Solution:

The vapor pressure at 80°F saturated is p_{vs} = 1.0321 in. Hg. The relative humidity $\phi = p_v/p_{vs}$, or $p_v = \phi \times p_{vs} = 0.50 \times 1.0321 =$ 0.516 in. Hg. The saturation temperature corresponding to a partial pressure $p_v = 0.516$ is 59.7°F dew point.

Illustrative Problem 4-9

What is the wet-bulb temperature corresponding to 72°F dry bulb and 56°F dew point?

Solution:

The heat content of dry air at 72°F is 17.28 Btu/lb. The heat content of the vapor at 56°F saturated is 10.36 Btu/lb. The total heat is 17.28 + 10.36 = 27.64 Btu/lb, which corresponds to 61.8 wet bulb by interpolation.

Illustrative Problem 4-10

What is the weight of dry air per cubic foot at 70°F saturated?

Solution:

From column 6 saturated air at 70°F has a volume of 13.67 ft^3/lb of dry air. The density is the reciprocal of specific volume and is then 1/13.67 = 0.0732 lb dry air/ft^3. Also, from column 7 511.9 grains/ft^3 divided by 7000 grains/lb is 0.0731 lb/ft^3.

Illustrative Problem 4-11

What is the weight of dry air per ft^3 of air at 70°F dry bulb and 60°F wet bulb?

Solution:

Total heat at 60°F wet bulb is 26.41 Btu/lb. Heat content of dry air at 70°F is 16.80 Btu/lb. Heat of the vapor is 26.41 − 16.80 = 9.61 Btu/lb. This corresponds to a dew-point temperature of 54°F, and a vapor pressure of 0.4203 in. Hg. The partial pressure of the dry air is 29.92 − 0.42 = 29.5 in. Hg. From the properties of dry air (Table A-4), at 70°F and 29.92 in. Hg, the density is 0.07489 lb/ft^3. From Boyles' law the density at 29.5 in. Hg and 70°F is (0.07489)(29.50)/29.92 = 0.07384 lb dry air/ft^3.

Illustrative Problem 4-12

How many Btu per hour of sensible heat is required to change 1000 ft^3 of air from 0°F to 70°F? (The 1000 ft^3 is measured at the entering condition.)

Solution:

From Table A-4 the air density at 0°F is 0.08629 lb/ft^3. The weight of air then is 1000 × 0.08629 = 86.3 lb. The sensible heat required is (86.3)(0.24)(70 − 0) = 1449.8 Btu.

Illustrative Problem 4-13

What is the total heat in Btu/lb of dry air in a mixture of air and water vapor at 70°F and 40 percent relative humidity?

Solution:

The heat of the dry air at 70°F from column 5 is 16.80 Btu/lb. The vapor pressure at 70°F saturated is 0.7392 in. Hg from column 2. At 40 percent relative humidity the vapor pressure

p_v = 0.40 × 0.7392 = 0.2957 in. Hg. By interpolation of column 2 and column 9, the heat of the vapor corresponding to 0.2957 in. Hg is 6.71 Btu/lb. Therefore, the total heat is 16.80 + 6.71 = 23.51 Btu/lb.

Alternate Solution:

This enthalpy may be closely approximated by using Eq. (4-7), $h = 0.240t + W(1061.2 + 0.45t)$. By Eq. (4-3), $W = 0.622\ p_v/(p_b - p_v) = 0.622 \times 0.2957/(29.92 - 0.2957) = 0.006208$ lb vapor per lb dry air. Substitution into Eq. (4-7) gives

$$h = 0.240(70) + 0.006208\,(1061.2 + 0.45 \times 70)$$
$$h = 23.59 \text{ Btu/lb da}$$

The preceding 11 examples were solved primarily by using data from Table 4-1, and Table A-4 in the appendix and were limited to the barometric pressure of 29.92 in. Hg. The following group of examples will make use of the tables, as well as equations presented in this chapter.

Illustrative Problem 4-14

Moist air exists at 90°F dry-bulb temperature and 40 percent relative humidity. The barometric pressure is 12.56 psia. Determine: (a) dew-point temperature and (b) enthalpy.

Solution:

(a) From Table A-1 at 90°F, find p_{vs} = 0.6988 psia; or, from Table 4-1 at 90°F, find p_{vs} = 1.4215 in. Hg. From Eq. (4-4) $p_v = \phi \times p_{vs}$ = 0.40 × 1.4215 = 0.5686 in. Hg. Table 4-1 interpolation of column 2 and column 1 with p_v = 0.5786, find dew-point temperature is 62.4°F *(ans)*

(b) By Eq. (4-7), $h = 0.240t + W(1061.2 + 0.45t)$

By Eq. (4-3), $W = 0.6220 \dfrac{p_v}{p_b - p_v}$

The barometric pressure p_b is given as 12.56 psia; for consistant units this is 12.56/0.491 = 25.58 in. Hg. Substitution into Eq. (4-3) gives

$$W = 0.622 \frac{0.5686}{25.58 - 0.5686}$$
$$W = 0.01414 \text{ lb water vapor/lb da}$$

Substitution into Eq. (4-7) gives

$$h = 0.240(90) + 0.01414\,(1061.2 + 0.45 \times 90)$$
$$h = 37.18 \text{ Btu/lb da} \quad \textit{(ans)}$$

Illustrative Problem 4-15

Air at 10°F dry bulb and 60 percent relative humidity after passing through a heating and humidifying device enters a space at 76°F dry bulb and 50 percent relative humidity. The barometric pressure is 29.92 in. Hg. How much water vapor and heat must be added to each pound of air entering the heating and humidifying device?

Solution:

The amount of water vapor added per pound of air must be the difference in humidity ratio between the initial and final air conditions. Also, the heat added must be the difference in enthalpy between the initial and final air condition.

At the initial condition (subscript 1):

Table 4-1 at 10°F, find p_{vs} = 0.0631 in. Hg. From Eq. (4-4) p_v = 0.60 × 0.0631 = 0.03786 in. Hg. By Eq. (4-3),

$$W_1 = 0.622 \left(\frac{0.3786}{29.92 - 0.03786}\right)$$
$$W_1 = 0.000788 \text{ lb water vapor/lb da}$$

By Eq. (4-7),

$$h_1 = 0.240t + W(1061.2 + 0.45t)$$
$$h_1 = 0.2402(10) + 0.000788(1060 + 0.45 \times 10)$$
$$h_1 = 3.24 \text{ Btu/lb da}$$

At final condition (subscript 2):
Table (4-1) at 76°F, find $p_{vs} = 0.9046$ in. Hg. From Eq. (4-4), $p_v = 0.50 \times 0.9046 = 0.4523$ in. Hg. By Eq. (4-3),

$$W_2 = 0.622\left(\frac{0.4523}{29.92 - 0.4523}\right)$$
$$W_2 = 0.00955 \text{ lb water vapor/lb da}$$

By Eq. (4-7),

$$h_2 = 0.240(76) + 0.00955 (1061.2 + 0.45 \times 76)$$
$$h_2 = 28.70 \text{ Btu/lb da}$$

Moisture addition $= W_2 - W_1 = 0.00955 - 0.000788 = 0.00876$ lb/lb da *(ans)*

Heat addition $= h_2 - h_1 = 28.70 - 3.24 = 25.46$ Btu/lb da *(ans)*

Illustrative Problem 4-16

Moist air exists at 80°F dry bulb, 60°F dew point, and 29.92 in. Hg barometric pressure. Determine the following:

a. humidity ratio
b. saturation ratio
c. relative humidity
d. enthalpy
e. specific volume of the dry air

<u>*Solution:*</u>

(a) Table 4-1 at a dew-point temperature of 60°F, find $p_v = 0.5218$ in. Hg. By Eq. (4-3),

$$W = 0.622\left(\frac{0.5218}{29.92 - 0.5218}\right)$$
$$W = 0.0110 \text{ lb water vapor/lb da}$$

(b) Table 4-1 at 80°F saturated, find $W_s = 155.52$ grains per lb da, or 155.52/7000 = 0.0222 lb/lb da (column 4). By Eq. (4-5),

$$\text{Saturation Ratio, } \mu = \frac{W}{W_s}$$
$$= \frac{0.110}{0.0222} - 0.495 \text{ or 49.5 percent.} \quad \textit{(ans)}$$

(c) From Table 4-1 at 80°F find $p_{vs} = 1.0321$ in. Hg. By Eq. (4-4),

$$\phi = P_v/p_{vs} = 0.5218/1.0321$$
$$= 0.506 \text{ or 50.6 percent.} \quad \textit{(ans)}$$

(d) By Eq. (4-7),

$$h = 0.240t + W(1061.2 + 0.45t)$$
$$h = 0.2402(80) + 0.0110 (1060 + 0.45 \times 80)$$
$$h = 31.27 \text{ Btu/lb da}$$

(e) By Eq. (2-19), $p_a v = RT$ where p_a is the partial pressure of the dry air in the mixture.

$$p_a = (0.491)(29.92 - 0.5218) = 14.43 \text{ psia}$$
$$144(14.43)v = 53.3\,(80 + 460)$$
$$v = 13.85 \text{ ft}^3\text{/lb} \quad \textit{(ans)}$$

Illustrative Problem 4-17

What is the dry-bulb temperature, humidity ratio, and total heat of a mixture of 2 lb of air at 30°F and 80 percent RH, and 3 lb of air at 70°F and 50 percent RH?

Solution:

Vapor pressure at 30°F saturated is 0.164 in. Hg. At 30°F and 80 percent RH, the vapor pressure is 0.80 × 0.164 = 0.131 in. Hg. This corresponds to a dew point of 25.2°F, a humidity ratio of 19.15 grains/lb da (column 4), and 2.93 Btu/lb of heat of the moisture (column 9). Heat of the dry air at 30°F is 7.20 Btu/lb (column 5). Total heat is 7.20 + 2.93 = 10.13 Btu/lb.

Vapor pressure at 70°F saturated is 0.7392 in. Hg. At 70°F and 50 percent RH, the vapor pressure is 0.50 × 0.7392 = 0.3696 in. Hg. This corresponds to a dew-point temperature of 50.5°F, 54.46 grains/lb da, and 8.43 Btu/lb heat of moisture. Dry air at 70°F has heat of 16.80 Btu/lb. The total heat is 16.80 + 8.43 − 25.23 Btu/lb da.

The dry-bulb temperature of the mixture is [(2 × 30) + (3 × 70)]/5 = 54°F (unless fog is formed). The humidity ratio is [(2 × 19.15) + (3 × 54.46)]/5 = 40.34 grains/lb da. At a temperature of 54°F saturation humidity ratio is 62.03 grains/lb da (column 4). Since the mixture humidity ratio is less than 62.03, fog will not form and 54°F is the dry-bulb temperature.

The total heat of the mixture is [(2 × 10.13) + (3 × 25.23)]/5 = 19.19 Btu/lb da.

4-17 THE PSYCHROMETRIC CHART

The illustrative problems of Section 4-16 indicate how problems in psychrometrics may be solved by using equations and the psychrometric table. As may be noted, this method of problem solving may be long and tedious, especially if more than one psychrometric process is involved. In addition, it is usually much more revealing to us to be able to see the process taking place by using some graphical representation.

The psychrometric chart is a plot of the psychrometric properties of moist air. Charts of various types have been designed and produced. Each has its usefulness, and selection of a chart for use is normally one of personal preference and the temperature range required.

For general use in this text we will be using the Trane Psychrometric Chart for normal temperatures (30°F to 115°F). Other charts are available for low temperatures (−40°F to 50°F) and for high temperatures (60°F to 250°F). A brief description of the Trane chart follows.

Figure 4-4 is a skeleton chart showing the various coordinates and identifies the various scales and lines for the psychrometric properties of air.

Dry-Bulb Temperature (Line 1). The temperature of air read on a standard thermometer. Lines of constant dry-bulb temperature are straight, vertical lines on the chart. The dry-bulb temperature scale is at the bottom of the chart. Units are °F.

Wet-Bulb Temperature (Line 2). The wet-bulb temperature above 32°F is the temperature indicated by a thermometer whose bulb is covered by a wet wick and exposed to a stream of air moving at a velocity of 1000 fpm. Wet-bulb temperatures below 32°F are obtained

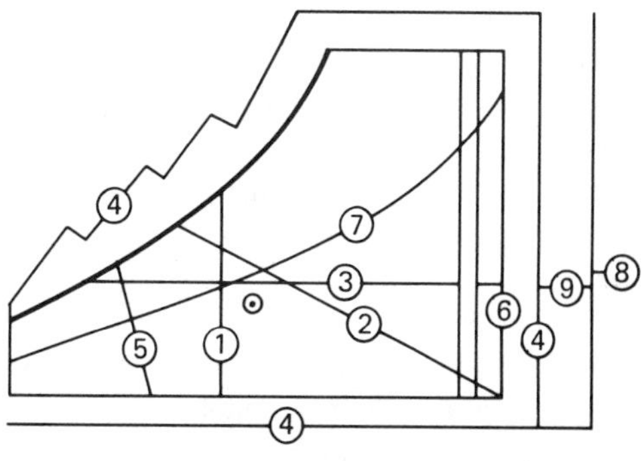

Figure 4–4

Skeleton psychrometric chart showing various coordinates, scales, and lines

from a thermometer on which the water in the wick has frozen to ice. This is the reason the slope of the wet-bulb temperature lines change below 32°F. The wet-bulb temperature scale is on the curved line at the left of the chart. Units are °F.

Humidity Ratio (Line 3). Lines of constant humidity ratio are straight, horizontal lines at right angle to the dry-bulb temperature lines. The humidity ratio scale is at the right of the chart. Units are grains of moisture per pound of dry air.

Enthalpy (Line 4). The thermodynamic property which serves as a measure of the heat energy in the air above some selected datum temperature. In this case it represents the enthalpy of 1 lb of dry air plus the enthalpy of W grains of moisture associated with it. The enthalpy scale, line 4, is shown on the upper left, and a duplication at the bottom and right side of the chart. When reading enthalpy values it will be necessary to use a straight-edge spanning the chart from side to side through the point in the chart for which the enthalpy is to be determined. Note that on this chart lines of constant enthalpy are not parallel to lines of constant wet-bulb temperature. The units of enthalpy are Btu/lb da.

Specific Volume (Line 5). Lines of constant specific volume. The units are ft^3/lb da.

Dew-Point Temperature (Line 6). Lines of constant dew-point temperature are horizontal, parallel to humidity ratio lines. Line 6 is the dew-point temperature scale. Units are °F.

Relative Humidity (Line 7). Lines of constant relative humidity curve upward from left to right.

Vapor Pressure (Line 8). Lines of constant partial pressure of water vapor in the air-water vapor mixture are horizontal and parallel to line of constant humidity ratio. Line 8 is the pressure scale in units of inches of mercury absolute.

Sensible Heat Ratio (Line 9). The ratio of sensible heat to total heat in a heating and humidifying process or a cooling and dehumidifying process. At 78°F dry-bulb temperature and 50 percent relative humidity is the alignment circle for the sensible heat ratio scale.

Figure 4-5 is the complete Trane Psychrometric chart. The chart had been developed for standard sea level barometric pressure of 29.92 in. Hg but is usable with little error for barometric pressures from 29.00 to 31.00 in. Hg.

4-18 BASIC PROCESSES INVOLVING AIR-WATER VAPOR MIXTURES

Figure 4-6 represents schematically the basic process lines for air-conditioning processes. Each process line shown, or combinations of process lines, may be used to show graphically the variation in the properties of an air-vapor mixture as it is heated, humidified, cooled, dehumidified, or treated in various combinations. Chapter 5 discusses in detail the various air-conditioning processes with illustrative problems commonly encountered in applied psychrometrics.

4-19 ILLUSTRATIVE EXAMPLES TO DETERMINE PSYCHROMETRIC PROPERTIES OF AIR USING THE PSYCHROMETRIC CHART

In Section 4-16 several problems were done to illustrate how the psychrometric properties of moist air could be calculated using equations and Table 4-1. We will now determine psychrometric properties of moist air using the psychrometric chart.

Illustrative Problem 4-18

Using the psychrometric chart, determine the psychrometric properties of moist air existing at 70°F DB and 60°F WB.

Solution:

Step 1: Referring to Fig. 4-7, locate the moist-air condition on the chart

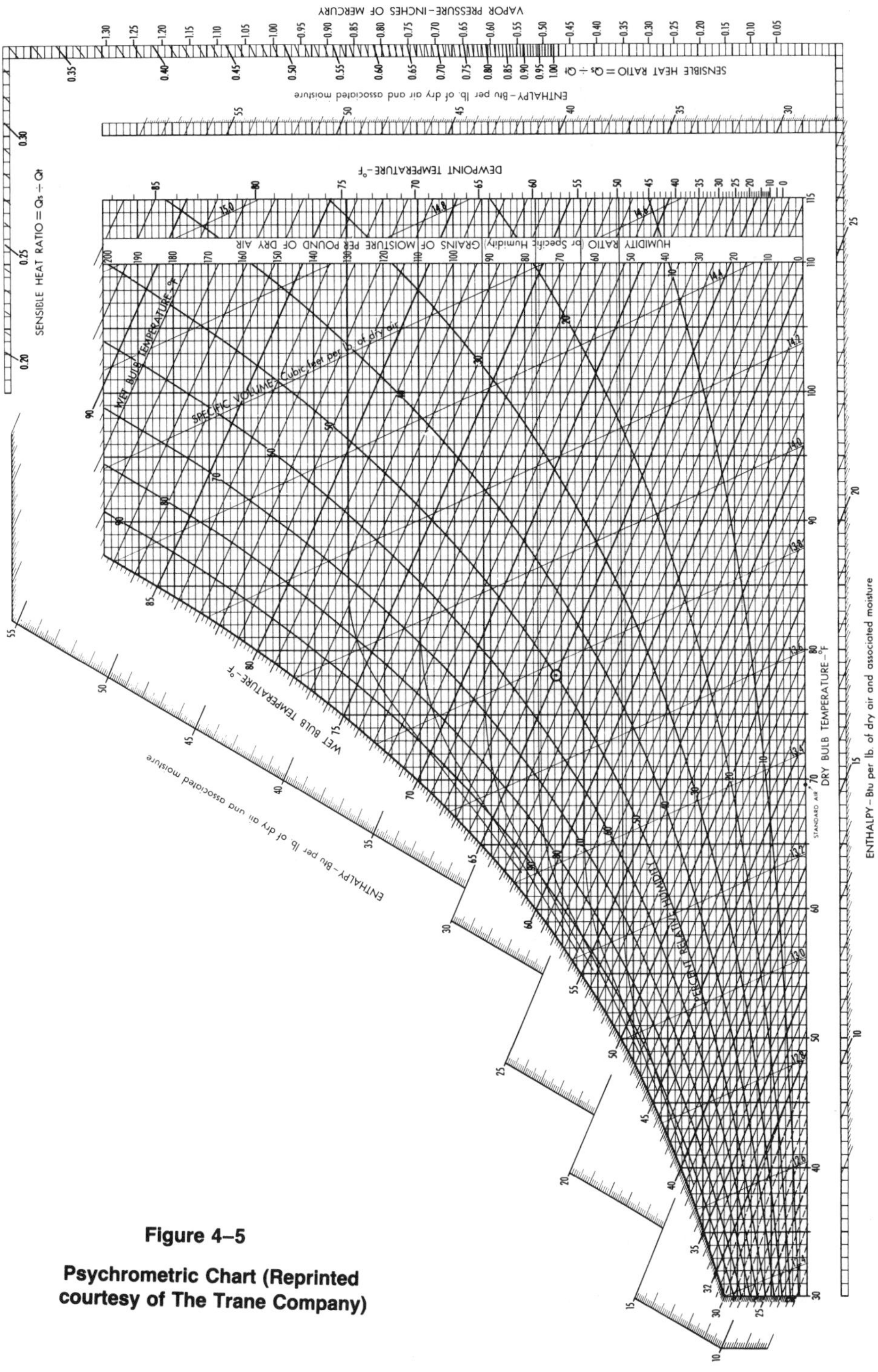

Figure 4–5

Psychrometric Chart (Reprinted courtesy of The Trane Company)

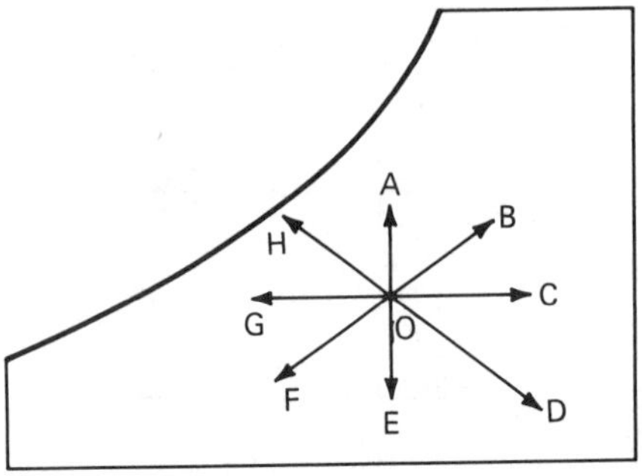

Figure 4–6

Schematic psychrometric chart representing various basic psychrometric processes:
OA = Humidifying only
OB = Heating and humidifying
OC = Sensible heating only
OD = Chemical dehumidifying
OE = Dehumidifying only
OF = Cooling and dehumidifying
OG = Sensible cooling only
OH = Cooling and humidifying

where the vertical line representing 70°F DB intersects the inclined line representing 60°F WB. This point is labeled *A*.

Step 2: At point *A* by interpolating between the inclined lines of constant specific volume, find v_A = 13.53 ft³/lb da.

Step 3: At point *A* by interpolating between the curved lines of constant relative humidity, find ϕ_A = 56 percent.

Step 4: Using a straight-edge, draw a straight horizontal line through point *A* to the right to intersect the humidity ratio, dew point, and vapor pressure scales, find:
Humidity ratio, W = 61.9 grains/lb da
Dew-point temperature, t_{dp} = 53.6°F
Vapor pressure, p_v = 0.420 in. Hg

Step 5: Using a straight-edge, span the chart from the enthalpy scale at upper left to the same enthalpy value on the lower enthalpy scale at bottom and right of chart through point *A*. Where the straight-edge intersects the enthalpy scale, find h_A = 26.4 Btu/lb da. (Note that lines of constant enthalpy are *not* parallel to lines of constant wet-bulb temperature in this area of the chart).

Illustrative Problem 4-19

Using the psychrometric chart, determine the following psychrometric properties of moist air existing at 85°F DB and 70 percent RH. Determine wet-bulb temperature, humidity ratio, enthalpy, dew-point temperature, specific volume, vapor pressure, and saturation ratio.

Solution:

Step 1: Referring to Fig. 4-7, locate the moist-air condition on the chart where the vertical line representing 85°F DB intersects the curved line representing 70 percent RH. This point is labeled *B*.

Step 2: Referring to the point *B*, locate the sloped wet-bulb line passing through the point and read from the 100 percent relative humidity curve the wet-bulb temperature of 77°F.

Step 3: Using a straight-edge, span the chart through point *B* to the enthalpy scales and read, h_B = 40.4 Btu/lb da.

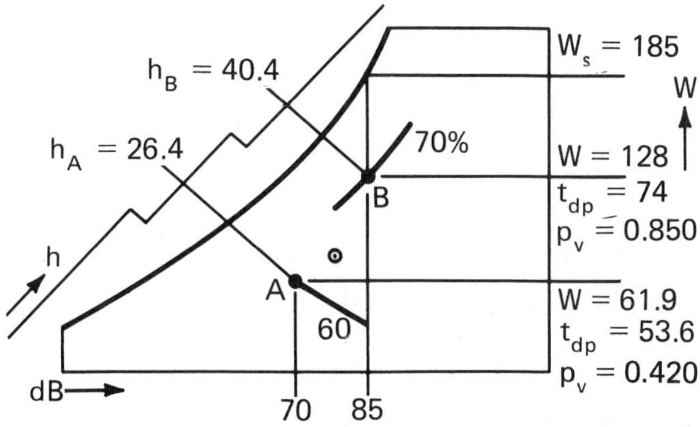

Figure 4–7

Solution to Examples 4-18 and 4-19

Step 4: Using a straight-edge, draw a straight horizontal line through point *B* to the right to intersect the humidity ratio, dew point, the vapor pressure scales; find:
Humidity ratio, W = 128 grains/lb da.
Dew-point temperature, t_{dp} = 74°F.
Vapor pressure, p_v = 0.850 in. Hg

Step 5: At point *B* by interpolating between the inclined lines of constant specific volume, find v_B = 14.12 ft^3/lb da.

Step 6: By Eq. (4-5), saturation ratio $\mu = W/W_s$. W_s is the humidity ratio of air at 85°F DB and 100 percent RH. Extend the vertical line representing 85°F DB upward from point *B* to 100 percent RH. Through this point draw a horizontal line to the right to find W_s = 185 grains/lb da. Therefore, μ = 128/185 = 0.692.

CHAPTER 4

REVIEW

4-1. Calculate the humidity ratio, enthalpy, and relative humidity of moist air at 90°DB and 80°WB. Barometric pressure is 29.92 in. Hg.

4-2. Determine the same quantities as Problem 4-1 using the psychrometric chart.

4-3. Moist air at a barometric pressure of 14.696 psia has a dry-bulb temperature of 80°F and a wet bulb temperature of 60°F.

(a) Calculate the air dew-point temperature.

(b) Calculate the humidity ratio.

(c) Calculate the enthalpy.

(d) Check the answers to parts (a), (b), and (c) by using the psychrometric chart.

4-4. Moist air exists at 70°DB and 65°DP, and a barometric pressure of 29.92 in. Hg.

(a) Calculate the relative humidity, humidity ratio, saturation ratio, and enthalpy.

(b) Determine the same properties from the psychrometric chart.

4-5. Air at 30°DB and 30°WB is heated to 80°DB without addition of moisture. Determine the following from the psychrometric chart:

(a) Initial and final humidity ratio.

(b) Initial and final relative humidity.

(c) Initial and final enthalpy.

(d) Initial and final specific volume.

(e) Initial and final dew-point temperature.

(f) Initial and final partial pressure of the water vapor in the air.

4-6. Moist air exists at 70° DB and 60 percent RH when the barometric pressure is 14.696 psia. Determine partial pressure of the water vapor in the air, wet-bulb temperature, specific volume, enthalpy, humidity ratio, and dew-point temperature.

4-7. Calculate the humidity ratio, enthalpy, and specific volume for saturated air at 14.696 psia total pressure for (a) t = 70°F and (b) t = 32°F. Check calculated values with those found in Table 4-1.

4-8. Moist air exists at 70°F and 50 percent relative humidity in a room containing several windows. If the inside surface temperature of the windows is at 45°, will condensation form on the windows?

4-9. Moist air exists at 32°F dry-bulb temperature and 80 percent relative humidity. How much moisture must be added to each pound of air if the final condition of the air is to be 80°F dry-bulb temperature and 40 percent relative humidity? Assume that total pressure is 14.696 psia.

BIBLIOGRAPHY

1. Strock, C., and Koral, R. L., *Handbook of Air Conditioning, Heating and Ventilating,* 2nd ed., The Industrial Press, New York, 1965.

2. Jordan, R. C., and Priester, G. B., *Refrigeration and Air Conditioning,* 2nd ed., Prentice-Hall, Inc., Englewood Cliffs, NJ, 1956.

3. ASHRAE *Handbook of Fundamentals,* American Society of Heating, Refrigerating and Air-Conditioning Engineers, Atlanta, GA, 1977.

Chapter 5

Applied Psychrometrics

5-1 INTRODUCTION

The psychrometric chart, introduced in Chapter 4, is one of the most useful tools of the practicing air-conditioning engineer and technologist. The chart assists in the solution of simple as well as complex processes involved in the conditioning of moist air. Most of the complex problems consist of various combinations of simple processes, such as sensible heating, sensible cooling, humidification, or dehumidification. The chart provides a means for showing these various processes graphically. In this chapter we will progress through the various types of basic processes and into the analysis of the more complex processes.

Although the psychrometric chart is most helpful, it must be understood that graphical solution of problems must be done with great care if satisfactory results are to be obtained. Therefore, we should stress careful and neat work on the chart to obtain as much accuracy as possible.

5-2 SENSIBLE HEATING OF MOIST AIR

If heat is added to moist air with no addition of moisture, the process is called *sensible heating*. Such a process may occur when moist air passes through or over a heat-transfer surface which causes a rise in the air dry-bulb temperature. The heat-transfer surface may be a tubular heating coil where a medium, such as steam or hot water, circulates within the coil tubes, or an electrically heated coil, or the heat-transfer surfaces of a warm-air furnace.

Figure 5-1 shows a typical heat-transfer coil for heating air by the use of steam. Figure 5-2 is a schematic drawing showing the heat-transfer coil and the skeleton psychrometric chart illustrating the sensible-heating process. As indicated in Fig. 5-2, during a sensible-heating process the dry- and the wet-bulb temperatures

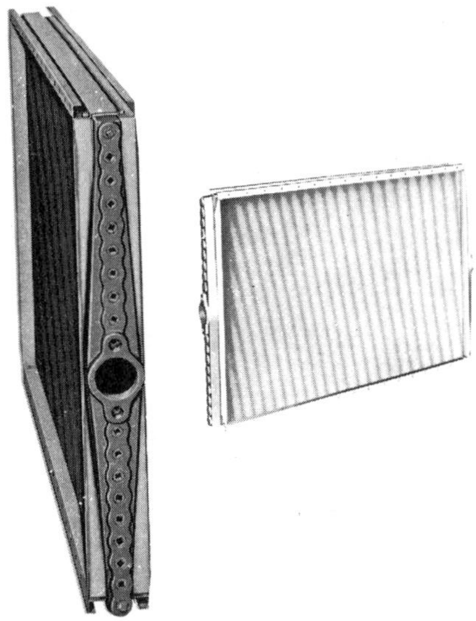

Figure 5–1

Single row heating coil (Courtesy of The Trane Company.)

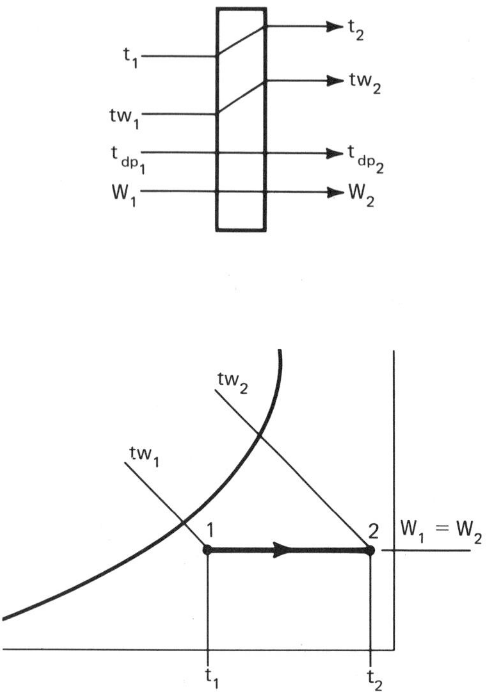

Figure 5–2

Schematic: Sensible heating process

increase, but the dew-point temperature and humidity ratio are constant.

The steady-flow and material-balance equations which apply to this process are

$$\dot{m}_{a_1} h_1 + {}_1\dot{q}_{s_2} = \dot{m}_{a_2} h_2 \qquad (5\text{-}1)$$

$$\dot{m}_{a_1} = \dot{m}_{a_2} \qquad (5\text{-}2)$$

$$\dot{m}_{a_1} W_1 = \dot{m}_{a_2} W_2 \qquad (5\text{-}3)$$

where

$\dot{m}_a$ = pounds of air flowing per unit of time (usually lb/hr)

${}_1\dot{q}_{s_2}$ = sensible heat added from condition 1 to condition 2 per unit of time (usually in Btu/hr)

W = humidity ratio (lb water vapor/lb dry air)

h = enthalpy of moist air (Btu/lb dry air)

Therefore, for sensible heating of moist air, and using Eq. (5-1) and (5-2), we have

$${}_1\dot{q}_{s_2} = \dot{m}_a(h_2 - h_1) \qquad (5\text{-}4)$$

Since the humidity ratio is constant during the process, it is closely true that the moist-air enthalpy is

$$h = 0.240t + W(1060 + 0.45t) \qquad (4\text{-}7)$$

Repeated

If t_1 and t_2 are substituted into Eq. (4-7) to obtain h_1 and h_2 and if the results are substituted in Eq. (5-4), we would have

$${}_1\dot{q}_{s_2} = \dot{m}_a(0.240 + 0.45W)(t_2 - t_1) \qquad (5\text{-}5)$$

where t_1 and t_2 are the initial and final dry-bulb temperatures, respectively, for the sensible-heating process. Equation (5-5) assumes a constant specific heat of superheated steam of 0.45. It is often convenient to combine the two specific heat values for an average value of humidity ratio, say 0.01 lb vapor/lb da (70 grains/lb da). This gives a single specific heat of moist air of 0.244, and Eq. (5-5) becomes

$${}_1\dot{q}_{s_2} = 0.244(\dot{m}_a)(t_2 - t_1) \qquad (5\text{-}6)$$

Air-handling components in systems, such as fans, ducts, and so forth, are selected on the basis of *volume* flow rather than mass flow of air. This introduces a little difficulty in problem solutions because the volume flow *varies* with temperature and pressure (or, the specific volume), whereas the mass flow rate is *constant*. Therefore, if the air volume flow rate is to be determined, it is necessary to specify the point in the system where flow volume rate is to be determined and find the specific volume of the air at that point. With the mass flow rate, $\dot{m}_a$, known and the specific volume of the air at the point, we may calculate the volume flow rate in cubic feet per minute from

$$\text{cfm} = \frac{(\dot{m}_a)(v)}{60} \qquad (5\text{-}7)$$

where $\dot{m}_a$ is the mass flow rate of air in lb/hr, 60 min/hr, and, v the specific volume of the air at the point in question in ft^3/lb da.

For uniformity in the manufacturing industry air-handling equipment is normally rated on the basis of "standard air" which has been defined as dry air at 70°F and 29.92 in. Hg barometric pressure, which gives a density of 0.075 lbm/ft^3. Using this density, equal to $1/v$, in Eq. (5-7) we have

$$\dot{m}_a = (60 \times 0.075)(\text{cfm}) = 4.5 \text{ cfm}$$

If this value of $\dot{m}_a$ is substituted in Eq. (5-4), we have

$$_1\dot{q}_{s_2} = (4.5 \text{ cfm})(h_2 - h_1) \qquad (5\text{-}8)$$

And, if we substitute the value of $\dot{m}_a$ into Eq. (5-6), we have

$$_1\dot{q}_{s_2} = 0.244(4.5 \text{ cfm})(t_2 - t_1)$$

or

$$_1\dot{q}_{s_2} = (1.10 \text{ cfm})(t_2 - t_1) \qquad (5\text{-}9)$$

Caution: You should recall that of the series of equations just developed for $_1\dot{q}_{s_2}$, Eq. (5-4) is the only one that may be called *exact*. The remaining equations, (5-5) through (5-9), can be called approximate because, as you have seen, we made certain assumptions about the air in their development. Regardless of this, Eqs. (5-5) through (5-9) are used quite generally in actual work because of their simplicity, and the dry-bulb temperatures are easy to read from a psychrometric chart. Another reason for using these equations instead of Eq. (5-4) is because of difficulty in reading accurate enthalpy values from the psychrometric chart, which is quite necessary in some problems. For example, assume that a system requires an air flow rate of 50,000 lb/hr of air and an enthalpy change $(h_2 - h_1)$ equal to 15.5 Btu/lb da. The resulting $_1\dot{q}_{s_2}$ would be 775,000 Btu/hr. However, if the enthalpy difference read from the psychrometric chart had been 15.7 Btu/lb da, the resulting $_1\dot{q}_{s_2}$ would be 785,000 Btu/hr, with an error of about 1.3 percent. This is well within the accuracy of Eqs. (5-5) through (5-9).

5-3 SENSIBLE COOLING OF MOIST AIR

If heat is removed from moist air with no change in moisture content, the process is called *sensible cooling*. Such a process occurs when warm air from a warm-air heating system enters a space and replaces the heat lost from the space. This is true if there are no sources within the space giving off moisture. This assumption is normally made when designing winter heating systems. Sensible cooling also occurs if moist air passes through a tubular cooling coil where a medium such as chilled water circulates through the coil tubes. The chilled water must be above the moist-air dew-point temperature because no condensation from the air is a requirement for sensible cooling.

Figure 5-3 shows a typical cooling coil for chilled water, and Fig. 5-4 shows the schematic drawing representing the sensible-cooling process. Equations (5-1) through (5-9) apply to sensible-cooling processes, however, the subscripts should be reversed if a positive value is wanted for $_1\dot{q}_{s_2}$. Remember that heat is being removed from the air in the sensible-cooling process.

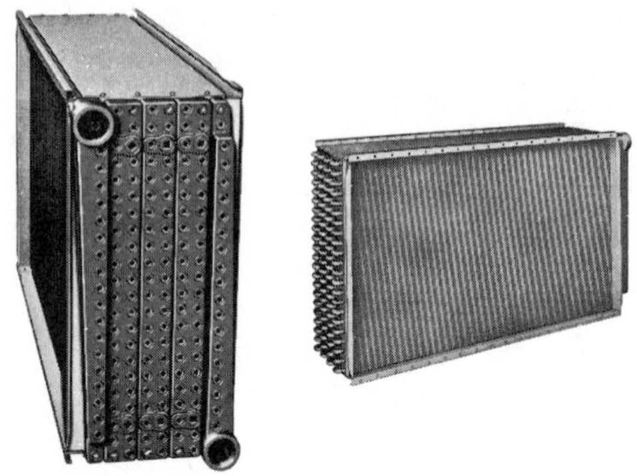

Figure 5–3

Drainable water cooling coil (Courtesy of The Trane Company.)

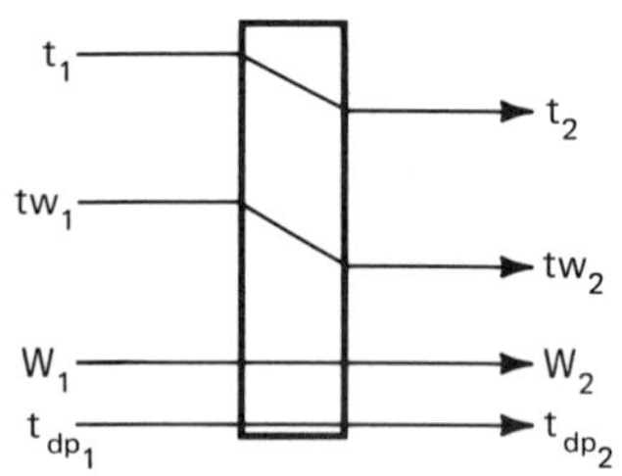

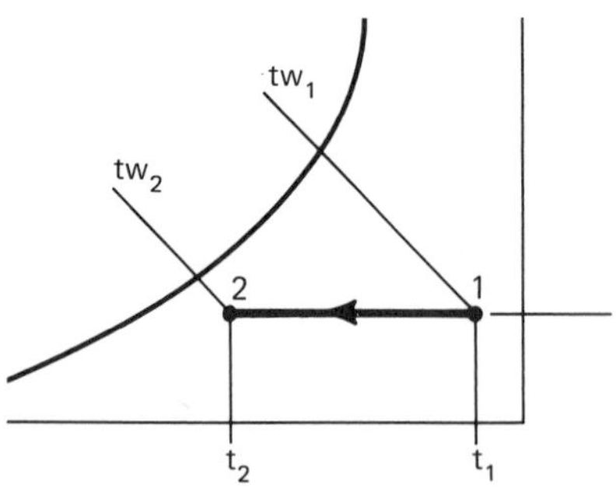

Figure 5–4

Schematic: Sensible cooling process

Illustrative Problem 5-1

How much heat is required to heat 3000 cfm of moist air at 45°F DB and 35°F WB to a final temperature of 90°F DB without change in humidity ratio?

Solution:

(Refer to Fig. 5-5). Since the problem states no change in humidity ratio, the process is a sensible-heating process.

Step 1: On the psychrometric chart locate point 1 at the intersection of 45°F DB and 35°F WB. At this point read W_1 = 14 grains/lb da, h_1 = 13.0 Btu/lb da and v_1 = 12.76 ft^3/lb da.

Step 2: On the psychrometric chart locate point 2 by drawing a horizontal line (constant humidity ratio) through point 1 and extend to point 2 at 90°F DB. At this point find W_2 = 14 grains/lb da, h_2 = 23.9 Btu/lb da, and v_2 = 13.90 ft^3/lb da.

Step 3: Determine the mass of air heated. From Eq. (5-7)

$$\dot{m}_a = \text{cfm} \times (60)/v_1 = 3000\ (60)/12.76$$
$$\dot{m}_a = 14{,}100 \text{ lb/hr}$$

Step 4: Determine sensible heat added. By Eq. (5-4)

$$_1\dot{q}_{s_2} = \dot{m}_a(h_2 - h_1) = 14{,}100\,(23.9 - 13.0)$$
$$_1\dot{q}_{s_2} = 153{,}700 \text{ Btu/hr}$$

(ans)

By Eq. (5-6)

$$_1\dot{q}_{s_2} = 0.244\dot{m}_a(t_2 - t_1)$$

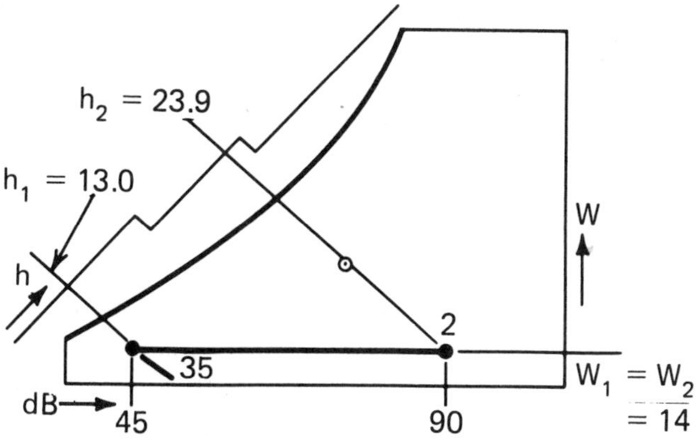

Figure 5–5

Solution for Example 5-1

$= 0.244\ (14{,}100)(90 - 45)$

$_1\dot{q}_{s_2} = 156{,}800$ Btu/hr *(ans)*

By Eq. (5-9)

$_1\dot{q}_{s_2} = (1.10 \text{ cfm})(t_2 - t_1)$

$= 1.10\ (3000)(90 - 45)$

$_1\dot{q}_{s_2} = 148{,}500$ Btu/hr *(ans)*

Illustrative Problem 5-2

(Refer to Fig. 5-6.) How much heated air must be supplied to a space within a building to offset a heat loss of 38,200 Btu/hr? The space air condition is to be maintained at 70°F DB and 50 percent RH, and the air supplied is at a temperature of 105°F DB. Assume no moisture addition within the space.

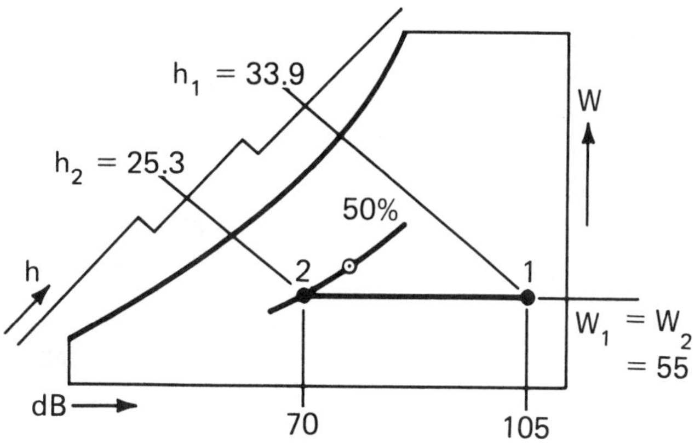

Figure 5–6

Solution for Example 5-2

Solution:

Since there is no moisture addition in the space, the supply air will have the same humidity ratio as the space air, and we have a sensible-cooling process. The supply air cools to the space condition as it gives up its sensible heat to overcome the space heat loss.

Step 1: On the psychrometric chart at 70°F DB and 50 percent RH, locate space air condition, point 2. At this point read W_2 = 55 grains/lb da, h_2 = 25.3 Btu/lb da, and v_2 = 13.51 ft^3/lb da.

Step 2: On the psychrometric chart locate the supply air condition by constructing a horizontal line through point 2 and extend to the supply air temperature, 105°F, point 1. At this point find W_1 = 55 grains/lb da, h_1 = 33.9 Btu/lb da, and v_1 = 14.41 ft^3/lb da.

Step 3: Determine the mass of heated air supplied. By Eq. (5-4),

$$_1\dot{q}_{s_2} = \dot{m}_a(h_1 - h_2)$$

Note that subscripts on h have been reversed to give a positive value of $_1\dot{q}_{s_2}$.

$$38{,}200 = \dot{m}_a(33.9 - 25.3)$$

$$\dot{m}_a = 4442 \text{ lb/hr} \qquad \textit{(ans)}$$

Step 4: Determine volume of supply air. By Eq. (5-7),

$$\text{cfm} = \frac{(\dot{m}_a \times v)}{60}$$

The specific volume of the supply air v_1 = 14.41 ft^3/lb da.

$$\text{cfm} = \frac{4442 \times 14.41}{60}$$

$$\text{cfm} = 1067 \text{ at supply air conditions} \qquad \textit{(ans)}$$

An approximate solution could be obtained by use of Eq. (5-9):

$$1\dot{q}_{s_2} = (1.10 \text{ cfm})(t_1 - t_2)$$

$$38{,}200 = (1.10 \text{ cfm})(105 - 70)$$

$$\text{cfm} = 992 \qquad \textit{(ans)}$$

5-4 ADIABATIC MIXING OF TWO STREAMS OF AIR

A frequently encountered process in air conditioning is the mixing of two or more streams of air having different psychrometric properties. Figure 5-7 represents the schematic drawing of the two flow streams mixing and the psychrometric chart for the process. The fundamental equations which apply to this mixing process are

$$\dot{m}_{a1}h_1 + \dot{m}_{a2}h_2 = \dot{m}_{a3}h_3 \qquad \textit{(5-10)}$$

$$\dot{m}_{a1} + \dot{m}_{a2} = \dot{m}_{a3} \qquad \textit{(5-11)}$$

$$\dot{m}_{a1}W_1 + \dot{m}_{a2}W_2 = \dot{m}_{a3}W_3 \qquad \textit{(5-12)}$$

If $\dot{m}_{a3}$ is eliminated from Eqs. (5-10), (5-11), and (5-12) we would have

$$\frac{\dot{m}_{a1}}{\dot{m}_{a2}} = \frac{h_2 - h_3}{h_3 - h_1} = \frac{W_2 - W_1}{W_3 - W_1} \qquad \textit{(5-13)}$$

This defines a straight line on the psychrometric chart between points 1 and 2. Point 3 lies on the line between points 1 and 2, and the line segments produced are proportional to the masses of air mixed, the temperature differences, humidity ratio differences, and the enthalpy differences.

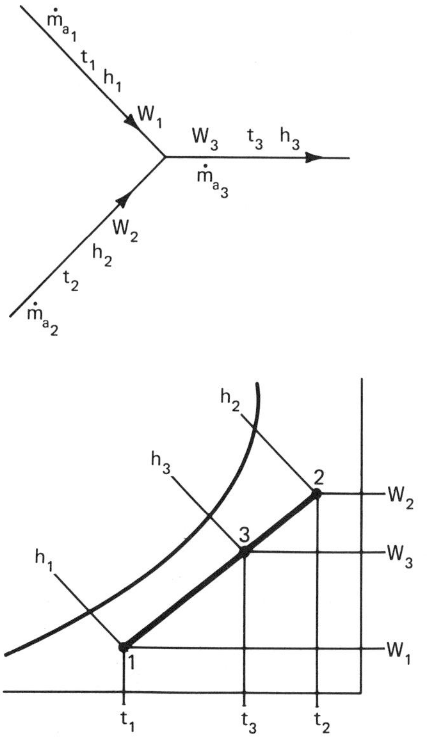

Figure 5–7
Schematic: Mixing process

Making use of Eqs. (5-10) through (5-13), the following three equations become useful in the evaluation of the mixed air properties:

$$t_3 = \frac{\dot{m}_{a1} \times t_1 + \dot{m}_{a2} \times t_2}{\dot{m}_{a1} + \dot{m}_{a2}} \qquad (5\text{-}14)$$

$$W_3 = \frac{\dot{m}_{a1} \times W_1 + \dot{m}_{a2} \times W_2}{\dot{m}_{a1} + \dot{m}_{a2}} \qquad (5\text{-}15)$$

$$\text{h}_3 = \frac{\dot{m}_{a1} \times h_1 + \dot{m}_{a2} \times h_2}{\dot{m}_{a1} + \dot{m}_{a2}} \qquad (5\text{-}16)$$

The solution of a mixed-air problem in psychrometrics normally makes use of one of Eqs. (5-14) through (5-16), and then the remaining mixed-air properties are determined by graphical means. This will be illustrated by Problem 5-3 to follow.

If the specific volumes of the two airstreams do not vary appreciably from each other (about 0.5 ft³/lb da) then Eq. (5-14) could be used, substituting cfm in place of $\dot{m}_a$ as follows:

$$t_3 = \frac{\text{cfm}_1 \times \text{t}_1 + \text{cfm}_2 \times t_2}{\text{cfm}_1 + \text{cfm}_2} \qquad (5\text{-}17)$$

Also, in some problems involving the mixing of two airstreams, data is given on the percent of the total flow entering at the two conditions. This is handled by substituting the percentage flow, expressed as a decimal, in place of the mass flow quantities.

Illustrative Problem 5-3

Three hundred cfm of air at 35°F DB and 100 percent RH are mixed with 600 cfm of air at 85°F DB and 50 percent RH. What will be the dry-bulb temperature, humidity ratio, and enthalpy of the mixture?

Solution:

(See Figure 5-8.)

Step 1: Determine the psychrometric properties of the two airstreams from psychrometric chart.

Condition 1	Condition 2
300 cfm at 35°F DB, 100 percent RH	600 cfm at 85°F DB, 50 percent RH
Specific volume	
v_1 = 12.55 ft³/lb da	v_2 = 14.02
Humidity ratio	
W_1 = 30 grains/lb da	W_2 = 90
Enthalpy	
h_1 = 13.1 Btu/lb da	h_2 = 34.5 Btu/lb da

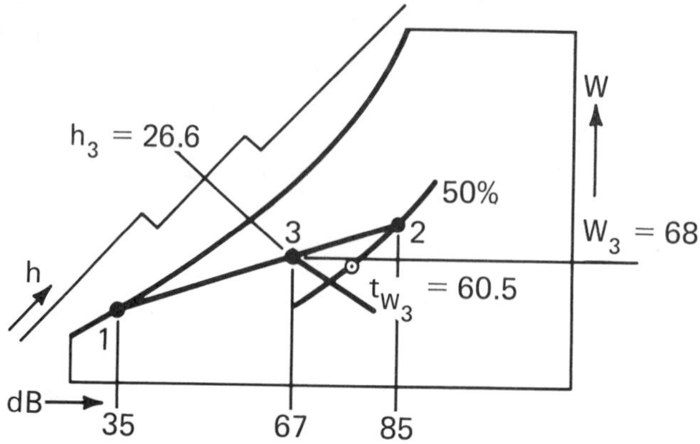

Figure 5–8

Solution for Example 5-3

Step 2: Determine mass of the two airstreams using Eq. (5-8) to determine lb per min, we have

$$\dot{m}_{a1} = \frac{\text{cfm}_1}{v_1} = \frac{300}{12.55}$$

$$\dot{m}_{a2} = \frac{\text{cfm}_2}{v_2} = \frac{600}{14.02}$$

$$\dot{m}_{a1} = 23.9 \text{ lb/min}$$

$$\dot{m}_{a2} = 42.8 \text{ lb/min}$$

Step 3: Determine psychrometric properties of the mixed air.

On the psychrometric chart, draw a straight line joining point 1 and point 2. Calculate the dry-bulb temperature of the mixture by using Eq. (5-14).

$$t_3 = \frac{\dot{m}_{a1} \times t_1 + \dot{m}_{a2} \times t_2}{\dot{m}_{a1} + \dot{m}_{a2}}$$

$$= \frac{23.9 \times 35 + 42.8 \times 85}{23.9 + 42.8}$$

$$t_3 = 67.08°\text{F} \quad (\text{say } 67°\text{F}) \qquad \textit{(ans)}$$

Where $t_3 = 67°$F intersects the mixture line between points 1 and 2 determines the mixed-air condition, point 3. From the chart at point 3 find $t_{w_3} = 60.5°$F, $W_3 = 68$ grains/lb da, $h_3 = 26.6$ Btu/lb da, dew point $t_{dp} = 56.4°$F, partial pressure, $p_v = 0.455$ in. Hg.

Step 4: Determine the dry-bulb temperature of the mixture by using approximate Eq. (5-17).

$$t_3 = \frac{300 \times 35 + 600 \times 85}{300 + 600}$$

$$t_3 = 68.3°\text{F} \qquad \textit{(ans)}$$

This varies 1.3°F from the previous value of $t_3 = 67°$F. The significance of the difference must be evaluated by the engineer as to the accuracy with which he should be working. The exact Eq. (5-14) should be used when possible.

5-5 HUMIDIFICATION OF AIR†

As stated previously, the purpose of environmental control is to make people more comfortable and the labor of people and equipment more efficient. Temperature, humidity, cleanliness, air movement, and thermal radiation interrelate to create conditions in which people are more comfortable or less comfortable. In a home, business, or industry, comfort or production efficiency is greatly affected by changes in these environmental variables.

Least evident of these variables to human perception is humidity. All of us will recognize and react more quickly to temperature changes, odors, or heavy dust in the air, drafts, or radiant heat from sunlight or a radiator, then we will to a change in relative humidity. However, as relative humidity interrelates with temperature and others of these variables, it becomes a vital ingredient in total environmental control.

Frequently, in air-conditioning work it is necessary to humidify the air by introducing moisture into the air. The moisture added may be already in the vapor state or it may be liquid. (It may also be solid, although infrequently.) Humidification is frequently required during cold weather conditions because the cold outside air infiltrating into a heated building, or being brought in by a mechanical ventilation system, is normally very dry (low humidity ratio, even though its relative humidity may be high). If the infiltration or ventilation air is not humidified, a low dew-point temperature, relative humidity, and humidity ratio could exist within the heated building.

Illustrative Problem 5-4

Assume that outside air at 30°F DB and 40 percent RH infiltrates into a building where the inside is being maintained at 72° F DB. What would be the theoretical relative humidity of the infiltrated air at 72°F?

Solution:

Locate on the psychrometric chart the condition of the outside air at 30°F DB and 40 percent RH. At that point find the humidity ratio of 9.5 grains/lb da. If there is no moisture addition in the building, the infiltration air will have the same humidity ratio at 72°F.

At a humidity ratio of 9.5 grains/lb da and 72°F, find the relative humidity to be slightly less than 9.0 percent. ***(ans.)***

Since one to three complete air changes occur every hour in most buildings through infiltration (and many more times with forced makeup or exhaust), cold outdoor air replaces the warm indoor air. The heating system heats this cold, moist outdoor air and it becomes warm, dry indoor air. This condition is unhealthy for people and may be troublesome for some industrial processes.

Indoor relative humidity calculated as we did in Problem 5-4 should be called the "theoretical" indoor relative humidity. This condition very seldom actually exists in a building. Relative humidity values observed on a hygrometer will almost always exceed this theoretical value. The reason is that this dry, heated air will absorb moisture from any source it can within the building. It will absorb moisture from any hygroscopic material in the building, as well as from nasal passages and skin of human beings. This is not "free" hu-

†Extracted from *The Armstrong Humidification Handbook,* Armstrong Machine Works, Three Rivers, Michigan p. 2–3, with permission.

midification; it is the most expensive when translated into terms of human health and comfort, material deterioration, and production difficulties. Moreover, it requires the same amount of energy whether the moisture is absorbed from people and materials or added to the air by an efficient humidification system.

Illustrative Problem 5-5

Let us examine the theoretical system of Problem 5-4 using enthalpy as the base and determine the energy exchanges taking place.

From the psychrometric chart at 30°F DB and 40 percent RH, find h_{oa} = 8.8 Btu/lb da. If this air is heated to 72°F without moisture addition, the enthalpy is h_{ra} = 18.7 Btu/lb da, and the resulting relative humidity of 9.0 percent.

If the actual inside relative humidity were 25 percent by actual measurement by a hygrometer, the enthalpy at 72°F and 25 percent relative humidity would be, h_{ra} = 21.7 Btu/lb da, with the additional moisture coming from hygroscopic materials and people in the area. The additional energy 21.7 − 18.7 = 3.0 Btu/lb da, which is 16 percent, must come from the heating system.

If a humidification system is used and moisture added to achieve a comfortable 35 percent relative humidity, the enthalpy at 72°F and 35 percent relative humidity is 23.6 Btu/lb da, which is only (23.6 − 21.7)/21.7 = 8.0 percent increase over the "inevitable" energy load of 21.7 Btu/lb da. This is substantially less than (23.6 − 18.7)/18.7 = 26 percent, which is the theoretical increase from 9.0 percent relative humidity and 35 percent relative humidity.

If the inside air temperature were only 68°F at 35 percent relative humidity (because people can be more comfortable at a lower temperature with higher humidity levels), the enthalpy is only 21.8 Btu/lb da, which is a very slight increase 21.8 − 21.7 = 0.1 Btu/lb da in energy. In many cases there will be a decrease in the required energy.

The introduction of water into the air for purposes of humidification is relatively easy and can be accomplished with rather simplified equipment. However, before we introduce this equipment, we will discuss briefly the principle of *vaporization* of a liquid.

Vaporization† of a liquid, such as water, may take place in two different ways: (1) by evaporation and (2) by boiling. *Evaporation* of a liquid can occur only at the free surface of the liquid and may take place at any liquid temperature between its saturation temperature and the dew-point temperature of the surrounding air. (*Free surface* means the interface between the liquid surface and the surrounding air.) *Boiling* takes place both at the liquid free surface and within the body of the liquid and can occur only at the saturation temperature corresponding to the absolute pressure on the liquid.

Water left standing in a pan open to the atmosphere (Fig. 5-9a) will eventually evaporate completely, even though the actual water temperature is below its saturation temperature. However, if the open pan is placed inside of a closed container, as shown in Fig. 5-9b, only a portion of the water will evaporate; the balance will remain in the pan for an indefinite length of time if the closed container remains sealed. The reason for this difference in behavior is found in the difference between the vapor pressure of the water in the pan and the pressure exerted by the vapor pressure of mois-

†Information extracted from *Trane Air Conditioning Manual*, The Trane Company, LaCrosse, Wisconsin, 1965, with permission.

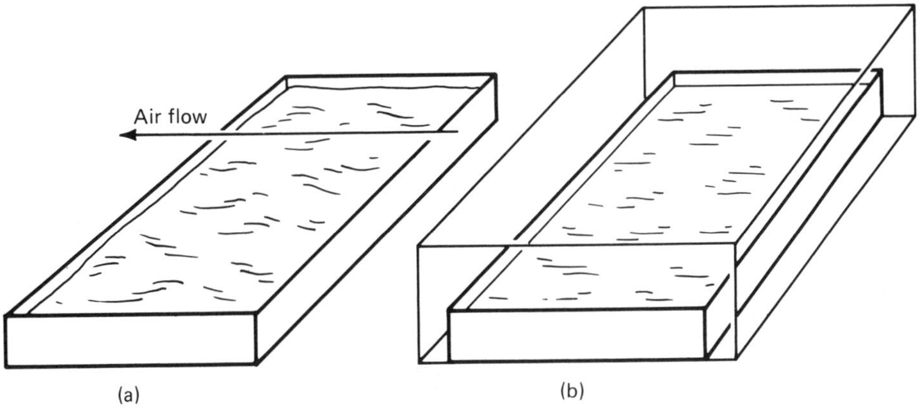

Figure 5–9

Evaporation of water

ture in the air in the sealed container. *Vapor pressure difference* is the controlling factor in evaporation and also condensation of water; or we may say that the dew-point temperature of the air above the liquid is the controlling factor.

Consider an open pan of water, such as Fig. 5-9a, setting in a room where the air temperature is 74°F and the relative humidity is 40 percent; the corresponding dew-point temperature is the saturation temperature of the moisture vapor at its vapor pressure, $p_v = \phi \times p_{vs} = 0.40 \times 0.8462 = 0.3385$ in. Hg. The dew-point temperature at 0.3385 in. Hg (Table 4-1) is 48.2°F. Although the water temperature in the pan could fall to the wet-bulb temperature ($t_w = 58.8$°F) of the surrounding air, it will in a case such as this probably assume a temperature that is closer to the air dry-bulb temperature ($t = 74$°F). Assume, for purposes of discussion, that the water assumes a temperature of 65°F. The vapor pressure corresponding to the air dew point was 0.3385 in. Hg. At the same time the vapor pressure of the water at 65°F is 0.6222 in. Hg. Since the vapor pressure of the surrounding air is less than that of the water, evaporation of the water into the air will occur.

This difference in vapor pressure is the driving force that causes the vapor to diffuse from the liquid surface into the air above. As long as the air currents above the liquid continually carry away the moisture being evaporated, evaporation will continue.

On the other hand, when the pan of water is placed inside the closed container (Fig. 5-9b), the pressure of the vapor mixed with the air in the container is continually increasing due to the water evaporating from the pan. Eventually, *vapor pressure equilibrium* will be established when the vapor pressure in the air and of the water are equal and no further evaporation takes place. When evaporation ceases, the dew-point temperature of the air inside the container will be equal to the temperature of the water in the pan. Under these conditions the air is said to be saturated.

From this, it is apparent that if the temperature of the water is greater than the air dew point, evaporation takes place; and when the air dew point is equal to the water temperature, evaporation ceases.

The evaporation of a liquid requires that the latent heat of vaporization be supplied. As a result of the water evaporation into the air, the water remaining tends to cool as it surrenders the necessary latent heat from vaporization. In the preceding discussion of the pan of water, it has been assumed that heat would flow from the surrounding air into the water, as no insulation was indicated. As a result, even though the water tends to cool slightly, its temperature remains nearly constant because of heat flow into it from the surrounding air.

5-6 HEAT EXCHANGE BETWEEN AIR AND WATER†

When air is brought in contact with water at a temperature different from the air wet-bulb temperature, an exchange of heat, as well as moisture, will take place between the air and water.

1. If the water temperature is higher than the air wet-bulb temperature, the water temperature will drop and the air wet-bulb temperature will rise because the water surrenders heat to the air.
2. If the water temperature is lower than the air wet-bulb temperature, the water temperature will rise and the air wet-bulb temperature will drop.
3. In any exchange of heat between water and air, the water temperature can never fall, or rise, to the initial air wet-bulb temperature.
4. Whenever an exchange of heat occurs between air and water, the temperature of both must change.

There is one *important exception* to item 4 above. In *air washers,* to be discussed later, where the water is continually recirculated being neither heated nor cooled externally, the temperature of the spray water does not change because there is no addition or removal of heat from the system. The latent heat required for evaporation of water comes from the sensible heat in the air, which causes a reduction in air dry-bulb temperature.

In order to humidify air with water sprays, the temperature of the spray water must always be higher than the required final dew point of the air. Such an amount of water must be used that, as the water cools down from its initial temperature, its final temperature will still be above the required final air dew point. It is also necessary to remember that, besides the relationship between the final air dew-point temperature and the water temperature, the exchange of heat and moisture between air and water depends upon the air dry-bulb and wet-bulb temperatures.

5-7 METHODS OF HUMIDIFICATION†

If you are humidifying a hospital operating room, obviously your design criteria are different than those for humidifying a textile mill, an office building, a laboratory, or a home. Different types of operations have substantially different requirements for the achievement of proper relative humidity. These requirements determine what means of humidification you should use.

Three basic humidifying media are available: steam, evaporative pan, and water spray. Each has particular advantages and limitations which determine its suitability for a particular

†Information extracted from *Trane Air Conditioning Manual,* The Trane Company, LaCrosse, Wisconsin, 1965, pp. 69–76, with permission.

†Extracted from *The Armstrong Humidification Handbook,* Armstrong Machine Works, Three Rivers, Michigan, pp. 12–15, with permission.

application. A fourth medium, wetted element humidifiers, applies primarily to private residences or similar small buildings where very low humidifying capacities are required (see Figs. 5-10, 5-11).

Steam is ready-made water vapor that needs only to be mixed with the air. With evaporative pan humidification, air flows across the surface of heated water in a pan and absorbs the water vapor. Both steam and evaporative pan humidification are essentially *isothermal* processes where little change in air temperature occurs.

Water spray humidification is an *adiabatic* process if evaporation of the water is caused by the water absorbing sensible heat from the air reducing the air dry-bulb temperature. The latent heat of evaporation reduces the sensible heat of the air by about 1000 Btu for each pound of water evaporated.

Let us examine the basic requirements for effective humidification and see how each of the humidifying media satisfies these requirements.

Capacity. Small-capacity requirements can be met most economically with evaporative pan or self-contained steam-generating unit humidifiers, installed either in the air-handling system or individually within the area(s) to be humidified. Larger-capacity needs, depending on the application, can best be met by either water spray or steam humidification systems, which may be installed either in nonducted industrial applications or in air-handling systems. As a rule, steam humidification can provide larger capacities than water spray units.

Maintenance. Steam humidification systems are largely maintenance free. Maintenance requirements for both water spray and evaporative pan humidification depend on mineral content of the water. With water spray equipment minerals dispersed in the water mist settle out as dust when the droplets evaporate. Spray nozzles can become clogged with mineral deposits and require regular cleaning to maintain capacity. Similarly, evaporative pan humidifiers are subject to "liming up," particularly on the heating coils in the pan, although chemical additives automatically or manually added to the water in the pan can reduce this problem by as much as 50 percent.

Response to control. Since steam is ready-made water vapor, it needs only to be mixed with the air to satisfy the demands of the system. Response to control is much faster than with either water spray or evaporative pan where evaporation must take place before humidified air can be circulated.

Control of output. Operation of the three basic types of systems is regulated by a humidity controller (hygrostat) located either in the space humidified or in the return air duct of an air duct system. Water spray and evaporative pan equipment operate intermittently in response to control. Output for these types of humidifiers is determined by size and number of spray nozzles or by water temperature and surface area, respectively. Steam humidifiers can meter the output by means of a modulating control valve in response to the humidity controller, by positioning the humidifier operator anywhere from closed to fully open. Steam humidifiers can thus respond more precisely to system demand than can the other two types.

Sanitation. The high temperatures inherent in steam humidification make it virtually a sterile medium. Assuming boiler makeup water is of satisfactory quality, and there is no condensation, dripping or spitting in the air ducts, no bacteria or odors will be disseminated with steam humidification. Evaporative pan humidifiers can sustain bacteria colonies in the reservoir and distribute them throughout the humidified space. High water temperatures, water treatment, and regular cleaning and

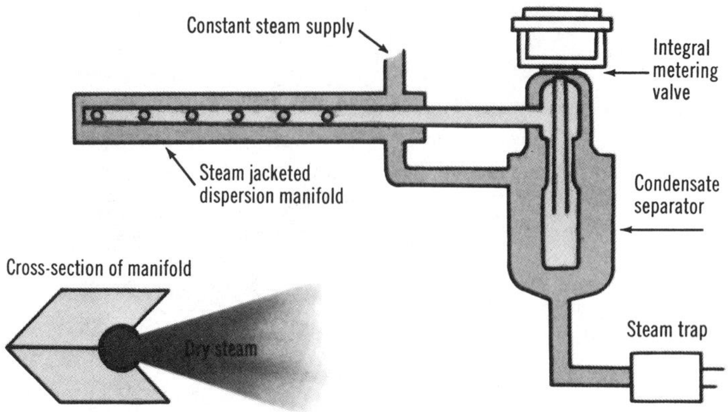

FIG. A-13 — STEAM METHOD

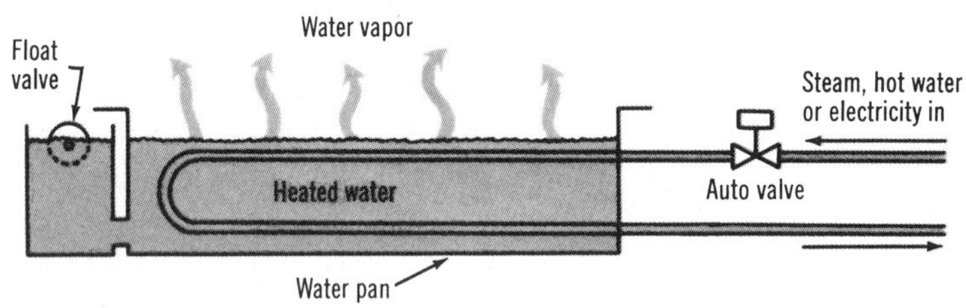

FIG. B-13 — EVAPORATIVE PAN METHOD

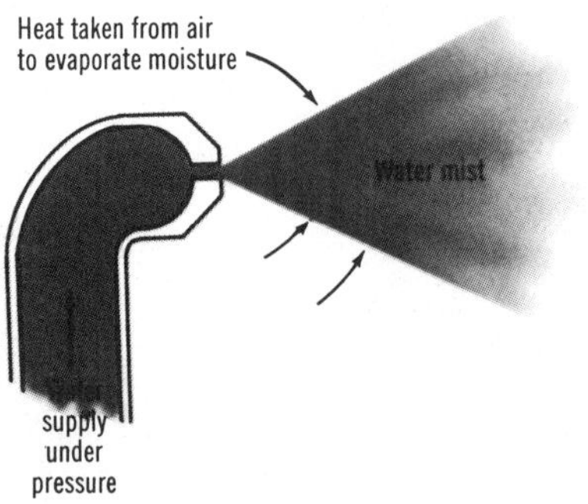

FIG. C-13 — WATER SPRAY METHOD

Figure 5–10

Humidification methods: (a) using dry steam (b) Using evaporated water (c) Using spray water

flushing of the humidifier help to minimize the problem, however. Water spray systems can distribute large amounts of bacteria and can present odor problems in air-handling systems. Drip from spray nozzles and unevaporated water discharge can collect in ducts, around drains and drip pans, and on eliminator plates, encouraging the growth of algae and bacteria.

Corrosion. Whenever liquid moisture is present in a humidification system, corrosion can be a problem. Since steam and evaporative pan humidifiers are designed to discharge only water vapor, corrosion in ductwork is not a problem. Evaporative pan units are subject to corrosion and deterioration, the severity of which depends on materials of construction and frequency of cleaning. Steam units are largely self-cleaning; scale or sediment, whether formed in the unit or entrained in the supply steam, are drained from the humidifier through the drain trap. Water spray humidifiers and all ductwork, eliminator plates, and so forth in proximity to them are highly susceptible to corrosion.

Costs. Evaluation of costs in selecting a humidification system should include installation, operating and maintenance costs, as well as first cost. However, total humidification costs should run far less than heating or cooling system costs.

First costs, of course, vary with the size of the units. Priced on a capacity basis, larger-capacity units are the most economical, regardless of the types of humidifier; that is, one humidifier capable of delivering 1000 lb of humidification per hour costs less than two 500-lb/hr units of the same type.

Of the three basic humidification media, steam humidifiers will provide the highest capacity per first-cost dollar; water spray is next, and evaporative pan is the least economical, assuming capacity needs are 75 lb/hr or more.

Installation costs for the three types cannot be accurately formulated, as availability of water, steam, electricity, and so forth, close to the humidifiers can vary greatly between jobs. Operating costs are similar between the three types, although they are insignificant compared to heating and cooling system costs.

Maintenance costs vary widely with water spray and to a lesser degree with evaporative pan equipment, requiring substantially more maintenance than steam humidification. Table 5-1 summarizes the capabilities of each type.

RECOMMENDED APPLICATIONS:

Steam. Recommended for virtually all commercial, institutional, and industrial applications. Where steam is not available, small-capacity needs up to 50 to 75 lb/hr can be met best using self-contained steam-generating units; above this capacity range, central system steam humidifiers are most effective and economical. Steam should be specified with caution where humidification is being used in small, confined areas to add large amounts of moisture to hygroscopic materials.

Evaporative pan. Recommended only as an alternative to self-contained steam-generating unit humidifiers for small commercial or institutional applications. Generally not recommended where load requirements exceed 50 to 75 lb/hr.

Water spray. Recommended for industrial applications where evaporative cooling is required; typical application is summer time humidification of textile mills in southern United States.

A fourth medium, mentioned previously, is wetted element humidifiers. These units are recommended for private residences, or similar small buildings where very low humidifying capacities are required. Figure 5-11 represents a humidifying device which exposes a large wetted surface to an airstream. The unit shown would be mounted in the warm airstream in a duct system. The unit consists of a series of

Table 5-1

Comparison of Humidification Methods*

	Steam	Evaporative Pan	Water Spray
Effect on air temperature	Virtually no change	Small temperature rise	Substantial temperature drop (if no heating of spray water)
Unit capacity per unit size	Small to very large	Small	Small
Maintenance frequency	Annual	Weekly to monthly	Weekly to bimonthly
Response to control	Immediate	Slow	Slow
Control of output	On-off or full range modulation	On-off	On-off
Sanitation/Corrosion	Sterile medium; corrosion free	Pan subject to corrosion; bacteria can be present	Subject to severe corrosion and bacteria problems
Cost: Price (per unit of capacity)	Low	High	Low to medium
Installation	Varies with availability of steam, water, electricity, etc.		
Operating	Low	Low	Low
Maintenance	Low	High	Very high

**Reproduced from* The Armstrong Humidification Handbook, *Armstrong Machine Works, Three Rivers, Michigan, with permission.*

rotating, alloy metal, screened disks rotated at slow speed (2 rpm) by a small electric motor drive. The disks rotate in a bath of water in the pan where water is picked up and held in the mesh of the screen by surface tension. The wetted disks rotate into the airstream, presenting a large surface area for water evaporation. The disks rotate only when humidification is called for by the humidity control system.

Figure 5-12 represents a type of humidifier which breaks up a stream of water into a fine mist by throwing it against a breaker. Operation is quite simple. The motor section, sealed against water, drives both disk and pump tube. Water is drawn from the reservoir through the pump tube and is placed on top of the rotating disk. The water is cast centrifugally against the breaker comb, which in turn atomizes the water. Primary air enters the disk and picks up the water particles and produces partial evaporation. This partially vaporized water is then introduced either into the space or the duct where the surrounding air completes the evaporation process.

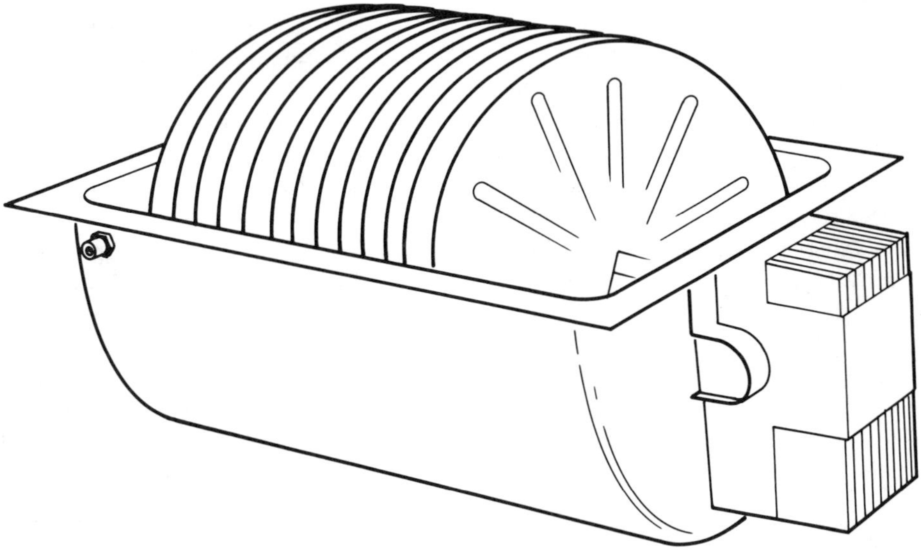

Figure 5–11

Wetted element humidifier, HUMID-AIRE (Courtesy of Hamilton Humidity, Inc.)

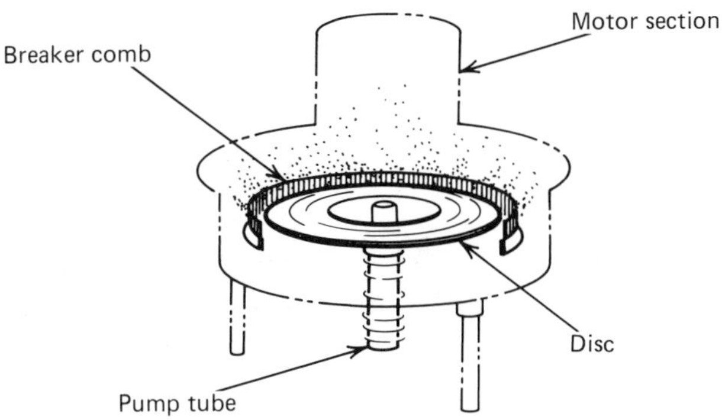

Figure 5–12

The Walton Atomizer (Courtesy Walton Laboratories, Inc.)

Illustrative Problem 5-6

An evaporative pan type humidifier is to be used in a duct system to humidify the air supplied to a building. The supply air upstream of the humidifier has a dry-bulb temperature of 90°F and a wet-bulb temperature of 58°F. The downstream air leaving the pan humidifier must have a humidity ratio of 70 grains/lb da. The quantity of air flowing through the duct upstream of the humidifier is 8000 cfm. If the water in the pan is maintained at 212°F by a steam immersion coil, determine the following: (a) moisture addition to the air and (b) resulting psychrometric properties of the air downstream from the humidifier.

Solution:

Step 1: Sketch the equipment.

Step 2: Determine initial condition of air upstream of humidifier. Locate point 1 on psychrometric chart, Fig. 5-14 at $t_1 = 90°F$, $t_{w1} = 58°F$. From the chart find $W_1 = 21$ grains/lb da, $h_1 = 24.9$ Btu/lb da, and $v_1 = 13.92$ ft³/lb da.

Step 3: Determine final condition of air downstream from humidifier. Given in the problem, $W_2 = 70$ grains/lb da. The final air condition must lie on the constant humidity ratio line W_2. The quantity of air passing through system is from Eq. (4-7),

$$\dot{m}_a = \frac{\text{cfm} \times 60}{v_1} = \frac{8000 \times 60}{13.92}$$

$$\dot{m}_a = 34{,}480 \text{ lb/hr}$$

The quantity of moisture added to *each pound* of air is $W_2 - W_1 = 70 - 21 = 49$ grains/lb da or 0.007 lb vapor/lb da. The heat added to each pound of air by the evaporating water is $(W_2 - W_1)h_w$, where h_w is the enthalpy of the evaporating water; in this case $h_w = h_g$ at 212°F and 14.696 psia and is 1150.5 Btu/lb steam.

$$\text{Heat added to air/lb air} = (49/7000)(1150.5)$$
$$= 8.05 \text{ Btu/lb air}$$

The enthalpy of air leaving the humidifier must be $h_2 = h_1 +$ enthalpy addition $= 24.9 + 8.05 = 32.15$ Btu/lb da.

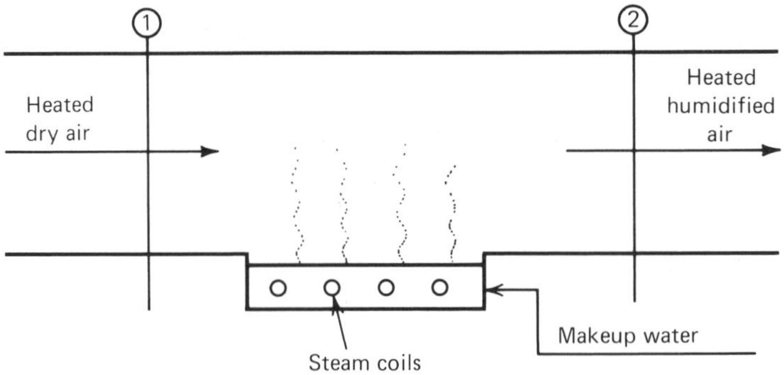

Figure 5–13

Schematic: Open-pan humidifier (Illustrative problem 5-6)

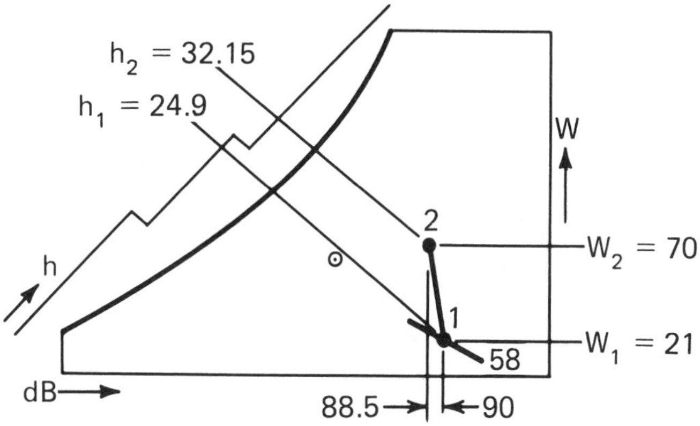

Figure 5-14

Solution for Example 5-6

With h_2 = 32.95 Btu/lb da and W_2 = 70 grains/lb da on Fig. 5-14, find t_2 = 91°F, t_{w_2} = 68.8°F. *(ans)*

A straight line on Fig. 5-14 joining points 1 and 2 represents the humidifying process.

Step 4: Total moisture added to the air. The total moisture added to the air determines the manufacturer's rating of the humidifier, and for this problem is $\dot{m}_a(W_2 - W_1)$.

Pounds/hr moisture

$$= 34{,}480\left(\frac{70 - 21}{7000}\right) = 241 \text{ lb/hr} \quad \textit{(ans)}$$

Illustrative Problem 5-7

Atomizing spray-type humidifiers (similar to Walton atomizers—Figure 5-12) are to be installed in a space within a building where the total air circulation rate is 8000 cfm and the inside air is 70°F DB and 50 percent RH. Fresh outside air is to be heated and introduced to the space at the rate of 1500 cfm. The heated outside air has a humidity ratio of 11 grains/lb da. The space is heated by steam convection units. Determine the amount of moisture to be added to the space air to maintain the desired space temperature and relative humidity.

Solution:

Step 1: Establish known conditions for points on psychrometric chart, Figure 5-15. The space air condition (ra) may be established at 70°F DB and 50 percent RH. At this point find $t_{w_{ra}}$ = 58.5°F WB, W_{ra} = 55 grains/lb da, and v_{ra} = 13.51 ft³/lb da. Also, establish the humidity ratio of the outside ventilation air (oa) as W_{oa} = 11 grains/lb da.

Step 2: Construct the space wet-bulb temperature line on the chart and extend downward and to the right until it intersects with the humidity ratio of the outside air (W_{oa} = 11 grains). This point of intersection is the dry-bulb temperature to

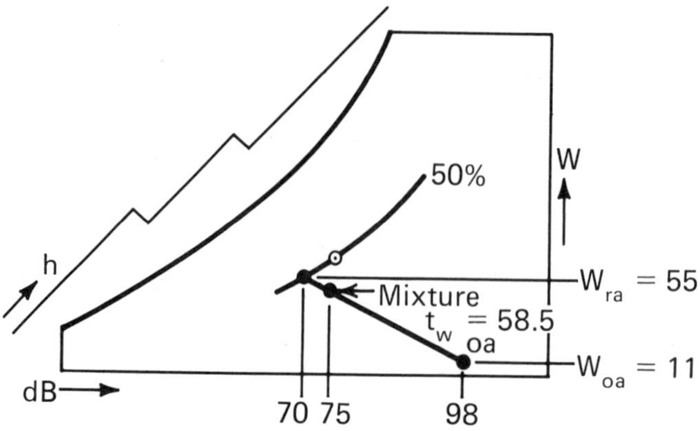

Figure 5–15

Solution for Example 5-7

which the outside air must be preheated before introducing to the space, t_{oa} = 98°F DB, $t_{w_{oa}}$ = 58.5 WB, and v_{oa} = 14.09 ft³/lb da.

Step 3: The outside air at the condition found in Step 2 is mixed with space air found in Step 1:

$$t_{mix} = \frac{\dot{m}_{a_{ra}} \times t_{ra} + \dot{m}_{a_{oa}} \times t_{oa}}{m_{a_{ra}} + m_{a_{oa}}}$$

$$\dot{m}_{a_{ra}} = \frac{6500 \times 60}{13.51} = 28{,}868 \text{ lb/hr}$$

$$\dot{m}_{a_{oa}} = \frac{1500 \times 60}{14.09} = 6388 \text{ lb/hr}$$

$$t_{mix} = \frac{28{,}868 \times 70 + 6388 \times 98}{28{,}868 + 6388}$$

$$t_{mix} = 75°\text{F}$$

In Fig. 5-15 find the point of intersection between t_{mix} = 75°F DB and $t_{w_{mix}}$ = 58.5 WB. This is the condition of the space without humidification. The dry-bulb temperature of the space is to be 70°F; therefore, the 5°F difference (75 − 70) cooling of the air represents the sensible heat required to evaporate the moisture from the humidifiers.

Step 4: The moisture addition to the space may be found from

$$\text{Moisture addition} = \dot{m}_{a_{oa}} (W_{ra} - W_{oa})$$

$$= 6388 \left(\frac{55 - 11}{7000}\right)$$

$$\text{Moisture addition} = 40.15 \text{ lb/hr}$$

Latent heat required to evaporate water is equal to the mass of water times the latent heat of evaporation, or h_{fg} at 58.5°F, which is 1060.5 Btu/lb (Table A-1).

$$\text{Latent heat } (q_L) = 40.15 \times 1060.5 = 42{,}580 \text{ Btu/hr}$$

This latent heat is supplied by the space air as it cools to space conditions from t_{mix} = 75°F.

$$q_s = q_L = 42{,}580 = 0.244\,(35{,}256)(\Delta t)$$
$$\Delta t = 4.95°\text{F}$$

This checks the 5°F value found in Step 3.

5-8 COMBINED HEATING AND HUMIDIFICATION OF AIR

During cold weather it may be necessary to provide for both heating and humidification of air to be circulated in a building to maintain comfort conditions. The heated air must be introduced into the space to be conditioned at a sufficiently high dry-bulb temperature and humidity ratio to offset heat losses and to maintain a satisfactory humidity condition.

Figure 5-16 is a typical heating and ventilating air-handling unit. The unit consists of several sections: a mixing box section where return air from the conditioned space is mixed with outside air for ventilation in a fixed or variable ratio which is adjusted by damper position, a heating coil and humidifier section, and a fan section. The unit shown is called a draw-through unit because of the location of the fan with respect to the other components.

Figure 5-17 is a generalized sketch of an air-handling system where moist air enters at point 1 and leaves at point 2. As the air passes through the system, it may have sensible heat and/or moisture added. The system is a steady-flow system and the following equations apply:

$$\dot{m}_{a_1}h_1 + {}_1q_{s_2} + \dot{m}_w h_w = \dot{m}_{a_2}h_2 \qquad \textit{(5-18)}$$

$$\dot{m}_{a_1} = \dot{m}_{a_2} \qquad \textit{(5-19)}$$

$$\dot{m}_{a_1}W_1 + \dot{m}_w = \dot{m}_{a_2}W_2 \qquad \textit{(5-20)}$$

where m_w is the pounds of water evaporated, or the pounds of vapor (steam) added per hour, and h_w is the corresponding enthalpy of the moisture added in Btu/lb of moisture.

Since $\dot{m}_{a_1} = \dot{m}_{a_2}$, we divide Eq. (5-18) by $\dot{m}_a$ and we have

$$h_1 + \frac{{}_1\dot{q}_{s2}}{\dot{m}_a} + \frac{\dot{m}_w h_w}{\dot{m}_a} = h_2 \qquad \textit{(5-21)}$$

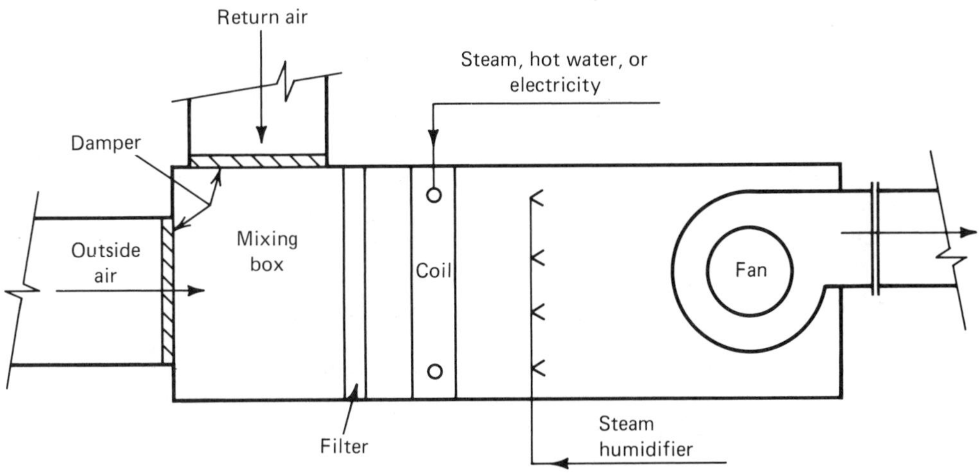

Figure 5–16

Schematic: Typical heating, humidifying unit

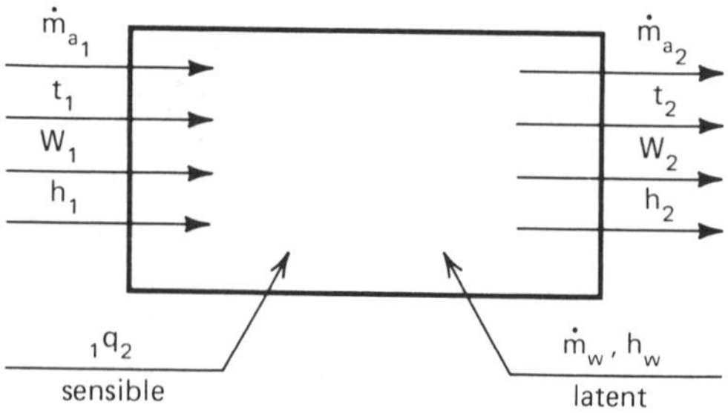

Figure 5–17

Schematic: Device for heating humidifying air

Also, from Eqs. (5-18) and (5-19) we have

$$ {}_1\dot{q}_{s_2} = \dot{m}_a(h_2 - h_1) - \dot{m}_w h_w \qquad (5\text{-}22)$$

and from Eq. (5-20) we have

$$\dot{m}_w = \dot{m}_a(W_2 - W_1) \qquad (5\text{-}23)$$

Substitution of Eq. (5-23) into Eq. (5-21) gives

$$h_1 + \frac{{}_1\dot{q}_{s2}}{\dot{m}_a} + (W_2 - W_1)h_w = h_2 \qquad (5\text{-}24)$$

Equation (5-24) is the general equation which states that the enthalpy of the air leaving, h_2, is equal to the enthalpy of the air entering, h_1, plus the sensible heat added plus the latent heat added by the moisture.

If the process were adiabatic (${}_1q_{s_2} = 0$), then Eq. (5-24) becomes

$$h_1 + (W_2 - W_1)h_w = h_2 \qquad (5\text{-}25)$$

In the process represented by Eq. (5-25), ideally the air entering at point 1 is brought to its "temperature of adiabatic saturation," commonly known as the "thermodynamic wet-bulb temperature" (see Section 4-14). The process represents cooling of the air at *constant wet bulb* as the water absorbs sensible heat from the air to provide the latent heat for vaporization. This is the process used to cool and humidify air with recirculated spray water in an air washer with no external heating or cooling of the spray water. The process is called *evaporative cooling,* to be discussed later.

By solving Eq. (5-23) for $\dot{m}_a$ and substituting into Eq. (5-22) and rearranging terms, we have

$$\frac{\Delta h}{\Delta W} = \frac{h_2 - h_1}{W_2 - W_1} = \frac{{}_1\dot{q}_{s2}}{\dot{m}_w} + h_w$$

$$= \frac{{}_1\dot{q}_{s2} + \dot{m}_w h_w}{\dot{m}_w} \qquad (5\text{-}26)$$

Equation (5-26) defines the enthalpy-humidity ratio where:

1. The rate of air flow has disappeared.
2. The ratio $\Delta h/\Delta W$ is the slope of a line on the psychrometric chart related only to the total space load and the moisture added.
3. If the condition of the air at point 1 or point 2 is set, the values of h and W for the other point may be allowed to vary but must always be on a straight line on the psychrometric chart having a slope of $\Delta h/\Delta W$.

Some psychrometric charts include an alignment quadrant which gives various values of $\Delta h/\Delta W$ which provides a means for determining the direction of heating and humidifying process line on the chart. The following example illustrates the process.

Illustrative Problem 5-8

Moist air at 40°F DB and 35°F WB enters a heating and humidifying device at the rate of

3000 cfm at a barometric pressure of 29.92 in. Hg. While passing through the chamber, the air absorbs sensible heat at the rate of 153,000 Btu/hr and picks up 83 lb/hr of saturated steam at 230°F. Determine the dry-bulb and wet-bulb temperatures of the air leaving the device.

Solution (Method 1):

(See Figure 5-18)

Step 1: Determine psychrometric properties of air at initial condition. On psychrometric chart (Fig. 5-18), at 40°F DB and 35°F WB find $v_1 = 12.64$ ft^3/da, $W_1 = 22$ grains/lb da, $h_1 = 13.0$ Btu/lb da.

Step 2: Determine mass of air flowing. From Eq. (5-7).

$$\dot{m}_{a_1} = \dot{m}_{a_2} = \frac{\text{cfm} \times 60}{v_1} = \frac{3000 \times 60}{12.64}$$

$$\dot{m}_a = 14{,}240 \text{ lb/hr} \quad \text{or} \quad \frac{14{,}240}{60}$$

$$= 237.3 \text{ lb/min}$$

Step 3: Determine psychrometric properties of air at the final condition. From Eq. (5-20)

$$W_2 = \frac{\dot{m}_a W_1 + \dot{m}_w}{\dot{m}_a} = W_1 + \frac{\dot{m}_w}{\dot{m}_a}$$

$$= \frac{22}{7000} + \frac{83}{14{,}240}$$

$$W_2 = 0.00898 \text{ lb vapor/lb da}$$

or

$$W_2 = 0.00898 \times 7000$$

$$= 62.86 \text{ grains/lb da} \qquad \textit{(ans)}$$

From Eq. (5-21)

$$h_2 = h_1 + \frac{{}_1\dot{q}_{s2}}{\dot{m}_a} + \frac{\dot{m}_w h_w}{\dot{m}_a}$$

From Table A-1 h_w at 230°F = h_g at 230°F = 1157.1 Btu/lb steam

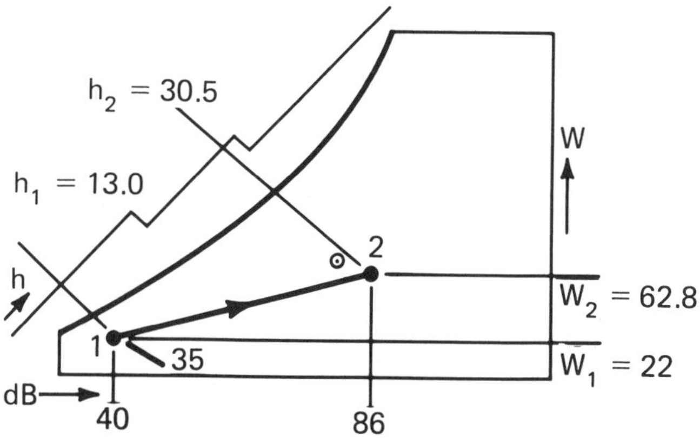

Figure 5–18

Solution for Example 5-8 (Method No. 1)

$$h_2 = 13.00 + \frac{153{,}000}{14{,}240} + \frac{83 \times 1157.1}{14{,}240}$$

$$h_2 = 13.00 + 10.74 + 6.74$$

$$h_2 = 30.48 \text{ Btu/lb da} \qquad \textit{(ans.)}$$

NOTE: In the above solution for h_2, the value 10.74 represents the sensible heat added, and the value 6.74 represents the latent heat added per pound of air passing through the device.

Step 4: Locate the final air condition on psychrometric chart. Draw a horizontal line on psychrometric chart for W_2 = 62.86 grains/lb da. Draw an inclined line on chart for h_2 = 30.48 Btu/lb da to intersect W_2 line. This point of intersection is the final air condition leaving the device, point 2. At point 2, read the dry-bulb temperature as 86°F and the wet-bulb temperature as 65.8°F.

Step 5: Connect point 1 and 2 by a straight line. This line represents the process of heating and humidification within the device.

<u>*Solution: (Method 2):*</u> (See Figure 5-19.)

Step 1: Determine the sensible-heat ratio (SHR) for the heating and humidifying process. From the solution of Method 1 above, it was determined that the sensible-heat addition to the air was q_s = 10.74 Btu/lb da and the latent heat addition was q_L = 6.74 Btu/lb da. The SHR is defined as the sensible heat divided by the total heat for a process. For this problem

$$\text{SHR} = \frac{q_s}{q_s + q_L} = \frac{10.74}{10.74 + 6.74} = 0.614$$

Step 2: Plot the SHR on the psychrometric chart. In Fig. 5-19 at a point where the 78°F DB line intersects with the 50 percent RH curve, you will note a small circle. This is the alignment circle for the SHR scale located on the upper right of the Fig. 5-19. Using a straight-edge, draw a construction

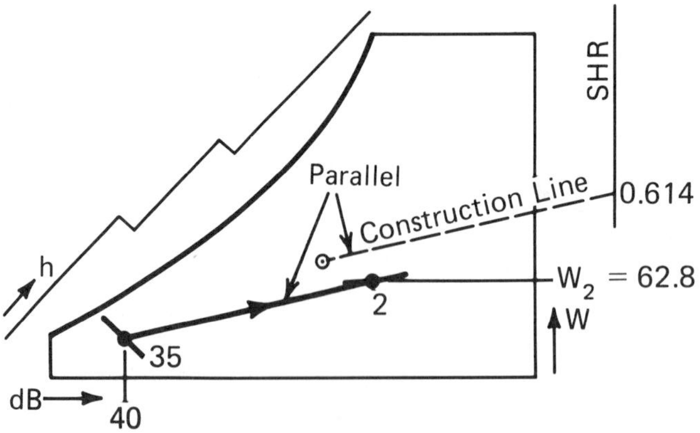

Figure 5–19

Solution for Example 5-8 (Method No. 2)

line between the alignment circle and a SHR value of 0.614. Parallel to this construction line draw the process line through the known point 1 at 40°F DB and 35°F WB. Point 2 must be on this process line.

Step 3; Determine location of point 2. Referring again to Method 1 solution, Step 3, the humidity ratio of the air leaving the device was W_2 = 62.86 grains/lb da. Determine the point of intersection of W_2 and the process line drawn in Step 2 above. This determines point 2 on the psychrometric chart, and we can read the dry-bulb temperature of 84°F and wet-bulb temperature of 65.2°F.

NOTE: The small variation in the results obtained is quite possible with the graphical-type solution. When using the SHR scale, extreme care must be used if a high degree of accuracy is to be expected.

<u>*Solution: (Method 3):*</u> (See Figure 5-20.)

After the development of Eq. (5-26) it was indicated that some psychrometric charts provide an alignment quadrant to identify the enthalpy-humidity ratio ($\Delta h/\Delta W$) for a heating and humidifying process. Figure 5-20 is a schematic drawing of such a chart and quadrant.

Step 1: Determine the enthalpy-humidity ratio for the process. By Eq. (5-26),

$$\frac{\Delta h}{\Delta W} = \frac{{}_1\dot{q}_{s2}}{\dot{m}_w} + h_w$$

$$= \frac{153{,}000}{83} + 1157.1$$

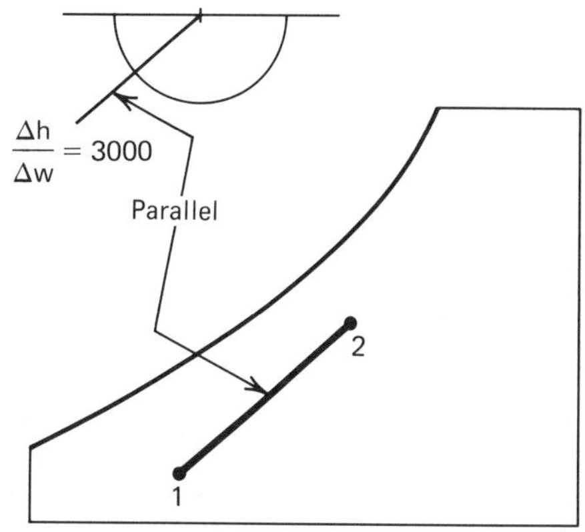

Figure 5–20

Alignment quadrant used with some psychrometric charts (See illustrative problem 5-8, Method 3).

$$\frac{\Delta h}{\Delta W} = 3004 \quad \text{(say 3000)} \qquad \textbf{\textit{(ans)}}$$

Step 2: On the quadrant draw a construction line from its center through 3000 on the scale. Parallel to this line, draw the heating and humidifying process line on the psychrometric chart through the known conditions at point 1.

Step 3: Determine the position of point 2 on the process line by using either Eq. (5-20) or (5-21)

5-9 HEATING AND HUMIDIFICATION USING SPRAY EQUIPMENT

Frequently, spray water equipment is used with or without various arrangements of heating coils to preheat and/or reheat the air passing through the spray chamber. The spray chamber is frequently called an *air washer* because it performs the secondary task of cleaning, removing dust and other particulate from the airstream. However, for our discussion at this point, we will be concerned primarily with the spray chamber as a device to humidify the air.

In Sections 5-5 and 5-6 there was a discussion of the behavior of air and water in contact with each other. Humidification of air by spraying water into it is a direct application of the principles of evaporation and heat exchange between air and water when in contact with each other.

In most instances air washers are used in industrial applications where the device performs the two objectives of humidification and cleaning of the air. Figure 5-21 shows a sectional view of a large, central station washer. It consists of a spray chamber in which a number of spray nozzles and risers are installed. The washer illustrated has one bank of spray nozzles. Sometimes two, and occasionally three, banks are installed. Nozzles used for air washers ordinarily have a capacity of approximately 1¼ to 2 gpm per nozzle. The quantity of water delivered by each bank can be varied by installing a larger or smaller number of nozzles. The water is forced through the nozzle by pumping at pressures from 20 psi to 40 psi to obtain good atomization of the water. The

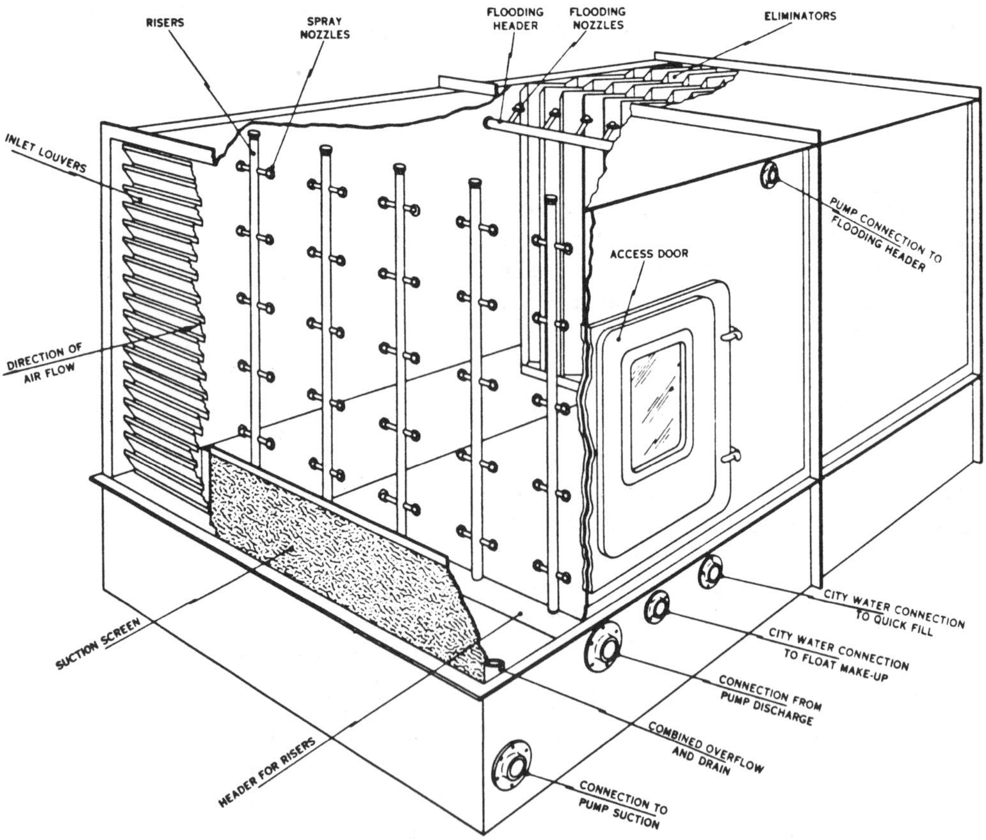

Figure 5–21

Industrial Tube Air-Washer (Courtesy of The Trane Company.)

washer housing is a rectangular steel casing, closed at top and sides and mounted on a shallow watertight tank base. Inlet baffles distribute the entering air uniformly through the chamber, and at the leaving end eliminators remove the entrained water droplets in the airstream preventing so-called carryover. The eliminators also pick up dirt particles. The eliminator plates are so designed and installed so that the direction of air flow is changed a number of times while flowing through them. In this way the air impinges against the wet surfaces of the plates and the dust particles and water droplets are deposited on the plates. The plates are washed down by a continuous stream of water from flooding nozzles. Air washers will not always remove greasy particles and soot. Tobacco smoke will ordinarily pass through an air washer.

Some odors can be removed from the air passing through an air washer. An odor is usually due to the vapor of some compound mixed with the air. Many of these vapors will dissolve in water. Of course, as the water becomes saturated with the soluble vapors, it becomes less and less able to remove odors from the air. Furthermore, the water itself acquires an odor due to the material in solution; therefore, under these conditions it must be changed frequently or continuous overflow used.

Air washers are designed to be installed on the suction side of the fan. They should never be installed on the discharge side unless they have been specially built for this purpose. There will be no difficulty with water leaking through the joints of the washer installed on the suction side of a fan. However, if a washer is installed on the discharge side of a fan, the air pressure developed by the fan is likely to cause water leakage through the joints, in spite of rubber gaskets usually installed.

If a set of eliminator plates is installed after

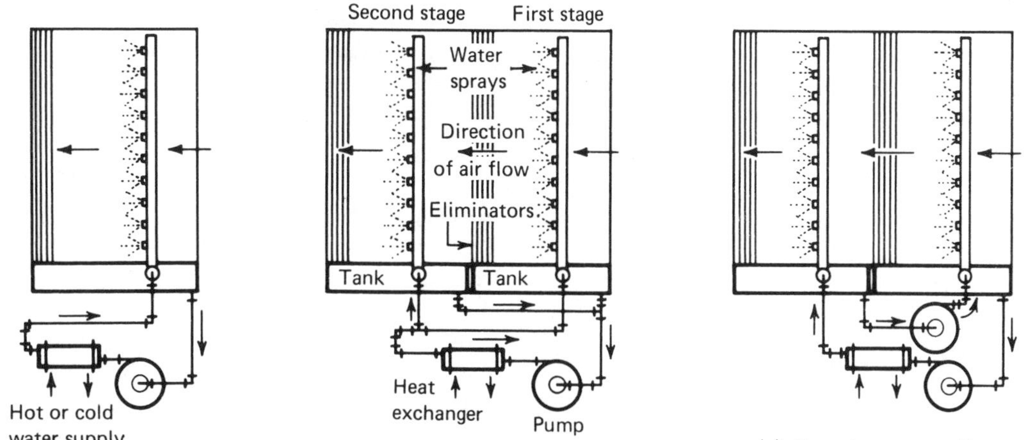

(a) Single bank air washer

(b) Two-stage air washer

(c) Two-stage counterflow air washer

Figure 5–22

Types of air washers

each bank of sprays, a washer having two banks is known as a two-stage washer. A two-stage washer is illustrated in Fig. 5-22b and c. Sometimes a two-stage washer may have two or three banks of spray nozzles in each stage. Ordinarily, the water in air washers is recirculated. When the air is to be humidified, either the air or the water, or sometimes both, are heated before entering the washer. If the air is to be cooled and dehumidified in the washer, the water must first be cooled. Heating or cooling of the water can be accomplished by supplying either hot or chilled water to the heat exchanger shown in the spray water piping of Fig. 5-22.

The total cross-sectional flow area of air washers is usually based on an air velocity of 500 fpm. If higher air velocities are used, trouble may be experienced with entrained moisture being carried over from the washer.

Some washers, smaller in size than the industrial type, are used in built-up air-handling units with various arrangements of heating and tempering coils, as well as means for heating or cooling the spray water used in the washer. Figure 5-23 is a schematic drawing of such a built-up air-handling unit. The arrangement is general, and only the components actually required to perform the heating and humidifying process would be included. On the sketch is shown the variation in air properties as the air passes through the unit.

There are several common arrangements that are used:

1. Preheat the air prior to its contact with recirculated spray water, followed by reheating of the air.
2. Heat the spray water only.
3. Moderately preheat the air then pass it through heated spray water.
4. Pass the air through spray washer using recirculated water with no preheating or reheating of air.

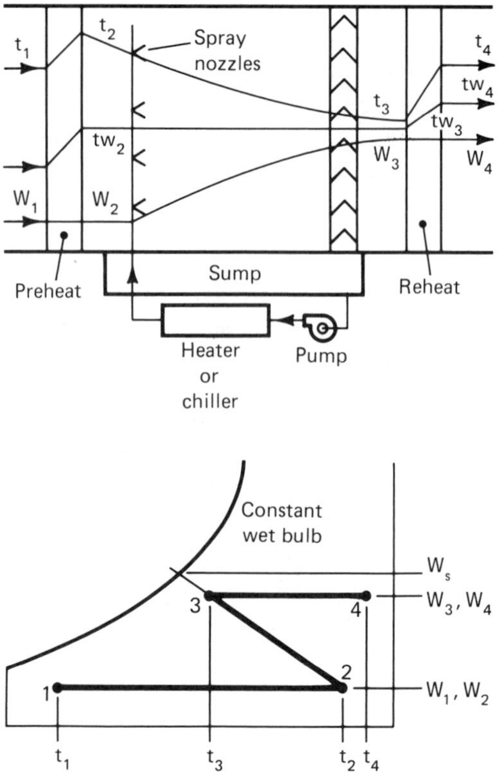

Figure 5–23

Schematic: Heating and humidifying devices using spray washer

Arrangement 1 normally uses recirculated spray water having no external heating and uses a reheating coil to heat the humidified air to a final temperature that may be required to provide sensible heating of a building. Arrangement 2 is seldom used if the incoming air temperature is expected to be below 32°F at any time. Arrangement 3 is used for special cases when arrangement 2 cannot meet the requirements. Since it involves additional equipment and piping, it is a more expensive system. Arrangement 4 is the familiar process of adiabatic saturation of air when the air dry-bulb temperature is lowered at constant wet bulb as some of the water evaporates to humidify the air. In hot, dry climates a device of this type becomes a very simple, useful unit for summer cooling.

Referring again to Fig. 5-23, as the air is preheated, its dry- and wet-bulb temperatures increase, but the humidity ratio remains constant. Passing through the spray chamber, the wet-bulb temperature remains constant as the dry-bulb temperature decreases and the humidity ratio increases. The reheating process increases the dry-bulb and wet-bulb temperatures at constant humidity ratio. Therefore, we have three elementary processes in series, a sensible-heating process, an adiabatic saturation process, and another sensible-heating process.

Humidifying efficiency (frequently called effectiveness) is a measure of the completeness of the humidifying process using a spray chamber. This efficiency is dependent on several factors:

a. number of spray nozzles, and banks of nozzles, and the direction of the sprays with respect to the air flow direction
b. production of a fine water mist by the nozzles, which depends on the nozzle design, size and water pressure
c. air velocity through the chamber
d. ratio of mass of air to water flow
e. time of contact between water and air, or actually the length of the unit

Under favorable conditions, nearly complete saturation of the air at the air wet-bulb temperature may be obtained. The amount of moisture added to the air depends on the entering air wet bulb and on the spray chamber design. Commercial air washers are less than 100 percent effective. That is, complete saturation of the air is not obtained. The saturation effectiveness will normally range from 80 to 90 percent. Referring to Fig. 5-23, the saturation efficiency is defined as

$$E_h = \frac{\text{Actual change in air dry-bulb temperature}}{\text{Maximum possible change in dry bulb}}$$

or

$$E_h = \frac{t_2 - t_3}{t_2 - t_w} \times 100 \qquad (5\text{-}27)$$

also

$$E_h = \frac{W_3 - W_2}{W_s - W_2} \times 100 \qquad (5\text{-}28)$$

where t_2 and t_3 are the dry-bulb temperatures of the air entering and leaving the spray chamber, respectively; t_w is the constant wet-bulb temperature for the process; W_2 and W_3 are the humidity ratios of the entering and leaving air, respectively; and W_s is the humidity ratio of the air saturated at the wet-bulb temperature.

Illustrative Problem 5-9

Five thousand cfm of air at 40°F DB and 30°F WB are to be heated and humidified to a final condition of 90°F DB and 30 percent RH. The process of heating will be done by using a preheating coil and a reheating coil arranged as shown in Fig. 5-23. The humidification process is to be done by using a spray chamber with recirculated unheated water. Assume that the humidifying process will have an effectiveness of 82 percent. Determine the following: (a) the psychrometric properties of the air entering and leaving the spray washer, (b) the amount of preheat and reheat required in

Btu/hr, and (c) the amount of water evaporated into the air in lb/hr.

Solution: (See Fig. 5-24.)

Step 1: Locate initial and final air condition on the psychrometric chart.

Initial Condition (1)	***Final Condition (4)***
t_1 = 40°F DB	t_4 = 90°F DB
t_{w_1} = 30°F WB	t_{w_4} = 67.2°F WB
W_1 = 10 grains/lb da	W_4 = 63 grains/lb da
h_1 = 11.2 Btu/lb da	h_4 = 31.6 Btu/lb da
v_1 = 12.62 ft^3/lb da	v_4 = 14.04 ft^3/lb da

Step 2: Construct the horizontal *preheat* line through point 1 and extend to the right. The air condition entering spray chamber must be on this line.

Step 3: Construct the horizontal *reheat* line through point 4 and extend to the left. The air condition leaving spray chamber must be on this line.

Step 4: Determine the psychrometric properties of the air entering and leaving the chamber. The humidifying process is assumed to have an effectiveness of 82 percent, therefore, with $W_4 = W_3$ and $W_2 = W_1$ Eq. (5-28) becomes

$$E_h = \frac{W_4 - W_1}{W_s - W_1}$$

where W_4 is the humidity ratio of the air leaving the spray chamber. Solving Eq. (5-28) for W_s, we have

$$W_s = \frac{W_4 - W_1}{E_h} + W_1$$

$$= \frac{63 - 10}{0.82} + 10$$

$$W_s = 74.6 \text{ grains/lb da}$$

In Fig. 5-24 locate W_s of 74.6 on humidity ra-

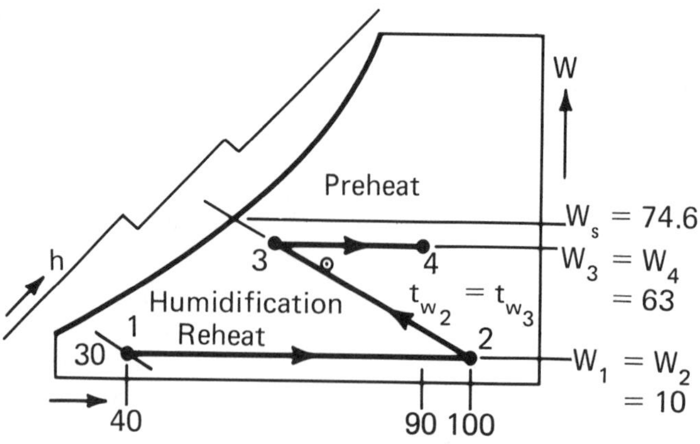

Figure 5–24

Solution for Example 5-9

tio scale and extend horizontal line to the left to intersect the 100 percent RH line. This point locates the wet-bulb temperature for the humidifying process as 59.2°F. Where this constant wet-bulb line intersects the preheat process line and the reheat process line will be the entering and leaving air condition, respectively, for the humidifying process, points 2 and 3. From Fig. 5-24 then

Point 2	Point 3
$t_2 = 100°F$ DB	$t_3 = 66.6°F$ DB
$t_{w_2} = 59.2°F$ WB	$t_{w_3} = 59.2°F$ WB
$W_2 = 10$ grains/lb da	$W_3 = 63$ grains/lb da
$h_2 = 25.6$ Btu/lb da	$h_3 = 25.8$ Btu/lb da
$v_2 = 14.13$ ft³/lb da	$v_3 = 13.44$ ft³/lb da

Step 5: Determine the quantity of preheat and reheat. From Eq. 4-7

$$\dot{m}_a = \frac{\text{cfm} \times 60}{v_1} = \frac{5000 \times 60}{12.62}$$

$$\dot{m}_a = 23{,}772 \text{ lb/hr} \qquad \textit{(ans)}$$

Remember this mass flow rate is *constant* throughout the flow system. The *volume* flow rate (cfm) changes because the specific volume is changing. By Eq. 5-4,

$$_1\dot{q}_{s_2} = \dot{m}_a\,(h_2 - h_1) = 23{,}772(25.6 - 11.2)$$

$$_1\dot{q}_{s_2} = 342{,}300 \text{ Btu/hr} \quad \text{(preheat)} \qquad \textit{(ans)}$$

By Eq. 5-4

$$_3\dot{q}_{s_4} = \dot{m}_a(h_4 - h_3) = 23{,}772(31.6 - 25.8)$$

$$_3\dot{q}_{s_4} = 137{,}800 \text{ Btu/hr} \quad \text{(reheat)} \qquad \textit{(ans)}$$

Step 6: Mass of water evaporated. Each pound of air absorbs $W_3 - W_2$ grains of moisture; therefore, by Eq. (5-23)

$$\dot{m}_w = \dot{m}_a(W_3 - W_2) = 23{,}772\left(\frac{63 - 10}{7000}\right)$$

$$\dot{m}_w = 179.9 \text{ lb/hr} \quad \text{(evaporated)} \qquad \textit{(ans)}$$

5-10 HEATING AND HUMIDIFYING AIR WITH HEATED SPRAY WATER

As noted in Section 5-9, air may be heated and humidified by passing it through a spray chamber where the spray water is heated externally. The initial spray water temperature must be high enough so that as it contacts the air and transfers its heat to the air and cools, the final air temperature must be above the initial air temperature. The water temperature can fall only to the wet-bulb temperature of the leaving air (ideally). The changes in the water temperature and air temperature within the spray chamber depend on the relative masses of air and water circulated.

If both the air and water temperature are changing, the process line on the psychrometric chart is not a straight line. The path of the process is a pursuit curve of the water temperature which is represented on the saturation curve (100 percent RH). To draw the pursuit curve on a psychrometric chart, it is necessary to use the following relationship.

Heat absorbed by air = heat given up by water

or

$$\dot{m}_a(h_B - h_A) = \dot{m}_w c_p(t_{A'} - t_{B'}) \qquad \textit{(5-29)}$$

where

$\dot{m}_a$ = lb/hr of air flow through spray chamber

$\dot{m}_w$ = lb/hr of water sprayed

h_A and h_B = air enthalpy values from psychrometric chart for two points, *A* and *B*

$t_{A'}$ and $t_{B'}$ = spray water temperatures (in °F)

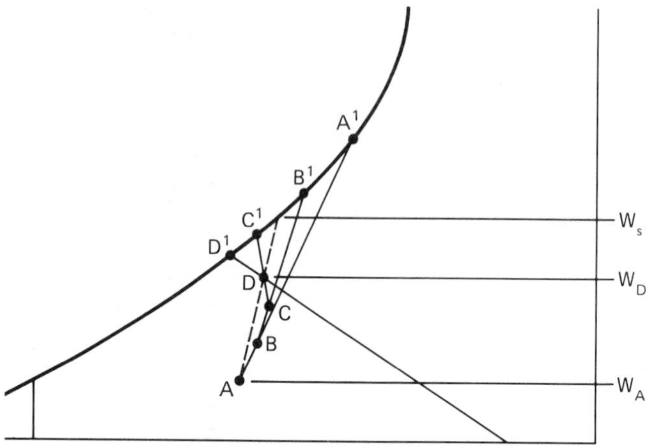

Figure 5–25

Construction of pursuit curve (heating and humidifying with heated spray water

c_p = specific heat of water, assumed constant and equal to unity

Figure 5-25 represents a procedure for the development of the pursuit curve on the psychrometric chart. To draw the curve, the initial air condition, point A, and the initial water temperature, point A', are plotted and then connected by a straight line. Select point B on this line a short distance from A. The water temperature at point B' may then be calculated from Eq. (5-29) with known values of $\dot{m}_a$ and $\dot{m}_w$. Connect points B and B' with a straight line. Next, select a point such as C a short distance from B and again use Eq. (5-29) to calculate the water temperature at point C'. The procedure is continued until the final water temperature, D', is approximately equal to the wet-bulb temperature of the air passing through D. A straight line drawn through point A and the final air dry bulb, point D, continued to intersect the saturation curve (shown dashed) defines the humidifying efficiency (effectiveness) of the process. For Fig. 5-25 the value of E_h would be $(W_D - W_A)/(W_s - W_A)$ expressed as a decimal.

This method of plotting the pursuit curve is time consuming, and a simpler solution of such problems using heated spray water may be obtained by the use of Fig. 5-26. This graph is based upon the assumption that the heat absorbed by the air is equal to the heat given up by the water. Also, the final water temperature and final air wet-bulb temperature are the same. Actually, the final water temperature will be higher than the final air wet-bulb temperature. Nevertheless, this chart enables many problems involving washers to be solved easily by first finding the theoretical solution and then applying a correction factor (CF).† The correction factor is determined by

†*Trane Air Conditioning Manual,* The Trane Company, LaCrosse, Wisconsin, 1965, p. 227.

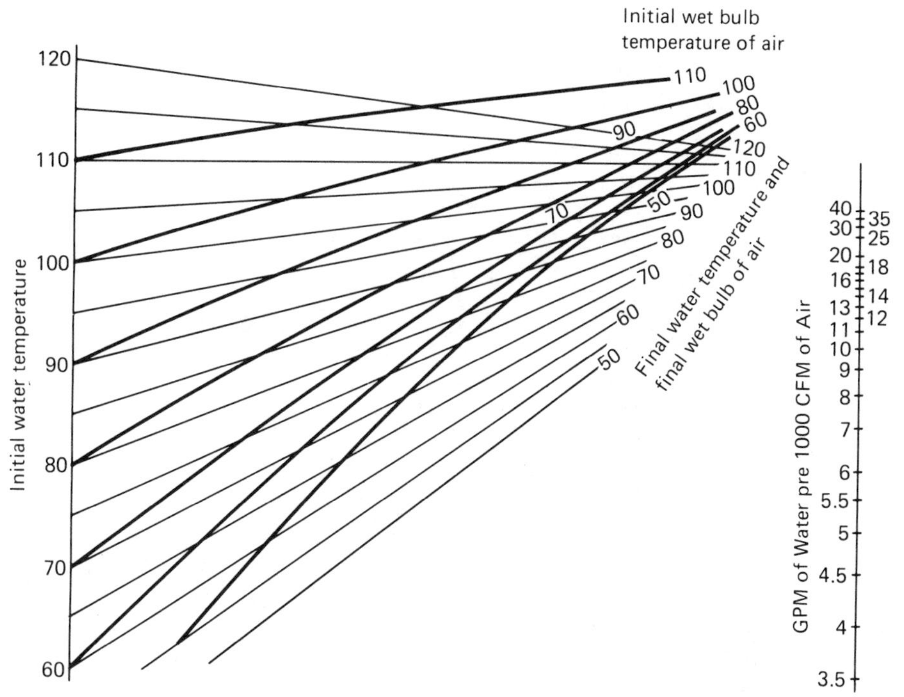

Figure 5–26

Spray Water Air-Washer Chart for Humidification (Courtesy of The Trane Company.)

$$CF = \frac{\text{Actual change in air enthalpy}}{\text{Theoretical change in air enthalpy}}$$

or

$$CF = \frac{\Delta h \text{ actual}}{\Delta h \text{ theoretical}} \qquad (5\text{-}30)$$

In effect, this correction factor corresponds to the humidifying efficiency, E_h, determined by Eqs. (5-27) and (5-28).

Illustrative Problem 5-10‡

An air quantity of 15,000 cfm enters a washer at 70°F DB and 50°F WB. The washer is supplied with 150 GPM of water at 90°F. For a correction factor of 0.85 determine the water temperature in the tank and the condition of the air leaving the washer.

Solution: (See Fig. 5-27.)

Step 1: Determine final water temperature

‡*Trane Air Conditioning Manual,* The Trane Company, LaCrosse, Wisconsin, 1965, p. 232.

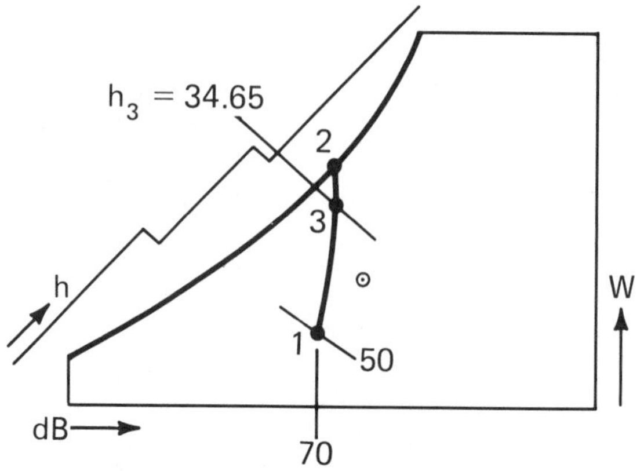

Figure 5–27

Solution to Problem 5-10

and theoretical final air wet-bulb temperature. The water-to-air ratio is 150/15 = 10 gpm/1000 cfm of air. Using Fig. 5-26, set the right end of the straight-edge on the right scale at 10 gpm/1000 cfm; set the left end of the straight-edge on the initial water temperature of 90°F on left scale. Where a line between these two points crosses the initial air wet bulb of 50°F, read the final water temperature of 73.5°F, which is also the *theoretical* final air wet-bulb temperature.

Step 2: Determine actual leaving air condition from washer. In Fig. 5-27 at entering air condition of 70°F DB and 50°F WB find $h_1 = 20.2$ Btu/lb da. Theoretically, the air could be heated and humidified to saturation at 73.5°F, and line 1-2 represents the process through the washer. The enthalpy of saturated air at 73.5°F is $h_2 = 37.2$ Btu/lb da. Theoretical change in enthalpy is $h_2 - h_1 = 37.2 - 20.2 = 17.0$ Btu/lb da. The actual change in enthalpy is the theoretical change times the correction factor. Actual enthalpy change $= 17.0 \times 0.85 = 14.45$ Btu/lb da. The actual enthalpy of air leaving washer will be $h_3 = 20.2 + 14.45 = 34.65$ Btu/lb da. In Fig. 5-27 construct a constant-enthalpy line at 34.65 and extend to intersect line 1-2 at point 3. At point 3 find $t_3 = 73.5$°F, $t_{w3} = 70.7$°F. ***(ans)***

A water-to-air ratio of 5 gpm per 1000 cfm is common for a single-bank washer, while approximately 10 to 15 gpm per 1000 cfm are usual ratios for two- and three-bank washers. Initial water temperature will vary within the range of 60°F to 120°F depending upon the inlet and outlet air temperatures. Ordinarily, the operator has the following items he can vary to control the performance of the washer:

a. inlet air temperature by changing the amount of heating medium to the preheat coil

b. inlet water temperature by controlling the heating capacity of heat exchanger

c. water quantity by taking a bank of nozzles out of service.

Illustrative Problem 5-11

An air-conditioning unit for an industrial building must provide 4500 cfm of outside air entering a preheat coil at 5°F DB and 50 percent RH. The preheat coils heat the air to 40°F DB. The outside air then mixes with 13,500 cfm of recirculated air at 70°F DB and 40 percent RH. The mixed air passes through a washer supplied with 90 GPM of water at 110°F and has a correction factor of 0.83. The humidified air is then reheated to a dry-bulb temperature of 115°F before entering the supply fan which delivers it to a duct system for distribution within the building. Determine (a) the psychrometric properties of the air as it flows through the conditioning equipment and (b) if the heat exchanger supplying heat to the spray water were turned off so that the water in the washer were recirculated without any heat being added, what would be the properties of the air through the conditioning equipment?

Solution: (Part a): (See Figs. 5-28 and 5-29.)

Step 1: Make a sketch of the mechanical arrangement of the equipment, Fig. 5-28. (Circled numbers establish points where air conditions are to be established.)

Step 2: Determine the properties of the preheated outside air (point 2).

The properties of the outside air from low-temperature psychrometric charts are listed in the table below. Since the preheating process is a sensible-heating process, it is carried out at constant humidity ratio $W_1 = W_2$.

Outside Air (1)	**Preheated Outside Air (2)**
$t_1 = 5°F$	$t_2 = 40°F$ (given)
$\phi_1 = 50$ percent	$\phi_1 = 10$ percent
$v_1 = 11.72$ ft³/lb da	$v_2 = 12.60$ ft³/lb da
$W_1 = 3.50$ grains/lb da	$W_2 = 3.50$ grains/lb da
$h_1 = 1.7$ Btu/lb da	$h_2 = 10.2$ Btu/lb da

Step 3: Determine the amount of preheat required. By Eq. 5-7,

$$\dot{m}_{a_1} = \dot{m}_{a_2} = \frac{60 \times \text{cfm}}{v_1} \text{ lb/hr}$$

$$\dot{m}_{a_2} = \frac{60 \times 4500}{11.72} = 23{,}038 \text{ lb/hr}$$

$$= 384 \text{ lb/min}$$

By Eq. 5-4,

$$_1\dot{q}_{s_2} = m_{a_2}(h_2 - h_1)$$

$$= 23{,}038(10.2 - 1.7)$$

$$_1\dot{q}_{s_2} = 195{,}820 \text{ Btu/hr}$$

NOTE: If approximate Eq. (5-9) had been used, the result would be

$$_1\dot{q}_{s_2} = (1.10 \text{ cfm})(t_2 - t_1)$$

$$= 1.10(4500)(40 - 5)$$

$$_1\dot{q}_{s_2} = 173{,}250 \text{ Btu/hr} \quad \text{(approximate)}$$

The difference of about 22,500 Btu/hr may be accounted for by the fact that the constant

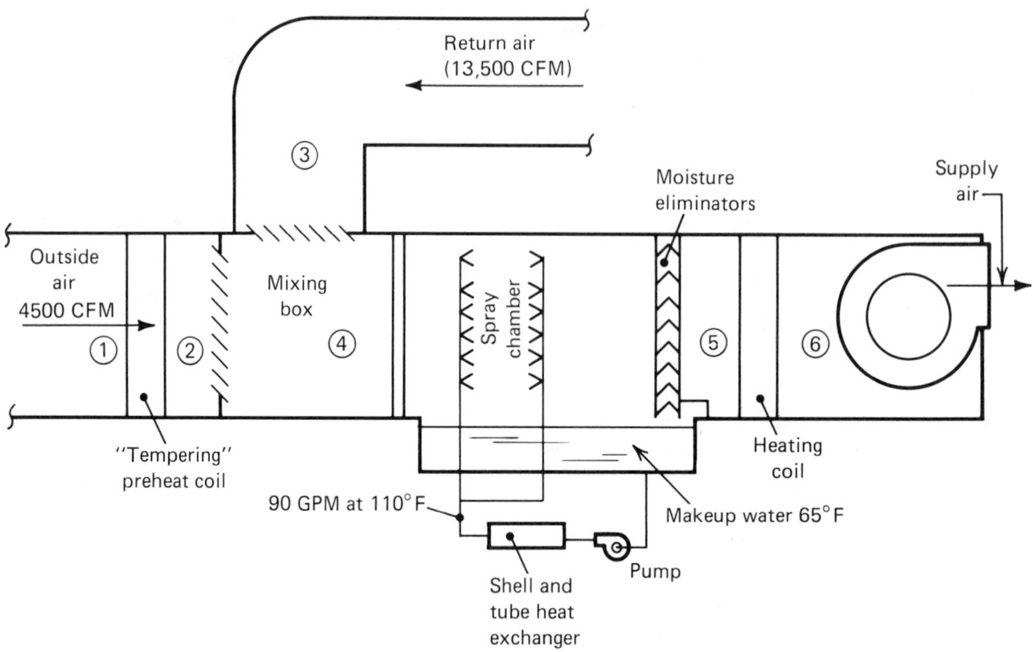

Figure 5–28

Schematic arrangement of one of the two units of Illustrative Problem 5-11

in Eq. (5-9) was determined by considering standard air ($v = 13.34$ ft^3/lb). The actual specific volume of the outside air in this problem is 11.72 ft^3/lb. The exact Eq. (5-4) should *always* be used when the actual specific volume is significantly different from standard.

Step 4: Determine mixture condition of preheated outside air and return air.

In Fig. 5-29 locate point 3 at 70°F DB and 40 percent RH, find $v_3 = 13.48$ ft^3/lb da. By Eq. (5-7),

$$\dot{m}_3 = \frac{60 \times \text{cfm}_3}{v_3} \text{ lb/hr}$$

$$= \frac{60 \times 13{,}500}{13.48}$$

$$\dot{m}_3 = 60{,}089 \text{ lb/hr}$$

$$= 1001.5 \text{ lb/min} \quad \text{(say 1000 lb/min)}$$

By Eq. (5-14)
Mixture temperature

$$(t_4) = \frac{\dot{m}_2 \times t_2 + \dot{m}_3 \times t_3}{\dot{m}_2 + \dot{m}_3}$$

$$t_4 = \frac{(384)(40) + (1000)(70)}{1384}$$

$$t_4 = 61.6°\text{F}$$

NOTE: The mixture temperature (t_4) could also be approximated by use of Eq. (5-17).

$$t = \frac{\text{cfm}_2 \times t_2 + \text{cfm}_3 \times t_3}{\text{cfm}_2 + \text{cfm}_3}$$

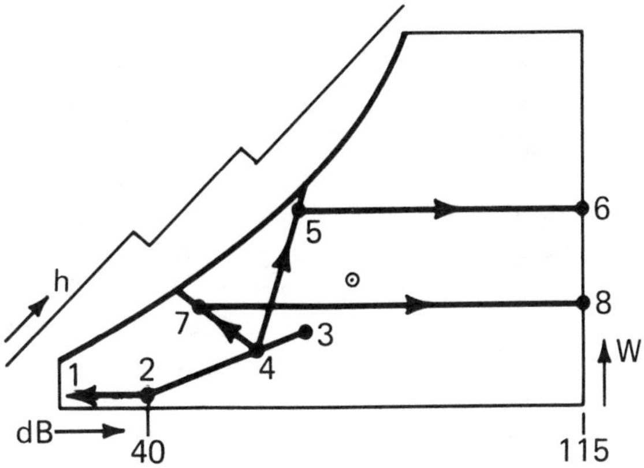

Figure 5–29

Solution of Example 5-11a and 5-11b

$$\text{cfm} = \dot{m} \times v = 384 \times 12.60 = 4838 \text{ cfm}$$

then

$$t_4 = \frac{(4838)(40) + (13{,}500)(70)}{18{,}338}$$

$$t_4 = 62.08°\text{F} \quad (\text{use } t_4 = 62°\text{F})$$

In Fig. 5-29 draw a straight line connecting points 2 and 3. Where this line intersects the dry-bulb temperature of 62°F locate point 4 and find $v_4 = 13.25$ ft^3/lb da, $W_4 = 33$ grains/lb da, $h_4 = 20.0$ Btu/lb da, and $t_{w4} = 49.8$°F.

Step 5: Determine the washer performance and calculate the leaving water temperature (t_{lw}), and the entering and leaving air conditions for the washer.

The entering air condition to the washer has already been determined, point 4, in Step 4. The volume of air entering the washer is

$$\text{cfm}_4 = \dot{m}_{a_4} \times v_4 = (1384)(13.25)$$
$$= 18{,}338 \text{ cfm}$$

The water-to-air ratio then is 90/18.338 = 4.9 gpm/1000 cfm.

Using Fig. 5-26 with water-to-air ratio of 4.9 and the inlet water temperature (t_{ew}) = 110°F, at $t_{w_4} = 49.8$°F, find

Final water temperature (t_{lw}) = 74°F

Theoretical final wet-bulb temperature of air leaving = 74°F.

The *actual* condition of the air leaving may be found by applying the correction factor 0.83.

On Fig. 5-29, locate t_{lw} = 74°F on the 100 percent relative humidity curve. At this point read the theoretical enthalpy of the air leaving as 37.6 Btu/lb da. The theoretical rise in air enthalpy is $37.6 - h_4 = 37.6 - 20.0 = 17.6$ Btu/lb da. The actual change in enthalpy would be (17.6)(0.83) = 14.6 Btu/lb da.

Therefore, the actual enthalpy of the air leaving would be 20.0 + 14.6 = 34.6 Btu/lb da, h_5.

Draw a straight line on Fig. 5-29 connecting point 4, the entering air condition, and the leaving water condition. Construct a constant-enthalpy line (h_5 = 34.6) to produce an intersection at point 5, the actual leaving air condition from the spray washer. Read t_5 = 72°F, t_{w_5} = 70.6°F, W_5 = 111 grains/lb da, and v_5 = 13.75 ft/lb da.

Step 6: Determine the final air properties after reheating (point 6).

The reheating process is a sensible-heating process at constant humidity ratio $W_6 = W_5$ = 111 grains/lb da. On Fig. 5-29 locate intersection of W = 111 and t_6 = 115°F and find the following properties of the reheated air: h_6 = 45.2 Btu/lb da, v_6 = 14.85 ft/lb da, and t_{w6} = 81.6°F.

Step 7: Determine the required amount of reheat. By Eq. (5-4),

$$_5\dot{q}_{s6} = \dot{m}_{a_5}(h_6 - h_5)$$

$$= (1384)(60)(45.2 - 34.6)$$

$$_5\dot{q}_{s6} = 880{,}200 \text{ Btu/hr}$$

Solution: (Part b) (Refer to Fig. 5-29)

All air properties up to and including point 4 are the same as found in part a.

Step 1: Determine air properties leaving spray washer when the spray water is unheated.

When no heat is added to the spray water, the water will quickly assume the wet-bulb temperature of the entering air (t_{w_4} = 49.8°F). The air condition as it passes through the spray chamber would follow the constant wet-bulb temperature of 49.8°F as the air is humidified and cooled.

Theoretically, the air would leave the spray washer saturated at a dry-bulb temperature of 49.8°F, or the drop in dry-bulb temperature would be 62 − 49.8 = 12.2 degrees. The correction factor, or humidifying effectiveness, must be applied in this case to the dry-bulb temperature change or the humidity ratio change. Therefore, the actual change in dry-bulb temperature would be (12.2)(0.83) = 10.1 degrees. Therefore, the actual dry-bulb temperature of the air leaving the washer would be 62 − 10.1 = 51.9°F = t_7.

On Fig. 5-29 locate the intersection of t_7 = 51.9°F and t_{w_7} = 49.8°F. Find h_7 = 20.1 Btu/lb da, W_7 = 50 grains /lb da, and v_7 = 13.04 ft^3/lb da.

Step 2: Determine final air properties after reheating.

The reheating process is a sensible-heating process at constant humidity ratio ($W_8 = W_7$ = 50 grains/lb da). On Fig. 5-29 draw a constant humidity ratio line from point 7 to intersect t_8 at 115°F. At point 8 find t_{w_8} = 72.2°F, h_8 = 35.5 Btu/lb da, and v_8 = 14.65 ft^3/lb da.

Step 3: Determine the amount of reheat required. By Eq. 5-4,

$$_7\dot{q}_{s8} = \dot{m}_{a_7}(h_8 - h_7)$$

$$= (1384)(60)(35.5 - 20.1)$$

$$_7\dot{q}_{s8} = 1{,}278{,}800 \text{ Btu/hr}$$

5-11 COOLING AND DEHUMIDIFICATION OF AIR

The preceding sections of this chapter have been concerned primarily with the heating and/or humidification of air to provide supply air to a conditioned space which would overcome heat losses and provide for satisfactory humidity conditions. We will now concern ourselves with cooling and dehumidification of air supplied to a space. This air would be used to overcome the heat gains and moisture gains of the space.

Most summer air-conditioning applications require the simultaneous removal of both sensible and latent heat quantities to maintain comfort conditions for people, or for maintaining stability of an industrial process. Determination of the required supply air quantity and its properties are the important steps in the satisfactory design of air-conditioning systems.

In Section 5-3 sensible cooling of moist air was discussed. However, as may be recalled, we were concerned at that time with the process of warm supply air providing the sensible heat to overcome heat loss from the conditioned space. We are now concerned with supplying air to a conditioned space at a sufficiently reduced dry-bulb temperature below the space design dry-bulb temperature to absorb the sensible heat gains to the space. The supply air must also be at a sufficiently reduced humidity ratio to absorb the latent heat gains to the space.

Figure 5-30a represents a typical cooling coil using chilled water in the tubes. If the chilled water circulated through the tubes is at a temperature low enough to maintain the air-side surface temperature of the coil below the dew-point temperature of the entering air, the properties of the air passing through the dehumidifying coil may be represented as shown in Fig. 5-30a. The dry- and wet-bulb temperatures, humidity ratio, and enthalpy of the entering air are reduced. The psychrometric process of cooling and humidification of the air from condition 1 to condition 2 is represented on Fig. 5-30b if all the air uniformly and perfectly contacts the chilled surface of the coil. The moisture could be condensed from the air at a variable temperature from the initial dew-point temperature (A) to the final saturation temperature (t_2). If the condensed moisture is brought to the final temperature (t_2) before draining from the system, we may write the following steady-flow energy and material-balance equations.

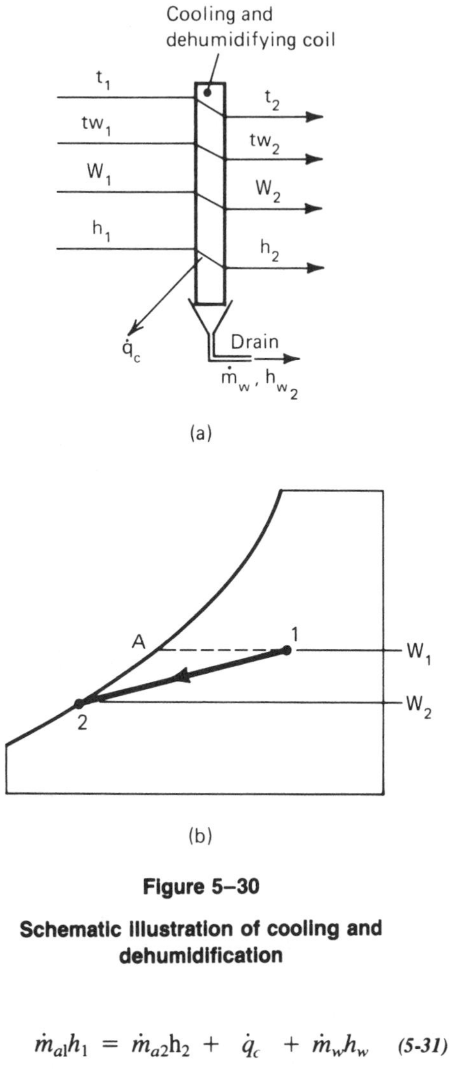

Figure 5–30

Schematic illustration of cooling and dehumidification

$$\dot{m}_{a1}h_1 = \dot{m}_{a2}h_2 + \dot{q}_c + \dot{m}_w h_w \quad (5\text{-}31)$$

$$\dot{m}_{a1} = \dot{m}_{a2} \quad (5\text{-}32)$$

$$\dot{m}_{a1}W_1 = \dot{m}_{a2}W_2 + \dot{m}_w \quad (5\text{-}33)$$

Note that these three equations are similar to Eqs. (5-18), (5-19), and (5-20) with the signs changed on $\dot{q}_s$ and $\dot{m}_w$. The process of cooling and dehumidification between points 1 and 2 on Fig. 5-30b may be represented by a straight line (1-2) on the psychrometric chart. There is no problem here because we are normally concerned with only the initial and final air condition. Combining Eqs. (5-31) and (5-32) we have

$$\dot{q}_c = \dot{m}_a (h_1 - h_2) - \dot{m}_w h_w \quad (5\text{-}34)$$

and from Eqs. (5-32) and (5-33) we have

$$\dot{m}_w = \dot{m}_a (W_1 - W_2) \quad (5\text{-}35)$$

In the above equations h_w would be the enthalpy of saturated water at the final air dew-point temperature or at a specified water temperature.

We may also look at the process of Fig. 5-30b and assume that air supplied to a space at condition 2 will absorb sensible and latent heat as it changes to condition 1. If condition 1 is the space design conditions to be maintained, then air may be supplied to the space at condition 2.

By using Eq. (5-9) with the temperature difference expressed as $t_1 - t_2$ and assuming various values of t_2, it is possible to determine the quantity of standard air required to absorb the sensible heat load; or the air quantity to be supplied may be arbitrarily selected and the required supply air temperature (t_2) may be calculated for a given sensible heat load. In either case it will be necessary to establish the humidity ratio (W_2) of the supply air if there is to be a simultaneous removal of latent heat as well as the sensible heat.

As may be seen in Fig. 5-30b, the humidity ratio of the supply air (W_2) is less than that of the conditioned space (W_1). At low partial pressures, approximately 1055 Btu are required to condense or evaporate 1 lb of water. Equation 5-35 gives the amount of moisture change in the air. Therefore, the latent heat ($\dot{q}_L$) associated with this moisture may be found by multiplying Eq. (5-35) by $h_{fg} = 1055$, or

$$\dot{q}_L = \dot{m}_w h_w = \dot{m}_a (W_1 - W_2) h_{fg}$$

or

$$\dot{q}_L = \dot{m}_a (W_1 - W_2)(1055) \quad (5\text{-}36)$$

If standard air is assumed with W_1 and W_2 expressed in units of grains/lb da, Eq. (5-36) becomes

$$\dot{q}_L = \frac{(1055)(0.075)(60)(\text{cfm})(W_1 - W_2)}{7000}$$

or

$$\dot{q}_L = 0.68\ \text{cfm}(W_1 - W_2) \quad (5\text{-}37)$$

where

$\dot{q}_L$ = latent heat absorbed by supply air from condition 2 to condition 1 (in Btu/hr)

cfm = supply air quantity at standard conditions

W_1 and W_2 = space and supply air humidity ratios, respectively, (grains/lb da)

Illustrative Problem 5-12

Assume that a given space has a calculated sensible heat gain of 50,000 Btu/hr and a latent heat gain of 12,500 Btu/hr with a space design condition of 78°F and 45 percent RH. Determine the supply air quantities and respective properties if the supply air temperature is assigned values of 55°F, 60°F, and 65°F.

<u>Solution:</u> (See Fig. 5-31)

Step 1: Using a supply air temperature $t_{sa} = 55$°F, determine the quantity of supply air. (Subscript

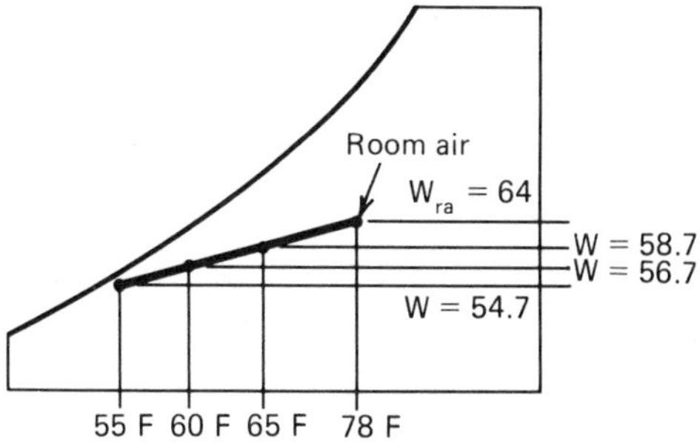

Figure 5–31

Schematic psychrometric chart for Illustrative Problem 5-12

"ra" refers to room air condition.) From Eq. (5-9)

$$\dot{q}_s = 1.10 \text{ CFM}_{sa} (t_{ra} - t_{sa})$$

or

$$\text{cfm}_{sa} = \frac{\dot{q}_s}{1.10(t_{ra} - t_{sa})} = \frac{50,000}{1.10(78 - 55)}$$

$$\text{cfm}_{sa} = 1976 \qquad \textit{(ans)}$$

Step 2: Determine the humidity ratio of the supply air.

Each pound of air supplied at 55°F must have a humidity ratio sufficiently depressed below the design humidity ratio of the space to absorb the latent heat load while rising to the space condition, or the vertical distance $(W_{ra} - W_{sa}) = \Delta W$. From Eq. (5-37)

$$\Delta W = W_{ra} - W_{sa} = \frac{q_L}{0.68 \text{ CFM}_{sa}}$$

$$= \frac{12,500}{0.68(1976)}$$

$$\Delta W = W_{ra} - W_{sa} = 9.3 \text{ grains/lb da}$$

From the psychrometric chart at t_{ra} = 78°F and RH_{ra} = 45 percent, find W_{ra} = 64 grains/lb da. Therefore, the required humidity ratio of the supply air will be

$$W_{sa} = W_{ra} - \Delta W = 64 - 9.3$$

$$W_{sa} = 54.7 \text{ grains/lb da}$$

Step 3: Plot the values of t_{sa} = 55°F and W_{sa} = 54.7 grains/lb da on Fig. 5-31.

Step 4: Perform similar calculations to determine the values of W_{sa} for t_{sa} = 60°F and 65°F. The results are shown in the table below.

Step 5: Plot the above corresponding values of t_{sa} and W_{sa} on Fig. 5-31.

By observing the plotted points on Fig. 5-31, it may be noted that a straight line may be drawn through them and, when extended to the right, will pass through the space design condition of 78°F DB and 45 percent RH. If this same line is extended to the left, it will intersect the 100 percent RH curve at a temperature of 49.6°F DB. This point is called the *room apparatus dew point* (t_{radp})

Supply Air Temperature	$t_{ra} - t_{sa}$	cfm_{sa}	$W_{ra} - W_{sa}$	W_{sa}
55	23	1976	9.3	54.7
60	18	2525	7.3	56.7
65	13	3496	5.3	58.7

It should be apparent that there could be an infinite number of possible supply air temperatures along this sloping line between 49.6°F and 78°F which would absorb the space sensible and latent heat loads simultaneously. Also, it may be noted by our calculations that, as the supply air temperature approaches the 100 percent RH curve, the required cfm_{sa} decreases and would reach a minimum when the air leaves the cooling device saturated at 49.6°F.

It should also be apparent from the preceding discussion that the supply air condition is dependent on the *slope* of the line on Fig. 5-31. While the method of solution used in this illustrative problem is quite workable, a more simple and direct method is normally used by air-conditioning engineers.

5-12 SENSIBLE HEAT RATIO (SHR)

In Section 5-8 and Problem 5-8 (Method 2) there was a brief discussion of the sensible heat ratio (sometimes called sensible heat factor, SHF). This ratio becomes a very useful tool for the solution of problems concerned with the cooling and dehumidification of air. In Problem 5-8 the sensible heat ratio was defined as

$$\text{SHR} = \frac{\dot{q}_s}{\dot{q}_s + \dot{q}_L} = \frac{\dot{q}_s}{\dot{q}_t} \qquad (5\text{-}38)$$

Using the same basic data of Problem 5-12 with the space condition at 78°F DB and 45 percent RH, and the supply air at a condition of 55°F DB, W_{sa} = 54.7 grains/lb da, we may replot these conditions on Fig. 5-32. Label these points 1 and 2, respectively. From the psychrometric chart the enthalpy values may be found as h_1 = 28.7 Btu/lb da and h_2 = 21.8 Btu/lb da. Construct the right triangle as shown, determine point 3, and find h_3 = 27.3 Btu/lb da.

The horizontal line 2-3 referred to the enthalpy scale defines the sensible heat capacity of the supply air. The vertical line 3-1 referred to the enthalpy scale defines the latent heat capacity of the supply air. The line 2-1 referred to the enthalpy scale defines total heat capacity of the supply air. Therefore,

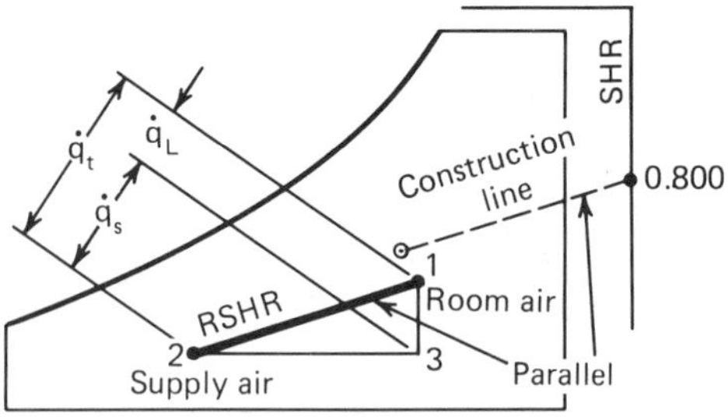

Figure 5–32

Schematic psychrometric chart illustrating sensible heat ratio (SHR)

The sensible heat capacity, $q_s = h_3 - h_2 = 27.3 - 21.8 = 5.5$ Btu/lb da

The latent heat capacity, $q_L = h_1 - h_3 = 28.7 - 27.3 = 1.4$ Btu/lb da

The total heat capacity, $q_t = h_1 - h_2 = 28.7 - 21.8 = 6.9$ Btu/lb da

The sensible heat ratio is

$$\text{SHR} = \frac{\dot{q}_s}{\dot{q}_s + \dot{q}_L} = \frac{5.5}{5.5 + 1.4}$$
$$= 0.797 \quad (\text{say } 0.800)$$

The same result is found by taking the ratio of the sensible and total heat loads of the space. Using room sensible heat (RSH) and room latent heat (RLH), we have

$$\text{RSHR} = \frac{\text{RSH}}{\text{RSH} + \text{RLH}}$$
$$= \frac{50{,}000}{50{,}000 + 12{,}500} = 0.800$$

where RSHR refers to the ratio of the room loads. The slight variation in the above two answers may be accounted for by the difficulty of reading accurate enthalpy values from the psychrometric chart.

Whenever the space sensible and latent heat loads are known, the slope of process line 1-2 may be determined by the sensible heat ratio. This *room sensible heat ratio* (RSHR) is plotted on the psychrometric chart through the room design conditions parallel to the reference line through the alignment circle (78°F DB and 50 percent RH) and the sensible heat ratio scale, as illustrated on Fig. 5-32.

From the above discussion, when the room sensible and latent loads are known, the numerical value of the room sensible ratio may be calculated. The sensible heat ratio scale and alignment circle on the psychrometric chart is used to establish the required slope of the RSHR line. The actual process line is drawn parallel to this reference line and through the room design conditions. *The required supply air conditions entering the space must be located on this line.*

Illustrative Problem 5-13

A room has a sensible heat gain of 76,000 Btu/hr and a RSHR of 0.85. The room is to be maintained at 80°F DB and 40 percent RH. If 3000 cfm of air is supplied to the room, find the required dry bulb and humidity ratio of the supply air.

<u>Solution:</u> (See Fig. 5-33.)

Step 1: Determine the location of the RSHR line on the psychrometric chart.

Construct the reference RSHR line on the psychrometric chart through the alignment circle and 0.85 on the SHR scale. Parallel to this reference line construct the RSHR line through the room design conditions of 80°F DB and 40 percent RH.

Step 2: Determine properties of the supply air. By Eq. 5-9,

$$\dot{q}_s = (1.10\ \text{cfm}_{sa})(t_{rm} - t_{sa})$$

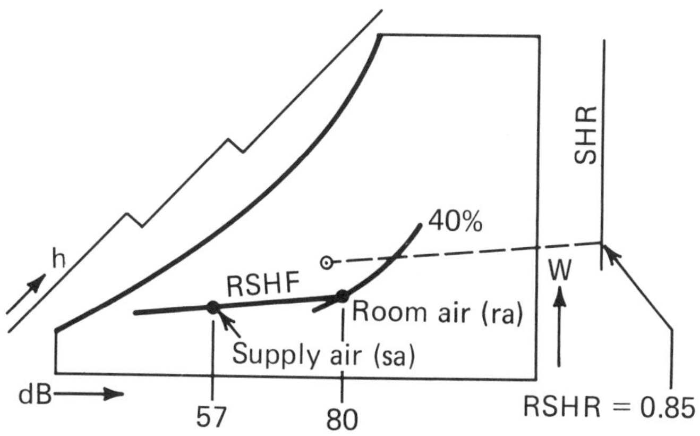

Figure 5–33

Solution to Problem 5-13

$$76{,}000 = 1.10(3000)(80 - t_{sa})$$

$$t_{sa} = 57°F$$

Locate the point where the dry-bulb temperature 57°F intersects the RSHR line. This is the condition of the supply air. Read W_{sa} = 55.5 grains/lb da, RH_{sa} = 80 percent, h_{sa} = 22.2 Btu/lb da, $t_{w_{sa}}$ = 53.4°F.

Illustrative Problem 5-14

A room within a building has a sensible heat gain of 42,000 Btu/hr and a latent heat gain of 14,000 Btu/hr. The room design conditions are 76°F DB and 63°F WB. Air is to be supplied to the room at 56°F DB. How much supply air must be used and what will be the supply air dew-point temperature?

Solution: (See Fig. 5-34.)

Step 1: Determine the RSHR and plot the RSHR line on the psychrometric chart. By Eq. 5-38,

$$\text{RSHR} = \frac{\dot{q}_s}{\dot{q}_s + \dot{q}_L}$$

$$= \frac{\text{RSH}}{\text{RSH} + \text{RLH}} = \frac{42{,}000}{42{,}000 + 14{,}000}$$

$$\text{RSHR} = 0.75$$

Step 2: Determine the dew-point temperature of the supply air. Determine the point where t_{sa} = 56°F intersects the RSHR line. From the chart at this point find t_{dp} = 50.5°F *(ans)*

Step 3: Determine supply air quantity. By Eq. 5-9,

$$\dot{q}_s = (1.10\ \text{cfm}_{sa})(t_{rm} - t_{sa})$$

$$42{,}000 = 1.10(\text{cfm}_{sa})(76 - 56)$$

$$\text{cfm}_{sa} = 1909 \quad \text{(say 1900 cfm)} \quad \textit{(ans)}$$

5-13 THE CONDITIONED AIR SUPPLY

Theoretically, the designer of an air-conditioning system may select for the condition of

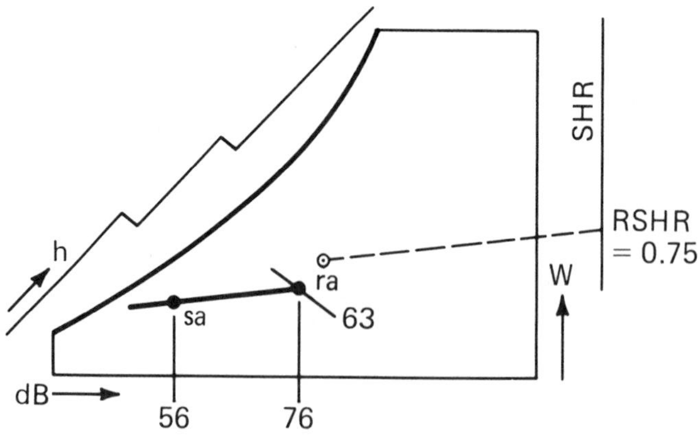

Figure 5–34

Solution to Problem 5-14

the supply air any combination of dry- and wet-bulb temperatures that intersect on the RSHR line. In practice, however, the combination of dry- and wet-bulb temperatures selected for the supply air must be one that can be obtained with the equipment used to cool the air. Air is ordinarily delivered to the conditioned room in the same state that it leaves the cooling equipment. The conditioning equipment selected must, therefore, be able to reduce the dry- and wet-bulb temperatures of the supply air to a point that will fall on the RSHR line for the room in question.

Air can be cooled to many different combinations of dry- and wet-bulb temperatures, the exact combination depending on the design of the cooling equipment. Equipment of the type ordinarily used tends to provide supply air having high relative humidities, in the order of 90 percent relative humidity, though lower or higher relative humidities are obtainable.

In any actual air-conditioning installation the sensible and latent heat gains can be calculated by methods presented in later chapters. The required dry- and wet-bulb temperatures to be maintained in the conditioned space are determined by the type of occupancy. So in every installation the problem resolves itself primarily into finding two quantities:

a. the amount of supply air required

b. the condition of the supply air

Illustrative Problem 5-15

A conditioned room is to be maintained at 75°F DB and 50 percent RH. The sensible heat gain of the room is 14,000 Btu/hr. The latent heat gain of the room is 3000 Btu/hr. If the characteristics of the cooling equipment are such that the air leaving it has a 90 percent RH, find the required dry- and wet-bulb temperatures of the supply air.

Solution:

Step 1: Determine RSHR

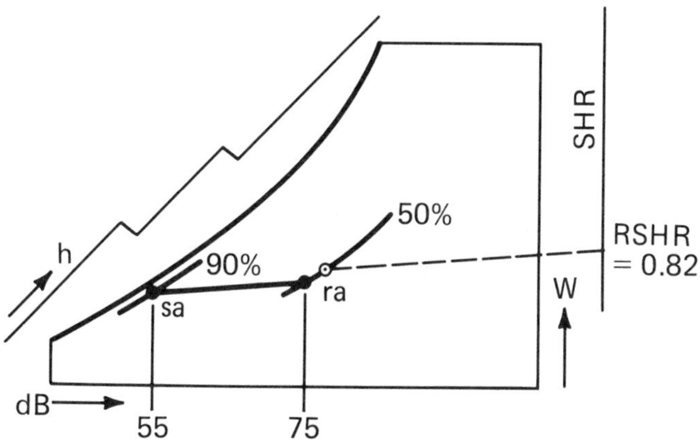

Figure 5–35

Solution to Problem 5-15

$$RSHR = \frac{RSH}{RSH + RLH} = \frac{14{,}000}{14{,}000 + 3000}$$

$$RSHR = 0.823 \quad \text{(say 0.82)}$$

Step 2: Construct RSHR line on psychrometric chart.

On Fig. 5-35 locate room design conditions at 75°F DB and 50 percent RH. Using the alignment circle and the SHR scale, construct the reference RSHR line. Parallel to this reference line construct the actual RSHR line through the room design condition and extend to the left to intersect the 100 percent RH curve. Where this RSHR line interesects the 90 percent RH curve, find the condition of the air leaving the cooling apparatus, which is also the supply air condition to the space. Read t_{sa} = 55°F, $t_{w_{sa}}$ = 53.4°F. ***(ans)***

Illustrative Problem 5-16

Air leaves a bank of cooling coils at 63°F DB and 90 percent RH. The room to which the air is delivered has a sensible heat gain of 25,000 Btu/hr and a latent heat gain of 8300 Btu/hr. If the room is to be maintained at 80°F DB by thermostat, find the resulting RH and humidity ratio in the conditioned room. How much air is being supplied to the room?

Solution:

Step 1: Determine the RSHR.

$$RSHR = \frac{RSH}{RSH + RLH} = \frac{25{,}000}{25{,}000 + 8300}$$

$$RSHR = 0.7507 \quad \text{(say 0.75)}$$

Step 2: Locate RSHR line on psychrometric chart.

On Fig. 5-36 locate the supply air condition at t_{sa} = 63°F, RH_{sa} = 90 percent. Using the alignment circle and the SHR scale, construct the reference RSHR line. Parallel to this reference line construct the actual RSHR line through the supply air condition and extend to the right to intersect the 80°F DB line. At this point find RH_{ra} = 57 percent,

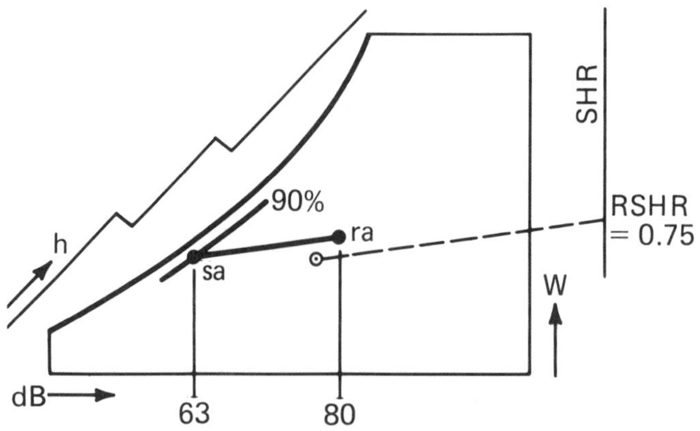

Figure 5–36

Solution to Problem 5-16

$$W_{ra} = 86 \text{ grains/lb da} \qquad \textit{(ans)}$$

Step 3: Determine the supply air quantity (cfm_{sa}). By Eq. 5-9,

$$q_s = (1.10\ cfm_{sa})(t_{ra} - t_{sa})$$
$$25{,}000 = (1.10\ cfm_{sa})(80 - 63)$$
$$cfm_{sa} = 1337 \qquad \textit{(ans)}$$

Problems 5-15 and 5-16 dealt with selecting the condition of the supply air; no reference has been made of selecting the volume of supply air. To begin by arbitrarily selecting the air volume is equivalent to fixing the dry- and wet-bulb temperatures of the supply air without first ascertaining whether air at these temperatures can be obtained by use of the conditioning equipment. This point is illustrated by the following problem.

Illustrative Problem 5-17

The sensible heat gain to a room is 18,200 Btu/hr and its latent heat gain is 4500 Btu/hr. A conditioned air supply of 1200 cfm is to be delivered to the room. If the room is to be maintained at 85°F DB and 69°F WB, find the required wet- and dry-bulb temperatures of the conditioned supply air.

Solution:

Step 1: Determine the dry-bulb temperature of the supply air (t_{sa}). By Eq. (5-19),

$$\dot{q}_s = (1.10\ cfm_{sa})(t_{ra} - t_{sa})$$
$$18{,}200 = 1.10(1200)(85 - t_{sa})$$
$$t_{sa} = 71.1°F \quad (\text{say } 71°F)$$

Step 2: Calculate the RSHR.

$$RSHR = \frac{RSH}{RSH + RLH} = \frac{18{,}200}{18{,}200 + 4500}$$
$$RSHR = 0.802 \quad (\text{say } 0.80)$$

Step 3: Locate RSHR line on psychrometric chart.

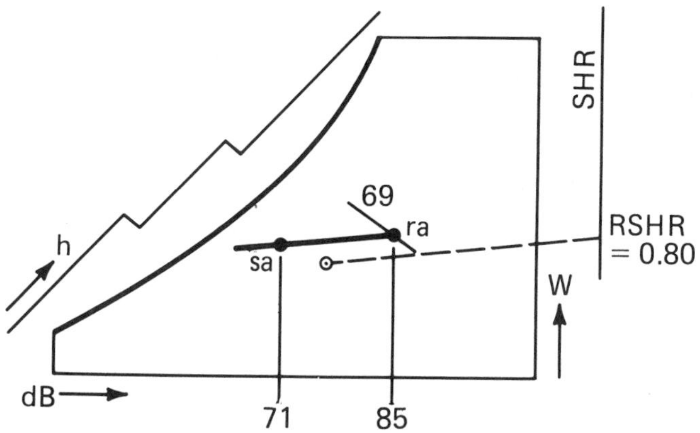

Figure 5–37

Solution to Problem 5-17

On Fig. 5-37 locate the room air condition at 85°F DB and 69°F WB. Using the SHR scale, construct the reference RSHR line. Parallel to this reference line and through the room design conditions draw the actual RSHR line and extend to the left. Where the dry-bulb temperature of the supply air t_{sa} = 71°F intersects the RSHR line find, $t_{w_{sa}}$ = 63.5°F, RH_{sa} = 68 percent.

This would be a difficult supply air condition to obtain without extensive modification of the system. It is always best to select the supply air condition before determining the required supply air quantity, as was done in Problem 5-16.

BYPASSED RETURN AIR: Figure 5-38 is a schematic drawing of a system where a portion of the air drawn from the conditioned room is admitted to the inlet of the supply fan without passing through the cooling equipment. The air supplied to the fan in this manner is known as "bypassed return air." The air entering the fan is a mixture of chilled air and the bypassed return air. Only the chilled air portion of the mixture can absorb sensible and latent heat in the conditioned space. The reason is that the dry-bulb and dew-point temperatures of the bypassed return air are equal to those of the room air, whereas the air chilled by the coil is at a lower dry-bulb and dew-point temperature. In problems involving mixtures of bypassed return air and conditioned air, the bypassed return air may be neglected entirely, and the problem figured as though chilled air alone were being delivered to the conditioned room.

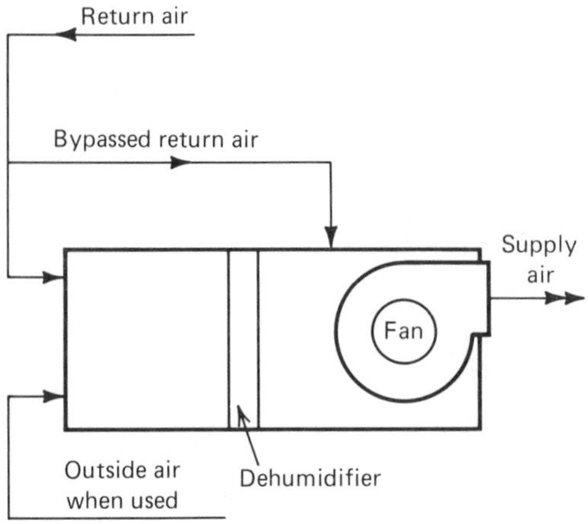

Figure 5–38

Schematic: Return air bypass system

Illustrative Problem 5-18†

A room has a sensible heat gain of 80,000 Btu/hr and a latent heat gain of 20,000 Btu/hr. The room is to be maintained at 78°F DB and 50 percent RH. The specifications require a total air volume flow of 6000 cfm to be circulated. Find the following:

a. the dry- and wet-bulb temperatures of the chilled air leaving the cooling coil if its RH is 90 percent

b. the volume of air to be cooled and dehumidified

c. the volume of return air to be bypassed

d. the dry- and wet-bulb temperatures of the mixture of chilled and bypassed return air delivered to the room

Solution:

Step 1: Determine RSHR line and locate on psychrometric chart.

$$\text{RSHR} = \frac{\text{RSH}}{\text{RSH} + \text{RLH}} = \frac{80{,}000}{80{,}000 + 20{,}000}$$

$$\text{RSHR} = 0.80$$

†Extracted from *Trane Air Conditioning Manual,* The Trane Company, LaCrosse, Wisconsin, p. 106, 1965, with permission.

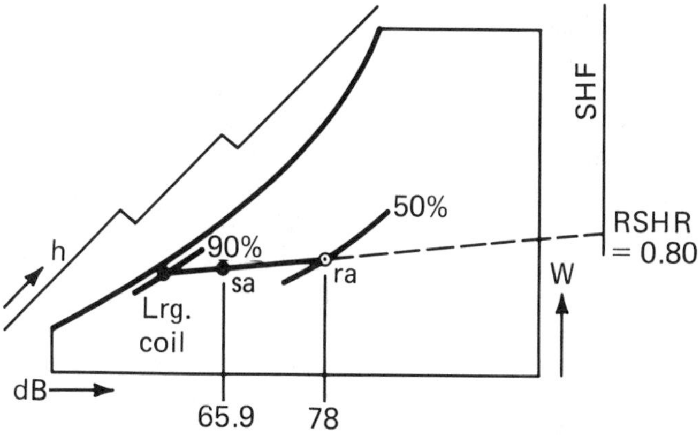

Figure 5–39

Solution to Problem 5-18

Draw RSHR line on chart through room design conditions and extend to the left to intersect the 90 percent RH curve. At this point find $t = 57.5°F$ DB and $t_w = 55.8°F$ WB. This is the condition of the chilled supply air.

Step 2: Find volume flow of chilled supply air. By Eq. (5-9),

$$\dot{q}_s = (1.10\ \text{cfm}_{sa})(t_{ra} - t_{sa})$$

$$80{,}000 = (1.10\ \text{cfm}_{sa})(78 - 57.5)$$

$$\text{cfm}_{sa} = 3548 \quad \text{(volume flow of chilled air)}$$

Step 3: Determine volume flow of return air.

The total volume flow of air was set at 6000 cfm. Therefore, the bypassed return air would be

$$\text{Bypassed Return Air } (\text{cfm}_{ra}) = 6000 - 3548$$

$$= 2452 \text{ cfm}$$

Step 4: The mixture condition of chilled air and bypassed return air may be found as follows. By Eq. 5-17,

$$t_{mix} = \frac{\text{cfm}_{sa} \times t_{sa} + \text{cfm}_{ra} \times t_{ra}}{\text{cfm}_{total}}$$

$$= \frac{3548 \times 57.5 + 2452 \times 78}{6000}$$

$$t_{mix} = 65.87 \quad \text{(say 65.9°F)}$$

The mixture temperature (t_{mix}) must be the air temperature entering the room and must fall on the RSHR line (0.80). At the point of intersection with $t_{mix} = 65.9°F$, find $t_{w_{mix}} = 59.8°F$, $W_{mix} = 67$ grains/lb da.

Old fallacies die hard.† One that has been most persistently long-lived is the belief that uncomfortable drafts result in a conditioned room if the temperature of the supply air is low. It is just as possible to have a drafty room with a warm air supply as with a cold air sup-

†Extracted from *Trane Air Conditioning Manual,* The Trane Company, LaCrosse, Wisconsin, p. 107, 1965.

ply. The whole matter of draft prevention is one of *properly introducing supply air into a room*. Poorly designed and improperly located air outlets will always cause drafts, regardless of the temperature of the supply air. With properly located air inlets of correct design, there is little danger of drafts when introducing air at temperatures usually needed to maintain the required temperature and humidity of a conditioned room. In spite of this, specifications fixing the lowest dry-bulb temperature at which an air supply may be introduced are occasionally encountered. Usually these specifications fix the minimum temperature of the supply air at a point 10 to 15 degrees below the dry-bulb temperature being maintained in the conditioned room. Usual temperature differences are 17 to 25 degrees.

Fixing both the dry-bulb temperature of the room and the lowest dry-bulb temperature at which the air supply may be introduced is equivalent to fixing both the temperature rise and the volume of the air supply. As the temperature rise of the air supply has been set, and because the sensible heat gain of the room is a given amount, it is apparent that by choosing the dry-bulb temperature of the air supply, the volume of chilled air that must be delivered is automatically determined. This has been illustrated in previous problems. Because of this, many specifications call for definite volumes of air to be supplied; these volumes have been calculated on some minimum assumed temperature which may, or may not, be stated.

Another situation that requires the circulation of larger air volumes than are desirable sometimes arises in the larger cities because of their ventilation ordinances. If larger volumes of air then are needed are cooled and dehumidified, the humidity in the conditioned space will frequently be too high. However, the requirement that the entire air supply be taken from outdoors is often waived, but the volume of air circulated according to these ventilation requirements must still be supplied, even though practically all return air is used.

Where the volume of air that must be circulated is larger than the volume of air that should be conditioned, the difference between the two—that is, the balance of the air needed to make up the total which must be circulated—should be withdrawn from the room and bypassed. (The use of bypass is suggested here not as a method of draft prevention or of temperature control but only as a method of meeting the requirements of specifications that call for needlessly large volumes of air.) This excess quantity of return air actually required to maintain the given room temperature and humidity should be conditioned.

Whenever possible, this excess return air should be eliminated because it contributes nothing toward the maintenance of the desired room conditions. The bypassed return air is carried around through the duct system as so much dead weight only in order to circulate the specified larger air volume. The circulation of larger air volumes than are needed has some disadvantages. First, there is an increase in the first cost of the installation because of the larger ducts and fans needed. Second, there is an increase in the operating cost because of the larger fan motor required to circulate the excess air.

The use of bypass return air is a means of controlling room design conditions at partial loads. Whenever the actual room sensible heat load is less than the maximum design load (latent load has a tendency to remain more constant than sensible load), the room temperature will be decreased if the design quantity of conditioned air is supplied. When the room loads are at the maximum design values, the return air bypass would be closed, but as these loads decrease, the bypass damper would be opened proportional to the decrease in the room load, thus preventing a decrease in the room temperature.

The same effect as results from the above arrangement can be accomplished by throttling the conditioned air supply by dampers or by varying the speed of the supply air fan. Both of these introduce air distribution problems if carried beyond certain limits, while the bypass system insures a constant rate of air supply.

Illustrative Problem 5-19

A room in which the sensible heat gain is 160,000 Btu/hr and the latent heat gain is 40,000 Btu/hr is to be maintained at 76°F DB and 50 percent RH. The specifications require that the air supply shall be delivered to the room at a temperature no lower than 61°F. The chilled air leaves the conditioner at 90 percent RH. Find the following:

a. the quantity of air to be conditioned by the cooling coils

b. the total air quantity to be delivered to the room

c. the quantity of return air to be bypassed.

Solution:

Step 1: Determine RSHR line and locate on psychrometric chart.

$$\text{RSHR} = \frac{160{,}000}{160{,}000 + 40{,}000} = 0.80$$

Step 2: Determine required quantity of chilled air (subscript "ca"). By Eq. (5-9),

$$\dot{q}_s = (1.10\ \text{cfm}_{ca})(t_{ra} - t_{ca})$$

$$160{,}000 = (1.10\ \text{cfm}_{ca})(76 - 55.3)$$

$$\text{cfm}_{ca} = 7027$$

Step 3: Determine total quantity of supply air.

According to the problem, the supply air to the conditioned space must have a temperature of 61°F. Therefore, the total quantity of supply air would be

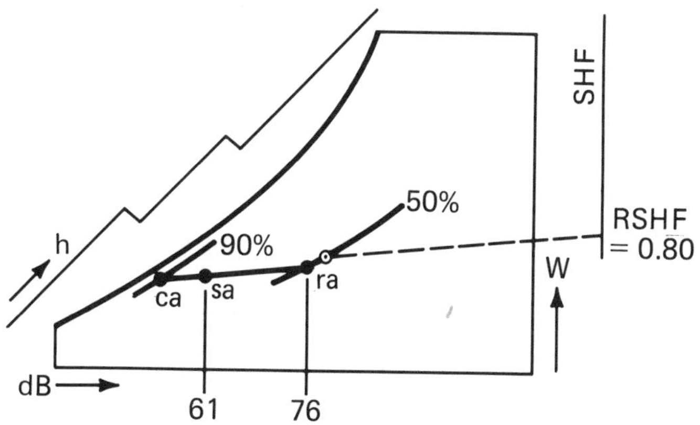

Figure 5–40

Solution to Problem 5-19

$$\dot{q}_s = (1.10\ \text{cfm}_{sa})(t_{ra} - t_{sa})$$
$$160{,}000 = (1.10\ \text{cfm}_{sa})(76 - 61)$$
$$\text{cfm}_{sa} = 9697$$

Step 4: Determine quantity of bypass air.

$$\text{Bypassed air} = \text{cfm}_{sa} - \text{cfm}_{ca}$$
$$= 9697 - 7027$$
$$\text{Bypassed air} = 2670\ \text{cfm}$$

5-14 COOLING COIL PERFORMANCE

In Section 5-13 certain conditions were assumed for the supply air. The major assumption was that the air leaving the cooling coil was at a relative humidity of 90 percent. Further, it was assumed that the air entering the cooling coil was recirculated return air from the conditioned space. The latter assumption would be correct only when no outside air was introduced to the conditioned space for proper ventilation. The actual condition of the air entering the cooling apparatus would normally be a mixture of return air from the conditioned space and outside fresh air for ventilation.

Several factors influence the air-side performance of a cooling and dehumidifying coil using a volatile refrigerant or chilled water. For a given coil configuration these factors are:

a. entering air condition, that is, dry- and wet-bulb temperatures

b. velocity of air flow, that is, usually established by the coil face velocity (FV)

c. coil depth in rows of tubes perpendicular to the airstream direction

d. refrigerant temperature

Figure 5-41 illustrates the actual performance of an extended surface (finned) cooling and dehumidifying coil. Line *AB* represents the RSHR for a particular situation. The air entering the cooling coil is at point *D*, which represents a mixture of outside air at condition *C*

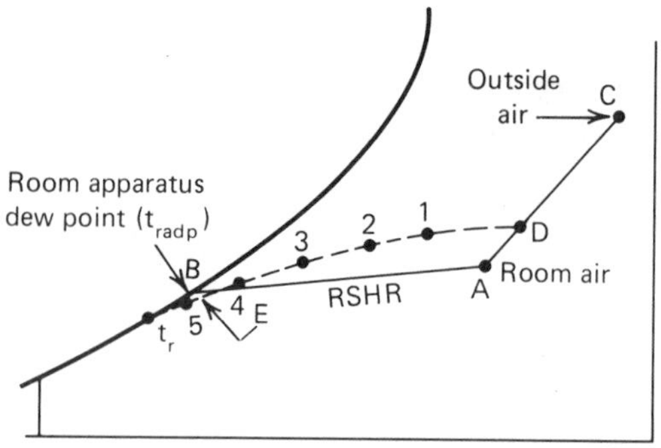

Figure 5–41

Actual performance of an extended surface cooling and dehumidifying coil

and room return air at condition *A*. *Whenever possible, outside air used for ventilation should be brought in through the dehumidifying equipment in order to reduce the effects of the outside air on the room load to a minimum.* This practice also increases the *effectiveness* of the coil because the higher temperature and moisture content of the entering mixture, compared to room air, raises the performance level of the coil and increases the rate of heat transfer.

In practice, the surface temperature of an extended surface dehumidifying coil is seldom uniform. The typical condition of the airstream as it passes through the coil is represented by the curved line $D12345t_r$, with the refrigerant temperature inside the coil at t_r. The numbered points (2, 3, 4, and 5) represent the number of rows of coil depth and indicate the air condition produced by coils of varying depth with the refrigerant temperature and coil face velocity constant. If the refrigerant temperature and/or coil face velocity is changed, a different curve would result. The point of intersection between the RSHR line *(AB)* and the coil performance curve, point *E*, represents the air condition required to satisfactorily absorb the room sensible and latent heat gains. If an exact number of rows of coil depth did not deliver

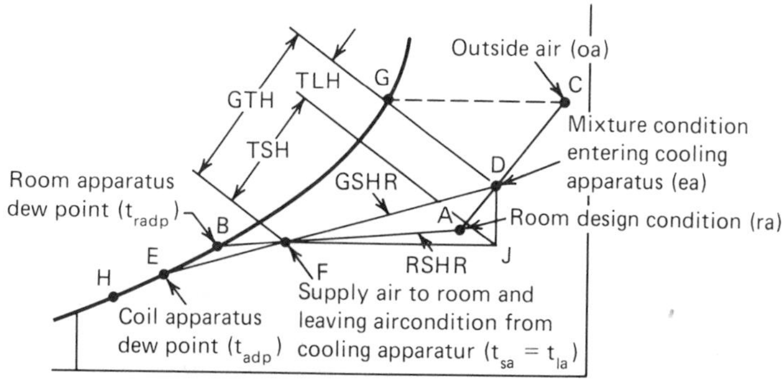

Figure 5–42

Relationship between RSHR and GSHR (cooling Coil Performance)

air at this point, the room loads could not be absorbed to produce the desired room conditions at point *A*. From the above discussion we may see that the design engineer may select coils of varying depth, vary the refrigerant temperature, or vary the face velocity to obtain the proper coil performance.

The actual surface temperature of the cooling coil could range from *G* to *H* on Fig. 5-42 and is not of primary importance in evaluating coil performance. If point *D* represents the condition of air entering the coil and point *F* the desired leaving condition, then a straight line from *D* through *F* extended to the 100 percent RH curve establishes a single uniform coil surface temperature at point *E* which would result in the same leaving air condition. Point *E* is called an equivalent or effective coil surface temperature, t_E, which can be considered as the uniform coil surface temperature which would produce the same leaving air conditions as the variable coil surface temperature that occurs when the coil is in operation.† It should be realized that the straight line *DFE* does not necessarily represent a condition line describing the air condition as the air passes through the coil. Rather, it represents the relationship between the entering and leaving air conditions and defines the effective coil surface temperature, t_E. It also defines the grand sensible heat ratio (GSHR) which represents the actual load conditions for the cooling coil. The point t_E is referred to as *the coil apparatus dew point* (t_{adp}). This point *E* should not be confused with point *B*, which is the *room apparatus dew point* (t_{radp}). When no outside air is used (an unusual situation), lines *AB* and *DE* coincide and the coil and room apparatus dew-point temperatures will be the same. They will also be the same when the air leaving the cooling coil is saturated at 100 percent relative humidity.

GRAND SENSIBLE HEAT RATIO (GSHR):‡ The *grand sensible heat ratio* is the ratio of the total sensible heat to the grand total heat load that the conditioning apparatus must handle, including the outdoor heat loads. This

†Extracted in part from *Carrier System Design Manual*, Part 1, *Load Estimating*, Carrier Air Conditioning Company, Syracuse, N.Y., 1972, pp. 1–120.

‡Extracted in part from *Carrier System Design Manual*, Part 1, *Load Estimating*, Carrier Air Conditioning Company, Syracuse, N.Y., 1972, pp. 1–118.

ratio is determined from the following equation:

$$\text{GSHR} = \frac{\text{TSH}}{\text{TSH} + \text{TLH}} = \frac{\text{TSH}}{\text{GTH}} \quad (5\text{-}39)$$

or, from Fig. 5-42

$$\text{GSHR} = \frac{h_J - h_F}{h_D - h_F}$$

In Eq. (5-39)

TSH = room sensible heat (RSH) + outside air sensible heat (OASH)

or

$$\text{TSH} = \text{RSH} + (1.10\ \text{cfm}_{oa})(t_{oa} - t_{ra}) \quad (5\text{-}40)$$

and

TLH = room latent heat (RLH) + outside air latent heat (OALH)

or

$$\text{TLH} = \text{RLH} + (0.68\ \text{cfm}_{oa})(W_{oa} - W_{ra}) \quad (5\text{-}41)$$

The slope of the GSHR line represents the ratio of sensible and latent heat that the cooling apparatus must handle. The GSHR can be plotted on the psychrometric chart without knowing the condition of the supply air, in much the same manner as the RSHR line, using the calculated GSHR, the mixture condition of air entering the apparatus, the sensible heat ratio scale, and the alignment circle on the chart. The resulting GSHR line is plotted through the mixture condition of the air entering the apparatus.

Referring again to Fig. 5-42, the point of intersection between the GSHR line and the RSHR line must represent the air condition leaving the apparatus and entering the conditioned space ($t_{la} = t_{sa}$). This neglects fan and duct heat gains, duct leakage losses, and so forth. In practice, these heat gains and losses are taken into account in estimating the cooling load. Therefore, the temperature of the air leaving the apparatus is not necessarily equal to the temperature of the air supplied to the space. This temperature difference is usually small (2 or 3 degrees); therefore, in our work in the following problems, we will assume ($t_{la} = t_{sa}$) unless otherwise stated.

REQUIRED AIR QUANTITY:† The quantity of air required to offset simultaneously the room sensible and latent heat loads and the quantity of air required through the cooling apparatus to handle the total sensible and latent heat loads may be calculated using the conditions on their respective RSHR and GSHR lines. For a particular application, when both the RSHR and GSHR lines are plotted on the psychrometric chart, the quantity of air required to satisfy the room loads may be calculated from

$$\text{cfm}_{sa} = \frac{\text{RSH}}{1.10(t_{ra} - t_{sa})} \quad (5\text{-}42)$$

The quantity of air required through the cooling apparatus to satisfy the total air conditioning load (including supplementary loads) may be calculated from

$$\text{cfm}_{da} = \frac{\text{TSH}}{1.10\ (t_{ea} - t_{la})} \quad (5\text{-}43)$$

†Extracted in part from *Carrier System Design Manual,* Part 1, *Load Estimating,* Carrier Air Conditioning Company, Syracuse, N.Y., 1972, pp. 1–119.

The required air quantity supplied to the space (cfm_{sa}) is equal to the air quantity required through the apparatus (cfm_{da}), neglecting leakage losses. In Eq. (5-43) the term t_{ea} is the mixture condition of the air entering the cooling apparatus. The value of t_{ea} can only be determined by trial and error, unless 100 percent outside air is specified.

BYPASSING OUTSIDE AIR:† Outdoor air should not be bypassed. Obviously, the effect of bypassing outdoor air is exactly the same as though this outdoor air were blown directly into the room by a separate fan. The fact that the outdoor air is mixed with the chilled air before being delivered to the conditioned space does not alter the truth of this statement.

If outdoor air is delivered directly to a conditioned room or, what amounts to the same thing, is bypassed through the cooling coil, the sensible and latent heat gains of the room will be increased by exactly the amount of heat that the outdoor air surrenders in cooling down to the room dry-bulb and dew-point temperatures. Therefore, an additional amount of chilled air supply will be required to offset the sensible and latent heat gains due to the delivery of outdoor air directly to the room.

Unconditioned outdoor air, when delivered to a conditioned room, will usually lower the sensible heat ratio of that room, thus making it necessary to chill the supply air to a lower temperature than would otherwise be needed. Hence, if outdoor air is bypassed, not only must a larger volume of chilled air be provided, but it must also be cooled to a lower temperature.

BYPASS FACTOR: Many practicing engineers take into account coil *bypass factor* (BF) when estimating the coil cooling-load and supply air quantity. Some manufacturers rate their cooling coils according to bypass factors. The theory‡ on which this practice is based is that all the air passing through a finned-tube cooling coil does not contact the chilled surface of the coil unless the coil is several rows deep normal to the direction of air flow. In other words the air leaving an extended surface cooling coil is a mechanical mixture of air which has contacted the chilled surface and chilled to that temperature and air which escapes (bypassed) through the coil in the same condition as when it entered.

The bypass factor is equal to the difference between the leaving air temperature and the mean surface temperature of the coil, divided by the difference between the entering air temperature and the mean surface temperature of the coil. Referring to Fig. 5-42, the bypass factor would be

$$\mathrm{BF} = \frac{t_F - t_E}{t_D - t_E}$$

where t_F = dry-bulb temperature of air leaving coil

t_E = coil mean (effective) surface temperature

t_D = dry-bulb temperature of air entering coil

More generally, we may use the following expression for the bypass factor:

$$\mathrm{BF} = \frac{t_{la} - t_{adp}}{t_{ea} - t_{adp}} \qquad (5\text{-}44)$$

†Extracted from *Trane Air Conditioning Manual,* The Trane Company, LaCrosse, Wisconsin, p. 109, 1965.
‡Carrier, W. H., "The Contact-Mixture Analogy Applied to Heat Transfer with Mixtures of Air and Water Vapor," *Transactions ASME,* Vol. 59, p. 49, 1937.

Since the line *DFE* is a straight line, it is closely true that

$$BF = \frac{W_{la} - W_{adp}}{W_{ea} - W_{adp}} \qquad (5\text{-}45)$$

The quantity 1 − BF is frequently referred to as the "contact factor" and may be considered to be that portion of the air leaving the cooling coil at the coil apparatus dew point.

$$1 - BF = \frac{t_{ea} - t_{la}}{t_{ea} - t_{adp}} = \frac{W_{ea} - W_{la}}{W_{ea} - W_{adp}} = \frac{h_{ea} - h_{la}}{h_{ea} - h_{adp}} \qquad (5\text{-}46)$$

In Eqs. (5-44), (5-45), and (5-46) subscripts used are as follows:

ea = entering air condition to cooling apparatus

la = leaving air condition from cooling apparatus

adp = air properties at the coil apparatus dew point

The bypass factor of a coil of *n* rows deep equals the bypass factor for one row raised to the nth power. Experimentation† has indicated that the approximate bypass factor for extended surface, ⅝-in. OD tube, ¹³⁄₃₂-in. high helical fins, spaced 14 per inch, may be predicted by the following empirical relationship:

$$BF = 0.56^n$$

where 0.56 is the bypass factor of a single row coil at a face velocity of 500 fpm and *n* is the number of rows of coil depth.

The bypass factor‡ is a function of the physical and operating characteristics of the air-conditioning apparatus and, as such, represents that portion of the air which is considered to pass through the conditioning apparatus completely unaltered.

The physical and operating characteristics affecting the bypass factor are as follows:

1. A decreasing amount of available apparatus heat-transfer surface results in an increase in the bypass factor; that is, less rows of coils, less coil surface area, wider spacing of coil tubes.
2. A decrease in the velocity of air through the conditioning apparatus results in a decrease in bypass factor; that is, more time for the air to contact the heat-transfer surface.

Decreasing or increasing the amount of heat-transfer surface has a greater effect on bypass factor than varying the velocity of air through the apparatus.

To properly maintain room design conditions, the air must be supplied to the space at some point along the RSHR line. Therefore, as the bypass factor varies, the relative position of the GSHR line and RSHR line changes, as shown by the dotted lines on Fig. 5-43. As the position of the GSHR line changes, the entering and leaving air conditions at the apparatus, the required air quantity, bypass factor, and the apparatus dew point also change.

†Carpenter, James H., "Fundamentals of Psychrometrics," Carrier Air Conditioning Company, Syracuse, New York, p. 17, 1962.

‡Extracted in part from *Carrier System Design Manual,* Part 1, *Load Estimating,* Carrier Air Conditioning Company, Syracuse, N.Y., 1972, pp. 1-121 to 1-123.

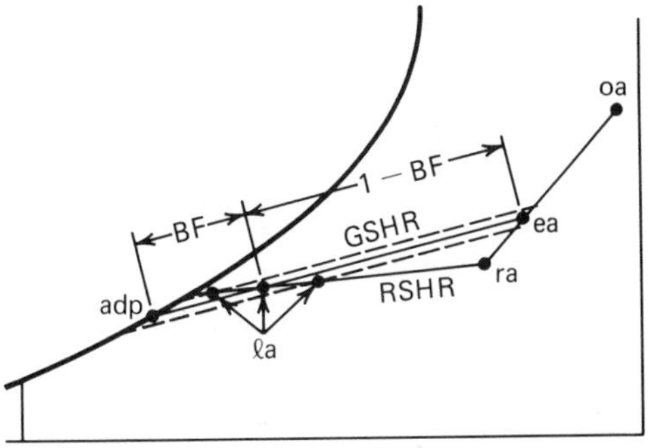

Figure 5–43

Effect of changing size of bypass factor (BF)

The effect of varying the bypass factor on the conditioning equipment is as follows:

1. Smaller bypass factor
   a. Higher adp—DX equipment selected for higher refrigerant temperature and chilled water equipment would be selected for less water flow or higher temperature chilled water. Possibly smaller refrigeration machine.
   b. Less air—smaller fan and fan motor.
   c. More heat-transfer surface—more rows of coil or more coil surface area available.
   d. Smaller piping if chilled water is used.
2. Larger bypass factor
   a. Lower adp—Lower refrigerant temperature to select DX equipment, and more water flow or lower water temperature for chilled water equipment. Possibly larger refrigeration machine.
   b. More air—larger fan and fan motor.
   c. Less heat-transfer surface—less rows of coil or less coil surface available.
   d. Larger piping if more chilled water is used.

It is, therefore, an economic balance of first cost and operating cost in selecting the proper bypass factor for a particular application. Table 5-2 gives suggested ranges of bypass factors for various applications.

Illustrative Problem 5-20

A conditioned room having sensible and latent heat gains of 40,000 and 6500 Btu/hr, respectively, is to be maintained at 77°F DB and 63°F WB. The outdoor air is at 95°F DB and 78°F WB. The specifications require that 2400 CFM of air be delivered to the room. Of this amount 25 percent is to be outdoor air and 75 percent return air from the room. Find the following:

a. the required dry- and wet-bulb temperatures of the supply air to the conditioned room

b. the dry- and wet-bulb temperatures of the air entering the cooling coils

c. the required condition of the chilled air supply leaving the cooling coils

d. the quantity of the air to be chilled by the cooling coils

e. the quantity of air to be bypassed

Solution: (See Fig. 5-44)

Step 1: Determine dry-bulb temperature of supply air and RSHR line. By Eq. 5-9,

$$\dot{q}_s = (1.10\ \text{cfm}_{sa})(t_{ra} - t_{sa})$$

$$40{,}000 = 1.10(2400)(77 - t_{sa})$$

$$t_{sa} = 61.8°\text{F}$$

$$\text{RSHR} = \frac{\text{RSH}}{\text{RSH} + \text{RLH}} = \frac{40{,}000}{40{,}000 + 6500}$$

$$\text{RSHR} = 0.86$$

Table 5-2

Typical Bypass Factors*

Coil Bypass Factor	Type of Application	Example
0.30 to 0.50	A *small* total load or a load that is somewhat larger with a low sensible heat ratio (high latent load)	Residence
0.20 to 0.30	Typical comfort application with a *relatively small* total load or a low sensible heat ratio with a somewhat larger load	Residence, small retail shop, factory
0.10 to 0.20	Typical comfort application	Department store, bank, factory
0.05 to 0.10	Applications with high internal sensible loads or requiring a large amount of outdoor air for ventilation	Department store, restaurant, factory
0.00 to 0.10	All outdoor air application	Hospital operating room, factory

**Extracted from* Carrier System Design Manual, *Part 1,* Load Estimating, *Carrier Air Conditioning Company, Syracuse, N.Y., 1972, p. 1-127.*

On psychrometric chart (Fig. 5-44) draw RSHR line (0.86) through room design conditions. Where t_{sa} = 61.8°F intersects the RSHR line, find $t_{w_{sa}}$ = 56.8°F, W_{sa} = 60 grains/lb da.

Step 2: Determine entering air conditions to cooling coil.

Locate outside air condition (95°F DB and 78°F WB) on Fig. 5-44. Connect outside air condition to room design conditions with a straight line. The mixture condition must lie on this straight line. By Eq. (5-14),

$$t_{mix} = \frac{0.25 \times 95 + 0.75 \times 77}{1.00}$$

$$t_{mix} = 81.5°\text{F} = t_{ea}$$

Locate t_{ea} = 81.5°F on straight line between outside air and room air and find $t_{w_{ea}}$ = 67.2°F, W_{ea} = 77 grains/lb da.

Step 3: Determine the GSHR line and the required condition of the chilled air leaving the cooling coil.

The chilled air leaving the cooling coil must be on the RSHR line and also on the GSHR line. This air condition has already been established at t_{sa} = 61.8°F, $t_{w_{sa}}$ = 56.8°F. Therefore,

$$t_{la} = t_{sa} = 61.8°\text{F}$$

$$t_{w_{la}} = t_{w_{sa}} = 56.8°\text{F}$$

$$W_{la} = W_{sa} = 60 \text{ grain/lb da}$$

Since the entering and leaving air conditions from the coil have been determined, the GSHR line is a straight line joining these two air conditions. On Fig. 5-44 draw the GSHR line and determine from the SHR scale and alignment circle that GSHR = 0.65.

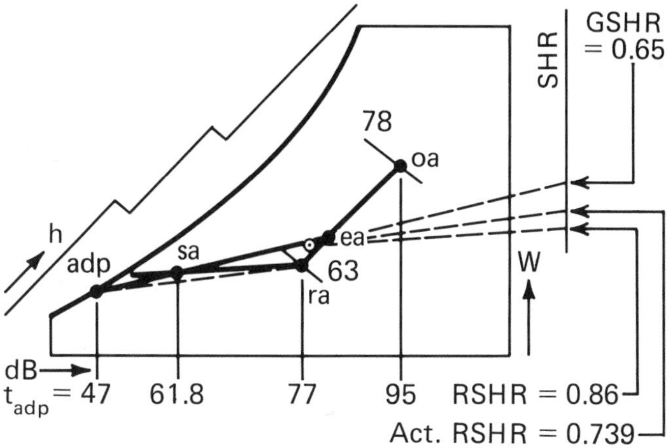

Figure 5–44

Solution to Problems 5-20 and 5-21

If the GSHR line is extended downward and to the left to intersect the 100 percent RH curve, the apparatus dew point is found, t_{adp} = 47°F.

Step 4: Determine bypassed air and chilled air.

By Eq. 5-44,

$$BF = \frac{t_{la} - t_{adp}}{t_{ea} - t_{adp}} = \frac{61.8 - 47}{81.5 - 47}$$

$$BF = 0.43$$

Bypassed air = BF × cfm_{sa} = 0.43 × 2400 = 1032 cfm

By Eq. 5-46

$$1 - BF = \frac{t_{ea} - t_{la}}{t_{ea} - t_{adp}} = \frac{81.5 - 61.8}{81.5 - 47}$$

$$1 - BF = 0.57$$

Chilled air = (1 − BF) (cfm_{sa}) = 0.57 × 2400 = 1368 cfm

Illustrative Problem 5-21

Using the same basic data of Problem 5-20, determine the actual RSHR caused by the bypassing of outside air.

Solution: (Refer to Fig. 5-44)

Step 1: Determine the sensible and latent heat quantities associated with the outside air which is bypassed.

Problem 5-20 stated that 25 percent of the air circulated was to be outside air, or 0.25 × 2400 = 600 cfm_{ca}. Of this amount, 43 percent was bypassed (BF = 0.43). Therefore, bypassed outside air is 0.43 × 600 = 258 cfm. The sensible and latent heat loads caused by this bypassed air may be found as follows:

Bypassed outside air

Sensible heat load = $1.10(258)(t_{oa} - t_{ra})$

= 1.10(258)(95 − 77)

Sensible heat load = 5100 Btu/hr

and

Bypassed outside air

$$\text{Latent heat load} = 0.68(258)(W_{oa} - W_{ra})$$
$$= 0.68(258)(117.5 - 64)$$
$$\text{Latent heat load} = 9390 \text{ Btu/hr}$$

New sensible heat for room
$$= 40{,}000 + 5100 = 45{,}100 \text{ Btu/hr}$$

New latent heat for room
$$= 6500 + 9390 = 15{,}890 \text{ Btu/hr}$$

Actual RSHR
$$= \frac{\text{New sensible heat}}{\text{New sensible heat} + \text{New latent heat}}$$
$$= \frac{45{,}100}{45{,}100 + 15{,}890}$$
$$\text{Actual RSHR} = 0.739$$

Draw actual RSHR line on psychrometric chart through room design conditions and extend to intersect the 100 percent RH curve. *At this point of intersection it may be noted that the actual RHSR line passes through the room design conditions and the coil apparatus dew point.*

EFFECTIVE SENSIBLE HEAT RATIO: From the results of Problems 5-20 and 5-21 it may be concluded that there must be psychrometric relationships between bypass factor (BF), apparatus dew point (t_{adp}), GSHR, and RSHR.

To relate BF and t_{adp} to load calculations, the *effective sensible heat ratio* (ESHR) term was developed. The ESHR is interwoven with BF and t_{adp} and thus greatly simplifies the calculation of air quantity and apparatus selection.†

The ESHR is the ratio of the effective room sensible heat to the sum of the effective room sensible and latent heat. Effective room sensible heat (ERSH) is the sum of room sensible heat plus that portion of outdoor air sensible heat which is considered as being bypassed, unaltered, through the conditioning apparatus. The effective room latent heat (ERLH) is the sum of the room latent heat plus that portion of outdoor air latent heat which is considered as being bypassed, unaltered, through the conditioning apparatus. The ESHR may be expressed in the following relationship:

$$\text{ESHR} = \frac{\text{ERSH}}{\text{ERSH} + \text{ERLH}} = \frac{\text{RSH} + (\text{OASH})(\text{BF})}{\text{RSH} + (\text{OASH})(\text{BF}) + \text{RLH} + (\text{OALH})(\text{BF})} \tag{5-47}$$

The bypassed outdoor air loads that are included in the calculation of ESHR are, in effect, loads imposed on the conditioned space in exactly the same manner as infiltration loads. The infiltration loads come through doors and windows; the bypassed outdoor air load is supplied to the space through the air distribution system.

It should be noted at this point that ESHR is equal to actual RSHR calculated in Problem 5-21.

Plotting the ESHR on the psychrometric chart locates the apparatus dew point (t_{adp}). The GSHR passes through the t_{adp} and the entering air condition to the cooling apparatus. Where the GSHR line and RSHR line intersect, locates graphically the leaving air condition from the apparatus and entering the space. Figure 5-45 illustrates the graphical arrangements of these conditions.

AIR QUANTITY USING ESHR, ADP,

†*Carrier System Design Manual,* Part 1, *Load Estimating,* Carrier Air Conditioning Company, Syracuse, N.Y., 1972, p. 1-127.

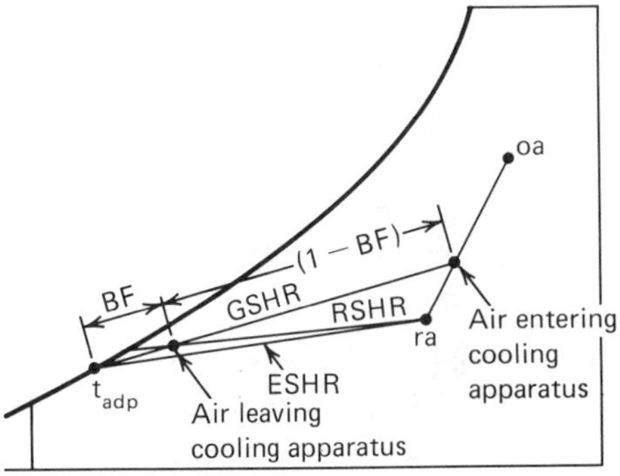

Figure 5–45

Schematic: Showing relationship between GSHR, ESHR, RSHR, and apparatus dew point

AND BF† A simplified approach for determining the required air quantity is to use the psychrometric correlation of ESHR, t_{adp}, and BF. The formula for calculating the air quantity, using BF and t_{adp} is

$$\text{cfm}_{da} = \frac{\text{ERSH}}{1.10(t_{ra} - t_{adp})(1 - \text{BF})} \qquad (5\text{-}48)$$

This air quantity simultaneously offsets the room sensible and room latent heat loads and also handles the total sensible and latent heat loads for which the conditioning apparatus is designed, including outdoor air loads and the supplementary loads.

Illustrative Problem 5-22

A small department store has a calculated sensible heat load of 150,000 Btu/hr and a latent heat load of 30,000 Btu/hr. The summer outside air design conditions are 95°F DB and 75°F WB. The inside design conditions are 80°F DB and 65°F WB. Outside air required for ventilation is 1500 cfm. Determine the following:

a. outdoor air heat loads
b. grand total heat load
c. effective sensible heat ratio
d. apparatus dew point
e. dehumidified supply air quantity
f. entering and leaving air conditions at the apparatus.

Solution:

Step 1: Determine outside air loads.

Outside air sensible heat

$$\begin{aligned}(\text{OASH}) &= (1.10\ \text{cfm}_{oa})(t_{oa} - t_{ra}) \\ &= 1.10(1500)(95 - 80) \\ \text{OASH} &= 24{,}750\ \text{Btu/hr} \qquad \textit{(ans)}\end{aligned}$$

Outside air latent heat

$$\begin{aligned}(\text{OALH}) &= (0.68\ \text{cfm}_{oa})(W_{oa} - W_{ra}) \\ &= 0.68(1500)(98.5 - 68.5) \\ \text{OALH} &= 30{,}600\ \text{Btu/hr} \qquad \textit{(ans)}\end{aligned}$$

Step 2: Determine grand total heat load.

Grand total heat load

$$\begin{aligned}(\text{GTH}) &= \text{RSH} + \text{RLH} + \text{OASH} + \text{OALH} \\ \text{GTH} &= 150{,}000 + 30{,}000 \\ &\quad + 24{,}750 + 30{,}600 \\ \text{GTH} &= 235{,}350\ \text{Btu/hr} \qquad \textit{(ans)}\end{aligned}$$

This is the total heat load the cooling apparatus must handle.

Step 3: Determine the ESHR and t_{adp}. Eq. 5-47,

$$\text{ESHR} = \frac{\text{ERSH}}{\text{ERSH} + \text{ERLH}} = \frac{\text{RSH} + \text{OASH(BF)}}{\text{RSH} + \text{OASH(BF)} + \text{RLH} + \text{OALH(BF)}}$$

From Table 5-2 select a bypass factor of 0.15.

$$\text{ESHR} = \frac{150{,}000 + 24{,}750(0.15)}{153{,}700 + 30{,}000 + 30{,}600(0.15)}$$

$$\text{ESHR} = \frac{153{,}700}{188{,}300}$$

$$\text{ESHR} = 0.816 \quad (\text{say } 0.82)$$

On the psychrometric chart (Fig. 5-46) lay off ESHR line through room design conditions and extend to intersect the 100 percent RH curve. At this point find

$$t_{adp} = 53°\text{F}$$

Step 4: Find dehumidified supply air quantity. By Eq. 5-48,

$$\text{cfm}_{da} = \frac{\text{ERSH}}{1.10(t_{rm} - t_{adp})(1 - \text{BF})}$$

$$= \frac{153{,}700}{1.10(80 - 53)(1 - 0.15)}$$

$$\text{cfm}_{da} = 6088$$ ***(ans)***

Step 5: Determine the entering and leaving air conditions at the apparatus.

The entering air dry-bulb temperature is the temperature of the mixture of 1500 cfm of outside air and 6088 − 1500 cfm of return air from the inside. By Eq. 5-17,

$$t_{ea} = \frac{(1500)(95) + (6088 - 1500)(80)}{6088}$$

$$t_{ea} = 83.7°\text{F}$$

Where t_{ea} = 85.7°F intersects with the straight line between outside air and room design conditions find $t_{w_{ea}}$ = 67.7°F, W_{ea} = 76 grains/lb da.

The condition of the air leaving the cooling apparatus must fall on the GSHR and RSHR lines. This point may be found graphically by

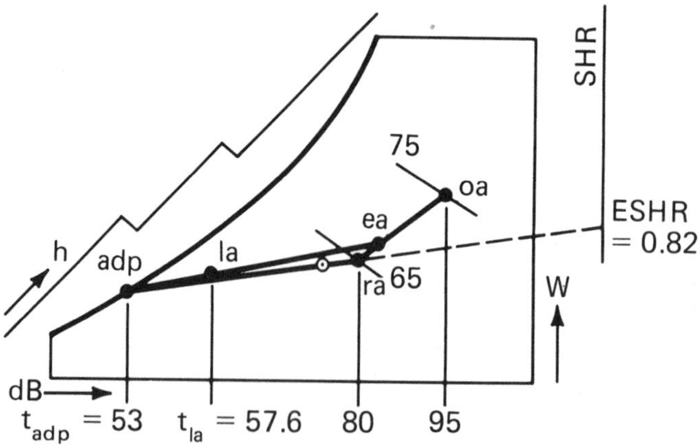

Figure 5–46

Solution to Problem 5-22

constructing these two lines and finding their point of intersection. This is not as accurate as determination by calculation. By calculation we have, by Eq. 5-44,

$$BF = \frac{t_{la} - t_{adp}}{t_{ea} - t_{adp}}$$

$$t_{la} = t_{adp} + BF(t_{ea} - t_{adp})$$

$$53 + 0.15(83.7 - 53)$$

$$t_{la} = 57.6°F$$

On the psychrometric chart (Fig. 5-46) construct the GSHR line by joining the point representing the coil entering air condition with the $t_{adp} = 53°F$. Where $t_{la} = 57.6°F$ intersects the GSHR line find $t_{w_{la}} = 55.6°F$, $W_{la} = 62$ grains/lb da. The leaving air condition from the cooling apparatus is equal to the supply air condition to the space. Therefore

$$t_{sa} = t_{la} = 57.6°F$$

$$t_{w_{sa}} = t_{w_{la}} = 55.6°F$$

$$W_{sa} = W_{la} = 62 \text{ grain/lb da}$$

SENSIBLE HEAT RATIO LINES: Solution of Problem 5-22 involved the determination of the apparatus dew point by finding the intersection between the ESHR line and the 100 percent RH curve on the psychrometric chart. When the ESHR line does not intersect the 100 percent RH curve, or, when the apparatus dew point is too low, a new procedure must be used to determine the apparatus dew point.

Low values of ESHR or RSHR occur when low relative humidities must be maintained in a conditioned space or when the room latent heat load is high in relationship to the sensible heat load. This latter condition frequently occurs at partial-load conditions because the latent load in a room tends to stay relatively constant, whereas the sensible heat load may vary widely (solar load decreasing).

When low values of ESHR and RSHR are encountered, it nearly always means that the chilled air leaving the cooling apparatus must be *reheated* before it is introduced to the conditioned room as supply air. Therefore, the leaving air condition from the cooling apparatus *will not* be equal to the supply air condition to the conditioned room.

The method used to solve a problem involving low values of ESHR and RSHR will be illustrated by the following problem.

Illustrative Problem 5-23

A room within a particular building has a sensible heat gain of 100,000 Btu/hr and a latent heat gain of 38,900 Btu/hr. The room is to be maintained at 80°F DB and 40 percent RH when the outside air at design conditions are 90°F DB and 74°F WB. Outside air for ventilation is to be 2000 cfm. Determine the following:

a. outdoor air heat load
b. effective sensible heat ratio
c. supply air condition to the space
d. supply air quantity
e. apparatus dew point
f. entering and leaving air conditions at the cooling apparatus
g. reheat quantity, if any.

<u>*Solution:*</u>

Step 1: Determine outside air loads.

$$OASH = (1.10\ cfm_{oa})(t_{oa} - t_{ra})$$

$$= (1.10)(2000)(90 - 80)$$

$$OASH = 22,000 \text{ Btu/hr}$$

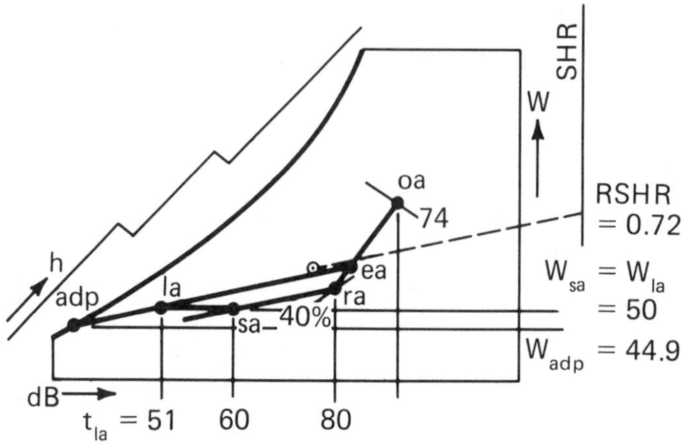

Figure 5–47

Solution to Problem 5-23

$$OALH = (0.68\ cfm_{oa})(W_{oa} - W_{ra})$$
$$= (0.68(2000)(101 - 62)$$
$$OALH = 53{,}000\ Btu/hr$$

Outside

$$\text{air heat load} = OASH + OALH$$
$$= 22{,}000 + 53{,}000$$
$$= 75{,}000\ Btu/hr$$

Grand total

$$\text{heat load} = RSH + RLH + OASH + OALH$$
$$= 100{,}000 + 38{,}900 + 22{,}000 + 53{,}000$$
$$GTH = 213{,}900\ Btu/hr \qquad \textit{(ans.)}$$

Step 2: Determine effective sensible heat ratio (ESHR).

$$ESHR = \frac{RSH + OASH(BF)}{RSH + OASH(BF) + RLH + OALH(BF)}$$

From Table 5-2, assume BF = 0.15.

$$ESHR = \frac{100{,}000 + 22{,}000(0.15)}{103{,}300 + 38{,}900 + 53{,}000(0.15)}$$
$$= \frac{103{,}300}{150{,}150}$$
$$ESHR = 0.688$$

When plotted on the psychrometric chart, this ESHR line intersects the 100 percent RH curve at approximately 35°F, which is a very low value and would give a low supply air temperature to the conditioned space.

Step 3: Assume a reasonable supply air temperature to the conditioned space and calculate supply air quantity.

The temperature difference between supply air and room air could be between 15 and 25 degrees normally. Select supply air temperature t_{sa} = 60°F. By Eq. 5-42,

$$\text{cfm}_{\text{sa}} = \frac{\text{RSH}}{1.10(t_{\text{ra}} - t_{\text{sa}})}$$

$$= \frac{100{,}000}{1.10(80 - 60)}$$

$$\text{cfm}_{\text{sa}} = 4545$$

$$\text{RSHR} = \frac{\text{RSH}}{\text{RSH} + \text{RLH}} = \frac{100{,}000}{100{,}000 + 38{,}900}$$

$$\text{RSHR} = 0.719 \quad \text{(say 0.72)}$$

Draw RSHR line on psychrometric chart (Fig. 5-47). Where $t_{\text{sa}} = 60°\text{F}$ intersects the RSHR line, find $t_{w_{\text{sa}}} = 53.3°\text{F}$, $W_{\text{sa}} = 50$ grains/lb da.

Step 4: Determine mixture condition of air entering the cooling apparatus.

The mixture entering the cooling apparatus consists of 2000 cfm at outside air conditions and 4545 − 2000 cfm of inside recirculated air. By Eq. 5-17

$$t_{\text{mix}} = t_{\text{ea}} = \frac{2000(90) + (4545 - 2000)(80)}{4545}$$

$$t_{\text{ea}} = 84.4°\text{F}$$

On a straight line connecting outside air conditions and inside air conditions, locate $t_{\text{ea}} = 84.4°\text{F}$. Find $t_{w_{\text{ea}}} = 68.5°\text{F}$, $W_{\text{ea}} = 79$ grains/lb da.

Step 5: Determine the coil apparatus dew point and the condition of the air leaving the cooling coil.

The air leaving the cooling coil must be at the same humidity ratio as that of the supply air, $W_{\text{la}} = W_{\text{sa}} = 50$ grains/lb da. By Eq. 5-46,

$$1 - \text{BF} = \frac{W_{\text{ea}} - W_{\text{la}}}{W_{\text{la}} - W_{\text{adp}}}$$

$$W_{\text{adp}} = W_{\text{ea}} - \frac{W_{\text{ea}} - W_{\text{la}}}{1 - \text{BF}}$$

$$= 79 - \frac{79 - 50}{1 - 0.15}$$

$$W_{\text{adp}} = 44.9 \text{ grains/lb da}$$

On psychrometric chart (Fig. 5-47) locate $t_{\text{adp}} = 45.4°\text{F}$ where $W_{\text{adp}} = 44.9$ intersects the 100 percent RH curve.

Join t_{ea} to t_{adp} by a straight line (GSHR line). Where $W_{\text{la}} = W_{\text{sa}} = 50$ intersects the GSHR line, find the leaving air condition from the cooling coil. At this point $t_{\text{la}} = 51°\text{F}$, $t_{w_{\text{la}}} = 49.5°\text{F}$.

Step 6: Determine amount of reheat required.

Since the air leaving the cooling apparatus is at a lower dry-bulb temperature than the required supply air dry-bulb temperature, reheat is required. This would be a sensible-heating process. By Eq. 5-9,

$$\text{Reheat} = (1.10\ \text{cfm}_{\text{sa}})(t_{\text{sa}} - t_{\text{la}})$$

$$= 1.10(4545)(60 - 51)$$

$$\text{Reheat} = 45{,}000 \text{ Btu/hr} \qquad \textit{(ans.)}$$

Reheating of the chilled air supply, for the cases when low sensible heat ratios are found, can be accomplished in several ways. Steam or hot water reheat coils or electric resistance coils are sometimes used. Hot water from the refrigeration condenser may be used in the reheat hot water coil. The so-called run-around system may be used. The run-around system uses a precooling water coil followed by a dehumidifying coil and the reheat water coil connected as indicated in Fig. 5-48.† The run-around cycle has the advantage of an operating cost that is small in comparison with other

---

[+]Extracted from *Trane Air Conditioning Manual,* The Trane Company, LaCrosse, Wisconsin, p. 115, 1965.

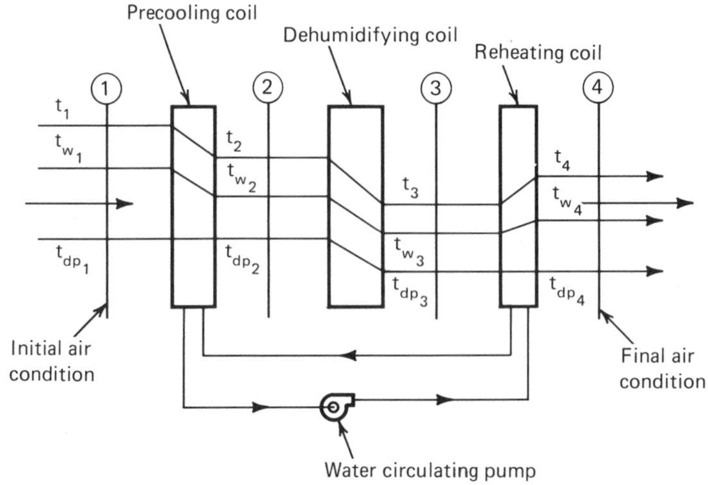

Figure 5–48

Schematic: Run-around reheat cycle

methods. This is because the refrigeration capacity needed to dehumidity the supply air is reduced by the amount of reheating done once the system is in operation. Water is recirculated continuously through the precooling and reheating coils. Sensible heat withdrawn from the warm air on its way to the dehumidifying coil is carried by the circulating water to the reheating coils. These reheating coils then return the sensible heat to the chilled air leaving the dehumidifying coil. Any sensible heat added to the flowing air by the reheating coils is equal to the heat removed by the precooling coils. Consequently, there is a decrease in the required refrigerating capacity when using the run-around cycle to reheat the chilled air supply.

When reheating, the smallest possible quantity of sensible heat should be added to the air supply in order to hold the required additional refrigerating capacity to a minimum. When using the run-around cycle, the quantity of sensible heat added during reheating should also be kept as small as possible. The larger the amount of heat transferred from the precooling coils to the reheating coils by the circulating water, the greater will be the required area of the coils, and the larger will be the amount of circulating water needed.

Theoretically, the chilled air supply leaving the cooling coils can be reheated by being mixed with air at a higher dry-bulb temperature. Actually, however, this method works out poorly. In the average air-conditioning installation air for reheating purposes can be obtained from only two sources: from outdoor air or from the conditioned room. Outdoor air is not always suitable because of its unstable temperature, and return air cannot be used for this purpose when the sensible heat percentage is low.

Outdoor air should not be introduced into a conditoned room without first being cooled and dehumidified. If outdoor air is used for reheating, it must of necessity be mixed directly with the chilled air supply, without being cooled and dehumified. The introduction lowers the sensible heat ratio of the room (see ESHR), thus aggravating the very condition that it is desired to correct by reheating.

5-15 COOLING OF MOIST AIR USING SPRAY WASHERS

Air spray washers were discussed generally in Section 5-9 and 5-10. However, the discussion there was concerned primarily with the heating and humidification of air. Problem 5-9 involved the process of humidification of air by use of recirculated spray water without heating or cooling of the spray water. This process of humidification also reduced the dry-bulb temperature of the air. Therefore, we may say that this represents a cooling and humidifying process. This process is commonly referred to as *evaporative cooling*.

The evaporative-cooling process is a simultaneous removal of sensible heat and addition of moisture which takes place at constant wet-bulb temperature, the wet-bulb temperature of the entering air. The spray water temperature remains constant at the air wet-bulb temperature, and the water is continually recirculated without external heating or cooling.

Evaporative cooling is commonly used for applications where inside relative humidity is to be controlled without, necessarily, any control of the inside temperature, except to hold it within certain limits. Evaporative cooling may be used in industrial applications where humidity alone is critical, and also in dry climates where evaporative cooling gives some relief by removing sensible heat.

Illustrative Problem 5-24

An industrial building has a calculated sensible heat load of 1,500,000 Btu/hr. The latent head load is considered negligible. The inside relative humidity is to be maintained at 50 percent. The outside air at design conditions is at 100°F DB and 65°F WB. Using all outdoor air passing through an evaporative cooler having a humidifying efficiency of 90 percent, determine the following:

a. the building inside dry- and wet-bulb temperatures at design conditions

b. the supply air quantity

Solution: (See Fig. 5-49)

Step 1: Determine entering and leaving air conditions at the evaporative cooler.

The entering air conditions are the outside air conditions of $t_{ea} = t_{oa} = 100°F$, and $t_{w_{ea}} = 65°F$. On the psychrometric chart (Fig. 5-49) locate the entering air condition. The wet-bulb temperature ($t_{w_{ea}} = 65°F$) is the constant wet-bulb temperature for the process.

The leaving air dry-bulb temperature may be found by use of Eq. (5-27) using new subscripts or

$$E_h = \frac{t_{ea} - t_{la}}{t_{ea} - t_w} \times 100$$

or

$$t_{la} = t_{ea} - \frac{E_h(t_{ea} - t_w)}{100}$$

$$= 100 - \frac{90(100 - 65)}{100}$$

$$t_{la} = 68.5°F$$

Locate the point of intersection where $t_{la} = 68.5°F$ crosses the wet-bulb temperature line of 65°F. This point represents the leaving air condition from the evaporative cooler; find $t_{la} = 68.5°F$, $t_{w_{la}} = 65°F$, and $W_{la} = 87$ grains/lb da.

Step 2: Determine inside air dry- and wet-bulb temperature.

The room sensible heat ratio (RSHR) is 1.00 because there is no latent heat gain. A RSHR of unity would be represented by a line

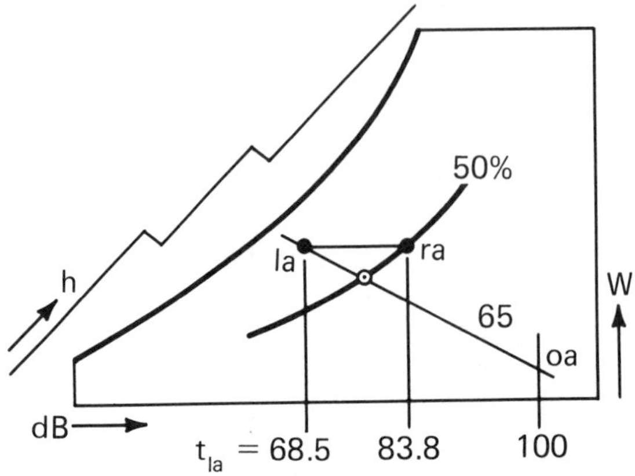

Figure 5–49

Solution to Problem 5-24

of constant humidity ratio or dew point. The air condition leaving the evaporative cooler is the condition of the supply air to the space. Construct a horizontal line through the air condition leaving the cooler and extend to the right to intersect the 50 percent RH curve. This point of intersection represents the inside conditions at design. Find t_{ra} = 83.8°F and $t_{w_{ra}}$ = 69.8°F.

Step 3: Determine the supply air quantity.

The air leaving the evaporative cooler and entering the building must absorb the sensible heat gain as the air increases in temperature from $t_{sa} = t_{la}$ = 68.5°F to t_{ra} = 83.5°F. This would be a sensible-heating process of the supply air. By Eq. 5-42,

$$\text{cfm}_{sa} = \frac{\text{RSH}}{1.10(t_{ra} - t_{sa})} = \frac{1{,}500{,}000}{1.10(83.5 - 68.5)}$$

$$\text{cfm}_{sa} = 89{,}100$$

If the RSHR had been less than unity for Problem 5-24, the method of solution would be the same. However, as may be observed by inspection, the inside dry- and wet-bulb temperatures would be higher, and the CFM_{sa} required would be less since the temperature rise of the supply air would be greater.

Air may be cooled and humidified by passing the air through water sprays when the water temperature is lower than the initial air wet-bulb temperature. To do so, the water temperature must be higher than both the initial and final dew-point temperatures of the air. If the water temperature could be maintained constant at some point between the initial wet-bulb and dew-point temperatures of the air, the condition of the air as it passes through the humidifier could be represented by a straight line such as *AC* in Fig. 5-50. Point *C* represents the constant water temperature and point *B* represents the final condition of the air leaving the humidifier. If the water temperature is not constant but changes as it contacts the air (the usual condition), the condition of the air follows a curved path such as *AD*, leaving the humidifier at point *E*. The final water temperature would be at point *D*. The water temper-

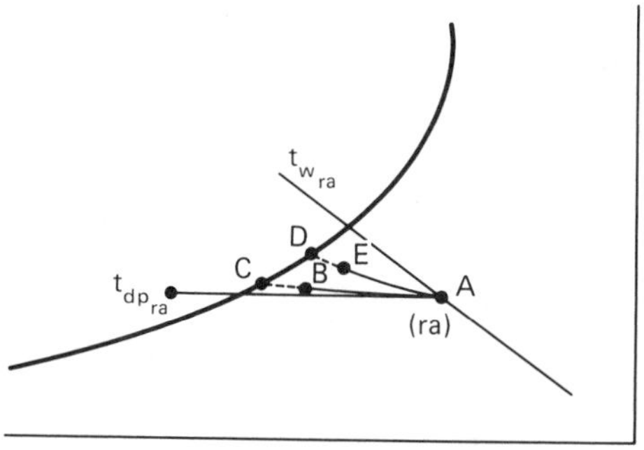

Figure 5–50

Cooling and humidifying with chilled water sprays

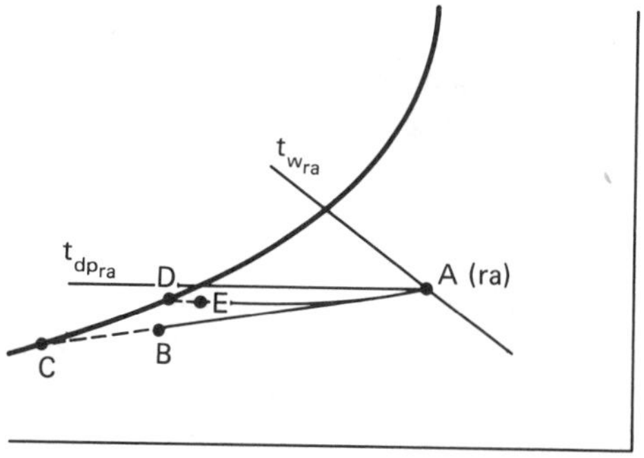

Figure 5–51

Cooling and dehumidifying with chilled water sprays

ature rises in spite of the fact that it is humidifying the air. In this case the air surrenders sufficient sensible heat to warm the spray water and also to evaporate a small portion of it. Because, in this case, the initial temperature of the water is less than the initial wet-bulb temperature of the air, the water temperature will rise and both the dry- and wet-bulb temperatures of the air will fall. The maximum final temperature to which the water can possibly rise is to some point that lies below the final wet-bulb temperature of the air.

If the water temperature is held constant at a point below the initial dew-point temperature of the air, *cooling and dehumidification* of the air will take place along a line such as *AC* in Fig. 5-51. However, if, as is the usual case, the water temperature rises while in contact with the air, the condition of the air would be represented by a curved line such as *AD*, with the air actually leaving at condition *E*. When the initial water temperature is below the initial dew-point temperature of the air, the water temperature rises as it is brought in contact with the warm air. In the case represented by Fig. 5-50 the water humidified the air. In the case represented by Fig. 5-51 the water will dehumidify the air if sufficient water is provided to hold the final water temperature down to a point below the initial dew-point temperature of the air. In the actual device sufficient water is provided to accomplish this.

As long as the final water temperature is kept below the initial dew-point temperature of the air, the air will be both cooled and dehumidified. However, if the final water temperature is allowed to rise above the initial air dew point, the air will still be cooled, but it will be humidified as well. Although the water temperature can rise above the initial air dew point, the water temperature cannot rise to the final wet-bulb temperature of the air.

In a single-stage air washer (Fig. 5-22a) the flow of water and air are parallel to each other, hence both the initial and final temperature of the water must be lower than the final dew-point temperature of the air if dehumidification is to take place. This is true only for a single-stage washer. When a two-stage washer (Fig. 5-22b) is used, the situation is somewhat different. In many such washers the chilled water is divided equally between each spray bank. The water is then supplied to each stage at the same temperature. Problems involving such two-stage washers can be solved by considering each stage as a separate, single-stage washer. Thus, for a two-stage washer, the problem is solved exactly as though the air flowed through two washers in series. The temperature of the water supplied to each stage is exactly the same, but the air temperature entering the second washer has been lowered because it has already been cooled in the first washer.

With a given amount of heat to be transferred from the air to the water, the temperature rise of the water is determined by the quantity of water supplied. Sufficient water should be supplied to a single-stage washer so that its final temperature will be lower than the final dew point to which the air is cooled. To provide a factor of safety, the final water temperature should be 1 degree lower than the required final dew-point temperature.

If the cooling water is obtained from a city main or a deep well, its initial temperature is fixed. If the water is cooled by a refrigeration system, the water should be cooled to such a temperature that the rise in water temperature in the spray washer will be 8 to 14 degrees. A rise of 10 degrees is usually a good average. The quantity of water to be circulated through an air washer for a given heat load can be computed from the following expression:

heat absorbed by spray water = heat given up by air

or

$$\text{gpm} = \frac{\dot{q}}{500(t_{lw} - t_{ew})} \qquad (5\text{-}49)$$

where

gpm = water flow rate (gal/min)

500 = (60 min/hr)(8.34 lb/gal)

$\dot{q}$ = heat transferred from air to water (Btu/hr).

$t_{lw} - t_{ew}$ = difference in water temperature, water temperature rise in spray washer

The reverse problem is also common, that is, determining the quantity of heat that can be removed from the air flowing through the washer when the water available is limited because its cost is high, or because the yield of a well is fixed.

Figure 5-52 is a chart which may be used to aid in the calculation of the final condition of water and air in a washer. Primarily the chart is based on the fundamental fact that the heat surrendered by the air is equal to the heat gained by the water. Also, when air and water are brought into intimate contact with each other in the washer, the final temperature of the water will be equal to the final wet-bulb temperature to which the air is cooled. Actually, the final water temperature will be lower than the final wet-bulb temperature of the air. Nevertheless, this chart enables many problems involving washers to be solved, by finding the theoretical solution and then applying a *correction factor* [see Eq. (5-30)]. The correction factor is normally about 0.85.

Illustrative problem 5-25

Outside air at 95°F DB and 72°F WB is to be cooled in a spray washer where the initial water temperature is 50°F. The quantity of air to be cooled is 16,000 cfm and the quantity of water available is 100 gpm. Determine the following:

1. Find the final temperature to which the air can theoretically be cooled.
2. If the washer correction factor is 0.85, estimate the final conditions of the air and final water temperature.

Solution: (See Fig. 5-53)

Step 1: Determine the water-to-air ratio.

$$\text{Water-to-air ratio} = \frac{100 \text{ gpm}}{16{,}000/1000}$$
$$= 6.25 \text{ gpm}/1000 \text{ cfm}$$

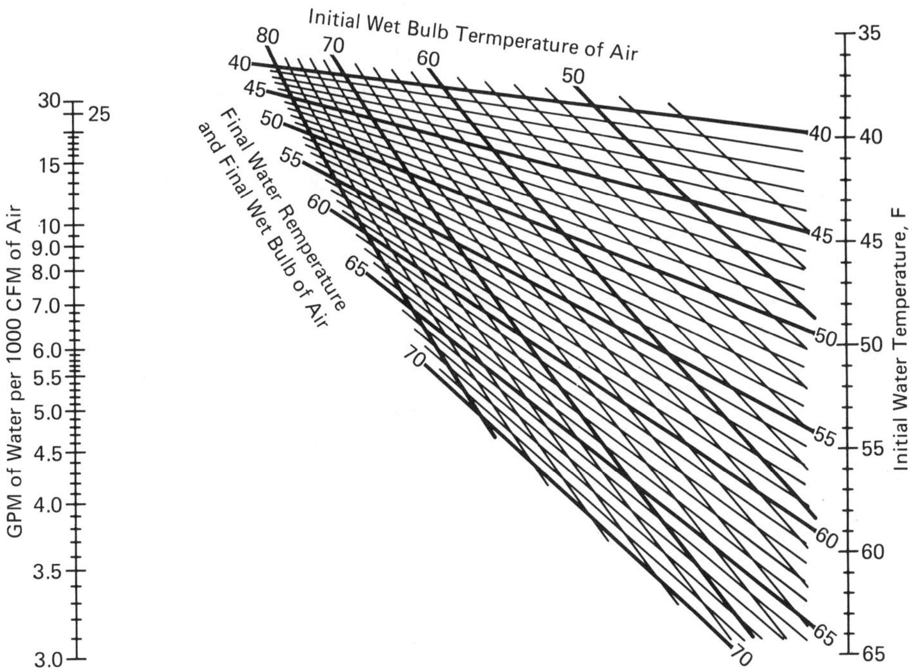

Figure 5–52

Spray Water Air-Washer Chart for Cooling and Dehumidification (Courtesy of The Trane Company.)

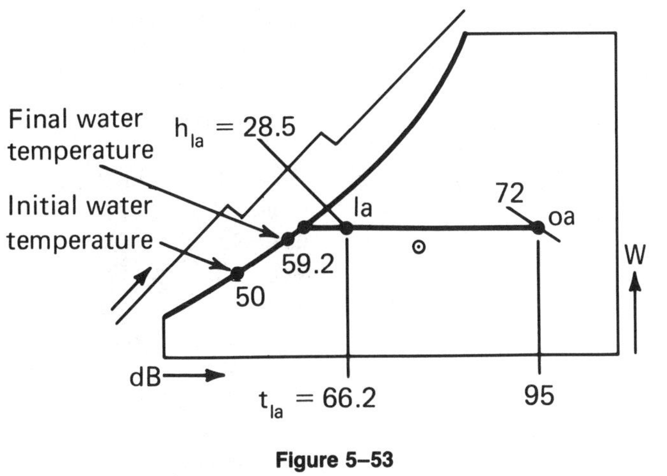

Figure 5–53

Solution to Problem 5-25

Step 2: Determine theoretical final air temperature.

Using Fig. 5-52 draw a straight line from 50°F initial water temperature on right scale to the water-to-air ratio 6.25 on left scale. Where straight line crosses 72°F, the initial air wet-bulb temperature, read theoretical final water temperature and wet-bulb temperature of 61.2 F. *(ans)*

On psychrometric chart (Fig. 5-53) draw a straight line from the initial air condition (95°F DB, 72°F WB) to the theoretical final air wet-bulb temperature of 61.2 located on 100 percent RH curve. This straight line represents the theoretical process of cooling the air from its initial conditions to saturation at 100 percent RH.

Step 3: Estimate the actual final air conditions leaving the washer.

From Fig. 5-53 determine the theoretical entering and leaving enthalpy of the air.

$$h_{ea} = 35.5 \text{ Btu/lb da}$$

$$h_{la} = 27.3 \text{ Btu/lb da} \quad \text{(theoretical)}$$

By Eq. 5-30

Correction factor

$$= \frac{\text{actual change in air enthalpy}}{\text{theoretical change in air enthalpy}}$$

$$CF = \frac{(h_{ea} - h_{la})_{act}}{(h_{ea} - h_{la})_{theor}}$$

$$(h_{ea} - h_{la})_{act} = 0.85(35.5 - 27.3)$$

$$= 6.97$$

Then

$$(h_{la})_{act} = h_{ea} - 6.97 = 35.5 - 6.97$$

$$= 28.53 \text{ Btu/lb}$$

On Fig. 5-53 where $(h_{la})_{act}$ intersects process line between theoretical entering and leaving air conditions, read actual leaving air condition as

$$t_{la} = 66.2°\text{F}$$

$$t_{w_{la}} = 63.0°F$$

Step 4: Estimate actual final water temperature.

Theoretically, the water would be heated by the air from 50°F to 61.2°F. Evidently, if it absorbs only 0.85 of the theoretical quantity of heat, its actual temperature rise would be only 0.85 (61.2 − 50) = 9.52°F. Therefore, the estimated final water temperature would be 50 + 9.52 = 59.52°F.

As may be noted in this problem, the initial and final dew-point temperature of the air are the same. Therefore, the air has been cooled from 95°F to 66.2°F, but no dehumidification has taken place.

Illustrative Problem 5-26†

An auditorium is supplied with 60,000 Cfm of conditioned air to maintain inside design conditions of 78°F DB and 50 percent RH. A total of 20,000 cfm of outside air at 95°F DB and 76°F WB is used. The auditorium is served by two air-handling units each having a precooling coil and a double-bank, single-stage air washer. Each washer operates at 0.87 correction factor. Figure 5-54 illustrates the mechanical arrangement of one of the two air-handling units. Well water used for condensing water in the refrigeration machine flows through the precooling coil and reduces the temperature of the outside air by 6 degrees. Each washer is supplied with 300 gpm of spray water at 42°F.

†Extracted from *Trane Air Conditioning Manual,* The Trane Company, LaCrosse, Wisconsin, pp. 228–230, with permission.

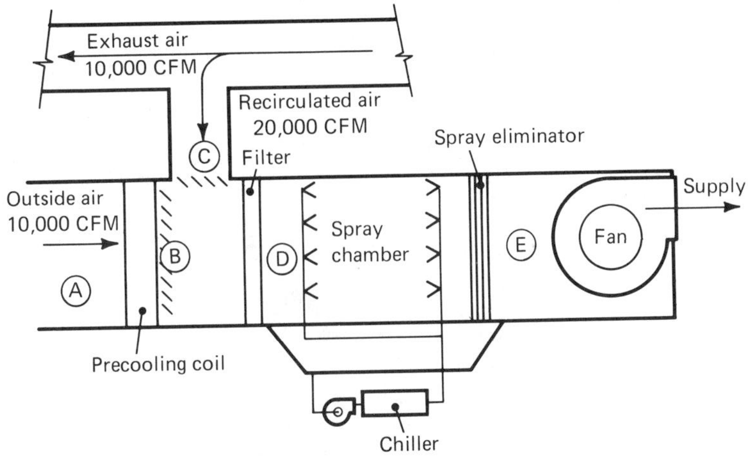

Figure 5–54

Schematic: Equipment arrangement for illustrative problem 5-26 (Circled letters identify corresponding points on Fig. 5-55.)

1. Determine the actual condition of the air leaving the air washer of each unit.
2. If the air is supplied directly to the auditorium from each of the units, how much sensible and latent heat could be absorbed by the supply air? What is the RSHR?

Solution:

Step 1: Determine the condition of the air leaving the air-handling unit.

On the psychrometric chart (Fig. 5-55) locate points A, B, and C for the conditions given. Connect A and B and B and C with straight lines.

Point D represents a mixture condition of precooled outside air at B and return air at C. The dry-bulb temperature of this mixture may be found as follows:
By Eq. 5-17

$$t_D = \frac{cfm_B \times t_B + \text{cfm}_C \times t_C}{\text{total cfm}}$$

$$= \frac{(10{,}000)(89) + (20{,}000)(78)}{30{,}000}$$

$$t_D = 81.67°\text{F}$$

Locate $t_D = 81.67°\text{F}$ on line BC. This represents the air condition entering the air washer. Find $t_{w_D} = 68.4°\text{F}$, $h_D = 32.6$ Btu/lb da, $W_D = 83$ grains/lb da.

$$\text{Water-to-air ratio} = \frac{300 \text{ GPM}}{30{,}000/1000}$$

$$= 10 \text{ gpm/1000 cfm air}$$

Using Fig. 5-52, connect initial water temperature of 42°F on right scale with 10 gpm/1000 cfm on left scale. Where this straight line intersects with initial air wet-bulb temperature ($t_{w_D} = 68.4°\text{F}$), read the theoretical final air wet-bulb and water temperatures as 52°F. Locate 52°F on the 100 percent RH curve and mark point G. Connect points D and G with a straight line. This is the theoretical process line for the air passing through the washer.

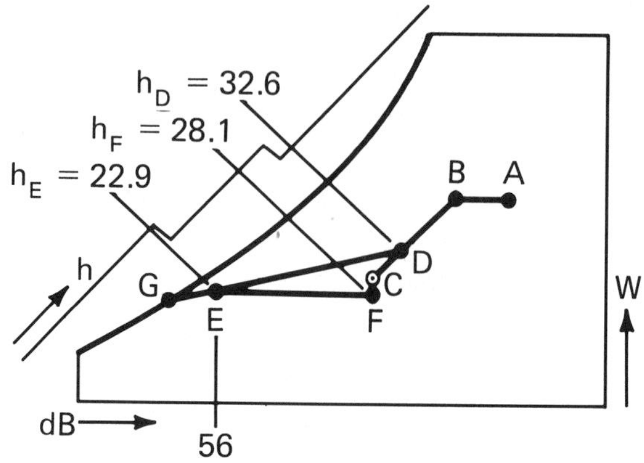

Figure 5–55

Solution to Problem 5-26

The washer has a correction factor of 0.87. Determine the theoretical change in air enthalpy as $h_D - h_G = 32.6 - 21.4 = 11.2$ Btu/lb da. The actual change in air enthalpy would be $(0.87)(11.2) = 9.7$ Btu/lb da. Therefore, the actual enthalpy of the air leaving the washer would be $32.6 - 9.7 = 22.9$ Btu/lb da $= h_E$. Locate $h_E = 22.9$ on line DG and find $t_E = 56°F$, $t_{w_E} = 54.6°F$, and $W_E = 61$ grains/lb da.

Step 2: Determine sensible and latent quantities that the air will absorb.

On Fig. 5-55 locate point F by drawing a horizontal line through E and a vertical line through C. Find $h_F = 28.1$ Btu/lb da.

$$\text{Sensible heat capacity} = h_F - h_E$$
$$= 28.1 - 22.9$$
$$= 5.2 \text{ Btu/lb da.}$$

$$\text{Latent heat capacity} = h_C - h_F$$
$$= 29.9 - 28.1$$
$$= 1.8 \text{ Btu/lb da.}$$

By Eq. 5-8

$$\dot{q}_s = (4.5 \text{ cfm}_{sa})(h_F - h_E)$$
$$= 4.5(30{,}000)(5.2)$$
$$\dot{q}_s = 702{,}000 \text{ Btu/hr}$$
$$\dot{q}_L = (4.5 \text{ cfm}_{sa})(h_C - h_F)$$
$$= 4.5(30{,}000)(1.8)$$
$$\dot{q}_L = 243{,}000 \text{ Btu/hr}$$
$$\text{RSHR} = \frac{5.2}{5.2 + 1.8} = 0.74$$

Illustrative Problem 5-27

Air initially at 80°F DB and 68°F WB is to be cooled to 55°F WB. Water at the rate of 15 gpm/1000 cfm of air is to be used in an air washer. Determine the following:

a. the initial temperature that the water must theoretically have

b. the actual initial water temperature, if a correction factor of 0.85 is assumed

c. an estimate of the final condition of the air leaving the washer

Solution:

Step 1: Determine the theoretical initial water temperature.

Using Fig. 5-52 locate where the initial wet-bulb temperature of 68°F and final wet-bulb temperature of 55°F intersect in the matrix. Lay a straight-edge through this point and through the water-to-air ratio of 15 gpm/1000 cfm. Read the required water temperature from the right scale at 50.0°F.

Step 2: Determine actual initial water temperature.

On psychrometric chart (Fig. 5-56) locate initial air condition (80°F DB, 68°F WB) at *A*. Find $h_A = 32.2$ Btu/lb da.

Theoretical final air wet bulb given is 55°F. At 55°F WB and 100 percent RH, read $h_B = 23.3$ Btu/lb da.

Actual heat to be surrendered by air is $h_A - h_B = 32.2 - 23.3 = 8.9$ Btu/lb da. Since the correction factor given is 0.85 and the actual heat lost by the air amounts to 8.9 Btu/lb da, it is evident that the problem must be solved as though the air were to surrender $8.9/0.85 = 10.47$ Btu/lb da.

The theoretical wet bulb temperature of the air may be found as follows:

The initial air enthalpy $h_A = 32.2$ Btu/lb da. The theoretical quantity of heat to be surrendered is 10.47 Btu/lb da. Therefore, the theoretical final enthalpy of the air should be $32.2 - 10.47 = 21.7$ Btu/lb da.

On Fig. 5-56 draw the line of constant enthalpy of 21.7 Btu/lb. Where this line intersects the 100 percent RH curve, find the wet-bulb temperature of 51.8°F, point *C*.

On Fig. 5-52 use a straight-edge between 15 gpm/1000 cfm and intersection of 68°F initial wet-bulb temperature and 51.8°F final wet

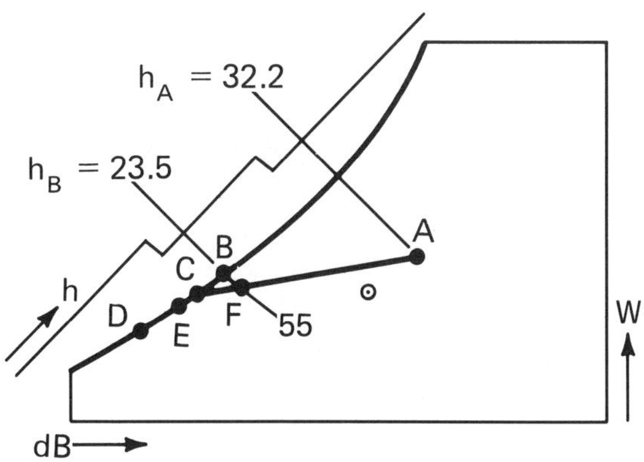

Figure 5–56

Solution to Problem 5-27

bulb. At right read the initial water temperature as 45.5°F, point D of Fig. 5-56. This is the initial water temperature that will actually cool the air to a final wet-bulb temperature of 55°F.

Theoretically, the rise in water temperature would be $51.8 - 45.5 = 6.3$ degrees. Since the correction factor is 0.85, the actual rise in water temperature would be $0.85 \times 6.3 = 5.36$ degrees. Therefore, the actual final water temperature would be $45.5 + 5.36 = 50.9$°F, point E of Fig. 5-56.

Step 3: Estimate final air condition leaving washer.

Draw line AC on Fig. 5-56. This represents the theoretical process line of the air passing through the washer. The actual final wet-bulb temperature of the air was stated as 55°F. Where 55°F WB intersects line AC, find point F, which is the approximate air leaving conditions. Find $t_F = 56.8$°F, $h_F = 23.2$ Btu/lb da, $W_F = 62$ grains/lb da.

CHAPTER 5

Review

5-1. Moist air at 35°F DB and 32°F WB is to be conditioned to 70°F DB and 50 percent RH. Using the psychrometric chart and any required calculations, determine the following:

(a) The initial and final values of humidity ratio, enthalpy, specific volume, and dew-point temperature

(b) If the initial air volume entering the conditioning device is 5000 CFM, determine:

(1) required heating capacity (Btu/hr)

(2) required moisture addition (lb/hr)

(3) pounds per hour of air flowing through the conditioner

5-2. Three thousand cfm of air at 40°F DB and 35°F WB are mixed with 5000 cfm of air at 90°F DB and 74°F WB in a steady-flow device. Determine the dry-bulb and wet-bulb temperatures, humidity ratio, and enthalpy of the resulting mixture.

5-3. Moist air at an initial condition of 40°F DB and 32°F WB is to be heated and humidified to a final condition of 90°F DB and 30 percent RH. The process of humidification is to be carried out in a spray chamber (air washer) using recirculated spray water with no external heating or cooling of the spray water. The spray washer has a humidifying effectiveness of 88 percent. Preheating and reheating of the air will be done by heat-transfer coils. If the initial air volume flow rate is 8000 cfm, determine:

(a) How much preheat and reheat capacity is required, Btu/hr?

(b) How much water must be added to the spray chamber as makeup, gpm?

5-4. Using the same initial and final air conditions of Problem 5-3, determine the range of preheated air temperatures that would permit the use of an open-pan type humidifier containing water maintained at 212°F by immersed heating coils. Assume that the air pressure is 14.7 psia.

5-5. A building interior is to be heated by forced circulation warm air supplied at 105°F DB. The calculated sensible heat loss from the building is 560,000 Btu/hr with the building interior maintained at 68°F DB and 60 percent RH when the outdoor air is at 30°F DB and 24°F WB. Proper ventilation requires that 25 percent of the air circulated must be outdoor air. If the building heat loss is all sensible heat, determine the following:

(a) the required supply air quantity in lb/hr and cfm

(b) if humidification is performed by spraying steam into the airstream after the mixing of outdoor and recirculated air, how much steam in lb/hr is required, assuming the steam is at 16 psia, dry and saturated?

(c) How much heat must be supplied by the heat-transfer coil after humidification?

5-6. Air at 60°F DB and 50°F WB is to be preheated and then humidified to a final condition of 100°F DB and 40 percent RH in a steady-flow device. Humidification is to be performed by spraying steam into the air. The steam pressure is 20 psia and has a quality of 80 percent.

(a) To what temperature must the air be preheated before humidification?

(b) If 4000 cfm of air at the initial condition enters the process, how much steam in lb/hr is required to humidify the air?

(c) What must be the heating capacity of the preheating coil in Btu/hr?

5-7. A building has a calculated heat loss of 400,000 Btu/hr and a sensible heat ratio of 0.80. The building interior is to be maintained at 70°F DB and 50 percent RH when the outdoor air is at 35°F DB and 40 percent RH. Outdoor air required for ventilation is 3000 cfm. Air supplied to the building interior is to be at 115°F DB. Humidification of the supply air to the building interior, if required, will be done by using dry saturated steam at 18 psia after the air has passed through the heating device.

(a) How much air must be supplied to the building interior, lb/hr and cfm?

(b) How much steam, if any, is required for humidification, lb/hr?

(c) Find the required temperature rise of the air through the heat-transfer device, and heat transfer in Btu/hr.

(d) Make a sketch of a psychrometric chart showing all process lines and points which pertain to the solution of the problem.

5-8. Moist air at 60°F DB and 30°F dew-point temperature enters a humidifying device at the rate of 4000 cfm. Dry saturated steam at 30 psia is sprayed into the airstream. Following humidification, the air is heated by a heat-transfer device to a final condition of 90°F DB and 55°F dew-point temperature.

(a) How much steam must be added to the air, lb/hr?

(b) How much heat must be supplied by the heat-transfer device, Btu/hr?

(c) What are the dry- and wet-bulb temperatures after humidification?

(d) Make a sketch of a psychrometric chart showing all process lines and points which pertain to the solution of the problem.

5-9. A building interior is to be maintained at 70°F DB ant 50 percent RH when outdoor design conditions are 30°F DB and 26°F WB. Heat losses from the building are 250,000 Btu/hr sensible and 44,100 Btu/hr latent. Latent heat transfer is due to infiltration of cold, dry outdoor air. Ventilation requires that 1000 cfm of outdoor air be introduced into the supply air. The supply air to the building interior is to be at 100°F DB. The conditioning equipment is shown in Fig. 5-57.

(a) Determine the quantity of supply air required, lb/hr and cfm.

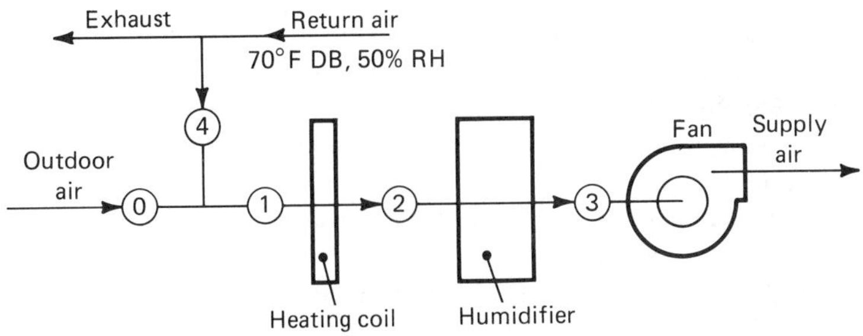

Figure 5–57

Sketch for Problem 5-9

(b) Determine the capacity of the heating device in Btu/hr, if

(1) The humidifier is a spray washer using recirculated spray water, without heating or cooling of the water.

(2) The humidifier is a steam humidifier using steam at 16 psia, dry and saturated.

(c) Make a sketch of a psychrometric chart showing all process lines and points which pertain to the solution of the problem.

5-10. A high humidity chamber is to be maintained at 75°F DB and 70 percent RH when outdoor design conditions are 30°F DB and 80 percent RH. The sensible heat loss from the chamber is 260,000 Btu/hr. Due to contamination within the chamber it will be necessary to use 75 percent outdoor air and 25 percent recirculated air for the supply air mixture.

The proposed equipment to condition the air is shown in Fig. 5-58. The preheat coil will heat the outdoor air to 70°F. The spray washer has a humidifying effectiveness of 90 percent and will use recirculated spray water with no heating or cooling of the spray water. The steam humidifier will use dry satu-

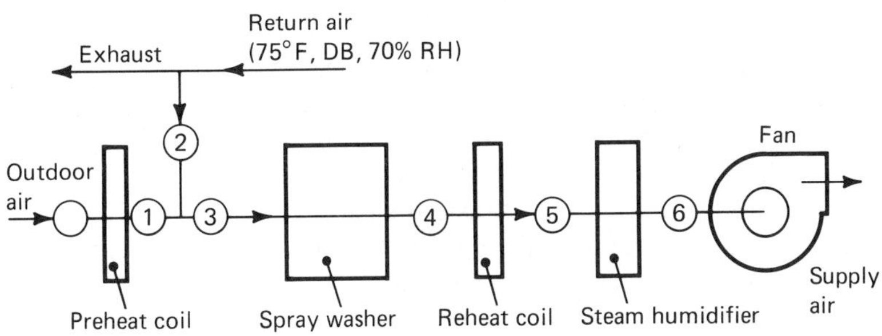

Figure 5–58

Sketch for Problem 5-10

rated steam at 20 psia. Supply air temperature to the chamber will be 110°F dry bulb.

(a) Determine the dry- and wet-bulb temperatures, humidity ratio, and enthalpy for the air at each numbered point in the system.

(b) Determine the required supply air quantity, lb/hr and cfm.

(c) Determine the heating capacity required for the preheat coil and the reheat coil, Btu/hr.

(d) Determine the amount of water absorbed by the air in the air washer, lb/hr.

(e) Determine the amount of steam absorbed by the air in the steam humidifier, lb/hr.

(f) Make a sketch of a psychrometric chart showing all process lines and points which pertain to the solution of the problem.

5-11. A given space within a building is to be maintained at 78°F DB and 50 percent RH. The calculated sensible heat gain to the space is 48,000 Btu/hr, and the latent heat gain is 12,000 Btu/hr. Supply air to the space is to have a dry-bulb temperature of 60°F.

(a) Calculate the space sensible heat ratio.

(b) Determine the wet-bulb and dew-point temperatures, humidity ratio, and enthalpy of the supply air.

(c) Calculate the quantity of supply air required, lb/hr and cfm.

5-12. A building has a calculated sensible heat gain of 130,000 Btu/hr and a latent heat gain of 32,500 Btu/hr. The inside design conditions are 78°F DB and 50 percent RH. The outside design conditions are 90°F DB and 76°F WB. Required outside air quantity for ventilation is 4000 cfm.

(a) For a cooling coil bypass factor of zero, what would be the apparatus dew-point temperature?

(b) If a cooling coil was selected that had a bypass factor of 0.20, determine

(1) Required supply air quantity, cfm.

(2) Entering and leaving air conditions at the cooling coil (dry- and wet-bulb temperatures).

(3) The refrigeration load on the cooling coil, Btu/hr and tons.

5-13. Given the following design information for a building:

Outside air conditions:	90°F DB, 75°F WB.
Inside air conditions:	80°F DB, 50 percent RH.
Sensible heat gain:	160,000 Btu/hr.
Latent heat gain:	40,000 Btu/hr.

If the total quantity of supply air to the building is to be 8000 cfm, 50 percent of which must be outdoor air for ventilation, determine:

(a) The required dry- and wet-bulb temperatures of the air leaving the cooling coil and entering the building.

(b) The dry- and wet-bulb temperatures of the air entering the cooling coil.

(c) The required bypass factor for the cooling coil.
(d) The effective sensible heat factor.

5-14. Given the following design information for a building:

Outside air conditions:	95°F DB, 75°F WB.
Inside air conditions:	80°F DB, 67°F WB.
Sensible heat gain:	226,000 Btu/hr.
Latent heat gain:	67,500 Btu/hr.

Required quantity of outdoor air for ventilation: 5400 cfm.
Inside building volume: 70,000 ft^3.

If 12 air changes per hour are desired for proper air circulation within the building, determine:

(a) Air volume to be circulated, cfm.
(b) The sensible heat factor for the building.
(c) Dry- and wet-bulb temperatures of the air supplied to the building.
(d) Dry- and wet-bulb temperatures of the air entering the cooling coil.

5-15. Given the following design information for a building:

Outside air conditions:	105°F DB, 70°F WB.
Inside air conditions:	50 percent RH.
Sensible heat gain:	180,000 Btu/hr.
Latent heat gain:	20,000 Btu/hr.

It has been decided to use 100 percent outdoor air to condition the building. The outdoor air will pass through a spray chamber using recirculated spray water. The spray chamber will have an estimated humidifying effectiveness of 80 percent. Determine:

(a) The resulting dry-bulb temperature in the building at design conditions.
(b) The required air flow, cfm.
(c) The dry- and wet-bulb temperatures of the air leaving the spray chamber at design conditions.

5-16. Given the following design information for a building.

Outside air conditions:	95°F DB, 80°F WB.
Inside air conditions:	80°F DB, 40 percent RH.
Sensible heat gain:	230,000 Btu/hr.
Latent heat gain:	55,000 Btu/hr.

Required outdoor air for ventilation: 5000 cfm.

(a) Using a cooling coil bypass factor of 0.15, determine:
(1) The apparatus dew-point temperature.
(2) The quantity of air supplied to the building interior to absorb the building heat gains, cfm.
(3) Entering and leaving air conditions at the cooling coil (dry- and wet-bulb temperature).
(4) Effective sensible heat factor.

(b) If the supply air dry-bulb temperature has to be 60°F and the same cooling coil bypass factor is used, determine:

(1) The quantity of supply air required, cfm.
(2) The required reheat capacity, Btu/hr.

5-17. Given the following design information for a building:

Outside air conditions:	92°F DB, 76°F WB.
Inside air conditions:	80°F DB, 50 percent RH.
Sensible heat gain:	68,500 Btu/hr.
Latent heat gain:	29,350 Btu/hr.

Ventilation requires that all air supplied to the building interior must be outdoor air.

Temperature difference between inside design dry bulb and supply air is to be 20 degrees.

Using a cooling coil bypass factor of 0.10, determine:

(a) Room sensible heat factor.
(b) The supply air quantity, Btu/hr.
(c) Apparatus dew-point temperature.
(d) Reheat quantity, Btu/hr.

5-18. Cooling-load calculations for a building indicate a sensible heat gain of 200,000 Btu/hr and a latent heat gain of 25,000 Btu/hr at design conditions. The outdoor air at design conditions is at 105° F DB and 65°F WB. A manufacturer can supply a spray water evaporative cooler using recirculated spray water which has a humidifying effectiveness of 90 percent when the air flow is 9000 cfm.

(a) If 100 percent outdoor air is used in the amount of 9000 cfm through the spray washer and enters the building as supply air, determine:
(1) The dry- and wet-bulb temperatures of the air leaving the spray washer as supply air.
(2) The resulting dry- and wet-bulb temperatures and the relative humidity of the air in the building.

5-19. Given the following design information for a building:

Outside air conditions:	95°F DB, 75°F WB.
Inside air conditions:	80°F DB, 50 percent RH.
Sensible heat gain:	150,000 Btu/hr.
Latent heat gain:	50,000 Btu/hr.

Minimum required outdoor air for ventilation: 3000 cfm.

Supply air for this application is to be all outdoor air.

For a cooling coil bypass factor of 0.10, determine:

(a) Heat loads caused by outdoor air, Btu/hr.
(b) Apparatus dew-point temperature.
(c) Supply air quantity to use, cfm.
(d) Supply dry- and wet-bulb temperatures.

BIBLIOGRAPHY

1. *The Armstrong Humidification Handbook,* Armstrong Machine Works, Three Rivers, Michigan.

2. *Trane Air Conditioning Manual,* The Trane Company, LaCrosse, Wisconsin.

3. *Carrier System Design Manual,* Part 1, *Load Estimating,* Carrier Air Conditioning Company, Syracuse, NY, 1972.

Chapter 6

Comfort in Air Conditioning

6-1 *INTRODUCTION*

Air conditioning has been defined as the simultaneous control of temperature, humidity, air movement, and the quality of air in a given space. The use of the conditioned space determines the temperature, humidity, air movement, and quality of the air to be maintained. Air conditioning is able to provide widely varying atmospheric conditions: from those necessary for drying processes to those necessary for high-humidity process application. Air conditioning can maintain any atmospheric condition regardless of variations in outdoor weather. The range of temperatures and humidities used in comfort cooling is a small band. The location of this band on the psychrometric chart depends on the season of the year.

Cleanliness and air movement must be considered. Air should be cleaned, that is, freed of dust and soot particles and odor. Air cleanliness is important from the standpoint of human health. Also, the walls and ceilings of rooms supplied with filtered air will need cleaning less frequently than those supplied with unfiltered air. Considerable dust is carried into a room by an unfiltered air supply. Some spaces, called "clean rooms," require special attention as to air filtration.

Air should circulate freely in the room to which it is delivered. This will allow it to absorb heat and moisture uniformly throughout the entire room during the cooling cycle and to deliver heat and moisture uniformly during the heating cycle. At the same time the air movement should be gentle, or it will cause objectional draft.

Comfort can be defined as any condition which when changed will make a person uncomfortable. Though this sounds paradoxical, it just means that a person is not aware of the best air-conditioning systems when comfortable. If a person is uncomfortably warm, then the temperature or humidity (or both) are too high. In a first-class air-conditioning system one is not really conscious of the temperature or humidity because one is comfortable. Neither is there any disturbance because of equipment noise or air movement.

Finding the best conditions for comfort and health has been the object of considerable research work. The *ASHRAE Handbook of Fundamentals 1981* is probably the most up-to-date and complete source of information relating to the physiological aspects of thermal comfort. Thus, much information is available concerning factors affecting human comfort.

6-2 HEAT BALANCE OF THE HUMAN BODY

Individual comfort depends on how fast the body is losing heat. The human body may be compared to a furnace using food as fuel. Food is largely carbon and hydrogen; energy contained in the fuel (food) is released by oxidation. The oxygen comes from the air, and the principal products of combustion are carbon dioxide and water vapor. The physician calls this process *metabolism*.

The human body is essentially a constant-temperature device. Its deep body temperature is kept at 98.6°F (37°C) by a delicate, complex temperature-regulating mechanism within the body.

The object of air conditioning is to assist the body in controlling its cooling rate. This is true for both winter and summer conditions. In summer the job is to increase the cooling rate of the body; in winter it is to decrease the cooling rate.

It is possible to compile a heat balance for the human body and study the ways the body's cooling rate can be controlled. As mentioned previously, metabolism is the process by which the body produces heat. This will be shown on the left side of the following equation. On the right side of the equation will be shown the ways the body loses heat.
Heat produced = Heat lost

$$M = \pm \dot{q}_s \pm \dot{q}_r + \dot{q}_L \qquad (6\text{-}1)$$

where

M = metabolism (Btu/hr or W)

$\dot{q}_s$ = sensible heat exchange (Btu/hr or W)

$\dot{q}_r$ = radiant heat exchange (Btu/hr or W)

$\dot{q}_L$ = latent heat exchange (evaporation) (Btu/hr or W)

Two things can cause the deep body temperature to rise: fever or continuous heavy activity. The body temperature will fall in freezing weather if it is not properly protected. If a rise or fall of deep body temperature were considered, Eq. (6-1) would have a "storage" term and a mechanical work term, and for a normal, healthy comfortable body it would be written as

$$S = M - (\pm W) \pm \dot{q}_L \pm \dot{q}_r \pm \dot{q}_s \qquad (6\text{-}2)$$

where

S = rate of heat storage and is proportional to the time rate of change in intrinsic body heat

W = mechanical work accomplished

The left side of Eq. (6-1) can be the basal metabolism. This is the amount of heat generated by an unclothed human at rest in an air temperature of about 70°F (21.1°C). For an individual of average weight and body build the basal metabolism is about 240 Btu/hr (70.3 W). About 60 Btu/hr (17.6 W) is used by the heart and lungs; the balance is used for oxidation. Except for clinics and hospitals, the air-conditioning engineer has only passing interest in basal metabolism. The people he must keep comfortable are clothed and have various degrees of activity. So the left side of Eq. (6-1), M, becomes the actual metabolic level of the individual. This is the heat generated by the individual and may vary widely dependent on activity levels. Naturally the higher the metabolic rate, the higher the heat-transfer rate from the body to the environment. A general knowledge of the amounts of energy expended are useful to engineering since body heat production increases in proportion to exercise intensity.

Table 6-1 extracted from *ASHRAE Handbook of Fundamentals 1981* presents probable metabolic rates (or the energy cost) for various typical activities. These values have been ex-

Table 6-1

Metabolic Rate at Different Typical Activities*

Activity	Metabolic Rates in Met Units[a]	Activity	Metabolic Rates in Met Units[a]
Resting		Teacher	1.6
Sleeping	0.7	Vehicle driving	
Reclining	0.8	Car	1.5
Seated, quiet	1.0	Motorcycle	2.0
Standing, relaxed	1.2	Heavy vehicle	3.2
Walking		Aircraft flying routine	1.4
On level *mph*		Instrument landing	1.8
2	2.0	Combat flying	2.4
3	2.6	Domestic Work, Women	
4	3.8	House cleaning	2.0-3.4
Miscellaneous Occupations		Cooking	1.6-2.0
Carpentry		Shopping	1.4-1.8
Machine sawing, table	1.8-2.2	Office Work	
Sawing by hand	4.0-4.8	Typing	1.2-1.4
Planing by hand	5.6-6.4	Drafting	1.1-1.3
Garage Work (i.e. replacing tires, raising cars by jack)	2.2-3.0	Leisure Activities	
General laboratory work	1.4-1.8	Stream fishing	1.2-2.0
Machine Work		Dancing, social	2.4-4.4
Light (electrical industry)	2.0-2.4	Tennis, singles	3.6-4.6
Heavy (steel work)	3.5-4.5		

[a]Ranges are for activities which may vary considerably from one place of work or leisure to another or when performed by different people. 1 met = 58.2 W/m^2; 50 kcal/hr m^2; 18.4 Btu/hr ft^2. Some activities are difficult to evaluate because of differences in exercise intensity and body position.

**Abridged from ASHRAE* Handbook of Fundamentals *1981*.

pressed in *met* units; 1 met is defined as 58.2 W/m^2; 50 kcal/m^2 hr; or 18.4 Btu/hr ft^2. For an average size man, the *met* unit corresponds to approximately 90 kcal/hr; 100 W; or 360 Btu/hr. The met represents the average heat produced by a sedentary man. In Table 6-1 all energy costs are expressed in multiples of this relative unit. Values given are typical and primarily used in engineering planning.

Activities chosen are those performed steadily rather than intermittently. The higher energy level a person can maintain for any continuous length of time is approximately 50 percent of his maximal capacity to utilize oxygen (maximum energy capacity). A normal, healthy man has a maximum energy capacity of approximately 12 mets at 20 years of age, which drops to 7 mets at 70 years of age. Women tend to have maximum levels about 30 percent lower. Well-trained athletes have a maximum as high as 20 mets. For an untrained person at 35 years of age, the overall average can be considered as 10 mets. Activities performed continuously above the 5-met level by an untrained 35-year-old man may prove exhausting and uncomfortable.

Table 6-1 conveniently expresses the effect of different body activities in terms of met units. Another table to be presented in Chapter 9 extends these data into a convenient form for making heat gain calculations.

The right side of Eq. (6-1) is the heat loss from the body. These terms represent the ways the body is able to keep itself at a constant temperature. This is quite a remarkable feat, considering that activity may vary from sleeping to heavy work and in an air-conditioned atmosphere or in the broiling sun. Now each of the heat loss terms will be discussed.

6-3 SENSIBLE HEAT LOSS

Note the plus-or-minus sign in front of the sensible heat term in Eq. (6-1). This means that the body can either gain or lose sensible heat. The key is air temperature. If the air temperature is below the skin temperature, the sensible heat term is "plus"; that is, the body is losing sensible heat to the air. If the air temperature is above the skin temperature, the sensible heat term is "minus" that is, the body is gaining sensible heat from the air.

Some of the heat liberated in a furnace is transferred to the air surrounding the steel furnace walls. Similarly, some of the heat liberated within the body is transmitted to the air in contact with the skin. Although the deep body temperature remains at 98.6°F, the skin temperature varies. The skin temperature will vary from 40°F to 105°F (4.5 °C to 40.6°C) according to the temperature, humidity, and velocity of the surrounding air. If the temperature of the surrounding air falls, the skin temperature will also fall.

The range of skin temperature, at any location on the body, is comparatively small. A change of 10 degrees in room temperature does not produce anything like a change of 10 degrees in skin temperature. For instance, one series of tests reported, the forehead temperature ranged from 93°F (33.9°C) at 64°F (17.8°C) to 100°F (37.8C) in air at 96°F (35.6°C). So, as the temperature of the air rises, the difference in temperature between the skin and air decreases. At normal indoor air temperature, there is a steady flow of sensible heat from the skin to the surrounding air. The amount of this sensible heat depends upon the temperature difference between the skin and air. Inasmuch as this difference decreases as the temperature of the surrounding air rises, less sensible heat will be transferred to the surrounding air. When the surrounding air is at about 70°F (21.1°C), most people lose sensible heat at such a rate that they are comfortable. If the air rises to 80°F (26.7°C), the sensible heat loss drops to zero. The reason is that 80°F (26.7°C) is an average skin surface temperature for an adult. This is when he is indoor during the heating season and wearing comfortable clothing. If the air temperature continues to rise, the body will gain sensible heat from the air.

6-4 RADIANT HEAT LOSS

The body gains or loses heat by radiation according to the difference between two temperatures: (a) the body surface (bare skin and clothing) temperature and (b) the mean radiant temperature. Mean radiant temperature (MRT) is a weighted average of the temperature of all the surfaces in direct line of sight of the body.

If the MRT is below the body temperature, the radiant heat loss term, $\dot{q}_r$, in Eq. (6-1) is "plus." Then the body is losing radiant heat to surfaces it can see. If the MRT is above the body temperature, $\dot{q}_r$, is "minus" and the body is gaining radiant heat. It should be kept in mind that the body loses (or gains) sensible and radiant heat according to its surface temperature. For a comfortable adult normally dressed, the weighted average between bare skin and clothed surfaces is 80°F (26.7°C).

Surface temperatures have a lot to do with comfort. Suppose, during the heating season, a person is working at a desk facing the center of a room with his back 5 ft (1.5 m) from an outside wall. The wall surface temperature is 59°F (15°C). Assume the air distribution in the room is almost perfect (74°F, 2 in. below ceiling; 73°F, 2 in. above the floor; and 73.5°F, all around including the space between the person's back and the wall). Will he be comfortable? Probably not. Because radiant heat loss from the body to the wall is so high, a chill may be caused. What can be done to correct this situation? The temperature of the wall surface cannot be conveniently changed. The position of the desk might be changed so the person's back was against an inside wall. The surface temperature of the inside wall will be near the air temperature, probably 73°F. So the radiant heat loss would be less because of the smaller temperature difference. If the desk cannot be moved, what else might be done? The temperature of the air might be increased by turning up the thermostat. Increasing the air temperature will decrease the sensible heat loss from the body. Suppose the air temperature is set at 77°F (25.0°C). This will decrease the sensible heat loss by the same amount the radiant heat loss was decreased by moving the desk from the outside wall. This balances the heat loss from the body. The trouble is that everyone else in the room is too warm.

Exactly the same thing might be true during the cooling season. One might be too warm because of the radiant heat the body gains from a warm outside wall or window. Now the sensible heat loss from the body should be *increased* by *decreasing* the air temperature. This puts the body heat loss in balance, but everyone else in the room is too cool.

There is an important point to this discussion. It was assumed that air distribution was excellent; it was also assumed that air temperature could be raised or lowered at will. Still everyone in the room will not be comfortable all the time. This means that there is more to comfort than mechanical equipment. Good equipment properly operated will provide comfort in most instances. But it must be correctly applied. (It just will not do the job in a sheet iron shack or a tent). The building construction and orientation and the type of occupancy all influence comfort.

6-5 LATENT HEAT LOSS (EVAPORATION)

The last term in Eq. (6-1) is latent heat loss or perspiration. Under all conceivable situations this term will always be positive. Even at rest the body requires about 100 Btu/hr (29.3 W) to evaporate moisture into the inhaled air to keep the lungs moist. If it were not for this fact, lung tissues would stick together and not enough air would enter the lungs to keep alive. It is possible to "see" the breath when exhaling on a frosty morning. This is evidence that the air leaving the lungs has a high moisture content.

Evaporation heat regulation is the body process that sustains life outside an air-conditioned space. Equation (6-1) explains this. Suppose one goes outside when the air temperature is 100°F (37.8°C). The sensible heat loss is "minus" because the body is *gaining* heat from the air. The MRT is much higher than the body surface temperature: the sidewalk, street, building walls, and everything the body "sees" is above body surface temperature. Thus, the second term, $\dot{q}_r$, is also "minus" because the body is *gaining* radiant heat. But as one walks down the sidewalk, the left side of Eq. (6-1) (metabolism) is about 700 Btu/hr (205 W). So to maintain its heat balance the body must lose about 1025 Btu/hr (300 W) by evaporation. Nature takes care of this by automatically opening the sweat glands. Moisture on the body surface then evaporates. Most of the heat required to convert (evaporate) the moisture to a vapor comes from the body. As long as the air will carry away the water vapor, the skin temperature will remain cool. This in turn keeps the deep body temperature at very nearly 98.6°F (37°C).

Moisture loss by the body, whether in the exhaled air or from the skin, is in the form of steam at very low pressures. The body supplies the necessary latent heat to evaporate the moisture that it loses.

What bearing does the loss of moisture have on comfort when the temperature of the surrounding air rises from the comfortable levels to uncomfortable levels? As the dry-bulb temperature of the air rises, less sensible heat is lost by the body, but the latent heat loss increases because the loss of moisture from the body increases. Thus at 70°F (21.1°C), 290 Btu/hr (85 W) of sensible heat and 110 Btu/hr (32 W) of latent heat is lost. At 80°F (26.7°C) the sensible heat loss drops to nearly zero; but the latent heat loss increases to nearly 400 Btu/hr (117 W). Still in both cases, the total loss is the same, approximately 400 Btu/hr (117 W). When the outside temperature rises to about 90°F (32.2°C), all the body heat is carried away as latent heat.

When people work under conditions of high temperature and high humidity, both the sensible heat loss and the evaporation of moisture from their skins are retarded. Under these conditions, the rate of sensible heat transfer and evaporation must be increased by blowing air at high velocity over the body. In this way the evaporation of moisture from the surface of the skin is greatly accelerated, with a consequent increase in the total heat transferred from the skin to the air.

6-6 COMFORT AND EFFECTIVE TEMPERATURE

There appears to be no rigid rule as to the best atmospheric conditions for comfort for all people. Under the same conditions of temperature and humidity, the healthy young man is slightly warm while the elderly woman is cool. The customer who steps into a conditioned store from the bright sun of the street experiences a sense of relief. The active clerk who has been in the store for several hours may be a bit too warm for perfect comfort. The dancers in a night club may feel somewhat warm, the patrons seated at tables are comfortable, or perhaps slightly cool.

The comfort of an individual is affected by many variables. Health, age, activity, clothing, sex, food, and acclimatization all play their part in determining the elusive "best comfort conditions" for any particular person or group of persons. Hard and fast rules that will apply to all conditions and to all people cannot be given. The best that can be done is to approximate those conditions under which a majority of the occupants of a room will feel comfortable.

The discussion of the comfort charts to follow is intended to show comparisons and the developments from numerous studies of the past 50 years.

Figure 6-1 is based on the results of tests in which people were subjected to air at various temperatures and humidities. The coordinates of this chart are dry-bulb temperature and wet-bulb temperature. The straight lines sloping upwards from left to right are lines of constant relative humidity corresponding to various combinations of dry- and wet-bulb temperatures. The short, curved lines drawn across the humidity lines, and approximately perpendicular to them, are lines of constant *effective temperature* (ET).

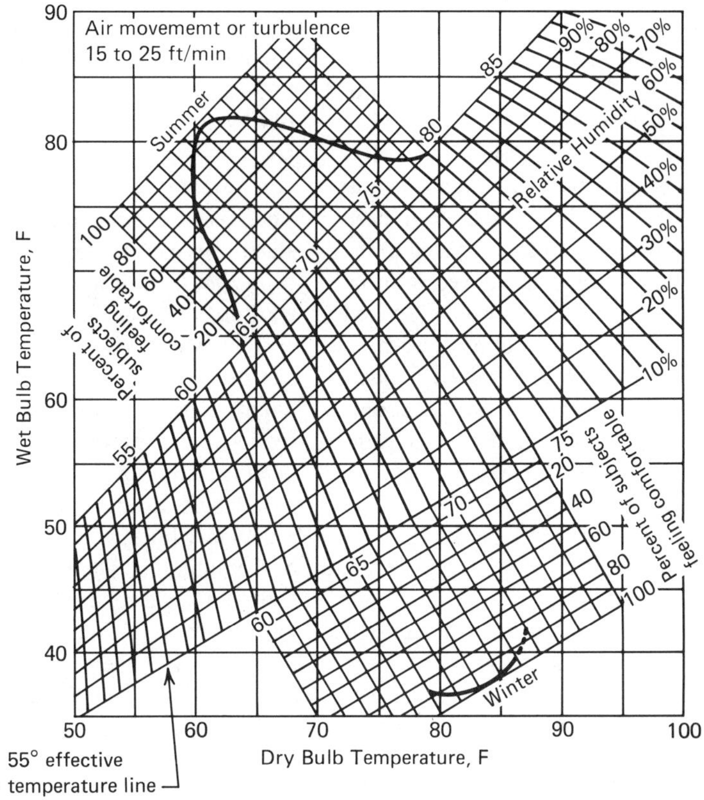

Figure 6–1

(Old) ASHRAE Comfort Chart for continuous occupancies of more than three hours duration (Reprinted by permission from ASHRAE Guide and Data Book 1961, Fundamentals and Equipment.)

Effective temperature is not an actual temperature in the sense that it can be measured by a thermometer. It is an experimentally determined index of the various combinations of temperature, humidity, and air movement which induce the same feeling of warmth. In summer most people who have been in a conditioned atmosphere for more than three hours will feel just about as cool at 75°F DB and 60 percent RH as at 79°F DB and 30 percent RH because both conditions fall on the line of 71 ET in Fig. 6-1. The curve in the upper left portion of Fig. 6-1 indicates the percent of subjects feeling comfortable during summer weather for conditions between 64 and 79 ET. Studies conducted by ASHRAE with relative humidities between 30 and 70 percent indicate that 98 percent of the subjects were comfortable when the wet- and dry-bulb temperatures fell on the 71 ET line. None of the test subjects were comfortable when the laboratory conditions were held at 64 ET and 79 ET.

Figure 6-1 must be used with care: Once the conditions under which the data were obtained are clearly understood, the reader's judgment will indicate whether or not the chart will apply to a particular condition. The original ASHRAE comfort chart (Fig. 6-1) was prepared in 1923 with an air movement of 15 to 25 fpm. Two test rooms were used. The dry- and wet-bulb temperatures were adjusted to give about the same feeling of warmth in each room. After a subject became acclimated to the conditions in one chamber he moved into the other chamber. The subject immediately voted his reaction to the new conditions: warmer, cooler, or same. Using several trained subjects and changing the conditions of the test rooms many times, a large mass of data was gathered. Thus, the lines of effective temperature (equal warmth) were drawn on Fig. 6-1. The 98 percent of subjects comfortable at 71 ET is applicable for summer air conditioning in offices and residences where the occupants will have a three-hour period to become acclimated to the air-conditioned space. Figure 6-1 was prepared for about 40° N latitude and for elevations below 1000 ft. An increase of about one degree effective temperature should be made for each five-degree reduction in North latitude.

Field studies by ASHRAE of large numbers of office workers led to some interesting observations: (a) Women of all age groups prefer an effective temperature one degree higher than men; (b) All men and women over 40 years of age prefer an effective temperature one degree higher than persons below this age; (c) For different geographical regions and age groups the most popular conditions vary from a low of 69 ET to a high of 73 ET.

While geographical location may account for a spread of four degrees in the desirable effective temperature, it should be pointed out that variations in the sensation of comfort among individuals may be greater in a particular environment than variations due to a difference in geographical location. Sometimes the engineer will be faced with psychological problems: The individuals who are not comfortable no matter what the effective temperature because they are in a building where glass blocks are used instead of windows and cannot see outside; or, the individuals who are not comfortable unless they have windows open. There is a valuable lesson in psychology in the statement that *a person who thinks he is uncomfortable is just as uncomfortable as if he were uncomfortable.*

The ASHRAE Comfort Chart of Fig. 6-1 can be used for occupancies of less than three hours, but the distribution curve of subjects feeling comfortable at a given effective temperature is no longer applicable. A conditioned space maintained at 71 ET may be too cool for a person who has just left the oppressive heat of the outdoors. The body functions having become adjusted to the higher outdoor temperature, a period of time is required for their readjustment to the lower temperature and humidity. Thus, the factor of acclimation plays a part in determining the best conditions for comfort.

For people who are going to be in a conditioned atmosphere for only a short period of time, the air conditions should be maintained above the 71 ET shown in Fig. 6-1 as optimum for a three-hour occupancy. Department stores, office buildings, theaters, restaurants, and many other commercial installations should maintain conditions higher than 71 ET. Even in homes, people are in and out several times a day, and 71 ET may be too cool for some people. In many commercial installations the temperatures and humidities to be maintained must be a compromise between those necessary to insure comfort of employees and those necessary to avoid too great a contrast between indoor and outdoor for the customers.

In some air-conditioned factories half-hour lunch periods, lunch rooms, and cafeteria service discourage employees from leaving the air-conditioned building. It has been found desirable in such instances to allow the effective temperature to slowly rise during the last hour of the working day in order to lower the contrast between indoor and outdoor conditions when the employees leave the air-conditioned building.

People doing physical work need a lower effective temperature for comfort. Naturally, the greater the activity and the more clothing worn, the lower must be the effective temperature for comfort. Although the latent heat liberated by people engaged in any physical activity rises markedly, the sensible heat liberated remains approximately the same. For example, an average man seated at rest in a room at 80°F DB gives off 180 Btu/hr (53 W) of sensible heat and 150 Btu/hr (44 W) of latent heat, or a total of 330 Btu/hr (97 W). Now, if he does light bench work, he will liberate 220 Btu/hr (64.5 W) as sensible heat and 530 Btu/hr (155 W) latent heat. Note that the sensible heat increased about 20 percent but the latent heat increased about 250 percent. Thus, the degree of activity must be taken into account in arriving at an effective temperature for comfort.

During the first 35 years of use of the ASHRAE comfort chart (Fig. 6-1) many changes were suggested. Some people thought the chart gave too much emphasis on the effect of humidity. From the viewpoint of body-heat dissipation alone, the range through which relative humidity is varied does not matter, but other considerations make the most desirable relative humidity range lie between 30 and 70 percent. For values below 30 percent the mucous membranes and the skin surface may become uncomfortably dry. For values above the 70 percent (or even 60 percent) range, there is a tendency for a clammy or sticky sensation to develop. The effect of increased air velocity is always to increase the convection and evaporative ($\dot{q}_s$ and $\dot{q}_L$) heat dissipation from the body except at high relative humidities combined with temperatures exceeding 100°F (37.8°C).

Other HVAC engineers believed that most people had about the same degree of comfort at the same dry-bulb temperature both winter and summer. However, over the 35 years of its use, the ASHRAE Comfort Chart remained essentially unchanged.

ASHRAE has supported research on thermal comfort for more than 50 years. The original ET scale (Houghton and Yaglou, 1923), Fig. 6-1, is still often used by physiologists and engineers as the standard in measuring comfort, although its shortcomings (overexaggeration of humidity at lower temperatures and underestimation of humidity at heat tolerance levels) were recognized by Yaglou as early as 1947. In 1962, Nevins[3] showed that temperature criteria for thermal comfort have risen steadily since 1900, from 18°C to 21°C (65°F to 70°F) DB range to 24°C to 26°C (75°F to 78°F) DB range in 1960. Effective temperature for comfort increased from 18°C (64°F) ET in 1923 to 20°C (68°F) ET in 1941. This increasing trend probably results from year-round use of lighter-weight clothing and from changing living patterns, diets, and *comfort expectations*.

In the 1950s ASHRAE decided to reevaluate its effective temperature scale. Studies showed humidity had negligible effect on comfort until 60 percent RH and 65°F (18°C) dry bulb were reached. Below these levels, dry-bulb temperature alone was the governing factor. Since 1956, comfort limits selected for field application have usually been expressed in terms of dry- and wet-bulb temperatures. The original ET scale is still used on a numerical basis for predicting extreme discomfort from high temperatures and humidities and man's heat tolerance while working at high humidities. But since 1956, it has seldom been used to specify comfort conditions.

Physiologists have recognized that sensations of comfort and of temperature may have different physiological and physical bases, and each should be considered separately. This dichotomy was recognized in ASHRAE Comfort Standard 55-74, where thermal comfort is defined as "that state of mind which expresses satisfaction with the thermal environment." Unfortunately, little practical research to date specifically fulfills this definition. Most current predictive charts are based on comfort defined as "a sensation that is neither slightly warm nor slightly cool."

ASHRAE research has usually been limited to *sedentary* man, lightly clothed. This approach is sound, since about 90 percent of man's indoor occupation and leisure time is spent at or near the *sedentary* activity level. Predictive methods available are believed accurate within stated limitations. During physical activity, a change occurs in man's physiology. Physiological thermal neutrality in the older (sedentary) sense does not exist. Some form of thermal regulation always occurs within man's body during exercise. Temperature and comfort during activity have different bases than during the sedentary condition. Sedentary skin and body temperatures used as indices of comfort, will likely prove false during moderate to heavy activity.

6-7 THE NEW ASHRAE COMFORT CHART

The most recent comfort chart based on ASHRAE research is shown on the standard ASHRAE psychrometric chart, Fig. 6-2. The comfort zone recommended in ASHRAE Comfort Standard 55-74 is shown. The new effective temperature (ET*) lines have been redefined with their reference base set on the 50 percent RH line.

Figure 6-2 generally applies to altitudes from sea level to 2134 m (7000 ft), and to the most common special case for indoor thermal environments in which the MRT is nearly equal to the dry-bulb air temperature and air velocity is less than 0.2286 m/s (45 fpm). For this special case the thermal environment is well specified by the two variables shown: dry-bulb air temperature and relative humidity.

The Institute for Environmental Research at Kansas State University under ASHRAE contracts conducted extensive research since 1963 on thermal comfort of clothed sedentary subjects. The KSU-ASHRAE envelope shown on Fig. 6-2 is the result of the data collected. The KSU data applies for occupants wearing slightly lighter clothing and activity level than used for the ASHRAE 55-74 data. These small differences in activity levels and clothing insulation account for the 1.5°C (3°F) displacement of the KSU-ASHRAE envelope above the standard 55-74 zone.

A wide range of environmental applications is covered by ASHRAE Comfort Standard 55-74. Offices, homes, schools, shops, theaters, and so forth can be approximated with these specifications. The standard specifies an environmental range for comfort based on available research data and field judgments of the standard's subcommittee members in 1974.

In place of the old effective temperature (ET) lines, shown on Fig. 6-1, the new effective temperature (ET*) lines are plotted on Fig. 6-2. The chart is primarily useful in the usual comfort range for sedentary, lightly clothed people, and secondarily at higher temperatures where sedentary heat stress is involved. Although the ET* scale is based on 1-hour exposure, previous data show no changes (of engineering importance) in response due to longer exposures unless the limits of heat stress (ET*32°C or 90°F) are approached.

The ET* lines show a moderate humidity effect on thermal comfort itself. Its effects on discomfort increases as both the thermal level of the environment and regulatory sweating in-

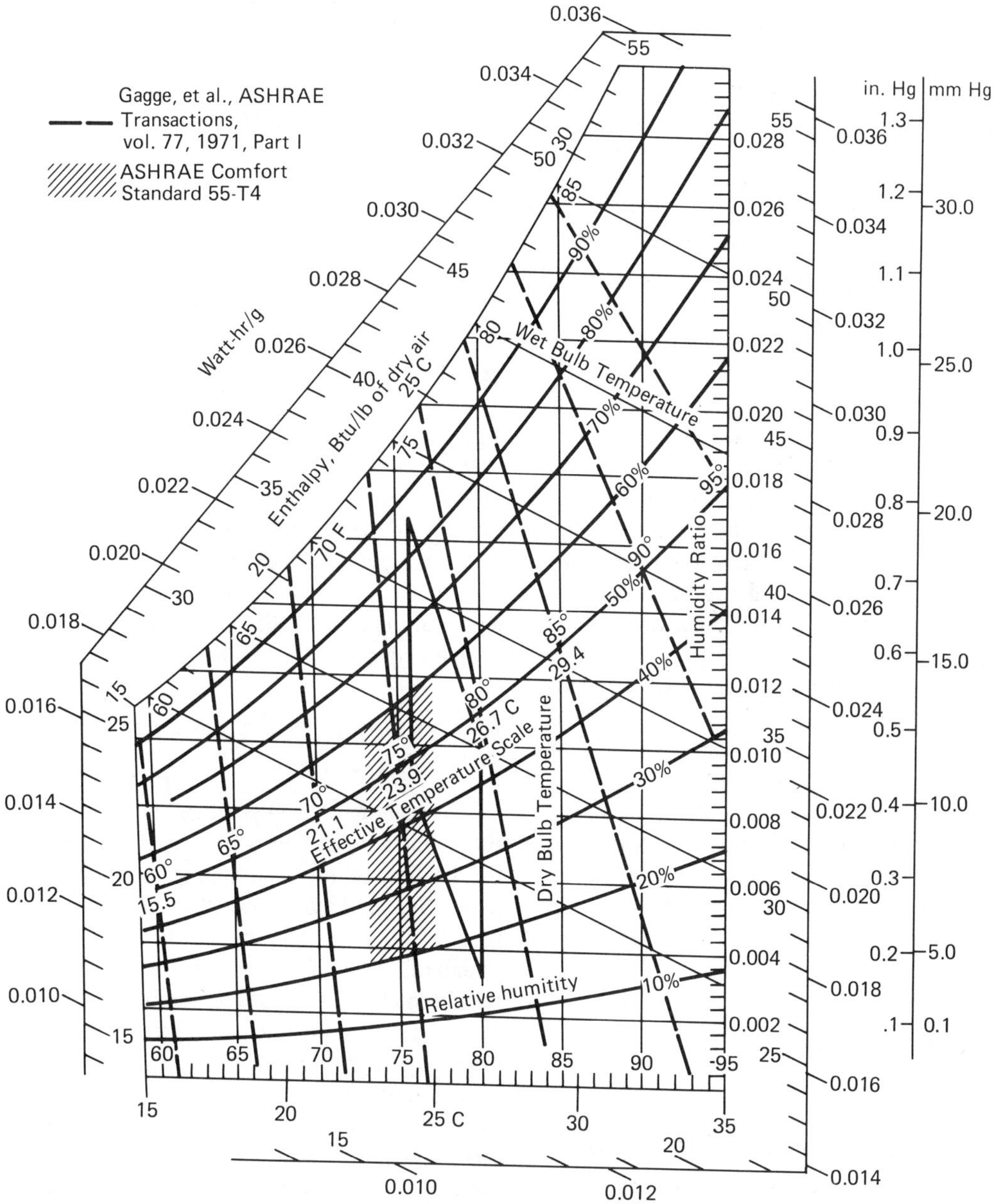

Figure 6–2

(New) Effective Temperature Chart (ET*) (Reprinted with permission from ASHRAE Handbook of Fundamentals 1981.)

crease. Man's total evaporative heat near the comfort range is only about 25 percent of total heat loss. As the thermal level increases, this percentage grows; evaporation accounts for 100 percent of the heat loss as the environmental temperature equals and rises above skin temperature of 35°C (or 95°F).

The new ET* scale is labeled as the dry-bulb temperature at the *intersection* of loci of constant physiological strain with the 50 percent relative humidity line. For dry-bulb temperatures above 24°C (75°F), this choice gives index temperature values which feel more accurate to engineers and the public, since most practical environments are usually much nearer 50 percent than 100 percent relative humidity. The old ET, indexed by the dry-bulb temperature at 100 percent relative humidity, resulted under practical conditions in a temperature always lower than the actual dry-bulb temperature. The new index, if used for laymen in reporting indoor and outdoor weather conditions, may result in better public acceptance of environmental control.

To summarize, data in Fig. 6-2 is useful for light clothing and seated or sedentary activity. ASHRAE Comfort Standard 55-74 generally applies for average clothing and activity. The most commonly recommended design conditions for comfort where the two zones overlap are therefore:

ET* = 24.5°C or 76°F

Dry bulb temperature = MRT

= 24.5°C or 76°F

Relative humidity 40 percent (20 to 60 range)

Air velocity less than 0.23 m/s (45 fpm)

With the increasing need to conserve energy, adjustments are being mandated by the federal government in the allowable winter and summer thermostat settings in public buildings.

A question then arises relative to temperatures as low as 68°F (20°C) in winter and as high as 80°F (26.7°C) in summer. For an individual to be comfortable, he has no choice but to wear substantially heavy clothing in winter and lighter clothing in summer.

Suggested indoor design dry-bulb temperature, wet-bulb temperature, and relative humidity are shown in Table 6-2. Note that each combination of these values gives values of ET* within the shaded area and diamond-shaped area of Fig. 6-2. The ET values shown in Table 6-2 are those found from Fig. 6-1.

Table 6-3 gives a range of winter indoor design dry-bulb temperatures which are most commonly found in practice. However, the most comfortable dry-bulb temperature to be maintained depends on the relative humidity and air motion. These three factors must be considered together to produce an effective temperature which will satisfy the majority of people occupying the conditioned space.

6-8 TEMPERATURE AND HUMIDITY FLUCTUATIONS

Environmental temperature and humidity often fluctuate as the control system operates the heating or air-conditioning system equipment. Studies have shown that allowable fluctuating limits stated in ASHRAE Comfort Standard 55-74 are conservative; and that in the thermal comfort range, the rate of change of temperature should not exceed 2.2°C/hr (4.0°F/hr) if the peak-to-peak variation in dry-bulb temperature cycle is 1.1°C (2.0°F) or greater. For the mean radiant temperature fluctuations, the rate of 1.7°F/hr should not be exceeded if the peak-to-peak variation in mean radiant temperature is 0.8°C (1.5°F) or greater. Humidity limits are quite broad, usually posing no problem for modern controls; the rate of RH change should not exceed 20 percent if the peak-to-peak variation in humidity is 10 percent or greater.

Table 6-2

Indoor Design Conditions for Comfort Cooling

Outside Design	Occupancy Over 40 Minutes				Occupancy Under 40 Minutes			
Dry Bulb Deg F	Dry Bulb °F	Wet Bulb °F	Rel. Hum. %	Eff. Temp E. T.	Dry Bulb °F	Wet Bulb °F	Rel. Hum. %	Eff. Temp E. T.
80	75	65	60	71	76	66	61	72
	76	64	53	71	77	65	54	72
	77	63	47	71	78	64	47	72
	78	62	40	71	79	63	41	72
	79	61	35	71	80	62	36	72
85	76	66	61	72	77	67	61	73
	77	65	54	72	78	66	54	73
	78	64	47	72	79	65	48	73
	79	63	41	72	80	64	42	73
	80	62	36	72	81	63	36	73
90	77	67	61	73	78	69	64	74
	78	66	54	73	79	68	58	74
	79	65	48	73	80	67	52	74
	80	64	42	73	81	66	46	74
	81	63	36	73	82	65	40	74
					83	64	35	74
95	78	69	64	74	79	70	65	75
	79	68	58	74	80	69	58	75
	80	67	52	74	81	68	52	75
	81	66	46	74	82	67	47	75
	82	65	40	74	83	66	41	75
	83	64	35	74	84	65	36	75
100	79	70	65	75	81	71	63	76
	80	69	58	75	82	70	56	76
	81	68	52	75	83	69	50	76
	82	67	47	75	84	68	44	76
	83	66	41	75	85	67	38	76
	84	65	36	75				
105	80	71	65	75.5	81	72	65	76.5
	81	70	58	75.5	82	71	59	76.5
	82	69	52	75.5	83	70	54	76.5
	83	68	47	75.5	84	69	47	76.5
	84	67	42	75.5	85	68	41	76.5
	85	66	37	75.5	86	67	36	76.5

Effective temperature (ET) values shown are those using Fig. 6-1.

6-9 AIR DISTRIBUTION WITHIN CONDITIONED SPACES

In Section 6-8 temperature and humidity fluctuations within a given conditioned space were discussed briefly. Air introduced into a conditioned space for heating or cooling plays an important role in maintaining equilibrium conditions within a space. Air introduced into a conditioned space should be distributed in such a way that within the occupied zone (floor level to 6 ft above floor) there are only minor horizontal or vertical temperature variations and the proper quantity of air is delivered to different sections of the space according to the heating or cooling requirements. Yet both of these requirements must be met without drafts. A *draft* may be defined as a noticeable air current. Drafts are objectionable. Yet there must be air motion or the occupants of a space will

Table 6-3

Winter Indoor Dry-Bulb Temperatures Usually Specified

TYPE OF BUILDING	DRY BULB °F.
SCHOOLS	
Classrooms	72–74
Assembly rooms	68–72
Gymnasiums	55–65
Toilets and baths	70
Wardrobe and locker rooms	65–68
Kitchens	66
Dining and lunch rooms	65–70
Playrooms	60–65
Natatoriums	75
HOSPITALS	
Private rooms	72–74
Private rooms (surgical)	70–80
Operating rooms	70–95
Wards	72–74
Kitchen and laundries	66
Toilets	68
Bathrooms	70–80
THEATERS	
Seating Space	68–72
Lounge rooms	68–72
Toilets	68

TYPE OF BUILDING	DRY BULB °F.
HOTELS	
Bedrooms and baths	75
Dining rooms	72
Kitchens and laundries	66
Ballrooms	65–68
Toilets and service rooms	68
HOMES	73–75
STORES	65–68
PUBLIC BUILDINGS	72–74
WARM AIR BATHS	120
STEAM BATHS	110
FACTORIES AND MACHINE SHOPS	60–65
FOUNDRIES AND BOILER SHOPS	50–60
PAINT SHOPS	80

feel uncomfortable. Heat and moisture must be carried away from the body as they are liberated or a stagnant film of warm, moist air would envelop each occupant. The type of occupancy, the physical arrangement of the space, the acceptable noise level, and the degree of activity of the occupants all have a bearing on the permissible air velocity in a conditioned space. Generally, a velocity of 15 to 25 fpm (0.08 to 0.13 m/s) is considered to be still air while air moving at 65 fpm would constitute a draft to most people.

In order to cool an air-conditioned space the supply air is usually introduced into the space at 12 to 30°F (6.7 to 16.7°C) below the required space air temperature at a velocity considerably above 15 fpm. Suppose a space were to be maintained at 78°F (25.6°C) and the conditioned supply air were introduced at 60°F (15.6°C) and 500 fpm (2.5 m/sec). A good air distribution system for this space must do the following:

1. Entrain enough room air with the supply air that upon reaching the occupied zone the airstream will be warm enough not to be objectionable.
2. Reduce the air velocity before reaching the occupied zone to a point that there will be freedom from drafts.
3. Provide a turbulent, eddying motion within the entire occupied zone.
4. Keep air noise from the supply outlets and return air inlets below the objectionable level.

While it may seem that these requirements are stringent, it should be pointed out that an air-conditioning job may have the very best mechanical components, properly sized for the load, and yet, from the owner's standpoint be unsatisfactory if the air distribution is not providing comfort for the occupants.

We will have more to say about air distribution in a later chapter.

6-10 QUALITY AND QUANTITY OF AIR SUPPLIED

The Value of Air (4). The quantity of air available remains constant though its quality does not. In a sense we breathe today the same air used and reused by countless preceding generations. Our air is continually regenerated for us by nature.

But the demands of modern life and industry are so complex and urgent that it is no longer practical to depend upon nature to provide air of the precise quality required, at a specific time or in a particular place. A climate neither too cold nor too hot, too dry nor too humid, all year-round is not man's heritage. To make our immediate atmosphere more suitable to our needs, we have learned to "condition" air mechanically. The process is not without cost.

Whether simple or intricate, an air-conditioning system represents a capital investment in equipment—in fans, ducts, dampers, diffusers, and grilles; in boilers, compressors, tempering and cooling coils, filters, air washers, and pumps; in thermostats and regulators of all types. It represents an operating charge for electrical energy, fuel, water, and operating and maintenance labor—all expended for the sole purpose of obtaining the precise quality and quantity of air desired and delivering it when and where required.

Once air is conditioned, contrary to being "free," it becomes a valuable commodity because of the conditioning energy that has been invested in it. The protection of that investment by conserving the conditioned air becomes important.

However, air becomes contaminated by "use," and we are as much concerned about air purity and freshness as we are about its temperature and humidity. Here we may anticipate nature by resorting to mechanical and chemical purification to restore air freshness while at the same time conserving its thermal and psychrometric value.

Air purity is relative. Except possibly on a mountain top immediately after a rainstorm, no air is entirely free of entrained impurities. Even mountain air, for example, will be found to contain a trace (though negligible) of sulfur dioxide. The extent to which impurities are present in air is termed their *concentration*. Concentration is normally expressed in one of two different ways; mass per unit volume of air (g/m^3) or volume per number of volumes of air (parts per million, ppm).

For social and economic reasons we have found it necessary to congregate in confined areas and consequently must meet the problem of obtaining uncontaminated air. All air which surrounds dwellings and industry accumulates foreign substances. These contribute no appreciable ill effects so long as their concentration is not excessive. There is, however, a "threshold" concentration of impurities which is considered optimal or tolerable for each contaminating substance under specified circumstances of exposure. Where the allowable threshold concentrations, as in the case of poisonous fumes, the criteria establishing threshold are often a compromise. The concentration tolerated in a foundry, for example, would not be considered suitable in a home. On the other hand the degree of air purity demanded in a factory devoted to the making of highly sensitive photographic films or precision instruments may be substantially higher than that which would be completely acceptable in the finest residence.

When air purification is employed to preserve the freshness of conditioned air so as to conserve its thermal and psychrometric value—its heat or lack of heat, its moisture or dryness—the process is called *air recovery*.

Regardless of the amount of outside air entering a building by natural or mechanical methods, the air must meet certain quality standards. ASHRAE Standard 62-81, *Ventilation for Acceptable Indoor Air Quality,* lists maximum allowable pollutant concentrations in air used for the ventilation of spaces used for human occupancy and are shown in Table 6-4.

The intrinsic worth or utility of an air-conditioning system is dependent on two factors: (a) the cost of the installation and (b) the cost of operation. Both of these factors are directly related to the amount of new, outdoor air required by the space to be conditioned. The internal heat gain or loss is a constant dictated by the locale, exposure, construction, occupancy, and function of the space. This constant represents the irreducible minimum of conditioning required. The air change, that is, the volume of outdoor air that must be supplied and conditioned, constitutes the variable and, when added to the internal load, determines the size and cost of the conditioning system.

The vitalizing quality in outdoor air is its oxygen content, which is almost exactly 21 percent by volume. The freshness of the air is entirely a function of its freedom from entrained impurities.

Ventilation standards, insofar as they are related to the above, establish the volume of air that must be supplied to a building or enclosure in order to (a) replenish the oxygen consumed and (b) dilute internally generated air-contaminating impurities.

This air needed to satisfy the first requirement must invariably be new outdoor air regardless of whether or not it is also fresh, that is, uncontaminated. The air required to serve the second purpose must obviously be fresh regardless of whether or not it is outdoor air. Far more air is required for dilution of impurities than for oxygen replenishment; the former may, in fact, be 10 to 20 times the latter.

All air expelled from an air-conditioned space carries with it cooling or heating energy. All air that can be recirculated conserves this conditioning energy. Whenever stale, vitiated, or otherwise contaminated conditioned air can be converted to its original freshness at a cost

Table 6-4

National Ambient Air Quality Standard[a]*

	Long Term		Short Term	
Contaminant	**Level**	**Time**	**Level**	**Time**
Carbon Monoxide			40 mg/m^3 10 mg/m^3	1 hr 8 hr
Lead	1.5 μg/m^3	3 mon		
Nitrogen Dioxide	100 μg/m^3	yr		
Oxidants (Ozone)			235 μg/m^3	1 hr
Particulates	75 μg/m^3	yr	260 μg/m^3	24 hr
Sulfur Dioxide	80 μg/m^3	yr	365 μg/m^3	24 hr

[a]U.S. Environmental Protection Agency, National Primary and Secondary Ambient Air Quality Standards, Code of Federal Regulations, Title 40 Part 50 (40 C.F.R. 50). Pertinent local regulations should also be checked. Some regulations may be more restrictive than given here, and additional substances may be regulated.

**Reprinted by permission from ASHRAE Standard 62-81, 1981.*

less than that required to replace it, the failure to employ such conversion adds to the cost and operation of the conditioning plant.

ASHRAE Standard 62-81 defines *ventilation* as "the process of supplying and removing air by natural or mechanical means to and from any space. Such air may or may not be conditioned." Further, *ventilation air* is defined as "that portion of supply air which is outdoor air plus any recirculated air that has been treated for the purpose of maintaining acceptable indoor air quality." Ventilation air becomes supply air after passing through the conditioning equipment. Ventilation air may be 100 percent outdoor air. The term *makeup* air may be used synonymously with *outdoor* air, and *return* and *recirculated* air are often used interchangeably.

6-11 OUTDOOR AIR REQUIREMENTS

Infiltration. The outdoor air makeup in any air-conditioning system must be at least equivalent to the anticipated natural air leakage of the building or structure. This leakage takes place through cracks around doors and windows, chimneys and other openings and, to a minor extent, through solid walls. The movement or exchange of air through leakage is caused mainly by (1) outside wind pressure, usually referred to as infiltration, and (2) the difference in density between inside and outside temperature, commonly called stack effect. Since the former predominates, except in high buildings, it is convenient to combine both factors under the term *infiltration effect.* In average practice this will be equivalent to from one-half to two air changes per hour. Accumulated experimental data, which enable us to estimate infiltration, will be found in Chapter 8.

The importance of infiltration in air conditioning is that it establishes the amount of new outdoor air that must be supplied to maintain an internal static air pressure adequate to prevent infiltration by leakage. This air supply to counteract infiltration is an independent demand and must be added to any outdoor air makeup required to replace mechanically exhausted air.

Oxygen Requirements. Aside from other considerations, such as infiltration and exhaust, the minimum of new outdoor air required in an air-conditioning or ventilating system is that needed to replenish the oxygen consumed. In the light of all research, the new air required to satisfy oxygen demands is so small as to be entirely negligible. The unavoidable infiltration factor in any type of above-ground structure will exceed the volume of outdoor air necessary for adequate oxygen replenishment. Except in overcrowded school rooms or, possibly, theaters and auditoriums, the space per occupant would be approximately 300 ft^3. The average infiltration will usually exceed one air change per hour. Thus, the infiltration per occupant in all but exceptional cases will be at least equivalent to 300 ft^3/hr or 5 cfm (2.4 l/s).

An average adult, even when engaged in physical activity will consume only 1.5 ft^3/hr (0.012 l/s) of oxygen. As air contains 21 percent oxygen by volume, the prevailing indoor oxygen concentration when 300 ft^3/hr of new air per occupant is supplied will be

$$\text{Oxygen concentration} = \frac{(300 \times 0.21) - 1.5}{300} \times 100 = 20.5 \text{ percent}$$

This small decrease in oxygen is no greater than that which is experienced naturally by going from sea level to an elevation of 700 ft. In fact, even smaller quantities of replenishment air are entirely tolerable as long as odor contamination levels are effectively controlled.

Dilution. Except for the minimum air volume

necessary to counteract infiltration and to supply oxygen, all ventilation is essentially dilution. This function of ventilation serves to displace contaminated indoor air with relatively uncontaminated air at a rate which will constantly maintain the desired threshold concentration of air-entrained impurities within the enclosure ventilated. Ventilation standards are thus governed primarily by the nature and the extent of internal air pollution. People, materials, products, and processes all contribute to air pollution. The greater the internal generation of air-contaminating impurities, or the lower the threshold concentration desired, the higher will be the rate of ventilation (fresh air) required. The principle involved is the same whether the ventilation is obtained by supplying outside air to the space or by purifying (recovering) and recirculating space air at an equivalent rate.

The amount of ventilation necessary for a given type of space for various kinds of occupancy and activity has evolved mainly through long experience. Where the findings of such experience has been sufficiently conclusive, ventilation requirements have been reduced to standardized practices, codes and, in some instances, statutes. These generally stipulate the ventilation (fresh air) requirements in such terms as cubic feet per minute (cfm) (liters per second) of fresh air per occupant; cfm/ft^2 of floor area or ft^3 of space; air changes per hour, or air velocity across the face of a ventilating hood.

Table 6-5 extracted from *ASHRAE Fundamentals Handbook 1977* gives a partial listing of minimum and recommended ventilation rates per certain types of occupancy. A more complete listing is found in ASHRAE Standard 62-73. The requirements listed in Table 6-5 are for 100 percent outdoor air when the outdoor air meets the specifications of Table 6-4. The outdoor air requirement may be reduced to 22 percent of the specified ventilation air quantity when adequate temperature control is provided in addition to filtering equipment so that contaminant levels given in Table 6-4 are not exceeded. If odor and gas removal equipment is also provided, the outdoor air requirement may be reduced to 15 percent of the specified ventilation air quantity. However, in no case can the outdoor air quantity be less than 5 cfm or 2.5 l/s per person. In some cases local codes and statutes may override the limits stated above. For energy conservation ASHRAE Standard 90-75, *Energy Conservation in New Building Design*[5], states that the minimum column values of Table 6-3 shall be used, and the outdoor air portion of the ventilation air shall be reduced as discussed above. However, the standard also specifies that allowance for 100 percent outside air shall be provided for cooling purposes when conditions are right (usually early fall and late spring). This is called an *economizer cycle* since it saves energy at off-design conditions.

What is significant is that the desired ventilation discussed is obtainable, without modifying current ventilation practices, through the proper application of air recovery. Ventilation air, over and above the infiltration or oxygen demand, need not be new outdoor air. Since the function of ventilation air is that of a dilution agent, it must only be fresh; that is, uncontaminated. When an air-conditioning system is used, this diluation agent may be obtained by purifying and thus converting a portion of the recirculated air to fresh air. Because such air is already conditioned, its recovery and use in lieu of outdoor air minimizes the conditioning load of the system.

It should be emphasized that the substitution of decontaminated recirculated air for outdoor air as a ventilating medium affects neither accepted ventilating standards nor the amount of ventilation that may be preferred by the design engineer.

6-12 AIR-ENTRAINED IMPURITIES

The impurities found in air may be divided into the following four general classifications:

Table 6-5

Ventilation Requirements for Occupants

	Estimated Persons Per 100 ft² Floor Space	Required Ventilation Air Per Human Occupant			
		Minimum		Recommended	
		cfm	l/s	cfm	l/s
Single Unit Dwellings					
General living areas, bedrooms, utility rooms	5	5	2.5	7-10	3.5-5
Kitchens, baths, toilet rooms	—	20	10	30-50	15-25
Commercial					
Public rest rooms	100	15	7.5	20-25	10-12.5
Department stores					
Sales floors	30	7	3.5	10-15	5-7.5
Dressing rooms	—	7	3.5	10-15	5-7.5
Malls and Arcades	40	7	3.5	10-15	5-7.5
Dining rooms	70	10	5.0	15-20	7.5-10
Kitchens	20	30	15	35	17.5
Cafeterias	100	30	15	35	17.5
Hotels, Motels, Resorts					
Bedrooms	5	7	3.5	10-15	5-7.5
Baths, toilets	—	20	10	30-50	15-25
Assembly rooms (large)	140	15	.5	20-25	10-12.5
Theaters					
Lobbies (Foyers & Lounges)	150	20	10	25-30	12.5-15
Auditorium (no smoking)	150	5	2.5	5-10	2.5-5
Auditoriums (smoking)	150	10	5	10-20	5-10
Gymnasiums and Arenas					
Playing floor	70	20	10	25-30	12.5-15
Locker rooms	20	30	15	40-50	20-25
Spectator area	150	20	10	25-30	12.5-15
Tennis, Squash, Handball Courts	—	20	10	25-30	12.5-15
Swimming Pools	25	15	7.5	20-25	10-12.5
Offices					
General office space	10	15	7.5	15-25	7.5-12.5
Conference rooms	60	25	12.5	30-40	15-20
Institutional					
Schools					
Classrooms	50	10	5	10-15	5-7.5
Multiple use rooms	70	10	5	10-15	5-7.5
Laboratories	30	10	5	10-15	5-7.5
Auditoriums	150	5	2.5	5-7.5	2.5-3.8
Gymnasiums	70	20	10	25-30	12.5-15
Lunch rooms, dining halls	100	10	5	15-20	7.5-10
Corridors	50	15	7.5	20-25	10-12.5

1. *Ducts and Smokes*—These terms define biological inert solid particles suspended in air. Dust particles are light enough to be blown about by air currents yet heavy enough to settle readily by gravity in comparatively still air. Smoke particles, being much smaller and lighter, may remain suspended for long periods even in still air.

2. *Mists and Fogs*—These terms define biologically, inert liquid particles (droplets) suspended in air. The migrations of such particles are determined by the same factors as those of dust and smoke.

3. *Microorganisms*—This term defines air-suspended bacteria, germs, viruses, and other living organisms which are light enough to "float" in air but more generally ride on inert suspended solid or liquid particles.

4. *Gases and Vapors*—These terms define matter consisting of dispersed molecules which are diffused rather than suspended in air. The motion of molecules is the result of their collisions and is influenced by temperature and pressure. For particles under 0.01 μm in size, the kinetic velocity becomes dominant, and this dimension is therefore sometimes conceived as the boundary between particles and gases. Gaseous impurities in air account for odor sensations, air "staleness" and, in many instances, irritating or toxic effects. Such impurities comprise many thousands of organic and inorganic substances.

Table 6-6 illustrates the relationships of size and characteristics of various air-entrained impurities.

The sources of atmospheric impurities are so diversified that enumeration would be endless. The principal sources of air-contaminating impurities in residential and commercial buildings are the occupants, their habits and services: body, apparel and respiratory emanations, tobacco smoke, cosmetics, liquor, edibles and their cooking or preparation, furnishings, painted surfaces, cleaning components, and so forth. In the average industrial establishment of a manufacturing nature, there are also contaminating sources such as cutting and lubricating oils and greases, solvents, fumes and metallic oxides, internally generated dust, soot, and smoke. Processing industries contribute vapors and fumes in almost infinite variety. Recently, outdoor air as a source of odors has increased in importance because in some areas it may contain a high percentage of automotive exhaust and electrical power generating stations emissions.

The air-conditioning system itself may contribute odors. Cooling coils collect dirt and lint both of which are moistened by the condensate; this helps mildew to form and leads to objectionable odors.

6-13 REMOVAL OF PARTICULATE MATTER

The means to be employed in the purification of air will depend upon the nature and concentration of the contaminants and the extent to which their elimination is necessary or desired. Dust, for example, is usually present to an extent warranting the use of *air filters*. Of these there are various types available, the selection of which will be dictated by the degree of efficiency demanded. Where extremely light and minute particles, such as those composing smoke and soot, prevail in objectionable concentration, elimination of these by the use of electrostatic precipitation will prove effective. Such filtration will likewise contribute to the control of pollen concentration. For those special instances where the hazardous accumulation of objectionable bacteria may be anticipated, the use of appropriate germicidal agents would be indicated.

Table 6-6

Air-Entrained Impurities, Relationship of Size and Characteristics.

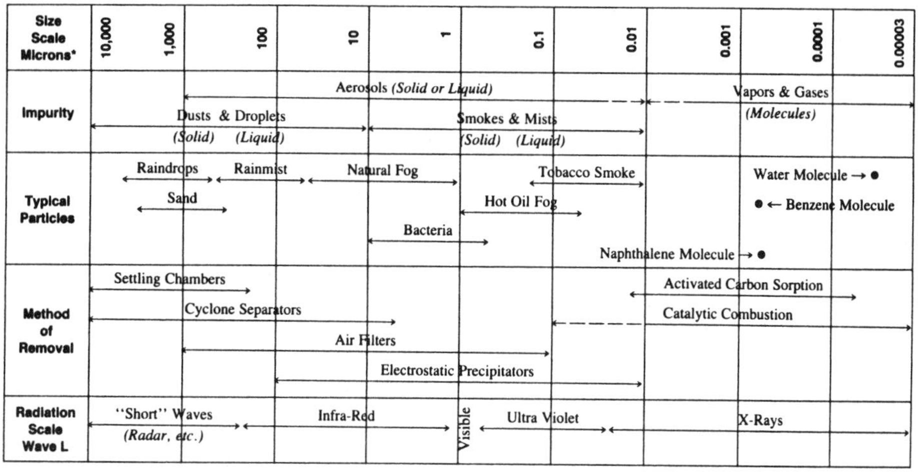

*1 Micron = 10^{-4} cm = $\frac{1}{25,400}$ in.

One or more of these devices may be required to produce the desired quality of air in a conditioning system, and technical data on their respective engineering features are available from manufacturers, and *ASHRAE Guide and Data Book, Systems 1976* has most valuable data. In addition, means of limiting the concentration of vaporous and gaseous impurities are necessary if air is to be recirculated indefinitely.

6-14 ELIMINATION OF VAPORS AND GASES

Various methods of eliminating air-entrained gaseous and vaporous contaminants have been employed. One persistent though futile approach to this problem has been the introduction of chemical agents either directly into the space or into the recirculated airstream to react with the entrained gaseous impurities in a manner tending to destroy, decompose, or otherwise alter their objectionable character. This method faces inescapable obstacles. On the one hand agents which will actually destroy or alter the chemical composition of the diffused substances are often toxic. On the other hand chemical agents which do not decompose or change the existing impurities are effective, if at all, merely as a screen; for example, masking a disagreeable odor with a stronger one of presumably more agreeable characteristics. Sometimes the masking effect is not merely sensory but pathological. Some so-called deodorizing substances may contain ingredients such as formaldehyde which deaden or anesthetize the olfactory nerves, thus preventing the detection of either the masking or the offending odors. In either event the result is one of adding to rather than reducing the prevailing contamination.

Of the methods to extract or separate gaseous impurities from air, two common processes cannot be applied practically to general ventilation. The first of these is *condensation* by a reduction in temperature. Most of the air-entrained gases and vapors are present in very low concentration even when highly objectionable. Hence their vapor pressures are extremely low and their condensation temperatures are far below zero Fahrenheit (−17.8°C). While the condensation method is useful in certain laboratory operations requiring the separation of gases by low-temperature distillation, it offers no possibility for large-scale air purification.

The second process is that of *air washing or scrubbing*. This method, if applied to general ventilation primarily for the purpose of air cleaning, is decidedly limited and generally costly. Its effectiveness in eliminating gaseous and vaporous impurities from air is confined to a relatively few gases and vapors which are water soluble. The vast majority of airborne odorants are organic substances which are insoluble in water and hence cannot be removed by water scrubbing. These insolubles include saturated and unsaturated hydrocarbons, sulfur and nitrogen compounds, esters, and most of the odorous acids, aldehydes, and ketones. It is sometimes possible to treat the water to obtain solubility of some gases; for example, an 8 percent caustic water solution in a spray-type washer effectively extracts airborne sulfur dioxide. This, however, makes necessary controlled maintenance of the strength of the water solution and the constant change of water to prevent accumulation and concentration of dissolved impurities. Even with the closest control, it is seldom possible to prevent some reevaporation and escape of dissolved impurities. The effectiveness of the washer in eliminating soluble substances alone will further depend upon the degree of contact obtained between the air and water. So-called scrubbers, imposing tortuous air passages, air turbulence, and counterflow features to increase contact are sometimes employed, but these add to flow resistance and, consequently, to the cost of the operation.

An added obstacle to air purification by washing is that the air becomes moisture saturated in the process and, unless supplementary dehumidification is provided, its application to ventilation is impractical. In air conditioning the air washer contributes the useful functions of cooling, humidification, or dehumidification (see Chapter 5) but is both inadequate and impractical as an air purifier.

Summarized, the preceding approaches to air purification are definitely limited, impractical, or uneconomical. There is, however, a simple and practical method of extracting nearly all odorous and objectionable gases and vapors from air, namely, the process of *adsorption*.

Adsorption (not to be confused with "absorption" which literally means the diffusion of one substance into the body of another) refers to the adhesion of fluid (gas or liquid) substances to the surfaces of solid substances. Because there are instances where the distinction between the two phenomena may not be very sharp, the term *sorption* and derivatives *sorbate* and *sorbent* are often used.

There are a number of substances that possess this unique physicochemical property of adsorption. The most common of these and the best adapted to practical air purification is activited carbon (charcoal). Other adsorbents include zeolite, silica gel, activated alumina, mica, and so forth.

Charcoal has a unique affinity for many gases and vapors, tenaciously holding them until they are removed by special means. While most charcoals in their natural state possess this adsorptive property to a degree, those intended for air purification are produced from particular raw materials and are specially processed or "activated." This purpose of the process is to create extensive surfaces on which adsorption can take place. This results in high adsorptive capacity for gases and vapors and distinguishes "activated carbon" from the more common varieties of charcoal. Activated carbons, in turn, vary in structural properties and are selected depending upon the purpose for which they are to be used, the most important distinction being between those adapted to deodorize or decolorize liquids and those suited to purify air and gas. The latter must be of much greater density and durability and a finer pore structure.

The air-purifying effectiveness of activated carbon depends in part upon the duration of contact between the air and carbon. Since the thickness of the carbon bed used is limited by the allowable resistance to air flow, the bed area must be large and the air velocity through the bed correspondingly low.

The amount of carbon employed must be sufficient to last a reasonable length of time before it becomes saturated. Upon saturation, the carbon can be reactivated to its original activity. Consequently, the total weight of carbon for any particular application is determined by the quantity of gaseous impurities to be adsorbed between reactivations.

Further information relating to activated carbon may be found in Ref. 4, and information concerning the use of other adsorbents may be found in the system volume of the ASHRAE Handbook series.

BIBLIOGRAPHY

1. ASHRAE *Fundamentals Handbook,* American Society of Heating, Refrigerating and Air-Conditioning Engineers, New York, 1977.

2. ASHRAE *Handbook of Fundamentals,* American Society of Heating, Refrigerating and Air-Conditioning Engineers, Atlanta, GA, 1981.

3. R. G. Nevins, Psychrometrics and Modern Comfort, presented at the joint ASHRAE-ASME meeting, November 28–29, 1961.

4. Sleik, H., and Turk, A., *Air Conservation Engineering,* Connor Engineering Corp., Danbury, CT, 1953.

5. ASHRAE Standard 90-75, *Energy Conservation in New Building Design,* American Society of Heating, Refrigerating and Air-Conditioning Engineers, New York, 1975.

6. ASHRAE Standard 62-1981 *Ventilation for Acceptable Indoor Air Quality,* American Society of Heating, Refrigerating and Air-Conditioning Engineers, Atlanta, GA.

Chapter 7

Heat Transfer in Building Sections

7-1 INTRODUCTION

Heat transfer is a basic engineering science of considerable magnitude and complexity. Many textbooks have been written in recent years concerning this important engineering science. Only the pertinent basic principles applicable to heat transfer through building sections will be discussed in this chapter.

The design of a satisfactory heating or cooling system requires that we be able to make a satisfactory estimate of the heat transfer, losses or gains, for a building. Also, in order to design or select practical, efficient, and economical heat exchangers to perform satisfactorily in heating or cooling buildings, a knowledge of the basic principles of heat transfer must be acquired.

In most cases heat is being transferred between two fluids which may be stationary or moving. For example, heat is transferred from the warm inside air to the cold outside air through a building structure. Or, heat is transferred from a warm fluid in a heat exchanger to the cooler surrounding air. Since we are concerned with stationary fluids or fluids in motion, we will find that heat transfer is closely related to the characteristic behavior and properties of fluids.

Heat is a form of energy, the transfer of which is subject to the first and second laws of thermodynamics. All the heat lost from a source or sources must equal that gained by the receiver or receivers involved. When unaided by mechanical means, heat flow is always from a higher to a lower temperature level.

In some cases complete and rapid heat transfer is desired, such as in heat-exchange devices. In other cases a minimum amount of heat transfer is desired, such as through building sections, or from piping or duct work carrying fluid to various heat-transfer devices in a heating or cooling system.

7-2 MODES OF HEAT TRANSFER

Essentially there are three methods (modes) of heat transfer, namely, *conduction, convection,* and *radiation*. In the usual situation all three modes occur simultaneously. In some instances the heat-transfer methods can be separated, but in others only the combined effect can be determined. The many variables involved in heat transfer have been the subject of much investigation in this field of science and great advances have been made in the past 30 years increasing the fund of basic knowledge.

In our study of heat transfer we will discuss separately the modes of heat transfer, and we will be concerned only with *steady-state heat transfer*. In steady-state heat transfer the various temperatures throughout a system remain constant with respect to time during heat transmission. In unsteady heat transfer temperature varies with time.

Thermal conduction is the term applied to

the mechanism of heat transfer whereby the molecules of higher kinetic energy transmit part of their energy to adjacent molecules of lower kinetic energy by direct molecular action. Since the temperature is proportional to the average kinetic energy of molecules, thermal transfer will occur in the direction of decreasing temperature. The motion of the molecules is random; there is no net material flow associated with the conduction mechanism. In the case of flowing fluids thermal conduction is significant in the region very close to a solid boundary or wall where the flow is laminar and parallel with the wall surface, and where practically no cross currents exist in the direction of the heat transfer across the solid-fluid boundary. In solid bodies the significant mechanism of heat transfer is always thermal conduction.

Thermal convection involves energy transfer by eddy mixing and diffusion in addition to conduction. Convection is the transfer of heat between a moving fluid (gas or liquid) and a surface, or the transfer of heat from one point to another within a fluid. In convection, if the fluid moves because of a difference in density resulting from temperature changes, the process is called *natural convection,* or *free convection;* if the fluid is moved by mechanical means (pumps or fans), the process is called *forced convection.*

In the conduction and convection mechanisms the transfer of heat is associated with matter. For *radiant heat transfer,* however, a change in energy form takes place, from internal energy at the source to electromagnetic energy for transmission, then back to internal energy at the receiver.

7-3 THERMAL CONDUCTION EQUATION

The theory of heat conduction was developed by a French mathematician, J. B. Fourier, and expresses steady-state conduction in one direction as

$$\dot{q} = -kA\frac{dt}{dx} \tag{7-1}$$

where

$\dot{q}$ = heat transfer rate (Btu/hr or W)

k = thermal conductivity [Btu/(hr ft °F) or W/(m°C)]

A = surface area normal to the direction of heat flow (ft^2 or m^2)

dt/dx = temperature gradient (°F/ft or °C/m) (Since t is decreasing as x increases, the negative sign gives a positive value of $\dot{q}$.)

When dt/dx is a constant negative value, Eq. (7-1) may be written as

$$\dot{q} = \frac{k}{x}A(t_1 - t_2) \tag{7-2}$$

$$= \frac{k}{x}A\Delta t$$

where t_1 and t_2 are the hot and cold surface temperatures, respectively, and $\Delta t = t_1 - t_2$, the temperature difference across the section. Figure 7-1 represents schematically the temperature variation through a homogeneous material.

It is important to know the units of thermal conductivity, k. These units may vary from one reference to another, and for Eq. (7-2) to yield the proper results, units must be consistant. The units of k in the foot-pound-second system are

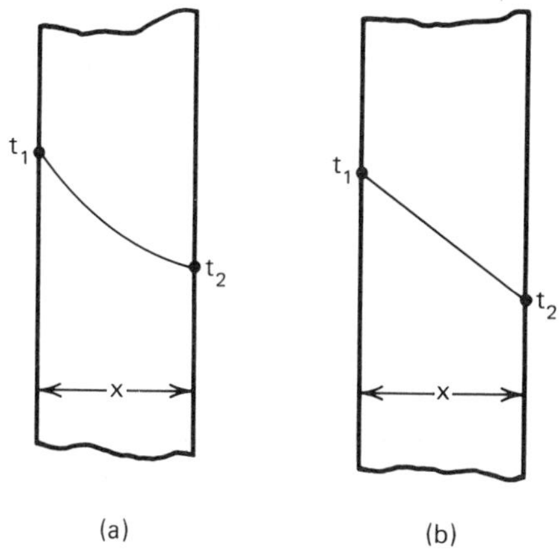

Figure 7–1

Temperature variation through homogeneous material (a) k, variable (b) k, constant mean value

$$k = \frac{\dot{q}}{A} \times \frac{x}{t} = \frac{(\text{Btu})(\text{ft})}{(\text{hr})(\text{ft}^2)(°\text{F})}$$

$$= \text{Btu/(hr ft °F)}$$

or

$$k = \frac{(\text{Btu})(\text{in.})}{(\text{hr})(\text{ft}^2)(°\text{F})} = \frac{\text{Btu}}{(\text{hr})(\text{ft}^2)(°\text{F/in.})}$$

$$= \frac{12(\text{Btu})(\text{ft})}{(\text{hr})(\text{ft}^2)(°\text{F})}$$

At the present time most reference data are expressed in the above units.

The units of k in SI are

$$k = \frac{\dot{q}}{A} \times \frac{x}{t} = \frac{(\text{W})(\text{m})}{(\text{m}^2)(°\text{C or K})} = \frac{\text{W}}{(\text{m})(°\text{C})}$$

For conversion we may use the following:

$$\frac{(\text{W})(\text{m})}{(\text{m}^2)(°\text{C})} = 6.938 \frac{(\text{Btu})(\text{in.})}{(\text{hr})(\text{ft}^2)(°\text{F})}$$

$$= 0.5782 \frac{(\text{Btu})(\text{ft})}{(\text{hr})(\text{ft}^2)(°\text{F})}$$

$$\frac{\text{Btu})(\text{ft})}{(\text{hr})(\text{ft}^2)(°\text{F})} = 1.73 \frac{\text{W}}{(\text{m})(°\text{C})}$$

$$= 0.0173 \frac{\text{W}}{(\text{cm})(°\text{C})}$$

$$\frac{(\text{Btu})(\text{in.})}{(\text{hr})(\text{ft}^2)(°\text{F})} = 0.1441 \frac{\text{W}}{(\text{m})(°\text{C})}$$

In general, the value of k varies with temperature, density, and type of material. Most tabular data give average or mean values of k for a specific temperature or range of temperature.

Symbols Used in Heat-Transfer Equations and Tables

U = overall heat transmission coefficient, in units of Btu/(hr)(ft^2)(°F) or W/(m^2)(K), existing between the air or other fluid on the two sides of a wall, floor, ceiling, roof, or heat-transfer section under consideration.

k = thermal *conductivity*, in units of (Btu)(in.)/(hr)(ft^2)(°F) or W/(m)(°C).

f = film or surface coefficient (conductance) for computing heat transfer through the gas or liquid film by conduction, convection, and radiation to the adjacent medium, in units of Btu/(hr)(ft^2)(°F) or W/(m^2)(°C).

C = thermal *conductance* is the heat transmitted through a nonhomogeneous or composite material of the thickness and type for which C is given, in units of Btu/(hr)(ft^2)(°F) or (W/m^2)(°C). Fluid film effects not usually considered.

R' = thermal resistance, equal to x/kA.

R = unit thermal resistance to heat flow, usually for 1 ft^2 of surface area. A general term for various kinds of resistance, in units of (hr)(ft^2)(°F)/Btu or (m^2)(°C)/W.
= $1/U$ = overall resistance, fluid to fluid on each side of composite section.

= 1/C = units resitance of a nonhomogeneous section or material, film effects usually not considered.
= x/k = unit resistance of a homogeneous material of x units thick.
= $1/k$ = unit resistance of a homogeneous material of unit thickness.
= $1/f$ = surface film resistance.
R_a = air space resistance.

Data for Table 7-1A and 7-1B have been selected from *ASHRAE Handbook and Product Directory, 1977 Fundamentals*. Table 7-1A presents unit thermal resistance values ($1/k$) and $1/C$) for some common building and insulating materials. Table 7-1B gives thermal conductivity values for a variety of materials.

Table 7-2A, also from ASHRAE, lists surface conductances *(f)* and unit resistance *(R)* for still air and for moving air. Note that the emittance (ϵ) of the surface is considered. Table 7-2B gives values of emittance and effective emittance *(E)* for various combinations of reflective surfaces.

Table 7-3, also from ASHRAE, lists air space thermal reistance, for ¾-in., 1½-in., and 3½-in. wide air spaces. Note that the thermal resistance (R_a) not only depends on the air space width but also on the position of the air space, the direction of the heat flow, the mean air temperature in the space, the temperature difference across the air space, and the effective emittance.

7-4 CONDUCTION THROUGH A COMPOSITE SECTION

Figure 7-2 represents a plane wall section made up of three homogeneous materials a, b, and c. The materials have mean conductivity values of k_a, k_b, and k_c, respectively, and thickness x_a, x_b, and x_c. The following assumptions are made:

1. Each of the materials is homogeneous.
2. The thermal conductivity of each material is constant with respect to temperature.
3. The surface of the different materials are in intimate contact so that no resistance to heat transfer exists at the interfaces.
4. The flow of heat is steady and perpendicular to the surface.

Equation (7-2) may be written as

$$\dot{q} = \frac{t_1 - t_2}{x/kA} = \frac{t_1 - t_2}{R'} \quad (7\text{-}3)$$

where $x/kA = R'$, called *thermal resistance*.

Applying Eq. (7-3) to the composite wall section of Fig. 7-2, for steady-state heat transfer we have

$$\dot{q} = \dot{q}_a = \dot{q}_b = \dot{q}_c$$

or

$$\dot{q} = \frac{t_1 - t_2}{x_a/k_aA_a} = \frac{t_2 - t_3}{x_b/k_bA_b} = \frac{t_3 - t_4}{x_c/k_cA_c} = \frac{t_1 - t_4}{R_t'} \quad (7\text{-}4)$$

where R_t' is the sum of all the thermal resistances in series. Considering 1 ft^2 of surface area, Eq. (7-4) may be written as

$$\frac{\dot{q}}{A} = \frac{t_1 - t_4}{x_a/k_a + x_b/k_b + x_c/k_c} = \frac{t_1 - t_4}{R_a + R_b + R_c} = \frac{t_1 - t_4}{R_t} \quad (7\text{-}5)$$

where $\dot{q}/A$ becomes the unit heat transfer in Btu/hr ft^2 or W/m^2, and R_a, R_b, and R_c are unit thermal resistances for the materials.

From Eq. (7-2) we may write

$$\Delta t = \left(\frac{\dot{q}}{A}\right)\left(\frac{x}{k}\right) \quad (7\text{-}6)$$

Table 7-1* (A)

Thermal Resistances *(R)* of Common Building and Insulating Materials*

	Material	Description	Density (Lb. per Cu. Ft.)	Thermal Resistance (R) ① Per Inch Thickness	Thermal Resistance (R) ① For Thickness Listed
1.	BUILDING BOARD Boards, Panels, Flooring, Sheathing, Etc.	a. Asbestos-cement board	120	0.25	--
		b. Asbestos-cement board ... 1/8 in.	120	--	0.033
		c. Gypsum or plaster board ... 3/8 in.	50	--	0.32
		d. Gypsum or plaster board ... 1/2 in.	50	--	0.45
		e. Plywood	34	1.25	--
		f. Sheathing, wood fiber (impreg. or coated)	20	2.63	--
		g. Sheathing, wood fiber (impreg. or coated)	22	2.44	--
		h. Sheathing, wood fiber (impreg. or coated)	25	2.27	--
		i. Wood fiber board, laminated or homogeneous	26	2.38	--
		j. Wood fiber board, laminated or homogeneous	33	1.82	--
		k. Wood fiber, hardboard type	65	0.72	--
		l. Wood fiber, hardboard type ... 1/4 in.	65	--	0.18
		m. Wood subfloor ... 25/32 in.	--	--	0.98
		n. Wood, hardwood finish ... 3/4 in.	--	--	0.68
2.	BUILDING PAPER	a. Vapor—permeable felt	--	--	0.06
		b. Vapor—seal, 2 layers of mopped 15 lb. felt	--	--	0.12
		c. Vapor—seal, plastic film	--	--	Negl.
3.	FINISH FLOORING MATERIALS	a. Carpet and fibrous pad	--	--	2.08
		b. Carpet and rubber pad	--	--	1.23
		c. Cork tile ... 1/8 in.	--	--	0.28
		d. Terrazzo ... 1 in.	--	--	0.08
		e. Tile—asphalt, linoleum, vinyl, rubber	--	--	0.05
4.	INSULATING MATERIALS Blanket and Batt	a. Cotton fiber	0.8-2.0	3.85	--
		b. Mineral wool, fibrous form processed from rock, slag, or glass	1.5-4.0	3.70	--
		c. Wood fiber	3.2-3.6	4.00	--
5.	INSULATING MATERIALS Board and Slabs	a. Cellular glass	9.0	2.50	--
		b. Corkboard	6.5-8.0	3.64	--
		c. Corkboard	12	3.28	--
		d. Glass fiber	4-9	4.00	--
		e. Expanded rubber (rigid)	4.5	4.55	--
		f. Expanded polyurethane (R-11 blown) (Thickness 1 in. and greater)	1.5-2.5	5.88	--
		g. Expanded polystyrene, extruded	1.9	3.85	--
		h. Expanded polystyrene, molded beads	1.0	3.57	--
		i. Mineral wool with resin binder	15	3.51	--
		j. Mineral fiberboard, wet felted			
		Core or roof insulation	16-17	2.94	--
		Acoustical tile	18	2.86	--
		Acoustical tile	21	2.73	--
		k. Mineral fiberboard, wet molded			
		Acoustical tile	23	2.38	--
		l. Wood or cane fiberboard			
		Acoustical tile ... 1/2 in.	--	--	1.19
		Acoustical tile ... 3/4 in.	--	--	1.78
		Interior finish (plank, tile)	15	2.86	--
		m. Insulating roof deck			
		Approximately ... 1-1/2 in.	--	--	4.17
		Approximately ... 2 in.	--	--	5.56
		Approximately ... 3 in.	--	--	8.33
		n. Wood shredded (cemented in preformed slabs)	22	1.67	--
6.	INSULATING MATERIALS Loose Fill	a. Macerated paper or pulp products	2.5-3.5	3.57	--
		b. Mineral wool (glass, slag, or rock)	2.0-5.0	3.51	--
		c. Sawdust or shavings	0.8-15	2.22	--
		d. Silica aerogel	7.6	6.07	--
		e. Vermiculite (expanded)	7.0-8.2	2.13	--
		f. Vermiculite (expanded)	4.0-6.0	2.28	--
		g. Wood fiber: redwood, hemlock, or fir	2.0-3.5	3.33	--
		h. Wood fiber: redwood bark	3	3.22	--
		i. Wood fiber: redwood bark	4	3.57	--
		j. Wood fiber: redwood bark	4.5	3.85	--

All resistance factors are based on 75°F. mean temperature.

Table 7-1 (A, Cont.)

Material	Description	Density (Lb. per Cu. Ft.)	Thermal Resistance (R)① Per Inch Thickness	Thermal Resistance (R)① For Thickness Listed
7. INSULATING MATERIALS Roof Insulation	a. Preformed, for use above deck			
	1. Approximately ... 1/2 in.	--	--	1.39
	2. Approximately ... 1 in.	--	--	2.78
	3. Approximately ... 1-1/2 in.	--	--	4.17
	4. Approximately ... 2 in.	--	--	5.26
	5. Approximately ... 2-1/2 in.	--	--	6.67
	6. Approximately ... 3 in.	--	--	8.33
	b. Cellular glass	--	2.56	--
8. MASONRY MATERIALS Concretes	a. Cement mortar	116	0.20	--
	b. Gypsum-fiber concrete 87½% gypsum, 12½% wood chips	51	0.60	--
	c. Lightweight aggregates including expanded shale, clay or slate; expanded slags; cinders; pumice; perlite; vermiculite; also cellular concretes	120	0.19	--
		100	0.28	--
		80	0.40	--
		60	0.59	--
		40	0.86	--
		30	1.11	--
		20	1.43	--
	d. Sand and gravel or stone aggregate (oven dried)	140	0.11	--
	e. Sand and gravel or stone aggregate (not dried)	140	0.08	--
	f. Stucco	116	0.20	--
9. MASONRY UNITS	a. Brick, common	120	0.20	--
	b. Brick, face	130	0.11	--
	c. Clay tile, hollow:			
	1 cell deep ... 3 in.	--	--	0.80
	1 cell deep ... 4 in.	--	--	1.11
	2 cells deep ... 6 in.	--	--	1.52
	2 cells deep ... 8 in.	--	--	1.85
	2 cells deep ... 10 in.	--	--	2.22
	3 cells deep ... 12 in.	--	--	2.50
	d. Concrete Blocks, three oval core:			
	Sand and gravel aggregate ... 4 in.	--	--	0.71
	... 8 in.	--	--	1.11
	... 12 in.	--	--	1.28
	Cinder aggregate ... 3 in.	--	--	0.86
	... 4 in.	--	--	1.11
	... 8 in.	--	--	1.72
	... 12 in.	--	--	1.89
	Lightweight aggregate (expanded shale, clay, slate or slag, pumice) 3 in.	--	--	1.27
	4 in.	--	--	1.50
	8 in.	--	--	2.00
	12 in.	--	--	2.27
	e. Concrete blocks, rectangular core:			
	Sand and gravel aggregate			
	2 core, 8 in. 36 lb.	--	--	1.04
	same with filled cores	--	--	1.93
	Lightweight aggregate (expandes shale, clay, slate or slag, pumice):			
	3 core, 6 in. 19 lb.	--	--	1.65
	Same with filled cores	--	--	2.99
	2 core, 8 in. 24 lb.	--	--	2.18
	Same with filled cores	--	--	5.03
	3 core, 12 in. 38 lb.	--	--	2.48
	Same with filled cores	--	--	5.82
	f. Stone, lime or sand	--	0.08	--
	g. Gypsum partition tile:			
	3 x 12 x 30 in. solid	--	--	1.26
	3 x 12 x 30 in. 4-cell	--	--	1.35
	4 x 12 x 30 in. 3-cell	--	--	1.67
10. PLASTERING MATERIALS	a. Cement plaster, sand aggregate	116	0.20	--
	1. Sand aggregate ... 1/2 in.	--	--	0.10
	2. Sand aggregate ... 3/4 in.	--	--	0.15
	b. Gypsum plaster:			
	1. Lightweight aggregate ... 1/2 in.	45	--	0.32
	2. Lightweight aggregate ... 5/8 in.	45	--	0.39
	3. Lightweight aggregate on metal lath ... 3/4 in.	--	--	0.47
	4. Perlite aggregate	45	0.67	--
	5. Sand aggregate	105	0.18	--
	6. Sand aggregate ... 1/2 in.	105	--	0.09
	7. Sand aggregate ... 5/8 in.	105	--	0.11
	8. Sand aggregate on metal lath ... 3/4 in.	--	--	0.10
	9. Sand aggregate on wood lath	--	--	0.40
	10. Vermiculite aggregate	45	0.59	--

①All resistance factors are based on 75°F. mean temperature.

Table 7-1 (A, Cont.)

Material	Description	Density (Lb. per Cu. Ft.)	Thermal Resistance (R)① Per Inch Thickness	Thermal Resistance (R)① For Thickness Listed
11. ROOFING	a. Asbestos-cement shingles	120	--	0.21
	b. Asphalt roll roofing	70	--	0.15
	c. Asphalt shingles	70	--	0.44
	d. Built-up roofing 3/8 in.	70	--	0.33
	e. Slate 1/2 in.	--	--	0.05
	f. Wood shingles	--	--	0.94
12. SIDING MATERIALS (On Flat Surface)	a. Shingles			
	Asbestos-cement	120	--	0.21
	Wood, 16 in., 7-1/2 in. exposure	--	--	0.87
	Wood, double, 16-in., 12-in. exposure	--	--	1.19
	Wood, plus insul. backer board 5/16 in.	--	--	1.40
	b. Siding			
	Asbestos-cement, 1/4 in., lapped	--	--	0.21
	Asphalt roll siding	--	--	0.15
	Asphalt insulating siding (1/2 in. bd.)	--	--	1.46
	Wood, drop, 1 x 8 in.	--	--	0.79
	Wood, bevel, 1/2 x 8 in., lapped	--	--	0.81
	Wood, bevel, 3/4 x 10 in., lapped	--	--	1.05
	Wood, plywood, 3/8 in., lapped	--	--	0.59
	c. Architectural glass	--	--	0.10
13. WOODS	a. Maple, oak, and similar hardwoods	45	0.91	--
	b. Fir, pine, and similar softwoods	32	1.25	--
	c. Fir, pine, and similar softwoods 25/32 in.	32	--	0.98
	d. Fir, pine, and similar softwoods 1-5/8 in.	32	--	2.03
	e. Fir, pine, and similar softwoods 2-5/8 in.	32	--	3.28
	f. Fir, pine, and similar softwoods 3-5/8 in.	32	--	4.55

① All resistance factors are based on 75°F. mean temperature.

**Copyright by ASHRAE. Extracted from* ASHRAE Handbook and Product Directory, 1977 Fundamentals. *With permission of the American Society of Heating, Refrigerating and Air-Conditioning Engineers, Atlanta, GA.*

and since $\dot{q}/A$ is the same for each material in steady flow, we may say that the temperature drop, Δt, through each material of the composite wall of Fig. 7-2 is

$$\Delta t_a = \frac{\dot{q}}{A}\left(\frac{x_a}{k_a}\right) = \frac{\dot{q}}{A}(R_a)$$

$$\Delta t_b = \frac{\dot{q}}{A}\left(\frac{x_b}{k_b}\right) = \frac{\dot{q}}{A}(R_b)$$

$$\Delta t_c = \frac{\dot{q}}{A}\left(\frac{x_c}{k_c}\right) = \frac{\dot{q}}{A}(R_c)$$

Illustrative Problem 7-1

Assume that a section of wall is made up of three different materials placed as shown in Fig. 7-2. Material a has a conductivity of 5.0 (Btu in.)/(hr ft^2°F); for material b, $k = 12.0$ (Btu in.)/(hr ft^2°F); and material c, $k = 0.80$ (Btu in.)/hr ft^2°F). Thicknesses are as follows: $x_a = 4.0$ in., $x_b = 6.0$ in., and $x_c = 0.5$ in. If the surface temperature $t_1 = 100$°F and the surface temperature $t_4 = 50$°F, what is the rate of heat transfer in Btu/(hr ft^2)? What are the theoretical interface temperatures in the section?

Solution:

By Eq. (7-5)

$$\frac{\dot{q}}{A} = \frac{t_1 - t_4}{x_a/k_a + x_b/k_b + x_c/k_c}$$

$$= \frac{100 - 50}{4.0/5.0 + 6.0/12.0 + 0.5/0.80}$$

$$= \frac{50}{0.80 + 0.50 + 0.625} = \frac{50}{1.925}$$

Table 7-1* B

Physical Properties of Material—Properties of Solids*

NAME OR DESCRIPTION	SPECIFIC HEAT BTU/lb. (°F.)	SPECIFIC GRAVITY	THERMAL CONDUCTIVITY ①	
			TEMP. °F.	k = 1/r
Aluminum	0.226	2.55–2.8	32	122.0
Asbestos	0.25	2.1–2.8	32	0.09
Cardboard	–			0.1–0.2
Charcoal (wood)	0.242	0.28–0.57	172	0.051
Earth (dry and packed)	–	1.5 (loose)	32	0.035
			100	0.039
Gold	0.0308	19.25–19.2	64	169.0
Iron (gray cast)	0.101	7.03–7.13	129	27.6
Iron (cast pig)	–	7.2		
Iron (wrought)	–	7.6–7.9	64	34.9
Lead	0.030	11.34	64	20.1
Limestone	0.217	2.1–2.8		0.3–0.75
Paper	0.324	0.70–1.15		0.075
Paraffin	0.6939	0.87–0.91	86	0.145
Platinum (cast)	–	21.5	64	40.2
Porcelain	0.22		329	0.945
Sawdust	–	0.21	68	0.042
Sand	0.191	1.4–1.9	68	0.188
Steel (cold-drawn)	0.12	7.83	32	28.0
Silver (cast)	–	10.4–10.6	64	244.0
Snow (fresh fallen)	–	0.125		
Tin (cast)	0.053	7.2–7.5	64	37.6
Wood (oak)	0.570	0.65–0.84		0.85–0.125
Wool	–	1.32	86	0.022
Zinc (cast)	–	7.1	32	63.0

① **Note:** k = Btu per (hr) (sq ft) (°F per ft. thickness)

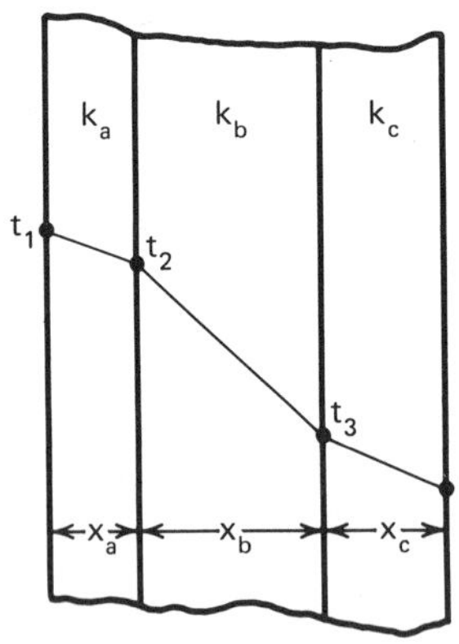

Figure 7–2

Temperature variation through a composite section

$$\frac{\dot{q}}{A} = 25.9 \text{ Btu/(hr ft}^2\text{)}$$

By Eq. (7-6)

$$\Delta t_a = \left(\frac{\dot{q}}{A}\right)(R_a) = (25.9)(0.80) = 20.72°\text{F}$$

$$\Delta t_b = \left(\frac{\dot{q}}{A}\right)(R_b) = (25.9)(0.50) = 12.95°\text{F}$$

$$\Delta t_c = \left(\frac{\dot{q}}{A}\right)(R_c) = (25.9)(0.625) = 16.19°\text{F}$$

The sum $\Delta t_a + \Delta t_b + \Delta t_c$ should be equal to $t_1 - t_4 = 50°\text{F}$. The calculated sum is

49.86°F which indicates a slight inaccuracy because of rounding-off numbers.

In the general case Eq. (7-5) becomes

$$\frac{\dot{q}}{A} = \frac{t_1 - t_m}{x_a/k_a + x_b/k_b + x_c/k_c + \cdot \cdot \cdot + x_m/k_m}$$

$$= \frac{t_1 - t_m}{R_t} \qquad (7\text{-}7)$$

The denominator of the above equation is the sum of all thermal resistances in series.

It is normally necessary to consider the resistance to heat flow caused by the fluid films which cling to the solid surfaces, as well as the resistance offered by air spaces within composite wall sections. The surface temperatures used in the development of Eqs. (7-5) and (7-7) are less frequently known than the bulk temperatures of the fluids on either side of the section. The method of heat transfer between a fluid (gas or liquid) and a solid surface, and through air spaces, is usually a combination of the three modes of heat transfer (conduction, convection, and radiation).

The thickness of the fluid film which clings to a solid surface varies, dependent on the flow conditions (laminar or turbulent), which in turn depends upon the fluid flow velocity, surface roughness, and surface orientation (horizontal, vertical, sloping). Thus, heat transfer through fluid films is influenced by the laws of fluid flow and by properties of fluids. Gas films offer much more resistance to heat transfer than do liquid films. Figure 7-3 represents the temperature profile through fluid films for both laminar and turbulent flow conditions. (The thickness of the films shown in the figure are greatly exagerated.) In both types of flow conditions shown there is a thin film of fluid at the surface that makes up what is called a *boundary layer*. For laminar flow the boundary layer is somewhat thicker and gives a greater temperature drop from the bulk fluid temperature (t_b) to the surface temperature (t_s). Adjacent to the surface, in both laminar and turbulent flow, the fluid is essentially stationary and heat is transferred by conduction. Further out from the surface convection currents are set up due to changes in fluid density. Natural or free convection is an important consideration in heat transfer through building sections. Forced con-

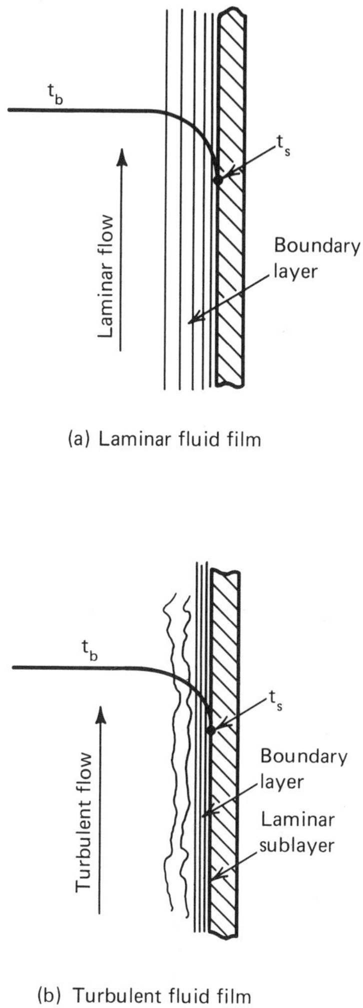

Figure 7–3

Temperature change through fluid films (not to scale)

Table 7-2* (A)

Surface Conductances and resistances for Air*[a]

Position of Surface	Direction of Heat Flow	Non-reflective $\epsilon = 0.90$		Reflective $\epsilon = 0.20$		Reflective $\epsilon = 0.05$	
Still air		f_i	R_a	f_i	R_a	f_i	R_a
Horizontal	Upward	1.63	0.61	0.91	1.10	0.76	1.32
Sloping (45)	Upward	1.60	0.62	0.88	1.14	0.73	1.37
Vertical	Horizontal	1.46	0.68	0.74	1.35	0.59	1.70
Sloping (45)	Downward	1.32	0.76	0.60	1.67	0.45	2.22
Horizontal	Downward	1.08	0.92	0.37	2.70	0.22	4.55
Moving Air (Any position)							
15-mph wind (for winter)	Any	6.00	0.17				
7.5-mph wind (for summer)	Any	4.00	0.25				

[a]All conductance values expressed in Btu per (hour) (square foot) (degree F temperature difference). A surface cannot take credit for both an air space resistance value and a surface resistance value. No credit for an air space value can be taken for any surface facing an air space of less than 0.5 in.

Table 7-2 (B)

Reflectivity and Emittance Values of Various Surfaces and Effective Emittances of Air Spaces

Surface	Reflectivity in Percent	Average Emittance ε	Effective Emittance E of Air Space: One Surface Emittance ε the Other 0.90	Effective Emittance E of Air Space: Both Surfaces Emittances ε
Aluminum foil, bright	92 to 97	0.05	0.05	0.03
Aluminum sheet	80 to 95	0.12	0.12	0.06
Aluminum coated paper, polished	75 to 84	0.20	0.20	0.11
Steel, galvanized, bright	70 to 80	0.25	0.24	0.15
Aluminum paint	30 to 70	0.50	0.47	0.35
Building materials: wood, paper, masonry, nonmetallic points	5 to 15	0.90	0.82	0.82
Regular glass	5 to 15	0.84	0.77	0.72

**Table 7-2 extracted from ASHRAE 1977* Fundamentals Handbook, *pg. 22.11, with permission of the American Society of Heating, Refrigerating and Air-Conditioning Engineers, Atlanta, Ga.*

vection frequently occurs in heat-transfer devices. In heat transfer through the metal surfaces of heat transfer devices, the fluid films offer the greatest resistance to heat flow and become an important consideration. In heat transfer through insulated sections, the fluid film resistances may become relatively unimportant.

Evaluation of heat transfer by convection is difficult because of the many variables involved. Mathematical expressions for convective heat transfer are best developed by an application of dimensional analysis, which is beyond the scope of this text. Textbooks on heat transfer cover this method of combining the variables to obtain practical working equations which may be used for predicting heat transfer by convection.

The usual simplified approach is to express the heat transfer by convection as

$$\dot{q}_c = h_c A(t_b - t_s) \quad (7\text{-}8)$$

where

$\dot{q}_c$ = convective heat-transfer rate (Btu/hr or W)

A = heat-transfer area (ft^2 or m^2)

t_b = bulk or mean temperature of main stream of fluid (°F or °C)

t_s = wall surface temperature (°F or °C)

h_c = coefficient of convective heat transfer [Btu/(hr ft^2 °F) or W/(m^2 °C)]

Dimensionally, h_c is equal to k/x, but since the fluid film thickness is so small, the value of h_c is usually given as a value independent of k

Table 7-3* (A)

Thermal Resistance (Ra) of ¾-in. Plane Air Spaces[a]*
All resistance values expressed in (hour) (square foot) (degree Fahrenheit temperature difference) per Btu.

Position of Air Space	Direction of Heat Flow	Mean Temp. (°F)	Temp. Diff. (°F)	¾-in. Air Space: Value of Effective Emittance *E* 0.03	0.05	0.20	0.82
		90	10	2.34	2.22	1.61	0.75
		50	30	1.71	1.66	1.35	0.77
Horizontal	Up	50	10	2.30	2.21	1.70	0.87
		0	20	1.83	1.79	1.52	0.93
		0	10	2.23	2.16	1.78	1.02
		90	10	2.96	2.78	1.88	0.81
		50	30	1.99	1.92	1.52	0.82
45° Slope	Up	50	10	2.90	2.75	2.00	0.94
		0	20	2.13	2.07	1.72	1.00
		0	10	2.72	2.62	2.08	1.12
		90	10	3.50	3.24	2.08	0.84
		50	30	2.91	2.77	2.01	0.94
Vertical	Horizontal	50	10	3.70	3.46	2.35	1.01
		0	20	3.14	3.02	2.32	1.18
		0	10	3.77	3.59	2.64	1.26
		90	10	3.53	3.27	2.10	0.84
		50	30	3.43	3.23	2.24	0.99
45° Slope	Down	50	10	3.81	3.57	2.40	1.02
		0	20	3.75	3.57	2.63	1.26
		0	10	4.12	3.91	2.81	1.30
		90	10	3.55	3.28	2.10	0.85
		50	30	3.77	3.52	2.38	1.02
Horizontal	Down	50	10	3.84	3.59	2.41	1.02
		0	20	4.18	3.96	2.83	1.30
		0	10	4.25	4.02	2.87	1.31

and x for each fluid considered, but dependent on flow conditions. The coefficient of convective heat transfer h_c is sometimes called the unit surface conductance, or simply *film coefficient*. Equation (7-8) may also be expressed in terms of thermal resistance:

$$\dot{q} = \frac{t_b - t_s}{R'} \qquad (7\text{-}9)$$

where

$$R' = \frac{1}{h_c A} \quad [(\text{hr °F})/\text{Btu or °C/W}] \qquad (7\text{-}10)$$

or

$$R = \frac{1}{h_c} = \frac{1}{C} \quad [(hr\,ft^2\,°F)/Btu \text{ or } (m^2\,°C)/W] \qquad (7\text{-}10a)$$

Table 7-3 (B)

Thermal Resistance (Ra) of 1½-In. Plane Air Spaces[a]*
All resistance values expressed in (hour) (square foot) (degree Fahrenheit temperature difference) per Btu.

Position of Air Space	Direction of Heat Flow	Mean Temp. (°F)	Temp. Diff. (°F)	1½-in. Air Space: Value of Effective Emittance *E* 0.03	0.05	0.20	0.82
		90	10	2.55	2.41	1.71	0.77
		50	30	1.87	1.81	1.45	0.80
Horizontal	Up	50	10	2.50	2.40	1.81	0.89
		0	20	2.01	1.95	1.63	0.97
		0	10	2.43	2.35	1.90	1.06
		90	10	2.92	2.73	1.86	0.80
		50	30	2.14	2.06	1.61	0.84
45° Slope	Up	50	10	2.88	2.74	1.99	0.94
		0	20	2.30	2.23	1.82	1.04
		0	10	2.79	2.69	2.12	1.13
		90	10	3.99	3.66	2.25	0.87
		50	30	2.58	2.46	1.84	0.90
Vertical	Horizontal	50	10	3.79	3.55	2.39	1.02
		0	20	2.76	2.66	2.10	1.12
		0	10	3.51	3.35	2.51	1.23
		90	10	5.07	4.55	2.56	0.91
		50	30	3.58	3.36	2.31	1.00
45° Slope	Down	50	10	5.10	4.66	2.85	1.09
		0	20	3.85	3.66	2.68	1.27
		0	10	4.92	4.62	3.16	1.37
		90	10	6.09	5.35	2.79	0.94
		50	30	6.27	5.63	3.18	1.14
Horizontal	Down	50	10	6.61	5.90	3.27	1.15
		0	20	7.03	6.43	3.91	1.49
		0	10	7.31	6.66	4.00	1.51

where C in Eq. (7-10a) is called the *unit conductance*. The thermal resistance shown in Eq. (7-10a) may be summed with other resistances in the conduction heat transfer Eq. (7-7).

Because of the low values of air film coefficients, especially with natural or free convection, the amount of heat transfer by thermal radiation may be equal to or greater than that by convection. This may be especially true in air spaces in building construction.

Thermal radiation is the transfer of thermal energy by electromagnetic waves and is an entirely different phenomenon from conduction and convection. In fact, thermal radiation can occur in a perfect vacuum and is actually

Table 7-3 (C)

Thermal Resistance (Ra) of 3½-in. Plane Air Spaces[a]*

All resistance values expressed in (hour) (square foot) (degree Fahrenheit temperature difference) per Btu.

Position of Air Space	Direction of Heat Flow	Mean Temp. (°F)	Temp. Diff. (°F)	3½-in. Air Space — Value of Effective Emittance E: 0.03	0.05	0.20	0.82
		90	10	2.84	2.66	1.83	0.80
		50	30	2.09	2.01	1.58	0.84
Horizontal	Up	50	10	2.80	2.66	1.95	0.93
		0	20	2.25	2.18	1.79	1.03
		0	10	2.71	2.62	2.07	1.12
		90	10	3.18	2.96	1.97	0.82
		50	30	2.26	2.17	1.67	0.86
45° Slope	Up	50	10	3.12	2.95	2.10	0.96
		0	20	2.42	2.35	1.90	1.06
		0	10	2.98	2.87	2.23	1.16
		90	10	3.69	3.40	2.15	0.85
		50	30	2.67	2.55	1.89	0.91
Vertical	Horizontal	50	10	3.63	3.40	2.32	1.01
		0	20	2.88	2.78	2.17	1.14
		0	10	3.49	3.33	2.50	1.23
		90	10	4.81	4.33	2.49	0.90
		50	30	3.51	3.30	2.28	1.00
45° Slope	Down	50	10	4.74	4.36	2.73	1.08
		0	20	3.81	3.63	2.66	1.27
		0	10	4.59	4.32	3.02	1.34
		90	10	10.07	8.19	3.41	1.00
		50	30	9.60	8.17	3.86	1.22
Horizontal	Down	50	10	11.15	9.27	4.09	1.24
		0	20	10.90	9.52	4.87	1.62
		0	10	11.97	10.32	5.08	1.64

[a]Value in Tables 7-3 apply only to air spaces of uniform thickness bounded by plane, smooth, parallel surfaces with no leakage of air to or from the space. Thermal resistance values for multiple air spaces must be based on careful estimates of mean temperature difference for each air space.

**Extracted* ASHRAE 1977 Fundamentals Handbook. *With permission of the American Society of Heating, Refrigerating and Air-Conditioning Engineers, Atlanta, Ga.*

impeded by an intervening medium between the two surfaces of unequal temperature.

The net transfer of energy by radiation from a warm to cool body is expressed as

$$\dot{q}_r = \sigma(T_1^4 - T_2^4)(A)(e_e)(F_a) \quad \text{(Btu/hr or W)} \tag{7-11}$$

where

σ = Boltzmann constant [0.1713 × 10^{-8} Btu/(hr ft^2 R^4) or 5.673 × 10^{-8} W/(m^2 − k^4)]

T_1 and T_2 = absolute temperatures of warm and cool surfaces, respectively (°R or K)

A = surface area (ft^2 or m^2)

e_e = effective absorptivity or emissivity factor, expressing the degree to which the two surfaces approach an "ideal black body" (an ideal black body is one which could absorb (or emit) all the radiant energy falling on it)

F_a = factor to account for the geometric configuration between the radiating surfaces. Fortunately, $F_a = 1$ for most applications

Values of effective emissivity, e_e, are dependent on the emissivity of each of the surfaces. For infinite parallel planes or completely enclosed body, large compared with enclosing body,

$$e_e = \frac{1}{1/e_1 + 1/e_2 - 1}$$

For a completely enclosed body, small compared with enclosing body, $e_e = e_1$.

Emissivity or absorptivity values depend somewhat on the temperature of the surface and may be found in heat-transfer references.

Although Eq. (7-11) is a suitable relationship for describing radiant heat exchange, it is not convenient for computation where other modes of heat transfer are in operation. For such cases it is convenient to define an *equivalent conductance for radiation* by the expression

$$\dot{q}_r = h_r A(t_1 - t_2) \qquad (7\text{-}12)$$

The unit conductance, h_r, thus defined is a function of the shape- or emissivity factor, as well as the temperatures of the radiator and receiver.

The calculation of heat transfer by convection and radiation may be made by use of equations and tables which apply; but because of uncertainties, theory and experiments have been combined to treat convection and radiation as a single combined process for fluid films and air spaces. The process is expressed by

$$\dot{q}_{rc} = h_{rc} A(t_1 - t_2) \qquad (7\text{-}13)$$

where

$\dot{q}_{rc}$ = heat transfer due to combined radiation and convection (Btu/hr or W)

h_{rc} = conductance of surface film or air space for combined radiation and convection [Btu/(hr ft^2 °F) or W/(m^2 °C]

In this text, as was noted in Section 7-3, the fluid film conductance will be used as f, and the air space conductance will be used as $1/R_a$ to conform with the usual data presented for HVAC work, and has been presented in Tables 7-2A, 7-2B, 7-3A, B, and C.

7-5 OVERALL HEAT TRANSMISSION COEFFICIENT *(U)*

Figure 7-4 represents a composite wall section similar to Fig. 7-2 except that fluid (air) films have been added and an air space has been included in the construction. We may write the overall heat-transfer equation now from the fluid on one side to the fluid on the other side using the same form as Eq. (7-7) and we have

$$\frac{\dot{q}}{A} = \frac{t_i - t_o}{1/f_i + x_2/k_2 + 1/C_a + x_1/k_1 + 1/C_1 + 1/f_o} = \frac{\Delta t}{R_t} \qquad (7\text{-}14)$$

where

t_i and t_o = inside and outside air temperatures respectively (°F or °C)

f_i and f_o = inside and outside film coefficients [Btu/(hr ft^2 °F) or W/(m^2 °C)]

C_a = conductance of the air space [Btu/(hr ft^2 °F) or W/(m^2 °C)]

C_1 = conductance of a nonhomogeneous material [Btu/(hr ft^2 °F) or W/(m^2 °C)]

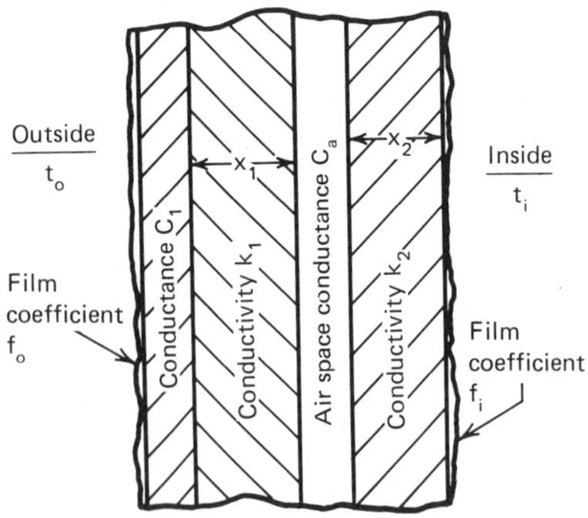

Figure 7–4

Composite wall section

As with Eq. (7-7), each term in the denominator of the right side is the unit thermal resistance offered by the individual components, the sum being the total unit thermal resistance, R_t.

It is convenient in calculating heat transfer to use the *overall heat transmission coefficient,* called *U factor,* for composite building sections. U is defined as $1/R_t$, and Eq. (7-14) becomes

$$\dot{q} = UA\Delta t \qquad (7\text{-}15)$$

where

$\dot{q}$ = heat transfer (Btu/hr or W)

U = overall heat transmission coefficient [Btu/(hr ft^2 °F) or W/(m^2 °C)]

A = surface area of section considered (ft^2 or m^2)

Δt = temperature change across the section from fluid to fluid (°F or °C)

7-6 CALCULATION OF HEAT TRANSMISSION COEFFICIENTS

The heat transmission coefficients (U values) for some common types of building construction are included in a following section. Occasionally, however, construction types not listed are encountered, and the U value must be calculated. As previously noted, the denominator of Eq. (7-14) represents the sum of the individual thermal resistance of the various component parts of a building section. Therefore, if we can evaluate all the individual resistances, they may be added to give total resistance. The reciprocal of this total resistance is the U value.

Illustrative Problem 7-2

Determine the total unit thermal resistance (R_t) and overall heat transmission coefficient (U) for the exterior vertical wall of a building which has the following components:

Solution	Unit Resistance (R)
(a) Interior surface air film. (Table 7-2A, still air, vertical surface, heat flow horizontal, nonreflective, f_i = 1.46)	$1/f_i$ = 0.68
(b) ½-in. cement plaster, sand aggregate. (Table 7-1A, item 10a.1)	= 0.10
(c) ⅜-in., gypsum plasterboard. (Table 7-1A, item 1 c)	= 0.32
(d) 2 by 4 nominal stud space, providing 3½-in. vertical air space, nonreflective, (Table 7-3C, assume mean temp. = 50°F and temp. diff. = 30°F, and E = 0.82)	= 0.91
(e) ⅝-in. fir plywood sheathing. (Table 7-1A, item 1e, resistance/inch = (1.25)	(1.25)(0.625) = 0.78
(f) Building paper.	= neg.
(g) Siding, asbestos cement. (Table 7-1A, item 12a.)	= 0.21
(h) Outside surface air film. (Table 7-2A, assume 15 mph wind)	= 0.17
	R_t = 3.17

$U = 1/R_t = 1/3.17$

$= 0.315$ say 0.32 Btu/(hr)(ft^2)(°F)

Illustrative Problem 7-3

Assume that the wall section of Problem 7-2 has a 2-in. thick, foil faced, mineral wool batt-type insulation installed on the inside portion of the air space between the 2 by 4 studs. The manufacturer of the insulation quotes an R factor of 7.0 for the insulation. Determine the total unit thermal resistance and U factor for the insulated wall.

Solution:

The air space between the studs has now been reduced to 1½-in. (approximately), and one side of the air space now has a reflective surface. From Table 7-2B, the effective emittance E is 0.05 because of aluminum foil on one surface. From Table 7-3B, with E = 0.05, we will assume that the mean temperature is now 0°F, and the temperature difference is 20°F. These assumptions give us R_a = 2.66. Therefore,

R_t from Problem 7-2	=	3.17
Less warm air space	=	−0.91
Plus cold air space	=	+2.66
Plus insulation	=	+7.00
R_t	=	11.92

$U = 1/R_t = 1/11.92 = 0.0839$ say 0.084 Btu/(hr)(ft^2)(°F).

Note that in Problems 7-2 and 7-3 it was necessary to assume an air space mean temperature and a temperature drop in the air space. To be accurate, these assumptions should be checked and corrected if necessary. We must know the inside and outside design temperatures and the manner in which the temperature changes through the wall. This will be shown in a later section.

7-7 SERIES AND PARALLEL HEAT CONDUCTION

In practical applications of heat conduction the heat may flow through the materials involved by series or parallel paths or by a combination of both. An analysis of heat conduction in this respect is similar to that for the conduction of electricity through series and parallel circuits. In a series heat flow path, the heat resistances of the materials involved are additive, or

$$R_t = R_1 + R_2 + R_3 + \ldots \quad (7\text{-}16)$$

Since unit conductance is the reciprocal of unit resistance,

$$1/C_t = 1/C_1 + 1/C_2 + 1/C_3 + \cdots \quad (7\text{-}17)$$

or for area A

$$1/C_t = x_1/k_1 + x_2/k_2 + x_3/k_3 + \cdots \quad (7\text{-}18)$$

When heat is conducted through parallel paths, the conductances are additive, or

$$C_t = C_1 + C_2 + C_3 + \cdots \quad (7\text{-}19)$$

For an area consisting of $A_1 + A_2 + A_3 + \ldots$

$$C_t = k_1A_1/x_1 + k_2A_2/x_2 + k_3A_3/x_3 \quad (7\text{-}20)$$

If C_t is desired on a per unit area basis, A_1, A_2, A_3, and so on are taken as percentages of the total typical area and expressed as decimals in Eq. (7-20).

For some situations each parallel path may be considered to extend from inside to outside, and the U factor of each path may be calculated. The average U factor is then

$$U_{\text{ave}} = aU_a + bU_b + \cdots + nU_n \quad (7\text{-}21)$$

where a, b, n, and so forth are respective fractions of a typical basic area composed of several different paths whose U factors are U_a, U_b, and so forth.

Illustrative Problem 7-4

Figure 7-5 shows a section through a stud wall of a building. The studs are nominal 2 by 6, 16-in. on center. The stud space is filled with loose-fill insulation having a conductivity $k = 0.27$ (Btu)(in.)/(hr)(ft^2)(°F). Calculate the average U factor for the wall.

<u>Solution:</u>

Step 1: Determine conductivity and conductance for each component of the wall section.

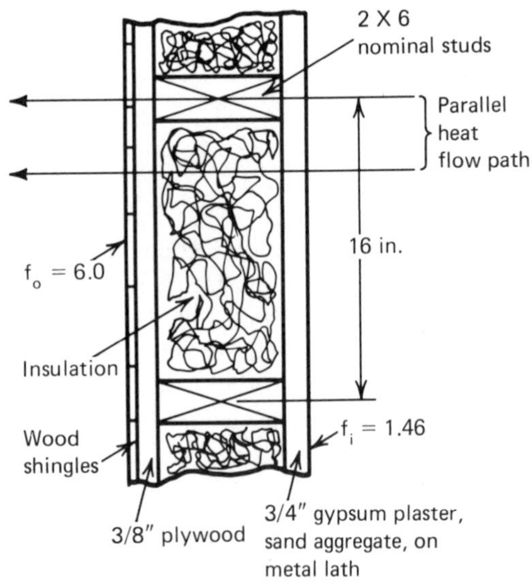

Figure 7–5

Illustrative problems 7-4 and 7-5

	Conductivity k, or Conductance C
Outside air film (Table 7-2A, f_o = 6.00)	f_o = 6.00
Inside air film (Table 7-2A, f_i = 1.46)	f_i = 1.46
Woodshingles, 16-in., 7½ in. exposure (Table 7-1A, item 12a, R = 0.87)	$C = 1/R = 1.15$
⅜-in.. fir plywood. (Table 7-1A, item 1e, R/in. = 1.25; $R = 0.375 \times 1.25 = 0.469$)	$C = 1/R = 2.13$
Insulation (k = 0.27 given)	k = 0.27
¾-in. gypsum plaster on metal lath, sand aggr. (Table 7-1A, item 10b.8, R = 0.10)	$C = 1/R = 10.00$
Spruce stud, nominal 2 by 6. (Table 7-1A, item 13b, R/in. = 1.25)	$k = 1/R = 0.80$

Step 2: Determine total unit conductance of stud space.

The 16-in., center-to-center, spacing of the studs gives 1.5/16 of the width occupied by one stud and 14.5/16 of the width occupied by insulation. By Eq. (7-20)

Stud space conductance (C_{sp})

$$= k_1A_1/x_1 + k_2A_2/x_2$$

Let k_1 = 0.80 for stud

k_2 = 0.27 for insulation

$$\text{Then } C_{sp} = (0.80)(1.5/16)/5.5 + (0.27)(14.5/16)/5.5 = 0.0136 + 0.0445$$

$$C_{sp} = 0.0581 \quad \text{(say 0.058)}$$

Step 3: Calculate the average U factor

Wall Component	Unit Resistance (R)
Outside air film	$1/f_o = 1/6 = 0.17$
Wood shingles	$1/C = 1/1.15 = 0.87$
⅜-in. plywood	$1/C = 1/2.13 = 0.47$
Stud space	$1/C_{sp} = 1/0.058 = 17.24$
¾-in. plaster	$1/C = 1/10 = 0.10$
Insider air film	$1/f_i = 1/1.46 = 0.68$
	$R_t = 19.53$

$$U = 1/R_t = 1/19.53$$

$$= 0.051 \text{ Btu/(hr ft}^2\text{ °F)} \quad \text{(say 0.05)}$$

If the conduction through the studs were neglected, the total unit resistance would be

$$R_t = 19.53 - 17.24 + 5.5/0.27 = 22.66$$

The corresponding U factor is

$$U = 1/22.66 = 0.044 \text{ Btu/(hr ft}^2\text{ °F)}$$

approximately 14 percent lower if the parallel heat flow path is neglected.

Illustrative Problem 7-5

Using the same Fig. 7-5 and data of Problem 7-4, calculate the U factor with the use of Eq. (7-21).

Solution:

Step 1: Determine U factor through insulation.

The total unit resistance through insulation was calculated in Step 3 of Problem 7-4 and found to be 22.66, and the corresponding U factor is 0.044.

Step 2: Determine U factor through studs.

The total unit resistance through the studs would be the total resistance found in step 3 of Problem 7-4 minus stud space resistance plus resistance of the stud, or

$$R_t = 19.53 - 17.24 + 5.5/0.80 = 9.20$$

$$U_{stud} = 1/R_t = 1/9.20$$

$$= 0.11 \text{ Btu/(hr)(ft}^2\text{)(°F)}$$

By Eq. (7-21)

$$U_{ave} = 0.044(14.5/16) + 0.11(1.5/16)$$

$$U_{ave} = 0.0502 \text{ Btu/(hr)(ft}^2\text{)(°F)} \quad \text{(say 0.05)}$$

Illustrative Problem 7-6

Determine the total unit thermal resistance (R) and U factor for a floor section (heat flow down) constructed as noted in the following tabulation.

	R
Top surface air film (Table 7-2, still air, horizontal, heat flow down)	0.92
Asphalt tile (Table 7-1A, item 3e)	0.05
Fir plywood, ½-in. (Table 701A, item 1e, R/in. = 1.25)	0.5 × 1.25 = 0.63
Air space, formed by nominal 2 by 4 wood laid flat, 1½-in. air space. (Table 7-3B, assume t_{mean} = 50°F, Δt = 10°F, E = 0.82)	1.15
Concrete slab, 4-in. sand and gravel aggr. dried. (Table 7-1A, item 8d, R/in. = 0.11)	4 × 0.11 = 0.44
Air space, 8-in. (Table 7-3C, assume equivalent to 3½-in. air space, t_{mean} = 50°F, Δt = 10°F)	1.24
Gypsum plaster, sand aggr. on metal lath, ¾-in. (Table 7-1A, item 10b.8.)	0.10
Bottom surface air film (Table 7-2, still air, horizontal, heat flow down)	0.92
	R_t = 5.45

$$U = \frac{1}{R_t} = \frac{1}{5.45} = 0.183$$
(say 0.18 Btu/hr ft^2 °F)

7-8 TABULATED HEAT TRANSMISSION COEFFICIENTS OF BUILDING SECTIONS

Tables 7-1, 7-2, and 7-3 give representative values of unit resistances for structural materials, air films, and air spaces, respectively. Because there are variations in commercially available materials of the same type, not all the values shown will be in exact agreement with data of individual manufacturers. The exact value for a certain manufacturer's product can be secured from unbiased tests or from guaranteed data of the manufacturer.

The most exact method of determining the heat transmission coefficient (U) for a given combination of building materials assembled as a building section is to test a representative section in a guarded hot box. However, it is not practical to test all such combinations which may be of interest in building construction. Experience has shown that U factors for many types of construction, using accurate values of resistance for the component materials, and when corrections are made for framing member heat loss, are in good agreement with values determined by guarded hot-box measurements.

For the convenience of the designer, tables have been constructed that give overall heat transmission coefficients *(U factors)* for many typical building sections.

Walls, floors and roofs. Table 7-4 is a typical table showing various combinations of structural materials used in walls, floors, and roofs. This particular table has been taken from *Load Calculation Digest, Commercial/Industrial Air Conditioning* published by the General Electric Company and was constructed from procedures and data contained in ASHRAE *Handbook and Product Directory, 1977 Fundamentals*. Table 7-4 is intended to have a great deal of flexibility. The tabulated U factors are based on conductivity and conductance values from Tables 7-1, 7-2, and 7-3 and the following conditions were used:

a. equilibrium or steady-state heat transfer, eliminating effects of heat capacity

b. surrounding surfaces at ambient air temperatures

c. exterior wind velocity of 15 mph for winter

d. surface emissivity of ordinary building materials is 0.90

e. stud space in frame construction not insulated

f. in construction involving air spaces, the U factors shown are calculated for areas between framing

g. air spaces are ¾-in. or more in width

h. variations in conductivity with mean temperature are neglected

It should be noted that the effects of poor workmanship in construction and installation have an increasingly greater percentage effect on heat transmission as the U factor becomes numerically smaller. A factor of safety may be applied as a precaution where it is judged desirable.

Illustrative Problem 7-7

1. Determine the U factor for a 2 by 4 nominal, brick veneer exterior wall built up as indicated. (Neglect parallel heat flow through studs.)
2. If the stud space contains paper-backed, batt-type fiberglass insulation that fills the stud space, 3 to 3¾-in. thickness ($R = 11.0$), determine the U factor.
3. Correct U factors of parts 1 and 2 for the presence of studs placed 16 in. on center.

Solution:

a. Total the unit thermal resistances, uninsulated wall, using the table at the bottom of the next page.
b. Total the unit thermal resistances for insulated wall.

$$\begin{aligned} \text{From part 1a } R_t &= 4.13 \\ \text{Less stud space} &= -1.01 \\ \text{Plus insulation} &= \underline{+11.00} \\ \text{New } R_t &= 14.12 \end{aligned}$$

$$\text{New } U = 1/R_t = 1/14.12 = 0.0708 \quad \text{(say 0.071)}$$

NOTE: Referring to Table 7-4, for construction detail 5d+ with 3-in. insulation, a U factor of 0.07 is noted.

c. Correct U factors for presence of studs.

For part 1 resistance through studs section is

$$R_s = 4.13 - 1.01 + 3.5/0.8 = 7.495$$

$$U_s = 1/R_s = 1/7.495 = 0.133$$

By Eq. (7-21), $U_{ave} = 0.24(14.5/16) + 0.133(1.5/16)$

$$U_{ave} = 0.2175 + 0.0125 = 0.23 \text{ Btu/(hr)(ft}^2\text{)(°F)}$$

For part 2 resistance through studs are the same, therefore, by Eq. (7-21),

$$U_{ave} = 0.071\ (14.5/16) + 0.133(1.5/16)$$

$$U_{ave} = 0.0643 + 0.0125$$

$$= 0.0768 \quad \text{(say 0.08)}$$

7-9 TEMPERATURE GRADIENT IN A WALL SECTION

The temperature in a wall section will vary from the inside to the outside surface. As indicated by Eq. (7-6), the temperature drop

	R
1. Outside air film (15 mph) (Table 7-2A)	= 0.17
2. Face brick (veneer). (Table 7-1A, item 9b., R/in. = 0.11)	4.0 × 0.11 = .44
3. Air space, ¾-in. (between brick and sheating) (Table 7-3A, assume mean temp. = 0°F, temp. diff. = 10°F, E = 0.82)	= 1.26
4. ⅜-in. fir plywood (sheating). (Table 7-1A, item 13, R/in. = 1.25)	⅜ × 1.25 = 0.47
5. Stud space (2 by 4 nominal wood studs). (Table 7-3C, 3½-in. air space, mean temp. = 50, temp. diff. = 10, E = 0.82)	= 1.01
6. Metal lath and gypsum plaster, sand aggr., ¾-in. (Table 7-1A, item 10b.8)	= 0.10
7. Inside air film (still air) (Table 7-2A)	= 0.68
	R_t = 4.13

$$U = 1/R_t = 1/4.13 = 0.242 \quad \text{say 0.24 Btu/(hr) (ft}^2\text{) (°F)}$$

NOTE: Referring to Table 7-4, the construction in our problem is similar to construction detail 5d, which gives a U factor of 0.31.

Table 7-4

Coefficients of Transmission *(U)* for Typical Building Sections*

CONSTRUCTION DETAIL	HEAT TRANSMISSION COEFFICIENT "U" Btuh per sq ft per °F. — INSULATION ADDED TO BASIC CONSTRUCTION			
	None	Blanket or Batt Type Thickness		
		1-1/2"	2"	3"
EXPOSED FRAME AND VENEER WALLS				
1 Wood Siding or Wood Shingle Exterior				
a. Building paper, 25/32 in. wood sheathing, with wood or 3/8 in. gypsum lath and 1/2 in. plaster interior	.24	.10	.09	.07
b. Same as (1a), but with 1/2 in. insulating board or 1/2 in. insulating lath and 1/2 in. plaster interior	.19	.10	.08	.06
c. 25/32 in. insulating sheathing with wood or 3/8 in. gypsum lath and 1/2 in. plaster interior	.19	.09	.08	.06
d. 3/8 in. plywood sheathing with wood or 3/8 in. gypsum lath and 1/2 in. plaster interior	.28	.11	.09	.07
2 Asbestos-Cement Siding or Shingles (1/4 in. thick, tapped) or Stucco (1" over bldg. paper) on Frame				
a. Building paper, 25/32 in. wood sheathing, with wood or 3/8 in. gypsum lath and 1/2 in. plaster interior	.29	.11	.09	.07
b. Same as (2a), but with 1/2 in. insulating board or 1/2 in. insulating lath and 1/2 in. plaster interior	.22	.10	.08	.06
c. 25/32 in. insulating sheathing with wood or 3/8 in. gypsum lath and 1/2 in. plaster interior	.22	.10	.08	.06
d. 3/8 in. plywood sheathing with wood or 3/8 in. gypsum lath and 1/2 in. plaster interior	.34	.12	.10	.07
3 Panel Walls, Steel or Aluminum Skin				
a. Blown polyurethane core		*.16	.08	.05
b. Polystyrene or fiber glass core		*.21	.12	.08
c. Cellular glass core		*.30	.17	.12
d. Corkboard, or, mineral wool with resin binder core		*.24	.14	.09
4 Insulating Siding (1/2 in.), or Wood Shingles over Insulating Backer Board (5/16 in.)				
a. Building paper, 25/32 in. wood sheathing, with wood or 3/8 in. gypsum lath and 1/2 in. plaster interior	.21	.10	.08	.06
b. Same as (4a), but with 1/2 in. insulating board or 1/2 in. insulating lath and 1/2 in. plaster interior	.18	.09	.08	.06
c. 25/32 in. insulating sheathing with wood or 3/8 in. gypsum lath and 1/2 in. plaster interior	.18	.09	.08	.06
d. 3/8 in. plywood sheathing with wood or 3/8 in. gypsum lath and 1/2 in. plaster interior	.24	.10	.09	.07

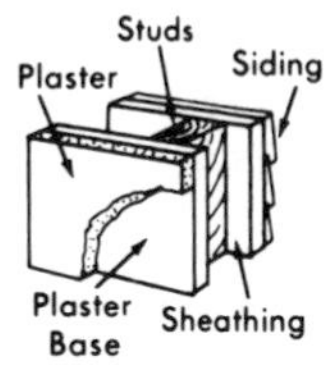

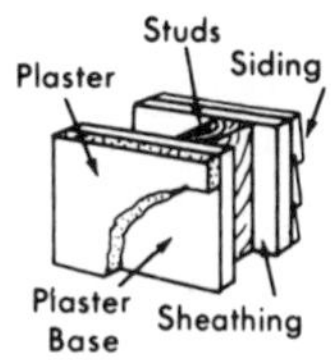

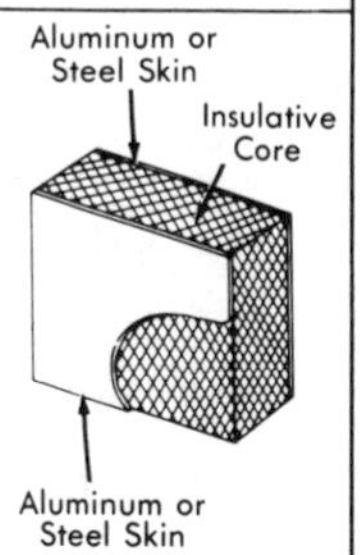

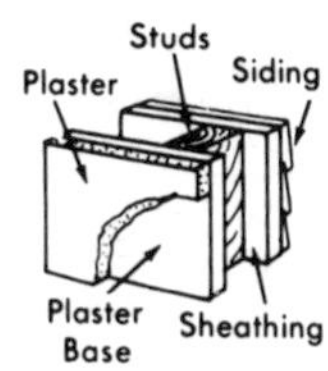

*1" Insulation Core

Table 7-4 (Cont.)

CONSTRUCTION DETAIL	HEAT TRANSMISSION COEFFICIENT "U" Btuh per sq ft per °F. INSULATION ADDED TO BASIC CONSTRUCTION — None	Blanket or Batt Type Thickness 1-1/2"	2"	3"
5 Veneer (4 in. Face Brick or 4 in. Stone)				
a. Building paper, 25/32 in. wood sheathing, with wood or 3/8 in. gypsum lath and 1/2 in. plaster interior	.27	.11	.09	.07
b. Same as (5a), but with 1/2 in. insulating board or 1/2 in. insulating lath and 1/2 in. plaster interior	.21	.10	.08	.06
c. 25/32 in. insulating sheathing with wood or 3/8 in. gypsum lath and 1/2 in. plaster interior	.21	.10	.08	.06
d. 3/8 in. plywood sheathing with wood or 3/8 in. gypsum lath and 1/2 in. plaster interior	.31	.11	.09	.07
6 Sheet Metal Siding				
a. Aluminum sheet on studs, no sheathing or interior finish	1.18	.15	.12	.08
b. Same as (6a), but with 1/2 in. insulating board interior expanded Polyesterene	.39	.12	.10	.07
c. Galvanized sheet (plain or corrugated) on studs, no sheathing or interior finish	1.18	.15	.12	.08
d. Same as (6c), but with 3/8 in. dry-wall interior	.47	.13	.11	.08
e. Aluminum or galvanized sheet on 25/32 in. insulating sheathing, no interior finish	.46	.13	.11	.08
7 Unsheathed Frame (old construction)				
a. Clapboard or wood siding, studs, lath and plaster interior	.33	.12	.10	.07
8 Frame Interior Partitions on Studs				
a. Wood or 3/8 in. gypsum lath and 1/2 in. plaster finish, one side	.57	.14	.11	.08
b. Metal lath and 3/4 in. plaster finish, one side	.68	.14	.11	.08
c. Same as (8b), but with finish, both sides	.40	.12	.10	.07
STANDARD MASONRY AND TILE WALLS (ABOVE GRADE ONLY)				
9 Brick, Face and Common				
a. 8 in. brick, no interior finish	.48			
b. Same as (9a), but with 5/8 in. plaster on interior surface of brick	.45			
c. Same as (9a), but with 3/8 in. gypsum lath and 1/2 in. plaster on furring	.29	.12	.11	.09
d. 10 in. brick, no interior finish	.40			
e. Same as (9d), but with 5/8 in. plaster on interior surface of brick	.39			
f. Same as (9d), but with 3/8 in. gypsum lath and 1/2 in. plaster on furring	.26	.11	.10	.09

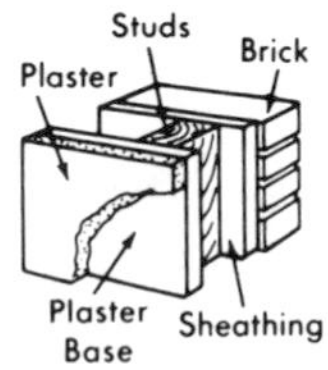

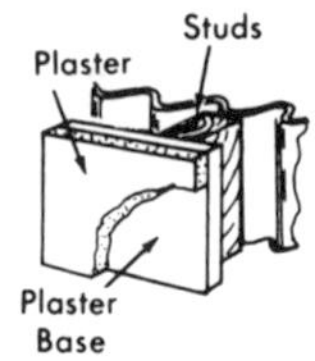

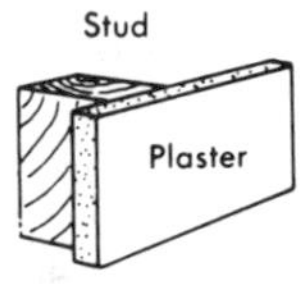

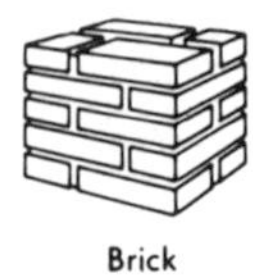

Table 7-4 (Cont.)

CONSTRUCTION DETAIL	HEAT TRANSMISSION COEFFICIENT "U" Btuh per sq ft per ^{o} F.				
	INSULATION ADDED TO BASIC CONSTRUCTION				
	None	Blanket or Batt Type Thickness			
		1-1/2"	2"	3"	
9 Brick, Face and Common (Cont'd)					
g. 12 in. brick, no interior finish	.35				
h. Same as (9g), but with 5/8 in plaster on interior surface of brick	.33				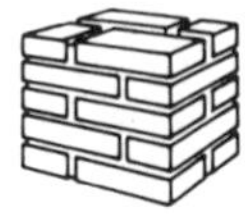
i. Same as (9g), but with metal lath and 3/4 in. plaster on furring	.25	.12	.10	.09	
j. Same as (9g), but with 3/8 in. gypsum lath and 1/2 in. plaster on furring	.24	.11	.10	.09	Brick
10 Poured Concrete – Gravel Aggregate (140 lb. per cu. ft. density)					
a. 6 in. poured concrete – gravel aggregate – no interior finish	.75				
b. Same as (10a), but with 5/8 in. plaster on interior surface of concrete	.70				
c. Same as (10a), but with 3/8 in. gypsum lath and 1/2 in. plaster on furring	.37	.13	.11	.10	
d. 8 in. poured concrete – gravel aggregate – no interior finish	.67				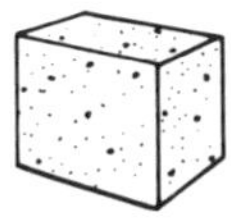
e. Same as (10d), but with 5/8 in. plaster on interior surface of concrete	.63				
f. Same as (10d), but with 3/8 in. gypsum lath and 1/2 in. plaster on furring	.35	.13	.11	.10	Concrete
g. 10 in. poured concrete – gravel aggregate – no interior finish	.61				
h. Same as (10g), but with 5/8 in. plaster on interior surface of concrete	.57				
i. Same as (10g), but with 3/8 in. gypsum lath and 1/2 in. plaster on furring	.33	.13	.11	.10	
11 Poured Concrete – Insulating (30 lb. per cu. ft. density)					
a. 6 in. poured concrete – insulating – no interior finish	.13				
b. Same as (11a), but with 5/8 in. plaster on interior surface of concrete	.13				
c. Same as (11a), but with 3/8 in. gypsum lath and 1/2 in. plaster on furring	.11	.09	.09	.08	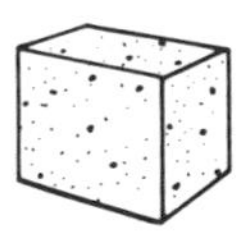
d. 8 in. poured concrete – insulating – no interior finish	.10				
e. Same as (11d), but with 5/8 in. plaster on interior surface of concrete	.10				
f. Same as (11d), but with 3/8 in. gypsum lath and 1/2 in. plaster on furring	.09	.09	.09	.08	Concrete
g. 10 in. poured concrete – insulating – no interior finish	.08				
h. Same as (11g), but with 5/8 in. plaster on interior surface of concrete	.08				
i. Same as (11g), but with 3/8 in. gypsum lath and 1/2 in. plaster on furring	.08	.08	.08	.08	

Table 7-4 (Cont.)

CONSTRUCTION DETAIL	HEAT TRANSMISSION COEFFICIENT "U" Btuh per sq ft per °F. INSULATION ADDED TO BASIC CONSTRUCTION				
	None	Blanket or Batt Type Thickness			
		1-1/2"	2"	3"	
12 Concrete Block – Gravel Aggregate					
a. 8 in. concrete block, no interior finish	.51				
b. Same as (12a), but with 5/8 in. plaster on interior surface of block	.48				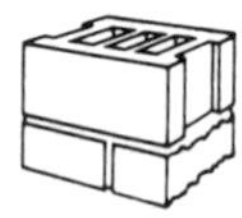
c. Same as (12a), but with 3/8 in. gypsum lath and 1/2 in. plaster on furring	.30	.13	.11	.09	
d. 12 in. concrete block, no interior finish	.47				Blocks
e. Same as (12d), but with 5/8 in. plaster on interior surface of blocks	.45				
f. Same as (12d), but with 3/8 in. gypsum lath and 1/2 in. plaster on furring	.29	.12	.11	.09	
13 Concrete Block – Cinder Aggregate					
a. 8 in. concrete block, no interior finish	.39				
b. Same as (13a), but with 5/8 in. plaster on interior surface of block	.38				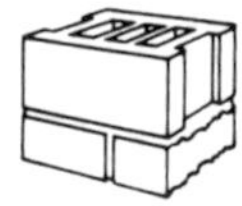
c. Same as (13a), but with 3/8 in. gypsum lath and 1/2 in. plaster on furring	.25	.12	.10	.09	
d. 12 in. concrete block, no interior finish	.36				
e. Same as (13d), but with 5/8 in. plaster on interior surface of block	.35				Blocks
f. Same as (13d), but with 3/8 in. gypsum lath and 1/2 in. plaster on furring	.24	.12	.10	.09	
14 Concrete Block – Light-Weight Aggregate					
a. 8 in. concrete block, no interior finish	.35				
b. Same as (14a), but with 5/8 in. plaster on interior surface of block	.34				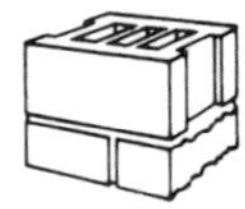
c. Same as (14a), but with 3/8 in. gypsum lath and 1/2 in. plaster on furring	.24	.12	.10	.09	
d. 12 in. concrete block, no interior finish	.32				
e. Same as (14d), but with 5/8 in. plaster on interior finish surface of block	.31				Blocks
f. Same as (14d), but with 3/8 in. gypsum lath and 1/2 in. plaster on furring	.22	.11	.10	.09	
15 4 in. Face Brick, Stone or Precast Concrete Backed by Gravel Aggregate Concrete Block					
a. 4 in. concrete block backing, no interior finish	.48				
b. Same as (15a), but with 5/8 in. plaster on interior surface of block	.46				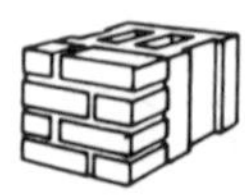
c. Same as (15a), but with 3/8 in. gypsum lath and 1/2 in. plaster on furring	.29	.12	.11	.09	
d. 8 in. concrete block backing, no interior finish	.40				Brick & Block or Tile
e. Same as (15d), but with 5/8 in. plaster on interior surface of block	.39				
f. Same as (15d), but with 3/8 in. gypsum lath and 1/2 in. plaster on furring	.26	.12	.10	.09	

Table 7-4 (Cont.)

CONSTRUCTION DETAIL	HEAT TRANSMISSION COEFFICIENT "U" Btuh per sq ft per ° F. INSULATION ADDED TO BASIC CONSTRUCTION				
	None	Blanket Batt Type Thickness 1-1/2"	2"	3"	
16 4 in. Face Brick, Stone, or Precast Concrete Backed by Cinder Aggregate Concrete Block					 Brick & Block or Tile
a. 4 in. concrete block backing, no interior finish	.40				
b. Same as (16a), but with 5/8 in. plaster on interior surface of block	.39				
c. Same as (16a), but with 3/8 in. gypsum lath and 1/2 in. plaster on furring	.26	.12	.10	.09	
d. 8 in. concrete block backing, no interior finish	.33				
e. Same as (16d), but with 5/8 in. plaster on interior surface of block	.32				
f. Same as (16d), but with 3/8 in. gypsum lath and 1/2 in. plaster on furring	.23	.11	.10	.09	
17 4 in. Face Brick, Stone, or Precast Concrete Backed by Light-Weight Aggregate Concrete Block					 Brick & Block or Tile
a. 4 in. concrete block backing, no interior finish	.35				
b. Same as (17a), but with 5/8 in. plaster on interior surface of block	.34				
c. Same as (17a), but with 3/8 in. gypsum lath and 1/2 in. plaster on furring	.24	.12	.10	.09	
d. 8 in. concrete block backing, no interior finish	.30				
e. Same as (17d), but with 5/8 in. plaster on interior surface of block	.29				
f. Same as (17d), but with 3/8 in. gypsum lath and 1/2 in. plaster on furring	.21	.11	.10	.09	
18 4 in. Face Brick, Stone, or Precast Concrete Backed by Gravel-Aggregate Poured Concrete					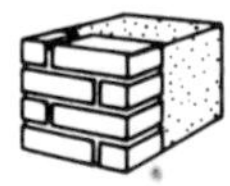 Brick, Stone or Precast Concrete Backed by Poured Concrete
a. 6 in. poured concrete backing, no interior finish	.54				
b. Same as (18a), but with 5/8 in. plaster on interior surface of concrete	.51				
c. Same as (18a), but with 3/8 in. gypsum lath and 1/2 in. plaster on furring	.31	.13	.11	.09	
d. 8 in. poured concrete backing, no interior finish	.49				
e. Same as (18d), but with 5/8 in. plaster on interior surface of concrete	.47				
f. Same as (18d), but with 3/8 in. gypsum lath and 1/2 in. plaster on furring	.30	.13	.11	.09	
MASONRY CAVITY WALLS					
19 4 in. Face Brick Exterior Construction Masonry Cavity Wall					
a. 4 in. gravel aggregate concrete block or 4 in. common brick inner section, no interior finish	.33				
b. Same as (19a), but with 5/8 in. plaster on interior surface of block or brick	.32				

Table 7-4 (Cont.)

CONSTRUCTION DETAIL	HEAT TRANSMISSION COEFFICIENT "U" Btuh per sq ft per °F.			
	INSULATION ADDED TO BASIC CONSTRUCTION			
	None	Blanket or Batt Type Thickness		
		1-1/2"	2"	3"
19 4 in. Face Brick Exterior Construction Masonry Cavity Wall (Cont'd)				
c. Same as (19a), but with 3/8 in. gypsum lath and 1/2 in. plaster on furring	.23	.11	.10	.09
d. 4 in. cinder aggregate concrete block or 4 in. clay tile inner section, no interior finish	.30			
e. Same as (19d), but with 5/8 in. plaster on interior surface of block or tile	.29			
MASONRY INTERIOR PARTITIONS				
20 Gravel-Aggregate Hollow Concrete Block Partitions				
a. 8 in. block, no finish either side	.40			
b. Same as (20a), but with 5/8 in. plaster finish one side	.39			
c. Same as (20a), but with gypsum lath and 1/2 in. plaster on furring one side	.26	.12	.10	.09
22 Cinder-Aggregate Hollow Concrete Block Partitions				
a. 8 in. block, no finish either side	.33			
b. Same as (21a), but with 5/8 in. plaster finish one side	.32			
c. Same as (21a), but with gypsum lath and 1/2 in. plaster on furring one side	.23	.11	.10	.09

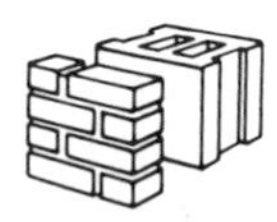

Masonry Cavity Wall

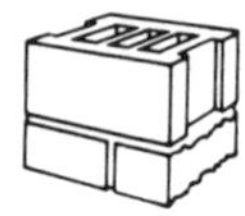

Blocks

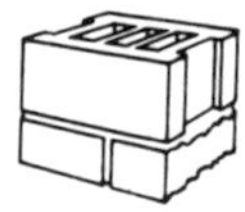

Blocks

CONSTRUCTION DETAIL	HEAT TRANSMISSION COEFFICIENT "U" Btuh per sq. ft. per °F.						
	INSULATION ADDED TO BASIC CONSTRUCTION						
			Blanket, Batt or Fill Type Thickness				
	None		2		3	4	6
	Clg.	Htg.	Clg.	Htg.	Clg. & Htg.	Clg. & Htg.	Clg. & Htg.
WOOD AND CONCRETE FLOORS							
22 WOOD FRAME FLOORS							
a. With subfloor, hardwood flooring. No ceiling. Over untreated, open, occupied, space, or kitchen, laundry, etc.	.28	.34	.09	.09	.07	.05	.04
b. Same as (20a) but with 1" accoustical tile on 3/8" gypsum board or ½" insulating board only.	.17	.19	.07	.08	.06	.05	.04
c. Same as (20a) but with metal lath and 3/8" plaster	.21	.26	.08	.09	.07	.05	.04
23 CONCRETE FLOORS							
a. 4" to 6" concrete floor, tile or linoleum, plaster below.	.42	.57	.10	.10	.08	.06	.04
b. Same as (23a) but with suspended ½" tile on 3/8" gypsum board ceiling.	.21	.25	.08	.09	.06	.05	.04
c. Concrete floor on ground with vertical edge insulation. (Btu/per linear ft. of perimeter/degree temperature difference.)		.83		.59			

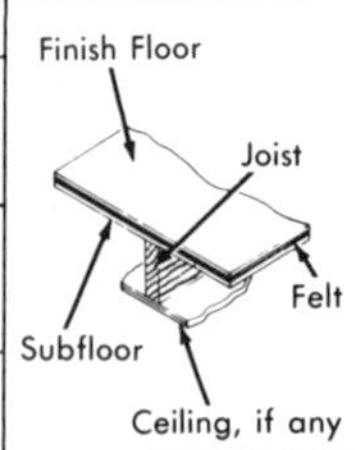

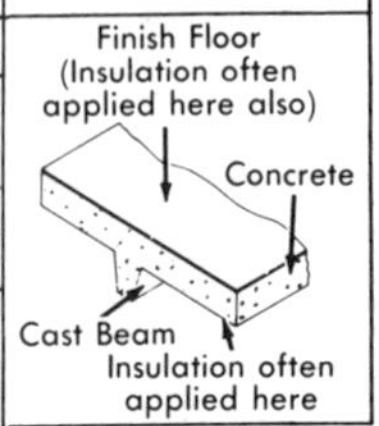

Table 7-4 (Cont.)

CONSTRUCTION DETAIL	HEAT TRANSMISSION COEFFICIENT "U" Btuh per sq ft per °F. INSULATION ADDED TO BASIC CONSTRUCTION							
	None		Blanket, Batt or Fill Type Thickness 2		3	4	6	
	Clg.	Htg.	Clg.	Htg.	Clg. & Htg.	Clg. & Htg.	Clg. & Htg.	
CEILINGS WITH ATTIC SPACE ABOVE (VENTED)								
24 Wood Joists, no floor								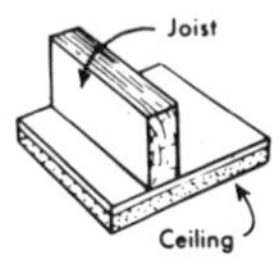
a. 3/8 in. dry-wall interior	.46	.65	.10	.11	.08	.06	.04	
b. 3/8 in. gypsum lath and 1/2 in. plaster interior	.44	.61	.10	.11	.08	.06	.04	
c. 1/2 in. acoustical tile on 3/8 in. gypsum board interior	.30	.37	.09	.10	.07	.06	.04	
d. 3/4 in. acoustical tile on 3/8 in. gypsum board interior	.25	.30	.09	.09	.07	.06	.04	
25 Wood Joists, with floor Above (Vented)								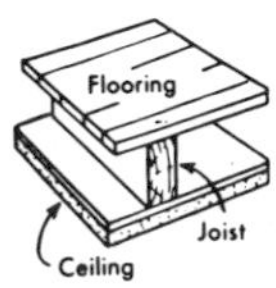
a. 25/32 in. wood or 3/4 in. plywood floor in attic, 3/8 in. dry-wall interior	.24	.30	.09	.09	.07	.06	.04	
b. Same as (25a), but with 3/8 in. gypsum lath and 1/2 in. plaster interior	.24	.29	.09	.09	.07	.06	.04	
c. Same as (25a), but with 1/2 in. acoustical tile on 3/8 in. gypsum board interior	.19	.22	.08	.08	.07	.05	.04	
d. Same as (25a), but with 1/2 in. acoustical tile on furring interior	.20	.24	.08	.09	.07	.05	.04	

CONSTRUCTION DETAIL	HEAT TRANSMISSION COEFFICIENT "U" Btuh per sq ft per °F. PREFORMED INSULATION BETWEEN DECK & ROOF Nominal Thickness, in.							
	None		1/2"		1"	2"	3"	
	Clg.	Htg.	Clg.	Htg.	Clg. & Htg.	Clg. & Htg.	Clg. & Htg.	
FLAT ROOFS – BUILT-UP ON DECK WITH & WITHOUT SUSPENDED CEILING								
26 4 in. Gravel-Aggregate Concrete Slab Deck Under Built-up Roof								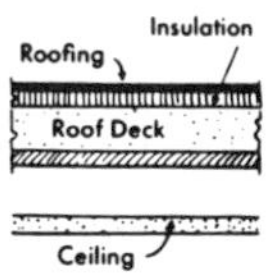
a. No ceiling	.55	.70	.31	.35	.22	.14	.10	
b. 3/8 in. dry-wall ceiling	.32	.38	.20	.22	.16	.10	.07	
c. 3/8 in. gypsum lath and 1/2 in. plaster ceiling	.31	.37	.20	.22	.15	.10	.07	
d. 1/2 in. acoustical tile ceiling on 3/8 in. gypsum board	.23	.26	.16	.18	.13	.09	.07	
e. 1/2 in. acoustical tile ceiling on furring or channels	.25	.29	.17	.19	.14	.09	.07	
27 8 in. Gravel-Aggregate Concrete Slab Deck Under Built-Up Roof								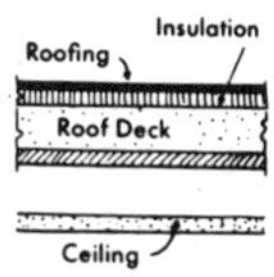
a. No ceiling	.47	.57	.28	.32	.20	.14	.10	
b. 3/8 in. dry-wall ceiling	.29	.34	.19	.21	.15	.10	.07	
c. 3/8 in. gypsum lath and 1/2 in. plaster ceiling	.28	.33	.18	.20	.15	.10	.07	
d. 1/2 in. acoustical tile ceiling on 3/8 in. gypsum board	.22	.24	.16	.17	.13	.09	.07	
e. 1/2 in. acoustical tile ceiling on furring or channels	.23	.26	.16	.18	.13	.09	.07	

Table 7-4 (Cont.)

CONSTRUCTION DETAIL	HEAT TRANSMISSION COEFFICIENT "U" Btuh per sq ft per °F. PREFORMED INSULATION BETWEEN DECK & ROOF Nominal Thickness, in.						
	None		1/2"		1"	2"	3"
	Clg.	Htg.	Clg.	Htg.	Clg. & Htg.	Clg. & Htg.	Clg. & Htg.
28 2 in. Light-Weight Aggregate Concrete Slab Deck Under Built-Up Roof							
a. Deck poured on corrugated metal, or 1/4 in. asbestos-cement board, no ceiling	.27	.30	.20	.21	.15	.11	.08
b. Same as (28a), but with 3/8 in. gypsum board ceiling	.20	.22	.15	.16	.12	.08	.06
c. Same as (28a), but with 3/8 in. gypsum lath and 1/2 in. plaster ceiling	.20	.22	.15	.16	.12	.08	.06
d. Same as (28a), but with 1/2 in. acoustical tile ceiling on 3/8 in. gypsum board	.16	.18	.12	.13	.11	.08	.06
e. Same as (28a), but with 1/2 in. acoustical tile ceiling on furring or channels	.17	.19	.13	.14	.11	.08	.06
f. Deck poured on 1 in. insulation board, no ceiling	.15	.16	.13	.13	.11	.09	.07
g. Same as (28f), but with 3/8 in. dry-wall ceiling	.13	.14	.10	.11	.09	.07	.05
h. Deck poured on 1 in. glass fiber board or 1-1/2 in. insulation board, no ceiling	.13	.14	.10	.11	.09	.07	.05
i. Same as (28h), but with 3/8 in. dry-wall ceiling	.11	.12	.09	.10	.08	.06	.05
29 4 in. Light-Weight Aggregate Concrete Slab Deck Under Built-Up Roof							
a. Deck poured on corrugated metal, or 1/4 in. asbestos-cement board, no ceiling	.17	.18	.14	.14	.11	.09	.07
b. Same as (29a), but with 3/8 in. gypsum board ceiling	.14	.15	.11	.12	.10	.07	.05
c. Same as (29a), but with 3/8 in. gypsum lath and 1/2 in. plaster ceiling	.14	.15	.11	.12	.10	.07	.05
d. Same as (29a), but with 1/2 in. acoustical tile ceiling on 3/8 in. gpysum board	.12	.13	.10	.10	.09	.07	.05
e. Same as (29a), but with 1/2 in. acoustical tile on ceiling furring or channels	.12	.13	.10	.10	.09	.07	.05
f. Deck poured on 1 in. insulation board, no ceiling	.11	.12	.10	.10	.09	.07	.06
g. Same as (29f), but with 3/8 in. dry-wall ceiling	.10	.11	.09	.09	.08	.06	.05
h. Same as (29f), but with 3/8 in. gypsum lath and 1/2 in. plaster ceiling	.10	.10	.09	.09	.08	.06	.05
i. Deck poured on 1 in. glass fiber board or 1-1/2 in. insulation board, no ceiling	.10	.10	.09	.09	.08	.07	.05
j. Same as (29i), but with 3/8 in. dry-wall ceiling	.09	.09	.08	.08	.07	.05	.04
k. Same as (29i), but with 3/8 in. gypsum lath and 1/2 in. plaster ceiling	.09	.09	.08	.08	.07	.05	.04
l. Same as (29i), but with 1/2 in. acoustical tile ceiling on 3/8 in. gpysum board	.08	.08	.07	.07	.06	.05	.04
m. Same as (29i), but with 1/2 in. acoustical tile ceiling on furring or channels	.08	.09	.07	.08	.07	.05	.04

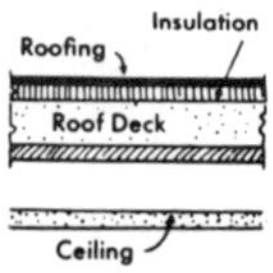

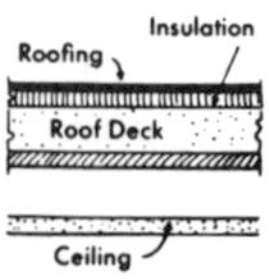

Table 7-4 (Cont.)

CONSTRUCTION DETAIL	HEAT TRANSMISSION COEFFICIENT "U" Btuh per sq ft per °F. PREFORMED INSULATION BETWEEN DECK & ROOF Nominal Thickness, in.						
	None		1/2"		1"	2"	3"
	Clg.	Htg.	Clg.	Htg.	Clg. & Htg.	Clg. & Htg.	Clg. & Htg.
30 2 in. Gypsum Deck Slab Under Built-up Roof							
a. Deck slab on 1/4 in. asbestos-cement board, no ceiling	.34	.40	.23	.26	.18	.12	.09
b. Same as (30a), but with 3/8 in. dry-wall ceiling	.25	.27	.17	.18	.13	.09	.07
c. Same as (30a), but with 3/8 in. gypsum lath and 1/2 in. plaster ceiling	.24	.26	.17	.18	.13	.09	.07
d. Same as (30a), but with 1/2 in. acoustical tile ceiling on 3/8 in. gypsum board	.19	.20	.14	.15	.12	.09	.06
e. Same as (30a), but with 1/2 in. acoustical tile ceiling on furring or channels	.20	.22	.15	.16	.12	.09	.06
f. Deck slab on 1 in. insulation board, no ceiling	.18	.20	.15	.15	.12	.09	.07
g. Same as (30f), but with 3/8 in. dry-wall ceiling	.15	.16	.11	.12	.10	.07	.06
h. Same as (30f), but with 3/8 in. gypsum lath and 1/2 in. plaster Ceiling	.15	.16	.12	.12	.10	.07	.06
i. Same as (30f), but with 1/2 in. acoustical tile ceiling on 3/8 in. gypsum board	.13	.13	.10	.10	.09	.07	.05
j. Same as (30f), but with 1/2 in. acoustical tile ceiling on furring or channels	.13	.14	.10	.11	.09	.07	.05
k. Deck slab on 1 in. glass fiber board or 1-1/2 in. insulation board, no ceiling	.15	.16	.12	.13	.11	.08	.07
l. Same as (30k), but with 3/8 in. dry-wall ceiling	.13	.13	.10	.10	.09	.07	.05
m. Same as (30k), but with 3/8 in. gpysum lath and 1/2 in. plaster ceiling	.12	.13	.10	.10	.09	.07	.05
n. Same as (30k), but with 1/2 in. acoustical tile ceiling on 3/8 in. gypsum board	.11	.12	.09	.10	.08	.06	.05
o. Same as (30K), but with 1/2 in. acoustical tile ceiling on furring or channels	.11	.12	.09	.10	.08	.06	.05
31 Wood Deck Under Built-Up Roof							
a. 25/32 in. wood deck, no ceiling	.40	.48	.26	.29	.19	.13	.09
b. Same as (31a), but with 3/8 in. dry-wall ceiling	.26	.31	.18	.19	.14	.09	.07
c. Same as (31a), but with 3/8 in. gypsum lath and 1/2 in. plaster ceiling	.26	.30	.18	.19	.14	.09	.07
d. Same as (31a), but with 1/2 in. acoustical tile ceiling on 3/8 in. gpysum board	.20	.22	.15	.16	.12	.08	.06

Table 7-4 (Cont.)

CONSTRUCTION DETAIL	HEAT TRANSMISSION COEFFICIENT "U" Btuh per sq ft per °F. — PREFORMED INSULATION BETWEEN DECK & ROOF Nominal Thickness, in.						
	None		1/2"		1"	2"	3"
	Clg.	Htg.	Clg.	Htg.	Clg. & Htg.	Clg. & Htg.	Clg. & Htg.
31 Wood Deck Under Built-Up Roof (Cont'd)							
e. Same as (31a), but with 1/2 in. acoustical tile ceiling on furring or channels	.22	.24	.16	.17	.13	.09	.07
f. 1-5/8 in. wood deck, no ceiling	.28	.32	.20	.22	.16	.11	.08
g. Same as (31f), but with 3/8 in. dry-wall ceiling	.21	.23	.15	.16	.12	.08	.06
h. Same as (31f), but with 3/8 in. gypsum lath and 1/2 in. plaster ceiling	.20	.23	.15	.16	.12	.08	.06
i. Same as (31f), but with 1/2 in. acoustical tile ceiling on 3/8 in. gypsum board	.17	.18	.13	.13	.11	.08	.06
j. Same as (31f), but with 1/2 in. acoustical tile ceiling on furring or channels	.18	.19	.13	.14	.11	.08	.06
32 Flat Metal Deck Built-Up Roof							
a. No ceiling	.67	.90	.35	.40	.23	.15	.10
b. 3/8 in. gypsum-board ceiling	.36	.44	.22	.24	.17	.10	.08
c. 3/8 in. gypsum lath and 1/2 in. plaster ceiling	.34	.42	.20	.23	.16	.10	.07
d. 1/2 in. acoustical tile ceiling on 3/8 in. gypsum board	.25	.29	.17	.19	.14	.09	.07
e. 1/2 in. acoustical tile ceiling on furring or channels	.27	.32	.18	.20	.14	.09	.07

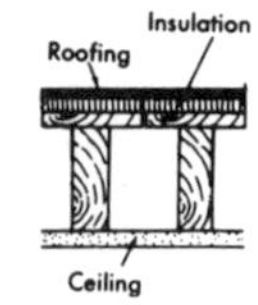

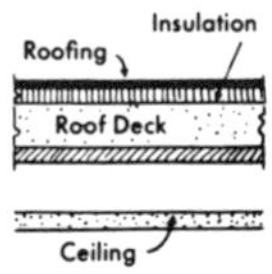

CONSTRUCTION DETAIL	HEAT TRANSMISSION COEFFICIENT "U" Btuh per sq ft per °F. — INSULATION ADDED TO BASIC CONSTRUCTION				
		RIGID INSULATION APPLIED TO INTERIOR OF FOUNDATION, THICKNESS			
		Insulating	Sheathing	Insulation Board	
	None	1/2"	25/32"	1"	2"
BASEMENT AND CRAWL-SPACE FOUNDATION WALLS					
33 Poured Concrete (No Interior Finish)					
a. 6 in. poured concrete — gravel-aggregate — above grade 140 lbs./cu. ft.	.75	.38	.30	.24	.16
b. Same as (33a), but with light-weight aggregate concrete 80 lbs./cu. ft.	.31	.22	.20	.16	.13
c. Same as (33a), but with insulating concrete 30 lbs./cu. ft.	.13	.13	.12	.11	.10
d. 8 in. poured concrete — gravel-aggregate — above grade 140 lbs./cu. ft.	.67	.36	.29	.23	.16
e. Same as (33d), but with light-weight aggregate concrete 8Q lbs./cu. ft.	.25	.19	.17	.15	.12
f. 6 in., 8 in., or 10 in. poured concrete below grade	.06				

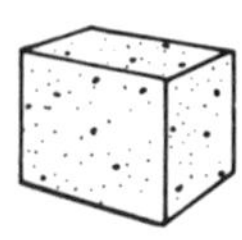

Concrete

Table 7-4 (Cont.)

CONSTRUCTION DETAIL	HEAT TRANSMISSION COEFFICIENT "U" Btuh per sq ft per °F.				
	INSULATION ADDED TO BASIC CONSTRUCTION				
	RIGID INSULATION APPLIED TO INTERIOR OF FOUNDATION, THICKNESS				
		Insulating	Sheathing	Insulation Board	
	None	1/2"	25/32"	1"	2"
34 Concrete Block (No Interior Finish)					
a. 8 in. concrete block – gravel-aggregate – above grade	.52	.31	.26	.21	.15
b. Same as (34a), but with cinder-aggregate block	.39	.26	.22	.18	.14
c. Same as (34a), but with light-weight aggregate block	.35	.25	.21	.18	.13
d. 10 in. concrete block – gravel-aggregate – above grade	.50				
e. Same as (34d), but with light-weight aggregate block	.33	.23	.20	.17	.13
f. 12 in. concrete block – gravel-aggregate – above grade	.47	.30	.25	.20	.14
g. Same as (34f), but with cinder-aggregate block	.36	.25	.22	.18	.13
h. 8 in., 10 in., or 12 in. block below grade	.06				

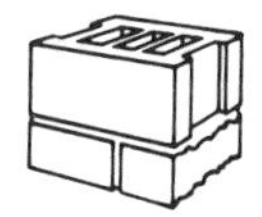

Blocks

**Extracted by permission.* Load Calculation Digest Commercial/Industrial Air Conditioning. *General Electric Company, 1971.*

through the various components of a building structure are proportional to the thermal resistance. For a wall section we may say

$$\frac{R_x}{R_t} = \frac{t_i - t_x}{t_i - t_o}$$

or

$$t_x = t_i - \frac{R_x}{R_t}(t_i - t_o) \qquad (7\text{-}22)$$

where

t_x = temperature at selected point in construction

t_i = inside air temperature

t_o = outside air temperature

R_x = sum of unit thermal resistance from inside air to selected point

R_t = total unit thermal resistance of section

The usual point of interest in a wall section would be the inside wall surface temperature. This surface temperature would be lower than the inside air temperature because of resistance offered by the inside air film. If the inside surface temperature is at or below the inside air dew-point temperature, condensation may occur on the surface. Condensation on interior surfaces may cause damage to plaster and woodwork, as well as constituting a nuisance. In winter insulated walls are effective in reducing condensation problems, but if walls cannot be effectively insulated, the inside air relative humidity must be reduced to lower its dew-point temperature. Even if the condensation problem does not exist, people may feel a chill when seated next to a cold exterior wall because of radiation heat loss from their body to the cold wall surface.

Illustrative Problem 7-8

An exterior wall of a building consists of 4-in. nominal face brick, ½-in. cement mortar, 8-in. concrete block, 3-oval core (sand and gravel aggregate), gypsum plaster (sand aggregate) ⅝-in. thick on inside surface. The inside air is a 70°F dry bulb and 58°F wet bulb. If the outside air temperature is −20°F,

1. What will be the inside surface temperature of the wall?
2. Will moisture condense on the wall surface?

Solution:

a. Determine resistance of wall components.

	R
1. Outside air film (15 mph) (Table 7-2A, f_o = 6.00)	0.17
2. Face brick (Table 7-1A, item 9b, $R/in.$ = 0.11)	4 × 0.11 = 0.44
3. Cement mortar (½-in.) (Table 7-1A, item 8a), R/in. = 0.20	0.5 × 0.20 = 0.10
4. Concrete block, 8-in., three oval core, sand and gravel aggr. (Table 7-1A, item 9d)	= 1.11
5. Gypsum plaster, ⅝-in., sand aggr. (Table 7-1A, item 10b.7)	= 0.11
6. Inside air film (still air) (Table 7-2A, f_i = 1.46)	= 0.68
	R_t = 2.61

Since the only resistance between the inside air and the interior wall surface is that due to the inside air film, $R_x = 0.68$. By Eq. (7-22),

$$t_x = t_i - \frac{R_x}{R_t}(t_i - t_o)$$

$$= 70 - \frac{0.68}{2.61}[70 - (-20)]$$

$$t_x = 46.6°\text{F} \quad \text{(inside wall surface temperature)}$$

b. Referring to the psychrometric chart, at 70°F DB and 58°F WB, the dew-point temperature (49°F) is above the wall surface temperature (46.6°F), condensation will form on the wall interior surface.

7-10 SERIES HEAT FLOW THROUGH UNEQUAL AREAS

Frequently, unheated and unventilated spaces exist in a structure where the construction surrounding the space may be made up of two or more layers (flat or of small curvature) of unequal area. Heat flows through the layers

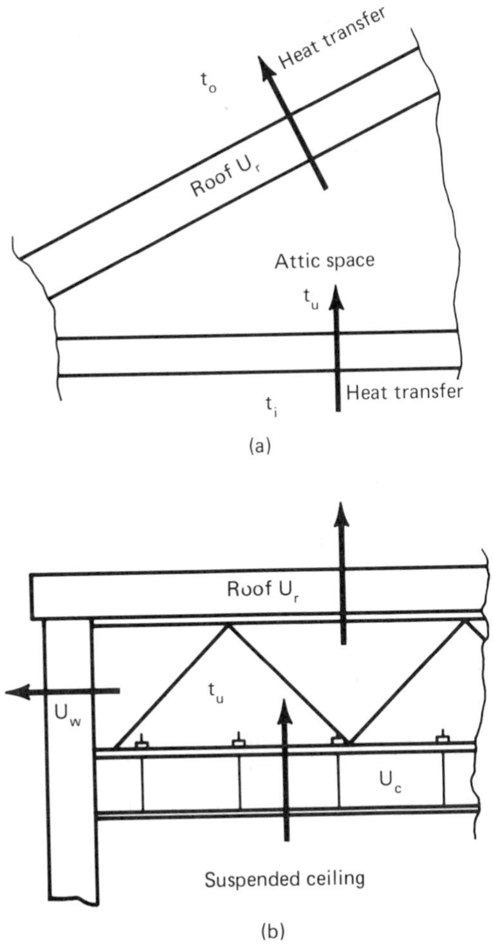

Figure 7–6

Typical unheated spaces in buildings. (a) pitched roof; (b) flat roof

in series. The most common such construction is a ceiling-roof combination where the attic space is unheated and unventilated. Figure 7-6 illustrates an attic space formed by a pitched roof and also a suspended ceiling below a flat roof. Another similar situation exists in the crawl space between the floor and ground when footing walls are used.

The air temperature of these unheated spaces would be some value between the inside and outside air temperature and may be estimated, assuming steady-state heat transfer, by

$$t_u = \frac{t_i\,(A_1U_1 + A_2U_2 + \cdots) + t_o(A_aU_a + A_bU_b + \cdots)}{(A_1U_1 + A_2U_2 + \cdots + (A_aU_a + A_bU_b + \cdots)} \qquad (7\text{-}23)$$

where

t_u = estimated air temperature in unheated space

t_i and t_o = inside and outside air temperature, respectively

Number subscripts refer to heated-to-unheated areas. Letter subscripts refer to unheated-to-outside areas.

Illustrative Problem 7-9

Estimate the temperature in the unheated space shown in Fig. 7-7. The structure is built on a slab. Exterior walls of the unheated space

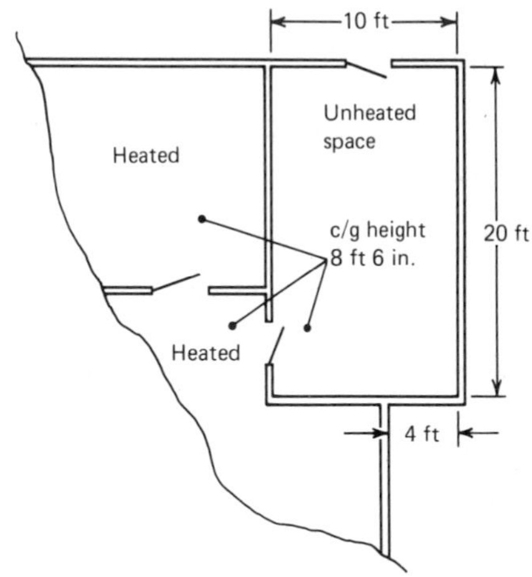

Figure 7–7

Illustrative problem 7-9

have a U factor of 0.25 Btu/hr ft^2 °F, roof-ceiling construction has a U factor of 0.10, interior walls have a U factor of 0.08. The inside heated spaces are maintained at 74°F, and the outside air temperature is −10°F.

Solution:

Assume no heat loss through slab floor because of expected low value of t_u. Neglect doors.

$$U_1 = 0.08 \text{ interior walls}$$

$$U_a = 0.25 \text{ exterior walls}$$

$$U_b = 0.10 \text{ ceiling-roof combination}$$

$$A_1 = (20 + 6)(8.5) = 221 \text{ ft}^2$$

$$A_a = (10 + 20 + 4)(8.5) = 289 \text{ ft}^2$$

$$A_b = (20)(10) = 200 \text{ ft}^2$$

By Eq. (7-23)

$$t_u = \frac{t_i(A_1U_1) + t_o(A_aU_a + A_bU_b)}{A_1U_1 + A_aU_a + A_bU_b}$$

$$= \frac{74(221)(0.08) + (-10)[(289)(0.25) + (200)(0.10)]}{(221)(0.08) + (289)(0.25) + (200)(0.10)}$$

$$t_u = \frac{385.82}{109.93} = 3.5°\text{F} \quad \text{(estimated space temperature)}$$

The original assumption of no heat transfer through the floor slab is justified because of the low space temperature determined.

It is convenient in some situations to determine a combined heat transmission coefficient for such spaces. The combined U factor would be based on the most convenient area from the air inside to the air outside and can be calculated from

$$R_t = \frac{1}{U_1} + \frac{1}{n_2U_2} + \frac{1}{n_3U_3} + \cdots + \frac{1}{n_pU_p} \tag{7-24}$$

The combined coefficient, U_t, is the reciprocal of R_t, or $U_t = 1/R_t$

where

U_t = combined coefficient to be used with A_1

R_t = total resitance of all elements in series

$U_1, U_2, \ldots, U_p$ = coefficients of transmission of $A_1, A_2, \ldots, A_p$, respectively

$n_2, n_3, \ldots, n_p$ = area ratios $A_2/A_1, A_3/A_1, \ldots, A_p/A_1$, respectively

Note the combined coefficient U_t should be multiplied by the area A_1 and the overall temperature difference $t_i - t_o$ to determine the heat loss.

Illustrative Problem 7-10

The top-story ceiling of a building measures 30 by 45 ft. An unventilated attic is formed by a pitched roof which has an area 1.6 times the ceiling area. The ceiling U factor is 0.09 Btu/hr ft^2 °F, the roof U factor is 0.58. The air temperature just below the ceiling is 78°F and the outside air temperature is 5°F. Find (a) the estimated attic temperature, (b) the ceiling-roof overall heat transmission coefficient based on ceiling area, and (c) the heat loss from the building through the ceiling-roof construction in Btu/hr.

Solution:

(a) Determine attic temperature. By Eq. (7-23),

$$t_u = \frac{t_i(A_1U_1) + t_o(A_aU_a)}{A_1U_1 + A_aU_a}$$

Letting

$$A_1 = A_c$$

$$A_a = A_r = 1.6A_c$$

$$U_1 = U_c \quad \text{(ceiling)}$$

$$U_a = U_r \quad \text{(roof)}$$

we have

$$t_u = \frac{t_i(A_c)(U_c) + t_o(1.6A_c)(U_r)}{A_cU_c + (1.6A_c)U_r}$$

A_c cancels and by substitution

$$t_u = \frac{(78)(0.09) + (5)(1.6)(0.58)}{(0.09) + (1.6)(0.58)}$$

$$t_u = 11.45°\text{F} \quad \text{(estimated attic temperature)}$$

(b) By Eq. (7-24)

$$R_t = \frac{1}{U_1} + \frac{1}{n_2U_2}$$

$$\text{Let } U_1 = 0.09$$

$$U_2 = 0.58$$

$$n_2 = 1.6$$

$$R_t = \frac{1}{0.09} + \frac{1}{(1.6)(0.58)}$$

$$R_t = 12.19$$

$U_t = 0.082$ Btu/hr ft^2 °F (combined U factor for ceiling-roof combination based on ceiling area)

(c) Heat loss through ceiling-roof construction based on ceiling area A_c.

$$\dot{q} = U_tA_c(t_i - t_o)$$
$$= 0.082(30 \times 45)(78 - 5)$$

$$\dot{q} = 8081 \text{ Btu/hr}$$

In Section 4 of ASHRAE *Standard 90-75, Energy Conservation in New Building Design,* requirements are stated in terms of U_o, where U_o is the combined thermal transmittance of the respective areas of gross exterior wall, roof-ceiling, and floor assemblies. The U_o equation for a wall could be

$$U_o = \frac{U_{\text{wall}}A_{\text{wall}} + U_{\text{window}}A_{\text{window}} + U_{\text{door}}A_{\text{door}}}{A_o} \tag{7-26}$$

where

U_o = average thermal transmittance of the gross wall area (Btu/hr ft^2 °F)

A_o = gross area of exterior walls (ft^2)

U_{wall} = thermal transmittance of all elements of the opaque wall area (Btu/hr ft^2 °F)

A_{wall} = opaque wall area (ft^2)

U_{window} = thermal transmittance of the window area (Btu/hr ft^2 °F)

A_{window} = window area (including sash) (ft^2)

U_{door} = thermal transmittance of the door area (Btu/hr ft^2 °F)

A_{door} = door area (ft^2)

Where more than one type of wall, window, and/or door is used, the $U \times A$ term for that

Table 7-5*

Overall Coefficients of Heat Transmission (*U* Factors) of Windows, Sliding Patio Doors, and Skylights for Use in Peak Load Determination and Mechanical Equipment Sizing Only and not in any Analysis of Annual Energy Usage, *W/m² °C (Btu/hr ft² °F)*.

Part A. Exterior[a] Vertical Panels

	No Storm Sash				Glass Outdoor Storm Sash 25-mm (1-in.) Air Space[b]			
	No Shade		Indoor Shade		No Shade		Indoor Shade	
	Winter	Summer	Winter	Summer	Winter	Summer	Winter	Summer
Flat Glass[c]								
Single Glass,	6.2(1.10)	5.9(1.04)	4.7(0.83)	4.6(0.81)	2.3(0.50)	2.8(0.50)	2.5(0.44)	2.8(0.49)
Insulating Glass, Double[c]								
5-mm (3/16-in.) air space[f]	3.5(0.62)	3.7(0.65)	3.0(0.52)	3.3(0.58)	2.1(0.37)	2.3(0.40)	1.7(0.29)	2.1(0.37)
6-mm (1/4-in.) air space[f]	3.3(0.59)	3.5(0.61)	2.7(0.48)	3.1(0.55)	2.0(0.35)	2.2(0.39)	1.6(0.28)	2.0(0.36)
13-mm (1/2-in.) air space[g]	2.8(0.49)	3.2(0.56)	2.4(0.42)	3.0(0.52)	1.8(0.32)	2.2(0.39)	1.4(0.25)	2.1(0.30)
13-mm (1/2-in.) air space low emittance coating[h]								
$e = 0.60$	2.4(0.43)	2.9(0.51)	2.2(0.38)	2.7(0.48)	1.7(0.30)	2.0(0.36)	1.4(0.24)	2.0(0.35)
$e = 0.40$	2.2(0.38)	2.6(0.48)	2.0(0.36)	2.5(0.43)	1.5(0.27)	1.9(0.39)	1.3(0.22)	1.8(0.35)
$e = 0.20$	1.8(0.32)	2.2(0.38)	1.7(0.30)	2.1(0.37)	1.4(0.24)	1.7(0.30)	1.1(0.20)	1.6(0.28)
Insulating Glass; Triple								
6-mm (1/4-in.) air space[f]	2.2(0.39)	2.5(0.44)	1.8(0.31)	2.3(0.40)	1.5(0.27)	1.8(0.32)	1.3(0.22)	1.7(0.30)
13-mm (1/2-in. air space[i]	1.8(0.31)	2.2(0.39)	1.5(0.26)	2.0(0.36)	1.3(0.23)	1.8(0.31)	1.1(0.19)	1.7(0.29)

	Glass Indoor Storm Sash 25-mm (1-in.) Air Space[b]				Acrylic Indoor Storm Sash 25-mm (1-in.) Air Space[b]			
	No Shade		Indoor Shade		No Shade		Indoor Shade	
	Winter*	Summer**	Winter*	Summer**	Winter*	Summer**	Winter*	Summer**
Flat Glass[c]								
Single Glass,	2.8(0.50)	2.8(0.50)	2.5(0.44)	2.8(0.49)	2.7(0.48)	2.7(0.48)	2.4(0.42)	2.7(0.47)
Insulating Glass; Double[e]								
5-mm (3/16-in.) air space[f]	2.1(0.37)	2.3(0.40)	1.7(0.29)	2.0(0.36)	2.0(0.35)	2.2(0.39)	1.6(0.28)	2.0(0.35)
6-mm (1/4-in.) air space[f]	2.0(0.35)	2.2(0.39)	1.6(0.28)	2.0(0.36)	1.9(0.34)	2.2(0.38)	1.5(0.27)	1.9(0.34)
13-mm (1/2-in.) air space[g]	1.8(0.31)	2.2(0.38)	1.4(0.25)	2.0(0.35)	1.7(0.30)	2.1(0.37)	1.4(0.24)	1.9(0.33)
13-mm (1/2-iin.) air space Low emittance costing[h]								
$e = 0.60$	1.7(0.29)	2.0(0.36)	1.4(0.24)	1.9(0.33)	1.6(0.28)	2.0(0.35)	1.3(0.23)	1.8(0.32)
$e = 0.40$	1.5(0.27)	1.9(0.33)	1.3(0.22)	1.8(0.31)	1.5(0.26)	1.8(0.32)	1.3(0.22)	1.7(0.30)
$e = 0.20$	1.4(0.25)	1.7(0.29)	1.1(0.20)	1.5(0.26)	1.4(0.24)	1.6(0.28)	1.1(0.20)	1.5(0.27)
Insulating Glass; Triple[e]								
6-mm (1/4-in.) air space[f]	1.5(0.27)	1.8(0.32)	1.3(0.22)	1.7(0.30)	1.5(0.26)	1.8(0.31)	1.3(0.22)	1.7(0.29)
13-mm (1/-in.) air space[i]	1.3(0.23)	1.7(0.30)	1.1(0.19)	1.6(0.28)	1.3(0.22)	1.7(0.29)	1.0(0.18)	1.6(0.28)

Part B. Exterior[a] Horizontal Panels (Skylights)

Description	Winter[i]	Summer[j]
Flat Glass[e]		
Single Glass	7.0(1.23)	4.7(0.83)
Insulating Glass; Double[c]		
5-mm (3/16-in.) air space[d]	4.0(0.70)	3.2(0.57)
6-mm (1/4-in.) air space[d]	3.7(0.65)	3.1(0.54)
13-mm (1/2-in.) air space[e]	3.4(0.59)	2.8(0.49)
13-mm (1/2-in.) air space, low emittance coating		
$e = 0.20$	2.7(0.48)	2.0(0.36)
$e = 0.40$	3.0(0.52)	2.4(0.42)
$e = 0.60$	3.2(0.56)	2.6(0.46)
Plastic Domes[k]		
Single Walled	6.5(1.15)	4.5(0.80)
Double Walled	4.0(0.70)	2.6(0.46)

[a]See Part C for adjustments for various windows and sliding patio doors.
[b]Emissivity of uncoated glass surface = 0.84.
[c]Double and triple refer to number of lights of glass.
[d]3-mm (1/8-in.) glass.
[e]6-mm (1/4-in.) glass.
[f]Coating on either glass surface facing air space; all other glass surfaces uncoated.
[g]Window design: 6-mm (1/4-in.) glass, 3-mm (1/8-in.) glass, 6-mm (1/4 in.) glass.
[h]Refers to windows with negligible opaque areas.
[i]For heat flow up.
[j]For heat flow down.
[k]Based on area of opening, not total surface area.
[m]Values will be less than these when metal sash and frame incorporate thermal breaks. In some thermal break designs *U* values will be equal to or less than those for the glass. Window manufacturers should be consulted for specific data.
*24 km/h (15 mph outdoor air velocity; − 18°C (0 F) outdoor air; 21°C (70 F) inside air temp. natural convection.
*12 km/h (7.5 mph) outdoor air velocity; 32°C (89 F) outdoor air; 21°C (75 F) inside air natural convection; solar radiation 782 W/m^2 (248.3 $Btuh/ft^2$).
The reciprocal of the above *U*-factors is the thermal resistance, *R* for each type of glazing. If tightly drawn drapes (heavy close weave), closed Venetian blinds or closely fitted roller shades are used internally, the additional *R* is approximately 0.05 $m^2 \cdot$ °C/W (0.29 $ft^2 \cdot$ F Btuh). If miniature louvered solar screens are used in close proximity to the outer fenestration surface, the additional *R* is approximately 0.04 $m^2 \cdot$ °C/W (0.24 $ft^2 \cdot$ F/Btuh).

Table 7-5 (Cont.)
Part C. Adjustment Factors for Various Window, Sliding Patio Door, and Skylight Types (Multiply *U*-Factors in Parts A and B by These Factors)

Product Description	Single Glass	Double Insulating Glass	Triple Insulating Glass	Storm Sash Applied Over Single Glass	Storm Sash Applied Over Double or Triple Insulating Glass
All Glass [m]	1.00	1.00	1.00	1.00	1.00
Wood Frame	0.85-0.95	0.90-1.00	0.95-1.00	0.90-1.00	0.95-1.00
Metal Frame	1.10-1.00	1.30-1.20	1.50-1.30	1.40-1.20	1.50-1.30
Thermally-Improved Metal Frame	0.90-1.00	0.95-1.15	1.00-1.25	0.90-1.20	0.95-1.25

[a]See Part *C* for appropriate adjustments for various windows and sliding patio doors. Window manufacturers should be consulted for specific data.
[b]3-mm (1/8-in.) glass or Acrylic as noted, 25 to 100-mm (1 to 4-in.) air space.
[c]Hemispherical emittance of uncoated glass surface = 0.84, coated glass surface as specified.
[d]Coating on second surface, i.e., room side of glass.
[e]Double and triple refer to number of lights of glass.
[f]3-mm (1/8-in.) glass.
[g]6-mm (1/4-in.) glass.
[h]Coating on either glass surface 2 or 3 for winter, and for surface 2 for summer *U*-factors.
[i]Window design 6-mm (1/4-in.) glass, 3-mm (1/8-in.) glass and 6-mm (1/4-in.) glass.
[j]For heat flow up.
[k]For heat flow down.
[l]Based on area of opening, not total surface area.
[m]Refers to windows with negligible opaque areas.
*24 km/h (15 mph) outdoor air velocity; −18°C (0 F) outdoor air; 21°C (70 F) inside air temperature, natural convection.

**12 km/h (7.5 mph) outdoor air velocity; 32°C (89 F) outdoor air; 24°C (75 F) inside air, natural convection; solar radiation 782 W/m^2 (248.3 Btu/h · ft^2).
The reciprocal of the above *U*-factors is the thermal resistance, *R* for each type of glazing. If tightly drawn drapes (heavy close weave), closed Venetian blinds, or closely fitted roller shades are used internally, the additional *R* is approximately 0.05 m^2 · °C/W (0.29 ft^2 · h · F/Btu). If miniature louvered solar screens are used in close proximity to the outer fenestration surface, the additional *R* is approximately 0.04 m^2 · °C/W (0.24 ft^2 · h · F/Btu).
Example: Find the winter *U*-factor for uncoated double insulating glass [13-mm (0.5-in.) air space] when (1) external miniature louvered sun screens are used; and (2) tightly woven drapes are added.
Solution: Winter *R* for 13-mm (0.5-in.) air space double insulating glass = 1/2.8 = 0.36 (2.04); added resistance for the miniature louvered sun screen = 0.04 (0.24); so total *R* = 0.40 (2.28) and *U*-factor = 2.5 W/m^2 · °C (0.44 Btu/h · ft^2 · F). Adding the tightly woven drape *R* = 0.05 (0.29), total R = 0.45 (2.57), so *U* = 2.2 W/m^2 · °C (0.39 Btu/h · ft^2 · F).

**Adapted from ASHRAE* Handbook of Fundamentals, *1981. With permission of the American Society of Heating, Refrigerating and Air-Conditioning Engineers, Atlanta, GA.*

exposure shall be expanded to include its subelements, as

$$U_{wall_1} \times A_{wall_1} + U_{wall_2} \times A_{wall_2} + \ldots$$

7-11 WINDOWS, SKYLIGHTS, AND LIGHT-TRANSMITTING PARTITIONS

Table 7-5 presents the *U* factors for windows, skylights, and light-transmitting partitions. Part C of Table 7-5 presents adjustment factors for various types of windows and sliding glass doors.

7-12 SLAB DOORS

Table 7-6 presents *U* factors for slab doors.

7-13 HEAT TRANSMISSION THROUGH BASEMENT WALLS, FLOORS, AND SLAB FLOORS NEAR GRADE

The allowance made for basement heat loss depends on whether or not the basement is heated. If the basement is heated to a specified temperature, heat loss is calculated in the usual manner, based on proper wall and floor *U* factors and outdoor air and ground temperatures. Heat loss through windows and walls above grade is based on outdoor air temperatures and proper air-to-air *U* factors. In addition, heat loss due to air leakage is calculated for this portion of the wall.

Heat loss through basement walls below grade and basement floors is based on floor and wall *U* factors for the surfaces in contact

Table 7-6A

Overall Coefficients of Heat Transmission (*U* Factors) for Wood Doors,* Btu/hr ft^2 °F

		Winter[b]			Summer[c]
Door Thickness, in.[d]	Description	No Storm Door	Wood Storm Door[e]	Metal Storm Door[f]	No Storm Door
1-3/8	Hollow core flush door	0.47	0.30	0.32	0.45
1-3/8	Solid core flush door	0.39	0.26	0.28	0.38
1-3/8	Panel door with 7/16-in. panels	0.57	0.33	0.37	0.54
1-3/4	Hollow core flush door	0.46	0.29	0.32	0.44
	with single glazing[g]	0.56	0.33	0.36	0.54
1-3/4	Solid core flush door	0.33	0.28	0.25	0.32
	With single glazing[g]	0.46	0.29	0.32	0.44
	With insulating glass[g]	0.37	0.25	0.27	0.36
1-3/4	Panel door with 7/16-in. panels[h]	0.54	0.32	0.36	0.52
	With single glazing[i]	0.67	0.36	0.41	0.63
	With insulating glass[i]	0.50	0.31	0.34	0.48
1-3/4	Panel door with 1-1/8-in. panels[h]	0.39	0.26	0.28	0.38
	With single glazing[i]	0.61	0.34	0.38	0.58
	With insulating glass[i]	0.44	0.28	0.31	0.42
2-1/4	Solid core flush door	0.27	0.20	0.21	0.26
	With single glazing[g]	0.41	0.27	0.29	0.40
	With insulating glass[g]	0.33	0.23	0.25	0.32

[a] Values for doors are based on nominal 3′8″ × 6′8″ door size. Interpolation and moderate extrapolation are permitted for glazing areas and door thicknesses other than those specified.
[b] 15 mph outdoor air velocity; 0 F outdoor air; 70 F inside air temp natural convection.
[c] 7.5 mph outdoor air velocity; 89 F outdoor air; 75 F inside air natural convection.
[d] Nominal thickness.
[e] Values for wood storm door are approximately 50% glass area.
[f] Values for metal storm door are for any percent of glass area.
[g] 17% exposed glass area; insulating glass contains 0.25 inch air space.
[h] 55% panel area.
[i] 33% glass area; 22% panel area; insulating glass contains 0.25 inch air space.

Table 7-6B

Overall Coefficients of Heat Transmission (*U*-Factors) for Steel Doors,* Btu/hr ft^2 °F

Btu/h · ft^2 · F				
Thickness	Steel Door[14]			No Storm Door
1.75 in.				
A[a]	0.59	—	—	0.58
B[b]	0.40	—	—	0.39
C[c]	0.47	—	—	0.46

[a] A = Mineral fiber core (2 lb/ft^3).
[b] B = Solid urethane foam core with thermal break.
[c] C = Solid Polystyrene core with thermal break.

**Reprinted from ASHRAE* Handbook of Fundamentals, *1981. With permission of the American Society of Heating, Refrigerating, and Air-Conditioning Engineers, Atlanta, GA.*

with the soil and on the proper ground temperature. The determination of heat transmission coefficients for below-grade structures is complicated because of the many variables involved. Rules of thumb have been established, by experiment and analysis, over the past years for heat loss through basements of residential buildings. The rate of heat transfer through a basement wall below grade varies with depth below the ground surface.

Walls below grade. The heat loss/ft^2/°F through walls below grade is given in Table 7-7 for uninsulated concrete walls as well as

Table 7-7

Heat Loss Below Grade in Basement Walls[a] [Btu/(hr ft^2 °F)]*

Depth (ft)	Path Length through Soil (ft)	Uninsulated	Heat Loss Insulation 1-in.	2-in.	3-in.
0–1(1st)	0.68	0.410	0.152	0.093	0.067
1–2(2nd)	2.27	0.222	0.116	0.079	0.059
2–3(3rd)	3.88	0.155	0.094	0.068	0.053
3–4(4th)	5.52	0.119	0.079	0.060	0.048
4–5(5th)	7.05	0.096	0.069	0.053	0.044
5–6(6th)	8.65	0.079	0.060	0.048	0.040
6–7(7th)	10.28	0.069	0.054	0.044	0.037

[a] k_{soil} = 9.6(Btu)(in.)/(ft^2)(°F); $k_{insulation}$ = 0.24(Btu)(in.)/(ft^2)(°F).

**Reproduced from* ASHRAE Handbook and Product Directory, 1977 Fundamentals. *With permission of the American Society of Heating, Refrigerating and Air Conditioning Engineers, Atlanta, Ga.*

walls to which 1, 2, or 3-in. of insulation have been added. An average value of 9.6 (Btu)(in.)/(hr)(ft^2)(°F) has been assumed for conductivity of the soil, and the insulation is assumed to have a thermal conductivity of 0.24 (Btu)(in.)/(hr)(ft^2)(°F).

Basement floors. Table 7-7 indicates that the heat loss from the seventh foot of the uninsulated basement wall is only a small fraction of the total heat loss through the wall; thus, it can be readily appreciated that (with the much longer heat flow path) the loss through each square foot of basement floor rapidly becomes a negligible part of the total basement heat loss. It is reasonable to take an average value for the heat loss through the basement floor. This value can be multiplied by the floor area to give the total floor heat loss.

The average rate of heat loss through the floor may be taken as equal to that from a point located one-quarter of the basement width from the side wall. Shallow, narrow basements will have higher heat loss per square foot than deep, wide basements. Typical values are given in Table 7-8.

Heat loss from the below-grade portion of the basement per °F temperature difference can be estimated using figures in Tables 7-7 and 7-8. For the wall the values of heat loss through each square foot are selected from Table 7-7, added together, and the total multiplied by the perimeter of the house. For the floor the average heat loss per square foot is estimated from Table 7-8 and multiplied by the floor area. The resulting two values may then be added together and multiplied by the appropriate design temperature difference to give the maximum rate of heat loss from that portion of the basement below grade.

Design temperature difference. Selecting the appropriate design temperature difference can be a problem. Although internal design temperature is given by basement air temperature, none of the usual external design air temperatures are applicable because of the heat capacity of the soil. However, ground surface temperature is known to fluctuate about a mean value by an amplitude A, which will vary with geographic location and surface cover. Thus, suitable external design temperatures can be

Table 7-8

Heat Loss Through Basement Floors*
[Btu/(hr ft² °F)]

Depth of Foundation Wall below Grade (ft)	Width of House 20 (ft)	24 (ft)	28 (ft)	32 (ft)
5	0.032	0.029	0.026	0.023
6	0.030	0.027	0.025	0.022
7	0.029	0.026	0.023	0.021

**Reproduced from* ASHRAE Handbook and Product Directory, 1977 Fundamentals. *With permission of the American Society of Heating, Refrigerating and Air-Conditioning Engineers, Atlanta, Ga.*

obtained by subtracting A for the location from the mean annual air temperature, $\bar{t}_a$. Values of $\bar{t}_a$ can be obtained from meteorological records; and A varies from 27°F through central Canada to 18°F through southern United States.

If a basement is completely below grade and unheated, its temperature normally will range between that in the rooms above and that of the ground. Of course, basement windows will lower basement temperature when it is cold outdoors, and heat given off by the heating plant will increase the basement temperature. The exact basement temperature is indeterminate if the basement is not heated. In general, heat from the heating plant sufficiently warms the air near the basement ceiling to make an allowance unnecessary for floor heat loss from rooms located over the basement.

Illustrative Problem 7-11

A full basement of a house measures 28 ft wide by 40 ft long and is 7 ft 6 in. deep from basement ceiling to floor. The top 1½ ft of the foundation wall is above grade, and the wall is 8-in. concrete with sand and gravel aggregate. The top 4½ ft of the wall has 1-in. insulation applied to its interior surface. Estimate the total heat loss from the basement if the inside air temperature is 70°F, the external design temperature is assumed to be −5°F, and the external ground temperature is assumed to be 15°F.

<u>Solution:</u>

Step 1: Determine U factor for above-grade basement wall and heat loss above grade.

	R
Inside air film	0.68
Insulation (1-in.) (Footnote Table 7-7)	1/0.24 = 4.17
Concrete (8-in.) (Table 7-1A, item 8e, R/in. = 0.08)	0.64
Outside air film	0.17
	R_t = 5.66

$$U = 1/5.66 = 0.177 \text{ (say 0.18 Btu/hr ft}^2\text{ °F)}$$

$$\text{Above grade } \dot{q} = UA(t_i - t_o)$$
$$= (0.18)(2 \times 28 + 2 \times 40)(1.5)[70 - (-5)]$$
$$\dot{q} = 2754 \text{ Btu/hr} \quad \text{(above grade)}$$

Step 2: Determine below-grade loss (wall and floor).

Applying 4½ ft of insulation down from the ceiling gives 3 ft of wall below grade insulated and 3 ft of wall uninsulated.

Wall (Using Table 7-7)	Btu/(hr ft° F)
1st ft below grade	0.152
2nd ft below grade	0.116
3rd ft below grade	0.094
4th ft below grade	0.119
5th ft below grade	0.096
6th ft below grade	0.079
Total per ft of wall	0.656 Btu/(hr ft° F)

Basement perimeter = 2(28 + 40) = 136 ft

Total wall loss = (0.656)(136)
= 89.2 Btu/hr °F

Floor (Using Table 7-8):

Average heat loss per ft^2 = 0.025 Btu/hr °F

Floor area (28 × 40) = 1120 ft^2

Total floor loss = (0.025)(1120)
= 28 Btu/hr °F

Total

Total basement heat loss below grade
= 89.2 + 28 = 117.2 Btu/hr °F

Design temperature difference below grade
= 70 − 15 = 55°F

Total heat loss below grade = (117.2)(55)
= 6446 Btu/hr

Step 3: Total loss from basement = 2754 + 6446 = 9200 Btu/hr

Floor slab at or near grade. Two types of concrete floors in basementless houses are (1) unheated, relying for warmth on heat delivered above floor level by the heating system; and (2) heated, containing heating ducts or pipes that constitute a radiant slab or portion thereof for complete or partial heating of the building.

For type 1 floors floor heat loss generally comprises only about 10 percent of the total heat loss from the building. For comfort, however, it may be the most important since houses with cold floors are difficult to heat. Not that a well-insulated floor does not in itself assure comfort if downdrafts from windows or exposed walls create pools of chilly air over considerable areas of the floor. Also, the cold perimeter of the floor may be below the inside air dew-point temperature and condensation may result. Therefore, a type 1 floor should not be used in a severe climate, except with a heating system that delivers enough heat near the floor to counteract the downdrafts of exterior walls and windows and heat transmission through the floor.

Heat loss from type 1 unheated floor slabs is more nearly proportional to the exposed perimeter than to the floor area. Table 7-9 presents data for calculating heat loss from type 1 floor slabs.

The insulation should extend under the floor horizontally 2 ft and can also be located along the foundation vertical wall with equal effectiveness if it extends 2 ft below floor level.

Floors of type 2, containing heating ducts or pipes, are now in common use. Heat loss downward into the ground and outward through the edges of the floor is called *reverse loss*. Table 7-10 values of perimeter heat loss are listed for heated slabs.

Desirability of edge insulation is apparent. The minimum thickness that should be used is 1 in. of water-resistant material, but 2 in. is recommended. The values of edge loss in Table 7-10 indicate that reverse heat loss of heated slabs is likely to be 20 percent of the total heat loss of many types of present-day houses, and may exceed 20 percent if only 1 in. of insulation is used at the floor's edge.

Table 7-9

Heat Loss of Concrete Floors at or Near-Grade level per Foot of Exposed Edge*

Outdoor Design Temperature, F	Heat Loss Per Foot of Exposed Edge, Btuh	
	Recommended 2-in. Edge Insulation	1-in. Edge Insulation
−20 to −30	50 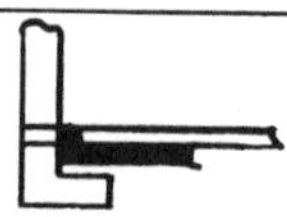	55 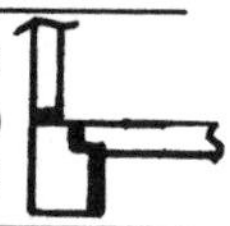
−10 to −20	45	50
0 to −10	40	45
Outdoor Design Temperature, F	**1-in. Edge Insulation**	**No Edge Insulation [a]**
−20 to −30	60 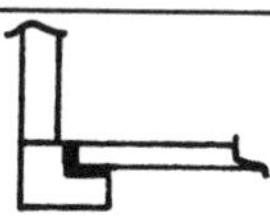	75
−10 to −20	55	65
0 to −10	50	60

[a]This construction not recommended; shown for comparison only.

**Reproduced from* ASHRAE Handbook and Product Directory, 1977 Fundamentals. *With permission of the American Society of Heating, Refrigerating and Air-Conditioning Engineers, Atlanta, GA.*

Table 7-10

Floor Heat Loss to Be Used When Warm Air Perimeter Heating Ducts Are Embedded in Slabs Btuh per (linear foot of heated edge)[a]*

Outdoor Design Temperature (°F)	Edge Insulation: 1-in. Vertical Extending Down 18 in. Below Floor Surface	1-in. L-Type Extending at Least 12 in. Deep and 12 in. Under	2-in. L-Type Extending at Least 12 in. Down and 12 in. Under
−20	105	100	85
−10	95	90	75
0	85	80	65
10	75	70	55
20	62	57	45

[a]Factors include loss downward through inner area of slab.

**Reproduced from* ASHRAE Handbook and Product Directory, 1977 Fundamentals. *With permission of the American Society of Heating, Refrigerating and Air-Conditioning Engineers, Atlanta, Ga.*

The concrete floor slab is usually placed on a gravel fill 4 in. thick or more to insulate the floor from the earth and to retard the rise of groundwater by capillary action. A waterproof membrane should be installed over the gravel fill. Obviously, such floors must be laid several inches above grade, and effective subsoil drainage provided to avoid slabs soaked by rain or melting snow, and consequent excessive heat loss.

CHAPTER 7

Review

7-1. An insulating material has a thermal conductivity of $K = 0.27$ Btu in./hr ft^2 °F. How many inches of the material should a contractor install if specifications call for a thermal resistance R-14?

7-2. The section of a masonry wall is shown in Fig. 7-8. Calculate the overall resistance and U factor for winter conditions. The wall consists of 4 in. of face brick, ½ in. of cement mortar, an 8-in. sand-and-gravel concrete block (three-hole oval-cored), 1½- by 2-in. furring strips on block surface, and metal lath and plaster (sand aggregate) attached to the furring strips. (Neglect resistance of furring strips.)

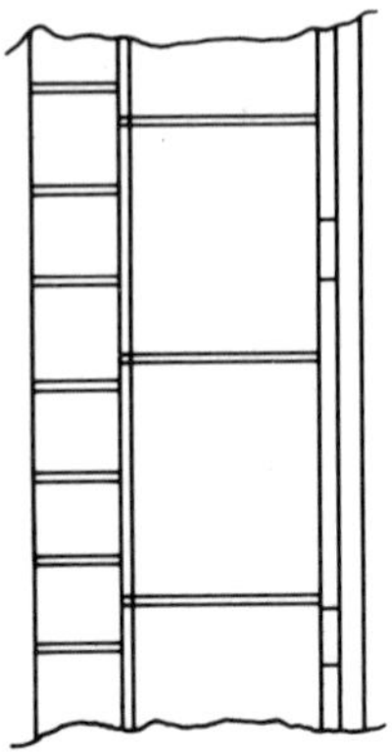

Figure 7–8

Sketch for Problem 2

7-3. A section of a wood frame wall is shown in Fig. 7-9. Calculate the overall resistance and U factor for winter conditions. The wall consists of lapped wood siding (½ by 8 in.), ½-in. fir plywood, 3½-in. stud space with 2-in. fiber glass batt insulation between studs, ½-in. thick gypsum wall board on inside surface. (Neglect resistance of wood studs.)

7-4. For the stud wall of Problem 3 correct the U factor for the presence of studs if studs are placed 16 in. on center.

7-5. The section of a ceiling-floor combination is shown in Fig. 7-10. The floor is asphalt tile on 1-in. plywood nailed to 2 by 6-in. joists. The ceiling is ¾-in. metal lath with lightweight aggregate plaster attached. Calculate the thermal resistance and U factor on the basis of heat flow up to unoccupied space above.

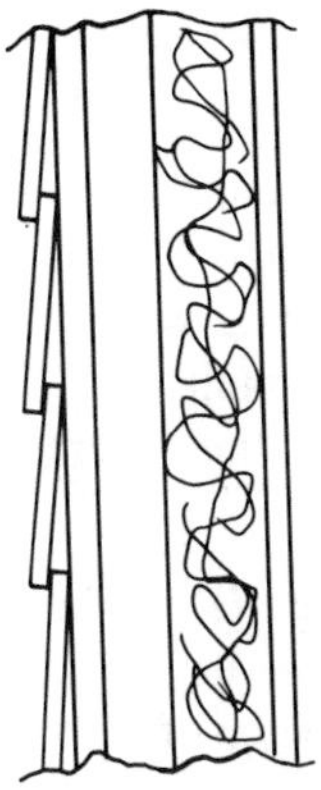

Figure 7–9

Sketch for Problems 3 and 4

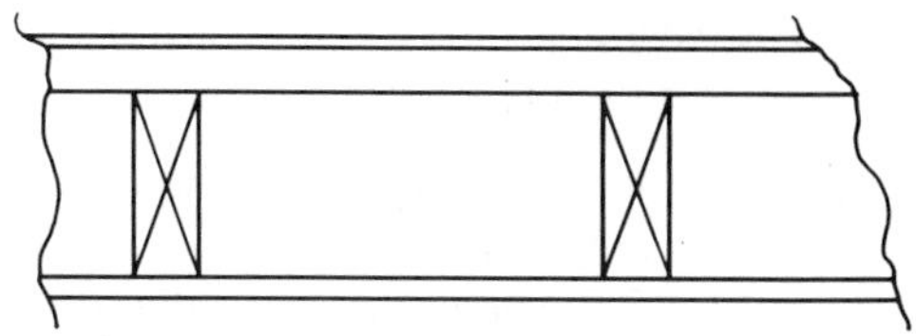

Figure 7–10

Sketch for Problems 5 and 6

7-6. Consider the ceiling-floor combination of Problem 5, except the ceiling is above an unheated basement and is ½-in. plywood nailed to the joists instead of the metal lath and plaster. Heat flow is downward. Calculate the thermal resistance and U factor.

7-7. A sloping (pitched) roof has asphalt shingles on felt laid on ⅝-in. plywood nailed to 2 by 6 nominal wood rafters placed 20-in. on center. Gypsum wallboard (½ in.) is fastened directly to the underside of the rafters. Both sides of the air space have reflective surfaces. Neglecting the effect of the 2 by 6 rafters, calculate the thermal resistance and U factor for the roof.

7-8. If the roof section of Problem 7 has ordinary building material surfaces in the air space and the affect of the rafters is included, calculate the thermal resistance and U factor for the roof.

7-9. A masonry wall consists of 4-in. face brick, ½-in. cement mortar, 8-in. sand-and-gravel concrete block (three-hole oval-core) finished on the inside surface with gypsum plaster, lightweight aggregate on metal lath on 2 by 2-in. furring strips (2-in. air space). If the interior of the room is maintained at 70°F and the outside air is 5 °F, calculate the estimated interior wall surface temperature. What relative humidity could be maintained in the interior space without condensation on the walls?

7-10. If the air space in the wall section of Problem 9 is filled with corkboard insulation (density 8 lb/ft^3) calculate the estimated interior wall surface temperature.

7-11. A pitched roof of a building has a calculated heat transmission coefficient $U_r = 0.43$ and the horizontal ceiling transmission coefficient $U_c = 0.15$.

The ratio of roof area to ceiling area is 1.4. If the attic space is unvented, calculate the heat transmission coefficient for the ceiling-roof combination based on the ceiling area.

7-12. A single-story home is built on a concrete floor slab laid at ground level. Waterproof 1-in.-thick insulation is placed at the edge of the slab where it joins the footing walls. The insulation extends downward along the footing walls a distance of 18 in. The house is rectangular in shape 50 by 30 ft. Compute the heat loss (edge loss) for the floor slab if the house is built in a location where the outside design temperature is −10°F.

7-13. Estimate the temperature in the unheated space shown in Fig. 7-11. The structure is built on a slab. The exterior walls of the unheated space have a U factor of 0.25 Btu/hr ft^2 °F, the ceiling-attic structure has a U factor of 0.09 Btu/hr ft^2 °F, and the interior walls have a U factor of 0.06 Btu/hr ft^2 °F. Design temperatures are 72°F inside and −10°F outside. Ceiling height is 9 ft.

7-14. The exterior wall of a room in a building has a U factor of 0.24. The wall measures 20 ft. long by 8 ft. high. It contains a solid wood door, 3 ft by 7 ft by 1½-in., and two 48 by 30 in. metal sash (80 percent glass) single-glass windows. Assuming parallel heat flow paths for the wall, door, and windows, determine the overall thermal resistance and overall U factor for the combination. Assume winter conditions.

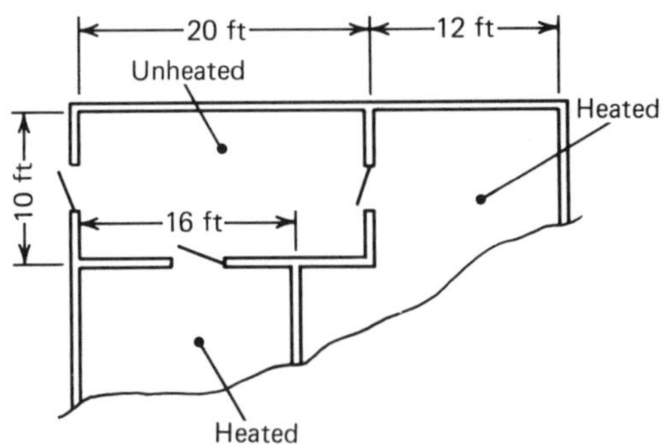

Figure 7–11

Sketch for Problem 13

BIBLIOGRAPHY

1. ASHRAE *Handbook and Product Directory, 1977 Fundamentals,* American Society of Heating, Refrigerating and Air-Conditioning Engineers, Inc., Atlanta, GA.

2. *Load Calculation Digest, Commercial/Industrial Air Conditioning,* General Electric Company, 1971.

3. *Trane Air Conditioning Manual,* The Trane Company, LaCrosse, Wisconsin, 1965.

Chapter 8

Heating and Ventilation Loads

8-1 INTRODUCTION

Regardless of the type of heating system to be used for a particular building, the first basic requirement is the determination of the heating and ventilating loads for the building. This calculation should be done with as much accuracy as possible. Rule of thumb methods should not be used, and the necessity for an accurate heat load calculation cannot be overemphasized. It is not just a question of the heating system being large enough but of selecting the size required to secure optimum performance. This will require an accurate determination of the building heat loads using accepted methods and good judgment on the part of the design engineer.

Most heating systems operate intermittently at a fixed rate, with short infrequent cycles in mild weather and nearly continuously during the most severe weather. An ideal arrangement is to have the heating system capable of operating continuously with varying output which exactly matches the building heat load with climate changes. This ideal can be approached in careful design if an accurate calculation is made for the building heat load.

8-2 ELEMENTS OF THE HEATING AND VENTILATION LOADS

The elements which enter into the calculation of the heating and ventilating loads for a building are as follows:

a. heat loss through all exposed areas to the outside, and heat loss through exposed areas to unheated spaces
b. heat required to warm infiltrated air which enters the building through cracks around outside doors and windows, and air entering through doorways when people enter or leave
c. heat load caused by controlled ventilation air mechanically brought into building. (Infiltration air should be distinguished from ventilation air)
d. miscellaneous loads such as humidification of outside air from infiltration (and ventilation), auxiliary heat sources, and intermittently heated buildings

Calculation of the heating and ventilating loads for a building must be done on a room-by-room basis, considering all spaces which are to be heated as a separate calculation. After all heat loads for the individual spaces have been calculated, the total system load is obtained by simple addition of the individual space loads at design conditions.

Calculation of heating loads for industrial and commercial buildings requires the additional consideration of power ventilation and exhaust equipment which may have an appreciable effect on the building load. When large quantities of air are exhausted from buildings, provision must be made for the introduction of heated makeup air to the building.

Some buildings, such as churches, public schools, and so forth, may be occupied only a portion of the time during the heating season. In these cases the building may be heated to comfort conditions only during occupied periods and controlled at a lower temperature at other times. This may place an additional load on the system, requiring consideration of reserve capacity to raise the building temperature rapidly to occupancy levels when required.

8-3 GENERAL PROCEDURE FOR CALCULATING HEAT LOSSES

The general procedure for calculating heat losses from a structure is

1. Select the design outdoor weather conditions: temperature, humidity, wind direction, and speed.
2. Select the design indoor air temperature to be maintained in each heated space when outside design conditions exist.
3. Estimate temperatures in adjacent unheated spaces.
4. Select or compute the heat transmission coefficient (*U* factors) for all building components through which heat losses are to be calculated.
5. Determine all surface areas through which the heat is lost.
6. Compute heat transmission losses for each kind of wall, glass, floor, ceiling, and roof in the building by multiplying the heat transmission coefficient in each case by the area of the surface and the temperature difference between indoor and outdoor air, or adjacent unheated space.
7. Compute heat losses from basement or grade-level slab floors by methods presented in Chapter 7.
8. Select infiltration air quantities and compute the heat load caused by infiltration around doors and windows.
9. When positive ventilation using outdoor air is provided by an air-heating or air-conditioning unit, the energy required to warm the outdoor air to room temperature must be provided by the unit. The principle for calculation of this load component is identical to that for infiltration. If mechanical exhaust from the room is provided in an amount equal to the outdoor air drawn in by the unit, the unit must also provide for the natural infiltration losses. If no mechanical exhaust is used, and the outdoor ventilation air supplied equals or exceeds the natural infiltration that would occur without ventilation, natural infiltration may be neglected.
10. The sum of the transmission losses or heat transmitted through the confining walls, floors, ceiling, glass, and other surfaces, plus the energy associated with the cold entering air by infiltration or required to replace mechanical exhaust, represents the total heating load.
11. In buildings that have a reasonably steady internal heat release of appreciable magnitude from sources other than the heating system, a computation of this heat release under design conditions should be made and deducted from the total of the heat losses computed above.
12. Additional heating capacity may be required for intermittently heated buildings to bring the temperature of the air,

confining surfaces, and building contents to the design indoor temperature within a specified time.

8-4 DESIGN OUTDOOR WEATHER CONDITIONS

Prior to the calculation of the design heat loss from a structure, it is necessary to establish the design outside air temperature and weather condition for the area in which the structure will be located. The ideal heating system would provide enough heat to match the heat loss from the structure. Studies of weather records show that severe weather conditions do not repeat themselves each year, and weather conditions vary over the heating season. If the heating system were designed for the most severe conditions, the system would have a great excess of capacity during most of the time.

In many cases occasional failure of a heating system to maintain a preselected indoor design temperature during brief periods of severe weather is not critical. However, the successful completion of some industrial or commercial processes may depend upon close regulation of indoor temperatures.

Table C-1 (Appendix C) contains weather data for selected locations in the United States. The ASHRAE *Fundamentals Handbook 1981* has a more complete tabulation of weather data.

Before selecting an outdoor design temperature from Table C-1, a heating engineer should consider each of the following factors:

1. Is the heat capacity of the structure high or low?
2. Is the structure insulated?
3. Is the structure exposed to high wind?
4. Is the ventilation or infiltration load high?
5. Is there more glass area than normal in the structure?
6. During what part of the day will the structure be used?
7. What is the nature of occupancy?
8. Will there be long periods of operation at reduced indoor temperature?
9. What is the amplitude between maximum and minimum daily temperature in the locality?
10. Are there local conditions which cause significant variation from temperatures reported by weather bureau?
11. What auxiliary heating devices will be in the building?
12. What is the expected cost of fuel or energy?

The designer must keep in mind, before reaching a final decision on outdoor design temperature, that if the indoor-outdoor design temperature difference is exceeded, the indoor temperature will probably fall. The question is to determine how large a drop in indoor temperature can be tolerated.

Finally, there is a factor, perhaps intangible, that should not be ignored. It is the performance expected by the owner of the system. The designer of the system should make clear to the owner the various factors he has had to consider in the design. In order to judge whether expected performance can be assured, the designer needs a full understanding of the basis on which the capacities of all the system components are derived or determined, the limits of accuracy of published performance data, and the accelerating capacity of certain types of equipment. There is no substitute for experienced engineering judgment in this type of problem.

Winter outside design dry bulb temperatures are presented in Table C-1 for selected cities. The 97.5 percent values shown represent the frequency level which would be equalled or exceeded by 97.5 percent of the total hours in the months of December, January and February (a total of 2160 hours) in the northern hemisphere. In a normal winter there would be approximately 54 hours at or below the 97.5 percent value. Reference 2 also lists a 99 percent value, and design temperatures for many more locations. In general, the 97.5 percent temperature gives satisfactory design heat loss calculations.

8-5 DESIGN INDOOR AIR TEMPERATURE

The temperature to be maintained inside a structure is understood to be the dry-bulb temperature at the breathing line, which is 3 to 5 ft above the floor and at a location where the temperature-sensing device (thermostat) is not exposed to a condition of abnormal heat gain or heat loss. Indoor design air temperatures usually specified vary in accordance with the intended use of the building.

The main purpose of Chapter 6 was to define indoor design conditions that would make most of the occupants comfortable. The proper dry-bulb temperature to be maintained for comfort depends upon the relative humidity and air motion.

Table 8-1 gives representative inside air design temperatures for particular types of spaces. As may be seen, a range of temperatures are given in some cases.

In making the actual heat loss calculations for a structure, it is sometimes necessary to modify the inside design temperature so that the air temperature at the proper level within a space will be used. The air temperature within a given space will usually vary from floor to ceiling. The allowance for air temperature variation from floor to ceiling may be somewhat difficult to predict as it depends on many

Table 8-1

Type of Building	(°F)	(°C)	Type of Building	(°F)	(°C)
Schools			*Theaters*		
Classrooms	70–72	21–23	Seating space	68–72	20–23
Assembly rooms	68–72	20–23	Lounge rooms	68–72	20–23
Gymnasiums	55–65	13–18	Toilets	68	20
Toilets and baths	70	21			
Locker rooms	65–68	18–20	*Hotels*		
Kitchens	66	19	Bedrooms and baths	70	21
Lunch rooms	65–70	18–21	Dining rooms	70	21
Playrooms	60–65	16–18	Kitchens	68	20
			Ballrooms	65–68	18–20
Hospitals			Public toilets	68	20
Private rooms	70–80	21–27			
Operating rooms	70–95	21–35			
Wards	68	20			
Kitchens	66	19			
Bathrooms	70–80	21–27			

factors, primarily the type of heating system; that is, whether the air moves through the space by natural convection or by forced circulation. Circulation by natural convection has a tendency to cause a greater temperature differential floor to ceiling than does force circulation.

It is impractical to set up rigid rules for determining temperature variations. However, normally if the floor-to-ceiling height is 12 ft or less, no variation is considered. For higher ceilings an allowance of approximately 1.0 percent per foot of height above the 5-ft level may be made for ceilings up to 15 ft and approximately one-tenth of 1.0 degree per foot above this level. The temperature at the floor level may be 3 to 6 degrees lower in most cases when calculating heat loss through floors in high-ceiling spaces.

The inside design air temperature may be different for various spaces within a given structure because of different requirements for these spaces. Normally, however, no distinction of 2 or 3 degrees should be made. Frequently a single inside design temperature, say 70°F, is used throughout the occupied areas of the structure (say a home). Most heating systems can be readily adjusted and controlled to deliver heat to various spaces to maintain them at the individual temperature desired, even if a single inside design air temperature has been used to calculate the heat losses.

Temperatures in unheated spaces may be estimated by use of Eq. (7-23). The heat loss from heated rooms to unheated rooms or spaces must be based on this estimated or assumed temperature.

8-6 MEASUREMENTS FOR HEAT TRANSMISSION

The basic equation for the calculation of heat transmission by conduction and convection through composite sections of a building structure was presented in Chapter 7 as

$$\dot{q} = UA\Delta t = UA(t_i - t_o) \qquad \textit{(7-15)}$$

Repeated

and we have discussed the methods used to determine the U factor, inside design air temperature t_i, and the outside design air temperature t_o. The remaining quantity in Eq. (7-15) is the area A, through which the heat is transferred. It is important to determine these areas with as much accuracy as possible.

The owner of the structure may indicate his intention of completing an unfinished space at some future time. In this case it is necessary to obtain specific information as to how this space will be constructed and to estimate the heat losses from such a space. Rooms below a future attic room should be considered as having a cold ceiling because the attic space is unheated, and it may be some time before it is finished and heated. The heat loss of a crawl space or basement should be calculated and included in the total heat loss of the building. Garages are seldom heated with the central heating system which heats the living areas of a house, unless certain precautions are taken in the system design. If the unheated garage is attached to the house, there exists a cold wall between the two building components.

When the heat loss of a new structure is to be calculated, the dimensions recorded on architectural plans of the building should be used to determine areas. If it is necessary to scale the dimensions, it is advisable to check at least one dimension for each direction on the sheet of the plans to be certain the scale is reasonably accurate. If errors are found, ask the architect whether the plans or the dimensions are correct.

It may be necessary to accurately measure spaces within an existing structure. Plans for existing structures should be used *only* after comparing them to the structure in event that alterations or remodeling has occurred since the plans were originally drawn.

Inside dimensions of spaces are usually

used in calculating areas. If measurements of an existing structure are being made, measure the distance from the surface of one wall to the surface of the opposite wall. If building plans are being used, it is sufficiently accurate to use the dimensions recorded on the plan, even though they show the distance from the center line of interior walls and to the outside of sheathing on exterior walls. Record these dimensions to the *nearest foot*. For example, if the distance is 14 ft 6 in., record 15 ft as the distance. Ceiling heights should be measured and recorded to the nearest *one-half foot*. For example, a ceiling height of 7 ft 9 in. should be recorded as 8 ft.

8-7 TYPES OF HEAT-TRANSMITTING AREAS

Exposed walls: Heat loss calculations should begin with the exposed walls. An exposed wall is considered to be one which faces the outside weather, or one which faces an unheated space. These areas should be determined as separate quantities. If a space has only one exposed wall, the length of the exposure is the linear feet of the exposed wall. If a space has two or more exposed walls, the total linear feet of exposed wall is the sum of the individual wall lengths.

The *gross exposed wall area* of a space is the length of exposed wall times the ceiling height. In many cases two or more different wall constructions are used for a particular space. The gross exposed wall area is calculated separately for each construction. The actual height of the particular construction would be used instead of the ceiling height if the division between construction is horizontal. Record wall areas to the nearest square foot.

A closet is considered a part of the room in which it opens. The length of exposed wall, if any, is included in the linear feet of exposed wall of the room. If the closet has a cold ceiling or floor, it is also included in the heat loss calculations. Large closets or dressing rooms should be considered as separate rooms and should have separate heat loss calculations.

The presence of cabinets, shelves, and bookcases on exposed walls does not affect the heat loss calculation. Calculate the exposed wall area as if no shelves, bookcases, or cabinets existed.

When a stairway is located adjacent to a cold wall, the exposed area up to the ceiling of the second floor is included with that of the first floor hall or room. Unless a heating unit is to be installed in the second floor hall, the heat loss through the entire ceiling of the stairwell and the second floor hall is included with that of the first floor hall or room opening to the stairway.

Basement wall area above grade is calculated separately from basement wall area below grade. The heat loss per square foot of basement wall area above grade is considerably more than that below grade (see Illustrative Problem 7-11).

Windows and outside doors and other glass surfaces such as skylights should be measured inside their casings. These measurements should be recorded to the nearest 0.1 ft and areas should be recorded to the nearest 0.1 ft^2. The *total* window, door, or skylight area should be recorded to the nearest square foot for heat loss calculations.

Most doors and windows come in stock sizes. Some architects specify the opening size for windows rather than glass size, but in most drawings the full opening dimensions are not specified. Instead, window sizes are specified as follows:

a. double-hung (DH), or casement, and so forth

b. number of lights or panes

c. dimensions in inches of each light or pane

For example, a window specification that reads 9 × 12-12LT-DH refers to a double-hung window having 12 lights (panes), each 9 in. wide and 12 in. high. A specification that reads, 26/24-2LT-DH indicates that there are two sashes in the double-hung window and the glass in each sash is 26 in. wide and 24 in. high. If there is no reference to the type of window, check with the architect or builder.

The opening size of a window is always greater than the glass size. The opening width is approximately 4 in. greater than the total glass width. The opening height is approximately 6 in. greater than the total glass height. For the 9 × 12-12LT-DH window noted above, the total glass width would be 32 in. The glass height would be 25 in. per sash or 50 in. total. The opening height would be 56 in. A 1-in. allowance is normally made for the mullions between the panes. The recorded dimensions of the above window should be: 2.7 ft × 4.7 ft (32 in. = 2 ft 8 in or 2.7 ft; 56 in. = 4 ft 8 in. or 4.7 ft). Similarly, the opening size of the 26/24-2-LT-DH window would be 30 in. (26 + 4) wide and 54 in. (24 + 24 + 6) high. The dimensions would be recorded as 2.5 ft × 4.5 ft.

When *storm sash* is used, record the dimensions of the inside windows, but indicate the presence of the storm sash. When other forms of double glazing are used, it should also be indicated.

The fit of a window does not affect the conducted heat loss, but the fit must be noted because it will greatly affect the calculation of infiltration, which we will discuss later in this chapter.

Net exposed wall area is the gross exposed wall area minus the exposed window and door areas. If two or more wall constructions are used in a space, care must be taken to subtract only areas of windows and doors which are actually located in each construction. Most designers distinguish between the heat loss through windows from that lost through doors. When this is done, obviously the door and window areas must be tabulated separately.

Cold partitions, floors, and ceilings separate heated spaces and unheated spaces. A room has a cold ceiling when it is located directly beneath an attic, roof, or any unheated space. Thus, all spaces in a one-story building have cold ceilings. In a two-story building all second-story spaces have cold ceilings. Always check elevation plans of a building to determine if all of the ceiling should be considered cold, or note which portion of the ceiling area should be considered.

When a basement or a crawl space is heated, the heat loss from these spaces should be calculated, and the floor above it is *not* considered to be a cold floor. It is recommended that these spaces be heated, particularly in colder climates. In buildings with unheated basements, crawl spaces, or slab floors, the first-story floor is considered to be a cold floor. If a room or bay extends over a foundation wall, porch, breezeway, or other unheated space, that part of the floor which is exposed is considered as a cold floor exposed to outside temperatures. Floors of spaces located over an unheated garage should be considered as exposed to outside temperature. Cold-storage areas in a basement located under a first-story space also create a cold floor for the space above.

In a heated basement the concrete floor is a cold floor. Although basement floors have a low heat loss per square foot, it is necessary to calculate this heat loss since it may be appreciable for a large area.

In areas which have a mild winter climate, heat is seldom provided in or underneath floors located over crawl spaces, and adequate crawl-space ventilation must be provided. Heat loss through these floors can be reduced by insulating them, resulting in a warmer floor.

8-8 HEAT LOSS BY INFILTRATION

Outside air will leak into a building no matter how well constructed it may be, and an equal amount of conditioned air will leak out. This leakage of outside air into a building is called *infiltration*. A certain amount of fresh outdoor air is desirable and necessary, of course, but when possible it should be placed under positive control. Fresh outside air should be introduced to a building through a duct or ducts, provided specifically for that purpose, and should be conditioned before it is distributed to the interior spaces (mechanical ventilation).

Infiltration heat loss differs from other forms of heat loss. It is not conducted through the structural components of the building. The amount of heat necessary to condition the incoming infiltrated air to inside conditions is made up of the sensible heat to raise the air temperature and the latent heat to evaporate the required moisture to humidify the air. The heat required to raise the temperature of the infiltrated air is given by

$$\dot{q}_s = \dot{m}_{oa} C_p (t_i - t_o) \qquad (8\text{-}1)$$

where

$\dot{m}_{oa}$ = mass of infiltrated air (lbm/hr or kg/s)

C_p = specific heat of moist air [Btu/(lbm °F) or J/(kg °C)]

Infiltration is normally estimated on the basis of volume flow of air at outside conditions. Equation (8-1) then becomes

$$\dot{q}_s = \frac{\dot{Q}_{oa} C_p (t_i - t_o)}{v_{oa}} \qquad (8\text{-}1a)$$

where

Q_{oa} = volume flow rate (ft^3/hr or m^3/s)

v_{oa} = specific volume (ft^3/lbm or m^3/kg)

The latent heat required to humidify the infiltrated air is given by

$$\dot{q}_L = \dot{m}_{oa} (W_i - W_o) h_{fg} \qquad (8\text{-}2)$$

where

$(W_i - W_o)$ = difference in design humidity ratio (lbv/lbda or kgv/kgda)

h_{fg} = latent heat of vaporization at indoor conditions (Btu/lbv or J/kgv)

In terms of volume flow of air, Eq. (8-2) becomes

$$\dot{q}_L = \frac{\dot{Q}_{oa}}{v_{oa}} (W_i - W_o) h_{fg} \qquad (8\text{-}2a)$$

We will see that the heat loss calculated by Eqs. (8-1a) and (8-2a) may account for a large portion of the heating load.

8-9 ESTIMATION OF INFILTRATION AIR QUANTITIES

Air leakage into a building is caused by the pressure of the wind against the structure and the temperature difference between inside air and outside air. The tightness of the construction is also a major factor. More infiltration occurs at higher wind velocities than at lower velocities; similarly, more leakage occurs at lower outside air temperatures than at higher outside temperatures. The warm inside air tends to rise and leak out of the upper portions of the structure, and it is replaced by colder, heavier air which leaks into the lower portions of the structure. This effect is more pronounced in high buildings than in one- or two-story buildings. Therefore, for residential-type buildings, only air leakage caused by wind pressure is considered.

The wind pressure may be calculated from

$$p_w = \frac{\rho V_w^2}{2\ g_c}$$

where ρ is the air density in lb/ft³ (kg/m³), V_w is wind velocity in ft/sec (m/s), and p_w is velocity pressure in units of force per unit area. The *velocity head* of a given wind speed for air at 0.075 lb/ft³ (1.20 kg/m³) density may be expressed as

$$p_w = 0.000482\ V_w^2 \qquad (8\text{-}3)$$

where

p_w = velocity head, inches of water gauge (= 249.1 Pa).

V_w = wind velocity, miles per hour (= 0.447 m/s).

(The coefficient 0.000482 becomes 0.601 when SI units are used.)

There are two methods that may be used to estimate the quantity of infiltration air: the air-change method and the crackage method. The air-change method assumes that the air volume within a building is replaced by outside air a certain number of times each hour. If this method is used, ASHRAE recommends the air changes shown in Table 8-2; however, experience and good judgment are required to obtain satisfactory results using the air-change method.

The crackage method for determining infiltration rates is most widely accepted by HVAC engineers because it is believed to be the more accurate method. This method is based on the fact that the quantity of air which leaks through a crack around a window or door is roughly proportional to the length of crack, width of crack, and the square root of the pressure difference across the crack. The pressure difference across the crack depends upon the amount of pressurization within the building. Pressures inside buildings due to wind action alone will depend on the resistance of cracks or openings and their location with respect to wind direction. In large buildings the tightness of internal space separations also may be a factor. If the openings are uniformly distributed around the walls of the building, inside pressures will usually be within plus or minus $0.2p_w$. If openings on the windward side predominate, inside pressures up to plus $0.8p_w$ may occur, approaching the positive values on the outside. Conversely, if openings on the leeward side predominate, inside pressures

Table 8-2

Air Changes Taking Place Under Average Conditions in Residences, Exclusive of Air Provided for Ventilation.*[a]

Kind of Room or Building	Air Changes per Hour
Rooms with no windows or exterior doors	0.5
Rooms with windows or exterior doors on one side only	1
Rooms with windows or exterior doors on two sides	1.5
Rooms with windows or exterior doors on three sides	2
Entrance halls	2

[a]For rooms with weather-stripped windows or storm sash, use ⅔ these values.

**Reprinted from ASHRAE* Handbook, *1977. With permission of the American Society of Heating, Refrigerating and Air Conditioning Engineers, Atlanta, Ga.*

may have negative values of $0.2p_w$ to $0.4p_w$ or greater.

The first step in using the crackage method consists of measuring the linear feet of crack around doors and windows for a particular building space. In frame construction measure also the crack length between foundation and sill. The second step is to determine "the fit" of the windows and doors. Windows of exactly the same type have vastly different air-leakage characteristics. The fit of the window (tight, average, or loose), the use of weather-stripping, and the use of storm sash have definite effects on the amount of air leakage. Even though it may be normal to assume that windows in a new building will be tight fitting, this is not always the case. In fact, all windows with movable sash cannot be tighter than average fit if they are to be opened or closed. Determining the fit of installed DH windows is relatively easy. On these windows lock the window, grasp the meeting rail of the window, and shake it. If the window rattles, it is loose fitting (or poorly fitted). Otherwise, consider it as average fit.

There is a difference between "length of crack" and "infiltration rate." Although it is true that air leakage through a crack depends upon its length, it also depends upon the width and shape of the crack. Furthermore, air leakage per foot of crack for doors differs from that per foot of crack for windows. It is also possible that various windows in a space will have different air-leakage rates per foot of crack. This will usually occur when different types of windows are used in a space, or when some windows are weather-stripped or equipped with storm sash while others are not.

Table 8-3 gives infiltration data for double-hung wood windows.

Infiltration through door cracks must be considered in relation to the type of door and room or building involved. For residences and small buildings where doors are used infrequently, infiltration can be based on the air leakage through cracks between door and frame. Such doors vary greatly in fit because of the tendency to warp. For a well-fitted door, the leakage values for a poorly fitted double-hung window may be used. If door is poorly fitted, twice this figure should be used. If weather-stripped, the values may be reduced one-half. A single door that is frequently opened, as might be the case in a small store, should have a value applied three times that for a well-fitted door. This extra allowance is for opening and closing losses, and is kept from being greater by the fact that doors are not used as much in the coldest and windiest weather. Large rates of leakage can occur with doors in use, and this must be taken into account in commercial establishments where high traffic rates are expected.

The amount of air entering each time the doors are opened depends on the type of door and whether there are doors in one wall or more than one wall. Tables 8-4 and 8-4A have been set up to give the amount of air per passage for single-swing doors, swing doors with vestibules, and revolving doors. Table 8-4A is for applications where the doors are in one wall only, while Table 8-4 covers applications with doors in more than one wall.

Table 8-5 represents the expected number of entrance passages per occupant per hour for various commercial type establishments.

The use of these tables may be explained by the following example.

Illustrative Problem 8-1

Assume that there is a drug store which has an average occupancy of 20 persons. The entrance is a single swinging door, and there are entrances in one wall only. Estimate the amount of air (ft^3/hr) which infiltrates due to door openings.

Table 8-3

Infiltration Through Window and Door Crack*
(Expressed in ft^3/hr per ft of crack)[d]

Type of Window or Door	Pressure Difference (inches of water)			
	0.05	0.10	0.20	0.30
A. Wood, Double-hung Window (Locked) (around sash crack only)	20	25	50	75
1. Non-weatherstripped, loose fit.[a]	28	77	122	150
2. Non-weatherstripped, average fit.[b]	18	27	43	57
3. Weatherstripped, loose fit.	19	28	44	58
4. Weatherstripped, average fit.	11	14	23	30
B. Frame Wall Leakage[c] (Leakage between frame of wood, double-hung window and wall)				
1. Around frame in masonry wall (not caulked).	8	17	26	34
2. Around frame in masonry wall (caulked).	2	3	5	6
3. Around frame in wood frame wall.	6	13	21	29
C. Double-hung metal windows				
1. Non-weatherstripped (unlocked)	47	74	104	137
2. Non-weatherstripped (locked)	45	70	96	125
3. Weatherstripped (unlocked)	19	32	46	60
D. Single-sash Metal				
1. Industrial (horizontally pivoted)	108	176	244	304
2. Residential casement	32	52	76	100
3. Residential (vertically pivoted)	88	145	186	221
E. Doors				
1. Well fitted	69	110	154	199
2. Poorly fitted	138	220	308	398

[a]A 0.094 in. crack and clearance represent a poorly fitted window, much poorer than average.

[b]The fit of the average double-hung wood window was determined as 0.0625 in. crack and 0.047 in. clearance by measurements on approximately 600 windows under heating season conditions.

[c]The values given for frame leakage are per foot of sash perimeter, as determined for double-hung wood windows. Some of the frame leakage in masonry walls originates in the brick wall itself, and cannot be prevented by caulking. For the additional reason that caulking is not done perfectly and deteriorates with time, it is considered advisable to choose the masonry frame leakage values for caulked frames as the average determined by the caulked and no caulked tests.

[d]Multiply table values by 0.0258 to obtain ℓ/s per meter of crack.

**Data from various sources. Primarily from research papers in Trans. ASHVE, vol. 30, 34, 36, 37, and 39. Also, from ASHRAE* Fundamentals Handbook, *1981. With permission of the American Society of Heating, Refrigerating and Air-Conditioning Engineers, Atlanta, Ga.*

Table 8-4

Infiltration Through Entrances with Doors in More Than One Wall* (Summer)[a]

No. of Passages per Hour, Up to	Single Swing Doors		Swing Doors-Vestibule		Revolving Doors	
	Infiltration per Passage, Cu. Ft.	Heat Gain per Passage per Deg. Differ.	Infiltration per Passage, Cu. Ft.	Heat Gain per Passage per Deg. Differ.	Infiltration per Passage, Cu. Ft.	Heat Gain per Passage per Deg. Differ.
300	168	3.02	125	2.25	48	0.86
500	168	3.02	125	2.25	46	0.83
700	168	3.02	125	2.25	43	0.78
900	168	3.02	125	2.25	39	0.70
1100	168	3.02	125	2.25	33	0.59
1200	168	3.02	125	2.25	30	0.54
1300	168	3.02	125	2.25	28	0.50
1400	168	3.02	125	2.25	25	0.45
1500	168	3.02	125	2.25	23	0.41
1600	167	3.01	125	2.25	21	0.38
1700	163	2.93	124	2.23	19	0.34
1800	159	2.86	122	2.20	18	0.32
1900	156	2.81	120	2.16	17	0.30
2000	152	2.74	118	2.12	16	0.29
2100	147	2.64	115	2.07	15	0.27

[a]Increase "Infiltration per passage, cu ft", by 50 percent for winter application.

**Reprinted with permission.* Handbook of Air Conditioning, Heating, and Ventilating. *The Industrial Press, New York, 1965.*

Solution:

Step 1: For drug stores Table 8-5 indicates eight entrance passages per occupant per hour for each occupant in the store. Therefore, the total passages per hour is 8 × 20 = 160/hr.

Step 2: Since there are doors in one wall only, refer to Table 8-4A with 160 passages per hour and find the infiltration rate is 110 ft^3/passage for a single-swing door. In other words each person entering or leaving would allow 110 ft^3 of air into the drug store. Since there are 160 passages/hour, the total infiltration due to door openings is 160 × 110 = 17,600 ft^3/hr.

Table 8-5 may not provide the necessary data for all required establishments. A relationship developed by ARI may be helpful to determine number of entrance passages per hour. Three of the more important factors which determine the rate of infiltration through frequently opened doors are

a. the number of door openings per hour

b. the type of door

c. the wind velocity

The number of door openings can be determined by actual count, or by the following relationship:

$$\text{Door openings/hour} = \frac{P \times F}{t_o \times n} \qquad (8\text{-}4)$$

Table 8-4A

Infiltration Through Entrances with Doors in One Wall Only* (Summer)[a]

No. of Passages per Hour. Up to	Single Swing Doors		Swing Doors-Vestibule		Revolving Doors†	
	Infiltration per Passage, Cu. Ft.	Heat Gain per Passage per Deg. Differ.	Infiltration per Passage. Cu. Ft.	Heat Gain per Passage per Deg. Differ.	Infiltration per Passage, Cu. Ft.	Heat Gain per Passage per Deg. Differ.
300	110	1.98	83	1.49	30	0.54
500	110	1.98	83	1.49	29	0.52
700	110	1.98	82	1.47	27	0.49
900	109	1.96	82	1.47	25	0.45
1100	109	1.96	82	1.47	23	0.41
1200	108	1.95	82	1.47	21	0.38
1300	108	1.95	82	1.47	19	0.34
1400	108	1.95	81	1.46	18	0.32
1500	108	1.95	81	1.46	17	0.31
1600	108	1.95	81	1.46	16	0.29
1700	107	1.93	81	1.46	15	0.27
1800	105	1.89	80	1.44	14	0.25
1900	104	1.87	80	1.44	13	0.23
2000	100	1.80	79	1.42	12	0.22
2100	96	1.73	79	1.42	11	0.20

[a]Increase "Infiltration per passage, cu ft", by 50 percent for winter application.

**Reprinted with permission.* Handbook of Air Conditioning, Heating and Ventilating. *The Industrial Press, New York, 1965.*

where

P = number of people in the conditioned space

F = factor for arrivals and departures. (Use 2 for light traffic and 1.33 for heavy traffic). The factor 1.33 assumes that one-third of the arrivals and departures will occur simultaneously

t_o = average time of occupancy (hr)

n = number of doors

When using the crackage method to determine infiltration rates, the basis of calculations recommended by ASHRAE is as follows: The amount of crack used for computing the infiltration heat loss should be not less than one-half the total length of crack in the outside walls of a room. For a building having no partitions air entering through the cracks on the windward side must leave through the cracks on the leeward side. Therefore, take one-half the total crack for computing each side and end of the building. In a room with one exposed wall take all the crack; with two exposed walls, take the wall having the most crack; and with three or four exposed walls, take the wall having the most crack; but in no case take less than half the total crack.

In small residences the total infiltration loss of the house is generally considered to be equal to the sum of the infiltration losses of the various rooms. However, this is not necessar-

Table 8-5

Entrance Passages per Occupant per Hour*

Banks	8	Dress shops	3	Office buildings	2
Barber shops	4	Drug stores	8	Professional offices	4
Brokers offices	8	Furriers	3	Public buildings	3
Candy and soda stores	6	Hospital rooms	4	Restaurants	3
Cigars and tobacco stores	25	Lunchrooms	6	Shoe stores	4
Department stores	8	Men's shops	4	Variety stores	12

**Reprinted with permission.* Handbook of Air Conditioning, Heating, and Ventilating. *The Industrial Press, New York, 1965.*

ily accurate as at any given time infiltration will take place only on the windward side or sides and not on the leeward side. Therefore, for determining the total heat requirements of larger buildings, it is more accurate to base the total infiltration loss on the wall having the most crack, but in no case take less than half the total crack in the building.

Measurement of crackage should be done with as much care as in determining other measurements for heat loss calculations. Determine the running feet of crackage around the edges of doors by adding the width and height of the door and multiplying the sum by two. A 3 by 7-ft door would have a crackage of $2(3 + 7) = 20$ ft. Determine the running feet of crackage for *each* window by adding the lengths of cracks in each sash. Crackage should be measured, even for *fixed* sash. Crack between the movable sections of a sash, like the meeting rail, is measured only *once*. It is impossible to give a general rule for calculating the crackage on all types of windows. For some common types, however, the following rules may be used:

1. Double-hung windows—use twice the height plus three times the width of the sash.
2. Single-sash casement windows—use twice the height plus twice the width.
3. Horizontal sliding windows—three times the height plus twice the width.
4. Jalousie, awning, and architectural projected—use twice the height plus the width multiplied by one more than the number of glass shutters.
5. Fixed windows—use twice the height plus twice the width.

All dimensions are to be expressed in feet. In new construction the dimensions of double-hung windows are usually specified on architectual drawings. The dimensions specified are usually the glass size, but they may be converted to opening sizes using the method previously outlined for areas. The opening size is used to calculate crack length. For other types of windows it is necessary to refer to the architect, builder, or to the manufacturer's catalog of the particular window to obtain the crackage length. These windows are often specified on building plans by the manufacturer's model number, and reference to his published material, or measurement of the installed window are the methods used to obtain the necessary information.

Air leakage through walls: A large amount of cold air leakage occurs through building walls if the wall is porous and poorly constructed. In severely cold climates buildings are usually built to prevent unwanted air leakage. Problems of excess air leakage through building construction are more common in warmer climates. As a result, the most difficult heating problem may occur in warmer climates. The necessity for good construction to reduce unwanted air leakage cannot be emphasized too strongly. Good construction practices include:

a. Tight siding and sheathing construction
b. Using a tightly fitting building paper between sheathing and siding and between subflooring and finish flooring
c. Insulating stud spaces or placing draft stops in the spaces
d. Using good vapor barrier beneath the plaster base
e. Making a tight seal around the floor mouldings and around window frames

Infiltration and air for combustion: Fuel-burning appliances (gas, oil, coal, wood) require combustion air. The presence of these appliances within the building can have a pronounced effect upon the rate of infiltration. It is recommended that combustion air from the outside be introduced directly to the fuel-burning appliance to reduce the effect on infiltration. This may not be practical nor desirable in the case of a fireplace.

When a fireplace is not operating, and the damper is open, large quantities of air will rise through the chimney flue. This air has to leak into the building through crackage. The amount of leakage depends upon the tightness of the building construction, design of the chimney, and wind velocity, but the infiltration rate could be doubled. If the damper is closed and is tight fitting (seldom the case), the problem is eliminated.

When the fireplace is in operation, the draft through the chimney is greatly increased. The air for the fireplace, when possible, should be introduced through the central heating system and not allowed to enter by infiltration. The air required when a fireplace is in operation could be as high as 200 cfm. It is normal to assume about 50 cfm of infiltration per fireplace if the fireplace is not operating and the damper is closed.

Direct-fired warm-air furnaces are often located within the confines of the building. Most codes require that outside air for combustion be introduced directly to the space where the furnace is located. Combustion air required is 12 to 16 ft^3 per 1000 Btu in furnace capacity.

8-10 AUXILIARY HEAT SOURCES

The heat supplied by persons, lights, motors, and machinery should always be determined in the case of theaters, assembly halls, industrial plants, and commerical buildings such as stores, office buildings, and so forth; but allowances for such heat sources must be made only after careful consideration of all local conditions. In many cases these heat sources may materially affect the size of the heating plant and may have a marked effect on the operation and control of the system. In any evaluation, however, the night, weekend, and any other unoccupied periods must be evaluated. In general, where audiences are present, the heating system must have adequate capacity to bring the building to a stipulated indoor temperature before the audience arrives. In industrial plants quite a different condition exists, and heat sources, if always available during occupancy, may be substituted for a portion of the heating requirements. In no case, however, should the actual heating installation (exclusive of heat sources) be reduced below that required to maintain at least 40°F in the building.

Inefficiencies in motors and machinery they

drive, if both are located in the heated space, appear as room heat. This heat is retained in the room, if the product manufactured is not removed, until its temperature is the same as the room temperature. If power is transmitted to the machinery from outside, only the heat equivalent to the brake horsepower supplied is used. In some industries this is the chief source of heating; it can frequently overheat the building even in zero weather, thus requiring year-round cooling. For information on heat supplied by people, and so forth refer to Chapter 9.

8-11 INTERMITTENTLY HEATED BUILDINGS

For intermittently heated buildings additional heat is required for raising the temperature of the air, building material, and material contents of the building to the specified indoor temperature. The rate at which this additional heat must be supplied depends on the heat capacity of the structure and its material contents, and the time in which these are to be heated.

Because design outdoor temperatures generally provide a substantial margin for outdoor temperatures typically experienced during operating hours, many engineers make no allowance for this additional heat in most buildings. However, if optimum equipment sizing is used (minimum safety factor), the additional heat should be computed and allowed for as conditions require. In the case of churches, auditoriums, and other intermittantly heated buildings, additional capacity of 10 percent should be provided.

8-12 EXPOSURE FACTORS

Some designers use empirical *exposure* factors to increase calculated heat loss from rooms or spaces on the side (sides) of the building exposed to prevailing winds. However, use of exposure factors should be unnecessary with the method of calculating heat loss described in this chapter. They may be considered as *safety factors* for the rooms or spaces exposed to prevailing winds, to allow for additional capacity for the spaces; or used to *balance the radiation,* particularly in the case of multistory buildings. Tall buildings may have severe infiltration heat losses, induced by the stack effect, that will require special consideration. Although a 15 percent exposure allowance frequently is assumed, the actual allowance, if any, must be largely a matter of experience and judgment; no authentic test data is available from which to develop rules for the many conditions encountered.

8-13 HEAT LOSS CALCULATIONS FOR A BUILDING

Since the results desired in evaluating heat losses is a sum of all losses, an orderly tabulation of calculations is most desirable. Such a tabulation gives a clear, orderly presentation of dimensions of components through which heat is being transmitted, and also provides a means for quickly checking results. There are many *heat load forms* available, and all have their advantages and disadvantages. You may find the tabulation work sheet at the top of the following page useful.

The above type of work sheet would be made up for each room or space to be heated. Adding all the total losses from each room or space provides the total building heat loss.

Illustrative Problem 8-2

Figure 8-1 represents the floor plan of a proposed school addition consisting of two classrooms, a cafeteria, kitchen, and auxilliary rooms. The scale of the floor plan is *one space equals one foot.* (The wall thicknesses are exaggerated.) Make a complete heat loss calculation for each room (excluding walk-in refrigerator, food storage area, and basement) using the design data at the bottom of the next page.

Tabulation Work Sheet for Heat Loss

Name of Room or Space:
Room Size: L = ; W = ; Ceiling Height = ; Volume =

	Area × U × Δt		$\dot{q}$(Btu/hr)
Gross exposed wall		=	
Windows		=	
Doors		=	
Net exposed wall		=	
Cold ceiling		=	
Cold floor		=	
Cold partition		=	
Floor edge loss		=	
Infiltration		=	
Miscellaneous		=	______
	Total loss	=	

Auxilliary calculations as required:

Design Conditions

	Summer	Winter
Outside air design	95°F/76°F	−10°F, 100 percent RH
Inside air design	78°F/50 percent RH	70°F, 40 percent RH
Outside air daily temperature range	20 degrees	15 mph wind
Location 40°N latitude		

1. Structure data:
   (a) *Full basement* below kitchen and cafeteria; slab-on-grade for remainder of building. Basement walls 10-in. concrete, sand and gravel aggregate; basement floor and floor slab 4-in. concrete, sand and gravel aggregate; edge insulation 1-in., 18-in. high. Basement ceiling height (floor to underside of floor joist) 9 ft; exterior grade level at top of foundation wall.
   (b) *Roof* (horizontal): 3-in. concrete, sand and gravel aggregate; 2-in. preformed polyurethane, expanded; built-up tar and gravel ⅜-in.; suspended metal lath and plaster (conventional) ceiling.
   (c) *Ceiling heights:* Classroom, 10 ft; corridors, 8 ft 6 in.; cafeteria, 10 ft, all other areas, 9 ft.
   (d) *Exterior Walls:*
      (1) *Section A-A* (Exterior wall of corridors 1 and 2, and classroom 1): 4-in. nominal face brick; ¾-in. air space; ¾-in. plywood sheathing; 2 by 6 nominal wood studs; 3 to 3½-in. paper-packed bat insulation in stud space; gypsum board, plain ⅜-in.; ¼-in. plywood panels.
      (2) *Section B-B* (Exterior wall of main entrance and cafeteria as far as kitchen partition): 4-in. nominal face brick; ¾-in. cement mortar; 8-in. concrete block, hollow, light-weight aggregate; 1½-in. corkboard, no added binder (density 10.6); ¼-in. plywood panels.
      (3) *Section C-C* (Exterior wall of kitchen): Same as Section B-B except corkboard and plywood panel is replaced with insulating board lath, 1-in. (and plaster), plaster thickness ½-in.
   (e) *Floor* (All areas): 4-in. concrete slab, sand and gravel aggregate, asphalt tile ⅛-in. thick, no ceiling in basement area.

(f) *Windows and Skylights:*

(1) *Corridor 1:* Wood sash, fixed panels, double glazing with ¼-in. air space; each panel 5 ft wide by 3 ft high; sill height 5 ft above floor.

(2) *Classroom 1:* Wood sash, double-hung, weather-stripped; double glazing with ¼-in. air space; each window 4 ft wide by 6 ft high; sill height 3 ft 6 in. above floor.

(3) *Cafeteria* (SE wall): Metal frame, fixed panels, double glazing with ¼-in. air space, 4 ft wide by 6 ft high; sill height 1 ft above floor.

(4) *Cafeteria* (NE wall): Glass block, 5¾ by 5¾ by 3⅞-in. thick; wall opening 2 ft wide floor to ceiling.

(5) *Skylights* (All areas): Metal frame, double glass, ½-in air space.

(g) *Doors:*

(1) *Main entrance:* ⅜-in. plate glass, 2-in. thick hollow steel frame (20 percent of area), size 3 ft wide by 7 ft high each door.

(2) *Cafeteria:* 1¾-in. steel, mineral fiber core, size 3 ft wide by 7 ft high. (Two doors, emergency exit only.)

(3) *Corridor 2 and classroom 1:* 2-in. solid wood doors, size 3 ft wide by 7 ft high. (Classroom 1 exterior door for emergency use only.)

2. Miscellaneous Information:

(a) *Lighting levels:*

Classrooms: 6 W/ft^2 of floor space

Cafeteria: 4 W/ft^2 of floor space

Kitchen: 7 W/ft^2 of floor space

Corridors: 3 W/ft^2 of floor space

(b) *Kitchen exhaust hood:*

Exhaust rate 2400 cfm; toilet rooms 10 changes/hr.

(c) *Design people density* (Maximum):

Classrooms: 40 per room

Cafeteria: 180

3. Requirements:

(1) Calculate the heat losses for each area designated (conduction and infiltration).

(b) Determine outside air heat loads caused by required ventilation air to each area.

<u>*Solution:*</u>

Step 1: Determine U factors.

Roof		**R**
(1) Outside air film (f_o = 6.0)		0.17
(2) Built-up tar and gravel (C = 3.0)		0.33
(3) 2-in. preformed polyurethane (k = 0.16)	x/k =	12.50
(4) 3-in. concrete (sand and gravel)(k = 12.0)	x/k =	0.25
(5) Air space (C = 1.15)		0.87
(6) Metal lath and plaster (C = 4.40)		0.23
(7) Inside air film (f_i = 1.63)		0.61
$U = 1/R_t = 1/14.96 = 0.0668$ (say 0.07)	R_t =	14.96

Walls **Section A-A**	**R**
(1) Outside air film (f_i = 6.0)	0.17
(2) Face brick (C = 2.27)	0.44
(3) Air space (¾-in.) (C = 0.99)	1.01
(4) Plywood (¾-in.) (C = 1.07)	0.93
(5) Air space (approx. 2-in) (C = 0.99)	1.01
(6) 3 to 3½-in. paper-backed bats	11.00
(7) Gypsum board, plain, ⅜-in. (C = 3.7)	0.27
(8) Plywood panels (¼-in.) (C = 3.21)	0.31
(9) Inside air film (f_i = 1.46)	0.68
	R_t = 15.82

$U = 1/R_t = 1/15.82 = 0.0632$

Correct for studs (16-in. on center)

Resistance through studs = 15.82 − 11.00 − 1.01 + 1.25(5.5) = 10.685

U factor through studs = 1/10.685 = 0.09359

Corrected U factor = 0.0632 (14.5/16) + 0.09359 (1/5/16)

Corrected U factor = 0.066 (say 0.07)

Section B-B	**R**
(1) Outside air film (f_o = 6.0)	0.17
(2) Face brick (C = 2.27)	0.44
(3) Cement mortar (¾-in.) (k = 5.0)	x/k = 0.15
(4) Concrete block (8-in) (C = 0.50)	2.00
(5) Corkboard (1½-in.) (k = 0.30)	x/k = 5.00
(6) Plywood panels (¼-in.) (C = 3.21)	0.31
(7) Inside air film (f_i = 1.46)	0.68
	R_t = 8.75

$U = 1/R_t = 1/8.75 = 0.114$ (say 0.12)

Section C-C

Insulating board 1-in. and plaster ½-in. (C = 0.31; R = 3.18)

Total resistance = 8.75 − 5.00 − 0.31 + 3.18 = 6.62

$U = 1/R_t = 1/6.62 = 0.151$ (say 0.15)

Floor (over basement)	R
(1) Air film (top surface) $C = 1.08$	0.92
(2) Asphalt tile (1/8-in.) $C = 24.8$	0.04
(3) Concrete floor (4-in.) $k = 12.0$	$x/k = 0.33$
(4) Air film (bottom surface) $C = 1.08$	0.92
	$R_t = 2.21$

Light Transmitting Areas

$U = 1/R_t = 1/2.21 = 0.452$ (say 0.45)
Window (Corridor 1): $U = (0.61)(0.95) = 0.579$ (say 0.58)
Window (Classroom No. 1): $U = (0.61)(0.95) = 0.579$ (say 0.58)
Window (Cafeteria) $U = (0.61)(1.20) = 0.732$ (say 0.73)
Glass blocks (Cafeteria) $U = 0.60$
Skylights $U = (0.66)(1.2) = 0.792$ (say 0.79)

Doors	R
Main Entrance	.17
(1) Outside air film ($f_o = 6.0$)	$x/k = 0.074$
(2) Plate glass (3/8-in.) ($k = 5.1$)	0.68
(3) Inside air film ($f_i = 1.46$)	$R_t = 0.924$

$U_{\text{glass}} = 1/0.924 = 1.08$
Correction for metal frame:
R through frame $= 0.924 - 0.074 + (0.00318)(2) = 0.856$
$U_{\text{frame}} = 1/0.856 = 1.168$
Corrected u factor $= (1.08)(0.80) + (1.168)(.20)$
$U = 1.097$ (say 1.10)
Door (Cafeteria) $U = 0.59$
Door (Corridor 2, Classroom 1) $U = 0.43$

Summary of *U* factors

	U factor		*U* factor
Roof	= 0.07	Window (Corridor 1)	= 0.58
Wall (A-A)	= 0.07	Window (Classroom 1)	= 0.58
Wall (B-B)	= 0.12	Window (Cafeteria)	= 0.73
Wall (C-C)	= 0.15	Glass blocks (Cafeteria)	= 0.60
Floor	= 0.45	Skylights	= 0.79
Door (Main entrance)	= 1.10		
Door (Cafeteria)	= 0.59		
Door (Corr. 2, CR 1)	= 0.43		

Step 2: Calculate heat loss classroom 1

Space volume = (28)(38)(10) = 10,640 ft^3

		A		U		Δt		Btu/hr
Gross wall (45 × 10)	=	450						
Windows (4)(4 × 6)	=	96	×	0.58	×	80	=	4454
Door (3 × 7)	=	21	×	0.43	×	80	=	722
Net wall (450 − 96 − 21)	=	333	×	0.07	×	80	=	1865
Ceiling-roof (28 × 38)	=	1064	×	0.07	×	80	=	5958
Floor edge (45 lin. ft × 50 Btu/hr per foot, Table 7-9)							=	2250
Infiltration (see calculations below)							=	4825
						Total loss	=	20,074 Btu/hr

Crackage (Infiltration)
Crack length windows = 4[(3 × 4) + (2 × 6)] = 96 ft
Crack length window frame = 4[(2 × 6) + (2 × 3)] = 72 ft
Crack length door = (2 × 3) + (2 × 7) = 20 ft

Wind pressure By Eq. 8-3, $p_w = 0.000482\,V_w^2$
$V_w^2 = 0.000482(15)^2 = 0.108$ (use 0.10)

Table 8-3 Infiltration windows (weather-stripped, average fit) = 14 ft^3/hr ft
Infiltration window frame = 13 ft^3/hr ft
Infiltration door (assume same as DH window, weatherstripped, loose fit) = 28 ft^3/hr ft
Total infiltration = (96)(14) + (72)(13) + (20)(28) = 2840 ft^3/hr
From psychrometric chart at −10°F and 100 percent RH, $v_{oa} = 11.3$
Infiltration heat load is

By Eq. 8-1a $$\dot{q}_{oa} = \frac{2840}{11.3}(0.240)[70 - (-10)]$$

$$\dot{q}_{oa} = 4825 \text{ Btu/hr}$$

Step 3: Calculate heat loss Classroom No. 2

Room volume = 38 × 29 × 10
= 11,020 ft^3

No exterior wall, and no floor loss, therefore,

		A		U		Δt		Btu/hr
Gross roof (38 × 29)	=	1102						
Skylights 4 (6 × 4)	=	96	×	0.79	×	80	=	6067
Net roof (1102 − 96)	=	1006	×	0.07	×	80	=	5634
Infiltration (assume zero)							=	0
						Total loss	=	11,701 Btu/hr

Step 4: Calculate heat loss Corridor No. 1

		A		U		Δt		Btu/hr
Gross wall (48 × 8.5)	=	408						
Windows 6 (3 × 5)	=	90	×	0.58	×	80	=	4176
Net wall (408 − 90)	=	318	×	0.07	×	80	=	1781
Gross roof (60 × 10)	=	600						
Skylights 2 (6 × 4)	=	48	×	0.79	×	80	=	3034
Net roof (600 − 48)	=	552	×	0.07	×	80	=	3090
Floor edge (48 ft × 50 Btu/hr per ft Table 7-9)							=	2400
Infiltration (see below)							=	1906
						Total loss	=	16,387 Btu/hr

Crackage (Infiltration): The windows located in corridor 1 are fixed sash wood windows in a masonry-wood stud wall. Infiltration is difficult to evaluate because of two reasons: (a) leakage around fixed windows is questionable, however, we will use Table 8-3 and (b) some infiltration may occur due to the air flow through the main entrance. We will disregard this.

Crackage around windows = (30 × 2) + (2 × 3) = 66 ft

Table 8-3 Around frame in masonry wall, not caulked = 17 ft^3/hr ft

Infiltration rate = 66 × 17 = 1122 ft^3/hr

By Eq. (8-1a)

$$\dot{q} = 1122/11.3\ (0.24)\ [70-(-10)]$$
$$= 1906 \text{ Btu/hr}$$

Step 5: Calculate heat loss corridor 2

Gross wall (10 × 8.5)	=	85						
Door 2 (3 × 7)	=	42	×	0.43	×	80	=	1445
Net wall (85 − 42)	=	43	×	0.07	×	80	=	241
Gross roof (51 × 10)	=	510						
Skylights 2 (2 × 10)	=	40	×	0.79	×	80	=	2528
Net roof (510 − 40)		470	×	0.07	×	80	=	2632
Floor edge								negligible
Infiltration (see below)							=	12,840
						Total loss	=	19,686 Btu/hr

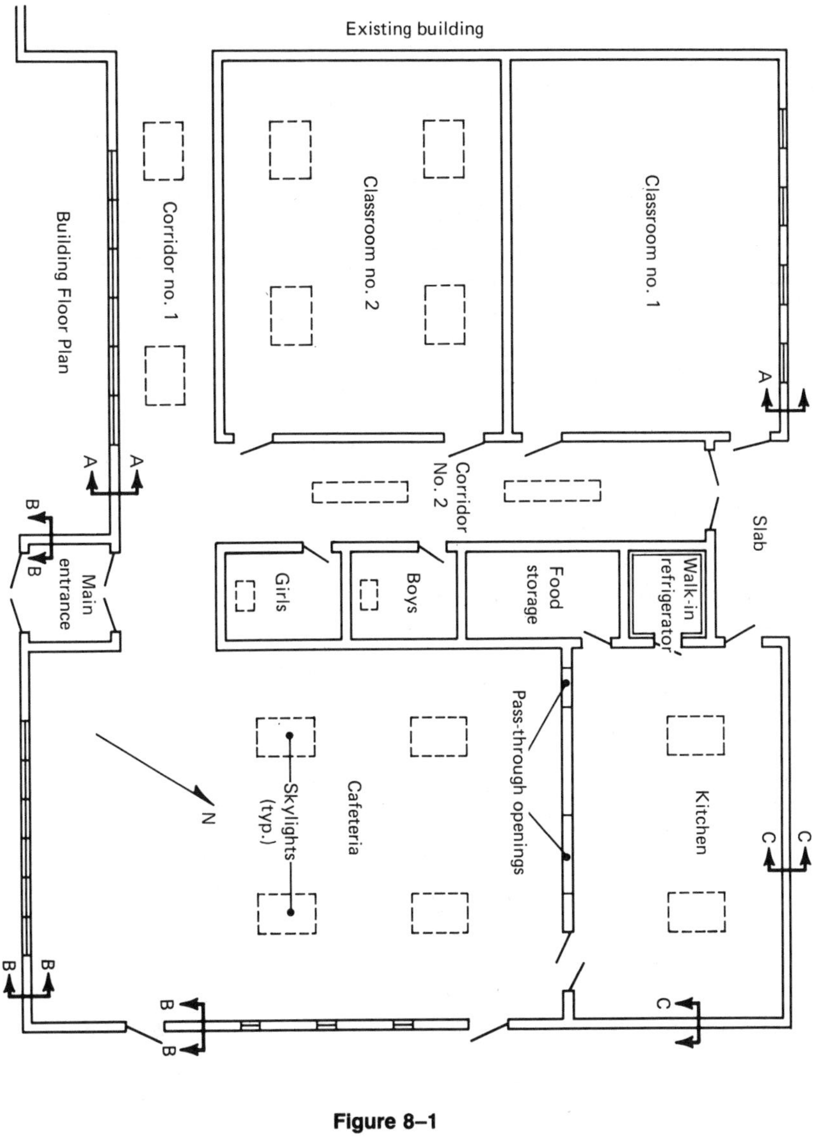

Figure 8–1

Illustrative Problem 8-2

Crackage: For the double doors, crackage length = (2 × 6) + (3 × 7) = 33 ft.
Infiltration: Since the doors would be frequently used, we shall assume the crack infiltration will be twice that for a loose fit, non-weather-stripped DH window. From Table 8-3 select 77 × 2 = 154 ft^3/hr per ft of crack. Therefore, infiltration through these doors will be 33 × 154 = 5082 ft^3/hr.

Infiltration (due to door openings): This portion of the infiltration load is one of the more difficult ones to evaluate. Tables 8-4, 8-4A, and 8-5 would normally be used. However, schools are not listed. Therefore, an alternate approach will be used with Eq. (8-4).

The total occupancy of the two classrooms is 80 students. We will increase this to 90 persons to account for teachers and other personnel who may be in this section of the building. We shall consider light traffic (F = 2) and time of occupancy (t_o = 4 hr). By Eq. 8-4

$$\text{Door openings /hr} = \frac{90 \times 2}{4 \times 2} = 22.5$$

From Table 8-4, for single-swing doors, find infiltration is 110 ft^3 per passage.

$$\text{Infiltration} = 22.5 \times 110 = 2475 \text{ ft}^3\text{/hr}$$

Total infiltration through doors = 5082 + 2475 = 7557 ft^3/hr

By Eq. (8-1a)

$$\dot{q} = 7557/11.3\ (0.24)\ [70 - (-10)]$$

$$\dot{q} = 12{,}840 \text{ Btu/hr}$$

Step 6: Calculate heat loss for main entrance

Space volume = 10 × 10 × 9 = 900 ft^3

(using the table at the bottom of the page)

Crackage: Assume same conditions exist as determined for exterior doors of corridor 2. Crackage infiltration is 5082 ft^3/hr.

Infiltration due to door openings will be affected by the presence of the vestibule. Assuming the same number of door openings per hour as for corridor 2 we have 22.5. From Table 8-4 we have 83 ft^3 per passage with a vestibule. Therefore, infiltration due to door openings is 22.5 × 83 = 1868 ft^3/hr.

$$\text{Total infiltration} = 5082 + 1868 = 6950 \text{ ft}^3\text{/hr}$$

By Eq. (8-1a)

$$\dot{q} = 6950/11.3\ (0.24)\ [70 - (-10)]$$

$$\dot{q} = 11{,}800 \text{ Btu/hr}$$

		A		***U***		**Δ*t***		**Btu/hr**
Gross Wall (19 × 9)	=	171						
Entrance doors (6 × 7)	=	42	×	1.10	×	80	=	3700
Net wall (171 − 42)	=	129	×	0.12	×	80	=	1240
Roof (10 × 10)	=	100	×	0.07	×	80	=	560
Floor edge								negligible
Infiltration (see below)							=	11,800
						Total loss	=	17,300 Btu/hr

Step 7: Calculate heat loss for cafeteria

Space volume = 38 × 56 × 10
= 21,280 ft^3

NOTE: Assume basement temperature is 55°F.

		A		U		Δt		Btu/hr
Gross wall (94 × 10)	=	940						
SE windows 6(4 × 6)	=	144	×	0.73	×	80	=	8400
Glass blocks 3(2 × 10)	=	60	×	0.60	×	80	=	2880
Doors 2(3 × 7)	=	42	×	0.59	×	80	=	1980
Net wall (940 − 144 − 60 − 42)	=	694	×	0.12	×	80		6660
Floor (38 × 56)	=	2128	×	0.45	×	(70 − 55)	=	14,360
Gross roof (38 × 56)	=	2128						
Skylights 4(4 × 6)	=	96	×	0.79	×	80	=	600
Net roof (2128 − 96)	=	2032	×	0.07	×	80	=	11,300
Infiltration (see below)	=							1900
						Total Loss	=	53,620 Btu/hr

Crackage: The south-east windows are fixed panels in masonry construction. We will assume that, with careful construction, these windows will have no infiltration.

The two emergency exit doors would have crackage amounting to 2 [(3 × 2) + (7 × 2)] = 40 ft. These doors are of the same type as found in classroom 1; therefore, infiltration is 28 ft^3/hr per foot of crack; or, infiltration is 28 × 40 = 1120 ft^3/hr.

By Eq. (8-1a)

$$\dot{q} = \frac{1120}{11.3} = (0.24)\,[70 - (-10)]$$

$$= 1900 \text{ Btu/hr}$$

Step 8: Calculate heat loss for kitchen

Space volume = 22 × 38 × 9 = 7524 ft^3

		A		U		Δt		Btu/hr
Gross wall (60 × 9)	=	540						
Door (3 × 7)	=	21	×	0.43	×	80	=	720
Net wall (540 − 21)	=	519	×	0.15	×	80	=	6230
Floor (22 × 38)	=	836	×	0.45	×	(70 − 55)	=	5640
Gross roof (22 × 38)	=	836						
Skylights 2(4 × 6)	=	48	×	0.79	×	80	=	3030
Net roof (836 − 48)	=	788	×	0.07	×	80	=	4400
Infiltration (see below)							=	negligible
						Total loss	=	20,020 Btu/hr

Infiltration: The kitchen presents a special problem because door infiltration would be negligible when consideration is made concerning the exhaust hood requirements. It is obvious that if 2400 cfm of air is to be exhausted (when the exhaust fan is operating), makeup air from the outside must be provided. If not, infiltration rates to the other areas (cafeteria, entrances, and so forth) would be greatly increased and difficult to evaluate. The presence of the exhaust hood in the kitchen is more a ventilation problem and will be discussed later under ventilation.

Step 9: Calculate heat loss for boys' room

Space volume = 9 × 11 × 9 = 891 ft^3

		A		*U*		Δ*t*		Btu/hr
Gross roof (11 × 9)	=	99						
Skylight (3 × 2)	=	6	×	0.79	×	80	=	380
Net roof (99 − 6)	=	93	×	0.07	x	80	=	520
					Total loss		=	900 Btu/hr

Step 10: Calculate heat loss for girls' room

Space volume = 9 × 12 × 9 = 972 ft^3

		A		*U*		Δ*t*		Btu/hr
Gross roof (9 × 12)	=	108						
Skylight (3 × 2)	=	6	×	0.79	×	80	=	380
Net roof (108 − 6)	=	102	×	0.07	×	80	=	570
					Total loss		=	950 Btu/hr

Step 11: Summarize heat losses by conduction and infiltration

Space	Heat Loss (Btu/hr)
Classroom 1	20,074
Classroom 2	11,701
Corridor 1	16,387
Corridor 2	19,686
Main Entrance	17,300
Cafeteria	53,620
Kitchen	20,020
Boys' Room	900
Girls' Room	950
Total loss =	160,638 Btu/hr

Step 12: Determine ventilation requirements

Ventilation air supplied to a building is *not* a heat load on the building. It is, however, a load on the air-conditioning system and must be evaluated to determine the required system capacity.

Recommended and minimum ventilation standards were presented in Chapter 6. Supplying outside air for ventilation is costly (from an energy standpoint), and in cold climates minimum ventilation air quantities are used at design conditions. In most well-designed HVAC systems provisions are made by the equipment design and control to admit more ventilation air as the outside air temperature rises, up to as much as 100 percent outside air when it may be necessary to cool the building during intermediate seasons, such as in the spring and fall.

For our present problem, we will design on the basis of *minimum* outside air, as follows:

a. For classroom 1 and 2, allow 5 cfm/student
b. For cafeteria, allow 10 cfm/student
c. For corridors 1 and 2 and main entrance, allow zero cfm. (Adequate ventilation is obtained by infiltration).
d. For kitchen, allow 4 cfm/ft^2 or floor area (maximum)
e. For boys' and girls' rooms, allow zero cfm. (This assumes that the doors to these facilities contain louvers through which corridor air will enter.)

Calculations of ventilation quantities:

Classroom 1 = 5 × 40 = 200 cfm

Classroom 2 = 5 × 40 = 200 cfm

Cafeteria 10 × 180 = 1800 cfm

Kitchen 4 × 22 × 38 = 3344 cfm (since this is greatly in excess of the exhaust fan capacity, 2400 cfm, and since the kitchen and cafeteria are connected by pass-through openings, the outside air required in the kitchen will be taken as 2400 − 1800 = 600 cfm.

Boys' room: From data presented in problem statement, exhaust rate is to be 10 changes/hr, or

Exhaust cfm

$$= \frac{\text{Room volume} \times \text{changes per hour}}{60}$$

$$\text{cfm} = \frac{891 \times 10}{60} = 148.5 \quad \text{(say 150 cfm)}$$

Girls' room: Again, allowing 10 changes/hr,

$$\text{Exhaust cfm} = \frac{972 \times 10}{60}$$

$$= 162 \quad \text{(say 160 cfm)}$$

The total outside air for building ventilation is

$$\text{Ventilation cfm} = 200 + 200 + 1800 + 600 = 2800 \text{ cfm}$$

The heating load placed on the equipment is by Eq. (8-1a)

$$\dot{q} = \frac{2800 \times 60}{11.3} \times (0.24)\,[70 - (-10)]$$

$$\dot{q} = 285{,}450 \text{ Btu/hr}$$

Step 13: Determine air balance for building at design conditions.

Infiltration:

Classroom 1	2840/60	=	47.3 cfm
Classroom 2		=	0 cfm
Corridor 1	1122/60	=	18.7 cfm
Corridor 2	7557/60	=	126.0 cfm
Main Entrance	6950/60	=	116.0 cfm
Cafeteria	1120/60	=	18.7 cfm
Total Infiltration		=	326.7 (say 327 cfm)

Ventilation: 2800 cfm

Exhaust Air:

Kitchen 2400 cfm

Boys' room	150 cfm
Girls' room	160 cfm
Total exhaust air =	2710 cfm

Total air entering building = 2800 + 327 = 3127 cfm
Total air leaving building = 2710 cfm

The difference (3127 − 2710 = 417 cfm) indicates a slight pressurization of the building, which helps to reduce the infiltration.

Step 14: Determine humidification load.

Previously in this problem we have omitted consideration of the heating load concerned with the humidification of the outside air which enters by infiltration and ventilation.

From the psychrometric chart the humidity ratio of the outside air W_{oa} = 0.00045 lb_v/lb_{da}. The humidity ratio of the inside air is W_{ra} = 0.0063 lb v/lb da. From step 13 above the total estimated volume of outside air entering the building is 3127 cfm. If the latent heat of evaporation of moisture at room conditions is assumed to have an average value of 1055, then by eq. (8-2a),

$$\dot{q} = \frac{\dot{Q}_{oa}}{v_{oa}} (W_i - W_o)\, h_{fg}$$

$$= \frac{(3127)(60)}{11.3} (0.0063 - 0.00045)(1055)$$

$$\dot{q}_L = 102{,}470 \text{ Btu/hr}$$

Step 15: Summarize the heating and ventilating loads at design conditions.

(a) Building heat loss (conduction and infiltration)	= 160,640
(b) Ventilation load	= 285,450
(c) Humidification load	= 102,470
Total equipment load	= 548,560 Btu/hr

8-14 AIR REQUIRED FOR SPACE HEATING

The computation of air supply quantities required for space heating was discussed in Chapter 5. The procedure used is always recommended when latent heat loss is involved or when a relatively large amount of outdoor ventilation air is used. In many cases, especially in residential or light commercial establishments, when latent heat loss is relatively small and may be neglected or included with the sensible heat loss, the air supply quantity may be calculated from

$$\dot{q}_t = \dot{m}_{sa} C_p (t_{sa} - t_{ra}] \qquad \textit{(8-5)}$$

or

$$\dot{q}_t = \frac{\dot{Q}_s}{v_{sa}} C_p (t_{sa} - t_{ra}) \qquad \textit{(8-5a)}$$

or (approximately)

$$\dot{q}_t = 1.10 \text{ cfm}_{sa} (t_{sa} - t_{ra}) \qquad \textit{(8-5b)}$$

where

$\dot{q}_t$ = sensible heat loss for a space, Btu/hr (including latent heat loss if negligible)

$\dot{m}_{sa}$ = mass of supply air (lb/hr)

v_{sa} = specific volume of supply air (ft^3/lb)

$\dot{Q}_s$ = volume of supply air (ft^3/hr)

t_{sa} = supply air temperature

t_{ra} = room air temperature

The temperature difference $t_{sa} - t_{ra}$ is normally less than 100°F. However, some small-capacity residential-type warm-air furnaces are rated on a 100-degree rise in air temperature through the heat exchanger. These units also have a rated capacity in cfm and a limited fan static pressure rise.

Most commercial warm-air systems are built-up units where the heat exchanger and fan may be selected to give performance levels which will meet any requirements.

If a system is being designed for year-round operation, the air flow quantities required for cooling are normally greater than for heating because of the differences in calculated heat loads and the $t_{ra} - t_{sa}$ difference used for cooling and $t_{sa} - t_{ra}$ used for heating. Usually, the air flow quantity determined by the cooling loads determines the supply air quantity.

As may be seen by Eqs. (8-5), (8-5a), and (8-5b), the quantity of supply air, for a given space heat load, is dependent on the supply air temperature difference $t_{sa} - t_{ra}$. A large temperature difference reduces the air supply quantity required. Another factor should also be considered. The total air circulated as supply air must be large enough to promote air motion in the space which is within comfort levels (20 to 40 fpm), or to provide for a reasonable number of air changes per hour.

Illustrative Problem 8-3

For the school addition of Problem 8-2 determine the supply air quantity required to each heated space at design conditions. Assume that the heating system is a forced circulation warm-air system using a central HV fan-coil unit located in the basement. Assume that the supply air temperature to each space is 90°F.

Solution:

Step 1: Determine supply air quantities to each space. For classroom 1: $\dot{q}_t$ = 20,074 Btu/hr. By Eq. (8-5b),

$$\dot{q}_t = 1.10 \text{ cfm}_{sa}(t_{sa} - t_{ra})$$

$$\text{cfm}_{sa} = \frac{\dot{q}_t}{1.10(t_{sa} - t_{ra})} = \frac{20{,}074}{1.10(90 - 70)}$$

$$\text{cfm}_{sa} = 912.5 \quad \text{(say 915 cfm)}$$

For classroom 2: $\dot{q}_t$ = 11,700 Btu/hr.

$$\text{cfm}_{sa} = \frac{11{,}700}{1.10(90 - 70)}$$

$$\text{cfm}_{sa} = 531.8 \quad \text{(say 530 cfm)}$$

For corridor 1: $\dot{q}_t$ = 16,387 Btu/hr.

$$\text{cfm}_{sa} = \frac{16{,}387}{1.10(90 - 70)}$$

$$\text{cfm}_{sa} = 744.8 \quad \text{(say 745 cfm)}$$

For corridor 2: $\dot{q}_t$ = 19,686 Btu/hr.

$$\text{cfm}_{sa} = \frac{19{,}686}{1.10(90 - 70)}$$

$$\text{cfm}_{sa} = 894.8 \quad \text{(say 895 cfm)}$$

For main entrance: $\dot{q}_t$ = 17,300 Btu/hr.

$$\text{cfm}_{sa} = \frac{17{,}300}{1.10(90 - 70)}$$

$$\text{cfm}_{sa} = 786.4 \quad \text{(say 786 cfm)}$$

For cafeteria: $\dot{q}_t$ = 53,620 Btu/hr.

$$\text{cfm}_{sa} = \frac{53{,}620}{1.10(90 - 70)}$$

$$\text{cfm}_{sa} = 2437 \quad \text{(say 2440 cfm)}$$

For kitchen: $\dot{q}_t$ = 20,020 Btu/hr.

$$\text{cfm}_{sa} = \frac{20.020}{1.10(90 - 70)}$$

$$\text{cfm}_{sa} = 910 \text{ cfm}$$

For boys' and girls' rooms combined: q_t = 900 + 950 = 1850 Btu/hr.

$$cfm_{sa} = \frac{1850}{1.10(90 - 70)}$$

$$cfm_{sa} = 84 \text{ cfm}$$

NOTE: All of these supply air quantities are relatively low, especially in the occupied classrooms. These values could be increased by supplying the air at a lower temperature, thus providing improved air circulation rates.

8-15 SEASONAL HEAT AND FUEL REQUIREMENTS

It is often necessary to estimate the energy requirements and fuel consumptions of HVAC systems for either short or long terms of operation. These quantities can be much more difficult to calculate than design heat loss and gain or required system capacity since they involve the integration over the period in question of the influence of many factors which may vary greatly with time. It is difficult to accurately foresee the way in which all the factors involved will vary throughout the time period. In addition to this, the calculations required to take all such variations into account become very involved. For these and other reasons, records of past operating experience of the building in question, when these are available, provide the most reliable and usually the most accurate basis for the prediction of future requirements.

Although the procedures used for estimating energy requirements vary considerably in degree of sophistication, they all have three common elements; that is, the calculation of (1) space load (heat loss of gain), (2) equipment load, and (3) equipment energy requirements.

The space load calculations determine the amount of energy which must be added to a space to maintain thermal comfort. The space load includes the effects of outdoor conditions and is often based upon no more than the calculated peak design heat load. More sophisticated methods include consideration of solar effects, internal heat gains, heat storage in building components, and the effects of wind.

The second step translates the space load into a load on the equipment. This can be a simple estimate of duct or piping losses, or a complex hour-by-hour analysis. The third step calculates the fuel and energy required to meet these loads. It gives consideration to efficiencies and part-load characteristics of the equipment.

The sophistication of the calculation procedure used can often be inferred from the number of separate ambient conditions and/or time increments used in the calculations. Thus, a simple procedure may use only one measure, such as annual degree-days, and would be appropriate only for simple systems of the residential and small commercial type. A detailed computer program may consider ambient conditions for each hour of the year and thus provide more realistic estimates for more complex building structures. Our discussion here will be limited to the simple procedure using the degree-day method, which is described in ASHRAE Fundamentals 1977, and is called the "single-measure method."

Records of past energy requirements and/or fuel consumption for a particular residence or small commercial building provides the best basis for estimating future energy use. However, in the absence of past records for a particular building, the data from similar buildings in the same locality may be used with caution. Living habits vary greatly, and apparently identical buildings may have very different energy-use patterns. Averages taken from many types of plants in many types of residences in various localities may prove to be inaccurate when applied to a particular residence. Thus, it is often necessary to estimate energy consumption from computed heat loads.

Any method of estimating energy use based primarily on calculated heat losses must, of necessity, also take into account consideration of the efficiency with which the energy is used. Efficiency can be defined in a variety of ways for different purposes, and the values in any given case can vary widely depending on the conditions. For estimating purposes the efficiency of utilization of the energy over the calculation period has been used. This is distinct, for example, from heating unit efficiency, which expresses the heat delivered by a furnace or boiler as a percentage of the heat available in the fuel, and which is the efficiency normally used in describing the performance of a unit under rated, or stated, load conditions.

The efficiency of utilization can vary quite widely even in similar buildings in the same area. A fixed value could be misleading because of the many factors affecting it. The efficiency of a system will depend not only on the operation of the equipment and control arrangement but also on the location of the heating unit and chimney and the net effect of infiltration. Efficiency may also vary with the time of year as the average load on the system varies. At times, people, lighting, appliance use, and solar gains all contribute substantial amounts of heat which results in frequent operation at part-load, and consequently, at lower than design efficiency.

In preparing energy estimates it should be realized that a more reliable estimate will result if operation over a long period is studied rather than a short period. Nearly all the methods in use give reasonable results over a full annual heating season, but estimates for shorter periods, such as a month, may produce inaccurate results. As the period of the estimate is shortened, there is more chance that some factor not directly taken into account in the estimating method will deviate from its long-term average value and thus lead to an error in the predicted energy requirement.

8-16 THE DEGREE-DAY METHOD

The traditional degree-day procedure for estimating heating energy requirements is based on the assumption that, on a long-term average, solar and internal heat gains will offset heat loss when the mean daily outdoor air temperature is 65°F, and that the fuel consumption will be proportional to the difference between the mean daily temperature and 65°F. In other words, on a day when the mean temperature is 20°F below 65°F, twice as much fuel is consumed as on days when the mean temperature is 10°F degrees below 65°F.

The number of degree-days (DD) per day is the difference between 65°F and the daily mean temperature when the latter is less than 65 degrees. Thus

$$DD = (65 - t_a)(1 \text{ day}) \qquad (8\text{-}6)$$

where t_a is the daily mean temperature. The number of degree-days for any longer period is the sum of all such products for as many days as the period covers.

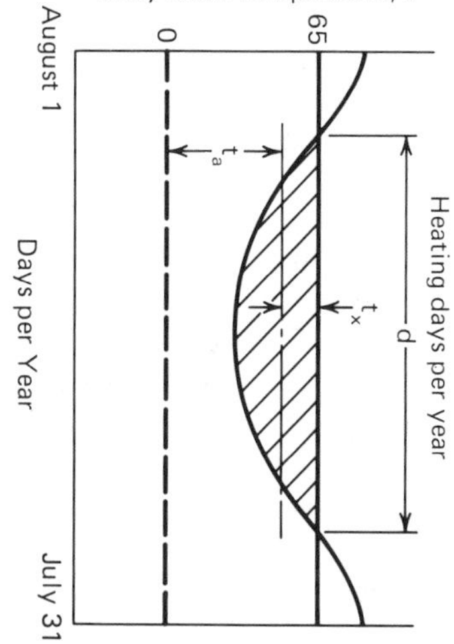

Figure 8–2

Degree-Day diagram

If the daily mean temperature for the days of a year, over a period of several years, were plotted against the number of days in a year, the result would be a diagram similar to Fig. 8-2. The diagram would be different for every section of the United States. The shaded area represents the average annual degree-days per year. The points where the curve crosses the 65°F degree line indicate the dates for the beginning and the end of the heating season. If we let t_x be the average height of the shaded area and d be the number of days in the heating season, we may say

$$\text{DD} = d \times t_x$$

or

$$t_x = \frac{\text{DD}}{d}$$

But, from the graph $t_x = 65 - t_a$, therefore

$$65 - t_a = \frac{\text{DD}}{d}$$

or

$$t_a = 65 - \frac{\text{DD}}{d} \tag{8-7}$$

where t_a is the average outside air temperature during the heating period of d number of days over which time DD degree-days accumulated.

Table C-2 (Appendix C) shows average degree-day per month and yearly totals. Figure C-2 indicates the date a normal heating season starts; Fig. C-3 indicates the number of days in a normal heating season; and Fig. C-4 indicates graphically the number of degree-days per year.

If the heat loss calculations for a building were accurate for the design conditions, and if the heat loss at any other outside temperature were proportional to that at design conditions, then the heat loss from the building for a whole heating season could be expressed by the equation

$$H = \frac{24\dot{q}d(t_x - t_a)}{t_i - t_o} \tag{8-8}$$

where

H = seasonal heat loss (Btu)

$\dot{q}$ = hourly heat loss from building for the design conditions (Btu/hr)

d = number of days in heating season

t_i = inside design temperature (°F)

t_o = outside design temperature (°F)

24 = hours per day

t_a = average outside air temperature during period (°F) calculated by Eq. (8-7)

t_x = weighted average inside air temperature (°F).

The weighted average inside air temperature may be used if the temperature control system is designed to change the inside temperature during certain periods of the day, such as using night set-back. It is calculated by

$$t_x = \frac{t_1 \times \text{hours at } t_1 + t_2 \times \text{hours at } t_2}{24} \tag{8-9}$$

The annual fuel consumption is estimated by

$$\text{Annual fuel consumption } F = \frac{H}{\text{Heat per sale unit} \times \text{Efficiency of Combustion}} \tag{8-10}$$

The calculated heat loss method represented

by Eq. (8-8) is theoretical, and constant temperatures are assumed for very definite hours each day throughout the entire heating season. It does not take into account factors which are difficult, if not impossible, to evaluate, such as

a. temporary opening of windows
b. abnormal heating above or below the inside design temperature
c. poor heating system design layout
d. heat gains from solar effects and interior sources
e. it is based on an average heating season
f. on-off operation of heating unit, and partial-load operation
g. does not consider wind chill factors

The seasonal heat loss, Eq. (8-8), and seasonal fuel consumption, Eq. (8-10), usually give rather high values and would be termed very conservative.

Until methods now under development at the National Bureau of Standards and other institutions are completed, a *modified degree-day* procedure is recommended by ASHRAE presented in ASHRAE Handbook of Fundamentals 1981.

Annual energy (fuel) consumption

$$(F) = \frac{24 \times \text{DD} \times \dot{q}}{\Delta t \times k \times \text{HHV}}(C_D) \tag{8-11}$$

where

F = fuel or energy consumption for the estimate period

$\dot{q}$ = design heat loss, including infiltration and ventilation (Btu/hr or W)

DD = number of degree-days for the estimate period

Δt = design temperature difference ($t_i - t_o$) (°F or °C)

k = a correction factor which includes the effects of rated full-load efficiency, part-load performance, oversizing, and energy conservation devices

HHV = heating value of fuel, units consistent with $\dot{q}$ and F.

C_D = emperical correction factor for heating effect versus degree-days

Although empirical, this equation has some basis in physical fact. In an approximate sense $\dot{q}/\Delta t$ may be thought of as an overall inside-to-outside thermal conductance (UA) for the building. In a similar sense the DD can be thought of as the inside-to-outside temperature difference averaged over the day so that the product yields the daily heating load. Division by the fuel heating value yields the ideal amount of fuel required for the day. The constant is for unit conversion (24 hr/day). The k and C_D are empirical, and attempt to account for various effects noted in their definitions.

Values of C_D may be taken from the following table, which has been adapted by permission from ASHRAE *Handbook, Fundamentals,* 1981.

Degree Days	C_D
1000	0.81
2000	0.75
3000	0.70
4000	0.65
5000	0.60
6000	0.61
7000	0.62
8000	0.64
9000	0.66

Correction factor k is empirical and should not be confused with any ratings for "seasonal efficiency." For electric resistance heating $k = 1.0$. For gas heating $k = 0.55$, for fuel oil $k = 0.80$, but values may vary between 0.55 and 0.80 for either.

Illustrative Problem 8-4

A large residence in the Portland, Maine, area has a calculated heat loss of 120,000 Btu/hr based on a design temperature $t_i = 70°F$ and $t_o = -5°F$. The oil-fired hot water boiler installed has a rated net output of 150,000 Btu/hr when fired with No. 2 fuel oil at the rate of 1.35 gallons per hour. If the heating system is to maintain the inside air temperature at the design value (70°F), estimate the fuel oil consumption (gal/season) using (a) Eqs. (8-8) and (8-10) and (b) Eq. (8-11). Assume No. 2 fuel oil has a heating value (HHV) of 140,000 Btu/gal.

Solution:

Step 1: Determine climate data.
From Table C-2: Annual average degree-days = 7681.
From Fig. C-3: Number of days in a normal heating season = 300.
From Table C-1: Design outside air temperature = −5°F.

Step 2: Determine t_a, t_x, k, C_D, and combustion efficiency.

By Eq. (8-7) $$t_a = 65 - \frac{\text{DD}}{d} = 65 - \frac{7681}{300} = 39.4°\text{F}$$

By Eq. (8-9) $t_x = 70°F$ because temperature is held constant 24 hr/day.

$$\text{Combustion Efficiency} = \frac{150{,}000}{1.35(140{,}000)} = 0.794$$

Assume $k = 0.60$

Assume $C_D = 0.64$

Step 3: Use Eqs. (8-8), (8-10), and (8-11).

By Eq. (8-8)

$$H = \frac{24\dot{q}d(t_x - t_a)}{t_i - t_o} = \frac{24(120{,}000)(300)(70 - 39.4)}{70 - (-5)}$$

$$H = 352{,}512{,}000 \text{ Btu/season}$$

By Eq. (8-10)

$$F = \frac{352{,}512{,}000}{140{,}000(0.794)} = 3171 \text{ gal/season}$$

By Eq. (8-11)

$$F = \frac{(24)(\text{DD})(\dot{q})}{(\Delta t)(k)(\text{HHV})}(C_D) = \frac{(24)(7681)(120{,}000)}{[70 - (-5)](0.60)(140{,}000)}(0.64)$$

$$F = 2247 \text{ gal/season}$$

As previously noted, Eqs. (8-8) and (8-10) give rather conservative estimates.

Illustrative Problem 8-5

For the residence of Problem 8-4, estimate the fuel consumption for the month of January.

Solution:

Step 1: Determine climate data.

From Table C-2: Average degree-days in January = 1373.
Number of days = 31.
Assume $k = 0.60$
Assume $C_D = 0.79$

Step 2: Determine fuel use.

By Eq. (8-7)

$$t_a = 65 - \frac{\text{DD}}{d} = 65 - \frac{1373}{31} = 20.7°\text{F}$$

By Eq. (8-8)

$$H = \frac{24(120{,}000)(31)(70 - 20.7)}{70 - (-5)}$$

$$= 58{,}686{,}720 \text{ Btu}$$

By Eq. (8-10)

$$\text{F} = \frac{58{,}686{,}720}{140{,}000(0.794)} = 528 \text{ gal}$$

By Eq. (8-11)

$$F = \frac{24(1373)(120{,}000)}{75(0.60)(140{,}000)}(0.79) = 496 \text{ gal}$$

8-17 COMPARISON OF ENERGY COST FOR VARIOUS TYPES OF FUELS

It is sometimes required in cost analyses to compare the relative unit cost of using alternate fuels such as gas, oil, coal, and electricity. The difficulty of such a comparison originates with the fact that fuels are sold on a different basis; such as cents/1000 ft^3 of gas; cents per gallon of oil; dollars per ton of coal, and cents per kW hr for electricity. Since a person should be interested in the unit of heat being purchased, the Btu, all costs should be reduced to utilize heat units as a basis of comparison. A convenient common basis is the *therm* (100,000 Btu), or in some cases one million Btu. The cost of utilized heat in cents/therm may be computed for any fuel by

$$\text{cents/therm} = \frac{\text{Price per sale unit (cents)} \times 100{,}000}{\text{Btu per sale unit} \times \text{utilization efficiency}} \quad \textit{(8-12)}$$

CHAPTER 8

Review

8-1. A room in a residence has two exposed walls and measures 20 ft long, 15 ft wide, and has a ceiling height of 8 ft 6 in. The 20-ft side contains three windows, and the 15-ft side contains two windows. The windows are double-hung, wood sash, single glazed, weather-stripped, and each measures 3 by 5 ft. The long side of the room faces the prevailing wind direction, which is assumed to be 15 mph. The overall heat transmission coefficient *(U)* for the walls is 0.28 Btu/hr ft^2 °F. Determine the heat loss from the room through the walls and windows, if the inside air is at 70°F and the outside air is at 0°F.

8-2. A large room in an old building having one exposed wall is fitted with 10 poorly fitted, double-hung, wood sash windows, 3 ft 6 in. wide by 6 ft high. Outside design conditions are −10°F and 15 mph wind. Inside temperature is maintained at 70°F. Estimate the reduction in heat loss from infiltration if the windows were weather-stripped. Assume one-half of the crack length contributes to infiltration.

8-3. Outside air at 30°F and 60 percent relative humidity enters a small building by infiltration at the rate of 2000 ft^3/hr. (a) How much moisture must be evaporated per hour to maintain an inside relative humidity of 40 percent at 70°F? (b) How much heat is required for humidification?

8-4. A work shop area in an industrial building has one exposed wall containing eight vertically pivoted, steel-sash windows each measuring 28 in. wide by 74 in. high. The wall is on the windward side of the building and also contains two 3 ft 6 in. by 7 ft 6 in. well-fitted doors. The shop area has interior well-fitted doors leading to the rest of the building. For a wind velocity of 10 mph, outside air temperature 10°F and inside air temperature 65°F, estimate the infiltration rate and heat loss when (a) interior doors are open and (b) interior doors are closed.

8-5. A small restaurant has a seating capacity of 50 people. The main entrance has two swinging double doors into a vestibule. There are two other doors in the restaurant, however, they are for emergency use only. Estimate the amount of air which infiltrates due to door openings.

8-6. A certain building in Boston has a calculated heat loss of 150,000 Btu/hr based on 70°F inside air temperature. Estimate the probable fuel oil consumption for an average heating system.

8-7. A large office building in Philadelphia has a calculated design heat load of 1,000,000 Btu/hr for a design indoor temperature of 70°F. The building is steam heated (1000 Btu/lb of steam).

(a) Estimate the steam consumption for a heating season.
(b) What is the average outside air temperature during the heating season (65°F base)?
(c) If the inside air temperature is reduced to 65°F from 6:00 PM to 7:00 AM, what is the weighted average inside air temperature?
(d) Estimate the probable savings in steam consumption that could be saved by reducing the inside air temperature during part of the 24-hr period (night set-back).

BIBLIOGRAPHY

1. ASHRAE *Handbook of Fundamentals,* American Society of Heating, Refrigerating, and Air-Conditioning Engineers, New York, 1977.

2. ASHRAE *Handbook of Fundamentals,* American Society of Heating, Refrigerating and Air-Conditioning Engineers, New York, 1981.

3. *Handbook of Air Conditioning, Heating and Ventilating,* The Industrial Press, New York, 1965.

Chapter 9

The Cooling Load

9-1 INTRODUCTION

The calculation for the heating load presented in Chapter 8 was made assuming steady-state heat transfer, which is the usual procedure, and this gives a reasonably accurate estimate of the heat necessary to maintain design indoor conditions when outside air conditions are stable at outdoor design conditions. The transient nature of the heat loss due to the variables of outdoor air conditions were neglected. The results obtained are normally quite adequate.

The cooling load, however, presents a much more complex problem because of the larger number of variables involved, and also, the transient nature of the heat gain must be considered. The instantaneous heat gain into a given space is variable with time of day, day of the month, and season of the year because of the transient nature of the solar radiation. There may be an appreciable difference between the instantaneous heat gain to a space and the actual instantaneous heat removal rate. The reason for this is the *heat storage effect* of a structure.

It will be necessary for us to distinguish between heat gain, cooling load, and heat extraction rate. The cyclic changes in cooling load components are often not in phase with each other; therefore, analysis is required to establish the resultant maximum cooling load for a building or particular zone within a building. A zoned system (a system serving several different areas within a structure each having its own temperature control) must handle cooling loads for each zone and for equipment serving more than one zone. At certain times of day during a heating season, some zones may require heating and others cooling.

In air-conditioning design it is important to differentiate between four related, but distinct, heat flow rates, each of which varies with time: (1) space heat gain; (2) space cooling load; (3) space heat extraction rate; and (4) cooling coil load. *ASHRAE 1981 Fundamentals* describes these heat flow rates as follows:

Space heat gain. Space heat gain (instantaneous rate of heat gain) is the rate at which heat enters into and/or is generated within a space at a given instant of time. Heat gain is classified by (1) the mode in which it enters the space and (2) whether it is sensible or latent heat gain.

The first classification is necessary because different fundamental principles and equations are used to calculate different modes of energy transfer. The heat gain occurs in the form of (1) solar radiation through transparent surfaces; (2) heat conduction through exterior walls and roofs; (3) heat conduction through interior partitions, ceilings, and floors; (4) heat generated within the space by occupants, lights, and appliances; (5) energy transfer as a result of ventilation and infiltration of outdoor air; and (6) miscellaneous heat gains.

The second classification, sensible or latent, is important for proper selection of cooling equipment. The heat gain is *sensible* when there is a direct addition of heat to a space by any or all mechanisms of conduction, convection, and radiation. The heat gain is *latent* when moisture is added to the space. To maintain a constant humidity ratio in a space, water vapor must be condensed out in the cooling apparatus at a rate equal to its rate of addition to the space. The amount of energy required to do this, the latent heat gain, essentially equals the product of the rate of condensation and the latent heat of condensation. The distinction between sensible and latent heat gain is necessary, because any cooling apparatus has a maximum sensible heat removal capacity and a maximum latent heat removal capacity for particular operating conditions.

Space cooling load. The space cooling load is the rate at which heat must be removed to maintain space temperature at a constant value. Note that the summation of all space instantaneous heat gains at a given time does not necessarily equal the cooling load for the space at the same time. The space heat gain by radiation is partially absorbed by the surfaces and contents of the space and does not affect the room air until some time later. The radiant energy must first be absorbed by surfaces that enclose the space (walls, floor, and ceiling), and the material in the space (furniture, etc.). As soon as these surfaces and objects become warmer than the space air, some of their heat will be transferred to the air in the room by convection. Since their heat storage capacity determines the rate at which their surface temperatures increase for a given radiant input, it governs the relationship between the radiant portion of heat gain and the corresponding part of the space cooling load. The thermal storage effect can be important in determining an economical equipment-cooling capacity.

Space heat extraction rate. The space heat extraction rate is the rate at which heat is removed from a conditioned space. It equals the space cooling load only when room air temperature is kept constant, which rarely occurs. Usually, the automatic control system, in conjunction with intermittent operation of the cooling equipment, will cause a ''swing'' in room temperature. Indeed, for a control system to function, there must be a change (swing) in temperature. Therefore, providing the control system is designed properly, computation of the space heat extraction rate results in a more realistic value of energy removal at the cooling equipment than merely using the values of the space cooling load.

Cooling coil load. Cooling coil load is the rate at which energy is removed at the cooling coil which serves one or more conditioned spaces in any central air-conditioning system. It is equal to the instantaneous sum of the space cooling load (or space heat extraction rate if the space temperature is assumed to ''swing'') for all the spaces served by the system, plus any additional load imposed on the system external to the conditioned spaces. Such additional load components include heat gain into the distribution system between the individual spaces and the cooling equipment (usually duct heat gain) and outdoor hot and moist air introduced into the system for ventilation.

Figures 9-1 and 9-2 show some of the relationships between instantaneous heat gain and cooling loads. The *delay* of peak heat gain is very evident, especially in heavy construction. The highly varying and relatively sharp peak of the instantaneous solar heat gain results in a large portion of it being stored at the time of peak solar heat gain, shown in Fig. 9-2a. The upper curve in Fig. 9-2a is typical of the *solar heat gain* for a west exposure, and

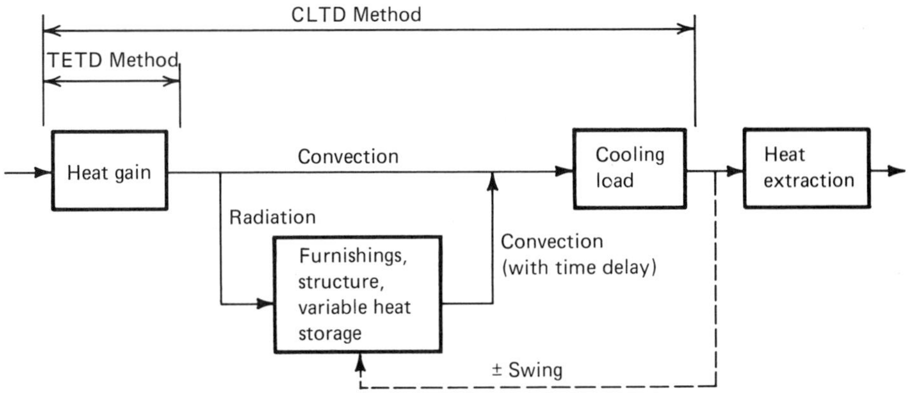

Figure 9–1

Origin of Difference between Magnitudes of Instantaneous Heat Gain and Instantaneous Cooling Load

the lower curve is the actual cooling load that results in an average construction application with the space temperature held constant. The reduction in the peak heat gain is approximately 40 percent and the peak cooling load lags the peak heat gain by approximately 1 hr. The cross-hatched areas (Fig. 9-2a) represent the heat stored and the stored heat removed from the construction. Since all the heat entering the space must be removed, these two areas are equal.

Figure 9-2b illustrates the relationship between the instantaneous heat gain and the actual cooling load in light, medium, and heavy construction. With light construction less heat is stored at the peak (less storage capacity available), and with heavy construction more heat is stored at the peak (more storage capacity available). This aspect affects the extent of zoning required in the design of a system for a given building; the lighter the building construction, the more attention should be given to zoning.

One more item that significantly affects the storage of heat is the *operating period* of the air-conditioning equipment. All the curves of Fig. 9-2 illustrate the actual cooling load for 24-hr operation. If the cooling equipment is shut down after a certain number of hours of operation (say 12 or 16 hr), some of the stored heat remains in the building construction. This heat must be removed (heat in must equal heat out) and will appear as a "pulldown load" when the equipment is turned on the next day (see Ref. 5).

The general procedure required to calculate the space cooling load is as follows:

1. Obtain characteristics of the building. Building materials, component size, external surface colors, and shape are usually determined from building plans and specifications.

2. Determine building location, orientation, and external shading. Plans and specifications should contain this information. Shading from adjacent buildings can be determined by site plans or visiting the building site.

3. Obtain appropriate weather data and select outdoor design conditions.

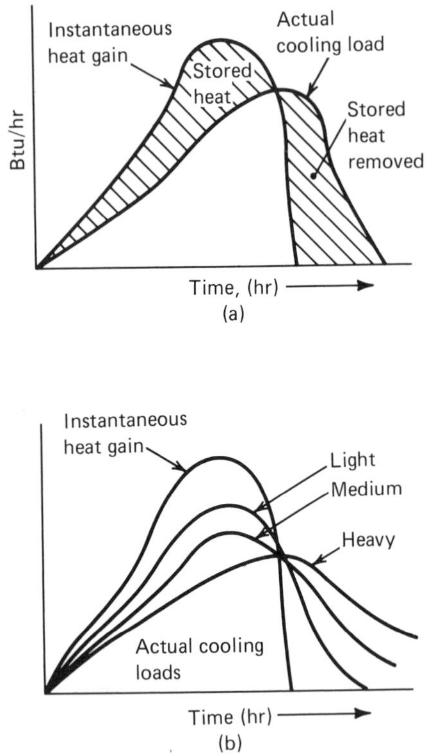

Figure 9–2

Actual cooling load compared with instantaneous heat gain. (a) West exposure, average construction; (b) light, medium, and heavy construction. (Constant space temperature)

4. Select indoor design conditions, such as indoor dry-bulb and wet-bulb temperatures, and ventilation rate. Include permissible variations and control limits.
5. Obtain a proposed schedule of lighting, occupants, internal equipment, appliances, and any processes that would contribute to the internal heat load.
6. Select the time of day and month of the year to do the cooling-load calculation. Frequently, several different times on a given day must be investigated. The particular day and month are often dictated by peak solar conditions.
7. Make instantaneous heat gain calculations of
   a. solar heat gain through walls, roofs, and fenestration (windows);
   b. conduction and convection through interior partitions, ceilings, and floors where a temperature difference exists across such structural components;
   c. heat sources within the conditioned space—people, lights, equipment, appliances;
   d. infiltration; and
   e. miscellaneous sensible and latent heat gains.
8. Calculate the space cooling load using the results of the instantaneous heat gain.

9-2 HEAT FLOW THROUGH BUILDING STRUCTURAL COMPONENTS BY CONDUCTION (Exclusive of Solar Effects)

In Chapter 7 we discussed the method for calculating the heat transferred through building structural components by conduction and convection. The relationship used for summer cooling would be

$$\dot{q} = UA \quad \text{(temperature difference)}$$

or

$$\dot{q} = UA(\Delta t) \tag{9-1}$$

where

$\dot{q}$ = rate of heat gain (Btu/hr or W)

U = overall heat transmission coefficient (Btu/hr ft^2 or W/m^{2}°C)

A = surface area normal to heat flow (ft^2 or m^2)

Δt = temperature difference air-to-air across structural component (°F or °C)

Note that the determination of the U factor for summer conditions normally uses an outside air wind velocity of 7½ mph. Also, direction of heat flow is important in determining the film coefficients (f_o and f_i). The Δt may be the overall temperature difference from outside air to inside air ($t_o - t_i$), or it may be the temperature difference across an interior floor, ceiling, or partition (see Table 9-1).

In Chapter 7 we also discussed a method for calculating the probable temperature at any point in a composite wall section. Let us apply the same concept to a wall such as is illustrated in Fig. 9-3 where the outside air temperature is at a constant 95°F (35°C) and the inside air temperature is a constant 80°F (26.7°C). The temperature profile would remain as shown as long as the outside air and inside air temperature were constant.

9-3 HEAT CONDUCTION THROUGH BUILDING STRUCTURAL COMPONENTS WITH VARYING OUTDOOR TEMPERATURES

Referring to Fig. 9-3, suppose the outdoor air temperature suddenly dropped to 85°F (29.4°C). For a short time heat would continue to flow from the inside face of the wall to the room air because all the temperatures in the wall would be caused by the higher outside air temperature. Now the temperature of the outer surface of the wall would temporarily be higher than the outdoor air at 85°F. So there would be heat flow from the outer face of the wall to the outdoor air, as well as from the inside face to the room air at 80°F. Under these conditions the heat of the wall would not be replenished. As a result all the temperatures at various points in the wall would fall as the wall surrenders its heat. This would continue

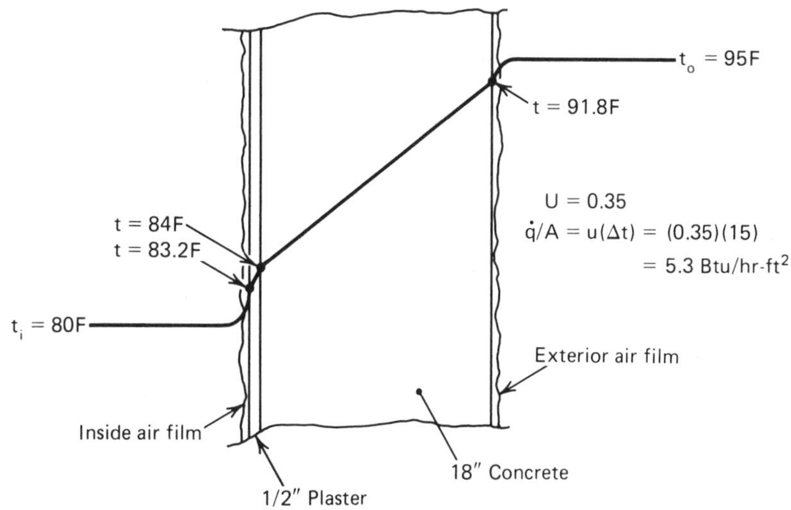

Figure 9–3

Temperature profile through a masonry wall

Table 9-1

Design Temperature Differences*

Item No.	Item	Temperature Differences [a] °F	°C
1	Walls, exterior	17	9.5
2	Glass in exterior walls	17	9.5
3	Glass in partitions (conditioned to unconditioned space)	10	5.6
4	Store show windows having a large lighting load	30	16.8
5	Partitions (conditioned to unconditioned space)	10	5.6
6	Partitions, or glass in partitions, adjacent to laundries, kitchens, or boiler rooms	25	14
7	Floors above unconditioned rooms	10	5.6
8	Floors on ground	0	0.0
9	Floors above basements	0	0.0
10	Floors above rooms or basements used as laundries, kitchens, or boiler rooms	35	19.6
11	Floors above vented spaces	17	9.5
12	Floors above unvented spaces	0	0.0
13	Ceilings with unconditioned rooms above	10	5.6
14	Ceilings with rooms above used as laundries, kitchens, etc.	20	11.2
15	Ceiling with roof directly above (no attic)	17	9.5
16	Ceiling with totally enclosed attic above	17	9.5
17	Ceiling with cross-ventilated attic above	17	9.5

[a]These temperature differences are based on the assumption that the air-conditioning system is being designed to maintain an inside temperature 17°F (9.5°C) lower than outdoor temperature. For air-conditioning systems designed to maintain a temperature differential greater than 17°F (9.5°C) between inside and outside, *add* the difference between the assumed design temperature difference and 17°F (9.5°C) to the values in the above table.

**Reprinted by permission from The Trane Company,* Trane Air Conditioning Manual, *1965.*

until these temperatures reached equilibrium for the outside air at 85°F and an inside air at 80°F. The point of steady temperatures within the wall would be reached only when the heat flowing out at the inner face of the wall was equal to the heat flowing in at the outer face. This may be a short or a long time depending on the heat capacity of the particular wall.

Now suppose that after several hours of 85°F the outdoor air temperature rose to 95°F. Then for a while the heat flow from the wall into the room would not increase. The inner surface temperature of the wall would still be very little above the room temperature due to the small heat flow when the outdoor temperature was 85°F. So the heat passing through the inside surface air film, from the inner face of the wall to the room air, would be no greater because of the outdoor air at 95°F.

On the outside, because of the higher outdoor temperature, the heat flow from the air through the outside surface film would be increased. The temperature of the outer surface of the wall would rise rapidly and the heat flow into the interior of the wall would increase. But the mass of the wall acts as a "trap" to soak up some of the heat pouring into the wall from the outside. As more heat flows into the wall than goes out, the temperature within the wall, as well as the inner surface temperature, will rise. This will continue until equilibrium heat flow is established at 95°F outside and 80°F inside air temperatures.

The inside surface temperature of the wall will not vary for some time after the change in outdoor air temperature. The time interval between the change in outdoor air temperature and the change in inside temperature is called *time lag*. Time lag is really the time required for the heat to travel through a wall from the outer to inner face. Massive walls with considerable insulation have a large time lag; substances such as glass or sheet metal have practically none. The more massive a wall the larger will be the quantity of heat needed to raise its temperature. So the inner surface of massive walls may never reach the maximum possible inner surface temperature if the high outdoor temperature does not last long.

Thin walls of lightweight construction require little heat to raise their temperature. So the maximum inner surface temperature is reached soon after the rise in outdoor temperature. Such walls store little heat and quickly lose it when the outdoor temperature drops. The time lag is short.

Heavy, thick walls have a large heat storage capacity. The time lag of such a wall is considerable. In hot climates buildings with massive walls are often cool inside. The heat of the day never penetrates to the inner surface. The wall may be so massive that it takes all day for the heat to go any distance into the wall and increase its temperature. By the time the temperature of the interior of the wall has been raised somewhat, night arrives. Then the relatively warm outer layers of the wall begin to lose heat to the cool night air. All night long heat flows from the hot outer layers of the wall to the cool night air. When morning arrives, the heat flow again reverses into the wall and the entire cycle is repeated. The wall acts as a buffer to the hot outdoor air temperatures, storing the heat in the daytime and delivering it back to the outdoors at night.

9-4 RADIANT HEAT

The discussion of heat transfer in Sections 9-2 and 9-3 involved some substance. The wall receives heat from the warm outside air, then surrenders it to the inside air. Another illustration in air-conditioning work would be the case of heat transferred from warm air to chilled water in an air-conditioning apparatus. After leaving the apparatus, the water flows back to a refrigerating machine where it surrenders this heat to the cold refrigerant. The heat has been conveyed from the warm air to the cold refrigerant by an actual substance—the circulating water.

Radiant energy or heat, on the other hand, needs no actual substance to carry it from one object to another. Radiant heat can and does travel through a vacuum. Radiant heat has many properties of light. It cannot pass through an opaque object. Radiant heat impinging on the face of an object causes a "heat shadow" to be formed behind the object. The area in the shadow of the object will remain cool, whereas the areas around the object exposed to the radiant heat will become warm. On the other hand, transparent materials allow both light and radiant heat to pass through them. Radiant heat can be reflected from a bright surface just as light is reflected by a mirror.

Suppose a furnace has a brisk fire in it. If one opens the door and looks at the fire, the full force of the radiant heat of the fire will be felt on the face. This sensation of heat cannot be due to a hot current of air because cool room air will be drawn into the furnace through the open door.

9-5 SOLAR RADIATION

The sun's rays pass through outer space and through the atmosphere to the earth. Any surface they strike (ground, roofs and walls of buildings) become warm. The heat reaching the surface of the earth varies considerably from hour to hour. Clouds, haziness of the atmosphere, and other factors cause large variations in the amount of solar heat reaching the surface of the earth.

Reflection. When light falls on a mirror, a white surface, or any other bright area, a large percentage of it is reflected. Similarly, if radiant heat strikes a light-colored surface, a large percentage of it will be reflected; only the balance will be absorbed by the surface. The following brief table shows the approximate solar radiation that is reflected by painted surfaces of various colors.

Color of Painted Surface	Percentage of Radiant Heat Reflected	Percentage of Radiant Heat Absorbed
Clean aluminum	72	28
Light Red	37	63
Black	6	94

The darker a surface, the more impinging solar radiation will be absorbed by the surface. Thus, dark surfaces always have higher temperatures than white surfaces exposed to the same sunlight.

Surface temperatures. Radiant energy striking any surface raises the surface temperature. A dark-colored roof may reach 150°F (65.6°C) during a summer day; yet the air temperature just above the roof may be only 90°F (32.2°C).

The surface temperature of a wall or roof depends on the clearness of the atmosphere and on the angle at which the sun's rays strike the surface. When a surface is perpendicular to these rays, the surface receives the full intensity of the sun. On the other hand, when the rays strike a surface at an angle, the intensity is much less. The earth rotates once each 24 hr, which causes day and night. The earth moves around the sun once a year, which causes the seasons. Because of these movements the angle at which the sun's rays strike a surface is always changing. This means the surface temperatures of a wall in the sun changes throughout the day.

In summary, the temperature of a wall or roof depends on (1) the angle of the sun's rays, (2) the type of construction, (3) the color and roughness of the surface, and (4) the reflectivity of the surface.

The direction that a vertical wall faces is important in finding the angle at which the rays of the sun strike it. The direction also determines the hours during which the wall will be exposed to the sun. A south wall at 35° N latitude will be in the sun at about eight in the morning. From then on the surface temperature of the outer face of the wall will steadily increase until noon. From noon on, the surface temperature will decrease until shortly after 4 PM when the wall will be in the shade. In the same latitude sunlight will not fall on a wall facing west until noon. The temperature of the outside surface of a west wall will reach a maximum at about 4 PM; when it will decline steadily. The temperature of a surface in the sun is usually above the outdoor air temperature. Thus, heat flows from the surface to the air through the outside surface film. Only a portion of the radiant heat striking a surface travels into the interior of the wall. Of the portion of the heat that starts the trip into the interior of the wall, only a part ever reaches the inside of the building. As we have mentioned previously, time is required for the heat to reach the inner face. Most of the radiant heat first striking a wall raises the temperature of only the outside portion of the wall, the outside surface temperature changes again because the sun changes position; the warm wall beings to surrender heat to the outdoor air. Nevertheless, a certain percentage of the heat will reach the inside surface of the wall and raise its temperature above room temperature, which adds heat to the interior.

9-6 DESIGN TEMPERATURE DIFFERENCES

The heat flow through walls and roofs depends upon the temperature difference across the wall or roof. The problem of selecting outdoor design conditions for calculating heat gain is similar to that for heat loss. Again, one should not select the worst conditions on record because a great excess of equipment capacity would result. The heat storage capacity and time lag previously discussed plays an important part in this instance.

Recommended outdoor design dry-bulb and wet-bulb temperatures and mean daily range are presented in Table C-1 (see Appendix C). The dry-bulb temperature represents values which have been equaled or exceeded 2.5 percent of the total hours during the months of June through September (a total of 2928 hours) in the northern hemisphere. The coincident wet-bulb temperature listed with each design dry-bulb temperature is the mean of all wet-bulb temperatures occurring at the specific dry-bulb temperature. The mean daily range is the difference between the average daily maximum and average daily minimum dry-bulb temperature in the warmest month. The 2.5 percent values are recommended for design purposes by ASHRAE Standard 90-75, Reference 4. One percent and 5 percent values of design outdoor temperatures are also quoted in Reference 1.

The indoor design dry-bulb and wet-bulb temperatures may be selected by applying the principles outlined in Chapter 6. However, it is common practice in the United States to use an inside design dry-bulb temperature of 78°F when the outside design dry bulb is 95°F. This gives a 17-degree difference between outdoor and indoor temperatures. In localities where the outdoor design dry bulb is as low as 85°F, an inside dry-bulb temperature of 72°F may be selected. This gives an agreeable contrast between the outdoor and indoor air. In most cases a temperature difference of about 17 degrees is used for design calculations.

The design temperature difference for a locality cannot be settled to everyone's satisfaction. Refinements in design temperature difference are not essential for several reasons: (1) seasonal weather cycles vary; (2) lack of agreement on what indoor temperature should match a given outdoor temperature; and (3) the highest 10 percent of the outdoor temperature are discarded. A temperature difference of 17 degrees agrees with most people. This will be satisfactory for all sections of the country except those having an outdoor temperature of 100°F or higher many hours each summer. Then a design temperature difference of 20 to 22 degrees is desirable. The indoor temperature will probably be maintained at 78°F.

There are many special conditions that must be considered for conduction heat gain, other than that for conduction through exterior structural components. For instance, in restaurants conditioned dining rooms are sometimes separated from the hot kitchen by a thin partition. Such kitchens may be as hot as 100°F (37.8°C) while the dining room is maintained at 78°F (25.6°C). So using a temperature difference of 17 degrees (95°F − 78°F) for such a partition would give too low a result. Therefore, various temperature differences should be used for different conditions encountered in actual work.

Table 9-1 lists the temperature differences generally found for various conditions. It covers only common situations found in figuring conduction heat gains to air-conditioned spaces, in the absence of solar heat gains.

9-7 SOLAR HEAT GAIN CALCULATIONS

The 1972 *ASHRAE Handbook of Fundamentals* described two methods for calculating space cooling load. In the first procedure, the total equivalent temperature differential (TETD) method, various components of space heat gain were added together to get an instantaneous total rate of space heat gain, which was then converted to an instantaneous space cooling load through the use of *weighting factors*. The second procedure, introduced for the first time in the 1972 volume, was the transfer function method. Although similar in principle to the first method, it employed entirely different weighting factors (called *coefficients of room transfer functions*) in converting heat gain to cooling load.

A brief clarification of the diverse terminology used in this field is in order. *Transfer function,* a term used because it relates to underlying mathematical principles, is nothing more than a set of coefficients that relate an output function at a given time to the value of one or more driving functions at the time and previous times. There is basically no difference between the *transfer function,* the *thermal response factor* used by some authors (for calculating wall and roof heat conduction), and the *weighting factors* (for obtaining cooling-load components from heat gain components) described by the ASHRAE Task Group (TG) on Energy Requirements.

To eliminate any discrepancy between TETD and transfer function methods, ASHRAE sponsored research RP-138. The final results of this project are extensively used in the preparation of Chapter 25 of *ASHRAE Handbook and Product Directory, 1977 Fundamentals*.

For RP-138, investigators used the methodology and basic equations of the transfer function method to generate *cooling-load factors* (CLF) for each component of the space cooling load.

We shall use the calculation procedure here where values of the space cooling-load components are calculated directly, through the use of CLFs which include the effect of time lag due to thermal storage. The CLTD and CLF tables which follow are still based on the original transfer functions appearing in the ASHRAE fundamentals volume, 1972, and are generally more conservative than those computed with the later room transfer functions.

9-8 SPACE COOLING LOAD DUE TO HEAT GAIN THROUGH ROOFS AND WALLS

The technique for calculating the space cooling-load component as a result of heat gain through exterior roofs and walls involves the concept of sol-air temperature. *Sol-air temperature* is that temperature of the outdoor air which, in the absence of all radiation exchanges, would give the same rate of heat entry into the surface as would exist with the actual combination of incident solar radiation, radiant energy exchange with the sky and other outdoor surroundings, and convective heat exchange with the outdoor air. Details for the calculation of sol-air temperatures may be found in Ref. 1.

The cooling-load temperature difference (CLTD) method makes use of temperature differences to determine the cooling load for walls and roofs. The CLTDs vary with time and are a function of environmental variations and construction characteristics. The CLTD values in Tables 9-2 and 9-3 are for fixed conditions of surface characteristics and environment. The cooling load is then calculated from

$$\dot{q} = UA(\text{CLTD}) \tag{9-2}$$

where

Table 9-2

Cooling-Load Temperature Differences (CLTD) for Calculating Cooling Load for Flat Roofs (°F).*[a,b]

Roof No	Description of Construction	Weight lb/ft²	U-value Btu/(h·ft²·°F)	1	2	3	4	5	6	7	8	9	10	11	12	13	14	15	16	17	18	19	20	21	22	23	24
				Solar Time, hr																							
				Without Suspended Ceiling																							
1	Steel sheet with 1-in. (or 2-in.) insulation	7 (8)	0.213 (0.124)	1	−2	−3	−3	−5	−3	6	19	34	49	61	71	78	79	77	70	59	45	30	18	12	8	5	3
2	1-in. wood with 1-in. insulation	8	0.170	6	3	0	−1	−3	−3	−2	4	14	27	39	52	62	70	74	74	70	62	51	38	28	20	14	9
3	4-in. l.w concrete	18	0.213	9	5	2	0	−2	−3	−3	1	9	20	32	44	55	64	70	73	71	66	57	45	34	25	18	13
4	2-in. h.w. concrete with 1-in. (or 2-in.) insulation	29	0.206 (0.122)	12	8	5	3	0	−1	−1	3	11	20	30	41	51	59	65	66	66	62	54	45	36	29	22	17
5	1-in. wood with 2-in. insulation	9	0.109	3	0	−3	−4	−5	−7	−6	−3	5	16	27	39	49	57	63	64	62	57	48	37	26	18	11	7
6	6-in. l.w. concrete	24	0.158	22	17	13	9	6	3	1	1	3	7	15	23	33	43	51	58	62	64	62	57	50	42	35	28
7	2.5-in. wood with 1-insu-ulation	13	0.130	29	24	20	16	13	10	7	6	6	9	13	20	27	34	42	48	53	55	56	54	49	44	39	34
8	8-in. l.w. concrete	31	0.126	35	30	26	22	18	14	11	9	7	7	9	13	19	25	33	39	46	50	53	54	53	49	45	40
9	4-in. h.w. concrete with 1-in. (or 2-in.) insulation	52 (52)	0.200 (0.120)	25	22	18	15	12	9	8	8	10	14	20	26	33	40	46	50	53	53	52	48	43	38	34	30
10	2.5-in. wood with 2-in. insulation	13	0.093	30	26	23	19	16	13	10	9	8	9	13	17	23	29	36	41	46	49	51	50	47	43	39	35
11	Roof terrace system	75	0.106	34	31	28	25	22	19	16	14	13	13	15	18	22	26	31	36	40	44	45	46	45	43	40	37
12	6-in. h.w. concrete with 1-in. (or 2-in.) insulation	(75) 75	0.192 (0.117)	31	28	25	22	20	17	15	14	14	16	18	22	26	31	36	40	43	45	45	44	42	40	37	34
13	4-in. wood with 1-in. (or 2-in) insulation	17 (18)	0.106 (0.078)	38	36	33	30	28	25	22	20	18	17	16	17	18	21	24	28	32	36	39	41	43	43	42	40
				With Suspended Ceiling																							
1	Steel Sheet with 1-in. (or 2-in.) insulation	9 (10)	0.134 (0.092)	2	0	−2	−3	−4	−4	−1	9	23	37	50	62	71	77	78	74	67	56	42	28	18	12	8	5
2	1-in. wood with 1-in. insulation	10	0.115	20	15	11	8	5	3	2	3	7	13	21	30	40	48	55	60	62	61	58	51	44	37	30	25
3	4-in. l.w. concrete	20	0.134	19	14	10	7	4	2	0	0	4	10	19	29	39	48	56	62	65	64	61	54	46	38	30	24
4	2-in. h.w. concrete with 1-in. insulation	30	0.131	28	25	23	20	17	15	13	13	14	16	20	25	30	35	39	43	46	47	46	44	41	38	35	32
5	1-in. wood with 2-in. insulation	10	0.083	25	20	16	13	10	7	5	5	7	12	18	25	33	41	48	53	57	57	56	52	46	40	34	29
6	6-in. l.w. concrete	26	0.109	32	28	23	19	16	13	10	8	7	8	11	16	22	29	36	42	48	52	54	54	51	47	42	37
7	2.5-in. wood with 1-in. insulation	15	0.096	34	31	29	26	23	21	18	16	15	15	16	18	21	25	30	34	38	41	43	44	44	42	40	37
8	8-in. l.w. concrete	33	0.093	39	36	33	29	26	23	20	18	15	14	14	15	17	20	25	29	34	38	42	45	46	45	44	42
9	4-in. h.w. concrete with 1-in. (or 2-in.) insulation	53 (54)	0.128 (0.090)	30	29	27	26	24	22	21	20	20	21	22	24	27	29	32	34	36	38	38	38	37	36	34	33
10	2.5-in. wood with 2-in. insulation	15	0.072	35	33	30	28	26	24	22	20	18	18	18	20	22	25	28	32	35	38	40	41	41	40	39	37
11	Roof terrace system	77	0.082	30	29	28	27	26	25	24	23	22	22	22	23	23	25	26	28	29	31	32	33	33	33	33	32
12	6-in. h.w. concrete with 1-in. (or 2-in) insulation	77 (77)	0.125 (0.088)	29	28	27	26	25	24	23	22	21	21	22	23	25	26	28	30	32	33	34	34	34	33	32	31
13	4-in. wood with 1-in (or 2-in.) insulation	19 (20)	0.082 (0.064)	35	34	33	32	31	29	27	26	24	23	22	21	22	22	24	25	27	30	32	34	35	36	37	36

[a]See text material for corrections applied to table values.
[b]CLTF may be converted to centigrade by multiplying by 5/9.

Reprinted from ASHRAE Handbook of Fundamentals, *1981. With permission of the American Society of Heating, Refrigerating and Air Conditioning Engineers, Atlanta, GA.*

Table 9-3

Cooling-Load Temperature Differentials (CLTD) for Calculating Cooling Load from Sunlit Walls (°F)*,a,b

North Latitude Wall Facing	Solar Time, hr 1	2	3	4	5	6	7	8	9	10	11	12	13	14	15	16	17	18	19	20	21	22	23	24
Group A Walls																								
N	14	14	14	13	13	13	12	12	11	11	10	10	10	10	10	10	11	11	12	12	13	13	14	14
NE	19	19	19	18	17	17	16	15	15	15	15	15	16	16	17	18	18	18	19	19	20	20	20	20
E	24	24	23	23	22	21	20	19	19	18	19	19	20	21	22	23	24	24	25	25	25	25	25	25
SE	24	23	23	22	21	20	20	19	18	18	18	18	18	19	20	21	22	23	23	24	24	24	24	24
S	20	20	19	19	18	18	17	16	16	15	14	14	14	14	14	15	16	17	18	19	19	20	20	20
SW	25	25	25	24	24	23	22	21	20	19	19	18	17	17	17	17	18	19	20	22	23	24	25	25
W	27	27	26	26	25	24	24	23	22	21	20	19	19	18	18	18	18	19	20	22	23	25	26	26
NW	21	21	21	20	20	19	19	18	17	16	16	15	15	14	14	14	15	15	16	17	18	19	20	21
Group B Walls																								
N	15	14	14	13	12	11	11	10	9	9	9	8	9	9	9	10	11	12	13	14	14	15	15	15
NE	19	18	17	16	15	14	13	12	12	13	14	15	16	17	18	19	19	20	20	21	21	21	20	20
E	23	22	21	20	18	17	16	15	15	15	17	19	21	22	24	25	26	26	27	27	26	26	25	24
SE	23	22	21	20	18	17	16	15	14	14	15	16	18	20	21	23	24	25	26	26	26	26	25	24
S	21	20	19	18	17	15	14	13	12	11	11	11	11	12	14	15	17	19	20	21	22	22	22	21
SW	27	26	25	24	22	21	19	18	16	15	14	14	13	13	14	15	17	20	22	25	27	28	28	28
W	29	28	27	26	24	23	21	19	18	17	16	15	14	14	14	15	17	19	22	25	27	29	29	30
NW	23	22	21	20	19	18	17	15	14	13	12	12	12	11	12	12	13	15	17	19	21	22	23	23
Group C Walls																								
N	15	14	13	12	11	10	9	8	8	7	7	8	8	9	10	12	13	14	15	16	17	17	17	16
NE	19	17	16	14	13	11	10	10	11	13	15	17	19	20	21	22	22	23	23	23	23	22	21	20
E	22	21	19	17	15	14	12	12	14	16	19	22	25	27	29	29	30	30	30	29	28	27	26	24
SE	22	21	19	17	15	14	12	12	12	13	16	19	22	24	26	28	29	29	29	29	28	27	26	24
S	21	19	18	16	15	13	12	10	9	9	9	10	11	14	17	20	22	24	25	26	25	25	24	22
SW	29	27	25	22	20	18	16	15	13	12	11	11	11	13	15	18	22	26	29	32	33	33	32	31
W	31	29	27	25	22	20	18	16	14	13	12	12	12	13	14	16	20	24	29	32	35	35	35	33
NW	25	23	21	20	18	16	14	13	11	10	10	10	10	11	12	13	15	18	22	25	27	27	27	26
Group D Walls																								
N	15	13	12	10	9	7	6	6	6	6	6	7	8	10	12	13	15	17	18	19	19	19	18	16
NE	17	15	13	11	10	8	7	8	10	14	17	20	22	23	23	24	24	25	25	24	23	22	20	18
E	19	17	15	13	11	9	8	9	12	17	22	27	30	32	33	33	32	32	31	30	28	26	24	22
SE	20	17	15	13	11	10	8	8	10	13	17	22	26	29	31	32	32	32	31	30	28	26	24	22
S	19	17	15	13	11	9	8	7	6	6	7	9	12	16	20	24	27	29	29	29	27	26	24	22
SW	28	25	22	19	16	14	12	10	9	8	8	8	10	12	16	21	27	32	36	38	38	37	34	31
W	31	27	24	21	18	15	13	11	10	9	9	9	10	11	14	18	24	30	36	40	41	40	38	34
NW	25	22	19	17	14	12	10	9	8	7	7	8	9	10	12	14	18	22	27	31	32	32	30	27
Group E Walls																								
N	12	10	8	7	5	4	3	4	5	6	7	9	11	13	15	17	19	20	21	23	20	18	16	14
NE	13	11	9	7	6	4	5	9	15	20	24	25	25	26	26	26	26	26	25	24	22	19	17	15
E	14	12	10	8	6	5	6	11	18	26	33	36	38	37	36	34	33	32	30	28	25	22	20	17
SE	15	12	10	8	7	5	5	8	12	19	25	31	35	37	37	36	34	33	31	28	26	23	20	17
S	15	12	10	8	7	5	4	3	4	5	9	13	19	24	29	32	34	33	31	29	26	23	20	17
SW	22	18	15	12	10	8	6	5	5	6	7	9	12	18	24	32	38	43	45	44	40	35	30	26
W	25	21	17	14	11	9	7	6	6	6	7	9	11	14	20	27	36	43	49	49	45	40	34	29
NW	20	17	14	11	9	7	6	5	5	5	6	8	10	13	16	20	26	32	37	38	36	32	28	24
Group F Walls																								
N	8	6	5	3	2	1	2	4	6	7	9	11	14	17	19	21	22	23	24	23	20	16	13	11
NE	9	7	5	3	2	1	5	14	23	28	30	29	28	27	27	27	27	26	24	22	19	16	13	11
E	10	7	6	4	3	2	6	17	28	38	44	45	43	39	36	34	32	30	27	24	21	17	15	12
SE	10	7	6	4	3	2	4	10	19	28	36	41	43	42	39	36	34	31	28	25	21	18	15	12
S	10	8	6	4	3	2	1	1	3	7	13	20	27	34	38	39	38	35	31	26	22	18	15	12
SW	15	11	9	6	5	3	2	2	4	5	8	11	17	26	35	44	50	53	52	45	37	28	23	18
W	17	13	10	7	5	4	3	3	4	6	8	11	14	20	28	39	49	57	60	54	43	34	27	21
NW	14	10	8	6	4	3	2	2	3	5	8	10	13	15	21	27	35	42	46	43	35	28	22	18
Group G Walls																								
N	3	2	1	0	−1	2	7	8	9	12	15	18	21	23	24	24	25	26	22	15	11	9	7	5
NE	3	2	1	0	−1	9	27	36	39	35	30	26	26	27	27	26	25	22	18	14	11	9	7	5
E	4	2	1	0	−1	11	31	47	54	55	50	40	33	31	30	29	27	24	19	15	12	10	8	6
SE	4	2	1	0	−1	5	18	32	42	49	51	48	42	36	32	30	27	24	19	15	12	10	8	6
S	4	2	1	0	−1	0	1	5	12	22	31	39	45	46	43	37	31	25	20	15	12	10	8	5
SW	5	4	3	1	0	0	2	5	8	12	16	26	38	50	59	63	61	52	37	24	17	13	10	8
W	6	5	3	2	1	1	2	5	8	11	15	19	27	41	56	67	72	67	48	29	20	15	11	8
NW	5	3	2	1	0	0	2	5	8	11	15	18	21	27	37	47	55	55	41	25	17	13	10	7

[a] See text material for corrections applied to table values.

[b] CLTD may be converted to centigrade by multiplying by 5/9.

Reprinted from ASHRAE Handbook of Fundamentals, *1981. With permission of the American Society of Heating, Refrigerating and Air Conditioning Engineers, Atlanta, GA.*

U = overall heat transmission coefficient (Btu/hr ft^2 °F or W/m^2 °C).

A = surface area of wall or roof (ft^2 for m^2)

CLTD = temperature difference which gives the cooling load at the designated time, corrected as noted in the following paragraphs.

Since tables 9-2 and 9-3 are for fixed conditions, the CLTD values may often need corrections. The following fixed conditions apply to these tables:

a. Dark surface roof (''dark'' for solar radiation absorption)

b. Indoor temperature of 78°F

c. Outdoor temperature of 95°F with outdoor mean temperature of 85°F and an outdoor daily temperature range of 21 degrees

d. Solar radiation typical of 40° N latitude on July 21

e. Outside surface film resistance of 0.333 ft^2 °F hr/Btu

f. Without and with suspended ceiling, but no attic fans or return air ducts in the suspended ceiling space

g. Inside surface film resistance of 0.685 ft^2 °F hr/Btu

The following equation makes adjustments for deviations of design and solar conditions from those listed above.

$$CLTD_{corr} = [(CLTD + LM)K + (78 - t_i) + (t_{oa} - 85)]f \tag{9-3}$$

where

CLTD = values in Table 9-2 or Table 9-3

LM = latitude-month correction from Table 9-7 for a horizontal surface (roof), or for a vertical surface (wall) at the designated orientation.

K = a color adjustment factor and is applied after first making month-latitude adjustments. Credit should not be taken for a light-colored roof or wall except where permanence of light color is established by experience, as in rural areas or where there is little smoke. K = 1.0 if dark colored; K = 0.5 if permanently light colored.

$78 - t_i$ = the indoor design temperature correction.

$t_{oa} - 85$ = the outdoor design temperature correction, where t_{oa} is the average outside temperature on the design day. t_{oa} = design temperature t_o − daily temperature range/2.

f = a factor for attic fan and/or ducts above ceiling and is applied after all other adjustments have been made. f = 1.0 for no attic or ducts and for walls. f = 0.75 for positive ventilation.

Values of CLTD for roofs (Table 9-2) were calculated without and with suspended ceiling but made no allowances for positive ventilation or return ducts through the space. If ceiling is insulated and a fan is used between the ceiling and roof, CLTD may be reduced by 25 percent (f = 0.75). Use of the suspended ceiling space for a return air plenum or with return air ducts should be analyzed separately.

The U values for roof constructions listed in Table 9-2 are to be used only as guides. The actual value of U should always be used. An actual roof construction not in Table 9-2 would be thermally similar to a roof in the table if it has similar mass (lbm/ft^3) and similar heat capacity (Btu/ft^2 °F). In such a case, use the CLTD from the table and correct it by Eq. (9-3). Roof construction codes are found in Table 9-4.

Adjustments for CLTD values for walls (Table 9-3) are made by Eq. (9-3), where LM is the latitude-month correction from Table 9-7 and K is a color adjustment factor, where $K = 1.0$ for dark color, 0.83 for permanently medium colored (rural areas), and 0.65 for permanently light colored (rural areas).

The U values listed are to be used only as guides. The actual value of U obtained from calculation should be used. Wall construction groups are listed in Table 9-5. The actual wall construction not listed in Tables 9-3 or 9-5 would be thermally similar to a wall in the table, if it has similar mass (lbm/ft^3) and similar heat capacity (Btu/ft^2 °F). In that case, use the CLTD from Table 9-3 and correct it by Eq. (9-3).

For each 7.0 increase in R value due to insulation added to wall structures in Table 9-5, use the wall group with the next higher letter in the alphabet. For example, move to a group B wall when the initial wall group is C. When the insulation is added to the exterior of the construction rather than the interior, use the CLTD for the wall group *two* letters higher. If this is not possible, due to having already se-

Table 9-4

Roof Construction Code*

Roof No.	Description	Code Numbers of Layers (see Table 9-6)
1	Steel Sheet with 25.4-mm (1-in.) insulation	A0, E2, E3, B5, A3, E0
2	25.4-mm (1-in.) wood with 25.4-mm (1-in.) insulation	A0, E2, E3, B5, B7, E0
3	101.6-mm (4-in.) l.w. concrete	A0, E2, E3, C14, E0
4	50.8-mm (2-in.) h.w. concrete with 25.4 (1-in.) insulation	A0, E2, E3, B5, C12, E0
5	25.4-mm (1-in.) wood with 50.8-mm (2-in.) insulation	A0, E2, E3, B6, B7, E0
6	152.4-mm (6-in.) l.w. concrete	A0, E2, E3, C15, E0
7	63.5-mm (2.5-in.) wood with 25.4-mm (1-in.) insulation	A0, E2, E3, B5, B8, E0
8	203.2-mm (8-in.) l.w. concrete	A0, E2, E3, C16, E0
9	101.6-mm (4-in.) h.w. concrete with 25.4-mm (1-in.) insulation	A0, E2, E3, B5, C5, E0
10	63.5-mm (2.5-in.) wood with 50.8-mm (2-in.) insulation	A0, E2, E3, B6, B8, E0
11	Roof terrace system	A0, C12, B1, B6, E2, E3, C5, E0
12	152.4-mm (6-in.) h.w. concrete with 25.4-mm (1-in.) insulation	A0, E2, E3, B5, C13, E0
13	101.6-mm (4-in.) wood with 25.4-mm (1-in.) insulation	A0, E2, E3, B5, B9, E0

**Reprinted from ASHRAE* Handbook of Fundamentals, *1981. With permission of the American Society of Heating, Refrigeration and Air-Conditioning Engineers, Atlanta, Ga.*

CLTD, Uncorrected, When Vertical Wall Structure is "Thermally" Heavier than Group *A* Due to Added Insulation

N	NE	E	SE	S	SW	W	NW
11	17	22	21	17	21	22	17

lected a wall as group *A*, use an effective CLTD in the load calculations as given in the above table.

Illustrative Problem 9-1

A building located in Boston, Massachusetts, has a flat roof without a suspended ceiling and has mass properties of 19 lb/ft^2. The calculated U value is 0.18 Btu/hr ft^2 °F. Determine the cooling load caused by the roof at 2:00 PM standard time during August (in Btu/hr ft^2).

Solution:

This roof corresponds to roof 13 of Table 9-2. At 1400 hours (2:00 PM) find $CLTD_{uncorr} = 21°F$.

From Table C-1 (see Appendix C):

N latitude = 42°

Design temperature outside = 88°F

Daily temperature range = 16

By Eq. (9-3)

$$CLTD_{corr} = [(CLTD + LM)K + (78 - t_i) + (t_{oa} - 85)]$$

$LM = -3°F$ (Table 9-7 for 40° N latitude, horizontal)

$K = 1.0$ (assumed dark)

$t_i = 78°F$ (assumed as standard)

$$t_{oa} = t_o - DR/2 = 88 - 16/2 = 80°F$$

$f = 1.0$ (assumed)

$$CLTD_{corr} = [(21 - 3)(1.0) + (78 - 78) + (80 - 85)](1.0)$$

$$CLTD_{corr} = 13°F$$

By Eq (9-2) $q/A = U(CLTD_{corr}) = (0.18)(13) = 2.34$ Btu/hr ft^2.

Illustrative Problem 9-2

A building measures 40 by 110 by 12 ft (floor to ceiling) and is located at 32° N latitude in an area where the outside air design dry-bulb temperature is 90°F and a daily range of 20°F. The inside design dry-bulb temperature is 78°F. Determine the cooling load as a result of heat gain through the roof and south and west walls at 1200, 1400, and 1600 hr for August 21. The south wall measures 110 by 12 ft; the west wall measures 40 by 12 ft. Construction details are as follows:

Roof	0.5-in. slag plus 0.375-in. felt and membrane, 2.0-in. heavyweight concrete, 2.0-in. insulation ($R = 6.7$), and a suspended ceiling of acoustic tile. Assume a dark surface ($K = 1.0$).
South Wall	8.0-in. heavyweight concrete with 2.0-in. insulation ($R = 6.7$) and interior finish.
West Wall	Stucco on 4-in. lightweight concrete block with 1-in. insulation ($R = 3.3$) and interior finish ($R = 0.15$).

Roof Table 9-2: No. 4 Table 9-4	Table 9-6	South Wall Table 9-5: Group *B* Table 9-5	Table 9-6	West Wall Table 9-5: Group *F* Table 9-5	Table 9-6
Code Numbers of Layers	***R***	**Code Numbers of Layers**	***R***	**Code Numbers of Layers**	***R***
A0	0.333	A0	0.333	A0	0.333
E2	0.050	A1	0.208	A1	0.208
E3	0.285	C10	0.667	C2	1.51
B6	6.68	B5/B6	6.68	B1/B2	3.32
C12	0.167	E1	0.149	E1	0.149
E0	0.685	E0	0.685	E0	0.685
	R_t = 8.20		R_t = 8.72		R_t = 6.205
	U_r = 0.122		U_{sw} = 0.115		U_{ww} = 0.161

Solution:

Step 1: Determine U values for roof and walls.

Step 2: Determine CLTD correction factors for Eq. (9-3).

Correction for Outdoor Air:

$$t_{oa} = 90 - 20/2 = 80$$

$$\text{Correction} = 80 - 85 = -5°\text{F}$$

Correction for Inside Air:

$$\text{Correction } (78 - 78) = 0°\text{F}$$

Correction for Latitude (Table 9-7):

$$\text{Roof} = -1°\text{F}$$

$$\text{S Wall} = +1°\text{F}$$

$$\text{W Wall} = 0°\text{F}$$

Correction for Surface Color

Dark roof: $K = 1.0$

Medium color walls: $K = 0.83$

$$\text{Roof, CLTD}_{\text{corr}} = [(\text{CLTD} - 1)(1.0) + 0 - 5)](1.0) = \text{CLTD} - 6$$

$$\text{S Wall, CLTD}_{\text{corr}} = [(\text{CLTD} + 1)(0.83) + 0 - 5] = (\text{CLTD} + 1)(0.83) - 5$$

$$\text{W Wall, CLTD}_{\text{corr}} = [(\text{CLTD} + 0)(0.83) + 0 - 5] = (0.83\ \text{CLTD}) - 5$$

Step 3: Determine CLTD, CLTD$_{\text{corr}}$, and heat gains by using the table at the bottom of page 383 and the top of page 384.

Step 4: Summary.

At 1200 hours: $\dot{q} = 10{,}200 + 750 + 320 = 11{,}270$ Btu/hr

At 1400 hours: $\dot{q} = 15{,}570 + 880 + 900 = 17{,}350$ Btu/hr

At 1600 hours: $\dot{q} = 19{,}860 + 1260 + 2120 = 23{,}240$ Btu/hr

Table 9-5

Wall Construction Group Description*[a]

Weight (kg/m²)	U-Value (W/m²·°C)	Group No.	Description of Construction	Weight (lb/ft²)	U-Value (Btu/h·ft²·F)	Code Numbers of Layers
405	2.033		101.6-mm (4-in.) Face Brick + (*Brick*)			
405	2.033	C	Air Space + 101.6-mm (4-in.) Face Brick	83	0.358	A0, A2, B1, A2, E0
439	2.356	D	101.6-mm (4-in.) Common Brick	90	0.415	A0, A2, C4, E1, E0
439	0.987-1.709	C	25.4-mm (1-in.) Insulation or Air Space + 101.6-mm (4-in.) Common Brick	90	0.174-0.301	A0, A2, C4, B1/B2, E1, E0
430	0.630	B	50.8-mm (2-in.) Insulation + 101.6-mm (4-in.) Common Brick	88	0.111	A0, A2, B3, C4, E1, E0
635	1.714	B	203.2-mm (8-in.) Common Brick	130	0.302	A0, A2, C9, E1, E0
635	0.874-1.379	A	Insulation or Air Space + 203.2-mm (8-in.) Common brick	130	0.154-0.243	A0, A2, C9, B1/B2, E1, E0
			101.6-mm (4-in.) Face Brick + (*H.W. Concrete*)			
459	1.987	C	Air Space + 50.8-mm (2-in.) Concrete	94	0.350	A0, A2, B1, C5, E1, E0
474	0.658	B	50.8-mm (2-in.) Insulation + 101.6-mm (4-in.) Concrete	97	0.116	A0, A2, B3, C5, E1, E0
698-928	0.625-0.636	A	Air Space or Insulation + 203.2-mm (8-in.) or more Concrete	143-190	0.110-0.112	A0, A2, B1, C10/11, E1, E0
			101.6-mm (4-in.) Face Brick + (*L.W. or H.W. Concrete Block*)			
303	1.811	E	101.6-mm (4-in.) Block	62	0.319	A0, A2, C2, E1, E0
303	0.868-1.397	D	Air Space or Insulation + 101.6-mm (4-in.) Block	62	0.153-0.246	A0, A2, C2, B1/B2, E1, E0
342	1.555	D	203.2-mm (8-in.) Block	70	0.274	A0, A2, C7, A6, E0
356-434	1.255-1.561	C	Air Space or 25.4-mm (1-in.) Insulation + 152.4-mm (6-in.) or 203.2-mm (8-in.) Block	73-89	0.221-0.275	A0, A2, B1, C7/C8, E1, E0
434	0.545-0.607	B	50.8-mm (2-in.) Insulation + 203.2-mm (8-in.) Block	89	0.096-0.107	A0, A2, B3, C7/C8, E1, E0
			101.6-mm (4-in.) Face Brick + (*Clay Tile*)			
347	2.163	D	101.6-mm (4-in.) Tile	71	0.381	A0, A2, C1, E1, E0
347	1.595	D	Air Space + 101.6-mm (4-in.) Tile	71	0.281	A0, A2, C1, B1, E1, E0
347	0.959	C	Insulation + 101.6-mm (4-in.) Tile	71	0.169	A0, A2, C1, B2, E1, E0
469	1.561	C	203.2-mm (8-in.) Tile	96	0.275	A0, A2, C6, E1, E0
469	0.806-1.255	B	Air Space or 25.4-mm (1-in.) Insulation + 203.2-mm (8-in.) Tile	96	0.142-0.221	A0, A2, C6, B1/B2, E1, E0
474	0.551	A	50.8-mm (2-in.) Insulation + 203.2-mm (8-in.) Tile	97	0.097	A0, A2, B3, C6, E1, E0
			H.W. Concrete Wall + (*Finish*)			
308	3.321	E	101.6-mm (4-in.) Concrete	63	0.585	A0, A1, C5, E1, E0
308	0.675-1.136	D	101.6-mm (4-in.) Concrete + 25.4-mm (1-in.) or 50.8-mm (2-in.) Insulation	63	0.119-0.200	A0, A1, C5, B2/B3, E1, E0
308	0.675	C	50.8-mm (2-in.) Insulation + 101.6-mm (4-in.) Concrete	63	0.119	A0, A1, B6, C5, E1, E0
532	2.782	C	203.2-mm (8-in.) Concrete	109	0.490	A0, A1, C10, E1, E0
537	0.653-1.061	B	203.2-mm (8-in.) Concrete + 25.4-mm (1-in.) or 50.8-mm (2-in.) Insulation	110	0.115-0.187	A0, A1, C10, B5/B6, E1, E0
537	0.653	A	50.8-mm (2-in.) Insulation + 203.2-mm (8-in.) Concrete	110	0.115	A0, A1, B3, C10, E1, E0
762	2.390	B	304.8-mm (12-in.) Concrete	156	0.421	A0, A1, C11, E1, E0
762	0.642	A	304.8-mm (12-in.) Concrete + Insulation	156	0.113	A0, C11, B6, A6, E0
			L. W. and H.W. Concrete Block + (*Finish*)			
142	0.914-1.493	F	101.6-mm (4-in.) Block + Air Space/Insulation	29	0.161-0.263	A0, A1, C2, B1/B2, E1, E0
142-181	0.596-0.647	E	50.8-mm (2-in.) Insulation + 101.6-mm (4-in.) Block	29-37	0.105-0.114	A0, A1, B3, C2/C3, E1, E0
229-249	1.669-2.282	E	203.2-mm (8-in.) Block	47-51	0.294-0.402	A0, A1, C7/C8, E1, E0
200-278	0.846-0982	D	203.2-mm (8-in.) Block + Air Space/Insulation	41-57	0.149-0.173	A0, A1, C7/C8, B1/B2, E1, E0
			Clay Tile + (*Finish*)			
190	2.379	F	101.6-mm (4-in.) Tile	39	0.419	A0, A1, C1, E1, E0
190	1.720	F	101.6-mm (4-in.) Tile + Air Space	39	0.303	A0, A1, C1, B1, E1, E0
190	0.993	E	101.6-mm (4-in.) Tile + 25.4-mm (1-in.) Insulation	39	0.175	A0, A1, C1, B2, E1, E0
195	0.625	D	50.8-mm (2-in.) Insulation + 101.6-mm (4-in.) Tile	40	0.110	A0, A1, B3, C1, E1, E0
308	1.681	D	203.2-mm (8-in.) Tile	63	0.296	A0, A1, C6, B1/B2, E1, E0
308	0.857-1.312	C	203.2-mm (8-in.) Tile + Air Space/25.4-mm (1-in.) Insulation	63	0.151-0.231	A0, A1, C6, B1/B2, E1, E0
308	0.562	B	50.8-mm (2-in.) Insulation + 203.2-mm (8-in.) Tile	63	0.099	A0, A1, B3, C6, E1, E0
			Metal Curtain Wall			
24-29	0.516-1.306	G	With/ withou t air Space + 25.4 -mm (1- in.)/50 .8-mm (2-in.) 76.2- mm (3-in.) Insulation	5-6	0.091-0.2 30	A0, A3, B5/B6/B12, A3, E0
			Frame Wall			
78	0.459-1.010	G	25.4-m m (1- in.) to 76.2- mm (3-in.) Insulation	16	0.081-0.1 78	A0, A1, B1, B2/B3/B4, E1, E0

*Reprinted from ASHRAE *Handbook of Fundamentals,* 1981. With permission of the American Society of Heating, Refrigerating and Air-Conditioning Engineers, Atlanta, GA.

[a]*Code number of layers are found in Table 9-6.*

Table 9-6

Thermal Properties and Code Numbers of Layers Used in Calculating Coefficients for Roofs and Walls*

Description	Code Number	Thickness and Thermal Properties[a] L	K	D	SH	R	Wt
Outside surface resistance	A0					0.333	
25.4-mm (1-in.) Stucco (asbestos cement or wood siding plaster, etc)	A1	0.0833	0.4	116	0.20	0.208	9.66
101.6-mm (4-in.) facebrick (dense concrete)	A2	0.3333	0.75	130	0.22	0.444	43.3
Steel siding (aluminum or other lightweight cladding)	A3	0.0050	26.0	480	0.10	0.000	2.40
Outside surface resistance							
12.7-mm (0.5-in.) slag, membrane	A4	0.0417	0.83	55	0.40		
9.5-mm, (0.375-in.) felt		0.0313	0.11	70	0.40		
Finish	A6	0.0417	0.24	78	0.26	0.174	3.25
101.6-mm (4-in.) facebrick	A7	0.3333	0.77	125	0.22	0.433	41.6
Air Space Resistance	B1					0.91	
25.4-mm (1-in.) insulation	B2	0.0833	0.025	2.0	0.2	3.32	0.17
50.8-mm (2-in.) insulation	B3	0.1667	0.025	2.0	0.2	6.68	0.33
76.2-mm (3-in.) insulation	B4	0.2500	0.025	2.0	0.2	10.03	0.50
25.4-mm (1-in.) insulation	B5	0.0833	0.025	5.7	0.2	3.33	0.47
50.8-mm (2-in.) insulation	B6	0.1667	0.025	5.7	0.2	6.68	0.95
25.4-mm (1-in.) wood	B7	0.0833	0.067	37.0	0.6	1.19	3.08
62.5-mm (2.5-in.) wood	B8	0.2083	0.067	37.0	0.6	2.98	7.71
14.6-mm (4-in.) wood	B9	0.3333	0.067	37.0	0.6	4.76	2.3
50.8-mm (2-in.) wood	B10	0.1667	0.067	37.0	0.6	2.39	6.18
76.2-mm (3-in.) wood	B11	0.2500	0.067	37.0	0.6	3.58	9.25
76.2-mm (3-in.) insulation	B12	0.2500	0.025	5.7	0.2	10.00	1.42
101.6-mm (4-in.) insulation	B13	0.3333	0.025	5.7	0.2	13.33	1.90
127.0-mm (5-in.) insulation	B14	0.4167	0.025	5.7	0.2	16.67	2.38
152.4-mm (6-in.) insulation	B15	0.5000	0.025	5.7	0.2	20.00	2.85
101.6-mm (4-in.) clay tile	C1	0.3333	0.33	70.0	0.2	1.01	23.3
101.6 mm (4-in.) l.w. concrete block	C2	0.3333	0.22	38.0	0.2	1.51	12.7
101.6-mm (4-in.) h.w. concrete block	C3	0.3333	0.47	61.0	0.2	0.71	20.3
101.6-mm (4-in.) common brick	C4	0.3333	0.42	120.0	0.2	0.79	40.0
101.6-mm (4-in.) h.w. concrete	C5	0.3333	1.00	140.0	0.2	0.333	46.6
203.2-mm (8-in.) clay tile	C6	0.6667	0.33	70.0	0.2	2.02	46.7
203.2 mm (8-in.) l.w. concrete block	C7	0.6667	0.33	38.0	0.2	2.02	25.4
203.2-mm (8-in.) h.w. concrete block	C8	0.6667	0.6	61.0	0.2	1.11	40.7
203.2-mm (8-in.) common brick	C9	0.6667	0.42	120.0	0.2	1.59	80.0
203.2-mm (8-in.) h.w. concrete	C10	0.6667	1.00	140.0	0.2	0.667	93.4
304.8-mm (12-in.) h.w. concrete	C11	1.0000	1.00	140.0	0.2	1.08	140.0
50.8-mm (2-in.) h.w. concrete	C12	1.6667	1.00	140.0	0.2	0.167	23.4
151.4-mm (6-in.) h.w. concrete	C13	0.5000	1.00	140.0	0.2	0.500	70.0
101.6-mm (4-in.) l.w. concrete	C14	0.3333	0.10	40.0	0.2	3.333	13.3
152.4-mm (6-in.) l.w. concrete	C15	0.5000	0.10	40.0	0.2	5.000	20.0
203.2-mm (8-in.) l.w. concrete	C16	0.6667	0.10	40.0	0.2	6.667	26.7
203.2-mm (8-in.) l.w. concrete block (filled insulation)	C17	0.6667	0.08	18.0	0.2	9.00	12.0
203.2-mm (8-in.) l.w. concrete block (filled insulation)	C18	0.6667	0.34	53.0	0.2	1.98	35.4
204.8-mm (12-in.) l.w. concrete (block filled insulation)	C19	1.0000	0.08	19.0	0.2	13.5	19.0
204.8-mm (12-in.) l.w. concrete block (filled insulation)	C20	1.0000	0.39	56.0	0.2	2.59	56.0
Inside surface resistance	E0					0.685	
19.0-mm (0.75-in.) plaster; 19.0-mm (0.75-in.) gypsum or other similar finishing layer	E1	0.0625	0.42	100	0.2	0.149	6.25
12.7-mm (0.5-in.) slag or stone	E2	0.0417	0.83	55	0.40	0.050	2.29
9.5-mm (0.375-in.) felt & membrane	E3	0.0313	0.11	70	0.40	0.285	2.19
Ceiling air space	E4					1.0	
Acoustic tile	E5	0.0625	0.035	30	0.20	1.786	1.88

[a]Units: L = ft; K = Btu/hr ft °F; D = lb/ft^3; SH = Btu/lb °F; R = ft^2 °F hr/Btu; WT = lb/ft^2.

Reprinted from ASHRAE Handbook of Fundamentals, *1981. With permission of the American Society of Heating, Refrigeration and Air-Conditioning Engineers, Atlanta, Ga.*

Table 9-7

CLTD Correction for Latitude and Month Applied to Walls and Roofs North Latitudes (°F)*

Latitude	Month	N	NNE NNW	NE NW	ENE WNW	E W	ESE WSW	SE SW	SSE SSW	S	HOR
24	Dec	−5	−7	−9	−10	−7	−3	3	9	13	−13
	Jan/Nov	−4	−6	−8	−9	−6	−3	3	9	13	−11
	Feb/Oct	−4	−5	−6	−6	−3	−1	3	7	10	−7
	Mar/Sept	−3	−4	−3	−3	−1	−1	1	2	4	−3
	Apr/Aug	−2	−1	0	−1	−1	−2	−1	−2	−3	0
	May/Jul	1	2	2	0	0	−3	−3	−5	−6	1
	Jun	3	3	3	1	0	−3	−4	−6	−6	1
32	Dec	−5	−7	−10	−11	−8	−5	2	9	12	−17
	Jan/Nov	−5	−7	−9	−11	−8	−4	2	9	12	−15
	Feb/Oct	−4	−6	−7	−8	−4	−2	4	8	11	−10
	Mar/Sept	−3	−4	−4	−4	−2	−1	3	5	7	−5
	Apr/Aug	−2	−2	−1	−2	0	−1	0	1	1	−1
	May/July	1	1	1	0	0	−1	−1	−3	−3	1
	Jun	1	2	2	1	0	−2	−2	−4	−4	2
40	Dec	−6	−8	−10	−13	−10	−7	0	7	10	−21
	Jan/Nov	−5	−7	−10	−12	−9	−6	1	8	11	−19
	Feb/Oct	−5	−7	−8	−9	−6	−3	3	8	12	−14
	Mar/Sept	−4	−5	−5	−6	−3	−1	4	7	10	−8
	Apr/Aug	−2	−3	−2	−2	0	0	2	3	4	−3
	May/July	0	0	0	0	0	0	0	0	1	1
	June	1	1	1	0	1	0	0	−1	−1	2
48	Dec	−6	−8	−11	−14	−13	−10	−3	2	6	−25
	Jan/Nov	−6	−8	−11	−13	−11	−8	−1	5	8	−24
	Feb/Oct	−5	−7	−10	−11	−8	−5	1	8	11	−18
	Mar/Sept	−4	−6	−6	−7	−4	−1	4	8	11	−11
	Apr/Aug	−3	−3	−3	−3	−1	0	4	6	7	−5
	May/July	0	−1	0	0	1	1	3	3	4	0
	Jun	1	1	2	1	2	1	2	2	3	2

[1]Corrections are in °F. The correction is applied directly to the CLTD for a roof or wall as well as given in Table 9-2 or 9-3.
[2]The CLTD correction is *not* applicable to Table 9-8.
[3]For South latitudes, replace Jan. through Dec. by July through June.
[4]CLTD corrections may be converted to centigrade by multiplying by 5/9.

**Reprinted from ASHRAE* Handbook of Fundamentals *1981. With permission of the American Society of Heating, Refrigerating and Air-Conditioning Engineers, Atlanta, Ga.*

Hour	CLTD (Table 9-2)	Roof $CLTD_{corr}$	$\dot{q} = UA(CLTD_{corr})$
1200	25	19	$\dot{q}$ = (0.122)(4400)(19) = 10,200 Btu/hr
1400	35	29	= (0.122)(4900)(29) = 15,570 Btu/hr
1600	43	37	= (0.122)(4400)(37) = 19,860 Btu/hr

Hour	CLTD (Table 9-3 (*B*))	South Wall $CLTD_{corr}$	$\dot{q} = UA(CLTD_{corr})$
1200	11	4.96	q = (0.115)(1320)(4.96) = 750 Btu/hr
1400	12	5.79	= (0.115)(1320)(5.79) = 880 Btu/hr
1600	15	8.28	= (0.115)(1320)(8.28) = 1260 Btu/hr

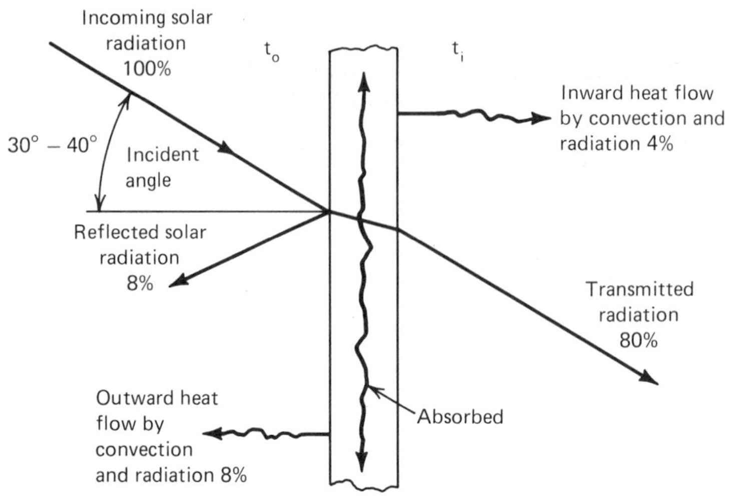

Figure 9–4

Instantaneous heat balance for Sunlit Glazing Material

Hour	CLTD (Table 9-3 (*F*))	West Wall $CLTD_{corr}$	$\dot{q} = UA(CLTD_{corr})$
1200	11	4.13	$\dot{q}$ = (0.161)(480)(4.13) = 320 Btu/hr
1400	20	11.6	= (0.161)(480)(11.6) = 900 Btu/hr
1600	39	27.37	= (0.161)(480)(27.37) = 2120 Btu/hr

9-9 SPACE COOLING LOAD DUE TO CONDUCTION AND SOLAR HEAT GAIN THROUGH FENESTRATION

Calculation of heat transfer through fenestration (any light-transmitting opening in a building) will be discussed in this section.

The ability of glazing materials to transmit solar radiation depends on radiation wave length, chemical composition and thickness of the material, and the incident angle.[1] Clear, single- or double-strength glass transmits 85 to 90 percent of the incident radiation, while bronze, gray and green heat-absorbing glass 0.25 in. (64 mm) thick transmits about 47 percent. Figure 9-4 illustrates an approximate heat balance for clear, double-strength glass.

Almost all types of architectural glass are completely opaque to the long-wave radiation emitted by surfaces at temperatures below 250°F (121°C). This characteristic produces a *greenhouse* effect, by which solar radiation entering through a window is trapped. When radiation is absorbed by the surfaces within a space and then emitted as long-wave radiation, heat cannot escape outward since the window is opaque to such wave lengths.

Of the many environmental factors affecting heat admission or loss through fenestration areas, the most significant are (1) solar radiation intensity and incident angle, (2) outdoor-indoor temperature difference, and (3) velocity and direction of air flow across the exterior and interior fenestration surfaces. The solar heat gain is the sum of the transmitted radiation and the portion of the absorbed radiation that flows inward. Heat is also transmitted through the glazing material by conduction whenever there is a temperature difference between the outdoor and indoor air. Therefore, the total rate of heat admission is

Total heat admission through glass

	Radiation		Inward Flow		Conduction
=	through	+	of absorbed	+	heat
	glass		radiation		gain

The first two quantities on the right are directly related to the solar radiation falling on the glass, and the third quantity is independent of solar radiation and exists even if the sun is not shining. Therefore, we may condense the above equation into

Total heat gain *(9-4)*
= solar heat gain + conduction heat gain

The *conduction heat gain* may be calculated using Eq. (9-1) where Δt is the overall temperature difference between outside and inside air ($t_o - t_i$) and the U value for the existing fenestration, which may be found in Table 7-5 of Chapter 7. The corresponding conduction *cooling load* is calculated by Eq. (9-2) using CLTD values from Table 9-8. The values in the table were calculated for an inside air temperature of 78°F (25.5°C), outdoor maximum temperature of 95°F (35°C), and outdoor daily range of 21°F (11.6°C). The footnote to Table 9-8 explains the technique for correcting the values to other indoor temperatures or outdoor temperature cycles, similar to the technique used with Tables 9-2 and 9-3 for roofs and walls.

The heat gain due to *transmitted* and *absorbed solar energy* is present in cooling-load calculations only when fenestration is irradiated. It is a direct function of the total shortwave radiation reaching the surface, the angle of incidence, and the characteristic of each type of fenestration.

To account for the different types of fenestration and shading devices used, the *shading coefficient* (SC) relating the solar heat gain through the glazing system under a specific set of conditions to the solar heat gain through the

Table 9-8

Cooling-Load Temperature Differences (CLTD) for Conduction Through Glass*[a]

Solar Time, h	1	2	3	4	5	6	7	8	9	10	11	12	13	14	15	16	17	18	19	20	21	22	23	24
CLTD																								
°C	1	0	−1	−1	−1	−1	−1	0	1	2	4	5	7	7	8	8	7	7	6	4	3	2	2	1
°F	1	0	−1	−2	−2	−2	−2	0	2	4	7	9	12	13	14	14	13	12	10	8	6	4	3	2

[a]Corrections: The values in the table were calculated for an inside temperature of 78°F (25.5°C) and an outdoor maximum temperature of 95°F (35°C) with an outdoor daily range of 21°F (11.5°C). The table remains approximately correct for other outdoor maximums, 93–102°F (33.8–38.8°C), and other outdoor daily ranges, 16–34°F (8.9–18.9°C), provided the outdoor daily average temperature remains approximately 85°F (29.4°C). If the room air temperature is different from 78°F (25.5°C) and the outdoor daily average temperature is different from 85°F (29.4°C), the following rules apply: (a) For room air temperature less than 78°F (25.5°C), add the difference between 78°F (25.5°C) and room temperature; if greater than 78°F (25.5°C), subtract the difference. (b) For outdoor daily average temperature less than 85°F (29.4°C), subtract the difference between 85°F (29.4°C) and the daily average temperature; if greater than 85°F (29.4°C), add the difference.

**Reprinted from ASHRAE* Handbook of Fundamentals *1981. With permission of the American Society of Heating, Refrigerating and Air Conditioning Engineers, Atlanta, Ga.*

reference material under the same conditions is defined:

$$SC = \frac{\text{Solar Heat Gain of Fenestration}}{\text{Solar Heat Gain of Double-Strength Glass}} \qquad (9\text{-}5)$$

Shading coefficients are found in Tables 9-10 through 9-14.

ASHRAE (1) generated a computer program to calculate solar heat gain factors (SHGFs) for double-strength sheet glass. Included are calculations for different latitudes (0°N to 64°N), months (January through December), orientations (17), and daylight hours. These were used as the "heat gain" input for calculating cooling-load factors (CLFs), considering type of interior construction (light, medium, or heavy), and presence or absence of an interior shading device for the glass.

Solar heat gain factors for 24, 32, 40, and 48°N latitude have been selected from Ref. 1 to be included here and are found in Table 9-9.

For calculation, the cooling load due to solar radiation must be analyzed in one of two cases: (1) presence of interior shading or (2) absense of indoor shading. In converting heat gain to cooling load, the time lag, due to the radiant solar energy entering the space, is variable; for example, it is different when energy is absorbed by interior draperies or venetian blinds from when it is absorbed by the floor. Interior shading devices cause the cooling load to track the solar heat gain profile more closely, while the case without the interior shading spreads the load out. The time lag difference in converting heat gain to cooling load will appear in the cooling load factors (CLFs) used to multiply the solar heat gains. CLFs are found in Tables 9-15 and 9-16.

Cooling load due to solar radiation through fenestration is calculated by

$$q = A(SC)(SHGF)(CLF) \qquad (9\text{-}6)$$

where

A = net glass area of the fenestration
SC = shading coefficient from Table 9-10, 9-11, 9-12, 9-13, or 9-14
SHGF = maximum solar heat gain from Table 9-9 for the appropriate latitude, month, and surface orientation
CLF = cooling load factor from Table 9-15 or 9-16 for the appropriate room thermal characteristic (light, medium, or heavy)

Solution:

Table Data:	***U* (Table 7-5)**	**SC (f_o = 3.0)**	**SHGF (Table 9-9)**
S Glass	0.61	0.82 (Table 9-10)	72
W Glass	0.81	0.53 (Table 9-11)	215

Conduction heat gain component: $q = UA$ (CLTD)

Time	CLTD (Table 9-8)	CLTD (Corrected)*	S Glass UA(CLTD)	W Glass UA(CLTD)
1200	9	9 − 5 = 4	(0.61)(150)(4) = 366	(0.81)(200)(4) = 648
1400	13	13 − 5 = 8	(0.61)(150)(8) = 732	(0.81)(200)(8) = 1296
1600	14	14 − 5 = 9	(0.61)(150)(9) = 824	(0.81)(200)(9) = 1458

*Footnote of Table 9-8: $t_{oa} = 90 - 20/2 = 80°F$. This is 5 degrees less than the value of 85°F. Therefore subtract 5 degrees from table values of CLTD to obtain corrected CLTD.

Solar heat gain component: $q = A$ (SC) (SHGF) (CLF)

South Glass:

Time	CLF (Table 9-15)	*A*(SC)(SHGF)CLF)
1200	0.52	(150)(0.82)(72)(0.52) = 4600
1400	0.58	(150)(0.82)(72)(0.58) = 5140
1600	0.47	(150)(0.82)(72)(0.47) = 4160

West Glass:

Time	CLF (Table 9-16)	*A*(SC)(SHGF)(CLF)
1200	0.17	200(0.53)(215)(0.17) = 3870
1400	0.53	200(0.53)(215)(0.53) = 12080
1600	0.82	200(0.53)(215)(0.82) = 18690

Total Cooling Load:

Time	South Glass	West Glass
1200	366 + 4,600 = 4,960 Btu/hr	648 + 3,870 = 4,518 Btu/hr
1400	732 + 5,140 = 5,872 Btu/hr	1,296 + 12,080 = 13,376 Btu/hr
1600	824 + 4,160 = 4,984 Btu/hr	1,458 + 18,690 = 20,148 Btu/hr

Illustrative Problem 9-3

Determine the cooling load due to windows on the south and west walls of a building at 1200, 1400, and 1600 hours in July. The building is located at 32°N latitude with outside design of 90°F dry-bulb temperature and a 20-degree daily range. The inside design dry-bulb temperature is 78°F. Assume the room construction is of medium weight. The south glass is insulating glass, 0.25 in. of air space having an area of 150 ft^2 with no interior shading. The west glass is a 0.21875-in. single tinted glass with an area of 200 ft^2 and with light-colored venetian blinds.

External Shading is more effective in reducing solar heat gain that internal shading and may produce reductions of up to 80 percent. Fenestration may be externally shaded by trees, shrubbery, another building, canvas awnings, or some architectural feature of the window openings. When such external shading

Table 9-9

Maximum Solar Heat Gain Factor (SHGF), Btu/hr ft² for Sunlit Glass, North Latitudes.*

32 Deg

	N (Shade)	NNE/ NW	NE/ NW	ENE/ WNW	E/ W	ESE/ WSW	SE/ SW	SSE/ SSW	S	HOR
Jan.	24	24	29	105	175	229	249	250	246	176
Feb.	27	27	65	149	205	242	248	232	221	217
Mar.	32	37	107	183	227	237	227	195	176	252
Apr.	36	80	146	200	227	219	187	141	115	271
May	38	111	170	208	220	199	155	99	74	277
June	44	122	176	208	214	189	139	83	60	276
July	40	111	167	204	215	194	150	96	72	273
Aug.	37	79	141	195	219	210	181	136	111	265
Sept.	33	35	103	173	215	227	218	189	171	244
Oct.	28	28	63	143	195	234	239	225	215	213
Nov.	24	24	29	103	173	225	245	246	243	175
Dec.	22	22	22	84	162	218	246	252	252	158

36 Deg

	N (Shade)	NNE/ NNW	NE/ NW	ENE/ WNW	E/ W	ESE/ WSW	SE/ SW	SSE/ SSW	S	HOR
Jan.	22	22	24	90	166	219	247	252	252	155
Feb.	26	26	57	139	195	239	248	239	232	199
Mar.	30	33	99	176	223	238	232	206	192	238
Apr.	35	76	144	196	225	221	196	156	135	262
May	38	107	168	204	220	204	165	116	93	272
June	47	118	175	205	215	194	150	99	77	273
July	39	107	165	201	216	199	161	113	90	268
Aug.	36	75	138	190	218	212	189	151	131	257
Sep.	31	31	95	167	210	228	223	200	187	230
Oct.	27	27	56	133	187	230	239	231	225	195
Nov.	22	22	24	87	163	215	243	248	248	154
Dec.	20	20	20	69	151	204	241	253	254	136

40 Deg

	N (Shade)	NNE/ NNW	NE/ NW	ENE/ WNW	E/ W	ESE/ WSW	SE/ SW	SSE/ SSW	S	HOR
Jan.	20	20	20	74	154	205	241	252	254	133
Feb.	24	24	50	129	186	234	246	244	241	180
Mar.	29	29	93	169	218	238	236	216	206	223
Apr.	34	71	140	190	224	223	203	170	154	252
May	37	102	165	202	220	208	175	133	113	265
June	48	113	172	205	216	199	161	116	95	267
July	38	102	163	198	216	203	170	129	109	262
Aug.	35	71	135	185	216	214	196	165	149	247
Sep.	30	30	87	160	203	227	226	209	200	215
Oct.	25	25	49	123	180	225	238	236	234	177
Nov.	20	20	20	73	151	201	237	248	250	132
Dec.	18	18	18	60	135	188	232	249	253	113

44 Deg

	N (Shade)	NNE/ NNW	NE/ NW	ENE/ WNW	E/ W	ESE/ WSW	SE/ SW	SSE/ SSW	S	HOR
Jan.	17	17	17	64	138	189	232	248	252	109
Feb.	22	22	43	117	178	227	246	248	247	160
Mar.	27	27	87	162	211	236	238	224	218	206
Apr.	33	66	136	185	221	224	210	183	171	240
May	36	96	162	201	219	211	183	148	132	257
June	47	108	169	205	215	203	171	132	115	261
July	37	96	159	198	215	206	179	144	128	254
Aug.	34	66	132	180	214	215	202	177	165	236
Sep.	28	28	80	152	198	226	227	216	211	199
Oct.	23	23	42	111	171	217	237	240	239	157
Nov.	18	18	18	64	135	186	227	244	248	109
Dec.	15	15	15	49	115	175	217	240	246	89

Table 9-9 (Cont.)

Maximum Solar Heat Gain Factor (SHGF), Btu/hr ft^2 for Sunlit Glass, North Latitudes.*

48 Deg										
	N (Shade)	NNE/ NNW	NE/ NW	ENE/ WNW	E/ W	ESE/ WSW	SE/ SW	SSE/ SSW	S	HOR
Jan.	15	15	15	53	118	175	216	239	245	.85
Feb.	20	20	36	103	168	216	242	249	250	138
Mar.	26	26	80	154	204	234	239	232	228	188
Apr.	31	61	132	180	219	225	215	194	186	226
May	35	97	158	200	218	214	192	163	150	247
June	46	110	165	204	215	206	180	148	134	252
July	37	96	156	196	214	209	187	158	146	244
Aug.	33	61	128	174	211	216	208	188	180	223
Sep.	27	27	72	144	191	223	228	223	220	182
Oct.	21	21	35	96	161	207	233	241	242	136
Nov.	15	15	15	52	115	172	212	234	240	85
Dec.	13	13	13	36	91	156	195	225	233	65

Reprinted from ASHRAE Handbook of Fundamentals, *1981. With permission of the American Society of Heating, Refrigerating and Air-Conditioning Engineers, Atlanta, GA.*

Table 9-10

Shading Coefficients for Single Glass and Insulating Glass*[a]

Type of Glass	Nominal Thickness[b] in.	mm	Solar Transmittance	Shading Coefficient: f_o = 4.0 Btu/(hr ft^2 °F) or 23 W/(m^2 °C)	Shading Coefficient: f_o = 3.0 Btu/(hr ft^2 °F) or 17 W/(m^2 °C)
A. Single Glass					
Clear	⅛	32	0.84	1.00	1.00
	¼	64	0.78	0.94	0.95
	⅜	95	0.72	0.90	0.92
	½	127	0.67	0.87	0.88
Heat Absorbing	⅛	32	0.64	0.83	0.85
	¼	64	0.46	0.69	0.73
	⅜	95	0.33	0.60	0.64
	½	127	0.24	0.53	0.58
B. Insulating Glass					
Clear Out–Clear In	⅛[c]	32	0.71[e]	0.88	0.88
Clear Out–Clear In	¼	64	0.61	0.81	0.82
Heat Absorbing[d] Out; Clear In	¼	64	0.36	0.55	0.58

[a]Refers to factory-fabricated units with 3⁄16, ¼, or ½ in. of air space or to prime windows plus storm sash.
[b]Refers to manufacturer's literature for values.
[c]Thickness of each pane of glass, not thickness of assembled unit.
[d]Refers to gray, bronze, and green tinted heat-absorbing float glass.
[e]Combined transmittance for assembled unit.

Reprinted from ASHRAE Handbook and Product Directory 1977 Fundamentals. *With permission of The American Society of Heating, Refrigerating and Air-Conditioning Engineers, Atlanta, Ga.*

Table 9-11

Shading Coefficients (SC) for Single Glass with Indoor Shading by Venetian Blinds or Roller Shades*

	Nominal Thickness[a] mm (in.)	Solar Trans.[b]	Type of Shading: Venetian Blinds: Medium	Venetian Blinds: Light	Roller Shade: Opaque: Dark	Roller Shade: Opaque: White	Roller Shade: Translucent: Light
Clear	2.5 to 6 (3/32 to 1/4)	0.87 to 0.80					
Clear	6 to 12 (1/4 to 1/2)	0.80 to 0.71					
Clear Pattern	3 to 12 (1/8 to 1/2)	0.87 to 0.79	0.64	0.55	0.55	0.25	0.39
Heat-Absorbing Pattern	3 (1/8)	—					
Tinted	5 to 5.5 (3/16, 7/32)	0.74, 0.71					
Heat-Absorbing[d]	5 to 6 (3/16, 1/4)	0.46					
Heat-Absorbing Pattern	5 to 6 (3/16, 1/4)	—	0.57	0.53	0.45	0.30	0.36
Tinted	3 to 5.5 (1/8, 7/32)	0.59, 0.45					
Heat-Absorbing							
or Pattern	—	0.44 to 0.30	0.54	0.52	0.40	0.28	0.32
Heat-Absorbing[d]	10 (3/8)	0.34					
Heat-Absorbing		0.29 to 0.15					
or Pattern	—	0.24	0.42	0.40	0.36	0.28	0.31
Reflective Coated Glass							
S.C.[c] = 0.30			0.25	0.23			
0.40			0.33	0.29			
0.50			0.42	0.38			
0.60			0.50	0.44			

[a]Refer to manufacturer's literature for values.
[b]For vertical blinds with opaque white and beige louvers in the tightly closed position, SC is 0.25 and 0.29 when used with glass of 0.71 to 0.80 transmittance.
[c]SC for glass with no shading device.
[d]Refers to grey, bronze, and green tinted heat-absorbing glass.

Reprinted from ASHRAE Handbook of Fundamentals, *1981. With permission of the American Society of Heating, Refrigerating and Air-Conditioning Engineers, Atlanta, GA.*

Table 9-12

Shading Coefficients (SC) for Insulating Glass[a] with Indoor Shading by Venetian Blinds or Roller Shades*

Type of Glass	Nominal Thickness, Each Light	Solar Trans.[b]: Outer Pane	Solar Trans.[b]: Inner Pane	Type of Shading: Venetian Blinds[c]: Medium	Venetian Blinds[c]: Light	Roller Shade: Opaque: Dark	Roller Shade: Opaque: White	Roller Shade: Translucent: Light
Clear Out	2.5, 3 mm (3/32, 1/8 in.)	0.87	0.87					
Clear In				0.57	0.51	0.60	0.25	0.37
Clear Out								
Clear In	6 mm (1/4 in.)	0.80	0.80					
Heat-Absorbing[d] Out	6 mm (1/4 in.)	0.46	0.80	0.39	0.36	0.40	0.22	0.30
Clear In								
Reflective Coated Glass								
SC[e] = 0.20				0.19	0.18			
0.30				0.27	0.26			
0.40				0.34	0.33			

[a]Refers to fabricated units with 5, 6, or 13-mm (3/16, 1/4, or 1/2-in.) air space, or to prime windows plus storm windows.
[b]Refer to manufacturer's literature for exact values.
[c]For vertical blinds with opaque white or beige louvers, tightly closed, SC is approximately the same as for opaque white roller shades.
[d]Refers to bronze, or green tinted, heat-absorbing glass.
[e]SC for glass with no shading device.

Reprinted from ASHRAE Handbook of Fundamentals, *1981. With permission of the American Society of Heating, Refrigerating and Air-Conditioning Engineers, Atlanta, GA.*

Table 9-13

Shading Coefficients for Domed Skylights* Curb (Fig. 9-5)

Dome	Light Diffuser (Translucent)	Height in.	Width to Height Ratio	Shading Coefficient (SC)	*U* factor Btu/(hr ft^2)
Clear	yes	0	∞	0.61	0.46
(Transmittance) 0.86	0.58	9	5	0.58	0.43
		18	2.5	0.50	0.40
Clear	none	0	∞	0.99	0.80
(Transmittance) 0.86		9	5	0.88	0.75
		18	2.5	0.80	0.70
Translucent	none	0	∞	0.57	0.80
(Transmittance) 0.52		18	2.5	0.46	0.70
Translucent	none	0	∞	0.34	0.80
(Transmittance) 0.27		9	5	0.30	0.75
		18	2.5	0.28	0.70

**Reproduced from* ASHRAE Handbook and Product Directory 1977 Fundamentals. *With permission of the American Society of Heating, Refrigerating and Air-Conditioning Engineers, Atlanta, Ga.*

Table 9-14

Shading Coefficients (SC) for Single and Insulating Glass with Draperies.*[a]

	SC for Index Letters in Fig. 9-6**											
Glazing	**Glass Trans.**	**Glass SC***	**A**	**B**	**C**	**D**	**E**	**F**	**G**	**H**	**I**	**J**
Single Glass												
6 mm (1/4 in.) Clear	0.80	0.95	0.80	0.75	0.70	0.65	0.60	0.55	0.50	0.45	0.40	0.35
12 mm (1/2 in.) Clear	0.71	0.88	0.74	0.70	0.66	0.61	0.56	0.52	0.48	0.43	0.39	0.35
6 mm (1/4 in.) Heat Abs.	0.46	0.67	0.57	0.54	0.52	0.49	0.46	0.44	0.41	0.38	0.36	0.33
12 mm (1/2 in.) Heat Abs.	0.24	0.50	0.43	0.42	0.40	0.39	0.38	0.36	0.34	0.33	0.32	0.30
Reflective Coated	—	0.60	0.57	0.54	0.51	0.49	0.46	0.43	0.41	0.38	0.36	0.33
(see manufacturers' literature	—	0.50	0.46	0.44	0.42	0.41	0.39	0.38	0.36	0.34	0.33	0.31
for exact values)	—	0.40	0.36	0.35	0.34	0.33	0.32	0.30	0.29	0.28	0.27	0.26
	—	0.30	0.25	0.24	0.24	0.23	0.23	0.23	0.22	0.21	0.21	0.20
Insulating Glass 12-mm(1/2-in.) Air												
Space Clear Out and Clear In	0.64	0.83	0.66	0.62	0.58	0.56	0.52	0.48	0.45	0.42	0.37	0.35
Heat Abs. Out and Clear In	0.37	0.55	0.49	0.47	0.45	0.43	0.41	0.39	0.37	0.35	0.33	0.32
Reflective Coated	—	0.40	0.38	0.37	0.37	0.36	0.34	0.32	0.31	0.29	0.28	0.28
(see manufacturers' literature	—	0.30	0.29	0.28	0.27	0.27	0.26	0.26	0.25	0.25	0.24	0.24
for exact values)	—	0.20	0.19	0.19	0.18	0.18	0.17	0.17	0.16	0.16	0.15	0.15

*For glass alone, with no drapery.

**Shading Coefficient values for the SC lines in Fig. 9-6 for representative glazings. Substitute for SC index letters in Fig. 9-6 values on the line of the glazing selected.

[1]*ASHRAE* Handbook of Fundamentals, *1981, with permission of The Society of Heating, Refrigerating and Air-Conditioning Engineers, Atlanta, Georgia.*

Table 9-15

Cooling-Load Factors (CLF) for Glass Without Interior Shading, North Latitudes*

Fenestration Facing	Room Construction	Solar Time, h																							
		1	2	3	4	5	6	7	8	9	10	11	12	13	14	15	16	17	18	19	20	21	22	23	24
N (Shaded)	L	0.17	0.14	0.11	0.09	0.08	0.33	0.42	0.48	0.56	0.63	0.71	0.76	0.80	0.82	0.82	0.79	0.75	0.84	0.61	0.48	0.38	0.31	0.25	0.20
	M	0.23	0.20	0.18	0.16	0.14	0.34	0.41	0.46	0.53	0.59	0.65	0.70	0.73	0.75	0.76	0.74	0.75	0.79	0.61	0.50	0.42	0.36	0.31	0.27
	H	0.25	0.23	0.21	0.20	0.19	0.38	0.45	0.49	0.55	0.60	0.65	0.69	0.72	0.72	0.72	0.70	0.70	0.75	0.57	0.46	0.39	0.34	0.31	0.28
NNE	L	0.06	0.05	0.04	0.03	0.03	0.26	0.43	0.47	0.44	0.41	0.40	0.39	0.39	0.38	0.36	0.33	0.30	0.26	0.20	0.16	0.13	0.10	0.08	0.07
	M	0.09	0.08	0.07	0.06	0.06	0.24	0.38	0.42	0.39	0.37	0.37	0.36	0.36	0.36	0.34	0.33	0.30	0.27	0.22	0.18	0.16	0.14	0.12	0.10
	H	0.11	0.10	0.09	0.09	0.08	0.26	0.39	0.42	0.39	0.36	0.35	0.34	0.34	0.33	0.32	0.31	0.28	0.25	0.21	0.18	0.16	0.14	0.13	0.12
NE	L	0.04	0.04	0.03	0.02	0.02	0.23	0.41	0.51	0.51	0.45	0.39	0.36	0.33	0.31	0.28	0.26	0.23	0.19	0.15	0.12	0.10	0.08	0.06	0.05
	M	0.07	0.06	0.06	0.05	0.04	0.21	0.36	0.44	0.45	0.40	0.36	0.33	0.31	0.30	0.28	0.26	0.23	0.21	0.17	0.15	0.13	0.11	0.09	0.08
	H	0.09	0.08	0.08	0.07	0.07	0.23	0.37	0.44	0.44	0.39	0.34	0.31	0.29	0.27	0.26	0.24	0.22	0.20	0.17	0.14	0.13	0.12	0.11	0.10
ENE	L	0.04	0.03	0.03	0.02	0.02	0.21	0.40	0.52	0.57	0.53	0.45	0.39	0.34	0.31	0.28	0.25	0.22	0.18	0.14	0.12	0.09	0.08	0.06	0.05
	M	0.07	0.06	0.05	0.05	0.04	0.20	0.35	0.45	0.49	0.47	0.41	0.36	0.33	0.30	0.28	0.26	0.23	0.20	0.17	0.14	0.12	0.11	0.09	0.08
	H	0.09	0.09	0.08	0.07	0.07	0.22	0.36	0.46	0.49	0.45	0.38	0.33	0.30	0.27	0.25	0.23	0.21	0.19	0.16	0.14	0.13	0.12	0.11	0.10
E	L	0.04	0.03	0.03	0.02	0.02	0.19	0.37	0.51	0.57	0.57	0.50	0.42	0.37	0.32	0.29	0.25	0.22	0.19	0.15	0.12	0.10	0.08	0.06	0.05
	M	0.07	0.06	0.06	0.05	0.05	0.18	0.33	0.44	0.50	0.51	0.46	0.39	0.35	0.31	0.29	0.26	0.23	0.21	0.17	0.15	0.13	0.11	0.10	0.08
	H	0.09	0.09	0.08	0.08	0.07	0.20	0.34	0.45	0.49	0.49	0.43	0.36	0.32	0.29	0.26	0.24	0.22	0.19	0.17	0.15	0.13	0.12	0.11	0.10
ESE	L	0.05	0.04	0.03	0.03	0.02	0.17	0.34	0.49	0.58	0.61	0.57	0.48	0.41	0.36	0.32	0.28	0.24	0.20	0.16	0.13	0.10	0.09	0.07	0.06
	M	0.08	0.07	0.06	0.05	0.05	0.16	0.31	0.43	0.51	0.54	0.51	0.44	0.39	0.35	0.32	0.29	0.26	0.22	0.19	0.16	0.14	0.12	0.11	0.09
	H	0.10	0.09	0.09	0.08	0.08	0.19	0.32	0.43	0.50	0.52	0.49	0.41	0.36	0.32	0.29	0.26	0.24	0.21	0.18	0.16	0.14	0.13	0.12	0.11
SE	L	0.05	0.04	0.04	0.03	0.03	0.13	0.28	0.43	0.55	0.62	0.63	0.57	0.48	0.42	0.37	0.33	0.28	0.24	0.19	0.15	0.12	0.10	0.08	0.07
	M	0.09	0.08	0.07	0.06	0.05	0.14	0.26	0.38	0.48	0.54	0.56	0.51	0.45	0.40	0.36	0.33	0.29	0.25	0.21	0.18	0.16	0.14	0.12	0.10
	H	0.11	0.10	0.10	0.09	0.08	0.17	0.28	0.40	0.49	0.53	0.53	0.48	0.41	0.36	0.33	0.30	0.27	0.24	0.20	0.18	0.16	0.14	0.13	0.12
SSE	L	0.07	0.05	0.04	0.04	0.03	0.06	0.15	0.29	0.43	0.55	0.63	0.64	0.60	0.52	0.45	0.40	0.35	0.29	0.23	0.18	0.15	0.12	0.10	0.08
	M	0.11	0.09	0.08	0.07	0.06	0.08	0.16	0.26	0.38	0.48	0.55	0.57	0.54	0.48	0.43	0.39	0.35	0.30	0.25	0.21	0.18	0.16	0.14	0.12
	H	0.12	0.11	0.11	0.10	0.09	0.12	0.19	0.29	0.40	0.49	0.54	0.55	0.51	0.44	0.39	0.35	0.31	0.27	0.23	0.20	0.18	0.16	0.15	0.13
S	L	0.08	0.07	0.05	0.04	0.04	0.06	0.09	0.14	0.22	0.34	0.48	0.59	0.65	0.65	0.59	0.50	0.43	0.36	0.28	0.22	0.18	0.15	0.12	0.10
	M	0.12	0.11	0.09	0.08	0.07	0.08	0.11	0.14	0.21	0.31	0.42	0.52	0.57	0.58	0.53	0.47	0.41	0.36	0.29	0.25	0.21	0.18	0.16	0.14
	H	0.13	0.12	0.12	0.11	0.10	0.11	0.14	0.17	0.24	0.33	0.43	0.51	0.56	0.55	0.50	0.43	0.37	0.32	0.26	0.22	0.20	0.18	0.16	0.15
SSW	L	0.10	0.08	0.07	0.06	0.05	0.06	0.09	0.11	0.15	0.19	0.27	0.39	0.52	0.62	0.67	0.65	0.58	0.46	0.36	0.28	0.23	0.19	0.15	0.12
	M	0.14	0.12	0.11	0.09	0.08	0.09	0.11	0.13	0.15	0.18	0.25	0.35	0.46	0.55	0.59	0.59	0.53	0.44	0.35	0.30	0.25	0.22	0.19	0.16
	H	0.15	0.14	0.13	0.12	0.11	0.12	0.14	0.16	0.18	0.21	0.27	0.37	0.46	0.53	0.57	0.55	0.49	0.40	0.32	0.26	0.23	0.20	0.18	0.16
SW	L	0.12	0.10	0.08	0.06	0.05	0.06	0.08	0.10	0.12	0.14	0.16	0.24	0.36	0.49	0.60	0.66	0.66	0.58	0.43	0.33	0.27	0.22	0.18	0.14
	M	0.15	0.14	0.12	0.10	0.09	0.09	0.10	0.12	0.13	0.15	0.17	0.23	0.33	0.44	0.53	0.58	0.59	0.53	0.41	0.33	0.28	0.24	0.21	0.18
	H	0.15	0.14	0.13	0.12	0.11	0.12	0.13	0.14	0.16	0.17	0.19	0.25	0.34	0.44	0.52	0.56	0.56	0.49	0.37	0.30	0.25	0.21	0.19	0.17
WSW	L	0.12	0.10	0.08	0.07	0.05	0.06	0.07	0.09	0.10	0.12	0.13	0.17	0.26	0.40	0.52	0.62	0.66	0.61	0.44	0.34	0.27	0.22	0.18	0.15
	M	0.15	0.13	0.12	0.10	0.09	0.09	0.10	0.11	0.12	0.13	0.14	0.17	0.24	0.35	0.46	0.54	0.58	0.55	0.42	0.34	0.28	0.24	0.21	0.18
	H	0.15	0.14	0.13	0.12	0.11	0.11	0.12	0.13	0.14	0.15	0.16	0.19	0.26	0.36	0.46	0.53	0.56	0.51	0.38	0.30	0.25	0.21	0.19	0.17
W	L	0.12	0.10	0.08	0.06	0.05	0.06	0.07	0.08	0.10	0.11	0.12	0.14	0.20	0.32	0.45	0.57	0.64	0.61	0.44	0.34	0.27	0.22	0.18	0.14
	M	0.15	0.13	0.11	0.10	0.09	0.09	0.09	0.10	0.11	0.12	0.13	0.14	0.19	0.29	0.40	0.50	0.56	0.55	0.41	0.33	0.27	0.23	0.20	0.17
	H	0.14	0.13	0.12	0.11	0.10	0.11	0.12	0.13	0.14	0.14	0.15	0.16	0.21	0.30	0.40	0.49	0.54	0.52	0.38	0.30	0.24	0.21	0.18	0.16
WNW	L	0.12	0.10	0.08	0.06	0.05	0.06	0.07	0.09	0.10	0.12	0.13	0.15	0.17	0.26	0.40	0.53	0.63	0.62	0.44	0.34	0.27	0.22	0.18	0.14
	M	0.15	0.13	0.11	0.10	0.09	0.09	0.10	0.11	0.12	0.13	0.14	0.15	0.17	0.24	0.35	0.47	0.55	0.55	0.41	0.33	0.27	0.23	0.20	0.17
	H	0.14	0.13	0.12	0.11	0.10	0.11	0.12	0.13	0.14	0.15	0.16	0.17	0.18	0.25	0.36	0.46	0.53	0.52	0.38	0.30	0.24	0.20	0.18	0.16
NW	L	0.11	0.09	0.08	0.06	0.05	0.06	0.08	0.10	0.12	0.14	0.16	0.17	0.19	0.23	0.33	0.47	0.59	0.60	0.42	0.33	0.26	0.21	0.17	0.14
	M	0.14	0.12	0.11	0.09	0.08	0.09	0.10	0.11	0.13	0.14	0.16	0.17	0.18	0.21	0.30	0.42	0.51	0.54	0.39	0.32	0.26	0.22	0.19	0.16
	H	0.14	0.12	0.11	0.10	0.10	0.10	0.12	0.13	0.15	0.16	0.18	0.18	0.19	0.22	0.30	0.41	0.50	0.51	0.36	0.29	0.23	0.20	0.17	0.15
NNW	L	0.12	0.09	0.08	0.06	0.05	0.07	0.11	0.14	0.18	0.22	0.25	0.27	0.29	0.30	0.33	0.44	0.57	0.62	0.44	0.33	0.26	0.21	0.17	0.14
	M	0.15	0.13	0.11	0.10	0.09	0.10	0.12	0.15	0.18	0.21	0.23	0.26	0.27	0.28	0.31	0.39	0.51	0.56	0.41	0.33	0.27	0.23	0.20	0.17
	H	0.14	0.13	0.12	0.11	0.10	0.12	0.15	0.17	0.20	0.23	0.25	0.26	0.28	0.28	0.31	0.38	0.49	0.53	0.38	0.30	0.25	0.21	0.18	0.16
HOR	L	0.11	0.09	0.07	0.06	0.05	0.07	0.14	0.24	0.36	0.48	0.58	0.66	0.72	0.74	0.73	0.67	0.59	0.47	0.37	0.29	0.24	0.19	0.16	0.13
	M	0.16	0.14	0.12	0.11	0.09	0.11	0.16	0.24	0.33	0.43	0.52	0.59	0.64	0.67	0.66	0.62	0.56	0.47	0.38	0.32	0.28	0.24	0.21	0.18
	H	0.17	0.16	0.15	0.14	0.13	0.15	0.20	0.28	0.36	0.45	0.52	0.59	0.62	0.64	0.62	0.58	0.51	0.42	0.35	0.29	0.26	0.23	0.21	0.19

L = Light construction: frame exterior wall, 50.8-mm (2-in.) concrete floor slab, approximately 146 kg (30 lb) of material/m^2 (ft^2) of floor area.
M = Medium construction: 101.6-mm (4-in.) concrete exterior wall, 101.6-mm (4-in.) concrete floor slab, approximately 341 kg (70 lb) of building material/m^2 (ft^2) of floor area.
H = Heavy construction: 152.4-mm (6-in.) concrete exterior wall, 152.4-mm (6-in.) concrete floor slab, approximately 635 kg (130 lb) of building materials/m^2 (ft^2) of floor area.

is considered, the solar heat gain through the windows must be considered in two parts. First, any shaded portion of the glass will have diffuse sky radiation; therefore, use a SHGF for North. Second, any portion in the sun should have the SHGF for the particular orientation of the window.

Domed Skylight—The solar and total heat gains for domed skylights can be determined by the procedure used for windows. The SHGFs for such calculations should be consistent with dome orientation. Values of SHGFs for horizontal roofs are given in Table 9-9. For sloping roofs, an approximate SHGF can be

Table 9-16

Cooling-Load Factors (CLF) for Glass with Interior Shading, North Latitudes*

(All Room Constructions)

Fenes-tration Facing	Solar Time, h																							
	1	2	3	4	5	6	7	8	9	10	11	12	13	14	15	16	17	18	19	20	21	22	23	24
N	0.08	0.07	0.06	0.06	0.07	0.73	0.66	0.65	0.73	0.80	0.86	0.89	0.89	0.86	0.82	0.75	0.78	0.91	0.24	0.18	0.15	0.13	0.11	0.10
NNE	0.03	0.03	0.02	0.02	0.03	0.64	0.77	0.62	0.42	0.37	0.37	0.37	0.36	0.35	0.32	0.28	0.23	0.17	0.08	0.07	0.06	0.05	0.04	0.04
NE	0.03	0.02	0.02	0.02	0.02	0.56	0.76	0.74	0.58	0.37	0.29	0.27	0.26	0.24	0.22	0.20	0.16	0.12	0.06	0.05	0.04	0.04	0.03	0.03
ENE	0.03	0.02	0.02	0.02	0.02	0.52	0.76	0.80	0.71	0.52	0.31	0.26	0.24	0.22	0.20	0.18	0.15	0.11	0.06	0.05	0.04	0.04	0.03	0.03
E	0.03	0.02	0.02	0.02	0.02	0.47	0.72	0.80	0.76	0.62	0.41	0.27	0.24	0.22	0.20	0.17	0.14	0.11	0.06	0.05	0.05	0.04	0.03	0.03
ESE	0.03	0.03	0.02	0.02	0.02	0.41	0.67	0.79	0.80	0.72	0.54	0.34	0.27	0.24	0.21	0.19	0.15	0.12	0.07	0.06	0.05	0.04	0.04	0.03
SE	0.03	0.03	0.02	0.02	0.02	0.30	0.57	0.74	0.81	0.79	0.68	0.49	0.33	0.28	0.25	0.22	0.18	0.13	0.08	0.07	0.06	0.05	0.04	0.04
SSE	0.04	0.03	0.03	0.03	0.02	0.12	0.31	0.54	0.72	0.81	0.81	0.71	0.54	0.38	0.32	0.27	0.22	0.16	0.09	0.08	0.07	0.06	0.05	0.04
S	0.04	0.04	0.03	0.03	0.03	0.09	0.16	0.23	0.38	0.58	0.75	0.83	0.80	0.68	0.50	0.35	0.27	0.19	0.11	0.09	0.08	0.07	0.06	0.05
SSW	0.05	0.04	0.04	0.03	0.03	0.09	0.14	0.18	0.22	0.27	0.43	0.63	0.78	0.84	0.80	0.66	0.46	0.25	0.13	0.11	0.09	0.08	0.07	0.06
SW	0.05	0.05	0.04	0.04	0.03	0.07	0.11	0.14	0.16	0.19	0.22	0.38	0.59	0.75	0.83	0.81	0.69	0.45	0.16	0.12	0.10	0.09	0.07	0.06
WSW	0.05	0.05	0.04	0.04	0.03	0.07	0.10	0.12	0.14	0.16	0.17	0.23	0.44	0.64	0.78	0.84	0.78	0.55	0.16	0.12	0.10	0.09	0.07	0.06
W	0.05	0.05	0.04	0.04	0.03	0.06	0.09	0.11	0.13	0.15	0.16	0.17	0.31	0.53	0.72	0.82	0.81	0.61	0.16	0.12	0.10	0.08	0.07	0.06
WNW	0.05	0.05	0.04	0.03	0.03	0.07	0.10	0.12	0.14	0.16	0.17	0.18	0.22	0.43	0.65	0.80	0.84	0.66	0.16	0.12	0.10	0.08	0.07	0.06
NW	0.05	0.04	0.04	0.03	0.03	0.07	0.11	0.14	0.17	0.19	0.20	0.21	0.22	0.30	0.52	0.73	0.82	0.69	0.16	0.12	0.10	0.08	0.07	0.06
NNW	0.05	0.05	0.04	0.03	0.03	0.11	0.17	0.22	0.26	0.30	0.32	0.33	0.34	0.34	0.39	0.61	0.82	0.76	0.17	0.12	0.10	0.08	0.07	0.06
HOR.	0.06	0.05	0.04	0.04	0.03	0.12	0.27	0.44	0.59	0.72	0.81	0.85	0.85	0.81	0.71	0.58	0.42	0.25	0.14	0.12	0.10	0.08	0.07	0.06

**Reprinted from ASHRAE* Handbook of Fundamentals, *1981. With permission of the American Society of Heating, Refrigerating Engineers, Atlanta, GA.*

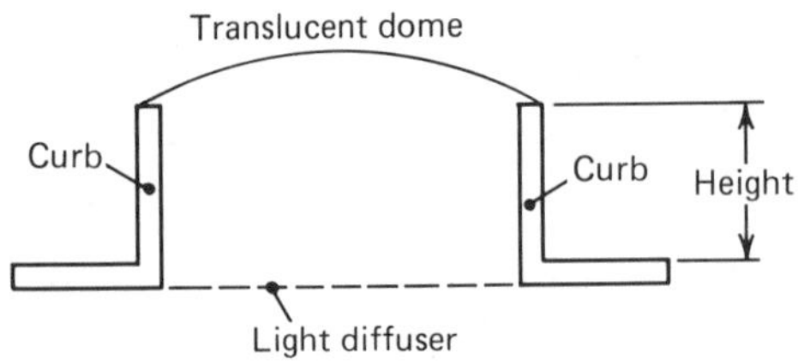

Figure 9–5

Terminology for Domed Skylights

found: SHGF = total solar irradiation on sloping roof divided by 1.15. Table 9-13 gives SC and U factors for plastic domed skylights. The terminology and dimensioning for typical skylights are shown in Fig. 9-5. For further details consult manufacturer's literature.

Shading coefficients for draperies in combination with single or insulating glass is a complex function of color and weave of the drapery fabric. ASHRAE 1 recommends that Table 9-14 in combination with Fig. 9-6 be

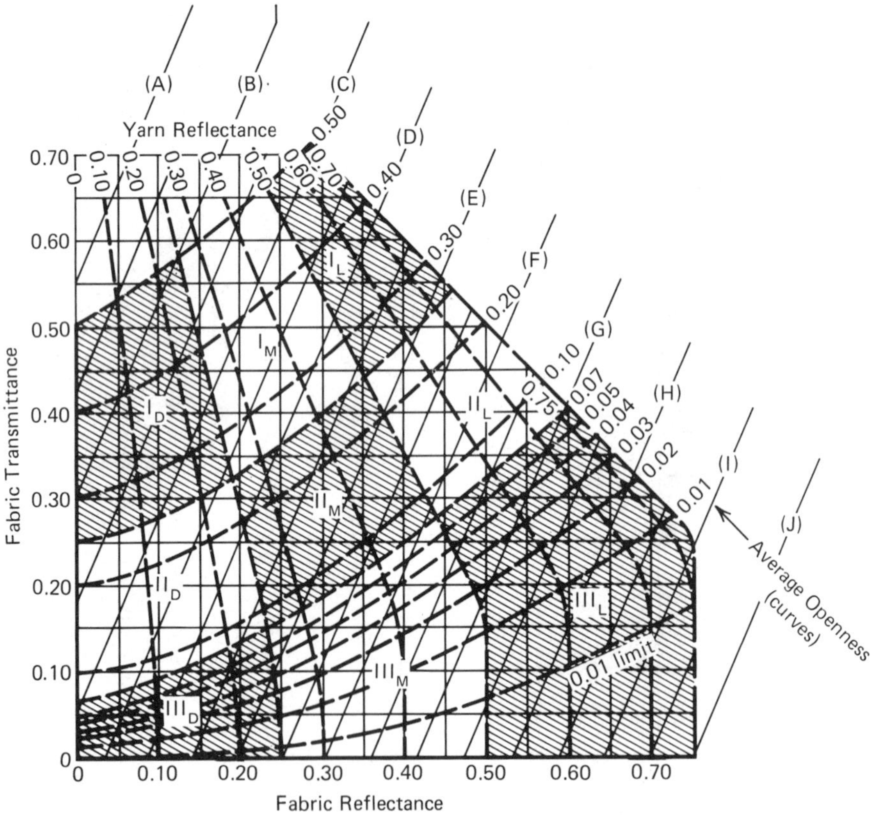

Figure 9–6

Indoor shading properties of drapery fabrics (Reprinted by permission from ASHRAE Handbook of Fundamentals 1981.)

Solution:

Table Data:

	U (Table 7-5)	SC (f_o = 3.0)	SHGF (Table 9-9)
E Glass	0.61	0.82	216
S Glass	0.81	0.70	109

used to determine the shading coefficient for these combinations. Many manufacturers of drapery materials furnish data on the reflectance and transmittance of their products. As we may see, Fig. 9-6 uses reflectance as the abscissa and transmittance as the ordinate. When this data is known we find the intersection of these values in Fig. 9-6 and determine the index letter (A through J). With this index letter we may find a SC in Table 9-14 for the combination of glass and drape. The data of Fig. 9-6 are for draperies of 100 percent fullness (fabric width two times draped width). If the reflectance and transmittance are not known, the letter index may be estimated from the openness of the weave and color of the material. Openness is classified as open weave (I); semiopen weave (II) and closed weave (III). Color is classified as dark (D), medium (M), and light (L). A dark-colored, closed-weave material would be classified as III_D. In Fig. 9-6 the index letter varies from G to I. Therefore, some judgment must be used in selecting the appropriate index letter.

Illustrative Problem 9-4

Determine the cooling load due to windows in the east and south walls of a building at 1000 and 1200 hours solar time in July by using the table at the top of the page. The building is located near 40°N latitude with outside design dry-bulb temperature of 98°F and a 22-degree daily range. The inside design dry-bulb temperature is 78°F. Assume the room construction is heavyweight. The east glass is insulating glass, 0.25 in. of air space having an area of 100 ft^2 with no interior shading. The south glass is ¼ in., single, clear, with inside drapes which have a reflectance of 20 percent and a transmittance of 50 percent. The south glass area is 200 ft^2.

To determine SC for the south glass refer to Fig. 9-6 with 20 percent reflectance and 50 percent transmittance. At intersection find index letter C. Using Table 9-14, index letter C, find SC = 0.70.

Conduction heat gain component: $q = UA(\text{CLTD})$

Time	CLTD Table 9-8	CLTD (Corrected)*	E Glass UA(CLTD)	S Glass UA(CLTD)
1000	4	4 + 2 = 6	(0.61)(100)(06) = 366	(0.81)(200)(06) = 972
1200	9	9 + 2 = 11	(0.61)(100)(11) = 671	(0.81)(200)(11) = 1782

*Footnote of Table 9-8: $t_{oa} = 98 - 22/2 = 87$°F. This is 2.0 degrees higher than the 85°F, therefore add 2 degrees to table values of CLTD to obtain corrected CLTD.

Solar heat gain component: $q = A(SC)(SHGF)(CLF)$

East Glass Time	CLF (Table 9-15)	A(SC)(SHGF)(CLF)
1000	0.49	(100)(0.82)(216)(0.49) = 8679 Btu/hr
1200	0.36	(100)(0.82)(216)(0.36) = 6376 Btu/hr
South Glass:	**CLF (Table 9-16)**	**A(SC)(SHGF)(CLF)**
1000	0.58	(200)(0.70)(109)(0.58) = 8,851 Btu/hr
1200	0.83	(200)(0.70)(109)(0.83) = 12,666 Btu/hr

Total cooling load:

Time	East Glass	South Glass
1000	366 + 8679 = 9045 Btu/hr	972 + 8851 = 9823 Btu/hr
1200	671 + 6376 = 7047 Btu/hr	1782 + 12666 = 14448 Btu/hr

9-10 HEAT GAIN FROM PEOPLE

The rate at which heat and moisture are given off by people in different states of activity, mode of dress, and environmental conditions was discussed in Chapter 6. The human body releases sensible heat by radiation and convection and latent heat by the evaporation of moisture from the skin surface. Table 9-17 gives some practical values of the heat release rates from people for commonly encountered conditions and activities.

The latent heat gain caused by people can be considered as instantaneous cooling load without delay. The radiant portion is first absorbed by the surroundings, then convected to the space air after some time delay. This time delay is taken into account by the CLF. The CLF is a function of the time people occupy the space and the time elapsed since first entering the space. The CLFs are shown in Table 9-18.

If the space temperature is not maintained constant during the 24-hr period (for example, if the cooling system is shut down during the night) a CLF of 1.0 should be used. This pull-down load results from not removing the stored sensible heat in the structure; thus, it reappears as cooling load when the system is started the next day. If there is a high occupant density, as in theaters and auditoriums, a CLF of 1.0 should be used.

Illustrative Problem 9-5

A secretarial office space in a building consists of a secretarial area with space for eight secretaries and a reception and waiting area where one receptionist works and there are spaces available for five people waiting. Determine the cooling load for the area at 1200, 1400, and 1600 hours.

Solution:

Step 1: In the absence of exact count of actual occupancy levels, the following assumptions will be made:

1. Assume secretarial area will have six secretaries on duty on any one day, 9:00 AM to 5:00 PM.

Table 9-17

Rate of Heat Gain from Occupants of Conditioned Spaces*[a]

Degree of Activity	Typical Application	Total Heat Adult, Male		Total Heat Adjusted[b]		Sensible Heat		Latent Heat	
		Watts	Btu/hr	Watts	Btu/hr	Watts	Btu/hr	Watts	Btu/hr
Seated at rest	theater, movie	115	400	100	350	60	210	40	140
Seated, very light work, writing	offices, hotels, apts.	140	480	120	420	65	230	55	190
Seated, eating	restaurant[c]	150	520	170	580[c]	75	255	95	325
Seated, light work typing	offices, hotels, apts.	185	640	150	510	75	255	75	255
Standing, light work, or walking slowly	retail store, bank	235	800	185	640	90	315	95	325
Light bench work	factory	255	880	230	780	100	345	130	435
Walking, 3 mph, light machine work	factory	305	1040	305	1040	100	345	205	695
Bowling[d]	bowling alley	350	1200	280	960	100	345	180	615
Moderate dancing	dance hall	400	1360	375	1280	120	405	255	875
Heavy work, heavy machine work, lifting	factory	470	1600	470	1600	165	565	300	1035
Heavy work, athletics	gymnasium	585	2000	525	1800	185	635	340	1165

[a]Note: Tabulated values are based on 78°F room dry-bulb temperature. For 80°F room dry bulb, the total heat remains the same, but the sensible heat value should be decreased by approximately 8 percent and the latent heat values increased accordingly.
[b]Adjusted total heat gain is based on normal percentage of men, women, and children for the application listed, with the postulate that the gain from an adult female is 85 percent of that of an adult male, and that the gain from a child is 75 percent of that of an adult male.
[c]Adjusted total heat value for eating in a restaurant, includes 60 Btu/hr for food per individual (30 Btu sensible, 30 Btu latent).
[d]For bowling figure one person per alley actually bowling, and all others are sitting (400 Btu/hr) or standing and walking slowly (790 Btu/hr).

**Reprinted from* ASHRAE Handbook and Product Directory 1977 Fundamentals. *With permission of the American Society of Heating, Refrigerating and Air Conditioning Engineers, Atlanta, Ga.*

Table 9-18

Sensible Cooling-Load Factors (CLF) for People*

Total Hours in Space	Hours after Each Entry Into Space																							
	1	2	3	4	5	6	7	8	9	10	11	12	13	14	15	16	17	18	19	20	21	22	23	24
2	0.49	0.58	0.17	0.13	0.10	0.08	0.07	0.06	0.05	0.04	0.04	0.03	0.03	0.02	0.02	0.02	0.02	0.01	0.01	0.01	0.01	0.01	0.01	0.01
4	0.49	0.59	0.66	0.71	0.27	0.21	0.16	0.14	0.11	0.10	0.08	0.07	0.06	0.06	0.05	0.04	0.04	0.03	0.03	0.03	0.02	0.02	0.02	0.01
6	0.50	0.60	0.67	0.72	0.76	0.79	0.34	0.26	0.21	0.18	0.15	0.13	0.11	0.10	0.08	0.07	0.06	0.06	0.05	0.04	0.04	0.03	0.03	0.03
8	0.51	0.61	0.67	0.72	0.76	0.80	0.82	0.84	0.38	0.30	0.25	0.21	0.18	0.15	0.13	0.12	0.10	0.09	0.08	0.07	0.06	0.05	0.05	0.04
10	0.53	0.62	0.69	0.74	0.77	0.80	0.83	0.85	0.87	0.89	0.42	0.34	0.28	0.23	0.20	0.17	0.15	0.13	0.11	0.10	0.09	0.08	0.07	0.06
12	0.55	0.64	0.70	0.75	0.79	0.81	0.84	0.86	0.88	0.89	0.91	0.92	0.45	0.36	0.30	0.25	0.21	0.19	0.16	0.14	0.12	0.11	0.09	0.08
14	0.58	0.66	0.72	0.77	0.80	0.83	0.85	0.87	0.89	0.90	0.91	0.92	0.93	0.94	0.47	0.38	0.31	0.26	0.23	0.20	0.17	0.15	0.13	0.11
16	0.62	0.70	0.75	0.79	0.82	0.85	0.87	0.88	0.90	0.91	0.92	0.93	0.94	0.95	0.95	0.96	0.49	0.39	0.33	0.28	0.24	0.20	0.18	0.16
18	0.66	0.74	0.79	0.82	0.85	0.87	0.89	0.90	0.92	0.93	0.94	0.94	0.95	0.96	0.96	0.97	0.97	0.97	0.50	0.40	0.33	0.28	0.24	0.21

**Reprinted from ASHRAE* Handbook of Fundamentals, *1981. With permission of the American Society of Heating, Refrigerating and Air-Conditioning Engineers, Atlanta, GA.*

2. Assume receptionist on duty 8:00 AM to 2:00 PM.
3. Assume an average of three people in waiting area entering at 10:00 AM.

Step 2: Assume secretaries and receptionist are seated, doing light work and typing.

Table 9-17: Sensible heat = 255 Btu/hr per person.
Latent heat = 255 Btu/hr per person.

Assume people in waiting area are seated, doing very light work, writing or reading.

Table 9-17: Sensible heat = 230 Btu/hr per person.
Latent heat = 190 Btu/hr per person.

Step 3: Determine CLF from Table 9-18.

Assume the three people in the waiting area spent a total of 2 hr in the space having entered at 10:00 AM. CLF = 0.58 at 1200 hours, 0.13 at 1400 hours, and 0.08 at 1600 hours. For the receptionist, total hours in space is 6, 4 hr after entry gives a CLF of 0.72, 6 hr after entry CLF = 0.79, 8 hr after entry CLF = 0.26. For the secretaries, total hours in space is 8, 3 hr after entry CLF = 0.67, 5 hr after entry CLF = 0.76, 7 hr after entry CLF = 0.82.

Step 4: Determine cooling loads

For people waiting:

$$1200 \text{ hr: } \dot{q}_t = 3(230)(0.58) + 3(190) = 970 \text{ Btu/hr}$$

$$1400 \text{ hr: } \dot{q}_t = 3(230)(0.13) + 0 = 90 \text{ Btu/hr}$$

$$1600 \text{ hr: } \dot{q}_t = 3(230)(0.08) + 0 = 55 \text{ Btu/hr}$$

For receptionist:

$$1200 \text{ hr: } \dot{q}_t = 1(255)(0.72) + 1(255) = 439 \text{ btu/hr}$$

$$1400 \text{ hr: } \dot{q}_t = 1(255)(0.79) + 1(255) = 456 \text{ Btu/hr}$$

$$1600 \text{ hr: } \dot{q}_t = 1(255)(0.26) + 0 = 66 \text{ Btu/hr}$$

For secretaries:

$$1200 \text{ hr: } \dot{q}_t = 6(255)(0.67) + 6(255) = 2555 \text{ Btu/hr}$$

$$1400 \text{ hr: } \dot{q}_t = 6(255)(0.76) + 6(255) = 2693 \text{ Btu/hr}$$

$$1600\ hr: \dot{q}_t = 6(255)(0.82) + 6(255) = 2785\ Btu/hr$$

At 1200 hr:

$$\dot{q}_s = 400 + 184 + 1025 = 1609 \text{ Btu/hr}$$

$$\dot{q}_L = 570 + 255 + 1530 = 2355 \text{ Btu/hr}$$

At 1400 hr:

$$\dot{q}_s = 90 + 202 + 1163 = 1455 \text{ Btu/hr}$$

$$q_L = 0 + 255 + 1530 = 1785 \text{ Btu/hr}$$

At 1600 hr:

$$\dot{q}_s = 55 + 66 + 1255 = 1376 \text{ Btu/hr}$$

$$q_L = 0 + 0 + 1530 = 1530 \text{ Btu/hr}$$

9-11 HEAT GAIN FROM INTERIOR LIGHTING

An accurate estimate of the space cooling load imposed by lighting, often a major cooling-load component, is essential. Calculation of this load component is not straightforward; the rate of heat gain to the air can be quite different from the power supplied to the lights. The instantaneous heat gain to a space may be estimated from

$$\dot{q}_s = 3.413 \times \text{total light wattage} \times \text{use factor} \times \text{special allowance factor}$$

or

$$\dot{q}_s = 3.413 \times \text{watts} \times \text{UF} \times \text{SA} \qquad (9\text{-}7)$$

where $\dot{q}_s$ is the sensible heat in Btu/hr. The total light wattage is obtained from the ratings of all light fixtures installed for general illumination or special use. The use factor (UF) is the ratio of wattage in use for the conditions under which the load estimation is being made to the total installed wattage. For commercial applications such as stores, the use factor is usually 1.0. The special allowance factor (SA) is used for fluorescent fixtures and other fixtures requiring more energy than their rated wattage. For fluorescent fixtures, the special allowance factor accounts for ballast losses which vary with the fixture type and voltage. In general, a special allowance factor of 1.2 is satisfactory for estimating work.

Since the heat gain from lighting fixtures is all sensible heat, the space cooling load due to lights does not immediately reflect the full energy output of the lights. Some of the energy emanating from lights in the form of radiation affects the air only after it has been absorbed by floors, walls, and furniture and then transferred to the air by convection after a time lag. This time lag effect must be taken into account in calculating the cooling load, and is done so by applying a cooling-load factor (CLF) to the results of Eq. (9-7).

Design values of CLF for lighting fixtures have been published in Ref. 1. This factor indicates that the cooling load caused by lighting is dependent on the type of fixture, type of air supply and return, space furnishings, and the thermal characteristics of the space. Also these CLFs depend on the total time the lights are on and the number of hours after the lights are turned on. Table 9-19 gives us CLFs for lights with the assistance of Tables 9-20 and 9-21 from which the *a* and *b* classifications are obtained.

Illustrative Problem 9-6

Determine the cooling load in a building at 1400 hours and at 1600 hours due to fluorescent lights, turned on at 8:00 AM and turned off at 6:00 PM. Total lamp wattage is 1200.

Table 9-19

Cooling-Load Factors (CLF) When Lights Are on for 10 Hours*

"a" Coefficients	"b" Classification	Number of hours after lights are turned on																							
		0	1	2	3	4	5	6	7	8	9	10	11	12	13	14	15	16	17	18	19	20	21	22	23
	A	0.03	0.47	0.58	0.66	0.73	0.78	0.82	0.86	0.88	0.91	0.93	0.49	0.39	0.32	0.26	0.21	0.17	0.13	0.11	0.09	0.07	0.06	0.05	0.04
	B	0.10	0.54	0.59	0.63	0.66	0.70	0.73	0.76	0.78	0.80	0.82	0.39	0.35	0.32	0.28	0.26	0.23	0.21	0.19	0.17	0.15	0.14	0.12	0.11
0.45	C	0.15	0.59	0.61	0.64	0.66	0.68	0.70	0.72	0.73	0.75	0.76	0.33	0.31	0.29	0.27	0.26	0.24	0.23	0.21	0.20	0.19	0.18	0.17	0.16
	D	0.18	0.62	0.63	0.64	0.66	0.67	0.68	0.69	0.69	0.70	0.71	0.27	0.26	0.26	0.25	0.24	0.23	0.23	0.22	0.21	0.21	0.20	0.19	0.19
	A	0.02	0.57	0.65	0.72	0.78	0.82	0.85	0.88	0.91	0.92	0.94	0.40	0.32	0.26	0.21	0.17	0.14	0.11	0.09	0.07	0.06	0.05	0.04	0.03
	B	0.08	0.62	0.66	0.69	0.73	0.75	0.78	0.80	0.82	0.84	0.85	0.32	0.29	0.26	0.23	0.21	0.19	0.17	0.15	0.14	0.12	0.11	0.10	0.09
0.55	C	0.12	0.66	0.68	0.70	0.72	0.74	0.75	0.77	0.78	0.79	0.81	0.27	0.25	0.24	0.22	0.21	0.20	0.19	0.17	0.16	0.15	0.14	0.14	0.13
	D	0.15	0.69	0.70	0.71	0.72	0.73	0.73	0.74	0.75	0.76	0.76	0.22	0.22	0.21	0.20	0.20	0.19	0.18	0.18	0.17	0.17	0.16	0.16	0.15
	A	0.02	0.66	0.73	0.78	0.83	0.86	0.89	0.91	0.93	0.94	0.95	0.31	0.25	0.20	0.16	0.13	0.11	0.08	0.07	0.05	0.04	0.04	0.03	0.02
	B	0.06	0.71	0.74	0.76	0.79	0.81	0.83	0.84	0.86	0.87	0.89	0.25	0.22	0.20	0.18	0.16	0.15	0.13	0.12	0.11	0.10	0.09	0.08	0.07
0.65	C	0.09	0.74	0.75	0.77	0.78	0.80	0.81	0.82	0.83	0.84	0.85	0.21	0.20	0.18	0.17	0.16	0.15	0.14	0.14	0.13	0.12	0.11	0.11	0.10
	D	0.11	0.76	0.77	0.77	0.78	0.79	0.79	0.80	0.81	0.81	0.82	0.17	0.17	0.16	0.16	0.15	0.15	0.14	0.14	0.14	0.13	0.13	0.12	0.12
	A	0.01	0.76	0.81	0.84	0.88	0.90	0.92	0.93	0.95	0.96	0.97	0.22	0.18	0.14	0.12	0.09	0.08	0.06	0.05	0.04	0.03	0.03	0.02	0.02
	B	0.04	0.79	0.81	0.83	0.85	0.86	0.88	0.89	0.90	0.91	0.92	0.18	0.16	0.14	0.13	0.12	0.10	0.09	0.08	0.08	0.07	0.06	0.06	0.05
0.75	C	0.07	0.81	0.82	0.83	0.84	0.85	0.86	0.87	0.88	0.89	0.89	0.15	0.14	0.13	0.12	0.12	0.11	0.10	0.10	0.09	0.09	0.08	0.08	0.07
	D	0.08	0.83	0.83	0.84	0.84	0.85	0.85	0.86	0.86	0.87	0.87	0.12	0.12	0.12	0.11	0.11	0.11	0.10	0.10	0.10	0.09	0.09	0.09	0.09

Reprinted from ASHRAE Handbook of Fundamentals, *1981. With permission of the American Society of Heating, Refrigerating and Air-Conditioning Engineers, Atlanta, GA.*

Table 9-20

Design Values of "a" Coefficient. Features of Room Furnishings, Light Fixtures, and Ventilation Arrangements[a]

a	Furnishings	Air Supply and Return	Type of Light Fixture
0.45	Heavyweight, simple furnishings, no carpet	Low rate; supply and return below ceiling [$V \leq 2.5$ (0.5)]*	Recessed, not vented
0.55	Ordinary furniture, no carpet	Medium to high ventilation rate; supply and return below ceiling or through ceiling grill and space [$V \geq 2.5$ (0.5)]*	Recessed, not vented
0.65	Ordinary furniture, with or without carpet	Medium to high ventilation rate or fan coil or induction type air-conditioning terminal unit; supply through ceiling or wall diffuser; return around light fixtures and through ceiling space. [$V \geq 2.5$ (0.5)]*	Vented
0.75 or greater	Any type of furniture	Ducted returns through light fixtures	Vented or free-hanging in air stream with ducted returns

*V is room air supply rate in L/s m^2 (CFM/ft^2) of floor area

[a]*Reprinted from ASHRAE* Handbook of Fundamentals, *1981. With permission of the American Society of Heating, Refrigerating and Air-Conditioning Engineers, Atlanta, GA.*

Assume an allowance factor of 1.20. The room has ordinary office furniture, tile flooring over a 3-in. concrete floor, and a medium air circulation rate. Lights are recessed and not vented.

Solution:

Step 1: Determine a and b classification.
Table 9-21, find b classification to be B.
Table 9-20, find a classification to be 0.55.

Step 2: Determine cooling-load factors.
Table 9-19, with a = 0.55 and b = B.
At 1400 hours, lights have been on for 6 hr, CLF = 0.78.
At 1600 hours, lights have been on for 8 hr, CLF = 0.82.

Step 3: Determine cooling load.
At 1400 hours:

$$\dot{q}_s = 3.413(1200)(1.0)(1.20)(0.78) = 3835 \text{ Btu/hr}$$

At 1600 hours:

Table 9-21

The "b" Classification Values Calculated for Different Envelope Construction and Room Air Circulation Rates[a]

Room Envelope Construction* [mass of floor area, (kg/m², lb/ft²)]	Room Air Circulation and Type of Supply and Return**			
	Low	Medium	High	Very High
50.8-mm (2-in.) Wood Floor (48.8, 10)	B	A	A	A
76.2-mm (3-in.) Concrete Floor (195.3, 40)	B	B	B	A
152.4-mm (6-in.) Concrete Floor (366.2, 75)	C	C	C	B
203.2-mm (8-in.) Concrete Floor (585.8, 120)	D	D	C	C
304.8-mm (12-in.) Concrete Floor (781.1, 160)	D	D	D	D

*Floor covered with carpet and rubber pad; for a floor covered only with floor tile take next classification to the right in the same row.

***Low:* Low ventilation rate—minimum required to cope with cooling load due to lights and occupants in interior zone. Supply through floor, wall, or ceiling diffuser. Ceiling space not vented and f = 2.27 W/m² °C (0.4 Btu/hr ft² °F)(where f = inside surface convection coefficient used in calculation of b classification).

Medium: Medium ventilation rate, supply through floor, wall, or ceiling diffuser. Ceiling space not vented and f = 3.41 W/m² °C (0.60 Btu/hr ft² °F).

High: Room air circulation induced by primary air of induction unit or by fan coil unit. Return through ceiling space and f = 4.54 W/m² °C (0.80 Btu/hr ft² °F).

Very High: High room air circulation used to minimize temperature gradients in a room. Return through ceiling space and f = 6.81 W/m² °C (1.2 Btu/hr ft² °F).

[a]*Reprinted from ASHRAE* Handbook of Fundamentals, *1981. With permission of the American Society of Heating, Refrigerating and Air-Conditioning Engineers, Atlanta, GA.*

$$\dot{q}_s = 3.413(1200)(1.0)(1.20)(0.82)$$
$$= 4030 \text{ Btu/hr}$$

9-12 HEAT GAIN FROM APPLIANCES AND OTHER EQUIPMENT

There is a wide variety of appliances and equipment used in buildings which may contribute a cooling load which needs to be considered.

The most common heat-producing appliances found in conditioned areas are those used for food preparation in commercial and industrial food service establishments such as restaurants, hospitals, school cafeterias, hotels, and in-plant cafeterias. Counter and back-bar appliances are frequently located in a dining or service area, while heavy-duty equipment is usually confined to the kitchen and installed under an exhaust hood. Table 9-22 contains data for heat gain from many appliances.

Laboratory tests[6] have shown that (a) appliance surfaces contribute most of the heat to commercial kitchens, (b) the heat is primarily radiant energy from appliance surfaces and cooking utensils, and (c) convected and latent heat are negligible when appliances are installed under an affective hood. It is reasonable to assume this applies to all hooded appliances.

A conservative estimate of the maximum heat due to radiation release into a kitchen is 32 percent of the hourly Btu input. Unfortunately for most appliances, the only information available is the total nameplate or catalog input rating, not the hourly input required to maintain typical appliance surface temperatures. Therefore, it is common practice to introduce a *usage factor* which when multiplied by the input rating approximates the actual hourly input to the appliances. Based on historical demand studies by utility companies, this usage factor is normally assumed to be 0.50. However, ASHRAE Ref. 1 recommends that the following be used to compute the maximum hourly heat gain for electric and steam appliances installed under a hood:

$$\dot{q} = q_i \times F_R = 0.32 q_i \qquad (9\text{-}8)$$

where

$\dot{q}_i$ = actual input energy rate to maintain typical appliance surface temperatures, watts (Btu/hr)

Table 9-22A

Recommended Rate of Heat Gain from Commercial Cooking Appliances Located in the Air-Conditioned Area*[1]

Appliance	Capacity	Overall Dim., Inches Width x Depth x Height	Miscellaneous Data (Dimensions in inches)	Manufacturer's Input Rating: Boiler hp or Watts	Manufacturer's Input Rating: Btu/Hr	Probable Max. Hourly Input Btu/Hr	Recommended Rate of Heat Gain, Btu/Hr — Without Hood: Sensible	Without Hood: Latent	Without Hood: Total	With Hood[2]: All Sensible
Gas-Burning, Counter Type										
Broiler-griddle		31 x 20 x 18			36,000	18,000	11,700	6,300	18,000	3,600
Coffee Brewer										
per burner			With *warm* position		5,500	2,500	1,750	750	2,500	500
Water heater burner			With storage tank		11,000	5,000	3,850	1,650	5,500	1,100
Coffee urn	3 gal.	12-inch dia.			10,000	5,000	3,500	1,500	5,000	1,000
	5 gal.	14-inch dia.			15,000	7,500	5,250	2,250	7,500	1,500
	8 gal. twin	25-inch wide			20,000	10,000	7,000	3,000	10,000	2,000
Deep fat fryer	15 lb fat	14 x 21 x 15			30,000	15,000	7,500	7,500	15,000	3,000
Dry food warmer										
per sq ft of top					1,400	700	560	140	700	140
Griddle, frying										
per sq ft of top					15,000	7,500	4,900	2,600	7,500	1,500
Short order stove,										
per burner			Open grates		10,000	5,000	3,200	1,800	5,000	1,000
Steam table										
per sq ft of top					2,500	1,250	750	500	1,250	250
Toaster, continuous	360 slices/hr	19 x 16 x 30	2 slices wide		12,000	6,000	3,600	2,400	6,000	1,200
	720 slices/hr	24 x 16 x 30	4 slices wide		20,000	10,000	6,000	4,000	10,000	2,000
Gas-Burning, Floor Mounted Type										
Broiler, unit		24 x 26 grid	Same burner heats oven		70,000	35,000	Exhaust hood required	Exhaust hood required	Exhaust hood required	7,000
Deep fat fryer	32 lb fat		14-inch kettle		65,000	32,500				6,500
	56 lb fat		18-inch kettle		100,000	50,000				10,000
Oven, deck, per sq ft			Same for 7 and 12							
of hearth area			high decks		4,000	2,000				400
Oven, roasting		32 x 32 x 60	Two ovens—24 x 28 x 15		80,000	40,000				8,000
Range, heavy duty		32 x 42 x 33								
Top section			32 wide x 39 deep		64,000	32,000				6,400
Oven			25 x 28 x 15		40,000	20,000				4,000
Range, Jr., heavy duty		31 x 35 x 33								
Top Section			31 wide x 32 deep		45,000	22,500				4,500
Oven			24 x 28 x 15		35,000	17,500				3,500
Range, restaurant type										
Per 2-burner sect.			12 wide x 28 deep		24,000	12,000				2,400
Per oven			24 x 22 x 14		30,000	15,000				3,000
Per broiler-griddle			24 wide x 26 deep		35,000	17,500				3,500
Electric, Counter Type										
Coffee brewer										
Per burner				625	2,130	1,000	770	230	1,000	340
Per warmer				160	545	300	230	70	300	90
Automatic	240 cups/hr	27 x 21 x 22	4-burner + water htr.	5,000	17,000	8,500	6,500	2,000	8,500	1,700
Coffee urn	3 gal.			2,000	6,800	3,400	2,550	850	3,400	1,000
	5 gal.			3,000	10,200	5,100	3,850	1,250	5,100	1,600
	8 gal. twin			4,000	13,600	6,800	5,200	1,600	6,800	2,100
Deep fat fryer	14 lb fat	13 x 22 x 10		5,500	18,750	9,400	2,800	6,600	9,400	3,000
	21 lb fat	16 x 22 x 10		8,000	27,300	13,700	4,100	9,600	13,700	4,300
Dry food warmer,				240	820	400	320	80	400	130
per sq ft of top										
Egg boiler	2 cups	10 x 13 x 25		1,100	3,750	1,900	1,140	760	1,900	600
Griddle, frying,				2,700	9,200	4,600	3,000	1,600	4,600	1,500
per sq ft of top										
Griddle-Grill		18 x 20 x 13	Grid, 200 sq in.	6,000	20,400	10,200	6,600	3,600	10,200	3,200
Hotplate		18 x 20 x 13	2 heating units	5,200	17,700	8,900	5,300	3,600	8,900	2,800
Roaster		18 x 20 x 13		1,650	5,620	2,800	1,700	1,100	2,800	900
Roll warmer		18 x 20 x 13		1,650	5,620	2,800	2,600	200	2,800	900
Toaster, continuous	360 slices/hr	15 x 15 x 28	2 slices wide	2,200	7,500	3,700	1,960	1,740	3,700	1,200
	720 slices/hr	20 x 15 x 28	4 slices wide	3,000	10,200	5,100	2,700	2,400	5,100	1,600
Toaster, pop-up	4 slice	12 x 11 x 9		2,540	8,350	4,200	2,230	1,970	4,200	1,300
Waffle iron		18 x 20 x 13	2 grids	1,650	5,620	2,800	1,680	1,120	2,800	900

Table 9-22A (Cont.)

Appliance	Capacity	Overall Dim., Inches Width x Depth x Height	Miscellaneous Data (Dimensions in Inches)	Manufacturer's Input Rating: Boiler hp or Watts	Manufacturer's Input Rating: Btu/Hr	Probable Max. Hourly Input Btu/Hr	Recommended Rate of Heat Gain, Btu/Hr — Without Hood: Sensible	Without Hood: Latent	Without Hood: Total	With[2] Hood: All Sensible
Electric, Floor Mounted Type										
Broiler, no oven			23 wide x 25 deep grid	12,000	40,900	20,500	Exhaust hood required	Exhaust hood required	Exhaust hood required	6,500
With oven			23 x 27 x 12 oven	18,000	61,400	30,700				9,800
Deep fat fryer	28 lb fat	20 x 38 x 36	14 wide x 15 deep kettle	12,000	40,900	20,500				6,500
	60 lb fat	24 x 36 x 36	20 wide x 20 deep kettle	18,000	61,400	30,700				9,800
Oven, baking, per sq ft of hearth			Compartment 8-in. high	500	1,700	850				270
Oven, roasting per sq ft of hearth			Compartment 12-in. high	900	3,070	1,500				490
Range, heavy duty		36 x 36 x 36								
Top section				15,000	51,100	25,600				8,200
Oven				6,000	20,400	10,200				3,200
Range, medium duty		30 x 32 x 36								
Top section				8,000	27,300	13,600				4,300
Oven				3,600	12,300	6,200				1,900
Range, light duty		30 x 29 x 36								
Top section				6,600	22,500	11,200				3,600
Oven				3,000	10,200	5,100				1,600
Steam Heated										
Coffee urn	3 gal.			0.2	6,600	3,300	2,180	1,120	3,300	1,000
	5 gal.			0.3	10,000	5,000	3,300	1,700	5,000	1,600
	8 gal twin			0.4	13,200	6,600	4,350	2,250	6,600	2,100
Steam table per sq ft of top			With insets	0.05	1,650	825	500	325	825	260
Bain marie per sq ft of top			Open tank	0.10	3,300	1,650	825	825	1,650	520
Oyster steamer				0.5	16,500	8,250	5,000	3,250	8,250	2,600
Steam kettles per gal. capacity			Jacketed type	0.06	2,000	1,000	600	400	1,000	320
Compartment steamer per compartment		24 x 25 x 12 compartment	Floor mounted	1.2	40,000	20,000	12,000	8,000	20,000	6,400
Compartment steamer	3 pans 12 x 20 x 2½		Single counter unit	0.5	16,500	8,250	5,000	3,250	8,250	2,600
Plate warmer per cu ft				0.05	1,650	825	550	275	825	260

Rate of Heat Gain from Miscellaneous Appliances

Electrical Appliances

Appliance	Miscellaneous Data	Manufacturer's Rating: Watts	Manufacturer's Rating: Btu/Hr	Recommended Rate of Heat Gain, Btu/Hr: Sensible	Latent	Total
Hair Dryer	Blower type	1580	5400	2300	400	2700
Hair dryer	Helmet type	705	2400	1870	330	2200
Permanent wave machine	60 heaters @25 W, 36 in normal use	1500	5000	850	150	1000
Neon sign, per linear ft of tube	½ in., dia.			30		30
	3/8 in., dia.			60		60
Sterilizer, instrument		1100	3750	650	1200	1850

Gas-Burning Appliances

Appliance	Miscellaneous Data	Manufacturer's Rating: Watts	Manufacturer's Rating: Btu/Hr	Recommended Rate of Heat Gain, Btu/Hr: Sensible	Latent	Total
Lab burners						
Bunsen	7/16 in. barrel		3000	1680	420	2100
Fishtail	1½ in. wide		5000	2800	700	3500
Meeker	1 in. diameter		6000	3360	840	4200
Gas light, per burner	Mantle type		2000	1800	200	2000
Cigar lighter	Continuous flame		2500	900	100	1000

[1]Heat gain from cooking appliances located in conditioned area (but not included in the table) should be estimated as follows: (1) Obtain *probable maximum hourly input* by multiplying the manufacturer's hourly input rating by the usage factor of 0.50. (2) If appliances are installed without an exhaust hood, the estimated latent heat gain is 34 percent of the *probable maximum hourly input* and the sensible heat gain is 66 percent. (3) If appliances are installed under an effective exhaust hood, the estimated heat gain is sensible heat and can be calculated from equations in Table 9-9B.

[2]For poorly designed or undersized exhaust systems the heat gains in this column should be doubled and half the increase assumed as latent heat.

**Reproduced from ASHRAE* Handbook and Product Directory, *1977* Fundamentals. *With permission of the American Society of Heating, Refrigerating and Air-Conditioning Engineers, Atlanta, Ga.*

Table 9-22B

This formula will help to determine how much heat will be exhausted and how much will remain in the conditioned space, with respect to cooking equipment under an effective exhaust hood.	
Gas Burning Appliance	QA = 0.10 x Qi
Electric or Steam Appliance	QA = 0.16 x Qi

QA = Heat Gain in conditioned space from appliance, Btuh. Qi = Rated input of appliance.

F_R = fraction of input energy of the hooded appliance released by radiation = 0.32

If the value of q_i is not available for the appliance in question, the following equation can be used as an estimate of the maximum hourly heat gain:

$$\dot{q} = q_r F_u F_R = 0.16\, q_r \qquad (9\text{-}9)$$

where

$\dot{q}_r$ = input rating of the hooded appliance, watts (Btu/hr)

F_u = usage factor = 0.50

Correct assessment of internal loads in laboratory spaces is essential for correct cooling-load calculations. As a general rule, in a laboratory with outdoor exposure, equipment load can be as high as four times the load from all other sources combined.[7]

In comparison to laboratories and computer areas heat gains from equipment in offices are generally very small; the only heat-generating equipment are electric typewriters, calculators, and sometimes posting machines, check writers, telewriters, and so on. The heat gain may average out to 3 to 4 Btu/hr ft^2 for general offices, and 6 to 7 Btu/hr ft^2 for purchasing and accounting departments. In offices provided with computer display units, it may be as high as 15 Btu/hr ft^2.

In computer areas the heat generated by electronic equipment ranges from 75 to 175 Btu/hr ft^2 for digital computers and 50 to 150 Btu/hr ft^2 for analog computers. Although transistors reduce the heat output of such equipment greatly, there seems to be much more electronic equipment installed per square foot of floor space, with the result that the decrease in heat dissipated per unit area amounts to only 25 percent less than the figures already mentioned.

Laboratory equipment heat output generally range from 15 to 70 Btu/hr ft^2. The equipment heat output in a manufacturing plant, on the other hand, ranges from 20 Btu/hr ft^2 for general assembly and stamping plants, up to 150 Btu/hr ft^2 in planting, foaming, and curing processes. In production, the trend today is towards automation; consequently, high concentrations of high heat output equipment are not uncommon.

The effect of the sensible heat gain from appliances and laboratory equipment on the cooling load is delayed in the same manner as the radiant fraction of the sensible heat gain of other load components. Table 9-23 provides the CLF for appliances. The sensible portion of the cooling load is obtained by multiplying the sensible heat gain by the appropriate CLF.

9-13 HEAT GAIN TO DUCTS AND PLENUMS

The cooling coils of any air-conditioning system must remove the heat gains to conditioned spaces. Also, they must remove any heat added to the air flowing through the ductwork which passes through unconditioned spaces.

If possible, the running of ducts through hot

Table 9-23

Sensible Heat Cooling Load Factors for Appliances*

Total Operational Hours	Hooded: Hours after appliances are on											
	1	2	3	4	5	6	7	8	9	10	11	12
2	0.27	0.40	0.25	0.18	0.14	0.11	0.09	0.08	0.07	0.06	0.05	0.04
4	0.28	0.41	0.51	0.59	0.39	0.30	0.24	0.19	0.16	0.14	0.12	0.10
6	0.29	0.42	0.52	0.59	0.65	0.70	0.48	0.37	0.30	0.25	0.21	0.18
8	0.31	0.44	0.54	0.61	0.66	0.71	0.75	0.78	0.55	0.43	0.35	0.30
10	0.33	0.46	0.55	0.62	0.68	0.72	0.76	0.79	0.81	0.84	0.60	0.48
12	0.36	0.49	0.58	0.64	0.69	0.74	0.77	0.80	0.82	0.85	0.87	0.88
Unhooded												
2	0.56	0.64	0.15	0.11	0.08	0.07	0.06	0.05	0.04	0.04	0.03	0.03
4	0.57	0.65	0.71	0.75	0.23	0.18	0.14	0.12	0.10	0.08	0.07	0.06
6	0.57	0.65	0.71	0.76	0.79	0.82	0.29	0.22	0.18	0.15	0.13	0.11
8	0.58	0.66	0.72	0.76	0.80	0.82	0.85	0.87	0.33	0.26	0.21	0.18
10	0.60	0.68	0.73	0.77	0.81	0.83	0.85	0.87	0.89	0.90	0.36	0.29
12	0.62	0.69	0.75	0.79	0.82	0.84	0.86	0.88	0.89	0.91	0.92	0.93

**Extracted from* ASHRAE Handbook and Product Directory 1977 Fundamentals. *With permission of the American Society of Heating, Refrigerating Engineers, Atlanta, Ga.*

spaces such as kitchens, laundries, boiler rooms, and unconditioned storage rooms should be avoided. When it is necessary, these ducts should be insulated.

Plenum chambers are spaces kept under pressure by a steady delivery of air to them. They are ordinarily located under the floor or above the ceiling of a conditioned space. The plenum chamber is connected to the space to be cooled by openings through either the floor or ceiling. Because of the pressure in the plenum chamber, there is a constant flow of cool air into the conditioned space.

Plenum chambers are often found under floors of theaters, though they are seldom used in new theaters. Also, plenum chambers may be used above the ceilings of some types of buildings. These chambers may be 2 to 4 ft deep and often extend under the entire floor or over the entire ceiling. Conditioned air is delivered to the plenum chamber by ducts from a supply air fan. Numerous small openings in the floor or ceiling allow the chilled air to flow into the conditioned space.

Ducts are preferred to plenum chambers. The delivery of exact quantities of air to given locations is always more positive with ducts than with plenum chambers.

Exhaust and return-air plenum chambers are also used occasionally. They are exactly like supply plenum chambers except that the pressure is slightly below atmospheric. So there is a constant flow of air from the room into the plenum chamber. An exhaust fan, or the supply air fan, continuously withdraws air from the plenum chamber. All the disadvantages of supply plenum chambers apply equally well to exhaust chambers.

Ordinarily, the heat gain through ducts may be neglected. It is usually small compared to the rest of the heat gains of the average installation. But the heat gains may be too large to be neglected. This could happen if there is a lot of ductwork or if it passes through hot spaces.

The heat gain through ducts cannot be determined until the ducts are sized. The ducts cannot be sized until the required air quantities are determined. The air quantities cannot be determined until the heat gains are found. So everything seems to depend upon something else and it appears there is no place to start.

The logical first step is to estimate the duct system required, as to probable size and extent, and where it may pass through unconditioned spaces. Next an estimate of the duct system's probable heat gain should be made, and all other heat gains determined. Next, the air quantities should be calculated and the duct system designed. Now, the preliminary heat gain estimate to the ducts may be checked.

Conduction and convection heat gain is calculated in a similar manner as other such heat transfer calculation from $q = UA\Delta t$, where Δt is the temperature difference between the air outside of the duct and the air inside the duct. Typical U factors may be used from Table 9-24.

Table 9-24

Overall Heat Transmission Coefficients for Ductwork*

Description of Ductwork	U Factor Btu/(hr ft^2 °F)
Sheet metal, not insulated	1.18
½-in.-thick insulation board with or without sheet metal	0.38
1-in. ditto	0.22
1½-in. ditto	0.15
2-in. ditto	0.12

Note: The heat gain or loss from a duct is dependent on the velocity of air flow through the duct. Therefore, the table values should be used for estimating work only.

**Reproduced by permission from* Trane Air Conditioning Manual, *The Trane Company, LaCrosse Wisconsin.*

Illustrative Problem 9-7

A 75-ft.-long section of sheet metal duct runs through a space where the temperature is expected to be 90°F. The air flowing through the duct is at 60°F. The duct dimensions are 36 in. wide and 24 in. deep. Find the heat gain to the air in the duct.

Solution:

The perimeter of the duct section is 2(3) + 2(2) = 10 ft. The surface area of the 75-ft.-long section is (75)(10) = 750 ft^2. From Table 9-24, $U = 1.18$, therefore $\dot{q} = UA\Delta t = 1.18(750)(90 - 60) = 26{,}550$ Btu/hr.

Sometimes ducts are installed with their top surface against a ceiling and one side against a wall. The heat gain is calculated as though all four sides of the duct were exposed, because it is difficult to hang a duct with its surfaces in tight contact with a ceiling or wall.

Insulation, including a vapor barrier, is always necessary on ducts passing through usually hot spaces.

Illustrative Problem 9-8

Using the same data as Problem 9-7, assume that the duct is covered with ½ in. of fiberglass duct insulation. Determine the percent savings in heat gain over the uninsulated duct.

Solution:

From Table 9-24, find $U = 0.38$. Therefore,

$$\dot{q} = (0.38)(750)(90 - 60) = 8550 \text{ Btu/hr}$$

$$\text{Savings} = \frac{26{,}550 - 8550}{26{,}550}(100)$$

$$= 67.8 \text{ percent}$$

9-14 HEAT GAIN FROM ELECTRIC MOTORS

Electric motors give off sensible heat while running. In Section 9-13 we found that heat added to the air stream in ductwork had to be removed by the cooling equipment. Likewise, the heat equivalent of the motor load must be removed by the cooling equipment. And this is true whether the motor load is in the conditioned room or in the air stream. Take a motor driving a supply air fan as an example. Assume that output of the motor is 5 hp and that the motor is outside the fan chamber. The 5 hp delivered to the fan is energy added to the air stream. The heat equivalent of 5 hp is 12,725 Btu/hr (5 × 2545 Btu/hp hr). The cooling equipment does not care where the 12,725 Btu/hr came from; it still must do the same cooling job—whether the heat gain was from the room, through the ductwork, or from the fan.

Of course, motors are less than 100 percent efficient. So in order for the motor to deliver 5 hp, its input must be larger than 5 hp. Assume that the efficiency of a 5-hp motor is 80 percent. Then the input to the motor is 6.25 hp (5.0/0.80 = 6.25). Of course this energy arrives at the motor in the form of electricity. But eventually it is converted to heat. Generally the heat equivalent of the input electrical energy is considered part of the space load. However, in some cases it may be desirable to consider the motor is outside the conditioned space while the driven device is inside the space.

Table 9-25 gives values of heat gain which may be used for estimating heat gain from electric motors.

Some electric-motor-driven appliances may be located in conditioned spaces. Heat gain to the space in which they are located may be estimated by use of the data in Tables 9-22 through 9-28.

Table 9-25

Heat Gain from Electric Motors*

Nameplate Rating of Motor in Horsepower	Average Motor Efficiency in Continuous Operation	Btu per Hr to Room Air per Rated Horsepower of Motor: Motor outside of Room, Driven Device Inside Room	Motor in Room, Driven Device Outside of Room	Motor and Driven Device both Inside Room
1/8 to 1/2	0.60	2545	1700	4246
1/2 to 3	0.69	2545	1100	3646
3 to 20	0.85	2545	400	2946

$$\text{General Rule for Motors: Btu/hr} = \frac{2545 \times \text{hp (connected load)}}{\text{Motor Efficiency}}$$

Note: Where possible obtain actual value of motor efficiency. Where not possible: for motors use average efficiencies as listed in table above; for motor generators use average efficiency sets up to 3 hp as 0.55; for larger sets use average efficiency of 0.80.

**Extracted with permission from* Handbook of Air Conditioning, Heating and Ventilating, *The Industrial Press, New York, 1965.*

Table 9-26

Electric-Motor-Driven Appliances*
(Motor and Driven Appliance both in Same Room)

Fans (Blade diameter in inches)		Btu/hr	Appliances	Btu/hr
Ceiling	32	340	Clock	7
	52	410	Hair dryer	1900
	56	600	Drink mixer	240
Desk or Wall	8	120	Sewing machine (domestic)	220
	10	140	Vacuum cleaner (domestic)	250
	12	200	Hair clipper	78
	16	300	Vibrator (beauty)	11

Note: Figures in table are thermal equivalent of nameplate rating corrected for motor efficiency.

**Extracted with permission from* Handbook of Air Conditioning, Heating and Ventilating, *The Industrial Press, New York, 1965.*

Table 9-27

Electric Refrigerators*
(With well insulated cabinet in 80°F air and 40°F inside cabinet)

Electric Motor and Air Cooled Condenser in Cabinet in Room Air		Cold Cabinet in Room Compressor and Condenser Remote	
Cabinet Volume (ft^3)	Btu/hr (Thermal equivalent of motor input)	Cabinet Volume (ft^3)	Btu/hr per Cu ft of Cabinet Volume
2–4	530	2–4	100 to 75
5	710	5–6	70 to 65
6–10	850	7–10	60 to 55
12–18	1060	12, 14, 16	55 to 50
		20, 25, 30	50, 45, 40

**Extracted with permission from* Handbook of Air Conditioning, Heating and Ventilating, *The Industrial Press, New York, 1965.*

9-15 HEAT GAIN FROM INFILTRATION AND OUTDOOR VENTILATION AIR.

Outdoor air must be introduced for ventilation of conditioned spaces. Chapter 6 suggests minimum outdoor air requirements for representative applications, although these requirements are not necessarily adequate for psychological attitudes and physiological responses. Also, it must be remembered that codes and ordinances must be satisfied. From a viewpoint of cooling-load calculations, occupants of a conditioned space should be provided with not less than 5 cfm per person when no smoking is involved, and 25 to 40 cfm when smoking is a factor.

Table 9-28

Miscellaneous Electric Appliances*

Type of Appliance	Btu/hr	Type of Appliance	Btu/hr
Domestic electric irons		Radio	340
small	—	Infrared lamp (heat)	850
medium	2300	Sun lamp	1350
large	3400	Neon lights (15 mm), per ft.	12
Tumbler water heaters	1200	Neon lights (11 mm), per ft.	18
Television set	680	Permanent waver (beauty)	4240

Note: Figures are thermal equivalent of nameplate rating of typical example of appliance listed.

**Extracted with permission from* Handbook of Air Conditioning, Heating and Ventilating, *The Industrial Press, New York, 1965.*

Infiltration air enters a building through cracks, through doors when they are opened, and through any other openings. The procedure for calculating infiltration air flow rates are the same as presented in Chapter 8, with one exception, the wind velocity for summer cooling-load calculations is 7½ mph instead of the 15 mph for heating-load calculations. Tables presented in Chapter 8 may be used for estimating infiltration air quantities.

In the case of air-conditioning design work, frequently the amount of ventilation air introduced to a building, with no exhaust fans operating, will be adequate to off-set infiltration through cracks and door openings. In other words, a slight positive pressure is maintained within the building. However, any infiltration air entering a building brings with it the high-temperature outside air with its associated high moisture content. The higher temperature outside air introduces a sensible heat gain to the space, and the high moisture content introduces a latent heat load to the space.

The relationships for calculating the sensible and latent heat loads are similar to those used in Chapter 8 and would be computed as follows:

$$\dot{q}_s = \dot{m}_{oa} c_p (t_o - t_i) = \frac{Q_{oa} C_p}{v_o} (t_o - t_i) \tag{9-10}$$

$$\dot{q}_l = \dot{m}_{oa} (W_o - W_i) h_{fg} = \frac{Q_{oa}}{v_o} (W_o - W_i) h_{fg}$$

These heat gains may also be evaluated by using the psychrometric chart and method presented in Chapter 5.

Outdoor air introduced into a conditioned space for purposes of ventilation should *always* be brought in through the cooling coil. It then becomes a part of the coil cooling load, not a load on the conditioned space.

9-16 DETERMINING THE TIME OF DAY WHEN THE COOLING LOAD IS AT MAXIMUM

The maximum heat gains from the many sources rarely occur at the same hour of the day. Assume a 100-seat restaurant on the first floor of an office building. The building is located at approximately 40° N latitude and has a glass front facing east (no other exposures). Referring to Tables 9-9, the maximum solar heat gain (224) occurs in April, and Table 9-16 gives a maximum CLF of 0.80 at 8:00 AM for interior shading (assumed). The restaurant would probably have the largest number of patrons between noon and 2:00 PM when the CLF has dropped to 0.27. It is evident that the maximum solar heat gain and the maximum people load do not occur at the same time.

The principle to follow in selecting the hour of the day at which to make the heat gain calculation is: *Select that hour of the day at which the sum of all the various heat gains is at a maximum*. This is entirely different from adding the maximum heat gains from each source. Considering the restaurant again, the solar heat gain peaks at 8:00 AM when there could be 25 people eating breakfast. Considering this, the solar heat gain at 8:00 AM and the body heat of 25 people should be added. Similarly, the solar heat gain at 1:00 PM and the body heat of 100 people should be added. The selection of the air-conditioning equipment would then be based on the larger of the two loads.

Selecting the hour at which the sum of the heat gains is likely to be a maximum comes with experience. There are some general principles that may be of aid.

For variable occupancy buildings (restaurants, theaters, auditoriums, etc.) the hour of maximum heat gains is generally the time of the maximum number of people present. However, some buildings may have large sunlit

Table 9-29*

	WINDOWS FACE IN FOLLOWING DIRECTIONS		SHADES BEHIND WINDOWS		AWNINGS ON WINDOWS	
	LARGEST GLASS AREAS IN THESE WALLS	SMALLEST GLASS AREAS IN THESE WALLS	FLOOR ABOVE	ROOF ABOVE	FLOOR ABOVE	ROOF ABOVE
WINDOWS IN 4 WALLS	N AND S	E AND W	10 OR 2	2	10 OR 2	2
	NE AND SW	SE AND NW	8 OR 4	3	8 OR 4	3
	E AND W	S AND N	9 OR 4	3	8 OR 4	3
	SE AND NW	SW AND NE	8 OR 4	4	8 OR 4	3

	WINDOWS FACE IN FOLLOWING DIRECTIONS		SHADES BEHIND WINDOWS		AWNINGS ON WINDOWS	
	LARGEST GLASS AREAS IN THESE WALLS	SMALLEST GLASS AREAS IN THESE WALLS	FLOOR ABOVE	ROOF ABOVE	FLOOR ABOVE	ROOF ABOVE
WINDOWS IN 3 WALLS	N	E AND W	8 OR 4	3	8 OR 4	3
	N AND S	W	2	2	2	2
	N AND S	E	10	1	10	1
	NE	SE AND NW	8	8 OR 3	8	3
	NE AND SW	NW	4	3	4	3
	NE AND SW	SE	8	3	8	2
	E	S AND N	8	10 OR 2	8	10 OR 2
	E AND W	N	8 OR 4	3	8 OR 4	3
	E AND W	S	9 OR 4	3	8 OR 4	3
	SE	SW AND NE	8	9 OR 2	8	2
	SE AND NW	NE	8	4	8	4
	SE AND NW	SW	4	4	4	4
	S	E AND W	10 OR 2	2	10 OR 2	2
	SW	SE AND NW	4	3	4	3
	W	S AND N	4	3	4	3
	NW	SW AND NE	4	4	4	3

	WINDOWS FACE IN FOLLOWING DIRECTIONS		SHADES BEHIND WINDOWS		AWNINGS ON WINDOWS	
	LARGEST GLASS AREAS IN THESE WALLS	SMALLEST GLASS AREAS IN THESE WALLS	FLOOR ABOVE	ROOF ABOVE	FLOOR ABOVE	ROOF ABOVE
WINDOWS IN 2 WALLS	N	E	8	2	8	2
	N	W	4	3	4	3
	NE	SE	8	2	8	2
	NE	NW	8	4	8	3
	E	N	8	8 OR 2	8	10 OR 2
	E	S	8	10 OR 1	8	10 OR 1
	SE	NE	8	9 OR 3	8	11 OR 1
	SE	SW	9	2	9	2
	S	E	10	1	10	1
	S	W	2	2	2	2
	SW	SE	3	3	3	3
	SW	NW	4	3	4	3
	W	N	4	3	4	3
	W	S	4	3	4	3
	NW	NE	4	4	4	4
	NW	SW	4	4	4	3

NOTES:

1. Figures set "10 or 2" mean one of two different things:
   (a) the excess solar heat gain may be a maximum and have approximately equal values at two different hours of the day.
   (b) The excess solar heat gain may have a maximum value at only one of the two hours given, and not at the remaining hour.
   Which of the two interpretations to use can be determined only for a particular problem.
2. In the above tables for windows in more than one wall, combinations of directions through which the sun cannot possibly shine at one time have been omitted. For example, such combinations as north and south, or east and west are not tabulated. For such combinations use either direction alone. Thus, if there are windows in both the east and west walls, use the time of day for the one wall having the largest window area. Similarly, for rooms having windows in north and south walls, use the time of day for windows in only the south wall.
3. Buildings having large areas of skylights in their roofs will probably have their maximum solar heat gain, at, or near noon.
4. If roofs are insulated, the better the insulation, the more closely the time of day when the cooling load is a maximum will tend to approach the time given in the columns headed "floor above."
5. The above table is for typical proportions of windows, walls and roof and for average wall, window and roof construction. For special conditions or unusual installations, a check of the heat gain at times which differ from those given in the tables, is recommended.

Reproduced with permission from Trane Air Conditioning Manual, *The Trane Company, LaCrosse, Wisconsin, 1965.*

glass areas during periods of low occupancy. The maximum solar heat gains plus the body heat of a few people may be larger than the body heat at peak occupancy plus a small solar heat gain. To check this, the total heat gains should be computed for both periods. Then the larger of the two totals is used for the design of the air-conditioning system.

Some buildings may have an unusually high lighting level during the evening hours and the sensible heat gain from the lights may be greater than solar heat gain during the day.

Table 9-30*

WINDOWS FACE IN	SHADES BEHIND WINDOWS		AWNINGS ON WINDOWS		WINDOWS IN 4 WALLS
FOLLOWING DIRECTIONS	FLOOR ABOVE	ROOF ABOVE	FLOOR ABOVE	ROOF ABOVE	
N, S, E AND W	8 OR 4	3	8 OR 4	3	
NE, SW, SE AND NW	8 OR 4	3	8 OR 4	3	

WINDOWS FACE IN	SHADES BEHIND WINDOWS		AWNINGS ON WINDOWS		WINDOWS IN 3 WALLS
FOLLOWING DIRECTIONS	FLOOR ABOVE	ROOF ABOVE	FLOOR ABOVE	ROOF ABOVE	
N, E AND W	8 OR 4	3	8 OR 4	3	
N, S AND W	4	3	4	3	
N, S AND E	8	10 OR 1	8	2	
NE, SE AND NW	8	9 OR 3	8	3	
NE, SW AND NW	4	3	4	3	
NE, SW AND SE	8	8 OR 3	8	2	
E, W AND S	8 OR 4	3	8 OR 4	3	
SE, NW AND SW	4	3	4	3	

WINDOWS FACE IN	SHADES BEHIND WINDOWS		AWNINGS ON WINDOWS		WINDOWS IN 2 WALLS
FOLLOWING DIRECTIONS	FLOOR ABOVE	ROOF ABOVE	FLOOR ABOVE	ROOF ABOVE	
N AND E	8	8 OR 2	8	2	
N AND W	4	3	4	3	
NE AND SE	8	9 OR 2	8	2	
NE AND NW	8 OR 4	3	8 OR 4	3	
E AND S	8	10 OR 1	8	1	
SE AND SW	9 OR 3	2	9 OR 3	2	
S AND W	4	3	4	3	
SW AND NW	4	3	4	3	

WINDOWS FACE IN	SHADES BEHIND WINDOWS		AWNINGS ON WINDOWS		WINDOWS IN 1 WALL
FOLLOWING DIRECTIONS	FLOOR ABOVE	ROOF ABOVE	FLOOR ABOVE	ROOF ABOVE	
N	-	2	-	2	
NE	8	2	8	2	
E	8	8 OR 2	8	2	
SE	9	10 OR 2	9	11 OR 2	
S	12	1	12	2	
SW	3	2	3	2	
W	4	3	4	3	
NW	4	4	4	3	

NOTES:

1. Figures set "10 or 2" may mean one of two different things:
   (a) the excess solar heat gain may be a maximum and have approximately equal values at two different hours of the day.
   (b) The excess solar heat gain may have a maximum value at only one of the two hours given, and not at the remaining hour.
   Which of the two interpretations to use can be determined only for a particular problem.
2. In the above tables for windows in more than one wall, combinations of directions through which the sun cannot possibly shine at one time have been omitted. For example, such combinations as north and south, or east and west are not tabulated. For such combinations use either direction alone. Thus, if there are windows in both the east and west walls, use the time of day for the one wall having the largest window area. Similarly, for rooms having windows in north and south walls, use the time of day for windows in only the south wall.
3. Buildings having large areas of skylights in their roofs will probably have their maximum solar heat gain, at, or near noon.
4. If roofs are insulated, the better the insulation, the more closely the time of day when the cooling load is a maximum will tend to approach the time given in the columns headed "floor above."
5. The above table is for typical proportions of windows, walls and roof and for average wall, window and roof construction. For special conditions or unusual installations, a check of the heat gain at times which differ from those given in the tables, is recommended.

**Reproduced with permission from* Trane Air Conditioning Manual, *The Trane Company, LaCrosse, Wisconsin, 1965.*

Such spaces ordinarily have their maximum occupancy at night. Hence, the total heat gain due to lighting and occupancy may be greater than the total heat gain due to solar heat and occupancy.

Numerous buildings have a fairly steady occupancy throughout the day—for example, office buildings and factories. Which hour gives the maximum heat gain in such buildings? Usually it is the hour when the sun shines on the walls having the largest window areas. This is true of buildings with large wall and window areas and small roof areas. All tall buildings are in this class.

The opposite of a tall building is the one- or two-story building spread over a large area. Such a building has a large roof area compared to its wall area. The solar heat gain through roofs is at a maximum around 2:00 PM. So the total heat gain will in all probability also be at its maximum at this hour. Suppose there is a lot of glass in the walls. Then it is a good idea to check the total heat gain at the hour when the sun shines on the walls having the greatest glass areas.

For a guide to find the approximate hour when the solar heat gain will be at maximum, use Tables 9-29 and 9-30. Use the tables *only* when the variation in the total heat gains is due to solar heat. In other words, these tables are for buildings where the lighting, motor, and people loads are approximately equal. If so, use Table 9-30. If not, use Table 9-29. When 2 hr are given for the probable maximum heat gain, it may mean one of two things: (1) the solar heat gain may be a maximum at 2 hr of the day; (2) the solar heat gain may have a maximum value at only one of the two hours listed. In order to know which to use, more facts about the building may be required.

CHAPTER 9

Review

9-1. Calculate the cooling load at 3:00 PM sun time on August 1 from 1000 ft^2 of horizontal flat roof constructed of 4.0 in. of heavyweight concrete with 2 in. of insulation covered with ⅜ in. of felt roofing and slag. The location is 40° N latitude. The outside air design dry-bulb temperature is 95°F with a daily range of 22 degrees, the inside design dry-bulb temperature is 80°F.

9-2. Calculate the cooling load at 12:00 noon sun time on June 1 from 800 ft^2 of horizontal flat roof constructed of 2.5 in. of wood with 1.0 in. of insulation covered with ⅜ in. of felt roofing and slag, with suspended ceiling. The location is 32° N latitude. The outside design dry-bulb temperature is 100°F with a daily range of 16 degrees; the inside design dry-bulb temperature is 78°F.

9-3. Calculate the cooling load in Btu/hr at 4:00 PM sun time on August 1 for a wall constructed of 4 in. of face brick, 8 in. of heavyweight concrete block, 1 in. of insulation, and ¾ in. of gypsum plaster on the interior. The wall is facing southwest at 40° N latitude and has a net area of 820 ft^2. The outside design dry-bulb temperature is 96°F with a 20-degree temperature range. The inside design dry-bulb temperature is 78°F.

9-4. Calculate the cooling load for a window facing southwest at 40° N latitude at 3:00 PM sun time in July. The window is regular insulating glass with ½ in. of air space. Inside drapes have a transmittance of 35 percent and a reflectance of 30 percent. The indoor design dry-bulb temperature is 78°F and the outdoor temperature is 98°F with a 22-degree daily range. Window dimensions are 3 ft wide by 6 ft high.

9-5. Compute the cooling load for the south windows of a building that has no external shading. The windows are insulating type with regular plate glass inside and out, ¼ in. of air space, and inside venetian blinds of light color. The building is located at 44° N latitude. Outside design dry-bulb temperature is 95°F with an 18-degree daily temperature range. Inside design dry-bulb temperature is 78°F. Make the calculations for 11:00 AM sun time in the month of August for a total window area of 450 ft^2.

9-6. Calculate the cooling load for a frame wall which has 3½ in. of insulation in the walls with no air space. The building is located at 32° N latitude and the wall faces south. Make calculations for 9:00 AM, 1:00 PM, and 5:00 PM sun time in July. Outdoor air design temperature is 100°F with a 20-degree daily range. Inside air design temperature is 75°F. Net wall area is 320 ft^2. Assume dark surface.

9-7. A meeting room in a building has a seating capacity of 30 people when fully occupied. Meetings are held on the average of two times a week and normally last an average of 2 hr starting at 8:00 AM. Following the meeting, ten of the people work in the room typing and doing light work such as filing papers, and so on. Estimate the cooling load caused by the people at 10:00 AM and at 4:00 PM. Inside design dry-bulb temperature is 80°F.

9-8. A medium-size restaurant located near 40° N latitude must be air conditioned. We wish to determine the time of day when the cooling load is at a maximum, and the cooling load caused by solar effects, people, lights and equipment, only in the month of July.

The east wall is 55 ft long and 10 ft high floor to ceiling and contains 240 ft^2 of ¼-in. plate glass (single thickness) including the main entrance. The south wall is 40 ft long and 10 ft high floor to ceiling and contains 180 ft^2 of ¼-in. plate glass (single thickness). The windows have interior drapes which have an estimated reflectance of 20 percent and transmittance of 50 percent. The east and south walls are similar to a Group D type with $U = 0.27$ Btu/hr ft^2 °F. The north wall is a partition leading to the kitchen. The west wall is a common wall with a conditioned store.

The roof is similar in construction to a No. 3 roof, with $U = 0.14$, and has a suspended ceiling. The floor is a 6-in. concrete slab covered with carpet and pad. Equipment located in the restaurant consists of two automatic coffee brewers (240 cups/hr), two toasters (4-slice), and two roll warmers. All units are electric. The total lighting is 6000 W of recessed, fluorescent lights.

The restaurant is open from 6:00 AM until 10:00 PM. The rush hours are from 7:30 to 8:30 AM, 12:00 noon to 2:00 PM, and 6:00 to 8:00 PM. The entire seating capacity (80 persons) may be used during the rush hours. One cashier and eight waitresses work in the serving area during these hours.

Outside design temperature is 95°F with a daily range of 20 degrees. Inside design temperature is 80°F.

9-9. For the building of Problem 9-8 determine the cooling load caused by solar effects at 10:00 AM sun time in the month of July.

9-10. A large office space in the administration building of a given industrial concern has an average occupancy of 25 people from 8:00 AM to 5:00 PM. The floor area is approximately 4600 ft^2. Assume a medium ventilation rate and a 6-in. concrete floor with carpet and pad. Recessed, unvented fluorescent lighting installed in the ceiling has a total wattage of 11,500 W. Electrical equipment located in the space operates intermittently and amounts to an average of 5.8 hp. Determine the cooling load at 3:00 PM sun time.

BIBLIOGRAPHY

1. ASHRAE *Handbook of Fundamentals,* American Society of Heating, Refrigerating and Air-Conditioning, Atlanta, GA, 1981.

2. ASHRAE *Handbook of Product Directory, Fundamentals,* American Society of Heating, Refrigerating and Air-Conditioning Engineers, New York, 1977.

3. ASHRAE *Handbook of Fundamentals,* American Society of Heating, Refrigerating, and Air-Conditioning Engineers, New York, 1972.

4. *Trane Air Conditioning Manual,* The Trane Company, La Crosse, Wisconsin, 1965.

5. ASHRAE Standard 90-75, *Energy Conservation in New Building Design,* American Society of Heating, Refrigerating and Air-Conditioning Engineers, New York, 1975.

6. *Handbook of Air Conditioning System Design,* Carrier Air Conditioning Company, McGraw-Hill Book Company, New York, 1965.

7. W. L. Marn, *Commercial Gas Kitchen Ventilation Studies,* (Research Bulletin No. 90, American Gas Association Laboratories, March 1962).

8. A. J. DeAlbuquerque: Equipment Loads in Laboratories, *ASHRAE Journal* 14(10), September 1972, p. 59.

Chapter 10

Heat-Transfer Devices

10-1 INTRODUCTION

Once the calculations have been made to determine the heating, ventilating, and cooling requirements, the next logical step in the development of the complete air-conditioning system would be to determine the type of heating and/or cooling system which will best serve the requirements. This step will determine the type of heat-transfer devices to be used in the various areas to be conditioned.

The types of systems that are in common use are forced circulation hot water, forced circulation warm air, steam, and electric for heating and forced circulation chilled water and forced circulation chilled air for cooling. It should be evident, then, that there can be combined heating and cooling systems using some of the same heat-transfer devices by simply changing the temperature of the circulating medium, such as water or air.

The number and types of available heat-transfer devices are many. This chapter will identify the most common types, present some performance data of these units, and describe some of the rating procedures used to establish standards for the equipment.

The numerous heat-transfer devices may be classified generally as (a) *energy conversion units* or (b) *terminal units*. Some devices encountered may be classified as combination units. One such type is the space heater located within the space to be heated. The space heater burns a fuel such as coal, gas, oil, or wood and some of the released heat energy is delivered by radiation and convection directly to the space to be heated. Another type is the electric resistance heater which, when located within the space, converts electrical energy into heat energy which is delivered directly to the space by radiation and convection. Basically, however, the energy conversion units convert some of the energy in fuels into heat energy which is transferred to a working fluid. The working fluid circulates to the various terminal units where the heat is transferred to the space. The most common energy conversion units are boilers and the hot-air furnaces.

10-2 BOILER TYPES

Steam and hot water boilers for heating and process are built of steel or cast iron in a wide variety of types and sizes for low- or high-pressure operation. The nationally recognized code governing the construction of low-pressure steel and cast-iron heating boilers is the ASME Code for Heating Boilers. Some states and municipalities have their own codes which apply locally, but they are usually patterned after the ASME Code.

The maximum allowable working pressures for heating boilers are limited by the ASME Code for Heating Boilers to 15 psig for steam and 160 psig for hot water boilers, with a maximum temperature limitation of 250°F (121°C). Hot water boilers are generally designed for a working pressure of 30 psig, and may be equipped for higher working pressures for either heating purposes or for hot water supply only when designed, tested, and stamped for higher pressures.

The nationally recommended code governing the design and construction of high-pressure steel boilers is the ASME Code for Power Boilers. This code is used almost universally in the design of steam boilers for operating pressures over 15 psig. High-pressure boilers used for combination heating and process loads, or process loads alone, are normally designed for 150 psig and higher, and most local codes require that boilers operating over 15 psig must have constant attendance by a licensed operator. Unusually large or specialized heating systems are sometimes equipped with boilers designed for pressures above the limits of the code for hot water heating boilers. These boilers fall under the jurisdiction of ASME Code for Power Boilers.

Boilers may be classified in a number of different ways, such as:

1. *Materials of construction*. Most low-pressure boilers are constructed of cast iron or steel, but nonferrous metals are used for some.

2. *Design pressures*. As previously noted, heating boiler pressures are 15 psig for steam and 160 psig for water. Boilers designed for higher pressures are power boilers and are normally constructed of steel.

3. *Raw energy used*. Coal (hand-fired or stoker-fired), oil, gas, or electricity. Most boilers are designed for just one type of fuel, but some may be designed for dual-fuel operation.

4. *Type of application*. Space heating or domestic hot water service.

5. *Construction*. Sectional, round, firetube, or watertube.

6. *Type of combustion air-handling system*. Natural draft, induced draft, or forced draft.

Cast-iron boilers are usually constructed with vertical sections and are square or rectangular. Large boilers are usually shipped in sections and assembled at the place of installation. However, some boilers are shipped factory-assembled as *boiler-burner units, or packaged-boilers* having all components in an assembled unit ready for connections to the piping, fuel, and electric power. In the majority of cast-iron boilers, the sections are assembled with push nipples and tie rods, but external headers and screw nipples are used in some. Many sectional boilers are provided with large push nipples to permit the circulation of water between adjacent sections at both the waterline and bottom of the boiler. This is also necessary to permit the use of an indirect water heater with the boiler for summer-winter domestic hot water supply. Many sectional-type boilers may be enlarged by the addition of sections and corresponding plate work.

Figure 10-1 shows one of the larger-capacity sectional-type cast-iron boilers. The No. 92 shown, manufactured by Weil-McLain Company, Inc., presents an impressive list of new and exclusive features. It meets all the basic requirements for high operating efficiency (large combustion volume for low heat release rate), a high percentage of primary heating surface for maximum radiant heat transfer and flue gas travel which utilizes secondary heat-

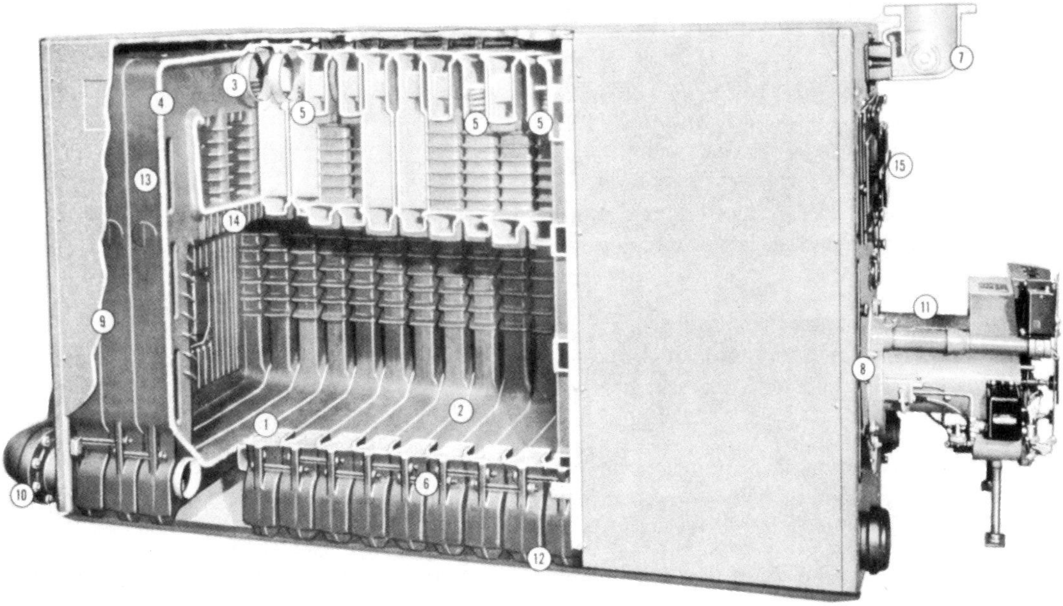

Figure 10–1

Sectional cast iron boiler for hot water or steam

transfer surfaces. Referring to Fig. 10-1, some of the features of this boiler are:

1. *Hydro-wall design*. Water backed combustion area with water circulating all around the firebox.
2. *Combustion chamber*. No chamber is required for any type of firing.
3. *Extra large top nipple*. Forms internal header for better water circulation. Generous steam space assures rapid generation of dry steam.
4. *Asbestos rope sealed*. Makes a completely gas-tight boiler, which permits forced-draft firing.
5. *Multiple tankless heaters*. Up to six heaters can be installed on either side of the larger size 92 boilers.
6. *Assembly with short draw rods*. Permits faster, easier erection of boiler and a strain-free assembly.
7. *Supply elbow*. Built-in trap separates air in hot water boilers.
8. *Burner-mounting plate*. Permits burner installation with minimum labor. Equipped with observation port.
9. *Insulated jacket*. Completely lined with 1-in. fiberglass.
10. *Return header*. Standard equipment on all sizes.
11. *Burner*. Available for light oil, gas, combination gas-light oil, heavy oil.
12. *Natural aeration*. 2½-in. space between floor and boiler is provided by cast-on legs of each section. No addi-

tional insulation of the boiler foundation is required.

13. *Stronger boiler construction and reversible sections.* *H*- and *T*-shaped cross sections provide maximum design strength, 80 lb, working pressure available. Reversible sections speed erection, permit right or left side piping for tankless heaters.

14. *Maximum radiant heat transfer.* Overhanging tongues and projecting fins on crown sheet and walls increase heating surface.

15. *Clean-out doors.* Two frameless doors, hinged to center plate, give complete access to center flueways for easy cleaning.

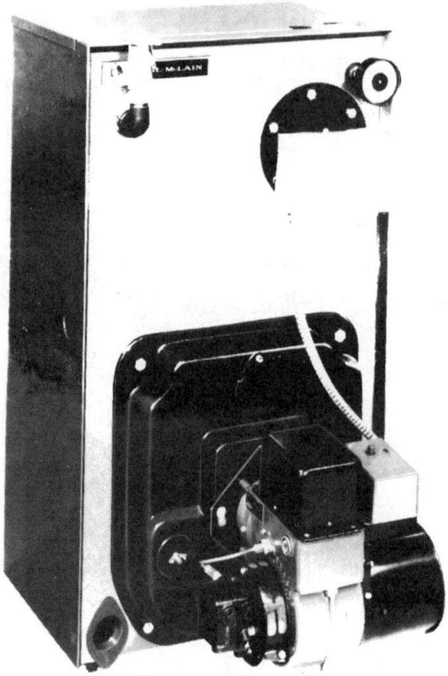

Figure 10–2
No. 66E Boiler-Burner Unit

Capacities of cast-iron heating boilers generally range from capacities required for small residences up to about 13,000 MBh (MBh means thousands of Btu per hour) gross output. Table 10-1 presents the engineering performance data for the 92 boiler shown in Fig. 10-1.

The Weil-McLain 66E boiler, or boiler-burner unit is shown in Fig. 10-2. It is a versatile, cast-iron sectional boiler designed specifically for oil-firing. It is very compact (35⅜ in. high and 20 in. wide) and is available with optional tankless domestic water heater. Sections are factory assembled and platework is mounted. The unit design includes wet base construction, provision for easy cleaning, and insulated jacket.

As may be seen in Table 10-2, the ratings vary as the number of boiler sections. Such a boiler would be used for domestic heating or small commercial heating system application.

Steel boilers may be of the firetube type in which the combustion gases pass through the tubes and boiler water circulates around them, or the watertube type in which the combustion

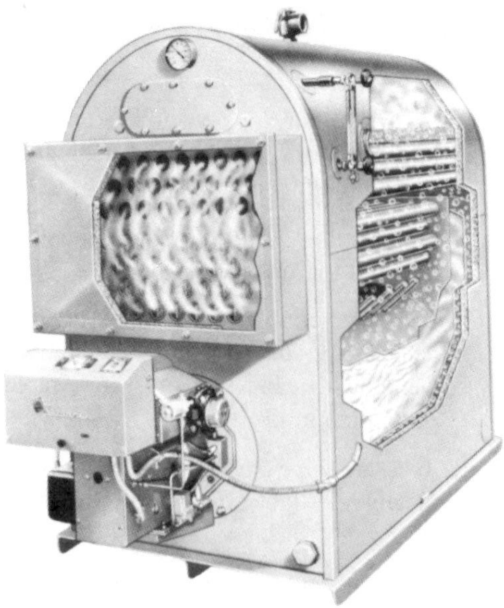

Figure 10–3
Firetube steel boiler for steam or hot water

Table 10-1

Performance Ratings of Boiler (Figure 10-1)*

Boiler Unit Number Steam or Water	I-B-R Burner Capacity ♦							Gross I-B-R Output		Net I-B-R Ratings ‡				
	Natural Draft			Forced Draft			Gas Pressure Required Inches W.C. ⊗	Natural Draft	Forced Draft	Natural Draft		Forced Draft		Sq. Ft. Water ****
	** Light Oil G.P.H.	Gas MBH	*** Heavy Oil G.P.H.	Light Oil G.P.H.	Gas MBH	Heavy Oil G.P.H.		BTU/hr	BTU/hr	Steam Sq. Ft.	Steam & Water BTU/hr ☆	Steam Sq. Ft.	Steam & Water BTU/hr ☆	
▲- 692*†•	10.45	1463	10.05	10.30	1442	10.05	6.5	1,180,000	1,180,000	3690	885,200	3690	885,200	5900
▲- 792*†•	12.50	1750	11.95	12.30	1722	11.95	6.5	1,406,000	1,406,000	4445	1,066,800	4445	1,066,800	7110
▲- 892*†•	14.50	2030	13.85	14.30	2002	13.85	6.5	1,632,000	1,632,000	5230	1,255,400	5230	1,255,400	8370
▲- 992*†•	16.50	2310	15.70	16.25	2275	15.70	6.5	1,856,000	1,856,000	5995	1,438,800	5995	1,438,800	9590
▲-1092*†•	18.55	2597	17.55	18.20	2548	17.55	6.5	2,080,000	2,080,000	6730	1,614,900	6730	1,614,900	10765
▲-1192*†•	20.60	2884	19.40	20.20	2828	19.40	6.5	2,304,000	2,304,000	7455	1,788,800	7455	1,788,800	11925
▲-1292*†•	22.65	3171	21.20	22.20	3108	21.20	6.5	2,532,000	2,532,000	8190	1,965,800	8190	1,965,800	13105
▲-1392*†•	24.70	3458	23.00	24.15	3381	23.00	6.5	2,756,000	2,756,000	8915	2,139,800	8915	2,139,800	14265
▲-1492*†•	26.75	3745	24.80	26.15	3661	24.80	6.5	2,980,000	2,980,000	9640	2,313,700	9640	2,313,700	15425
▲-1592*†•	28.80	4032	26.70	28.15	3941	26.70	6.5	3,210,000	3,210,000	10385	2,492,200	10385	2,492,200	16615
▲-1692*†•	30.90	4326	28.55	30.10	4214	28.55	6.5	3,430,000	3,430,000	11095	2,663,000	11095	2,663,000	17755
▲-1792*†•	32.90	4606	30.40	32.10	4494	30.40	6.5	3,660,000	3,660,000	11840	2,841,600	11840	2,841,600	18945
▲-1892*†•	34.95	4893	32.25	34.05	4767	32.25	6.5	3,880,000	3,880,000	12550	3,012,400	12550	3,012,400	20085
▲-1992*†•	36.95	5173	34.10	36.10	5054	34.10	7	4,110,000	4,110,000	13295	3,191,000	13295	3,191,000	21275
▲-2092*†•	39.00	5460	35.95	38.05	5327	35.95	7	4,330,000	4,330,000	14010	3,361,800	14010	3,361,800	22410

Table 10-1 (Cont.)

Performance Ratings of Boiler (Figure 10-1)*

Boiler Unit Number Steam or Water	Boiler H.P.	Net Firebox Volume Cu. Ft.	Stack Gas Volume CFM***** Light Oil and Gas Natural Draft	Stack Gas Volume CFM***** Light Oil and Gas Forced Draft	Stack Gas Volume CFM***** Heavy Oil Natural & Forced Draft	Draft Loss Thru Boiler in H_2O □ Light Oil and Gas Natural Draft	Draft Loss Thru Boiler in H_2O □ Light Oil and Gas Forced Draft	Draft Loss Thru Boiler in H_2O □ Heavy Oil Natural Draft	Draft Loss Thru Boiler in H_2O □ Heavy Oil Forced Draft	I-B-R Chimney Size ★Natural Draft Light Oil and Gas Size Inches	I-B-R Chimney Size ★Natural Draft Light Oil and Gas Height Feet	I-B-R Chimney Size ★Natural Draft Heavy Oil Size Inches	I-B-R Chimney Size ★Natural Draft Heavy Oil Height Feet	I-B-R Chimney Size Forced Draft Vent Dia. Inches
▲- 692*†●	35.3	33.24	784	639	754	.042	.032	.053	.043	16 x 16	19	16 x 16	23	12
▲- 792*†●	42.0	39.32	938	763	896	.048	.034	.057	.046	16 x 20	20	16 x 20	24	12
▲- 892*†●	48.8	45.40	1,088	887	1,039	.054	.037	.060	.049	16 x 20	21	16 x 20	24	12
▲- 992*†●	55.4	51.48	1,238	1,008	1,178	.060	.039	.064	.052	20 x 20	22	20 x 20	25	14
▲-1092*†●	62.1	57.56	1,392	1,128	1,316	.066	.041	.068	.054	20 x 20	23	20 x 20	26	14
▲-1192*†●	68.8	63.64	1,545	1,253	1,455	.072	.043	.071	.057	20 x 20	24	20 x 20	26	14
▲-1292*†●	75.6	69.72	1,699	1,377	1,590	.078	.046	.075	.060	20 x 24	25	20 x 24	27	16
▲-1392*†●	82.3	75.80	1,853	1,497	1,725	.083	.048	.078	.063	20 x 24	26	20 x 24	28	16
▲-1492*†●	89.0	81.88	2,007	1,622	1,860	.088	.050	.082	.066	24 x 24	27	20 x 24	28	16
▲-1592*†●	95.9	87.96	2,160	1,746	2,003	.094	.052	.085	.069	24 x 24	28	24 x 24	29	16
▲-1692*†●	102.5	94.04	2,318	1,866	2,141	.099	.055	.089	.072	24 x 24	29	24 x 24	30	18
▲-1792*†●	109.3	100.12	2,468	1,990	2,280	.105	.057	.093	.075	24 x 24	30	24 x 24	30	18
▲-1892*†●	115.9	106.20	2,622	2,111	2,419	.110	.059	.096	.078	24 x 28	30	24 x 24	31	18
▲-1992*†●	122.8	112.28	2,772	2,238	2,558	.115	.062	.100	.080	24 x 28	31	24 x 28	32	18
▲-2092*†●	129.4	118.36	2,925	2,359	2,696	.120	.064	.103	.083	24 x 28	32	24 x 28	32	18

▲ Substitute "BL" for light oil, "BGL" for gas-light oil, "BH" for heavy oil, "BG" for gas.

* Substitute "S" for steam, "W" for water.

† Substitute "N" for natural draft, "F" for forced draft.

● For T-Intermediate section (s) and tankless heater (s) add suffix "(number required) TIH"; for T-Intermediate section (s) with cover plate (s) only add suffix "(number required) TIP."

♦ Burner input based on maximum of 2,000 ft. altitude—for other altitudes consult factory.

** No. 2 fuel oil—Commercial Standard Spec. CS12-48, Heat value of oil = 140,000 BTU/G.

*** No. 4 or 5 fuel oil—Commercial Standard Spec. CS12-48, Heat value of oil = 150,000 BTU/G.

⊗ Gas pressure required at control inlet for maximum burner input, based on 1,000 BTU, 0.60 specific gravity gas. Burners for different pressures and gases are available—consult factory.

‡ Net I-B-R Ratings are based on net installed radiation of sufficient quantity for the requirements of the building and nothing need be added for normal piping and pick up.

☆ An additional allowance should be made for gravity hot water systems or for unusual piping and pick up loads.

**** Based on average water temperature of 170° F. in radiators.

***** Stack gas volume at outlet temperature.

□ Draft over fire must be added to obtain draft required at smoke collar.

★ When chimney is lined with the largest standard clay chimney tile, the equivalent area is considered the same as the unlined chimney area.

NOTE: Water boilers available upon special request at 80 P.S.I. working pressure.

TANKLESS WATER HEATER CAPACITIES*

HEATER NUMBER	**Intermittent Draw GPM 100° Av. Temp. Rise	***Continuous Draw GPM 100° Temp. Rise	Inlet and Outlet Tappings
92-K-34	**9 GPM**	**11 GPM**	**3/4"**

*Weil-McLain Ratings

**Gallons of Water per minute heated from 40° to 140° with 200°F. boiler water temperature

***Continuous Draw—no recovery period

Courtesy of Weil-McLain Company, Inc.

Table 10-2

Ratings for No. 66E Boiler, Boiler Burner Unit*

RATINGS

Boiler Number	I-B-R Burner Capacity GPH	Gross I-B-R Output BTU/Hr.†	Net I-B-R Ratings**			Net Water Sq. Ft.***	I-B-R Chimney	
			Steam Sq. Ft.	Steam BTU/Hr.	Water BTU/Hr.		Size Inches	Height Feet
▲-366E-*	.95	106,000	330	79,500	92,200	615	8 x 8	15
▲-466E-*	1.25	138,000	430	103,500	120,000	800	8 x 8	15
▲-566E-*	1.50	170,000	530	127,500	147,800	985	8 x 8	15
▲-666E-*	1.80	203,000	635	152,300	176,500	1175	8 x 8	15
▲-766E-*	2.05	236,000	740	177,000	205,200	1370	8 x 8	15

▲Substitute "B" for Boiler-Burner Unit or "A" for Oil-Fired Boiler for use with approved burners (see page 2).
*For your convenience when ordering, add to boiler number "W" for water, "S" for steam, without tankless heater. Add "WT" for water or "ST" for steam, with tankless heater.
†At combustion condition of $12\frac{1}{4} \pm \frac{1}{4}\%$ CO_2.
**Net I-B-R ratings are based on net installed radiation of sufficient quantity for the requirements of the building and nothing need be added for normal piping and pickup. Water ratings are based on a piping and pickup allowance of 1.15, steam ratings on an allowance of 1.333. An additional allowance should be made for unusual piping and pickup loads. Consult Customer Services Department.
***Based on average water temperature of 170°F. in heat distributing units.

Table 10-2A

Tankless Water Heater Capacities for No. 66E Boiler, Boiler Burner Unit*

Boiler No.	Heater No.	*Intermittent Draw GPM 100 F. Average Temp. Rise	**Continuous Draw GPM 100 F. Temp. Rise	Inlet and Outlet Tappings	Temp. Control Tapping
WATER					
366E	E-624	3.00	2.00	½"	¾"
466E	E-624	3.25	2.70	½"	¾"
566E	E-624	3.25	3.30	½"	¾"
666E	E-626	3.50	4.00	½"	¾"
766E	E-632	4.25	4.60	½"	¾"
STEAM					
366E	35-S-29	3.00	2.00	¾"	¾"
466E	35-S-29	3.25	2.70	¾"	¾"
566E	35-S-29	3.50	3.30	¾"	¾"
666E	35-S-29	3.75	4.00	¾"	¾"
766E	35-S-29	4.00	4.60	¾"	¾"

Weil-McLain ratings based on 60 PSIG domestic water pressure at heater.
*Gallons of water heated per minute from 40°F. to 140°F. with 200°F. boiler water temperature.
**Continuous draw, no recovery period.

**Courtesy of Weil-McLain Company, Inc.*

Table 10-3

Performance and Rating of Firetube Boiler (Fig. 10-3)*

UNIT NUMBER		3R1-F	3R2-F	3R3-F	3R4-F	3R5-F	3R6-F	3R7-F	3R8-F	3R9-F	3R10-F	3R11-F	3R12-F
SBI gross output	MBh.	324	396	468	540	648	792	936	1080	1260	1440	1620	1800
horsepower		10	12	14	16	19	24	28	32	38	43	48	54
steam per hour 212°F.	lbs.	334	408	482	557	668	816	965	1113	1299	1484	1670	1855
steam	sq. ft.	1350	1650	1950	2250	2700	3300	3900	4500	5250	6000	6750	7500
SBI net rating—steam	sq. ft.	1010	1240	1460	1690	2020	2480	2930	3380	3940	4500	5060	5620
Firing rate—gas 1000 Btu/cu. ft.		405	495	585	675	810	990	1170	1350	1575	1800	2025	2250
oil 140,000 Btu/gal.		2.9	3.5	4.2	4.8	5.8	7.1	8.4	9.6	11.2	12.9	14.5	16.1
Heating surface—Total	sq. ft.	53	65	77	88	106	129	153	177	206	236	265	294
primary	sq. ft.	20.5	22.8	25.2	27.5	33.0	36.6	40.4	44.5	49.5	54.1	58.4	62.6
Furnace vol. net	cu. ft.	7.8	9.1	10.5	11.7	15.4	17.9	20.4	23.3	29.5	33.4	37.1	40.8
Safety valve capacity steam per hour	lbs.	424	520	616	704	848	1032	1224	1416	1648	1888	2120	2352
Approximate dry weight	lbs.	1650	1800	1950	2100	2350	2700	3000	3300	3700	4150	4500	4900
Vent diameter—forced draft	in.	6	6	6	6	8	8	8	8	10	10	10	10
Stack diameter—natural draft	in.	12	12	12	12	15	15	15	15	18	18	18	18
Stack height—natural draft	ft.	30	35	35	40	35	40	40	45	35	40	45	50

**Courtesy of Kewanee Boiler Corp.*

gases circulate around the tubes and the water passes through the tubes. Either the firetube or watertube type may be designed with integral water-jacketed furnaces or arranged for refractory lined brick or refractory lined jacketed furnaces. Those with integral water-jacketed furnaces are called firebox boilers and are the most common type. They are usually shipped in one piece, ready for piping connections. A typical firetube boiler is shown in Fig. 10-3 with performance data presented in Table 10-3.

A *packaged boiler* (also called steam generator or hot water generator) is a boiler unit having all components, including burner, boiler, controls, and auxilliary equipment, assembled as a unit. Medium-sized packaged boilers or generators are usually a modified Scotch type. Smaller units are usually of the firebox type.

Capacities of steel boilers range from those required by small residences up to about 23,500.0 MBh gross output.

Electric boilers generally can be classified in one of the following categories:

1. Electric steam boiler with (a) immersion heater type consisting of resistance heaters immersed below the water line, or (b) electrode type consisting of bare electrodes immersed in the water. Heat is generated by resistance of the water to the flow of electricity between the electrodes.
2. Electric hot water boilers which are (a) conventional boilers (large internal volume) with immersed electric resistance heaters, (b) instantaneous type (minimum internal volume) with immersed electric resistance heaters, or (c) instantaneous type with wrap-around electric heaters. The heater is bonded to the outside of the water shell.

Residential and small commercial model electric boilers are generally quite small and compact and are often mounted on the wall to save floor space. Some of the smaller models are designed to be mounted behind a finned-tube enclosure.

Large commercial electric boilers of the electrode type are available for floor mounting with steam capacities from 1070 to 9500 lb steam per hour or 200 to 300 kW, and hot water capacities from 1440 to 6520 MBh or 400 to 20,000 kW. Pressure selection of these boilers are from low pressure to 2000 psig.

Boilers for special applications. Boilers for *domestic hot water supply* are classified as *direct* if the water heated passes through the boiler, and as *indirect* if the water heated does not come in contact with the water or steam in the boiler.

Direct heaters are built to operate at the pressures found in city water mains and are tested at pressures from 200 to 300 psig. The life of direct water heaters depends almost entirely on the scale-forming properties of the water supplied and the temperatures maintained. If low water temperatures are maintained, the life of the heater will be much longer due to decreased scale formation and minimized corrosion. Direct water heaters may be designed to burn refuse and garbage.

Indirect heaters generally consist of steam boilers in connection with heat exchangers of the coil or tube types which transfer the heat from the steam to the water. This type of installation has the following advantages:

1. Steam boiler can operate at low pressure.
2. The boiler is protected from scale and corrosion.
3. The scale formed on the heat exchanger parts may be removed by cleaning, or by replacing the heat exchanger components. The accumulation of scale does not affect efficiency, although it will affect the capacity of the heat exchanger.

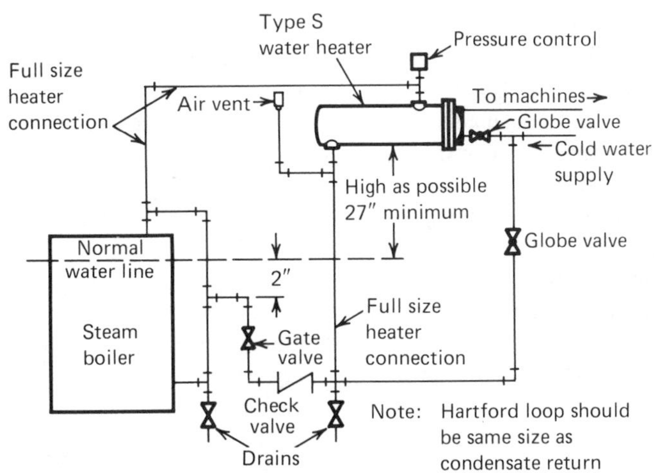

Figure 10–4

Indirect Hot Water Heater showing connections to steam boiler

4. Discoloration of water may be prevented if the water supply comes in contact with only nonferrous metal.

Where a steam or forced circulation hot water heating boiler system is installed, the domestic hot water may be heated by an indirect heater attached to the boiler. For most satisfactory performance in a steam system, the heater is placed just below the water line of the boiler. In a forced circulation hot water system, it should be located as high as practical with respect to the boiler, see Figure 10-4.

10-3 BOILER RATING AND TESTING CODES[1]

Heating boilers are usually rated according to codes developed by the Hydronics Institute (formally the Institute of Boiler and Radiator Manufacturers), the American Gas Association, and the Packaged Firetube Branch of the American Boiler Manufacturers Association. These test codes have been prepared specifically for the purpose of obtaining information required for establishing acceptable ratings.

A word about the Hydronics Institute may be in order at this point.[4] The Hydronic Institute is a nonprofit organization maintained by its membership. Members of the institute are manufacturers of equipment used in hydronic heating systems (boilers, radiators, baseboards, convectors, and commercial finned-tube radiation). Associate members are manufacturers of water heaters, valves and controls, and other accessory equipment.

In addition to its I=B=R rating program for cast-iron boilers, baseboards, and commercial finned-tube radiation, and its SBI rating program for steel boilers, the institute is actively engaged in other projects. All are designed to create a better understanding and appreciation of the superior comfort qualities of hydronic systems and to assure proper calculation, design, and installation of these systems so that the customer may obtain the maximum of comfort at a minimum of cost.

Some of the major projects of the institute are

a. Technical investigations including the development of codes and standards

b. Research investigations on hydronic systems.

c. Publications embodying the combined results of research and practical experience. (A list of I=B=R guides and other publications may be obtained by writing the institute.)

d. I=B=R Hydronic Heating School—a three-day course designed to keep heating contractors and wholesalers abreast of the latest methods for designing and installing more efficient and economical residential hydronic heating systems.

e. I=B=R Commercial Heating School—a course designed to keep heating contractors and wholesalers abreast of the latest methods for designing and installing hydronic heating systems in small commercial and industrial applications.

f. Compilation of facts and data useful to manufacturers and the heating industry in general.

The I=B=R emblem and the SBI emblem are the property of The Hydronics Institute, registered in its name in the United States Patent Office and with the Registrar of Trademarks of Canada.

The I=B=R and SBI emblems were designed and are to be used only to indicate I=B=R and SBI approval of ratings in specific units. Therefore, the use of the emblems by a manufacturer is restricted to those portions of his literature where I=B=R or SBI ratings are shown. No one has the right to state that his boilers have been tested under the Institute Boiler Standard, or to publish I=B=R or SBI ratings or to use the I=B=R or SBI emblem in connection therewith, unless (a) the manufacturer has executed a license with the institute and (b) the institute has advised the manufacturer in writing the exact I=B=R or SBI ratings which are applicable to the specific unit.

The current edition of the *Testing and Rating Standard for Cast Iron and Steel Heating Boilers*[3,4] is used for determining boiler ratings. This standard specifies the conditions under which the tests are conducted. All tests are conducted by a qualified representative of the institute at the manufacturer's laboratory, using institute instruments. Each boiler series is tested prior to approval. Approval is limited to a maximum of five years, but may be renewed following successful retesting. In addition, a qualified representative of the institute makes annual unannounced inspections of each manufacturing plant to determine that rated boilers are being produced according to the specifications on file with the institute.

All I=B=R or SBI rated boilers have been manufactured in accordance with the latest edition of Section IV of the American Society of Mechanical Engineers' (ASME) Boiler and Pressure Vessel Code.

I=B=R and SBI Ratings are expressed in terms of:

1. "Gross I=B=R or SBI Output" which is the total quantity of Btu's which the boiler will deliver per hour and at the same time meet all the limitations of the code.
2. "Net I=B=R or SBI Rating" which is the amount of installed radiation that will be served by the boiler based on the normal allowance for piping and pickup.

NOTE: Net ratings equal gross output divided by the piping and pickup factor. All net I=B=R and SBI ratings are based on piping and pickup factors of 1.15 for water, 1.333 for steam up to 1254 MBh gross output, decreasing regularly from 1.333 to 1.288 between gross outputs of 1256 to 1936 MBh, and 1.288 for larger steam sizes. The manufacturer should be consulted before selecting a boiler for installation on a job having unusual piping and pickup requirements. It is essential that advice of the manufacturer of the boiler be solicited before selecting a boiler for replacement on an existing installation which has a larger amount of piping than would normally be installed, or where other unusual conditions exist such as intermittent operation, temperature conditions other than normal, and unbalanced systems.

The net I=B=R or SBI ratings usually show steam ratings in square feet (square feet of equivalent direct radiation, EDR) and Btu/hr, and water ratings in Btu/hr.

10-4 SELECTION OF BOILERS

In addition to the obvious choices of materials of construction, steam or hot water, type of fuel, and so forth, the selection of a boiler for a particular application depends upon the load the boiler must handle.

The *maximum* or *gross load* on the boiler is the sum of the following four load components:

1. *Radiation Load.* The estimated heat emission in Btu/hr of the connected radiation. The connected radiation is determined by calculating the heat losses for each room. The sum of these heat losses is the total required heat emission of connected radiation, expressed in Btu/hr.

2. *Domestic Hot Water Load.* The estimated maximum heat in Btu/hr required to heat water for domestic use. (Obviously, if the domestic hot water is supplied by a separate heating device, this boiler load does not exist.)
   I = B = R recommends that: (a) no allowance is necessary for domestic hot water service unless there are more than two bathrooms to be served, or if the estimated use of hot water will not exceed 75 gal per day; and (b) if the estimated use of domestic hot water exceeds these limits, the following allowances should be made: (1) for a storage-type heater allow 120 Btu/hr for each gallon of water storage tank capacity and (2) for a tankless heater allow 12,000 Btu/hr for each bathroom in excess of two.
3. *Piping Tax.* The estimated heat loss from all sections of piping in the system.
4. *Warm-Up or Pickup Allowance.*

The *net boiler load* is the sum of items 1 and 2.

10-5 WARM-AIR FURNACES

If a warm-air furnace is defined as a heat-transfer device in which heat is released on one side of a heat exchanger surface and heat is absorbed by circulating air on the other side, a large number of devices can be included in this category, some of which are illustrated in schematic form in Fig. 10-5. The early types were:

1. *Stove.* The earliest ancester of the present-day warm-air furnace is the parlor stove, sometimes referred to as the pot-bellied stove. Aside from the cheery aspects of the flame that could be viewed through the isinglass openings, the stove

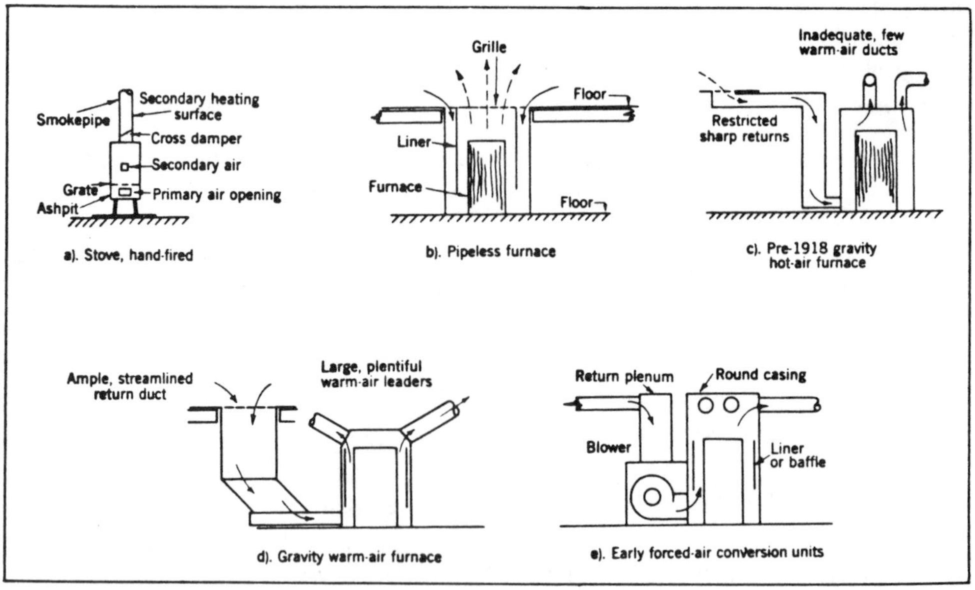

Figure 10–5

Development stages of warm-air furnaces

was a utility that was tolerated but not aesthetically admired.

2. *Pipeless Furnace.* Warmed air rose by gravity action and was delivered into the room above through a large grilled opening in the floor. Return air from the room passed through the same floor grilled but into the cooler downflow concentric passage.

3. *Gravity Hot-Air Furnaces (Pre-1918).* These early systems were installed without benefit of engineering knowledge. Air flow was restricted, and the furnaces were aptly described as ''hot-air'' furnaces since the air temperatures were in excess of 200°F.

4. *Gravity Warm-Air Furnace (Pre-1930).* The early experimental work at the University of Illinois by Professors A. C. Willard and A. P. Kratz in the period following 1918 showed that, for example, markedly improved results could be obtained by use of streamlined, amply sized return-air ducts, large warm-air stacks, and registers with ample free area. In these improved systems the actual operating register-air temperature was much less than the assumed design value of 175°F, even under design heater loads.

5. *Early Forced-Air Conversion Models (Pre-1940).* Even in the greatly improved gravity warm-air furnace system, the surfaces had to be extensive and the furnaces were large in size. The obvious step to increase the heat-transfer rate was to use some mechanical means for circulating the air. Propeller fans located in the return-air boot or in the supply bonnet were found to be lacking in pressure and capacity. When centrifugal fans were placed in a compartment attached to the casing, it was found that furnace casings had to be made smaller to prevent bypassing the heating surface. Later, casings became rectangular and smaller; fans, filters, burners, and controls were integrally built into the casing, and a whole new line of products were introduced. In the trade the centrifugal fans were referred to as ''blowers.''

Current Forced-Air Furnace Types. The current types of furnaces, some of which are illustrated in Fig. 10-6, show a great diversity of shapes and arrangements. Among the principal types used for residences and small buildings, the following arrangements are common:

1. Low-boy arrangement with furnace and blower in separate compartments, and with air discharged upwards. Used for basement installation, but also usable in furnace closets located on the first story.

2. High-boy arrangement with up-flow of air in which the blower is located at the bottom of the casing and the heat ex-

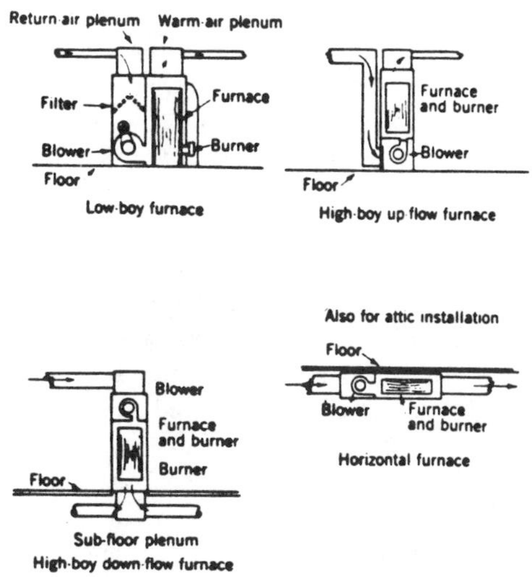

Figure 10–6

Current types of warm-air furnaces

changer is located in the top. Used for either basement installation or for first-story installation.

3. High-boy arrangement with down-flow of air in which the blower is located at the top of the casing and the heat exchanger is located below the blower. Used mainly for first-story installation in which warm air is discharged downward.

4. Horizontal arrangement in which the blower and heat exchanger are located side by side and the air is discharged horizontally. Used for basement, attic, or crawl-space installation.

The current types are integrally designed and coordinated units in which the blower and control equipment are selected and arranged for the specific unit. In fact, in the smaller-capacity units, the furnace package consists of a factory-wired and factory-assembled heat exchanger, blower, blower motor, filter, controls, and humidifier.

Special Types of Warm-Air Furnaces. The dividing line between a warm-air furnace and a device which is not a warm-air furnace is difficult to define sharply. The following are included here as special types of warm-air furnaces.

1. *Space Heater*. This reminder of the early parlor stove still persists. Most space heaters are gravity circulation systems, but some are provided with fans. No ducts are attached. Oil-fired and gas-fired space heaters are available. Those equipped to burn wood or coal are again becoming popular because of the high cost and short supplies of gas and fuel oil. Modern versions of the space heaters may include thermostatic temperature control equipment and operate at very reasonable efficiencies.

2. *Floor Furnace*. This device is reminiscent of the earlier pipeless furnace, but on a much smaller scale and with refinements of automatic firing and controls. The unit is usually attached to the ceiling joists of the basement and is ductless. The circulating air temperature is extremely high with gravity circulating units.

3. *Wall Furnace*. This device is a space-saving unit that is installed in the wall of a room and is usually a gravity circulation, pipeless unit. Some are tall and some are low enough to be installed just above the baseboard.

4. *Direct-Fired Unit Heater*. Unit heaters can be direct-fired, either with or without a circulating fan behind the heat exchanger surface. Used mainly for industrial or commercial installation.

5. *Direct-Fired Floor-Model Unit Heaters*. Extremely large unit heaters are available either with or without duct connections. Used for commercial and industrial buildings. Capacities in excess of 1000 MBh are available.

6. *Industrial Warm-Air Furnaces*. Industrial warm-air furnaces, or heavy-duty furnaces, are available with capacities in excess of 1000 MBh. Used for schools, churches, commercial buildings, and industrial buildings.

The styles, models, and types of warm-air furnaces are extremely diverse. Models are available for burning wood, coal, coke, fuel oil, kerosene, natural gas, bottled gas, and manufactured gas. The diversity was the result of demand: the demand of special building

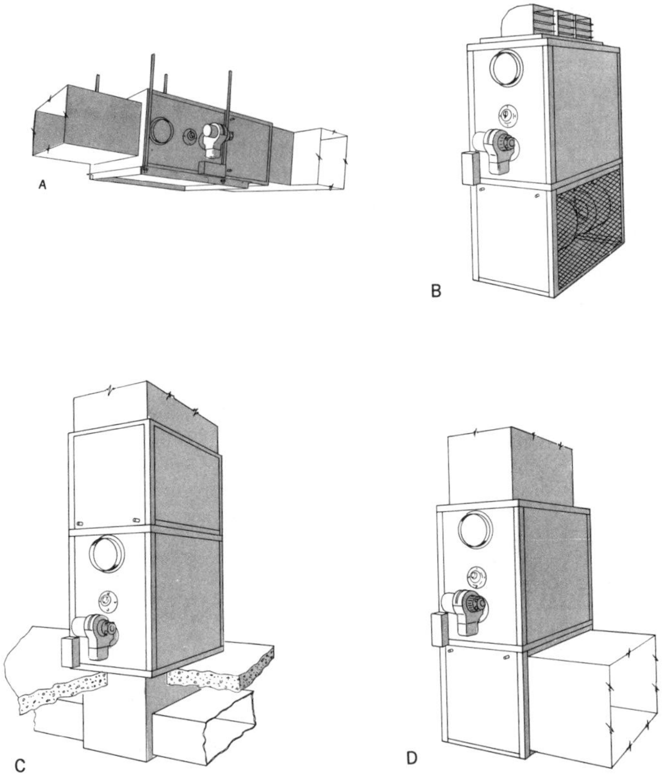

Figure 10–7 Types of warm-air furnaces
(a) Suspended horizontal for use with distribution system
(b) Floor upright unit with air delivery nozzles for direct space heating
(c) Inverted vertical unit for use with distribution system
(d) Floor-mounted upright unit for use with distribution system

types and the demands of both the builder and the public for small, efficient, and economical units.

Furnace Performance Terminology. A few terms appear in catalog description of warm-air furnaces which require definition (see also Fig. 10-8).

1. *Input, or heat input* is the rate at which heat is released inside the furnace and is expressed in Btu/hr units.

For gas:

$$\begin{matrix}\text{Input,} \\ \text{Btu/hr}\end{matrix} = \begin{matrix}\text{Cubic ft of} \\ \text{gas per hr}\end{matrix} \times \begin{matrix}\text{Heating value,} \\ \text{Btu/ft}^3\end{matrix}$$

For fuel oil:

$$\begin{matrix}\text{Input,} \\ \text{Btu/hr}\end{matrix} = \begin{matrix}\text{Gallons of oil} \\ \text{per hour}\end{matrix} \times \begin{matrix}\text{Heating value,} \\ \text{Btu/gal}\end{matrix}$$

For coal or solid fuel:

$$\begin{matrix}\text{Input,} \\ \text{Btu/hr}\end{matrix} = \begin{matrix}\text{Pounds of fuel} \\ \text{per hour}\end{matrix} \times \begin{matrix}\text{Heating value} \\ \text{Btu/lb}\end{matrix}$$

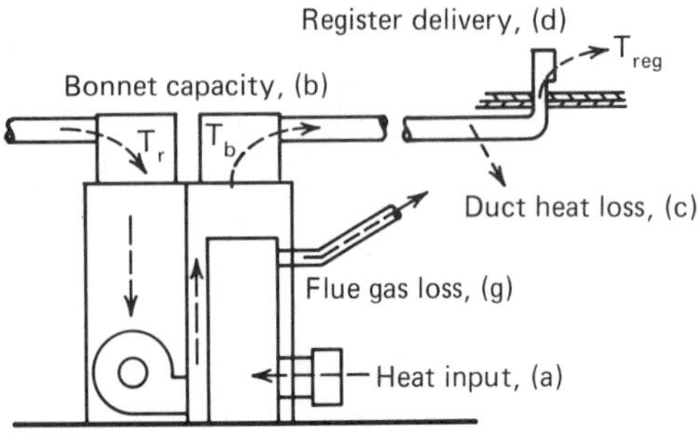

Figure 10–8

Terminology related to furnace performance

2. *Bonnet capacity* is the heat available in the air at the furnace supply air bonnet, in units of Btu/hr.

Bonnet capacity

$$= (\text{cfm})_b(60)(0.24)(\rho_b)(t_b - t_r) \quad (10\text{-}1)$$

where

$(\text{cfm})_b$ = air flow rate in cfm measured at bonnet temperature, t_b.

ρ_b = air density at atmospheric pressure and bonnet temperature, t_b.

t_b = bonnet supply air temperature, °F.

t_r = bonnet return air temperature, °F.

3. *Duct heat loss* is the heat loss from the warm-air supply ducts to the spaces surrounding the ducts expressed in Btu/hr. (Sometimes expressed as a percentage, say 5 to 10 percent of the bonnet capacity).

4. *Register delivery* is the heat available in the supply air that is delivered from the registers into the spaces to be heated, expressed in Btu/hr.

Register delivery =

$$(\text{cfm})_{reg}(60)(0.24)(\rho_{reg})(t_{reg} - t_r) \quad (10\text{-}2)$$

where

$(\text{cfm})_{reg}$ = air flow rate in cfm measured at register air temperature, t_{reg}.

ρ_{reg} = air density at atmospheric pressure and register air temperature, t_{reg}.

t_{reg} = register supply air temperature, °F.

5. *Bonnet efficiency* is the ratio of the bonnet capacity to the input.

6. *Duct transmission efficiency* is the ratio of register delivery to the bonnet capacity.

7. *Flue gas loss* is the heat loss of the flue gases, expressed as a percentage of the heat input, and includes both sensible and latent heat losses.

8. *Combustion efficiency* is 100 minus the flue gas loss, expressed as a percentage.

Testing and rating of furnaces. A given furnace in combination with a given blower can operate in a number of given ways and provide almost an indefinitely large number of different capacities and efficiencies. For example, the rate of heat input to the furnace could be successively increased with a fixed speed of the blower and a set of performance data taken. After this, the whole set of tests could be repeated with another blower speed. From the standpoint of a commercial rating, it becomes necessary, therefore, to hold some of the variables at a constant value and vary only one item. This is shown by a typical performance curve for a furnace, such as illustrated in Fig. 10-8. For this furnace the rate of fuel input was successively increased after each test, but each time the blower speed was so adjusted that the air temperature rise through the furnace was maintained at a constant 100°F. The typical curves in Fig. 10-9 show some interesting trends, common to all furnace performance tests:

1. In common with most heat-transfer devices, the maximum efficiency occurs at a low value of input. In fact, the efficiency gradually decreases as the input increases.
2. The capacity increases with the increase in input, both not in a linear relationship. Theoretically, if the input could be made large enough, the capacity could be extended indefinitely.
3. The flue gas temperatures constantly increase with an increase in input since the heat exchanger surface is fixed in amount.
4. Since the tests were conducted with a constant rise in temperature of the air passing through the furnace (100°F), the air flow rate also increased. (Note that it would be possible to conduct tests with a constant air flow rate and a constantly increasing air temperature rise.)

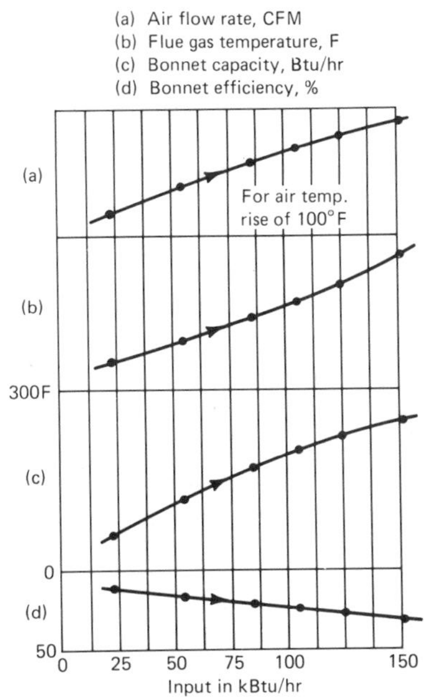

Figure 10–9

Typical trends of furnace performance tests

From the standpoint of the engineer a complete set of performance curves of the type shown in Fig. 10-9 is much more informative and useful than a single-point rating; but, in general a single-point rating is the only kind of rating that is acceptable to the trade. Every manufacturer is interested in a rating of the single-point type. The user is also interested in a single-point rating so that a furnace can be selected for a given application with reasonable assurance that it is adequate but not too large.

Acceptable limits. Theoretically, the furnace whose performance curves are shown in Fig. 10-9 could be rated at any value between say 10,000 Btu/hr up to 160,000 Btu/hr input. As a matter of fact if no limits of any kind were imposed, the furnace could have been tested with higher and higher inputs until finally the furnace disintegrated. It becomes obvious that in order for a manufacturer to specify a single-point rating it is necessary for the testing laboratory to impose limits, or boundaries, as to what is acceptable. The single-point rating that is finally chosen would not necessarily be a point of maximum efficiency since the maximum efficiency was obtained at an abnormally low input and capacity. On the other hand the single-point rating would not be one for which the equipment was strained to the last notch. The arbitrary limits that happen to be selected by industry are not important for this discussion. What is important is that industry has sensibly and wisely agreed upon certain limits not only for the safeguard of the consumer but primarily for protection of the manufacturer who hopes to stay in business. Industry could mutually agree upon a value of only 75 percent as a minimum bonnet efficiency that would be acceptable. Or, a flue-gas temperature of say 800°F could be considered a maximum, or the surface temperature of the heat exchanger surface could be limited to say 900°F, or the draft could be limited, or all of these limits could be imposed simultaneously. Whatever capacity that corresponded with the most stringent limit would then be the capacity for rating purposes. In practice, an extremely stringent set of limits have been imposed by The American National Standard Approval Requirements for Gas-Fired Gravity and Forced Air Central Furnaces (ANSI Z21.47) which is used universally in testing and rating. All gravity gas furnaces certified by the American Gas Association (AGA) under these requirements are assigned a rating based on 75 percent efficiency for gravity furnaces, and 80 percent efficiency for forced-air furnaces.

Oil-fired furnaces equipped with pressure-type or rotary burners should bear the Underwriters' Laboratory (UL) label showing compliance with UL 296, which is basically a safety standard, and Commercial Standard label (CS 75), which is a performance standard. In addition, the complete furnace should bear the UL 727 label and be so listed.

Selection of furnace for given installation. The selection of a furnace for a given installation is simple in principle since it consists of selecting a furnace whose capacity is adequate to take care of the heat losses for the building under design winter weather conditions. In practice, a number of minor points arise that require clarification. For example, if the register delivery is to be made equal to the design heat loss, the manufacturer's catalog should either state the delivery or give some means of estimating the duct transmission efficiency. In this connection, the design manuals of the National Environmental Systems Contractors Association (NESCA) arbitrarily assume that register delivery will be 0.85 times the bonnet capacity or, in other words, that the duct transmission efficiency will be 85 percent. This assumption may be greatly in error in large installations. Another question that arises in connection with design heat loss is that of determining what portion of the house should be included in the heat loss calculations. For example, should the basement or crawl-space heat loss be considered?

For the sake of uniformity and consistency, one set rule is offered for the selection of furnaces as follows:[9]

> Calculate the design heat loss for the entire structure, including the basement or crawl-space, and select a furnace whose bonnet capacity is equal to this heat loss.

Another very important consideration in furnace selection is to realize that a great many installations are for year-round application where cooling equipment is installed in the duct work for summer air conditioning. The blower must have adequate capacity to handle the required flow rate of air for summer cooling and to overcome the additional resistance of the cooling coil located in the air duct.

Controls.[1] Controls on a warm-air furnace usually consist of a low-voltage room thermostat which starts and stops the gas or oil burner (open or close draft ports on coal or wood furnaces) in response to room temperature changes, combustion controls, and thermostatic controls which respond to the temperature of the discharge air to prevent overheating and to start and stop the furnace blower.

The limit control is used to shut off the burner if the temperature of the air leaving the furnace is excessive. This temperature for furnaces listed for installation with reduced clearances from combustible materials may be 150°F, 175°F, or 200°F, depending on the range of temperature rise, and for furnaces listed for installation with standard clearances, 250°F. The limit control usually has a fixed stop at its maximum permissible setting and a differential of 25°F or more to permit resumption of burner operation when the temperature has dropped to a safe value.

The limit control performs only a safety function to prevent overheating in the event of a reduction in the proper air flow through the furnace which can occur with a clogged air filter, restrictions at the inlet or outlet grills of the duct system, broken blower drive belts, or a defective blower motor.

A fan control starts the blower when the temperature of the air in the furnace cabinet reaches a predetermined level and keeps it running until the heat in the furnace cabinet has been dissipated. This control is independent of burner controls. It usually has a "Fan On" temperature setting and may either have a fixed differential for turning off the fan or may also incorporate a "Fan Off" adjustment.

The fan control settings and the speed or air delivery of the blower must be carefully adjusted to give as continuous air circulation as possible. Whenever possible, it is recommended that the blower be turned on manually and be operated continuously throughout the heating season. Publications of the National Environment Systems Contractors Association (NESCA) give details of setting blower speeds and fan controls.

The fan and limit controls are often incorporated in the same housing and are sometimes operated by the same thermostatic element, through separate switches. The actuating elements of the controls may be located in the warm-air plenum or duct or within the furnace cabinet at some point in the circulating airstream where the temperature is the same as the outlet air temperature.

10-6 HEAT TRANSFER THROUGH HOLLOW CYLINDERS

One of the simplest terminal heat-transfer devices could be considered to be a circular hollow cylinder (say a pipe), shown schematically in Fig. 10-10. If a fluid is flowing inside the cylinder at temperature t_i and the fluid surrounding the outside of the cylinder was at temperature t_o, there would be a flow of heat from the higher to the lower temperature. If the two temperatures t_i and t_o were considered constant throughout the length L, we may express the steady-flow heat transfer by

$$\dot{q} = UA(t_i - t_o) \tag{10-3}$$

where U is the overall heat transmission coefficient in Btu/(hr ft^2 °F) or W/(m^2 °C), A is the

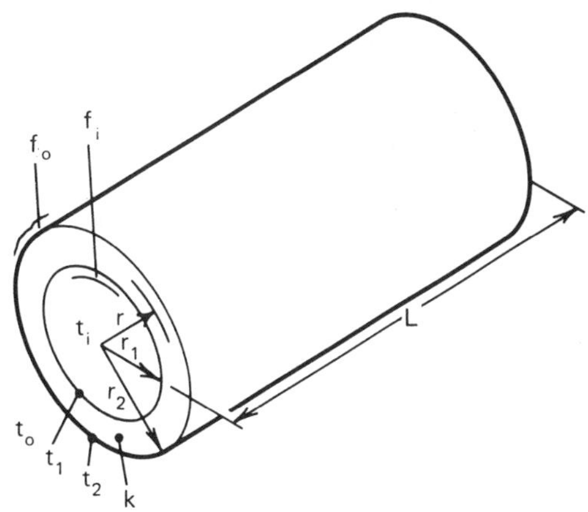

Figure 10–10

Radial heat flow through hollow cylinder

surface area of the cylinder in ft^2 or m^2, and $t_i - t_o$ is the overall temperature difference (fluid to fluid) in °F or °C.

Since we are dealing here with a surface area which is different on the outside than on the inside, we must refer to this area as A_o (outside surface area) or A_i (inside surface area); also, the U factor is dependent on which surface area we are referring to.

The U factor of Eq. (10-3) requires further explanation. This is the reciprocal of sum of all the thermal resistances in the heat flow path. These thermal resistances are the inside fluid film resistance, f_i, the resistance of the metal wall, and the resistance of the outside fluid film, f_o. The fluid film resistances are dependent on the type and rate of fluid flow, the properties of the fluids, the size of fluid-flow path, the nature of state change (evaporation, condensation), if any, that is produced by the heat transfer, and whether convection or radiation, or both, are involved. Heat transfer books and *ASHRAE Handbook 1977 Fundamentals* present many relationships and much data concerned with evaluating film coefficients.

The thermal resistance of the cylinder wall itself must take into account the fact that the inner and outer surface areas are not equal. At the inner surface the area is $2\pi r_1 L$, and at the outer surface the area is $2\pi r_2 L$. Under steady-flow conditions the heat transfer through the two areas must be equal. It can be shown that the heat transferred in Btu/hr is given by

$$\dot{q} = \frac{2\pi kL}{\ln(r_2/r_1)}(t_1 - t_2) = \frac{2\pi kL}{2.3 \log_{10}(r_2/r_1)}(t_1 - t_2) \quad (10\text{-}4)$$

where t_1 and t_2 are the inner and outer surface temperatures, L is the cylinder length measured along the cylinder axis, k is the thermal conductivity of the material, and r_1 and r_2 are the inner and outer radii, respectively. The thermal current is $\dot{q}/A$ and may be referred to either the inside or outside cylinder surface. Based on the inside surface area,

$$\dot{q}/A_1 = \frac{\dot{q}}{2\pi r_1 L} = \frac{k}{r_1 \ln(r_2/r_1)}(t_1 - t_2) \quad (10\text{-}5)$$

From Eq. (10-5) we may see that the thermal resistance of the cylinder wall must be

$$R = \frac{r_1 \ln(r_2/r_1)}{k} \quad (10\text{-}6)$$

or, based on the outside surface area, the thermal resistance would be expressed as

$$R = \frac{r_2 \ln(r_2/r_1)}{k} \quad (10\text{-}7)$$

Now we may combine the resistance of fluid films and the cylinder wall to obtain the total thermal resistance. Based upon the outside surface area, the total resistance is

$$R_{t_o} = \frac{1}{f_i}\left(\frac{r_2}{r_1}\right) + \frac{r_2 \ln(r_2/r_1)}{k} + \frac{1}{f_o} \quad (10\text{-}8)$$

Based upon the inner surface area, the total resistance is

$$R_{t_i} = \frac{1}{f_i} + \frac{r_1 \ln(r_2/r_1)}{k} + \frac{1}{f_o}\left(\frac{r_1}{r_2}\right) \quad (10\text{-}9)$$

The U factor now for Eq. (10-3) is simply the reciprocal of either Eq. (10-8) or Eq. (10-9), depending upon which surface area is being considered; usually we are concerned with the outside surface area.

For thin-wall cylinders such as pipe, the thermal resistance of the pipe wall, center term of Eq. (10-8), is very small in comparison to the resistances offered by the fluid films, especially if the fluids are gases. Therefore, frequently it is assumed that the inner and outer surface temperatures of the pipe wall are equal ($t_1 = t_2$). This is the assumption made for Table D-3 (Appendix D) and discussed in the following sections.

10-7 HEAT LOSSES FROM BARE PIPE†

In the preceding section we referred to the outside fluid film heat transmission coefficient, f_o, as being dependent on both radiation and convection conditions. Heat is lost or gained by pipes by both radiation and convection. Radiation was discussed briefly in Chapter 7 where the Stephan-Boltzman equation was introduced (Eq. 7-11). It is presented here in a slightly different form as

$$\frac{\dot{q}_r}{A} = 0.1713e_e\left[\left(\frac{T_1}{100}\right)^4 - \left(\frac{T_2}{100}\right)^4\right] \qquad (10\text{-}10)$$

where

$\dot{q}_r/A$ = Btu transferred by radiation per hour per square foot of pipe surface.

e_e = effective emissivity of the pipe.

T_1 = absolute temperature of the pipe, °F + 460.

T_2 = absolute temperature of the surroundings, °F + 460.

Heat transferred by convection can be determined by the Rice-Heilman formula, which resulted from work at Mellon Institute, as follows:

$$\frac{\dot{q}_c}{A} = C\left(\frac{1}{d}\right)^{0.2}\left(\frac{1}{T_{\text{ave}}}\right)^{0.181}(t_1 - t_2)^{1.266} \qquad (10\text{-}11)$$

where

$\dot{q}_c/A$ = Btu transferred by convection per hour per square foot of pipe surface.

C = constant with a value of 1.016 for horizontal pipe and 1.394 for vertical plates.

d = outside diameter of pipe (in.)

T_{ave} = absolute average temperature of hot body and surrounding air (°F + 460)

t_1 = temperature of pipe surface (°F)

t_2 = temperature of surrounding air (°F)

The emissivity in the radiation formula is the "effective emissivity," taking into account the absorbtivity of the bodies receiving the radiation. An emissivity of 0.94 was used for oxidized steel pipe to determine the values in Table D-1; a value of 0.44 is used for tarnished copper tube to determine the values of Table D-2.

No convection formula is directly available for vertical pipes. However, the emission can be assumed to be closely approximating that of a steel plate. In this connection note that C in Eq. (10-11) becomes 1.394. In the case of vertical pipes d in the convection formula is not the diameter of the pipe but the height of the plate (pipe) in inches, in which the value of $(1/d)^{0.20}$ becomes constant when d = 24 inches. Therefore, the convection formula for vertical pipes or vertical surfaces becomes

$$\dot{q}_c/A = 1.39 \times 0.53\left(\frac{1}{T_{\text{ave}}}\right)^{0.181}(t_1 - t_2)^{1.266} \qquad (10\text{-}12)$$

†This section is extracted with permission from *"Handbook of Air Conditioning, Heating, and Ventilating,"* 2nd ed., The Industrial Press, 1965.

where the terms are as defined previously. For vertical pipe, the radiation is the same as for horizontal pipe.

The tables which follow in this section are based on the assumption that the outside surface of the pipe is at the same temperature as the fluid flowing in the pipe; this is not strictly true, but is close enough for all practical estimating purposes.

Also, the tables assume an air temperature of 70°F and, in the case of radiation, a temperature of 70°F for the surrounding walls, machinery, and so forth. It can be seen, then, that if the surroundings are at temperatures appreciably above or below 70°F, the heat transferred by both radiation and convection would be appreciably lower than or above, respectively, the values given in the tables. Consequently, the tables should not be used for cases where the air temperatures and surrounding bodies' temperatures are lower than 60°F or higher than 80°F.

The convection formulas are both based on free convection with no appreciable air motion from fans or open doors; in other words, the tables apply to "still" air condition.

In the formula for radiation, values for e_e other than those given for steel and iron pipe and copper tube are

Surface	Emissivity, e_e
Aluminum, polished	0.08
Aluminum paint	0.40
Brass	0.05
Cast iron	0.20
Lead	0.08
Nickel	0.06
Paint	0.94
Tin	0.08
Nonmetallic surfaces	0.90

Small iron pipe of ½-in. size, frequently left uninsulated, can be profitably painted with aluminum paint which, with little effort, serves to reduce the emissivity of the pipe and, consequently, the radiation.

Although the emissivity of copper pipe is substantially lower than that for steel, so that the radiation loss is less, the convection loss is the same for copper as for iron where the conditions are the same. Therefore, it is not true that there is no reason to insulate hot lines because they are made of copper.

Many tables are available on heat losses from bare pipe; most of them, however, are on the basis of either Btu per square foot of pipe surface per hour "or Btu per linear foot of pipe per degree temperature difference (between pipe surface and air) per hour." Note, then, that Table D-1 is in Btu per linear foot per hour, with no additional calculation necessary.

Illustrative Problem 10.1

A steel pipe, nominal size 3 in., carries hot water at 210°F under 10 psig of pressure. If the air temperature surrounding the pipe is 70°F, what is the heat loss per foot of pipe? If the pipe is 100 ft long, what would be the hourly heat loss?

Solution:

From Table D-1, if the pipe temperature is assumed equal to the water temperature, 210°F, and the air temperature is 70°F, then the heat loss per linear foot is 303 Btu/(hr ft).

Therefore, a 100-ft section would have a heat loss of (303 × 100) = 30,300 Btu/hr.

Heating units made up of bare pipe coils were used some number of years ago mainly in industrial applications. The coils were built-up in a serpentine arrangement and hung vertically against a wall or horizontally at the ceiling. They are seldom, if ever, installed at the present time. However, for estimating replacement of any existing units, it is desirable to

Table 10-4*

Heat Emission of Pipe Coils Placed Vertically on a Wall (Pipes Horizontal) Containing Steam at 215°F and Surrounded with Air at 70°F Btu per linear foot of coil-hour (not linear feet of pipe)

Size of Pipe	1 in.	1¼ in.	1⅓ in.
Single row	132	162	185
Two	252	312	348
Four	440	545	616
Six	567	702	793
Eight	651	796	907
Ten	732	907	1020
Twelve	812	1005	1135

**Reprinted from ASHRAE* Handbook and Product Directory, 1977, Fundamentals, *with permission of the American Society of Heating, Refrigerating and Air-Conditioning Engineers, Atlanta, Georgia.*

have capacity values of such old pipe coils. (See Table 10-4.)

10-8 HEAT LOSS FROM INSULATED PIPE

In most cases for heating and cooling piping systems, we are concerned with reducing the heat loss or gain in the piping. Insulation installed on pipes and fittings is an effective method of saving energy.

Figure 10-11 illustrates a pipe section covered with a single thickness of insulation. Again, if we concern outselves with only the heat transfer from the inner surface of the pipe to the outer surface of the insulation, we will be able to write a heat-transfer conduction equation similar in form to Eq. (10-4), and it would appear as

$$\dot{q} = \frac{t_1 - t_3}{R_t}$$

$$= \frac{2\pi L(t_1 - t_3)}{\ln(r_2/r_1)/k_1 + \ln(r_3/r_2)/k_2} \qquad (10\text{-}13)$$

where

$\dot{q}$ = Btu/hr heat loss through the composite structure for L feet of length

t_1 and t_3 = inner surface temperature of pipe and outer surface temperature of the insulation, respectively

k_1 = conductivity of metal pipe

k_2 = conductivity of insulation (see Table 10-5)

Again, to simplify our calculations we will assume the resistance offered by the metal pipe is very small and concern ourselves with only the insulation.

For pipe covering in single thicknesses the heat loss from hot pipes can be expressed as

$$\dot{q}_1 = \frac{k\,(t_2 - t_3)}{r_3 \ln(r_3/r_2)} \qquad (10\text{-}14)$$

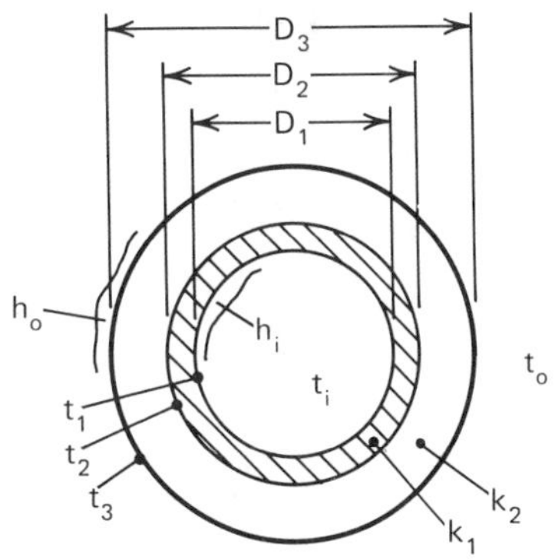

Figure 10–11

Radial heat flow through composite cylinder

Table 10-5

Heat Conductivity of Pipe Insulating Materials*
(Value to be used in the absence of specific data for the exact material and brand name used)*

Insulation	Approx. Use Range (°F)	Approx. Density lb/ft³	100	200	300	400	500	600	700	800
			Mean Temperature (°F) — Conductivity, k, (Btu in.)/(hr ft² °F)							
85% Magnesia	600	11	0.35	0.38	0.42	0.46	—	—	—	—
Laminated asbestos	700	30	0.40	0.45	0.50	0.55	—	—	—	—
4-ply corrugated asbestos	300	12	0.57	0.62	—	—	—	—	—	—
Molded asbestos	1000	16	0.33	0.38	0.43	0.48	0.53	0.58	—	
Mineral fiber[a] wire-reinforced	1000	10	0.29	0.35	0.42	0.49	0.56	0.63	—	—
Diatomaceous silica	1600	22	—	—	—	0.64	0.66	0.68	0.70	0.72
Calcium silicate	1200	11	0.32	0.37	0.42	0.46	0.51	0.56	—	—
Mineral fiber,[a] molded	350	9	0.26	0.31	0.39	—	—	—	—	—
Mineral fiber,[a] fine fiber, molded	350	3	0.23	0.27	0.31	—	—	—	—	—
Wood felt	225	20	0.33	0.37	—	—	—	—	—	—

[a]Includes glass, mineral wool, rock wool, etc.

**Reprinted with permission from* Handbook of Air Conditioning, Heating, and Ventilating, *2nd ed., The Industrial Press, 1965.*

where

$\dot{q}_1$ = heat loss (Btu/hr ft^2 of outer surface of insulation)

k = conductivity of insulation

t_2 = temperature of inner surface of insulation (assumed equal to pipe temperature)

t_3 = temperature at the outer surface of the insulation

r_2 = inner radius of the insulation

r_3 = outer radius of the insulation

The objection to the practical application of Eq. (10-14) is that the temperature of the outer surface of the insulation, t_3, must be known.† A more useful formula is

$$\dot{q}_2 = \frac{t_p - t_a}{\dfrac{r_2 \ln(r_2/r_1)}{k} + \dfrac{1}{f_o}} \qquad \textit{(10-15)}$$

where

$\dot{q}_2$ = heat loss (Btu/hr ft^2 of outer surface of pipe insulation)

t_p = temperature of pipe surface (°F)

t_a = temperature of ambient air (°F)

r_2 = radius of outer surface of insulation (in.)

r_1 = radius of outer surface of pipe (in.)

k = conductivity of insulation (Btu/ft^2 hr °F in. of thickness).

f_o = surface conductance of outer surface of insulation (Btu/ft^2 hr °F).

Since in most cases one knows or can assume the temperature of the ambient air, Eq. (10-15) is more readily applicable than Eq. (10-14).

For very low air movements and low rates of heat transmission, the value of $1/f_o$ can be taken as 0.6.

The heat loss figure is usually, in actual problems, desired in Btu per linear foot of pipe rather than in terms of Btu per square foot of outer surface. The outer surface of insulation per linear foot of pipe is $(2\pi r_2)/12$, so Eq. (10-15) becomes

$$\dot{q} = \frac{0.523 r_2 (t_p - t_a)}{\dfrac{r_2 \ln(r_2/r_1)}{k} + 0.6} \qquad \textit{(10-16)}$$

where

$\dot{q}$ = Btu loss per (hr)(linear ft of pipe)

r_2 = outer radius of insulation (in.)

r_1 = outer radius of pipe (in.)

t_p = temperature of pipe (°F)

t_a = temperature of ambient air (°F)

Equation (10-16) was used to calculate Table D-3 (Appendix D) and shows the heat loss for a range of values of k and a range of values of $t_p - t_a$.

The problem becomes more complicated when two or more separate layers of pipe insulation are applied. The equation for this condition would be

$$\dot{q}_2 = \frac{t_p - t_a}{\dfrac{r_n \ln(r_1/r_p)}{k_1} + \dfrac{r_n \ln(r_2/r_1)}{k_2} + \dfrac{r_n \ln(r_n/r_{n-1})}{k_n} + \dfrac{1}{f_o}} \qquad \textit{(10-17)}$$

†This section is extracted with permission from *Handbook of Air Conditioning, Heating, and Ventilating*, The Industrial Press, p. 8-173 to 8-174.

where q_2, t_p, t_a, and f_o are as given in Eq. (10-15) and r_1, r_2, . . ., r_n are the outer radius, in inches, of the first, second, and nth layer of insulation, respectively; r_p is the outer radius of pipe, in inches; k_1, k_2, . . ., k_n are conductivities of first, second, and nth layer of insulation, respectively.

The problem of two or more layers of insulation occurs more frequently in high-temperature work where, for example, the temperature may be too high for 85 percent magnesia, so the first layer may be calcium silicate or diatomaceous silica.

The values of thermal conductivity k for specific insulation is available from manufacturers, and it is suggested that calculations be based on data which apply to the material and brand name of the insulation to be used. In the absence of such data Table 10-5 has been included as a rough guide. Mean temperatures in Table 10-5 are the arithmetic mean between the inside surface and the outside surface of the insulation.

Tables D-3 (Appendix D) were calculated on the basis of still air with a $1/f_o = 0.60$. Actually the surface resistance decreases as the air movement increases and as the rate of heat transfer is increased. Table 10-6 gives the effect of air movement on surface resistance.

When insulated piping is located where people or animals may come in contact with it, the criterion of economical heat loss in selecting insulation thickness may be outweighed by the need to maintain the outside surface temperature of the insulation at a safe level. The outside surface temperature may be estimated from

$$t_o = t_p - \left[\frac{x/k}{x/k + 1/f_o}\right](t_p - t_a) \qquad (10\text{-}18)$$

where t_o is the outside surface temperature of the insulation in terms of t_p, t_a, k, and f_o as previously defined and x is the insulation thickness in inches.

Table 10-6

Effect of Air Velocity on Surface Resistance*

Heat Transmitted Btu/(hr ft²)	Velocity of Air, fpm: 0	100	200	400
	Value of $1/f_o$			
0	—	0.56	0.50	0.41
50	0.60	0.52	0.45	0.39
100	0.55	0.48	0.41	0.36
150	0.50	0.45	0.39	0.34
200	0.48	0.42	0.37	0.32
300	0.43	0.38	0.33	0.29
500	0.36	0.33	0.28	0.25

**Reprinted with permission from* Handbook of Air Conditioning, Heating, and Ventilating, *2nd ed., The Industrial Press, New York, 1956.*

Illustrative Problem 10.2

A 4-in. nominal steel pipe carries water at 300°F and is insulated with 2 in. of 85 percent magnesia whose density is 12 lb/ft³. What will be the estimated outside surface temperature of the insulation if $t_a = 70$°Ff?

Solution:

Assume $t_p = 300$°F and the approximate mean temperature in the insulation is 200°F. From Table 10-5 we find $k = 0.38$, and for still air conditions $1/f_o = 0.60$.
By Eq. (10-18),

$$t_o = 300 - \left[\frac{2/0.38}{2/0.38 + 0.6}\right](300 - 70)$$

$$t_o = 94°\text{F}$$

This result of 94°F for the outside surface temperature is close to the 100°F used to estimate the arithmetic mean temperature in the insulation of 200°F. If there had been a significant difference between the calculated t_o and the estimated t_o, then a new value should be assumed to get a new arithmetic mean temperature, a new k, and the calculation repeated.

10-9 RADIATORS, CONVECTORS, BASEBOARDS, AND FINNED-TUBE UNITS

Radiators, convectors, baseboard, and finned-tubes are types of terminal heat-transfer units commonly used in hot water heating and steam heating systems. They transfer heat from the system fluid (water or steam) to the surroundings by a combination of radiation and natural convection, and their function is to maintain the desired air temperature in the spaces where they are located. In general, these types of heat-transfer devices should be placed at points of greatest heat loss from the space in which they are located. For example, such units are located under windows, along exposed walls, and at door openings.

In general, those heat-transfer terminal units which have a large portion of their heated surface exposed to the space (radiators and cast-iron baseboard) emit a larger portion of heat by radiation than do units having completely or partially concealed heating surfaces (convectors, finned-tube, finned-pipe). Also, finned-pipe heating elements constructed of steel emit a larger portion of heat by radiation than do finned-tube elements constructed of nonferrous materials. The heat output ratings of these various units are normally expressed in units of Btu/hr (Btuh), 1000 Btu/hr (MBh), or in square feet equivalent direct radiation (EDR).

NOTE: 240 Btuh = 1 ft^2 EDR with 1 psig steam condensing in the heat-transfer unit. Ratings in EDR are largely being abandoned by many manufacturers in the industry.

Radiators. The small-tube-type radiators, with a length of only 1¾ in. per section, occupy less space than the older column and large tube units and are particularly suited for installation in recesses (see Table 10-7).

After a study of the demand for various sizes of radiators, the Institute of Boiler and Radiator Manufactures (I=B=R), in cooperation with the Division of Simplified Practice of the National Bureau of Standards, established the Simplified Practice Recommendation R 174-65, Cast Iron Radiators, for small-tube radiators. Table 10-7 shows the dimensions and ratings of the *wall-type,* and *large-tube* radiators are no longer manufactured, but since many of these units are still in use, refer to reference (1) Chapter 28 for dimensions and capacities.

Convectors. Convectors consist of a finned cast iron, or finned steel, or finned copper heating element installed within a cabinet-type steel enclosure. The same finned heating element will produce different capacities depending upon the height of the enclosure and upon the location of the inlet and outlet air openings in the enclosure. Air enters the enclosure below the heating element and leaves the enclosure through the outlet grille located above the heating element. The heat output of convectors is largely due to convection heating of the air as it passes over the heating element. Since convectors are manufactured in a wide variety of enclosure types, height, width, and depth of enclosure a wide range of capacities are available. The units may be floor mounted (free standing), wall hung, or recessed, and may have outlet grilles and arched inlets or inlet grilles as desired.

Capacities of convectors are quoted for different cabinet enclosure types, height, width, and depth of cabinet, average water tempera-

Table 10-7

Small-Tube Cast-Iron Radiators (Current)*

Number of Tubes per Section	Catalog Rating per Section[a]		Section Dimensions				
			A	B Width		C	D
			Height[c]	Min	Max	Spacing[b]	Leg Height[c]
	Sq Ft	Btuh	In.	In.	In.	In.	In.
3[d]	1.6	384	25	3¼	3½	1¾	2½
4[d]	1.6	384	19	4 7/16	4 13/16	1¾	2½
	1.8	432	22	4 7/16	4 13/16	1¾	2½
	2.0	480	25	4 7/16	4 13/16	1¾	2½
5[d]	2.1	504	22	5⅝	6 5/16	1¾	2½
	2.4	576	25	5⅝	6 5/16	1¾	2½
6[d]	2.3	552	19	6 13/16	8	1¾	2½
	3.0	720	25	6 13/16	8	1¾	2½
	3.7	888	32	6 13/16	8	1¾	2½

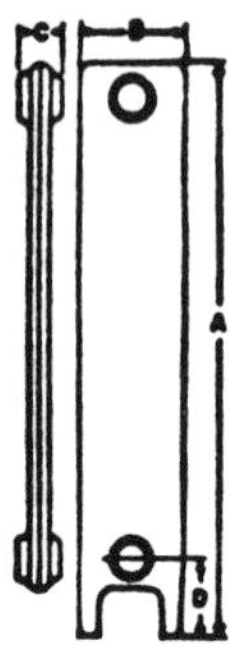

[a] These ratings are based on steam at 215 F and air at 70 F. They apply only to installed radiators exposed in a normal manner, not to radiators installed behind enclosures, grilles, or under shelves. For Btu per hour ratings at other temperatures, divide table values by factors found in Table 7.

[b] Length equals number of sections times 1¾ in.

[c] Overall height and leg height, as produced by some manufacturers, are one inch greater than shown in Columns *A* and *D*. Radiators may be furnished without legs. Where greater than standard leg heights are required this dimension shall be 4½ in.

[d] Or equal.

**Reprinted from ASHRAE* Handbook and Product Directory, 1977, Fundamentals, *with permission of the American Society of Heating, Refrigerating and Air-Conditioning Engineers, Atlanta, GA.*

ture in the heating element, water temperature drop in the unit, and are normally expressed in MBh.

Figure 10-12 illustrates one type of convector, and Table 10-8 quotes only a partial listing of capacities available from this type of unit. Capacity ratings are also available for other average water temperatures and water temperature drops, as well as ratings using low-pressure steam.

To illustrate the use of Table 10-8, assume that a heating capacity of 6200 Btu/hr is needed for a particular space which is to be heated with a convector of the type illustrated in Fig. 10-12. The average water temperature in the convector is 190°F. The space for the convector limits its height to 32 in. and a maximum depth of 6 in. How long must the unit be to deliver the required heating capacity? Referring to Table 10-8, a 32-in. long unit deliv-

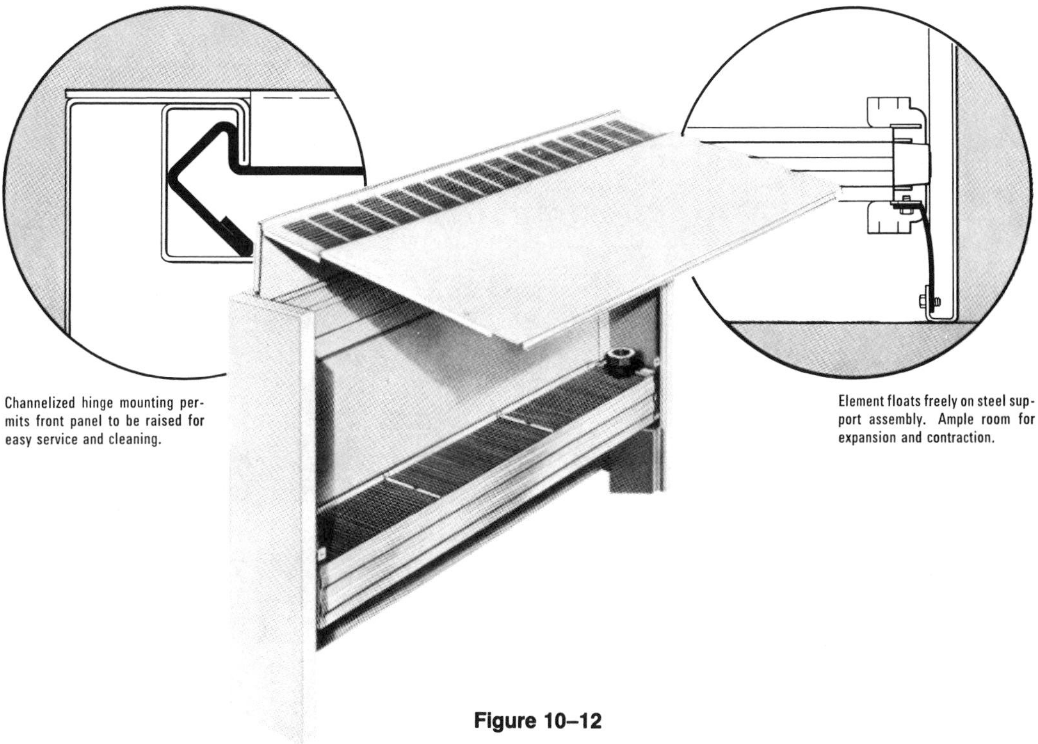

Channelized hinge mounting permits front panel to be raised for easy service and cleaning.
Element floats freely on steel support assembly. Ample room for expansion and contraction.

Figure 10–12

Front discharge, floor-mounted convector with front panel in raised position to show heating element (steam or hot water)

ers 6.3 MBh with a 10-degree temperature drop, 5.6 MBh with a 20-degree drop, and 5.4 MBh with a 30-degree drop. *Decision:* Select the 32-in. unit and a 10-degree water temperature drop.

Baseboard units. Baseboard heat-transfer units are commonly available in two types: cast iron and finned pipe. The cast-iron units may be either radiant type or radiant convector type. Both are provided with covered passages for the flow of water or steam. The radiant type has a plain front with no openings and produces mostly radiant heat. The radiant convector type is open at the bottom, grille work or openings at the top, and integral fins cast in the back providing openings for convection air flow over the unit. Finned-pipe baseboard units consist of a finned pipe or tube mounted within a sheet metal enclosure which is open top and bottom for air circulation. Nearly all the heat output from such units is by convection. The finned heating elements may be steel pipe with steel fins, or copper tubes with aluminum fins.

The capacity of baseboard heating units varies over a wide range depending on the physical dimensions and the materials used. Ratings are usually expressed in Btuh per linear foot of unit, at a specific average water temperature or a standard steam pressure.

Figure 10-13 illustrates a type of baseboard heating unit, and Table 10-9 gives typical performance ratings for such units.

Table 10-8

Capacity Table for Convectors (Trane Models SKF and AFG) for Various AWT and 65F EAT.*[a,b]

180 F Average Water

HEIGHT	DEPTH	LENGTH 26			32			38			44			50			56			62		
14	4	2.5	2.2	2.1	3.2	2.8	2.7	3.9	3.4	3.3	4.6	4.0	3.9	5.3	4.6	4.5	6.0	5.2	5.1	6.7	5.8	5.7
16	6	3.4	3.0	2.9	4.5	3.9	3.8	5.8	5.0	4.9	7.0	6.2	6.0	8.2	7.4	7.1	9.4	8.6	8.2	10.6	9.8	9.3
18	4	2.6	2.3	2.2	3.2	2.8	2.7	4.1	3.6	3.4	5.0	4.4	4.1	5.9	5.2	4.8	6.8	6.0	5.5	7.7	6.8	6.2
20	6	3.9	3.4	3.3	4.9	4.3	4.2	6.2	5.4	5.2	7.4	6.5	6.2	8.6	7.6	7.2	9.8	8.7	8.2	11.0	9.8	9.2
	8	4.8	4.2	4.0	6.0	5.3	5.1	7.2	6.4	6.2	8.6	7.5	7.3	9.9	8.6	8.4	11.2	9.8	9.5	12.5	10.9	10.5
24	4	2.7	2.3	2.2	3.4	3.0	2.9	4.2	3.7	3.6	5.0	4.4	4.3	5.8	5.1	5.0	6.6	5.8	5.7	7.4	6.5	6.4
26	6	4.3	3.8	3.6	5.4	4.8	4.6	6.6	5.8	5.5	7.7	6.7	6.5	8.8	7.7	7.5	9.9	8.7	8.5	11.0	9.7	9.5
	8	5.1	4.5	4.3	6.5	5.7	5.5	7.8	6.8	6.6	9.2	8.0	7.7	10.5	9.2	8.9	11.9	10.4	10.1	13.2	11.6	11.2
30	4	2.8	2.5	2.4	3.6	3.2	3.1	4.4	3.8	3.7	5.2	4.5	4.4	6.0	5.2	5.1	6.8	5.9	5.8	7.6	6.6	6.5
32	6	4.5	4.0	3.8	5.7	5.0	4.8	6.9	6.0	5.8	8.0	7.0	6.8	9.2	8.0	7.8	10.4	9.0	8.8	11.6	10.0	9.8
	8	5.4	4.8	4.6	6.7	5.9	5.7	8.0	7.0	6.8	9.4	8.2	7.9	10.8	9.4	9.1	12.2	10.7	10.3	13.5	11.9	11.4
36	4	3.1	2.6	2.5	3.8	3.3	3.2	4.5	4.0	3.8	5.3	4.7	4.5	6.2	5.4	5.2	7.0	6.1	5.9	7.8	6.8	6.6
38	6	4.8	4.2	4.0	5.8	5.0	4.9	7.0	6.2	6.0	8.3	7.3	7.1	9.6	8.3	8.1	10.8	9.4	9.1	12.0	10.5	10.1
	8	5.7	5.1	4.9	7.2	6.3	6.1	8.7	7.6	7.3	10.1	8.9	8.6	11.7	10.2	9.9	13.1	11.5	11.1	14.6	12.8	12.3

190 F Average Water

HEIGHT	DEPTH	LENGTH 26			32			38			44			50			56			62		
14	4	2.8	2.5	2.4	3.5	3.1	3.0	4.3	3.9	3.7	5.1	4.7	4.4	5.9	5.5	5.1	6.7	6.3	5.8	7.5	7.1	6.5
16	6	3.8	3.4	3.2	5.0	4.4	4.2	6.4	5.7	5.5	7.9	7.0	6.7	9.4	8.3	7.9	10.9	9.6	9.1	12.4	10.9	10.3
18	4	2.9	2.6	2.5	3.6	3.2	3.1	4.5	4.0	3.9	5.4	4.8	4.7	6.3	5.6	5.5	7.2	6.4	6.3	8.1	7.2	7.1
20	6	4.3	3.9	3.7	5.5	4.9	4.7	6.9	6.1	5.9	8.2	7.3	7.0	9.5	8.5	8.1	10.8	9.7	9.2	12.1	10.9	10.3
	8	5.3	4.7	4.5	6.7	6.0	5.7	8.1	7.2	6.9	9.6	8.5	8.2	11.0	9.8	9.4	12.5	11.1	10.6	13.9	12.4	11.9
24	4	3.0	2.7	2.5	3.8	3.4	3.2	4.7	4.2	4.0	5.6	5.0	4.8	6.5	5.8	5.6	7.4	6.6	6.4	8.3	7.4	7.2
26	6	4.8	4.3	4.1	6.1	5.4	5.2	7.3	6.5	6.2	8.6	7.6	7.3	9.9	8.7	8.4	11.2	9.8	9.5	12.5	10.9	10.6
	8	5.7	5.1	4.9	7.2	6.4	6.2	8.7	7.7	7.4	10.2	9.1	8.7	11.8	10.5	10.0	13.3	11.8	11.3	14.8	13.1	12.6
30	4	3.2	2.8	2.7	4.1	3.6	3.5	4.9	4.3	4.2	5.8	5.2	4.9	6.7	6.1	5.6	7.6	7.0	6.3	8.5	7.9	7.0
32	6	5.1	4.5	4.3	6.3	5.6	5.4	7.7	6.8	6.5	9.0	8.0	7.6	10.3	9.1	8.8	11.6	10.3	9.9	12.9	11.5	11.1
	8	6.1	5.4	5.2	7.5	6.7	6.4	9.0	8.0	7.6	10.5	9.3	8.9	12.0	10.7	10.2	13.6	12.0	11.6	15.1	13.4	12.9
36	4	3.5	3.1	2.9	4.3	3.8	3.6	5.1	4.5	4.3	6.0	5.3	5.1	6.9	6.1	5.9	7.8	6.9	6.6	8.7	7.7	7.4
38	6	5.3	4.7	4.5	6.4	5.7	5.5	7.9	7.0	6.7	9.3	8.3	7.9	10.7	9.5	9.1	12.0	10.7	10.2	13.3	11.9	11.4
	8	6.5	5.7	5.5	8.1	7.2	6.9	9.7	8.6	8.2	11.3	10.1	9.6	13.0	11.6	11.1	14.7	13.0	12.5	16.3	14.5	13.9

200 F Average Water

HEIGHT	DEPTH	LENGTH 26			32			38			44			50			56			62		
14	4	3.1	2.8	2.6	3.9	3.5	3.3	4.8	4.3	4.1	5.7	5.1	4.9	6.6	5.9	5.7	7.5	6.7	6.5	8.4	7.5	7.3
16	6	4.2	3.8	3.6	5.5	5.0	4.7	7.1	6.4	6.0	8.7	7.8	7.4	10.3	9.2	8.8	11.9	10.6	10.2	13.5	12.0	11.6
18	4	3.2	2.9	2.7	4.0	3.6	3.4	5.0	4.5	4.3	6.0	5.4	5.2	7.0	6.3	6.1	8.0	7.2	7.0	9.0	8.1	7.9
20	6	4.8	4.3	4.1	6.1	5.5	5.2	7.6	6.8	6.5	9.1	8.2	7.7	10.6	9.6	9.0	12.1	11.0	10.3	13.6	12.4	11.6
	8	5.9	5.3	5.0	7.4	6.7	6.3	9.0	8.1	7.7	10.6	9.5	9.0	12.2	11.0	10.4	13.8	12.4	11.7	15.4	13.9	13.1
24	4	3.3	2.9	2.8	4.2	3.8	3.6	5.2	4.7	4.4	6.2	5.6	5.2	7.2	6.5	6.0	8.2	7.4	6.8	9.2	8.3	7.6
26	6	5.3	4.8	4.5	6.7	6.0	5.7	8.1	7.3	6.9	9.5	8.6	8.1	10.9	9.9	9.3	12.3	11.2	10.5	13.7	12.5	11.7
	8	6.3	5.7	5.4	8.0	7.2	6.8	9.6	8.6	8.2	11.3	10.2	9.6	13.0	11.7	11.1	14.7	13.2	12.5	16.3	14.7	13.9
30	4	3.5	3.2	3.0	4.5	4.1	3.8	5.4	4.9	4.6	6.4	5.8	5.4	7.4	6.7	6.2	8.4	7.6	7.0	9.4	8.5	7.8
32	6	5.6	5.0	4.8	7.0	6.3	6.0	8.5	7.7	7.2	9.9	8.9	8.4	11.4	10.3	9.7	12.8	11.5	10.9	14.2	12.7	12.1
	8	6.7	6.0	5.7	8.3	7.5	7.1	9.9	8.9	8.4	11.6	10.4	9.9	13.3	12.0	11.3	15.0	13.5	12.7	16.7	15.0	14.2
36	4	3.8	3.4	3.2	4.7	4.2	4.0	5.6	5.0	4.8	6.6	5.9	5.6	7.6	6.8	6.5	8.6	7.7	7.3	9.6	8.6	8.2
38	6	5.8	5.2	5.0	7.1	6.4	6.0	8.7	7.8	7.4	10.3	9.3	8.8	11.8	10.6	10.0	13.3	12.0	11.3	14.8	13.3	12.6
	8	7.0	6.4	6.0	8.9	8.0	7.6	10.7	9.6	9.1	12.5	11.3	10.6	14.4	13.0	12.2	16.2	14.6	13.8	18.0	16.2	15.3

210 F Average Water

HEIGHT	DEPTH	LENGTH 26			32			38			44			50			56			62		
14	4	3.3	3.1	2.9	4.2	3.9	3.6	5.2	4.8	4.4	6.2	5.7	5.2	7.2	6.6	6.0	8.2	7.5	6.8	9.2	8.4	7.6
16	6	4.5	4.2	3.9	5.9	5.5	5.1	7.7	7.1	6.6	9.4	8.7	8.0	11.1	10.3	9.4	12.8	11.9	10.8	14.5	13.5	12.2
18	4	3.5	3.2	3.0	4.3	4.0	3.7	5.4	5.0	4.6	6.5	6.0	5.5	7.6	7.0	6.4	8.7	8.0	7.3	9.8	9.0	8.2
20	6	5.2	4.8	4.4	6.6	6.1	5.6	8.2	7.6	7.0	9.8	9.1	8.4	11.4	10.6	9.8	13.0	12.1	11.2	14.6	13.6	12.6
	8	6.4	5.9	5.5	8.0	7.4	6.8	9.7	9.0	8.3	11.4	10.6	9.8	13.1	12.2	11.3	14.9	13.8	12.8	16.6	15.4	14.2
24	4	3.6	3.3	3.1	4.5	4.2	3.9	5.6	5.2	4.8	6.7	6.2	5.7	7.8	7.2	6.6	8.9	8.2	7.5	10.0	9.2	8.4
26	6	5.7	5.3	4.9	7.2	6.7	6.2	8.7	8.1	7.5	10.3	9.5	8.8	11.9	10.9	10.1	13.5	12.3	11.4	15.1	13.7	12.7
	8	6.8	6.3	5.8	8.6	8.0	7.4	10.3	9.6	8.9	12.2	11.3	10.5	14.0	13.0	12.0	15.9	14.7	13.6	17.6	16.3	15.1
30	4	3.8	3.5	3.2	4.9	4.5	4.2	5.8	5.4	5.0	6.9	6.4	5.9	8.0	7.4	6.8	9.1	8.4	7.7	10.2	9.4	8.6
32	6	6.0	5.6	5.2	7.6	7.0	6.5	9.2	8.5	7.9	10.7	9.9	9.2	12.3	11.4	10.5	13.8	12.8	11.8	15.3	14.2	13.1
	8	7.2	6.7	6.2	8.9	8.3	7.7	10.7	9.9	9.2	12.5	11.6	10.7	14.4	13.3	12.3	16.2	15.0	13.9	18.0	16.7	15.4
36	4	4.2	3.8	3.4	5.1	4.7	4.3	6.0	5.6	5.2	7.1	6.6	6.1	8.2	7.6	7.0	9.3	8.6	8.0	10.4	9.6	8.9
38	6	6.3	5.8	5.5	7.7	7.1	6.6	9.4	8.7	8.0	11.1	10.3	9.5	12.7	11.8	10.9	14.4	13.3	12.3	16.0	14.8	13.7
	8	7.7	7.0	6.6	9.6	8.9	8.2	11.6	10.7	9.9	13.5	12.5	11.6	15.5	14.4	13.3	17.5	16.2	15.0	19.4	18.0	16.7

[a]Capacities expressed in MBh for 10, 20, and 30°F. Temperature drops are located below each cabinet length.
[b]See manufacturer's catalog for complete data.

Reproduced in part (Courtesy of The Trane Company, LaCrosse, Wisconsin).

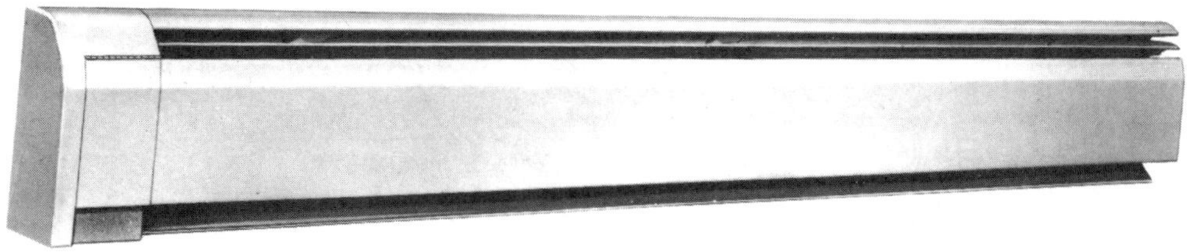

Figure 10–13

Model 75 WL-3 Therma Trim Baseboard

When selecting the type of baseboard, the designer should avoid using a unit with too high an output per linear foot.[1] Optimum comfort for room occupants is usually obtained when units are so selected that they are installed along as much of the exposed wall as possible.

The basic advantage of the baseboard units is that its normal placement is along cold walls and under areas where the greatest heat loss occurs. Other advantages claimed are: it is inconspicuous; it offers a minimum of interference with furniture placement; and it distributes the heat near the floor. This last characteristic reduces the floor-to-ceiling temperature gradient to about 2 to 4°F, and tends to produce more uniform temperature throughout the room.[1]

Commercial finned tube or finned pipe. These heating units are similar, as to component parts, to baseboard units. The main difference is in the size and enclosure design, resulting in a higher capacity per unit length than baseboard units.

Finned-tube heating elements are available in several pipe sizes, either steel or copper; 1 to 2 in. I.P.S. for steel, and 3/4 to 1 1/4 nominal tube size for copper. Fins are usually mechanically bonded to the pipe and tube and are steel or aluminum, either square or rectangular depending on the manufacturer.

The enclosures are sheet metal and have an open or grilled air inlet below the heating element and normally a grilled air outlet at the top or top front above the heating element.

The ratings of commercial finned-pipe units, like baseboard units, are quoted in Btuh per linear foot at a specific average water temperature or specific steam pressure. Figure 10-14 illustrates typical shapes of finned tube enclosures and Table 10-10 gives ratings of one enclosure type with different finned tube (or pipe) installed.

To standardize rating procedures for convectors, baseboard, and commercial finned pipe, various codes have been established and used by industry. The generally accepted method of testing and rating of both ferrous and nonferrous convectors is given in *Commercial Standard CS140-47, Testing and Rating Convectors,* which has been developed cooperatively by the Convector Manufacturer Association, The Institute of Boiler and Radiator Manufacturers, other members of the trade and the National Bureau of Standards.

Baseboard testing and rating is based upon *I-B-R Testing and Rating Code for Baseboard Radiation,* and finned-pipe testing and rating is based upon *I-B-R Testing and Rating Code for Finned Tube (Commercial) Radiation.*

Table 10-9*[a]

MODEL 75WL-3 (¾″ TUBE) THERMA TRIM
I-B-R APPROVED WATER RATINGS
Capacities in BTU/Hr. per Linear Foot with 65° F. Entering Air

Number of Linear Feet	Water Flow Rate—500 lbs./hr. or 1 GPM Average Water Temperature (°F.)					
	170	180	190	200	210	220
1	530	600	670	740	810	870
2	1,060	1,200	1,340	1,480	1,620	1,740
3	1,590	1,800	2,010	2,220	2,430	2,610
4	2,120	2,400	2,680	2,960	3,240	3,480
5	2,650	3,000	3,350	3,700	4,050	4,350
6	3,180	3,600	4,020	4,440	4,860	5,220
7	3,710	4,200	4,690	5,180	5,670	6,090
8	4,240	4,800	5,360	5,920	6,480	6,960
9	4,770	5,400	6,030	6,660	7,290	7,830
10	5,300	6,000	6,700	7,400	8,100	8,700
11	5,830	6,600	7,370	8,140	8,910	9,570
12	6,360	7,200	8,040	8,880	9,720	10,440
13	6,890	7,800	8,710	9,620	10,530	11,310
14	7,420	8,400	9,380	10,360	11,340	12,180
15	7,950	9,000	10,050	11,100	12,150	13,050
16	8,480	9,600	10,720	11,840	12,960	13,920
17	9,010	10,200	11,390	12,580	13,770	14,790
18	9,540	10,800	12,060	13,320	14,580	15,660
19	10,070	11,400	12,730	14,060	15,390	16,530
20	10,600	12,000	13,400	14,800	16,200	17,400

Number of Linear Feet	Water Flow Rate—2000 lbs./hr. or 4 GPM Average Water Temperature (°F.)					
	170	180	190	200	210	220
1	560	630	710	780	860	920

[a]I-B-R approved water ratings are based on the active (finned) length and include the 15 percent addition for heating effect allowed by the I-B-R Testing and Rating Code for Baseboard Type of Radiation. The active length is 3 in. less than the enclosure length. Ratings apply to the assembly with the damper installed and adjusted to the normal open position.

The heating elements are constructed of ¾-in. nominal copper tubing expanded into 2⅛ × 2$\frac{5}{16}$-in. flanged aluminum fins, which are spaced 52 to the foot; fin thickness is 0.007-in. Elements are unpainted.

Use of I-B-R ratings at 4 gpm flow rate is limited to installations where the water flow rate is not known, the I-B-R ratings at the standard flow rate of 1 gpm must be used. Flow rates exceeding 6 gpm should be avoided because of possible noise.

Pressure drop through Therma Trim Baseboard is 0.047 inches of water per linear foot at the 1 gpm flow rate. At 4 gpm flow rate, pressure drop is 0.525 in. water per linear foot.

Courtesy of Weil-McLain Company, Inc.

Figure 10–14

Diagrammatic view of Trane Wall Fin convection radiation provided with heating element (steam or hot water)

10-10 UNIT HEATERS, UNIT VENTILATORS, AND MAKEUP AIR UNITS

These units are grouped together because they all have the same basic component elements, which are a finned coil heating element, a motor-driven fan, and usually a filter, all enclosed in a housing (cabinet) equipped with inlet and outlet air grilles or diffusers, and, in some cases, duct collars for attachment of inlet and outlet air ducts. However, each may have special components and functions which accounts for the variation in the names.

Unit heaters. These units have the primary function of heating. Ventilation is sometimes provided by introducing outside air, but the majority of these units have 100 percent air recirculation. Cooling may be provided by these units if chilled water is circulated through the coil, and some means of draining condensate is provided.

There are three general types of unit heaters which may be identified: (a) cabinet heater, (b) horizontal projection heater, and (c) vertical projection heater. (See Figs. 10-15, 10-16, and 10-17.)

The *cabinet heater* is basically similar, in external appearance, to a convector. A centrifugal fan and filters have been added as well as a different heating coil design. The fan is employed to force air across the heating element which greatly increases the capacity of the unit, and in addition, provides desirable air circulation within the space to be conditioned. These units are available in a wide variety of cabinet styles and may be floor mounted, wall hung, ceiling mounted, or recessed. The air flow may be upward (normal) or downward through the unit (called inverted air flow). These units are used in building entrances, below windows, and in any location where a high capacity is required, and where the low-level fan noise can be tolerated.

Table 10-11 gives a typical capacity table using hot water for a few of the sizes available from one manufacturer.

Figures 10-16 and 10-17 represent two styles of propeller fan unit heaters. These types of unit heaters are normally employed in factories and warehouses where the units are suspended from the ceiling or mounted at a high level on a wall. The propeller fan forces air across the heat-transfer coil, and this air is directed outward or downward by means of adjustable louvers (Fig. 10-16). Figure 10-17 shows a vertical projection unit where the fan, mounted vertically, draws the air horizontally across the heat-transfer coil and blows the heated air downward. These unit heaters are available in a wide range of sizes, varying in capacity from 5 to 500 MBh, and use either steam or hot water for the heating medium.

Tables 10-12 and 10-13 give partial capacity ratings of the units shown in Fig. 10-16 and Fig. 10-17, respectively.

Table 10-10

Ratings of Trane Wall Fin Elements with Types *S* Enclosure*[a]

TABLE 11—Ratings of Wall Fin Elements With Type S Enclosure

Element	Rows	Enclo-sure	Installed Height	Steam Capacity/Ft. 1 PSI at 65 F Air		Hot Water Capacity—Btu/Hr./Ft. at 65 F Air, Average Water Temp. (IBR Factor—Steam to Hot Water)					
				EDR Sq. Ft.	Btu/Hr.	220F (1.05)	210F (0.95)	200F (0.86)	190F (0.78)	180F (0.69)	170F (0.61)
Steel—1¼" Series—40 Fins—2½" x 5¼" x .027" Finish—Black	1	12S	15 7/32	5.45	1310	1380	1240	1130	1020	900	800
		16S	18 7/32	5.60	1340	1410	1270	1150	1050	920	820
		20S	21 7/32	5.70	1370	1440	1300	1180	1070	950	840
	2*	16S	18 7/32	8.75	2100	2210	2000	1810	1640	1450	1280
		20S	21 7/32	9.00	2160	2270	2050	1860	1680	1490	1320
	2**	20S	21 7/32	9.40	2260	2370	2150	1940	1760	1560	1380
Steel—1¼" Series—52 Fins—2½" x 5¼" x .027" Finish—Black	1	12S	15 7/32	6.40	1530	1610	1450	1320	1190	1060	930
		16S	18 7/32	6.90	1650	1730	1570	1420	1290	1140	1010
		20S	21 7/32	7.25	1740	1830	1650	1500	1360	1200	1060
	2*	16S	18 7/32	9.15	2200	2310	2090	1890	1720	1520	1340
		20S	21 7/32	9.90	2380	2500	2260	2050	1860	1640	1450
	2**	20S	21 7/32	10.25	2460	2580	2340	2120	1920	1700	1500
Copper-Aluminum—1¼" Series—60 Fins—2½" x 5¼" x .015" Finish—Black	1	12S	15 7/32	7.65	1840	1930	1750	1580	1440	1270	1120
		16S	18 7/32	8.55	2050	2150	1950	1760	1600	1410	1250
		20S	21 7/32	9.35	2240	2350	2130	1930	1750	1550	1370
	2*	16S	18 7/32	10.15	2440	2560	2320	2100	1900	1680	1490
		20S	21 7/32	11.25	2700	2840	2570	2320	2110	1860	1650
	2**	20S	21 7/32	11.45	2750	2890	2610	2370	2150	1900	1680
Copper-Aluminum—1" Series—68 Fins—2" x 3¼" x .011" Finish—Black	1	10S	14 3/32	4.70	1130	1190	1070	970	880	780	690
		14S	17 3/32	5.30	1290	1330	1210	1090	990	880	790
		18S	20¾	5.90	1410	1480	1340	1210	1100	970	860
	2*	14S	17 3/32	6.15	1470	1540	1400	1260	1150	1010	900
		18S	20¾	6.75	1620	1700	1540	1390	1260	1120	990
	2**	18S	20¾	7.25	1740	1830	1650	1500	1360	1200	1060

*4" CENTERS **8" CENTERS

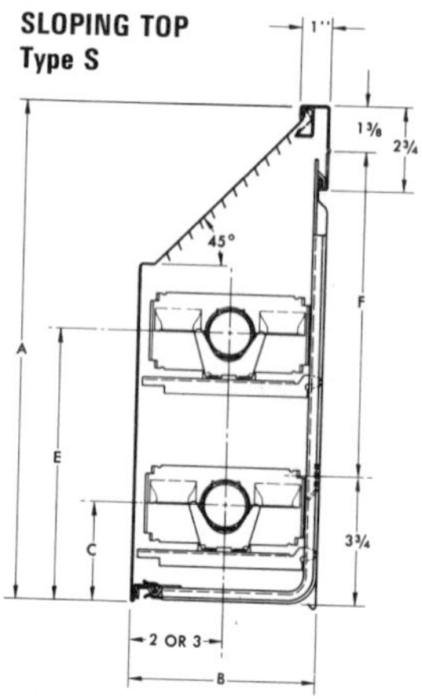

TABLE 12—Types S and F Enclosure Dimensions

Dimensions	3¼" Fin			5¼" Fin		
A	10	14	18	12	16	20
B	4	4	4	6	6	6
C Min.	2¼	2¼	2¼	2½	2½	2½
C Max.	3⅝	3⅝	3⅝	3⅞	3⅞	3⅞
D*	1⅝	1⅝	1⅝	1⅞	1⅞	1⅞
E Min.	—	6¼	10¼	—	6½	10½
E Max.	—	7⅝	11⅝	—	7⅞	11⅞
F	4⅞	8⅞	12⅞	6⅞	10⅞	14⅞

[a]See complete catalog data for other enclosure types.

**Reproduced by permission (Courtesy of The Trane Company).*

Ratings of unit heaters are standardized by various codes. For *steam* the code[7] uses the following items for the basis of rating: dry saturated steam at 2 psig at the heater coil, air entering at 60°F (29.92 in. Hg), and unit operating free of external resistance to air flow. For *hot water* the code[8] uses the following items for the basis of rating: entering water temperature to coil at 200°F, water temperature drop of 20°F, entering air temperature of 60°F, and unit operating free of external resistance to air flow. The code also prescribes a method for translating the heating capacity, as obtained under test, to other conditions of air and water temperatures.

Other available types of unit heaters are gas fired, oil fired, and electric.

Unit ventilators. Unit ventilators (sometimes called classroom unit ventilators because of their frequent use in school classrooms) are a specially designed fan-coil unit which provides for a variety of applications. ASHRAE[1] identifies two general types of unit ventilators.

A *heating unit ventilator* denotes an assembly, the principal functions of which are to heat, ventilate, and cool a space by the introduction of outside air in quantities up to 100 percent. The heating medium may be steam, hot water, gas, or electricity. The essential

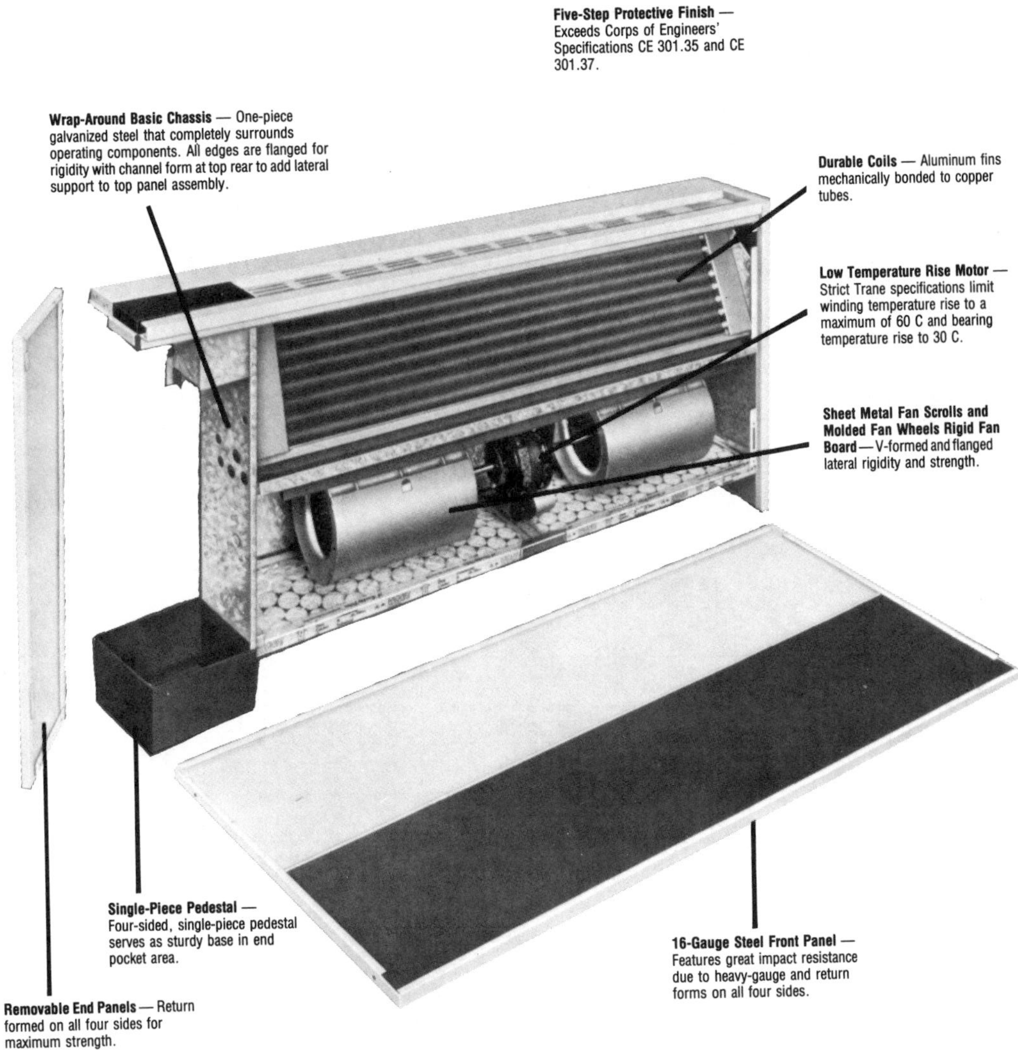

Five-Step Protective Finish — Exceeds Corps of Engineers' Specifications CE 301.35 and CE 301.37.
Wrap-Around Basic Chassis — One-piece galvanized steel that completely surrounds operating components. All edges are flanged for rigidity with channel form at top rear to add lateral support to top panel assembly.
Durable Coils — Aluminum fins mechanically bonded to copper tubes.
Low Temperature Rise Motor — Strict Trane specifications limit winding temperature rise to a maximum of 60 C and bearing temperature rise to 30 C.
Sheet Metal Fan Scrolls and Molded Fan Wheels Rigid Fan Board — V-formed and flanged lateral rigidity and strength.
Single-Piece Pedestal — Four-sided, single-piece pedestal serves as sturdy base in end pocket area.
16-Gauge Steel Front Panel — Features great impact resistance due to heavy-gauge and return forms on all four sides.
Removable End Panels — Return formed on all four sides for maximum strength.

Figure 10–15

Trane Force-Flow Cabinet Heater, front and end panel removed to show fans and heating element (steam or hot water)

components of a heating unit ventilator are: electric motor-driven centrifugal fans, heating element, dampers, filters, inlet grilles, and outlet grilles (diffusers), all encased in a housing. Some or all of the automatic controls are also included in the unit housing.

An *air-conditioning unit ventilator* is similar to the heating unit ventilator, but, in addition to the normal winter season function of heating, ventilating, and cooling with outdoor air, it is equipped to cool and dehumidify during the summer season. It is usually arranged

Table 10-11

A-Coil Hot Water Capacity, 60°F EAT and 180°F EWT*[a]

UNIT SIZE	HIGH SPEED (1100 RPM)					LOW SPEED (700 RPM)				
	GPM	WPD	MBH	WTD	FAT	GPM	WPD	MBH	WTD	FAT
02	0.50	0.07	11.1	44.4	102.8	0.50	0.07	9.6	38.3	118.2
	0.75	0.15	13.6	36.5	112.5	0.75	0.15	11.3	30.2	128.6
	1.00	0.24	15.3	30.9	119.2	1.00	0.24	12.4	24.9	135.3
	1.50	0.50	17.6	23.8	128.0	1.50	0.50	13.7	18.5	143.5
	2.00	0.83	19.0	19.3	133.5	2.00	0.83	14.5	14.7	148.3
	3.00	1.70	20.7	14.0	140.0	3.00	1.70	15.4	10.4	153.7
	5.00	4.20	22.3	9.0	146.0	5.00	4.20	16.2	6.6	158.6
	1.90	0.76	18.8	20.0	132.6	1.35	0.41	13.4	20.0	141.5
	0.63	0.11	12.5	40.0	108.1	0.46	0.06	9.2	40.0	116.1
03	0.50	0.08	14.9	54.9	100.2	0.50	0.08	13.0	51.5	115.1
	0.75	0.17	18.4	49.2	109.7	0.75	0.17	15.4	40.9	125.3
	1.00	0.28	20.9	42.0	116.4	1.00	0.28	17.0	34.0	132.1
	1.50	0.58	24.1	32.5	125.2	1.50	0.58	18.9	25.4	140.5
	2.00	0.97	26.2	26.5	130.8	2.00	0.97	20.1	20.2	145.5
	3.00	1.99	28.6	19.3	137.4	3.00	1.99	21.5	14.4	151.2
	5.00	4.90	31.0	12.6	143.7	5.00	4.90	22.7	9.1	156.3
	2.87	1.84	28.4	20.0	136.8	2.03	0.99	20.2	20.0	145.7
	1.08	0.33	21.5	40.0	118.2	0.78	0.18	15.6	40.0	126.2
04	0.50	0.11	17.0	68.6	95.5	0.50	0.11	15.2	61.1	108.7
	0.75	0.23	21.4	57.8	104.7	0.75	0.23	18.4	49.5	119.1
	1.00	0.38	24.6	49.9	111.4	1.00	0.38	20.6	41.7	126.1
	1.50	0.77	28.9	39.2	120.5	1.50	0.77	23.4	31.6	135.1
	2.00	1.28	31.7	32.3	126.4	2.00	1.28	25.1	25.5	140.7
	3.00	2.63	35.2	23.9	133.5	3.00	2.63	27.1	18.4	147.1
	5.00	6.49	38.5	15.7	140.5	5.00	6.49	28.9	11.8	153.0
	3.75	3.90	36.7	20.0	136.9	2.71	2.19	26.6	20.0	145.6
	1.46	0.73	28.6	40.0	119.9	1.07	0.42	21.0	40.0	127.6
06	0.50	0.13	21.3	85.4	91.9	0.50	0.13	19.2	77.0	104.3
	0.75	0.27	27.2	73.1	100.8	0.75	0.27	23.7	63.6	114.6
	1.00	0.45	31.6	64.0	107.5	1.00	0.45	26.8	54.1	121.9
	1.50	0.93	37.8	51.1	116.8	1.50	0.93	30.9	41.7	131.4
	2.00	1.54	41.9	42.6	123.0	2.00	1.54	33.5	34.0	137.3
	3.00	3.16	47.1	31.9	130.7	3.00	3.16	36.6	24.8	144.4
	5.00	7.80	52.2	21.3	138.4	5.00	7.80	39.5	16.0	151.1
	5.38	8.89	52.8	20.0	139.4	3.88	4.97	38.2	20.0	148.1
	2.19	1.82	43.2	40.0	124.8	1.60	1.03	31.5	40.0	132.7

WPD = WATER PRESSURE DROP.
WTD = WATER TEMPERATURE DIFFERENCE.
FAT = FINAL AIR TEMPERATURE.

[a]This is only a partial listing of capacity and sizes. See the complete data for other sizes in manufacturer's catalog.
**Reprinted courtesy of the Trane Company.*

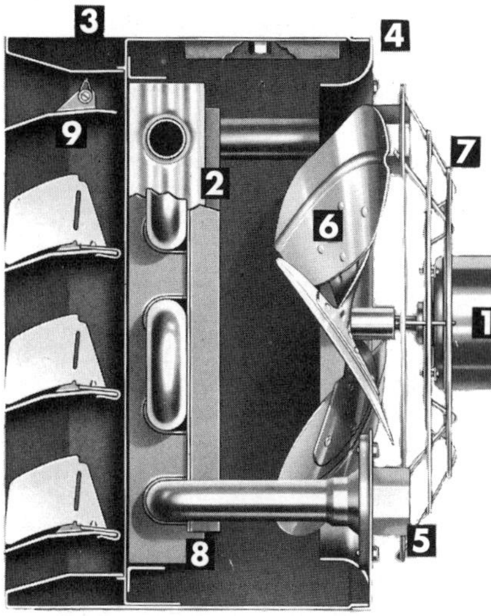

Figure 10–16

Model S unit heater by Trane Company

and controlled to introduce a fixed quantity of outdoor air for ventilating during warm periods in mild weather. The air-conditioning unit ventilator may be provided with a variety of combinations of heating and air-conditioning elements. Some of the more common arrangements include: (a) combination hot and chilled water coil (two-pipe), (b) separate hot and chilled water coils (four-pipe), hot water or steam coil and direct expansion refrigerant coil, (d) electric heating coil and chilled water or direct expansion coil, and (e) gas-fired furnaces with direct expansion coil.

Unit ventilators are used primarily for schools, meeting rooms, offices, and other areas where density of occupancy indicates the need for controlled ventilation. The typical unit is equipped with a system of control that permits the heating, ventilating, and cooling effect to be varied while the fans are operating continuously. In normal operation the discharge air temperature from a unit is varied in accordance with the room requirements. When heating is required, the delivered air is above room temperature. When heat is generated within the room from people, lights, solar radiation, and so forth, in sufficient quantity to exceed the heat losses, the temperature of the delivered air is below that of the room. With the heating unit ventilator this can be done

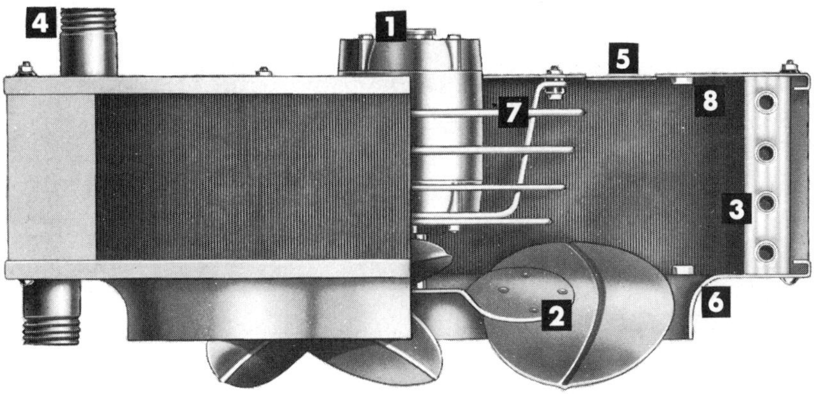

Figure 10–17

Model P unit heater by Trane Company

Table 10-12

Model S Unit Heater Hot Water Capacities — 60°F EAT*[a]

UNIT SIZE	WATER TEMP. DROP °F	180 F MBH	180 F GPM	180 F FT	180 F PD S.W.T.	180 F PD H.W.T.	200 F MBH	200 F GPM	200 F FT	200 F PD S.W.T.	200 F PD H.W.T.	220 F MBH	220 F GPM	220 F FT	220 F PD S.W.T.	220 F PD H.W.T.	240 F MBH	240 F GPM	240 F FT	240 F PD S.W.T.	240 F PD H.W.T.
18-S CFM 280 RPM 1050 HP 1/25	5	9.9	4.07	92.4	.17	.21	12.0	4.94	99.4	.24	.28	14.1	5.81	106.4	.31	.37	16.2	6.69	113.3	.39	.47
	10	7.0	1.45	83.1	.03	.03	9.2	1.89	90.1	.04	.05	11.3	2.32	97.0	.06	.07	13.4	2.76	104.0	.08	.10
	15	4.2	.58	73.8	.01	.01	6.3	.87	80.8	.01	.01	8.4	1.16	87.7	.02	.02	10.6	1.45	94.7	.03	.03
20-S2 CFM 315 RPM 1050 HP 1/25	5	11.8	4.86	94.5	.24	.28	14.3	5.90	101.9	.33	.39	16.9	6.95	109.4	.43	.51	19.4	8.00	116.8	.54	.64
	10	8.3	1.71	84.2	.04	.04	10.8	2.23	91.7	.06	.07	13.4	2.76	99.1	.08	.10	15.9	3.28	106.6	.11	.13
	15	4.8	.66	74.0	.01	.01	7.3	1.01	81.5	.01	.02	9.9	1.36	88.9	.02	.03	12.4	1.71	96.4	.04	.04
38-S	5	25.0	10.30	102.4	1.23	1.48	29.8	12.27	110.5	1.64	1.96	34.5	14.23	118.6	2.09	2.50	39.3	16.19	126.7	2.58	3.09
CFM 543	10	21.4	4.42	96.4	.28	.33	26.2	5.40	104.5	.38	.46	31.0	6.38	112.5	.51	.61	35.7	7.36	120.6	.64	.77
RPM 1550	15	17.9	2.46	90.3	.10	.12	22.6	3.11	98.4	.14	.17	27.4	3.76	106.5	.20	.24	32.2	4.42	114.6	.26	.31
HP 1/20	20	14.3	1.47	84.3	.04	.05	19.1	1.97	92.4	.06	.08	23.8	2.46	100.5	.09	.11	28.6	2.95	108.5	.13	.15
	25	10.8	.89	78.2	.02	.02	15.5	1.28	86.3	.03	.04	20.3	1.67	94.4	.05	.06	25.0	2.06	102.5	.07	.08
	30	—	—	—	—	—	11.9	.82	80.3	.01	.02	16.7	1.15	88.4	.02	.03	21.5	1.47	96.4	.04	.06
42-S	5	26.2	10.81	100.9	1.34	1.61	31.2	12.87	108.7	1.78	2.13	36.2	14.93	116.5	2.28	2.72	41.2	16.98	124.3	2.81	3.36
CFM 591	10	22.5	4.64	95.1	.30	.36	27.5	5.67	102.9	.42	.50	32.5	6.69	110.7	.55	.66	37.5	7.72	118.4	.70	.84
RPM 1550	15	18.8	2.58	89.3	.11	.13	23.8	3.26	97.1	.16	.19	28.8	3.95	104.8	.22	.26	33.7	4.64	112.6	.28	.34
HP 1/20	20	15.0	1.55	83.5	.04	.05	20.0	2.06	91.2	.07	.08	25.0	2.58	99.0	.10	.12	30.0	3.09	106.8	.14	.17
	25	11.3	.93	77.6	.02	.02	16.3	1.34	85.4	.03	.04	21.3	1.76	93.2	.05	.06	26.3	2.17	101.0	.07	.09
	30	—	—	—	—	—	12.6	.86	79.6	.02	.02	17.6	1.21	87.4	.03	.03	22.6	1.55	95.2	.04	.05
60-S	5	39.0	16.09	104.1	3.40	4.07	46.2	19.03	112.2	4.46	5.33	53.3	21.98	120.3	5.65	6.76	—	—	—	—	—
CFM 815	10	35.2	7.25	99.8	.83	1.00	42.3	8.72	107.8	1.12	1.34	49.5	10.19	115.9	1.45	1.74	56.6	11.67	124.0	1.81	2.17
RPM 1550	15	31.3	4.30	95.4	.33	.40	38.4	5.28	103.5	.46	.55	45.6	6.26	111.6	.62	.74	52.7	7.25	119.6	.78	.94
HP 1/20	20	27.4	2.83	91.0	.16	.19	34.6	3.56	99.1	.23	.28	41.7	4.30	107.2	.32	.38	48.9	5.04	115.3	.41	.49
	25	23.6	1.94	86.7	.08	.10	30.7	2.53	94.7	.13	.15	37.9	3.12	102.8	.18	.22	45.0	3.71	110.9	.24	.29
	30	19.7	1.35	82.3	.04	.05	26.9	1.84	90.4	.07	.09	34.0	2.34	98.5	.11	.13	41.1	2.83	106.5	.15	.18
	40	12.0	.62	73.5	.01	.01	19.1	.99	81.6	.02	.03	26.3	1.35	89.7	.04	.05	33.4	1.72	97.8	.06	.07
70-S	5	45.7	18.84	98.3	4.49	5.38	54.1	22.29	105.3	5.90	7.06	—	—	—	—	—	—	—	—	—	—
CFM 1100	10	41.1	8.48	94.5	1.10	1.32	49.5	10.21	101.5	1.48	1.77	57.9	11.93	108.5	1.92	2.30	66.3	13.66	115.5	2.40	2.87
RPM 1550	15	36.6	5.03	90.7	.44	.52	45.0	6.18	97.7	.61	.73	53.3	7.33	104.7	.81	.97	61.7	8.48	111.7	1.03	1.24
HP 1/8	20	32.1	3.30	86.9	.21	.25	40.4	4.17	93.9	.30	.36	48.8	5.03	100.9	.42	.50	57.2	5.89	107.9	.54	.65
	25	27.5	2.27	83.0	.11	.13	35.9	2.96	90.1	.17	.20	44.3	3.65	97.1	.24	.28	52.6	4.34	104.1	.32	.38
	30	23.0	1.58	79.2	.06	.07	31.3	2.15	86.2	.10	.11	39.7	2.73	93.3	.14	.17	48.1	3.30	100.3	.20	.23
	40	13.9	.71	71.6	.01	.02	22.2	1.15	78.6	.03	.04	30.6	1.58	85.6	.05	.06	39.0	2.01	92.7	.08	.10
90-S	10	53.9	11.10	100.9	2.14	2.56	64.3	13.25	108.8	2.84	3.40	74.7	15.39	116.7	3.64	4.36	85.1	17.54	124.6	4.50	5.39
CFM 1214	15	49.6	6.81	97.6	.91	1.08	60.0	8.24	105.5	1.23	1.47	70.4	9.67	113.4	1.60	1.92	80.8	11.10	121.3	2.01	2.40
RPM 1550	20	45.3	4.67	94.4	.47	.56	55.7	5.74	102.3	.65	.78	66.1	6.81	110.2	.86	1.03	76.5	7.88	118.1	1.10	1.32
HP 1/8	25	41.0	3.38	91.1	.26	.32	51.4	4.24	99.0	.38	.45	61.8	5.09	106.9	.52	.62	72.2	5.95	114.8	.67	.80
	30	36.7	2.52	87.9	.16	.19	47.1	3.24	95.8	.24	.28	57.5	3.95	103.7	.33	.40	67.9	4.67	111.6	.44	.52
	40	28.1	1.45	81.3	.06	.07	38.5	1.98	89.2	.10	.12	48.9	2.52	97.1	.15	.18	59.3	3.06	105.0	.21	.25
	50	19.5	.80	74.8	.02	.03	29.9	1.23	82.7	.04	.05	40.3	1.66	90.6	.07	.09	50.7	2.09	98.5	.11	.13
100-S	10	60.6	12.48	96.4	2.63	3.15	72.3	14.90	103.4	3.50	4.19	84.0	17.32	110.5	4.49	5.37	95.8	19.74	117.5	5.55	6.64
CFM 1535	15	55.7	7.65	93.4	1.11	1.33	67.4	9.26	100.5	1.51	1.81	79.1	10.87	107.5	1.97	2.36	90.9	12.48	114.6	2.47	2.96
RPM 1550	20	50.8	5.23	90.5	.57	.68	62.5	6.44	97.5	.80	.95	74.2	7.65	104.6	1.06	1.27	86.0	8.86	111.6	1.35	1.62
HP 1/8	25	45.9	3.78	87.5	.32	.39	57.6	4.75	94.6	.47	.56	69.3	5.72	101.6	.63	.76	81.1	6.68	108.7	.82	.98
	30	41.0	2.82	84.6	.19	.23	52.7	3.62	91.6	.29	.35	64.4	4.43	98.7	.40	.48	76.2	5.23	105.7	.53	.64
	40	31.2	1.61	78.7	.07	.09	42.9	2.21	85.8	.12	.15	54.6	2.82	92.8	.18	.22	66.4	3.42	99.9	.25	.30
	50	21.4	.88	72.8	.03	.03	33.1	1.37	79.9	.05	.06	44.8	1.85	86.9	.09	.10	56.6	2.33	94.0	.13	.15
	60	—	—	—	—	—	23.3	.80	74.0	.02	.02	35.0	1.20	81.0	.04	.05	46.8	1.61	88.1	.07	.08

Table 10-12 (Continued)

Model S Unit Heater Hot Water Capacities – 60°F EAT*[a]

UNIT SIZE	WATER TEMP. DROP °F	ENTERING WATER TEMPERATURE 180 F MBH	180 F GPM	180 F FT	180 F PD S.W.T.	180 F PD H.W.T.	200 F MBH	200 F GPM	200 F FT	200 F PD S.W.T.	200 F PD H.W.T.	220 F MBH	220 F GPM	220 F FT	220 F PD S.W.T.	220 F PD H.W.T.	240 F MBH	240 F GPM	240 F FT	240 F PD S.W.T.	240 F PD H.W.T.
126-S	10	80.0	16.48	101.9	5.29	6.33	95.0	19.58	109.7	6.97	8.34	110.0	22.68	117.6	8.88	10.62	— —	— —	— —	— —	— —
CFM 1760	15	74.9	10.29	99.2	2.31	2.76	89.9	12.36	107.1	3.09	3.70	104.9	14.42	115.0	3.99	4.78	120.0	16.48	122.8	4.97	5.95
RPM 1100	**20**	**69.8**	**7.20**	**96.6**	**1.23**	**1.47**	**84.9**	**8.74**	**104.4**	**1.68**	**2.01**	**99.9**	**10.29**	**112.3**	**2.20**	**2.64**	**114.9**	**11.84**	**120.2**	**2.77**	**3.32**
HP 1/6	25	64.8	5.34	93.9	.73	.87	79.8	6.58	101.8	1.02	1.22	94.8	7.82	109.6	1.36	1.62	109.8	9.05	117.5	1.73	2.07
	30	59.7	4.10	91.3	.46	.55	74.7	5.13	99.1	.66	.79	89.7	6.16	107.0	.89	1.07	104.7	7.20	114.9	1.15	1.38
	40	49.5	2.55	85.9	.20	.24	64.6	3.33	93.8	.31	.37	79.6	4.10	101.7	.44	.52	94.6	4.87	109.5	.58	.70
	50	39.4	1.62	80.6	.09	.11	54.4	2.24	88.5	.15	.19	69.4	2.86	96.4	.23	.28	84.4	3.48	104.2	.32	.39
	60	29.2	1.00	75.3	.04	.05	44.3	1.52	83.2	.08	.09	59.3	2.04	91.0	.13	.15	74.3	2.55	98.9	.19	.22
168-S	15	101.4	13.93	99.2	4.60	5.51	121.2	16.66	106.9	6.12	7.33	141.1	19.39	114.6	7.87	9.41	161.0	22.12	122.3	9.75	11.67
CFM 2381	**20**	**95.5**	**9.84**	**97.0**	**2.50**	**2.99**	**115.3**	**11.88**	**104.6**	**3.37**	**4.04**	**135.2**	**13.93**	**112.3**	**4.39**	**5.26**	**155.0**	**15.98**	**120.0**	**5.50**	**6.58**
RPM 1100	25	89.6	7.38	94.7	1.51	1.81	109.4	9.02	102.4	2.08	2.48	129.3	10.66	110.0	2.74	3.28	149.1	12.29	117.7	3.47	4.15
HP 1/6	30	83.6	5.75	92.4	.97	1.17	103.5	7.11	100.1	1.37	1.63	123.4	8.47	107.7	1.83	2.19	143.2	9.84	115.4	2.34	2.80
	40	71.8	3.70	87.8	.45	.54	91.7	4.72	95.5	.67	.80	111.5	5.75	103.2	.92	1.11	131.4	6.77	110.9	1.21	1.45
	50	60.0	2.47	83.2	.22	.27	79.8	3.29	90.9	.36	.43	99.7	4.11	98.6	.51	.61	119.5	4.93	106.3	.69	.83
	60	48.1	1.65	78.6	.11	.13	68.0	2.34	86.3	.20	.23	87.9	3.02	94.0	.30	.36	107.7	3.70	101.7	.42	.50
186-S	15	110.2	15.14	96.2	5.33	6.38	131.7	18.10	103.2	7.09	8.48	153.2	21.05	110.3	9.10	10.89	— —	— —	— —	— —	— —
CFM 2808	**20**	**103.9**	**10.70**	**94.1**	**2.90**	**3.47**	**125.4**	**12.92**	**101.2**	**3.91**	**4.68**	**146.9**	**15.14**	**108.2**	**5.09**	**6.09**	**168.4**	**17.36**	**115.3**	**6.37**	**7.62**
RPM 1100	25	97.5	8.04	92.0	1.76	2.10	119.1	9.81	99.1	2.41	2.88	140.6	11.59	106.1	3.18	3.80	162.1	13.36	113.2	4.02	4.81
HP 1/4	30	91.2	6.27	89.9	1.14	1.36	112.7	7.74	97.0	1.59	1.90	134.3	9.22	104.1	2.13	2.54	155.8	10.70	111.1	2.72	3.25
	40	78.6	4.05	85.8	.53	.63	100.1	5.16	92.8	.78	.93	121.6	6.27	99.9	1.08	1.29	143.1	7.37	107.0	1.41	1.69
	50	65.9	2.72	81.6	.26	.32	87.4	3.60	88.7	.42	.50	109.0	4.49	95.8	.60	.72	130.5	5.38	102.8	.81	.97
	60	53.3	1.83	77.5	.13	.16	74.8	2.57	84.5	.23	.28	96.3	3.31	91.6	.35	.42	117.8	4.05	98.7	.49	.59
230-S	15	142.7	19.61	99.9	1.98	2.37	171.3	23.53	107.8	2.65	3.18	199.8	27.46	115.8	3.43	4.10	228.4	31.38	123.8	4.26	5.10
CFM 3299	**20**	**133.2**	**13.72**	**97.2**	**1.06**	**1.27**	**161.7**	**16.67**	**105.2**	**1.44**	**1.73**	**190.3**	**19.61**	**113.2**	**1.89**	**2.26**	**218.8**	**22.55**	**121.1**	**2.38**	**2.85**
RPM 1100	25	123.6	10.19	94.5	.63	.75	152.2	12.54	102.5	.88	1.05	180.7	14.90	110.5	1.17	1.40	209.3	17.25	118.5	1.49	1.78
HP 1/4	30	114.1	7.84	91.9	.40	.47	142.6	9.80	99.8	.57	.68	171.2	11.76	107.8	.77	.92	199.7	13.72	115.8	.99	1.19
	40	95.0	4.89	86.5	.17	.21	123.5	6.36	94.5	.27	.32	152.1	7.84	102.5	.38	.45	180.6	9.31	110.5	.50	.60
	50	75.8	3.13	81.2	.08	.10	104.4	4.30	89.2	.13	.16	133.0	5.48	97.1	.20	.24	161.5	6.66	105.1	.28	.33
	60	56.7	1.95	75.9	.03	.04	85.3	2.93	83.8	.07	.08	113.9	3.91	91.8	.11	.13	142.4	4.89	99.8	.16	.19
	70	— —	— —	— —	— —	— —	66.2	1.95	78.5	.03	.04	94.8	2.79	86.5	.06	.07	123.3	3.63	94.5	.10	.11
260-S	15	160.0	21.98	96.0	2.43	2.91	192.1	26.40	103.2	3.25	3.89	224.3	30.81	110.4	4.20	5.03	256.4	35.23	117.7	5.23	6.26
CFM 4099	**20**	**149.0**	**15.35**	**93.5**	**1.29**	**1.55**	**181.1**	**18.67**	**100.7**	**1.77**	**2.11**	**213.3**	**21.98**	**108.0**	**2.32**	**2.77**	**245.4**	**25.29**	**115.2**	**2.92**	**3.49**
RPM 1100	25	138.0	11.38	91.0	.76	.92	170.2	14.03	98.3	1.07	1.28	202.3	16.68	105.5	1.42	1.70	234.5	19.33	112.7	1.82	2.17
HP 1/2	30	127.1	8.73	88.6	.48	.58	159.2	10.94	95.8	.69	.82	191.4	13.15	103.0	.94	1.12	223.5	15.35	110.3	1.21	1.45
	40	105.1	5.42	83.6	.21	.25	137.3	7.07	90.9	.32	.38	169.4	8.73	98.1	.46	.55	201.6	10.39	105.3	.61	.73
	50	83.2	3.43	78.7	.09	.11	115.3	4.75	85.9	.16	.19	147.5	6.08	93.2	.24	.29	179.6	7.40	100.4	.34	.40
	60	61.3	2.10	73.8	.04	.05	93.4	3.21	81.0	.08	.10	125.6	4.31	88.2	.13	.16	157.7	5.42	95.5	.19	.23
	70	— —	— —	— —	— —	— —	71.5	2.10	76.1	.04	.05	103.6	3.05	83.3	.07	.09	135.8	4.00	90.5	.11	.14

[a]This is only a partial listing of available sizes and capacities. See manufacturer's catalog for complete data. Abbreviations: MBH = 1000 Btu/hr; GPM = gal/min; FT = final air temperature, F; PD-S.W.T. = feet of water pressure drop - standard wall tubes; PH-H.W.T. = feet of water pressure drop - heavy wall tubes.

Reproduced courtesy of the Trane Company.

Table 10-13

Model P Unit Heater Hot Water Capacities – 60°F EAT*[a]

UNIT SIZE	WATER TEMP. DROP °F	ENTERING WATER TEMPERATURE 180 F					200 F					220 F					240 F				
		MBH	GPM	FT	PD S.W.T.	PD H.W.T.	MBH	GPM	FT	PD S.W.T.	PD H.W.T.	MBH	GPM	FT	PD S.W.T.	PD H.W.T.	MBH	GPM	FT	PD S.W.T.	PD H.W.T.
42-P	10	23.8	4.90	96.9	.26	.34	28.8	5.93	104.6	.37	.49	33.7	6.95	112.2	.49	.65	38.7	7.98	119.9	.64	.84
CFM 595	15	20.8	2.86	92.2	.09	.12	25.7	3.54	99.9	.14	.18	30.7	4.22	107.6	.19	.25	35.7	4.90	115.3	.25	.33
RPM 1550	20	17.8	1.83	87.5	.04	.05	22.7	2.34	95.2	.06	.08	27.7	2.86	102.9	.09	.12	32.7	3.37	110.6	.12	.16
HP 1/25	25	14.8	1.22	82.9	.02	.02	19.7	1.63	90.6	.03	.04	24.7	2.04	98.3	.05	.06	29.7	2.45	105.9	.07	.09
	30	—	—	—	—	—	16.7	1.15	85.9	.02	.02	21.7	1.49	93.6	.03	.03	26.7	1.83	101.3	.04	.05
	40	—	—	—	—	—	—	—	—	—	—	—	—	—	—	—	20.6	1.06	91.9	.01	.02
42-P	10	18.9	3.90	100.0	.17	.22	22.9	4.71	108.3	.24	.31	26.8	5.52	116.6	.32	.42	30.7	6.33	125.0	.41	.54
CFM 436	15	16.6	2.27	95.0	.06	.08	20.5	2.82	103.3	.09	.12	24.4	3.36	111.6	.12	.16	28.4	3.90	120.0	.16	.21
RPM 1150	20	14.2	1.46	90.0	.03	.03	18.1	1.87	98.3	.04	.05	22.1	2.27	106.7	.06	.08	26.0	2.68	115.0	.08	.10
HP 1/25	25	—	—	—	—	—	15.8	1.30	93.3	.02	.03	19.7	1.63	101.7	.03	.04	23.7	1.95	110.0	.04	.06
	30	—	—	—	—	—	13.4	.92	88.4	.01	.01	17.4	1.19	96.7	.02	.02	21.3	1.46	105.0	.02	.03
	40	—	—	—	—	—	—	—	—	—	—	—	—	—	—	—	16.6	.85	95.0	.01	.01
64-P	10	40.0	8.25	97.3	.75	1.00	48.1	9.92	104.8	1.05	1.40	56.2	11.58	112.4	1.40	1.86	64.3	13.25	119.9	1.80	2.39
CFM 989	15	35.8	4.91	93.3	.28	.37	43.9	6.03	100.9	.41	.54	51.9	7.14	108.4	.56	.74	60.0	8.25	115.9	.73	.97
RPM 1550	20	31.5	3.25	89.4	.13	.17	39.6	4.08	96.9	.19	.26	47.7	4.19	104.4	.27	.36	55.8	5.75	112.0	.36	.49
HP 1/20	25	27.3	2.25	85.4	.06	.08	35.3	2.91	92.9	.10	.14	43.4	3.58	100.5	.15	.20	51.5	4.25	108.0	.21	.27
	30	23.0	1.58	81.4	.03	.04	31.1	2.14	89.0	.06	.08	39.2	2.69	96.5	.09	.12	47.3	3.25	104.0	.12	.16
	40	—	—	—	—	—	22.6	1.16	81.0	.02	.02	30.7	1.58	88.6	.03	.04	38.8	2.00	96.1	.05	.07
	50	—	—	—	—	—	—	—	—	—	—	22.2	.91	80.7	.01	.02	30.2	1.25	88.2	.02	.03
64-P	10	31.7	6.53	101.4	.48	.64	38.1	7.85	109.7	.67	.90	44.5	9.17	118.1	.90	1.19	50.9	10.48	126.4	1.15	1.53
CFM 706	15	28.4	3.90	97.1	.18	.24	34.8	4.78	105.4	.26	.35	41.2	5.66	113.7	.36	.48	47.5	6.53	122.1	.47	.62
RPM 1150	20	25.1	2.58	92.7	.08	.11	31.5	3.24	101.1	.13	.17	37.8	3.90	109.4	.18	.23	44.2	4.56	117.7	.23	.31
HP 1/20	25	21.8	1.79	88.4	.04	.06	28.2	2.32	96.7	.07	.09	34.5	2.85	105.1	.10	.13	40.9	3.37	113.4	.13	.18
	30	18.5	1.27	84.1	.02	.03	24.8	1.71	92.4	.04	.05	31.2	2.15	100.8	.06	.08	37.6	2.58	109.1	.08	.11
	40	—	—	—	—	—	—	—	—	—	—	24.6	1.27	92.1	.02	.03	31.0	1.60	100.5	.03	.04
	50	—	—	—	—	—	—	—	—	—	—	—	—	—	—	—	24.4	1.00	91.8	.01	.02
80-P	10	48.9	10.07	97.5	.70	.90	58.7	12.11	105.1	.98	1.27	68.6	14.14	112.7	1.31	1.69	78.5	16.17	120.3	1.68	2.16
CFM 1200	15	43.7	6.01	93.6	.26	.34	53.6	7.36	101.2	.38	.49	63.5	8.72	108.7	.52	.67	73.3	10.07	116.3	.68	.87
RPM 1550	20	38.6	3.97	89.6	.12	.15	48.4	4.99	97.2	.18	.23	58.3	6.01	104.8	.25	.33	68.2	7.02	112.3	.34	.44
HP 1/20	25	33.4	2.75	85.7	.06	.08	43.3	3.57	93.2	.10	.12	53.1	4.38	100.8	.14	.18	63.0	5.19	108.4	.19	.25
	30	28.2	1.94	81.7	.03	.04	38.1	2.62	89.3	.05	.07	48.0	3.30	96.8	.08	.10	57.8	3.97	104.4	.11	.15
	40	—	—	—	—	—	27.8	1.43	81.3	.02	.02	37.7	1.94	88.9	.03	.04	47.5	2.45	96.5	.05	.06
	50	—	—	—	—	—	—	—	—	—	—	—	—	—	—	—	37.2	1.53	88.6	.02	.02
80-P	10	38.7	7.99	101.6	.45	.58	46.5	9.59	110.0	.63	.81	54.3	11.20	118.4	.84	1.08	62.1	12.81	126.7	1.07	1.38
CFM 858	15	34.7	4.77	97.3	.17	.22	42.5	5.84	105.7	.24	.31	50.3	6.91	114.1	.33	.43	58.1	7.99	122.4	.43	.56
RPM 1150	20	30.7	3.16	93.0	.08	.10	38.5	3.97	101.4	.12	.15	46.3	4.77	109.7	.16	.21	54.1	5.57	118.1	.22	.28
HP 1/20	25	26.7	2.20	88.7	.04	.05	34.5	2.84	97.0	.06	.08	42.3	3.49	105.4	.09	.12	50.1	4.13	113.8	.12	.16
	30	22.7	1.56	84.4	.02	.03	30.5	2.09	92.7	.03	.04	38.3	2.63	101.1	.05	.07	46.1	3.16	109.5	.07	.10
	40	—	—	—	—	—	—	—	—	—	—	30.2	1.56	92.5	.02	.03	38.0	1.96	100.9	.03	.04
	50	—	—	—	—	—	—	—	—	—	—	—	—	—	—	—	30.0	1.24	92.2	.01	.02

Table 10-13 (Continued)

UNIT SIZE	WATER TEMP DROP °F	ENTERING WATER TEMPERATURE																			
		180 F					200 F					220 F					240 F				
		MBH	GPM	FT	PD S.W.T.	PD H.W.T.	MBH	GPM	FT	PD S.W.T.	PD H.W.T.	MBH	GPM	FT	PD S.W.T.	PD H.W.T.	MBH	GPM	FT	PD S.W.T.	PD H.W.T.
102-P CFM 1528 RPM 1070 HP 1/8	10	64.9	13.38	99.1	1.50	1.96	77.2	15.91	106.6	2.06	2.68	89.5	18.45	114.0	2.69	3.50	101.8	20.98	121.4	3.40	4.42
	15	60.4	8.30	96.4	.61	.80	72.7	9.99	103.9	.85	1.11	85.0	11.68	111.3	1.13	1.48	97.3	13.38	118.7	1.45	1.88
	20	**56.0**	**5.77**	**93.8**	**.31**	**.40**	**68.3**	**7.03**	**101.2**	**.44**	**.58**	**80.6**	**8.30**	**108.6**	**.59**	**.78**	**92.9**	**9.57**	**116.0**	**.77**	**1.00**
	25	51.5	4.24	91.1	.17	.23	63.8	5.26	98.5	.25	.33	76.1	6.27	105.9	.35	.46	88.4	7.29	113.3	.46	.60
	30	47.0	3.23	88.4	.10	.14	59.3	4.08	95.8	.16	.21	71.6	4.92	103.2	.22	.29	83.9	5.77	110.6	.30	.39
	40	38.1	1.96	83.0	.04	.05	50.4	2.60	90.4	.07	.09	62.7	3.23	97.8	.10	.13	75.0	3.86	105.2	.14	.18
	50	—	—	—	—	—	41.5	1.71	85.0	.03	.04	53.8	2.22	92.4	.05	.07	66.1	2.72	99.9	.07	.09
	60	—	—	—	—	—	—	—	—	—	—	44.8	1.54	87.0	.03	.03	57.1	1.96	94.5	.04	.05
102-P CFM 1208 RPM 850 HP 1/8	10	53.6	11.04	100.9	1.04	1.36	63.7	13.13	108.6	1.43	1.86	73.8	15.22	116.3	1.87	2.43	84.0	17.31	124.1	2.36	3.07
	15	50.0	6.86	98.1	.43	.56	60.1	8.26	105.8	.59	.78	70.2	9.65	113.6	.79	1.03	80.4	11.04	121.3	1.01	1.31
	20	**46.3**	**4.77**	**95.4**	**.22**	**.28**	**56.5**	**5.82**	**103.1**	**.31**	**.40**	**66.6**	**6.86**	**110.8**	**.41**	**.54**	**76.7**	**7.91**	**118.6**	**.53**	**.70**
	25	42.7	3.52	92.6	.12	.16	52.9	4.36	100.3	.18	.23	63.0	5.19	108.1	.24	.32	73.1	6.03	115.8	.32	.42
	30	39.1	2.69	89.8	.07	.10	49.2	3.38	97.6	.11	.15	59.4	4.08	105.3	.16	.20	69.5	4.77	113.0	.21	.27
	40	31.9	1.64	84.3	.03	.04	42.0	2.16	92.0	.05	.06	52.1	2.69	99.8	.07	.09	62.3	3.21	107.5	.10	.13
	50	—	—	—	—	—	34.8	1.43	86.5	.02	.03	44.9	1.85	94.2	.04	.05	55.0	2.27	102.0	.05	.07
	60	—	—	—	—	—	—	—	—	—	—	37.6	1.29	88.7	.02	.02	47.8	1.64	96.5	.03	.04
122-P CFM 1790 RPM 1070 HP 1/6	10	79.7	16.43	101.1	2.22	2.89	94.9	19.55	108.9	3.04	3.96	110.0	22.68	116.7	3.99	5.18	125.2	25.80	124.5	5.04	6.55
	15	74.1	10.19	98.2	.90	1.17	89.3	12.27	106.0	1.26	1.64	104.4	14.35	113.8	1.67	2.18	119.6	16.43	121.6	2.14	2.78
	20	**68.6**	**7.06**	**95.3**	**.45**	**.59**	**83.7**	**8.63**	**103.1**	**.65**	**.85**	**98.9**	**10.19**	**110.9**	**.87**	**1.14**	**114.0**	**11.75**	**118.7**	**1.13**	**1.48**
	25	63.0	5.19	92.4	.25	.33	78.1	6.44	100.2	.37	.49	93.3	7.69	108.0	.51	.67	108.4	8.94	115.8	.68	.88
	30	57.4	3.94	89.5	.15	.20	72.5	4.98	97.3	.23	.30	87.7	6.02	105.1	.32	.42	102.8	7.06	112.9	.43	.57
	40	46.2	2.38	83.8	.06	.08	61.4	3.16	91.6	.10	.13	76.5	3.94	9.94	.15	.19	91.7	4.72	107.2	.20	.27
	50	—	—	—	—	—	50.2	2.07	85.8	.04	.06	65.3	2.69	93.6	.07	.09	80.5	3.32	101.4	.10	.14
	60	—	—	—	—	—	39.0	1.34	80.1	.02	.03	54.2	1.86	87.9	.04	.05	69.3	2.38	95.7	.06	.07
146-P CFM 2220 RPM 1070 HP 1/6	10	99.0	20.41	101.1	3.18	4.27	117.6	24.24	108.8	4.32	5.81	136.2	28.07	116.5	5.61	7.54	154.8	31.89	124.2	7.05	9.47
	15	92.8	12.76	98.5	1.32	1.77	111.4	15.31	106.3	1.83	2.45	130.0	17.86	114.0	2.40	3.23	148.6	20.41	121.7	3.04	4.09
	20	**86.7**	**8.93**	**96.0**	**.68**	**.91**	**105.2**	**10.84**	**103.7**	**.96**	**1.29**	**123.8**	**12.76**	**111.4**	**1.28**	**1.72**	**142.4**	**14.67**	**119.1**	**1.64**	**2.20**
	25	80.5	6.63	93.4	.39	.52	99.0	8.16	101.1	.56	.76	117.6	9.70	108.8	.77	1.03	136.2	11.23	116.5	.99	1.33
	30	74.3	5.10	90.8	.24	.32	92.8	6.38	98.5	.36	.48	111.4	7.65	106.3	.49	.66	130.0	8.93	114.0	.65	.87
	40	61.9	3.19	85.7	.10	.14	80.5	4.15	93.4	.16	.22	99.0	5.10	101.1	.23	.31	117.6	6.06	108.8	.31	.42
	50	49.5	2.04	80.6	.04	.06	68.1	2.81	88.3	.08	.10	86.6	3.57	96.0	.12	.16	105.2	4.34	103.7	.17	.23
	60	—	—	—	—	—	55.7	1.91	83.1	.04	.05	74.3	2.55	90.8	.06	.09	92.8	3.19	98.5	.10	.13
166-P CFM 2620 RPM 1070 HP 1/6	10	111.5	22.98	99.2	2.70	3.59	132.4	27.29	106.6	3.67	4.88	153.3	31.59	113.9	4.77	6.34	174.2	35.89	121.3	6.00	7.96
	15	104.6	14.38	96.8	1.12	1.49	125.5	17.24	104.1	1.55	2.06	146.4	20.11	111.5	2.04	2.71	167.3	22.98	118.8	2.59	3.44
	20	**97.7**	**10.07**	**94.4**	**.58**	**.77**	**118.6**	**12.22**	**101.7**	**.81**	**1.08**	**139.5**	**14.38**	**109.1**	**1.09**	**1.45**	**160.4**	**16.53**	**116.4**	**1.39**	**1.85**
	25	90.8	7.49	92.0	.33	.44	111.7	9.21	99.3	.48	.64	132.6	10.93	106.6	.65	.87	153.5	12.65	114.0	.85	1.12
	30	84.0	5.77	89.5	.20	.27	104.8	7.20	96.9	.30	.40	125.7	8.64	104.2	.42	.56	146.6	10.07	111.6	.55	.79
	40	70.2	3.62	84.7	.09	.11	91.1	4.69	92.0	.14	.18	111.9	5.77	99.4	.20	.26	132.8	6.84	106.7	.27	.36
	50	56.4	2.32	79.8	.04	.05	77.3	3.19	87.2	.07	.09	98.2	4.05	94.5	.10	.14	119.0	4.91	101.9	.14	.19
	60	—	—	—	—	—	63.5	2.18	82.3	.03	.04	84.4	2.90	89.7	.06	.07	105.3	3.62	97.0	.08	.11

[a]This is only a partial listing of available sizes and capacities. See manufacturer's catalog for complete data. Abbreviations: See Table 10-12.

**Reproduced courtesy of the Trane Company.*

Airflow path engineered to handle full rated CFM quietly, efficiently.

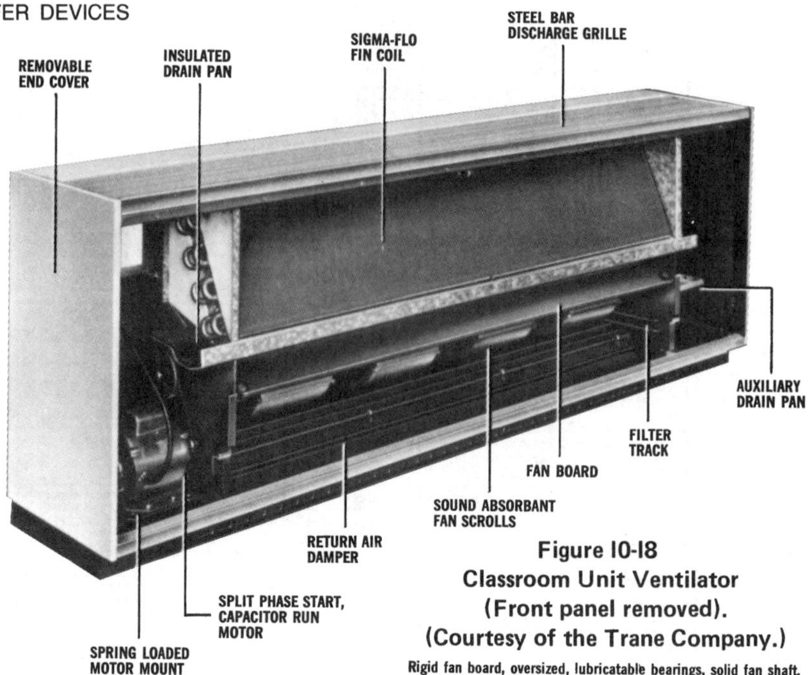

Figure 10-18
Classroom Unit Ventilator (Front panel removed). (Courtesy of the Trane Company.)

Rigid fan board, oversized, lubricatable bearings, solid fan shaft, durable drive provide high reliability, extend unit life.

whenever the outdoor air is below the room temperature by bringing in more outdoor air, thereby providing what is called ventilation cooling. Air-conditioning unit ventilators can provide mechanical cooling in cases where the outdoor air is too high to be used effectively for ventilation cooling.

The standard air rating of unit ventilators is obtained with the *ASHRAE Standard Method of Testing for Rating Unit Ventilators*.[6] The *Standard Air Rating* of a unit is the air delivery, in cfm, converted to standard air at 70°F.

The heating and cooling capacities of unit ventilators are published by manufacturers. Each manufacturer may have a different method for selection of a unit ventilator for a particular application, therefore catalog data should be used as indicated by illustrative examples in such catalogs. In general, when a unit ventilator is to be selected we must consider the following: (a) unit air capacity, cfm; (b) minimum percent outdoor air; (c) heating and/or cooling requirements; (d) control cycle required; and (e) location of unit.

The three basic *control cycles* commonly used are defined by ASHRAE as cycle I, cycle II, and cycle III, which are briefly defined as follows:

Cycle I: 100 percent outdoor air admitted at all times, except during warm-up.

Cycle II: A minimum amount of outdoor air (normally 25 to 50 percent) is admitted during the *heating and ventilating* stage. The percentage is gradually increased to 100 percent, if needed, during the *ventilation cooling* stage.

Cycle III: Except during the warmup stage, a variable amount of outdoor air is admitted as needed to maintain a fixed temperature of the air entering the heating element. This is controlled by the airstream thermostat, which is set low enough (often 55°F) to provide ventilation cooling when needed.

Air-conditioning unit ventilators can utilize

any of the cycles above, but in addition have a mechanical cooling stage. In this stage a fixed amount of outdoor air is introduced, and the cooling capacity is controlled by the room thermostat.

In all cases energy savings may be realized by using what is called "night-setback," which is used during periods of no occupancy. Normally, the unit ventilators are arranged to provide for gravity air circulation with no outdoor air combined with fan cycling.

Air makeup units. ASHRAE[1] identifies these units as an assembly of elements, the principal function of which is to properly condition makeup air being introduced into a space, customarily to make up for air being exhausted from a building. The air exhausted may be from a particular process or from general area exhaust.

Whatever the method of exhaust, a volume of air is removed from a structure and must be replaced with new outdoor air. If the temperature, humidity, or both, within a structure are controlled, the environmental control system must provide capacity to condition the replacement air if the space condition is to be maintained. The most practical way to condition the replacement air is through the use of a makeup air unit.

Certain symptoms of an "air-starved" building may be recognized, for example: (a) gravity stacks will back-vent; (b) exhaust systems do not perform at rated volume, resulting in poor control of contaminants; (c) the perimeter of the building will be cold due to a high rate of infiltration; (d) there will be severe in-draft at exterior doors; (e) exterior doors will be difficult to open; and (f) heating systems will not be able to maintain uniform comfort conditions throughout the building. The center core will become overheated.

Makeup air units may be required to do heating, cooling, humidifying, dehumidifying, and filtering. They may be required to replace the air at space conditions, or they may be used for part or all of the heating and cooling load.

Various types of air makeup units may be identified, such as *blow-through* or *draw-through* referring to the relative location of the fan with respect to the heat-transfer device. The fans may be either centrifugal (single or double width), forward or backward inclined, flat plate, or air foil blades), axial flow (propeller, ducted propeller, vane axial, tube axial, or axial centrifugal). *Heating sources* include gas (direct- or indirect-fired), oil (direct- or indirect-fired), steam, hot water, electric blast coils, heat-transfer fluids, partial recirculation (heat conservation), and summer supply only. *Cooling sources* include direct expansion (DX) refrigerant or chilled water coils, evaporative cooling, sprayed coils, or chilled water air washers. *Filters* may be automatic or manual roll type, throw away or cleanable panels, bag type, electrostatic, or a combination of these types.

10-11 HEAT-TRANSFER COILS

Finned-tube heat-transfer coils are widely used for forced convection cooling or heating of air. Cooling is normally accomplished with chilled water or refrigerants circulating through the coil tubes, while heating is accomplished by circulating hot water or steam through the coil tubes. Finned-tube coils owe their popularity to their relative compactness and to the ease with which they can be installed. Finned-tube coils are used extensively as the heat-transfer device in central station air-handling units, room air terminals, and factory-assembled self-contained air conditioners. The application of each type of coil is limited to the field within which it is rated. Other limitations are imposed by code regulations, choice of materials, the fluids used, and the condition of the air handled, or by an economic analysis of the possible alternatives for each installation.

Selection of cooling coils. ASHRAE[1] recommends that the following items be considered when selecting a cooling coil for a particular application:

a. The duty required, cooling, dehumidification, and capacity required to maintain balance with other system components such as compressor equipment in the case of direct expansion coils
b. Temperature of the entering air dry bulb only if there is no dehumidification; dry-bulb and wet-bulb temperatures if moisture is to be removed
c. Available cooling medium and operating temperatures
d. Space and dimensional limitations
e. Air quantity limitations
f. Allowable frictional limitations in air circuit (including coils)
g. Allowable frictional limitations in cooling media piping system (including coils)
h. Characteristics of individual coil designs
i. Individual installation requirements, such as the type of automatic temperature control to be used, corrosive atmosphere, and so forth
j. Coil face velocity. Selection of the face velocity is determined by an economical evaluation of initial and operating costs for the complete installation as influenced by (1) heat-transfer performance of the specific coil surface type for various combinations of face areas and row depths as a function of air velocity and (2) air side frictional resistance for the complete air circuit (including coils) which affects fan size, horsepower, and sound level requirements
k. The total heat capacity of the coil(s) should be in balance with the other refrigerant system components, such as the compressor(s), water chiller(s), condenser(s), and refrigerant liquid metering device(s)
l. In the case of dehumidifying coils it is also important that the proper amount of face area be installed to obtain the ratio of air side sensible-to-total (SHF) which is required for maintaining the dry-bulb and wet-bulb temperatures in the conditioned space

Ratings of cooling coils. Based on information in *ARI Standard 410-64*, dry surface (sensible cooling) coils and dehumidifying coils (which accomplish both cooling and dehumidification), particularly for field-assembled coil banks or factory-assembled central station type air conditioners using different combinations of coils, are usually rated within the following limits:

a. Entering air dry-bulb temperature = 65°F to 100°F
b. Entering air wet-bulb temperature = 60°F to 85°F
c. Air face velocity = 300 to 800 fpm (sometimes as low as 200 and as high as 1500 fpm)
d. Volatile refrigerant saturation temperature = 30°F to 55°F at coil suction outlet (refrigerant vapor superheat at coil suction outlet is 6°F or higher)
e. Entering chilled water temperature = 35°F to 65°F
f. Chilled water flow quantity = 1.2 to 6.0 gpm per ton of refrigeration (equivalent to a water temperature rise of from 4°F to 20°F)
g. Water velocity = 1 to 8 fps

Selection of heating coils. Heating coil selection is relatively simple because it involves dry-bulb temperature and sensible heat only, without the complication of simultaneous latent heat loads, as in cooling coils. Heating coils are usually selected from charts and tables giving performance data such as the final air temperature at various air flow velocities (face velocities), entering air temperature, and entering saturated steam pressure or entering water temperature.

The selection of hot water heating coils is slightly more complex than steam because of the added water velocity variable and water temperature drop as it flows through the coil circuits. Therefore, in selecting hot water coils, the water velocity and the mean temperature difference (between hot water flowing in the tubes and the air passing over the fins) must be taken into consideration.

Ratings of heating coils. Steam and hot water heating coils are usually rated within the following limits, which may be exceeded for special applications:

a. Air face velocity = 200 to 1500 fpm, based on air at standard density of 0.075 lb/ft^3

b. Entering air temperature = −20°F to 100°F for steam coils; 0°F to 100°F for hot water coils

c. Steam pressure = 2 to 250 psig at the coil supply connection (pressure drop through the steam control valve must be considered)

d. Hot water supply temperatures = 120°F to 250°F

e. Water flow velocities = 0.5 to 8 fps

The delivered air temperature from heating coils may range from 72°F (for ventilation) to 150°F for full heating. Steam pressures used for steam heating coils are usually 2 to 10 psig, with 5 psig most common. If the entering air to the coil is below freezing, steam pressure should be 5 psig minimum. Entering water temperatures normally used for hot water coils are between 180°F and 200°F, with water flow velocities between 4 and 6 fps, and a water temperature drop of 20°F.

Performance rating tables. Most coil manufacturers have their own methods of producing performance rating tables from a suitable number of coil performance tests. A method of testing air heating and/or cooling coils is given in *ASHRAE Standard 33-64, Method of Testing for Rating Forced-Circulation Air-Cooling and Air-Heating Coils.* A basic method of rating to provide a fundamental means for establishing thermal performance of air heating coils by extension of test data, as determined from laboratory tests on prototypes, to other operating conditions, coil sizes, and row depths of a particular surface design and arrangement is given in *Air-Conditioning and Refrigeration Institute Standard 410-72 for Forced-Circulation Air-Cooling and Air-Heating Coils.*

10-12 CHILLED WATER COILS

Water coils are used for cooling air with or without accompanying dehumidification of the air. When the chilled water or refrigerant in the coil tubes is at a lower temperature than the dew-point temperature of the entering air, moisture will condense out of the air onto the tube surfaces. Thus, such coils can be used for dehumidification as well as cooling. Examples of cooling applications without dehumidification are precooling coils using well water or other relatively high temperature water to reduce the load on refrigerating machinery, chilled water coils removing sensible heat in connection with chemical moisture-absorption apparatus, or where the air-conditioning load in a building is all sensible heat. The major portion of chilled water cooling coil equipment is designed to provide, simultaneously, both air sensible cooling and dehumidification. Proper provision is made in all well-designed systems to clean the air before it enters the coil, protecting the coil from accumulation of dirt and keep dust and foreign matter out of the conditioned spaces. Coils are sometimes purposely wetted with spray water or a hygroscopic liquid to aid in air cleaning, odor absorption, or prevention of frost buildup.

Brines may be used instead of chilled water through the coil tubes. Brines are seldom required in air-conditioning work, although there are cases with nonvolatile refrigerants where low entering air temperatures with large latent heat loads require a refrigerant temperature so low that use of chilled water becomes impractical. Sometimes, also, the brine from an industrial system already installed is the only convenient source of refrigeration.

Several types of water coils using chilled water as the cooling medium are illustrated in Fig. 10-19. In practically all such coils which use water as the cooling medium, the flow of water and air through the coil are in opposite directions, as illustrated in Fig. 10-20b. Such an arrangement is known as *counterflow* because the air and water flow in opposite directions. *Parallel flow,* in which the water and air flow through the coil in the same direction, illustrated in Fig. 10-20a, is seldom used in commercial work because of the additional heat-transfer surface area required for a given set of conditions. Notice that in *counterflow* (Fig. 10-20b) the cold water enters the coil at the point where the coldest air is leaving the coil. In *parallel flow,* on the other hand, the cold water enters the coil at the same end where the warm air is entering.

Let us recall at this point Eq. (10-3) which is repeated here. The t_i is the temperature of one fluid and t_o is the temperature of the other fluid.

$$\dot{q} = UA\,(t_i - t_o) \qquad \textit{(10-3) Repeated}$$

To make the value of $\dot{q}$ positive for a cooling coil, t_o and t_i should be reversed in the equation. Equation (10-3) depends upon t_i and t_o being constant. As we can see by the diagrams of Fig. 10-20, the temperature of the outside fluid (the air) and the temperature of the inside fluid (the water) are both changing as the two fluids flow through the system. Therefore, to make Eq. (10-3) usable, we must establish *an average temperature difference* between the two fluids. This average temperature difference is called the *logarithmic mean temperature difference* (LMTD). The LMTD usually gives the most reliable value of mean temperature difference to use in heat transfer relationships for heat exchangers. The LMTD is defined as

$$\text{LMTD} = \frac{\Delta t_{\max} - \Delta t_{\min}}{\ln\left(\dfrac{\Delta t_{\max}}{\Delta t_{\min}}\right)} \qquad \textit{(10-19)}$$

where

$\Delta t_{\max}$ = maximum temperature difference (°F or °C)

$\Delta t_{\min}$ = minimum temperature difference (°F or °C)

Now Eq. (10-3) can be written as

$$\dot{q} = UA(\text{LMTD}) \qquad \textit{(10-20)}$$

Let us now look at the heat-transfer situation illustrated in Fig. 10-20 for parallel flow and counterflow. We will assume for this that the entering air temperature is 90°F and the air leaves at 60°F, and water enters at 40°F and leaves at 55°F.

For parallel flow (Fig. 10-22a)

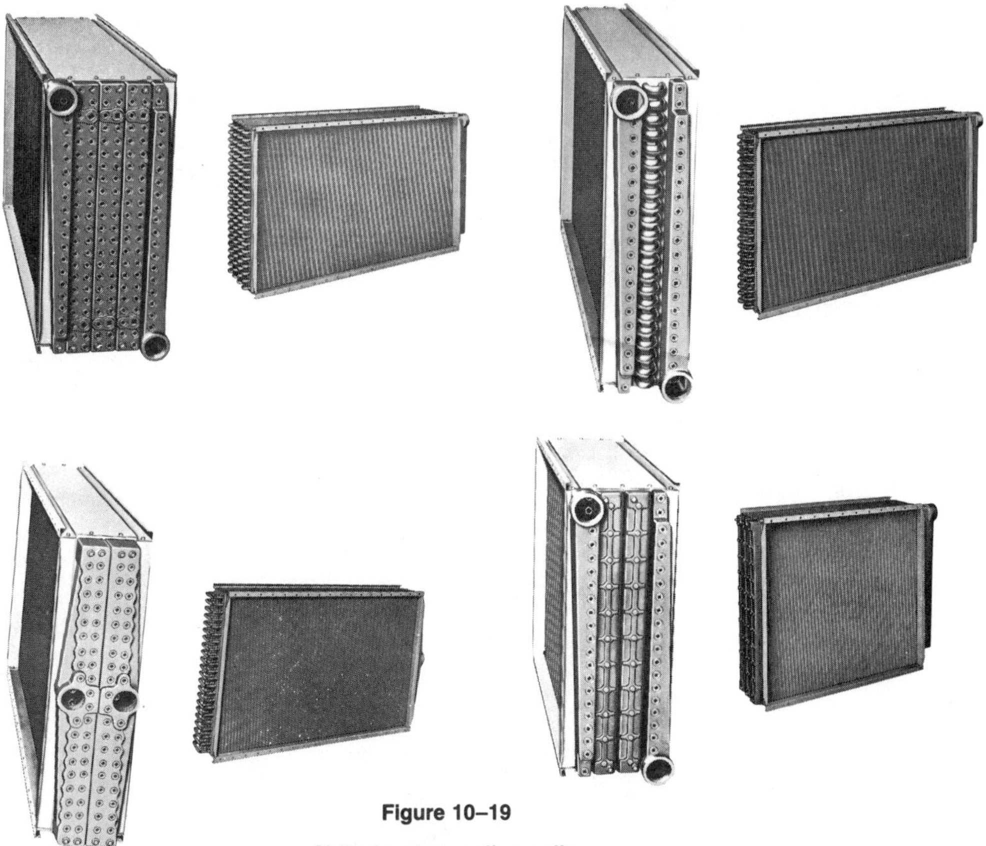

Figure 10–19

Chilled water cooling coils. (a) Drainable water coil (b) Cleanable water coil (c) Single-row water coil (d) Drainable, double-row serpentine water coil

$$\text{LMTD} = \frac{(90 - 40) - (60 - 55)}{\ln 50/5}$$

$$= 19.5°\text{F}$$

For counterflow (Fig. 10-22b)

$$\text{LMTD} = \frac{(90 - 55) - (60 - 40)}{\ln 35/20}$$

$$= 26.8°\text{F}$$

Now referring to Eq. (10-20), for a given amount of heat transfer $\dot{q}$ and a given overall heat transmission coefficient U, a greater surface area would be required using parallel flow because the LMTD is smaller than that calculated for counterflow conditions.

Inlet water connections to water cooling coils are usually made at the bottom tapping so that the water flow is up through the coil and out the top tapping. There are two basic rea-

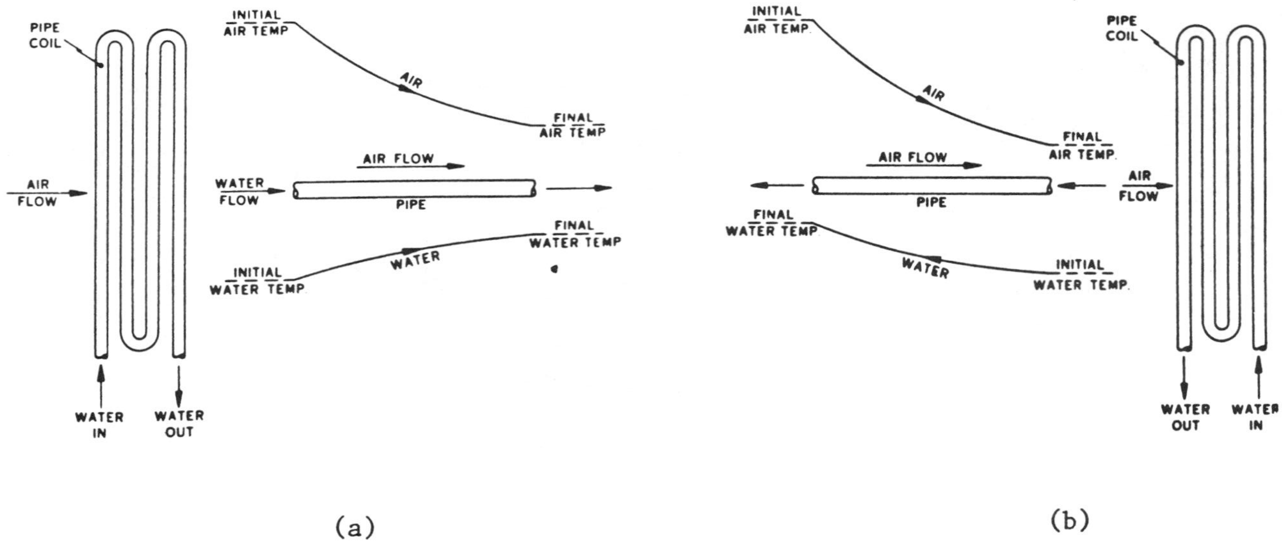

Figure 10–20

Schematic illustrating (a) Parallel flow (b) Counterflow heat transfer

sons for connecting water coils in this manner. First, all the air in the coil will be pushed ahead of the water and will accumulate in the top part of the coil, where it can be easily vented. Second, the coil will remain completely filled with water even though the control valve is closed. Figure 10-21 illustrates methods which are used to pipe-up coils used for cooling with cold water.

10-13 EVAPORATORS

When the refrigerant in the cooling device boils and evaporates as it absorbs heat, the device is called an *evaporator*. Evaporator coils are constructed the same as water cooling coils, except the tubes are serpentined, or circuited, differently. The tubes in the coil are arranged in circuits so that the proper amount of refrigerant is metered to each circuit. The flow of refrigerant to the coil tubes (dry evaporator) is controlled by an automatic expansion valve. This valve meters the amount of refrigerant to the coil in order to maintain the proper amount of superheat in the refrigerant gas as it leaves the evaporator and enters the compressor.

The evaporator coil is perhaps the most commonly used cooling device in air-conditioning systems. A typical coil is shown in Fig. 10-22. These units are known by various names such as direct expansion coil, DX coil, or DE coil. This type of coil was discussed quite generally in Chapter 3.

Regarding coil performance, heat transfer from the air passing over the exterior finned surfaces to the chilled water or boiling refrigerant inside the tubes involves a great many variables. Customarily, manufacturers test their coils and publish performance data. Typical of such published performance data is illustrated in Tables 10-14, 10-15, and 10-16. The Trane Company publish their tabulated data for dehumidifying water cooling coils and for refrigerant coils in the manner represented in these tables, where multiple interpolation may be used.

For given entering conditions within the ranges noted below, the MBh/ft^2 of face area capacity, leaving dry-bulb air temperature and leaving wet-bulb air temperatures are tabulated for 2-, 4-, 6-, 8-, and 10-row water coils and 3-, 4-, 6-, and 8-row refrigerant coils for fin series 15, 16, and 18 (fin series means fins per inch). These capacities and leaving air temperatures are tabulated for the following air and water or refrigerant conditions.

(a) Air wet- and dry-bulb temperatures (both water and refrigerant coils).

EWB(°F)	EDB (°F)
58	75 and 85
60	75 and 85
62	75 and 85
64	75 and 85
65	75 and 85
66	75 and 85
67	80 and 90
68	80 and 90
69	80 and 90
70	80 and 90
75	95
78	95

(b) For each of the above entering wet bulb (EWB) and entering dry bulb (EDB) conditions, capacities are given in the tabulated data section for any combination of the following variables:

(1) *Water Coils*

Entering water temperature (EWT) 42, 44, 45°F.

Water Temperature rise 10, 12, 16, 20°F.

Water velocity 1, 2, 4, 8 fps.

Air velocity 400, 500, 600, 700 fpm.

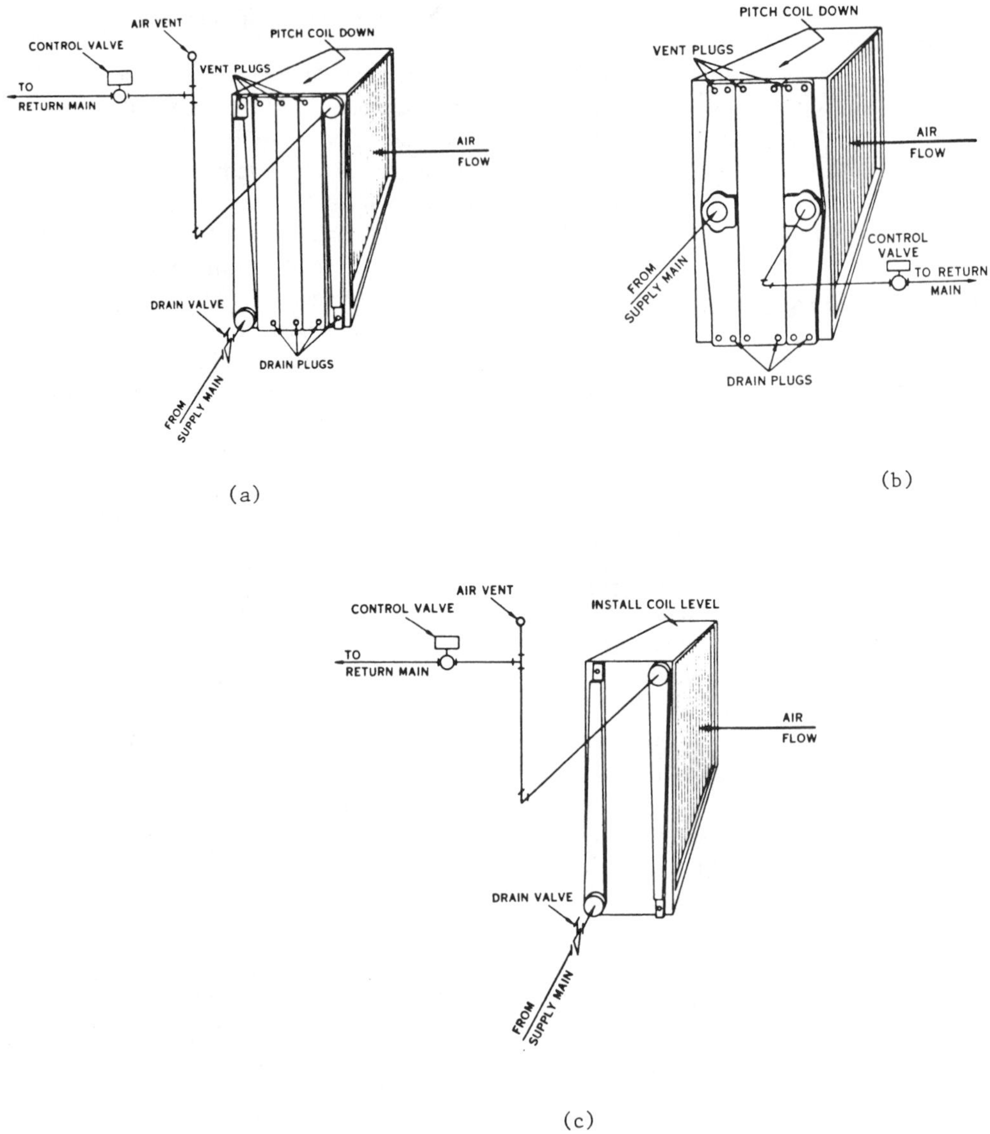

Figure 10–21

Methods used to pipe-up water cooling coils (a) Drainable coils (b) Water coil, two-row serpentine (c) Water coil, one row serpentine

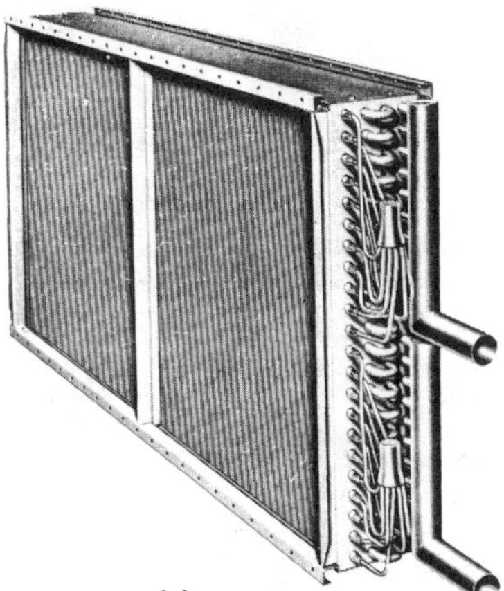

(a)

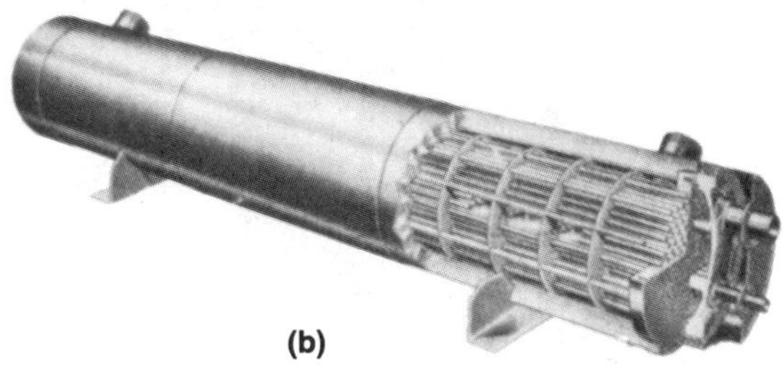

(b)

Figure 10–22

Evaporations (a) Direct expansion coil (b) Dry expansion chiller

(2) Refrigerant Coils:
Suction temperatures 35, 40, 45, 50°F.
Air velocity 400, 500, 600, 700 fpm.

Series 18 fins produce maximum capacity and are recommended for the most economical and compact selections. Fin Series 15, 16, and 18 are finished in the Sigma-Flow II design (a Trane Company design feature). Trane cooling coils dehumidify at face velocities up to 750 fpm for Series 15 and 16, 650 fpm for Series 18. The characteristics of the Sigma-Flow II fin design allow condensed moisture to flow to the bottom channel of the coil without being carried into the airstream. For any given face area a long narrow coil will give the most economical selection. Dehumidifying cooling coils are not recommended for vertical airflow applications, but may be supplied as a special. Vertical tube installation is not recommended for dehumidifying applications because horizontal fins do not provide a continuous condensate flow path to the bottom channel.

The Trane Company recommends the following when selecting water cooling coils and refrigerant coils from their catalog:

1. Water Cooling Coils:
   a. Capacities are based on counterflow of air and water through the coil. Counterflow is obtained when the cooling medium enters the coil bank on the leaving air side and leaves on the entering air side.
   b. The minimum economical difference between the leaving air dry bulb and entering water temperatures for counterflow operation is 6 to 8°F. Lower differentials are possible but usually require excessive coil surface.
   c. Maintain as high a water velocity (up to 8 fps) as pressure drop will permit.
   d. Where water flow rate is low (below 4 fps), the use of turbulators is recommended to obtain maximum heat transfer. This gives capacities equivalent to those with double the water velocity. Therefore, when selecting a coil with turbulators, double the actual water velocity for selection purposes.
   e. Standard water cooling coils may be used in hot water applications for a water temperature drop up to 60°F if

80 90 EDB
67 EWB

Table 10-14

Chilled Water-Cooling Coil Capacities*[a]

WTR	FPS	ROW	FIN	80/67 EDB/EWB												90/67 EDB/EWB												FIN	ROW	FPS
				400 FPM			500 FPM			600 FPM			700 FPM			400 FPM			500 FPM			600 FPM			700 FPM					
				MBH	LDB	LWB	MBH	LDB	LWB	MBH	LDB	LWB	MBH	LDB	LWB	MBH	LDB	LWB	MBH	LDB	LWB	MBH	LDB	LWB	MBH	LDB	LWB			
10	1	2	15	5.6	67.5	62.8	6.1	68.7	63.3	6.6	70.0	63.7	7.0	70.9	64.0	7.3	73.4	61.4	8.1	75.2	62.1	8.8	76.7	62.6	9.4	77.8	63.0	15	2	1
			16	6.6	65.0	62.0	7.2	66.9	62.7	7.7	68.2	63.1	8.1	69.3	63.5	8.6	70.5	60.4	9.5	72.7	61.2	10.3	74.4	61.8	10.9	75.8	62.3	16		
			18	7.3	63.2	61.5	8.0	65.3	62.2	8.6	66.9	62.7	9.1	68.2	63.1	9.5	68.3	59.6	10.6	70.7	60.5	11.4	72.6	61.2	12.1	74.2	61.7	18		
		4	15	9.8	61.2	59.5	10.8	62.4	60.4	11.6	63.4	61.1	12.3	64.2	61.7	11.8	63.0	57.6	13.5	65.5	58.6	14.8	67.5	59.3	16.0	69.2	59.9	15	4	
			16	11.1	59.1	58.4	12.2	60.3	59.5	13.2	61.2	60.3	14.0	62.0	60.9	13.3	59.7	56.4	15.2	62.3	57.4	16.9	64.4	58.1	18.2	66.3	58.9	16		
			18	12.0	57.8	57.6	13.3	59.1	58.7	14.4	60.0	59.6	15.3	60.8	60.3	14.3	57.3	55.4	16.3	60.1	56.5	18.3	62.3	57.3	19.8	64.2	58.1	18		
		6	15	12.9	57.4	56.8	14.4	58.7	58.0	15.7	59.9	58.9	16.8	60.7	59.6	14.8	56.3	55.0	17.0	58.9	56.0	19.0	61.1	56.9	20.8	62.8	57.5	15	6	
			16	14.3	55.7	55.5	16.0	57.1	56.9	17.6	58.1	57.8	18.8	59.0	58.7	16.3	53.9	53.6	18.8	55.7	54.8	21.1	57.9	55.6	23.0	59.8	56.4	16		
			18	15.4	54.7	54.6	17.2	56.1	56.0	18.9	57.2	57.1	20.3	58.1	58.0	17.3	52.7	52.6	20.0	54.0	53.9	22.5	55.8	54.8	24.7	57.9	55.6	18		
		8	15	15.4	54.8	54.6	17.4	56.2	55.9	19.0	57.4	57.0	20.5	58.3	57.8	17.1	53.3	52.9	19.8	54.6	54.1	22.1	56.4	55.0	24.4	58.2	55.8	15	8	
			16	16.8	53.4	53.3	19.1	54.8	54.7	21.0	55.9	55.8	22.6	56.9	56.8	18.5	51.6	51.5	21.5	52.9	52.8	24.3	53.8	53.7	26.7	55.2	54.6	16		
			18	17.8	52.4	52.3	20.3	53.9	53.8	22.4	55.1	55.0	24.2	56.1	56.0	19.4	50.7	50.6	22.7	51.9	51.8	25.6	53.0	52.9	28.3	53.8	53.7	18		
		10	15	17.3	[illegible]	[illegible]	[illegible]	[illegible]	[illegible]	[illegible]	[illegible]	[illegible]	[illegible]	[illegible]	[illegible]	[illegible]	[illegible]	[illegible]	21.9	[illegible]	[illegible]	24.7	[illegible]	[illegible]	27.3	[illegible]	[illegible]	15	10	
			16	18.7	51.6	51.5	21.4	53.0	52.9	23.9	54.2	54.1	25.9	55.2	[illegible]	20.0	50.1	50.0	23.5	51.3	51.2	26.7	52.3	52.2	29.6	53.1	53.0	16		
			18	19.6	50.7	50.6	22.6	52.1	52.0	25.2	53.4	53.3	27.5	54.4	[illegible]	20.9	49.3	49.2	24.6	50.4	50.3	28.0	51.4	51.3	31.1	52.3	52.2	18		
	2	2	15	6.7	66.6	61.9	7.4	67.6	62.6	7.9	68.5	63.0	8.4	69.0	63.4	8.3	71.2	60.6	9.3	73.0	61.3	10.2	74.5	61.8	11.0	75.7	62.2	15	2	2
			16	8.0	63.8	60.9	8.8	65.0	61.7	9.5	65.8	62.2	10.1	66.8	62.7	9.9	67.4	59.3	11.2	69.6	60.1	12.4	71.2	60.7	13.3	72.7	61.2	16		
			18	9.1	61.9	60.0	10.0	63.1	60.9	10.8	63.9	61.5	11.4	65.0	62.1	11.2	64.5	58.2	12.6	66.9	59.1	14.0	68.8	59.8	15.1	70.4	60.4	18		
		4	15	11.5	59.7	58.0	12.8	61.0	59.1	13.8	62.2	59.9	14.8	63.0	60.5	13.1	60.3	56.5	15.1	62.4	57.4	16.8	64.3	58.2	18.4	65.9	58.7	15	4	
			16	13.2	57.2	56.6	14.8	58.6	57.8	16.2	59.6	58.6	17.3	60.5	59.4	15.0	55.9	54.8	17.2	58.6	55.9	19.3	60.7	56.7	21.2	62.2	57.3	16		
			18	14.5	55.6	55.4	16.3	57.0	56.7	17.9	58.1	57.7	19.2	59.0	58.5	16.3	54.0	53.6	18.8	55.7	54.7	21.2	57.8	55.6	23.3	59.7	56.3	18		
		6	15	14.9	55.6	55.0	16.8	57.1	56.3	18.5	58.2	57.3	19.9	59.2	58.1	16.3	54.7	53.6	18.9	56.0	54.7	21.2	57.8	55.6	23.4	59.6	56.3	15	6	
			16	16.7	53.5	53.4	19.0	55.0	54.8	21.0	56.1	55.8	22.7	57.1	56.7	18.1	52.1	51.9	21.1	53.4	53.0	23.9	54.4	53.9	26.4	55.7	54.8	16		
			18	18.0	52.2	52.1	20.6	53.6	53.5	22.9	54.8	54.7	24.9	55.7	55.6	19.4	50.8	50.7	22.7	51.9	51.8	25.7	52.9	52.8	28.5	53.8	53.6	18		
		8	15	17.4	52.9	52.7	19.9	54.3	54.1	22.1	55.5	55.2	24.0	56.5	56.1	18.6	51.8	51.4	21.8	53.1	52.5	24.6	54.2	53.5	27.2	55.1	54.3	15	8	
			16	19.1	51.2	51.1	22.1	52.5	52.4	24.8	53.7	53.6	27.0	54.6	54.5	20.2	49.9	49.8	23.8	51.0	50.9	27.2	52.0	51.9	30.2	52.9	52.7	16		
			18	20.3	50.0	49.9	23.6	51.3	51.2	26.6	52.5	52.4	29.2	53.5	53.4	21.2	48.9	48.8	25.2	49.9	49.8	28.9	50.9	50.8	32.2	51.7	51.6	18		
		10	15	19.2	51.0	50.9	22.4	52.3	52.2	25.0	[illegible]	[illegible]	27.4	54.5	[illegible]	20.2	50.0	[illegible]	[illegible]	[illegible]	[illegible]	27.2	52.2	[illegible]	30.3	[illegible]	[illegible]	15	10	
			16	20.8	49.5	49.4	24.4	50.8	50.7	27.8	51.9	51.8	30.4	52.8	52.7	21.5	[illegible]	[illegible]	25.7	[illegible]	[illegible]	29.6	50.4	50.3	33.1	51.2	51.1	16		
			18	21.7	48.5	48.4	25.7	49.7	49.6	29.3	50.7	50.6	32.4	51.7	[illegible]	22.4	47.6	47.5	26.9	48.5	[illegible]	31.1	[illegible]	[illegible]	34.9	50.2	50.1	18		
	4	2	15	7.7	65.8	61.2	8.4	66.9	61.9	9.1	67.9	62.4	9.7	68.5	62.8	8.9	69.5	60.0	10.2	71.3	60.7	11.2	72.9	61.3	12.3	74.0	61.7	15	2	4
			16	9.3	62.7	59.8	10.3	64.0	60.7	11.2	64.8	61.3	11.9	65.7	61.8	10.9	65.0	58.4	12.5	67.1	59.2	13.9	68.8	59.8	15.1	70.3	60.3	16		
			18	10.7	60.4	58.7	11.9	61.8	59.7	13.0	62.7	60.4	13.8	63.5	61.0	12.4	61.6	57.1	14.3	63.8	57.9	15.9	65.7	58.6	17.4	67.3	59.2	18		
		4	15	12.8	58.5	56.9	14.4	59.9	58.0	15.7	61.1	58.9	16.9	62.0	59.6	14.1	58.6	55.6	16.2	60.4	56.6	18.2	62.2	57.4	20.0	63.8	58.0	15	4	
			16	14.9	55.6	55.0	16.9	57.1	56.3	18.7	58.1	57.2	20.2	59.1	58.0	16.3	54.7	53.6	18.9	56.1	54.7	21.4	57.6	55.5	23.5	59.4	56.2	16		
			18	16.6	53.7	53.5	18.9	55.1	54.8	21.0	56.3	55.9	22.7	57.2	56.7	17.9	52.5	52.1	20.9	53.8	53.2	23.7	54.8	54.1	26.1	56.0	54.9	18		
		6	15	16.3	54.3	53.7	18.7	55.7	55.0	20.6	56.9	56.0	22.5	57.9	56.9	17.4	53.6	52.5	20.3	55.0	53.6	22.9	56.2	54.6	25.3	57.2	55.3	15	6	
			16	18.4	51.9	51.7	21.3	53.2	53.0	23.9	54.3	54.1	26.1	55.3	55.0	19.5	50.8	50.6	22.9	52.0	51.7	26.2	53.0	52.5	29.0	53.9	53.3	16		
			18	19.9	50.4	50.3	23.3	51.6	51.5	26.2	52.7	52.6	28.8	53.7	53.6	20.8	49.3	49.2	24.7	50.3	50.2	28.3	51.3	51.1	31.6	52.1	51.9	18		
		8	15	18.8	51.6	51.4	21.8	52.9	52.7	24.4	54.1	53.8	26.8	55.1	54.7	19.6	50.8	50.4	23.2	52.0	51.5	26.4	53.1	52.4	29.4	54.0	53.1	15	8	
			16	20.6	49.6	49.5	24.3	50.8	50.7	27.6	51.9	51.8	30.5	52.8	52.7	21.4	48.7	48.6	25.5	49.7	49.6	29.4	50.6	50.4	32.8	51.4	51.2	16		
			18	21.9	48.4	48.3	26.0	49.5	49.4	29.7	50.5	50.4	33.0	51.4	51.3	22.4	47.6	47.5	27.0	48.5	48.4	31.2	49.3	49.2	35.2	50.0	49.9	18		
		10	15	20.5	49.8	49.7	24.3	51.0	50.9	27.5	52.1	[illegible]	30.2	53.0	[illegible]	[illegible]	49.0	[illegible]	25.2	50.0	[illegible]	28.9	51.0	50.7	[illegible]	[illegible]	51.5	15	10	
			16	22.0	48.2	48.1	26.3	49.2	49.1	30.2	50.1	50.0	33.7	51.0	50.9	22.6	47.5	47.4	27.2	48.3	[illegible]	31.6	49.1	49.0	[illegible]	[illegible]	49.7	16		
			18	23.0	47.2	47.1	27.7	48.0	47.9	32.0	48.9	48.8	36.0	49.7	[illegible]	23.3	46.7	46.6	28.4	47.3	47.2	33.2	48.0	47.9	[illegible]	[illegible]	[illegible]	18		
	8	2	15	8.3	65.2	50.6	9.2	66.4	61.4	10.0	67.4	62.0	10.7	68.1	62.4	9.4	68.4	59.6	10.8	70.2	60.3	11.9	71.7	50.8	13.0	72.9	61.2	15	2	8
			16	10.3	61.8	59.0	11.5	63.1	59.9	12.6	64.1	60.6	13.5	64.9	61.1	11.7	63.2	57.7	13.4	65.5	58.6	15.0	67.0	59.1	16.4	68.5	59.7	16		
			18	12.1	59.3	57.5	13.6	60.6	58.6	14.8	61.7	59.4	16.0	62.5	60.0	13.5	59.3	56.2	15.5	61.6	57.1	17.5	63.3	57.8	19.1	64.9	58.4	18		
		4	15	13.7	57.7	56.1	15.6	59.1	57.2	17.1	60.3	58.1	18.5	61.2	58.8	14.8	57.9	55.0	17.1	59.4	56.0	19.2	60.8	56.8	21.2	62.3	57.4	15	4	
			16	16.2	54.5	53.9	18.5	55.9	55.1	20.6	57.0	56.1	22.4	58.0	56.9	17.3	53.8	52.7	20.1	55.2	53.8	22.9	56.2	54.6	25.2	57.3	55.3	16		
			18	18.0	52.3	52.1	20.9	53.7	53.3	23.4	54.8	54.4	25.6	55.8	55.3	19.0	51.4	51.0	22.4	52.6	52.0	25.5	53.6	52.9	28.4	54.6	53.7	18		
		6	15	17.3	53.4	52.8	19.9	54.8	54.1	22.2	56.0	55.1	24.3	57.0	55.9	18.2	52.9	51.8	21.3	54.2	52.9	24.1	55.4	53.8	26.8	56.4	54.5	15	6	
			16	19.6	50.7	50.6	22.9	52.0	51.8	25.9	53.0	52.8	28.6	54.0	53.7	20.3	49.9	49.7	24.1	51.1	50.7	27.7	52.0	51.5	30.9	52.9	52.3	16		
			18	21.1	49.1	49.0	25.0	50.2	50.1	28.6	51.2	51.1	31.7	52.1	52.0	21.7	48.3	48.2	26.1	49.3	49.2	30.1	50.1	50.0	33.8	50.9	50.7	18		
		8	15	19.6	50.7	50.5	23.0	52.0	51.7	26.0	53.1	52.8	28.7	54.1	53.6	20.3	50.1	49.7	24.1	51.2	50.7	27.6	52.3	51.6	30.9	53.2	52.3	15	8	
			16	21.5	48.7	48.6	25.7	49.7	49.6	29.5	50.6	50.5	32.8	51.5	51.4	22.1	48.0	47.9	26.6	48.9	48.8	30.8	49.6	49.5	34.7	50.4	50.2	16		
			18	22.8	47.4	47.3	27.4	48.3	48.2	31.8	49.1	49.0	35.7	49.9	49.8	23.1	46.9	46.8	28.1	47.6	47.5	32.8	48.2	48.1	37.1	48.9	48.8	18		
		10	15	21.2	49.0	48.9	25.2	50.1	50.0	28.8	51.1	51.0	32.1	52.0	51.9	21.7	[illegible]	[illegible]	[illegible]	49.4	[illegible]	[illegible]	[illegible]	[illegible]	[illegible]	[illegible]	[illegible]	15	10	
			16	22.8	47.4	47.3	27.5	48.3	48.2	31.8	49.0	48.9	35.6	49.8	49.7	23.1	[illegible]	[illegible]	[illegible]	47.6	[illegible]	[illegible]	[illegible]	[illegible]	[illegible]	[illegible]	[illegible]	16		
			18	23.7	46.5	46.3	28.9	47.1	47.0	33.7	47.7	47.6	38.2	48.4	48.3	[illegible]	[illegible]	46.0	[illegible]	[illegible]	[illegible]	[illegible]	47.1	47.0	[illegible]	[illegible]	[illegible]	18		

45 EWT

Table 10-14 (Continued)

Chilled Water-Cooling Coil Capacities*[a]

12	1	2	15	5.4	67.5	62.9	5.9	69.1	63.4	6.3	70.3	63.8	6.7	71.2	64.1
			16	6.3	65.5	62.2	6.9	67.3	62.8	7.5	68.6	63.3	7.9	69.7	63.6
			18	7.1	63.8	61.6	7.7	65.9	62.3	8.3	67.3	62.8	8.9	68.5	63.2
		4	15	9.4	61.5	59.8	10.4	62.7	60.7	11.2	63.6	61.3	11.9	64.3	61.8
			16	10.7	59.4	58.7	11.8	60.6	59.7	12.8	61.5	60.5	13.6	62.2	61.1
			18	11.6	58.2	57.9	12.9	59.4	59.0	14.0	60.3	59.9	14.9	61.0	60.5
		6	15	12.5	57.8	57.2	13.9	59.1	58.3	15.1	60.2	59.2	16.2	61.0	59.9
			16	13.8	56.1	56.0	15.5	57.4	57.2	17.0	58.4	58.2	18.2	59.3	59.0
			18	14.9	55.2	55.1	16.7	56.5	56.4	18.3	57.5	57.4	19.6	58.4	58.3
		8	15	14.9	55.3	55.1	16.8	56.7	56.4	18.4	57.8	57.4	19.8	58.7	58.2
			16	16.3	53.9	53.8	18.4	55.3	55.2	20.3	56.3	56.2	21.9	57.2	57.1
			18	17.3	52.9	52.8	19.6	54.4	54.3	21.7	55.5	55.4	23.4	56.5	56.4
		10	15	[illegible]	[illegible]	[illegible]	[illegible]	[illegible]	[illegible]	[illegible]	[illegible]	[illegible]	[illegible]	[illegible]	[illegible]
			16	[illegible]	[illegible]	[illegible]	[illegible]	[illegible]	[illegible]	[illegible]	[illegible]	[illegible]	[illegible]	[illegible]	[illegible]
			18	[illegible]	[illegible]	[illegible]	[illegible]	[illegible]	[illegible]	[illegible]	[illegible]	[illegible]	[illegible]	[illegible]	[illegible]
	2	2	15	6.4	66.8	62.1	7.1	67.8	62.7	7.6	68.7	63.2	8.1	69.4	63.5
			16	7.7	64.1	61.1	8.5	65.2	61.9	9.2	66.0	62.4	9.7	67.3	62.8
			18	8.8	62.1	60.3	9.6	63.3	61.1	10.4	64.0	61.7	11.1	65.4	62.2
		4	15	11.0	60.1	58.4	12.3	61.4	59.4	13.3	62.5	60.2	14.2	63.3	60.8
			16	12.7	57.6	57.0	14.2	59.0	58.1	15.6	59.9	59.0	16.7	60.8	59.7
			18	14.0	56.1	55.8	15.7	57.4	57.1	17.3	58.4	58.0	18.6	59.3	58.8
		6	15	14.3	56.1	55.6	16.2	57.5	56.8	17.8	58.7	57.7	19.2	59.5	58.5
			16	16.1	54.1	53.9	18.3	55.5	55.3	20.3	56.5	56.3	21.9	57.5	57.1
			18	17.4	52.8	52.7	19.9	54.2	54.1	22.1	55.3	55.2	24.0	56.2	56.1
		8	15	16.8	53.5	53.3	19.2	54.9	54.6	21.3	56.1	55.7	23.2	57.0	56.5
			16	18.5	51.7	51.6	21.4	53.1	53.0	23.9	54.2	54.1	26.1	55.1	55.0
			18	19.7	50.6	50.5	22.9	51.9	51.8	25.7	53.0	52.9	28.2	54.0	53.9
		10	15	[illegible]	[illegible]	[illegible]	[illegible]	[illegible]	[illegible]	[illegible]	[illegible]	[illegible]	[illegible]	[illegible]	[illegible]
			16	[illegible]	[illegible]	[illegible]	[illegible]	[illegible]	[illegible]	[illegible]	[illegible]	[illegible]	[illegible]	[illegible]	[illegible]
			18	[illegible]	[illegible]	[illegible]	[illegible]	[illegible]	[illegible]	[illegible]	[illegible]	[illegible]	[illegible]	[illegible]	[illegible]
	4	2	15	7.3	66.1	61.5	8.0	67.2	62.1	8.7	68.1	62.6	9.3	68.7	63.0
			16	8.9	63.1	60.2	9.9	64.3	61.0	10.8	65.1	61.6	11.5	65.9	62.0
			18	10.3	60.8	59.0	11.5	62.1	60.0	12.5	63.0	60.7	13.3	63.8	61.2
		4	15	12.3	59.0	57.4	13.8	60.3	58.4	15.0	61.5	59.3	16.2	62.3	59.9
			16	14.3	56.2	55.5	16.2	57.5	56.7	18.0	58.5	57.6	19.4	59.5	58.4
			18	16.0	54.3	54.0	18.2	55.6	55.3	20.2	56.7	56.3	21.9	57.6	57.1
		6	15	15.7	54.9	54.3	17.9	56.3	55.5	19.8	57.4	56.5	21.6	58.3	57.3
			16	17.8	52.5	52.3	20.5	53.8	53.6	23.0	54.9	54.6	25.1	55.8	55.5
			18	19.3	51.0	50.9	22.5	52.2	52.1	25.3	53.3	53.2	27.8	54.2	54.1
		8	15	18.1	52.2	52.0	21.0	53.5	53.3	23.5	54.7	54.3	25.8	55.6	55.2
			16	20.0	50.2	50.1	23.5	51.4	51.3	26.7	52.4	52.3	29.4	53.4	53.3
			18	21.3	48.9	48.8	25.3	50.0	49.9	28.8	51.0	50.9	32.0	51.9	51.8
		10	15	[illegible]	[illegible]	[illegible]	[illegible]	[illegible]	[illegible]	[illegible]	[illegible]	[illegible]	[illegible]	[illegible]	[illegible]
			16	[illegible]	[illegible]	[illegible]	[illegible]	[illegible]	[illegible]	[illegible]	[illegible]	[illegible]	[illegible]	[illegible]	[illegible]
			18	[illegible]	[illegible]	[illegible]	[illegible]	[illegible]	[illegible]	[illegible]	[illegible]	[illegible]	[illegible]	[illegible]	[illegible]
	8	2	15	7.9	65.6	61.0	8.8	66.7	61.7	9.5	67.6	62.2	10.2	68.3	62.6
			16	9.9	62.3	59.4	11.0	63.5	60.3	12.1	64.4	60.9	12.9	65.2	61.4
			18	11.6	59.7	58.0	13.0	61.0	59.0	14.2	62.0	59.7	15.3	62.8	60.3
		4	15	13.1	58.2	56.6	14.9	59.6	57.7	16.3	60.7	58.5	17.7	61.6	59.2
			16	15.5	55.1	54.5	17.7	56.5	55.7	19.8	57.5	56.6	21.5	58.4	57.3
			18	17.4	52.9	52.7	20.1	54.3	53.9	22.5	55.3	54.9	24.6	56.3	55.8
		6	15	16.6	54.0	53.5	19.1	55.4	54.7	21.3	56.5	55.7	23.3	57.5	56.4
			16	18.9	51.4	51.2	22.1	52.6	52.5	25.0	53.6	53.4	27.5	54.6	54.3
			18	20.5	49.7	49.6	24.3	50.8	50.7	27.6	51.8	51.7	30.7	52.7	52.6
		8	15	19.0	51.4	51.2	22.2	52.6	52.4	25.0	53.7	53.4	27.6	54.7	54.2
			16	21.0	49.2	49.1	24.9	50.3	50.2	28.5	51.2	51.1	31.7	52.1	52.0
			18	22.3	47.9	47.8	26.7	48.9	48.8	30.9	49.7	49.6	34.6	50.5	50.4
		10	15	[illegible]	[illegible]	[illegible]	[illegible]	[illegible]	[illegible]	[illegible]	[illegible]	[illegible]	[illegible]	[illegible]	[illegible]
			16	[illegible]	[illegible]	[illegible]	[illegible]	[illegible]	[illegible]	[illegible]	[illegible]	[illegible]	[illegible]	[illegible]	[illegible]
			18	[illegible]	[illegible]	[illegible]	[illegible]	[illegible]	[illegible]	[illegible]	[illegible]	[illegible]	[illegible]	[illegible]	[illegible]

7.2	73.7	61.5	7.9	75.5	62.2	8.6	77.0	62.7	9.1	78.1	63.1	15	2	1
8.4	70.8	60.5	9.3	73.0	61.3	10.1	74.7	61.9	10.7	76.1	62.4	16		
9.3	68.7	59.7	10.4	71.1	60.6	11.2	72.9	61.3	11.9	74.6	61.9	18		
11.6	63.4	57.8	13.2	66.0	58.7	14.5	67.9	59.5	15.7	69.5	60.0	15	4	
13.0	60.3	56.6	14.9	62.8	57.5	16.5	64.9	58.3	18.0	66.6	59.0	16		
14.1	57.8	55.6	16.0	60.5	56.7	18.0	62.7	57.5	19.5	64.6	58.2	18		
14.5	56.8	55.2	16.7	59.4	56.2	18.7	61.6	57.1	20.6	63.3	57.7	15	6	
16.0	54.1	53.8	18.5	56.2	55.0	20.8	58.5	55.9	22.7	60.3	56.6	16		
17.1	53.0	52.9	19.7	54.2	54.1	22.2	56.3	55.0	24.4	58.3	55.8	18		
16.8	53.6	53.2	19.4	54.8	54.3	21.8	56.9	55.2	24.0	58.7	56.0	15	8	
18.2	51.9	51.8	21.2	53.1	53.0	23.9	54.1	53.9	26.3	55.7	54.8	16		
19.1	51.0	50.9	22.3	52.2	52.1	25.3	53.2	53.1	27.9	54.0	53.9	18		
[illegible]	[illegible]	[illegible]	[illegible]	[illegible]	[illegible]	[illegible]	[illegible]	[illegible]	[illegible]	[illegible]	[illegible]	15	10	
[illegible]	[illegible]	[illegible]	[illegible]	[illegible]	[illegible]	[illegible]	[illegible]	[illegible]	[illegible]	[illegible]	[illegible]	16		
[illegible]	[illegible]	[illegible]	[illegible]	[illegible]	[illegible]	[illegible]	[illegible]	[illegible]	[illegible]	[illegible]	[illegible]	18		
8.1	71.5	60.8	9.2	73.3	61.4	10.0	74.9	62.0	10.7	76.0	62.4	15	2	2
9.7	67.8	59.4	11.0	70.0	60.2	12.1	71.6	60.8	13.0	73.0	61.3	16		
10.9	65.1	58.4	12.4	67.3	59.2	13.7	69.1	59.9	14.8	70.8	60.5	18		
12.8	60.8	56.8	14.8	62.8	57.6	16.4	64.9	58.4	17.9	66.5	58.9	15	4	
14.7	56.5	55.1	16.9	59.1	56.1	19.0	61.1	56.9	20.7	62.8	57.5	16		
16.0	54.3	53.8	18.5	56.2	55.0	20.8	58.3	55.8	22.9	60.2	56.5	18		
16.0	54.9	53.9	18.6	56.3	54.9	20.8	58.4	55.8	22.9	60.2	56.5	15	6	
17.8	52.4	52.2	20.8	53.7	53.3	23.5	54.6	54.2	25.9	56.2	55.0	16		
19.1	51.1	51.0	22.3	52.2	52.1	25.3	53.2	53.0	28.1	54.0	53.9	18		
18.2	52.1	51.8	21.4	53.4	52.8	24.1	54.5	53.8	26.8	55.4	54.5	15	8	
19.9	50.2	50.1	23.4	51.4	51.3	26.8	52.3	52.1	29.7	53.1	53.0	16		
20.9	49.2	49.1	24.9	50.2	50.1	28.4	51.2	51.1	31.7	52.0	51.9	18		
[illegible]	[illegible]	[illegible]	[illegible]	[illegible]	[illegible]	[illegible]	[illegible]	[illegible]	[illegible]	[illegible]	[illegible]	15	10	
[illegible]	[illegible]	[illegible]	[illegible]	[illegible]	[illegible]	[illegible]	[illegible]	[illegible]	[illegible]	[illegible]	[illegible]	16		
[illegible]	[illegible]	[illegible]	[illegible]	[illegible]	[illegible]	[illegible]	[illegible]	[illegible]	[illegible]	[illegible]	[illegible]	18		
8.7	70.0	60.2	10.0	71.7	60.8	11.1	73.2	61.4	12.0	74.3	61.8	15	2	4
10.7	65.5	58.6	12.2	67.6	59.3	13.7	69.2	59.9	14.8	70.7	60.5	16		
12.2	61.9	57.2	14.0	64.3	58.1	15.7	66.1	58.8	17.1	67.7	59.4	18		
13.8	58.8	55.9	15.9	60.9	56.8	17.8	62.7	57.6	19.6	64.3	58.1	15	4	
16.0	55.0	53.9	18.6	56.4	54.9	21.0	58.1	55.7	23.0	60.1	56.5	16		
17.6	52.8	52.4	20.5	54.1	53.5	23.2	55.1	54.3	25.7	56.6	55.1	18		
17.1	53.9	52.9	19.9	55.3	53.9	22.5	56.5	54.8	24.9	57.6	55.5	15	6	
19.1	51.2	50.9	22.5	52.4	52.0	25.7	53.3	52.8	28.5	54.2	53.6	16		
20.5	49.7	49.6	24.3	50.7	50.6	27.8	51.6	51.5	31.1	52.4	52.2	18		
19.3	51.1	50.8	22.8	52.3	51.8	25.9	53.4	52.7	28.8	54.3	53.4	15	8	
21.0	49.1	49.0	25.1	50.1	50.0	28.9	50.9	50.8	32.3	51.7	51.6	16		
22.1	48.0	47.9	26.6	48.8	48.7	30.8	49.6	49.5	34.6	50.3	50.2	18		
[illegible]	[illegible]	[illegible]	[illegible]	[illegible]	[illegible]	[illegible]	[illegible]	[illegible]	[illegible]	[illegible]	[illegible]	15	10	
[illegible]	[illegible]	[illegible]	[illegible]	[illegible]	[illegible]	[illegible]	[illegible]	[illegible]	[illegible]	[illegible]	[illegible]	16		
[illegible]	[illegible]	[illegible]	[illegible]	[illegible]	[illegible]	[illegible]	[illegible]	[illegible]	[illegible]	[illegible]	[illegible]	18		
9.2	68.9	59.8	10.5	70.6	60.4	11.7	72.2	61.0	12.8	73.3	61.4	15	2	8
11.4	63.8	58.0	13.1	66.0	58.8	14.7	67.5	59.3	16.0	69.0	59.8	16		
13.2	60.0	56.4	15.2	62.0	57.3	17.1	63.8	58.0	18.8	65.4	58.5	18		
14.5	58.2	55.3	16.8	59.7	56.2	18.8	61.4	57.0	20.8	62.7	57.6	15	4	
16.9	54.2	53.0	19.7	55.5	54.1	22.4	56.5	54.8	24.8	57.7	55.6	16		
18.7	51.8	51.3	22.0	53.0	52.4	25.1	53.9	53.2	27.8	54.9	54.0	18		
17.8	53.3	52.2	20.9	54.5	53.2	23.6	55.8	54.1	26.2	56.7	54.8	15	6	
20.0	50.3	50.1	23.7	51.4	51.1	27.2	52.3	51.9	30.3	53.2	52.6	16		
21.4	48.7	48.6	25.6	49.6	49.5	29.6	50.4	50.3	33.2	51.2	51.0	18		
19.9	50.5	50.1	23.7	51.6	51.1	27.0	52.6	52.0	30.3	53.5	52.7	15	8	
21.7	48.4	48.3	26.1	49.2	49.1	30.3	50.0	49.9	34.1	50.7	50.6	16		
22.8	47.2	47.1	27.7	47.9	47.8	32.3	48.6	48.5	36.6	49.2	49.1	18		
[illegible]	[illegible]	[illegible]	[illegible]	[illegible]	[illegible]	[illegible]	[illegible]	[illegible]	[illegible]	[illegible]	[illegible]	15	10	
[illegible]	[illegible]	[illegible]	[illegible]	[illegible]	[illegible]	[illegible]	[illegible]	[illegible]	[illegible]	[illegible]	[illegible]	16		
[illegible]	[illegible]	[illegible]	[illegible]	[illegible]	[illegible]	[illegible]	[illegible]	[illegible]	[illegible]	[illegible]	[illegible]	18		

[a]Only a partial data listing. See manufacturer's catalog for complete data.

Note: (1) MBH = MBH/ft^2 coil area; WTR = water temperature rise, degrees F; FPS = water velocity, ft/sec; LDB = leaving air dry bulb, degrees F; LWB = leaving air wet bulb, degrees F; EWT = entering water temperature, degrees F.

(2) When using turbulators, make selection based on double the actual water velocity.

Reproduced courtesy of The Trane Company.

Table 10-15

Type F1 Refrigerant Cooling Coil Capacities*[a]
(Freon-12)

80 / 90 EDB — 67 EWB

80/67 EDB/EWB

ST	ROW	FIN	400 FPM MBH	LDB	LWB	Nc	500 FPM MBH	LDB	LWB	Nc	600 FPM MBH	LDB	LWB	Nc	700 FPM MBH	LDB	LWB	Nc
35	3	15	14.8	57.5	55.1	1.02	16.6	59.3	56.5	1.12	18.1	60.7	57.5	1.13	19.5	61.8	58.3	1.16
		16	17.6	53.7	52.5	1.05	20.0	55.5	54.0	1.17	22.0	56.9	55.2	1.28	23.8	58.0	56.2	1.36
		18	19.9	50.8	50.3	1.10	22.8	52.6	51.9	1.20	25.4	54.0	53.1	1.31	27.7	55.2	54.2	1.42
	4	15	17.7	53.7	52.4	1.19	20.2	55.6	53.9	1.30	22.2	57.1	55.1	1.42	24.0	58.3	56.1	1.53
		16	20.6	50.1	49.6	1.26	23.7	51.9	51.2	1.33	26.4	53.3	52.5	1.46	28.8	54.5	53.6	1.58
		18	22.7	47.6	47.4	1.31	26.5	49.2	49.0	1.44	29.8	50.6	50.3	1.55	32.8	51.8	51.4	1.66
	6	15	22.0	48.6	48.2	1.58	25.4	50.4	49.8	1.71	28.5	51.9	51.2	1.83	31.2	53.2	52.3	1.94
		16	24.5	45.6	45.5	1.65	28.7	47.2	47.1	1.80	32.5	48.6	48.4	1.93	36.0	49.8	49.6	2.07
		18	26.2	43.7	43.6	1.65	31.1	45.2	45.1	1.88	35.6	46.4	46.3	2.01	39.7	47.5	47.4	2.11
	8	15	24.7	45.4	45.3	1.80	29.0	47.1	46.8	2.05	32.9	48.5	48.2	2.20	36.8	49.7	49.3	2.35
		16	26.8	43.0	42.9	1.80	31.9	44.5	44.4	2.05	36.6	45.7	45.6	2.29	40.9	46.8	46.7	2.36
		18	28.1	41.5	41.4	1.80	33.9	42.7	42.6	2.05	39.2	43.8	43.7	2.29	44.1	44.8	44.7	2.51
40	3	15	12.2	60.0	57.4	0.89	13.7	61.5	58.5	0.91	14.9	62.7	59.3	0.94	16.0	63.7	60.0	0.98
		16	14.5	56.6	55.4	0.94	16.4	58.1	56.6	1.04	18.0	59.3	57.6	1.09	19.4	60.3	58.3	1.08
		18	16.4	54.2	53.6	0.95	18.8	55.7	54.9	1.06	20.8	56.9	56.0	1.16	22.5	57.9	56.8	1.22
	4	15	14.8	56.7	55.2	1.04	16.7	58.2	56.4	1.16	18.3	59.5	57.4	1.27	19.7	60.6	58.2	1.33
		16	17.1	53.6	53.0	1.06	19.6	55.1	54.3	1.19	21.7	56.3	55.4	1.30	23.6	57.3	56.3	1.40
		18	18.9	51.4	51.2	1.12	21.9	52.8	52.6	1.21	24.6	54.0	53.6	1.32	27.0	55.0	54.6	1.43
	6	15	18.4	52.2	51.7	1.36	21.2	53.8	53.1	1.48	23.7	55.0	54.2	1.58	25.8	56.1	55.1	1.65
		16	20.6	49.7	49.6	1.39	24.0	51.1	50.9	1.54	27.1	52.3	52.1	1.65	29.9	53.3	53.0	1.74
		18	22.1	48.1	48.0	1.43	26.2	49.3	49.2	1.57	29.8	50.4	50.3	1.72	33.1	51.3	51.2	1.83
	8	15	20.9	49.5	49.3	1.64	24.4	50.8	50.6	1.77	27.5	52.1	51.8	1.91	30.4	53.2	52.7	2.04
		16	22.7	47.5	47.4	1.64	26.9	48.7	48.6	1.75	30.7	49.8	49.7	1.97	34.3	50.7	50.6	2.11
		18	24.0	46.1	46.0	1.64	28.7	47.2	47.1	1.87	33.1	48.2	48.1	2.08	37.2	49.0	48.9	2.16
45	3	15	9.6	62.4	59.6	0.70	10.7	63.6	60.5	0.74	11.6	64.6	61.1	0.75	12.4	65.4	61.6	0.78
		16	11.4	59.5	58.1	0.78	12.8	60.7	59.1	0.80	14.1	61.6	59.8	0.83	15.2	62.4	60.3	0.86
		18	12.9	57.5	56.8	0.83	14.7	58.6	57.8	0.89	16.3	59.5	58.6	0.90	17.6	60.3	59.2	0.94
	4	15	11.6	59.5	57.9	0.91	13.1	60.8	58.9	0.98	14.3	61.9	59.6	0.98	15.5	62.7	60.2	1.01
		16	13.5	56.9	56.3	0.93	15.4	58.1	57.3	1.03	17.0	59.1	58.1	1.11	18.6	59.9	58.8	1.11
		18	15.0	55.2	54.9	0.94	17.3	56.3	56.0	1.05	19.3	57.2	56.8	1.15	21.1	58.0	57.5	1.24
	6	15	14.7	55.8	55.2	1.13	16.9	57.0	56.3	1.21	18.7	58.1	57.2	1.33	20.4	59.0	57.9	1.43
		16	16.5	53.7	53.6	1.17	19.2	54.8	54.6	1.28	21.6	55.8	55.5	1.35	23.7	56.6	56.2	1.46
		18	17.8	52.4	52.3	1.21	21.0	53.4	53.3	1.33	23.8	54.3	54.2	1.43	26.3	55.0	54.9	1.53
	8	15	16.8	53.4	53.2	1.37	19.8	54.6	54.3	1.49	22.0	55.6	55.2	1.59	24.2	56.5	56.0	1.69
		16	18.4	51.8	51.7	1.37	21.7	52.8	52.7	1.55	24.7	53.7	53.6	1.66	27.4	54.8	54.3	1.76
		18	19.5	50.7	50.6	1.42	23.3	51.6	51.5	1.55	26.7	52.4	52.3	1.72	29.8	53.2	53.1	1.84
50	3	15	7.1	64.4	61.6	0.52	8.0	65.4	62.2	0.55	8.8	66.3	62.6	0.61	9.5	67.3	62.9	0.62
		16	8.4	62.0	60.6	0.58	9.6	62.8	61.2	0.61	10.6	63.5	61.6	0.64	11.6	64.5	62.0	0.66
		18	9.6	60.3	59.6	0.62	11.0	61.1	60.3	0.66	12.2	61.8	60.8	0.70	13.3	62.3	61.2	0.73
	4	15	8.6	62.1	60.4	0.68	9.8	63.0	61.0	0.72	10.8	63.8	61.5	0.75	11.7	64.4	61.9	0.78
		16	10.0	59.9	59.3	0.76	11.5	60.8	59.9	0.79	12.8	61.5	60.5	0.82	14.0	62.0	60.9	0.86
		18	11.2	58.5	58.3	0.80	12.9	59.4	59.0	0.87	14.4	60.0	59.6	0.89	15.8	60.6	60.1	0.93
	6	15	11.0	59.1	58.4	0.92	12.6	60.0	59.2	1.02	14.0	60.8	59.8	1.11	15.3	61.4	60.3	1.16
		16	12.4	57.4	57.3	0.94	14.4	58.2	58.0	1.04	16.1	58.9	58.6	1.14	17.8	59.5	59.1	1.23
		18	13.4	56.4	56.3	0.97	15.7	57.2	57.1	1.06	17.8	57.8	57.7	1.16	19.7	58.3	58.2	1.25
	8	15	12.9	57.2	57.0	1.08	14.7	58.1	57.8	1.20	16.5	58.8	58.4	1.27	18.2	59.5	59.0	1.37
		16	14.0	56.0	55.9	1.13	16.4	56.7	56.6	1.24	18.6	57.4	57.3	1.34	20.6	57.9	57.8	1.40
		18	14.9	55.2	55.1	1.16	17.6	55.9	55.8	1.20	20.1	56.5	56.4	1.38	22.4	57.0	56.9	1.47

90/67 EDB/EWB

400 FPM MBH	LDB	LWB	Nc	500 FPM MBH	LDB	LWB	Nc	600 FPM MBH	LDB	LWB	Nc	700 FPM MBH	LDB	LWB	Nc	FIN	ROW	ST
14.8	59.6	55.0	1.01	16.7	61.6	56.3	1.06	18.4	63.2	57.2	1.06	20.0	64.5	58.0	1.11	15	3	35
17.8	54.5	52.2	1.03	20.3	56.4	53.7	1.15	22.6	57.9	54.8	1.23	24.7	59.1	55.6	1.28	16		
20.1	51.0	49.9	1.05	23.3	52.7	51.4	1.17	26.1	54.2	52.6	1.29	28.7	55.4	53.5	1.39	18		
17.8	54.9	52.2	1.14	20.3	56.9	53.6	1.28	22.5	58.6	54.8	1.40	24.5	59.9	55.7	1.48	15	4	
20.7	50.3	49.3	1.21	24.0	52.1	50.8	1.30	27.0	53.6	52.0	1.43	29.7	54.8	53.0	1.55	16		
23.0	47.4	47.0	1.27	26.9	49.0	48.5	1.38	30.5	50.4	49.7	1.50	33.7	51.6	50.7	1.58	18		
22.0	48.9	48.0	1.53	25.6	50.7	49.5	1.66	28.8	52.3	50.8	1.77	31.7	53.7	51.9	1.87	15	6	
24.7	45.4	45.1	1.57	29.1	47.0	46.6	1.74	33.1	48.3	47.9	1.87	36.9	49.4	48.9	1.98	16		
26.4	43.3	43.2	1.59	31.6	44.6	44.5	1.77	36.3	45.8	45.7	1.91	40.6	46.8	46.7	2.07	18		
24.8	45.3	45.0	1.80	29.2	47.0	46.5	1.98	33.2	48.4	47.8	2.14	36.9	49.7	48.9	2.28	15	8	
27.0	42.6	42.5	1.80	32.3	44.0	43.9	2.05	37.1	45.1	45.0	2.20	41.7	46.2	46.1	2.37	16		
28.4	41.0	40.9	1.80	34.3	42.1	42.0	2.05	39.8	43.2	43.1	2.29	45.0	44.1	44.0	2.51	18		
12.6	61.7	56.9	0.81	14.4	63.2	57.9	0.85	15.9	65.2	58.6	0.89	17.4	66.8	59.2	0.93	15	3	40
15.1	57.1	54.7	0.92	17.3	58.7	55.8	0.98	19.4	59.8	56.6	1.00	21.4	61.5	57.3	1.04	16		
17.1	54.0	52.9	0.94	19.8	55.4	54.0	1.00	22.3	56.6	54.9	1.13	24.6	57.6	55.7	1.18	18		
15.1	57.5	54.7	1.02	17.3	59.2	55.8	1.14	19.3	60.6	56.7	1.20	21.2	61.8	57.4	1.21	15	4	
17.6	53.5	52.3	1.04	20.5	54.9	53.5	1.16	23.0	56.1	54.5	1.27	25.4	57.1	55.2	1.38	16		
19.6	50.9	50.5	1.08	22.9	52.2	51.7	1.18	26.0	53.4	52.6	1.30	28.8	54.3	53.5	1.41	18		
18.8	52.2	51.2	1.30	21.9	53.8	52.5	1.41	24.6	55.1	53.5	1.48	27.1	56.2	54.3	1.61	15	6	
21.1	49.2	49.0	1.35	24.8	50.5	50.2	1.48	28.3	51.6	51.2	1.59	31.5	52.5	52.0	1.70	16		
22.6	47.4	47.3	1.37	27.0	48.5	48.4	1.54	31.0	49.5	49.4	1.66	34.7	50.4	50.2	1.76	18		
21.2	49.1	48.8	1.58	25.0	50.5	50.0	1.71	28.4	51.7	51.1	1.84	31.6	52.8	52.0	1.95	15	8	
23.2	46.9	46.8	1.64	27.6	48.0	47.9	1.79	31.8	49.0	48.8	1.93	35.6	49.8	49.7	2.05	16		
24.4	45.5	45.4	1.64	29.4	46.5	46.4	1.80	34.1	47.3	47.2	1.98	38.4	48.1	48.0	2.12	18		
10.8	64.9	58.5	0.68	12.5	66.8	59.2	0.76	14.0	68.3	59.7	0.79	15.4	69.5	60.2	0.83	15	3	45
12.7	60.4	56.8	0.76	14.8	62.3	57.6	0.81	16.8	63.9	58.1	0.88	18.7	65.1	58.6	0.92	16		
14.3	56.6	55.4	0.83	16.8	58.7	56.2	0.89	19.0	60.4	56.9	0.95	21.2	61.8	57.4	0.99	18		
12.7	60.3	56.8	0.90	14.8	62.5	57.6	0.97	16.7	64.1	58.2	1.01	18.4	65.5	58.7	1.08	15	4	
14.7	56.2	55.0	0.91	17.3	57.8	55.9	1.02	19.6	59.6	56.5	1.13	21.8	60.9	57.1	1.19	16		
16.3	54.1	53.6	0.92	19.2	55.1	54.5	1.04	21.9	56.0	55.2	1.14	24.4	57.5	55.8	1.24	18		
15.7	55.2	54.2	1.08	18.4	56.5	55.1	1.18	20.8	57.7	55.8	1.30	23.1	59.3	56.4	1.42	15	6	
17.5	52.7	52.5	1.13	20.7	53.7	53.3	1.24	23.7	54.5	54.1	1.32	26.5	55.2	54.7	1.44	16		
18.8	51.3	51.2	1.17	22.4	52.1	52.0	1.28	25.8	52.9	52.7	1.39	29.0	53.5	53.4	1.45	18		
17.7	52.7	52.3	1.32	20.9	53.8	53.2	1.44	23.8	54.7	54.0	1.53	26.6	55.5	54.6	1.59	15	8	
19.2	50.9	50.8	1.35	23.0	51.7	51.6	1.49	26.5	52.4	52.3	1.61	29.8	53.1	52.9	1.72	16		
20.3	49.8	49.7	1.37	24.4	50.6	50.5	1.53	28.3	51.2	51.1	1.67	32.0	51.8	51.7	1.78	18		
9.4	68.0	59.6	0.61	10.9	69.6	60.2	0.67	12.2	71.0	60.7	0.73	13.4	72.1	61.1	0.74	15	3	50
11.0	64.3	58.3	0.69	12.9	65.8	58.9	0.72	14.7	67.1	59.3	0.77	16.3	68.3	59.7	0.82	16		
12.3	61.4	57.2	0.73	14.5	62.9	57.8	0.80	16.6	64.3	58.3	0.86	18.4	65.4	58.7	0.91	18		
11.0	64.3	58.3	0.79	12.9	66.0	58.9	0.87	14.6	67.4	59.4	0.93	16.1	68.5	59.8	0.94	15	4	
12.6	60.8	57.0	0.80	14.9	62.3	57.5	0.90	17.0	63.5	58.0	0.99	19.0	64.6	58.4	1.07	16		
13.7	58.3	56.0	0.80	16.3	59.6	56.5	0.90	18.8	60.8	57.0	1.00	21.1	61.9	57.4	1.08	18		
13.2	59.4	56.4	0.91	15.7	60.9	57.0	1.03	17.9	62.2	57.5	1.13	20.1	63.3	57.9	1.23	15	6	
14.4	56.4	55.3	0.91	17.3	57.7	55.8	1.03	20.1	58.8	56.2	1.14	22.7	59.8	56.6	1.24	16		
15.3	54.6	54.5	0.95	18.5	55.5	55.0	1.03	21.5	56.6	55.4	1.14	24.4	57.6	55.8	1.25	18		
14.5	56.2	55.2	1.11	17.4	57.6	55.8	1.19	20.1	58.9	56.2	1.31	22.7	58.9	56.6	1.36	15	8	
15.6	54.3	54.2	1.12	18.9	54.9	54.7	1.26	21.9	55.9	55.1	1.32	24.9	55.8	55.5	1.43	16		
16.4	53.6	53.5	1.13	19.9	54.1	54.0	1.27	23.2	54.5	54.4	1.40	26.4	54.8	54.7	1.52	18		

NOTE: 1) MBH=MBH/FT.^{2} Coil Face Area
2) ST=Suction Temp.°F
3) Refrigerant liquid temperature=110 F

[a]Only a partial data listing. See Manufacturer's catalog for complete data.

Reproduced courtesy of The Trane Company.

LENGTH CORRECTION FACTOR (LF)

Length (Inches)	20	40	60	80	100	120
Correction Factor	0.950	0.990	1.010	1.020	1.025	1.030

NOTE COIL CAPACITY (MBH) $= \frac{MBH}{FT^2}$ x FACE AREA x LF

R 12

R 22

Table 10-16

Type F2 Refrigerant Cooling Coil Capacities*[a] (Freon-22)

80/67 EDB/EWB

ST	ROW	FIN	400 FPM				500 FPM				600 FPM				700 FPM			
			MBH	LDB	LWB	Nc	MBH	LDB	LWB	Nc	MBH	LDB	LWB	Nc	MBH	LDB	LWB	Nc
35	3	15	15.9	56.4	54.1	0.72	17.9	58.3	55.5	0.80	19.6	59.7	56.6	0.85	21.2	60.9	57.5	0.85
		16	19.0	52.3	51.2	0.74	21.7	54.1	52.8	0.82	24.0	55.6	54.0	0.90	26.1	56.8	55.0	0.97
		18	21.4	49.2	48.7	0.78	24.8	51.0	50.3	0.87	27.8	52.4	51.6	0.92	30.4	53.7	52.7	1.00
	4	15	19.0	52.5	51.2	0.85	21.7	54.4	52.8	0.93	24.0	55.9	54.0	0.99	26.0	57.2	55.0	1.07
		16	22.0	48.6	48.1	0.90	25.5	50.4	49.8	0.99	28.6	51.9	51.1	1.06	31.3	53.1	52.2	1.10
		18	24.2	45.9	45.8	0.93	28.5	47.6	47.3	1.04	32.2	49.0	48.7	1.12	35.7	50.2	49.8	1.19
	6	15	23.2	47.2	46.8	1.11	27.1	49.0	48.5	1.23	30.5	50.6	49.9	1.31	33.5	51.9	51.0	1.41
		16	25.8	44.1	44.0	1.15	30.6	45.7	45.5	1.30	34.8	47.0	46.9	1.39	38.7	48.2	48.0	1.49
		18	27.5	42.2	42.1	1.15	33.0	43.5	43.4	1.30	38.0	44.7	44.6	1.44	42.6	45.7	45.6	1.57
	8	15	[illegible]	[illegible]	[illegible]	[illegible]	[illegible]	[illegible]	[illegible]	[illegible]	[illegible]	[illegible]	[illegible]	[illegible]	[illegible]	[illegible]	[illegible]	[illegible]
		16	[illegible]	[illegible]	[illegible]	[illegible]	[illegible]	[illegible]	[illegible]	[illegible]	[illegible]	[illegible]	[illegible]	[illegible]	[illegible]	[illegible]	[illegible]	[illegible]
		18	[illegible]	[illegible]	[illegible]	[illegible]	[illegible]	[illegible]	[illegible]	[illegible]	[illegible]	[illegible]	[illegible]	[illegible]	[illegible]	[illegible]	[illegible]	[illegible]
40	3	15	13.3	59.0	56.5	0.66	14.9	60.6	57.6	0.69	16.3	61.8	58.5	0.70	17.6	62.8	59.2	0.72
		16	15.9	55.4	54.1	0.67	18.0	56.9	55.5	0.75	19.9	58.2	56.5	0.81	21.5	59.2	57.3	0.85
		18	17.9	52.8	52.2	0.68	20.7	54.3	53.5	0.78	23.0	55.5	54.6	0.83	25.1	56.6	55.5	0.89
	4	15	15.9	55.5	54.1	0.74	18.1	57.2	55.4	0.82	20.0	58.5	56.4	0.90	21.6	59.6	57.3	0.97
		16	18.5	52.2	51.6	0.78	21.3	53.7	53.0	0.86	23.8	55.0	54.1	0.92	26.0	56.0	55.0	0.99
		18	20.4	49.9	49.7	0.82	23.9	51.4	51.1	0.84	26.9	52.5	52.2	0.97	29.7	53.6	53.1	1.01
	6	15	19.7	51.0	50.5	0.98	22.8	52.5	51.9	1.07	25.6	53.8	53.0	1.14	28.1	54.9	54.0	1.20
		16	21.9	48.3	48.2	1.00	25.8	49.7	49.5	1.10	29.3	50.8	50.6	1.22	32.5	51.8	51.6	1.29
		18	23.5	46.7	46.6	1.02	28.0	47.8	47.7	1.13	32.1	48.9	48.8	1.24	35.9	49.8	49.7	1.32
	8	15	[illegible]	[illegible]	[illegible]	[illegible]	[illegible]	[illegible]	[illegible]	[illegible]	[illegible]	[illegible]	[illegible]	[illegible]	[illegible]	[illegible]	[illegible]	[illegible]
		16	[illegible]	[illegible]	[illegible]	[illegible]	[illegible]	[illegible]	[illegible]	[illegible]	[illegible]	[illegible]	[illegible]	[illegible]	[illegible]	[illegible]	[illegible]	[illegible]
		18	[illegible]	[illegible]	[illegible]	[illegible]	[illegible]	[illegible]	[illegible]	[illegible]	[illegible]	[illegible]	[illegible]	[illegible]	[illegible]	[illegible]	[illegible]	[illegible]
45	3	15	10.6	61.5	58.8	0.52	11.8	62.8	59.7	0.55	12.9	63.9	60.4	0.58	13.8	64.7	61.0	0.59
		16	12.6	58.4	57.1	0.59	14.2	59.7	58.1	0.63	15.7	60.7	58.9	0.63	17.0	61.5	59.5	0.65
		18	14.3	56.2	55.6	0.61	16.4	57.4	56.6	0.67	18.2	58.4	57.5	0.71	19.8	59.3	58.2	0.72
	4	15	12.7	58.5	57.0	0.66	14.4	59.9	58.0	0.73	15.8	61.0	58.8	0.77	17.1	61.9	59.5	0.77
		16	14.8	55.7	55.1	0.67	17.0	57.0	56.2	0.74	18.9	58.0	57.1	0.81	20.6	58.8	57.8	0.87
		18	16.4	53.8	53.6	0.69	19.1	55.0	54.7	0.75	21.5	56.0	55.5	0.82	23.6	56.8	56.3	0.89
	6	15	15.9	54.7	54.1	0.83	18.3	56.0	55.2	0.90	20.5	57.0	56.2	0.97	22.4	58.0	56.9	1.01
		16	17.8	52.4	52.3	0.86	20.8	53.6	53.4	0.94	23.6	54.5	54.3	1.00	26.1	55.3	55.0	1.07
		18	19.2	51.1	51.0	0.87	22.7	52.1	52.0	0.97	25.9	52.9	52.8	1.05	28.9	53.6	53.5	1.12
	8	15	[illegible]	[illegible]	[illegible]	[illegible]	[illegible]	[illegible]	[illegible]	[illegible]	[illegible]	[illegible]	[illegible]	[illegible]	[illegible]	[illegible]	[illegible]	[illegible]
		16	[illegible]	[illegible]	[illegible]	[illegible]	[illegible]	[illegible]	[illegible]	[illegible]	[illegible]	[illegible]	[illegible]	[illegible]	[illegible]	[illegible]	[illegible]	[illegible]
		18	[illegible]	[illegible]	[illegible]	[illegible]	[illegible]	[illegible]	[illegible]	[illegible]	[illegible]	[illegible]	[illegible]	[illegible]	[illegible]	[illegible]	[illegible]	[illegible]
50	3	15	7.9	63.8	61.0	0.40	8.8	64.8	61.6	0.42	9.7	65.6	62.1	0.44	10.5	66.3	62.5	0.47
		16	9.4	61.2	59.8	0.43	10.7	62.1	60.5	0.46	11.8	62.9	61.0	0.48	12.9	63.5	61.4	0.50
		18	10.7	59.4	58.7	0.47	12.3	60.3	59.4	0.50	13.7	61.0	60.0	0.52	14.9	61.6	60.5	0.56
	4	15	9.5	61.3	59.7	0.51	10.8	62.4	60.4	0.54	11.9	63.2	61.0	0.57	12.9	63.9	61.4	0.59
		16	11.1	59.0	58.4	0.57	12.7	60.0	59.1	0.62	14.2	60.7	59.7	0.63	15.6	61.3	60.2	0.65
		18	12.4	57.5	57.3	0.58	14.3	58.4	58.0	0.65	16.1	59.1	58.7	0.70	17.7	59.7	59.2	0.73
	6	15	12.0	58.2	57.6	0.66	13.8	59.2	58.4	0.72	15.4	60.0	59.0	0.80	16.9	60.7	59.6	0.86
		16	13.5	56.4	56.3	0.69	15.8	57.3	57.0	0.75	17.8	58.0	57.7	0.82	19.7	58.6	58.2	0.88
		18	14.6	55.4	55.3	0.71	17.3	56.1	56.0	0.79	19.6	56.7	56.6	0.86	21.8	57.3	57.2	0.90
	8	15	[illegible]	[illegible]	[illegible]	[illegible]	[illegible]	[illegible]	[illegible]	[illegible]	[illegible]	[illegible]	[illegible]	[illegible]	[illegible]	[illegible]	[illegible]	[illegible]
		16	[illegible]	[illegible]	[illegible]	[illegible]	[illegible]	[illegible]	[illegible]	[illegible]	[illegible]	[illegible]	[illegible]	[illegible]	[illegible]	[illegible]	[illegible]	[illegible]
		18	[illegible]	[illegible]	[illegible]	[illegible]	[illegible]	[illegible]	[illegible]	[illegible]	[illegible]	[illegible]	[illegible]	[illegible]	[illegible]	[illegible]	[illegible]	[illegible]

90/67 EDB/EWB

400 FPM				500 FPM				600 FPM				700 FPM				FIN	ROW	ST
MBH	LDB	LWB	Nc	MBH	LDB	LWB	Nc	MBH	LDB	LWB	Nc	MBH	LDB	LWB	Nc			
15.8	58.6	54.0	0.72	17.9	60.7	55.4	0.78	19.7	62.4	56.5	0.81	21.4	63.7	57.3	0.81	15	3	35
19.0	53.3	51.0	0.73	21.8	55.2	52.5	0.81	24.3	56.8	53.7	0.89	26.6	58.0	54.6	0.95	16		
21.5	49.5	48.5	0.77	25.0	51.3	50.0	0.83	28.1	52.9	51.2	0.91	31.0	54.2	52.3	0.98	18		
18.9	53.7	51.1	0.84	21.7	55.8	52.6	0.90	24.1	57.6	53.8	0.98	26.2	59.0	54.8	1.06	15	4	
22.0	49.0	47.9	0.88	25.6	50.8	49.5	0.96	28.9	52.3	50.8	1.00	31.8	53.6	51.8	1.08	16		
24.3	45.9	45.5	0.91	28.7	47.5	47.0	1.01	32.6	48.9	48.2	1.08	36.2	50.1	49.3	1.14	18		
23.2	47.6	46.7	1.09	27.1	49.5	48.3	1.20	30.5	51.1	49.7	1.29	33.7	52.5	50.8	1.35	15	6	
25.9	44.0	43.8	1.15	30.7	45.5	45.2	1.24	35.1	46.9	46.5	1.35	39.2	48.1	47.6	1.44	16		
27.6	41.9	41.8	1.15	33.2	43.1	43.0	1.30	38.3	44.3	44.1	1.38	43.1	45.3	45.2	1.49	18		
[illegible]	[illegible]	[illegible]	[illegible]	[illegible]	[illegible]	[illegible]	[illegible]	[illegible]	[illegible]	[illegible]	[illegible]	[illegible]	[illegible]	[illegible]	[illegible]	15	8	
[illegible]	[illegible]	[illegible]	[illegible]	[illegible]	[illegible]	[illegible]	[illegible]	[illegible]	[illegible]	[illegible]	[illegible]	[illegible]	[illegible]	[illegible]	[illegible]	16		
[illegible]	[illegible]	[illegible]	[illegible]	[illegible]	[illegible]	[illegible]	[illegible]	[illegible]	[illegible]	[illegible]	[illegible]	[illegible]	[illegible]	[illegible]	[illegible]	18		
13.5	60.9	56.2	0.61	15.3	62.7	57.2	0.63	17.0	64.1	58.0	0.66	18.5	65.4	58.7	0.68	15	3	40
16.2	56.1	53.7	0.66	18.6	57.7	54.9	0.72	20.8	59.0	55.8	0.76	22.9	60.0	56.5	0.77	16		
18.3	52.8	51.7	0.67	21.3	54.3	52.9	0.75	24.0	55.5	53.9	0.81	26.5	56.6	54.7	0.86	18		
16.1	56.6	53.8	0.73	18.5	58.4	55.0	0.81	20.6	59.8	55.9	0.88	22.5	61.0	56.7	0.92	15	4	
18.8	52.3	51.2	0.76	21.9	53.8	52.5	0.82	24.7	55.1	53.5	0.90	27.3	56.1	54.3	0.97	16		
20.8	49.6	49.2	0.78	24.5	51.0	50.4	0.86	27.9	52.1	51.4	0.91	31.0	53.1	52.3	0.99	18		
19.8	51.2	50.2	0.94	23.2	52.7	51.5	1.02	26.2	54.1	52.5	1.09	28.9	55.3	53.4	1.16	15	6	
22.2	48.0	47.8	0.96	26.3	49.3	49.0	1.08	30.1	50.4	50.0	1.16	33.6	51.4	50.8	1.23	16		
23.8	46.2	46.1	0.99	28.5	47.2	47.1	1.10	32.9	48.2	48.1	1.20	36.9	49.1	48.9	1.29	18		
[illegible]	[illegible]	[illegible]	[illegible]	[illegible]	[illegible]	[illegible]	[illegible]	[illegible]	[illegible]	[illegible]	[illegible]	[illegible]	[illegible]	[illegible]	[illegible]	15	8	
[illegible]	[illegible]	[illegible]	[illegible]	[illegible]	[illegible]	[illegible]	[illegible]	[illegible]	[illegible]	[illegible]	[illegible]	[illegible]	[illegible]	[illegible]	[illegible]	16		
[illegible]	[illegible]	[illegible]	[illegible]	[illegible]	[illegible]	[illegible]	[illegible]	[illegible]	[illegible]	[illegible]	[illegible]	[illegible]	[illegible]	[illegible]	[illegible]	18		
11.4	63.5	58.0	0.49	13.1	65.5	58.7	0.54	14.7	67.1	59.3	0.59	16.2	68.4	59.7	0.63	15	3	45
13.5	58.5	56.1	0.56	15.8	60.6	56.9	0.58	17.9	62.3	57.5	0.63	19.8	63.6	58.0	0.67	16		
15.3	55.7	54.5	0.60	17.9	56.9	55.4	0.65	20.3	58.4	56.1	0.70	22.6	59.9	56.7	0.74	18		
13.5	59.1	56.2	0.64	15.6	60.9	57.0	0.70	17.6	62.7	57.7	0.75	19.4	64.1	58.2	0.78	15	4	
15.7	55.4	54.2	0.65	18.4	56.5	55.1	0.73	20.9	57.6	55.8	0.80	23.2	59.1	56.4	0.86	16		
17.3	53.1	52.6	0.66	20.5	54.1	53.5	0.74	23.4	55.0	54.3	0.81	26.1	55.8	54.9	0.87	18		
16.5	54.4	53.4	0.78	19.4	55.7	54.3	0.86	22.0	56.7	55.1	0.92	24.4	57.5	55.7	0.99	15	6	
18.5	51.8	51.5	0.82	21.9	52.8	52.4	0.90	25.2	53.6	53.2	0.97	28.2	54.3	53.8	1.04	16		
19.8	50.3	50.2	0.84	23.7	51.1	51.0	0.93	27.4	51.9	51.7	1.01	30.8	52.5	52.4	1.07	18		
[illegible]	[illegible]	[illegible]	[illegible]	[illegible]	[illegible]	[illegible]	[illegible]	[illegible]	[illegible]	[illegible]	[illegible]	[illegible]	[illegible]	[illegible]	[illegible]	15	8	
[illegible]	[illegible]	[illegible]	[illegible]	[illegible]	[illegible]	[illegible]	[illegible]	[illegible]	[illegible]	[illegible]	[illegible]	[illegible]	[illegible]	[illegible]	[illegible]	16		
[illegible]	[illegible]	[illegible]	[illegible]	[illegible]	[illegible]	[illegible]	[illegible]	[illegible]	[illegible]	[illegible]	[illegible]	[illegible]	[illegible]	[illegible]	[illegible]	18		
9.9	66.9	59.2	0.44	11.5	68.5	59.8	0.48	12.9	69.9	60.3	0.52	14.2	71.0	60.7	0.54	15	3	50
11.6	63.0	57.8	0.49	13.7	64.5	58.4	0.53	15.6	65.8	58.8	0.57	17.4	66.9	59.2	0.60	16		
12.9	60.0	56.7	0.52	15.3	61.4	57.2	0.58	17.6	62.7	57.7	0.62	19.6	63.8	58.1	0.66	18		
11.5	63.1	57.9	0.57	13.5	64.7	58.5	0.62	15.4	66.1	59.0	0.67	17.0	67.3	59.4	0.71	15	4	
13.1	59.5	56.5	0.58	15.7	60.9	57.0	0.64	18.0	62.1	57.5	0.70	20.2	63.2	57.9	0.76	16		
14.3	56.7	55.4	0.58	17.1	58.1	56.0	0.65	19.8	59.3	56.4	0.71	22.3	60.3	56.8	0.77	18		
13.7	58.1	56.0	0.65	16.3	59.6	56.5	0.73	18.8	60.9	57.0	0.80	21.0	62.0	57.4	0.87	15	6	
15.1	54.9	54.7	0.65	18.1	56.2	55.3	0.73	21.0	57.4	55.7	0.80	23.8	58.3	56.1	0.87	16		
16.0	54.0	53.9	0.68	19.4	54.5	54.4	0.76	22.5	55.0	54.8	0.84	25.6	56.0	55.2	0.87	18		
[illegible]	[illegible]	[illegible]	[illegible]	[illegible]	[illegible]	[illegible]	[illegible]	[illegible]	[illegible]	[illegible]	[illegible]	[illegible]	[illegible]	[illegible]	[illegible]	15	8	
[illegible]	[illegible]	[illegible]	[illegible]	[illegible]	[illegible]	[illegible]	[illegible]	[illegible]	[illegible]	[illegible]	[illegible]	[illegible]	[illegible]	[illegible]	[illegible]	16		
[illegible]	[illegible]	[illegible]	[illegible]	[illegible]	[illegible]	[illegible]	[illegible]	[illegible]	[illegible]	[illegible]	[illegible]	[illegible]	[illegible]	[illegible]	[illegible]	18		

80 / 90 EDB — 67 EWB

AVAILABLE CIRCUITS FOR EACH COIL WIDTH

Finned Width (Inches)	12	18	24	30	33
Available Circuits (C)	1,2,4,8*	1,2,3,6,12*	2,3,8,16*	2,4*5,10,20*	3,7*11,22*

*OPPOSITE END CONNECTIONS FOR ODD NO ROWS.

NOTE OC=Nc x FA

OC=OPTIMUM NO OF CIRCUITS FOR MAX CAPACITY

NC=OPTIMUM NO OF CIRCUITS PER FT^{2}FA (FROM TABULATED DATA)

FA=COIL FACE AREA (FT^2)

C=AVAILABE CIRCUITS CLOSEST TO CIRCUITS (OC)

IF C/OC < 0.7 SPECIAL VERTICAL SPLIT IS REQUIRED

NOTE: 1) MBH=MBH/FT.^{2} Coil Face Area
2) ST=Suction Temp.°F
3) Refrigerant liquid temperature=110 F

[a]Only a partial data listing. See manufacturer's catalog for complete data.

**Reproduced courtesy of The Trane Company.*

water velocity is 1.0 fps or more. Entering water temperature should not exceed 220°F. Refer to the appropriate design selection curves and tables for capacities.

1. Refrigerant Cooling Coils:
   a. Selecting the proper number of coil circuits for a given refrigerant is extremely important. The number of circuits required depends upon the coil capacity, coil size (width and length), number of rows, and suction and condensing temperatures. When these are determined, the optimum number of circuits can be obtained by multiplying N_c (obtained from the tabulated data) times the face area. The selected number of circuits would then be the number of available circuits which is closest to the optimum circuits.
   b. Avoid selecting refrigerant coils for dehumidifying applications with suction temperatures between 32°F and 35°F. Reduction in load or entering air conditions will cause suction temperature to drop, which may result in coil frosting. Hot-gas bypass (see Chapter 3), properly applied, may permit satisfactory operation at lower than normal suction temperatures.
   c. Proper expansion valve selection and installation is necessary for optimum performance of refrigerant coils. The Trane Refrigeration Manual[11] presents a thorough discussion of expansion valve selection and installation. In calculating pressure difference available for expansion valve selection, use 20 psi pressure drop for refrigerant coil, including distributor.

Figure 10-23 illustrates a typical piping hookup for a refrigerant cooling coil.

Another type of evaporator is called a *flooded evaporator* (see Fig. 10-24). In this evaporator a float valve mechanism maintains a constant level of liquid refrigerant in the coil. In other words, as fast as the liquid evaporates,

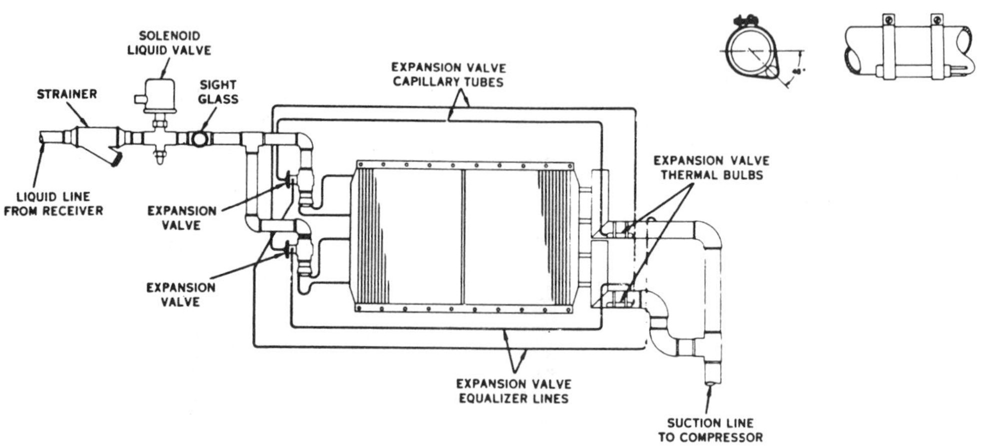

Figure 10–23

Typical piping detail of dry evaporator coil

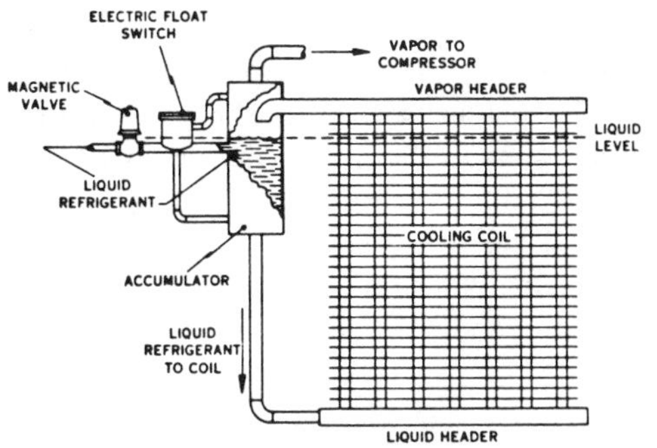

Figure 10–24

Flooded evaporator coil

more liquid is admitted by the float valve. As a result, the entire interior of the evaporator is filled with liquid refrigerant up to the level determined by the float. In contrast to this, the "dry evaporator" has a mixture of liquid and gas throughout the coil until it reaches the outlet where all the liquid turns to gas and is superheated.

Although flooded evaporators are more efficient (better heat transfer because the entire inside surface of the tubes is covered with liquid), they have found little use in air-conditioning installations. One of the more important reasons is the expense of the additional refrigerant required.

10-14 SHELL-AND-TUBE EVAPORATORS

There are two common types of shell-and-tube evaporators used to provide chilled water for air-conditioning systems. They are the same two types as discussed previously with fin and tube coil evaporators and also in Chapter 3. The two types are the *flooded* type and the *direct expansion* (dry) type. In the flooded type the shell contains a tube bundle through which the water to be chilled is pumped. About half or three-fourths of the tube bundle is immersed in liquid refrigerant, which boils because of heat received from the water being cooled.

A modification of the flooded evaporator is the semiflooded type with only the bottom row of tubes being submerged in the liquid refrigerant. For this type special consideration must be given to the uniform distribution of refrigerant inside the shell to insure wetting of all heat-transfer surfaces. Otherwise, a larger shell must be used for a given capacity. In the full-flooded type, this is not a problem as the violent boiling of the refrigerant thoroughly wets the outside surface of all tubes. With the lower refrigerant level in the semiflooded type, a trough running the full length of the shell is commonly employed to insure good distribution of liquid refrigerant underneath the tube bundle.

Laboratory tests have shown that finned tubes give higher rates of heat transfer than smooth tubes. Thus, for a given capacity, an evaporator having finned tubes will have a smaller shell and fewer tubes than an evaporator with smooth tubing.

Liquid carryover from the flooded evaporator to the compressor may be a problem at some load conditions. It is customary to install eliminator plates near the refrigerant vapor outlet to block liquid carryover. In some designs the eliminator plates vary in resistance in order to equilize the pressure drop throughout the evaporator length. Such an arrangement promotes uniform vapor flow and a high rate of heat transfer.

In the direct expansion (dry) evaporator of the shell-and-tube type, see Fig. 10-22, liquid refrigerant boils and evaporates *inside* the tubes while water is circulated over the tube bundle. The shell contains several baffles so the water passes over the tube bundle numerous times. Tube bundles are of U tube or straight tube design. The advantage of the U-tube bundle is that it may be removed from the shell as a unit for cleaning the water side of the tubes. The advantage of the straight tube design is that individual tubes may be easily replaced if necessary. The straight tube design is currently more widely used.

In building air-conditioning and refrigeration units, manufacturers usually supply the direct expansion type for their smaller equipment, say, up to 175 tons, and the flooded type for larger units. The direct expansion type has the advantage of a smaller pressure drop in the chilled water circuit and a smaller charge of refrigerant. All evaporators are provided with controls to prevent freezing, but should a control fail to function, the dry expansion evaporator is less likely to be damaged by temperatures a few degrees below 32°F. The water freezing on the outside of tubes will not likely crush the tubes nor rupture the shell, whereas, in the flooded evaporator, the expansion of water freezing inside the tubes may cause them to burst.

Performance of Shell-and-Tube Evaporators. Manufacturers of shell-and-tube evaporators run laboratory tests under carefully controlled conditions to determine the performance of their particular design configuration. If they sell the evaporators separately as water chillers, then the usual form of the capacity data is a series of curves. One form of presentation is shown in Fig. 10-25 where the capacity in tons is plotted against gpm/nominal ton for various size chillers at varying ITD/Δt_w, where ITD is the initial temperature difference between the entering water to the chiller and the evaporating refrigerant and Δt_w is the difference in temperature between the water entering the chiller and leaving it.

If the chiller is to be combined with a compressor and sold as a package, Fig. 10-26, to produce chilled water, the capacity may be presented in tabular form. Table 10-17 is a capacity table for a nominal 50-ton reciprocating compressor water chiller with a water-cooled condenser. As can be seen, all the parameters of selection are indicated. The designer of the system into which this equipment is to be incorporated merely refers to the table and finds the capacity. Pressure drop, through the water side of the chiller of the package, can be presented in tabular or chart form.

10-15 HEATING COILS

Finnned-tube coils are also used for heating air in air-conditioning systems. The heating medium is nearly always steam or hot water which circulates through the tubes with the air circulating over the exterior finned surfaces.

Finned-tube coils for use with hot water as the heating medium are similar in construction to those used for chilled water. The chief differences is that coils used for hot water usually require fewer coil banks (banks of tubes normal to the direction of air flow) then do those used for chilled water. The reason for this is that the difference between the water temperature and the air temperature is much greater. Figure 10-27 illustrates two typical types of hot water heating coils. Figure 10-28 illustrates some methods used for piping connections to hot water coils used for heating air.

Heating coils, using steam as the heating medium, are made in two general types. First are those used for general heating applications where no attempt is made to obtain equal air temperatures over the entire length of the coil or which are *not* used in airstreams which drop below 32°F in temperature. Second are those called steam-distributing heating coils, made with perforated inner tubes to distribute the steam along the full length of the inner surface of the outer tubes.

The first type, illustrated in Fig. 10-29, is most often used in air-handling systems where face and bypass dampers may be used, or where comparatively short tubes can be used.

Figure 10–25

Shell-and-tube evaporator performance

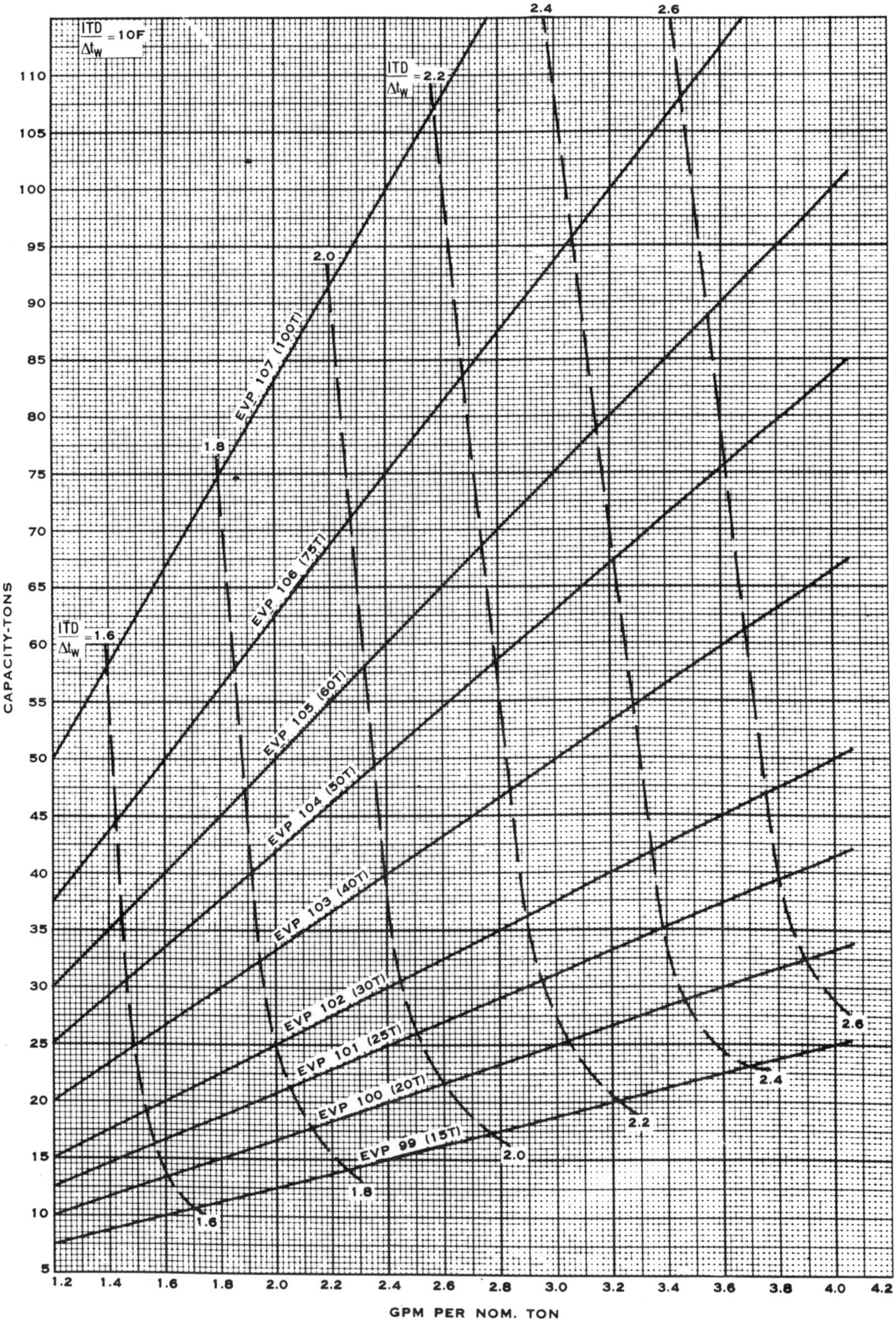

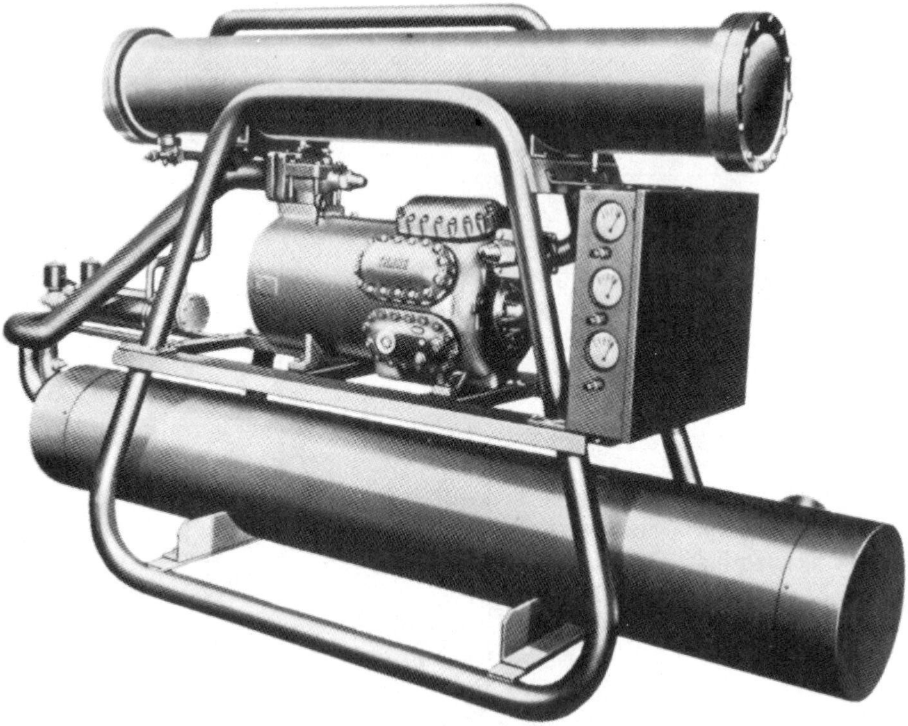

Figure 10–26

Packaged water chiller

Table 10-17

Capacities of 50-Ton Water Chiller*

LVG. WATER TEMP.	WATER TEMP. DROP	COLD GENERATOR — CONDENSING TEMPERATURE 95° CAP.	95° GPM	95° KW	95° COND. GPM	105° CAP.	105° GPM	105° KW	105° COND. GPM	115° CAP	115° GPM	115° KW	115° COND GPM
CG50B													
40°	8°	47.2	142	36.9	141	44.2	133	40.3	132	40.4	121	43.2	123
	10°	47.6	117	36.9	144	44.6	108	40.4	134	40.8	98	43.3	126
	12°	48.1	96	37.0	148	45.1	90	40.4	137	41.2	82	43.4	129
42°	8°	49.2	148	37.2	152	46.1	138	40.6	141	42.2	127	43.7	132
	10°	49.7	119	37.2	155	46.5	111	40.7	143	42.6	102	43.8	134
	12°	50.1	100	37.3	157	46.8	93	40.8	145	43.0	86	43.9	137
44°	8°	51.2	154	37.4	161	48.0	144	40.9	149	43.9	132	44.1	140
	10°	51.6	124	37.4	164	48.4	116	41.0	151	44.3	106	44.2	142
	12°	52.1	104	37.5	166	48.8	97	41.0	154	44.7	89	44.3	144
45°	8°	52.7	158	37.6	166	48.9	146	41.2	153	45.1	135	44.5	144
	10°	52.9	127	37.7	167	49.0	117	41.2	154	45.3	109	44.6	146
	12°	53.0	106	37.7	168	49.2	98	41.3	155	45.4	91	44.8	148
46°	8°	53.3	160	37.8	167	49.5	148	41.4	156	45.6	137	45.0	147
	10°	53.7	129	37.8	168	50.0	120	41.4	158	46.1	111	45.1	148
	12°	54.1	108	37.9	169	50.5	101	41.5	162	46.6	93	45.2	152
48°	8°	55.4	166	38.0	170	51.6	155	41.7	167	47.7	143	45.5	157
	10°	55.8	134	38.1	175	52.0	125	41.7	171	48.2	115	45.6	160
	12°	56.1	112	38.1	179	52.4	105	41.8	175	48.7	97	45.7	163
50°	8°	57.4	172	38.3	187	53.6	161	41.9	177	49.9	150	46.0	168
	10°	57.8	139	38.3	191	54.0	129	42.0	180	50.3	121	46.1	172
	12°	58.1	116	38.4	196	54.4	109	42.1	182	50.7	101	46.2	175

**Courtesy of The Trane Company.*

Figure 10–27

Typical hot water heating coils: (a) Single-row (b) Double-row

The so-called steam-distributing type coils are usually used where the steam flow to the coil must be throttled in order to obtain control, where even temperatures are required along the length of the coil, or where freezing temperatures are frequently encountered with the entering air. This type of coil, illustrated in section in Fig. 10-30, shows the inner steam-distributing tubes. Steam condenses in the outer tubes where heat transfer takes place and drains to the return header.

Particular care in piping, controls, and installation is necessary to protect the coils from freeze-up due to incomplete drainage of condensate. Figure 10-31 illustrates several methods frequently used to connect steam coils into the steam piping system.

10-16 HEAT REJECTION DEVICES AND SYSTEMS

In Chapter 3 we discussed the various items of equipment used in vapor compression refrigeration systems. The ratings and selection of evaporators and chilled water coils has been discussed in this chapter. For the refrigeration cycle to be complete, the heat absorbed in the evaporator and the heat equivalent of the work required to raise the pressure of the refrigerant gas must be removed and dissipated. This is the function of the heat rejection equipment. The heat may be dissipated by sensible heat transfer or by a combination of sensible heat transfer and latent heat transfer (mass transfer). The means of heat rejection is the basis of equipment classification. In general, since the objective of the heat rejection equipment is to convert the hot refrigerant gas to a liquid, such a device is called a *condenser*.

There are three types of heat rejection devices commonly used:

1. Air-cooled condenser in which the heat is rejected directly to the air by sensible heat transfer.
2. Evaporative condenser in which sprayed

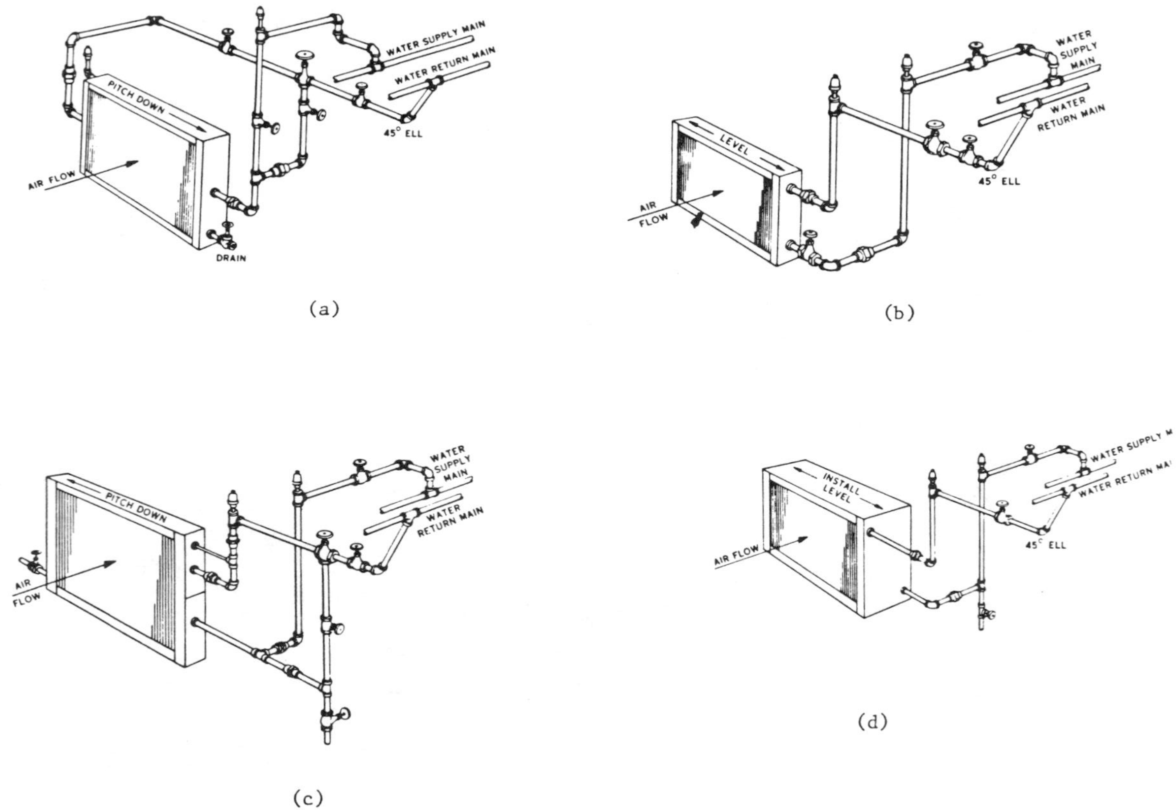

Figure 10–28

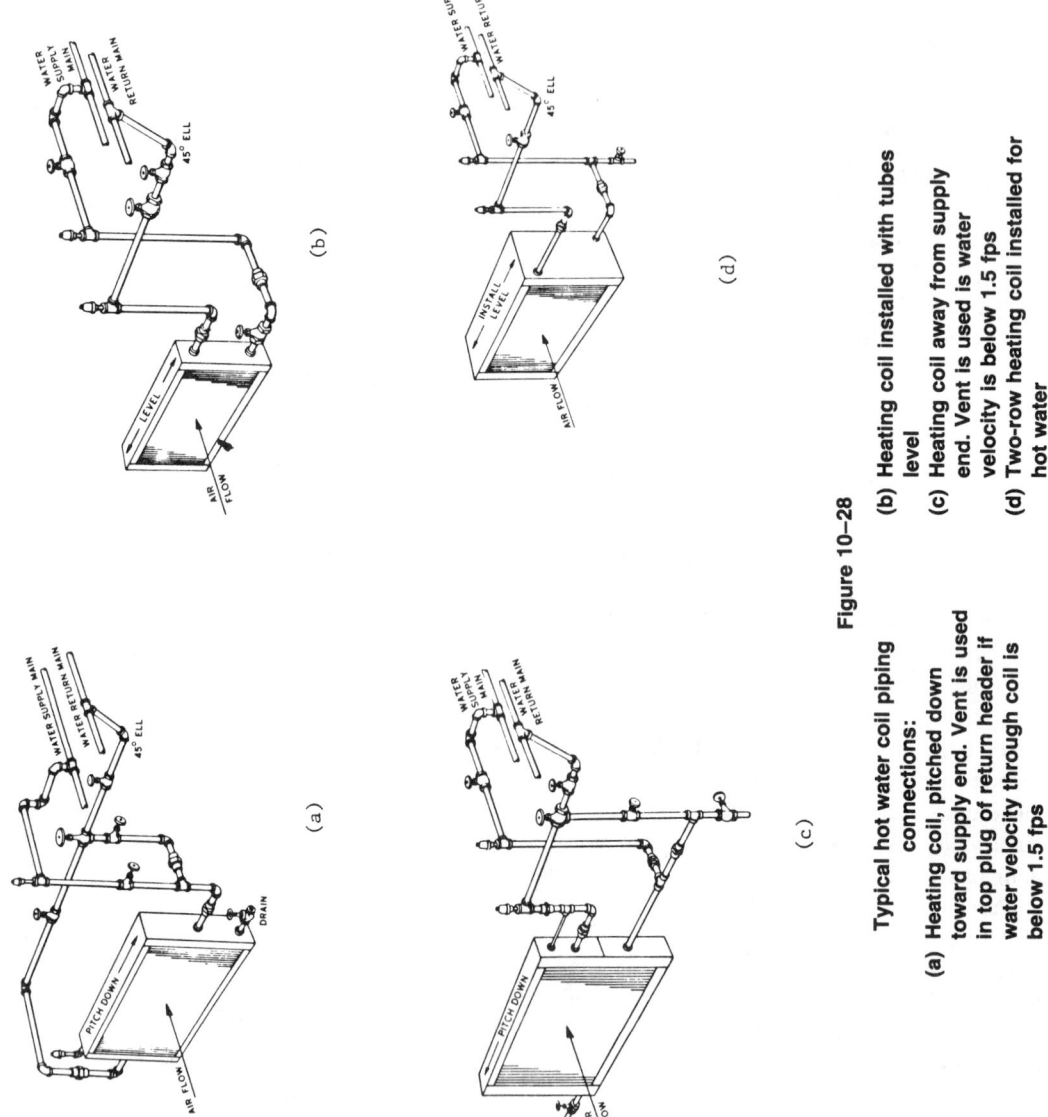

Typical hot water coil piping connections:

(a) Heating coil, pitched down toward supply end. Vent is used in top plug of return header if water velocity through coil is below 1.5 fps

(b) Heating coil installed with tubes level

(c) Heating coil away from supply end. Vent is used is water velocity is below 1.5 fps

(d) Two-row heating coil installed for hot water

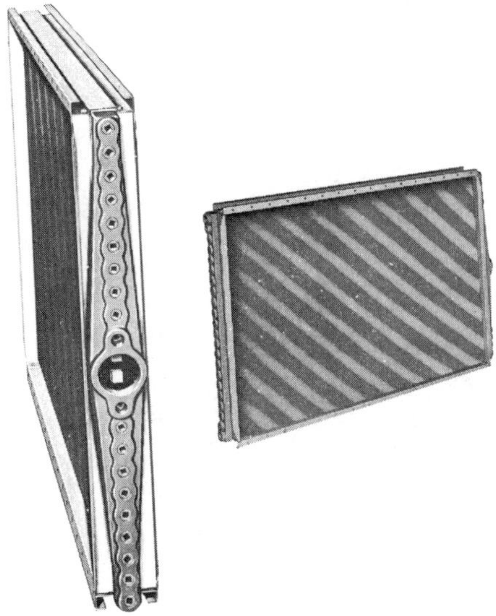

Figure 10–29

Single-row steam heating coil with opposite end connections

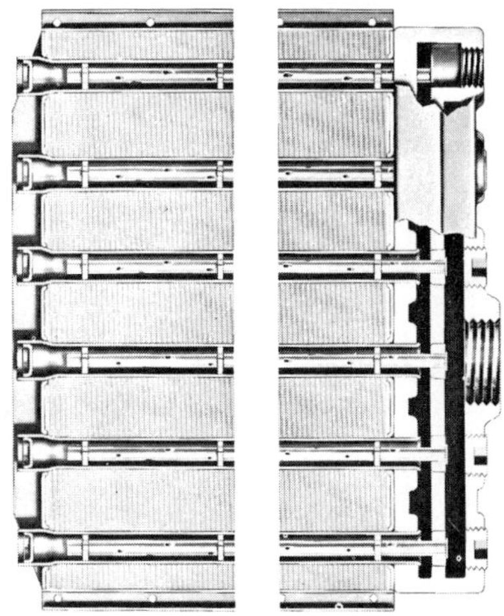

Figure 10–30

Section, steam heating coil with supply and return headers, and steam distributing tubes

coils are used to dissipate heat to the air by sensible and latent heat transfer.

3. Water-cooled condenser in which heat is sensible transferred to water. Although this water may be wasted, it is usually conserved by a process of sensible and latent heat transfer in a cooling tower. The resulting cooled water is then recirculated to the condenser.

An evaluation of the owning and operating costs is usually the basis for selection of a means of heat rejection. Customer preference and provision for future conditions may influence the choices as well. Local design factors such as air and water conditions and system application affect the selection insofar as they affect the economics.

In an economic analysis system size is important since the installed cost per ton of the various condensing methods decrease at different rates with increasing size. All factors being equal, air-cooled condensing is often chosen for capacities up to 75 tons. Evaporative condensing is a primary alternative in the 50- to 150-ton range. Above 100 tons, water-cooled condensing, in conjunction with a mechanical draft cooling tower, is the most common choice.

In the capacity range where all three condensing methods are alternatives, air-cooled condensing is usually the highest in first cost. However, maintenance costs for air-cooled condensers are considerably lower for a given capacity. Therefore, air-cooled condensing is well suited to systems where service is infrequent or incomplete. Similarly, long operating hours at light loads favor air-cooled condensing. Overall operating costs of this method for the commonly applied capacity range are less than for water-cooled condensing and compare favorably with the evaporative condenser. Other advantages and disadvantages in the use

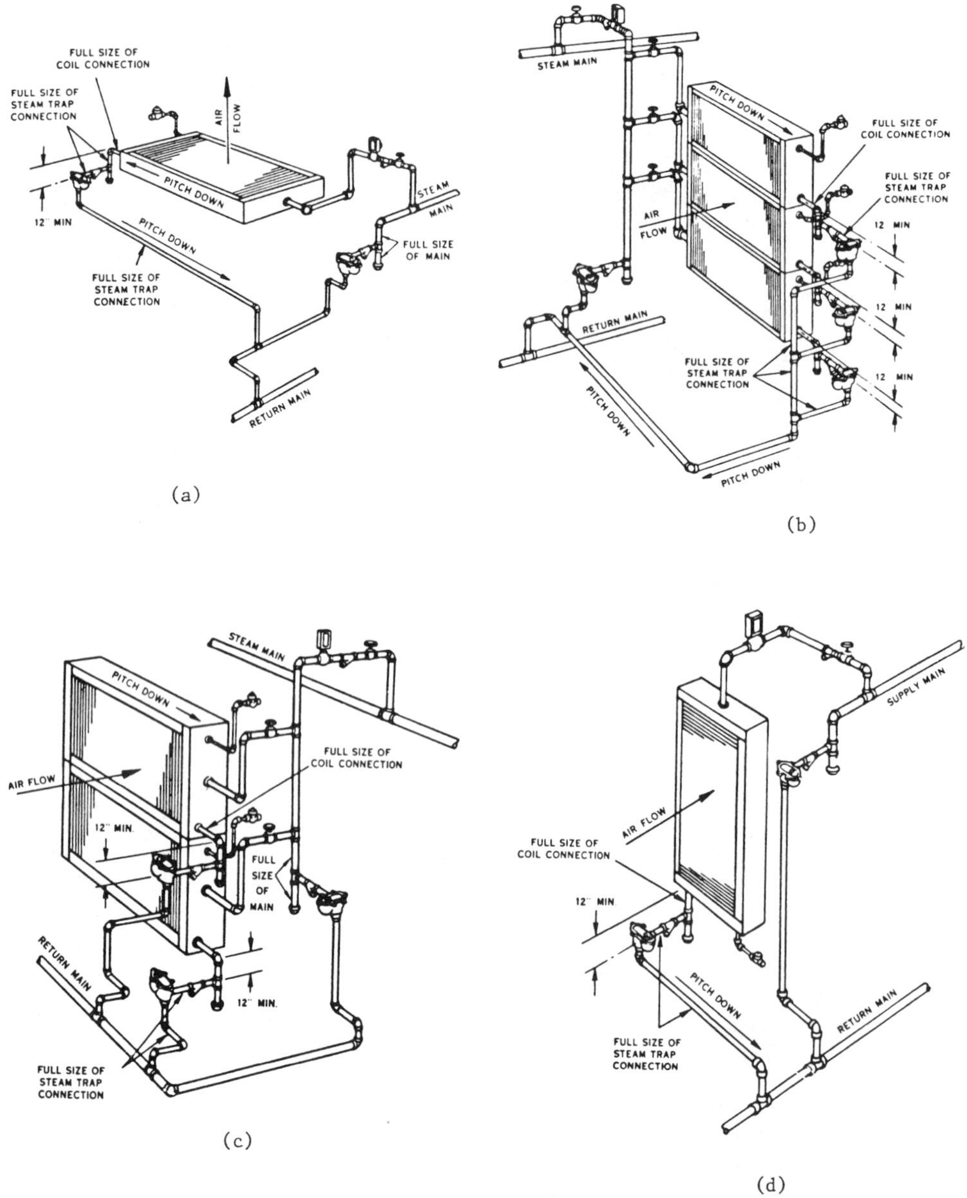

Figure 10–31

Some methods of making piping connections to steam heating coils

of air-cooled condensing were discussed in Chapter 3.

In the 50- to 150-ton capacity range evaporative condensing usually has the lowest first cost. Some other factors which encourage its use are low wet-bulb temperatures and high dry-bulb temperatures, or the availability of inexpensive water of adequate quality. Operating costs may be below those of air-cooled condensing, particularly if the condensing temperature considered is lower, with consequently smaller compressor power input requirements.

In general, conditions favoring the use of evaporative condensing also favor water-cooled condensing in combination with a cooling tower. When the heat rejection equipment is located further away from the other refrigeration system components, the use of a close-coupled water-cooled condenser and a remote cooling tower becomes economically more attractive. This is because the refrigerant piping necessary with air-cooled or evaporative condensing is more costly than water piping for a given capacity.

The application and installation of heat rejection equipment should conform to codes, laws, and regulations applying at the location.

Methods of testing and rating mechanical draft cooling towers are prescribed in the ARI standards, as are similar procedures for evaporative and air-cooled condensers. For water-cooled condensers design, testing, and installations should be in accordance with the ASME Unfired Pressure Vessel Code and the ASA B9.1. Code for Mechanical Refrigeration.

Sensible Heat Transfer. The heat transfer which takes place in an air-cooled or water-cooled condenser between the hot refrigerant gas and the cooling medium (air or water) is a sensible heat-transfer process. The temperature of the condensing refrigerant gas remains constant (at constant pressure) and the temperature of the cooling medium increases. This may be illustrated by Fig. 10-32, where temperature is plotted against area of condenser surface transversed by the cooling medium.

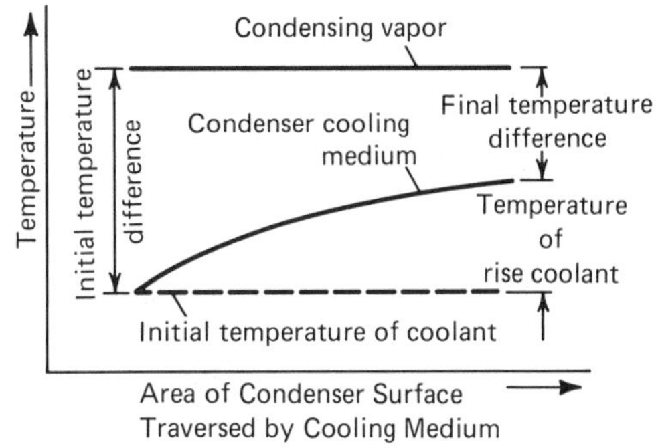

Figure 10–32

Temperature rise of cooling medium in a condenser

The capacity of a condenser is affected by the temperature of the cooling medium, the quantity of cooling medium used, and the temperature of the condensing refrigerant gas. The capacity of the condenser will increase whenever the temperature difference between the refrigerant gas and the cooling medium is increased. This temperature difference may be increased by raising the condenser pressure (temperature), by lowering the temperature of the entering cooling medium, or by increasing the quantity of cooling medium in order to maintain a lower average temperature of the cooling medium.

At a constant-suction pressure any increase in head pressure (condensing pressure) will require an increase in power input because of the additional work to be done in compressing the refrigerant gas to the higher pressure. On the other hand the higher the condensing pressure (temperature), the smaller the quantity of cooling medium required. In the case of water cooling, then, this means it is possible to save power at the expense of water or save water at the expense of power. The most economical conditions of operation can be determined only by a study of the local power and water costs.

In order to obtain a high rate of heat transfer through the heat-transfer surface of a condenser, it is necessary for the cooling medium to pass through the condenser at a fairly high velocity. For the shell-and-tube type water condensers, Fig. 10-33, this means that the tubes are separated into several groups, called passes, the same water traveling in series through each of the passes. A condenser having four groups of water tubes would be a four-pass unit because the water flows back and forth along the condenser length four times. Four-pass condensers are common, although any reasonable number of passes may be used. The greater the number of tube passes the greater will be the pressure drop (friction) through the tubes, and consequently the greater will be the power required to pump the condenser water.

The allowable temperature rise of the cooling medium is an important factor in circuiting the cooling medium through the condenser. Knowing the allowable temperature rise in the condenser-cooling medium, the quantity of coolant may be determined.

In general, the quantity of heat to be rejected in the condenser is given by Eq. (3-7). The quantity of cooling medium required to carry away this heat is dependent on the allowable temperature rise of the coolant. The following relationship may be used to determine the required quantity of cooling medium if the allowable temperature rise is known:

$$\dot{q}_{\text{cond}} = \dot{m} c_p \Delta t$$

or

$$\dot{m} = \frac{\dot{q}_{\text{cond}}}{c_p \Delta t} \tag{10-21}$$

where

$\dot{m}$ = weight of cooling medium required (lb/hr)

q_{cond} = heat rejected by refrigerant (Btu/hr)

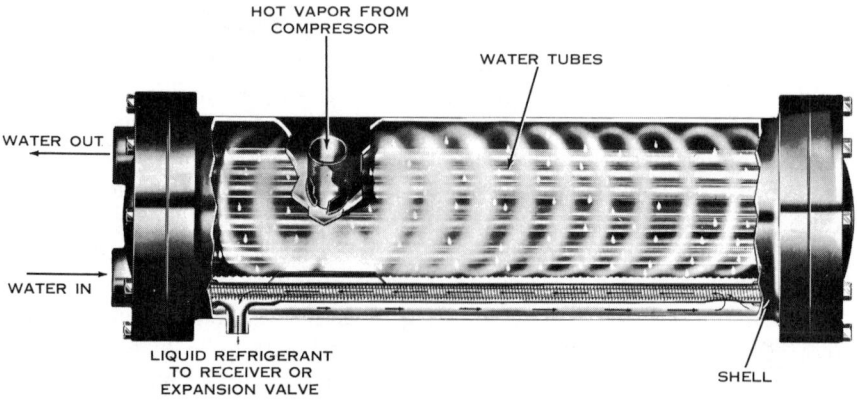

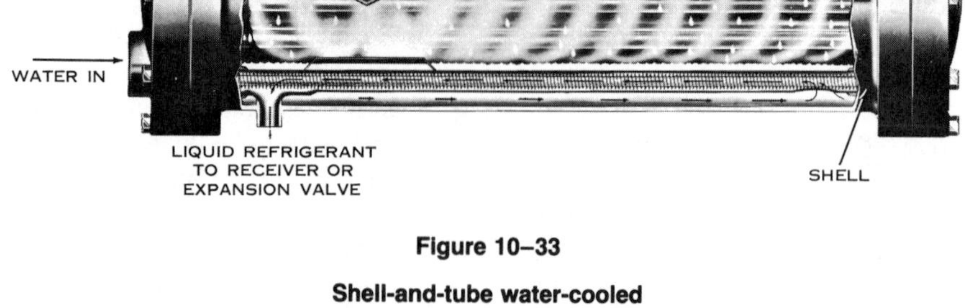

Figure 10–33

Shell-and-tube water-cooled condenser with subcooling circuit

c_p = specific heat of cooling medium (Btu/lb °F)

Δt = allowable temperature rise of cooling medium (°F)

If we consider the water-cooled condenser, the quantity of condenser water needed is normally expressed in gallons per minute (gpm). There are 8.33 lb water per gallon of water at normal condenser temperatures, and c_p is usually taken as unity. Therefore, Eq. (10-21) may be easily modified to give the required water flow in gpm. Since $\dot{m} = (60)(8.33)(\text{gpm}) = 500$ gpm,

$$\text{gpm} = \frac{\dot{q}_{\text{cond}}}{500\Delta t} \qquad \textit{(10-21a)}$$

Illustrative Problem 10-3

A compressor pumping Freon-22 for a 10-ton refrigeration plant is operating with a suction temperature of 40°F and a condensing pressure of 220 psig. If the condenser water is allowed to rise 15 degrees, find the quantity of water required. Assume that saturated vapor leaves the evaporator and enters the compressor, and saturated liquid leaves the condenser and enters the expansion valve.

Solution:

Make a sketch of the *p-h* diagram for the problem.

Referring to Table B-5:

At $t_3 = 40°\text{F}$, $h_3 = h_g = 108.142$ Btu/lb, and $p_2 = p_3 = 83.206$ psia

At $p_1 = 220$ psig (234.71 psia), $h_1 = h_f = 41.804$ Btu/lb, and

$t_1 = 108°\text{F}$ condensing temperature.

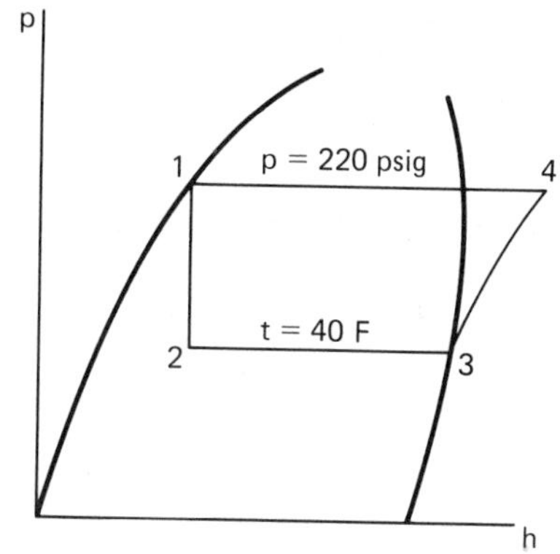

Figure 10–34

Sketch of p-h diagram for Illustrative Problem 10-3

Referring to Fig. B-3, at $p_4 = 234.71$ psia and $s_3 = s_4$ find $h_4 = 118.5$ Btu/lb.

$$\text{Compression ratio} = \frac{P_d}{p_s} = \frac{220 + 14.7}{83.206}$$

$$= 2.82$$

Referring to Fig. 3-18, the overall compressor efficiency is approximately 77.5 percent. By Eq. (3-6) the theoretical work of compression is $h_4 - h_3 = 118.5 - 108.142$, or

$$\text{iwk}_c = 10.36 \text{ Btu/lb}$$

Then, the actual work of compression is

$$\text{wk}_c = 10.36/.775 = 13.37 \text{ Btu/lb}$$

$$\text{Mass flow of refrigerant } (\dot{m}_r) = \frac{(200)(\text{tons})}{h_3 - h_1}$$

$$\dot{m}_r = \frac{200(10)}{108.142 - 41.804}$$

$$\dot{m}_r = 30.15 \text{ lb/min}$$

Heat rejected to condenser coolant equals the refrigeration effect plus the heat of compression, or

$$\dot{q}_{\text{cond}} = (10)(200) + (13.37)(30.15)$$

$$\dot{q}_{\text{cond}} = 2000 + 403 = 2403 \text{ Btu/min}$$

$$= 2403(60) = 144{,}180 \text{ Btu/hr}$$

By Eq. (10-19a) $\text{gpm} = \dfrac{\dot{q}_{\text{cond}}}{500\Delta t_w} = \dfrac{144{,}180}{500(15)}$

$\text{gpm} = 19.2$ water flow required

The enthalpy of the vapor leaving a compressor is generally close to the value computed by theoretical methods (see Chapter 3) because the relatively warm vapor is cooled by the cylinder head at the end of the compression stroke. In spite of this, the power required to run a compressor is always greater than the amount computed by theoretical methods. This was discussed in Chapter 3 under the sections discussing volumetric efficiency and power requirements of compressors.

For the most practical comfort air-conditioning applications, a value of $(200 \times 1.2) = 240$ Btu/min per ton of refrigeration (14,400 Btu/hr) may be used as the total quantity of heat transferred to the condenser coolant for Freon-12 and Freon-22 systems. A variation of ± 10 Btu/min from 240 Btu/min amounts to an error of approximately 4 percent in the computed quantity of water. If the figure of 14,400 Btu/hr is substituted into Eq. (10-21a) we would have

$$\text{gpm/ton} = \frac{14{,}400}{500\Delta t}$$

$$= \frac{28.8}{\Delta t} \qquad \textit{(10-21b)}$$

Equation (10-21b) gives the required condenser water rate in gpm per ton of refrigeration capacity.

Illustrative Problem 10-4

Find the quantity of condenser water required for a 60-ton refrigeration plant if the allowable temperature rise of the condenser water is 12 degrees.

Solution:

By Eq. (10-21b)

$$\text{gpm/ton} = \frac{28.8}{\Delta t} = \frac{28.8}{12}$$

$$= 2.4 \text{ gpm per ton}$$

Total condenser water required is

$$\text{gpm} = (2.4)(60 \text{ tons}) = 144 \text{ gpm}$$

The preceding discussion has shown that, for a given heat load, the quantity of condenser-cooling medium (in this case water) needed depends upon its allowable temperature rise. However, before the allowable temperature rise can be determined, both the head pressure (temperature) to be maintained and the initial temperature of the cooling water must be known. Although the initial temperature of the water is usually fixed by conditions beyond control, the head pressure to be maintained may, within reasonable limits, be selected.

The head pressure that will result is determined by the temperature at which the refrigerant vapor is condensing. As will be shown in Problem 10-5, it is desirable to maintain as low a condensing temperature as is practical.

With the initial temperature of the condenser water known, the number of degrees by which the temperature of the condensing vapor will exceed the initial temperature of the water is determined by two factors: (1) the quantity of condenser water available and (2) the amount of heat-transfer surface installed in the condenser.

The smaller the quantity of water used for condensing purposes, the larger will be its temperature rise. But, the temperature rise of the condenser water is always less than the difference between the temperature of the condensing vapor and the initial water temperature (see Fig. 10-32). The left side of the chart shows a large temperature difference, and the right side of the chart shows that the rise in water temperature is about one-half the initial temperature difference for the particular operating conditions of a condenser. The initial temperature difference represents the theoretical maximum temperature rise of the water.

In a given condenser the temperature of the condensing vapor will decrease as the water quantity increases. This is illustrated in the following example.

Illustrative Problem 10-5

Two identical shell-and-tube condensers are condensing Freon-22, initially at the same condensing pressure. Condenser *A* is provided with cooling water at the rate of 1.92 gpm per ton and condenser *B* at the rate of 4.8 gpm per ton. The cooling water enters both condensers at the same temperature. Compute the temperatue rise of the cooling water in each condenser.

<u>*Solution:*</u>

From Eq. (10-21b) $\Delta t = \dfrac{28.8}{\text{gpm/ton}}$

For Condenser *A* $\Delta t = \dfrac{28.8}{1.92}$

$= 15$ degrees

For Condenser *B* $\Delta t = \dfrac{28.8}{4.8}$

$= 6$ degrees

If the initial temperature difference between the condensing vapor and entering water were 30 degrees, the results may be shown as in Fig. 10-35. As calculated and as illustrated in Fig. 10-35a, the 1.92-gpm/ton water rate results in a 15 degree rise in water temperature for condenser *A*. This means a final temperature difference between condensing refrigerant vapor and leaving water of 15 degrees.

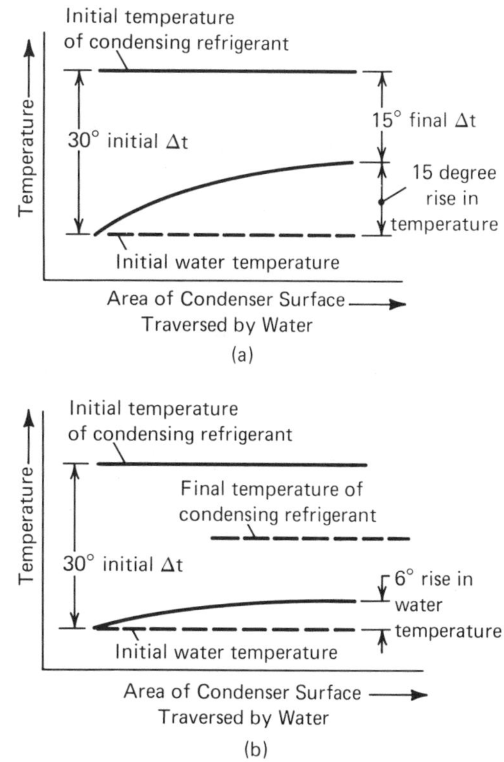

Figure 10–35

An increase in water flow rate decreases head pressure (See Illustrative Problem 10-3)

Condenser *B*, as shown in Fig. 10-35b, has the same initial temperature difference of 30 degrees. As computed, however, the water rate of 4.8 gpm per ton gives a temperature rise of only 6 degrees. Because of the greater LMTD between the refrigerant and water, more heat will be transferred to the water in a given time. This means that after operating for a short time the temperature of the condensing vapor will drop. The temperature drop will depend on the amount of heat-transfer surface, but it might be as much as 12 or 15 degrees, as indicated in Fig. 10-35b.

A 12 or 15 degree drop in condensing temperature has a beneficial effect upon the operating conditions of the refrigeration system (see Chapter 3). From Table B-5 it will be noted that if the condensing saturation temperature decreases from 108°F to 96°F (12 degrees), the corresponding drop in condenser pressure is (234.71 − 199.26) = 35.5 psi. For the same suction pressure at the compressor, this means a lower compression ratio. From Table 3-4 and Fig. 3-18 it will be noted that a lower compression ratio means a higher volumetric efficiency results in an increase in refrigeration capacity and a higher overall efficiency results in a lower brake horsepower. Thus, the condensing temperature, and therefore the head pressure, should be kept as low as feasible.

The amount of condenser water required for any specific condensing unit must be determined from the manufacturer's data. While well-designed condensers of different makes will probably have approximately the same capacity for a given water rate, this must not be assumed to hold true for all makes. The water requirements may vary widely due to difference in water velocity through the condenser tubes and differences in the amount of condensing surface supplied.

For example, the quantity of water required for The Trane Company Model HFC-225 hermetic condensing unit using Freon-22 may be found by means of Tables 10-18 and 10-19, taken from the Trane Hermetic Compressor Bulletin DS-361 H. With Tables 10-18 and 10-19 the following formula is used:

$$Z = \frac{NT}{D}$$

where

D = temperature difference in °F between the condensing temperature of the refrigerant and the temperature of the condenser water as it *enters* the condenser

Table 10-18

Values of *N[a]**

COND. TEMP.	-10	0	10	20	30	35	40	45	50
85	1.44	1.36	1.29	1.23	1.19	1.17	1.16	1.14	1.12
95	1.48	1.40	1.33	1.27	1.22	1.21	1.19	1.17	1.15
105	1.53	1.45	1.38	1.31	1.26	1.24	1.22	1.20	1.18
115	1.59	1.51	1.43	1.36	1.30	1.28	1.26	1.24	1.22
125	1.66	1.57	1.49	1.41	1.35	1.32	1.29	1.27	1.25
135	1.73	1.64	1.56	1.47	1.41	1.36	1.33	1.31	1.29

* FOR TRANE HERMETIC COMPRESSORS.

[a]*Reproduced courtesy of The Trane Company.*

Table 10-19

Condenser Capacity[a]

CDS147—R-22

NOMINAL 25 TON			
Z (0 0005)	Z (0 001)	GPM	P D
0 360	0 337	10	0 41
C 504	0 459	15	0 55
0 632	0 563	20	0 92
0 747	0 655	25	1 38
0 853	0 733	30	1 91
1 040	0 867	40	3 22
1 197	0 973	50	4 85
1 328	1 060	60	6 68
1 458	1 132	70	8 75
1 549	1 198	80	11 1
1 639	1 249	90	13 8
1 718	1 294	100	16 6

Note: All pressure drops (PD) shown in feet of water.

[a]*Reproduced courtesy of The Trane Company.*

N = a constant found in Table 10-18. The value N depends upon the condensing unit. It will be higher than shown if the compressor is the hermetic type

T = the refrigerating capacity, in tons, of the condensing unit as determined from the rating tables

Z = a constant found in Table 10-19. The value of Z depends upon the size and type of the condenser and the quantity of condenser water supplied

To determine the quantity of water required under any definite condition of operation, the value of Z is first found by means of the forgoing formula. Having determined Z for the selected conditions, the required gpm and the resulting pressure drop may be read from Table 10-19. The following example will serve to illustrate this method.

Illustrative Problem 10-6

Assume a Trane HFC-225 hermetic condensing unit has been selected for an operating condition of 40°F suction temperature and 105°F condensing temperature. At this operating condition it has a capacity of 27.5 tons. Condenser water is available at 75°F. How much condenser water will be required with a standard CDS-147 condenser if the fouling factor is 0.0005?

Solution:

$D = 105 - 75$

$= 30$ (initial temperature difference)

$N = 1.22$ (from Table 10-18, for 40 and 105 degrees)

$T = 27.5$ (load in tons)

$$Z = \frac{NT}{D}$$

$$= \frac{1.22(27.5)}{30}$$

$$Z = 1.12$$

From Table 10-19 by interpolation with $Z = 1.12$, find 45.1 gpm and a pressure drop of 4.04 ft of water or 1.75 psi for this flow through the condenser.

The Z factors are given in Table 10-19 for fouling factors of 0.0005 and 0.001. Selection of the proper fouling factor depends on the quality of the condenser water.

As the Z factor for any condenser is actually a capacity factor, this can frequently be used to advantage in determining other conditions of operation when certain factors are known or limited. For example, it may be necessary to know the maximum tonnage available from a given condenser when the quantity and temperature of the condensing water is limited. By rearranging the Z formula we have $T = (ZD)/N$.

In applying this formula, Z is determined from Table 10-19 for the maximum water flow available and for the proper fouling factor. D is determined by subtracting the temperature of the condensing water from the maximum allowable condenser temperature. N is found in Table 10-18 for the desired suction temperature and the maximum allowable condensing temperature.

Quality of Condenser Water. The condenser water supply is extremely important. The water must be reasonably cool, available in adequate quantity, at sufficient pressure, and must not excessively foul or scale the water passages.

The temperature of the water determines the quantity to be used, so it is advisable to guard against an unnecessarily warm water supply. If city water is used, the service connection from the street main should be as close to condenser as possible. This piping should not run through any hot rooms or near any heat-generating equipment.

The amount of water available and the pressure of the water supply must be carefully checked. Sometimes the pressure in the building system will fall so low, at the time when many fixtures are in use, that the condenser will receive insufficient water. To avoid this, it may be necessary to install a new and larger service from the street main and increase the size of the water meter. Before doing this, a cooling tower to provide condenser water or an evaporative condenser should be considered.

The *fouling factor* is a measure of the buildup of impurities on the water side of the condenser tubes that decrease the overall heat-transfer rate. Water with a high degree of hardness or impurities will form scales on the tubes. It is necessary to select a condenser with sufficient surface or water flow so that there will be a reasonable time before the buildup of impurities will seriously affect capacity and require cleaning of the tubes. Such selection does not reduce the buildup of impurities, however.

Normal fouling factors used are:

City water	0.0005
Cooling tower water	0.0005
Well water	0.001
River water	0.001

There is a wide variation in the chemical content of water and it is often necessary to consider local conditions which may cause the selection to vary from the normal factors above.

Referring to Problem 10-6 again, for a Z factor of 1.12, Table 10-19 shows that the water requirement is 45.1 gpm based on 0.0005 fouling factor or 68.3 gpm based upon 0.001 fouling factor. This increase is 51 percent and can be more severe in larger-sized units. It is often necessary to use a larger condenser when basing the selection upon 0.001 rather than 0.0005 fouling factor.

As we can see from the preceding four illustrative problems, the quantity of cooling water required is a relatively high quantity when we consider the tonnage capacity of the systems. The disposal of such large quantities of water may present problems, to say nothing of the cost of supplying it.

As municipalities grew in population and air-conditioning systems employing mechanical refrigeration increased in number, it became increasingly more difficult to supply water for all needs. The sewage systems also became overtaxed. Many localities now prohibit the use of sanitary sewers for the disposal of waste water from condensing equipment because the large volumes of clean water upset the bacterial action at the sewage treatment plant. When confronted with these problems, the local authorities restricted more and more the use of once-through city water for condensing purposes. This forced the consideration of cooling towers and evaporative condensers, where water is reused, and air-cooled condensers that completely eliminate water from the condensing process.

10-17 OTHER HEAT REJECTION METHODS

In Chapters 4 and 5 considerable attention was given to the case of air moving through a spray of water, as in a spray chamber. There the emphasis was on the *increase* in humidity ratio of the air as it passed through the spray chamber. In the cooling tower, on the other hand, the emphasis is on the *decrease* in the water temperature. Therefore, a cooling tower is arranged to give the maximum drop in water temperature for the design inlet air conditions. (Cooling towers and spray ponds were discussed generally in Chapter 3.) Only a negligible amount of heat is transferred through the walls of the cooling tower. Since the temperature of the water decreases in falling through the cooling tower, the heat transfer (loss of heat from the water) required to bring about this temperature change is called a sensible heat loss. The heat lost in cooling the water, as it falls through the tower, is transferred to the air passing through the tower. The water is cooled slightly by the air, but most of the cooling is because of a small part of water evaporating into the moving airstream. The amount of water evaporated is known as the *evaporation loss*.

The latent heat for the evaporation process is supplied by the sensible heat from the droplets of water falling through the air. The cool water in the sump is then available for another pass through the condenser and, in this way, all heat transferred from the condensing refrigerant vapor to the water is, in turn, transferred to the atmosphere. If all the heat lost by the water became latent heat, the evaporation would theoretically not exceed 1.0 percent of the water supply to the tower for every 10 degrees the water is cooled.

The cooling tower will cool water below the dry-bulb temperature of the ambient air. The cooling effect by evaporation is completely dependent upon the ability of the surrounding air to absorb moisture; that is directly related to the air enthalpy, and this is dependent on the wet-bulb temperature. No method of heat transfer is 100 percent efficient so water is never cooled to its lowest potential—the wet-bulb temperature. The difference between actual cold water temperature and existing wet-bulb temperature is termed *approach to the wet bulb,* commonly shortened to *approach.* The actual number of degrees the water temperature is reduced is known as the *cooling range.* ''Range'' and ''approach'' are the yardsticks of performance for any device cooling water by evaporation and are illustrated in Fig. 10-36.

Drift is the percentage of the warm water entrained by the air and carried away from the cooling device. It is the function of a device which cools water by evaporation to speed up the evaporation process and to cause the heat to transfer rapidly from the water being cooled to the air. The evaporation of 1 percent of a quantity of water will theoretically reduce the temperature of the remaining 99 percent by approximately 10°F. Cooling water devices can be smaller and still achieve a given range for a certain volume of water if the cooling effect is accelerated. The following factors, taken singly or in combination, can be used to accelerate the cooling effect:

a. breaking up the flow of water into droplets for exposure of greater water surface.

b. increasing the wetted surface to expose more water to air

c. increasing the heat-transfer potential by increasing the air flow

There are several major considerations that should be kept in mind when applying cooling towers:

1. When selecting the location, sufficient clearance should be allowed for the free flow of air to the inlet of the tower and for its discharge from the tower. Obstructions will reduce the air flow causing a reduction in capacity.

2. Cooling towers should be located so that noise created by air or water is not a source of annoyance.

3. The cooling tower location should be such that the air discharge will not cause condensation on nearby surfaces or wetting because of normal drift.

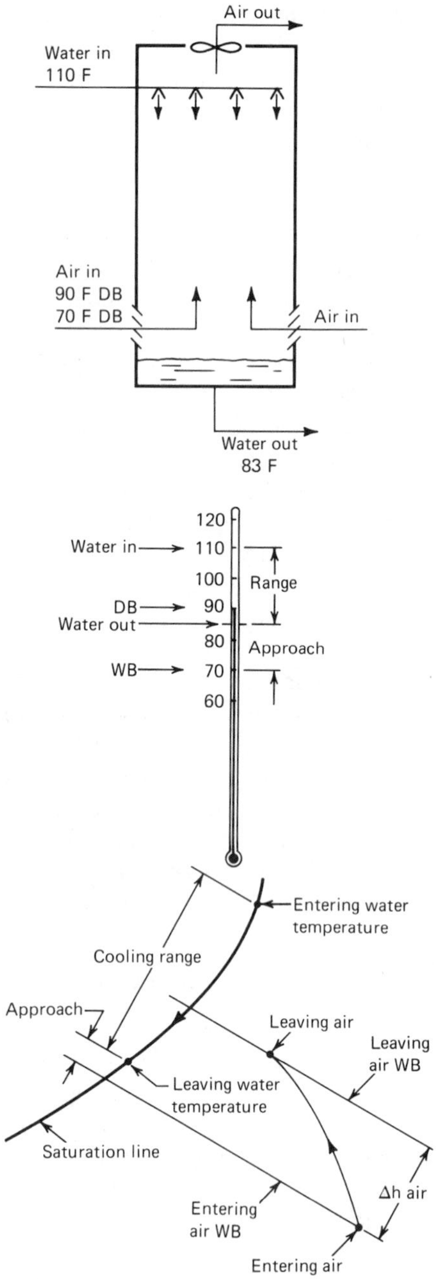

Figure 10–36

Cooling tower psychrometrics illustrating "range" and "approach"

4. The tower should be located away from sources of exhaust heat and contaminated air.

Cooling tower ratings are usually given for various entering and leaving water temperatures and entering air wet-bulb temperatures, and water quantities of 3 to 4 gpm per ton of refrigeration. The ratings may be given as tons or Btu/hr of gross heat rejection or refrigeration effect.

The *evaporative condenser* is a water-conserving device which may be used to replace the functions performed by the water-cooled condenser and the cooling tower (see Chapter 3 and Fig. 3-37). In an evaporative condenser water is pumped from a sump or base pan to a spray header or perforated distributing pan. The droplets of water completely wet the condenser coil and fall to the sump. Air is drawn in through the unit over the coil and discharged to the outside. Forced-air circulation is needed for moving the air to overcome the airway resistance of the coil. Eliminators are sometimes used on the discharge to limit carryover of water from the unit. The hot refrigerant gas from the compressor is piped to the condenser coil where the condenser heat rejection load is removed and the refrigerant is condensed and subcooled.

Heat is removed in the evaporative condenser by evaporation of water from the coil's exterior surface. The heat to evaporate the water is supplied by the condensing refrigerant inside the coil. Each pound of water evaporated removes approximately 1000 Btu from the refrigerant. The rate of evaporation is related to the enthalpy of the entering air, which in turn is dependent on the entering air wet-bulb temperature.

To minimize scale and facilitate cleaning, the condenser coil is usually fabricated of prime surface tubing rather than a finned surface. Hot-dipped galvanized steel tubing or copper tubing are used in making the coil. The copper coil is more resistant to corrosion but has a significantly higher first cost. The choice of steel or copper tube has virtually no effect on the heat-transfer rate because the thermal resistance of the tube wall is only a small part of the total resistance of heat transfer.

As in the cooling tower, water is lost by evaporation, drift, and bleed-off. These water losses are made up by a float valve assembly which maintains a fixed water level in the sump. Water consumption for an evaporative condenser normally runs about 5 percent of the water that would be required by a once-through waste water condensing system.

Application of Evaporative Condensers. In a refrigerant-condensing system using a cooling tower all the water required for the condenser (approximately 3 gpm/ton) must be pumped through the cooling tower-condenser circuit. As a result, the pump horsepower is relatively high. In an evaporative condenser only enough water is circulated within the condenser casing to insure constant wetting of the condenser coil tubes. The pumping horsepower will be much less than for a cooling tower of the same capacity. The fan horsepower will be comparable for cooling towers and evaporative condensers of equal capacity.

Evaporative condensers are designed for outdoor installation and are available in horizontal and vertical component arrangements. The sizes offered by various manufacturers will vary, but units are available in a capacity range of 10 to 200 tons of gross heat rejection. The primary use of evaporative condensers is to condense refrigerants, but they may be used to cool engine jacket water, oil-cooled transformers, or process fluids. Evaporative condensers are usually floor mounted on roofs or on concrete pads at grade level, although they may be suspended or hung.

If winter operation of the unit is required,

consideration must be given to freeze-up problems. The unit can be drained of water and run as a dry coil (air-cooled condenser). The capacity of an evaporative condenser with a dry coil in winter with 30°F entering air and 105°F condensing temperature is about 50 percent of the unit's capacity in summer with a wet coil, 75°F entering wet bulb and 105°F condensing temperature. If more than 50 percent capacity is required in winter, it will be necessary to select the unit on its dry coil capacity. Then the unit will likely be oversized in summer and control of head pressure with air volume dampers will be necessary to reduce the unit's capacity. As a second possibility, serious consideration should be given to locating the unit indoors in a heated space where freezing during off cycles will not be a problem. Duct work is usually required on both the inlet and discharge of the unit in such cases. Dampers should be provided to close during off cycles to prevent gravity flow of outdoor air and some type of head pressure control is usually required.

When evaporative condensers are mounted outdoors, the liquid refrigerant receiver should be located so the direct rays of the sun can never strike it. The heat added by the sun may cause excessive pressure or create trouble due to too warm liquid. Furthermore, the water supply and drain lines must be designed for perfect drainage so the lines will not hold water in pockets when drained for winter shutdown.

Selection of Evaporative Condenser. The selection of an evaporative condenser is a simple matter. Table 10-20 is typical of most evaporative condenser-rating tables. The capacities of the various evaporative condensers are usually given in Btu/hr of refrigerating capacity. Heat of compression, which must be dissipated through the condenser, may be ignored as this is taken into account in preparing the tables. The capacity of the evaporative condenser, as given in the normal capacity table, can be matched directly to the capacity of the compressor.

Illustrative Problem 10-7

Assume that a job requires a compressor delivering 205,000 Btu/hr at 30°F suction temperature and 105°F condensing temperature. The average maximum air wet-bulb temperature is 78°F. Select an evaporative condenser by using Table 10-20.

Solution:

Table 10-20 is based upon 40°F suction temperature; therefore, the compressor load must be adjusted before selecting the evaporative condenser. From the lower portion of Table 10-20, the multiplier for 30°F suction temperature is 1.035. Therefore, the compressor load, for the purpose of this table, is (1.035)(205,000) = 212,000 Btu/hr. The average maximum wet-bulb temperature was given as 78°F; therefore, the 78°F wet-bulb column is read, considering only those values for 105°F condensing temperature. A 20-ton evaporative condenser is found to have a capacity of 240,000 Btu/hr. The 20-ton evaporative condenser will be satisfactory because the slight surplus capacity will provide a margin of safety and insure proper operation when the wet-bulb temperature slightly exceeds the average maximum.

Naturally, the wet-bulb temperature of the air will not stay at the average maximum, except for short intervals, but will fluctuate at lower levels. This will affect the operation of the evaporative condenser in that as the wet-bulb temperature falls the capacity of the condenser will rise. As the capacity of the condenser rises the condensing temperature will fall to compensate. This can be seen from Ta-

Table 10-20

Refrigerating Capacity of a Typical Evaporative Condenser*

Btu Per Hour Based on Refrigerant-22 at 40 Degrees Suction Temperature

UNIT SIZE	CONDENSING PRESSURE (PSIG) REF-22	COND. TEMP.	ENTERING AIR WET BULB TEMPERATURE 85°	82°	80°	78°	76°	75°	72°	70°	67°	64°
20 TONS	168.40	90	47.600	68.700	81.900	95.100	108.000	114.000	131.000	142.000	158.000	172.000
	181.80	95	92.800	114.000	128.000	141.000	154.000	160.000	178.000	188.000	204.000	219.000
	195.91	100	140.000	162.000	176.000	190.000	203.000	209.000	227.000	238.000	254.000	269.000
	210.75	105	189.000	211.000	225.000	240.000	254.000	259.000†	277.000	289.000	305.000	320.000
	226.35	110	240.000	262.000	277.000	291.000	306.000	311.000	330.000	341.000	358.000	373.000
	242.72	115	293.000	316.000	331.000	345.000	360.000	366.000	385.000	396.000	414.000	429.000

FOR SUCTION TEMPERATURES OTHER THAN 40 DEGREES, MULTIPLY COMPRESSOR LOAD BY THE PROPER MULTIPLIER FROM THE FOLLOWING BEFORE SELECTING EVAPORATIVE CONDENSER.

SUCTION TEMPERATURE	-20	-10	0	10	20	30	40	50
MULTIPLY COMP. LOAD BY	1.27	1.21	1.155	1.10	1.07	1.035	1.000	0.970

†RATINGS AT ASRE STANDARD CONDITIONS

*Reproduced courtesy of The Trane Company.

ble 10-20. By assuming that the load remains constant as figured in the preceding problem, the condensing temperature at 78°F wet bulb is found (by interpolation) to be approximately 102°F, at 75°F wet bulb to be approximately 100°F, and at 72°F wet bulb to be approximately 98°F, based on the use of the 20-ton evaporative condenser. This reduction in condensing temperature within certain limits is not objectionable as it has been pointed out that a lower head pressure is more favorable to economical compressor operation.

Air-Cooled Condensers. Prior to World War II, air-cooled condensers were limited primarily to small commercial refrigeration systems and room air conditioners. Following this period, the air-cooled condenser began to gain acceptance for small central residential and commercial air-conditioning equipment. Today they are commonplace and are being applied to installations of ever increasing size. Air-cooled condensers are easy to install, requiring only power and refrigerant connections. Maintenance is simple, and they do not have to be winterized in the fall and started up in the spring. Their primary disadvantage is that they usually must operate at higher condensing temperatures than water-cooled condensers or evaporative condensers to keep their physical size reasonable.

The higher condensing temperatures, of course, increase compressor horsepower input and increases operating costs. It should also be pointed out that the first cost of air-cooled condensers usually equals or exceeds that of an evaporative condenser of comparable capacity. Against these factors, one must consider potentially higher maintenance, water, and water treatment costs for evaporative condensers or water-cooled condensers used with cooling towers.

The circulation of air over the air-cooled coil may be provided by natural convection or forced convection produced by a fan. The natural-draft air-cooled condenser is limited to applications of relatively small capacity for the rate of heat removal is slow. Large surface areas are required for each unit of capacity. A common application is household refrigerators and freezers.

A typical forced-draft air-cooled condenser is shown in Fig. 10-37. This condenser used a propeller-type fan to force the air across the heat-transfer surfaces. Such condensers are available in sizes as large as 125 tons in a single unit. Multiple units may be used for higher tonnages, but it is always desirable to use a single condenser, if possible, because the piping for a multiple unit is more involved and requires greater care in design and installation.

Rating data are presented in two general ways by manufacturers of air condensers. One way is to present the condensing capacity at various temperature differences between entering air dry bulb and condensing. This data then is compared with compressor performance at the same condensing temperature, and the designer of the system determines the balance of the two components.

Recently, manufacturers have assumed the obligation of making this balance for components that they manufacture. By doing so, the benefit of subcooling in the condenser can be utilized to the advantage of the entire system. The resulting combinations of components thus matched will result in the lowest first cost as well as the lowest operating cost for normal air-conditioning applications. In Table 10-21 is found a nominal 30-ton compressor and two different size propeller fan condensers. (Ratings of centrifugal fan condensers is done in the same manner.) The use of Table 10-21 is illustrated in the following example.

Illustrative Problem 10-8

Given the following data:

Cooling load = 29.4 tons

Figure 10–37

Typical Air-Cooled Condenser with propeller type fan and automatic shutter

Outdoor design temperature = 95°F
Suction temperature at
compressor = 40°F to 45°F

Determine:

a. Condenser-compressor

b. actual suction temperature at compressor.

<u>Solution:</u>

a. Part a of the solution is obtained by the appropriate rating table for the load range, in this case Table 10-21.

b. Through interpolation, the actual suction temperature is found to be 41.7°F. Therefore, and HF230 compressor and a CA3008 condenser will produce the required capacity of 29.4 tons at 41.7°F suction temperature and 95°F ambient air temperature.

Where a manufacturer provides matched equipment combinations in published literature, they generally will have individual component data available on request so a system designer can do his own component balancing if desired.

10-18 SHELL-AND-TUBE HEAT EXCHANGERS (HEATING)

We have already discussed the use of the shell-and-tube heat exchanger used in refrigeration systems and have referred to them as

Table 10-21

Trane CA Condenser Rating*

COMPRESSOR MODEL HF-230 R-22			CONDENSER MODEL CA-3008				CONDENSER MODEL CA-4008			
AMBIENT TEMPERATURE			85°	95°	105°	115°	85°	95°	105°	115°
SUCTION TEMP.	20°	NET TONS	20.4	18.7	16.9	15.1	21.0	19.2	17.4	15.7
		KILOWATTS	22.7	23.6	24.7	25.1	22.1	23.1	24.1	24.8
	30°	NET TONS	25.3	23.5	21.5	19.4	26.3	24.3	22.2	20.2
		KILOWATTS	25.0	26.4	27.6	28.5	24.2	25.6	27.0	28.1
	40°	NET TONS	31.1	28.5	26.0	...	32.4	29.8	27.2	24.6
		KILOWATTS	27.6	29.3	30.9	...	26.1	28.1	29.8	31.5
	50°	NET TONS	36.5	33.8	31.3	...	37.9	35.3	32.6	...
		KILOWATTS	29.7	32.0	33.9	...	27.7	30.2	32.5	...

**Courtesy of The Trane Company.*

evaporators or condensers dependent upon their specific function in the refrigeration system.

Shell-and-tube heat exchangers are frequently used in heating systems. One type of heat exchanger is illustrated in Fig. 10-38. This type of unit is designed to heat water with steam. The water flows through the *U*-bent tubes while the steam condenses inside the shell and on the tubes. The shell, tube sheets, spacer plates, tie rods are normally made of steel. The *U*-tubes are normally of ¾-in. OD seamless copper and may serpentine through the length of the shell two, four, or six times (call two-pass, four-pass, and six-pass).

Another type of shell-and-tube heat exchanger used when there is a source of high temperature water is a water-to-water unit. These units are constructed in the same general way as shown in Fig. 10-38, except the inside of the shell has baffles spaced along its length to force the water flow in a definite path through the shell.

The usual installation arrangement is to have the medium to be heated passing through the tubes and the medium being cooled passing through the shell. The steam-to-water heat exchanger is an instantaneous-type heater (water enters the tubes at one temperature and leaves at the outlet at the desired final temperature). Most manufacturers' catalogs list ratings for the most commonly required temperature rises for the water through a wide range of steam pressures. They are available in two-, four- or six-pass construction and usually are cataloged for lengths up to 10 ft and shell diameters from 4 to 30 in.

The most common application for steam-to-water, shell-and-tube units is to use existing steam from a process or existing steam heating system to heat water for use in a auxilliary heating system, to heat domestic hot water and so forth.

For heating water with steam, the following conditions must be known: (a) water flow rate through the tubes, gpm; (b) water temperature in and out, °F; and (c) the steam condition in the heater shell. When selecting a steam-to-water heat exchanger a few miscellaneous hints may be in order, such as:

1. Water flow velocity through the tubes should be limited to less than 7.5 ft/sec, because:

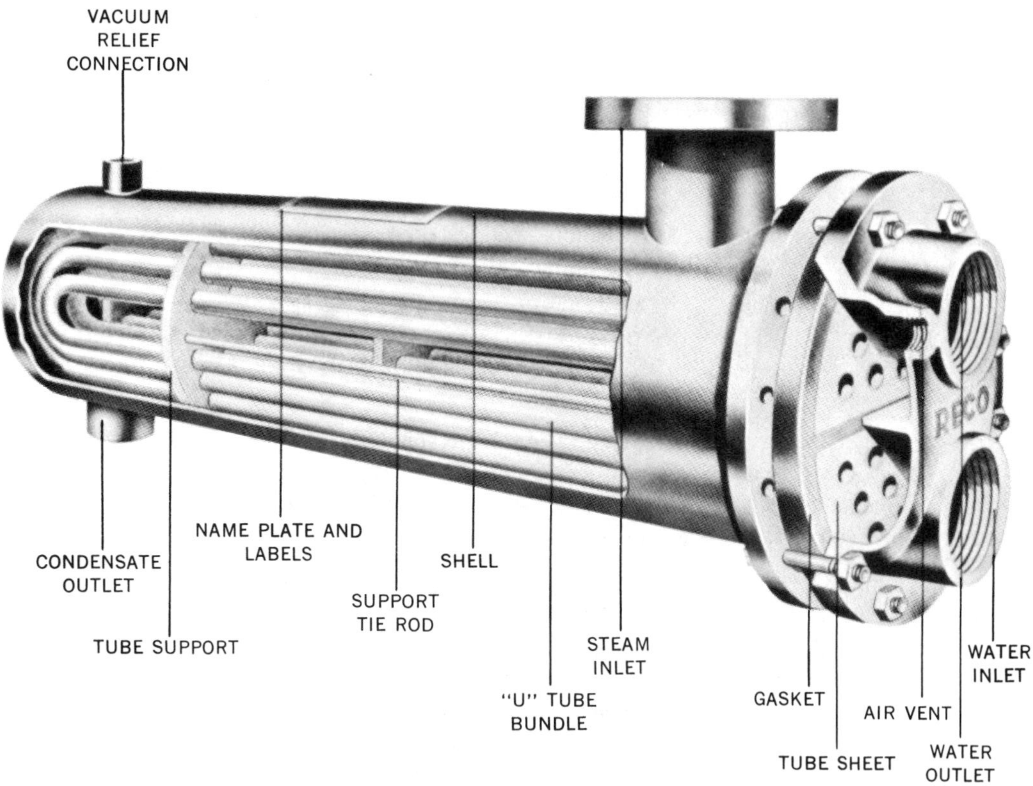

Figure 10–38

Steam-to-water, sheel-and-tube heat exchanger

a. Water velocities of 7.5 ft/sec and above can become erosive. Rapid wear of the tubing is the result.
b. Any small accumulation of scale in the unit that has been rated at high velocity causes a very sharp drop-off in heating capacity.
c. The high pressure drop resulting from very high velocities can make pump selections difficult and costly.

2. Fouling factors as set forth by TEMA (Tubular Exchanger Manufacturers Association) should always be considered. Water from different localities varies in mineral content. In the process of being heated, the minerals are precipitated in the form of lime, scale, and so forth. They then collect on the tube walls, and the ability of the unit to transfer heat is reduced. To offset a loss in heater capacity from fouling, the size of the heater should be increased so that after scale has collected, the unit will still operate at its rated capacity.

3. When using a steam-to-water heat exchanger where human contact with any part of the system is possible, insulation of the heater shell and piping should always be used. Also, a low steam pressure (15 psig or less) should be used.

CHAPTER 10

REVIEW

10-1. A hot water heating system in a building had installed radiation capacity of 119,000 Btu/hr. Domestic hot water requirements are 3 gpm of 140°F water. Select a boiler for the system.

(a) What is the piping and pickup factor?

(b) What is the overall efficiency using No. 2 fuel (heating value = 140,000 Btu/gal)?

10-2. An oil-fired boiler uses 2.1 gph of No. 2 fuel oil (heating value = 140,000 Btu/gal) to give a gross output of 220,000 Btu/hr. What efficiency of operation was obtained? If the net I-B-R rating was 145,000 Btu/hr, what total pickup and piping-loss factor was allowed?

10-3. A commercial building, which is to be heated continuously, has a calculated heat loss of 1,950,000 Btu/hr. The piping will be typical of such buildings. Select by number the boiler best suited to this installation using data from Table 10-1 and assuming forced draft.

(a) What reserve capacity will the unit have in percent of calculated load?

(b) What overall efficiency is indicated by the tabular data when the unit is fired with (1) gas, (2) No. 2 fuel oil, and (3) No. 5 oil?

10-4. Estimate the hourly heat loss per linear foot of 2 in. schedule 40 steel pipe covered with canvas-wrapped 85 percent magnesia insulation 1½ in. thick when 200°F water flows in the pipe and the surrounding air temperature is 70°F.

10-5. A 4-in. schedule 40 steel pipe carries steam at 400°F and is insulated with 2½ in. of 85 percent magnesia insulation, density 11 lbs/ft^3. What will be the estimated outside surface temperature of the insulation, if the surrounding air is at 80°F?

10-6. A space within a building has a heat loss of 8200 Btu/hr at design conditions. The heating system is forced circulation hot water with an average water temperature of 200°F. Select a convector to heat the space using Table 10-8. The convector height is limited to 26 in. maximum.

10-7. For the same data of Problem 10-6 how many linear feet of Model 75W-3 baseboard should be used? Assume (a) 1.0 gpm and (b) 4.0 gpm.

10-8. A space in a building has an exterior wall which is 32 ft long. The heat loss from the space has been calculated to be 48,000 Btu/hr. It has been determined that Trane Type S wall fin shall be used to provide the heating. Forced circulation hot water at an average temperature of 190°F is used in the system. Specify the size and type of heating element, enclosure height, and length to be used.

10-9. A cabinet heater is to be installed in the entrance of a building where the calculated heat loss is 30,000 Btu/hr at design conditions. Using Table 10-11, determine the unit size required, water flow rate, water pressure drop, and final air temperature assuming 180°F entering water temperature and 60°F entering air temperature.

10-10. A machine shop in an industrial building has a calculated heat loss of 120,000 Btu/hr. The decision of the heating engineer is to use Trane Model S unit heaters in the space. The forced circulating hot water system circulates 220°F water with a system water temperature drop of 30°F. Select the unit size and number of unit heaters to use and the capacity, in gpm, final air temperature, water pressure headdrop for standard wall tubes.

10-11. Air supplied to a building is to be cooled with a chilled water coil from 80°F/67°F to a leaving wet-bulb temperature of 56°F using chilled water entering the coil at 45°F. The building sensible heat load is 280,000 Btu/hr and the sensible heat ratio is 0.85. Inside design conditions are 78°F DB and 50 percent RH. Select a cooling coil from Table 10-14 for a 12-degree water temperature rise and a maximum coil face velocity of 500 fpm.

10-12. Using the same basic data as Problem 10-11, select a DX cooling coil from Table 10-15 for Freon-12. Specify the suction temperature, number of rows of coil, fin series and actual capacity, leaving dry-bulb and wet-bulb temperatures for 500 fpm face velocity.

10-13. Find the amount of heat in Btu/hr which must be rejected in the water-cooled condenser of a 30-ton system using Freon-12 if the following conditions are assumed: saturation temperature in the evaporator is 20°F, vapor leaves the evaporator with 10 degrees of superheat, saturation temperature in the condenser is 110°F, no subcooling of the liquid.

10-14. For the conditions of Problem 10-13, how much cooling water in gpm must be used through the condenser, if a 10-degree water temperature rise is assumed?

BIBLIOGRAPHY

1. ASHRAE *Handbook and Product Directory, 1975 Equipment*. American Society of Heating, Refrigerating and Air-Conditioning Engineers, Atlanta, GA.

2. *SBI Testing and Rating Code for Steel Boilers,* The Hydronics Institute, Berkeley Heights, N.J. 10th ed., November 1967.

3. *I=B=R Testing and Rating Code for Low Pressure Cast-Iron Heating Boilers,* The Hydronics Institute, Berkeley Heights, N.J. November, 1967.

4. *I=B=R Ratings for Cast-Iron Boilers,* The Hydronics Institute, Berkeley Heights, NJ. May 1978.

5. *Steel Boiler Ratings,* The Hydronics Institute, Berkeley Heights, NJ. May 1978.

6. *ASHRAE Standard Method of Testing for Rating Unit Ventilators* (ASHRAE Standard 71-73, 1973). American Society of Heating, Refrigerating and Air Conditioning Engineers, Atlanta, GA.

7. *Standard Code for Testing and Rating Steam Unit Heaters,* Air Moving and Conditioning Association, AMCA Bulletin No. 20, 1950.

8. *Standard Code for Testing and Rating Hot Water Unit Heaters* (Adopted jointly by The American Society of Heating and Ventilating Engineers and Industrial Unit Heater Association, IUHA Bulletin No. 11, April 1953, 2nd ed.).

9. Strock, C., and Koral, R. L., *Handbook of Air Conditioning, Heating, and Ventilating,* 2nd ed., The Industrial Press, 1965.

10. *Trane Air Conditioning Manual,* The Trane Company, LaCrosse, Wisconsin, 1965.

11. *The Trane Refrigeration Manual,* The Trane Company, LaCrosse, Wisconsin, 1966.

12. *Cooling and Heating Coils, Catalog D-Coil 1,* The Trane Company, LaCrosse, Wisconsin, 1973.

Chapter 11

Fluid Flow Fundamentals and Piping Systems

11-1 INTRODUCTION

This chapter deals with basic fluid mechanics, and our discussion will be limited primarily to the flow of fluids. Design of heating and cooling systems normally involves fluids that carry the heat to or from various heat-exchange devices. These fluids must be transported from one place to another in pipes and ducts. The proper design and layout of a piping system or a duct system involves consideration of the behavior of fluids as they flow through these conduits. Much of the design sizing of pipes and ducts are done by use of tables and charts which are convenient and save much valuable time. However, before such charts and tables are used, it is desirable to study and understand the basic principles of fluid flow through closed conduits. This should lead us to a more complete understanding of the charts and tables, and we should be able to use them with a clearer understanding of their worth and limitations.

11-2 FLUID MECHANICS

Fluid mechanics is the study of the physical behavior of fluids and fluid systems and the laws describing this behavior. In general, the behavior of fluids follows laws similar to those that describe the behavior of solid bodies, which are familiar from studies in statics and dynamics. The major difference lies in the interpretation that must be placed on the laws due to the nature of the substance with which we are dealing.

A *fluid* is a substance which has a definite mass and volume but has no definite shape. A fluid cannot sustain a shear stress under equilibrium conditions. To be more precise, there is one additional restriction that must be made to the foregoing statement to cover all possible cases. That is, a fluid has a definite mass and volume at constant pressure and temperature. There are two basic classes of fluids: *liquids* and *gases*.

Liquids are fluids which have definite volumes independent of the size of the container. Under conditions of constant temperature and pressure, liquids will assume the shape of the container and fill a part of it which is equal in volume to the volume of the liquid. Liquids are generally considered to be incompressible in air-conditioning work; that is, their volumes do not change with a change in pressure. The specific volume of a liquid *does* change appreciably with temperature, however. Therefore, we must be careful to account for this in our calculations. For instance, in a closed hot water heating system, if the system is initially filled with cold water and the water is heated to its operating temperature, the volume of water in the system may increase by 5 or 6 percent. Since the liquid is confined, excessive pressures would be encountered which could easily burst pipes or connections in the system. Liquids, when exposed to atmospheric pressure are said to have a "free surface."

Gases are fluids which are compressible. Gases will vary in volume to fill the vessel containing them. Unlike liquids, which have a definite volume for a given mass, the volume of a given mass of gas will change to fill the container. The behavior of gases was described generally by the gas laws presented previously. Gases cannot have a "free surface" which liquids can display. Although gases are compressible, they may be treated as incompressible in many problems in air-conditioning work because pressure and temperature changes may be small during a gas flow process.

11-3 FLUID PROPERTIES

Most of the fluid properties have been discussed in Chapter 1. These properties were specific volume, density, specific weight, internal energy, enthalpy, entropy, temperature, pressure, and specific gravity. One additional property of fluids is necessary in our study of fluid flow. This property is called *viscosity*.

11-4 VISCOSITY

Viscosity is that physical property of a fluid that determines its resistance to a shearing force. It is due primarily to the interaction between fluid molecules (cohesion and molecular momentum). Viscosity expresses the readiness with which the fluid flows when it is acted upon by an external force. The coefficient of absolute viscosity or, simply, the absolute viscosity of a fluid, is a measure of its resistance to internal deformation or shear. Molasses is a highly viscous fluid; water is comparatively much less viscous; and the viscosity of gases is quite small compared with water.

Fluids used in air-conditioning systems have viscosities which are predictable. Some fluids not commonly encountered have viscosities which depend upon previous working of the fluid. Printer's ink, wood pulp slurries, and catsup are examples of fluids possessing such thixotropic properties of viscosity.

Considerable confusion exists concerning the units used to express viscosity. Proper units must be used whenever substituting values of viscosity into formulas.

Absolute viscosity. (Symbol: μ). In the English Engineering System of units (pound-force, foot, second) the units of absolute viscosity are

$$\frac{\text{lbf sec}}{\text{ft}^2} = \frac{\text{slugs}}{\text{ft sec}} \quad \text{or} \quad \frac{\text{lbm}}{\text{ft sec}}$$

It should be recalled that 1 slug is equal to 32.174 lbm, therefore, 1 slug/ft sec is equal to 32.174 lbm/ft sec.

The most commonly used unit of absolute viscosity is in the cgs (centimeter, gram, second) system. The unit is called *poise*. The units would be

$$\mu \text{ (poise)} = \frac{\text{dyn sec}}{\text{cm}^2} = \frac{\text{gm}}{\text{cm sec}}$$

The poise (P) is a relatively large number, therefore, *centipoise* (cP) is frequently quoted in the literature. One centipoise is equal to 0.01 poise.

Conversion of the unit poise to SI units gives an equivalent term called *pascal-second* (Pa·s). One poise is equal to 0.1 Pa·s. The conversion can be made with the following SI unit equivalents.

1 pascal = 1 newton/square meter (N/m^2)

Therefore, 1 Pa·s = 1 $(N \cdot s)/m^2$

Summary of Units and Conversions (Absolute Viscosity)

Symbol: μ (mu)

Units: English Engineering lbf sec/ft^2, slugs/ft sec, and lbm/ft sec

cgs: gm/cm sec (poise)

SI: newton-sec/meter^2 ($N \cdot s/m^2$) = Pa·s

Conversion: To convert lbf sec/ft^2 to poise multiply by 478.8
To convert lbf sec/ft^2 to $N \cdot s/m^2$ multiply by 47.88
To convert lbm/ft sec to $N \cdot s/m^2$ multiply by 1.488
To convert centipoise to $(N \cdot s)/m^2$ multiply by 0.0001
To convert poise to $(N \cdot s)/m^2$ multiply by 0.1

Kinematic Viscosity (Symbol: ν). Kinematic viscosity is the ratio of the absolute viscosity to the mass density of the fluid.

In the English Engineering System of units we found that absolute viscosity, μ may have units of lbf sec/ft^2, slugs/ft sec, or lbm/ft sec. Therefore, the kinematic viscosity units would be defined as

$$\nu = \frac{\mu}{\rho} = \frac{\mu}{\gamma/g} = \frac{\mu \times g}{\gamma}$$

$$= \frac{(\text{lbf sec/ft}^2)(\text{ft/sec}^2)}{\text{lbf/ft}^3} = \frac{\text{ft}^2}{\text{sec}}$$

where γ, the specific weight, is equal to $\rho \times g$. Also

$$\nu = \frac{\mu}{\rho} = \frac{\text{slugs/ft sec}}{\text{slugs/ft}^3} = \text{ft}^2/\text{sec}$$

Also,

$$\nu = \frac{\mu}{\rho} = \frac{\text{lbm/ft sec}}{\text{lbm/ft}^3} = \text{ft}^2/\text{sec}$$

In the cgs system of units, the unit of kinematic viscosity is called *stoke*-(S) which has dimensions of cm^2/sec. Centistoke (cS) is frequently quoted, and 1 cS is equal to 0.01 S. Occasionally it is desirable to convert stoke to poise. It should be recognized that stoke (cm^2/sec) multiplied by the mass density (gm/cm^3) will yield the dimensions of gm/cm sec, which is poise. Mass density in gm/cm^3 is equal numerically to the fluid specific gravity (SG). Therefore,

$$\mu \text{ (poise)} = \nu \text{ (stokes)} \times \text{SG} \qquad (11\text{-}1)$$

In SI units, kinematic viscosity has units of m^2/s.

Summary of Units and Conversions (Kinematic Viscosity)

Symbol: ν (nu)

Units: English Engineering: ft^2/sec

cgs: cm^2/sec (stoke)

SI: m^2/s

Conversion: To convert ft^2/sec to stoke multiply by $(30.48)^2$
To convert ft^2/sec to m^2/s multiply by 0.0929
To convert stoke to m^2/s multiply by 0.0001
To convert centistoke to m^2/s multiply by 0.000001

Illustrative Problem 11-1

At 100°F the absolute viscosity of water is 0.680 cP. Express the absolute viscosity in (a) English Engineering Units and (b) SI Units.

Solution: (a) 0.680 cP = 0.0068 P

$$\mu\ (\text{lbf sec/ft}^2) = 0.0068/478.8 = 1.42 \times 10^{-5}$$ *(ans)*

(b) 1 P = 0.1 Pa·s

$$\mu\ (\text{Pa·s}) = (0.0068)(0.1)\ \text{Pa·s} = 0.00068\ \text{Pa·s}$$ *(ans)*

Illustrative Problem 11-2

The density of water at 100°F is 61.996 lbm/ft². Determine its kinematic viscosity in (a) ft²/sec, (b) stokes, and (c) m²/sec.

Solution: (a)

$$SG_{100} = \frac{\rho_{100}}{\rho_{60}} = \frac{61.996}{62.371} = 0.994$$

By Eq. (11-1) μ (poise) = ν (stokes) × SG
From Problem 11-1, μ (poise) = 0.0068 P

Substituting

$$\nu\ (\text{stokes}) = \frac{0.0068}{0.994} = 0.00684\ \text{S}$$ *(ans)*

(b)

$$\nu\ (\text{ft}^2/\text{sec}) = \frac{\nu\ (\text{stokes})}{(30.48)^2} = \frac{0.00684}{(30.48)^2} = 7.36 \times 10^{-6}\ \text{ft}^2/\text{sec}$$ *(ans)*

$$\nu\ (\text{m}^2/\text{s}) = \nu\ (\text{stokes}) \times 0.0001$$

$$\nu\ (\text{m}^2/\text{s}) = (0.00684)(0.0001) = 6.84 \times 10^{-7}\ \text{m}^2/\text{s}$$ *(ans)*

Saybolt Universal Viscosity. The measurement of the absolute viscosity of fluids (especially gases and vapors) requires elaborate equipment and considerable experimental skill. On the other hand a rather simple instrument can be used for measuring the kinematic viscosity of oils and other viscous liquids. The instrument adopted as a standard in this country is the Saybolt Universal Viscosimeter. In measuring kinematic viscosity with this instrument, the time required for a small volume of liquid (60 cm³) to pass through a calibrated orifice at a specified temperature is determined. (Viscosity is a measure of the "flowability" at a given temperature.) The elapsed time is measured in seconds with the fluid sample maintained at a constant temperature, usually at 100°F or 210°F. The elapsed time in seconds is the Saybolt seconds universal (SSU) viscosity at the particular temperature.

If the elapsed time, t, is between 32 and 100 sec for a control temperature of 100°F, conversion to other viscosity units is as follows:

$$\nu\ (\text{centistokes}) = 0.226t - \frac{195}{t} \qquad (11\text{-}2)$$

If the elapsed time, t, is greater than 100 sec for a control temperature of 100°F,

$$\nu\ (\text{centistokes}) = 0.220t - \frac{135}{t} \qquad (11\text{-}3)$$

In Eqs. (11-2) and (11-3), t is the Saybolt seconds universal (SSU) obtained by test.

Illustrative Problem 11-3

The viscosity of a fluid measured in a Saybolt Viscosimeter is 140 sec at 100°F, and the specific gravity measured at 60°F is 0.875. Determine the fluid viscosity in (a) centistokes, (b) centipoise, (c) ft²/sec, (d) lbf sec/ft², (e) pascal-second, at a temperature of 100°F.

Solution:

(a) Since elapsed time (SSU) is greater than 100 sec, use Eq. (11-3).

$$\nu \text{ (centistokes)} = 0.220t - \frac{135}{t}$$

$$= (0.220)(140) - \frac{135}{140}$$

$$\nu = 30.8 - 0.964 = 29.84 \text{ cS} \qquad \textit{(ans)}$$

(b) Specific gravity of the fluid at 100°F may be found from Eq. (1-19).

$$SG_t = SG_{60/60} - 0.00035(t - 60)$$

$$SG_{100} = 0.875 - 0.00035(100 - 60)$$

$$SG_{100} = 0.861$$

From Eq. (11-1)

$$\mu \text{ (centipoise)} = \nu \text{ (centistokes)} \times SG$$

$$= (29.84)(0.861)$$

$$\mu \text{ (centipoise)} = 25.69 \text{ cP} \qquad \textit{(ans)}$$

(c) $(\text{ft}^2/\text{sec})(30.48)^2 = \text{stokes}$

or

$$\nu \text{ (ft}^2\text{/sec)} = \frac{29.84/100}{(30.48)^2}$$

$$\nu = 0.000321$$

$$= 3.21 \times 10^{-4} \text{ ft}^2\text{/sec} \qquad \textit{(ans)}$$

(d) $(\text{lbf sec/ft}^2)(478.8) = \text{poise}$

or

$$\mu \text{ (lbf sec/ft}^2) = \frac{\text{poise}}{478.8} = \frac{25.69/100}{478.8}$$

$$\mu = 0.000537$$

$$= 5.37 \times 10^{-4} \text{ lbf sec/ft}^2$$

(ans)

(e) $1 \text{ Pa·s} = 1 \text{ (N·s)/m}^2$

$$(\text{lbf sec/ft}^2)(47.88) = (\text{N·s})/\text{m}^2$$

$$\text{N·s/m}^2 = (5.37 \times 10^{-4})(47.88)$$

$$\text{N·s/m}^2 = 2.57 \times 10^{-2} \text{ Pa·s} \qquad \textit{(ans)}$$

11-5 FLUID FLOW

In general there are two types of fluid flow: (1) free, natural, or gravity flow and (2) forced flow. Fluid flow is further subdivided into three classifications: (a) viscous, laminar, or streamlined flow, (b) turbulent flow, and (c) supersonic flow. Supersonic flow does not occur as yet in air-conditioning systems.

Gravity flow is the type of flow caused by an unbalanced difference in density of portions of a fluid that are attempting to find their equilibrium level. This unbalance may result from an increase in density of an upper portion of a fluid due to cooling, or a decrease in density of the lower portion of a fluid due to heating. Cooling or heating of the fluid occurs when the fluid is in contact with a cool or warm surface.

A common example of free flow occurs surrounding a heating element such as fin pipe. Air in contact with the heated pipe and fins is heated, which decreases its density. The heated air, less dense than the air in the surroundings, rises and is replaced by the more dense air in the surroundings. The resulting "draft" can be appreciable under good conditions. Gravity flow may be either laminar or turbulent.

Forced flow occurs when a fluid is forced to flow by some mechanical means, such as a fan or a pump, rather than by thermal means. Forced flow may be either laminar or turbulent. Turbulent flow is most common in air-conditioning systems.

Laminar flow (viscous, streamline) implies that all portions of the fluid move in paths parallel to each other and parallel to the walls of confining surfaces, if any. When heat is being transferred to or from the fluid flowing, natural convection currents within the fluid have a tendency to distort the streamline pattern. In such cases the flow is classified as nonisothermal or modified laminar flow.

Figure 11-1a represents the "velocity profile" for laminar flow of a fluid through a pipe. It may be seen that the velocity varies from maximum (V_{max}) at the center of the pipe to zero velocity at the pipe walls. In fluid flow work we are concerned with *average flow velocity* (V_{av}), which is used in the continuity equation. For laminar flow in a pipe, $V_{av} = \frac{1}{2}V_{max}$. Laminar flow may occur in some air-conditioning equipment and also may be encountered when handling viscous fluids such as fuel oil flowing in a pipe.

Turbulent flow is the most common type of flow encountered in practice. In turbulent flow there is random motion of the fluid particles in directions transverse to the direction of the main flow. The velocity distribution in turbulent flow is more uniform across a pipe diameter than in laminar flow. This may be observed in Fig. 11-1b. Even though a turbulent motion of the fluid exists throughout the greater portion of the pipe diameter, there is always a thin layer of fluid at the pipe wall, known as the "boundary layer" or "laminar sublayer" which is moving in laminar flow.

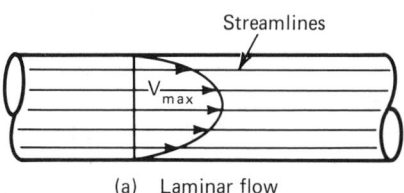

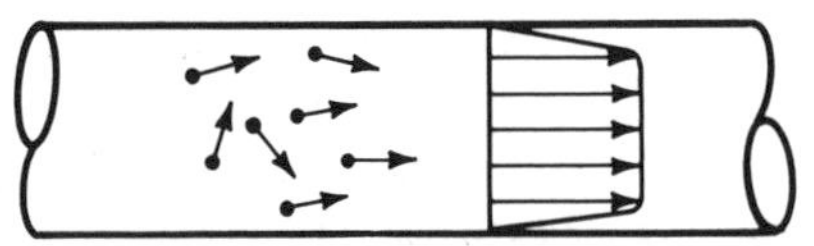

Figure 11–1

Laminar and Turbulent Flow in a Pipe

The average velocity in turbulent flow cannot be found by a simple relationship as for laminar flow. It is normally determined from the continuity equation first presented in Chapter 2, Eq. (2-5).

$$\dot{m} = \rho AV \tag{2-5}$$

Repeated

where

$\dot{m}$ = mass flow rate (lbm/sec or kg/s)

A = cross-sectional flow area (ft^2 or m^2)

V = average flow velocity (ft/sec or m/s)

When the fluid is incompressible, Eq. (2-5) becomes

$$\dot{Q} = AV \tag{11-4}$$

where $\dot{Q}$ = volume flow rate (ft^3/sec or m^3/s).

Frequently, the flow rate of a liquid is given in gallons per minute (gpm), the flow cross-sectional area in in.^{2}, or the diameter of flow path in inches. It is convenient to recognize that, using appropriate conversion units, Eq. (11-4) may be rearranged as

$$V = \frac{0.408 \times \text{gpm}}{d^2}$$

or

$$V = \frac{0.3208 \times \text{gpm}}{a} \tag{11-5}$$

where

V = average flow velocity (ft/sec)

gpm = gallons per minute

d = internal pipe diameter (in.)

a = internal pipe flow area (in.^{2})

11-6 REYNOLDS NUMBER

The work of Osborne Reynolds (1842–1912) has shown that the nature of fluid flow, that is, whether it is laminar or turbulent, depends upon the ratio of the inertia forces to the viscous forces, and may be defined as

Reynolds Number

$$(N_R) = \frac{\text{velocity} \times \text{diameter} \times \text{density}}{\text{absolute viscosity}}$$

which is a dimensionless combination of the four variables. In symbol form, Reynolds number becomes

$$N_R = \frac{VD\rho}{\mu} = \frac{VD}{\mu/\rho}$$

$$= \frac{VD}{\nu} \quad \text{(dimensionless ratio)} \qquad (11\text{-}6)$$

where

V = average flow velocity (ft/sec or m/sec)

D = characteristic dimension of system (length of surface, diameter of pipe, hydraulic radius) (ft or m)

ρ = fluid mass density (lbm/ft^3 or kg/m^3)

μ = absolute fluid viscosity [lbm/ft sec or Pa·s (N·s/m^2) or, if ρ is in slugs/ft^3, then μ must be in slugs/ft sec]

ν = kinematic viscosity (ft^2/sec or m^2/s)

Caution must be used when substituting into Eq. (11-6) to use units to produce a dimensionless ratio.

For engineering work flow in conduits is usually considered to be laminar if N_R is less than 2000 and turbulent if N_R is greater than 4000. Between the two values lies the "critical zone" where the flow may be laminar, turbulent, or a combination of both.

Reynolds number may also be calculated from the following relationships using different, and sometimes more convenient, units in the ratio.

$$N_r = \frac{123.9dV\rho}{\mu} = \frac{7740dV}{\nu} = \frac{50.6(\text{gpm})(\rho)}{d\mu} \qquad (11\text{-}7)$$

where

d = inside diameter of the pipe (in.)

V = flow velocity (ft/sec)

ρ = density (lbm/ft^3)

μ = absolute viscosity (centipoise)

ν = kinematic viscosity (centistokes)

gpm = gallons per minute flow rate

Occasionally a conduit of noncircular cross section is encountered or a conduit partially filled with flowing fluid. To calculate Reynolds number for these conditions, an equivalent diameter (four times the hydraulic radius) is substituted for the dimension D. The hydraulic radius is defined as

Hydraulic radius (R_H)

$$= \frac{\text{cross-sectional flow area in ft}^2 \text{ or m}^2}{\text{wetted perimeter in ft or m}}$$

This applies to any ordinary conduit (circular conduit not flowing full, oval, square, or rectangular) but not to extremely narrow shapes, such as annular or elongated openings, where the width is small relative to the length. In such cases the hydraulic radius is approximately equal to one-half the width of the passage.

11-7 THE STEADY-FLOW ENERGY EQUATION

In Chapter 2, the steady-flow energy equation was discussed and is repeated here with the heat and work terms omitted.

$$\left(\frac{g}{g_c}\right)Z_1 + p_1v_1 + \frac{V_1^2}{2g_c} + JU_1 = \left(\frac{g}{g_c}\right)Z_2 + p_2v_2 + \frac{V_2^2}{2g_c} + JU_2$$

The change in internal energy $J(U_2 - U_1)$ is usually very small in fluid flow work, and letting $v = 1/\rho$ = a constant (incompressible flow), we would have

$$\left(\frac{g}{g_c}\right)Z_1 + \frac{P_1}{\rho} + \frac{V_1^2}{2g_c} = \left(\frac{g}{g_c}\right)Z_2 + \frac{P_2}{\rho} + \frac{V_2^2}{2g_c} \qquad (11\text{-}8)$$

Each term in Eq. (11-8) has units of ft lbf/lbm or J/kg, or *energy per unit mass*.

If each term of Eq. (11-8) is multiplied by the ratio g_c/g(lbf/lbm), we would have

$$Z_1 + \frac{g_cP_1}{\rho g} + \frac{V_1^2}{2g} = Z_2 + \frac{g_cP_2}{\rho g} + \frac{V_2^2}{2g} \qquad (11\text{-}9)$$

Since $\rho(g/g_c) = \gamma$ the specific weight, Eq. (11-9) frequently appears as

$$Z_1 + \frac{P_1}{\gamma} + \frac{V_1^2}{2g} = Z_2 + \frac{P_2}{\gamma} + \frac{V_2^2}{2g} \qquad (11\text{-}9a)$$

Each term in Eqs. (11-9) has units of ft lbf/lbf, or head (ft), and is *energy per unit weight.* This form of the steady-flow equation is Bernoulli's equation, frequently used in the solution of many problems involving incompressible fluid flow. The three terms on each side of the equation are known respectively as static elevation head, static pressure head, and velocity head.

With the flow of real fluids, losses due to fluid friction must be accounted for. Therefore, Eq. (11-9) must include a term to represent these losses, as follows:

$$Z_1 + \frac{P_1}{\gamma_1} + \frac{V_1^2}{2g} = Z_2 + \frac{P_2}{\gamma_2} + \frac{V_2^2}{2g} + h_L \qquad (11\text{-}10)$$

where h_L is the *lost head.*

Figure 11-2 represents a section of pipe through which a fluid is flowing under steady-flow conditions. Since Eqs. (11-9) and (11-10)

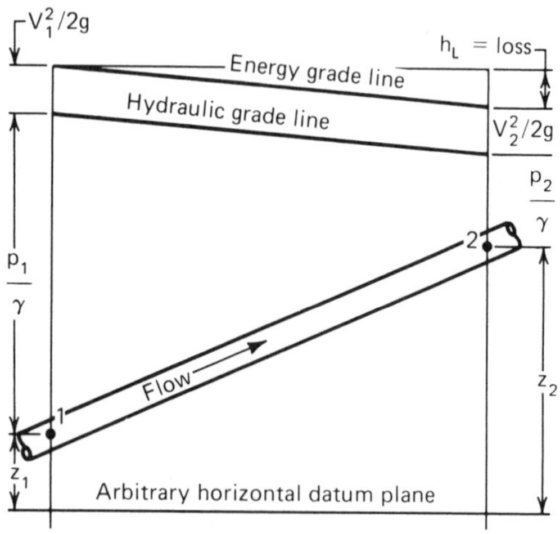

Figure 11–2

Steady flow of fluid through pipe section

represent energy levels at points 1 and 2, in units of feet (or meters) of fluid, it is convenient to represent graphically the change in energy levels as fluid flows through the pipe. A vertical scale may be established to represent these energy levels above a conveniently selected datum plane, and values of Z, P/γ, $V^2/2g$, and h_L may be plotted above the datum plane.

The sum $(Z + P/\gamma)$ plots the *hydraulic grade line*, and the sum $(Z + P/\gamma + V^2/2g)$ plots the *energy grade line*. The energy grade line represents graphically the available energy in the flow stream at various points between section 1 and 2, and always slopes downward in the direction of fluid flow. The loss term, h_L, represents the loss of energy due to friction.

The various terms in Eq. (11-10) are frequently grouped and referred to by particular names.

Elevation head change $= Z_2 - Z_1$

Pressure head change $= \dfrac{P_2 - P_1}{\gamma}$

($\gamma_1 = \gamma_2$ for incompressible flow)

Velocity head change $= \dfrac{V_2^2 - V_1^2}{2g}$

Head loss $= h_L$

Illustrative Problem 11-4

If water is flowing through the section of pipe of Fig. 11-3 at the rate of 100 gpm and the static pressure at point 1 is 100 psi, what is the static pressure at point 2. The inside diameter at point 1 is 2.5 in. and at point 2 is 1.5 in. Assume lost head h_L is 3 ft of water.

Solution:

By Eq. (11-5) $V_1 = \dfrac{0.408 \text{ gpm}}{d_1^2} = \dfrac{0.408(100)}{(2.5)^2}$

$= 6.528$ ft/sec

$V_2 = \dfrac{0.408 \text{ gpm}}{d_2^2} = \dfrac{0.408(100)}{(1.5)^2}$

$= 18.133$ ft/sec

By Eq. (11-10) $Z_1 + \dfrac{P_1}{\gamma} + \dfrac{V_1^2}{2g} = Z_2 + \dfrac{P_2}{\gamma} + \dfrac{V_2^2}{2g} + h_L$

Since the datum plane is selected at the pipe centerline and the pipe is horizontal, $Z_1 = Z_2$ and are canceled from the equation. By substitution

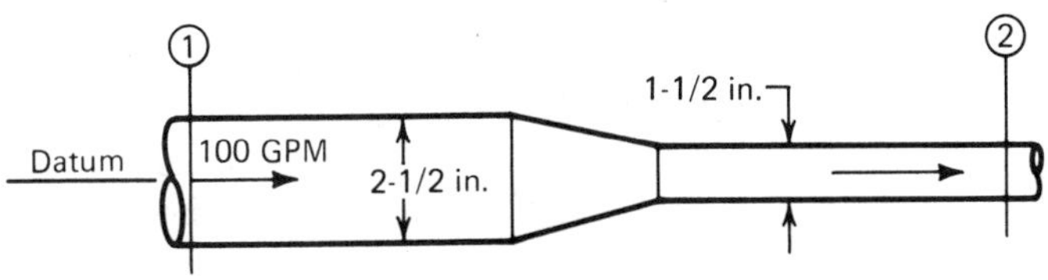

Figure 11–3

Illustration for Problem 11-4

$$\frac{144(100)}{62.4} + \frac{(6.528)^2}{2(32.2)} = \frac{144P_2}{62.4} + \frac{(18.133)^2}{2(32.2)} + 3$$

$$230.77 + 0.66 = 2.31p_2 + 5.11 + 3$$

$$p_2 = 96.7 \text{ psi} \qquad \textbf{\textit{(ans)}}$$

Note in the above calculations the magnitude of the $V^2/2g$ terms in comparison to p/γ terms. In some cases velocity head terms may be omitted in calculations because of their low value. However, it is recommended that they always be included in calculations to determine if they are of importance.

11-8 FRICTION LOSSES IN A CLOSED CONDUIT

Flow of real fluids in a conduit is always accompanied by friction, which represents a loss of available energy; or we may say, there is a loss in pressure in the direction of fluid flow. This loss is expressed in Eq. (11-10) as h_L. The general equation for calculating this loss is known as the *Darcy equation*. It is an empirical equation, meaning that it is based on the results of tests rather than theory. The equation may be expressed in feet (or meters of fluid flowing or in lbf/ft^2 (or N/m^2) by use of Eqs. (11-11) and (11-12), respectively. For incompressible flow in closed conduits

$$h_L = f\left(\frac{L}{D}\right)\left(\frac{V^2}{2g}\right) \quad \text{(ft or m)} \qquad \textit{(11-11)}$$

$$\Delta p_L = f\left(\frac{L}{D}\right)\left(\frac{\gamma V^2}{2g}\right) \quad (\text{lbf/ft}^2 \text{ or N/m}^2) \qquad \textit{(11-12)}$$

where

h_L = head loss due to fluid friction (ft or m)

Δp_L = pressure loss due to fluid friction (lbf/ft^2 or N/m^2)

f = friction factor (dimensionless)

L = length of conduit (ft or m)

D = conduit inside diameter (ft or m)

V = mean flow velocity (ft/sec or m/s)

γ = specific weight of the fluid (lbf/ft^3 or N/m^3)

g = acceleration of gravity (ft/sec^2 or m/s^2)

The Darcy equation is valid for laminar or turbulent flow of any liquid in a closed conduit of cross-sectional diameter D. However, at high velocities with warm liquids flowing in a pipe, the downstream pressure may drop to the vapor pressure of the liquid. This may cause a phenomenon called *cavitation* to occur and calculated flow rates will be inaccurate. Cavitation may be described as the rapid formation and collapse of vapor bubbles in the flowing stream. It will be discussed in more detail in the chapter on pumps.

The Darcy equation may be used, with suitable restrictsions, for the flow of gases and vapors (compressible fluids) in closed conduits. Equations (11-11) and (11-12) give the head loss and pressure loss, respectively, due to fluid friction in a closed conduit of constant diameter carrying fluids of reasonably constant specific weight in straight pipe, whether horizontal, vertical, or sloping. For inclined conduits, vertical conduits, or conduits of varying diameter, the change in pressure due to changes in elevation, velocity, and specific weight of the fluid must be accounted for with Eq. (11-10).

The Darcy equation may be developed from the methods of dimensional analysis, with the exception of the *friction factor, f*. The friction factor must be determined experimentally. Numerous investigators over past years have determined that, for laminar flow (N_R less than 2000), the friction factor is a function of Reynolds number only; but, for turbulent flow (N_R greater than 4000), the friction factor depends upon Reynolds number and also the surface roughness of the conduit. In the region known as the "critical zone" (N_R between 2000 and 4000), the friction factor is indeterminate. It depends on various factors such as changes in pipe section or direction of flow and obstructions such as valves and fittings in the upstream piping.

For laminar flow the friction factor f is

$$f = 64/N_R \qquad \textit{(11-13)}$$

If appropriate units are used for N_R and the value of f resulting from Eq. (11-13) are substituted into Eq. (11-12), we would have

$$\Delta p_L = 0.000668 \frac{\mu L V}{d^2} \quad \text{(laminar flow)} \qquad \textit{(11-14)}$$

where

Δp_L = pressure loss (lbf/in.^{2})

μ = absolute viscosity (centipoise)

L = conduit length (ft)

V = average flow velocity (ft/sec)

d = internal pipe diameter (in.)

When flow is turbulent (N_R greater than 4000), flow conditions become quite stable and the friction factor depends upon Reynolds number and the *relative roughness* (ϵ/D) of the pipe surface. The term ϵ/D is the absolute surface roughness divided by the internal pipe diameter, both expressed in feet. For very smooth pipes, such as drawn tubing, the friction factor decreases more rapidly with increasing Reynolds number than for pipe with a comparatively rough surface. Since the character of the internal surface of commercial pipe is practically independent of the diameter, the roughness of the pipe walls has a greater effect on the friction factor in the smaller pipe sizes. Consequently, pipe of small diameter will approach the very rough condition, and, in general, will have higher friction factors than large pipe of the same material.

The most useful and widely accepted data of friction factors for use with the Darcy equation has been presented by L. F. Moody, and are reproduced here as Figs. 11-4 and 11-5. Figure 11-4 shows the relative roughness of pipe of various materials and inside diameters. It also indicates the friction factor, f, for complete turbulence.

Figure 11-5 shows a plot of the friction factor against Reynolds number for laminar flow and for turbulent flow for pipe of various relative roughness conditions. The friction factor is normally determined by the following procedure:

1. Calculate Reynolds number (N_R).
2. Calculate the relative roughness (ϵ/D).
3. Using ϵ/D, enter Fig. 11-5 at the right, and approximate a curve of relative roughness parallel to an existing curve.
4. Follow the curve of ϵ/D to the left to intersect with the vertical line for the calculated N_R.
5. From the point of intersection of ϵ/D and N_R, move horizontally to the left and read the friction factor f.

Use of Figs. 11-4 and 11-5 will be illustrated by the solution of three common *classes* of problems involving flow in pipes.

Illustrative Problem 11-5 (Class I)

In this type of problem the head loss or pressure loss is to be determined when the

Relative Roughness of Pipe Materials and Friction Factors For Complete Turbulence

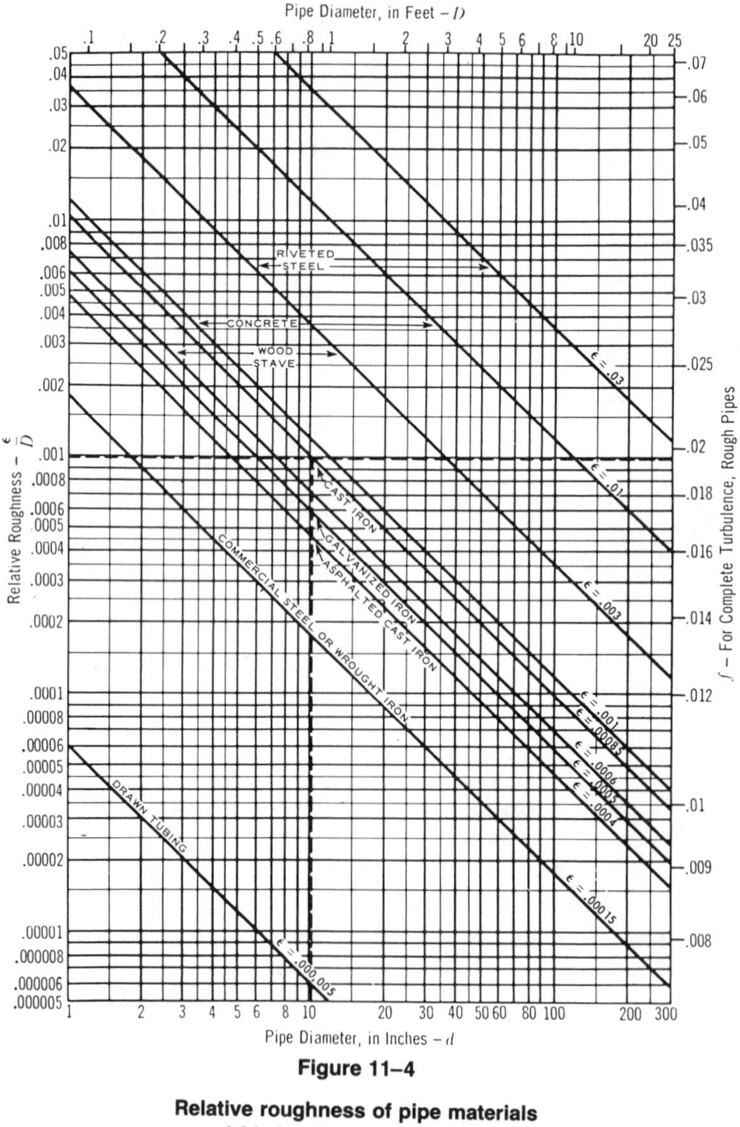

Figure 11–4

Relative roughness of pipe materials and friction factors for complete turbulence.

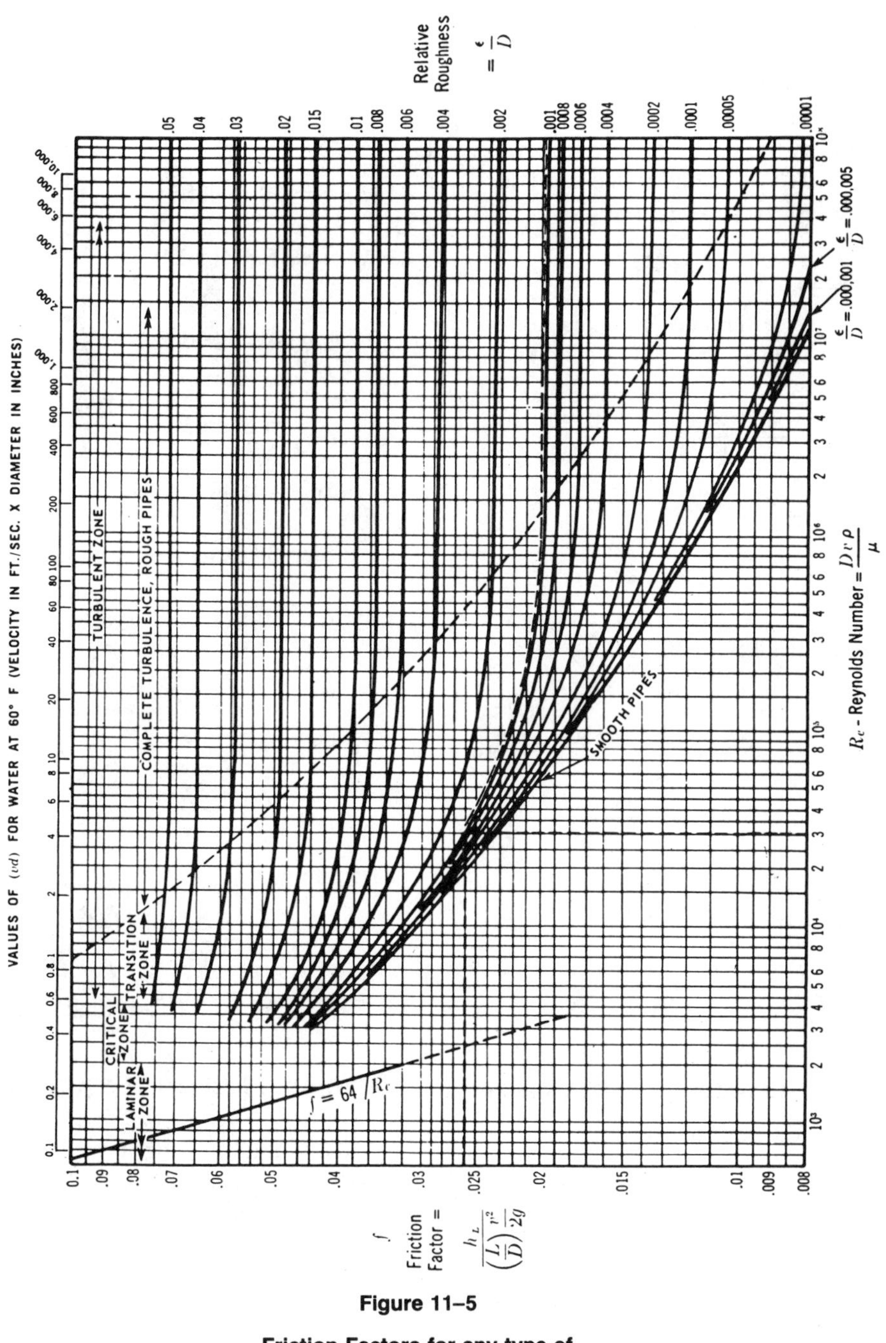

Figure 11–5

Friction Factors for any type of commercial pipe

quantity of flow, length of pipe, size of pipe, pipe material, and fluid viscosity are known.

Problem statement: A section of 2-in. Schedule 40 commercial steel pipe is 50 ft long and carries 20 gpm of fluid having a specific gravity of 0.85 and a viscosity of 150 SSU at 100°F. Determine the head loss and pressure loss in the section of pipe.

Solution:

Step 1: Determine Reynolds number for the flow. By Eq. (11-5)

$$V = \frac{0.408 \times \text{gpm}}{d^2} = \frac{(0.408)(20)}{(2.067)^2}$$

$$V = 1.910 \text{ ft/sec}$$

By Eq. (11-3)

$$\nu \text{ (centistokes)} = 0.220t - \frac{135}{t}$$

$$\nu = 0.220(150) - \frac{135}{150}$$

$$\nu = 32.1 \text{ cS}$$

By Eq. (11-7)

$$N_R = \frac{7740dV}{\nu}$$

$$= \frac{(7740)(2.067)(1.910)}{32.1}$$

$N_R = 952$ (less than 2000, therefore flow is laminar)

Step 2: Determine pressure loss. By Eq. (11-13), for laminar flow,

$$f = 64/N_R = 64/952 = 0.0672$$

By Eq. (11-12)

$$\Delta p_L = f\left(\frac{L}{D}\right)\left(\frac{\gamma V^2}{2g}\right)$$

$$= (0.0672)\left(\frac{50}{2.067/12}\right)\left[\frac{(62.4)(0.85)(1.91)^2}{2(32.2)}\right]$$

$$\Delta p_L = 58.61 \text{ lbf/ft}^2 = 0.407 \text{ lbf/in.}^2 \qquad \textit{(ans)}$$

Also, by Eq. (11-14)

$$\Delta p_L = 0.000668 \frac{\mu L V}{d^2}$$

$$= 0.00668 \frac{(32.1)(0.85)(50)(1.91)}{(2.067)^2}$$

$$\Delta p_L = 0.407 \text{ lbf/in.}^2$$

Step 3: Determine head loss.

By Eq. (11-11)

$$h_L = f\left(\frac{L}{D}\right)\left(\frac{V^2}{2g}\right)$$

or, note that

$$h_L = \frac{\Delta p_L}{\gamma}$$

Therefore,

$$h_L = \frac{(0.407)(144)}{(62.4)(0.85)}$$

$$h_L = 1.1 \text{ ft of fluid} \qquad \textit{(ans)}$$

Illustrative Problem 11-6 (Class I)

If the fluid flowing in the pipe of Problem 11-5 was water at 60°F with a viscosity of 1.15 cP, determine the pressure loss and head loss.

Solution:

Step 1: Determine Reynolds number.

$V = 1.910$ ft/sec (from before)

By Eq. (11-7)

$$N_R = \frac{123.9 dV\rho}{\mu}$$

$$= \frac{(123.9)(2.067)(1.910)(62.4)}{1.15}$$

$N_R = 26{,}540$ (greater than 4000, therefore flow is turbulent)

Step 2: Determine friction factor.

From Fig. 11-4 find ϵ/D for 2-in., commercial steel pipe to be 0.0009.
On Fig. 11-5 find point of intersection of $\epsilon/D = 0.0009$ and $N_R = 2.654 \times 10^4$ and read $f = 0.026$.

Step 3: Determine head loss and pressure loss.

By Eq. (11-11)

$$h_L = f\left(\frac{L}{D}\right)\left(\frac{V^2}{2g}\right)$$

$$= (0.026)\left(\frac{50}{2.067/12}\right)\left[\frac{(1.910)^2}{2(32.2)}\right]$$

$h_L = 0.428$ ft of water *(ans)*

$$\Delta p_L = \gamma \times h_L = (62.4)(0.428)$$

$\Delta p_L = 26.71 \text{ lbf/ft}^2 = 0.185 \text{ lbf/in}^2$ *(ans)*

or, use Eq. (11-12) directly.

Illustrative Problem 11-7 (Class II)

In this type of problem the flow quantity is to be determined when the head loss or pressure loss, pipe size and material, and fluid properties are known. This type of problem requires a trial-and-error type of solution because the flow velocity and friction factor are unknown.

Problem statement: Water of 60°F flows through a 12-in. inside diameter riveted steel pipe ($\epsilon = 0.01$) with a measured head loss of 40 ft of water in a section of pipe 1000 ft long. Estimate the flow rate through the pipe in ft^3/sec if the viscosity is assumed to be 2.35×10^{-5} lbf sec/ft^2.

Solution:
Procedure:

Step 1: Solve for relative roughness ϵ/D and assume a trial value of f from Fig. 11-4, assuming complete turbulence.

Step 2: Substitute trial value of f in Eq. (11-11) and solve for velocity V.

Step 3: Determine Reynolds number.

Step 4: With this Reynolds number and ϵ/D, find f from Fig. 11-5.

Step 5: If f does not agree with the value assumed in step 1, use new value of f, and repeat steps 1 through 4. Repeat until f does not change.

Step 6: Calculate Q using Eq. (11-4).

1. $\epsilon/D = 0.01/1 = 0.01$. From Fig. 11-4, assume f for complete turbulence is 0.038.

2. By Eq. (11-11)

$$h_L = f\left(\frac{L}{D}\right)\left(\frac{V^2}{2g}\right)$$

$$40 = (0.038)\left(\frac{1000}{1}\right)\left(\frac{V^2}{2 \times 32.2}\right)$$

$$V = 8.23 \text{ ft/sec}$$

3. By Eq. (11-6)

$$N_R = \frac{VD\rho}{\mu} = \frac{(8.23)(1)(62.4/32.2)}{2.35 \times 10^{-5}}$$

$$N_R = 6.79 \times 10^5 \quad \text{(turbulent)}$$

4. With $N_R = 6.79 \times 10^5$ and $\epsilon/D = 0.01$ enter Fig. 11-5 for an improved value of f. As may be seen, f is in the zone of complete turbulence; therefore, $f = 0.038$ and the flow velocity is 8.23 ft/sec found in step 2.
5. Not required.
6. By Eq. (11-4)

$$\dot{Q} = AV$$

$$= \frac{\pi D^2 V}{4}$$

$$= \frac{\pi(1)^2(8.23)}{4}$$

$$\dot{Q} = 6.46 \text{ ft}^3\text{/sec} \qquad \textit{(ans)}$$

Illustrative Problem 11-8 (Class III)

In this type of problem the pipe inside diameter is to be determined if the head loss or pressure loss, pipe material, flow quantity, and fluid properties are known.

This type of problem requires a trial-and-error solution because D, V, and f are all unknown. The solution of such a problem is sometimes made easier by recognizing the following relationships.

By Eq. (11-11) $$h_L = f\left(\frac{L}{D}\right)\left(\frac{V^2}{2g}\right)$$

By Eq. (11-4) $$V^2 = \frac{\dot{Q}^2}{A^2} = \frac{\dot{Q}^2}{\left(\frac{\pi D^2}{4}\right)^2}$$

Substituting into Eq. (11-11) we have

$$h_L = f\left(\frac{L}{D}\right)\left[\frac{\dot{Q}^2}{\left(\frac{\pi D^2}{4}\right)^2(2g)}\right]$$

Rearranging and simplifying

$$D^5 = \left[\frac{8L\dot{Q}^2}{h_L g \pi^2}\right]f$$

All terms within the brackets are constant for a given problem; therefore, we may say

$$D^5 = C_1 \times f \qquad \textit{(a)}$$

By Eq. (11-6) $$N_R = \frac{VD}{\nu}$$

From Eq. (11-4) $$\dot{Q} = \frac{\pi D^2 V}{4}$$

or $$V = \frac{4\dot{Q}}{\pi D^2}$$

Substituting into Eq. (11-6), we have

$$N_R = \frac{4\dot{Q} \times D}{\pi D^2 \times \nu}$$

or

$$N_R = \left[\frac{4\dot{Q}}{\pi\nu}\right]\left(\frac{1}{D}\right)$$

Again, all terms within the brackets are constant; therefore, we may say

$$N_R = \frac{C_2}{D} \qquad \textit{(b)}$$

Problem statement: What nominal size commercial steel Schedule 40 pipe should be used to carry 40 gpm of water with a pressure loss of not more than 5 lbf/in.^{2} per 100 ft of length. Assume that the kinematic viscosity is 7.61×10^{-6} ft^2/sec.

Solution:

Procedure:

Step (a): Assume a value of f and solve Eq. (a) for D.

Step (b): Using this value of D, solve Eq. (b) for N_R, and determine ϵ/D for the pipe.

Step (c): With N_R and ϵ/D enter Fig. 11-5 and determine f. If this f is the same as assumed value in step (a), the problem is solved.

Step (d): If f found in step (c) is different from assumed value in Step (a), use the new value of f and repeat steps (a), (b), and (c), until f does not change.

(a) Assume $f = 0.02$. By Eq. (a)

$$D^5 = \left[\frac{8L\dot{Q}^2}{h_L g \pi^2}\right] f = C_1 \times f$$

$$\dot{Q} = \frac{40 \text{ gpm}}{(60 \text{ sec/min})(7.48 \text{ gal/ft}^3)} = 0.0892 \text{ ft}^3\text{/sec}$$

$$h_L = \frac{(5 \text{ lbf/in.}^2)(144 \text{ in.}^2\text{/ft}^2)}{62.4 \text{ lbf/ft}^3} = 11.54 \text{ ft of water}$$

Substitution $$D^5 = \left[\frac{(8)(100)(0.0892)^2}{(11.54)(32.2)(\pi^2)}\right] f$$

$$D^5 = 0.00173f$$

For assumed value of $f = 0.02$

$$D^5 = 0.00173(0.02)$$

$$D = 0.128 \text{ ft}$$

(b) By Eq. (b)

$$N_R = \frac{C_2}{D} = \left[\frac{4\dot{Q}}{\pi\nu}\right]\left(\frac{1}{D}\right) = \left[\frac{(4)(0.0892)}{\pi(7.61 \times 10^{-6}}\right]\left(\frac{1}{D}\right)$$

$$N_R = \frac{1.49 \times 10^4}{D}$$

Substituting $D = 0.128$ ft

$$N_R = \frac{1.49 \times 10^4}{0.128} = 1.164 \times 10^5 \quad \text{(turbulent)}$$

From Fig. 11-4, find $\epsilon = 0.00015$ for commercial steel pipe. Therefore, $\epsilon/D = 0.00015/0.128 = 0.0012$.

(c) With $\epsilon/D = 0.0012$ and $N_R = 1.164 \times 10^5$ enter Fig. 11-5 and find $f = 0.023$. This value of f is different from the assumed value of 0.02, use $f = 0.023$ and repeat steps (a), (b), and (c).

Step (a) $D^5 = 0.00173f = 0.00173(0.023)$

$D = 0.132$ ft

(b) $$N_R = \frac{1.49 \times 10^4}{D} = \frac{1.49 \times 10^4}{0.132}$$

$N_R = 1.129 \times 10^5$ (turbulent)

$\epsilon/D = 0.00015/0.132 = 0.0011$

(c) With $\epsilon/D = 0.0011$ and $N_R = 1.129 \times 10^5$ enter Fig. 11-5 and find f approximates the value of 0.023. Therefore, since f does not change, the pipe inside diameter should be 0.132 ft or 1.584 in.

From pipe size table, the nominal size pipe having an inside diameter closest to this value is 1½ in. Schedule 40, which has an actual inside diameter of 1.610 in.

Most of the pipe flow problems encountered in heating and cooling systems are of the class III type. The trial-and-error solution is very cumbersome as may be noted by studying Problem 11-8. Pipe-sizing methods commonly used in system design usually involve charts and tables that greatly simplify the task. These methods will be demonstrated in following chapters.

11-9 TYPES OF VALVES AND FITTINGS USED IN PIPE SYSTEMS

The preceding sections have considered fluid flow through closed conduits (pipes or ducts). The friction loss calculated by Eqs. (11-11) and (11-12) assumes that the loss occurred in a straight section of conduit L units long. Any conduit system also contains devices, such as fittings and valves, that cause losses (sometimes called minor losses in fluid mechanics textbooks). Piping designed to carry fluids in heating and cooling systems normally contains a relatively large number of fittings, valves, and other devices. The losses due to these fittings may be quite high in comparison to the loss in the straight pipe. Therefore, it is necessary to determine the types of fittings in the system and evaluate the losses associated with them in order to accurately determine the total loss in the system.

Extensive work has been done by Crane Company to develop comprehensive methods of evaluating flow resistances of valves and fittings. The results of that company's work has been published in Crane Company, Technical Paper No. 410, Flow of Fluids.[3]

Valves. The great variety of valve designs available makes it impossible to make a thorough classification. Figure 11-6 shows only a portion of the various types available. If valves were classified according to the resistance they offer to flow, those exhibiting a straight-through flow path such as gate, ball, plug, and butterfly valves would fall in the low-resistance class, and those having a change in flow path direction such as globe, and angle valves would fall in the high-resistance class.

Fittings. Fittings may be classified as branching, reducing, expanding, or deflecting. Such fittings as tees, crosses, side outlet elbows, and so forth, may be called branching fittings.

Reducing or expanding fittings are those that change the area of the flow passageway. In this class are reducers and bushings. Deflecting fittings such as bends, elbows, return bends, and so forth, are those which change the direction of flow.

Some fittings, of course, may be combinations of any of the foregoing general classifications. In addition, there are types, such as couplings and unions, that offer no appreciable resistance to flow and, therefore, need not be considered.

11-10 LOSSES IN VALVES AND FITTINGS

When fluid is flowing steadily in a long straight pipe of uniform diameter, the flow pattern, as indicated by the velocity distribution across the pipe diameter (see Fig. 11-1), will assume a certain characteristic form. Any obstruction in the pipe that changes the direction of the whole stream, or even part of it, will alter the characteristic flow pattern and create turbulence, causing an energy loss greater than that normally accompanying flow in a straight

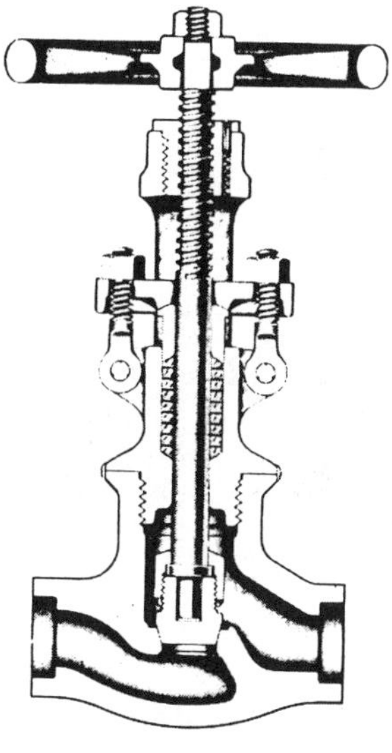

Conventional Globe Valve

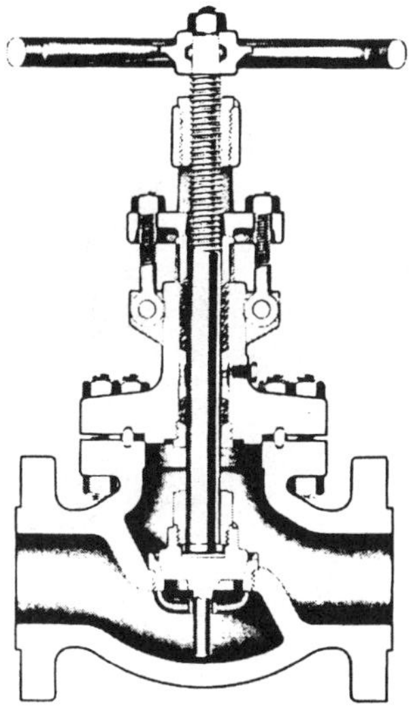

Conventional Globe Valve
With Disc Guide

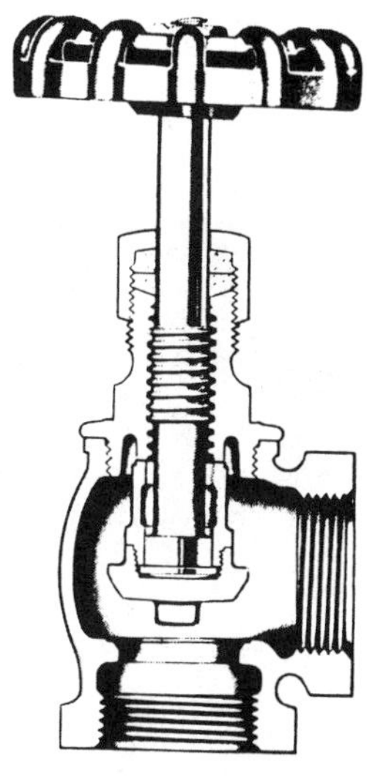

Conventional Angle Valve

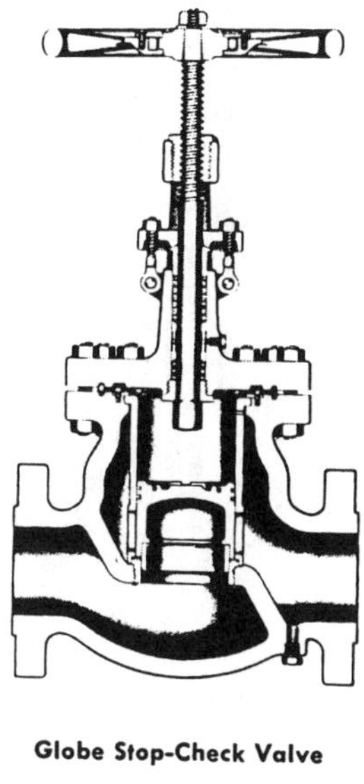

Globe Stop-Check Valve

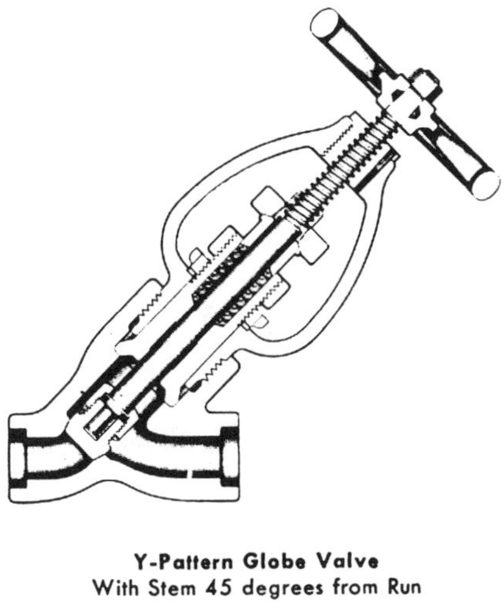

Y-Pattern Globe Valve
With Stem 45 degrees from Run

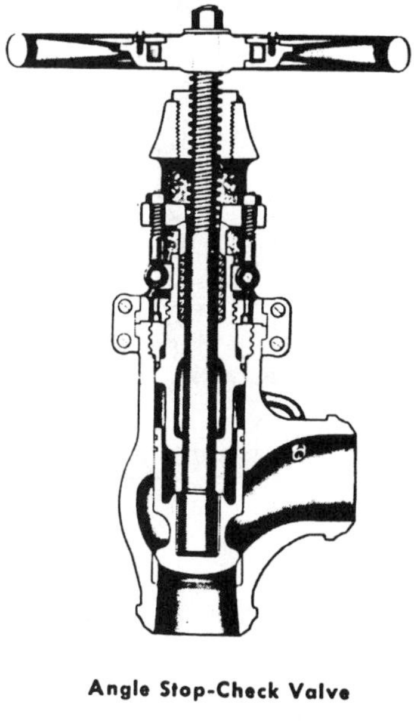

Angle Stop-Check Valve

Figure 11–6

Types of Valves

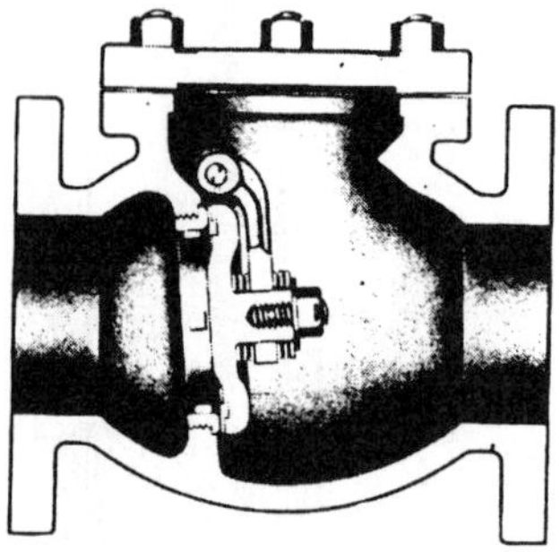

Conventional Swing Check Valve

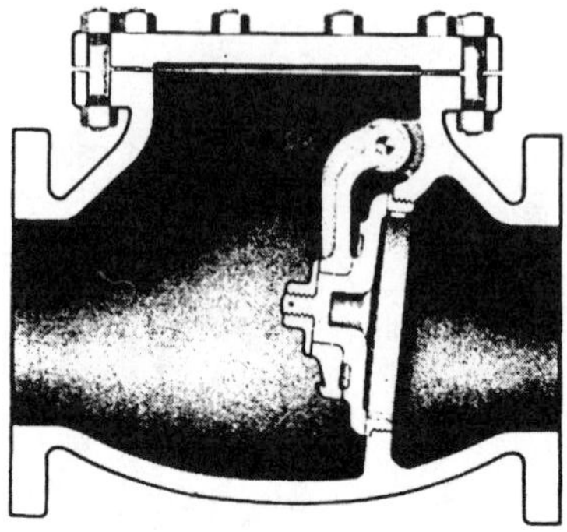

Clearway Swing Check Valve

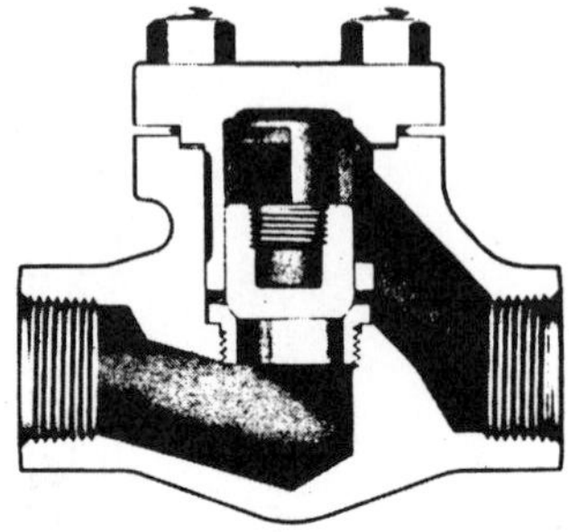

Globe Type Lift Check Valve

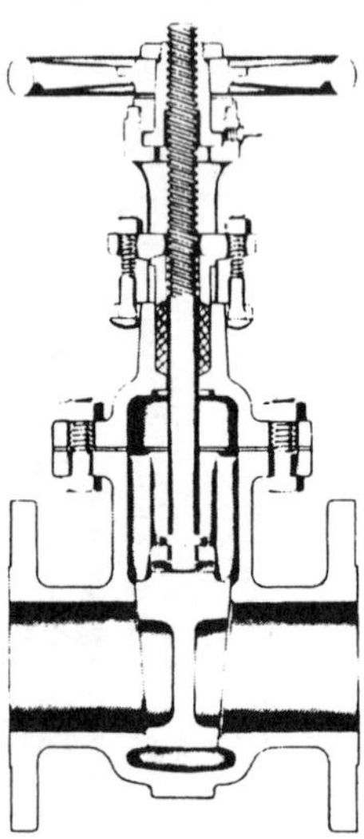

Wedge Gate Valve
(Bolted Bonnet)

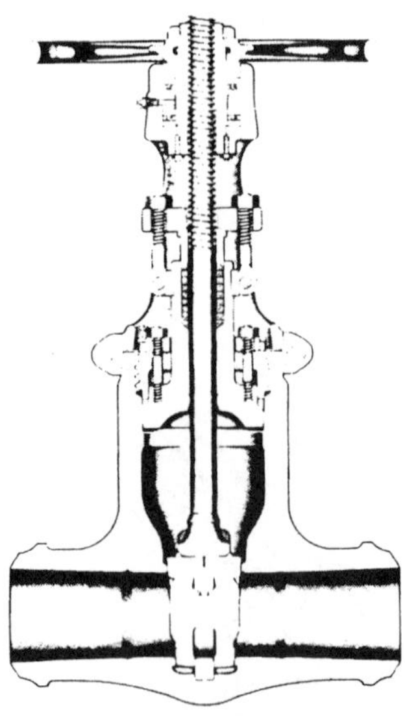

Flexible Wedge Gate Valve
(Pressure-Seal Bonnet)

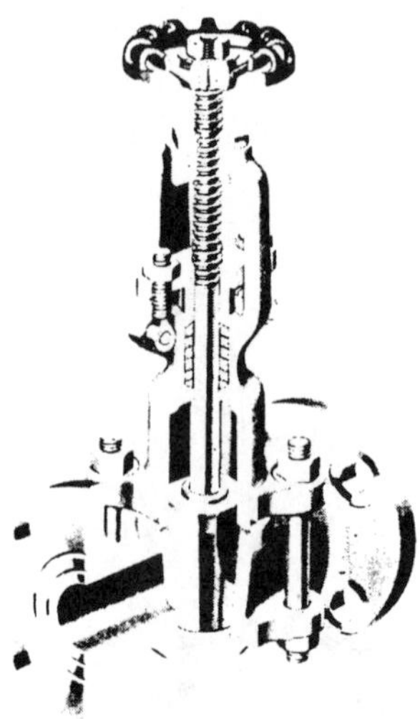

Plug Gate Valve
(Bolted Bonnet)

Figure 11–6 (cont)

Types of Valves

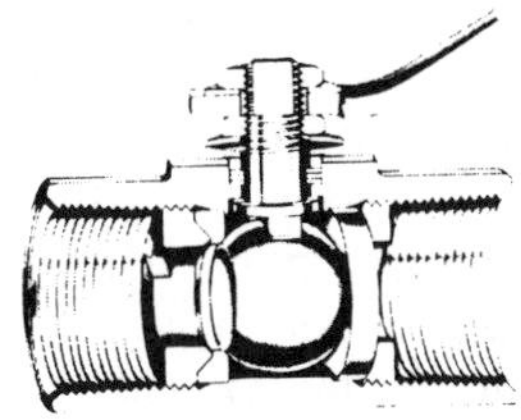
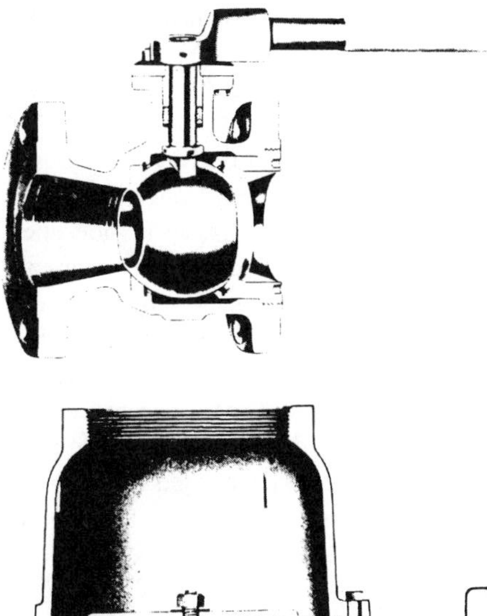

Ball Valves

Butterfly Wafer Valve

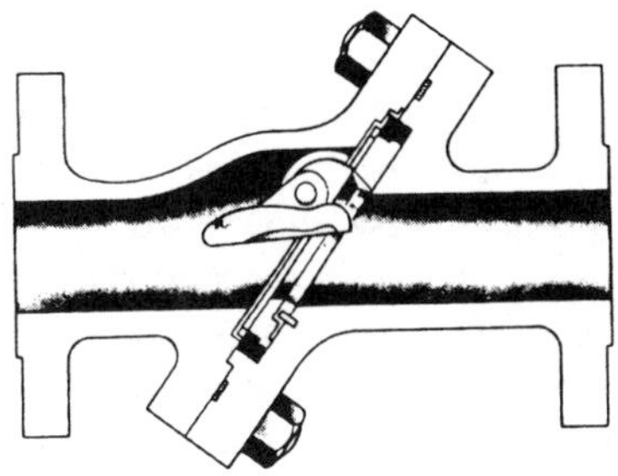

Tilting Disc Check Valve

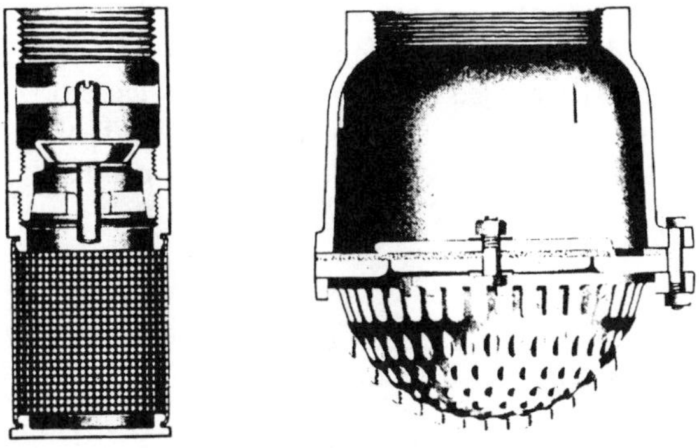

Foot Valves
Poppet and Hinged Types

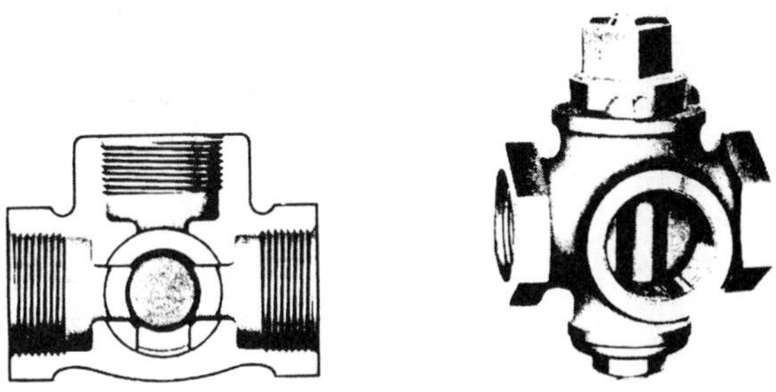

Three-Way Cock
Sectional and Outside Views

Figure 11–6 (cont)

Types of Valves

pipe. Because valves and fittings in a pipe line disturb the flow pattern, they produce an additional pressure drop. This loss of pressure by a valve (or fitting) consists of:

a. the pressure drop within the valve itself

b. the pressure drop in the upstream piping in excess of that which would normally occur if there were no valve in the line. This effect is small

c. the pressure drop in the downstream piping in excess of that which would normally occur if there were no valve in the line. This effect may be comparatively large

From an experimental point of view it is difficult to measure the three items separately. Their combined effect is the desired quantity, however, and this can be accurately measured by well-known methods.

Figure 11-7 shows two sections of a pipe line of the same diameter and length. The upper section contains a globe valve. If the pressure drops, Δp_1 and Δp_2, were measured between the points indicated, it would be found that Δp_1 is greater than Δp_2. Actually, the loss chargeable to the globe valve of length c is Δp_1 minus the loss in a section of pipe of length $a + b$. The losses, expressed in terms of resistance coefficient K of various valves and fittings as given in the K factor table (Table 11-1), include the loss due to the length of the valve or fitting.

Many experiments have shown that the pressure loss due to valves and fittings is mainly dynamic and is proportional to some power of the flow velocity. When pressure drop or head loss is plotted against velocity on logarithmic coordinates, the resulting curve is therefore a straight line. In the turbulent flow range the value of the power of V has been found to vary from about 1.8 to 2.1 for different designs of valves and fittings. However, for all practical purposes, it can be assumed that the pressure drop or head loss due to the flow of fluids in the turbulent range through valves and fittings varies as the square of the velocity.

A special situation must be considered for check valves because of their construction. The relationship of pressure drop to velocity of flow is valid only if there is sufficient flow to hold the disc in a wide-open position. Most of the difficulties encountered with check valves, both lift and swing types, have been found to be due to oversizing, which results in noisy operation and premature wear of the moving parts. The minimum velocity required to lift the disc to the full-open and stable position has been determined by tests of numerous types of check and foot valves and is given in Table 11-1. It is expressed in terms of a constant

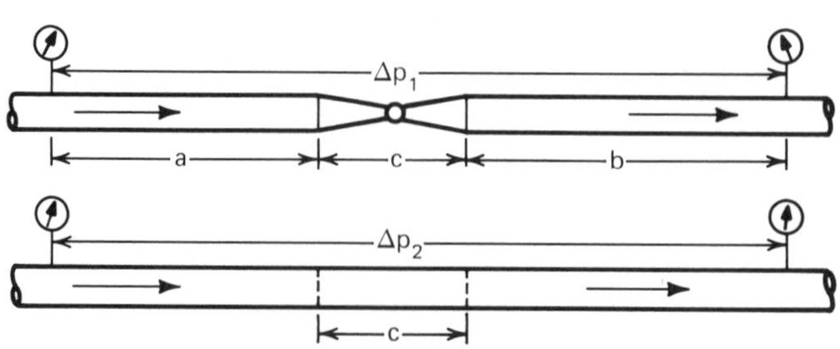

Figure 11–7

Flow through value or fitting

Table 11-1

K Factor Table*

Representative Resistance Coefficients (*K*) for Valves and Fittings

PIPE FRICTION DATA FOR CLEAN COMMERCIAL STEEL PIPE WITH FLOW IN ZONE OF COMPLETE TURBULENCE

Nominal Size	½"	¾"	1"	1¼"	1½"	2"	2½, 3"	4"	5"	6"	8-10"	12-16"	18-24"
Friction Factor (f_T)	.027	.025	.023	.022	.021	.019	.018	.017	.016	.015	.014	.013	.012

FORMULAS FOR CALCULATING "K" FACTORS FOR VALVES AND FITTINGS WITH REDUCED PORT

• **Formula 1**

$$K_2 = \frac{0.8 \sin\frac{\theta}{2}(1-\beta^2)}{\beta^4}$$

• **Formula 2**

$$K_2 = \frac{0.5\,(1-\beta^2)\sqrt{\sin\frac{\theta}{2}}}{\beta^4}$$

• **Formula 3**

$$K_2 = \frac{2.6 \sin\frac{\theta}{2}(1-\beta^2)^2}{\beta^4}$$

• **Formula 4**

$$K_2 = \frac{(1-\beta^2)^2}{\beta^4}$$

• **Formula 5**

$$K_2 = \frac{K_1}{\beta^4} + \text{Formula 1} + \text{Formula 3}$$

$$K_2 = \frac{K_1 + \sin\frac{\theta}{2}\left[0.8\,(1-\beta^2) + 2.6\,(1-\beta^2)^2\right]}{\beta^4}$$

• **Formula 6**

$$K_2 = \frac{K_1}{\beta^4} + \text{Formula 2} + \text{Formula 4}$$

$$K_2 = \frac{K_1 + 0.5\sqrt{\sin\frac{\theta}{2}}\,(1-\beta^2) + (1-\beta^2)^2}{\beta^4}$$

• **Formula 7**

$$K_2 = \frac{K_1}{\beta^4} + \beta\,(\text{Formula 2} + \text{Formula 4}) \text{ when } \theta = 180°$$

$$K_2 = \frac{K_1 + \beta\left[0.5\,(1-\beta^2) + (1-\beta^2)^2\right]}{\beta^4}$$

$$\beta = \frac{d_1}{d_2}$$

$$\beta^2 = \left(\frac{d_1}{d_2}\right)^2 = \frac{a_1}{a_2}$$

Subscript 1 defines dimensions and coefficients with reference to the smaller diameter.

Subscript 2 refers to the larger diameter.

SUDDEN AND GRADUAL CONTRACTION

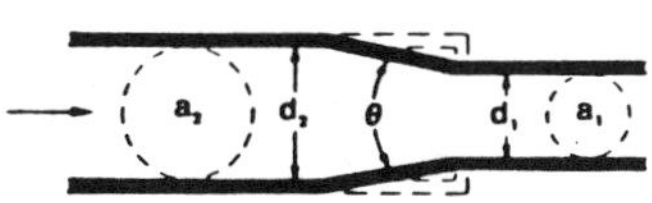

If: $\theta \le 45°$ .... K_2 = Formula 1

$\theta > 45° \le 180°$ ... K_2 = Formula 2

SUDDEN AND GRADUAL ENLARGEMENT

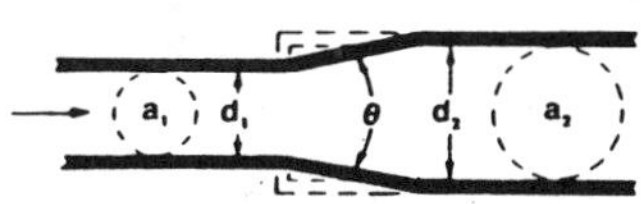

If: $\theta \le 45°$ .... K_2 = Formula 3

$\theta > 45° \le 180°$ ... K_2 = Formula 4

**Reprinted from Technical Paper No. 410,* Flow of Fluids, *with permission of the publisher, Crane Company.*

Table 11-1

K (Cont.)

GATE VALVES
Wedge Disc, Double Disc, or Plug Type

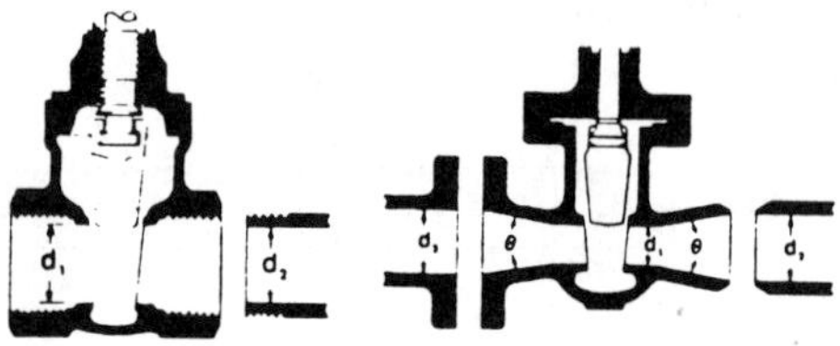

If: $\beta = 1$, $\theta = 0$. . . . . $K_1 = 8\,f_T$

$\beta < 1$ and $\theta \leqq 45°$. . . . . K_2 = Formula 5

$\beta < 1$ and $\theta > 45° \leqq 180°$. . . K_2 = Formula 6

SWING CHECK VALVES

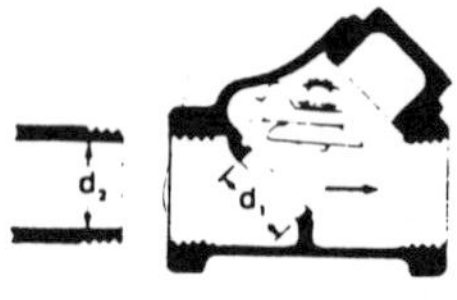

$K = 100\,f_T$

Minimum pipe velocity (fps) for full disc lift $= 35\sqrt{\bar{V}}$

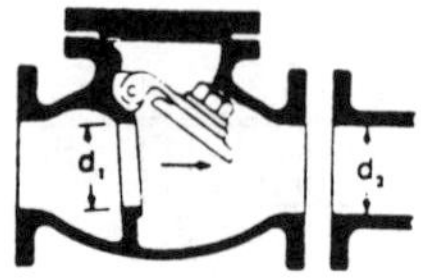

$K = 50\,f_T$

Minimum pipe velocity (fps) for full disc lift $= 48\sqrt{\bar{V}}$

GLOBE AND ANGLE VALVES

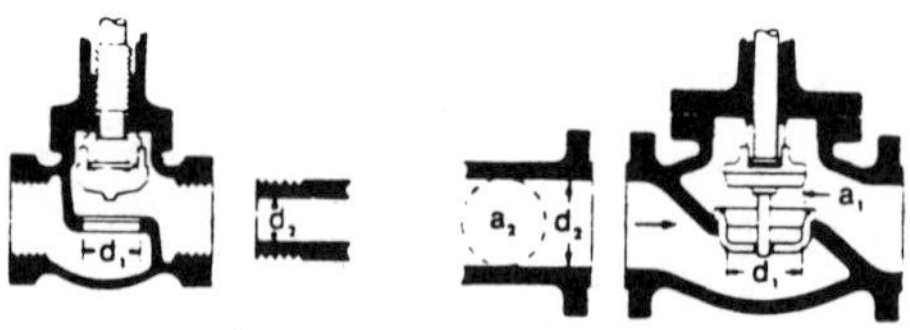

If: $\beta = 1$. . . $K_1 = 340\,f_T$

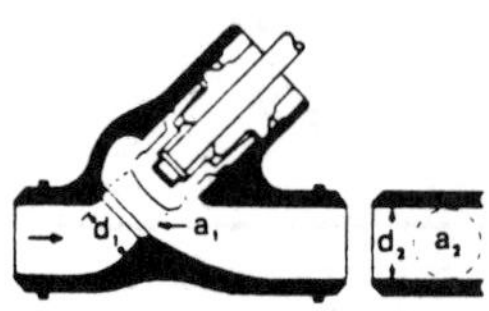

If: $\beta = 1$. . . $K_1 = 55\,f_T$

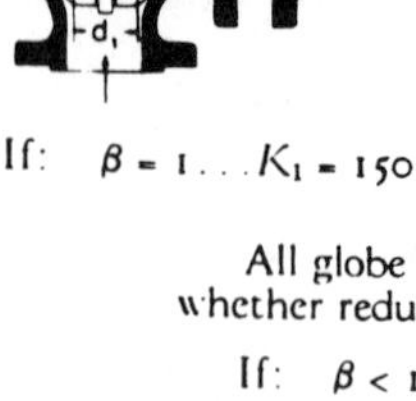

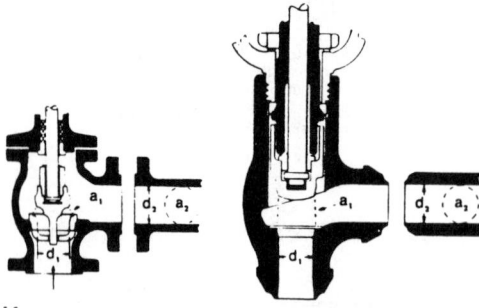

If: $\beta = 1$. . . $K_1 = 150\,f_T$

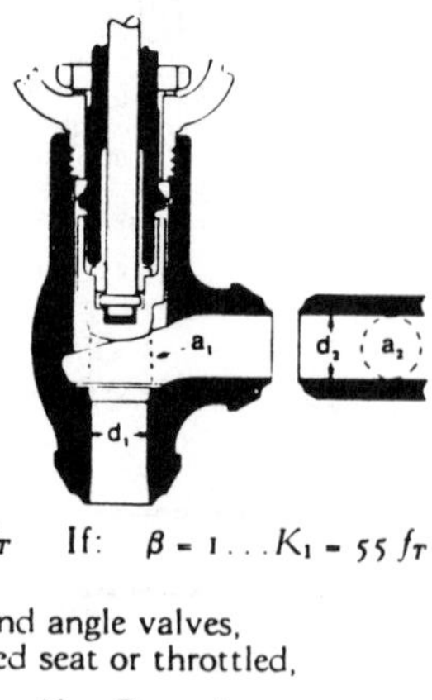

If: $\beta = 1$. . . $K_1 = 55\,f_T$

All globe and angle valves, whether reduced seat or throttled,

If: $\beta < 1$. . . K_2 = Formula 7

LIFT CHECK VALVES

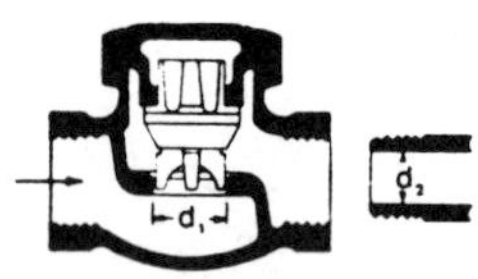

If: $\beta = 1$. . . $K_1 = 600\,f_T$

$\beta < 1$. . . K_2 = Formula 7

Minimum pipe velocity (fps) for full disc lift $= 40\,\beta^2\sqrt{\bar{V}}$

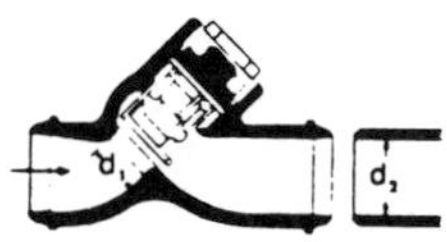

If: $\beta = 1$. . . $K_1 = 55\,f_T$

$\beta < 1$. . . K_2 = Formula 7

Minimum pipe velocity (fps) for full disc lift $= 140\,\beta^2\sqrt{\bar{V}}$

TILTING DISC CHECK VALVES

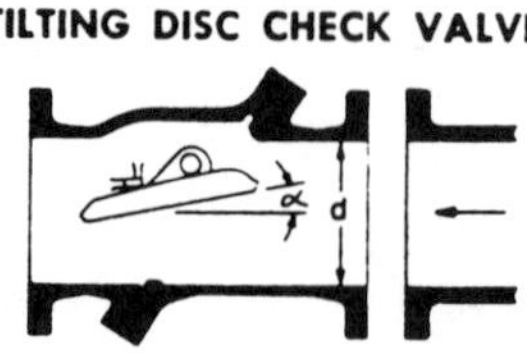

	$\alpha = 5°$	$\alpha = 15°$
Sizes 2 to 8". . . K =	$40\,f_T$	$120\,f_T$
Sizes 10 to 14". . . K =	$30\,f_T$	$90\,f_T$
Sizes 16 to 48". . . K =	$20\,f_T$	$60\,f_T$
Minimum pipe velocity (fps) for full disc lift =	$80\sqrt{\bar{V}}$	$30\sqrt{\bar{V}}$

Table 11-1

K (Cont.)

STOP-CHECK VALVES
(Globe and Angle Types)

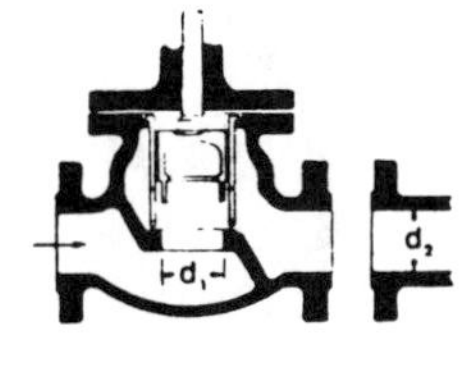

If:
$\beta = 1 \ldots K_1 = 400\, f_T$
$\beta < 1 \ldots K_2 =$ Formula 7

Minimum pipe velocity for full disc lift
$= 55\, \beta^2 \sqrt{V}$

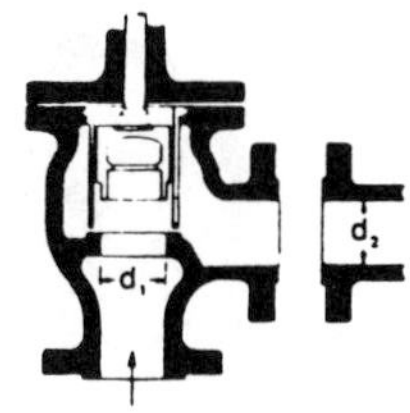

If:
$\beta = 1 \ldots K_1 = 200\, f_T$
$\beta < 1 \ldots K_2 =$ Formula 7

Minimum pipe velocity for full disc lift
$= 75\, \beta^2 \sqrt{V}$

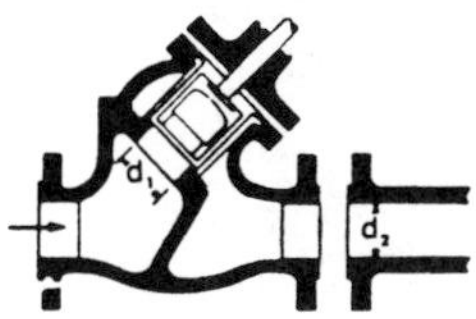

If:
$\beta = 1 \ldots K_1 = 350\, f_T$
$\beta < 1 \ldots K_2 =$ Formula 7

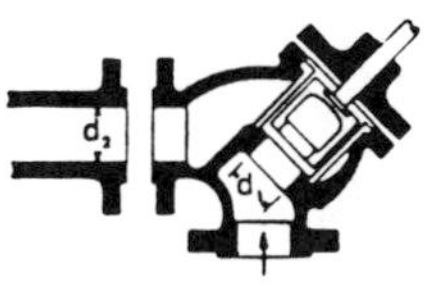

If:
$\beta = 1 \ldots K_1 = 300\, f_T$
$\beta < 1 \ldots K_2 =$ Formula 7

Minimum pipe velocity (fps) for full disc lift
$= 60\, \beta^2 \sqrt{V}$

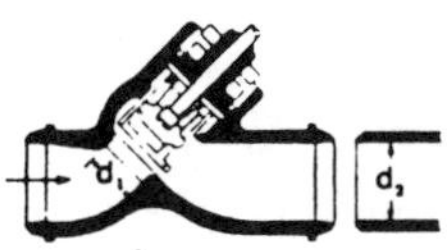

If:
$\beta = 1 \ldots K_1 = 55\, f_T$
$\beta < 1 \ldots K_2 =$ Formula 7

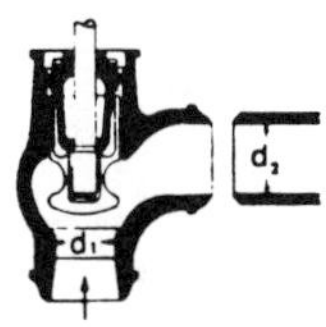

If:
$\beta = 1 \ldots K_1 = 55\, f_T$
$\beta < 1 \ldots K_2 =$ Formula 7

Minimum pipe velocity (fps) for full disc lift
$= 140\, \beta^2 \sqrt{V}$

FOOT VALVES WITH STRAINER

Poppet Disc

$K = 420 f_T$

Minimum pipe velocity (fps) for full disc lift
$= 15 \sqrt{V}$

Hinged Disc

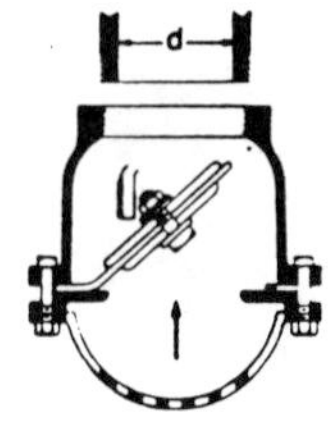

$K = 75 f_T$

Minimum pipe velocity (fps) for full disc lift
$= 35 \sqrt{V}$

BALL VALVES

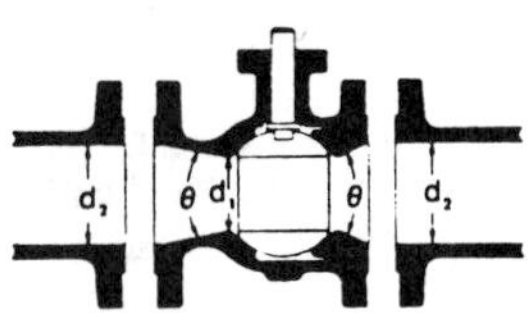

If: $\beta = 1,\ \theta = 0 \ldots\ldots\ldots\ldots\ldots K_1 = 3\, f_T$
$\beta < 1$ and $\theta \leq 45° \ldots\ldots\ldots K_2 =$ Formula 5
$\beta < 1$ and $\theta > 45° \leq 180° \ldots K_2 =$ Formula 6

BUTTERFLY VALVES

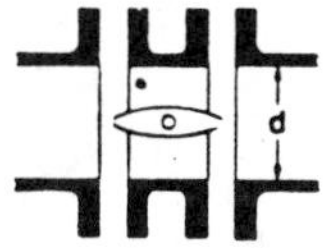

Sizes 2 to 8″ . . . $K = 45\, f_T$
Sizes 10 to 14″ . . . $K = 35\, f_T$
Sizes 16 to 24″ . . . $K = 25\, f_T$

Table 11-1
K (Cont.)

PLUG VALVES AND COCKS

Straight-Way

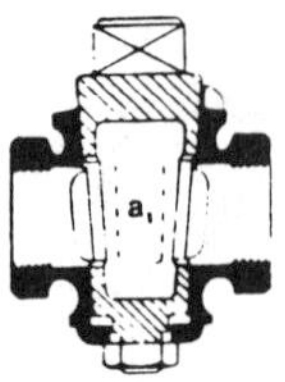

If: $\beta = 1$,
$K_1 = 18 f_T$

3-Way

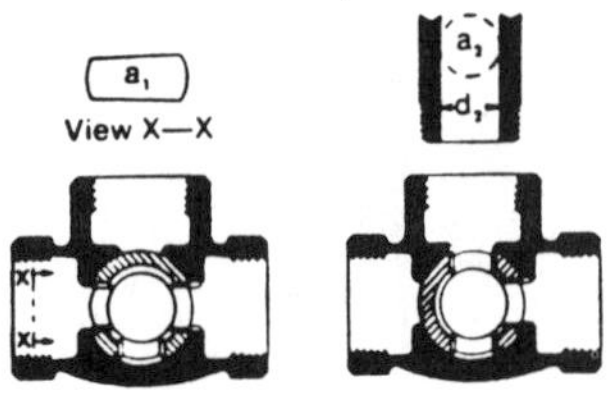

If: $\beta = 1$,
$K_1 = 30 f_T$

If: $\beta = 1$,
$K_1 = 90 f_T$

If: $\beta < 1 \ldots K_2$ = Formula 6

MITRE BENDS

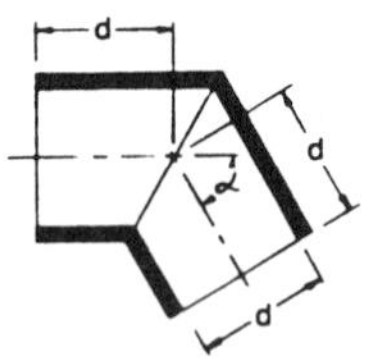

α	K
0°	2 f_T
15°	4 f_T
30°	8 f_T
45°	15 f_T
60°	25 f_T
75°	40 f_T
90°	60 f_T

90° PIPE BENDS AND FLANGED OR BUTT-WELDING 90° ELBOWS

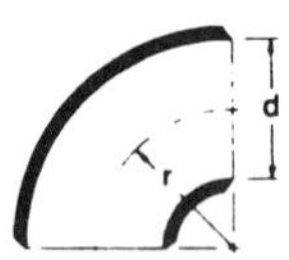

r/d	K	r/d	K
1	20 f_T	10	30 f_T
2	12 f_T	12	34 f_T
3	12 f_T	14	38 f_T
4	14 f_T	16	42 f_T
6	17 f_T	18	46 f_T
8	24 f_T	20	50 f_T

The resistance coefficient, K_B, for pipe bends other than 90° may be determined as follows:

$$K_B = (n - 1)\left(0.25\ \pi\ f_T \frac{r}{d} + 0.5\ K\right) + K$$

n = number of 90° bends
K = resistance coefficient for one 90° bend (per table)

CLOSE PATTERN RETURN BENDS

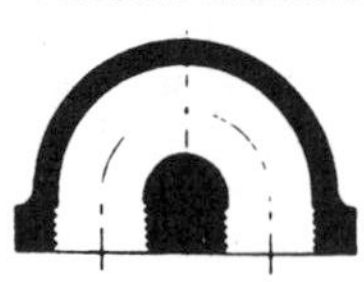

$K = 50 f_T$

STANDARD ELBOWS

90°

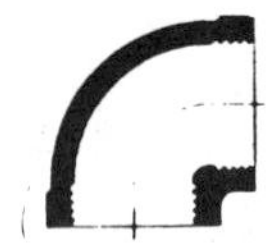

$K = 30 f_T$

45°

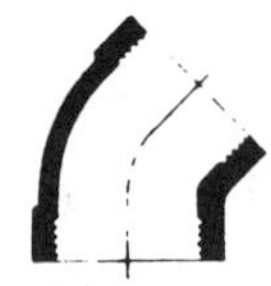

$K = 16 f_T$

STANDARD TEES

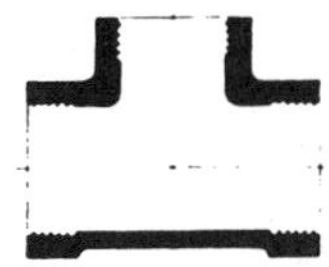

Flow thru run. . . . . . . $K = 20 f_T$
Flow thru branch. . . . $K = 60 f_T$

PIPE ENTRANCE

Inward Projecting

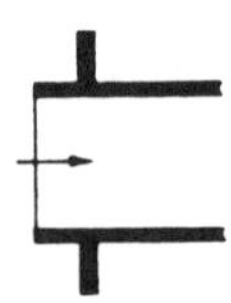

$K = 0.78$

r/d	K
0.00*	0.5
0.02	0.28
0.04	0.24
0.06	0.15
0.10	0.09
0.15 & up	0.04

*Sharp-edged

Flush

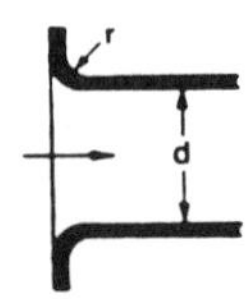

For K, see table

PIPE EXIT

Projecting

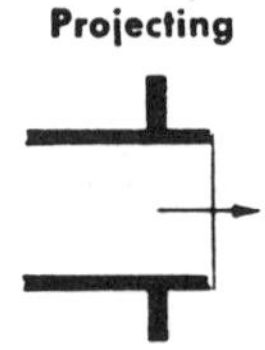

$K = 1.0$

Sharp-Edged

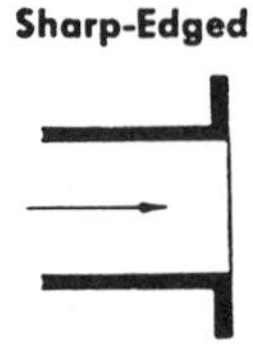

$K = 1.0$

Rounded

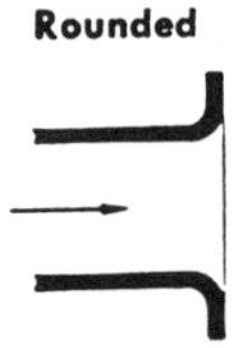

$K = 1.0$

times the square root of the specific volume of the fluid being handled, making it applicable for use with any fluid.

Sizing of check valves in accordance with the specified minimum velocity for full disc lift will often result in valves smaller in size than the pipe in which they are installed: however, the actual pressure drop will be small, if any, and higher than that of a full-size valve which is used in other than the wide-open positions. The advantages are longer valve life and quieter operation. The losses due to sudden or gradual contraction and enlargement, which will occur in such installations with bushings, reducing flanges, or tapered reducers, can be readily calculated from the date given in the Table 11-1.

As previously stated, pressure drop or head loss through fittings are due primarily to dynamic effects; that is, they are due to changes in flow direction and velocity (change in flow area). To a lesser extent, they are frictional, which depends upon surface roughness and Reynolds number.

Pressure drop test data for a wide variety of valves and fittings are available from numerous investigators. However, due to the time-consuming and costly nature of the tests, it is virtually impossible to obtain test data for every size and type of valve and fitting. It is therefore desirable to provide a means of reliably extrapolating available test data to envelop those items which have not been or cannot readily be tested. Manufacturer's tests are usually made under fully turbulent flow conditions and sometimes include entrance and exit losses. Commonly used methods of reporting resistances of valves and fittings are:

a. the equivalent length, L, of straight pipe of the same nominal pipe size having the same loss as the fitting

b. the equivalent length in pipe diameters, L/D, where D equals the pipe inside diameter in feet

c. a resistance coefficient, K

d. a flow coefficient, C_v

The relationship between L and L/D may be readily seen by stating that

$$L = \left(\frac{L}{D}\right)\left(\frac{d}{12}\right) \qquad (11\text{-}15)$$

where d is the internal pipe diameter in inches. The relationship of the resistance coefficient K, equivalent length L/D, and flow coefficient C_v, requires special attention.

11-11 RESISTANCE COEFFICIENT K, EQUIVALENT LENGTH L/D, AND FLOW COEFFICIENT C_v

Velocity of fluid flow in a pipe is obtained at the expense of static head, and decrease in static head due to velocity is

$$h_v = V^2/2g \qquad (11\text{-}16)$$

which is defined as "velocity head" in feet (meters) of fluid. Flow of fluid through a valve or fitting in a pipe line also causes a reduction in static head, which may be expressed in terms of velocity head. The resistance coefficient K in the equation

$$h_L = K\left(\frac{V^2}{2g}\right) \qquad (11\text{-}17)$$

is defined as the number of velocity heads lost due to a valve or fitting. *It is always associated with the diameter in which the velocity occurs.* In most valves and fittings the losses due to friction resulting from the length of flow path are minor in comparison to the dynamic loss. The resistance coefficient K is therefore considered as being independent of friction factor or Reynolds number, and may be considered constant for any given obstruction (valve or fitting) in a piping system under all conditions of flow.

Equation (11-11) may be arranged as

$$h_L = \left(f\frac{L}{D}\right)\left(\frac{V^2}{2g}\right) \qquad \textit{(11-11) Repeated}$$

Comparing Eqs. (11-11) and (11-17), it follows that

$$K = f\left(\frac{L}{D}\right) \qquad \textit{(11-18)}$$

The ratio L/D is the equivalent length, in pipe diameters of straight pipe, that will cause the same head loss as the obstruction under the same flow conditions. Since the resistance coefficient K is constant for all conditions of flow, the value of L/D for any given valve or fitting must necessarily vary inversely with the change in friction factor f for different flow conditions.

The resistance coefficient K would theoretically be a constant for all sizes of a given design or line of valves and fittings if all sizes were geometrically similar. However, geometric similarity is seldom, if ever, achieved because the design of valves and fittings is dictated by manufacturing economies, standards, structural strength, and other considerations. Based upon evidence from numerous investigations, it can be said that the resistance coefficient K, for a given line of valves or fittings, tends to vary with size as does the friction factor f, for straight clean commercial steel pipe at flow conditions resulting in a constant-friction factor and that the equivalent length L/D tends toward a constant for the various sizes of a given line of valves or fittings at the same flow conditions.

On the basis of this relationship the resistance coefficient K for each illustrated type of valve and fitting is presented in Table 11-1. These coefficients are given as the product of the friction factor f_t for the desired size of clean commercial steel pipe with flow in the zone of complete turbulence and a constant, which represents the equivalent length L/D, for the valve or fitting in pipe diameters for the same flow conditions, on the basis of test data. This equivalent length, or constant, is valid for all sizes of the valve or fitting type with which it is identified.

The friction factor f_t for clean commercial steel pipe with flow in the zone of complete turbulence, for nominal sizes from ½- to 24-in., are tabulated at the beginning of Table 11-1.

There are some resistances to flow in piping, such as sudden and gradual contractions and enlargements, and pipe entrances and exits, that have geometric similarity between sizes. The resistance coefficients K for these items are therefore independent of size, as indicated by the absence of a friction factor in their values given in Table 11-1.

As previously stated, the resistance coefficient K is always associated with the diameter in which the velocity in the term $V^2/2g$ occurs. The values in the K factor table (Table 11-1) are associated with the internal diameter of the following pipe schedule numbers for the various ANSI classes of valves and fittings:

Class 300 and lower	Schedule 40
Class 400 to 600	Schedule 80
Class 900	Schedule 120
Class 1500	Schedule 160

When the resistance coefficient K is used in flow equations such as Eqs. (11-17) or (11-18), the velocity and internal diameter dimensions used in the equation must be based on the dimensions of these schedule numbers regardless of the pipe in which the valve may be installed.

An alternate procedure, which yields identical results for Eq. (11-17), is to adjust K in proportion to the fourth power of the diameter ratio and to base values of velocity or diameter on the internal diameter of the connecting pipe:

$$K_a = K_b\left(\frac{d_a}{d_b}\right)^4 \tag{11-19}$$

where subscript a defines K and d with reference to the internal diameter of the connecting pipe. Subscript b defines K and d with reference to the internal diameter of the pipe for which the values of K were estabished, as given in the foregoing list of pipe schedule numbers.

When a piping system contains more than one size of pipe, valves, or fittings, Eq. (11-19) may be used to express all resistances in terms of one size. For this, subscript a relates to the size with reference to which all resistances are to be expressed, and subscript b relates to any other size in the system.

Flow Coefficient C_v. It has been found convenient in some branches of the valve industry, particularly in connection with control valves, to express the valve capacity and valve flow characteristics in terms of a flow coefficient, C_v. The coefficient is numerically equal to the flow rate in gpm of 60°F water, which will give a pressure drop of 1 lb/in.^{2} (2.31 ft of water) across the device.

By substitution of appropriate English Engineering System units into the Darcy equation, it can be shown that

$$C_v = \frac{29.9d^2}{\sqrt{K}} \tag{11-20}$$

where d is the internal diameter in inches. In SI units a flow coefficient $C_{v\mathrm{SI}}$ is defined as the flow rate in m^3/s of water at 15°C with a pressure loss of one kPa given by

$$C_{v\mathrm{SI}} = \frac{1.11D^2}{\sqrt{K}} \tag{11-20a}$$

where D is in meters.

The quantity in gallons per minute (gpm) of any liquid that will flow through the device can be determined from

$$\text{gpm} = C_v\sqrt{\Delta P_L\left(\frac{62.4}{\rho}\right)} \tag{11-21}$$

and the pressure drop can be computed from the same relationship arranged as

$$\Delta p_L = \frac{\rho}{62.4}\left(\frac{\text{gpm}}{C_v}\right)^2 \tag{11-22}$$

In Eq. (11-21) and (11-22), Δp_L is the pressure drop across the valve in lbf/in.^{2} and ρ is the fluid density in lbm/ft^3.

Since the pressure loss is proportional to the square of the flow velocity, the pressure drop or lost head may be calculated at other flow rates:

$$\frac{h_{L1}}{h_{L2}} = \left(\frac{\text{gpm}_1}{\text{gpm}_2}\right)^2 \tag{11-23}$$

In terms of the flow coefficient C_v:

$$\frac{h_L}{2.31} = \left(\frac{\text{gpm}}{C_v}\right)^2$$

or,

$$h_L = 2.31\left(\frac{\text{gpm}}{C_v}\right)^2 \tag{11-24}$$

where C_v is in gpm and h_L is in feet of water.

Heating and cooling devices usually have

pressure or head loss information furnished by the manufacturer. In fact, the flow rate may be adjusted by measurement of the pressure drop across the device. This is frequently done to balance the flows in the system. Equation (11-23) may be used to estimate the flow rate at other than specified conditions.

Laminar Flow Conditions. Equation (11-17) is valid for computing the head loss due to valves and fittings for all conditions of flow, including laminar flow, using the resistance coefficient K as given in Table 11-1. When this equation is used to determine the losses in straight pipe, it is necessary to compute the Reynolds number in order to establish the friction factor f to be used to determine the resistance coefficient K for the pipe in accordance with Eq. (11-18), ($K = fL/D$).

Conversions between K, L/D, and L can be obtained for various pipe sizes by use of Fig. 11-8. When using SI units, it is suggested that the L/D ratio be determined from Fig. 11-8, using the nominal pipe size. The equivalent length in meters may then be determined using the inside diameter D in meters.

The relationship between resistance coefficient K and flow coefficient C_v is shown in Fig. 11-9.

Illustrative Problem 11-9

A 6-in. class 125 iron Y-pattern globe valve has a flow coefficient C_v of 600. Determine the resistance coefficient K and equivalent lengths L/D and L for flow in the zone of complete turbulence.

Solution:

K, L/D, and L should be calculated in terms of 6-in. Schedule 40 pipe. Nominal 6-in. Schedule 40 pipe has an internal diameter (ID) of 6.065 in.

By Eq. (11-20)

$$C_v = \frac{29.9d^2}{\sqrt{K}}$$

$$K = \frac{(29.9)^2(d^2)^2}{(C_v)^2} = \frac{894d^4}{(C_v)^2} = \frac{894(6.065)^4}{(600)^2}$$

$$K = 3.36 \qquad \textit{(ans)}$$

Also, determine from Fig. 11-9.
By Eq. (11-18)

$$K = f\left(\frac{L}{D}\right)$$

From Table 11-1 $f = 0.015$, therefore,

$$3.36 = (0.015)(L/D)$$

$$L/D = 224 \qquad \textit{(ans)}$$

Also, determine from Fig. 11-8, By Eq. (11-15)

$$L = (L/D)(d/12) = 224(6.065/12)$$

$$L = 113.2 \text{ ft} \qquad \textit{(ans)}$$

Also, from Fig. 11-8:

Illustrative Problem 11-10

For 4-in. class 600 steel conventional angle valve with full seat area, find the resistance coefficient K, flow coefficient C_v, and equivalent lengths L/D and L for flow in zone of complete turbulence.

Solution:

K, L/D, and L should be given in terms of 4-in. Schedule 80 pipe. Nominal 4-in. Schedule 80 pipe has an ID of 3.826 in.

Equivalent Lengths *L* and *L/D* and Resistance Coefficient *K*

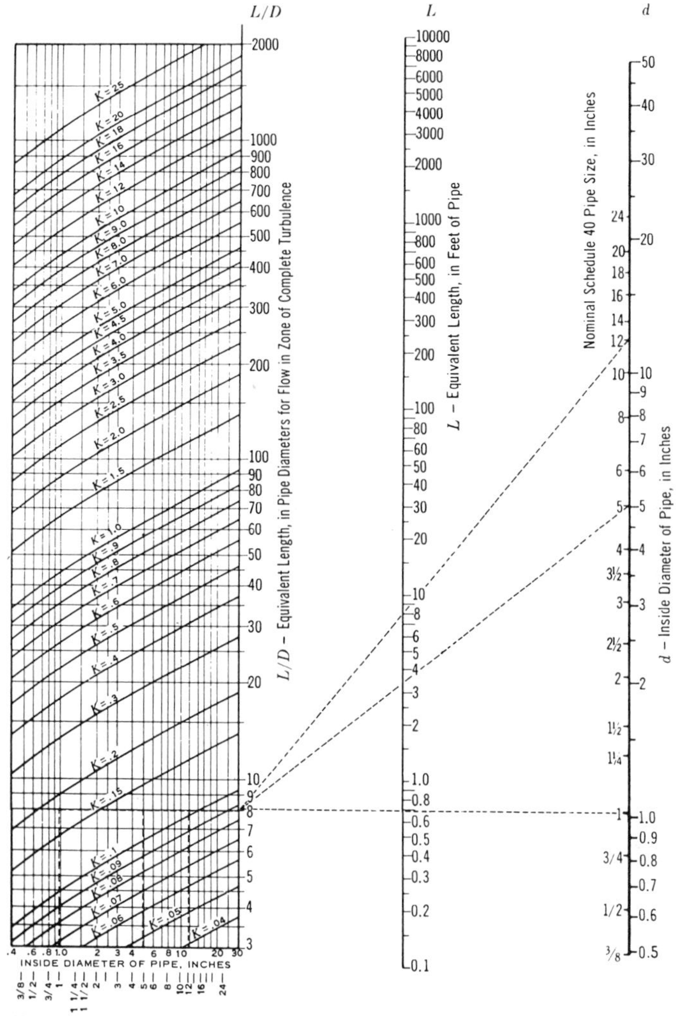

Figure 11–8

Reprinted from Technical Paper No. 410, *Flow of Fluids* with permission of the publisher, Crane Company

Equivalents of Resistance Coefficient K And Flow Coefficient C_v

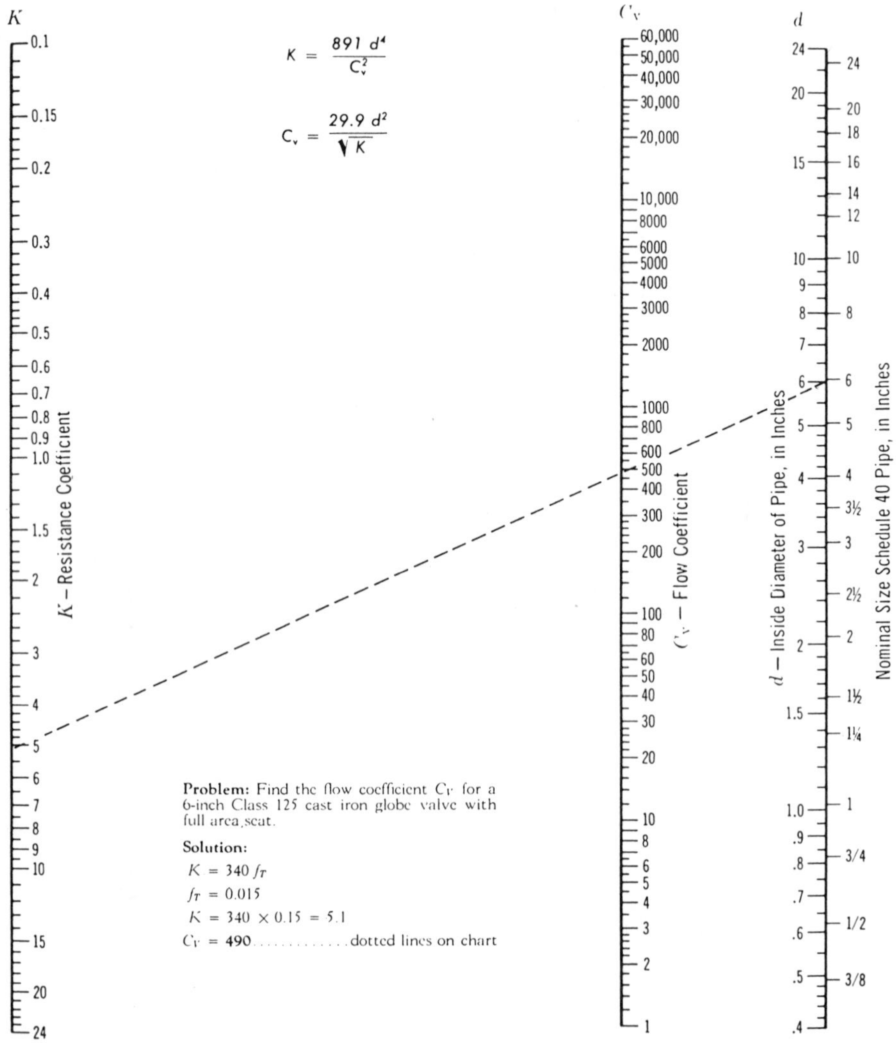

Figure 11–9

Reprinted from Technical Paper No. 410, *Flow of Fluids* with permission of the publisher, Crane Company

From Table 11-1, $K = 150\ f_t$, and $f_t = 0.017$

By Eq. (11-20) $$C_v = \frac{29.9d^2}{\sqrt{K}}$$

By Eq. (11-18) $K = f_t(L/D)$

or $L/D = K/f_t$ (for complete turbulence)

By Eq. (11-15) $L = (L/D)(d/12)$

Substituting: $K = 150(0.017) = 2.55$ *(ans)*

$$C_v = \frac{29.9(3.826)^2}{\sqrt{2.55}} = 274$$ *(ans)*

$$\frac{L}{D} = \frac{2.55}{0.017} = 150$$ *(ans)*

$$L = 150\left(\frac{3.826}{12}\right) = 47.8 \text{ ft}$$ *(ans)*

11-12 SUDDEN CONTRACTIONS AND ENLARGEMENTS

The resistance to flow due to sudden enlargements may be expressed by

$$K_1 = \left(1 - \frac{d_1^2}{d_2^2}\right)^2 \qquad (11\text{-}25)$$

and the resistance due to sudden contractions by

$$K_1 = 0.5\left(1 - \frac{d_1^2}{d_2^2}\right) \qquad (11\text{-}26)$$

where subscripts 1 and 2 define the internal diameters of the small and large pipes, respectively. It is convenient to identify the ratio of diameters of the small to large pipes by the Greek letter β (beta). Using this notation, we would have

$$K_1 = (1 - \beta^2)^2 \quad \text{(sudden enlargements)} \qquad (11\text{-}25a)$$

and

$$K_1 = (1 - \beta^2) \quad \text{(sudden contractions)} \qquad (11\text{-}26a)$$

Equation (11-25) is derived from the momentum equation together with the Bernoulli equation. Equation (11-26) uses the derivation of Eq. (11-25) together with the continuity equation and a close approximation of the contraction coefficients determined by Julius Weisbach.

The value of the resistance coefficient in terms of the larger pipe is determined by dividing Eqs. (11-15) and (11-16) by β^4,

$$K_2 = \frac{K_1}{\beta^4} \qquad (11\text{-}27)$$

The losses due to gradual enlargements in pipes were investigated by A. H. Gibson and may be expressed as a coefficient, C_e applied to Eq. (11-25). Approximate averages of Gibson's coefficients for different included angles of divergence θ are defined by the following equations:

For $\theta \leqslant 45°$ $$C_e = 2.6 \sin\frac{\theta}{2} \qquad (11\text{-}28)$$

For $\theta > 45° \leqslant 180°$ $$C_e = 1 \qquad (11\text{-}28a)$$

The losses due to gradual contractions in pipes were established by analysis of Crane test data, using the same basis as that of Gibson for gradual enlargements, to provide a contraction coefficient C_c to be applied to Eq. (11-26). The approximate averages of these coefficients for different included angles of convergence θ are defined by the following equations:

For $\theta \leq 45°$ $$C_c = 0.8 \sin\frac{\theta}{2} \qquad (11\text{-}29)$$

For $\theta > 45° \leq 180°$ $$C_c\ 0.5 \sin\frac{\theta}{2} \qquad (11\text{-}29a)$$

The resistance coefficient K for sudden and gradual enlargements and contractions, expressed in terms of the large pipe, is established by combining Eqs. (11-25) through (11-29). The formulas are summarized in Table 11-1.

Valves with reduced seats. Valves are often designed with reduced seats, and the transition from seat to valve ends may be either abrupt or gradual. Straight-through types, such as gate and ball valves, so designed with gradual transition are sometimes referred to as venturi valves. Analysis of tests on such straight-through valves indicate an excellent correlation between test results and calculated values of K. See summation of formulas in Table 11-1.

The procedure for determining K for reduced-seat globe and angle valves is also applicable to *throttled* globe and angle valves. For this case the value of β must be based on the square root of the ratio of areas, $\beta = \sqrt{a_1/a_2}$, where a_1 defines the area at the most restricted point in the flow path and a_2 defines the internal area of the connecting pipe. See formula summation in Table 11-1.

Illustrative Problem 11-11

Given a 6 by 4-in. class 600 steel gate valve with inlet and outlet ports conically tapered from back to body rings to valve ends. Face-to-face dimension is 22 in. and back of seat ring to back of seat ring is about 6 in. Determine K_2 for any flow condition, and L/D and L for flow in the zone of complete turbulence.

Solution:

2 For class 600, K_2, L/D, and L should be given in terms of 6-in. Schedule 80 pipe. From Table 11-1 $K_1 = 8f_t$

$$K_2 = \frac{K_1 + \sin\frac{\theta}{2}[0.8(1-\beta^2) + 2.6(1-\beta^2)^2]}{\beta^4} \qquad (Formula\ 5)$$

By Eq. 11-18 $K = f(L/D)$ or $L/D = K/f_t$

and $\beta = d_1/d_2$

For 4-in. Schedule 80 pipe, ID = 3.826 in.
For 6-in. Schedule 80 pipe, ID = 5.761 in.
From Table 11-1 $f_t = 0.015$ for 6-in. size

Substitution: $$\beta = \frac{3.826}{5.761} = 0.664$$

$$\beta^2 = 0.441 \qquad \beta^4 = 0.194$$

$$\tan\frac{\theta}{2} = \frac{0.5(5.761 - 3.826)}{0.5(22 - 6)}$$

$$\tan\frac{\theta}{2} = 0.121$$

$$= \sin\theta/2 \quad \text{(approximate)}$$

$$K_2 = \frac{(8)(0.015) + (0.121)[0.8(1 - 0.441) + 2.6(1 - 0.441)^2]}{0.194} \qquad (ans)$$

$$K_2 = 1.40$$

$$\frac{L}{D} = \frac{K_2}{f_t} = \frac{1.40}{0.015} =$$

93.3 diameters of 6-in. Schedule 80 pipe *(ans)*

$$L = 93.3\left(\frac{5.761}{12}\right) =$$

44.8 ft of 6-in. Schedule 80 pipe *(ans)*

Illustrative Problem 11-12

A globe-type lift check valve with a wing-guided disc is required in a 3-in. Schedule 40 horizontal pipe carrying 70°F water at the rate of 80 gpm. Determine the proper size check valve and its pressure drop. The valve is to be sized so that the disc is fully lifted at the specified flow.

Solution:

From Table 11-1

$$V_{min} = 40\sqrt{v}$$

$$K_1 = 600f_t$$

$$K_2 = \frac{K_1 + \beta[0.5(1 - \beta^2) + (1 - \beta^2)^2]}{\beta^4} \quad \text{(Formula 7)}$$

$f_t = 0.018$ (for either 2 1/2- or 3-in. pipe)

$\beta = d_1/d_2$

For 2½-in. Schedule 40 pipe, ID = 2.469 in. For 3-in. Schedule 40 pipe, ID = 3.068 in. At 70°F, specific volume $v = 0.016051$ ft³/lb, and $\rho = 1/v = 62.30$ lb/ft³.

Substitution:
Assume that check valve will be same size as pipe, or $\beta = 1$.

$$V_{min} = 40\sqrt{v} = 40\sqrt{0.016051}$$
$$= 5.07 \text{ ft/sec}$$

By Eq. (11-15)

$$V_{actual} = \frac{0.408(\text{gpm})}{d^2} = \frac{0.408(80)}{(3.068)^2}$$
$$= 3.47 \text{ ft/sec}$$

Since V_{actual} is less than V_{min} required, try 2½-in. check valve size.

$$V_{actual} = \frac{0.408(80)}{(2.469)^2} = 5.35 \text{ ft/sec}$$

Use 2½-in. check valve size with reducers from 3-in. Schedule 40 pipe. *(ans)*

$$\beta = \frac{d_1}{d_2} = \frac{2.469}{3.068} = 0.805$$

$$\beta^2 = 0.648 \qquad \beta^4 = 0.420$$

$$K_2 = \frac{(600)(0.018) +}{0.420} \frac{8[0.5(1 - 0.648) + (1 - 0.648)^2]}{0.420}$$

$K_2 = 26.3$ (based on 3-in. size)

By Eq. 11-11

$$h_L = f\left(\frac{L}{D}\right)\left(\frac{V^2}{2_g}\right) = K\left(\frac{V^2}{2_g}\right) = K\left(\frac{V^2}{2_g}\right)$$

$$h_L = 26.3\left[\frac{(3.47)^2}{2(32.2)}\right] = 4.9 \text{ ft of fluid}$$

or

$$\Delta p = h_L \times \gamma = 4.9 \times 62.3$$
$$= 305.7 \text{ lb/ft}^2$$
$$= 2.1 \text{ lb/in.}^2$$

(ans)

Illustrative Problem 11-13

Water at 60°F discharges from a tank with 22-ft average head to atmosphere through:

200 ft of 3-in. Schedule 40 pipe

6, 3-in. standard 90 degree threaded elbows

1, 3-in. flanged ball valve having a 2⅜-in. diameter seat, 16-degree conical inlet, and 30-degree conical outlet and sharp-edged entrance is flush with the inside of the tank.

Determine the flow velocity in the pipe and rate of discharge in gallons per minute.

Solution:

Step 1: Determine K values for fittings and pipe.

By Eq. (11-17) $h_L = K\left(\frac{V^2}{2g}\right)$

or $V = \sqrt{\frac{2_g h_L}{K}}$

By Eq. (11-15) $V = \frac{0.408(\text{gpm})}{d^2}$

or $\text{gpm} = \frac{Vd^2}{0.408}$

From Table 11-1

For pipe entrance, $K = 0.5$

For pipe exit, $K = 1.0$

$f_t = 0.018$

For K (ball valve) use formula (Table 11-1) 5. However, when inlet and outlet angles (θ) differ, formula 5 must be expanded to

$$K_2 = \frac{K_1 + 0.8 \sin\frac{\theta}{2}(1 - \beta^2) + 2.6 \sin\frac{\theta}{2}(1 - \beta^2)^2}{\beta^4}$$

$$K_1 = 3\ f_t$$

and $d_1 = 2\tfrac{3}{8}$ in. = 2.375 in.

$d_2 = 3.068$ in. (3-in. nominal, Schedule 40 pipe)

$$\beta = \frac{d_1}{d_2} = \frac{2.375}{3.068} = 0.774$$

$$\beta^2 = 0.599 \qquad \beta^4 = 0.359$$

$$\text{Inlet} - \sin\frac{\theta}{2} = \sin 8° = 0.139$$

$$\text{Outlet} - \sin\frac{\theta}{2} = \sin 15° = 0.259$$

$$K_2 = \frac{(3)(0.018) + (0.8)(0.139)(1 - 0.599)}{0.359} + \frac{(2.6)(0.259)(1 - 0.599)^2}{0.359}$$

$$K_2 = \frac{0.054 + 0.045 + 0.108}{0.359}$$

$$K_2 = 0.577$$

For six pipe elbows, $K = 6 \times 30f_t = 6(30)(0.018) = 3.24$

For pipe $K = f\left(\frac{L}{D}\right) = \frac{(0.018)(200)}{3.068/12} = 14.08$

For entire system, $K = 0.5 + 14.08 + 0.577 + 3.24 + 1.0 = 19.4$

Step 2: Calculate velocity and flow rate

$$V = \sqrt{\frac{2gh_L}{K}} + \sqrt{\frac{2(32.2)(22)}{19.4}}$$

$V = 8.54$ ft/sec ***(ans)***

$$\text{gpm} = \frac{Vd^2}{0.408} = \frac{(8.54)(3.068)^2}{0.408}$$

$$\text{gpm} = 197 \qquad \textit{(ans)}$$

Step 3: At this point the Reynolds number should be checked to verify that the friction factor ($f_t = 0.018$) is in the zone of complete turbulence for the existing flow conditions. When the N_R and ϵ/D values are determined, find f from Fig. 11-5. It will be found that f is approximately 0.019, which is close enough to the assumed value of 0.018 to make it unnecessary to recalculate the K values for pipe and fittings.

Illustrative Problem 11-14

Determine the head loss for 100 ft of 2-in. Schedule 40 commercial steel pipe carrying 100 gpm of 60°F water. The pipe contains three standard 90 degree elbows, a globe valve, and a swing check valve ($K = 100 f_t$). Assume that the head loss due to friction is 7.1×10^{-2} ft of water per foot of pipe length.

Solution:

Determine equivalent length of fittings.
Using Table 11-1, $f_t = 0.019$

Elbow: $K = 30f_t = 30(0.019) = 0.57$

$L = 4.8$ ft (Fig. 11-8)

Globe valve: $K = 340f_t = 340(0.019) = 6.45$

$L = 30$ ft

Check valve: $K = 100f_t = 100(0.019) = 1.9$

$L = 12$ ft

Total equivalent length (TEL) then is

Actual straight length	100 ft
Three elbows (3 × 4.8)	14.4 ft
One globe valve	30 ft
One check valve	12 ft
TEL =	156.4 ft

$$h_L = (7.1 \times 10^{-2} \text{ ft of water/ft of pipe})(\text{TEL})$$

$$h_L = (7.1 \times 10^{-2})(156.4)$$

$$= 11.1 \text{ ft of water} \qquad \textit{(ans)}$$

11-13 PIPING SYSTEMS

Heating and cooling systems may use one or more fluids enclosed in conduits and flowing to various locations in the system. Typical fluids are water, water-glycol mixtures, brine, steam, air, refrigerants, and so on. The flow of these fluids within the system and system components must be completely controlled and coordinated. This leads to a variety of integrated piping and duct systems that must be installed throughout the building to connect all the heat-transfer devices. Some of these systems include:

a. hot water piping systems
b. chilled water piping systems
c. water-glycol piping systems
d. steam supply and condensate return systems
e. refrigerant liquid and vapor systems
f. air piping systems for pneumatic control systems
g. air duct systems to carry heated or cooled air to and from spaces to be conditioned.

In the following sections we shall discuss piping systems for the distribution of liquids.

11-14 METHODS OF DISTRIBUTING HEATED AND CHILLED WATER

Water piping systems used in HVAC industry may be classified as *open systems* or *closed systems*. An open system refers to that type of system where one or both ends of the piping system is open to atmospheric pressure. Figure 11-10 is an example of an open system. This system represents a cooling tower used to cool water that is heated as it flows through the refrigerant condenser. Water entering the cooling tower and also that leaving the tower is exposed to atmospheric pressure. Another open piping system would be that used in a spray washer for conditioning air. This was discussed in Chapter 5.

Most heating and cooling water systems that are installed at the present time are referred to as closed systems. The water is sealed in the system of pipes and in the heating and cooling heat-transfer devices under some pressure above atmospheric pressure. One of the reasons for this is that the system water temperature may be above its boiling point (212°F at 14.7 psia) to obtain higher heat output with less water flow. A second reason is that elimination of air from the system is made easier and reduces the tendency of air to leak into the system. Air trapped in a water system produces a restriction to water flow, noisy operation, and provisions must be included in the system for its elimination.

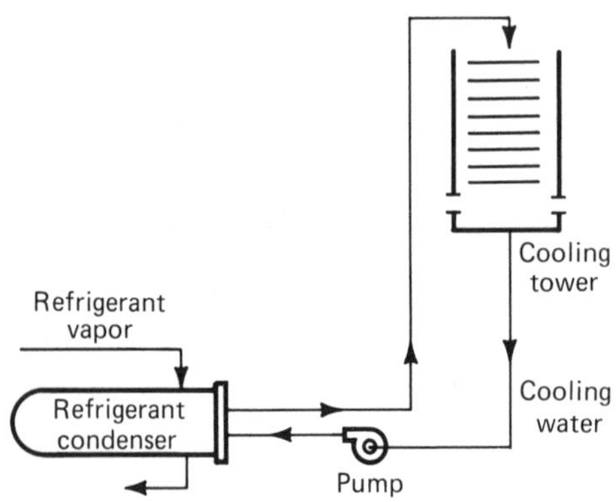

Figure 11–10

Cooling Tower (Open water system)

When water is used in a closed piping system, the arrangement takes the form of supply and return pipes along with the necessary pump, or pumps, to circulate the water through the pipes and heat-transfer devices. Water system piping arrangements, for heating and cooling, can be divided into four general classifications: (1) one-pipe, (2) two-pipe, (3) three-pipe, and (4) four-pipe systems.

The first two are arranged so that either heated or chilled water can be circulated through the piping. However, when hot water is being circulated in one-pipe or two-pipe arrangements, *only* hot water is available to each heating unit connected to that circuit. Conversely, when chilled water is being circulated, *only* chilled water is available to each unit.

The three-pipe arrangement has two supply pipes and one return pipe connected to each heat-transfer device. One supply pipe carries hot water; the other supply pipe carries chilled water. The return carries a mixture of hot and cold water back to the chilling or heating equipment where it is diverted to the chiller or heater, depending on its temperature.

The four-pipe arrangement is actually two, two-pipe arrangements. One of the two supply pipes carries chilled water, and the other supply pipe carries hot water to the heat-transfer devices connected to the circuit. There are also two return lines connected to each device. One carries the hot water back to the boiler or heat exchanger, and the other carries the cold water back to the chiller. Thus, hot and cold water do not mix.

11-15 THE SERIES LOOP HEATING SYSTEM

Perhaps the simplest of all heating systems is called a *series loop system*. It is used almost

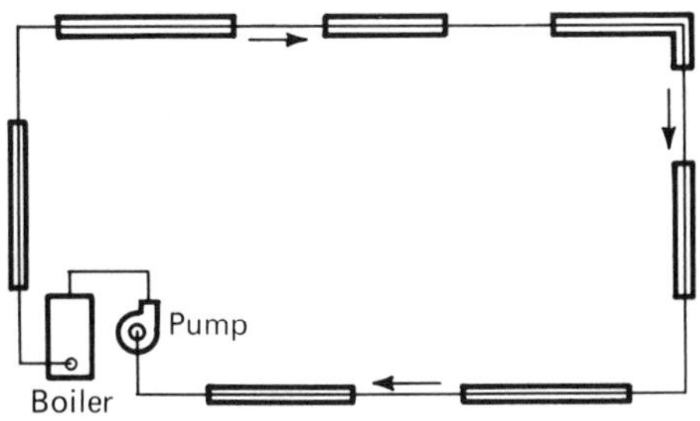

Figure 11–11

Series Loop Heating System (Plan)

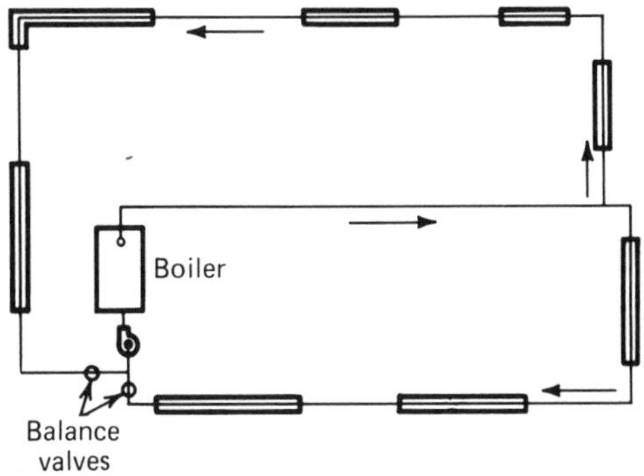

Figure 11–12

Two-Circuit Series Loop System (Plan)

exclusively for residential heating systems because of its limited heating capacity. Figure 11-11 represents a schematic layout of a series loop system. It may be noted that hot water leaving the boiler passes through each consecutive heating unit (usually baseboard heating panels) in series and the cooled water returns to the boiler. The main advantages of such a system are its simplicity and low cost. However, certain conditions exist which limit its use. If one heating unit is shut off, it prevents flow in the remainder of the loop. The pipe size is usually limited to a maximum of 1 in., with ¾ in. more common, which limits its heat output. Also, as the water flows through each heater, its temperature drops because of the heat output. This means that each heater downstream of the first heater receives successively cooler water and may require larger heating units downstream to obtain a given heat output.

Sometimes, if a larger-capacity system is involved, a two-circuit series loop system or a zoned series loop system is used as shown in Figures 11-12 and 11-13

11-16 CONVENTIONAL ONE-PIPE SYSTEM

Figure 11-14 illustrates the most common type of one-pipe systems for water distribution. The system name sometimes depends upon the manufacturer's tradename for the *special Tee fitting* required for each of the heating or cooling units. This special fitting is designed to divert a portion of the flow in the main through the heat transfer device. This special Tee fitting will be discussed in detail in the chapter on piping design. The system generally, consists of one pipe, looped around the building, which is both the supply and return main. This pipe is maintained the same size throughout its length because all the water circulated goes through this one main. At each heat-transfer device a special Tee is installed in the main. The Tee causes a pressure drop which is equal to or greater than the pressure loss through the risers plus the heater, thus causing some water to flow through the short branch circuits. The water returning to the main is at a lower temperature than that in the main. Since this cooled water mixes with the water in the main, the resulting lower temperature water will enter the next heat-transfer device. This water temperature change in the main has the same effect on system design as that discussed with the series loop system.

The heat-transfer devices must be selected for a low pressure drop because the special Tee fittings are designed to create a small pressure

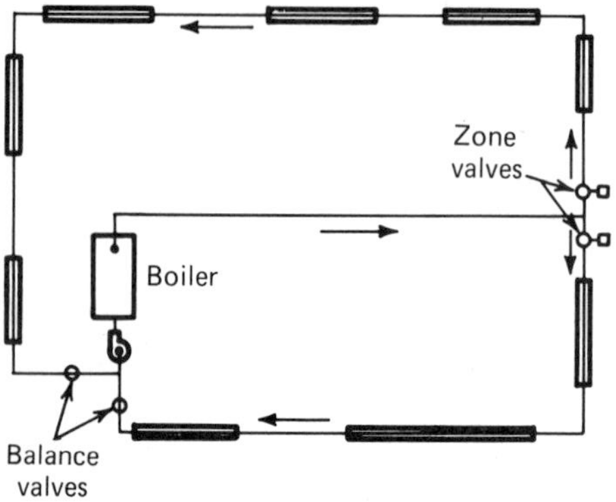

Figure 11–13

Zoned Series Loop System (Plan)

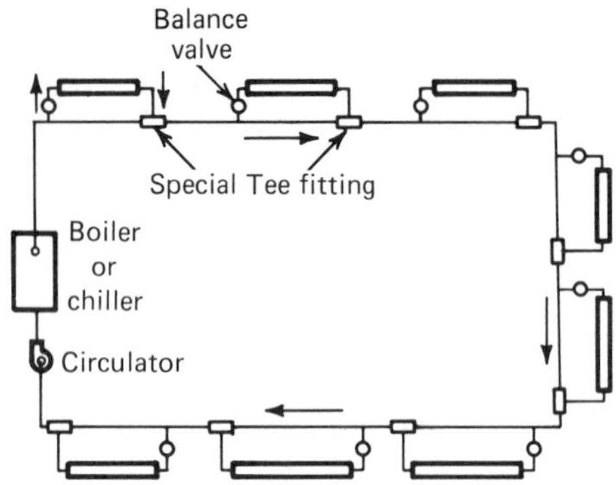

Figure 11–14

One-Pipe Heating or Cooling System (Single Circuit) (Plan)

drop in order to keep the pumping head required for the system within reasonable limits.

The one-pipe system of piping has been used extensively for heating in residences and small commercial establishments. For larger-capacity systems the one-pipe arrangement may be designed in two or more circuits, or combined with two-pipe systems as we shall see.

Some use has been made of the one-pipe arrangement for cooling systems, although this arrangement is not used as extensively for cooling as for heating. However, by proper design and application, the one-pipe arrangement for cooling represents a considerable saving in first cost.

11-17 TWO-PIPE DIRECT-RETURN, AND REVERSE-RETURN SYSTEMS

Figure 11-15 represents the basic *direct-return* piping arrangement. In this layout a supply main carries supply water to each heating or cooling device and a return main returns the water to the boiler or chiller. You may note that the distance along the piping through each heat-transfer device is different. The water flowing from the boiler or chiller through the first unit and back has a much shorter distance to travel than the water flowing through the last unit. Therefore, balancing the flows may be a problem. The pressure loss from the supply main, through the first unit, and back to the return main must be increased by means of some restriction. Usually, an adjustable valve, orifice, or change in pipe size is used; otherwise most of the water would flow through the first unit.

Figure 11-16 illustrates what is called the *reverse-return* piping system. In this system the distance traveled by the water from the boiler or chiller, through any unit, and back is essentially the same. This has been accomplished by making the unit nearest the boiler or chiller on the supply main the farthest on the return main. In other words the water from the supply main to the first unit must flow through the full length of the return main before it returns to the boiler or chiller. From the diagram it would appear that this arrangement would be self-balancing. In practice, it is necessary to install balancing valves or orifices in the piping to obtain the exact water flow required through each unit.

The two-pipe arrangements require more pipe for a given system than the one-pipe system; therefore, they are more expensive. Pipe size along the mains changes because less wa-

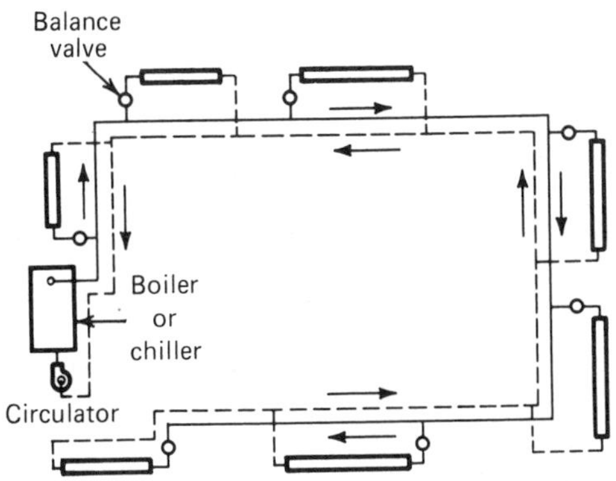

Figure 11–15

Two-Pipe Direct Return System
(Plan)

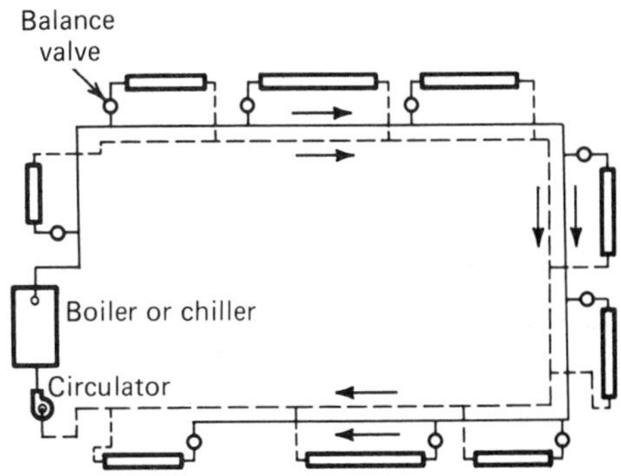

Figure 11–16

Two-Pipe Reversed Return System
(Plan)

ter is flowing after each successive unit (reverse on return main), requiring more fittings and more labor. However, the flexibility of water distribution and temperature control, permissible capacity, and simplicity make these systems well suited to the larger systems.

11-18 COMBINATION ONE-PIPE AND TWO-PIPE SYSTEMS

Frequently, in system design, advantage is taken of the simplicity of both the one-pipe and two-pipe arrangements and their individual advantages are combined without much complication. This is particularly true of multistory structures. Figures 11-17, 11-18, and 11-19 illustrate schematically three such arrangements.

Two-pipe arrangements may also be combined to obtain certain advantages in multistory buildings. Figures 11-20, 11-21, and 11-22 illustrate schematically three such arrangements.

11-19 THREE-PIPE ARRANGEMENT

Figure 11-24 illustrates a simple three-pipe water distribution arrangement. This arrangement makes use of two supply mains and one return main. One supply pipe carries hot water and the other chilled water. A special three-way valve is used for each unit. This valve has two inlets and one outlet. One inlet is connected to the chilled water supply, one inlet to the hot water supply, and the outlet is connected to the coil of the unit. In operation the valve allows either chilled water or hot water to enter the unit coil in variable quantities, but does not allow a mixture of hot and chilled water to flow into the coil. Because some units may be using hot water and some chilled water, there will be a mixing of hot and cold water in the single return line. In order to avoid excessive mixing, and reduce operating costs, it is usually necessary to zone the return lines so that all units that would normally be using cold water will then discharge into the other return. The cooler water is discharged through the chiller and the warmer water through the boiler or heat exchanger. Some method of diverting the return water must be used. One method is to use three pumps, as shown in Figure 11-23. Another method is to use a three-way diverting valve in the return line. This valve is then controlled from the temperature of the return water.

The use of three pumps shown is satisfac-

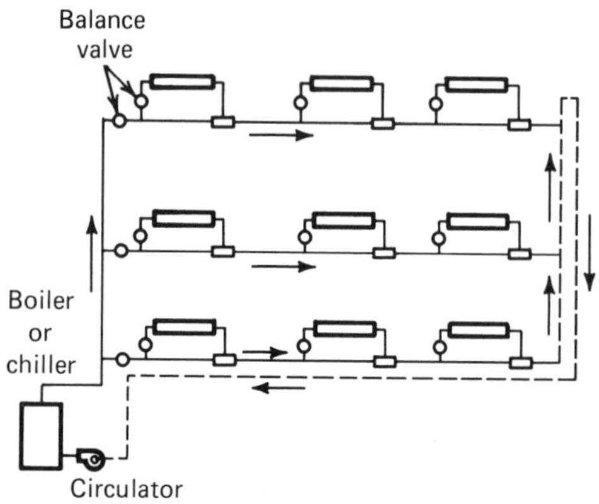

Figure 11–17

Reversed Return System with Horizontal One-Pipe Mains (Elevation)

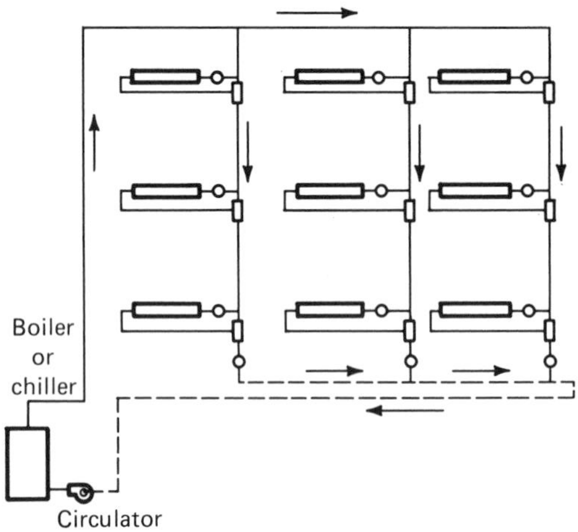

Figure 11–19

Overhead Reversed Return System with Downfeed One-pipe Circuits (Elevation)

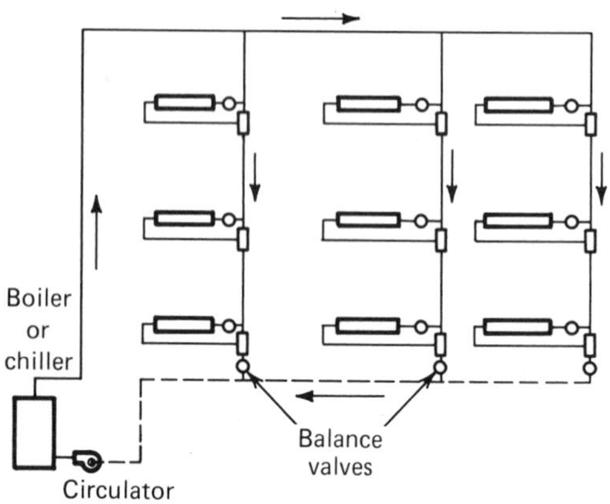

Figure 11–18

Overhead Direct Return System with Downfeed One-pipe Circuits (Elevation)

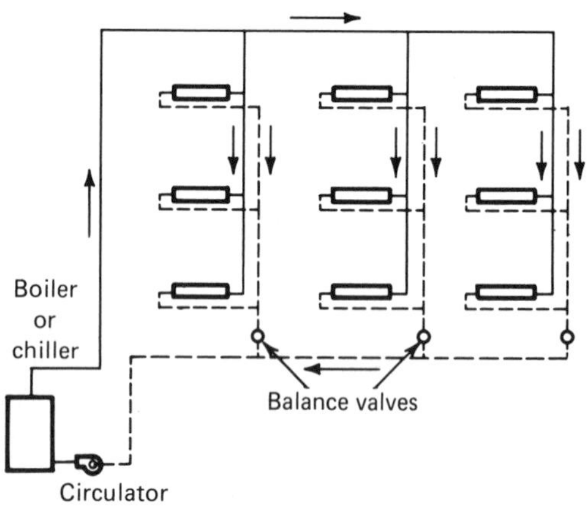

Figure 11–20

Overhead Two-Pipe Direct Return with Reversed Return Radiation Circuits (Elevation)

tory for two zones only. If more than two zones are used, two pumps should be used for each zone. In some cases a diverting valve in the return main from each zone is used to reduce the number of pumps required. Although the use of diverting valves reduces the number of pumps required, additional control valves are required to prevent excessive flow of water through the chiller or boiler.

Three-pipe water circulation arrangements must be designed very carefully to prevent excessive pressure differentials, back flow

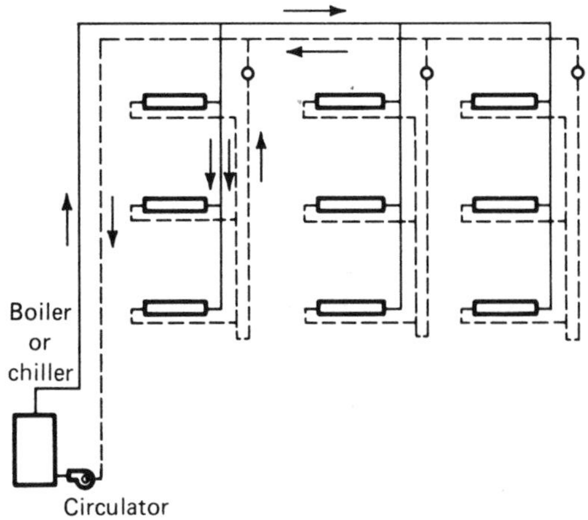

Figure 11–21

Overhead Two-Pipe Downfeed Direct Return with Reverse Return Radiation Circuits (Eelevation)

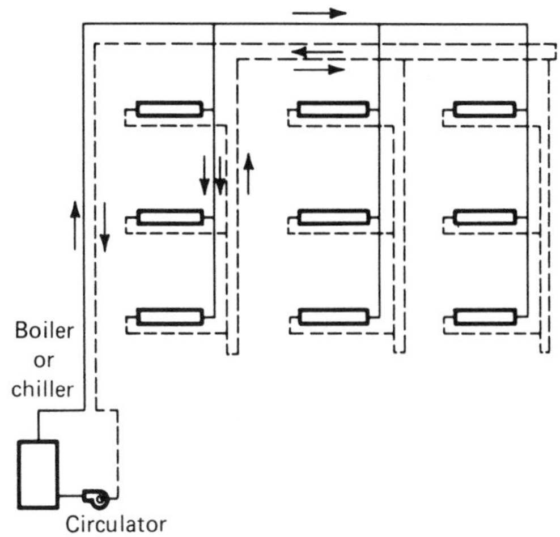

Figure 11–22

Overhead Downfeed Reverse Return System with Reverse Return Radiation Circuits (Elevation)

through the control valves at the units, and balancing problems. The designer should have extensive design experience before attempting this type of system arrangement.

11-20 FOUR-PIPE ARRANGEMENT

The four-pipe arrangement is illustrated in Figure 11-25. This arrangement is essentially two, two-pipe circuits with one two-pipe circuit circulating chilled water and one circulating hot water. There is one distinct difference. Because the circuit through which the hot water is flowing handles only hot water, the temperature of the water can be adjusted upward to reduce the size of piping in this circuit. In this arrangement no water is mixed since the chilled water returns to the chiller and hot water returns to the boiler, or heat exchanger.

By the use of separate coils for heating and cooling, it is possible to dehumidify and reheat with this type of system. Thus, cooling and heating can be made available at all units simultaneously. When one coil is used for cool-

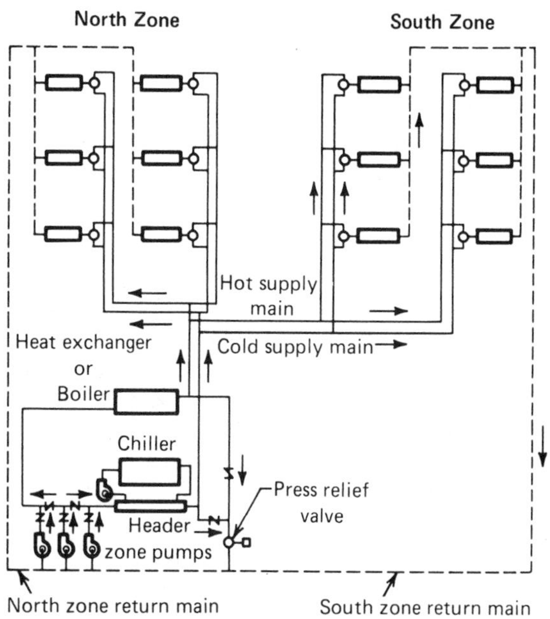

Figure 11–23

Three-Pipe, Three-Pump Distribution System

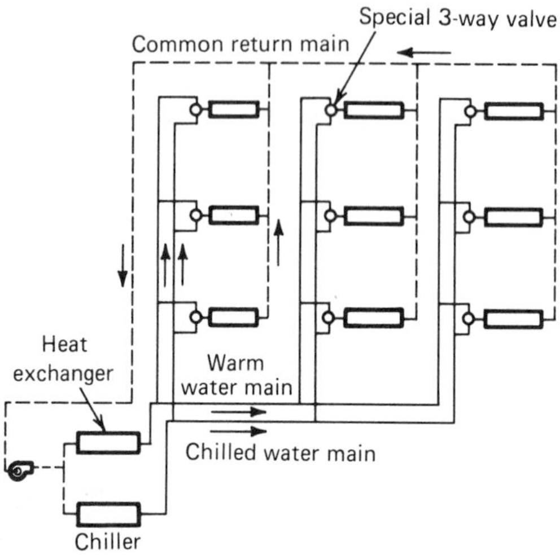

Figure 11–24

Simple Three-Pipe Water Distribution System (Elevation)

ing and heating as shown in Fig. 11-25, the same type of valve is used on the unit supply lines, and in addition, another valve is used on the return from each unit coil to divert the water to the proper return line.

In many large buildings it is customary to use two different hot water circuits—one for the cooling and heating units and one at a higher temperature for the convectors or fin pipe radiation in toilets, stairwells, storage rooms, and so forth. With the four-pipe arrangement it is unnecessary to use two heating circuits since the hot water circuit is entirely separate from the chilled water circuit. Neither is it necessary to zone the piping because each unit has become a zone by itself, furnishing either cooling or heating as required. If a unit, such as a convector, is not to be used for cooling, it is only necessary to omit any cold water connections to it, and merely connnect it to the hot water supply and return lines.

With the four-pipe arrangement, the ultimate in simplicity of design, room temperature control and economy of operation is accomplished with only a slight increase in first cost over the three-pipe arrangement.

11-21 STEAM PIPING SYSTEMS

Steam piping for steam heating systems is normally sized for low-pressure steam (0–15 psig). Steam at these low pressures and usually at saturation conditions create some special problems in piping layout because of condensation of some of the steam within the pipes. This means that in most sections of the steam supply mains both steam and water are flowing.

Steam will flow through a pipe only when the pressure at one end is greater than at the other end. The higher the velocity, the greater the pressure difference must be between the in-

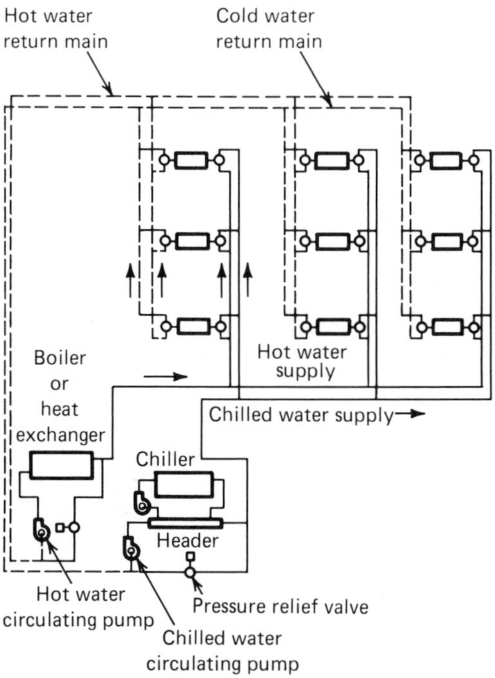

Figure 11–25

Four-Pipe Distribution System

let and outlet. The problems in layout and sizing of a steam piping system then resolve themselves into striking a balance between high velocity with large pressure drop for small pipe and a low velocity with a low pressure drop, requiring a large pipe, and the problem of two-phase (steam and condensate) flow in the same pipe which usually means larger pipes must be used.

The pressure drop caused by friction does not cause a loss in energy because this energy appears as heat. If the steam leaving the boiler is wet or saturated, the heat generated due to friction tends to evaporate the condensate in the supply lines and in some cases may even superheat the steam. Whether the heat gained at the expense of pressure drop is utilized or wasted depends on the equipment that is using the steam. Naturally, if the heat is not utilized, then the pressure used to overcome friction is a total loss to the system. For the average application in heating work, the loss in pressure is useful and not wasted.

Piping arrangements for steam systems are usually the two-pipe direct-return or two-pipe reverse-return arrangements similar to those for water shown in Figs. 11-15 and 11-16. The primary differences are the presence of steam traps at the return side of each heating unit and at the end of each supply main to permit the escape of air and condensate but prevent the passage of steam.

11-12 REFRIGERATION PIPING SYSTEMS

There are no clearly distinguishing piping arrangements for refrigeration systems as we have discussed in the preceding sections for water and steam. However, refrigeration system components must be connected with piping elements, as with any other system. Piping that is too small or improperly installed will cause operating problems. The piping for all refrigeration systems consists of four main sections:

a. the liquid line connecting the liquid receiver and the cooling coil

b. the suction line connecting the cooling coil to the compressor

c. the discharge line connecting compressor to the condenser

d. the condenser drain line connecting the condenser (that does not have an integral subcooler) to the liquid receiver.

11-23 PIPE EXPANSION

It is important in the design of any piping system to give consideration to the expansion and contraction of the piping because of temperature changes in the system. If a pipe is not free to expand and contract, pipe joints may leak, pipe fittings may crack or buckle, and damage may be done to the building structure. Where pipes pass through holes in floors, walls, or beams, the holes must be of ample size to prevent rubbing. The expansion and contraction which occur in other components, such as the boiler and heat distributing units, usually does not need to be considered because they are designed to absorb these stresses. However, careful consideration of the expansion in sections of baseboard and other fin pipe heating elements must be treated as if they were sections of pipes.

The problem, therefore, relates to the changes that occur in the piping system in going from a cold start to the operating temperatures. When the temperature of the pipe in the system is raised or lowered in relation to the surrounding air temperature, there is a corresponding expansion or contraction in the length and diameter of the pipe. Usually the changes in diameter are of minor importance and can be ignored.

The change in length of a pipe may be calculated from the following expression:

$$\Delta L = L_o\alpha(t - t_o) \qquad (11\text{-}30)$$

where ΔL = change in length of pipe for the temperature change $(t - t_o)$.

L_o = original length of pipe at t_o

α = linear coefficient of expansion of the pipe material

Table 11-2 lists the linear expansion of pipe for various temperature increases based on Eq. (11-30), an average coefficient of linear expansion, and a pipe length of 100 ft for steel, brass, or copper. The linear expansion for any length of pipe may be found by dividing the actual pipe length by 100 and multiplying by the linear expansion for the temperature increase under consideration from Table 11-2. For example, consider a 2-in. steel pipe 150 ft long which has an increase in temperature from 70°F to 200°F. From Table 11-2 for a temperature increase of 200 − 70 = 130°F, the expansion per 100 ft is 0.99 in. Therefore, the linear expansion for 150 ft would be (150/100)(0.99) = 1.485 in.

The increase in pipe length due to temperature change must be absorbed either by the design of the system or by the installation of devices designed for this purpose. Packless expansion joints and slip joints in a large variety of forms are commercially available to absorb pipe expansion. Expansion joints composed of pipe fittings and commercially available expansion bends may also be used.

Table 11-2

Linear Expansion of Pipe in Inches Per 100 Feet

Temperature Change,* (°F)	Steel	Brass and Copper
50	0.38	0.57
60	0.45	0.64
70	0.53	0.74
80	0.61	0.85
90	0.68	0.95
100	0.76	1.06
110	0.84	1.17
120	0.91	1.28
130	0.99	1.38
140	1.07	1.49
150	1.15	1.60
160	1.22	1.70
170	1.30	1.81
180	1.37	1.91
190	1.45	2.02
200	1.57	2.13
250	1.99	2.66
300	2.47	3.19
350	2.94	3.72
400	3.46	4.25

*Between fluid in pipe and surrounding air.

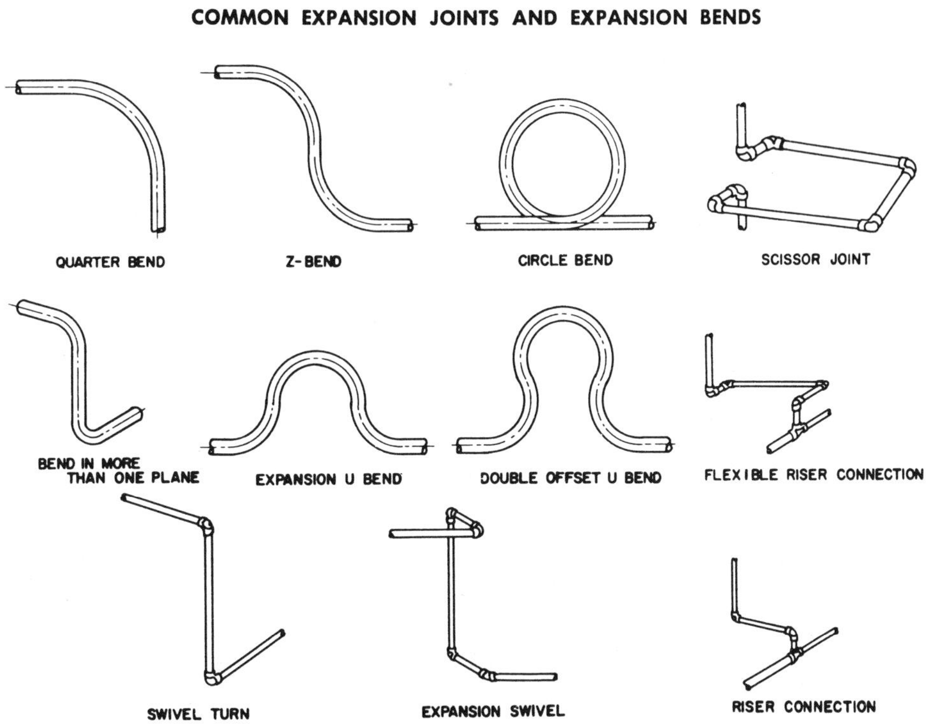

Figure 11–26

Common expansion joints and expansions bends

Figure 11-26 shows a variety of the more common types.

11-24 PIPE ANCHORS AND SUPPORTS

Pipes must be securely anchored and properly supported. Anchoring of the pipe causes the expansion of the pipe to be absorbed at the proper point in the system. Adequate support must be provided by specifically designed hangers. The pipe is heavy and if it contains liquid, such as water, it will tend to sag between even short distances such as 10 ft. The proper spacing for supports for the pipe is dependent primarily on the pipe size and the fluid handled. Manufacturers should always be consulted for proper support spacing and anchoring techniques for pipe systems.

11-25 STEEL PIPE AND COPPER TUBE DATA

The most common pipe material for industrial use is wrought steel. Cast iron is also used, but much less frequently. Table 11-3 gives dimensional data for Schedule 40 (sometimes called standard) wrought steel pipe, as well as Schedule 80 (sometimes called extra strong) pipe. For steam and condensate lines and some hot water systems, steel pipe is used with threaded fittings. In some cases, especially for the larger pipe sizes, pipes are connected by welding. When threaded fittings are

Table 11-3

Dimensions and Properties of Steel Pipe

Nominal[a] Size	ASTM[b] Schedule	Diameter OD In.	Diameter ID In.	Wall Thickness In.	Surface Area Sq Ft/Lin Ft OD	Surface Area Sq Ft/Lin Ft ID	Section Area Sq In. OD	Section Area Sq In. ID	Area of Metal Sq In.	Volume Gal/Lin Ft	Weight[c] (plain end) Lb/Lin Ft	Working Pressure Psia
1/8	40 (s)	0.405	0.269	0.068	0.106	0.0704	0.129	0.0568	0.0720	0.00295	0.244	314[d]
	80 (x)	0.405	0.215	0.095	0.106	0.0563	0.129	0.0363	0.0925	0.00189	0.314	1084
1/4	40 (s)	0.540	0.364	0.088	0.141	0.0953	0.229	0.104	0.125	0.00541	0.424	649
	80 (x)	0.540	0.302	0.119	0.141	0.0791	0.229	0.0716	0.157	0.00372	0.535	1353
3/8	40 (s)	0.675	0.493	0.091	0.177	0.129	0.358	0.191	0.167	0.00992	0.567	574
	80 (x)	0.675	0.423	0.126	0.177	0.111	0.358	0.140	0.217	0.00730	0.738	1191
1/2	40 (s)	0.840	0.622	0.109	0.220	0.163	0.554	0.304	0.250	0.0158	0.850	697
	80 (x)	0.840	0.546	0.147	0.220	0.143	0.554	0.234	0.320	0.0122	1.09	1266
	XX	0.840	0.252	0.294	0.220	0.0660	0.554	0.0499	0.504	0.00259	1.71	3824
3/4	40 (s)	1.050	0.824	0.113	0.275	0.216	0.886	0.533	0.333	0.0277	1.13	604
	80 (x)	1.050	0.742	0.154	0.275	0.194	0.866	0.432	0.434	0.0225	1.47	1078
	XX	1.050	0.434	0.308	0.275	0.114	0.866	0.148	0.718	0.00768	2.44	3134
1	40 (s)	1.315	1.049	0.133	0.344	0.275	1.36	0.864	0.494	0.0449	1.68	651
	80 (x)	1.315	0.957	0.179	0.344	0.251	1.36	0.719	0.639	0.0374	2.17	1083
	XX	1.315	0.599	0.358	0.344	0.157	1.36	0.282	1.08	0.0146	3.66	2963
1¼	40 (s)	1.660	1.380	0.140	0.435	0.361	2.16	1.50	0.669	0.0777	2.27	440
	80 (x)	1.660	1.278	0.191	0.435	0.335	2.16	1.28	0.881	0.0666	3.00	805
	XX	1.660	0.896	0.382	0.435	0.235	2.16	0.630	1.53	0.0328	5.21	2318
1½	40 (s)	1.900	1.610	0.145	0.497	0.421	2.84	2.04	0.800	0.1058	2.72	417
	80 (x)	1.900	1.500	0.200	0.497	0.393	2.84	1.77	1.07	0.0918	3.65	756
	XX	1.900	1.100	0.400	0.497	0.288	2.84	0.950	1.89	0.0494	6.41	2122
2	40 (s)	2.375	2.067	0.154	0.622	0.541	4.43	3.36	1.07	0.174	3.65	376
	80 (x)	2.375	1.939	0.218	0.622	0.508	4.43	2.95	1.48	0.153	5.02	690
	XX	2.375	1.503	0.436	0.622	0.393	4.43	1.77	2.66	0.0922	9.03	1861
2½	40 (s)	2.875	2.469	0.203	0.753	0.646	6.49	4.79	1.70	0.249	5.79	505
	80 (x)	2.875	2.323	0.276	0.753	0.608	6.49	4.24	2.25	0.220	7.66	806
	XX	2.875	1.771	0.552	0.753	0.364	6.49	2.46	4.03	0.128	13.7	2048

Table 11-3 (Continued)

Dimensions and Properties of Steel Pipe

Nominal[a] Size	ASTM[b] Schedule	Diameter OD In.	Diameter ID In.	Wall Thickness In.	Surface Area Sq Ft/Lin Ft OD	Surface Area Sq Ft/Lin Ft ID	Section Area Sq In. OD	Section Area Sq In. ID	Area of Metal Sq In.	Volume Gal/Lin Ft	Weight[c] (plain end) Lb/Lin Ft	Working Pressure Psia
3	40 (s)	3.500	3.068	0.216	0.916	0.803	9.62	7.39	2.23	0.384	7.57	454
	80 (x)	3.500	2.900	0.300	0.916	0.759	9.62	6.61	3.02	0.343	10.3	734
	XX	3.500	2.300	0.600	0.916	0.602	9.62	4.15	5.47	0.216	18.5	1829
3½	40 (s)	4.000	3.548	0.226	1.05	0.929	12.6	9.89	2.68	0.514	9.11	425
	80 (x)	4.000	3.364	0.318	1.05	0.881	12.6	8.89	3.68	0.462	12.5	692
	XX *	4.000	2.728	0.636	1.05	0.714	12.6	5.85	6.72	0.304	22.9	1699
4	40 (s)	4.500	4.026	0.237	1.18	1.05	15.9	12.7	3.17	0.661	10.8	403
	80 (x)	4.500	3.826	0.337	1.18	1.00	15.9	11.5	4.41	0.597	14.9	663
	XX	4.500	3.152	0.674	1.18	0.825	15.9	7.80	8.10	0.405	27.5	1602
5	40 (s)	5.563	5.047	0.258	1.46	1.32	24.3	20.0	4.30	1.04	14.6	498[d(2)]
	80 (x)	5.563	4.813	0.375	1.46	1.26	24.3	18.2	6.11	0.945	20.8	825
	XX	5.563	4.063	0.750	1.46	1.06	24.3	13.0	11.3	0.673	38.6	1951
6	40 (s)	6.625	6.065	0.280	1.73	1.59	34.5	28.9	5.58	1.50	18.0	467
	80 (x)	6.625	5.761	0.432	1.73	1.51	34.5	26.1	8.40	1.35	28.6	825
	XX	6.625	4.897	0.864	1.73	1.28	34.5	18.8	15.6	0.978	53.1	1912

*3½ double extra strong is no longer considered in ASTM specification but some pipe of this size is still manufactured.

[a]The sizes of wrought iron are approximately the same except wall thickness is slightly heavier.

[b]*American Society for Testing and Materials Schedule*. The numbers 30, 40, etc., refer to ASTM Schedule; the letter (s) refers to the former designation *Standard Weight;* the letter (x) refers to the former designation *Extra Strong;* the letters (XX) refer to the former designation *Double Extra Strong*.

[c]Weight per foot is based on plain end pipe. Threaded and coupled (T and C) pipe is slightly heavier.

[d]Working pressure for welded joints.

Table 11-4

Dimensions and Properties of Copper Tube

Nominal Size	Type	Diameter OD In.	Diameter ID In.	Wall Thickness In.	Surface Area Sq Ft/Lin Ft OD	Surface Area Sq Ft/Lin Ft ID	Section Area Sq In. OD	Section Area Sq In. ID	Area of Metal Sq In.	Volume Gal/Lin Ft	Weight[a] Lb/Lin Ft	Working Pressure[b] Psia
¼	K•	0.375	0.305	0.035	0.0982	0.0798	0.110	0.0730	0.0374	0.00379	0.145	918
	L•	0.375	0.315	0.030	0.0982	0.0825	0.110	0.0779	0.0324	0.00404	0.126	764
⅜	K	0.500	0.402	0.049	0.131	0.105	0.196	0.127	0.0695	0.00660	0.269	988
	L	0.500	0.430	0.035	0.131	0.113	0.196	0.145	0.0512	0.00753	0.198	677
½	K	0.625	0.527	0.049	0.164	0.138	0.306	0.218	0.0887	0.0113	0.344	779
	L	0.625	0.545	0.040	0.164	0.143	0.306	0.233	0.0735	0.0121	0.285	625
⅝	K	0.750	0.652	0.049	0.193	0.171	0.441	0.334	0.108	0.0174	0.418	643
	L	0.750	0.666	0.042	0.193	0.174	0.441	0.348	0.0934	0.0181	0.362	547
¾	K	0.875	0.745	0.065	0.229	0.195	0.601	0.436	0.165	0.0227	0.641	747
	L	0.875	0.785	0.045	0.229	0.206	0.601	0.484	0.117	0.0250	0.455	497
1	K	1.125	0.995	0.065	0.295	0.260	0.994	0.778	0.216	0.0405	0.839	574
	L	1.125	1.025	0.050	0.295	0.268	0.994	0.825	0.169	0.0442	0.655	432
1¼	K	1.375	1.245	0.065	0.360	0.326	1.48	1.22	0.268	0.0634	1.04	466
	L	1.375	1.265	0.055	0.360	0.331	1.48	1.26	0.228	0.0655	0.884	387
	M	1.375	1.291	0.042	0.360	0.338	1.48	1.31	0.176	0.0681	0.682	293
	DWV	1.375	1.295	0.040	0.360	0.339	1.48	1.32	0.163	0.0684	0.650	—
1½	K	1.625	1.481	0.072	0.425	0.388	2.07	1.72	0.351	0.0894	1.36	421
	L	1.625	1.505	0.060	0.425	0.394	2.07	1.78	0.295	0.0925	1.14	359
	M	1.625	1.527	0.049	0.425	0.400	2.07	1.83	0.243	0.0950	0.940	289
	DWV	1.625	1.541	0.042	0.425	0.403	2.07	1.86	0.205	0.0969	0.809	—
2	K	2.125	1.959	0.083	0.556	0.513	3.56	3.01	0.532	0.157	2.06	376
	L	2.125	1.985	0.070	0.556	0.520	3.56	3.10	0.452	0.161	1.75	316
	M	2.125	2.009	0.058	0.556	0.526	3.56	3.17	0.377	0.164	1.46	255
	DWV	2.125	2.041	0.042	0.556	0.534	3.56	3.27	0.288	0.142	1.07	—
2½	K	2.625	2.435	0.095	0.687	0.638	5.41	4.66	0.755	0.242	2.93	352
	L	2.625	2.465	0.080	0.687	0.645	5.41	4.77	0.640	0.247	2.48	295
	M	2.625	2.495	0.065	0.687	0.653	5.41	4.89	0.523	0.254	2.03	234

Table 11-4 (Continued)

Dimensions and Properties of Copper Tube

Nominal Size	Type	Diameter OD In.	Diameter ID In.	Wall Thickness In.	Surface Area Sq Ft/Lin Ft OD	Surface Area Sq Ft/Lin Ft ID	Section Area Sq In. OD	Section Area Sq In. ID	Area of Metal Sq In.	Volume Gal/Lin Ft	Weight[a] Lb/Lin Ft	Working Pressure[b] Psia
3	K	3.125	2.907	0.109	0.818	0.761	7.67	6.64	1.03	0.345	4.00	343
	L	3.125	2.945	0.090	0.818	0.771	7.67	6.81	0.858	0.354	3.33	278
	M	3.125	2.981	0.072	0.818	0.780	7.67	6.98	0.691	0.362	2.68	220
	DWV	3.125	3.035	0.045	0.818	0.796	7.67	7.23	0.435	0.376	1.69	—
3½	K	3.625	3.385	0.120	0.949	0.886	10.3	9.00	1.32	0.468	5.12	324
	L	3.625	3.425	0.100	0.949	0.897	10.3	9.21	1.11	0.478	4.29	268
	M	3.625	3.459	0.083	0.949	0.906	10.3	9.40	0.924	0.489	3.58	218
4	K	4.125	3.857	0.134	1.08	1.01	13.3	11.7	1.68	0.607	6.51	315
	L	4.125	3.905	0.110	1.08	1.02	13.3	12.0	1.39	0.623	5.38	256
	M	4.125	3.935	0.095	1.08	1.03	13.3	12.2	1.20	0.634	4.66	217
	DWV	4.125	4.009	0.058	1.08	1.05	13.3	12.6	0.67	0.656	2.87	—
5	K	5.125	4.805	0.160	1.34	1.26	20.7	18.1	2.50	0.940	9.67	307
	L	5.125	4.875	0.125	1.34	1.28	20.7	18.7	1.96	0.971	7.61	234
	M	5.125	4.907	0.109	1.34	1.29	20.7	18.9	1.72	0.981	6.66	203
6	K	6.225	5.741	0.192	1.60	1.50	29.4	25.9	3.50	1.35	13.9	308
	L	6.125	5.845	0.140	1.60	1.53	29.4	26.8	2.63	1.39	10.2	221
	M	6.125	5.881	0.122	1.60	1.54	29.4	27.2	2.30	1.42	8.92	190
	DWV	6.125	5.959	0.083	1.60	1.56	29.4	27.9	1.58	1.55	6.10	—

[a]Weight per foot is based on tube without couplings.

[b]Working pressure is based on the ANSI *Code for Pressure Piping,* published by the ASME, 1968 edition, for plain end tubing (sweat joints).

[c]Type K and L furnished in both hard and soft temper, Type M is hard only. Standard straight lengths are 20 ft. Standard coils (¼ to ½ in.) are 60 ft.

[d]3½ in. copper tube is not readily available.

used, it is necessary to ream the ends of the pipe to eliminate burrs and to avoid the possibility of a partial closure of the internal bore. If this is not done, design computations for friction loss are uncertain. When steel pipe is used for hot water systems, it may be desirable to use galvanized pipe, which is steel pipe covered with a corrosion-resistant zinc coating. Such coatings can prevent rusting for long periods of time.

Copper tubing has been used extensively for hot water and chilled water systems (less frequently for steam systems), for refrigeration systems. Table 11-4 gives the characteristic data for commercial copper tubing. Copper tubing is seamless, deoxidized copper tube supplied in three wall thicknesses. Type *K*, the heaviest, is usually supplied in coils for the smaller sizes, and in 12- and 20-ft lengths for the larger sizes. It is furnished in either hard or soft temper, and is usually specified for underground service.

Type *L* has a lesser wall thickness. It is supplied in coils up to and including 1½ in. nominal and in 12- and 20-ft lengths for sizes above 1½ in. It is supplied in either hard or soft temper.

Type *M* tubing, which has the thinnest wall, is furnished in hard temper in 12- and 20-ft lengths. Hard temper tubing, which is rigid, is pleasing in appearance and is most often used where runs of pipe are exposed.

Copper water tubing, such as types *K, L,* and *M,* is not intended to be joined by threaded fittings. A most common type of connection is the solder joint, or sweated joint. In assembly the tube ends are cut square to length, the end mating surfaces are cleaned with steel wool, and a flux is applied to the surfaces. The tube is inserted into the fitting, and the joint is then heated. Solder, usually of lead-tin or tin-antimony, melts on the hot surface and is drawn into the space between the fitting and tube by capillary action. After cooling and solidification, a tight and permanent joint results. Where temperatures may exceed 250°F for lead-tin solder, or 312°F for tin-antimony, it is necessary to use a brazing alloy or silver solder for the joints. Type *K* and type *L* copper water tubes are also joined by compression fittings.

CHAPTER 11

Review

11-1. A fluid tested in a Saybolt viscosimeter has a viscosity of 93 sec at 100°F and has a specific gravity measured at 60°F of 0.850. Calculate the viscosity of the fluid at 100°F in centistokes, centipoise, ft^2/sec, and lb sec/ft^2.

11-2. The fluid of Problem 11-1 is flowing through a 2-in. nominal Schedule 40 steel pipe at the rate of 40 gpm. What is the flow velocity and the Reynolds number of the flow? Is it laminar or turbulent?

11-3. A 30 percent solution of ethylene glycol and water has a specific gravity of 1.049 and a viscosity of 5.6 cP at 20°F. It is flowing through a 2½-in. Schedule 40 steel pipe at the rate of 100 gpm. What is the flow velocity and Reynolds number of the flow?

11-4. Water flows in a 2-in. copper tube (type *L*) at the rate of 120 gpm. The 2-in. tube reduces to a 1½-in. tube at a point in the flow. A pressure gauge attached to the 2-in. tube indicates a pressure of 50 psig. Downstream, in the 1½-in. tube, a pressure gauge is reading 46 psia. What is the lost pressure head (in feet of water) between the two gauges? Assume the system is horizontal.

11-5. No. 3 fuel oil at 60°F flows through a 2-in. Schedule 40 steel pipe at the rate of 40,000 lb/hr. Determine the flow rate in gallons per minute and the average flow velocity. Assume the specific gravity at 60°F is 0.898.

11-6. The maximum flow rate of a liquid will be 300 gpm with a maximum average velocity of 12 ft per second through a Schedule 40 steel pipe. Determine the smallest suitable nominal pipe size required and the actual average flow velocity in the pipe.

11-7. Water at 150°F flows steadily through a 4-in. nominal Schedule 40 steel pipe at the rate of 400 gpm. Determine the flow rate in lb/hr, the Reynolds number and the friction factor. Assume the density of water is 61.19 lb/ft^3 and the viscosity is 0.43 cP at 150°F.

11-8. Calculate the head loss and pressure loss for the flow conditions of Problem 11-7 if the pipe is 125 ft long.

11-9. Water at 100°F flows at the rate of 100 gpm through a 2-in. type *L* copper tube. Determine the head loss and pressure loss per 100 ft of tube. Assume the density and viscosity are 61.99 lb/ft^3 and 0.69 cP, respectively.

11-10. Water at 80°F flows through a section of 2-in. commercial steel pipe (Schedule 40) with a measured pressure loss of 4.5 psi in a section 250 ft long. Estimate the flow rate through the pipe in ft^3/sec and gpm if the absolute viscosity is 0.85 cP and the water density is 62.22 lb/ft^3.

11-11. What nominal, Schedule 40 steel pipe should be used for a flow rate of 150 gpm with a head loss of not more than 20 ft of water per 100 ft of pipe length? Assume the water has a density of 62.4 lb/ft^3 and a viscosity of 7.6×10^{-6} ft^2/sec.

11-12. Calculate the head loss for 260 ft of 2-in. nominal Schedule 40 steel pipe. The pipe contains four standard 90-degree elbows, one globe valve, one gate valve, and one swing check valve ($K = 50\ f_t$). Water is flowing at the rate of 75 gpm and has a density of 62.4 lb/ft^3 and a viscosity of 1.3 cP.

11-13. Water at 60°F is flowing through the piping system shown in Fig. 11-27 at the rate of 400 gpm. Determine the average flow velocity in the 4- and 5-in. pipe sections and the pressure difference between gauges p_1 and p_2. Assume water density of 62.37 lb/ft^3, viscosity 1.1 cP.

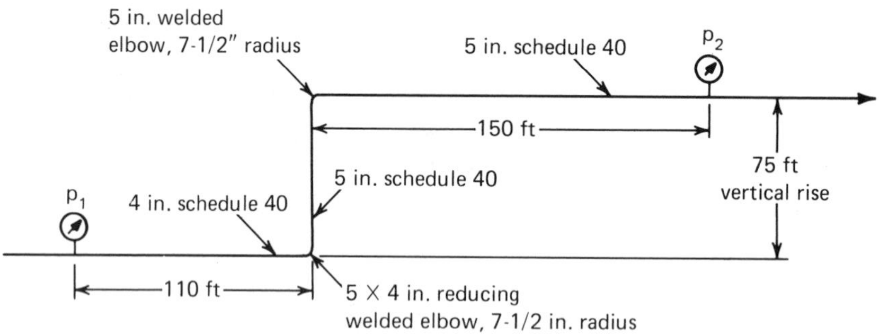

Figure 11–27

Piping Diagram for Problem 11-13 (Elevation)

11-14. Water at 80°F is flowing through the 3-in. Schedule 40 pipe at the rate of 200 gpm. The valve shown is a gate valve, the check valve is a swing check valve ($K = 100\ f_t$, $\beta = 1$), and all elbows are standard threaded 90-degree elbows. Determine the pressure difference between gauges p_1 and p_2. Assume water density is 62.22 lb/ft^3 and viscosity is 0.85 cP.

Figure 11–28

Piping Diagram for Problem 11-14 (Elevation)

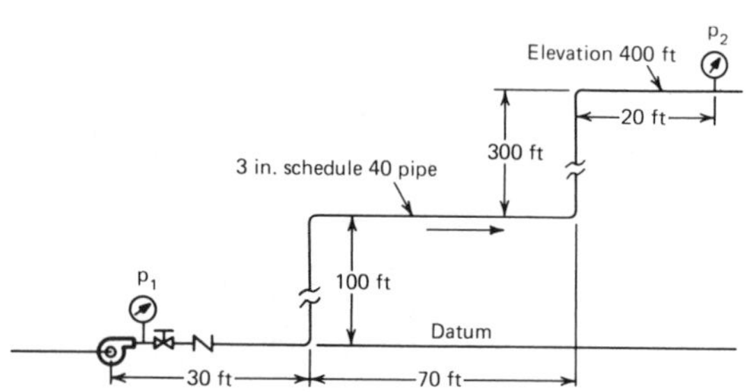

BIBLIOGRAPHY

1. ASHRAE *Handbook of Fundamentals,* American Society of Heating, Refrigerating and Air-Conditioning Engineers, Atlanta, GA, 1981.

2. *Pipe Friction Manual,* Hydraulic Institute, Cleveland, Ohio, 1961.

3. Technical Paper No. 410, *Flow of Fluids Through Valves, Fittings and Pipes,* Crane Company, Chicago, IL, 1976.

4. *ASHRAE Handbook and Product Directory—Systems,* American Society of Heating, Refrigerating, and Air-Conditioning Engineers, Atlanta, GA, 1980.

5. ASHRAE, *Guide and Data Book—Equipment,* American Society of Heating, Refrigerating, and Air-Conditioning Engineers, Atlanta, GA, 1979.

Chapter 12

Pumps and Piping System Design

12-1 INTRODUCTION

Movement of liquids through piping systems against system resistance requires the use of some form of pump. Generally, pumps are classified as *positive displacement* or *nonpositive displacement*.

The reciprocating piston pump is a positive-acting type, which means it is a displacement pump which creates lift and pressure by displacing liquid with a piston moving in a cylinder. The chamber or cylinder is alternately filled and emptied by forcing and drawing the liquid by mechanical action. This type is called *positive* inasmuch as the only limitation on pressure that may be developed is the strength of the structural parts. Volume or capacity delivered is constant (except for slippage) regardless of pressure and is varied only by speed changes. This type of pump has limited use in the HVAC industry; however, one may find such pumps used for pumping hot and cold water, pneumatic pressure systems, feeding small boilers, steam condensate return, gasoline and light-oil pumping, and small sprinkler irrigation jobs.

Another type of positive displacement pump is a rotary pump. The rotary pump is simple in design, has few moving parts and, like a reciprocating pump, is positive acting. It consists primarily of two cams or gears, spur or herringbone, in mesh, the so-called idler gear driven by the "driving" gear which is driven from an outside source of power. A close fitting casing surrounds the cams and contains the suction and discharge connections. Liquid fills the spaces between the cam teeth and casing. The cams rotate carrying the trapped liquid which is discharged (squeezed) out the discharge. Such operation produces a very even and continuous flow. As in the case of reciprocating pumps, the capacity delivered is constant (except for slippage) regardless of pressure. Due to the necessary close clearance and metal-to-metal contact, such units naturally work best and last longest when pumping liquids that have good lubricating qualities. These pumps are generally used where the capacity required is small but the pressure required may be medium to high. One particular application is the oil pump installed in an oil burner.

The centrifugal pump is by far the most frequently used pump in industry, and in particular, the HVAC industry. It is a nonpositive displacement pump, which means that its capacity is dependent on the resistance offered to flow by the system in which the pump is installed.

12-2 CENTRIFUGAL PUMPS

The name of this type of pump comes from the type of force exerted by a moving body in a circular path—centrifugal force. In such a pump the fluid is forced to revolve, and it exerts a centrifugal force on the fluid in the case surrounding the revolving wheel (impeller) which is equal to the discharge pressure or head. The fluid enters the pump inlet (''eye'' of the impeller) and is thrown by centrifugal force into the volute casing surrounding the impeller. The fluid leaving the impeller tip has high kinetic energy that is partially converted to static pressure in the volute and diffuser. The fluid forced out of the casing creates a partial vacuum permitting atmospheric pressure to force more fluid into the pump suction, and the operation is continuous. There are numerous types of impellers and casings; however, the principal of operation is the same.

With no pistons, valves, or close tolerances, a centrifugal pump is not positive acting. In other words a particular impeller is good for so much pressure (at a given speed) or head, and if actual pressure or head is higher than this, the impeller will churn the fluid in the casing.

The features of the centrifugal pump that makes it desirable for so many applications are simple construction, no close clearances, small floor space, low first cost, inexpensive and easy to maintain, no excessive pressures even with discharge valve closed, maximum practical suction lift (15 ft), smooth nonpulsating flow, quiet, no valves or reciprocating parts, high-speed operation, and may be belt driven, direct motor driven, or driven by turbines or engines.

The essential parts of a centrifugal pump are the rotating member (impeller) and the surrounding volute casing. The impeller is usually driven by a constant-speed electric motor and may be *inline, close-coupled,* or *base-mounted type*.

Inline pumps are normally used with fractional horsepower electric motors up to 1 hp operating at 1750 rpm. They are relatively light in weight and, because they are supported by the piping, are designed for extremely quiet operation. For this reason sleeve bearings are used in both the motor and pump bearing bracket (see Fig. 12-1). This type of pump is generally equipped with mechanical seals, and the motor is connected to the impeller shaft with a flexible coupling. Because of their quiet operation, these pumps are used in residences and small commercial and apartment buildings.

Close-coupled pumps are normally used with motors from ¼ to 40 hp and at speeds of 1750 to 3450 rpm. Motors are furnished with sleeve or ball bearings, and the impeller is connected directly to the motor shaft. Seals

Figure 12–1

Inline centrifugal water pumps. (See Fig. 12-28 for performance curves.) (Courtesy of Taco, Inc., Cranston, Rhode Island.)

may be either mechanical or packed type. These pumps are somewhat less expensive than the base mounted type (Fig. 12-2b) and are used in larger installations where noise is not a factor. However, when quiet operation is required, an extra quiet sleeve-bearing motor can be supplied. For high-temperature operation, cooling jackets are available.

Base-mounted type of pump, Fig 12-2b, is normally used with motors from ¼ hp on up and at speeds of 1750 or 3450 rpm. (Different speeds are obtained by using pulleys and belt drive.) Motors are furnished with sleeve or ball bearings and connected to the impeller shaft by a flexible coupling.

12-3 HEAD DEVELOPED BY A CENTRIFUGAL PUMP

The pressure at any point in a liquid can be thought of as being caused by a vertical column of liquid which, due to its weight, exerts a pressure equal to the pressure at the point in question. The height of this column is called *static head* and is expressed in feet or meters. The static head corresponding to any specific pressure is dependent upon the weight of liquid according to the following relationship:

$$\text{Head in feet} = \frac{\text{Pressure in psi} \times 2.31}{\text{Specific gravity}} \qquad (12\text{-}1)$$

A centrifugal pump imparts velocity to a liquid (adds kinetic energy). This velocity energy is transformed largely into pressure energy in the diffuser as the liquid leaves the pump. Therefore, the head developed is approximately equal to the velocity energy at the periphery of the impeller. This relationship is expressed by the following well-known relationship:

$$\text{Pump Head } (h_p) = \frac{V_p^2}{2g} \qquad (12\text{-}2)$$

where h_p = total developed pump head (ft)

V_p = linear velocity at periphery of impeller (ft/sec)

g = 32.2 ft/sec^2

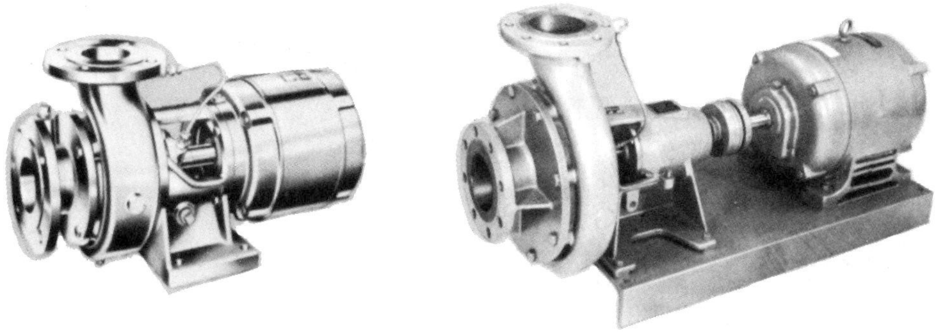

Figure 12–2

Centrifugal water pumps: (a) Close-coupled (b) Base-mounted (Courtesy of Taco, Inc., Cranston, Rhode Island.)

We may predict the approximate head developed by a centrifugal pump by calculating the peripheral velocity of the impeller and substituting into Eq. (12-2). A convenient formula for peripheral velocity is

$$V = \frac{\text{rpm} \times d}{229} \tag{12-3}$$

where V = peripheral velocity (ft/sec)

d = impeller diameter (inches)

The above demonstrates why we must always think in terms of feet of liquid rather than pressure when working with centrifugal pumps. A given pump with a given impeller diameter and speed will raise a liquid to a certain height regardless of the weight of liquid, as shown in Fig. 12-3.

12-4 SYSTEM HEAD ON A PUMP

Selection of the proper pump for a particular application is of primary importance. It is necessary to determine the required flow rate through the pump and the head or pressure against which the pump is to operate.

Equation (11-8) was written to represent the energy levels in a steady-flow fluid stream at two different points in the system. Equation (11-8) may be rewritten to include the energy added to the fluid between two points in a fluid system by the device we have been calling a pump. The resulting equation would be

$$\frac{g}{g_c}(Z_1) + p_1v_1 + \frac{v_1^2}{2g_c} + Ju_1 = \frac{g}{g_c}(Z_2) + p_2v_2 + \frac{V_2^2}{2g_c} + Ju_2 + {}_1\text{wk}_2 + h_L \tag{12-4}$$

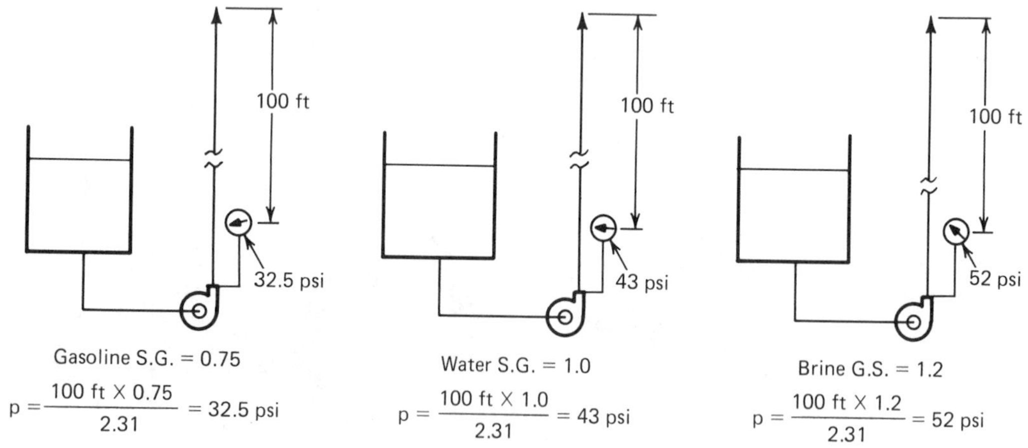

Figure 12–3

Identical pumps handling liquids of different specific gravities (adopted from GPM, Goulds Pump Manual, Goulds Pumps, Inc., by permission.)

Again, recall that each term has units of ft lbf/lbm or J/kg, which are energy units. The work term, ${}_1wk_2$ represents the energy added to the fluid by the pump between points 1 and 2, per pound-mass of fluid. Figure 12-4 represents a schematic drawing of a steady-flow system where Eq. (12-4) applies. Assuming the same conditions as in Section 11-7, we may write

$$\frac{g}{g_c}(Z_1)\left(\frac{g_c}{g}\right) + \left(\frac{g_c}{g}\right)(p_2v_1) + \left(\frac{g_c}{g}\right)\left(\frac{V^2}{2g_c}\right)$$
$$= \frac{g}{g_c}(Z_2)\left(\frac{g_c}{g}\right) + \left(\frac{g_c}{g}\right)(p_2v_2) + \left(\frac{g_c}{g}\right)\left(\frac{V_2^2}{2g_c}\right)$$
$$+ \left(\frac{g_c}{g}\right)({}_1wk_2) + h_L$$

Recalling that $v = 1/\rho$, $\gamma = \rho(g/g_c)$, and for incompressible flow $\gamma_1 = \gamma_2$, we would have

$$Z_1 + \frac{P_1}{\gamma} + \frac{V_1^2}{2g}$$
$$= Z_2 + \frac{P_2}{\gamma} + \frac{V_2^2}{2g} + h_p + h_L \qquad (12\text{-}5)$$

where $h_p = (g_c/g)({}_1wk_2)$ and will be called pump head having units of length, feet or meters. Rearranging, combining terms, and changing signs we have

$$h_p = (Z_2 - Z_1) + \left(\frac{P_2 - P_1}{\gamma}\right)$$
$$+ \left(\frac{V_2^2 - V_1^2}{2g}\right) + h_L \qquad (12\text{-}6)$$

This arrangement gives a positive value for pump head for convenience. The various combined terms are frequently referred to as

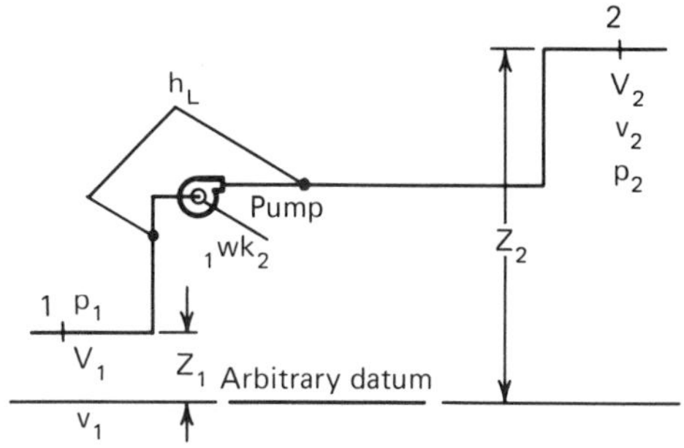

Figure 12–4

General pumping system, elevation (Not to scale.)

$(Z_2 - Z_1)$ = net static elevation head

$\left(\frac{P_2 - P_1}{\gamma}\right)$ = net static pressure head

$\left(\frac{V_2^2 - V_1^2}{2g}\right)$ = net velocity head

In Fig 12-4 an arbitrary datum was selected in such a way that both Z_1 and Z_2 were positive (above datum). It would also have been completely satisfactory to have the datum passing through the centerline of the pump. This is the usual practice to evaluate pump head. To determine the *total dynamic head* (TDH) against which a pump must operate, certain terms and definitions are used. Referring to Fig. 12-5a and 12-5b, these definitions are as follows:

Suction Lift exists when the source of supply is below the centerline of the pump. Thus, the *static suction lift* is the vertical distance in feet from the centerline of the pump to the free level of the liquid to be pumped.

Suction head exists when the source of supply is above the centerline of the pump. Thus, the *static suction head* is the vertical distance in feet from the centerline of the pump

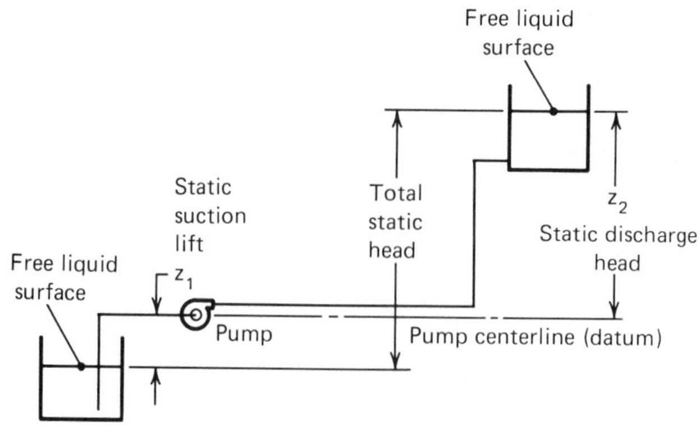

Figure 12–5a

General pumping system showing suction lift (Elevation, no scale)

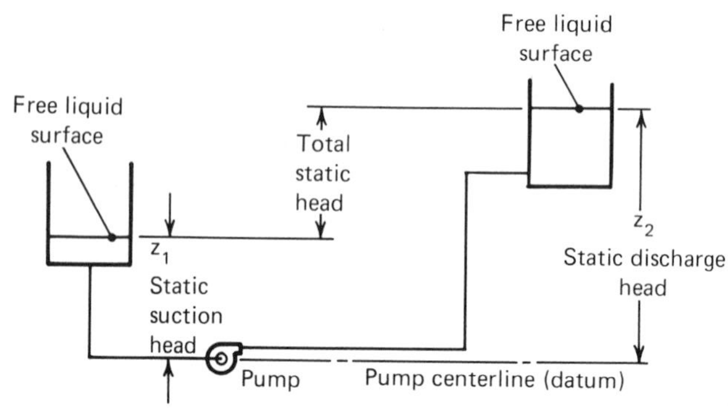

Figure 12–5b

General pumping system showing suction head (Elevation, no scale)

to the free level of the liquid to be pumped.

Static discharge head is the vertical distance in feet between the pump centerline and the point of free discharge or the surface of the liquid in the discharge tank.

Total static head is the vertical distance in feet between the free level of the source of supply and the point of free discharge or the free surface of the discharge liquid.

Friction head (h_L) is the head required to overcome the resistance to flow in the pipe and fittings. It is dependent upon the size and type of pipe, flow rate, and nature of the liquid (see Chapter 11).

Velocity head (h_v) is the energy of the liquid as a result of its motion at some velocity V. It is the equivalent head in feet through which the fluid would have to fall to acquire the same velocity; or in other words, the head necessary to accelerate the water ($h_v = V^2/2g$) (see Chapter 11). The velocity head is usually insignificant and can be ignored in most high-head systems. However, it can be a large factor and must be considered in low-head systems.

Pressure head must be considered when a pumping system either begins or terminates in a tank which is under some pressure other than atmospheric. The pressure in such a tank must first be converted into feet of liquid, Eq. (12-1). A vacuum in the suction tank or a positive pressure in the discharge tank must be added to the system head, whereas a positive pressure in the suction tank or vacuum in the discharge tank would be subtracted. Conversion of inches of mercury vacuum to feet of liquid may be accomplished by,

$$\text{Feet of liquid} = \frac{\text{Vacuum, in. Hg} \times 1.13}{\text{SG}}$$

The above forms of head, namely static, friction, velocity, and pressure, are combined to make up the total system head at any particular flow rate. Following are definitions of these combined or dynamic head terms as they apply to the pump.

Total dynamic suction lift (TDSL) is the static suction lift plus the velocity head at the pump suction flange plus the total friction head in the suction line. The TDSL, as determined on a pump test, is the reading of a gauge on the suction flange, converted to feet of liquid and corrected to the pump centerline, minus the velocity head at the point of gauge attachment.

Total dynamic suction head (TDSH) is the static suction head minus the velocity head at the pump suction flange minus the total friction head in the suction line. The total dynamic suction head, as determined on a pump test, is the reading of a gauge on the suction flange, converted to feet of liquid and corrected to pump centerline, plus the velocity head at the point of gauge attachment.

Total dynamic discharge head (TDDH) is the static discharge head plus the velocity head at the pump discharge flange plus the total friction head in the discharge line. The total dynamic discharge head, as determined on a pump test, is the reading of a gauge at the discharge flange, converted to feet of liquid and corrected to pump centerline, plus the velocity head at the point of gauge attachment.

Total head or total dynamic head (TDH) is the total dynamic discharge head minus the total dynamic suction head or plus the total dynamic dynamic suction lift.

$$h_p = \text{TDH} = \text{TDDH} - \text{TDSH}$$

(with a suction head)

or

$$h_p = \text{TDH} = \text{TDDH} + \text{TDSL}$$

(with a suction lift)

12-5 NET POSITIVE SUCTION HEAD (NPSH) AND CAVITATION

The Hydraulic Institute defines NPSH as the total suction head in feet of fluid absolute, determined at the suction nozzle of the pump and corrected to datum, less the vapor pressure of the liquid in feet absolute. Simply stated, it is an analysis of energy conditions on the suction side of a pump to determine if the liquid will vaporize at the lowest pressure point in the pump.

The pressure which a liquid exerts on its surroundings is dependent upon its temperature. This pressure, called vapor pressure, is a unique characteristic of every fluid and increases with increasing temperature. When the vapor pressure within the fluid reaches the pressure of the surrounding medium, the fluid begins to vaporize or boil. The temperature at which this vaporization occurs will decrease as the pressure of the surrounding medium decreases.

A liquid increases greatly in volume when it vaporizes. One cubic foot of water at room temperature becomes 1700 ft^3 of vapor at the same temperature. It is obvious from this that if we are to pump a liquid effectively, we must keep it in liquid form. NPSH is simply a measure of the amount of suction head present to prevent this vaporization at the lowest pressure point in the pump.

NPSH required ($NPSH_R$) is a function of the pump design. As the liquid passes from the pump suction to the eye of the impeller, the velocity increases and the pressure decreases. There are also pressure losses due to shock and turbulence as the liquid strikes the impeller. The centrifugal force of the impeller vanes further increases the velocity and decreases the pressure of the liquid. The $NPSH_R$ is the positive head in feet absolute required at the pump suction to overcome these pressure drops in the pump and maintain the liquid above its vapor pressure. The $NPSH_R$ varies with speed and capacity within any given pump. Pump manufacturers' performance curves normally provide this information.

NPSH available ($NPSH_A$) is a function of the system in which the pump operates. It is the excess pressure of the liquid in feet absolute over its vapor pressure as it arrives at the pump suction. The $NPSH_A$ may be calculated for any system by

$$NPSH_A = \frac{P_B - P_v}{\gamma} + (S - h_L) \qquad (12\text{-}7)$$

where P_B = pressure in lbf/ft^2 absolute on the liquid surface in the suction tank

P_v = vapor pressure of the liquid at the liquid temperature (lbf/ft^2 absolute)

γ = specific weight of the liquid (lbf/ft^3)

S = vertical distance from pump centerline to level of liquid in the suction tank (ft). (S is positive if level is above pump centerline, negative if below)

h_L = total friction loss in pump suction pipe and fittings (ft of fluid)

In an existing system the $NPSH_A$ can be determined by a gauge reading on the pump suction. The following relationship may be used:

$$NPSH_A = \frac{P_B - P_v}{\gamma} \pm GR + h_v \qquad (12\text{-}8)$$

where GR = gauge reading at pump suction expressed in feet (plus if above atmospheric, minus if below atmospheric) corrected to the pump centerline

h_v = velocity head in the suction pipe at the gauge connection (ft)

Cavitation is a term used to describe the phenomenon that occurs in a pump when there is insufficient $NPSH_A$. The pressure of the liquid is reduced to a value equal to or below its vapor pressure, and small vapor bubbles or pockets begin to form. As these vapor bubbles move along the impeller vanes to a higher-pressure area, they rapidly collapse. The collapse, or implosion is so rapid that it may be heard as a rumbling noise. The forces during the collapse are generally high enough to cause minute pockets of fatigue failure on the impeller vane surfaces. This action may be progressive, and under severe conditions can cause serious pitting damage to the impeller. The accompanying noise is the easiest way to recognize cavitation. Besides impeller damage, cavitation normally results in reduced capacity due to vapor present in the pump. Also, the head may be reduced and unstable, and the power consumption may be erratic. Vibration and mechanical damage such as bearing failure can also occur as a result of operating in cavitation. The only way to prevent the undesirable effects of cavitation is to insure that $NPSH_A$ in the system is greater than $NPSH_R$ by the pump.

Illustrative Problem 12-1

Figure 12-6 represents an elevation drawing of a pumping system. The pump receives water at 100°F from a storage reservoir and transfers the water through a shell and tube heat exchanger to the inlet of a closed storage tank where the pressure is 20 psig. The flow rate required is 99,400 lb/hr. Determine: (a) pump capacity in gpm, (b) pump total dynamic head, and (c) available net positive suction head for the pump. (Assume commercial steel Schedule 40 pipe)

Solution:

(a) Determine flow rate gpm.

Since the flow rate is given in lb/hr, the specific gravity must be calculated at 100°F and at 180°F to determine flows in gpm:

$$SG_{100} = \frac{\rho_{100}}{\rho_{60}} = \frac{v_{60}}{v_{100}} = \frac{0.016035}{0.016130} = 0.994$$

$$SG_{180} = \frac{v_{60}}{v_{180}} = \frac{0.016035}{0.016509} = 0.971$$

By Eq. (12-11)

$$gpm_{100} = \frac{\text{lb/hr}}{500 \times SG} = \frac{99{,}400}{(500)(0.994)}$$
$$= 200 \text{ gpm}$$

$$gpm_{180} = \frac{99{,}400}{(500)(0.971)} = 204.7 \text{ gpm}$$

This shows a relatively small change in flow rate with the temperatures involved. Therefore, we will assume a flow rate constant throughout system of 200 gpm. As far as the pump is concerned, it handles 200 gpm of 100°F water.

(b) Step 1: Determine velocity of flow in the pump suction pipe and pump discharge pipe, and velocity heads.

3-in. NPS, ID = 3.068 in.

2 1/2-in. NPS, ID = 2.469 in.

By Eq. (11-5)

$$V_{3\text{ in.}} = \frac{\text{gpm} \times 0.408}{d^2} = \frac{(200)(0.408)}{(3.068)^2}$$
$$= 8.69 \text{ ft/sec}$$

$$V_{2^{1/2}\text{ in.}} = \frac{(200)(0.409)}{(2.469)^2} = 13.42 \text{ ft/sec}$$

$$\text{Suction velocity head} = \frac{V^2}{2g} = \frac{(8.69)^2}{2 \times 32.2}$$
$$= 1.2 \text{ ft}$$

$$\text{Discharge velocity head} = \frac{V^2}{2g} = \frac{(13.42)^2}{2 \times 32.2}$$
$$= 2.8 \text{ ft}$$

Step 2: Determine pressure heads

By Eq. (12-1)

$$\text{Suction pressure head} = \frac{\text{psi} \times 2.31}{SG}$$

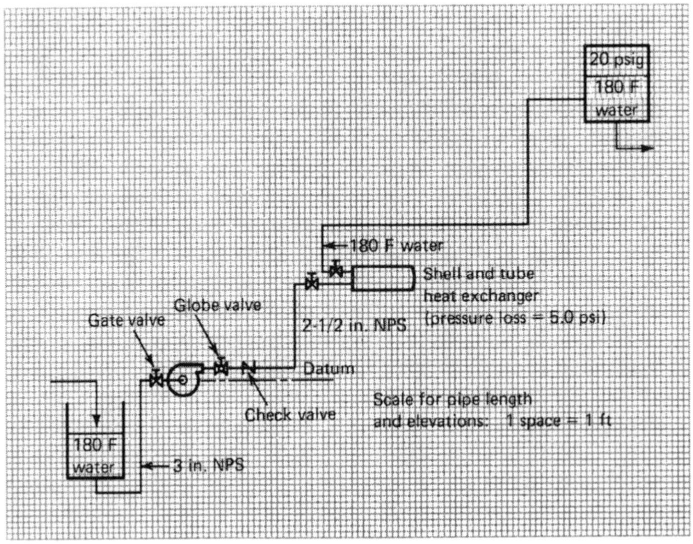

Figure 12–6

Diagram for Illustrative Problem 12-1

$$= \frac{(0)(2.31)}{0.944}$$

$$= 0 \text{ ft gauge}$$

$$\text{Discharge pressure head} = \frac{(20)(2.31)}{0.971}$$

$$= 47.6 \text{ ft gauge}$$

Step 3: Determine static elevation heads

Suction = 15 ft below pump centerline to lowest level in tank (lift)

Discharge = 51 ft above pump centerline to water level in discharge tank

Step 4: Determine losses in pipe and fittings. By Eq. (12-1)

$$\text{Heater head loss} = \frac{\text{Psi} \times 2.31}{\text{SG}}$$

Average specific gravity of water in heater =

$$\frac{0.994 + 0.971}{2} = 0.983$$

$$\text{Heater head loss} = \frac{(5.0)(2.31)}{0.983} = 11.75 \text{ ft}$$

Suction pipe and fitting loss (refer to Chapter 11):

From Table 11-1 f_t = 0.018 for 2½- or 3 in.-pipe, assuming complete turbulence.

$$\text{Gate valve, } K_1 = 8f_t = 8(0.018) = 0.144$$

$$\text{90-degree elbow, } K_1 = 30f_t = 30(0.018) = 0.54$$

Pipe entrance, K = 0.5.

From Fig. 11-8:

Gate valve = 2 ft

Three 90-degree elbow = 24 ft

Pipe entrance = 7 ft

Length of pipe = 38 ft (approximately)

Total equivalent length = 71 ft (approximately)

By Eq. (11-11) Pipe loss suction (h_L) = $f_t\left(\frac{L}{D}\right)\left(\frac{V^2}{2g}\right)$

$$h_L = (0.018)\left(\frac{71}{3.068/12}\right)\left[\frac{(8.69)^2}{2 \times 32.2}\right]$$

$$h_L = 5.86 \quad \text{(say 6 ft)}$$

Discharge pipe and fitting loss (from Table 11-1):

globe valve $K_1 = 340f_t = 340(0.018) = 6.12$

check valve $K_1 = 100f_t = 100(0.018) = 1.80$

gate valve $K_1 = 8f_t = 8(0.018) = 0.144$

pipe exit $K = 1.0$

elbow $K = 30f_t = 30(0.018) = 0.54$

From Fig. 11-8:

globe valve	70 ft	
check valve	21 ft	
2 gate valves	1.8 ft	
pipe exit	11.0	
6 elbows	36.0	
length of pipe	119.0	(approximately)
Total equivalent length =		258.8
		(say 259 ft)

By Eq. (11-11) Pipe loss discharge (h_L) = $f_t\left(\frac{L}{D}\right)\left(\frac{V^2}{2g}\right)$

$$h_L = (0.018)\left(\frac{259}{2.469/12}\right)\left[\frac{(13.42)^2}{2 \times 32.2}\right]$$

$$h_L = 63.36 \quad \text{(say 63 ft)}$$

TDSL = static suction lift + pressure head + velocity head + losses

= 15 + 0 + 1.2 + 6

TDSL = 22.2 ft gauge

TDDH = static discharge head + pressure head + velocity head + losses

TDDH = 51 + 47.6 + 2.8 + 11.75 + 63

TDDH = 176.2 ft

Pump TDH = TDDH + TDSL

= 176.2 + 22.2

198.4 (say 198 ft) ***(ans)***

Or Eq. (12-6)

$$h_p = (Z_2 - Z_1] + \left(\frac{P_2 - P_1}{\gamma}\right) + \frac{V_2^2 - V_1^2}{2g} + h_L$$

$$= [51 - (-15)] + \left[\frac{144(20 - 0)}{1/0.016509}\right] + \left[\frac{(13.42)^2 - (8.69)^2}{2 \times 32.2}\right] + (6 + 11.75 + 63)$$

$$= 66 + 47.6 + 1.6 + 80.75$$

$$h_P = 195.9 \quad \text{(say 196 ft)}$$

(c) Determine net positive suction head available.

By Eq. (12-7)

$$\text{NPSH}_A = \frac{P_B - P_v}{\gamma} + (S - h_L)$$

Assume P_B = 14.7 psia

P_v = vapor pressure at 100°F

= 0.9503 psia

γ = specific weight at 100°F

= $1/v = 1/0.01613 = 62$ lb/ft^3

S = −15 ft (lift)

h_L = 6 ft (suction loss)

$$\text{NPSH}_A = \frac{(14.7 - 0.9503)(144)}{62.0} - 15 - 6$$

NPSH_A = 10.9 ft (this must be equal to or greater than NPSH_R by pump)

12-6 SPECIFIC SPEED AND PUMP TYPE

Specific speed (N_s) is a nondimensional design index used to classify centrifugal pump impellers as to their type and proportions. It is defined as the speed in revolutions per minute at which a geometrically similar impeller would operate if it were of such a size as to deliver 1 gpm against a 1-ft head. The understanding of this definition is of design engineering significance only, however, and specific speed should be thought of only as an index used to predict certain pump characteristics. The following formula is used to determine specific speed:

$$N_s = \frac{N\sqrt{\text{gpm}}}{H^{3/4}} \qquad (12\text{-}9)$$

where N = pump speed in rpm

gpm = pump capacity in gpm at the best efficiency point

H = total head per stage at best efficiency point

For double suction impellers the total flow should be divided by 2 in calculating specific speed.

The specific speed determines the general shape or class of the impeller, as shown in Fig. 12-7. As the specific speed increases, the ratio of the impeller outlet diameter, D_2, to the inlet or eye diameter, D_1, decreases. This ratio becomes 1.0 for a true axial-flow impeller.

Radial flow impellers develop head principally through centrifugal force. Pumps of higher specific speed develop head partly by centrifugal force and partly by axial force. A higher specific speed indicates a pump design with head generation more by axial forces and less by centrifugal forces. An axial flow or propeller pump with a specific speed of 10,000 or greater generates its head exclusively through axial forces.

Generally, impellers with low specific speeds are used where head requirements are high and capacity requirements are low. Higher specific speed impellers are for lower head requirements and higher capacities.

12-7 CENTRIFUGAL PUMP OPERATING CHARACTERISTICS AND PERFORMANCE

We are concerned primarily with the capacity, power, and efficiency of centrifugal pumps. The capacity $(\dot{Q})$ is normally expressed in gallons per minute (gpm). Since liquids are essentially incompressible, there is a direct relationship between the capacity in a pipe and the velocity of flow discussed in Chapter 11, or $\dot{Q} = A \times V$. Equation (11-5) expresses the flow in gpm.

Equation (12-4) expressed the energy added (work) to a unit mass of fluid. The total work done per unit of time on the fluid would be

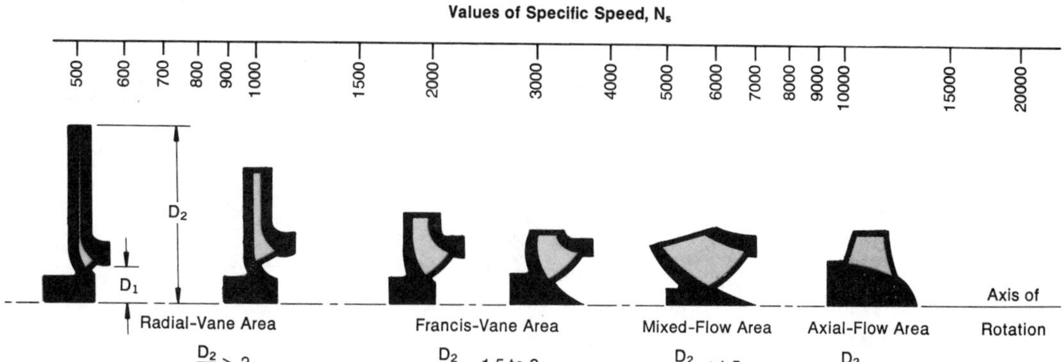

Figure 12–7

Impeller design versus specific speed. (*Goulds Pump Manual,* Goulds Pumps, Inc. Reproduced with permission.)

found by multiplying by the total mass per unit of time. Work per unit of time is power. Therefore, we may say that power added to the fluid passing through a pump is a function of the mass flow per unit of time and the total head developed, expressed as

$$\text{WHP} = \frac{\dot{m} \times \text{TDH}}{33000} \qquad (12\text{-}10)$$

where WHP = horsepower added to the fluid by the pump, frequently called water horsepower, or hydraulic horsepower.

$\dot{m}$ = mass flow (lb/min)

TDH = total dynamic head (ft)

33000 = ft lbf/min per horsepower

Pump capacity in gpm and liquid specific gravity are normally used in formulas rather than the mass flow of fluid.

$$\text{gpm} = \frac{\text{lb/hr}}{500 \times \text{SG}} \qquad (12\text{-}11)$$

Water horsepower then becomes

$$\text{WHP} = \frac{\text{gpm} \times \text{TDH} \times \text{SG}}{3960} \qquad (12\text{-}12)$$

The brake horsepower (BHP) or input to a pump is greater than the water horsepower due to the mechanical and hydraulic losses incurred in the pump. Therefore, the pump efficiency is the ratio WHP/BHP, or

$$\eta_p = \frac{\text{gpm} \times \text{TDH} \times \text{SG}}{3960 \times \text{BHP}} \qquad (12\text{-}13)$$

Although centrifugal pumps can be selected from rating charts (tables), performance curves give a much clearer picture of the characteristics at a given speed. Various forms of curves are used; however, the typical characteristic curve shows the total dynamic head, brake horsepower, efficiency, and net positive suction head all plotted over the capacity range of the pump. Figures 12-8, 12-9, and 12-10 are nondimensional curves that indicate the general shape of the characteristic curves for the various types of pumps. They show the head, brake horsepower, and efficiency plotted as a percent of their values at the design or best efficiency point of the pump.

Figure 12-8 shows that the head curve for a radial-flow pump is relatively flat, and the head decreases gradually as the flow increases. Note that the brake horsepower increases gradually over the flow range with the maximum normally at the point of maximum flow.

Mixed-flow centrifugal pumps and axial-flow or propeller pumps have considerably different characteristics as shown in Figs. 12-9 and 12-10. The head curve for a mixed-flow pump is steeper than for a radial-flow pump. The shutoff head is usually 150 percent to 200 percent of the design head. The brake horsepower remains fairly constant over the flow range. For a typical axial-flow pump the head and brake horsepower both increase sharply near the shutoff as shown in Fig. 12-10.

The distinction between the above three classes of pumps is not absolute, and there are many pumps with characteristics falling somewhere between the three shown. For instance, the Francis vane impeller would have characteristics between the radial and mixed flow classes. Most turbine pumps are also in this range depending upon their specific speeds.

Figure 12-11 shows typical characteristic curves resulting from actual test on a radial-flow centrifugal pump operating at constant speed and for a specific impeller diameter. On the test at 200 gpm the gauges indicated a total head of 90 ft and 6.6 BHP was required. At 120 gpm the gauges indicated a total head of 120 ft and 6.1 BHP was required, and so on for the complete test. Drawing curves through the various plotted points from test data gives the head curve and BHP curve. The efficiency curve is next determined by calculating the efficiency using Eq. (12-13), at various flow rates. The efficiency curve results when the calculated efficiencies are plotted and joined by a curved line.

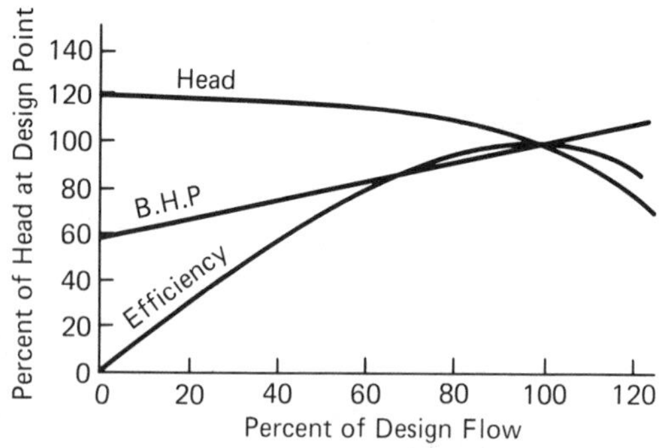

Figure 12–8

Radial Flow Pump

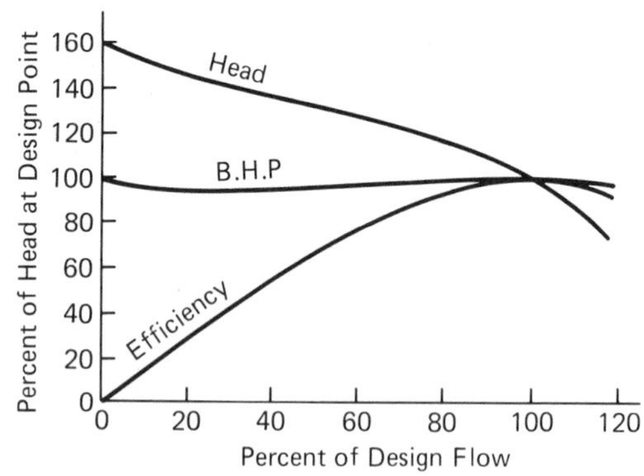

Figure 12–9

Mixed Flow Pump

The curves of Fig. 12-11 show what the pump will do with one specific impeller diameter. However, centrifugal pumps are flexible in that one casing or volute can be used with various impeller diameters cut down from the maximum. To make separate curves of several impeller diameters of each pump would result in multiplicity of sheets, which would be awkward and confusing. Therefore, a curve as shown in Fig. 12-12 has been evolved which tells at a glance what the pump will do at a specified speed with various impeller diameters, from the maximum to the minimum. Each impeller is actually tested, and separate test curves made up similar to Fig. 12-11. All the head curves are then correlated on a single sheet. It can readily be seen, however, that if an attempt was made to include also all the horsepower and efficiency curves, there would be such a conglomeration of lines that it would be impossible to pick out the individual curves. This is greatly simplified by taking several points of efficiency on each of the head curves and connecting them to make a composite efficiency curve; the same is done with the horsepower curves. For instance, in Fig. 12-12, 70 percent efficiency was obtained with a 5⅞-in. impeller at 161 gpm and 119 ft head and at 302 gpm and 82 ft head. The same efficiency was obtained with a 5⅜-in. impeller at 150 gpm and 98 ft and at 264 gpm and 68 ft head. Connecting all the points of the same efficiency results in the various constant efficiency curves.

This is done similarly to make the various horsepower curves. Note that, arbitrarily, specific diameter impellers have been tested. This is done as a check but does not mean that intermediate diameter impellers cannot be furnished. For instance, if a particular application should require 180 gpm at 105 ft head, referring to Fig. 12-12, the impeller would be cut to approximately 5½-in. in diameter, the efficiency would be approximately 74 percent, and a 7½-hp motor would be required.

Referring to the horsepower curves of Fig. 12-12, whenever a point of operation falls above a certain horsepower curve, use the next larger horsepower motor. Note that in a centrifugal pump the capacity increases as the head falls, with the resultant greater horsepower requirement. If on a particular application the head in feet should drop for any reason, use the next larger horsepower motor to be safe if the rating point is too close to the motor rating curve. As an example, 120 gpm at 103 ft head is required. At the rating point on Fig. 12-12 it shows a 5.0 hp motor is adequate. However, it is possible the head may drop to 96 ft. At this point the capacity goes to 160 gpm; also this causes an increase in power to above 5.0 hp. Therefore, a 7½-hp motor should be selected.

Figure 12-11 shows the performance curves

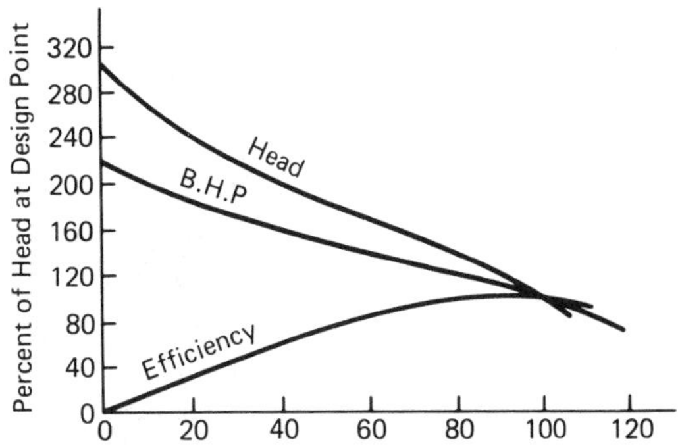

Figure 12–10

Axial Flow Pump

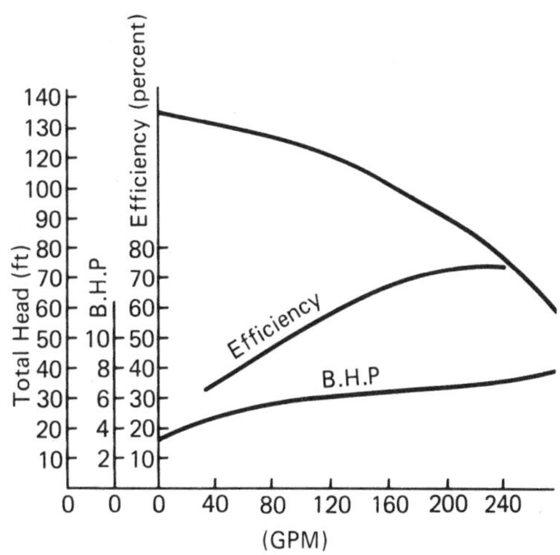

Figure 12–11

Typical performance curves for a centrifugal pump

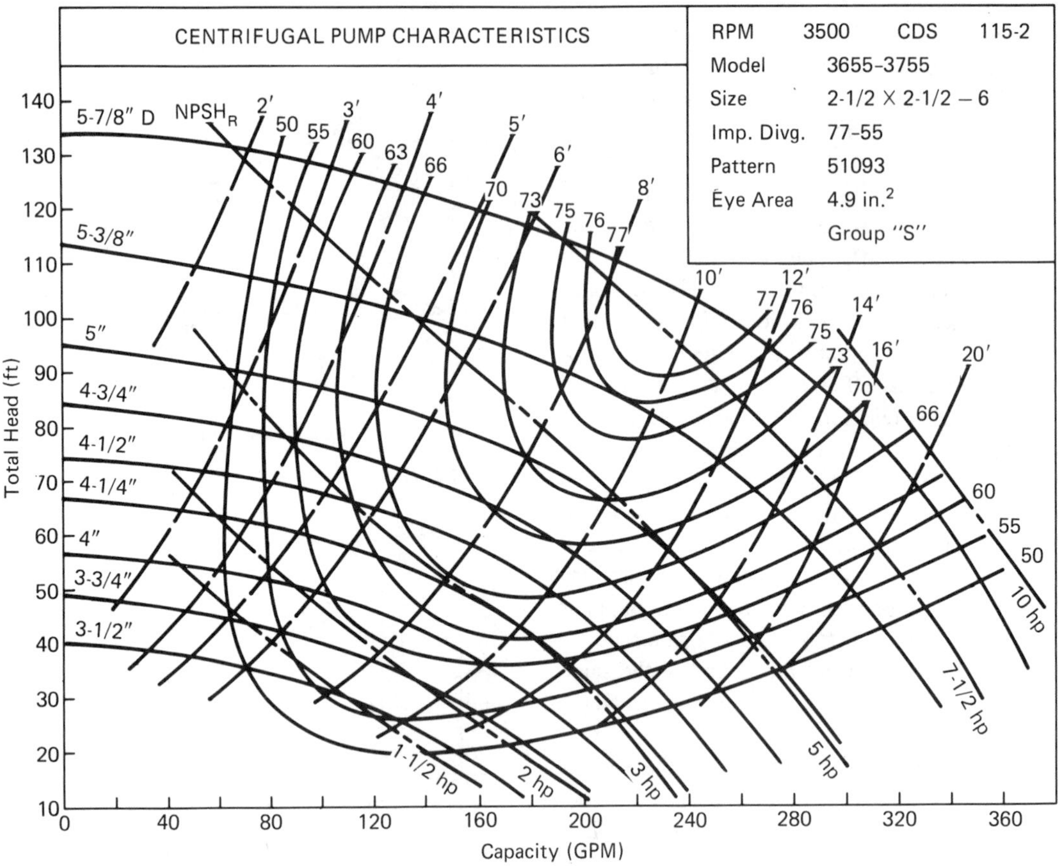

Figure 12–12

Courtesy of Goulds Pumps, Inc.

for a specific impeller diameter of a centrifugal pump operating at constant speed. Figure 12-12 gives the performance of a particular pump at constant speed but with various impeller diameters. However, a particular design of pump frequently consists of a multiplicity of pump sizes, and for selecting a pump for a specific application, a composite curve of the entire line in question is the quickest and simplest method.

The practical usable limit of a centrifugal pump with its various impeller diameters is approximately inside the boundary efficiency curves (center area of Fig. 12-12). This does not cover the entire range of the pump because the area outside this boundary is better covered by other sizes of the same line of pumps. If such an area of ratings were taken for each size of a pump line, a composite of the entire line could be made. Such a composite is shown in Fig. 12-13 with the heavy line denoting the pump size of Fig. 12-12. (Vertical scale in Fig. 12-13 is reduced, which accounts for the difference in appearance of these identical curves). On such a composite it is impractical to try to show efficiencies, but for the purpose

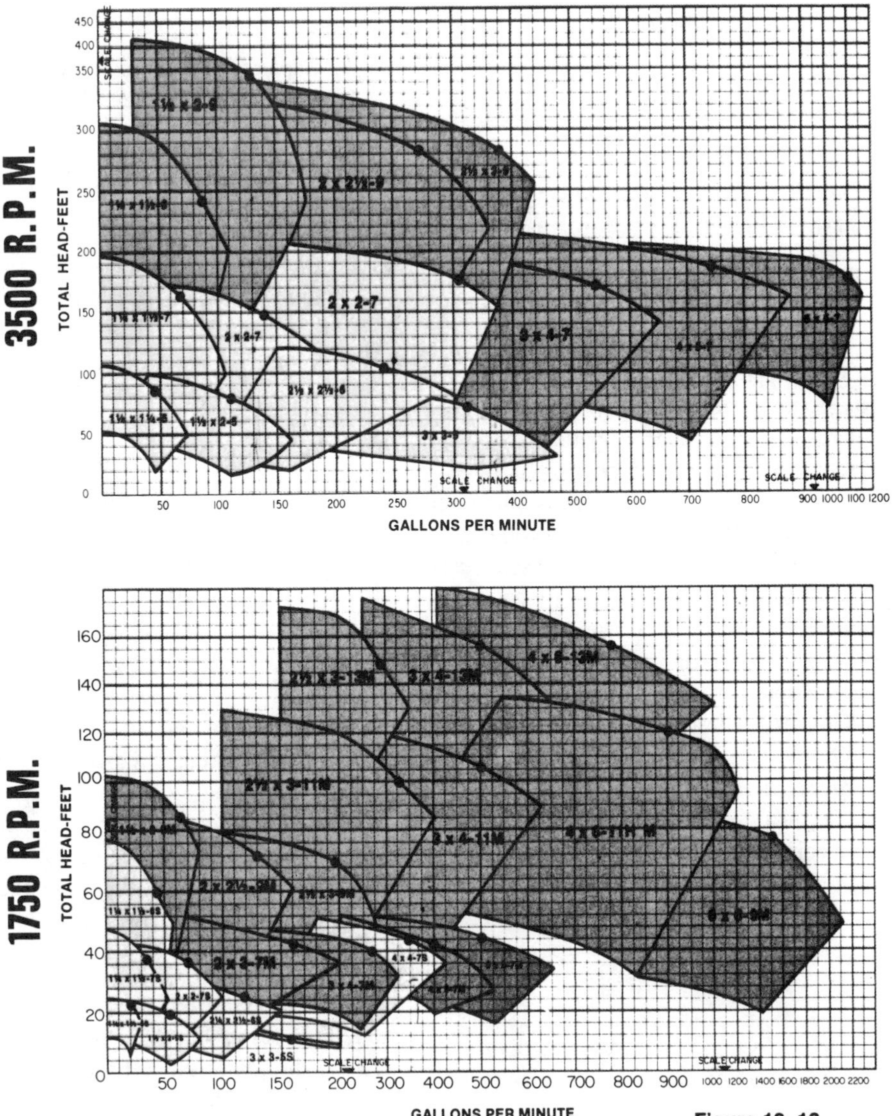

Figure 12–13

Typical composite centrifugal pump curves for various pump sizes. *Goulds Pump Manual,* reproduced courtesy of Goulds Pumps, Inc.

of ease of selection, such a composite gives information concerning gpm, pump head, and impeller size. Curves such as shown in Figs. 12-11 and 12-12 are always available for each pump size. Such curves may be referred to after a pump has been selected in order to obtain pump efficiency, impeller sizes, and so forth.

Several factors govern the rating of a centrifugal pump. Generally, these factors include:

a. *speed* (rpm)—the higher the speed, the higher the capacity and head

b. *impeller diameter*—the larger the diameter, the higher the head in feet

c. *impeller width*—the wider the impeller, the higher the capacity in gpm

In a multistage pump the head builds up in each stage, taking advantage of the head developed by each preceding impeller. In other words the total head developed by a three-stage pump, where each impeller is rated at 100 ft, would be 100 × 3 or 300 ft total, the capacity remaining the same, regardless of the number of stages.

Within limits (assuming constant efficiency) it is possible to calculate pump characteristics at different operating speeds or different impeller diameters. These relationships are commonly referred to as *pump laws*.

For speed change:

1. Capacity varies in direct proportion to speed:

$$\frac{\text{gpm}_1}{\text{gpm}_2} = \frac{\text{rpm}_1}{\text{rpm}_2} \tag{12-14}$$

2. Head in feet varies in direct proportion to the square of the

$$\frac{\text{TDH}_1}{\text{TDH}_2} = \left(\frac{\text{rpm}_1}{\text{rpm}_2}\right)^2 \tag{12-15}$$

3. Horsepower varies in direct proportion to the cube of the speed:

$$\frac{\text{BHP}_1}{\text{BHP}_2} = \left(\frac{\text{rpm}_1}{\text{rpm}_2}\right)^3 \tag{12-16}$$

For impeller diameter change:

1. Capacity varies in direct proportion to the impeller diameter:

$$\frac{\text{gpm}_1}{\text{gpm}_2} = \frac{d_1}{d_2} \tag{12-17}$$

2. Head in feet varies in direct porportion to the square of the impeller diameter:

$$\frac{\text{TDH}_1}{\text{TDH}_2} = \left(\frac{d_1}{d_2}\right)^2 \tag{12-18}$$

3. Horsepower varies in direct proportion to the cube of the diameter:

$$\frac{\text{BHP}_1}{\text{BHP}_2} = \left(\frac{d_1}{d_2}\right)^3 \tag{12-19}$$

Illustrative Problem 12-2

To illustrate the use of the pump laws, refer to Fig. 12-12 where it shows the performance of a particular pump at 3500 rpm for various impeller diameters. Assume that the pump with an impeller diameter of 5⅞ in. is to be driven at 3000 rpm. Determine the capacity head and brake horsepower at 3000 rpm.

Solution:

Select a point of best efficiency for illustration. This point is at 240 gpm and 104 ft head at 3500 rpm, with the BHP approximately 8.0. Therefore, let $\text{gpm}_1 = 240$, $\text{TDH}_1 = 104$ ft, $\text{rpm}_1 = 3500$, and $\text{rpm}_2 = 3000$.

By Eq. (12-14) $$\frac{\text{gpm}_1}{\text{gpm}_2} = \frac{\text{rpm}_1}{\text{rpm}_2}$$

or

$$\text{gpm}_2 = \text{gpm}_1 \left(\frac{\text{rpm}_2}{\text{rpm}_1}\right)$$

$$= 240 \left(\frac{3000}{3500}\right)$$

$$\text{gpm}_2 = 206 \text{ gpm} \qquad \textit{(ans)}$$

By Eq. (12-15)
$$\frac{\text{TDH}_1}{\text{TDH}_2} = \left(\frac{\text{rpm}_1}{\text{rpm}_2}\right)^2$$

or
$$\text{TDH}_2 = \text{TDH}_1 \left(\frac{\text{rpm}_2}{\text{rpm}_1}\right)^2$$

$$= 104 \left(\frac{3000}{3500}\right)^2$$

$$\text{TDH}_2 = 76.4 \text{ ft} \qquad \textit{(ans)}$$

By Eq. (12-16)
$$\frac{\text{BHP}_1}{\text{BHP}_2} = \left(\frac{\text{rpm}_1}{\text{rpm}_2}\right)^3$$

or
$$\text{BHP}_2 = \text{BHP}_1 \left(\frac{\text{rpm}_2}{\text{rpm}_1}\right)^3$$

$$= 8.0 \left(\frac{3000}{3500}\right)^3$$

$$\text{BHP}_2 = 5.0 \text{ hp} \qquad \textit{(ans)}$$

These results are for the best efficiency point operating at 3000 rpm. By performing similar calculations for several other points on the 3500-rpm curve, a new curve can be drawn that will approximate the pump performance at 3000 rpm.

Chapter 11 and the preceding sections have presented flow of fluids in piping, general piping systems, and pumps. Much work has been done in past years to simplify the required calculations for sizing pipe and for selecting pumps in the design of forced-circulation water heating and cooling systems. Although the simplifications introduce minor inaccuracies in our calculations, much time is saved, which reduces the engineering costs. We will now proceed with accepted procedures for the design of forced-circulation water heating and cooling systems.

12-8 FORCED-CIRCULATION HOT WATER HEATING SYSTEMS

Heating systems using hot water as a means of carrying heat to points of utilization are called hot water heating systems or, in the heating industry at present, *hydronic heating systems*. In hydronic heating systems water usually under static pressure is heated in a heat exchanger (boiler or shell and tube) and distributed to the various space-heating units through a system of pipes. Then, the water, reduced in temperature, is returned to the heat exchanger for reheating. Circulation of the water is accomplished by a pump (sometimes called circulator).

Forced-circulation hot water heating systems have become very popular in recent years. Some of their advantages are:

1. Small size tubing or pipe may be used, which may be easily concealed in a building structure.
2. Control versatility; the temperature of the water being circulated may be automatically reduced or increased to change the output of heating units to more closely match the demand variation as outdoor climate changes.
3. Small power requirements of the pump, which circulates the water through the system.
4. Various piping arrangements are possible to meet any system demand.

Hot water heating systems are flexible in operating temperature. For example, if a system is designed for 220°F supply water temperature, it is easy to use a lower supply water temperature in the spring and fall by setting back the boiler controls to a control temperature such as 160°F or lower. This lower temperature reduces the system output. The temperature of the water circulating through the system may also be controlled by bypassing some of the water around the boiler and recirculating the return water.

Most hydronic heating systems for residences and small commercial establishments use supply water temperature at the design point between 180°F and 250°F, with the most common at 200°F. Design water temperature drop may be 10°F, 20°F, or 30°F, with 20°F most common.

High-temperature hot water (HTW) systems use supply water temperatures between 300°F and 400°F with system temperature drops of 100°F to 150°F.

12-9 DESIGN PROCEDURE FOR HOT WATER HEATING SYSTEMS

The design of any type of heating system is sometimes made easier if we recognize that there are certain definite steps in the design process. For a forced-circulation hot water heating system the following steps may be recognized in the overall design process:

1. Determine the heat losses for the building on a room-by-room basis.
2. Determine type and size of the hot water heating units required for each space to be heated and locate these heating units within these spaces.
3. Determine the type of piping arrangement to use to connect all heating units.
4. Determine and locate all system piping components, including special fittings.
5. Determine water flow rates in all parts of the system.
6. Determine pipe sizes for the system.
7. Determine system pumping head; select and locate pump in the system.
8. Determine boiler size required.
9. Determine the expansion (compression) tank size.

Heat loss calculations. The first essential step in the design of any heating system is to accurately calculate the heat losses of the building. This determines the amount of heat the system will be required to supply and thus permits the designer to select the correct size and type of heating units and other system components. The procedures outlined in Chapters 7 and 8 may be used to calculate the heat losses from a building, or the procedure found in the ASHRAE guides may be used. The heat loss must be calculated for each individual space within a building where heating is to be provided.

Types of terminal heat-transfer devices. These units were discussed generally in Chapter 10. Some of these units are selected on the basis of the average temperature of the water circulating through the unit, others are selected on the basis of entering water temperature (EWT), entering air temperature (EAT), and a specified water temperature change through the unit. Therefore, the selection of the terminal heat-transfer unit depends on the water temperature in the system.

The *average water temperature in the system* (one-pipe) is the average between the temperature leaving the boiler (chiller for cooling system) and the temperature returning to boiler (chiller). The *average water temperature in an individual terminal heat-transfer device* may, or may not be, the same as the average water temperature in the system because of the particular water temperature changes selected during the system design.

The most commonly used terminal heat-transfer units used for hydronic heating systems are:

a. baseboard units
b. convectors

c. commercial finned tube and finned pipe

d. fan coil units

e. heating coils

f. radiators (infrequently)

The most commonly used terminal heat-transfer units used for hydronic cooling systems are fan coil units, and cooling coils, where a fan is available for force convection of the air.

Location of heat distributing units. Extensive research has been conducted by ASHRAE, I-B-R, and other agencies to determine the best location for the heating unit within the space. The research has established that convector heating units should be located under windows whenever possible, or along outside walls between windows if window sill heights are low.

Baseboard units and commercial fin tube units are located preferably along outside walls and should extend under as much of the window area as possible. It is not necessary for these finned units to extend the full length of the outside wall if a lesser length will supply the required capacity. Portions of the wall not used for finned units may be filled with blank enclosure sections if a continuous enclosure is desired for appearance. If there is insufficient outside wall space available, the inside partitions may also be used.

Fan coil units should be located under windows whenever possible. Since these units have a fan to circulate the air, their exact location is less critical.

12-10 PIPING SYSTEMS LAYOUT AND SYSTEM COMPONENTS

After the heating units have been selected and located in the various spaces within the building, the next step in the design is to determine the piping layout to connect all heating units and the boiler.

One of the factors affecting the size of pipe required for a hydronic system is the *length* of pipe involved. It is therefore necessary to make a piping layout of the system so that pipe lengths may be measured with reasonable accuracy. The piping layout is not intended to be a detailed drawing of the entire system; however, the drawing should be made to scale, showing the piping connections to boiler and to all heating units. This scale drawing should be a single-line drawing representing the centerline of the pipe. It should be drawn so that pipes follow the structural layout of the building, parallel or perpendicular to structural components using 90-degree or 45-degree elbow connections in all changes of direction. All special fittings and devices, such as special venturi tees, gate and globe valves, balancing valves, flow control valves, circulating pump, and so forth should be located, symbolized on the piping sketch, and identified. Some of these special fittings offer considerable resistance to fluid flow, and this resistance must be evaluated to properly determine the overall system flow resistance.

The most common classification of hydronic systems is according to the piping arrangement employed. These piping arrangements were discussed generally in Chapter 11. The four most common are one pipe, series loop, two-pipe reverse return, and two-pipe direct return.

12-11 SPECIAL FITTINGS AND DEVICES REQUIRED IN HYDRONIC SYSTEMS

Certain special fittings and devices are required for the proper functioning of hydronic systems. These include air elimination devices, flow control devices, pressure-relief valves, pressure-reducing valves, compression or expansion tank, and for one-pipe systems the specially designed venturi tee is required. Provision must be made for pipe expansion and, in some cases, elimination of vibration.

Venturi Pipe Fittings. Figures 12-14 through 12-16 represent venturi Tee fittings as manufactured by Taco, Inc. These specially constructed tees are engineered to produce a conversion of pressure head to velocity head in the venturi throat section. The reduction of pressure head at the throat causes the higher static pressure at the upstream standard tee to force a portion of the main flow through the branch supply riser, the heating unit, and the return riser. The flow through the return riser, reduced in temperature, joins the main flow in the venturi tee. Figure 12-17 shows a sketch of typical connections for an upfeed and a downfeed heating unit using venturi tees. Generally, only one venturi tee fitting is required per heating unit whether the unit is above or below the main. The same type of venturi may be used for upfeed or downfeed.

Taco *super venturi* fittings create a higher available pressure change and cause a substantially higher percentage of water diversion capacity than the *standard* fittings. These super venturis may be used with standard venturis in any one-pipe system. They are particularly well suited for downfeed heating units and for combination heating and cooling systems.

The operating principle of the venturi fitting may be summarized by referring to Fig. 12-16. At point 1 water enters from the upstream main; at 2 the restriction of the venturi nozzle causes an increase in flow velocity with a resultant lowering of system pressure at this point; at 3, as a pressure differential will always cause a flow from the high-pressure to the low-pressure region, water continually enters at this point as long as flow exists in the main; and, at point 4, velocity is converted back to pressure in the main with minimum overall shock loss. This action results in (1) a constant percentage of water diversion to the heating unit at any rate of water flow through the main and fitting and (2) a jet action with its highest efficiency (maximum diversion with minimum over all loss) within the flow rates recommended.

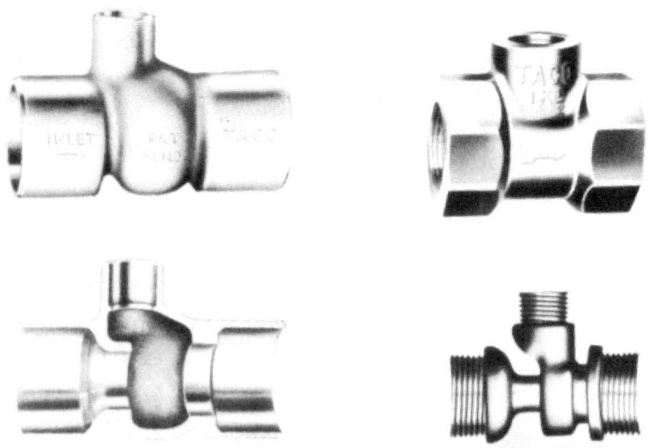

Figure 12–14

Taco venturi fittings (Courtesy of Taco, Inc., Cranston, Rhode Island.)

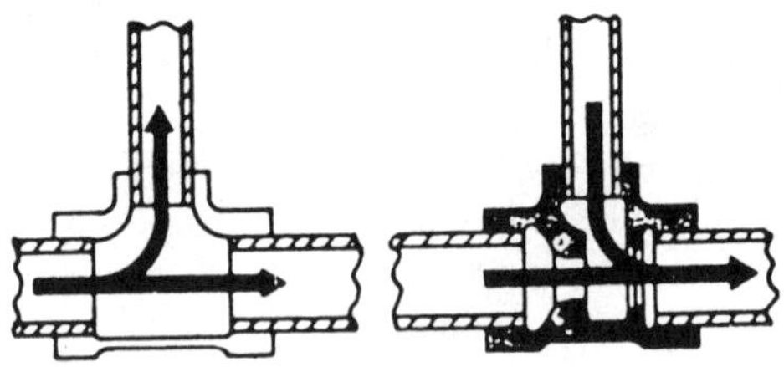

Figure 12–15

Typical water flow through main and riser connections using Taco venturi fittings (Courtesy of Taco, Inc., Cranston, Rhode Island.)

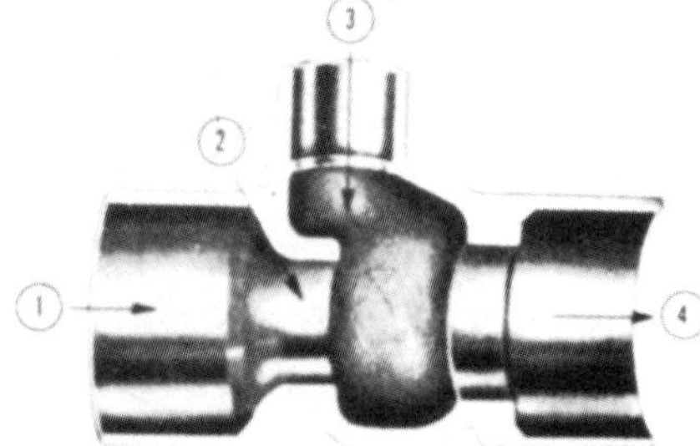

Figure 12–16

Operating principle of Taco venturi fittings (Courtesy of Taco, Inc., Cranston, Rhode Island.)

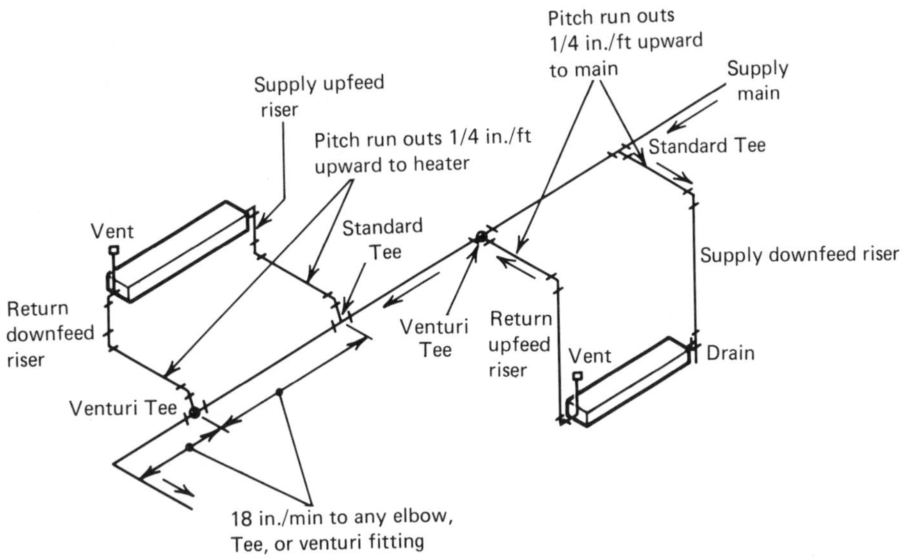

Figure 12–17

Venturi connections for heaters above and below main (one-pipe system)

Taco, Inc., was the originator of the application of the Bernoulli principle to the forced-flow one-pipe hydronic systems.

Air Elimination. All water contains air. As the cold water is heated, the air is driven out of solution and has a tendency to collect in heating units and at high points in the system piping. This may cause uneven heating from the heating unit, noisy operation, and may completely block water circulation. The air must be eliminated from the system to provide satisfactory performance.

Figures 12-18 through 12-20 illustrate typical devices used in hydronic systems to control the air in the system. The "air-vent" (Taco-Vent, Fig. 12-18) normally would be located on all heating units. The "hy-vent," Fig. 12-19, would be located in the piping system at all system high points. The operating principle of the "air-vent" is as follows: air passes through the special fiber discs and vent holes to the atmosphere. Water, following the air, wets the fiber discs causing them to swell, completely sealing the valve. As more air accumulates, the fiber discs dry and shrink, thus permitting the valve to repeat the cycle.

The operation of the "hy-vent" is as follows: when the valve shell is full of water, the valve is closed. When sufficient air accumulates, the mechanical float drops and the valve opens. As air passes out, water again fills the shell closing the valve. As fast as air accumulates the action is repeated.

Figure 11-20 shows one type of air eliminator. The unit shown is manufactured by Taco, Inc. It is located in the supply outlet pipe from the boiler and operates as follows:

1. Air, being less dense than water, will generally tend to travel along the upper portion of a horizontal pipe at velocities commonly used in hydronic systems.

Figure 12–18

Automatic hot water air vent (Courtesy of Taco, Inc., Cranston, Rhode Island.)

Figure 12–19

High vent (Courtesy of Taco, Inc., Cranston, Rhode Island.)

2. As the air and water enter the "air-scoop," the air bubbles are scooped up by the first baffle and rise into the upper chambers. Any bubbles that get through the first baffle are scooped up by the second and third.
3. Air that accumulates in the first chamber is removed from the system by an air-vent valve. Air that accumulates in the second chamber passes into the expansion tank to act as a system air cushion.
4. Should the air completely fill the expansion tank and back down into the "air-scoop," the excess will be removed by the air vent without disturbing the system operation.

Flow Control Device. When the circulating pump in a hydronic system is not operating,

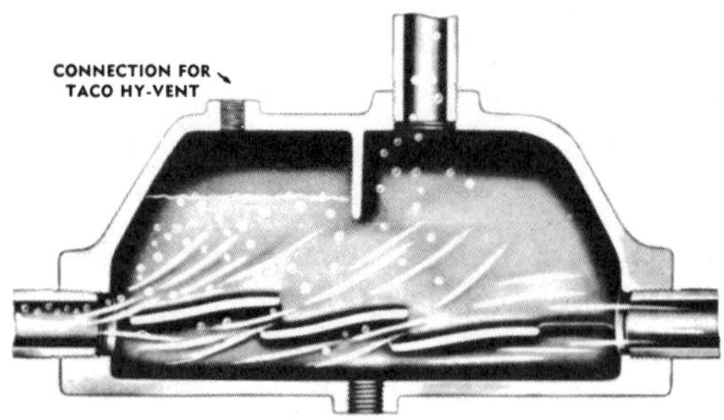

Figure 12–20

Air stoop (Courtesy of Taco, Inc., Cranston, Rhode Island.)

water flow in the system should stop. However, since there is always a temperature difference in the various parts of the system, gravity circulation of the water may be expected. This could cause overheating of the building during periods when no heat is required. Also, during the summer season, if domestic hot water is supplied by a heating coil in the boiler, the boiler water must be maintained at 160°F. This may cause gravity circulation in the piping of the heating system.

A specially designed check valve, Fig. 12-21, is installed in the supply main, close to the boiler outlet, which closes when the pump is not in operation to prevent gravity circulation. These flow control devices may be installed in each zoned circuit of a system that is not controlled by a motorized valve.

Relief Valve. Relief valves are designed to prevent the boiler from rupturing, should the pressure exceed its working pressure. These valves operate on thermal expansion pressures and/or steam pressures.

Relief valves that carry the ASME symbol must be tested and approved under the ASME Boiler and Pressure Vessel Code and rated by their discharge capacity in Btu per hour or pounds per hour. They are installed close to or directly into the top of the boiler without any shutoff valve between the relief valve and the

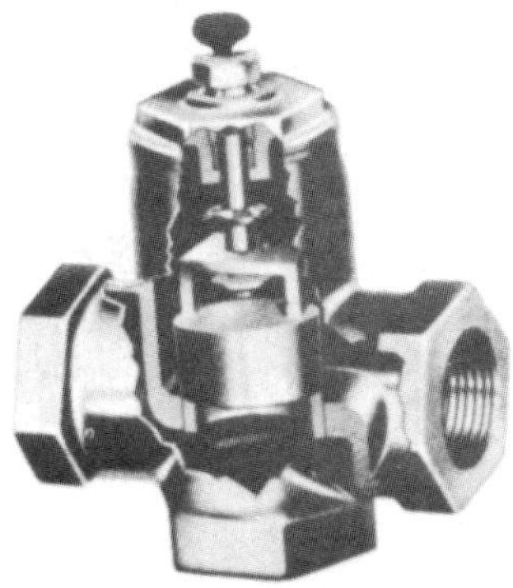

Figure 12–21

Flow control valve. Horizontal Flowcheck (Courtesy of Taco, Inc., Cranston, Rhode Island.)

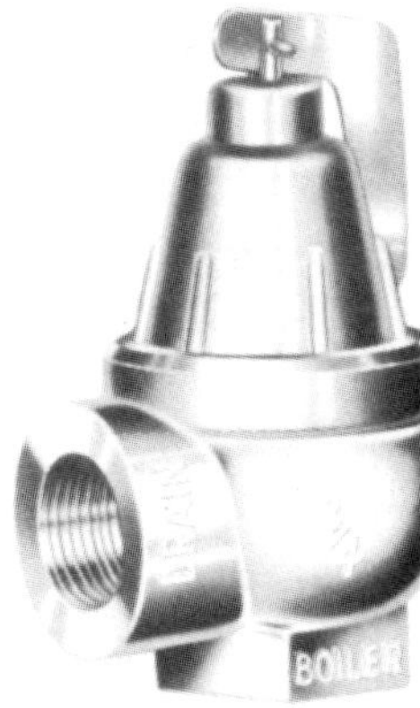

Figure 12–22

Relief valve (Courtesy of Taco, Inc., Cranston, Rhode Island.)

boiler. The drain line from the discharge of the relief valve must be full size and a shutoff valve must never be installed in this drain. Figure 12-22 shows a typical ASME pressure-relief valve.

Reducing Valve. The pressure-reducing valve for a hydronic system is designed to automatically feed makeup water into the system when the system water pressure drops below the setting of the reducing valve. The pressure-reducing valve must be located at a point in the piping system where there is no pressure change, as explained in a later section. Figure 12-23 shows a typical pressure-reducing valve.

Figure 12–23

Reducing valve (Courtesy of Taco, Inc., Cranston, Rhode Island.)

Figure 12-24 shows the normal placement of the various special fittings required in hydronic systems discussed in preceding sections. The locations shown are typical of a residential installation. These locations may be different in other installations, as will be noted later.

Expansion Tank. The expansion tank (sometimes called compression tank or air cushion tank) for most hydronic systems is an airtight cylindrical tank located above the boiler, connected in such a way that when the system is initially filled with water, air is trapped within the tank. When the system water temperature changes, it expands or contracts (depending on whether it is a heating or cooling system). The air trapped in the tank acts as a cushion providing space for this expansion or contraction and maintenance of reasonably stable pressure in the system.

If a hydronic system is initially filled with 50°F water (v_f = 0.016024 ft^3/lb) and is then heated to a temperature of 200°F (v_f =

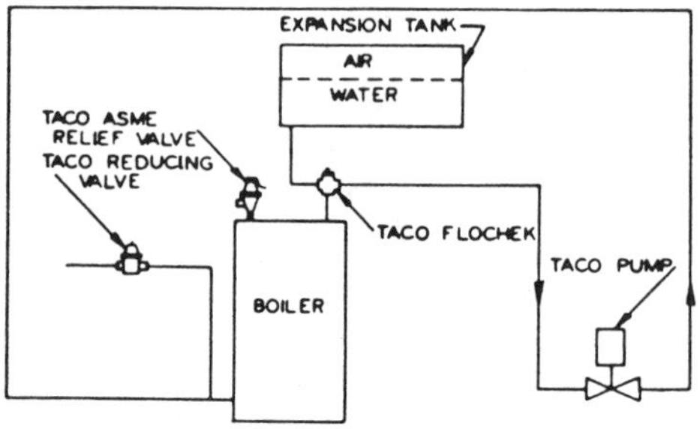

(a)

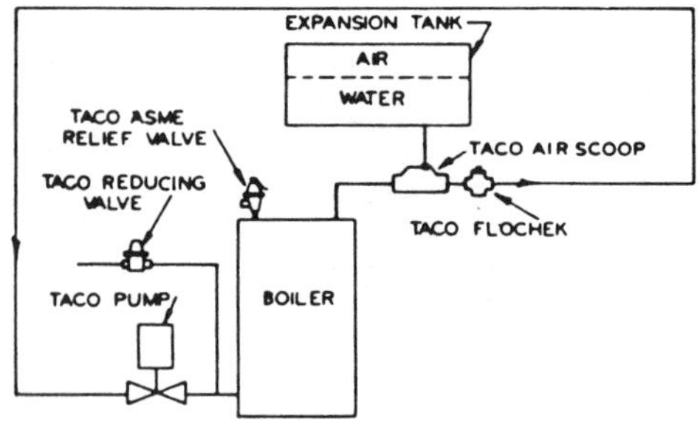

(b)

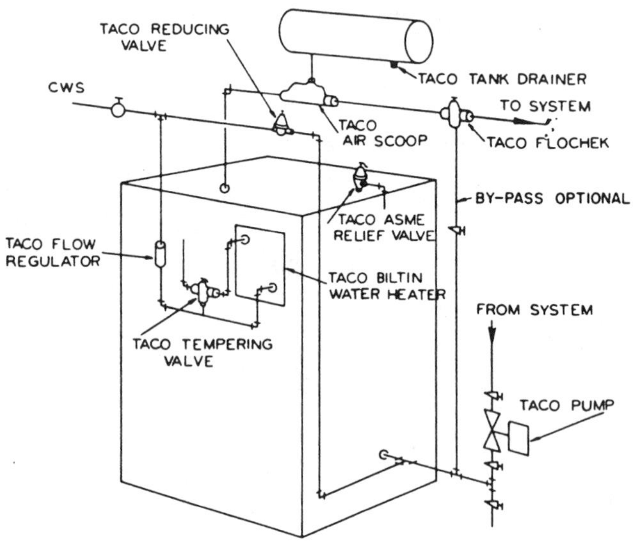

(c)

Figure 12–24

Hookups for forced circulation hot water system:

(a) Recommended hookup when expansion tank is connected to suction side of pump

(b) Recommended hookup when expansion tank is connected to discharge side of pump

(c) Typical boiler layout for residential installation (Courtesy of Taco, Inc., Cranston, Rhode Island.)

0.016634 ft^3/lb), the system water volume increase may be found from

$$\Delta V_s = V_{sc}\left(\frac{v_{fh}}{v_{fc}} - 1\right)$$

where

ΔV_s = change in system volume

V_{sc} = volume of cold water in system

v_{fc} and v_{fh} = specific volumes of cold water and hot water, respectively

Substitution of the values of specific volume given into the expression gives

$$\Delta V_s = V_{sc}\left(\frac{0.016634}{0.016024} - 1\right) = V_{sc}(0.038)$$

or, the system volume increases by 3.8 percent when heated from 50°F to 200°F.

The use of an expansion tank permits system operation at various water temperatures, including temperatures above the normal boiling point of water (212°F at 14.7 psia) because the water is under pressure, usually 12 to 20 psig.

Satisfactory operation of a hydronic system requires a properly sized expansion tank. Commercial sizes of tanks may range from 15 to 300 gal or more, dependent on the manufacturer. Frequently, in medium and large systems, it is more convenient to use two or more smaller tanks than to use one large tank. All expansion tanks should bear the ASME label signifying that the design has been approved by the ASME Construction Code.

For residential and small commercial installations the expansion tank size in gallons may be estimated by dividing the total system radiation load in Btu/hr by a factor of 5000.

A more accurate method for sizing the expansion tank is to use equations presented by various authorities such as ASME and ASHRAE. The following ASME formula is applicable for operating temperatures between 160°F and 280°F:

$$V_t = \frac{(0.00041t - 0.0466)V_s}{\left(\frac{h_a}{h_f} - \frac{h_a}{h_o}\right)}$$

where V_t = minimum expansion tank volume (gal)

V_s = system volume (gal)

t = maximum average operating temperature (°F)

h_a = pressure head in tank when water first enters, feet of water absolute (usually atmospheric = 34 ft)

h_f = initial fill or minimum head in tank, feet of water absolute (normally the static system head)

h_o = maximum operating head in tank, feet of water absolute (atmospheric pressure head plus relief valve setting, or working pressure

System volume, V_s, can be determined from capacities of boilers and heat-distributing devices in manufacturer's catalog data and from tables of dimensional and capacity data for piping.

12-12 BOILERS

An essential component of any hydronic heating system is an efficient, correctly sized boiler or heat exchanger. Boilers were discussed in some detail in Chapter 10. If the boiler is undersized, it is quite obvious that it will not supply sufficient heat for the system. Although, not quite so obvious, is that an oversized boiler may prove to be inefficient in its operation as well as being more costly. Research conducted by I-B-R indicates that there is no appreciable reduction in the warmup period when an oversized boiler is installed.

The boiler capacity required for a hydronic system is normally the *net rating,* or the boiler must have a net rating equal to the calculated heat loss from a building at design conditions and, also, the domestic hot water service if such water is heated by the boiler.

At times it may be desirable to use two or more boilers instead of one in order to provide greater flexibility and higher boiler efficiency. A boiler operates most efficiently if it can operate steadily at its rated capacity. Splitting the system load between boilers will depend upon the judgment of the design engineer. An accepted practice is to size each boiler at two-thirds of the total system load so that in the event of breakdown, the remaining boiler will carry the total load, except under extremely cold weather conditions.

Boiler location. The location of the boiler in the hydronic system is normally dictated by the chimney location if the boiler is gas, coal, wood or oil fired. An electric boiler, requiring no exhaust flue, may be located anywhere. If the building has a basement, the boiler would normally be located there. If the building has a slab floor without a basement or crawl space, the boiler is normally placed on a reinforced section of the first floor. Multistory buildings may have the boiler located on any floor if the piping arrangement is proper, but most often is located in the basement or on the first floor.

Careful selection of the boiler location will result in better operation and usually lower installation costs. I-B-R recommends the following as a guide in selecting a boiler location:

1. Locate the boiler as near to a chimney as possible. Keep smoke pipe as short and direct as possible, using the least number of turns. The cross-section area of the smoke pipe and flue should not be smaller than recommended by the manufacturer.
2. Avoid connecting boiler to a fireplace flue, or to a flue used for other appliances.
3. If a choice exists between using an inside or an outside chimney, make connections to the inside one. This provides better draft conditions as well as fuel economy.
4. Provide ample space around boiler and domestic hot water heater for proper installation and servicing of equipment.
5. Provide air supply for combustion in accordance with local ordinances or other regulations applicable to method of firing the boiler. When installation of firing devices is not governed by local codes, it is recommended that the American Gas Association's requirements be followed for gas-fired units, and the National Board of Fire Underwriters' requirements be followed for other fuels.
6. Provide a substantial and level foundation. If it is necessary to use waterproof concrete for the foundation, take necessary precautions to prevent cracking of the concrete due to heat from the combustion chamber.
7. If the boiler is to be located on a combustible floor, provide adequate insulation in accordance with the boiler manufacturer's recommendations and any local codes or ordinances.
8. The boiler location should conform to all local codes or ordinances governing the installation of heating equipment and firing devices.

Boiler connections. When connecting the supply and return mains to the boiler, connections should be selected that will permit the longest flow path of water through the boiler. This will provide better balanced water temperature conditions in the boiler and improve the overall efficiency of the boiler and system. Short-circuiting of the water from return connection to outlet leads to poor operation and danger of heat stress in the boiler.

12-13 WATER FLOW RATES IN HYDRONIC SYSTEMS

The water flow rate through any portion of a hydronic system may be determined after a design water temperature change has been established. For a heating system the water temperature drop may be established for the entire system or may be selected for any heat-distributing unit within the system to determine the flow rate through respective parts of the system.

The quantity of water circulated may be found from

$$\dot{q} = \dot{m}_w C_p (t_1 - t_2) \qquad (12\text{-}20)$$

where $\dot{q}$ = the total system heat load, or the heat load for a particular heat distributing unit (Btu/hr)

$\dot{m}_w$ = mass of water circulated (lb/hr)

C_p = specific heat of water usually taken as unity

$(t_1 - t_2)$ = design water temperature difference (TD) for the entire system or for a single heat-distributing unit within the system (°F)

Equation (12-20) may be easily rearranged to yield the mass flow of water ($\dot{m}_w$), which is normally done for high-temperature water (HTW) systems. Most calculations for low-temperature water (LTW) systems use flow rate in gallons per minute. Therefore $\dot{m}_w$ in lb/hr may be converted to gpm by the following relationship:

$$\dot{m}_w = \text{gpm} \times 60 \times 8.17 = 490 \text{ gpm}$$

where 60 is minutes per hour and 8.17 is the weight in pounds of 1 gal of water at 180°F. If the average water temperature in the system is much different from 180°F, a new value should replace the 8.17 in the above expression to adjust for water density. Equation (12-20) now becomes

$$\dot{q} = 490 \text{ gpm } (t_1 - t_2)$$

or

$$\text{gpm} = \frac{\dot{q}}{490(t_1 - t_2)} \qquad (12\text{-}21)$$

12-14 PIPE SIZING AND PRESSURE LOSSES

Chapter 11 presented means for accurately determining pipe sizes and pressure (head losses) in pipes. We will now discuss the methods used by many engineers in the design of hydronic piping systems.

The selected water flow velocity through a pipe determines the pipe size for a given gpm flow rate and also establishes the pressure loss in the pipe. At a given flow rate, higher flow velocities result in smaller pipe sizes but will also mean greater pressure loss, which will increase the power cost for pumping. It is therefore a matter of an economic balance between the cost of pipe and installation and the operating cost of the system that must be determined.

It has been found that water flow velocities in excess of 4 ft per second in 1¼-in. pipe, or smaller, are likely to cause noise in residential systems. In industrial applications, which normally have larger pipes, velocities as high as 8 ft per second are not uncommon. Table 12-1 shows recommended flow velocities through pipes of sizes normally encountered in hydronic systems.

Water flowing through a pipe produces friction, with a resulting loss in available energy as was discussed in Chapter 11. In hydronic systems this loss of available energy is called "pressure loss" or "head loss" and is normally expressed in *feet of water per linear foot of pipe*, or *millinches of water per linear foot of pipe* (1000 millinches of water head = 1 in. of water head) or (12,000 millinches of water head = 1 ft of water head). The pressure loss or head loss per foot of pipe will be called a *friction rate (f)*.

The relationships between pipe size, flow velocity, and friction rate in LTW systems is shown in Figs. 12-25 and 12-26. Such charts greatly simplify the determination of friction rates at various flow rates in different sizes of pipe.

Figure 12-25 shows head loss in feet of water per foot of pipe and millinches per foot of pipe plotted against gpm flow rate in various nominal sizes of black iron or steel pipe. The flow velocity in feet per second is also shown. Figure 12-26 shows the same quantities for water flow in type *L* copper tubing. Type *L* copper tube is the most common type used in hydronic systems.

In order to calculate the head loss through any section of piping, it is necessary to know the friction rate, f, and the total equivalent length (TEL) of pipe and fittings in the section of pipe. The head loss, h_L, will be

$$h_L = f \times \text{TEL} \qquad \textit{(12-22)}$$

where f = friction rate (ft of water/ft of pipe).

The TEL of pipe is the measured length of straight pipe plus an allowance for the straight-pipe length equivalents of all the fittings in the particular section of pipe. In Chapter 11 it was shown that the head loss through a pipe fitting could be represented by the head loss through an equivalent length of straight pipe. Tables 12-2 and 12-3 show typical equivalent lengths of straight pipe for various fittings commonly found in hydronic systems.

NOTE: As may be noted in Tables 12-2 and 12-3, there are some conflicting data concerning the equivalent lengths of some fittings. This is often encountered when taking data from different sources. The differences are relatively minor, and values from either table would be considered satisfactory in our design work. The following equation may be helpful in making conversions.

To convert head loss in feet of water to feet of pipe equivalents (EL):

$$\text{EL} = \frac{h_L \text{ (ft of water)}}{f \text{ (ft of water/ft of pipe)}} \qquad \textit{(12-23a)}$$

To convert head loss in inches of water to feet of pipe equivalents:

$$\text{EL} = \frac{h_L \text{ (in. of water)}}{12 \times f \text{ (ft of water/ft of pipe)}} \qquad \textit{(12-23b)}$$

To convert pressure loss in psi to feet of pipe equivalents:

$$\text{EL} = \frac{\text{psi} \times 2.31}{f \text{ (ft of water/ft of pipe)}} \qquad \textit{(12-23c)}$$

Illustrative Problem 12-3

A section of pipe in a hydronic heating system carries 30 gpm of water. The pipe is 2-in. nominal steel pipe and scales 50 ft in length. Contained in this section of pipe are two gate valves, two 45-degree elbows, four 90-degree

Table 12-1*

QUICK PIPE SIZING TABLE
RECOMMENDED FLOW RATES IN GPM
FOR ALL PIPING EXCEPT BRANCH CONNECTIONS ON TACO VENTURI SYSTEMS

Nominal Pipe or Tubing Size In.	STEEL PIPE				COPPER TUBING			
	Minimum GPM	Velocity Ft./Sec.	Maximum GPM	Velocity Ft./Sec.	Minimum GPM	Velocity Ft./Sec.	Maximum GPM	Velocity Ft./Sec.
½"	—	—	1.8	1.9	—	—	1.5	2.1
¾"	1.8	1.1	4.0	2.4	1.5	1.0	3.5	2.3
1"	4.0	1.5	7.2	2.7	3.5	1.4	7.5	3.0
1¼"	7.2	1.6	16.	3.5	7.	1.9	13.	3.5
1½"	14.	2.2	23.	3.7	12.	2.1	20.	3.6
2"	23.	2.3	45.	4.6	20.	2.1	40.	4.1
2½"	40.	2.7	70.	4.7	40.	2.7	75.	5.1
3"	70.	3.0	120.	5.2	65.	3.0	110.	5.2
3½"	100.	3.2	170.	5.4	90.	3.1	150.	5.2
4"	140.	3.5	230.	5.8	130.	3.5	210.	5.6
5"	230.	3.7	400.	6.4	—	—	—	—
6"	350.	3.8	610.	6.6	—	—	—	—
8"	600.	3.8	1200.	7.6	—	—	—	—
10"	1000.	4.1	1800.	7.4	—	—	—	—
12"	1500.	4.3	2800.	8.1	—	—	—	—

Courtesy of Taco, Inc.

Table 12-2*

EQUIVALENT RESISTANCE OF VALVES AND FITTINGS

Nominal Pipe Diameter, Inches	Valve or Fitting									
	Globe Valve, Open	Gate Valve				Angle Valve, Open	Close Return Bend	Tee		Ordinary Entrance‡
		¾ Closed	½ Closed	¼ Closed	Open			Through Run	Through Side†	
	Equivalent Resistance, Feet of Pipe									
½	16	40	10	2	0.3	9	4	1.0	4	0.9
¾	22	55	14	3	0.5	12	5	1.4	5	1.2
1	27	70	17	4	0.6	15	6	1.7	6	1.5
1¼	37	90	22	4	0.8	18	8	2.3	8	2.0
1½	44	110	28	6	0.9	21	10	2.7	9	2.4
2	55	140	35	7	1.2	28	13	3.5	12	3.0
2½	65	160	40	8	1.4	32	15	4.2	14	3.3
3	80	200	50	10	1.6	41	18	5.0	17	4.5
3½	100	240	60	12	2.0	50	21	6.0	19	5.0
4	120	275	69	14	2.2	55	25	7.0	21	6.0
5	140	325	81	16	2.9	70	30	8.5	27	7.5
6	160	400	100	20	3.5	80	36	10.1	34	9.0
8	220	525	131	26	4.5	110	50	14.0	44	12.0
10	—	700	175	35	5.5	140	60	17.0	55	15.0
12	—	800	200	40	6.5	160	72	19.0	65	16.5
14	—	950	238	48	8.0	190	85	23.0	75	20.0
16	—	1100	272	52	9.0	220	100	26.0	88	22.0
18	—	1300	325	65	10.0	250	115	30.0	110	25.0

Nominal Pipe Diameter, Inches	Valve or Fitting									
	Sudden Contraction§			Borda Entrance	Reducing Tee*		90° Elbow			45° Elbow
	D/d=4	D/d=2	D/d=4/3		D/d=2	D/d=4/3	Standard	Medium Sweep	Long Sweep	
	Equivalent Resistance, Feet of Pipe									
½	0.8	0.6	0.3	1.4	1.5	1.3	1.5	1.3	1.0	0.8
¾	1.0	0.8	0.5	1.9	2.0	1.8	2.0	1.8	1.4	1.0
1	1.3	1.0	0.6	2.5	2.6	2.4	2.6	2.4	1.7	1.3
1¼	1.6	1.3	0.8	3.5	3.5	3.1	3.5	3.1	2.3	1.6
1½	2.0	1.5	0.9	4.0	4.5	3.7	4.5	3.7	2.7	2.0
2	2.5	1.9	1.2	5.0	5.3	4.5	5.3	4.5	3.5	2.5
2½	3.0	2.2	1.4	6.0	6.3	5.5	6.3	5.5	4.2	3.0
3	3.7	2.8	1.6	7.5	8.0	6.9	8.0	6.9	5.0	3.7
3½	4.4	3.3	2.0	9.0	9.5	8.0	9.5	8.0	6.0	4.4
4	5.0	3.7	2.2	11.0	11.5	9.5	11.0	9.5	7.0	5.0
5	6.0	4.7	2.9	12.0	12.5	12.0	13.0	12.0	8.5	6.0
6	7.5	5.6	3.5	15.0	16.0	14.0	16.0	14.0	10.1	7.5
8	11.0	7.2	4.5	19.0	20.0	18.0	20.0	18.0	14.0	10.0
10	13.0	9.5	5.5	24.0	25.0	22.0	25.0	22.0	17.0	13.0
12	15.0	11.0	6.5	29.0	31.0	26.0	31.0	26.0	19.0	—
14	17.0	12.5	8.0	34.0	36.0	30.0	36.0	30.0	23.0	—
16	19.0	15.0	9.0	38.0	40.0	35.0	40.0	35.0	26.0	—
18	21.0	16.5	10.0	43.0	45.0	40.0	45.0	40.0	30.0	—

† For flow making 90° turn. ‡ Into pipe of given diameter from tank, etc.

* For run of tee. Use pipe size of small diameter.

§ Use pipe size of small diameter. For Sudden Enlargements, d/D=¼, values are same as for Reducing Tee, D/d=2; d/D=½, values are same as for Tee, Through Run; d/D=¾, values are same as for Sudden Contraction, D/d=4/3.

Table 12-3*

FEET OF PIPE EQUIVALENTS FOR VARIOUS FITTINGS USED IN MAIN OR TRUNK CONNECTIONS

	½		¾		1		1¼		1½		2		2½		3		4		5	6	8	10	12
	C.I.	COP.	C.I.	COP.	C.I.	COP.	C.I.	COP.	C.I.	COP.	C.I.	COP	C.I.	COP.	C.I.	COP.	C.I.	COP.	C.I.	C.I	C.I.	C.I	C.I.
45° Elbow	<				1.0			>	2.2	2.2	2.8	2.8	3.3	3.3	4.0	4.0	5.5	5.5	6.6	8.0	11.0	13.2	16.0
90° Elbow or Union Elbow	<				2.5			>	4.3	4.3	5.5	5.5	6.5	6.5	8.0	8.0	11.0	11.0	13.0	16.0	22.0	26.0	32.0
90° Elbow Long Turn	<				1.5			>	2.7	2.7	3.5	3.5	4.2	4.2	5.2	5.2	7.0	7.0	8.4	10.4	14.0	16.8	20.8
Gate Valve Open	<				.5			>	1.0	1.0	1.2	1.2	1.4	1.4	1.7	1.7	2.3	2.3	2.8	3.4	4.6	5.6	6.8
Globe Valve Open	17	17	22	22	27	27	36	36	43	43	55	55	67	67	82	82	110	110	134	164	220	268	328
Radiator Valve Angle	<				5.0			>															
Tee	<				5.0			>	9	9	12	12	14	14	17	17	22	22	28	34	44	56	68
Boiler	4.0	5.0	5.0	7.0	6.5	9.0	9.0	12.0	9.0	12.0	12.6	16.8	15.6	20.8	18.6	24.8	25.2	33.6	31.2	37.2	50.4	62.4	74.4
Square Head Cock Open	<				1.5			>	2.5	2.5	3.2	3.2	3.9	3.9	4.7	4.7	6.1	6.1					
Radiator C.I.	<				7.5			>															
Std. Venturi (1)			22.5	18	13.5	14	13.5	9.5	13.5	10.5	14.5		14		14								
Super Venturi (1)			37.0	31.5	27.5	28.5	33	21	32.5	26	30		29.5		28.5								
Air Scoop			2.0	2.0	2.7	2.7	4.0	4.0	4.8	4.8	6.8	6.8	8.0	8.0	13.3	13.3	15.0	15.0					
Flow Check			27	27	42	42	60	60	63	63	83	83	104	104	125	125	166	166					
3 Way Mix Valve	Refer To Manufacturers Catalog																						
Convector	Refer To Manufacturers Catalog																						
Baseboard	One Lineal Foot is equal to One Lineal Foot of Pipe																						

(1) Figures shown include Venturi Fitting and Standard Tee

Reproduced by permission. Courtesy of Taco, Inc.

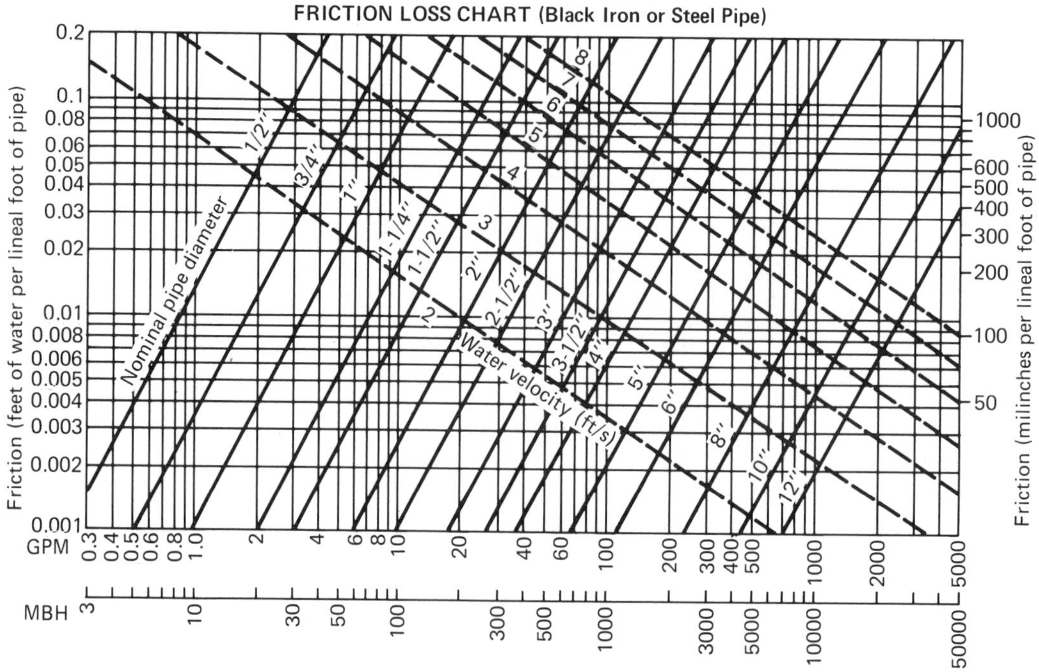

Figure 12–25

Pipe sizing chart. Water flow through black iron or steel pipe (Courtesy of Taco, Inc., Cranston, Rhode Island)

elbows, six flow-through tees, and one square head cock. Determine the head loss in the section of pipe.

Solution:

Step 1: Determine the friction rate.

Using Fig. 12-25, enter the bottom of the graph at 30 gpm and proceed vertically upward until a point of intersection is found with the 2-in. pipe size. From this point of intersection proceed horizontally to the left scale and read $f = 0.02$ ft of water/ft of pipe, or horizontally to the right and read $f = 230$ millinches/ft of pipe.

Step 2: Determine total equivalent length of pipe and fittings.

Type of Fitting	(Table 12-3) Equivalent Length (ft)
2-gate values	2 × 1.2 = 2.4
2-45 elbows	2 × 2.8 = 5.6
4-90 elbows	4 × 5.5 = 22.0
6-flow-through tees	6 × 3.5 = 21.0 (Table 12-2)
1-square head cock	1 × 3.2 = 3.2
	EL Fittings = 54.2 ft
	SL Pipe = 50.0 ft
	TEL = 104.2 ft

Step 3: Determine head loss. By Eq. (12-22)

$$h_L = f \times \text{TEL}$$
$$= (0.02)(104.2)$$
$$h_L = 2.08 \quad \text{(say 2.1 ft of water)}$$

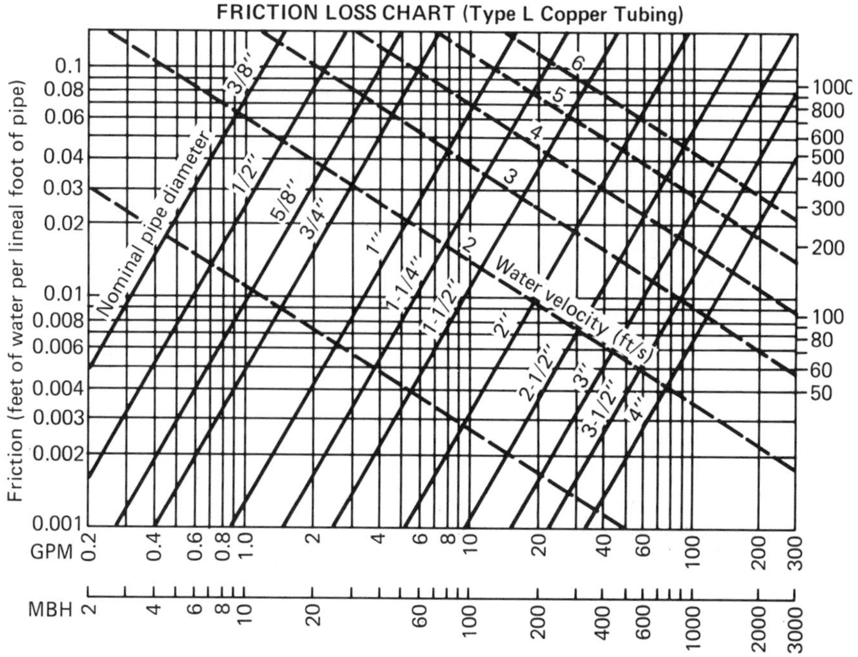

Figure 12–26

Tube sizing chart. Water flow through Type L copper tubing (Courtesy of Taco, Inc., Cranston, Rhode Island)

Illustrative Problem 12-4

Determine a tentative size of type *L* copper tube to carry a flow of 15 gpm of water within recommended limits of velocity. What will be the corresponding friction rate.

Solution:

Step 1: Determine recommended flow velocity range.

Referring to Table 12-1, for a 1½-in. tube a minimum flow of 12 gpm at a velocity of 2.1 ft/sec or a maximum flow of 20 gpm at a velocity of 3.6 ft/sec. Also, note that a 1¼-in. tube would carry 13 gpm at a velocity of 3.5 ft/sec. (Recall that 4 ft/sec is a maximum recommended velocity in residential applications.)

Step 2: Determine tube size.

Refer to Fig. 12-26 and locate a flow rate of 15 gpm at bottom of the figure. Proceed upward to locate a pipe size where the velocity would be between approximately 2.1 to 3.5 ft/sec. A choice is evident between a 1½- and 1¼-in. tube.

For a 1½-in. tube at 15 gpm the velocity is approximately 2.7 ft/sec with a corresponding $f = 0.022$ ft/ft of pipe. For a 1¼-in. tube at 15 gpm the velocity is approximately 4 ft/sec with a corresponding $f = 0.053$ ft/ft of pipe.

For the limited data presented in the problem statement a 1¼-in. tube size would be selected. However, since the friction rate is relatively high, the total head loss for a system in which the pipe is located may be too high to be reasonable.

12-15 LOSSES IN HYDRONIC SYSTEMS

Section 12-4 discussed in a general way the system heads that must be produced by a pump. These system heads were combined in Eq. (12-6). Most hydronic heating and cooling systems are *closed systems* (no part of the system in exposed to atmospheric pressure). This fact removes all head terms from Eq. (12-6) except for the head loss, h_L. Therefore, for a closed system, the pump must overcome *only* the friction losses in the system, or

$$h_p = h_L$$

To determine the head loss, h_L, due to friction for an entire hydronic system, it is necessary to determine the measured length of straight pipe, add to it the equivalent length of pipe fittings, and multiply by the friction rate per foot of pipe. This was partially illustrated in Problem 12-3.

To determine the correct pump size for a *single-size* hydronic piping system, such as a series system, or conventional one-pipe system, it may be necessary to plot a system resistance curve. This may be accomplished by selecting the friction rate at several different flow rates for the pipe size involved and multiplying by the total equivalent length of pipe. These resulting head losses may be plotted at the respective flow rates. The resulting curve is the system resistance curve. Where this curve intersects the constant-speed pump characteristic curve will be the point of operation for the system and that pump.

Illustrative Problem 12-5

Assume that a hydronic system has a total equivalent length of 875 ft, and the pipe size is 1½-in. nominal steel. Determine the system resistance curve between a flow of 10 and 40 gpm.

Solution:

Referring to Fig. 12-25, determine the friction rate at flows of 10, 20, 30, and 40 gpm.

Flow Rate (gpm)	Friction Rate f (ft/ft of pipe)	Total System Resistance (ft) ($f \times$ TEL)
10	0.0085	7.4
20	0.030	26.3
30	0.065	56.9
40	0.120	105.0

Figure 12-27 shows the plot of the system resistance curve. If the design flow rate for the system were assumed to be 25 gpm, the required pump head would be 40 ft.

In piping systems where a number of different pipe sizes in series are involved, such as in a two-pipe reverse return or a two-pipe direct return, the determination of the system resistance curve becomes a complicated procedure. This procedure may be greatly simplified if it is assumed that the resistance through all the various sized piping will be increased or decreased proportionally depending on the flow. As the resistance through a pipe increases as the square of the velocity, it becomes a simple matter to determine the system resistance at various flow rates.

If the calculated system resistance of a piping system is h_L and the gpm requirement is $\dot{Q}$, then, if the flow is increased 10 percent, $1.10\dot{Q}$ equals the new flow rate and the new head loss would be $(1.10)^2 h_L$ or $1.21 h_L$. If the flow rate is decreased 10 percent, then $0.90\dot{Q}$ equals the new flow rate and the new head loss would be $(0.90)^2 h_L$, or $0.81 h_L$. Using similar calculations, the system resistance curve may be constructed.

12-16 PUMP SELECTION FOR HYDRONIC SYSTEMS

Centrifugal pumps were discussed in the beginning sections of this chapter. The large ma-

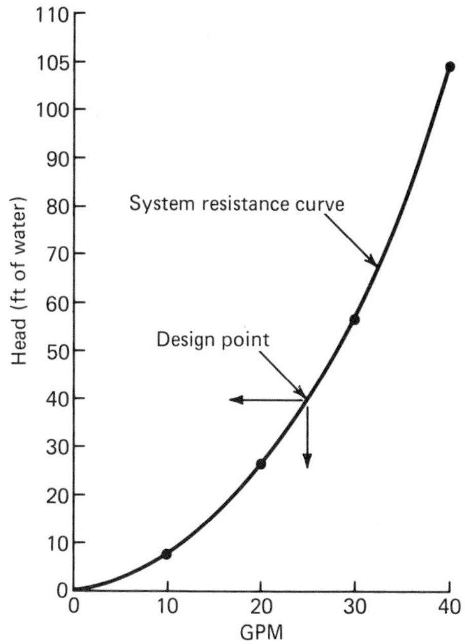

Figure 12–27

System Resistance Curve (Illustrative Problem 12-5).

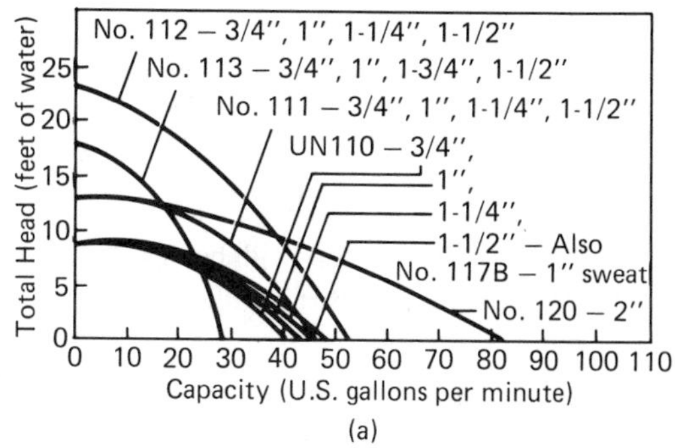

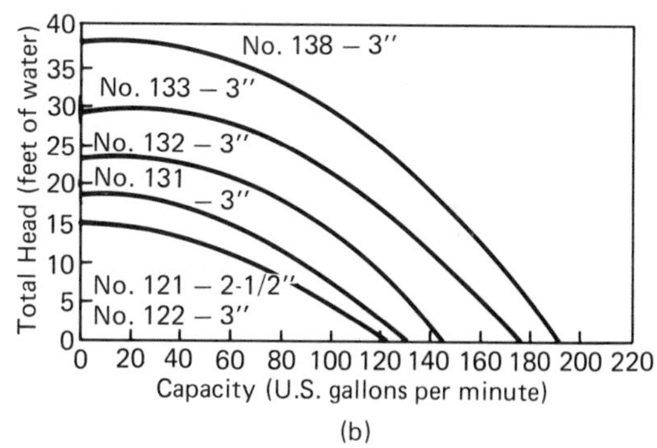

Figure 12–28

Circulator performance curves (a) Circulators Nos. 110, 111, 112, 113, 117B and 120 (b) Circulators Nos. 1212, 122, 131, 132, 133 and 138 (Reproduced courtesy of Taco, Inc., Cranston, Rhode Island)

jority of centrifugal pumps used in residential and small commercial hydronic systems are in-line pumps (Fig. 12-1) with their relatively low heads and capacities, light weight, and ease of installation. Due to their availability in a large number of sizes, a selection can be made very close to the actual system requirements. However, since this type of pump is furnished with one diameter impeller and one operating speed, it will have only one performance curve. Therefore, it is usually recommended that a valve be installed in the discharge line to adjust the flow to the required gpm. Inline pumps are generally designed with maximum heads of approximately 40 ft and maximum flows of approximately 170 gpm. Figures 12-28a and 12-28b are typical performance curves of inline pumps.

Where system flow rates and heads are larger, close-couple or base-mounted centrifugal pumps are used. It is always good practice to select them as close as possible to their highest efficiency curves. Rarely will it be possible to select a pump with a performance curve that will intersect the design operating point of the system curve. It is therefore necessary to select a pump with more or less capacity. As a recommendation, set this limit at plus or minus 10 percent of the design gpm.

If the flow deviates from the design gpm, the operation of the system will be affected as follows:

For Flows Below Design gpm:

a. higher temperature drop (2°F higher at 10 percent lower flow with a 20°F design system temperature drop)

b. power consumption will be less, which may in some cases result in a lower-horsepower motor

c. lower velocity in main

d. lower gpm side diversion in venturi fittings in one-pipe systems

e. in some cases a smaller size pump may be used

For Flows Above Design gpm:

a. lower temperature drop (2°F lower at 10 percent higher flow with a 20°F design system temperature drop)

b. power consumption may be higher, which may in some cases result in a higher-horsepower motor

c. higher main velocity with possibility of creating noise in the system. The flow should not exceed the maximum recommended values shown in Table 12-1

d. higher gpm diversion of venturi fittings in one-pipe systems

e. in some cases a larger size pump may be required

Figure 12-12 illustrates typical performance curves for close-coupled and base-mounted pumps.

12-17 LOCATION OF SYSTEM COMPONENTS

The proper location of a pump in forced-flow hydronic systems, as well as the expansion tank, relief valve, and pressure-reducing valve, has an important bearing on the system operation. Improper location of these system components can often lead to problems of pump cavitation and the production of negative pressures (below atmospheric) in parts of the system.

Point of no pressure change. Figure 12-29a illustrates the main of a closed hydronic system with a pump and compression tank shown and without means of feeding additional water. The expansion tank contains a certain volume of air, which is compressible, and the balance is water, which is essentially incompressible. To increase the pressure in the expansion tank, the air must be compressed. To compress the air requires more water. Since this is a closed system without means to receive more water, the pressure at the expansion tank cannot change, regardless of whether the pump is operating or not. Therefore, the point in the system where the expansion tank is connected will be known as the "point of no pressure change."

Figure 12-29b illustrates the main of a closed hydronic system with the expansion tank connected *on the suction side of the pump* with the pump not operating. At the initial filling of the system with cold water, assume the pressure gauges read 12 psig, as reducing valves are normally set to fill the system at that pressure. After the water in the system is heated to operating temperature, the gauges might read 18 psig because of the water expansion. Note that all points in the system indicate the same pressure (18 psig).

Figure 12-29c indicates the same system except the pump is now operating, and it is assumed that it develops a differential pressure of 8.0 psi. Since the expansion tank is located on the suction side of the pump and this is the "point of no pressure change," the pressure at this point will remain at 18 psig. The pressure differential developed by the pump will therefore appear as a positive pressure of 26 psig

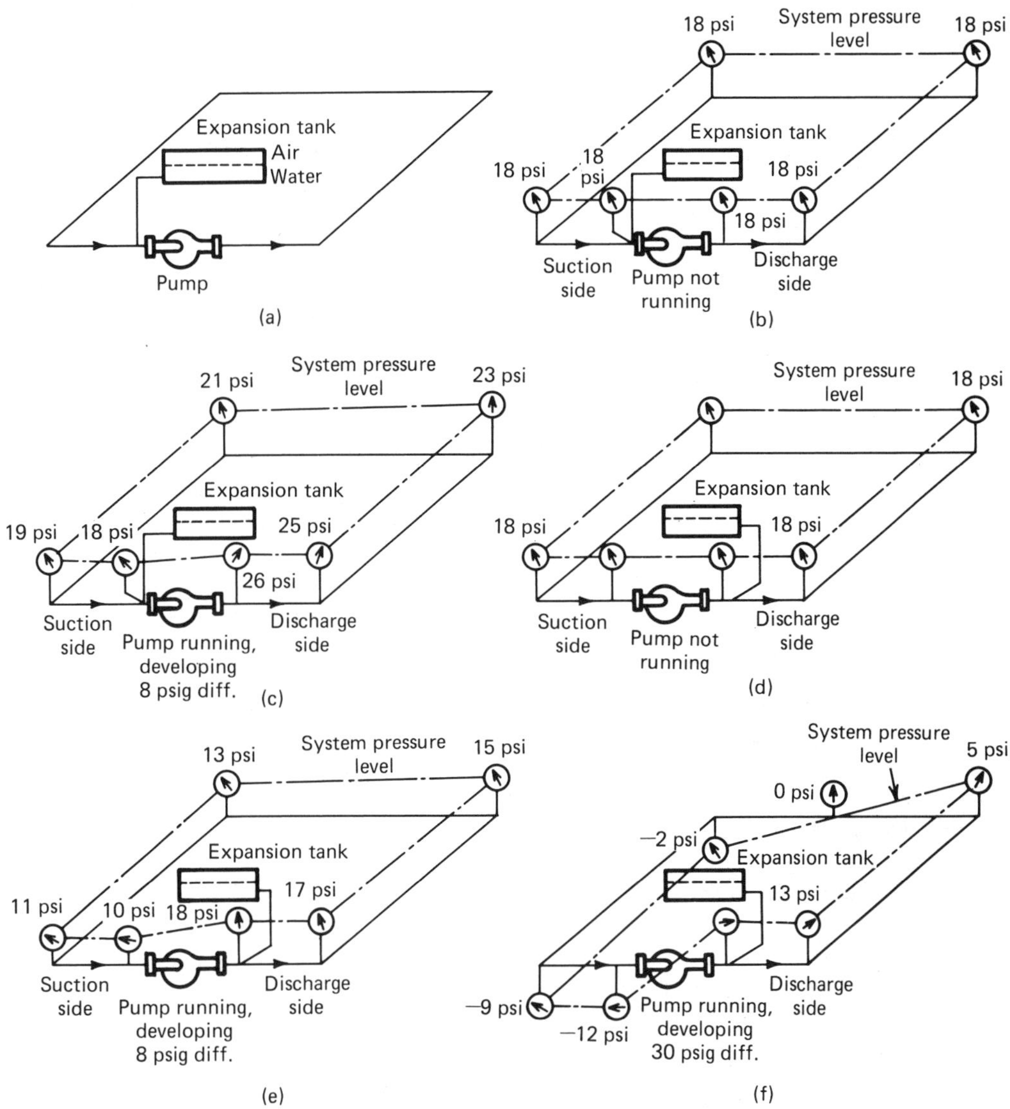

Figure 12–29

"Point of no pressure change" in hydronic system (Reproduced with permission from Taco, Inc., Cranston, Rhode Island.)

(18 + 8) at the discharge of the pump. The 8.0 psi will be dissipated by system friction as indicated by the reduced gauge readings around the system.

Figure 12-29d illustrates the same system as 12-29b and all conditions are the same, except the expansion tank is now connected *on the discharge side of the pump*. Note that the pump is not running and all pressure gauges read the same.

Figure 12-29e is the same system as 12-29d except the pump is now running. Since the expansion tank is located on the discharge side of the pump, this point now becomes the "point of no pressure change," and the pressure at this point remains at 18 psig. The pressure differential developed by the pump now appears as a negative pressure, and the pressure on the suction side of the pump will be reduced by 8.0 psi to 10 psig (18 − 8). As the friction of the system produces a drop of 8.0 psi, the pressure at various points in the system will become progressively lower.

Figure 12-29f illustrates the same system as 12-29d and e with the expansion tank connected on the discharge side of the pump, but with a pump that develops a pressure differential of 30 psi when running. This pressure differential produces a negative gauge pressure of −12 psig (18 − 30) on the suction side of the pump. Water at −12 psig (2.7 psia) will boil and produce vapor (steam) if the water temperature is above 137°F. This vapor formed on the suction side of the pump will cause the pump to cavitate. Also, due to part of the system being subjected to vacuum, air may leak into the system creating additional problems.

Both the pressure-relief valve and pressure-reducing valves should be connected to the boiler on the expansion tank side, which is the "point of no pressure change." Figure 12-24a illustrates the hookup when the expansion tank is on the suction side of the pump. Figure 12-24b illustrates the hookup when the expansion tank is connected to the discharge side of the pump.

12-18 STATIC PRESSURE AT THE HIGHEST POINT IN A SYSTEM

For proper operation of a hydronic system a positive pressure should be maintained at the highest point in the piping or heating device at all times. This pressure should be at least 1.5 psig or about 4 ft of water head for LTW systems. On boilers equipped with altitude gauges the red hand should be set manually to the minimum desired altitude in feet. This would be the distance from the top of the boiler to the highest point in the system plus 4 ft. The black hand, which indicates the actual system pressure, should never be permitted to fall below the position of the red hand. The following relationships will be helpful in determining the minimum pressure to be maintained at the highest point in the system.

For systems with expansion tank connected *on suction side of pump:*

$$p_s = \frac{h + 4}{2.31}$$

For systems with expansion tank connected *on discharge side of pump:*

$$p_s = \frac{h + 4 + h_p}{2.31}$$

where p_s = minimum static pressure (psi)

h = distance from top of boiler to highest point in system (ft)

h_p = operating head of pump (ft of water)

12-19 HYDRONIC SYSTEMS DESIGN EXAMPLES

Series loop baseboard heating systems. When designing a series loop baseboard system, the maximum load carried by a loop or circuit is dependent upon the size of the baseboard connections. A baseboard unit with ¾-in. connections will naturally carry a smaller load than one with 1-in. connections. It is the selection of the baseboard size that will determine in all cases whether one or more circuits are necessary.

Because all the water flowing through the main also flows through each baseboard, care must be taken to limit the amount of baseboard connected in series, otherwise velocity noises may occur. Table 12-4 shows the maximum recommended loop or circuit capacities to be used with various sizes of baseboard units. These capacities will insure satisfactory velocities, as indicated in Table 12-4, and also will result in reasonable friction losses.

If it is planned to use 1-in. steel pipe baseboard, the total circuit capacity should not exceed 72,000 Btu/hr. A circuit of 1-in. copper tube baseboard should not exceed a capacity of 75,000 Btu/hr. If the system load were to exceed the values, two or more circuits should be used.

It should be recognized at this point that, as the water flows through each heater in series, the water temperature decreases in proportion to the heat output from the individual units. This decrease in water temperature in the main may be readily calculated by

Table 12-4*

MAXIMUM RECOMMENDED LOOP OR CIRCUIT CAPACITIES USING BASEBOARDS

Baseboard Size	Total Load Btu/Hr.	Velocity Based on 20F. Temp. Drop
Cast Iron Baseboard — ¾″	40,000	2.4
Steel Pipe Baseboard — ½″	18,000	1.9
Steel Pipe Baseboard — ¾″	40,000	2.4
Steel Pipe Baseboard — 1″	72,000	2.7
Steel Pipe Baseboard — 1¼″	160,000	3.5
Steel Pipe Baseboard — 1½″	230,000	3.7
Steel Pipe Baseboard — 2″	450,000	4.6
Copper Tube Baseboard — ½″	15,000	2.1
Copper Tube Baseboard — ¾″	35,000	2.3
Copper Tube Baseboard — 1″	75,000	3.0
Copper Tube Baseboard — 1¼″	130,000	3.5

**Courtesy of Taco, Inc.*

$$\Delta t_{\text{main}} = \Delta t_{\text{circuit}} \left(\frac{\dot{q}_{\text{heater}}}{\dot{q}_{\text{circuit}}} \right) \qquad (12\text{-}24)$$

where Δt_{main} = temperature drop in the main caused by the heat output of an installed heater

$\Delta t_{\text{circuit}}$ = design temperature drop for the circuit

$\dot{q}_{\text{heater}}$ = heat output of a particular heater

$\dot{q}_{\text{circuit}}$ = total heat output of circuit

Illustrative Problem 12-6

Heat loss calculations for a residence gives the following room-by-room heat losses:

Room	*Heat Loss (Btu/hr)*
1	17,400
2	6,700
3	13,400
4	6,700
5	10,000
6	5,400
7	10,700
	Total = 70,300 Btu/hr

Design a series loop hydronic system for the residence using 1-in. steel baseboard units and steel pipe in the main. Assume average water temperature in first heating unit is 190°F and a system water temperature drop of 20 degrees.

Solution:

Step 1: Determine number of circuits.

Referring to Table 12-4, a 1-in. steel baseboard circuit should not have a load greater than 72,000 Btu/hr. Since the system load is 70,300, the system will consist of one circuit.

Step 2: Determine average water temperatures to use in evaluating baseboard lengths.

Manufacturers rate baseboard heating elements in Btu/hr per linear foot at an average water temperature in the unit. These average water temperature values are normally quoted in 10°F steps; that is, 190°F, 180°F, 170°F, and so forth. Interpolation can be used to obtain ratings at other average temperatures. To simplify the process, we will assume that the capacity of the baseboard will be evaluated at 5-degree intervals starting at 190°F leaving the boiler. This means that the circuit load will be divided into essentially four equal sections.

Using Eq. (12-24), the heater load to cause a 5-degree drop in temperature would be

$$\Delta t_{\text{main}} = \Delta t_{\text{circuit}} \left(\frac{\dot{q}_{\text{heater}}}{\dot{q}_{\text{circuit}}} \right)$$

$$5 = 20 \left(\frac{\dot{q}_{\text{heater}}}{70{,}300} \right)$$

$$\dot{q}_{\text{heater}} = 17{,}600 \text{ Btu/hr}$$

Therefore, we can assume that a load of 17,600 Btu/hr causes a 5-degree drop in water temperature in the main.

The approximate temperature of the water leaving the boiler then should be 192.5°F. The sum of heat loads for rooms 2 and 3 is 20,100 Btu/hr (approximately 17,600), and we will assume another 5-degree drop in temperature, giving an average water temper of 185°F for rooms 2 and 3. In a similar manner rooms 4 and 5 would have an average water temperature of 180°F, and rooms 6 and 7 an average water temperature of 175°F. With these average temperatures, we may now determine the unit capacity for each heating unit and the linear feet of baseboard required for each room (see work sheet on the following page).

Work Sheet
Series Loop Baseboard System

Room	Btu/hr Load	Average Water Temperature	Baseboard Rating Btu/hr/lin. ft	Lin. ft Baseboard Required
1	17,400	190	670	26
2	6,700	185	635	11
3	13,400	185	635	21
4	6,700	180	600	11
5	10,000	180	600	17
6	5,400	175	560	10
7	10,700	175	560	19
Total	70,300			115 ft

Step 3: Draw the sketch for the piping system.

Figure 12-30 shows a scaled layout of the piping and baseboard units. Pipe will be 1-in. NPS, same as baseboard.

Step 4: Calculate flow rate and system resistance.

The flow rate may be calculated by use of Eq. (12–21) and is

$$\text{gpm} = \frac{q}{490(t_1 - t_2)} = \frac{70.000}{490(20)} = 7.17$$

Length of straight pipe	= 131 ft (scaled from sketch)
Length of baseboard	= 115 ft (work sheet)
Fittings (1 in.)	
Boiler	= 6.5 ft (Table 12-3)
Air scoop	= 2.7 ft (Table 12-3)
Flow control valve (flow check)	= 42.0 ft (Table 12-3)
90-degree elbows (14 estimated at 2.5)	= 35.0 ft (Table 12-3)
Side-flow tee	= 6.0 ft (Table 12-2)
Gate values (2 at 0.5)	= 1.0 ft (Table 12-3)
Total equivalent length (TEL)	= 339.2 ft of pipe

From Fig. 12-25, f = 0.04 ft of water per ft of pipe length. By Eq. (12-22)

$$h_L = f \times \text{TEL} = 0.04(339.2)$$
$$= 13.568 \quad \text{(say 13.6 ft)}$$

Step 5: Select pump and expansion tank.

From a pump manufacturer's catalog select a pump having a capacity of 7.17 gpm against a head of 13.6 ft. Referring to Fig. 12-28a, one selection would be No. 113, 1-in. size. However, the discharge would have to be throttled to produce the required flow. Expansion tank minimum capacity would be 70,300/5000 = 14 gal.

Illustrative Problem 12-7

Heat loss calculations for a residence gives the following room-by-room heat losses:

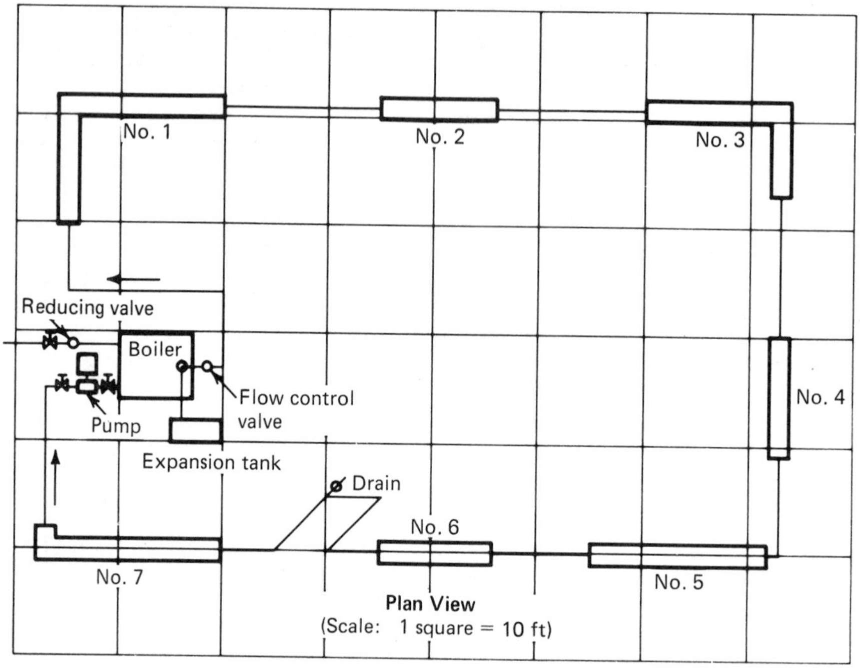

Figure 12–30

Series Loop Baseboard System (Illustrative Prob. 12-6)

Room	Heat Loss (Btu/hr)
1	18,500
2	4,000
3	12,300
4	10,000
5	28,000
6	6,500
7	15,000
	Total = 94,300 Btu/hr

Design a series loop heating system for the residence using 1-in. copper tube baseboard units with copper tube main. Assume an average water temperature in the first heating units in 190°F and a system water temperature drop of 20 degrees.

Solution:

Step 1: Determine number of circuits.

Referring to Table 12-4, a 1-in. copper tube baseboard circuit should not have a load exceeding 75,000 Btu/hr. Since the system load is 94,300 Btu/hr, we will use two circuits.

As nearly as possible, split the system load equally between the two circuits, (approximately 47,000 Btu/hr per circuit). Combining Nos. 1, 2, 3, and 4 gives a circuit load of 44,800 Btu/hr (circuit 1). Combining Nos. 5, 6, and 7 gives a circuit load of 49,500 Btu/hr (circuit 2).

Step 2: Determine average water temperatures to use in evaluating baseboard unit capacity and required length.

Each circuit will have a total temperature drop of 20 degrees. Assume that evaluation of average water temperature will be based on a 10-degree drop. Therefore the baseboard capacity for a 10-degree drop would be, By Eq. (12-24),

$$\Delta t_{\text{main}} = \Delta t_{\text{circuit}} \left(\frac{\dot{q}_{\text{heater}}}{\dot{q}_{\text{circuit}}} \right)$$

For circuit 1 $$10 = 20 \left(\frac{\dot{q}_{\text{heater}}}{44{,}800} \right)$$

$$\dot{q}_{\text{heater}} = 22{,}400 \text{ Btu/hr}$$

For circuit 2 $$10 = 20 \left(\frac{\dot{q}_{\text{heater}}}{49{,}500} \right)$$

$$\dot{q}_{\text{heater}} = 24{,}750 \text{ Btu/hr}$$

The average water temperatures for evaluating baseboard capacities then, are shown in the following work sheet.

Step 4: Determine flows in each part of system.

By Eq. (12-21) $$\text{gpm} = \frac{\dot{q}}{490(t_1 - t_2)}$$

For system main: $$\text{gpm} = \frac{94{,}300}{490(20)} = 9.6 \text{ gpm}$$

For circuit 1: $$\text{gpm} = \frac{44{,}800}{490(20)} = 4.57 \text{ gpm}$$

For circuit 2: $$\text{gpm} = \frac{49{,}500}{490(20)} = 5.03 \text{ gpm}$$

Step 5: Determine system resistance. See Table page 602.

The pump would be selected for a capacity of 9.6 gpm against a head of 7.32 ft of water.

Work Sheet
Series Loop Baseboard System (2-Circuit)

Room	Btu/hr Load	Ave. Water Temp.	Baseboard Rating Btu/hr/lin. ft.	Lin. ft. Baseboard Required
Circuit 1				
1	18,500	190	830	22
2	4,000	190	830	5
3	12,300	180	730	18
4	10,000	180	730	14
	Total = 44,800			Total = 58 ft
Circuit 2				
5	28,000	190	830	34
6	6,500	180	730	9
7	15,000	180	730	21
	Total = 49,500			Total = 64 ft

Step 3: Draw sketch of piping system.

Figure 12-31 shows a scaled layout of the piping and baseboard units. Pipe will be 1-in. copper tubes for each circuit.

One-pipe hydronic system using venturi fittings. These systems have become very popular in recent years. They overcome some of the difficulties of the series loop systems and are limited in capacity only by the available size

	Main	Circuit 1	Circuit 2
Scaled pipe length	73.0 ft	85.0 ft	73.0 ft
Baseboard length		58.0	64.0
Boiler (1¼ in.)	12.0		
Air scoop (1¼ in.)	4.0		
Flow control valve (flow check, 1¼ in.)	60.0		
Flow-through tee (1 in.)	2 × 1.7 = 3.4		
Gate values (1¼ in.)	2 × 0.8 = 1.6		
90-degree elbows	3 × 3.5 = 10.5	7 × 2.6 = 18.2	5 × 2.6 = 13.0
Side-flow tee (1 in.)		1 × 6 = 6.0	3 × 6 = 1.8
Square head cock		1 × 1.5 = 1.5	1 × 1.5 = 1.5
Total equivalent length (TEL)	164.5 ft	168.7 ft	169.5 ft
gpm flow	9.6	4.57	5.03
Friction rate (Fig. 12-26)	0.025 ft/lin. ft	0.0175	0.019
Resistance ($h_L = f \times$ TEL)	4.1 ft water	2.95	3.22
Circuit having greatest resistance (2)	3.22		
Pump head required	7.32 ft of water		

and diverting capacity of the special venturi tees used on each heating unit.

The design of one-pipe venturi systems will involve the use of tables and figures already presented and also Figs. 12-32 and 12-33. Figure 12-32 presents performance data of various sizes of venturi tees. For a given flow in the main in gpm, each size venturi within its operating range will cause a pressure drop that causes a small portion of the main flow to be diverted through a heater circuit. For example, if the main size is 1¼ in. carrying a flow of 15 gpm, a *a standard venturi* will cause a pressure head drop of 6.0 in. of water. A *super venturi* would cause a pressure head drop of 15.0 in. of water under the same flow conditions. This pressure head drop caused by the venturi fittings is available to overcome flow resistance in the heater branch circuits.

Figure 12-33 is a graph showing the available gravity heads to cause flow in risers. This is a positive assistance when considering upflow risers but is often not considered during system design unless the risers are long. For heaters located *below* the main, the gravity effect is negative in the downfeed risers and must be considered in the proper selection of the venturi tee fittings.

Illustrative Problem 12-8

Assume that a section of a circuit main has three heaters connected as shown in Fig. 12-34. The first heater is a second-floor heater having a capacity of 10,000 Btu/hr, the second heater is a first-floor heater with a capacity of 8,000 Btu/hr, and the third heater is a basement heater located below the main having a capacity of 6,000 Btu/hr. Assume that manufacturers' data on the heaters states that the pressure head loss through the heaters is 2.0 in. water/gpm flow through heater. Also, assume that the circuit main is 1¼ in. steel pipe with water flow at 16 gpm, 200°F water exists upstream of heater 1, and 20-degree water temperature drop occurs in each heater.

Determine steel pipe size for the risers on each heater, required venturi tee size, and the

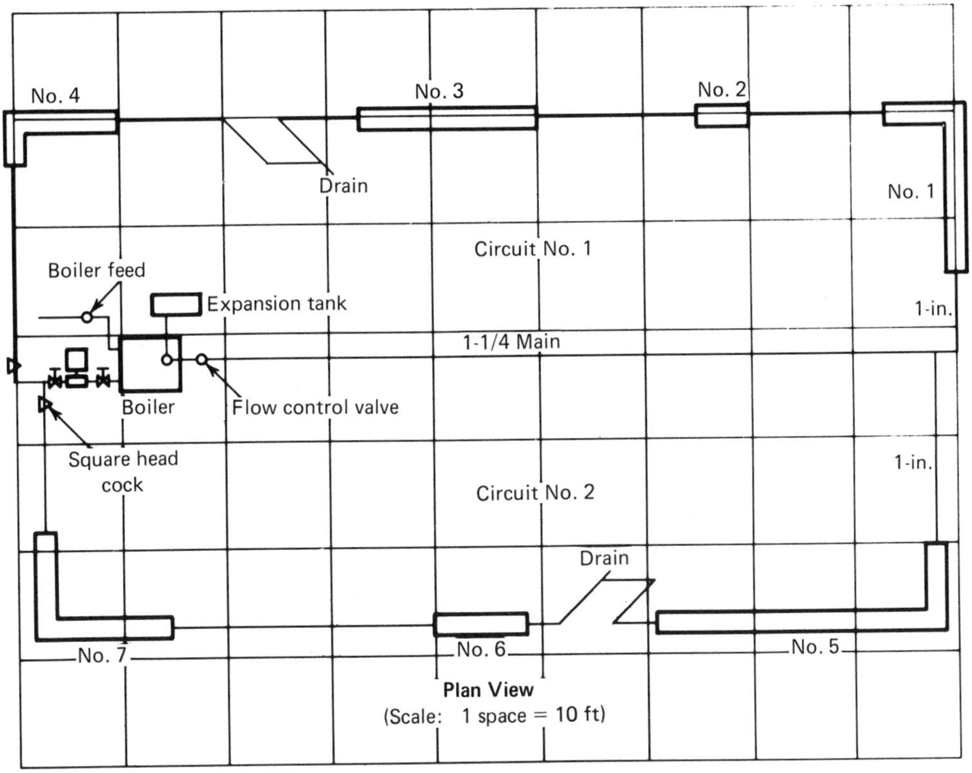

Figure 12–31

Series Loop Baseboard System—Two Circuit (Illustrative Prob. 12-7)

water temperature in the main after each heater.

Solution: Heater 1

Step 1: Determine flow rate in heater circuit. By Eq. (12-21)

Flow through heater (gpm)

$$= \frac{\dot{q}_{\text{heater}}}{490(t_1 - t_2}$$

$$= \frac{10{,}000}{490(20)}$$

$$\text{gpm}_{\text{heater}} = 1.02 \text{ gpm}$$

Step 2: Estimate pipe size for heater circuit and calculate circuit resistance.

From Table 12-1, a ½-in. steel pipe should carry this flow within given velocity limits. Therefore, we shall assume that ½-in. pipe will be used for risers.

Equivalent length of pipe and fittings in the heater circuit must be determined in order to calculate circuit resistance, using the table on the next page.

From Fig. 12-25, for a flow of 1.02 gpm in a ½-in. pipe, the friction rate f is 0.015 ft of water/ft of pipe. Therefore, the equivalent

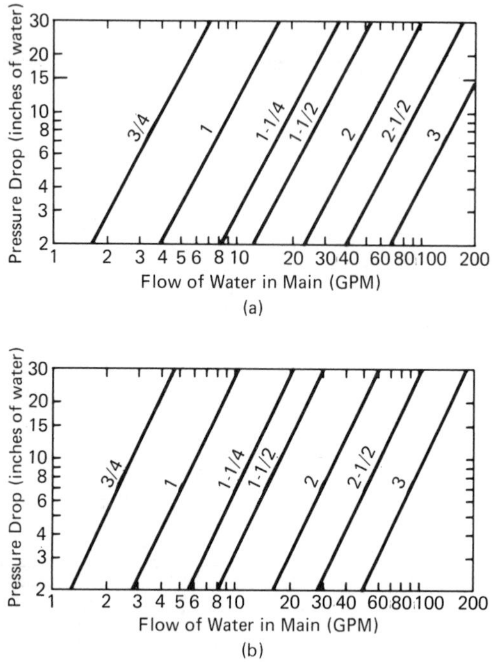

Figure 12–32

Pressure drop through Taco venturi fittings (a) *Standard* venturi, bronze and cast iron. (b) *Super* venturi, bronze and cast iron. (*Taco Guide for Hydronic Engineers*, with permission from Taco, Inc., Cranston, Rhode Island.)

Fittings		Equivalent Length	
Riser length (total)		= 48.0 ft	
Sq. hd. Cock (balance valve)	(1 at 1.5 ft)	= 1.5 ft	(Table 12-3)
Angle radiator valve	(1 at 5 ft)	= 5.0	
90-degree elbow	(8 at 2.5 ft)	= 20.0	
Side-flow tee (upstream tee used as elbow)		= 5.0	
Equivalent length pipe and fittings		= 79.5 ft	

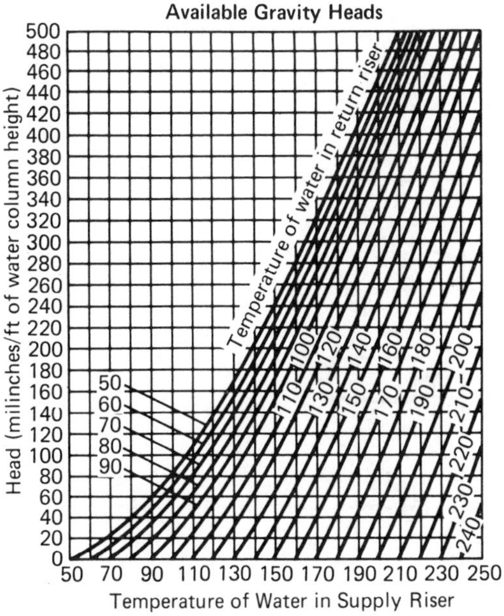

Figure 12–33

Available gravity heads in hydronic systems, from *Taco Guide for Hydronic Engineers.* (Used by permission, courtesy of Taco, Inc., Cranston, Rhode Island)

length of the heater is, by Eq. (12-23b),

$$EL_{heater} = \frac{2.0(1.02)}{12(0.015)} = 11.3 \text{ ft}$$

Total equivalent length of heater circuit = 79.5 + 11.3 = 90.8 ft. By Eq. (12-22)

Heater circuit resistance (h_L) = $f \times$ TEL

h_L = 0.015(90.8) = 1.36 ft water = 16.3 in. water

Step 3: Select venturi tee size.

From Fig. 12-32, a 1¼ *standard* venturi tee with a flow of 16 gpm in the main produces a head drop of 6.7 in. water. For a 1¼ *super* venturi tee the head drop is 17.0 in. water.

The required drop is 16.3 in. water (step 2), therefore a super venturi is required when ½-in. risers are used.

At this point, we could check to see the effect of using ¾-in. risers.

Step 4: Assume ¾-in. risers and recalculate steps 2 and 3.

Referring to Table 12-3, we find that the equivalent length of pipe and fittings does not change (79.5). However, the equivalent length of the heater does change because the friction rate changes.

Referring to Fig. 12-25, 1.02 gpm flowing in ¾-in. pipe has a friction rate of 0.0035 ft water/ft of pipe. Therefore,

Equivalent length heater 1 =

$$\frac{2.0(1.02)}{12(0.0035)} = 48.6 \text{ ft.}$$

Total equivalent length of heater circuit =

$$79.5 + 48.6 = 128.1 \text{ ft.}$$

Heater circuit loss (h_L) =

$$0.0035(128.1) = 0.448 \text{ ft} = 5.38 \text{ in. water}$$

From before (step 3), a 1¼-in. standard venturi produces a head drop of 6.7 in. water, which is satisfactory being greater than 5.38 in. water required.

We may now conclude that a 1¼ by ½ super venturi could be used with ½-in. risers, or 1¼ by ¾ standard venturi could be used with ¾-in. risers.

Step 5: Determine temperature drop of water in the main caused by heater 1.

If the flow of water in the main is 16 gpm and a 20-degree drop in temperature is assumed for the main, the total system heat load is, by Eq. (12-21)

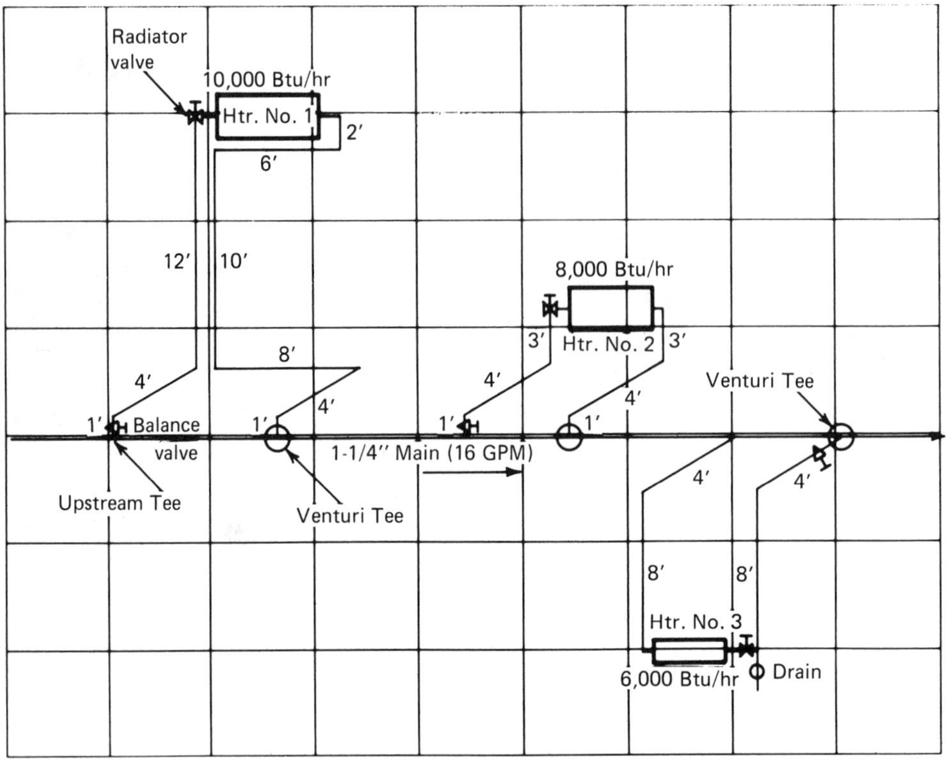

Figure 12–34

Section of a One-Pipe Venturi System (Illustrative Prob. 12-8)

$$\text{gpm} = \frac{\dot{q}_{\text{system}}}{490(t_1 - t_2)}$$

$$16 = \frac{\dot{q}_{\text{system}}}{490(20)}$$

$$\dot{q}_{\text{system}} = 156{,}800 \text{ Btu/hr}$$

The water temperature drop in the main caused by heater 1 is, by Eq. (12-24),

$$\Delta t_{\text{main}} = \Delta t_{\text{circuit}} \left(\frac{\dot{q}_{\text{heater}}}{\dot{q}_{\text{system}}} \right)$$

$$= 20\left(\frac{10{,}000}{156{,}800} \right)$$

$$\Delta t_{\text{main}} = 1.27 \text{ degrees}$$

Therefore, the water temperature in the main following heater 1 and entering heater 2 would be 200 − 1.27 = 198.7°F.

Heater 2

Step 1: Determine flow and assume pipe size for heater circuit.

$$\text{gpm} = \frac{8000}{490(20)} = 0.82 \text{ gpm}$$

From Table 12-1, a ½-in. steel pipe should carry this flow within given velocity limits.

Step 2: Determine resistance of heater circuit.

From Fig. 12-25, for a flow of 0.82 gpm, f = 0.0095 ft/ft of length.

Fittings		Equivalent Length	
Riser length		16.0 ft	
Sq. hd. cock	(1 at 1.5)	1.5	(Table 12-3)
Angle rad. valve	(1 at 5.0)	5.0	(Table 12-3)
90-degree elbow	(5 at 2.5)	12.5	(Table 12-3)
Side-flow tee	(1 at 5)	5.0	(Table 12-3)
Equivalent length pipe and fittings =		40 ft	

Equivalent length of heater $= \dfrac{2.0(0.82)}{12(0.0095)} = 14.4$ ft

Total equivalent length of heater circuit = 40 + 14.4 = 54.4 ft

Heater circuit resistance, $h_L = 0.0095(54.4) = 0.51$ ft = 6.2 in. water

Step 3: Determine venturi size.

From before a 1¼- by ½-in. standard venturi produces a head drop of 6.7 in., which is adequate.

Step 4: Determine temperature drop in main.

$$\Delta t_{\text{main}} = 20\left(\frac{8000}{156{,}800}\right) = 1.02 \text{ degrees}$$

Water in main after heater 2 is 198.7 − 1.02 = 197.6°F.

Heater 3

Step 1: Determine gravity head to be overcome.

When the heating unit is below the main, it is necessary to provide sufficient head by the venturi to force water down the risers against the gravity head. This gravity head is caused by the higher-temperature water in the main and the lower-temperature water in the heater (usually assumed to be 60°F, or room temperature).

From preceding calculations the water temperature in the main at the downfeed supply riser is 197.6°F, and the water in the return riser will be assumed at 60°F. Referring to Fig. 12-33, the gravity head to be overcome is approximately 430 millinches per foot of height. Since the riser height is 8 ft, the gravity head is 8 × 430 = 3440 millinches, or 3.44 in. water.

Step 2: Select a tentative venturi size.

From Fig. 12-32 we have found previously that a 1¼-in. standard venturi tee produces a head drop of 6.7 in. and a super venturi tee produces a head drop of 17.0 in. water.

Step 1 determined that 3.44-in. gravity head must overcome, therefore the remaining available head to overcome friction would be

$$\begin{aligned}\text{Standard venturi} &= 6.7 - 3.44\\ &= 3.26 \text{ in. water (very low)}\end{aligned}$$

$$\begin{aligned}\text{Super venturi} &= 17.0 - 3.44\\ &= 13.56 \text{ in. water}\end{aligned}$$

At this point, we may select a 1¼-in. super venturi, which gives us a head of 13.56 in. water to overcome friction losses in risers and heater 3.

Step 3: Determine riser size, using the table at the top of the next page.

From Fig. 12-25, for a flow rate of 0.612 gpm, the friction rate f is 0.0056 for a ½-in. pipe and 0.0014 for a ¾-in. pipe.

For ½-in. pipe risers:

Fittings		Equivalent Length	
Riser runouts	(2 at 4)	= 8.0 ft	
Risers	(2 at 8)	= 16.0	
Sq. hd. cock	(1 at 1.5)	= 1.5	(Table 12-3)
90-degree elbows	(4 at 2.5)	= 10.0	(Table 12-3)
Radiator valve	(1 at 5)	= 5.0	(Table 12-3)
Side-flow tee	(1 at 5)	= 5.0	(Table 12-3)
Equivalent length pipe and fittings		= 45.5 ft	
Flow through heater 3 (gpm)		$= \frac{6000}{490(20)} = 0.612$ gpm	

Equivalent length of heater is $\frac{2.0(0.612)}{12(0.0056)} =$ 18.2 ft

Total equivalent length = 45.5 + 18.2 = 63.7 ft

Head loss = $f \times$ TEL = 0.0056(63.7) = 0.357 ft = 4.28 in.

For ¾-in. pipe risers:

Equivalent length of heater is $\frac{2.0(0.612)}{12(0.0014)} =$ 72.8 ft

Total equivalent length = 45.5 + 72.8 = 118.3 ft

Head loss = 0.0014 (118.3) = 0.166 ft = 1.98 in.

Step 4: Determine venturi size.

We may select a 1¼- × ½-in. super venturi with ½-in. pipe risers. or, we may select a 1-¼- × ¾-in. standard venturi with ¾-in. pipe risers.

Step 5: Temperature in main after heater 3

$$\Delta t_{main} = \Delta t_{circuit}\left(\frac{\dot{q}_{heater}}{\dot{q}_{circuit}}\right)$$

$$= 20\left(\frac{6000}{156,800}\right)$$

$$\Delta t_{main} = 0.76 \text{ degrees}$$

Temperature in main after heater is 197.6 − 0.76 = 196.8°F.

Illustrative Problem 12-9

Heat loss calculations for a residence gives the following room-by-room heat losses:

Room	Heat Loss (Btu/hr)	Room	Heat Loss (Btu/hr)
1	6,000	6	9,000
2	10,000	7	14,000
3	12,000	8	7,000
4	8,000	9	6,000
5	15,000	10	9,000

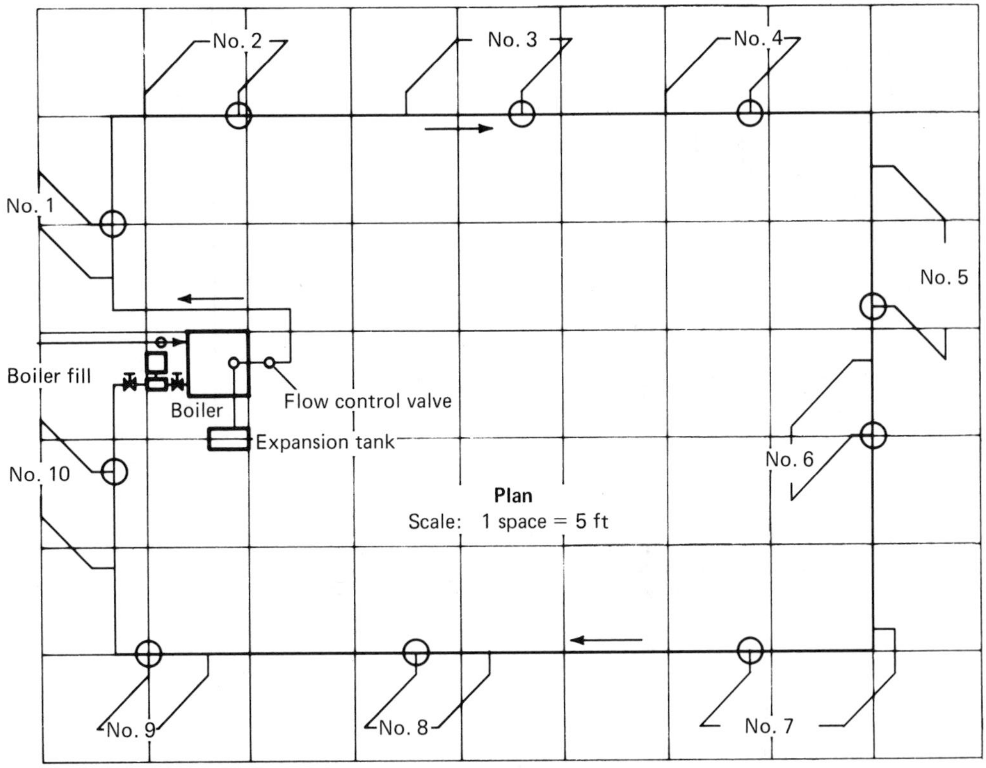

Figure 12–35

One-Pipe Venturi System (Illustrative Prob. 12-9)

Design a one-pipe venturi fitting hydronic system for the residence using a 20-degree design water temperature drop for the system, and a 20-degree water temperature drop through each heater (Fig. 12-35). All heaters shall be considered convectors having an internal resistance to flow of 2.0 in. water per gpm water flow. Assume the water temperature leaving the boiler is at 200°F. Use copper tube for the main and all heater risers.

Solution:

Assume that the risers to each of heater 1, 2, 3, 4, 5, 8, 9, and 10 contain 10 linear feet of tubing, two 45-degree elbows, three 90-degree elbows, one angle radiator valve, and one square head cock (used as balance valve).

For heater 6 (basement, below main) assume risers contain 18 linear feet of tubing (8 ft risers), three 90-degree elbows, one angle radiator valve, and one square head cock.

For heater 7 assume risers contain 14 linear feet of tube, two 45-degree elbows, three 90-degree elbows, one angle radiator valve, and one square head cock.

Step 1: Numerous calculations are involved in problems of this type, therefore, it is convenient to tabulate results on a results sheet.

Results Sheet
(Illustrative Problem 12-9)

Room or Heater	Heat Loss (Btu/hr)	Required Flow (gpm)	Branch Size (in.)	Fig. 12-27 Friction Rate (ft/ft pipe)	Equivalent Length TEL (ft)	Venturi	
						Size	Std. or Super
1	6,000	0.612	1/2	0.011	40.3	1 × 1/2	Std.
2	10,000	1.020	1/2	0.027	37.3	1 × 1/2	Super
3	12,000	1.220	1/2	0.035	36.8	1 × 1/2	Super
4	8,000	0.816	1/2	0.017	39.0	1 × 1/2	Std.
5	15,000	1.531	1/2	0.052	35.9	1 × 1/2	Super
6	9,000	0.918	1/2	0.021	44.3	1 × 1/2	Super
7	14,000	1.428	3/4	0.009	61.4	1 × 3/4	Std.
8	7,000	0.714	1/2	0.014	39.5	1 × 1/2	Std.
9	6,000	0.612	1/2	0.011	40.3	1 × 1/2	Std.
10	9,000	0.918	1/2	0.021	38.3	1 × 1/2	Std.
Total	96,000	9.789 (9.8)					

Use 1-in. copper tube for main, $f = 0.065$ ft/ft tube.

Step 2: Calculate flows in various parts of system. By Eq. (12-21)

$$\text{gpm} = \frac{\dot{q}}{490(t_1 - t_2)} = \frac{\dot{q}}{490(20)} = \frac{\dot{q}}{9800}$$

For heater 1 (Typical for all heaters and system)

$$\text{gpm} = \frac{6000}{9800} = 0.612 \text{ gpm}$$

Step 3: Determine equivalent length of heater circuits and select venturi sizes.

Heater circuits, except for 6 and 7, are all the same. The equivalent length of pipe and fittings will be equal for each; however, the equivalent length of the heaters depends upon flow rate, tube size, and friction rate.

Assume that all risers for heaters, except 6 and 7, will be ½-in. tube. The equivalent length of tube and fittings will be

Fittings		Equivalent Length	
Risers and runouts		10.0 ft	
45-degree elbows	(2 at 1.0) =	2.0	(Table 12-3)
90-degree elbows	(3 at 3.5) =	7.5	(Table 12-3)
Angle radiator valve	(1 at 5.0) =	5.0	(Table 12-3)
Sq. hd. cock	(1 at 1.5) =	1.5	(Table 12-3)
Side-flow tee	(1 at 5.0) =	5.0	(Table 12-3)
Equivalent length pipe and fittings	=	31.0 ft	

By Eq. (12-23b) $\quad \text{EL} = \dfrac{\text{inches water}}{12 \times f}$

For heater 1

From Fig. 12-26, with 0.612 GPM in 1/2-in. copper tube, f = 0.011 ft/ft of tube.

$$\text{EL} = \frac{2.0(0.612)}{12(0.011)} = 9.3 \text{ ft}$$

Total equivalent length heater circuit 1 = 31.0 + 9.3 = 40.3 ft. By Eq. (12-24)

$$\text{Head loss } (h_L) = f \times \text{TEL} = 0.011(40.3)$$
$$= 0.443 \text{ ft}$$
$$= 5.3 \text{ in. water}$$

Referring to Fig. 12-32, for a flow of 9.8 gpm in the 1-in. main, a 1-in. standard venturi produces a head drop of 11.0 in. water, and a 1-in. super venturi produces a head drop of 26.0 in. water.

A 1- by ½-in. standard venturi may be used on heater circuit 1. For the remaining similar heater circuits calculations would produce the results shown on the work sheet.

For heater 6

From Fig. 12-33, with an estimated water temperature in the main of 190°F and 60°F in heater, the gravity head to be overcome is 390 millinches/ft of riser height. If the riser height is 8 ft, the total gravity head is (8 × 390) = 3120 millinches, or 3.12 in. water.

From before, a 1-in. standard venturi produces a head drop of 11.0 in. water. Therefore, 11.0 − 3.12 = 7.88 in. water available to overcome flow resistance. (Assume riser will be ½-in. tube.)

Fitting			Equivalent Length
Downfeed risers and runouts			18.0 ft
90-degree elbows	(3 at 2.5 ft)	=	7.5
Angle radiator valve	(1 at 5.0 ft)	=	5.0
Sq. head cock	(1 at 1.5 ft)	=	1.5
Side-flow tee	(1 at 5.0 ft)	=	5.0
Equivalent length tube and fittings		=	37.0 ft

From Fig. 12-26, with 0.918 gpm flowing in ½-in. tube, $f = 0.021$ ft/ft of tube.

$$\text{Equivalent length of heater 6} = \frac{(2.0)(0.918)}{12(0.021)} = 7.3 \text{ ft.}$$

Total equivalent length heater circuit 6 = 37.0 + 7.3 = 44.3 ft.

$$h_L = 0.021(44.3) = 0.93 \text{ ft} = 11.2 \text{ in. water}$$

This is higher than the 7.88 in. available, therefore, we may use a super venturi, or we may increase the riser tube size to ¾-in. We will use a super venturi.

Heater 7

Assume riser size to be ¾-in. tube.

Fittings			Equivalent Length
Risers and runouts			14.0 ft
45-degree elbows	(2 at 1.0 ft)	=	2.0
90-degree elbows	(3 at 2.5 ft)	=	7.5
angle radiator valve	(1 at 5.0 ft)	=	5.0
Sq. hd. cock	(1 at 1.5 ft)	=	1.5
Side-flow tie	(1 at 5.0 ft)	=	5.0
Equivalent length tube and fittings		=	35.0 ft.

From Fig. 12-26, with 1.428 gpm flowing in ¾-in. tube, $f = 0.009$ ft/ft of tube.

$$\text{Equivalent length heater 7} = \frac{2.0(1.428)}{12(0.009)} = 26.4 \text{ ft}$$

Total equivalent length heater circuit 7 = 35 + 26.4 = 61.4 ft.

$$h_L = (0.009)(61.4) = 0.55 \text{ ft} = 6.63 \text{ in. water}$$

Available head drop from standard venturi is 11.0 in. water. Use a 1- by ¾-in. standard venturi.

Step 4: Determine head loss in main.

The flow in the main, recorded on work sheet, is 9.789 gpm, or say 9.8. From Fig. 12-26 with 9.8 gpm flowing in 1-in. tubing, the friction rate f is 0.065 ft/ft of tube.

NOTE: The required pump head (26.2 ft) for this system may seem excessive. Many hydronic systems are designed with the required pump head between 10 and 15 ft. The reason for the high head loss for this system is that the 1-in. main was selected to obtain higher diversion capacities for the venturi fittings. If the main size was 1¼ in. (recommended in Table 12-1) to reduce the flow velocity, ¾-in. risers would be required on all heater circuits. These changes would increase the installed cost of the system but would decrease slightly the operating cost because of possibly using a smaller pump motor. Normally, it is a decision made by the design engineer as to which procedure to follow.

Illustrative Problem 12-10

Figure 12-36 shows the plan layout of a two-circuit, one-pipe hydronic system. The scale of 1 in. = 10 ft applies only to the horizontal mains. The circled numbers identify the heaters in the system, which have the following capacities in Btu/hr:

Fitting (1-in.)			Equivalent Length	
Length of main			125.0 ft	
Flow control valve	(1 at 42 ft)	=	42.0	(Table 12-3)
Boiler	(1 at 9.0 ft)	=	9.0	(Table 12-3)
Gate valves	(2 at 0.5 ft)	=	1.0	(Table 12-3)
90-degree elbows	(11 at 2.5 ft)	=	27.5	(Table 12-3)
Venturi (standard)	(6 at 14.0 ft)	=	84.0	(Table 12-3)
Venturi (super)	(4 at 28.5 ft)	=	114.0	(Table 12-3)
Total equivalent length pipe and fittings		=	402.5 ft	

Main $h_L = 0.065(402.5) = 26.2$ ft water

Pump must pump 9.8 gpm against a head of 26.2 ft water.

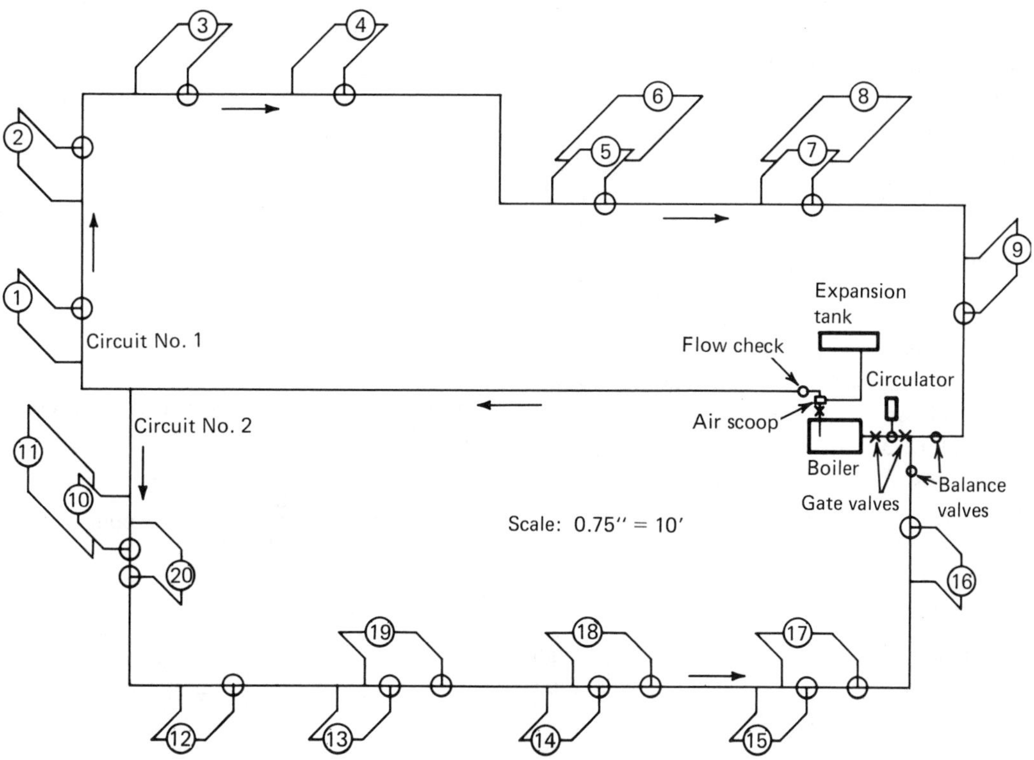

Figure 12–36

One-Pipe Hydronic System

First-Floor Heaters	Second-Floor Heaters	Basement Heaters
1— 8,000	6—11,000	17— 8,000
2—10,000	8—10,000	18— 8,000
3— 6,000	11— 8,000	19—10,000
4— 6,000		20—12,000
5— 7,000		
7— 8,000		
9—15,000		
10— 9,000		
12—10,000		
13—11,000		
14— 8,000		
15—11,000		
16—12,000		

The friction loss through the individual heaters will be assumed at 250 millinches per 1000 Btu/hr capacity at a water temperature drop of 20°F through the heater.

Figure 12-37 shows the riser diagram which may be assumed typical of all first-floor heaters, except heaters 5, 7, and 10. Figure 12-38 shows the riser diagram typical for each basement heater. Figure 12-39 shows the riser diagram for heater combinations 5 and 6, 7 and 8, and 10 and 11.

Problem Statement: Determine the venturi type and size and riser sizes required for the heater circuits illustrated in Figs. 12-37, 12-38, and 12-39, assuming a water temperature drop through each heater of 20°F and using type *L* copper tube. Also, assume a system design water temperature drop of 20°F.

<u>*Solution:*</u>

Step 1: Determine gpm flow rate through each heater circuit and circuit main. By Eq. (12-21)

$$\text{gpm} = \frac{\dot{q}}{490(t_1 - t_2)}$$

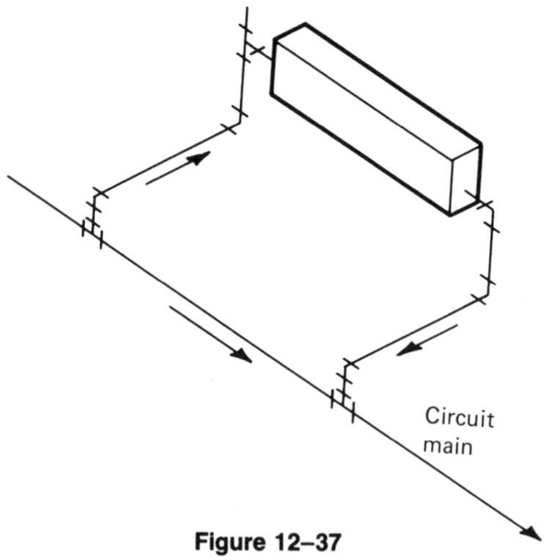

Figure 12–37

Illustrative Problem 12-10. Riser diagram for 1st floor heaters except heaters No. 5, 7, and 10. Straight length of tube = 18 feet

Since $t_1 - t_2$ is 20 degrees for each heater circuit and circuit main, then

$$\text{gpm} = \frac{\dot{q}}{490(20)} = \frac{\dot{q}}{9800}$$

$$= \frac{\dot{q}}{10{,}000} \quad \text{(approximately)}$$

For heater 1 circuit (consisting of heaters 1 through 9)

$$\text{gpm} = \frac{8{,}000}{10{,}000} = 0.8 \text{ gpm}$$

For circuit 1

$$\text{gpm} = \frac{81{,}000}{10{,}000} = 8.1 \text{ gpm}$$

The same calculations should be made for each heater, circuit, and main and enter results on the results sheet.

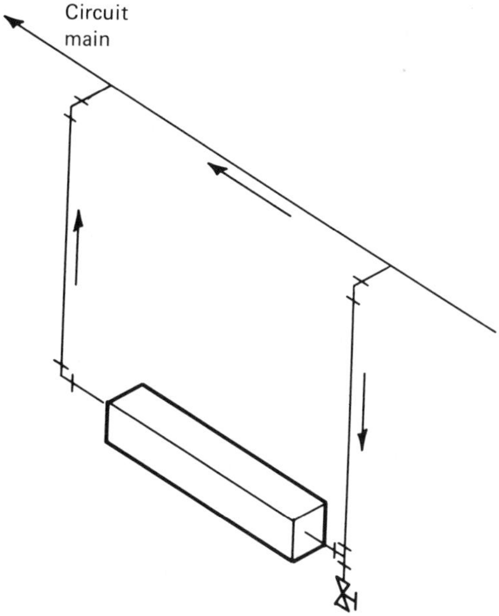

Figure 12–38

Illustrative Problem 12-10. Riser diagram, typical of each basement heater Straight length of tube = 24 feet.

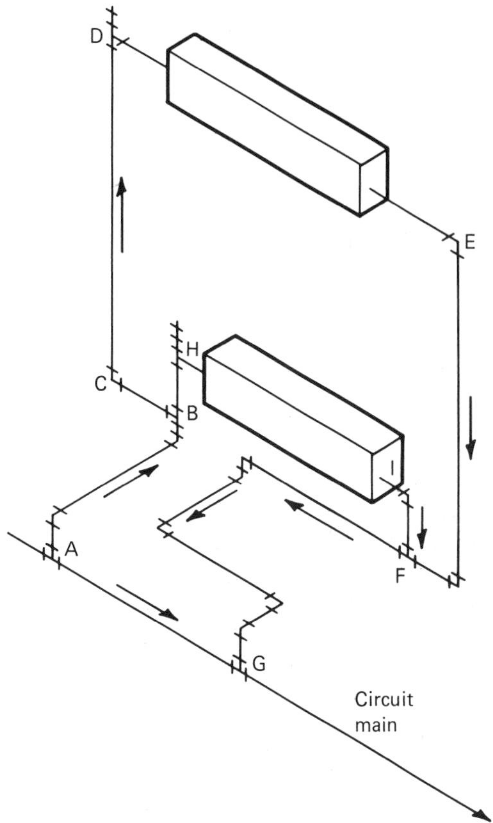

Figure 12–39

Illustrative Problem 12-10. Riser diagram for heater combinations 5 and 6, 7 and 8, 10 and 11. Estimated tube length A B C D E F G = 35 feet

Step 2: Determine tube size.

Using Fig. 12-26, determine *tentative* tube sizes for the various pipes, maintaining flow velocities within recommended velocity limits of Table 12-1. Enter tube sizes on results sheet, with the indicated friction rate.

Step 3: Determine venturi size and type (standard or super) for each first-floor heater circuit (Fig. 12-37).

The largest first-floor heater is 9 having a capacity of 15,000 Btu/hr. All other riser circuits for first-floor heaters will be assumed to have the same equivalent length for pipe and fittings.

For heater 9 riser circuit, the tube size has been tentatively sized at 3/4-in. with a friction rate of 120 millinches per foot of tube. Using Table 12-3, determine the equivalent length of fittings in the riser circuit.

Fitting		EL (Table 12-3)
90-degree elbows	(5 at 2.5)	= 12.5 ft
Angle radiator valve	(1 at 5.0)	= 5.0 ft
EL of fittings		= 17.5 ft
Straight length of pipe		= 18.0 ft
	TEL	= 35.5 ft

By Eq. (12-22) circuit head Loss (h_L) = $f \times$ TEL

$$h_L = 120(35.5) = 4260 \text{ millinches}$$
$$= 4.26 \text{ in. water}$$

$$\text{Heater head loss} = 250 \times 15$$
$$= 3750 \text{ millinches}$$
$$= 3.75 \text{ in. water}$$

$$\text{Total heater circuit loss} = 4.26 + 3.75$$
$$= 8.01 \text{ in. water}$$

Referring to Fig. 12-32, we find that with circuit 1 main sized at 1 1/4-in. and a flow rate of 8.1 gpm, the standard venturi produces a head drop of less than 2 in. water, and the super venturi produces a head drop of 4.25 in. water. Since this is less than required, we must reduce the resistance of the heater circuit or increase the head drop available from the venturi. The latter choice is accomplished by reducing the circuit main tube size to 1 in. Again, referring to Fig. 12-32, a 1-in. standard venturi produces a head drop of 7.8 in. water at 8.1 gpm, and a super venturi produces a head drop of 18.0 in. water.

If we consider reducing the resistance of the heater circuit, we would increase the tube size to 1 in. Referring to Fig. 12-26, for a flow rate of 1.5 gpm in a 1-in. tube, the friction rate is 30 millinches per foot of tube. The circuit head loss would be 30 × 35.5 = 1065 millinches for piping plus 3750 millinches for the heater, or a total of 4815 millinches. This is 4.8 in. water, which is still higher than the available head drop of a 1 1/4-in. super venturi (4.25).

Therefore, we will say that circuit 1 main will be sized at 1-in. with a friction rate of 550 millinches per foot.

Since the 1-in. standard venturi produces a head drop of 7.8 in. water, this may be increased slightly by considering the length of main between the upstream tee and the venturi. Assume this length if 1 1/2 ft (minimum recommended distance). The head drop in this section would be 1.5 × 550 = 825 millinches, or 0.825 in. When this is added to the 7.8 available from the venturi, the total is 8.6 in., which exceeds the required 8.01 in. Therefore, at this point we will say that each first-floor heater on circuit 1 requires a 1-in. standard venturi.

Step 4: Determine venturi size for basement heater circuits (Fig. 12-38).

The basement heater having the greatest capacity is 20 with a capacity of 12,000 Btu/hr. The heater is 9 ft below the circuit main. If we assume that water temperature in the heater is 60°F and the water in the circuit 2 main is 200°F, Fig. 12-33 indicates that the gravity head which must be overcome is 430 millinches per foot of height. Since the column height is 9 ft, the total gravity head is 9 × 430 = 3780 millinches, or 3.78 in. water.

Referring to Fig. 12-32, a 1 1/4-in. standard venturi for a flow of 10.7 gpm produces a head drop of 3.2 in., and a super venturi produces a head drop of 7.6 in. water. Using a super venturi, the head available to overcome resistance of the heater circuit is 7.6 − 3.78 = 3.73 in. water.

Fittings (3/4in.)			
90-degree elbows	(4 at 2.5)	=	10.0 ft
Angle radiator valve	(1 at 5.0)	=	5.0 ft
EL fittings		=	15.0 ft
Straight length of pipe		=	24.0
	TEL	=	39.0 ft
h_L = 75 × 39.0 = 2925 millinches		=	2.9 in.
Heater loss = 250 × 12 = 3000 millinches		=	3.0 in.
Total heater circuit loss		=	5.9 in.

This is greater than the available head drop of 3.73 in. Again we are confronted with the same situation as in step 3. We require a drop of 5.9 in. water and with 3.73 in. available, the difference 5.9 − 3.73 = 2.17 in. could be made up considering the distance between the upstream tee and the venturi. To find the required distance, each foot of the 1 1/4-in. circuit main has a drop of 350 millinches. Therefore, 2.17 in., or 2170 millinches, drop requires 2170/350 = 6.2 ft of tube. Specify then that the venturi be placed 6.2 ft downstream of the supply tee.

Step 5: Determine venturi size for combination heater circuit, Fig. 12-39.

Combinations of heater 5 and 6 or 7 and 8 have the same capacity of 18,000 Btu/hr, so Fig. 12-39 will apply to either. It will be necessary to calculate the equivalent length and head loss through the longer circuit *ABCDEFG*.

Fittings	EL (Table 12-3)
Tees	(2 at 5.0) = 10.0 ft
90-degree elbows	(10 at 2.5) = 25.0 ft
Angle radiator valve	(1 at 5.0) = 5.0 ft
EL	= 40.0 ft
Straight length of tube	= 35.0 ft (est.)
	TEL = 75.0 ft

Sections *AB* and *FG* carry 1.8 gpm at f = 160 millinches per foot, and section *BCDEF* carries 1.1 gpm at f = 65 millinches per foot. Assume EL of section *BCDEF* is 50 ft and EL of section *AB* plus *FG* is 25 ft.

$$h_L = 25 \times 160 = 4000$$

$$h_L = 50 \times 65 = \underline{3250}$$

$$\text{Total pipe} = 7250 \text{ millinches}$$

$$\text{Heater loss} = 250 \times 11 = 2750 \text{ millinches}$$

Total circuit loss = 7250 + 2750 = 10,000 millinches, or 10.0 in. water. From step 3, a 1-in. super venturi produces a head drop of 18.0 in. water.

The heater combination 10 and 11 on circuit 2 main presents a new situation and must be checked. Referring to Fig. 12-32, a 1 1/4-in. super venturi produces a head drop of 7.6 in. water. The estimated head loss through the heater circuit is

$$25 \times 130 = 3250$$

$$50 \times 45 = 2250$$

$$250 \times 8 = 2000$$

$$\text{Circuit loss} = 7500 \text{ millinches} = 7.5 \text{ in.}$$

This is slightly less than that available from the 1 1/4-in. super venturi.

Step 6: Check venturi type and size for first-floor heaters on circuit 2.

Heater 16 is the largest first-floor heater on circuit 2. Referring to step 3, the total equivalent length of the heater circuit is 35.5 ft. Flow through heater circuit is 1.2 gpm at friction rate of 80 millinches per foot (Fig. 12-26).

$$h_L = 75 \times 35.5 = 2660 \text{ millinches}$$
$$= 2.66 \text{ in.}$$

$$\text{Heater loss} = 250 \times 12 = 3000 \text{ millinches}$$
$$= 3.00 \text{ in.}$$

$$\text{Total circuit loss} = 5.66 \text{ in. water}$$

From step 4, a 1 1/4-in. super venturi should be used with its head drop of 7.6 in. water (see table at top of next page).

Illustrative Problem 12-11

Determine the size of the supply main for circuit mains 1 and 2 of Fig. 12-36 and Problem 12-10. Calculate the required pump head and capacity.

Solution:

Referring to Fig. 12-36, circuit 1 and 2 are in parallel. Therefore the pump head required

Results Sheet
(Illustrative Problem 12-10)
Circuit 1

Heater	Capacity (Btu/hr)	gpm	Tube Size	Friction Rate (millinches/ft)	Venturi Size	Venturi Type
1	8,000	0.8	1/2	200	1 × 3/4	Standard
2	10,000	1.0	1/2	300	1 × 3/4	Standard
3	6,000	0.6	1/2	130	1 × 3/4	Standard
4	6,000	0.6	1/2	130	1 × 3/4	Standard
5 & 6	18,000	1.8	3/4	160	1 × 3/4	Super
7 & 8	18,000	1.8	3/4	160	1 × 3/4	Super
9	15,000	1.5	3/4	120	1 × 3/4	Standard
	81,000	8.1				
Circuit Main			1	550		

Circuit 2

Heater	Capacity (Btu/hr)	gpm	Tube Size	Friction Rate (millinches/ft)	Venturi Size	Venturi Type
10 & 11	17,000	1.7	3/4	130	1 × 3/4	Super
20	12,000	1.2	3/4	75	1 × 3/4	Super
12	10,000	1.0	3/4	60	1 × 3/4	Super
13	11,000	1.1	3/4	65	1 × 3/4	Super
19	10,000	1.0	3/4	60	1 × 3/4	Super
14	8,000	0.8	3/4	45	1 × 3/4	Super
18	8,000	0.8	3/4	45	1 × 3/4	Super
15	11,000	1.1	3/4	65	1 × 3/4	Super
17	8,000	0.8	3/4	45	1 × 3/4	Super
16	12,000	1.2	3/4	75	1 × 3/4	Super
	107,000	10.7				
Circuit Main			1 1/4	350		

will be that head required to overcome the circuit having the greater resistance. Since the greater resistance circuit is usually not obvious by inspection, the resistance of each circuit must be determined separately.

Step 1: Determine the resistance of the system main.

The system main is common to both circuit 1 and 2 and extends from the junction of the returns at the pump inlet, through the boiler, and to the point where the circuits separate. This section carries the total flow (8.1 + 10.7), or 18.8 gpm. From Table 12-1, we may see that this flow could be carried in a 1 1/2-in. tube within recommended velocity limits. However, since this is a large capacity system with probable high resistance in the two circuits, we will use a 2-in. supply main to reduce the overall system resistance.

From Fig. 12-26, with 18.8 gpm flowing in a 2-in. copper tube, the friction rate is 120 millinches per foot of tube.

Fittings (2-in. main)		EL (Table 12-3)
Gate valve	(3 at 1.2) =	3.6 ft
Boiler	(1 at 16.8) =	16.8
Air scoop	(1 at 6.8) =	6.8
Flow control valve	(1 at 83.0) =	83.0
90-degree elbows	(4 at 5.5) =	22.0
EL fittings	=	132.0 ft
Scaled length of tube	=	73.0
TEL	=	205.2 ft

Head loss in main = 205.2 × 120
= 24,600 millinches

Step 2: Determine resistance of circuit 1.

From Problem 12-10, the circuit main is 1 in. carrying 8.1 gpm with a friction rate of 550 millinches per foot of tube.

Fittings (1-in.)		EL (Table 12-3)
Flow-through tee	(2 at 1.7) =	3.4 ft (Table 12-2)
90-degree elbows	(6 at 2.5) =	15.0
Super venturi	(2 at 28.5) =	57.0
Standard venturi	(5 at 14.0) =	70.0
Sq. hd. cock	(1 at 1.5) =	1.5
EL fittings	=	146.9 ft
Scaled length of tube	=	150.0 ft
TEL	=	296.9 ft

Head loss circuit 1 = 550 × 296.9
= 163,3000 millinches

Step 3: Determine resistance of circuit 2

From Problem 12-10, the circuit main is 1 1/4-in. carrying 10.7 gpm with a friction rate of 350 millinches per foot of tube.

Fittings (1 1/4-in.)		EL (Table 12-3)
Side-flow tee	(2 at 5.0) =	10.0 ft
90-degree elbows	(2 at 1.5) =	3.0
Super venturi	(10 at 21) =	210.0
Sq. hd. cock	(1 at 1.5) =	1.5
EL fittings	=	224.5 ft
Scaled length of tube	=	130.0 ft
TEL	=	354.5 ft

Head loss circuit 2 = 350 × 354.5
= 124,000 millinches

Step 4: Select circuit having greater resistance.

From steps 2 and 3 we may see that circuit 1 has the greater resistance. Therefore, the pump head required would be

$$h_p = \text{Resistance of main} + \text{resistance circuit}$$

$$h_p = 24{,}600 + 163{,}300$$

$$h_p = 187{,}900 \text{ millinches} = 15.7 \text{ ft of water}$$

The pump would be required to pump 18.8 gpm against a head of 15.7 ft of water.

Illustrative Problem 12-12 (Two-pipe, reverse-return hydronic system)

The two-pipe, reverse-return hydronic system is used most frequently for medium- and large-capacity systems. Figure 12-40 represents a typical system layout. The circled numbers represent the heating units having the following capacities:

The solid line on the piping layout represent the supply main and risers; the dashed lines represent the return main and risers. The system has been designed for two circuits. For this illustrative problem we will be concerned with circuit 1 only. The same solution procedure would be used for circuit 2.

Problem Statement: From Fig. 12-40 determine the required steel pipe size for mains and risers for circuit 1 and the required pump capacity and head. Assume that each first-floor

Heater	Capacity (Btu/hr)	Heater	Capacity (Btu/hr)
1	6,000		
2	8,000	11	10,000
3	8,000	12	16,000
4	12,000	13	14,000
5	4,000	14	8,000
6	6,000	15	10,000
7	10,000	16	9,000
8	8,000	17	8,000
9	12,000	18	12,000
10	9,000		

heater circuit contains five 90-degree elbows, one angle radiator valve, and 8 linear feet of pipe. Second-floor heater 4 has an additional two side-flow tees, and three 90-degree elbows and 16 ft pipe. Friction loss through any heater will be considered as 2.0 in. water per 1000 Btu/hr capacity. Assume a system water temperature drop of 20 degrees.

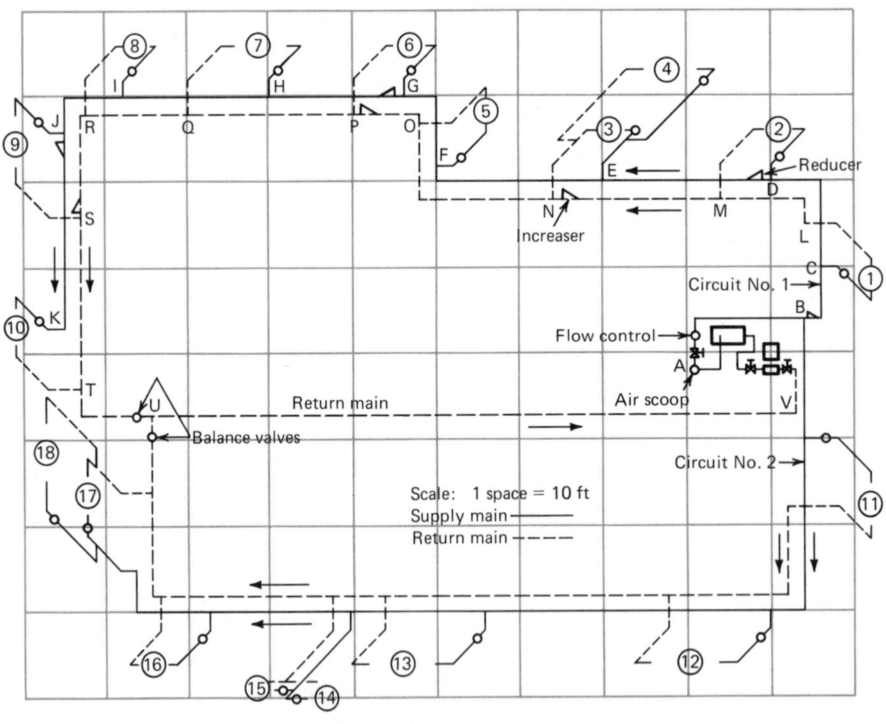

Figure 12–40

Two-Pipe Reverse Return System

Solution:

Step 1: Construct a results sheet for tabulation of data and results. Refer to results sheet:

Column 1: pipe section under consideration.

Column 2: heat load in Btu/hr carried by each section of pipe. Note that section *AB* carries the total system load, section *BC* carries the load for circuit 1, section *CD* carries circuit 1 load minus heater 1 capacity, and so on.

Column 3: flow rate in gpm calculated from Eq. (12-21), assuming gpm = $\dot{q}/10{,}000$.

Column 4: pipe size using Table 12-1 as a guide to maintain reasonable velocities.

Column 5: friction rate obtained from Fig. 12-25 for the flow rate and pipe size of columns 3 and 4.

Column 6: scaled length of straight pipe from the piping layout.

Column 7: calculated equivalent length of the fittings in each individual section of pipe. See step 2 for typical calculations.

Column 8: total equivalent length; the sum of columns 6 and 7.

Column 9: product of the friction rate (column 5) and TEL (column 8).

Column 10: identifies the friction loss in the highest resistance run.

Step 2: Determine equivalent lengths of various pipe sections.

Fittings—Section *AB* (1½ in.)		**EL**	
Air scoop	(1 at 4.8 ft)	= 4.8	(Table 12-3)
Flow control	(1 at 63.0 ft)	= 63.0	(Table 12-3)
Gate valve	(1 at 1.0 ft)	= 1.0	(Table 12-3)
90-degree elbow	(2 at 4.3 ft)	= 8.6	(Table 12-3)
EL fittings		= 77.4 ft.	

Fittings—Section *BC* (1¼ in.)		**EL**	
Reducing tee	(1 at 3.1 ft)	= 3.1	(Table 12-2)
90-degree elbow	(1 at 3.5 ft)	= 3.5	(Table 12-2)
Flow-through tee	(1 at 2.3 ft)	= 2.3	(Table 12-2)
EL fittings		= 8.9 ft.	

Fittings—Section *CD* (1¼ in.)		**EL**	
90-degree elbow	(1 at 3.5 ft)	= 3.5	(Table 12-2)
Reducing tee	(1 at 2.4 ft)	= 2.4	(Table 12-2)
EL fittings		= 5.9	

Fittings—Section *EF* (1 in.)		**EL**	
90-degree elbow	(1 at 2.6 ft)	= 2.6	(Table 12-2)
Flow-through tee	(1 at 1.7 ft)	= 1.7	(Table 12-2)
EL fittings		= 4.3 ft	

Results Sheet
(Illustrative Problem 12-12)

Supply Pipe Section (1)	Heat Load (Btu/hr) (2)	Flow Rate (gpm) (3)	Pipe Size (4)	Friction Rate (millinches/ft) (5)	Str. Length of Pipe (ft) (6)	EL Fittings (7)	TEL (ft) (8)	Friction Loss (Millinches) (9)	Resistance Longest Circuit (millinches) (10)
AB	170,000	17.0	1 1/2	275	22	77.4	99.4	27,330	27,300
BC	83,000	8.3	1 1/4	160	8	8.9	16.9	2,700	2,700
CD	77,000	7.7	1 1/4	130	16	5.9	21.9	2,800	2,800
DE	69,000	6.9	1	490	20	1.7	21.7	10,600	10,600
EF	49,000	4.9	1	250	22	4.3	26.3	6,600	
FG	45,000	4.5	1	200	12	4.4	16.4	3,300	
GH	39,000	3.9	3/4	500	16	1.4	17.4	8,700	
HI	29,000	2.9	3/4	300	18	1.4	19.4	5,800	
IJ	21,000	2.1	3/4	150	11	3.3	14.3	2,100	
JK	9,000	0.9	1/2	140	23	1.5	24.5	3,400	
Return									
LM	6,000	0.6	1/2	65	13	4.0	17.0	1,100	
MN	14,000	1.4	1/2	300	20	1.3	21.3	6,400	
NO	34,000	3.4	3/4	350	25	3.4	28.4	9,900	9,900
OP	38,000	3.8	3/4	490	9	3.3	12.3	6,000	6,000
PQ	44,000	4.4	1	180	20	1.7	21.7	3,900	3,900
QR	54,000	5.4	1	300	12	1.7	13.7	4,100	4,100
RS	62,000	6.2	1	350	13	5.0	18.0	6,300	6,300
ST	74,000	7.4	1 1/4	140	20	2.3	22.3	3,100	3,100
TU	83,000	8.3	1 1/4	180	12	7.3	19.3	3,500	3,500
UV	170,000	17.0	1 1/2	275	77	4.5	81.5	22,400	22,400
VA	170,000	17.0	1 1/2	275	20	19.8	39.8	10,900	10,900
								Total =	113,530

Fittings—Section *FG* (1 in.)		EL	
90-degree elbow	(1 at 2.6 ft)	= 2.6	(Table 12-2)
Reducing tee	(1 at 1.8 ft)	= 1.8	(Table 12-2)
EL fittings		= 4.4 ft	

Fittings—Section *IJ* (¾ in.)		EL	
90-degree elbow	(1 at 2.0 ft)	= 2.0	(Table 12-2)
Reducing tee	(1 at 1.3 ft)	= 1.3	(Table 12-2)
EL fittings		= 3.3 ft	

Fittings—Section *NO* (¾ in.)		EL	
90-degree elbow	(1 at 2.0 ft)	= 2.0	(Table 12-2)
Flow-through tee	(1 at 1.4 ft)	= 1.4	(Table 12-2)
EL fittings		= 3.4 ft	

Fittings—Section *OP* (¾ in.)		EL	
90-degree elbow	(1 at 2.0 ft)	= 2.0	(Table 12-2)
Reducing tee			(Table 12-2)
(Increaser)	(1 at 1.3 ft)	= 1.3	
EL fittings		= 3.3 ft	

Fittings—*RS* (1 in.)		EL	
90-degree elbow	(1 at 2.6 ft)	= 2.6	(Table 12-2)
Reducing tee			
(Increaser)	(1 at 2.4 ft)	= 2.4	(Table 12-2)
EL fittings		= 5.0 ft	

Fittings—*TU* (1¼ in.)		EL	
90-degree elbow	(1 at 3.5 ft)	= 3.5	(Table 12-2)
Sq. hd. cock	(1 at 1.5 ft)	= 1.5	(Table 12-3)
Flow-through tee	(1 at 2.3 ft)	= 2.3	(Table 12-2)
EL fittings		= 7.3 ft	

Fittings—Section *VA* (1½ in.)		EL	
90-degree elbow	(2 at 4.5 ft)	= 9.0	(Table 12-2)
Gate valve	(2 at 0.9 ft)	= 1.8	(Table 12-2)
Boiler	(1 at 9.0 ft)	= 9.0	(Table 12-3)
EL fittings		= 19.8 ft	

Step 3: Identify circuit having greatest resistance.

A study of the piping layout will show that all heaters are connected in parallel. Therefore, if we are able to identify the heater circuit (boiler to heater and back to boiler) having the greatest resistance, this is the resistance that must be overcome by the pump.

It would seem, by inspection, that the circuit to heater 4 would have the greatest resistance, and we will select it for illustration. This heater circuit starts at *A* continues through *B, C, D, E,* up supply riser, down return riser to *N*, then continues through *O, P, Q, R, S, T, U, V* and *A*.

Fittings—Risers to Heater 4 (¾ in.)		EL	
Side-flow tee	(4 at 5.0 ft)	= 20	(Table 12-2)
90-degree elbows (¾-in.)	(8 at 2.0 ft)	= 16	(Table 12-2)
Angle radiator valve	(1 at 12 ft)	= 12	(Table 12-2)
EL fittings		= 48 ft	
Straight length of pipe		= 24 ft	
TEL		= 72 ft	

From Fig. 12-25, average friction rate is approximately 150 millinches per foot. Pipe loss (h_L) = 72 × 150 = 10,800 millinches. Heater loss = 2.0 × 12 = 24 in. or 24,000 millinches. Head loss (point *E* through heater 4 to *N*) is 10,800 + 24,000 = 34,800 millinches.

On the result sheet identify the section of pipe and the calculated values of friction. List these friction values in column 10 and add. This sum is then combined with the riser loss to give the total resistance that must be overcome by the pump. This would be

Results sheet (column 10)	= 113,530
Riser loss (heater 4)	= 10,800
Heater loss	= 24,000
Total resistance longest circuit	= 148,330 millinches
	= 12.36 ft (say 12.4 ft)

Step 4: Select the pump capacity and head.

The pump required must have a flow capacity of 17.0 gpm at a head of 12.4 ft water.

NOTE: The selection of the highest resistance circuit (heater 4) seemed to be the most obvious choice for evaluation. However, depending on pipe sizes selected, heater 18 on circuit 2 could also be a high-resistance system and would be checked for a complete system design. In most cases the resistance from the boiler to each heater and back to the boiler is approximately the same because the distance traveled by the water is approximately the same. However, due to the limited number of pipe sizes available, the calculated resistance of each heater circuit may be different. Therefore, a balance valve (square head cock, or equivalent) should be installed in each heater riser for final balancing of flow.

Illustrative Problem 12-13

Another popular two-pipe system is called a direct-return piping system. Figure 12-41 shows a typical layout of such a system. As may be seen, the distance through the piping to each heater and back to the boiler is different in each case. The pump head required is controlled by the circuit having the greatest resistance, which is usually the longest circuit. When the pump is selected for this longest circuit, it would cause too much flow through the shorter, or low resistance, circuits. To balance this flow, higher resistance must be built into the shorter circuits by reducing pipe sizes or by use of balance valves or orifice devices. In many cases a square head cock valve is used in each heater supply riser to control the flow.

Problem Statement: Figure 12-41 illustrate a two-pipe, direct-return hydronic heating system. The various heater capacities are as follows:

Circuit 1 Heater	Capacity (Btu/hr)	Circuit 2 Heater	Capacity (Btu/hr)
1	20,000		
2	15,000	9	15,000
3	10,000	10	18,000
4	12,000	11	8,000
5	14,000	12	12,000
6	8,000	13	20,000
7	10,000	14	16,000
8	22,000		

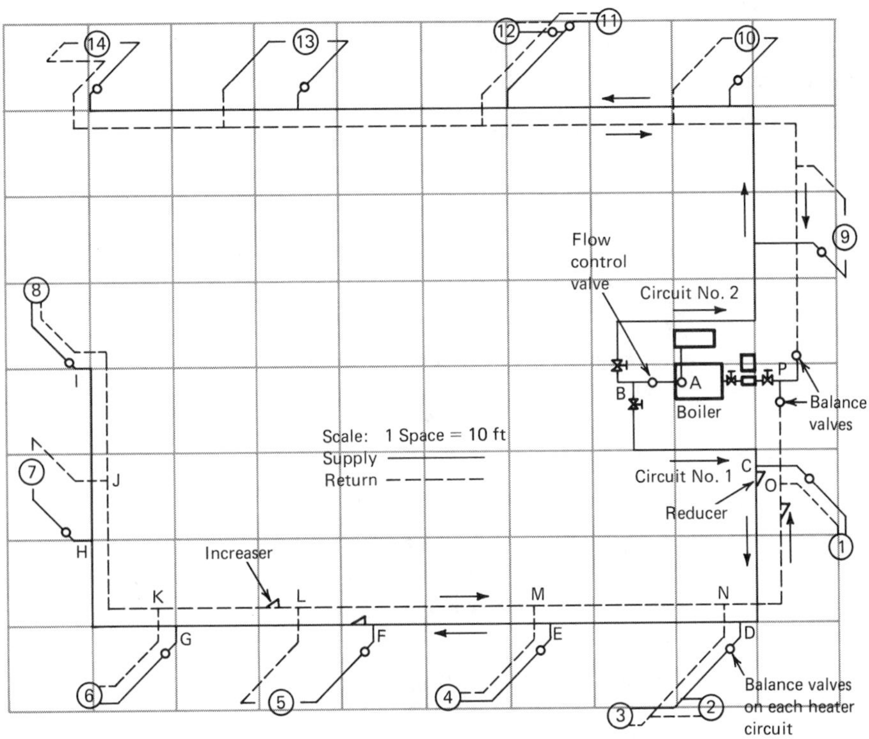

Figure 12–41

Two-Pipe Direct Return System

Assume that all the heaters are small, fan coil unit heaters. Manufacturer's data indicate the head loss through the heater coils is 0.30 ft of water per gpm flow rate. The design temperature drop for the system and for each of the heaters will be 30°F with the water temperature leaving the boiler at 215°F. Assume that each heater riser system contains 22 ft of pipe, one square head cock, two globe valves, one union elbow, and six 90-degree elbows.

Size the steel pipe on circuit 1 and determine the capacity and head required for the pump.

Solution:

Step 1: Construct a results sheet for data tabulation.

Tabulate data and results as determined from the following steps.

Step 2: Determine flow rates in each section of piping.

The design water temperature drop has been selected at 30 degrees, therefore from Eq. (12-21)

$$\text{gpm} \frac{\dot{q}}{490(t_1 - t_2)} = \frac{\dot{q}}{(490)(30)} = \frac{\dot{q}}{14{,}700}$$

Calculate gpm for each section of pipe and enter values on results sheet, column 3.

Step 3: Determine pipe sizes and friction rate for each section of pipe.

Using Fig. 12-25 and Table 12-1, determine the pipe size and friction rate maintaining velocities within acceptable limits. Enter values on results sheet in columns 4 and 5.

Step 4: Determine equivalent lengths of fittings in the various sections of pipe. (Use Table 12-2 and 12-3.) Enter values in column 7 of results sheet.

Fittings—Section *AB* (1½ in.)		**EL**	
Air scoop	(1 at 4.8 ft)	= 4.8 ft	(Table 12-3)
Flow control valve	(1 at 63 ft)	= 63.0	(Table 12-3)
Side-flow tee	(1 at 9.0 ft)	= 9.0	(Table 12-2)
90-degree elbow	(1 at 4.5 ft)	= 4.5	(Table 12-2)
EL of fittings		= 81.3 ft	

Fittings—Section *BC* (1¼ in.)		**EL**	
Gate valve	(1 at 0.8 ft)	= 0.8 ft	(Table 12-2)
90-degree elbow	(2 at 3.5 ft)	= 7.0	(Table 12-2)
Reducing tee (1-in.)	(1 at 2.4 ft)	= 2.4	(Table 12-2)
EL of fittings		= 10.2 ft	

Fittings—Section *CD* (1 in.)		**EL**	
90-degree elbow	(1 at 2.6 ft)	= 2.6 ft	(Table 12-2)
Run-through tee	(1 at 1.7 ft)	= 1.7	(Table 12-2)
EL of fittings		= 4.3 ft	

Fittings—Section IJ (¾ in.)		EL	
Sq. hd. cock	(1 at 1.5 ft)	= 1.5	(Table 12-3)
Globe valves	(2 at 22 ft)	= 44.0	(Table 12-3)
Union elbow	(1 at 2.5 ft)	= 2.5	(Table 12-3)
90-degree elbow	(6 at 2.0 ft)	= 12.0	(Table 12-2)
EL of fittings		= 60.0 ft	

Heater 8 loss (h_L) = 0.30 × 1.5 = 0.45 ft water or, by Eq. (12-23a),

$$\text{Heater EL} = \frac{0.45}{0.007} = 64 \text{ ft of pipe}$$

TEL of section IJ including heater 8, (60 + 64) = 124 ft.

Fittings—Section *PA* (1½ in.)		EL	
Gate valves	(2 at 0.9 ft)	1.8	(Table 12-2)
90-degree elbow	(2 at 4.5 ft)	9.0	(Table 12-2)
Boiler	(1 at 9.0 ft)	9.0	(Table 12-3)
EL of fittings		19.8 ft	

Step 5: Determine total equivalent length of pipe sections and calculate friction loss in each section.

The total equivalent length (TEL) is the sum of columns 6 and 7 of the results sheet, enter values in column 8.

The resistance of each pipe section is the product of the friction rate (column 5) and the TEL (column 8). Enter these products in column 9 of results sheet.

Step 6: Determine pump head and capacity.

The pump must have sufficient capacity to flow the entire gpm for the system, which is 13.6 gpm. The required pump head is the head required to overcome the losses in the longest circuit, which is the sum of column 9 of the results sheet. The loss is 100,300 millinches, or 8.36 ft of water.

Step 7: Determine riser sizes for other heaters on circuit 1.

The distance from boiler to a heater and back to boiler is different for each heater, which means the resistance is different, and the system is not in balance. Some balancing of the individual heater circuits may be accomplished by reducing the size of the risers as we work back toward the boiler. This method is limited because of the standard pipe sizes available, and a ½-in. size is about the minimum recommended pipe size.

As an illustration of this method we will use heater 5 circuit. The resistance of flow from boiler to supply riser takeoff and from return riser back to boiler is the sum of the resistances in the various pipe sections already determined and shown on results sheet. They are

Section	Resistance	Section	Resistance
AB	16,400	*LM*	4,500
BC	4,900	*MN*	5,000
CD	8,500	*NO*	9,100
DE	4,900	*OP*	3,000
EF	3,400	*PA*	6,800

The sum is 66,500 millinches. The pump has a head of 100,300 millinches. This leaves

Results Sheet
(Illustrative Problem 12-13)

Supply Pipe Section (1)	Circuit 1 Heat Load (Btu/hr) (2)	Flow Rate (gpm) (3)	Pipe Size (4)	Friction Rate (millinches/ft) (5)	Str. Length of Pipe (ft) (6)	EL Fittings (ft) (7)	TEL (ft) (8)	Friction Loss (millinches) (9)
AB	200,000	13.6	1½	180	10	81.3	91.3	16,400
BC	111,000	7.6	1¼	140	25	10.2	35.1	4,900
CD	91,000	6.2	1	350	20	4.3	24.3	8,500
DE	66,000	4.5	1	200	23	1.7	24.7	4,900
EF	54,000	3.7	1	150	21	1.8	22.8	3,400
FG	40,000	2.7	¾	250	24	1.4	25.4	6,400
GH	32,000	2.2	¾	180	20	3.4	23.4	4,200
HI	22,000	1.5	¾	80	20	2.0	22.0	1,800
Return								
IJ	22,000	1.5	¾	80	37	124.0	161.0	12,900
JK	32,000	2.2	¾	180	21	2.0	23.0	4,100
KL	40,000	2.7	1¾	250	17	0.5	17.5	4,400
LM	54,000	3.7	1	150	28	1.7	29.7	4,500
MN	66,000	4.5	1	200	23	1.7	24.7	5,000
NO	91,000	6.2	1¼	350	21	5.1	26.1	9,100
OP	111,000	7.6	1½	140	12	9.5	21.5	3,000
PA	200,000	13.6		180	18	19.8	37.8	6,800
						Total loss longest circuit		= 100,300 millinches
								= 8.36 ft. water

100,300 − 66,500 = 33,800 millinches to be used in the risers and heater 5.

The flow rate through heater 5 is 14,000/14,700 = 0.95 gpm. The head loss through heater is 0.30 × 0.95 = 0.285 ft of water = 3400 millinches. The available head to overcome friction in the risers is then, 33,800 − 3400 = 30,400 millinches. From step 3 we found that the EL of the riser fittings was 60 ft. The straight length of pipe in the risers was given as 22 ft, and the TEL would be 60 + 22 = 82 ft, exclusive of the heater loss.

The friction rate in the risers should be approximately 30,400/82 = 370 millinches/ft. Referring to Fig. 12-25 with f = 370 and 0.95 gpm, a pipe size less than ½ in. would be required. Use ½ in., and the remaining friction head would be obtained by a balancing valve, or an orifice could be used.

12-20 COMBINATION HYDRONIC HEATING-COOLING SYSTEMS

Hydronic cooling systems can be combined with hydronic heating systems using the same piping and circulating pump if the terminal equipment is fan-equipped heating and cooling units. These units may be connected to a one-pipe venturi system, a two-pipe direct-return, a two-pipe reverse-return system, a three-pipe, or four-pipe system. These fan coil units contain a finned-type heat-transfer coil; when heating is required, hot water from the boiler is circulated through the piping system and unit coil by the system pump. When cooling is required, chilled water from a water chiller is circulated through the piping and unit coil by means of the same pump used for heating or by a separate cooling pump.

Fan coil units used with one-pipe venturi piping systems must be carefully selected for low pressure drop through the unit coil since the special venturi tees are designed to produce small pressure drop in order to keep the system pumping head within reasonable limits. The maximum loss through the unit coil, runouts, and risers is fixed, once a specific venturi fitting is selected. The only way this maximum pressure loss can be altered is by changing the venturi fitting or by changing the water flow rate in the main. The water flow through the branch takeoffs and unit coils can be changed by increasing or decreasing the size of piping to and from the unit, by selecting a unit coil with a greater or lesser pressure drop, or by installing flow balancing valves in the runout piping.

Location of water chiller. The water chiller should be located as close to the boiler as practical to simplify the required piping. It should have a level foundation of adequate strength to support the unit. If necessary, place cork pads or other vibration isolators under the frame to prevent the transmission of noise to the system. It is good practice not to locate the chiller directly under or adjacent to sleeping areas. It is important that ample clearances be provided on all sides of any unit requiring servicing.

Insulation of piping. The temperature of water circulating through hydronic systems may range from 160°F to 220°F for heating, and 40°F to 50°F for cooling. It is essential that all piping or tubing, including fittings, between the water chiller and the fan coil units be completely insulated with a vapor or moisture proof insulation. While this insulation reduces heat gain to the piping, its most important function is to prevent condensation on the piping. The insulation material must therefore be vapor and moisture proof.

12-21 OTHER TYPES OF COOLING UNITS USED WITH HYDRONIC SYSTEMS: *Self-contained room cooling units.* In this type of system baseboard or any other type of heat-distributing units are connected to the boiler for heating, while self-contained room units provide summer cooling. These can be of the console or window types. Each self-contained unit has cooling, filtering, ventilating, and dehumidifying apparatus. These units may be installed when the building is constructed or at any later time.

Self-contained central cooling unit. In this system baseboard or any other type of heat-distributing units are connected to the boiler for heating, while the self-contained central unit provides summer cooling. The cooling unit may be installed in an attic, a closet, or a dropped ceiling chamber and connected by short sections of duct to ceiling diffusers or highwall registers located in rooms to be cooled.

Fan-equipped cooling coil unit. In this system baseboard or any other type of heat-distributing units are connected to the boiler for heating. For cooling, a water chiller supplies chilled water, which is circulated through piping by means of a pump to one or more fan-equipped cooling coil units. These units are connected to simple duct work located either in a dropped ceiling or an attic crawl space. Cool, filtered, dehumidified air is supplied through this duct work to nearby rooms through ceiling diffusers or high wall registers.

12-22 DESIGN CONSIDERATIONS FOR ONE-PIPE COMBINATION HEATING-COOLING SYSTEMS

The calculated heat loss from and heat gain to the various spaces to be conditioned must be determined as a first step in the design process. From manufacturers' catalogs fan coil units may be selected and sized to meet the required loads. The catalog data should include the gpm flow rate through a unit and the head loss in feet of water. Remember that for one-pipe venturi systems the head loss through the fan coil unit should be as low as possible.

Make a piping layout. It should be understood that in a one-pipe system, the temperature of water in the main decreases for heating and increases for cooling, after each heat-transfer device is supplied. This means the temperature of the supply water to each successive unit becomes slightly lower in heating and slightly higher in cooling. This gradual change in supply water temperature is not critical for heating because (a) outdoor design temperatures are experienced only a few days during the heating season, (b) of the high temperature difference that exists between the heating medium and the room air, and (c) a reduction in the supply water temperature of 1°F will change the unit output less than 1 percent.

However, in cooling, an increase of 1°F will change the output of a unit 4 to 5 percent, which may have an effect on comfort conditions. Therefore, in a one-pipe cooling system, units should be selected on the temperature of water that will be supplied to each unit rather than on a constant temperature.

The design water temperature increase for a chilled water system is normally 10 degrees, rather than the 20- or 30-degree drop for hot water systems. If we were to assume that the average increase supply water temperature after heat unit were 2 degrees, that would indicate that no more than five cooling units should be installed on any one circuit. This is recommended practice for one-pipe heating-cooling systems. If the supply water temperature to the first cooling unit on the circuit were 40°F, the fifth heater would have a supply water temperature of 48°F.

Selecting units, determining gpm requirements, and pressure drop through units. Based on the supply water temperatures selected, the size of the heating-cooling units, the gpm flow rate for heating and cooling, and the pressure drop through the units may be determined from manufacturers' catalogs. In most cases the flow rate in gpm through the unit will be different for heating than for cooling. Both flows should be recorded for data, but the higher flow will be used to determine system flow and pump size.

Pressure drop in feet of water through each heating-cooling unit should be converted to feet of pipe equivalents (EL). In all cases the pressure drop through the unit expressed in feet of pipe equivalents will be constant regardless of the flow through it. The feet of pipe equivalent is determined by use of Eq. (12-23) using the friction rate f in feet of water of the branch pipe size to and from the unit.

Selecting the type and size of venturi required, the main flow, and main size. The procedure used is the same as that for designing a one-pipe hot water system. The flow through the main will be the sum of the maximum individual flows to the various units.

Pump capacity and head would be calculated in the same manner as used in a one-pipe hot water heating system, using the maximum flow in the main and calculating the friction loss for this flow rate.

12-23 DESIGN CONSIDERATIONS FOR TWO-PIPE COMBINATION HEATING-COOLING SYSTEMS

In a two-pipe system the supply water temperature to each heating-cooling unit is constant. With this supply water temperature, possibly 200°F for heating and 40°F for cooling, refer to manufacturers' catalogs for unit size, flow rate (gpm), and head loss for the terminal units for the system. The flow rate (gpm) for heating and cooling may be different. They should be recorded as data, and the higher individual values should be added to give the required flow for the system. The only limit to the number of units on a particular circuit is the availability of a pump to overcome the head loss. The head loss should be kept at or below 20 ft of water in most small and medium-sized systems.

The normal connections and fittings for the boiler and chiller in combination heating-cooling systems is shown in Fig. 12-42. The circulator (pump) is normally located in the supply main after the boiler and chiller. The three-way control valves may be manual or automatic and are used to isolate either the boiler or chiller depending upon whether heating or cooling is being called for. Only one three-way valve is required if the boiler is not used for year-round hot water.

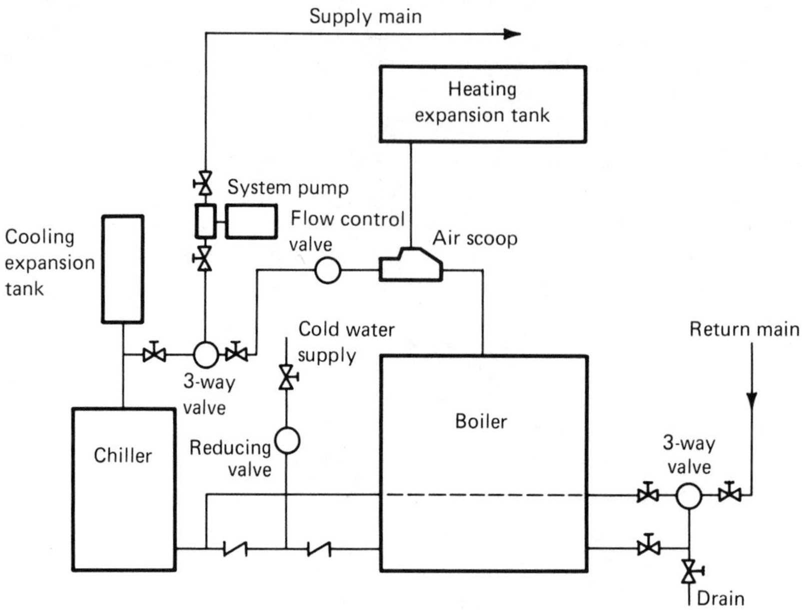

Figure 12–42

Combination heating-cooling connections

CHAPTER 12

Review

12-1. Referring to Illustrative Problem 12-1, what is the pump horsepower for the system, if the pump has an efficiency of 70 percent?

12-2. A piping system handling 60°F water requires a pump to flow 200 GPM against a head of 86 ft. Select a pump impeller size from Fig. 12-12, and state the pump efficiency, pump motor horsepower required, and required NPSH.

12-3. A pump operating at 1750 RPM at its best efficiency point produces a flow rate of 30 GPM against a head of 25 ft and requires a power input of 0.30 BHP. A drive arrangement is available which will increase the pump speed to 2000 RPM. What would be the flow rate, head and BHP at the speed of 2000 RPM, assuming constant efficiency?

12-4. Figure 12-43 shows a pump installed above a reservoir containing water at 80°F. Normally the 3-inch suction pipe extends 6 ft below the reservoir water surface. Assume barometric pressure is 14.7 psia, head loss in suction pipe 10 ft of water, and the water temperature is 80°F. (a) If the water level in the reservoir is 15 ft below the pump centerline, what is the available net positive suction head? (b) Determine available net positive suction head if the reservoir water level dropped to 20 ft below the pump centerline.

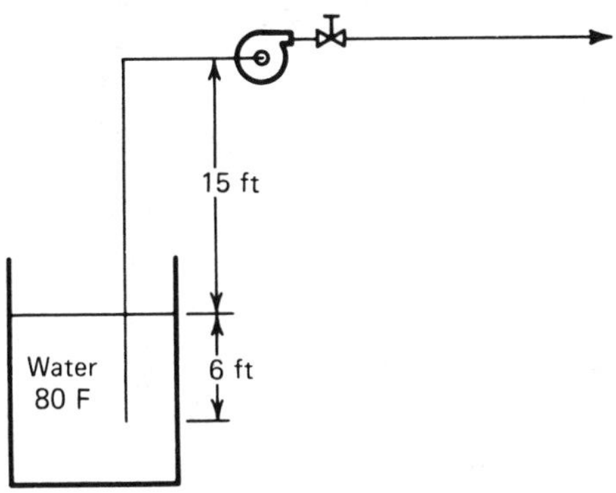

Figure 12–43

12-5. A section of pipe in a hydronic heating system carries 20 GPM of water. The pipe is 1½-inch nominal steel pipe and has a scaled length of 50 ft. Contained in this section of pipe are two gate valves, 3 flow-through tees, four 90-degree elbows, two 45-degree elbows, and one square head cock valve. Determine the head loss in the section of pipe, using (a) Tables 12-2 and 12-3; and (b) using methods of Chapter 11.

12-6. Determine a tentative size of steel pipe to carry a flow of 20 GPM of water within recommended limits of velocity. What will be the corresponding friction rate?

12-7. Figure 12-44 shows a plan layout for a one-pipe, two-circuit series loop hydronic heating system using baseboard terminal heating units. The baseboard units have been selected from Table 10-9 of Chapter 10. The baseboard units have the following capacities and lengths:

Baseboard Unit No.	Capacity (Btu/hr)	Length (ft)
1	18,000	22
2	8,000	10
3	10,000	16
4	14,000	20
5	6,000	8
6	9,000	16

If the supply water leaving the boiler is at 200°F and a system water temperature drop of 20 degrees is allowed, determine: (a) the average water temperature in each heating unit; (b) the required copper tube size for the supply main and each loop; (c) required pump capacity and head to flow the system. (Include resistance of all required fittings and accessories that would normally be installed in the system.)

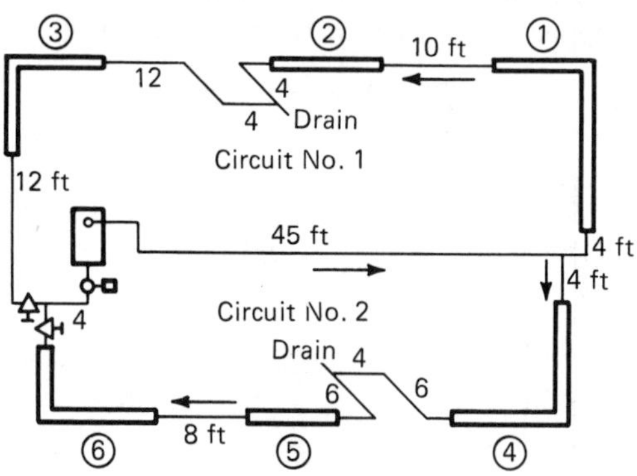

Figure 12–44

12-8. Figure 12-45 represents part of a one-pipe hot water heating system. The longer circuit shown carries a total heating load of 80,000 Btu/hr and has a measured length (AB) of 150 ft. The total system heating load is 150,000 Btu/hr supplied by the main (BA) which has a measured length of 75 ft. One terminal heat transfer unit on the longer circuit is shown. Assume that there are eight such terminal units each having a capacity of 10,000 Btu/hr and an internal pressure head loss of 2000 millinches of water. Also, assume each

THTU circuit contains seven 90-degree standard elbows, one angle valve, one balance valve, and twelve feet of straight pipe.
The system main (BA) contains one boiler, one flow control valve, four 90-degree standard elbows, one air scoop, and two gate valves.

(a) Determine the steel pipe size required for the longer circuit (AB) and supply main (BA) for a system water temperature drop of 20°F.

(b) Determine the venturi fitting size required for the THTU shown, and the pipe size for the heater circuit.

(c) Determine the average water temperature in the THTU shown, if the supply water at point A is 200°F.

(d) Determine the pump capacity and heat required for the system.

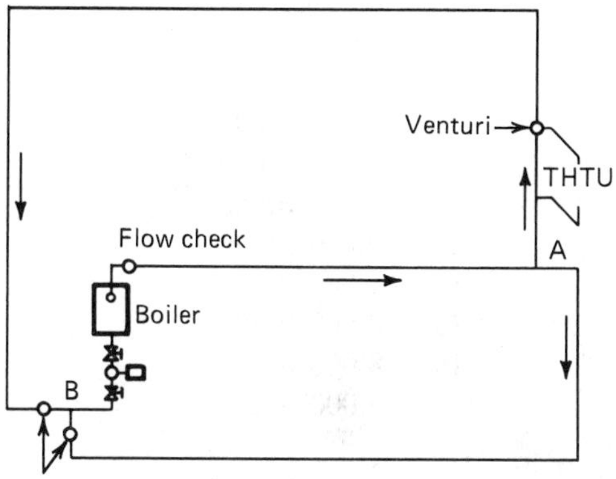

Figure 12–45

12-9. Figure 12-46 illustrates a two-pipe, hot water system supplying four Model S unit heaters (see Table 10-12). Heater A is a model 100-S, B and C are both model 70-S, and D is a model 126-S. For a system design water

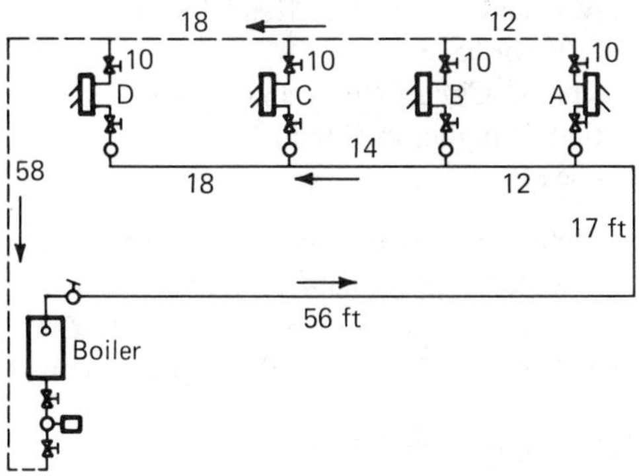

Figure 12–46

temperature drop of 20 degrees, and supply water temperature of 200°F, determine:

(a) Heating capacity, water flow rate, head loss (S.W.T.), final air temperature for each unit heater assuming a 20°F temperature drop.
(b) Steel pipe size for each section of piping.
(c) Required pump capacity and head. Assume that the supply and return mains contain one air scoop, one flow check valve, one boiler, two gate valves, one balance valve, and 90-degree standard elbows as shown. Assume the fittings at each unit heater include one balance valve, two globe valves, and four 90-degree standard elbows.

12-10 Figure 12-47 represents a two-pipe, reverse return hydronic system. The heater loads are as follows:

Circuit No. 1		Circuit No. 2	
No. 1	12,000 Btu/hr	No. 10	10,000 Btu/hr
2	8,000 Btu/hr	11	12,000 Btu/hr
3	10,000 Btu/hr	12	8,000 Btu/hr
4	12,000 Btu/hr	13	9,000 Btu/hr
5	9,000 Btu/hr	14	15,000 Btu/hr
6	10,000 Btu/hr	15	10,000 Btu/hr
7	8,000 Btu/hr	16	8,000 Btu/hr
8	15,000 Btu/hr	17	6,000 Btu/hr
9	9,000 Btu/hr		

Supply water temperature is 200°F and a system design temperature drop is 20 degrees. Each first floor heater branch circuit contains eight feet of pipe, one balance valve, two side-outlet tees, six 90-degree standard elbows, and one angle valve. Each second floor heater branch circuit contains twenty feet of pipe, one balance valve, two side-outlet tees, eight 90-degree standard elbows, and one angle valve. The system main contains one air scoop, one flow check, three gate valves, six 90-degree elbows, and one boiler. Head loss through each heater is one ft of water per 5000 Btu/hr capacity.
Consider that the greatest head loss occurs in Circuit No. 1 and determine:

(a) Steel pipe sizes for each section of piping in Circuit No. 1 .
(b) Required pump capacity and head.

12-11. Figure 12-48 represents an isometric view of the two first floor heaters and one second floor heater connected to the supply main of a one-pipe hot water heating system. Heating unit capacities are as follows: No. 1 = 8,000 Btu/hr, No. 2 = 6,000 Btu/hr, and No. 3 = 10,000 Btu/hr. The entire system load is 120,000 Btu/hr.
Copper tube lengths are shown on each section of the diagram, and fittings are identified. Friction loss through each heater is estimated at 200 millinches per 1000 Btu/hr capacity.

System design water temperature drop is 20 degrees:

(a) Select copper tube sizes for each section of piping.

(b) Determine the venturi sizes required.

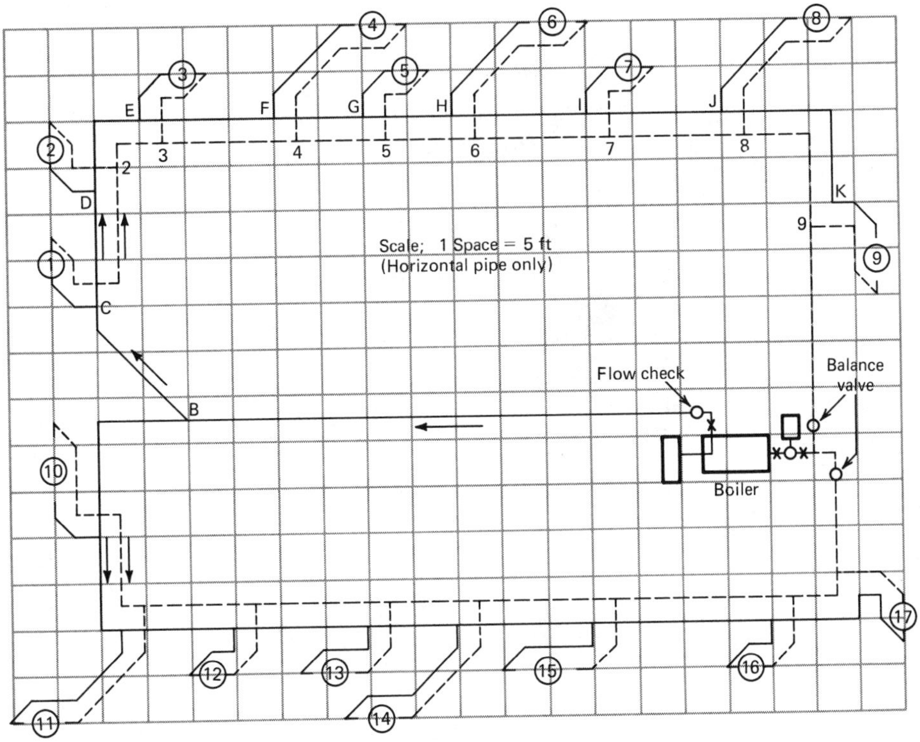

Figure 12–47

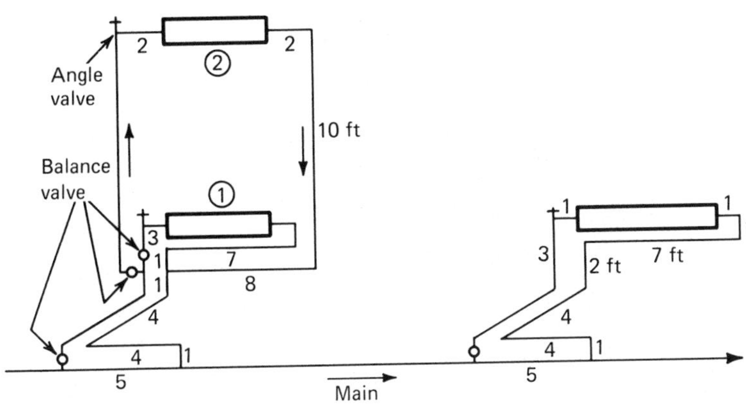

Figure 12–48

BIBLIOGRAPHY

1. *Taco Guide for Hydronic Engineers,* Taco, Inc. Cranston, Rhode Island.

2. *Goulds Pump Manual,* Goulds Pumps, Inc. Seneca Falls, NJ.

3. Strock, C., and Koral, R. L., *Handbook of Air Conditioning, Heating and Ventilating,* 2nd ed., The Industrial Press, New York, 1965.

Chapter 13

Principles of Room Air Distribution

13-1 INTRODUCTION

Air distribution is one of the most important branches of the engineering science of HVAC. The purpose of this chapter is to provide a source of accurate engineering information in the field of air distribution within conditioned spaces. An attempt is made to reduce somewhat intricate formulas to simple equations and to introduce a concept of the subject based on authentic facts discussed in simple language.

Air mechanically supplied to interior spaces of a building to provide for ventilation, heating, and/or cooling requires careful selection of terminal air distribution devices, a well-designed duct system, and fan or fans. This chapter deals with some fundamentals of terminal air distribution devices, a very important link in the chain of load estimating, duct design, fan equipment selection, and control. Chapter 14 will deal with duct system design and fan performance and selection.

The success of a good heating, ventilating, or air-conditioning system depends upon its air distribution. Most of the complaints arising out of faulty room air distribution can be classified as either *drafts* or *stuffiness*. It is well to analyze these two sensations because they illustrate the inter-relationship of the accepted standards of room temperatures, humidity, air motion, and air flow direction.

Possibly a third source of complaints, *noise,* should be added. We will certainly be concerned with taking the necessary precautions to prevent noise problems when we select outlets. However, for purposes of this discussion, we will consider that outlet velocities, which are the biggest factor in noisy air distribution, are within the recommended limits.

A draft is a current of air which due to its temperature, humidity, rate of movement, or any combination of these factors removes more heat from the surface of a body than that surface normally dissipates. The complaint of drafts may arise whenever there is a local or general sensation of feeling too cool.

Conversely, stuffiness is usually thought of as a stagnant condition of the air, but actually the complaint of stuffiness generally arises when less heat is removed from the surface of the body than it normally generates.

This indicates that drafts and stuffiness are not functions of the single property of air velocity but are due to the properties of motion, temperature, and humidity in combination, or in other words, the effective temperature. To these properties may be added a fourth factor, air direction, whenever the sensation is localized, such as a draft on the back of the neck.

This concept explains difficulties frequently encountered in practice. Comparatively high air motion may be tolerated with temperature and humidity adjusted upward to result in the optimum effective temperature. This same higher temperature and humidity with lower air motion may be considered stuffy even though the residual velocity may be what is normally considered optimum. The effect of inequities and fluctuations in either temperature or velocity serves to call the occupant's attention to differences in sensations. Good room air distribution must avoid temperature stratification, fluctuating gusts of air, warm or cold spots, and local high velocities.

13-2 TERMINOLOGY

Air distribution is one of the least understood branches of HVAC systems. This is true as the result of many factors, not the least of which is the lack of standard terminology and methods of rating the performance of similar air distribution devices in the industry. The following definitions are largely from ASHRAE *Handbook and Product Directory,* 1977 *Fundamentals*.

Air distribution outlet: Denotes any mechanical device for distributing air within an enclosure. It may be a grille, register, ceiling outlet, plaque, slot, etc.

Aspect ratio: The ratio of length to width of an opening, or core of a grille. Also, with regard to rectangular sheet metal ducts, it is the ratio of the long dimension to the short dimension.

Attenuation: The process of absorbing or dissipating sound waves between point of origin and the human ear.

Axial-flow jet: A stream of air whose motion is approximately symmetrical along a line, although some spreading and drop or rise may occur due to diffusion and buoyancy effects.

Ceiling diffuser: A circular, square, or rectangular outlet located in the ceiling of an enclosure through which supply air is discharged on a plane approximately parallel to the ceiling.

Coefficient of discharge: Ratio of area of vena contractor to area of opening.

Core area: The total plane area of that portion of a grille included within lines tangent to the outer edges of the opening through which air can pass.

Damper: A device used to vary the volume of air passing through a confined cross section by varying the cross-sectional area.

Diffuser: An outlet discharging supply air in various directions and planes.

Diffusion: Distribution of air within a space by an outlet discharging supply air in various directions and planes.

Drop: The vertical distance that the lower edge of the horizontally projected airstream drops between the outlet and the end of its throw. It is affected by the length of throw and by the temperature differential between the primary and secondary air.

Duct area: The area of the cross section of the duct based on its inside measurements at the point of installation of the device.

Effective Area: The net area of an outlet or inlet device through which air can pass, equal to the free area times the coefficient of discharge. In field practice it is the area which when multiplied by the average jet velocity gives the volume flow rate (cfm) passing through the device.

Entrainment: The entrainment of room air by the airstream discharged from the outlet (called secondary air motion, also induction).

Entrainment ratio: The total air divided by the discharged air. *See* Induction ratio.

Envelope: The outer boundary of an airstream moving at a perceptible velocity.

Exhaust opening or inlet: Any opening through which air is removed from a space.

Free area: The total minimum area of the openings in the air outlet or inlet through which air can pass.

Grille: A covering for an opening through which air passes.

High-pressure system: A supply or return air system where the maximum static pressure at any outlet on the system exceeds 3.5 in. WG but does not exceed 6.0 in. WG.

Induction: The induction of room air drawn

into an outlet by the primary airstream (commonly called aspiration).

Induction ratio: A value that may be established for any air distribution device by measuring the relative quantities of primary and secondary air in the mixed airstream at a predetermined distance from the device.

Jet velocity: The velocity of the air in fpm measured at a point in the vena contracta between the bars of a grille, register, or at the lips of a ceiling outlet. *See* Outlet velocity.

Low-pressure system: A supply or return air system where the maximum static pressure at any outlet on the system does not exceed 0.5 in. WG.

Medium-pressure system: A supply or return air system where the maximum static pressure on the system does not exceed 3.5 in. WG.

Multi-louver damper: A damper having a number of blades, used for the purpose of throttling supply or exhaust air passing through a duct.

Opposed-blade damper: A multi-louver-type damper with center-pivoted blades, where action moves adjacent blades in opposite directions in closing and opening.

Outlet velocity: The average velocity of air emerging from the outlet, measured in the plane of the opening.

Plaque: A ceiling outlet in which the supply air impinges against a flat plate or series of parallel plates and is deflected horizontally.

Primary air: The air delivered to the outlet by the supply duct.

Radial-flow jet: A stream of air distributed outwardly from a center so that the area of jet increases approximately in proportion to the distance from the center. Ceiling diffusers usually produce this type of jet.

Radius of diffusion: The horizontal axial distance an airstream travels after leaving an air outlet before the maximum stream velocity is reduced to a specified terminal level, i.e., 200, 150, or 100 fpm.

Register: A grille equipped with a damper or control valve.

Residual velocity: The average air velocity in fpm within an occupied zone of a room being conditioned by a distribution device. *See* Room velocity.

Rise: The converse of drop.

Room velocity: The residual air velocity level in the occupied zone of the conditioned space, i.e., 65, 50, or 35 fpm.

Secondary air: The air which the primary airstream entrains by the passage of the primary air through the enclosure.

Spread: The divergence of an airstream in a horizontal or vertical plane (measure in degrees) after it leaves the outlet.

Supply opening or outlet: Any opening through which air is delivered into a space which is being heated, cooled, humidified, dehumidified, or ventilated.

Temperature differential: The temperature difference between primary and room air.

Temperature variation: The temperature difference between points within the space.

Terminal velocity: The maximum airstream velocity at the end of the throw. Normally taken as 50 fpm for grilles and registers, and 100, 150, or 200 fpm for circular ceiling outlets, depending on application.

Throw: The horizontal or vertical axial distance an airstream travels after leaving an air outlet before the maximum stream velocity is reduced to a specified terminal level. *See* Terminal velocity.

Total air: The mixture of primary air and entrained air.

Vane: A thin plate in the opening of a grille.

Vane ratio: The ratio of depth of vane to minimum width between two adjacent vanes.

Ventilating ceilings: Multiple ceiling supply air openings with vertical discharge, in close proximity with each other, covering a significant part of the ceiling area and acting as a whole, not as individual units.

13-3 STATIC, VELOCITY, AND TOTAL PRESSURES

It is desirable at this point to briefly discuss some of the characteristics of air flow. This should help us to better understand the design, performance, and selection of the various air distribution devices for different applications.

Flow of any fluid is produced by a pressure difference. It is not important whether the regions in which the pressure difference occurs are above or below atmospheric pressure; it is the magnitude of the difference in pressure which determines the characteristics of flow. This phenomenon of fluid flow from one point to another under pressure difference is common throughout heating, ventilating, and air-conditioning system design. In air-conditioning and industrial ventilation applications the pressure differences to be dealt with are usually quite low, and as a result air is considered incompressible. Therefore, Bernoulli's equation for incompressible fluid flow applies. Bernoulli's equation was developed in Chapter 11 from the steady-flow energy equation (Eq. 11-8), which is repeated here:

$$\frac{g}{g_c} z_1 + \frac{p_1}{\rho} + \frac{V^2}{2g_c} = \frac{g}{g_c} z_2 + \frac{p_2}{\rho} + \frac{V_2^{\,2}}{2g_c} \qquad \textit{(11-8)Repeated}$$

Recall that each term of Eq. (11-8) has units of ft lbf/lbm, or J/kg, which is *energy per unit mass*.

If each term of Eq. (11-8) is multiplied by the fluid density ρ, we would have

$$\frac{g}{g_c}(\rho z_2) + p_1 + \frac{\rho V_1^{\,2}}{2g_c} = \frac{g}{g_c}(\rho z_2) + p_2 + \frac{\rho V_2^{\,2}}{2g_c} \qquad \textit{(13-1)}$$

Each term in Eq. (13-1) has units of ft lbf/ft^3, or pressure in lbf/ft^2, or pascals, and is *energy per unit volume*.

Recall that the symbols and dimensions used in the steady-flow equation, Eqs. (11-8), (11-9), and (13-1) are:

g_c = dimensional constant (32.2 lbm ft/lbf sec^2)

g = acceleration due to gravity (ft/sec)

z = elevation (ft)

V = mean flow velocity (ft/sec)

ρ = mass density (lbm/ft^3)

p = pressure (lbf/ft^2)

$\gamma = \rho(g/g_c)$ = specific weight (lbf/ft^3)

The first equation, Eq. (11-8), is the general form of Bernoulli's equation and states that in ideal flow the *total energy* remains constant. The second form, Eq. (11-9), states that the *total head* remains constant. And, the third form, Eq. (13-1), states that the *total pressure* remains constant.

In gas flow and air-conditioning duct system design the pressure form, Eq. (13-1), is usually used; however, Eq. (11-8) should be used when significant density variations ($\rho_1 \neq \rho_2$) occur. For liquid flow the head form, Eq. (11-9), is commonly used. Identical results are obtained no matter which of the three forms are used if proper care is taken with units and if the fluid is homogeneous.

Head and *pressure* are often used interchangeably, but these terms have specific meanings. *Head* is the height of a column of fluid supported by fluid pressure, while *pressure* is the normal force per unit area. With a liquid it is frequently convenient to measure the head of a fluid in terms of "feet of fluid flowing." With a gas or air, however, "feet of gas" would be a large number, and it is practical and customary to measure pressure by a column of liquid. In the case of air flow the liquid is water at a constant specified condition. Water at 62°F has a weight per unit volume of 62.36 lbf/ft^3; thus, a 1-in. column of water creates a pressure of 5.19 lbf/ft^2.

In Eq. (13-1) the combined term $(g/g_c)(\rho)(z_2 - z_1)$ represents elevation pressure change, which in most cases is negligibly small. Therefore, it is frequently omitted from consideration in air flow problems. The terms p and $\rho V^2/2g_c$ are the static pressure (p_s) and velocity pressure (p_v), respectively. The sum of static pressure and velocity pressure is called total pressure (p_t).

As stated previously, it is practical and customary in air flow work to express pressures in terms of the height of a column of water, called "inches water gauge" or in. WG. Conversion of units is accomplished simply by stating that

$$p_a = p_w$$

or

$$h_a(\rho_a)\left(\frac{g}{g_c}\right) = \frac{h_w}{12}(\rho_w)\left(\frac{g}{g_c}\right)$$

where p_a and p_w = the pressures of air and water, respectively, in lbf/ft^2

h_a = pressure head (ft of air)

h_w = pressure head (inches of water)

ρ_a = air density (lbm/ft^3)

ρ_w = water density [lbm/ft^3 (usually assumed to be 62.36 lbm/ft^3)]

Rearrangement gives,

$$h_a = \left(\frac{h_w}{12}\right)\left(\frac{62.36}{\rho_a}\right) = 5.19\left(\frac{h_w}{\rho_a}\right) \qquad (13\text{-}2)$$

For velocity head we know that $h_a = V^2/2g$, therefore,

$$\frac{V^2}{2g} = 5.19\left(\frac{h_w}{\rho_a}\right)$$

or

$$h_w = \frac{\rho_a V^2}{2g(5.19)} = \rho_a\left(\frac{V}{18.28}\right)^2 \qquad (13\text{-}3)$$

If V is to be expressed in feet per minute (fpm), then

$$h_v = h_w = \rho_a\left(\frac{V/60}{18.28}\right)^2 = \rho_a\left(\frac{V}{1096.5}\right)^2 \qquad (13\text{-}4)$$

where h_v = velocity pressure head (in in. WG)

ρ_a = air density at given conditions of pressure and temperature (lbm/ft^3)

V = mean flow velocity ($V = Q/A$) (feet per minute)

By rearrangement of Eq. (13-4), we would have

$$V = 1096.5\sqrt{\frac{h_v}{\rho_a}} \qquad (13\text{-}5)$$

In much design work "standard air" is assumed. Standard air is assumed to be at 70°F and 29.92 in. Hg, and has a density of 0.07494 (say 0.075 lbm/ft^3); therefore, Eq. (13-4) would become

$$h_v = \left(\frac{V}{4005}\right)^2 \qquad (13\text{-}6)$$

and Eq. (13-5) becomes

$$V = 4005\sqrt{h_w} \qquad (13\text{-}7)$$

It must be noted that in Eqs. (13-4) through (13-7) that flow velocity is in fpm, and we have now called h_v the velocity pressure head in inches water (in. WG). Therefore, we should refer to the static pressure head as h_s in inches water (in. WG). Again, the sum of static and velocity pressure heads will be the total pressure head, h_t, or

$$h_t = h_s + h_v = h_s + \left(\frac{V}{1096.5}\right)^2 \rho_a \qquad (13\text{-}8)$$

with all terms expressed in inches water gauge. The total pressure head in a duct system remains constant except for friction and shock losses. These losses will be discussed in detail in the following sections and in Chapter 14.

If fluid flow through a sharp-edge, round orifice is analyzed, much can be learned about the principles of air flow through air distribution devices. Let Fig. 13-1 represent such an orifice with a total pressure in the airstream at section 1 of say 1.0 in. WG higher than section 2, which is in free space. The flow will produce a convergence toward the orifice opening, and the air will reach a theoretical maximum velocity of about 4000 fpm at section 3 (Eq. 13-7), the smallest area of the stream, technically called the "vena contractor." The area of the stream at the vena contractor will be approximately 0.6 times the area of the orifice opening. Its location will be a distance downstream equal to about 0.5 times the diameter of the pipe, depending on the air velocity and size of the orifice.

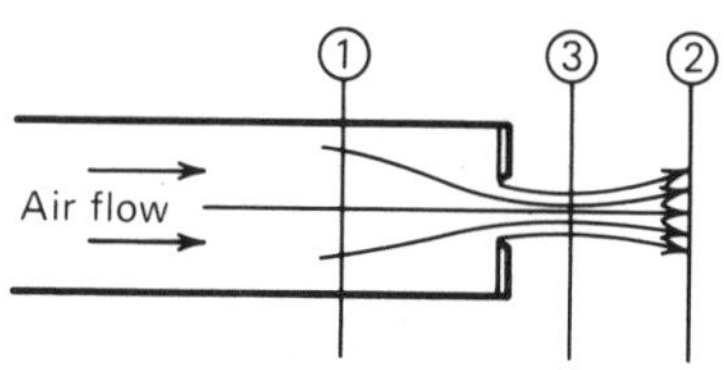

Figure 13–1

Vena contractor in sharp-edged orifice

The analogy of air flow through a sharp-edged orifice located at the end of a duct can be made to demonstrate the performance of air distribution devices. Any diffuser or grille performs in exactly the same way as in Fig. 13-2, where air flow is shown through the members of a ceiling diffuser and the bars of a grille. Note that the only difference between Figs. 13-1 and 13-2 is that the latter shows the action of multiple orifices. In other words there is a vena contractor existing in every ring of a circular ceiling diffuser and between pairs of bars of a grille instead of a single vena contractor as in the sharp-edged orifice.

The sum of the areas of the vena contractors existing in a circular ceiling diffuser is sometimes referred to as the "effective area", A_e. In the case of a grille the effective area would be the sum of the vena contractors existing between the bars. When considering a sharp-edged orifice, the effective area will be the area of a single stream at the vena contractor. The shape, size, and number of bars or rings of a distribution device will affect the effective area and location of the vena contractors, but the basic concept applies regardless of the individual design.

Bernoulli's equation is applicable to any air diffuser in that the total pressure in the airstream behind the outlet is converted to velocity pressure at the points of the vena contractors since the friction loss by air in contact with the rings and bars is negligible.

13-4 MEASUREMENT OF STATIC, VELOCITY, AND TOTAL PRESSURES

Any fluid, including air, will exert pressure on the walls of the duct in which it is confined. Thus, a gauge connected to a tank of compressed air will indicate what is called *static pressure;* that is, the outward push of the air against the walls. In a duct in which the fluid is at rest because a damper at its outlet has been shut tightly, the static pressure of the air

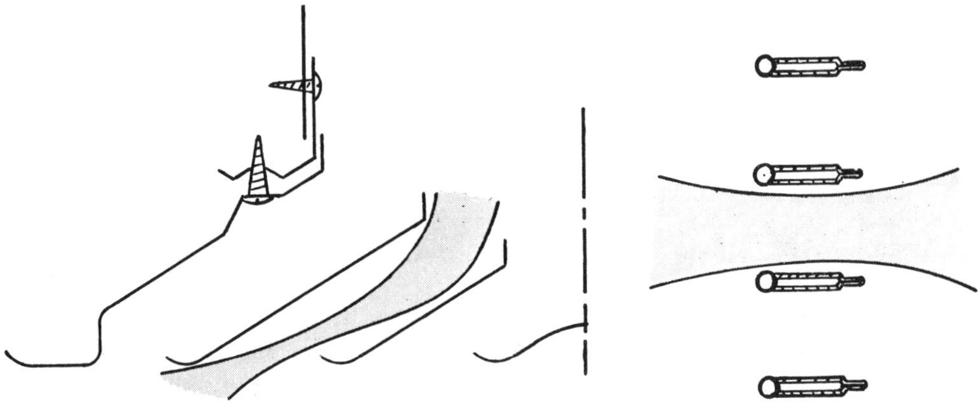

Figure 13–2

Vena contractor in diffusers
(a) Ceiling diffuser (b) Grille

in the duct can be measured by means of a gauge connected to a tube which is inserted through the wall of the duct as shown in Fig. 13-3.

In this same figure a hinged vane is shown in the airstream. If the air were at rest, the vane would hang vertically. However, if the damper at the duct outlet is opened and the air starts to move, the vane will be tipped back due to the impact of the moving air against its face. The pressure on the face of the vane, which holds it up at an angle, is due altogether to the velocity of the moving stream of air.

If the vane in Fig. 13-3 were hung parallel to the flow of air, it would not be affected by the movement of the air. The pressure on the two faces of the vane would be equal because only the static pressure of the air would be exerted against these two faces. Similarly, even though the air is in motion, a tube with its opening perpendicular to the air movement as in Fig. 13-3 will not be affected by the movement of air. A gauge connected to it will indicate only static pressure or sidewards push of the air exactly as it would when the air was at rest.

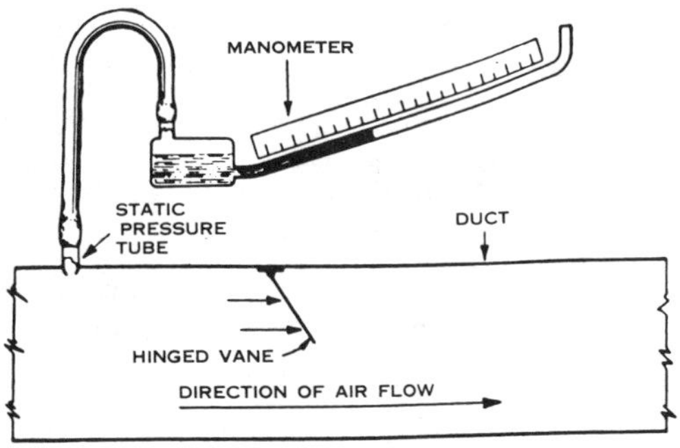

Figure 13–3

Measuring static pressure

On the other hand, if a tube is inserted into the duct with its opening pointing upstream as in Fig. 13-4, the pressure indicated by the gauge will be higher than when the static tube was used. The reason for this is apparent when the cause for the tilt of the vane is considered. The vane swings up at an angle because of the pressure exerted upon its face by the moving stream of air. Similarly, a gauge connected to a tube with its open end facing the stream of air would indicate not only the static pressure

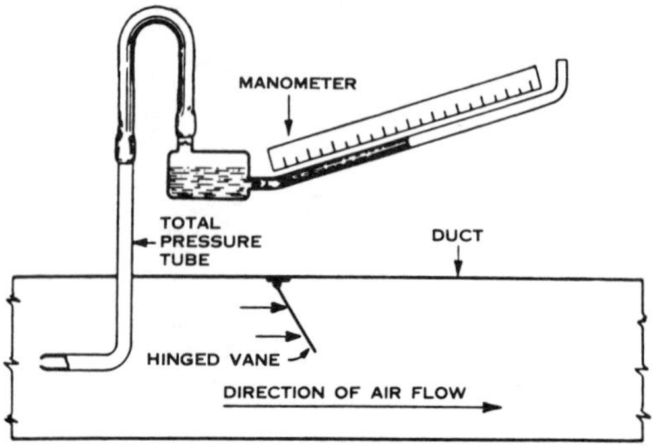

Figure 13–4

Measuring total pressure

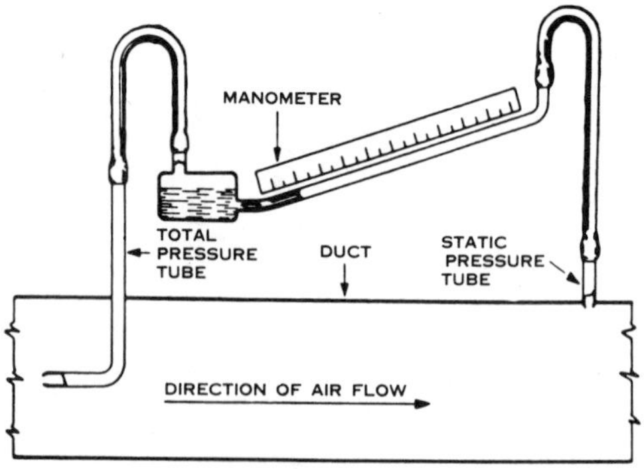

Figure 13–5

Measuring velocity pressure

but, in addition, a pressure due to the moving stream of air. The greater the velocity, the greater the pressure exerted against the tube opening and the face of the vane.

The action of the moving stream of air offers a means of measuring the velocity of the air. The pressure measured in this way is known as the total pressure. It is greater than the static pressure because of the added pressure due to the moving stream of air.

However, if measurements are made of the total and static pressures by means of two separate tubes, the difference in the measurements of the two gauges would give the excess pressure which is due only to the velocity. This excess pressure above the static pressure is known as velocity pressure.

As indicated by Eq. (13-8), we are interested in three pressures—static, velocity, and total. Actually, only the total and static pressures can be measured directly, and the velocity pressure is found from their difference. After the velocity pressure has been found in this way, the air velocity can be computed by using Eqs. (13-5) or (13-7).

In actual test work the manometer gauges used in Figs. 13-3 and 13-4 may be used to indicate the velocity pressure directly by connecting the total and static tubes to the opposite ends of a manometer, as shown in Fig. 13-5. In this way the displacement of the fluid in the manometer is due only to the *difference* between the total and static pressures, or it indicates the velocity pressure only ($p_v = p_t - p_s$).

The arrangement in Fig. 13-5 is unsatisfactory for actual work. Both for convenience and accuracy, the static and total pressure tubes are combined in one instrument called a Pitot tube. Such an instrument consists of two tubes, one inside the other, as shown in Fig. 13-6. The inner tube measures the total pressure. Eight holes 0.04 in. in diameter are drilled into the side of the outer tube for measuring the static pressure.

The Pitot tube is satisfactory for measuring velocities in excess of 700 fpm. Below this velocity, the deflection indicated on the inclined manometer scale is too small to read accurately. For laboratory work there have been developed ultrasensitive manometers that permit accurate velocity measurements below 700 fpm. These manometers are called *micromanometers*.

In conclusion, then, we may say that static pressure is the amount of compressive or expansive energy contained in a fluid or gas and is a measure of its potential energy. Static pressure may exist in a fluid or gas at rest or in motion and is the means of producing and maintaining flow against resistance.

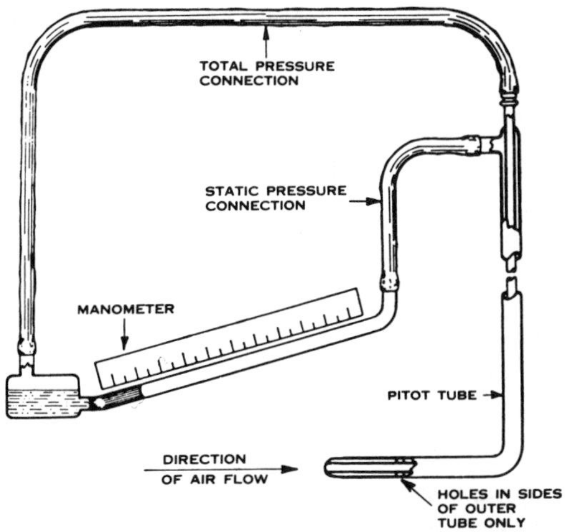

Figure 13–6

Pitot tube used to measure velocity pressure

Velocity pressure is the pressure corresponding to the velocity of flow and is a measure of the kinetic energy in the fluid or gas.

Total pressure is the sum of the static and velocity pressures and is a measure of the total energy within the fluid or gas. Static pressure may be transformed into velocity pressure and vice versa, but only at the expense of total energy.

13-5 PRESSURE CHANGES IN A DUCT SYSTEM

An air duct of varying cross-sectional flow area is illustrated in Fig. 13-7. If the air flow through the duct were ideal, there would be no friction, turbulence, or dynamic losses, and the total pressure (energy) is constant everywhere along the duct regardless of the changes in flow area. On the other hand the static pressure and the velocity pressure will change in magnitude with each change in flow area. Figure 13-7 shows graphically the changes in static and velocity pressures (heads) as the air flows through the ideal system.

Figure 13-8 illustrates an actual air flow system that includes a fan, intake duct, and an exit duct, again illustrating the changes in static and velocity pressures. Illustrated also is the decrease in total pressure in the direction of fluid flow caused by friction and dynamic (shock) losses. The following points should be noted:

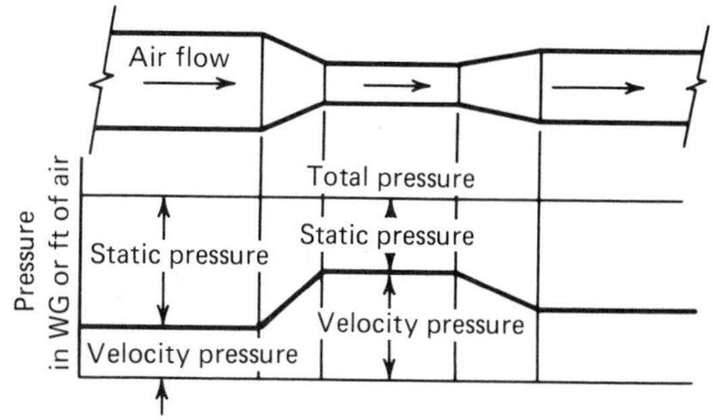

Figure 13–7

Theoretical changes in pressures with changes in duct flow area

1. The total pressure, h_t, at any point in the system is a measure of the total mechanical energy at that point.
2. In any actual flow system the total pressure decreases in the direction of flow.
3. Static and velocity pressures are mutually convertible and can either increase or decrease in the direction of flow.
4. In section *AB* the static pressure and total pressure decrease because of fan entrance losses and inlet duct friction loss. (Duct friction loss is dependent on Reynolds number; entrance losses *do not* depend directly on Reynolds number, but are functions of velocity pressure or head.)
5. At *B* the fan adds mechanical energy to the fluid increasing both its velocity and static pressure heads.

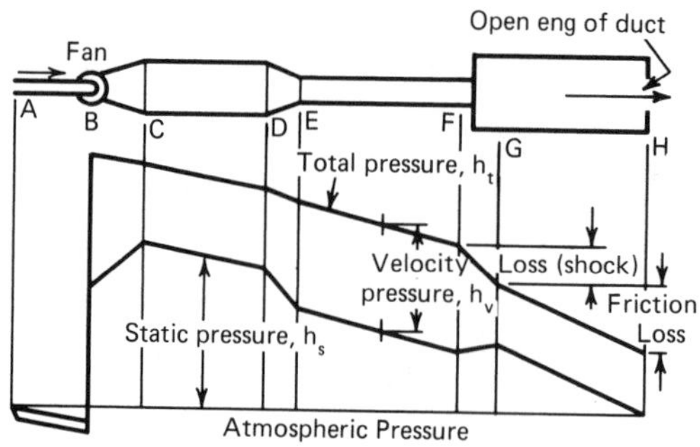

Figure 13–8

Actual changes in pressure with changes in duct flow area, and energy supplied by fan

6. In section *BC* the expansion of the duct causes a deceleration of the flow. The reduction of velocity causes a *conversion* of velocity pressure head to static pressure head. The loss in total head is caused by shock losses associated with the change in flow area.

7. In section *CD* the decrease in total head is caused by duct friction. The decrease in static head follows at the same rate because the velocity head is constant with the uniform cross section of the duct.

8. In section *DE* the flow is accelerated by the reduction in cross section of the duct. In this section there is a *conversion* of static to velocity head. The change in total head is due to shock losses associated with the change in flow area.

9. In section *EF* there is a decrease in total and static pressure head due to duct friction. The velocity head is constant.

10. In section *FG* there is a sudden expansion of the duct. There is a *conversion* of velocity head to static head with a drop in total head due to shock losses.

11. In section *GH* there is a uniform decrease in total and static heads due to duct friction.

12. The total pressure drops continually until at the very outlet of the duct, the static pressure is zero gauge (atmosphere) and the total and velocity pressures are equal. The reason for this is that at the outlet of the duct the pressure of the air is the same as the surrounding atmospheric pressure, that is, the static pressure is zero. However, the velocity pressure is still the same because the air issues from the mouth of the duct with the same velocity that it had while flowing in the duct. The velocity pressure is finally dissipated after the moving air issues from the duct and diffuser through the surrounding atmosphere. If the area at the outlet end of the duct is the same as the area at its inlet end, the velocity pressure will have the same value at both ends of the duct.

It is for this reason that the term *static pressure loss* is often used in practical work. While the total pressure loss is the actual loss in the system, the change in static pressures from inlet to outlet is numerically equal to the friction loss when the duct inlet and outlet areas are identical.

In many installations the area at the outlet end of the duct is larger than at the inlet. Hence, the velocity at the outlet end of the duct is smaller than at the inlet. Because of this, a small part of the initial velocity pressure is converted into static pressure, which may then be lost in friction. As a result, not only the initial static pressure but a portion of the initial velocity pressure will be lost in friction. However, the initial velocity pressure is usually small compared to the static pressure. For this reason the small part of the velocity pressure converted into static, which is lost in friction, can usually be disregarded. It is evident that for ordinary work the use of static pressure is of greater utility than total pressure because its magnitude is a very close measure of the pressure available to deliver the air against the friction of a given duct system.

Although Fig. 13-8 shows a simple duct with one outlet, the preceding discussion is true regardless of the number and size of the outlets connected to the duct system.

13-6 LAWS GOVERNING OUTLET PERFORMANCE

The exchange of thermal and kinetic energy, in the form of heating, cooling, and air motion, from the primary to the secondary air, may be done by induction only. The problem of properly distributing air to a conditioned space lies largely in completing the induction exchange above the occupied zone in a space.

Air emerging from an outlet at once commences to entrain room air. This induction effect increases the volume and decreases the velocity of the mixture stream according to Newton's second law, or the conservation of momentum theorem, which may be stated as,

$$m_{a_1}V_1 + m_{a_2}V_2 = (m_{a_1} + m_{a_2})V_3 \qquad (13\text{-}9)$$

where m_{a_1} = mass of primary air

m_{a_2} = mass of secondary air

V_1 = velocity of primary air

V_2 = velocity of secondary air

V_3 = velocity of mixture

The term V_2 may for practical purposes be considered zero, and since the density is essentially constant, the volume rate may be substituted for mass and Eq. (13-9) becomes

$$Q_1V_1 = (Q_1 + Q_2)V_3 \qquad (13\text{-}10)$$

where Q_1 = cfm of primary air

Q_2 = cfm of secondary air

Induction decreases gradually at lower velocities giving way to turbulence and eddies.

The induction ratio *(R)* is defined as the ratio of the total air to primary air, or

$$R = \frac{\text{total air}}{\text{primary air}} = \frac{\text{primary + secondary air}}{\text{primary air}} \qquad (13\text{-}11)$$

The practical operating induction ratio of a diffuser or grille may be determined by velocity or temperature measurement. Readings of either temperature or velocity should be taken as follows:

t_1 or V_1 = temperature or velocity of the primary airstream at the last confines of the diffuser or grille

t_2 or V_2 = average temperature or velocity of the room air

t_3 or V_3 = average mixed airstream temperature or velocity at a reference point

The readings must be taken within the confines of the envelope of the mixed airstream. The accuracy of determining the induction rate of a diffuser or grille depends upon taking temperature or velocity measurements at a multiplicity of points in a plane at right angles to the mixed airstream at a reference point. Figure 13-9 illustrates a schematic of a setup for determining the induction rate for a wall-mounted rectangular diffuser. The plane grid network is placed at a reference distance from the face of the diffuser (say 3 ft). The grid network is composed of a series of squares, the centers of which serve as a velocity-measuring station or the point location of a thermocouple for temperature measurement. Readings are taken in the center of each square, either temperature or velocity, and averaged. Then the induction ratio is determined from

$$R = \frac{t_1 - t_3}{t_3 - t_2} = \frac{V_1 - V_3}{V_3 - V_2} \qquad (13\text{-}12)$$

Where a potentiometer and thermocouples are available, the temperature method is preferred since it gives a more stable reading than a velocity meter, due to inertia and other characteristics generally inherent with velocity meters.

It must be borne in mind that an induction ratio must be referred to a given point in the stream travel. For example, assume a diffuser discharging air under conditions illustrated in Fig. 13-9. Assume the reference point is established at 3.0 ft from the diffuser. The following average temperatures were recorded: t_1 = 65°F, primary air; t_2 = 74.5°F, secondary air; t_3 = 73.5°F, mixed-air temperature (average) at reference point. By Eq. (13-12)

$$R = \frac{t_1 - t_3}{t_3 \quad t_2} = \frac{65 - 73.5}{73.5 - 74.5} = 8.5 \text{ to } 1$$

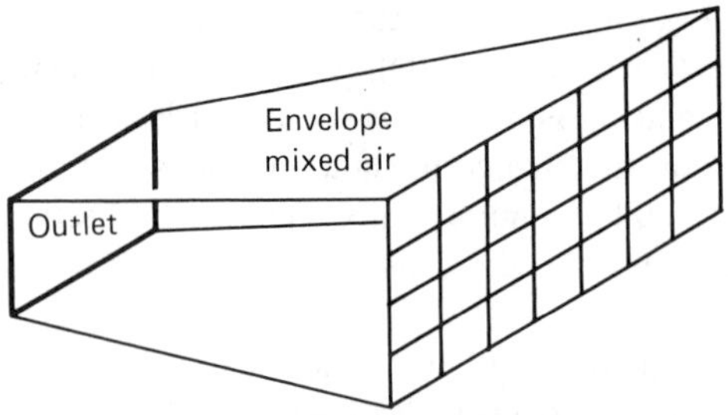

Figure 13–9

Setup for determining induction ratio of a rectangular diffuser

If the reference point is taken at 6.0 ft from the diffuser, the temperatures are: t_1 = 65°F; t_2 = 74.5°F; t_3 = 74.2°F. Then,

$$R = \frac{65 - 74.2}{74.2 - 74.5} = 30.7 \text{ to } 1$$

The difference in density between the primary air and the secondary air for extreme differences of temperature between the two will cause some error due to gravity effects. Therefore, such tests should be run with normal difference of temperature between the primary and secondary air. The above example illustrates the necessity of establishing a common reference point when comparing the induction rates of competitive equipment or of different types of equipment.

It will be seen that a given volume of air discharged at a given velocity from a low induction ratio device will carry a greater distance before reaching a specified mixture velocity as compared with the same volume discharged at the same velocity from a device with a high induction ratio characteristic.

Various types of air distribution devices have different induction characteristics and vary quite widely in the rapidity with which induction takes place. Table 13-1 shows the

Table 13-1

Relative Induction Characteristics of Grilles and Diffusers*

Type	Mounting	Induction
Rectangular or linear diffuser	On ceiling	Low
Straight-flow grille	In sidewall at ceiling	Low
Straight-flow grille	In sidewall below ceiling	Low
Side deflection grille	In sidewall at ceiling	Medium
Half-round ceiling diffuser	In ceiling	Medium
Full-round ceiling diffuser	In ceiling	High
Full-round ceiling diffuser	Located down from ceiling	High
Perforated ceiling	—	High

**From Ref. 3. Courtesy of Tuttle & Bailey, Division of Allied Thermal Corporation.*

relative induction characteristics of present-day equipment.

Each of these types has its application, and the success of an air distribution system depends on the selection of equipment with induction characteristics best suited to the installation. It is obvious that an air outlet with a low induction characteristic must not be used where the volume to be distributed is high and the throws are short. Conversely, a high induction device should not be used where the volume to be distributed is low and the throws long. Ceiling height and type of occupancy will affect the selection, of course, but the general statement may be made that the higher the cfm per square foot of area to be conditioned, the higher the induction characteristic of the air distribution device should be.

Effective area (A_e). One of the most important things to know about any air distribution outlet is its effective area. The entire air distribution performance and any pressure calculations involving the device are predicted upon its effective area. The effective area may be defined by the following expression:

$$A_e = A_g \times C \qquad (13\text{-}13)$$

where A_g = geometrically measured area

C = combined coefficient of entry and discharge

The value of C can only be determined from test for various outlets, many of which are quite complicated and difficult to measure geometrically. A typical laboratory test arrangement for such a test is shown in Fig. 13-10.

The term *effective area* applies to any air distribution outlet. An analogy between air distribution equipment and a sharp-edged orifice located at the end of a duct may be drawn. As air passes down the duct, it contains both static pressure and velocity pressure, the sum of which is, of course, total pressure. As the air approaches the sharp-edged orifice, it assumes a stream shape which, at some point in its travel, is smaller than the aperture through which the air passes (see Fig. 13-1). This area is called the vena contractor, as previously discussed. It may be said then, that the vena contractor area is the effective area of the sharp-edged orifice.

Exactly the same comparison may be made with a ceiling outlet. Air does not completely fill the passages of a ceiling outlet and, at some point within the passages, the air assumes a stream shape smaller than the actual opening of the passage (see Fig. 13-2). The sum of the vena contractors existing between

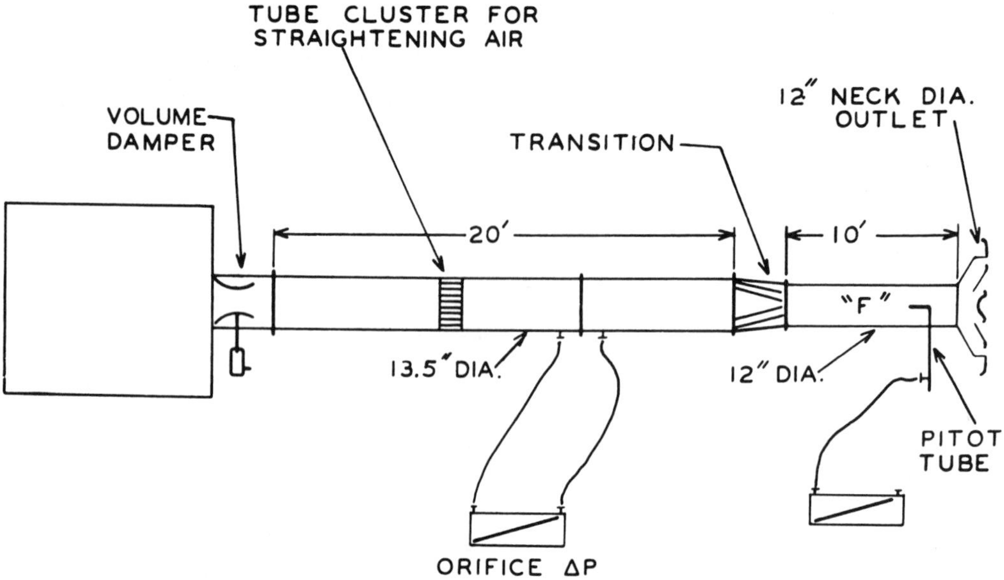

Figure 13–10

Equipment hookup for testing effective area of a ceiling diffuser

each ring is called the *effective area* of the complete diffuser.

The same fact applies to a grille. Air does not completely fill the space between the bars of a grille and, at some point within the passages, the air assumes a stream shape smaller than the actual aperture opening, no matter how streamlined the bars may be. Again, the sum of the vena contractors is called the *effective area* of the grille.

Referring to Fig. 13-10, the effective area may be found as follows:

1. Determine the average flow velocity in duct section F, from $Q = AV$, with Q through orifice.
2. Calculate velocity pressure head by Eq. (13-6).
3. Given the static pressure head at F, find total pressure head from Eq. (13-8).
4. Assume total pressure head is converted 100 percent to velocity pressure head at vena contractor.
5. Determine jet velocity in vena contractor by Eq. (13-7).
6. Determine effective area A_e by $A_e = Q/V_{jet}$.

Illustrative Problem 13-1

Assume that a test setup as shown in Fig. 13-10 is used to test a ceiling diffuser connected to a 12-in. round duct. The orifice Δp indicates a flow rate of 300 cfm. The static pressure head in section F is 0.096 in. WG.

Determine the effective area of the ceiling diffuser.

Solution:

Velocity at Section F = 300/0.7854
= 382 fpm

Velocity pressure head at F = $(V/4005)^2$
= $(382/4005)^2$
= 0.009 in. WG

Static pressure head at F = 0.096 in. WG (given)

Total pressure head at F = 0.096 + 0.009
= 0.105 in. WG

Velocity in jets = $4005(0.105)^{1/2}$
= 1298 fpm

$A_e = 300/1298 = 0.231\ \text{ft}^2$

Since many outlets are of complicated shapes, this method is often the only way possible to determine the effective area of the diffuser. The above procedure is primarily restricted to laboratory use due to the equipment required but is applicable to all types of air distribution equipment.

Average room velocity. Eq. (13-11) defines induction ratio R as the total air divided by the primary air. By definition, the entire airstream is composed of primary and secondary air, which is called total air. Primary air is supplied from the outlet and, by induction, the secondary air is entrained in the airstream.

Average room velocity is defined as the total air divided by the area of the wall opposite to the outlet. To take into consideration room velocity at approximately the level of the head and shoulders, the area of the wall is multiplied by 0.7. Essentially, this means 0.7 of the wall height times the length, or

$$\text{Average room velocity} = \frac{\text{Total Air}}{0.7 \times \text{Area of wall opposite outlet}} = \frac{\text{Total air} \times 1.4}{\text{Area of wall opposite outlet}} \qquad (13\text{-}14)$$

Some manufacturers of supply outlets utilize a room circulation factor K in the rating of such devices. The K factor is expressed as primary air cfm per square foot of wall area opposite the outlet. Note how this factor takes into account the average room velocity, primary and secondary air flows, and the vertical area opposite the outlet. From Eq. (13-11) we see that total air equals primary air times R. Substituting into Eq. (13-14) we have

$$\text{Average room velocity} = \frac{1.4 \times R \times \text{Primary air}}{\text{Area of wall opposite outlet}}$$

or

$$K = \frac{\text{Average room velocity}}{1.4 \times R}$$

$$K = \frac{\dfrac{1.4 \times \text{Total Air}}{\text{Area of wall opposite outlet}}}{1.4 \times \dfrac{\text{Total Air}}{\text{Primary Air}}} = \frac{\text{Primary Air}}{\text{Area of wall opposite outlet}} \qquad (13\text{-}15)$$

Once the required primary air is determined, it may be divided by the wall area opposite the outlet to determine the K factor. Knowing the application, we will be able to determine the outlet velocity that will be acceptable for a given average room velocity.

13-7 TYPES OF OUTLETS

An outlet is designed to distribute air that is supplied to it with velocity, pressure, and direction within limits that enable it to completely perform its function. However, an outlet is not designed to correct unreasonable conditions of flow in the air supplied to it.

Where an outlet without vanes is located directly against the side of the duct, the direction of blow of the air from the outlet is the vector sum of the duct velocity and the outlet velocity (Fig. 13-11a). This may be modified by the peculiarity of the duct opening.

Where an outlet is applied to the face of the duct, the resultant velocity V_c can be modified by adjustable vanes behind the outlet. Whether they should be applied or not depends on the amount of divergence from straight blow that is acceptable.

Often outlets are mounted on short extension collars away from the face of the duct (Fig. 13-11b). Whenever the duct velocity exceeds the outlet discharge velocity V_b (velocity due to pressure difference across the outlet), vanes should be used where the collar joins the duct (Fig. 13-11c).

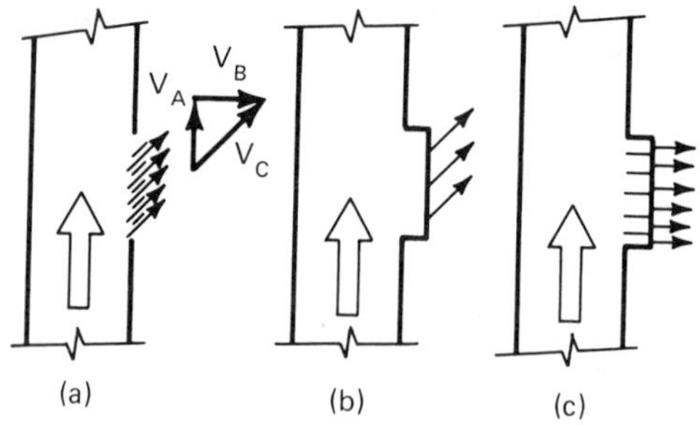

Figure 13–11

Supply outlet grilles mounted on face of duct

Importance of correct blow. Normally, it is not necessary to blow the entire length or width of a room. A good rule of thumb to follow is to blow three-fourths of the distance to the opposite wall. Exceptions occur, however, when there are local sources of heat at the end of the room opposite the outlet. These sources can be equipment heat and open doors. Under these circumstances, overblow may be required and caution must be exercised to prevent draft conditions.

Supply air temperature differential. The allowable supply temperature difference that can be tolerated depends to a great extent on (1) outlet induction ratio, (2) obstructions in the path of the primary air, and (3) the ceiling effect (see Fig. 13-12).

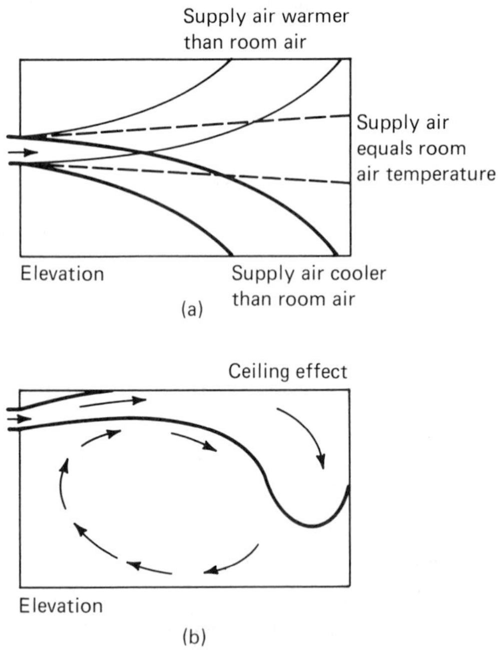

Figure 13–12

(a) Effect of supply air temperature differences
(b) Ceiling effects

There are basically three different types of supply air outlets: (1) ceiling diffusers, (2) grilles and registers, and (3) slot diffusers.

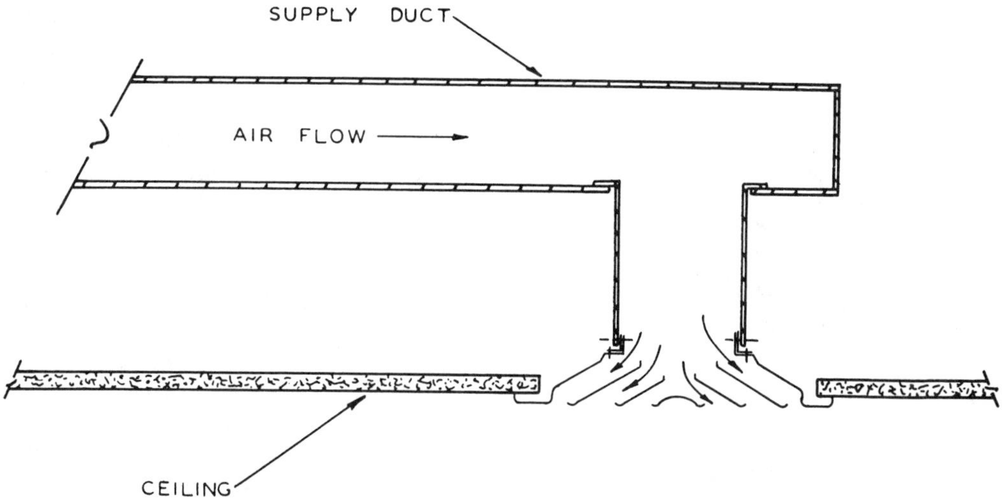

Figure 13–13

Multi-passage type ceiling diffuser

13-8 CEILING DIFFUSERS

By definition, a ceiling diffuser is a round, square, or rectangular air distribution device mounted in or near the ceiling, discharging air in a plane approximately parallel with the ceiling. The first ceiling diffusers were single-plane plaques mounted under ducts or conduits projecting through the ceiling. Some single-passage plaques are commercially available today, although most ceiling diffusers are of the multi-passage type to provide greater efficiency and better appearance (see Fig. 13-13).

The variable pattern diffuser, the linear and rectangular diffuser, and the perforated ceiling tile diffuser must be placed in the category of ceiling diffusers, although the discharge air patterns of each does not necessarily conform to the above definition.

In a homogeneous group of diffusers of any manufacturer, the effective areas of the diffusers in the group usually have a definite ratio to the neck areas. However, in comparing the diffusers of competitive manufacturers the same ratios do not apply, and comparison must be made on the basis of effective areas. Diffusers of the same effective area will produce equivalent performance regardless of the neck area.

Full-round diffuser with fixed effective area and fixed pattern (see Fig. 13-14) is a multi-passage diffuser mounted on the ceiling or suspended down from the ceiling. The neck of the unit is round and connects to round duct work by means of a duct ring to permit leveling and ease of installation, or by sheet metal screws attaching the neck directly to the duct. The discharge air pattern of the airstream is approximately parallel to the ceiling. Some diffusers use different blade angles when the diffuser is specified for use on the ceiling as compared with an exposed installation in order to get greater effective area. ''Flush''-type diffusers usually should not be used on exposed duct work since they will blow downward 10 to 20 degrees when so mounted. The air pattern is fixed and the effective area is not adjustable in the field.

The type PS supply diffusers shown in Fig. 13-14 are recommended for heating, ventilating, and cooling and are equipped with fully

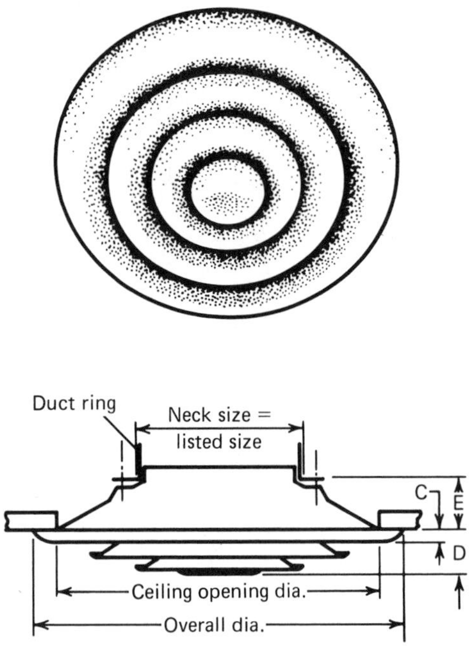

DIMENSIONS

Listed Size	Ceiling opening dia.	Overall dia.	C	D	E ± 1/8
6	11	12-5/8	3/8	1-3/8	1-3/4
8	14-3/4	16-15/16	1/2	1-13/16	2-5/16
10	18-3/8	21-1/4	5/8	2-7/16	2-15/16
12	21-15/16	25-3/8	5/8	2-13/16	3-1/2
14	25-1/2	29-7/16	11/16	3-1/16	4-1/4
16	29-1/8	34	7/8	3-7/8	4-3/4
18	32-3/4	38-1/2	1-1/16	4-5/16	5-3/8
20	36-3/8	43	1-3/16	4-15/16	6
24	43-1/2	51	1-3/8	5-11/16	7-1/8
30	54-3/8	63-5/8	1-5/8	7-5/8	9-1/8
36	54-3/8	63-5/8	1-5/8	8-3/4	7-1/2

All dimensions in in.

Figure 13–14

Type PS Round Ceiling Diffuser with Fixed Pattern for ceiling or exposed duct installation. (Courtesy of Tuttle & Bailey, Division of Allied Thermal Corporation.)

stepped-down cores for maximum range of performance (see Fig. 13-15). High induction rates result in rapid temperature and velocity equalization of the mixed-air mass well above the zone of occupancy. Horizontal performance assures confident use of cooling temperature differentials of 30°F and greater, at predicted low air motion (35 fpm) in the zone of occupancy. PS supply diffusers perform efficiently with air loadings of 6 to 30 air changes per hour (based on 10-ft ceiling height) and sound level range of NC 25 to 35.

The PS supply diffusers feature:

1. Fixed air pattern with stepped-down core. Horizontal air pattern of 360 degrees with maximum performance range.
2. Removable center core affords access to accessories. All necessary adjustments are easily made without disturbing the ceiling or removing the diffuser.
3. Cores are interchangeable with PF (flush face) and PA (adjustable pattern) diffusers in field.
4. Diffusers are equipped with margins designed to minimize smudging. Minimum dirt development on ceiling in normal application.
5. Duct rings for mounting furnished as standard. Diffuser outer shell fastens directly to duct ring with concealed fasteners; margins fit tight to ceiling for optimum ceiling appearance. Cost savings result from less installation time.
6. Diffusers and accessories are constructed of steel. Aluminum diffusers are available up through 18-in.

Round Diffusers with Fully Adjustable Pattern. The type PA round ceiling diffuser (see Fig. 13-16) with mechanical adjustments enables the angle of discharge pattern to be

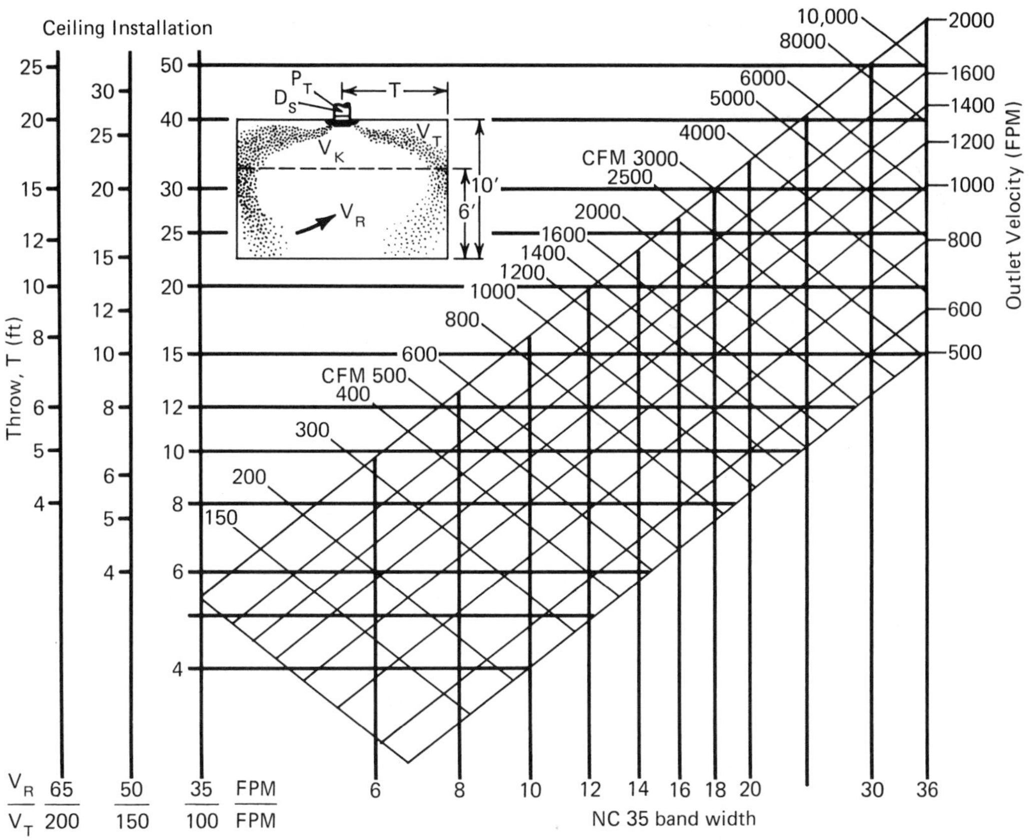

Listed size

Listed size	6	8	10	12	14	16	18	20	24	30	36
A_K outlet area	0.16	0.28	0.44	0.66	0.91	1.2	1.5	1.9	2.8	4.3	6.2
A_N neck area	0.20	0.35	0.55	0.79	1.1	1.4	1.8	2.2	3.1	4.9	7.1

V_K outlet velocity	500	600	700	800	900	1000	1200	1400	1600	1800	2000
P_T w/#4 damper	0.02	0.03	0.04	0.05	0.07	0.09	0.12	0.17	0.22	0.28	0.34
P_S w/#4 damper	0.01	0.01	0.02	0.02	0.03	0.04	0.06	0.08	0.11	0.14	0.17
P_T w/o #4 damper	0.02	0.02	0.03	0.04	0.05	0.06	0.09	0.12	0.16	0.20	0.25
P_S w/o #4 damper	< 0.01	< 0.01	0.01	0.01	0.02	0.02	0.03	0.04	0.05	0.07	0.08

Pressure accuracy ±0.01 in. or 10%, whichever is greater
When diffusers are used on exposed duct, multiply the throw (T) by 0.07.

SYMBOLS

V_T	Terminal velocity, FPM	T	Throw, feet	P_T	Total pressure, in. H_2O
V_R	Room velocity, FPM	A_K	Outlet area in ft^2	P_S	Static pressure, in. H_2O
V_K	Outlet velocity, FPM	A_N	Neck area in ft^2	NC re 8 dB room attenuation	

Figure 13–15

Engineering performance data for Type PS Round Ceiling Diffuser with Fixed Pattern. (Courtesy of Tuttle & Bailey, Division of Allied Thermal Corporation.)

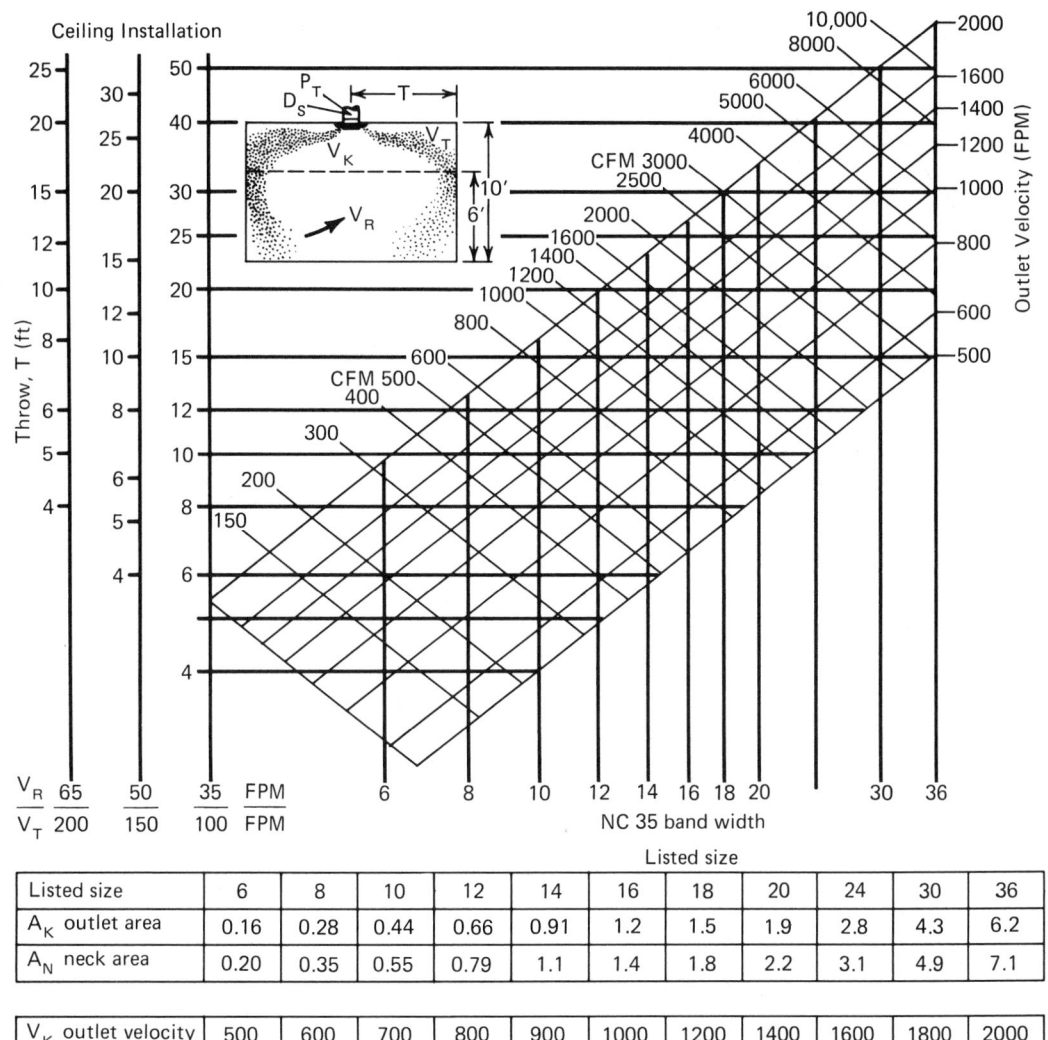

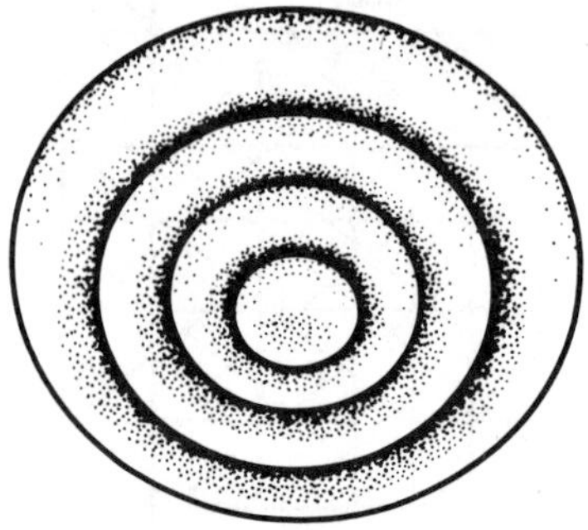

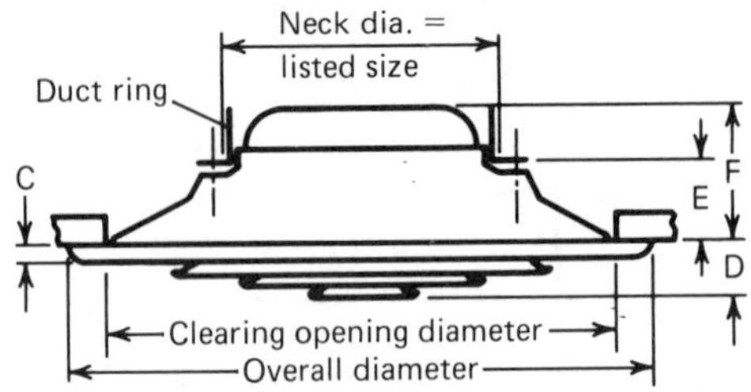

Dimensions

Listed Size	Clearing opening diameter	Overall diameter	C	D	E ± 1/8	F
6	11	12-5/8	3/8	1-3/16	1-3/4	3-3/16
8	14-3/4	16-15/16	1/2	1-9/16	2-5/16	3-3/4
10	18-3/8	21-1/4	5/8	1-7/8	2-15/16	4-5/8
12	21-15/16	25-3/8	5/8	2-9/16	3-1/2	5-9/16
14	25-1/2	29-7/16	11/16	3-1/8	4-1/4	6-5/8
16	29-1/8	34	7/8	3-1/8	4-3/4	7-3/8
18	32-3/4	38-1/2	1-1/16	3-5/8	5-3/8	8-5/16
20	36-3/8	43	1-3/16	4-3/16	6	9-1/2
24	43-1/2	51	1-3/8	4-3/4	7-1/8	11-9/16
30	54-3/8	63-5/8	1-5/8	6-7/16	9-1/8	14-15/16
36	54-3/8	63-5/8	1-5/8	7	7-1/2	14

All dimensions in inches

Figure 13–16

Type PA Round Ceiling Diffuser with Fully Adjustable Pattern. (Courtesy of Tuttle & Bailey, Division of Allied Thermal Corporation.)

changed. This type of diffuser has become popular. Users feel that such adjustability enables the air distribution pattern of a given system to be varied to suit conditions. In general, this is true *if* the diffuser is mounted on exposed duct work and not closer than 1 1/2 neck diameters to the ceiling. If diffuser is mounted on a ceiling, it has only two patterns: (1) a horizontal blow along the ceiling typical of type PS diffuser due to ceiling effect or (2) a straight columnar discharge downward into the occupied zone. The columnar downblow pattern rarely accomplishes good air distribution, and the use of this style is primarily restricted to spot heating, ventilating, or cooling where high ceiling heights are involved. A downblow diffuser changes its pattern by mechanically changing the components of force of the airstream. When the air pattern is changed from horizontal to vertical, the total effective area is decreased and the velocity for a given flow rate (cfm) is increased by maybe 30 percent. When mounted on exposed duct work, however, intermediate patterns may be obtained.

Type PA diffusers are recommended for heating, ventilating, and cooling and for conditions requiring adjustable air diffusion. High diffusion induction rates result in rapid temperature and velocity equalization of the mixed-air mass well above the zone of occupancy when adjustable cores are positioned for horizontal air diffusion. Horizontal performance assures confident use of cooling temperature differentials of 30°F and greater, at predicted low air velocity (35 fpm) in the zone of occupancy. Vertical downblow is effective for high ceiling installation and for dissipation of concentrated heating and cooling loads. PA supply diffusers perform efficiently at horizontal air diffusion with air loadings of 6 to 30 air changes per hour (based on 10 ft ceiling heights) and sound level range of NC 25 to 30.

Square and rectangular ceiling diffusers. Architectural demand has produced many variations in the design of ceiling diffusers. Due to structural conditions, it is sometimes impossible to mount a ceiling diffuser in the center of spaces to be conditioned. Also, the need to harmonize with acoustic tile ceilings had to be satisfied. Figures 13-17 through 13-21 illustrate three types of square and rectangular ceiling diffusers designed to meet architectural requirements.

Type DA diffusers (Fig. 13-17) with integral round necks are recommended for heating, ventilating, and cooling and for conditions requiring adjustable air diffusion. Round neck-to-square face construction results in a 360 degree air diffusion pattern similar to a full round diffuser. Horizontal pattern and vertical pattern air flow is similar to the type PA round diffuser.

Type M AM square and rectangular directional pattern diffusers (Fig. 13-18) are designed for ceiling and high sidewall installation. They are recommended for heating, ventilating, and cooling and for irregularly shaped space conditions requiring directional air diffusion patterns. Removable snap-in directional cores provide four-, three-, two-, one-way horizontal diffusion patterns on ceiling applications (see Fig. 13-19). Type M supply diffusers with three-way and one-way directional cores are applicable for high sidewalls applications. High diffusion induction rates result in rapid temperature and velocity equalization of the mixed-air mass well above the zone of occupancy. The reliable performance of the multidirectional pattern air diffusers assures confident use of cooling temperature differentials of 25°F and greater, at predicted low air motion (35 fpm) in the zone of occupancy. Type M supply diffusers perform efficiently with air loadings of 6 to 20 air changes per hour (based on 10 ft ceiling height) and sound level range of NC 25 to 35.

Type AVP ceiling diffusers (Fig. 13-20) are designed with a louvered face appearance. The louvered face design consists of four individual modules which can be positioned in four directions to provide the ultimate in pattern control and flexibility. They are recommended for heating, ventilating, and cooling applications where horizontal air patterns, along with the flexibility of air pattern, changes to meet specific job condition requirements of four-, three-, two-, or one-way horizontal patterns with a simple repositioning of the module (see Fig. 13-21).

High diffusion rates result in rapid temperature and velocity equalization of the mixed-air mass well above the occupied zone. The reliable performance of the horizontal pattern assures confident use of cooling temperature differentials of 25°F and greater, at predictable low air motion (35 fpm) in the zone of occu-

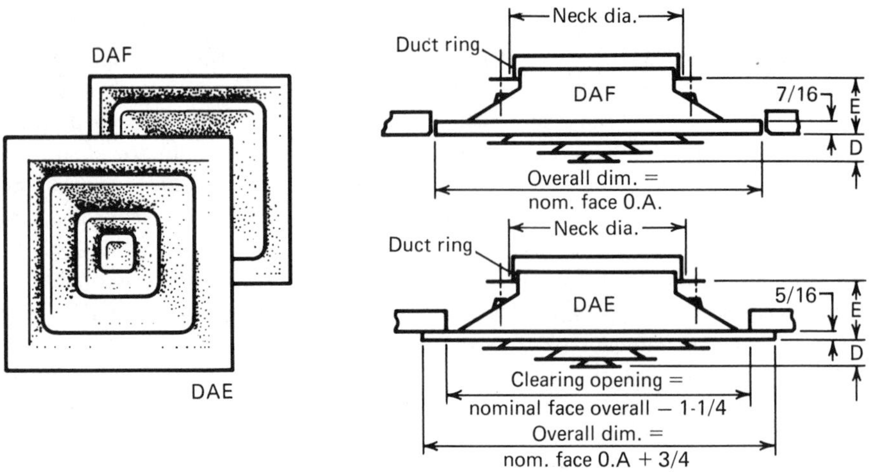

Dimensions

Listed Size	Neck Dia.	Nominal Face Overall	D		E ± 1/8	
			DAE	DAF	DAE	DAF
1206	6	12 X 12	1-7/16	1-3/16	1-11/16	1-15/16
1208	8	12 X 12	1-7/16	1-3/16	1-3/16	1-7/16
1608	8	16 X 16	1-13/16	–	2-7/16	–
2010	10	20 X 20	2-1/8	–	3-1/8	–
2406	6	24 X 24	1-7/16	1-3/16	1-11/16	1-15/15
2408	8	24 X 24	1-7/16	1-3/16	1-3/16	1-7/16
2410	10	24 X 24	2-1/8	1-7/8	3-1/8	3-3/8
2412	12	24 X 24	2-1/2	2-1/4	3-13/16	4-1/16
2415	15	24 X 24	2-1/2	2-1/4	3-1/8	3-3/8

All dimensions in inches

Figure 13–17

Type DA Square Diffuser with Fully Adjustable Pattern for ceiling or exposed duct installation. (Courtesy of Tuttle & Bailey, Division of Allied Thermal Corporation.)

pancy. AVP diffusers perform efficiently with air loading of 1 to 3 cfm per square foot of floor area or 6 to 20 air changes per hour (based on 10 ft ceiling height) and sound level range of NC 25 to 35.

13-9 PERFORMANCE OF CEILING DIFFUSERS

In predicting the performance of a ceiling diffuser, it is first necessary to determine its

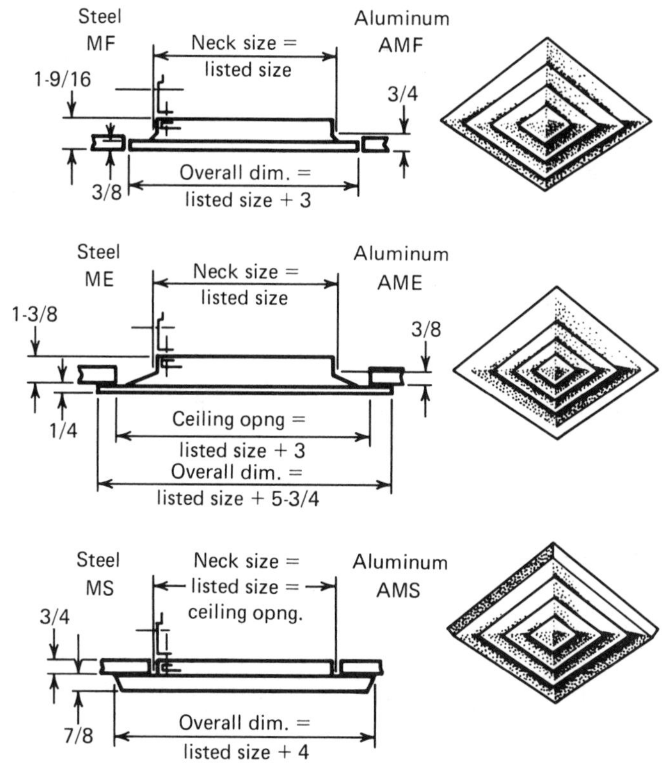

LISTED SIZES AVAILABLE W × H										
6 × 6	9 × 9	12 × 12	15 × 15	18 × 18	21 × 21	24 × 24	27 × 27	30 × 30	33 × 33	36 × 36
9 × 6	12 × 9	15 × 12	18 × 15	21 × 18	24 × 21	27 × 24	30 × 27	33 × 30	36 × 33	
12 × 6	15 × 9	18 × 12	21 × 15	24 × 18	27 × 21	30 × 24	33 × 27	36 × 30		
15 × 6	18 × 9	21 × 12	24 × 15	27 × 18	30 × 21	33 × 24	36 × 27			
18 × 6	21 × 9	24 × 12	27 × 15	30 × 18	33 × 21	36 × 24				
21 × 6	24 × 9	27 × 12	30 × 15	33 × 18	36 × 21					
24 × 6	27 × 9	30 × 12	33 × 15	36 × 18						
27 × 6	30 × 9	33 × 12	36 × 15							
30 × 6	33 × 9	36 × 12								
33 × 6	36 × 9									
36 × 6										

All dimensions in inches.

Figure 13–18

Type M AM Square and Rectangular Directional Pattern Diffusers for ceiling and high sidewall installation (Courtesy of Tuttle & Bailey, Division of Allied Thermal Corporation.)

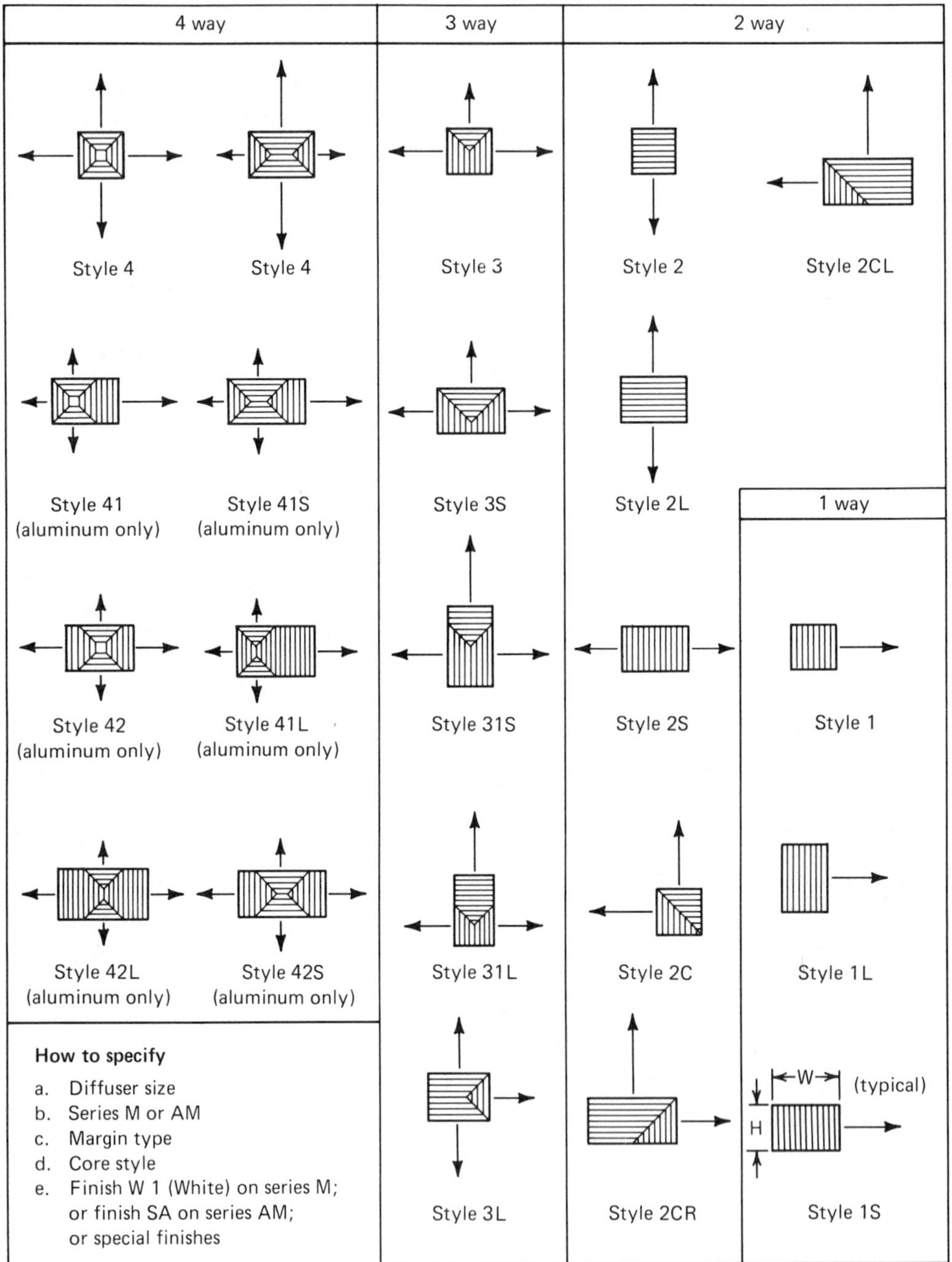

How to specify

a. Diffuser size
b. Series M or AM
c. Margin type
d. Core style
e. Finish W 1 (White) on series M; or finish SA on series AM; or special finishes

Examples:

18 × 12 MS-2CR finish W1
18 × 12 AMS-2CR finish SA

Figure 13–19

Type M AM Core Styles Plan or Reflected Ceiling View (Courtesy of Tuttle & Bailey, Division of Allied Thermal Corporation.)

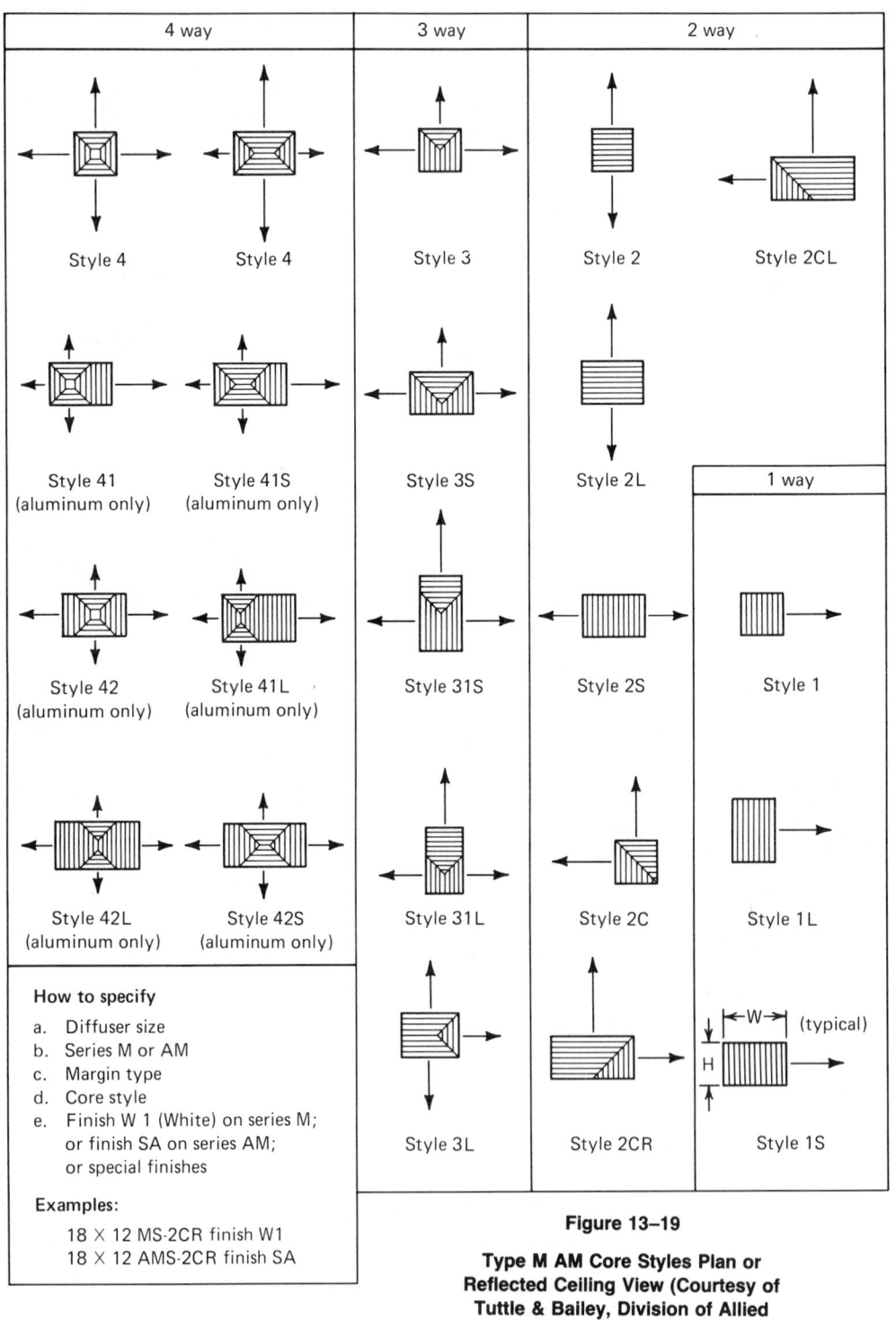

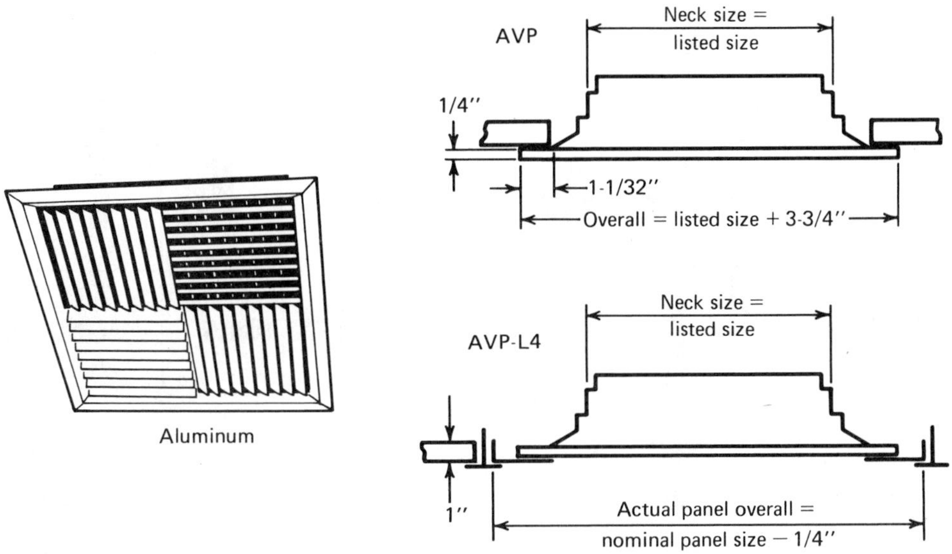

Listed sizes available

6 X 6	8 X 8	10 X 10	12 X 12	14 X 14	16 X 16	18 X 18	20 X 20	22 X 22	24 X 24

All dimensions in inches.

Figure 13–20

Type AVP Square Face, Square Neck Adjustable Module Pattern Diffusers (Courtesy of Tuttle & Bailey, Division of Allied Thermal Corporation.)

effective area A_e (see Section 13-6). Certain empirical formulas have been derived[1] by which to predict the performance of a ceiling diffuser after its effective area has been determined. Exhaustive tests have been run by various laboratories to establish and substantiate these formulas. Some investigators write the formulas in slightly different forms, but practically they are the same. For ceiling diffusers the throw (sometimes referred to as "radius of diffusion") can be predicted from

$$\text{Throw } (T) = \frac{K \times \text{cfm}}{\sqrt{A_e}} \qquad (13\text{-}16)$$

where T = the distance in feet the airstream travels to a point at which its motion falls to a predetermined value (terminal velocity V_T)

cfm = the flow volume through the diffuser

K = a constant determined by test and based on the induction characteristics of the diffuser (see Table 13-2)

A_e = effective area of the diffuser in square feet

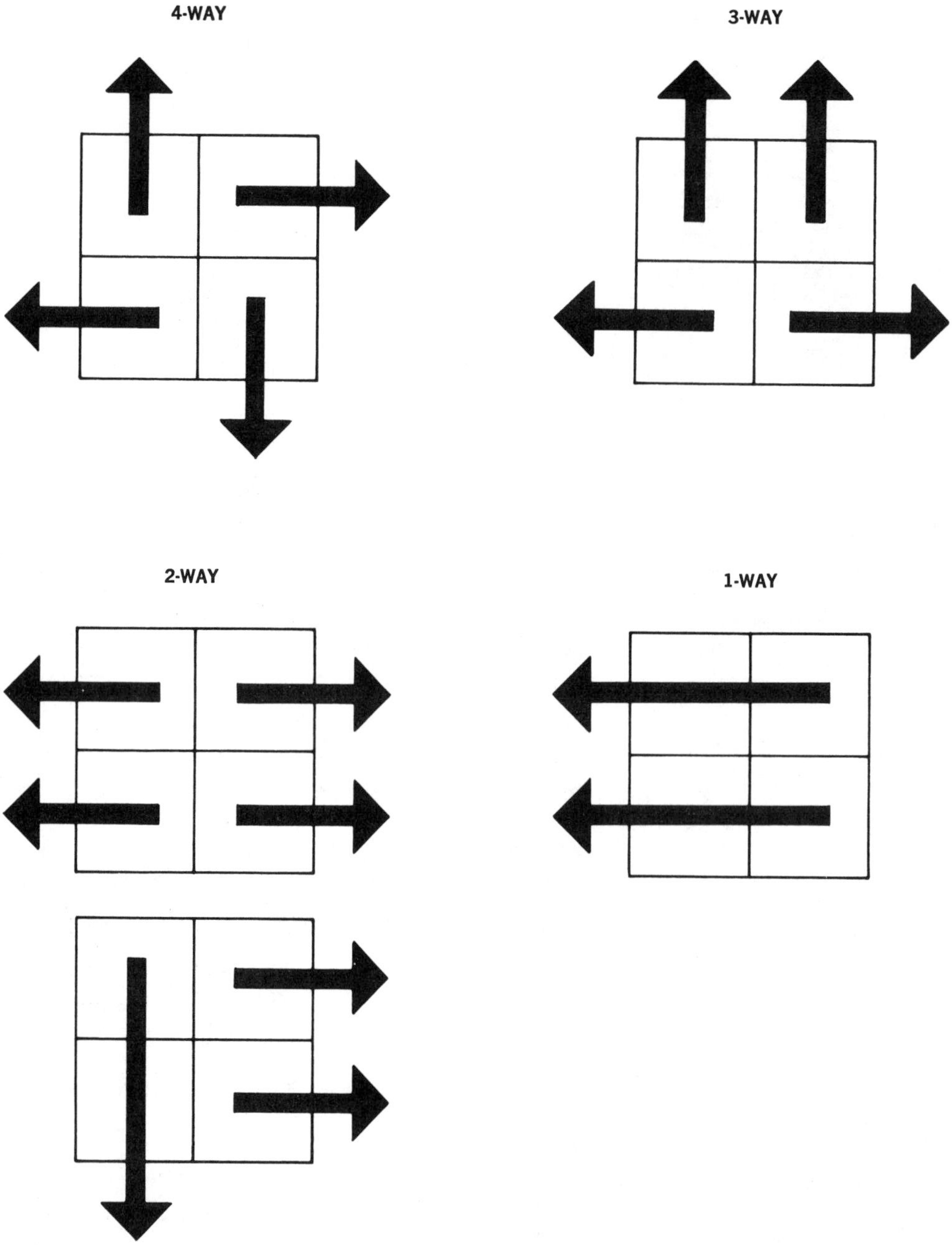

Figure 13–21

Type AVP Module Patterns (Courtesy of Tuttle & Bailey, Division of Allied Thermal Corporation.)

Table 13-2

K Values for Diffusers Discharging 360° Circular Air Pattern*

Terminal Velocity V_T (fpm)	Average Residual Velocity V_R (fmr)	Flush Ceiling Mounting	Exposed Mounting
100	25–40	0.0120	0.0080
150	40–60	0.0100	0.0066
200	60–85	0.0075	0.0050

**From Ref. 3. Courtesy of Tuttle & Bailey, Division of Allied Thermal Corporation.*

Illustrative Problem 13-2

Find the throw (T) of a ceiling diffuser discharging air on a 360-degree periphery with an effective area of 0.229 ft^2 when installed in a ceiling and handling 300 cfm, with 100 fpm terminal velocity required. By Eq. (13-16) and from Table 13-2, $K = 0.0120$, we have

$$T = \frac{K \times \text{cfm}}{\sqrt{A_e}} = \frac{(0.0120)(300)}{\sqrt{0.229}} = 7.5 \text{ ft}$$

If the diffuser were mounted on an exposed duct, Table 13-2 gives $K = 0.0080$. Then

$$T = \frac{(0.008)(300)}{\sqrt{0.229}} = 5.0 \text{ ft}$$

By Eq. (13-16), the required effective area could be found, if cfm, K, throw and terminal velocity are given.

The illustrative problem shows that a diffuser of the same size handling the same cfm when installed on a ceiling will throw further than one mounted on exposed duct work. The airstream after discharge is slowed and its energy absorbed by inducing room air into the primary airstream. Since, with a ceiling no air is induced on the ceiling side of the stream, the kinetic energy is not dissipated as rapidly and the airstream throws further. This is true, not only of a ceiling diffuser, but of any air distribution device located at or near the ceiling.

It will also be readily seen that if a ceiling diffuser with a given effective area has that effective area reduced by some means while still discharging the same volume of air, the throw will be longer. In problem 13-2, if the effective area of 0.229 ft^2 were reduced to 0.150 ft^2 by some mechanical adjustment while the air flow remained at 300 cfm and K was 0.0120, the throw would be

$$T = \frac{(0.0120)(300)}{\sqrt{0.150}} = 9.3 \text{ ft}$$

Therefore, it is possible to secure a variation in throw while using the same diffuser and the same cfm, if the effective area is changed. The total pressure required at discharge will vary, of course.

In the examples above, a constant volume was discharged through a varying effective area. Now consider a constant effective area and a varying air volume. If the volume is reduced from 300 to 150 cfm by dampering, the throw would be

$$T = \frac{(0.0120)(150)}{\sqrt{0.229}} = 3.75 \text{ ft}$$

Next, it must be understood that restricting the volume flow by dampering in the collar will not change the effective area of the diffuser. Dampering does nothing but reduce the cfm. The number of passages through which the air is discharged and their size is unaffected.

It is impossible to get the same throw, and hence the same performance, when the cfm is reduced by dampering unless the effective area is also changed. It is also apparent that on many systems where the fan cfm is reduced, the performance of the diffusers on the job is indiscriminately altered.

Illustrative Problem 13-3

Assume an example of an actual installation where the design conditions for a diffuser are 300 cfm, 7.5 ft throw, 100 fpm terminal velocity, and effective area of 0.229 ft^2. The jet velocity would be 300/0.229 = 1310 fpm.

After installation, it is found that the jet velocity is 1960 fpm, giving a cfm of (0.229)(1960) = 450 cfm, which does not correspond to 300 cfm required. The throw is

$$T = \frac{(0.0120)(450)}{\sqrt{0.229}}$$

$$= 11.9 \text{ ft at 100 fpm terminal}$$

instead of 7.5 ft required. If the diffuser effective area is reduced to 0.153 ft^2 by mechanical means, then the cfm becomes (0.153)(1960) = 300 cfm, which is the volume required; but the throw becomes $T = (0.012)(300)/\sqrt{0.153} = 9.2$ ft, which does not meet the throw requirement. However, the cfm can be reduced to 300 cfm by dampering in the neck or the diffuser so that the total pressure at this point corresponds to 1310 fpm instead of 1960 fpm. Therefore, the effective area of 0.229 ft^2 can be made to produce the required flow, or

$$T = \frac{(0.012)(300)}{\sqrt{0.229}} = 7.5 \text{ ft}$$

Thus, all requirements of the design are satisfied. A damper in the collar allows the total pressure available at the diffuser to be adjusted to the proper degree; changing the effective area only does not.

It can also be seen that volume control means must be available *in addition* to effective area control in order for a diffuser to perform all the requirements for which it may be called upon. It is also rarely possible to calculate the friction loss in any system closely enough to eliminate the need for volume control dampers to equalize the friction losses between the diffusers on various branch ducts (Chapter 14).

13-10 GRILLES AND REGISTERS

The oldest form of commercially manufactured air distribution device is a grille. Despite this fact, there is less general understanding of the laws governing performance of grilles than of ceiling diffusers. Actually, there are more variables to be considered when selecting grilles than any other air distribution device, particularly since the adjustable-bar grille (see Fig. 13-22) has practically supplanted the fixed-bar type. Inasmuch as the adjustable-bar grille contains all the variables to be considered with other types, this discussion will apply to the adjustable-bar grille.

Single-deflection grille with vertical bars. This is perhaps the most popular type in use. The vertical bars may be adjusted to permit maximum side spread of the airstream by deflecting the air as it strikes the blades (vanes). The grille may be combined with an opposed-blade damper to form a register (see Fig. 13-23). Opposed-blade dampers may be constructed with their blades running horizontally or vertically. Ideally, damper blades should run horizontally with a vertical front-bar grille and vertically with a horizontal-bar grille so that all blade passages will fill with air.

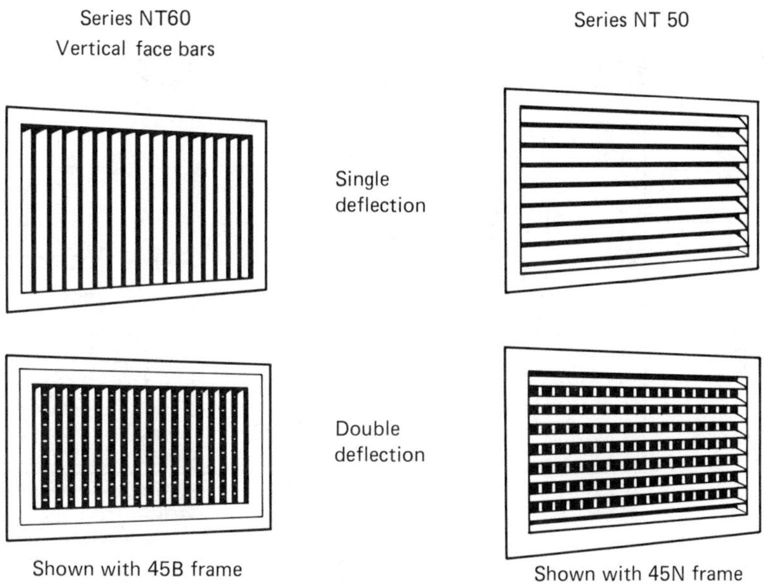

Listed sizes available in one piece

Min. W X H	Max. W X H
4 X 3	36 X 36
One inch increments of width and height.	
Multiple sections furnished for sizes greater than maximum width; maximum height remains 36.	

All dimensions in inches.

Figure 13–22

Series AN60, AN50 Supply Air Grilles and Registers. See manufacturer's catalog for complete data (Courtesy of Tuttle & Bailey, Division of Allied Thermal Corporation.)

Single-deflection grille with horizontal bars. This is popular from an appearance standpoint because it often harmonizes with the architectural motif of the building. However, functionally, it does not provide throw control to the extend that is possible with the vertical-bar grille since it cannot obtain horizontal spread and is therefore relegated to straight-flow applications. An opposed-blade damper may be combined with it to form a register. The damper blades should be vertical.

Double-deflection grille with vertical front bars. This type is the most versatile and functionally satisfactory combination, particularly if cooling is used. The vertical front bars allow throw control by obtaining variable side spread of the airstream. The rear blades (called deflectors) permit the airstream to be arched to control the drop of the airstream due to temperature differential in cooling installations. The assembly may be combined with a vertical opposed-blade damper to form a register.

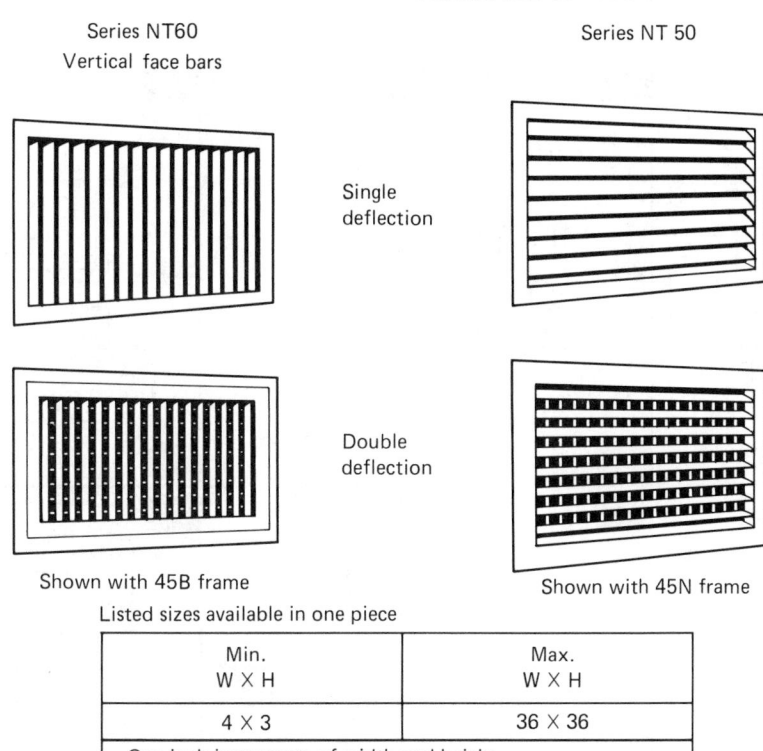

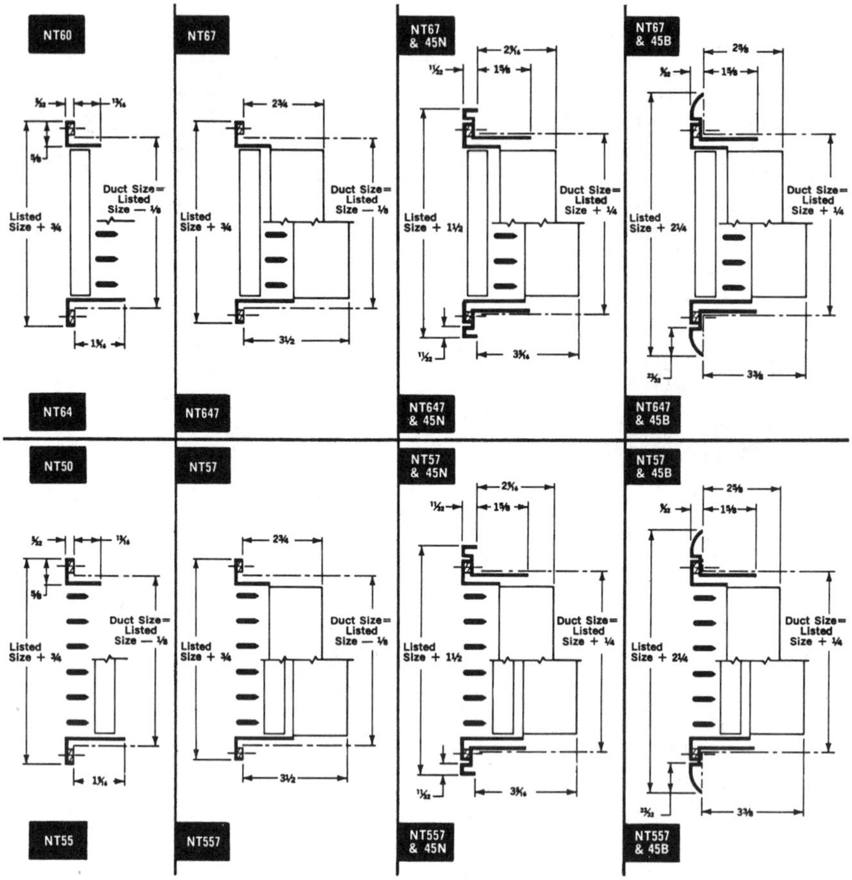

Figure 13–23

Series AN60, AN50 Supply Grilles and Registers (Sections) (Courtesy of Tuttle & Bailey, Division of Allied Thermal Corporation.)

Double-deflection grille with horizontal front bars. While this combination is pleasing architecturally, it is less effective than the vertical front bar grille because maximum side spread cannot be secured from the rear (deflector) bars. At maximum deflection of the rear bars, the airstream will strike the sides of the grille frame, thus reducing the control of throw. The application of this style of grille is therefore limited to medium and straight-deflection settings. The grille may be combined with a horizontal opposed-blade damper to form a register.

The Series AN 50 and AN 60 single-deflection supply air grilles and registers (Figs. 13-22 and 13-23) are recommended for applications requiring pattern adjustability in a single horizontal or vertical plane—sill grilles, sidewall location at ceiling line, or heating applications only. They perform efficiently with temperature differentials of up to 18–20°F cooling, 20–50°F heating, and for ventilating while handling 0.75 to 1.75 cfm per square foot of room area with draftless distribution.

The Series AN 55 and AN 64 double-deflection supply grilles and registers (see Figs. 13-22 and 13-23) are recommended for application in systems requiring optimum flexibility of pattern change to accommodate changing job conditions. They perform efficiently with temperature differentials of 20–22°F cooling, 20–50°F heating, and for ventilating while handling 1.0 to 2.0 cfm per square foot with draftless distribution.

The combination of streamlined foil-shaped bars and 3/4-in. bar spacing maintains a high effective area capacity of greater than 80 percent, which minimizes outlet velocity, reduces pressure drop requirement, and assures quiet operation. Individually adjustable bars are capable of shortening throw up to one-half with a widespread deflection requiring only a 20 percent increase in outlet velocity at a fixed volume.

Series A30-1 and A40-1 supply ceiling grilles and registers have individually adjustable curved vanes arranged to provide a one-way ceiling air pattern (see Figs. 13-24 and 13-25). They are recommended for applications in ceilings and sidewall locations for heating and cooling systems handling 0.75 to 1.75 cfm per square foot of room area.

Series A30-2 and A40-2 supply ceiling grilles and registers have individually adjustable curved vanes arranged to provide a two-way ceiling air pattern. They are recommended for applications in ceiling locations for heating and cooling systems handling 1.0 to 2.0 cfm per square foot of room area.

Both one-way and two-way pattern grilles can be adjusted to a full or partial downblow position. The curved streamlined vanes are adjusted to a uniformly partially closed position to deflect the air path horizontally while retaining an effective area capacity of 35 percent of the neck area. In the full downblow position grille effective area is increased to 75 percent.

Return air grilles and registers. These types are usually designed to match the appearance of the supply grilles that are used in the same area. The bars (or vanes) of such grilles do not need to be adjustable since no directional control of the airstream is required (see Section 13-21). Where such grilles are to be mounted on the side of the wall, fixed horizontal face bars, deflected downward, are often used to provide a reasonable sight-proof installation. A horizontal opposed-blade damper should be attached to form a register since balancing of the air flow through the returns is usually necessary.

Series A80, A70, and A110 return air grilles and registers, manufactured by Tuttle & Bailey, are shown in Fig. 13-26. The A70 return grilles and registers have fixed horizontal bars spaced on 3/4-in. centers with 0-degree straight or 40-degree face deflection. Series A80 return grilles have fixed vertical bars spaced on 3/4-in. centers with 0-degree straight deflection. The appearance matches the supply grille and register Series A50 and A60. Series A110 return grilles and registers have fixed horizontal bars spaced on 1/2-in. centers with 0-degree straight or 40-degree face deflection.

The streamlined foil-shaped bars and open bar spacing maintains an effective area capacity of greater than 75 percent, which minimizes intake velocity, reduces inlet pressure, and provides quiet operation. The smooth bar shapes do not accumulate lint and plug up as with sharp edge core-type returns. Deflected bar grilles installed in a low or high sidewall location are vision-proof with the grille deflection facing away from the line of sight. See Section 13-21 and Tables 13-14 and 13-15 for performance data.

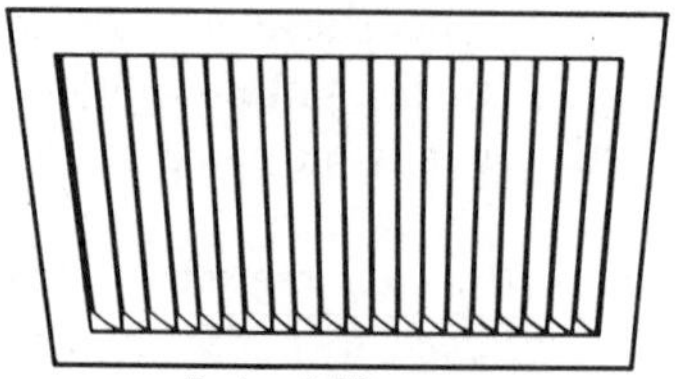

Series A40
Vertical face vanes
One way

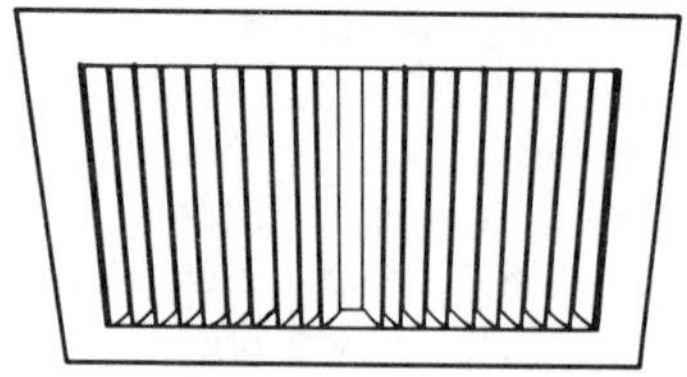

Series A40
Vertical face vanes
One way

Series A30
Horizontal face vanes
One way

Series A30
Horizontal face vanes
One way

Listed sizes available in one piece

Min. W × H	Max. W × H
6 × 6	48 × 48
One inch increments of width and height.	
Multiple sections furnished for sizes greater than maximum.	

All dimensions in inches.

Figure 13–24

Series A30, A40 Ceiling Supply Grilles and Registers (Curved vane). See manufacturer's catalog for complete data (Courtesy of Tuttle & Bailey, Division of Allied Thermal Corporation.)

13-11 PERFORMANCE OF GRILLES

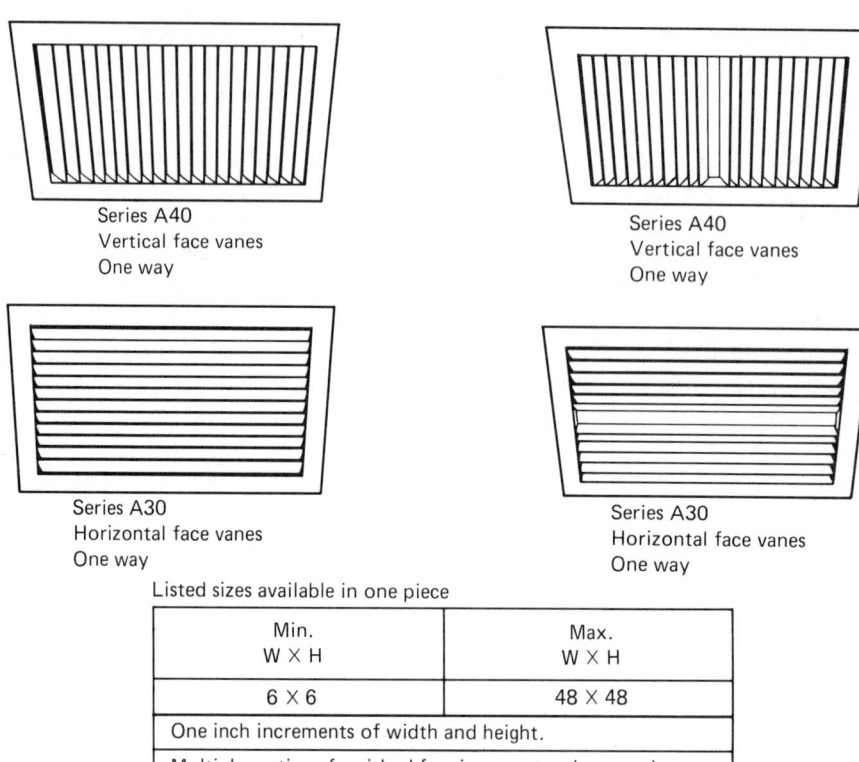

In order to predict the performance of grilles and registers, the effective area must be determined. In principle, the procedure for determining the effective area for a grille is the same as for ceiling diffusers except that the test installation is made with square or rectangular duct work instead of round (see Fig. 13-27). A difference exists between ceiling diffusers and grilles in that ceiling diffusers usually have a fixed effective area, whereas adjustable-bar grilles theoretically have an infinite number of effective areas. Any time a grille bar is deflected one degree the effective area is minutely changed. Practically, such a small change does not affect performance. In order to rate a grille properly, however, it is necessary to test at several deflection settings. Four to six such tests will establish the effective area limits of any grille line. Tuttle & Bailey normally determine effective area of supply

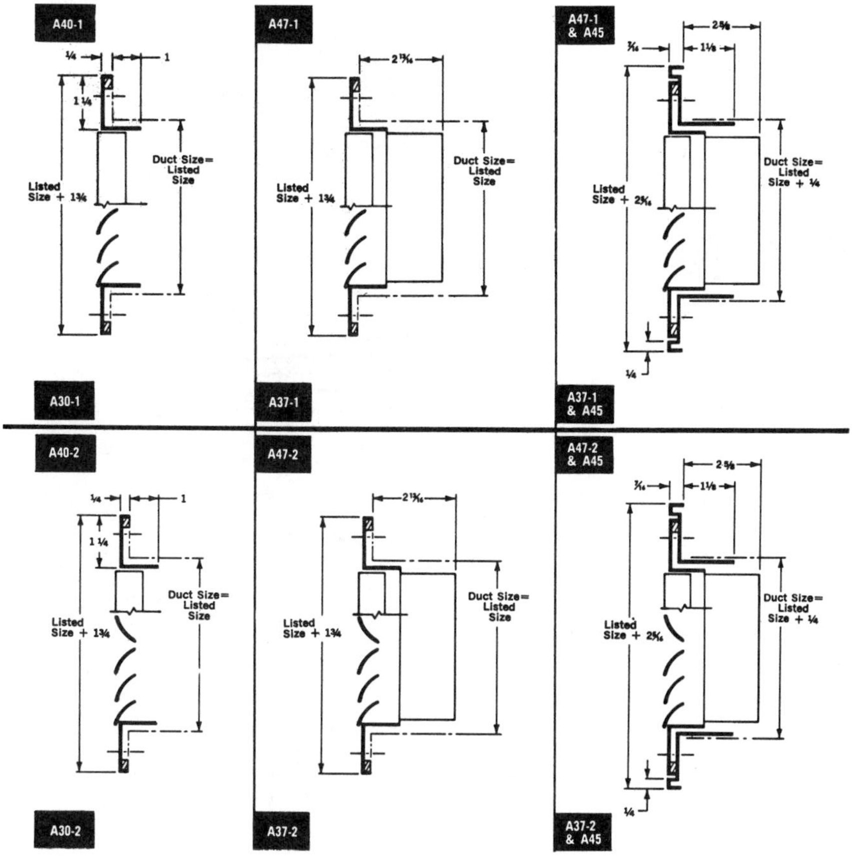

Figure 13–25

Series A30, A40 Ceiling Supply Grilles and Registers (Sections) (Courtesy of Tuttle & Bailey, Division of Allied Thermal Corporation.)

grilles at "straight", 20°–0°–20°, 40°–20°–0°–20°–40°, and 55°–40°–20°–0°–20°–40°–55°.

Illustrative Problem 13-4

As an example of a laboratory procedure for determining the effective area of a grille, assume it is desired to find the effective area of a 14-in. by 6-in. adjustable-bar grille with the bars set for straight flow, and using a test rig like Fig. 13-27.

Given: cfm through orifice = 500 cfm

Static pressure at F = 0.0359 in. WG

Duct size at F = 14 in. by 6 in.

<u>*Solution:*</u>

1. Duct area at $F = 14(6)/144 = 0.583\ \text{ft}^2$

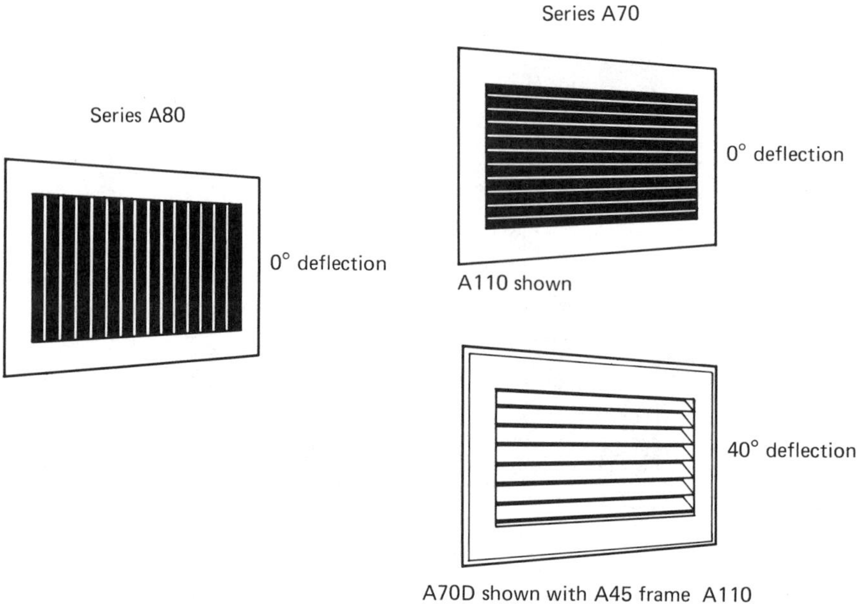

Listed sizes available in one piece

Min. W X H	Max. W X H
4 X 3	72 X 48
One inch increments of width and height.	
Multiple sections furnished for sizes greater than maximum.	

All dimensions in inches.

Figure 13–26

Return air grilles and registers. Series A80, A70, A110. See Manufacturer's catalogs for complete data (Courtesy of Tuttle & Bailey, Division of Allied Thermal Corporation.)

2. Velocity at F = 500/0.583 = 858 fpm
3. Velocity pressure at F = $(858/4005)^2$ = 0.0459 in. WG
4. Static pressure (read at F) = 0.0359 in. WG
5. Total pressure at F = 0.0359 + 0.0459 = 0.0818 in WG
6. Velocity in vena contractor between grille bars = 4005 $\sqrt{0.0818}$ = 1145 fpm
7. Effective area (A_e) = 500/1145 = 0.437 ft^2

The core area (A_c) of the adjustable-bar grille shown in Fig. 13-22 may be determined

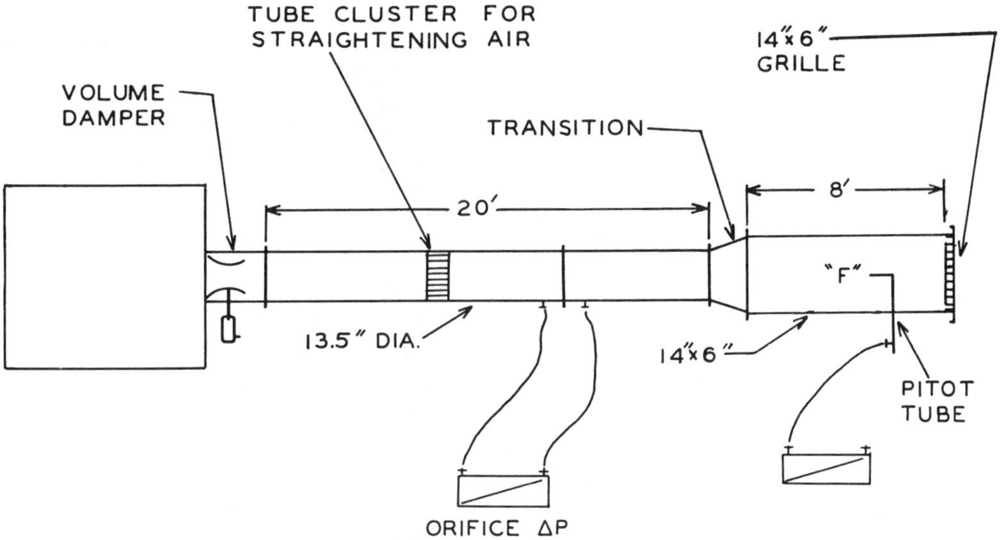

Figure 13–27

Rig for testing effective area of grilles

by multiplying the grille length, minus ½ in., times the grille height, minus ½ in. This is the total plane area within the limits of the grille frame.

If it is desired to find the ratio of the effective area to the core area of a given grille, the core area must be expressed in square feet, or

$$A_c = \frac{(14.0 - 0.5)(6.0 - 0.5)}{144} = 0.516 \text{ ft}^2$$

The effective area by test was 0.437 ft². Therefore,

$$\frac{A_e}{A_c} = \frac{0.437}{0.516} = 0.85$$

This means that the effective area of the grille, when the bars are set for straight flow, is 85 percent of the core area. The relationship between core area and effective area can be found for any other deflection setting by a similar computation. In the grille lines of most manufacturers the same relative proportion of effective area to core area is usually maintained for all sizes.

So far, the discussion of the effective area of grilles and registers has centered around straight-flow or "no deflection" vane setting. It can readily be seen that as the vanes are deflected, the effective area of any grille or register is reduced, depending on the extent of the deflection. For every deflection setting, it follows that there is a definite effective area factor (A_{ef}).

For grilles as shown in Fig. 13-22 the effective area factors for various deflection settings are shown in Table 13-3.

Therefore, the jet velocity (V_k) for any grille with a given cfm passing through it may be obtained from

$$V_k = \frac{\text{cfm}}{A_c \times A_{ef}} \qquad (13\text{-}17)$$

It should be noted that the effective area values in Table 13-3 are based on single-de-

Table 13-3

Effective Area Factors for Single-Deflection Grilles*

Angle of Deflection	Effective Area Factor A_{ef}
Straight	0.850
20°–0°–20°	0.750
40°–20°–0°–20°–40°	0.675
55°–40°–20°–0°–20°–40°–55°	0.630

**From Ref. 3. Courtesy of Tuttle & Bailey, Division of Allied Thermal Corporation.*

flection grilles (also, Tuttle & Bailey). The use of a double-deflection core reduces the given effective area factors by 5 percent. The use of an opposed-blade damper in the fully open position with a single-deflection grille reduces the effective area by 4 percent. Usually, these reductions are not significant unless the installation is a very critical one. Most rating tables do not differentiate between the various combinations of component register parts, but it may be calculated if necessary.

13-12 PARAMETERS AFFECTING GRILLE PERFORMANCE

The application and selection of a grille or register involves job condition requirements, selection judgment, and performance data analysis. Grilles and registers should be selected and sized for:

a. type and style

b. function

c. air volume requirement

d. throw, spread, and drop requirement

e. sound requirement

f. pressure requirement

We have previously discussed the various types and styles of grilles and registers, and how they function. We shall now discuss the remaining items above.

13-13 AIR VOLUME REQUIREMENT

The air volume requirement is usually accepted as the prime requisite following the grille type and style selection. The air volume per grille is that which is necessary for design heating, cooling, or ventilation requirements of the specific area served by the grille. The air volume required, when related to throw, sound, or pressure design limitations, determines the proper grille size. Single-deflection grilles and registers perform efficiently with temperature differentials of up to 18–20°F cooling, 20–50°F heating, and for ventilating while handling 0.75 to 1.75 cfm per square foot of room area. Double-deflection supply grilles and registers perform efficiently with temperature differentials of 20–22°F cooling, 20–50°F heating, and for ventilating while handling 1.0 to 2.0 cfm per square foot with draftless distribution.

13-14 THROW REQUIREMENT

As with ceiling diffusers, an empirical formula substantiated by tests [see Eq. (13-18)] may be used to determine the distance a grille or register will throw. Values of the K factor are shown in Table 13-4.

$$T = \frac{K \times \text{cfm}}{\sqrt{A_c}} \tag{13-18}$$

where T = throw in feet

K = a constant, developed from experiment

cfm = volume air flow from device

A_c = core area in square inches

Table 13-4

K Values for Adjustable-Bar Grilles and Registers*

Angle of Deflection	K Values
Straight	0.80
20°–0°–20°	0.68
40°–20°–0°–20°–40°	0.55
55°–40°–20°–0°–20°–40°–55°	0.43

**From Ref. 3. Courtesy of Tuttle & Bailey, Division of Allied Thermal Corporation.*

Illustrative Problem 13-5

Determine the throw for a 24- by 12-in. adjustable-bar grille (core area = 23.5 by 11.5 in.) discharging 400 cfm with maximum deflection.

Solution:

By Eq. (13-18)

$$T = \frac{K \times \text{cfm}}{\sqrt{A_c}} = \frac{(0.43)(400)}{\sqrt{(23.5)(11.5)}}$$

$$T = 10.5 \text{ ft}$$

If the same grille is used to discharge the same cfm with straight-bar deflection,

$$T = \frac{(0.80)(400)}{\sqrt{(23.5)(11.5)}} = 19.5 \text{ ft}$$

This illustrates the fact that a grille of a given size handling a given cfm, can be made to satisfactorily have a throw from 10.5 to 19.5 ft by simply adjusting the face bars. This explains why the adjustable-bar grille has become so popular. By manipulation of Eq. (13-18), the required core area can be found when the cfm, deflection required, and throw is given. Also, the deflection setting may be found when cfm, throw, and grille size are known, by solving for the value of K.

Throw and air motion in the occupied area are closely related, and both should be considered in the analysis of specific area requirements. The rated throw (T) for a given condition of operation is based on a stated terminal velocity (V_T) at that distance (T) from the grille. The residual room velocity (V_R) is a function of throw, terminal velocity, and the deflection setting for the outlet discharge pattern. The rated V_R is obtained with an unobstructed air path along the throw dimension. Ceiling protrusions, such as soffits, beams, surface-mounted luminairs, and so forth, may prematurely divert the stream into the occupied area.

Room air motion (V_R) is rated in the occupied zone of the room and is based on measured residual velocities at a given condition of supply air volume, terminal velocity (V_T) and temperature difference (ΔT). A definite relationship exists between terminal velocity and occpuied zone room velocity. An increase in V_T at the rated throw (T) dimension will result in an increased V_R. Figure 13-28 illustrates throw and room air motion.

Terminal velocity (V_T) is a value in fpm that exists at the end of the throw (T) from the grille. When the V_T occurs at the juncture of a ceiling and wall or at another specified location or throw dimension, its value determines the

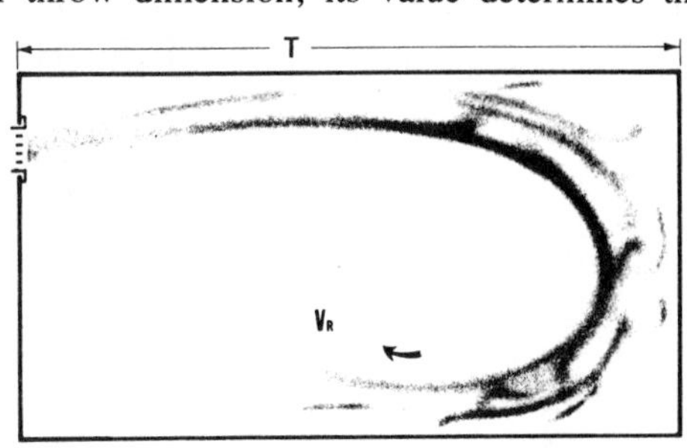

Figure 13–28

Relationship of throw (T) and room velocity (V_R)

Table 13-5

Engineering Performance Data, Supply Air Grilles and Registers*
Single and Double Deflection, Adjustable Bar Type

(Series T50, T60, NT50, NT60)
(Series A50, A60, AN50, AN60, AB50, AB60)

LISTED WIDTH

LISTED HEIGHT				Outlet Velocity 450-600 FPM					Outlet Velocity 550-700 FPM					Outlet Velocity 600-800 FPM					Outlet Velocity 700-1000 FPM				
				CFM	Deflection				CFM	Deflection				CFM	Deflection				CFM	Deflection			
					0°	20°	40°	55°		0°	20°	40°	55°		0°	20°	40°	55°		0°	20°	40°	55°
					V_K					V_K					V_K					V_K			
					450	500	550	600		550	600	650	700		600	700	750	800		700	800	900	1000
					P_T					P_T					P_T					P_T			
					.01	.01	.02	.02		.02	.02	.03.	.03		.02	.03	.04	.04		.03	.04	.05	.06
4 5 6	8 10 12	14 16 18	20 24		Throw (T) V_T 75					Throw (T) V_T 75					Throw (T) V_T 75					Throw (T) V_T 75			
6 5				40	6	5	4	3	45	8	7	6	4	60	10	9	7	5	65	11	10	8	6
8 6				50	7	6	5	4	60	9	11	6	5	70	11	10	8	6	80	12	11	9	7
10 8 6				75	9	7	6	5	90	10	9	7	6	100	12	11	9	7	125	13	12	10	8
12 10 8				100	10	8	6	5	125	11	9	8	7	150	13	12	9	8	175	14	13	11	8
14 12 10	8			125	11	9	7	6	150	12	11	9	7	175	14	13	10	8	200	16	14	11	9
18 14 12				150	11	10	8	7	175	13	11	10	8	200	15	14	11	9	250	17	15	12	10
20 16 14	10			175	12	11	9	7	200	15	13	11	9	250	17	15	12	9	275	19	17	14	11
24 20 16	12 10			225	13	12	10	8	275	16	14	12	10	300	19	16	13	10	350	21	18	15	12
30 24 20	14 12			300	14	13	11	9	350	18	15	13	10	425	20	17	15	11	475	24	20	17	13
38 30 24	18 14 12			350	17	14	12	9	425	20	17	13	11	500	22	20	17	12	550	26	22	18	14
40 32 26	20 16 14			400	18	16	13	10	475	21	18	15	12	550	24	21	18	13	650	28	24	20	15
44 36 30	24 18 16	14		450	19	17	13	11	550	22	19	17	13	625	26	22	19	14	725	30	25	22	17
44 36	26 22 18	16		500	20	17	14	12	600	24	20	18	13	700	27	24	20	15	800	32	27	23	18
48 40	30 24 20	18 16		600	21	18	15	12	700	25	22	18	14	850	29	25	22	16	950	34	29	25	19
48	38 30 24	22 18		700	23	20	17	13	850	28	24	20	15	975	33	28	24	18	1125	37	31	27	21
	40 32 28	24 20 18		800	25	21	18	14	950	31	26	22	17	1125	35	30	26	20	1275	39	34	29	22
	42 36 30	26 24 22	20	900	27	22	19	15	1100	32	27	23	18	1250	36	31	27	21	1450	42	36	31	23
	46 42 36	30 26 24	22	1000	29	25	21	16	1200	34	29	25	19	1400	40	34	29	22	1600	45	39	34	25
	44	38 34 30	28 24	1250	32	27	23	18	1500	36	30	27	23	1750	43	37	31	24	2000	48	43	36	27
		48 42 36	34 28	1500	34	29	25	19	1800	39	32	29	25	2100	45	40	34	26	2400	52	46	39	30
		44 40	36 30	1750	37	32	27	21	2100	42	35	32	27	2450	48	44	37	28	2800	55	49	42	32
		48	42 36	2000	39	34	29	22	2400	45	38	34	29	2800	52	46	40	30	3200	58	52	44	34
			48 40	2250	41	36	30	24	2700	48	40	36	31	3150	54	48	42	32	3600	61	54	46	37
			42	2500	43	38	32	25	3000	51	42	38	33	3500	57	51	44	33	4000	64	57	49	39
			48	3000	46	40	34	27	3600	54	45	40	35	4200	60	53	47	35	4800	68	60	52	42

≤NC30 NC35 NC40 ≥NC45

SYMBOLS V_T Terminal Velocity in FPM
V_R Room Velocity in FPM
V_K Outlet Velocity in FPM
A_K Outlet Area in Sq. Ft.
P_T Total Pressure in. H_2O
P_S Static Pressure in. H_2O
NC re 18db room attenuation
T Throw in Feet

**Courtesy of Tuttle & Bailey, Division of Allied Thermal Corporation. See manufacturer's catalog for complete data and description.*

Table 13-5 (Cont.)

LISTED HEIGHT				Outlet Velocity 900-1200 FPM					Outlet Velocity 1100-1450 FPM					Outlet Velocity 1250-1700 FPM					Outlet Velocity 1450-1900 FPM				
					Deflection					Deflection					Deflection					Deflection			
					0°	20°	40°	55°		0°	20°	40°	55°		0°	20°	40°	55°		0°	20°	40°	55°
					V_K					V_K					V_K					V_K			
				CFM	900	1000	1100	1200	CFM	1100	1200	1300	1450	CFM	1250	1400	1550	1700	CFM	1450	1600	1750	1900
					P_T					P_T					P_T					P_T			
					.05	.06	.08	.09		.08	.09	.10	.13		.09	.12	.15	.18		.13	.16	.19	.23
4 5 6	8 10 12	14 16 18	20 24		Throw (T) V_T 75					Throw (T) V_T 75					Throw (T) V_T 75					Throw (T) V_T 75			
6 5				80	13	11	9	7	100	15	13	11	8	125	19	16	13	10	150	22	19	15	12
8 6				100	15	13	11	8	125	18	15	13	10	150	21	18	14	11	175	25	21	17	13
10 8 6				150	17	14	12	9	175	20	17	14	11	200	24	20	17	13	225	27	23	19	15
12 10 8				200	18	15	13	10	250	22	19	15	12	275	25	21	18	14	325	29	25	21	17
14 12 10	8			250	21	18	15	11	300	25	21	18	14	350	28	25	20	15	400	33	28	24	18
18 14 12				300	24	20	16	13	350	27	24	20	15	425	31	27	22	17	475	36	32	26	20
20 16 14	10			350	25	21	18	14	425	30	25	21	17	500	34	30	25	19	550	39	32	27	21
24 20 16	12 10			450	27	23	19	15	550	32	28	24	18	625	37	32	27	22	725	42	36	31	24
30 24 20	14 12			600	29	25	21	16	725	35	30	26	20	850	41	35	30	23	950	45	40	34	26
38 30 24	18 14 12			700	32	28	24	18	850	39	33	28	21	975	45	39	33	25	1125	49	43	38	28
40 32 26	20 16 14			800	34	30	25	19	950	41	35	30	22	1125	48	41	35	27	1275	52	46	40	31
44 36 30	24 18 16	14		900	36	31	26	20	1075	43	37	32	24	1250	51	43	37	28	1450	55	49	42	32
44 36	26 22 18	16		1000	39	33	28	22	1200	45	40	34	26	1400	53	45	39	30	1600	58	51	44	34
48 40	30 24 20	18 16		1200	42	36	31	24	1450	48	42	36	28	1675	57	48	42	32	1925	61	54	46	36
48	38 30 24	22 18		1400	45	39	34	26	1675	52	45	40	31	1950	60	51	45	35	2250		58	49	40
	40 32 28	24 20 18		1600	47	43	37	28	1925	56	48	43	34	2250		54	47	39	2500		62	53	43
	42 36 30	26 24 22	20	1800	49	44	40	31	2150	59	51	46	35	2500		57	50	41	2750			56	46
	46 42 36	30 26 24	22	2000	52	47	42	32	2400		54	49	38	2800		61	53	43	3200			59	48
	44	38 34 30	28 24	2500	56	50	45	34	3000		59	53	41	3500			58	47	4000				54
		48 42 36	34 28	3000	60	54	48	36	3600			58	44	4200				52	4800				59
		44 40	36 30	3500		59	53	40	4200				49	4900				56					
		48	42 36	4000			57	43	4800				54	5600				60					
			48 40	4500			60	45	5400				58										
			42	5000				49															
			48	6000				54															

LISTED WIDTH

The indicated air paths may be easily obtained by using the appropriate notch on the bar deflecting key. The bars should be adjusted as shown for the required pattern.

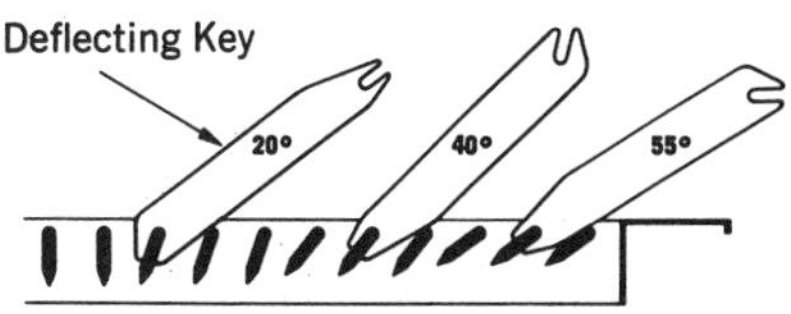

resultant V_R. Room air motion values of sidewall grilles (A50-60, see Fig. 13-22) in performance data ratings range from 35 to 40 fpm at corresponding terminal velocities of 75 to 100 fpm.

The specific values of V_T and V_R of each grille type are stated on the respective engineering performance data page (see Table 13-5). Performance data is based on an 18°F and 25°F cooled air supply temperature differential which, for the rated condition, results in a maximum ± 1F variation in the occupied zone.

Throw values in feet for terminal velocities (V_T) other than the base of V_T = 75 fpm performance table rating can be obtained from Table 13-6.

Sidewall supply grilles have horizontal throw *(T)* ratings based on a grille mounting

Table 13-6

Throw Factors*[a]
(V_T = 75 fpm Base)
Multiply Throw (*T*) by Factor

	V_T 50	V_T 75	V_T 100	V_T 150
Factor	1.5	1.0	0.75	0.50

[a]Throw factors applies to data of Table 13-5.

**Courtesy of Tuttle & Bailey, Division of Allied Thermal Corporation.*

height of 8 to 10 ft. For a given listed throw, the room air motion V_R will increase or decrease inversely with the mounting height; or for given conditions of size, capacity, deflection, and V_R, the listed throw can be decreased 1 ft for each 1-ft increase in mounting height above 10 ft.

The rated throw value is based on the full distance the airstream is to travel to a wall or opposing airstream. When grilles are located on opposite walls discharging toward each other, they should be sized for equal capacities and to throw one-half the distance between them. Selecting for throws up to one-third greater than these dimensions will provide slightly increased V_R values up to 50 fpm for noncritical occupied areas.

Sill supply grilles find use in perimeter application involving moderate-to-large glass areas. The standard single- or double-deflection grille will perform satisfactorily in these applications. Grilles should generally be selected for outlet velocities (V_K) in the range of 700 to 1200 fpm and for throws (*T*) equal to the vertical distance to the ceiling plus the required horizontal projection (normally 10 to 15 ft). The rated throw value can be decreased 1 ft for each 1 ft increase in ceiling height above 10 ft.

Ceiling supply grilles have horizontal throw (*T*) ratings based on a ceiling height of 8 to 10 ft. For a given listed throw, the room air motion (V_R) will increase or decrease inversely with the ceiling height; or for given conditions of size, capacity, pattern, and V_R, the listed throw can be decreased 1 ft for each 1 ft increase in mounting height above 10 ft.

The minimum throw results in room air motion higher than that for the maximum throw, generally 65 fpm versus 35 fpm. The minimum throw is one-half the recommended minimum distance between ceiling grilles discharging toward each other, and the maximum throw is one-half the maximum recommended distance between grilles. The minimum-maximum throw is also the recommended distance to a wall or other air path obstruction.

13-15 SPREAD REQUIREMENT

The spread of an unrestricted airstream is determined by the grille bar deflection and is expressed in feet at a given throw dimension (*T*) from the face of the grille. Air discharged from a grille with vertical face bars at zero-degree deflection will spread about 7 degrees per side in both horizontal and vertical planes or about 1.0 ft in every 8.0 ft of throw due only to the natural expansion of the free airstream.

As the deflection setting of the vertical bars is increased, the air stream covers a wider area and the throw (*T*) decreases. The shorter throw results from the greater stream surface that entrains more room air, dissipating stream energy and equalizing temperatures more rapidly. For the same size, capacity, V_T, and V_R, a 55-degree deflection setting will throw about one-half the distance of a zero-degree deflection setting. The actual spread for several standard deflections is shown in Fig. 13-29.

The distance between the centers of individual grilles mounted in the same wall (see Fig. 13-30) should not be less than the indicated spread of a single grille at a design throw (*T*) dimension. The distance from the center of an individual grille to an adjoining side wall

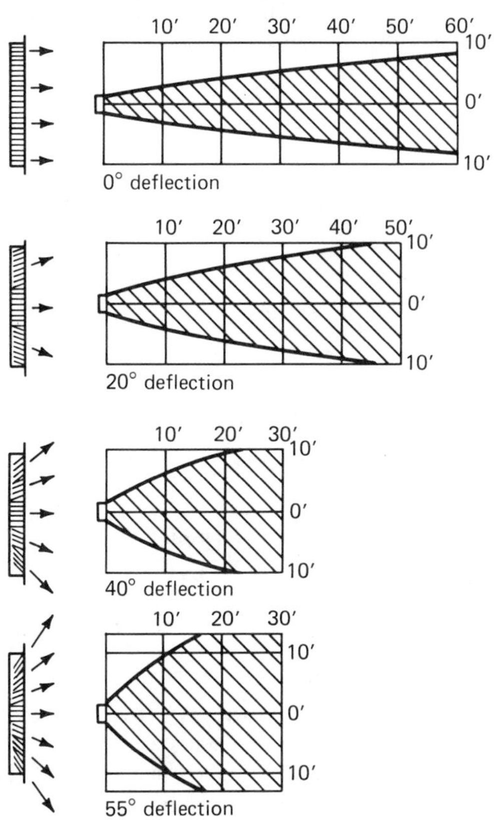

Figure 13–29

Spread related to throw and deflection setting. Required settings are made after installation using a key which gauges the bars as they are deflected (see Table 13–6)

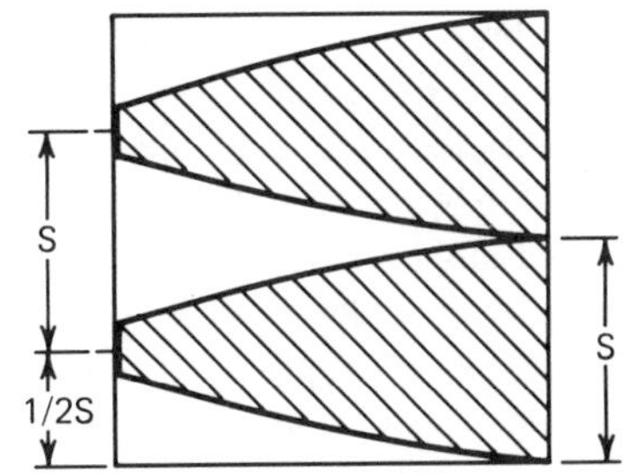

Figure 13–30

Location of sidewall outlets considering spread (S) and throw (T)

should not be less than one-half the indicated spread at the required throw.

The vertical face bar grille is generally recommended for applications where spread deflections are required. When vertical rear bars are used, a 40-degree deflection setting is the maximum available because of the confining action of the side margins. When very narrow grilles are used, this may be reduced to 20 degrees maximum.

Airstream adjustment in the vertical plane (trajectory) can be obtained with horizontal face or rear bars to avoid direct impingement with beams, pendant lighting fixtures, or other obstructions.

13-16 DROP REQUIREMENT

If the air supplied from a sidewall grille is cooled below room temperature, the airstream will tend to drop into the occupied zone due to the density differential between primary and secondary air. When supplying heated air at a temperature above that of the room air, the drop due to temperature is a negative one, as the primary warm airstream will rise, again due to density differences. Therefore, both of these conditions must be evaluated to correctly design a grille installation (see Fig. 13-12).

If there is no significant temperature differential between the primary and secondary air (ventilating), the airstream still exhibits a minor drop due to the vertical expansion of the stream as it entrains room air. This drop is approximately 1.0 ft for every 8.0 ft of throw.

The drop of an airstream is normally computed only for low induction devices such as grilles, registers, rectangular diffusers, and slots. The circular air pattern with which most ceiling outlets discharge air provide a high induction rate between primary and secondary air. This high rate of induction eliminates the temperature differential before the airstream can fall into the occupied zone.

The drop varies directly with the temperature differential (ΔT) and the throw (T), and inversely with outlet velocity (V_K). It can be calculated by an empirical formula substantiated by the tests of many investigators.

For drop due to temperature differentials, we have

$$D_t = \frac{(N)(t_r - t_a)(T)^{1.2}}{V_k} \qquad \textit{(13-19)}$$

where D_t = drop in feet due to temperature differential

N = a constant, numerically equal to 5

t_r = temperature of room air (°F)

t_a = temperature of supply air (°F)

T = throw in feet

V_K = grille jet velocity measured in vena contractor

For drop due to spread angle, we have

D_s = Tangent of spread angle times throw

or

$$D_s = \tan 7° \times T \qquad \textit{(13-20)}$$

where D_s = drop in feet due to spread angle

T = throw in feet

The total drop is additive, therefore

$$\text{Total drop } D = D_t + D_s \quad \text{(for cooling)} \qquad \textit{(13-21)}$$

$$\text{Total drop } D = D_s - D_t \quad \text{(for heating)} \qquad \textit{(13-22)}$$

Illustrative Problem 13-6

Assume that we wish to estimate the total drop for a 36- by 8-in. grille with straight-vane deflection. The volume flow rate is 1000 cfm, cooling differential $t_r - t_a$ = 15 degrees, and throw is 49 ft.

Solution:

$$V_k \text{ (jet velocity)} = \frac{\text{cfm}}{A_e}$$

$$A_e = \text{(Core area)(Effective area factor)}$$

$$= \frac{(35.5)(7.5)(0.85)}{144}$$

$$A_e = 1.57 \text{ ft}^2$$

$$V_k = \frac{1000}{1.57} = 637 \text{ fpm}$$

$$\text{By Eq. (13-19) } D_t = \frac{5(15)(49)^{1.2}}{637} = 12.6 \text{ ft}$$

$$\text{By Eq. (13-20) } D_s = \tan 7° \times 49 = 6.0 \text{ ft}$$

$$\text{By Eq. (13-21) } D = D_t + D_s = 12.6 + 6.0 = 18.6 \text{ ft} \quad \text{(for cooling)}$$

The fact that air supplied at the top of the room will fall by density differential towards the occupied zone of the room is advantageous in aiding general room air distribution in one respect, but extremely disadvantageous if the airstream falls into the occupied zone.

Table 13-7 shows tabulated values of total drop for a cooled airstream (including the vertical expansion) for two typical conditions of cooling temperature differentials and a range of throws and outlet velocities. The table values correspond reasonably well with calculation from Eqs. (13-19) through (13-21).

Figure 13-31a illustrates the calculated drop at certain conditions. It is obvious that unsatisfactory conditions will result in the occupied zone (below the 6-ft level). The minimum separation between mounting height and ceiling height must equal 60 percent of the total air drop (minimum separation of 3 ft). The minimum mounting height should be 7 ft. The rec-

Table 13-7
Total Air Drop from Sidewall Outlet (ft)[a]*

V_x in fpm	Sidewall Throw In Feet													
	10		15		20		25		30		40		50	
	—18F	—25F	—18F	—25F	—18F	—25F	—18F	—25F	—18F	—25F	—18F	—25F	—18F	—25F
500	3.5	4.0	5.5	6.0	7.5	8.5	9.0	10.0	10.5	13.5	15.5	18.0	18.5	23.0
750	2.5	3.5	4.0	5.5	6.0	6.5	7.0	8.0	8.5	10.5	11.5	14.5	15.0	18.5
1000	2.0	3.0	3.5	4.0	5.0	5.5	6.0	6.5	7.0	8.5	10.0	12.0	12.5	16.0
1250	2.0	2.5	3.0	3.5	4.5	5.0	5.5	6.0	6.5	7.5	9.0	11.0	11.5	13.5
1500	1.5	2.0	3.0	3.5	4.0	4.5	5.0	5.5	6.0	7.0	8.5	9.5	10.5	12.5
1750	1.0	2.0	2.5	3.0	3.5	4.0	4.5	5.0	5.5	6.5	8.0	9.0	10.0	11.5
2000	1.0	1.5	2.5	3.0	3.5	4.0	4.0	4.5	5.0	6.0	7.5	8.5	9.5	10.5

[a]Total Drop = drop due to temperature differential plus airstream expansion.

**Courtesy of Tuttle & Bailey, Division of Allied Thermal Corporation. See manufacturer's catalog for complete data.*

ommended mounting height is equal to the ceiling height minus the minimum separation distance.

There are two ways of locating the sidewall grille to reduce an excessive drop, which will create an objectionable draft in the occupied zone. One way is to locate a single-deflection grille high enough on the sidewall so that most of the velocity is lost, and the airstream, utilizing the ceiling effect, reaches the opposite wall before it falls into the occupied zone (see Fig. 13-31b). The more flexible way is to locate a double-deflection grille just above the 6-ft level and deflect the air upward 15 to 20 degrees toward the ceiling (see Fig. 13-31c). This is a somewhat longer induction travel and causes the air to be brought up to room temperature on the way up to the ceiling and also on the way down again. In effect, there is twice as much height above the occupied zone as actually exists. This allows an amount of drop equal to twice the distance between the ceiling and the grille to be absorbed plus the distance from the grille to the 6-ft level of the occupied zone.

When the air is arched in this manner, there is no appreciable effect on the throw of the airstream. Generally, an upward setting of 15 to 20 degrees of the deflector bars will suffice. Table 13-8 shows values of drop correction to

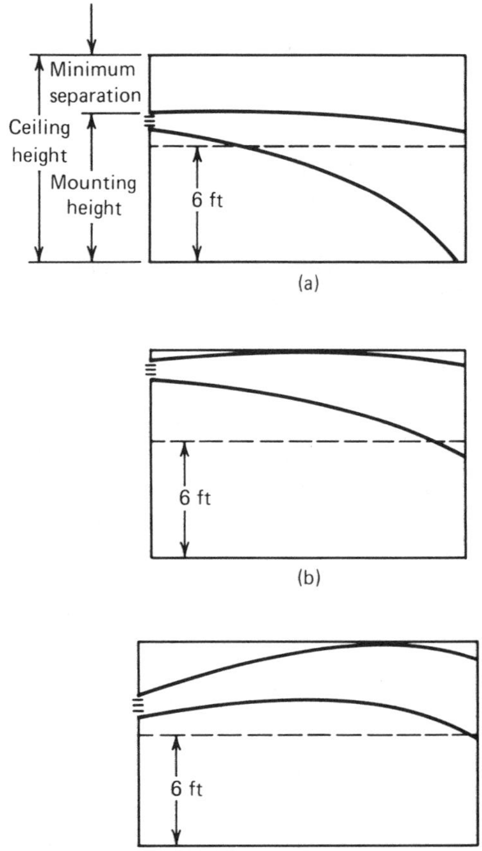

Figure 13–31

Air patterns for sidewall grilles showing effect of air stream drop

Table 13-8

Drop Correction with Deflected[a] Grille Bars*

	Throw in Feet						
	10	15	20	25	30	40	50
Drop reduction in feet, apply to Table 13-7.	2.5	3.5	4.5	6.0	7.0	9.0	11.5

[a]15 to 20 degrees upward deflection.

**Courtesy of Tuttle & Bailey, Division of Clevepak Corporation.*

be applied to the values of drop from Table 13-7.

13-17 SOUND REQUIREMENT

Acoustical control is as important as temperature and humidity control. To cope with noise problems, it is necessary to have a basic understanding of sound, for noise is simply unwanted sound.

Sound may be defined in two ways, physically and physiologically. Physical sound is a particular form of wave motion taking place in matter due to an original vibration or disturbance set up in the surrounding body. Physiologically, or psychologically, it is a sensory interpretation in the brain, conveyed from the ear, and differing to a great extent with each individual. It is in the attempt to rationalize phsyical sound with the psychological aspect that the complicated phases of sound study arise.

Sound is invariably produced by a compressive and expansive motion set up in a gaseous, liquid, or solid body. If the motion is sharp and discontinuous, such as a pistol shot, or if it is continuous, as when the fingernail is rubbed across a rough surface, the sound may be harsh and unmusical. On the other hand, if the vibrations are more or less defined and regularly spaced and have one or more definite frequencies, the sound presented may be musical in nature, such as the tones of a violin or a piano. In the language of the layman, the former represents noise to the human ear, whereas the latter represents a musical tone, although in the technical sense both represent sound.

Waveforms where the motion of particles in a gaseous medium are parallel to the direction in which the wave travels are considered to be longitudinal waves. Transverse waves, which occur perpendicular to longitudinal waves, cannot exist in a gaseous atmosphere, as the molecule structure is widely spaced and individual particles cannot transmit the force magnitude required to generate such motion. As the molecule structure vibrates with simple harmonic motion, the particles vibrate about their equilibrium point, creating a configuration or arrangement of particles moving in a given direction. A series of compression and rarefaction areas occur which propagate this type of longitudinal or compressional wave motion. Sound is created by a vibrating body and is transferred through a medium having mass and elasticity where compressional waves can be generated. Sound cannot be distributed in a vacuum.

When discussing sound, one must consider magnitude and frequency, measurement of sound magnitude and frequency, and human reaction to sound of varying frequency and magnitude.

Sound, for practical purposes, is any vibration that is audible to the human ear. As noted, sound waves may be transmitted through solids, liquids, or gases. Any pure sound is completely described by two terms, frequency and magnitude. The frequency of a sound corresponds to the pitch of the sound. The frequency of a sound is the same as the frequency of the vibrating source. The magnitude of a sound relates to the energy contained in the sound wave per unit area perpendicular to the direction of travel.

The human ear responds to sound due to the variation of pressure from the normal atmospheric pressure. This regular pattern of increasing and decreasing air pressure is the result of energy expended to the air. The amount of energy contained in a sound is the factor used in measuring the magnitude of the sound. The magnitude of a sound is expressed as the ratio of a measured sound level above an energy reference level.

The unit of sound measurement is the dimensionless decibel (db). The decibel expresses the ratio of magnitude of measured sound to magnitude of reference level sound. The sound level in decibels is a logarithmic function of that ratio:

$$\text{Level in db} = 10 \log_{10} \left(\frac{\text{measured quantity}}{\text{reference value}} \right) \tag{13-23}$$

In sound measurement the decibel is used to narrow the range of values of measured sound for ease of mathematical handling. The range of audible sound in directly measured units varies from 0.000 000 001 W for a whisper to 100,000 W for a jet plane. The corresponding units of decibels varies from 40 db for a whisper to 180 db for a jet plane.

In sound-level measurement there are two types of sound levels used. These types are (1) sound pressure level, L_p, and (2) sound power level, L_w. These two sound levels use the same unit of measurement, the decibel, measuring sound magnitude, but they are two different quantities.

To understand the definitions of sound pressure and sound power, first assume a given source of sound. Assume this source is a fan located in a room. If the magnitude of the sound produced by this fan is measured throughout the room, this magnitude will obviously vary with distance from the sound source and the acoustical properties of the room. This sound magnitude, which is affected by acoustical environment and distance from the source, is designated as sound pressure level, L_p. Sound pressure level is the sound level a person hears from a source of sound at the given point of reference. This sound level is the value of the ratio of the measured sound level to the reference sound level. This value is expressed in decibels by the following relationship:

$$L_p = 10 \log_{10} \left(\frac{p}{0.0002} \right)^2$$

$$L_p = 20 \log_{10} \left(\frac{p}{0.0002} \right) \tag{13-24}$$

where p is sound pressure in microbars. Energy reference for sound pressure level is 0.0002 microbar, which is the same as 0.0002 dyn per square centimeter. The sound pressure level is not used to rate the sound level produced by a given source because sound pressure varies with the distance from the source and the acoustical environment.

Sound power level, L_w, is used to rate the sound output of a sound source. The sound output or sound power level of a given sound source is a constant. Its sound pressure level at a given observation point may be varied by introducing varying acoustical arrangements. For example, for a given sound source, a bare hard-walled room will give a higher sound pressure level than a carpeted room with walls covered by drapes. However, the sound power level of this sound source is unchanged regardless of this change in room environment. A sound power level may be determined for any sound source such as a fan. Sound power levels must be measured under ideal acoustical conditions, and in this instance the measured sound pressure level equals the sound power level. This sound power rating may then be used to determine the resulting sound pressure level in a given acoustical environment and distance from the source.

When a manufacturer of air-conditioning equipment rates equipment noise, this rating must be in sound power, rather than sound pressure, to have specific meaning. Sound power level, L_w, in decibels is defined as

$$L_w = 10 \log_{10} \left(\frac{W}{10^{-12}} \right) \qquad (13\text{-}25)$$

where W is the measured sound power level in watts, and the reference energy level is 10^{-12} W.

In order to provide a comfortable acoustical atmosphere, the frequency as well as the magnitude of the sound must be given careful study. Sound frequency is directly involved in noise control for such reasons as (1) the human ear is more sensitive to high frequencies than low frequencies and (2) the design of material used to attenuate sound depends on the frequency of the sound.

The audible frequency range include frequencies from 20 to 10,000 cycles per second. For ease of mathematical handling, this vast frequency range has been subdivided into eight separate frequency ranges called octaves. The frequency of the upper limit of each octave band is twice that of the lowest frequency in each octave band. Frequency measurement and application of frequency criteria to noise level problems is done in octave bands. Since sounds of the same magnitude but of different frequencies will have different degrees of loudness to the human ear, a graph of sound level versus frequency is often used. This type graph is called a frequency spectrum.

Sound measurement. The reception of sound by the human ear varies with each individual. This fact increases the difficulty of establishing absolute values for sounds generated by air-conditioning equipment in order to predict what the environmental noise level of a given system design will be.

Modern sound measurement is obtained by means of a sound-level meter used with a microphone-amplifier circuit. A sound-level meter reading is the actual sound pressure level in decibels when the frequency network of the meter is set for "flat response", which is the "C" network of the meter. A flat response rating means that the meter reads the actual sound pressure level experienced by the meter and not a "loudness rating" simulating the reaction of the human ear, which emphasizes high-frequency sound.

When it is necessary to know the magnitude of sound at certain octaves, an octave band analyzer is used in conjunction with a sound-level meter. By using the octave band analyzer to electronically filter out all frequencies except those in a certain octave band, a sound pressure level of this certain octave band can be determined in decibels. This reading will be an octave band sound pressure level in decibels when the meter is set for flat response.

Sound-level meters may electronically deemphasize the low-frequency end of the frequency spectrum to approximate the "loudness-level" reaction of the human ear in decibels. This type of measurement is done on the A-scale of a standard meter and is therefore called the *A-scale noise level.*

The response of the human ear to a sound level can be approximated by direct measurement or by a combination of octave band sound pressure measurements and mathematical calculation of the loudness level. The A-scale noise level, explained above, is a direct measurement estimate of the human ear's response. Two other methods of loudness-level determination of human ear response uses the unit of sone and phon for total loudness and loudness level, respectively. These methods use a combination of sound pressure level measurement and an equation to calculate the loudness response. In addition to these methods, a widely used method of representing loudness criteria is the "noise criteria curve" (NC curve), which defines the maximum allowable sound pressure level for each frequency band.

However, for sound control problems where the noise level in certain octave bands is especially important, design based on octave band sound levels is necessary. Noise criteria (NC) curves provide this information. NC curves plot octave band frequencies versus sound level in decibels. NC curves are numbered and have increasing NC numbers as the allowable magnitude of the sound spectrum increases. Areas such as a concert hall have a low NC number and factories a relatively high NC number.

Absorptive characteristics of rooms. When a sound wave strikes a solid barrier, a part of its energy is reflected, part is absorbed, and a part is transmitted through the barrier. If the sound wave originates within the room, the portions absorbed and transmitted by the walls are not returned to the room so that the two may be taken together under the single heading of absorption. The fraction that is returned to the room is the absorption coefficient.

It is obvious that when a given sound pressure is released within a room, the construction of the room, the depth of its walls, the acoustic properties of the walls, ceiling and floor, and the furnishings and occupants within the space all affect the ratio of the reflective fraction to the absorption fraction. Absorption of sound concerns the dissipation in the form of heat of the vibrational energy of sound waves. Materials that are absorbant to sound waves to any degree are either porous, inelastically flexible or inelastically compressible, or they may possess two or more of these properties in varying degrees. Therefore, in any given room, the sound waves returned to the ear of an occupant from an air distribution device operating at a given outlet velocity will be different than if that same device operating at the same velocity were placed in a room whose absorption coefficient was different.

An accurate determination of the sound level in any given room would involve the calculation of the sound-absorbing rate of that room. For a complete air-conditioning installation, this would obviously be a long and tedious procedure. Inasmuch as the majority of air-conditioned spaces for human occupancy concern themselves with relatively similar building materials and construction procedures, it is much simpler to avoid the complexity of noise problems by selecting the air distribution devices to operate at velocities that are known to be quiet or at NC levels that are acceptable. Practical experience indicates that noise from fan apparatus, ducts, and accessories such as dampers is more likely to cause sound problems than from outlets conservatively selected. Table 13-9 gives recommended outlet velocities for various applications and applies to both ceiling diffusers and grilles.

In summary, the intensity of sound created by a grille or diffuser for a given condition of operation is not generally the sound that is heard by an observer in a typical building environment. The sound created by a grille is the total sound generated, independent of environment and is termed *sound power level* (L_w). The sound actually heard in a given environment is the power level value minus room at-

Table 13-9

Recommended Maximum Jet Velocities for Diffusers and Grilles

Application	Recommended Maximum Jet Velocities (fpm)
Broadcast studios	500
Residences	750
Apartments	750
Churches	750
Hotel bedrooms	750
Legitimate theaters	1000
Private offices, acoustically treated	1000
Motion picture theaters	1250
Private offices, not treated	1250
General offices	1500
Stores, upper floors	1500
Stores, main floors	1500
Industrial buildings	2000

tenuation and is termed *sound pressure level* (L_p). The amount of reduction from L_w to L_p is determined by the acoustic absorption of the environment. An 18-db room reduction is considered typical, and performance data ratings are values of sound pressure based on L_w − 18 db with the power level referenced to 10^{-13} W. A change in the 18-db room factor will inversely change the apparent NC rating by the same amount. Room factors commonly range from 15 to 22 db.

A given power level contains various intensity levels in prescribed frequencies ranging from 75 to 10,000 cps; this range is commonly divided in seven octave bands for convenience. The power level intensity within frequency-limited octave bands is numerically reduced by 18 db and plotted on NC charts. The point on the plot tangent to the highest numerical NC curve designation is the NC rating for a specific condition of operation (see Fig. 13-32).

Performance ratings of grilles in NC values are levels that will not be exceeded for a given condition of operation (not including system noise) in an environment having an 18-db absorption. Noise created by or transmitted through the duct system to the grille can add to the total noise being introduced or heard within the room, falsely indicating an increased grille sound level. Removal of the grille quickly establishes whether the offending noise is duct system noise, which will persist, or grille generated sound, which will cease. The noise in the duct system at the grille collar should be at least 5 db lower than the grille NC rating.

The effect of combining two sound levels. The decibel scale is a logarithmic function and decibel values, of course, cannot be added directly. The values to be added must be reverted back to the actual intensities, added, and then converted back to the decibel values. For example, in a space with a background sound level of 42 db (perhaps a private office), the use of a diffuser (grille) selected at 40 db will produce a combined noise level of 44 db (see Table 13-10).

Table 13-10

Effect of Combining Two Sound Levels

Difference between two sound levels, db	0	2	5	8
db addition to higher level to obtain combined level	3	2	1	0

Grille sound ratings will be increased by integral or localized air volume dampering, by nonuniform air flow in the grille collar, and by addition of inherent system noise. Table 13-11 indicates register throttling sound correction.

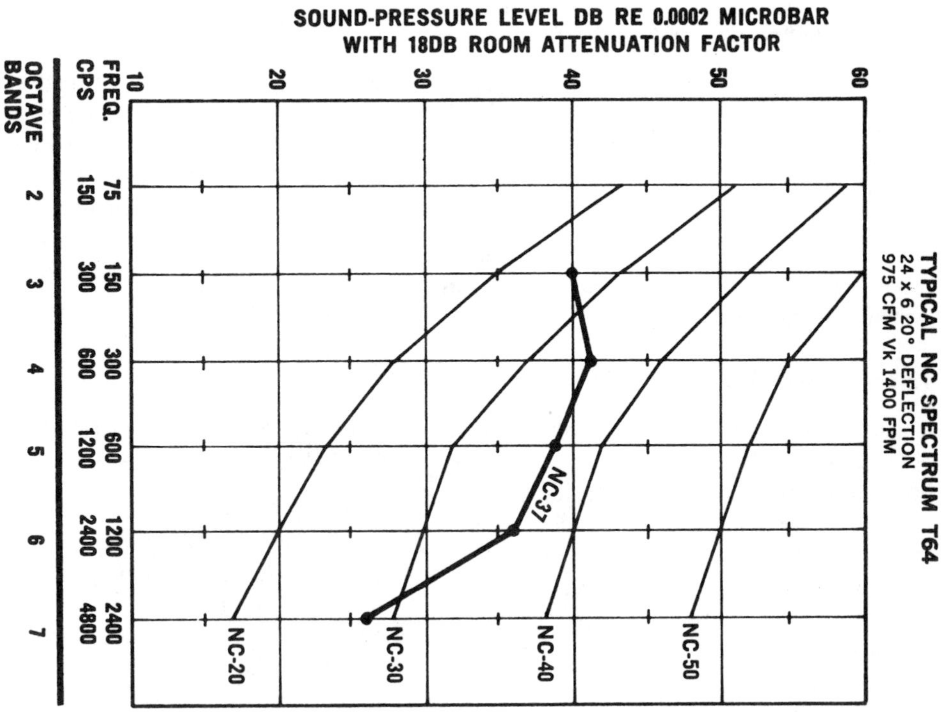

Figure 13–32

Typical NC Spectrum T64. 24 in. by 6 in. -20 degree deflection, 975 CFM, V_K 1400 FPM (Courtesy of Tuttle & Bailey, Division of Allied Thermal Corporation.)

Table 13-11

Register Throttling Sound Correction

	Pressure Drop in WG		
Damper Throttling Effect	**0.05**	**0.15**	**0.25**
Approximate damper opening	¾	⅔	½
NC addition to single outlet sound rating	5	10	15

The grille NC rating will also be increased if two or more outlets are located in an area of 200 ft^2 or less (see Table 13-12).

Grilles should be selected for the recommended NC rating for a specific application (see Table 13-13), or for recommended velocities (see Table 13-9) dependent on a particular manufacturer's method of rating. Grilles may also be selected for sound levels higher than low ambient background levels to mask extraneous and occupant-disturbing noises. A grille

Table 13-12

Multiple-Outlet Sound Correction

Number of outlets in 200 ft² area	2	3	4 to 6
NC addition to single-outlet sound to obtain total sound	3	5	6

with a low sound rating is not necessarily a properly selected grille for a good "sound-conditioned" application.

13-18 PRESSURE REQUIREMENT

The *grille minimum pressure* for a given air volume reflects itself in ultimate system fan horsepower requirements. A grille with a lower pressure rating requires less total energy than a grille with higher pressure rating for a given air volume and effective area. Grilles of a given size having lower pressure ratings usually have a low sound-level rating at a specified air volume.

Velocity pressure (p_v), discussed in Section 13-3, exerts a force only in the direction of air motion. The only energy available for diverting the air at right angles to the direction of motion into the connecting duct and through the grille is static pressure (p_s) in the main duct, which exerts a force equally in all directions. The static pressure in the main duct must be generally equal to or greater than the total pressure (p_t) required in the connecting duct $(p_t = p_v + p_s)$.

The engineering performance data (typical of Table 13-5) provide both total pressure and static pressure required in the collar behind the grille to effect rated performance. The additional static pressure loss (turning loss) encountered in transferring flow from the duct into the grille collar is approximately 20 percent of the rated total pressure. This "turning loss" must be *added* to the *grille-only total pressure* (p_t) in order to establish the total static pressure required in the branch duct connected to the collar.

The static pressure loss encountered in the duct system (see Chapter 14) between the outlet closest to the fan and the outlet on the longest equivalent duct run is important. The last outlet on the duct run must have the minimum pressures as determined above. Of necessity then, the outlet closest to the fan will have available a pressure equal to the sum of (1) minimum total static pressure at last outlet and (2) duct friction loss between last grille and first grille.

13-19 GENERAL APPLICATION DATA (CEILING DIFFUSERS)

The application of ceiling diffusers is a matter of judgment and careful analysis of job condition. As diffusers project air equally in all directions, they best serve a square area. Therefore, they should be located as nearly as possible in the center of equal squares within the room. Divide the ceiling area as nearly as possible into squares, the sides of which should not exceed three times the ceiling height.

In many cases, because of structural conditions or other reasons, it is impossible to locate the diffuser in the exact center of equal squares. When the ceiling area cannot be divided into squares, divide it into rectangles; the long side of such a rectangle should not exceed 1 1/2 times the short side.

The ceiling diffusers are then located in the center of these squares or rectangular areas. When this is not possible, the distance from the diffuser center to one side of the square or rectangle should not exceed 1½ times the distance to the other side.

Selecting the throw. Throw is the linear distance from the center of the diffuser to the point at which a desired terminal velocity (V_T) is reached. A terminal velocity of 100 fpm has

Table 13-13

Recommended NC Criteria*

NC Curve	Communication Environment	Typical Occupancy
Below NC 25	Extremely quiet environment, suppressed speech is quite audible, suitable for acute pickup of all sounds.	Broadcasting studios, concert halls, music rooms.
NC 30	Very quiet office, suitable for large conferences; telephone use satisfactory.	Residences, theaters, libraries, executive offices, directors' rooms.
NC 35	Quiet office; satisfactory for conference at a 15 ft. table; normal voice 10-30 ft.; telephone use satisfactory.	Private offices, schools, hotel rooms, courtrooms, churches, hospital rooms.
NC 40	Satisfactory for conferences at a 6-8 ft. table; normal voice 6-12 ft.; telephone use satisfactory.	General offices, labs, dining rooms.
NC 45	Satisfactory for conferences at a 4-5 ft. table; normal voice 3-6 ft.; raised voice 6-12 ft.; telephone use occasionally difficult.	Retail stores, cafeterias, lobby areas, large drafting & engineering offices, reception areas.
Above NC 50	Unsatisfactory for conferences of more than two or three persons; normal voice 1-2 ft.; raised voice 3-6 ft.; telephone use slightly difficult.	IBM rooms, stenographic pools, print machine rooms, process areas.

**Reprinted courtesy of Tuttle & Bailey, Division of Clevepak Corporation.*

been generally accepted as a conservative standard for ceiling diffuser installation. However, it is obvious that the nature of the occupancy in the space to be conditioned should determine the terminal velocity and may allow the use of considerably higher terminal velocities, such as 150 or 200 fpm (see Fig. 13-15), with associated changes in room velocity (V_R). Satisfactory results will be obtained because of the rapid rate of temperature and energy equalization provided by the circular air pattern issuing from the diffuser. This performance differs from that of a sidewall grille or linear diffuser, where the primary airstream, because of its mass and slower induction rate, may still be several degrees below room temperature at the end of its throw (cooling).

The following table shows some suggested maximum terminal velocities varying with length and time of occupancy for installation using diffusers providing a circular air pattern.

It is apparent that, all other factors being equal, the use of different terminal velocities will affect the throw of a given size diffuser

V_T = 100 fpm V_R = 35 fpm	Recommended where people are located adjacent to walls of structures for extended periods of time at sedentary occupations; private offices, apartments, residences, hotel bedrooms, hospital (private rooms, and wards).
V_T = 150 fpm V_R = 50 fpm	Recommended where people are not located adjacent to walls of structure for extended periods of time: general offices, restaurants, theaters, department stores, clothing stores, churches, operating rooms.
V_T = 200 fpm V_R = 65 fpm	Recommended where people are not located adjacent to walls of structures at any time; industrial plants, corridors, process areas.

and also the room velocity (V_R) in the occupied space. For example, the manufacturer's selection data may indicate the use of a diffuser that provides a 9-ft throw at 100 fpm terminal velocity when the measured throw is 12 ft. In this case the remaining 3 ft will still be properly conditioned, although the terminal velocity at the 12-ft point will be lower than that originally chosen for the installation. The room velocity will also be somewhat reduced, but not to a critical degree. In cases where the measured throw is somewhat less than that provided by the selected diffuser, satisfactory results will still be obtained, even though the terminal and room velocities at the end of the measured throw will be somewhat higher than that originally selected.

Most ceiling diffusers are installed in rooms with ceiling heights of 8 to 12 ft. No correction need be made for this variation in selecting the diffuser. However, when ceiling heights are above 12 ft, a corrected procedure should be used, as follows:

1. Measure the distance from the diffuser to the nearest wall or opposing airstream.
2. Measure the distance from the ceiling to a point 12 ft from the floor. Add to the plan throw (step 1) 75 percent of the difference between the actual ceiling and the 12-ft height.
3. Use this value as the actual throw and select the diffuser for the desired terminal velocity.

For installations with excessive ceiling heights, it is better to increase the cfm per diffuser and use fewer diffusers. For ceiling heights below 9 ft, it is better to reduce the cfm per diffuser to a minimum and use more diffusers.

Collars should always be used to connect the branch duct to the diffuser (see Fig. 13-33). A duct ring should be used to provide means for leveling the diffuser against the ceiling and to allow for variations in collar length due to construction.

Adjustable turning grids should be used at the point where the branch duct connects to the collar in order to correct improper conditions of approach; or in the case of short collars, a grid should be located between the damper and the diffuser. On extremely short collars two grids placed with blades at right angles to each other should be used. Functionally, a diffuser is designed and rated to distribute air with the supposition that the air supplied to it has a total pressure and a direction within the limits that will enable it to completely perform its function. No diffuser can be expected to correct unreasonable conditions of flow in the air supplied to it.

Dampers should be installed in the collar of each diffuser in order to maintain the required total pressure back of each diffuser and to balance unequal losses due to duct friction. Figure 13-33 shows an ideal diffuser assembly. Note cushion head, branch take-off, connecting collar, grid, damper, and diffuser. Omission of any of these component parts will result in less than ideal air distribution.

Since diffusers mounted on a ceiling produce a relatively thin, flat airstream moving parallel with the ceiling, they will produce a direct impingement on suspended light fixtures, exposed beams, and so forth. The airstreams will be deflected down into the occupied zone, creating objectionable drafts. Manufacturers' catalogs should be studied and manufacturers should be consulted when these problems exist.

The location of lighting fixtures and other ceiling obstructions in relation to high sidewall grilles should also be checked. This is often a more difficult problem than with ceiling diffusers since a greater mass of air is likely to strike

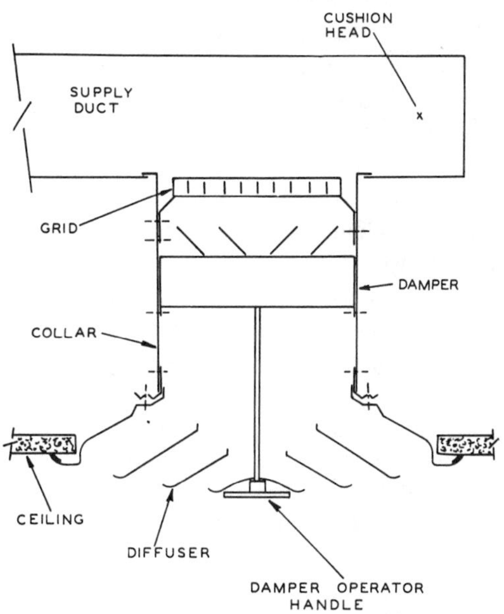

Figure 13–33

Ideal diffuser assembly including grid and damper

a given obstruction. With the lower induction characteristics of grilles, the airstream may lack several degrees of being up to room temperature at the point of impingement and may fall rapidly into the occupied zone after striking the obstruction.

13-20 GENERAL APPLICATION DATA (GRILLES AND REGISTERS)

As with ceiling diffusers, the selection and application of grilles and registers is a matter of judgment and analysis. There are a few rules which should be borne in mind when applying grilles and registers. They are:

1. The throw *(T)* from a straight-flow grilles varies with the square root of the core area of the grille and with the face velocity.
2. The aspect ratio (the ratio of the height to the width) has no appreciable effect on the distance of the throw from grilles whose aspect ratio is less than 25 to 1.
3. If the airstreams from a grille are converged, it results only in cutting down the effective area of the grille.
4. Breaking the airstream into jets has no effect on either the rate of the mixing or the throw.
5. Deflecting the airstream by turning the vanes outward to increase the spread shortens the throw, depending on the amount of deflection (see Fig. 13-29).
6. The drop for a given throw varies inversely as the face velocity for an airstream below room temperature and varies directly as the temperature differential.

Unlike ceiling diffusers, grilles have a comparatively low induction rate. Moreover, the airstream moving in just one direction induces room air into the primary airstream in such a manner that this room air must be replaced by noninduced room air. This causes a counterflow room air movement about one-half the terminal velocity. Since ideal room air motion is limited to 50 fpm or less, the terminal velocity should not exceed 100 fpm (see Tables 13-5, 13-6, and refer to Section 13-14).

Collars should be used to connect the branch duct and the grilles or registers. The collar should not be less than 12 in. in length. A cushion head should be used at the end of a branch run to air distribution in the last grille.

Dampers should be used to provide the correct total pressure behind the grille so that the required volume is discharged. Santrols (Tuttle & Bailey), or other air turning devices with adjustable blades, should be used to control volume and correct imperfections in the condition of approach of air to the grille. Figure 13-34 shows an ideal register assembly. Note collar, santrol, grille face, and damper.

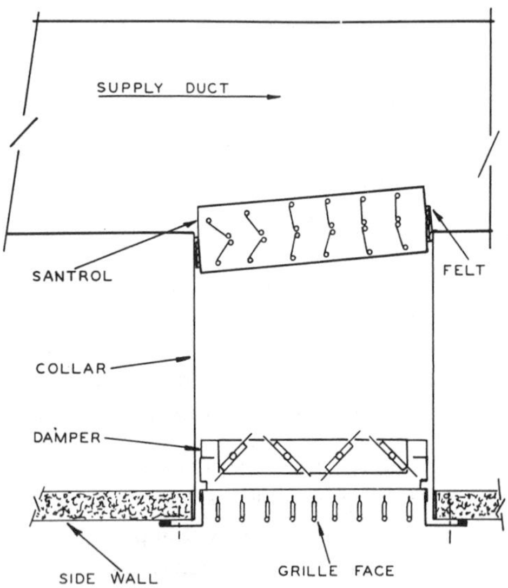

Figure 13–34

Ideal register assembly consisting of grille, damper, and santrol (Courtesy of Tuttle & Bailey, Division of Allied Thermal Corporation.)

In the location of grilles within an enclosure, it is well to visualize the actual air pattern that various deflection settings will produce. Refer to Section 13-15 and Figs. 13-29 and 13-30 for proper spacing of sidewall grilles. If the airstreams from adjacent grilles strike too soon, undesirable turbulence in the occupied zone may occur.

Grilles versus ceiling outlets. The question often arises in the designer's mind, when considering cooling application, as to when to use ceiling outlets and when to use sidewall grilles. Frequently, for architectural reasons, one is preferred over the other. However, there is no clear line of demarcation, but it may be said that for distributing large quantities of air (over 1 1/2 cfm per square foot of floor space) into rooms with low ceilings or short throws, ceiling diffusers work somewhat better than sidewall grilles. Other than these points, properly designed installations using either method will produce satisfactory results.

For heating applications floor registers located at the base of cold walls and below windows are normally considered best. The air should be blown upward with sufficient velocity to obtain distribution across the ceiling and downward along the opposite wall. This is particularly true if the floor registers are used for cooling as well as heating to prevent stratification of air within the conditioned space. Double-deflection, high sidewall grilles may also be satisfactory in relatively mild climates. However, for more severe winter conditions auxiliary radiation may be required, located under windows and along cold walls. Ceiling diffusers are seldom used in heating applications.

13-21 GENERAL APPLICATIONS (RETURN AIR GRILLES)

Let us now consider some design criteria for return air inlets. Return air grilles may be connected to a return air duct or they may be simple vents which transfer air from one space to another.

Return air grilles are usually selected for the required air volume at a given sound level or pressure value. The intake air velocity at the face of the grille depends mainly on the grille size and air volume. The grille style and damper setting have a small effect on this intake velocity. The grille style, however, has a very great effect on the pressure drop. And this, in turn, directly influences the sound effect.

The intake velocity is evident only in the immediate vicinity of the return grille and cannot influence room air distribution. Recent ASHRAE research projects have developed a scientific computerized method of relating intake grille velocities, measured 1 in. out from the face of the grille, to air volume with a high degree of precision. Figure 13-35 illustrates a typical velocity fall-off chart. If the general room air velocity 2 ft from the face of the grille is 10 fpm, one-half that distance would indicate a velocity of only 40 fpm, and just 6 in. away the velocity is only 160 fpm, even though the velocity through the outlet is 500 fpm. Also, as long as the general drift of the room air toward the return air grille is not more than 50 fpm, then no objectionable drafts would arise.

Return air grille location. It still remains the function of the supply outlets to establish proper coverage, air motion, and thermal equilibrium. It is erroneous to feel that the supply air can be "pulled" from the supply opening to the return grille location. A return grille creates a low-pressure area and atmospheric pressure "pushes" air into the grille. Because of this, the location of return grilles is not critical, and their placement can be largely a matter of convenience in many cases. Specific locations in the ceiling may be desirable for local heat loads or smoke exhaust, or a location in the perimeter sill may be desirable for an exterior zone intake under a window wall section. Despite the localized effects of a return grille (see Fig. 13-35), it is not advisable to locate large centralized return grilles in an occupied area because the large mass of air moving toward the grille can cause objectionable air motion for nearby occupants.

A wall return near the floor is the best location for both heating and cooling applications. Stratification due to poor mixing in winter is counteracted by a low return since the cool air near the floor is withdrawn first and is replaced by the warmer upper air strata. Also, a return near the floor makes it possible for the supply air to release its full capacity without any possibility of short circuiting to the return outlet.

On a heating installation return and supply ceiling diffusers should never be used unless baseboard heating or strip radiation is used to raise the mean radiant temperature of the outside wall. In this way cold air falling down the wall will not pass across the feet and legs of the occupants when it is returning to the diffuser to be reinduced. Returns may then be installed as for a cooling installation.

In some cases, for economic reasons, such radiation must be omitted. Several return grilles must then be located at the base of the exposed wall to trap as much cold air as possible. This installation does not compare in quality of results, however, with an installation where radiation is placed along the outside walls.

Floor returns should be avoided whenever possible because they tend to be a catch-all for dirt. This imposes a severe handicap on central system filters and coils. When there are common returns in corridors, and so forth, return air grilles are frequently located in the bottom of doors, or many times the doors are undercut.

In the interest of symmetry it is often desirable to furnish return units that look the same as the supply units. Round or square ceiling

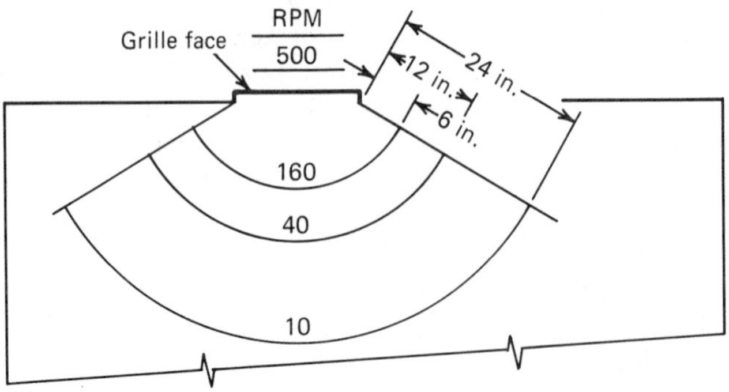

Figure 13–35

Velocity fall-off with distance from return grille

diffusers may be used for the purpose of matching returns. However, the coefficient of entry changes rather drastically, and the effective area is therefore reduced so that for a given volume and given size diffuser, the velocity is considerably higher when used as a return than it would be when used for supply. Therefore, additional units may be necessary to withdraw the required amount of air. A ceiling diffuser used as a return has an effective area about 65 percent of its supply effective area. Therefore, in order to avoid excessive pressure requirements, more diffusers may have to be used.

In summary, it should be remembered that even though many locations of return grilles are possible, wall returns located near the floor are usually the best. Also, objectionable drafts may result if return air velocities through the outlet become excessive.

Return air velocities. From the standpoint of results to be obtained in the conditioned space, return air velocities in air distribution devices may be considerably higher than present practice indicates. The following conditions should be observed. When return air grilles are located at or adjacent to the floor, they should be sized for low air volumes per grille, and the velocity should be between 400 and 600 fpm so that localized high-velocity areas in the occupied zone adjacent to the return grilles may be avoided. If returns are mounted high on the sidewall or in the ceiling, they may operate from 600 to 1200 fpm with no undue difficulty from localized high-velocity or objectionable noise generation. Since the grilles mounted in ceilings and high sidewalls are some distance from the occupied zone, they may successfully handle larger air quantities per unit than return grilles mounted in the occupied zone.

Return grille sound requirement. Return grilles should be selected for static pressures that will provide the required NC rating and conform to the return system performance characteristics. Since fan sound power is transmitted through the return air system as well as the supply system, fan silencing may be necessary or desirable in the return side particularly if silencing is being considered on the supply side.

When excessive pressure must be throttled through the grille damper, the resulting NC rating increases sharply. Large pressure drops should be handled with remote duct dampers with intervening sound attenuation treatment in the duct. Return grilles up to 3 ft^2 in nominal size can be close-coupled to PRV pressure reducing valves (normally used in high-pressure supply duct system) to produce an intake unit capable of providing low noise levels at high duct pressures without supplementary duct attenuation.

Transfer grilles venting into a ceiling plenum should be located remote from plenum noise sources. The use of a lined sheet metal elbow can reduce transmitted sound as much as 12 to 15 db in octave bands 5 to 7 and 3 to 5 db in bands 3 to 4. Lined elbows on vent grilles and lined common ducts on ducted return grilles can minimize "cross-talk" between private offices.

Return grille pressure requirement. The design of a return air system and the selection of return grilles require the same consideration as a supply system. The pressure drop across return grilles will be least at the end of a duct system and greatest near the riser and fan due to friction and turning losses as air flows toward the fan. It does no good to select a return grille size to operate at a low pressure drop and low sound level and install it in a system where a large negative static pressure exists. This will result in a large pressure drop across the grille causing excess return air flow and high sound levels. If registers are installed, the damper may be throttled to control the air flow, but the large pressure drop and accompanying high sound level still exist.

It is extremely important that the negative static pressure be controlled by designing for minimum friction and turning losses. This is done by use of appropriate friction factors, radius elbows or turning vanes in square elbows, static pressure regain, and the use of branch duct balancing dampers.

Tables 13-14 and 13-15 show performance data for return air grilles and registers of the type shown in Fig. 13-26. Performance of any size grille not shown in the tables will be the same as for the size shown having the same listed size area. For example, a 36 by 16 (not shown) has an area of 576 $in.^2$ Its performance would be the same as a 24 by 24 grille which is listed.

Table 13-14

Engineering Performance Data, Return Air Grilles and Registers*

(Fixed Bar Type, 0-degree Straight Bars)
(Series T70, NT70, T70G, T80, NT80)
(Series A70, AN70, AB70, A80, AN80, AB80)

RETURN AIR CAPACITY

		NC 20—25 Application Non-Ducted		NC 25—30 Application Ducted		NC 30—40 Application Ducted	
		P_s		P_s		P_s	
Listed Size W x H	A_K	—.02″ CFM	—.03″ CFM	—.08″ CFM	—.10″ CFM	—.15″ CFM	—.20″ CFM
8 x 4	.26	85	105	170	190	230	265
8 x 6	.34	110	135	220	245	300	345
10 x 6	.42	140	170	275	310	380	435
12 x 6	.50	170	205	335	375	455	525
10 x 8	.53	195	235	385	430	525	605
12 x 8	.63	240	290	470	525	640	735
10 x 10	.64	245	300	490	550	670	770
18 x 6	.75	265	325	530	595	725	835
12 x 12	.89	375	455	740	830	1010	1160
18 x 12	1.3	570	695	1130	1265	1545	1775
22 x 10	1.4	615	750	1220	1365	1665	1915
24 x 12	1.7	770	940	1530	1715	2090	2405
18 x 18	1.9	880	1075	1750	1960	2390	2750
34 x 10	2.1	940	1145	1865	2090	2550	2930
30 x 12	2.2	975	1190	1940	2170	2645	3040
24 x 18	2.5	1190	1450	2365	2650	3235	3720
22 x 22	2.8	1300	1585	2585	2895	3530	4060
30 x 18	3.2	1495	1825	2975	3330	4060	4670
24 x 24	3.3	1610	1965	3200	3585	4375	5030
36 x 18	3.8	1810	2210	3600	4030	4915	5650
30 x 24	4.1	2020	2465	4015	4495	5485	6305
34 x 22	4.3	2150	2625	4280	4795	5850	6725
36 x 24	4.9	2440	2975	4850	5430	6625	7620
46 x 22	5.9	2980	3635	5925	6635	8095	9310
36 x 30	6.1	3090	3770	6145	6880	8395	9655
48 x 24	6.6	3250	3965	6460	7235	8825	10150
48 x 30	8.1	4170	5085	8290	9285	11325	13025
48 x 36	9.7	4950	6040	9845	11025	13450	15465

**Courtesy of Tuttle & Bailey, Division of Allied Thermal Corporation. See manufacturer's catalog for complete data.*

Table 13-15

Engineering Performance Data, Return Air Grilles and Registers*

(Fixed Bar Type, 40-degree deflected Bars)
(Series A70D, NT70d, T70DG)
(Series A70D, AN70D, AB70D)
(Series A70C, AN70C, AB70C)

RETURN AIR CAPACITY

		NC 20—25 Application Non-Ducted		NC 25—30 Application Ducted		NC 30—40 Application Ducted	
		P_s		P_s		P_s	
Listed Size W x H	A_K	—.02" CFM	—.03" CFM	—.08" CFM	—.10" CFM	—.15" CFM	—.20" CFM
8 x 4	.26	55	65	105	115	140	160
8 x 6	.34	75	90	145	160	195	225
10 x 6	.42	90	110	180	200	245	280
12 x 6	.50	110	135	220	245	300	345
10 x 8	.53	125	150	245	275	335	385
12 x 8	.63	155	190	310	345	420	485
10 x 10	.64	160	195	315	350	425	490
18 x 6	.75	170	205	335	375	455	525
12 x 12	.89	240	290	470	525	640	735
18 x 12	1.3	370	450	735	825	1005	1155
22 x 10	1.4	380	465	755	845	1030	1185
24 x 12	1.7	505	615	1000	1120	1365	1570
18 x 18	1.9	575	700	1140	1275	1555	1790
34 x 10	2.1	600	732	1195	1340	1635	1880
30 x 12	2.2	630	770	1255	1405	1715	1970
24 x 18	2.5	770	940	1530	1715	2090	2405
22 x 22	2.8	880	1075	1750	1960	2390	2750
30 x 18	3.2	970	1185	1930	2160	2635	3030
24 x 24	3.3	1040	1270	2070	2320	2830	3255
36 x 18	3.8	1170	1425	2320	2600	3170	3645
30 x 24	4.1	1320	1610	2625	2940	3585	4120
34 x 22	4.3	1355	1655	2695	3020	3685	4235
36 x 24	4.9	1580	1925	3135	3510	4280	4920
46 x 22	5.9	1940	2365	3855	4315	5265	6055
36 x 30	6.1	2010	2450	3995	4475	5460	6280
48 x 24	6.6	2130	2600	4240	4750	5795	6665
48 x 30	8.1	2700	3295	5370	6015	7340	8440
48 x 36	9.7	3220	3930	6405	7175	8755	10065

**Courtesy of Tuttle & Bailey, Division of Allied Thermal Corporation. See manufacturer's catalog for complete data.*

BIBLIOGRAPHY

1. *Handbook and Product Directory—Fundamentals*. American Society of Heating, Refrigerating and Air-Conditioning Engineers, New York, 1977.

2. *ASHRAE Handbook and Product Directory—Equipment,* American Society of Heating, Refrigerating and Air-Conditioning Engineers, Atlanta, GA, 1979.

3. Richard D. Tutt, *Principles of Air Distribution,* Tuttle & Bailey, Division of Clevepak Corporation, 1955.

4. Grilles and Registers (GR8-67), Catalog, Tuttle & Bailey, Division of Clevepak Corporation.

5. Ceiling Diffusers (CD 81), Catalog, Tuttle & Bailey, Division of Clevepak Corporation.

Chapter 14

Air Flow in Duct Systems and Fans

14-1 INTRODUCTION

Air mechanically supplied to a building for ventilation, heating, and/or cooling requires a well-designed air duct system extending from the fan or fans to the ultimate points of discharge at distribution diffusers and grilles, as well as returning all or part of the air to the conditioning apparatus from the return grilles. This chapter describes the methods used in the design of sheet metal duct systems and the performance characteristics and selection of fans used to force air through such systems.

The purpose of an air duct system is to convey conditioned air from one location to another within prescribed limits of space, noise, and cost. Good duct design takes into consideration not only the fundamentals of fluid flow in closed conduits but also the aesthetics and economics of the finished job.

All the necessary steps in the design of a duct system are summarized below. Different listed steps may overlap and would be combined in practice. With experience, each HVAC engineer will usually develop his own technique, but the following breakdown will serve as a convenient checklist.

1. Calculate the volume of air required to each space within the building enveloped to meet the heat losses, heat gains, and/or ventilation loads.
2. Determine the sizes and location of all supply air outlets and return inlets in each space or area to be conditioned using the techniques presented in Chapter 13.
3. Study the building plans and draw a tentative system of supply air ducts connecting all supply air outlets with the central-fan unit. (This would normally be a single-line drawing, the line representing the centerline of the proposed supply duct system.) In this layout remember to avoid obstructions and any direct contact with steel work and concrete. The best system is usually the simplest and most direct.
4. Layout a return air duct system using a single-line drawing. Do not return air from spaces where contaminated air may exist, such as bathrooms, kitchens, laundries, and so forth. Provide for suitable direct exhaust or vents for such areas.
5. Size the supply and return air duct systems using an accepted method as directed in this chapter.
6. Determine the friction loss of the duct offering the greatest resistance to air flow. To the duct static pressure friction loss, add an amount equal to the *total* pressure required by the last outlet, plus an amount required for the turning loss of the air into the diffuser. The sum of all these pressures represents the static discharge pressure the fan must maintain

to insure distribution of air in all parts of the system. To this must be added the suction static pressure requirements from a similar analysis of the return air system. The sum of these static pressure requirements is the total static pressure that the fan must produce. (Note that the duct having the longest linear footage of run will usually have the greatest resistance, but not necessarily always.)

7. Select a fan to flow the system.

14-2 BASIC PRINCIPLES OF GOOD DUCT DESIGN AND CONSTRUCTION

Keep the following principles in mind when designing any sheet metal, air duct system.

1. Get the air where it is wanted as directly as possible, using the least material, power, space, and at the least possible cost.
2. Stay within the maximum recommended velocities of Table 14-1 unless there is good reason for a deviation. The recommended velocities provide the quietest, most trouble-free operations.
3. Avoid sharp elbows and bends in the duct. Splitters and turning vanes should be used to reduce elbow and outlet pressure losses.
4. Avoid both sudden enlargements, and abrupt contractions. The angle of divergence of enlargements should not exceed 20 degrees. The angle of convergence in contractions should not be greater than 60 degrees. Special care should be taken to avoid restriction of flow.
5. For the *greatest* air-carrying capacity per square foot of sheet metal, use *round* ducts. In installations where rectangular ducts must be used, make the ducts as square as possible, and select the dimensions on the basis of equal friction with the round ducts from the table of circular equivalents (Table 14-2). Rectangular ducts with aspect ratio (ratio of long side to short side) greater than 6 to 1 should be avoided.
6. Make ducts as tight as possible. All laps should be in the direction of air flow. Avoid raw edges on splitters and vanes.
7. Provide expansion joints in long duct runs.
8. Provide adequate bracing and secure supports for all duct work.
9. Use plated sheet metal screws, rivets, or bolts. Hardware for aluminum air ducts should be made of aluminum, or should be zinc or cadmium-plated.
10. Use duct insulation where necessary to prevent excessive heat loss, heat gain, or condensation.
11. Provide dampers in all duct branches for final balancing.
12. Provide adequate access openings where apparatus and components may be serviced.

14-3 LOSS IN PRESSURE DUE TO FRICTION

As noted in Section 13-5, there is a decrease in the total and static pressures as air flows through a system of ducts. This loss is due to fluid friction and to fluid turbulence. In this section we shall confine our discussion to fluid friction losses. This loss is due to internal fluid friction and movement of the air over the interior surfaces of the duct.

Table 14-1

Recommended and Maximum Duct Velocities*

Designation	Recommended Velocity (feet per minute)			Maximum Velocity (feet per minute)		
	Residences	Schools Theaters Public Buildings	Industrial Buildings	Residences	Schools Theaters Public Buildings	Industrial Buildings
Outside air intakes[a]	500	500	500	800	900	1200
Filters[a]	250	300	350	300	350	350
Heating coils[a]	450	500	600	500	600	700
Air washers	500	500	500	500	500	500
Suction connections	700	800	1000	1000	1400	1400
Fan outlets	1000–1600	1300–2000	1600–2400	1500–2000	1700–2800	1700–2800
Main ducts	700–900	1000–1300	1200–1800	800–1200	1100–1600	1300–2200
Branch ducts	600	600–900	800–1000	700–1000	800–1300	1000–1800
Branch risers	500	600–700	800	650–800	800–1200	1000–1600

[a]The velocities are for total face area; not the net free area. Other velocities are for net free area.

**Used by permission, Reynolds Metal Company.*

Table 14-2

Circular Equivalents of Rectangular Ducts for Equal Friction and Capacity* (Dimension in inches, feet, or meters.)

Side Rectangular Duct	4.0	4.5	5.0	5.5	6.0	6.5	7.0	7.5	8.0	8.5	9.0	9.5	10.0	10.5	11.0	11.5	12.0	12.5	13.0	13.5
3.0	3.8	4.0	4.2	4.4	4.6	4.8	4.9	5.1	5.2	5.4	5.5	5.6	5.7	5.9	6.0	6.1	6.2	6.3	6.4	6.5
3.5	4.1	4.3	4.6	4.8	5.0	5.2	5.3	5.5	5.7	5.8	6.0	6.1	6.3	6.4	6.5	6.7	6.8	6.9	7.0	7.1
4.0	4.4	4.6	4.9	5.1	5.3	5.5	5.7	5.9	6.1	6.3	6.4	6.6	6.8	6.9	7.1	7.2	7.3	7.5	7.6	7.7
4.5	4.6	4.9	5.2	5.4	5.6	5.9	6.1	6.3	6.5	6.7	6.9	7.0	7.2	7.4	7.5	7.7	7.8	8.0	8.1	8.2
5.0	4.9	5.2	5.5	5.7	6.0	6.2	6.4	6.7	6.9	7.1	7.3	7.4	7.6	7.8	8.0	8.1	8.3	8.4	8.6	8.7
5.5	5.1	5.4	5.7	6.0	6.3	6.5	6.8	7.0	7.2	7.4	7.6	7.8	8.0	8.2	8.4	8.6	8.7	8.8	9.0	9.2

Side Rectangular Duct	6	7	8	9	10	11	12	13	14	15	16	17	18	19	20	22	24	26	28	30
6	6.6																			
7	7.1	7.7																		
8	7.5	8.2	8.8																	
9	8.0	8.6	9.3	9.9																
10	8.4	9.1	9.8	10.4	10.9															
11	8.8	9.5	10.2	10.8	11.4	12.0														
12	9.1	9.9	10.7	11.3	11.9	12.5	13.1													
13	9.5	10.3	11.1	11.8	12.4	13.0	13.6	14.2												
14	9.8	10.7	11.5	12.2	12.9	13.5	14.2	14.7	15.3											
15	10.1	11.0	11.8	12.6	13.3	14.0	14.6	15.3	15.8	16.4										
16	10.4	11.4	12.2	13.0	13.7	14.4	15.1	15.7	16.3	16.9	17.5									
17	10.7	11.7	12.5	13.4	14.1	14.9	15.5	16.1	16.8	17.4	18.0	18.6								
18	11.0	11.9	12.9	13.7	14.5	15.3	16.0	16.6	17.3	17.9	18.5	19.1	19.7							
19	11.2	12.2	13.2	14.1	14.9	15.6	16.4	17.1	17.8	18.4	19.0	19.6	20.2	20.8						
20	11.5	12.5	13.5	14.4	15.2	15.9	16.8	17.5	18.2	18.8	19.5	20.1	20.7	21.3	21.9					
22	12.0	13.1	14.1	15.0	15.9	16.7	17.6	18.3	19.1	19.7	20.4	21.0	21.7	22.3	22.9	24.1				
24	12.4	13.6	14.6	15.6	16.6	17.5	18.3	19.1	19.8	20.6	21.3	21.9	22.6	23.2	23.9	25.1	26.2			
26	12.8	14.1	15.2	16.2	17.2	18.1	19.0	19.8	20.6	21.4	22.1	22.8	23.5	24.1	24.8	26.1	27.2	28.4		
28	13.2	14.5	15.6	16.7	17.7	18.7	19.6	20.5	21.3	22.1	22.9	23.6	24.4	25.0	25.7	27.1	28.2	29.5	30.6	
30	13.6	14.9	16.1	17.2	18.3	19.3	20.2	21.1	22.0	22.9	23.7	24.4	25.2	25.9	26.7	28.0	29.3	30.5	31.6	32.8
32	14.0	15.3	16.5	17.7	18.8	19.8	20.8	21.8	22.7	23.6	24.4	25.2	26.0	26.7	27.5	28.9	30.1	31.4	32.6	33.8
34	14.4	15.7	17.0	18.2	19.3	20.4	21.4	22.4	23.3	24.2	25.1	25.9	26.7	27.5	28.3	29.7	31.0	32.3	33.6	34.8
36	14.7	16.1	17.4	18.6	19.8	20.9	21.9	23.0	23.9	24.8	25.8	26.6	27.4	28.3	29.0	30.5	32.0	33.0	34.6	35.8
38	15.0	16.4	17.8	19.0	20.3	21.4	22.5	23.5	24.5	25.4	26.4	27.3	28.1	29.0	29.8	31.4	32.8	34.2	35.5	36.7
40	15.3	16.8	18.2	19.4	20.7	21.9	23.0	24.0	25.1	26.0	27.0	27.9	28.8	29.7	30.5	32.1	33.6	35.1	36.4	37.6
42	15.6	17.1	18.5	19.8	21.1	22.3	23.4	24.5	25.6	26.6	27.6	28.5	29.4	30.4	31.2	32.8	34.4	35.9	37.3	38.6
44	15.9	17.5	18.9	20.2	21.5	22.7	23.9	25.0	26.1	27.2	28.2	29.1	30.0	31.0	31.9	33.5	35.2	36.7	38.1	39.5
46	16.2	17.8	19.2	20.6	21.9	23.2	24.3	25.5	26.7	27.7	28.7	29.7	30.6	31.6	32.5	34.2	35.9	37.4	38.9	40.3
48	16.5	18.1	19.6	20.9	22.3	23.6	24.8	26.0	27.2	28.2	29.2	30.2	31.2	32.2	33.1	34.9	36.6	38.2	39.7	41.2
50	16.8	18.4	19.9	21.3	22.7	24.0	25.2	26.4	27.6	28.7	29.8	30.8	31.8	32.8	33.7	35.5	37.3	38.9	40.4	42.0
52	17.0	18.7	20.2	21.6	23.1	24.4	25.6	26.8	28.1	29.2	30.3	31.4	32.4	33.4	34.3	36.2	38.0	39.6	41.2	42.8
54	17.3	19.0	20.5	22.0	23.4	24.8	26.1	27.3	28.5	29.7	30.8	31.9	32.9	33.9	34.9	36.8	38.7	40.3	42.0	43.6
56	17.6	19.3	20.9	22.4	23.8	25.2	26.5	27.7	28.9	30.1	31.2	32.4	33.4	34.5	35.5	37.4	39.3	41.0	42.7	44.3
58	17.8	19.5	21.1	22.7	24.2	25.5	26.9	28.2	29.3	30.5	31.7	32.9	33.9	35.0	36.0	38.0	39.8	41.7	43.4	45.0
60	18.1	19.8	21.4	23.0	24.5	25.8	27.3	28.7	29.8	31.0	32.2	33.4	34.5	35.5	36.5	38.6	40.4	42.3	44.0	45.8
62	18.3	20.1	21.7	23.3	24.8	26.2	27.6	29.0	30.2	31.4	32.6	33.8	35.0	36.0	37.1	39.2	41.0	42.9	44.7	46.5
64	18.6	20.3	22.0	23.6	25.2	26.5	27.9	29.3	30.6	31.8	33.1	34.2	35.5	36.5	37.6	39.7	41.6	43.5	45.4	47.2
66	18.8	20.6	22.3	23.9	25.5	26.9	28.3	29.7	31.0	32.2	33.5	34.7	35.9	37.0	38.1	40.2	42.2	44.1	46.0	47.8
68	19.0	20.8	22.5	24.2	25.8	27.3	28.7	30.1	31.4	32.6	33.9	35.1	36.3	37.5	38.6	40.7	42.8	44.7	46.6	48.4
70	19.2	21.1	22.8	24.5	26.1	27.6	29.1	30.4	31.8	33.1	34.3	35.6	36.8	37.9	39.1	41.3	43.3	45.3	47.2	49.0
72															3916	41.8	43.8	45.9	47.8	49.7
74															40.0	42.3	44.4	46.4	48.4	50.3
76															40.5	42.8	44.9	47.0	49.0	50.8
78															40.9	43.3	45.5	47.5	49.5	51.5
80															41.3	43.8	46.0	48.0	50.1	52.0
82															41.8	44.2	46.4	48.6	50.6	52.6
84															42.2	44.6	46.9	49.2	51.1	53.2
86															42.6	45.0	47.4	49.6	51.6	53.7
88															43.0	45.4	47.9	50.1	52.2	54.3
90															43.4	45.9	48.3	50.6	52.8	54.8
92															43.8	46.3	48.7	51.1	53.4	55.4
94															44.2	46.7	49.1	51.6	53.9	55.9
96															44.6	47.2	49.5	52.0	54.4	56.3

* From R. G. Huebscher, *Trans. ASHVE*, Vol. 54 (1948), pp. 112–13. Reprinted by permission.

Table 14-2 (Cont.)

14.0	14.5	15.0	15.5	16
6.6	6.7	6.8	6.9	7.0
7.2	7.3	7.4	7.5	7.6
7.8	7.9	8.1	8.2	8.3
8.4	8.5	8.6	8.7	8.9
8.9	9.0	9.1	9.3	9.4
9.4	9.5	9.6	9.8	9.8

32	34	36	38	40	42	44	46	48	50	52	56	60	64	68	72	76	80	84	88	Side Rectangular Duct
																				6
																				7
																				8
																				9
																				10
																				11
																				12
																				13
																				14
																				15
																				16
																				17
																				18
																				19
																				20
																				22
																				24
																				26
																				28
																				30
35.0																				32
36.0	37.2																			34
37.0	38.2	39.4																		36
38.0	39.2	40.4	41.6																	38
39.0	40.2	41.4	42.6	43.8																40
39.9	41.1	42.4	43.6	44.8	45.9															42
40.8	42.0	43.4	44.6	45.8	46.9	48.1														44
41.7	43.0	44.3	45.6	46.8	47.9	49.1	50.3													46
42.6	43.9	45.2	46.5	47.8	48.9	50.2	51.3	52.6												48
43.5	44.8	46.1	47.4	48.8	49.8	51.2	52.3	53.6	54.7											50
44.3	45.7	47.1	48.3	49.7	50.8	52.2	53.3	54.6	55.8	56.9										52
45.0	46.5	48.0	49.2	50.6	51.8	53.2	54.3	55.6	56.8	57.9										54
45.8	47.3	48.8	50.1	51.5	52.7	54.1	55.3	56.5	57.8	58.9	61.3									56
46.6	48.1	49.6	51.0	52.4	53.7	55.0	56.2	57.5	58.8	60.0	62.3									58
47.3	48.9	50.4	51.8	53.3	54.6	55.9	57.1	58.5	59.8	61.0	63.3	65.7								60
48.0	49.7	51.2	52.6	54.2	55.5	56.8	58.0	59.4	60.7	62.0	64.3	66.7								62
48.7	50.4	52.0	53.4	55.0	56.4	57.7	59.0	60.3	61.6	62.9	65.3	67.7	70.0							64
49.5	51.1	52.8	54.2	55.8	57.2	58.6	59.9	61.2	62.5	63.9	66.3	68.7	71.1							66
50.2	51.8	53.5	55.0	56.6	58.0	59.5	60.8	62.1	63.4	64.8	67.3	69.7	72.1	74.4						68
50.9	52.5	54.2	55.8	57.3	58.8	60.3	61.7	63.0	64.3	65.7	68.3	70.7	73.1	75.4						70
51.5	53.2	54.9	56.5	58.0	59.6	61.1	62.6	63.9	65.2	66.6	69.2	71.7	74.1	76.4	78.8					72
52.1	53.9	55.6	57.2	58.8	60.4	61.9	63.3	64.8	66.1	67.5	70.1	72.7	75.1	77.4	79.9					74
52.7	54.6	56.3	57.9	59.5	61.2	62.7	64.1	65.6	67.0	68.4	71.0	73.6	76.1	78.4	80.9	83.2				76
53.3	55.2	57.0	58.6	60.3	62.0	63.4	64.9	66.4	67.9	69.3	71.8	74.5	77.1	79.4	81.8	84.2				78
53.9	55.8	57.6	59.3	61.0	62.7	64.1	65.7	67.2	68.7	70.1	72.7	75.4	78.1	80.4	82.8	85.2	87.5			80
54.5	56.4	58.2	60.0	61.7	63.4	64.9	66.5	68.0	69.5	71.0	73.6	76.3	79.0	81.4	83.8	86.2	88.6			82
55.1	57.0	58.9	60.7	62.4	64.1	65.7	67.3	68.8	70.3	71.8	74.5	77.2	79.9	82.4	84.8	87.2	89.6	91.9		84
55.7	57.6	59.5	61.3	63.0	64.8	66.4	68.0	69.5	71.1	72.6	75.4	78.1	80.8	83.3	85.8	88.2	90.6	92.9		86
56.3	58.2	60.1	62.0	63.7	65.4	67.0	68.7	70.3	71.8	73.4	76.3	79.0	81.6	84.2	86.8	89.2	91.6	93.9	96.3	88
56.9	58.8	60.7	62.6	64.4	66.0	67.8	69.4	71.1	72.6	74.2	77.1	79.9	82.5	85.1	87.8	90.2	92.6	94.9	97.3	90
57.4	59.4	61.3	63.2	65.0	66.8	68.5	70.1	71.8	73.3	74.9	77.8	80.8	83.4	86.0	88.7	91.2	93.6	95.9	98.3	92
57.9	60.0	61.9	63.8	65.6	67.5	69.2	70.8	72.5	74.1	75.6	78.6	81.7	84.3	86.9	89.6	92.1	94.6	96.9	99.3	94
58.4	60.5	62.4	64.4	66.2	68.2	69.8	71.5	73.2	74.8	76.3	79.4	82.6	85.2	87.8	90.5	93.0	95.6	97.9	100.3	96

In Chapter 11 we developed an equation for head loss due to friction in a round, closed conduit for any fluid as

$$h_L = f\left(\frac{L}{D}\right)\left(\frac{V^2}{2g}\right) \text{ ft of fluid} \qquad \textit{(11-11)}$$

Repeated

Since $V^2/2g$ has been defined a velocity head, we may substitute Eq. (13-4) into Eq. (11-11) and obtain

$$h_L = f\left(\frac{L}{D}\right)\left(\frac{V}{1096.5}\right)^2 \rho_a \text{ in. WG} \qquad \textit{(14-1)}$$

or, if standard air is considered, we may use Eq. (13-6) and obtain

$$h_L = f\left(\frac{L}{D}\right)\left(\frac{V}{4005}\right)^2 \text{ in. WG} \qquad \textit{(14-1a)}$$

The symbols and units of Eqs. (14-1) and (14-1a) are as follows:

h_L = friction loss (in. WG)

f = dimensionless friction factor

L = duct length (ft)

D = duct diameter (ft)

V = mean air flow velocity (fpm)

ρ_a = air density (lbm/ft^3)

As was the case with water, the greater the quantity of air flowing through a given sized duct, the greater will be the loss in pressure due to friction, which means, greater operating costs. If the size of a duct is increased to reduce the flow velocity and decrease the friction loss, the first cost increases. Therefore, the principles of good duct design must be known in order to secure satisfactory operation of the system with the proper balance existing between first cost and final operating cost. It will be readily seen that a poorly designed system with inherently large friction losses will cost considerably more to operate than a well-designed system where the pressure losses are kept to a minimum. From the fan laws (discussed later) it can be seen that a fan of a given size selected for a given pressure must be speeded up if it is to overcome an additional resistance. Since the power consumption varies as the cube of the speed change, it is apparent that pressure losses must be kept to a minimum if economical operation is to be obtained.

In practice, this balancing of costs is a lengthy process and sometimes a useless one because of the lack of reliable cost data. Duct designers invariably depend upon past experience with air-conditioning and ventilating systems in selecting the pressure losses to be allowed in the duct system. In addition to this, in conventional low velocity duct design, the problem is further complicated in commercial buildings by the fact that air velocities in ducts must be maintained below some definite point if quiet operation is to be obtained. Recommended practice is shown in Table 14-1. If the maximum velocity is thus set, this immediately fixes the maximum friction loss for a given air quantity without any other considerations of cost. In industrial systems where noise is not as much a factor, the problem is only one of balancing the operating cost against the first cost.

The friction loss in straight sound ducts may be determined from charts such as Figs. 14-1 and 14-1a, which are based on Eq. (14-1a). Therefore, these charts are based on standard air (ρ = 0.0750 lb/ft^3) flowing through average, clean, round aluminum ducts. Ducts are also constructed of galvanized sheet metal. Friction loss in galvanized metal ducts is slightly more than that in aluminum ducts. However, considering the degree of accuracy with which we can read the charts and the accuracy in fabrication of the ducts, we may use the design charts (Figs. 14-1 and 14-1a) for

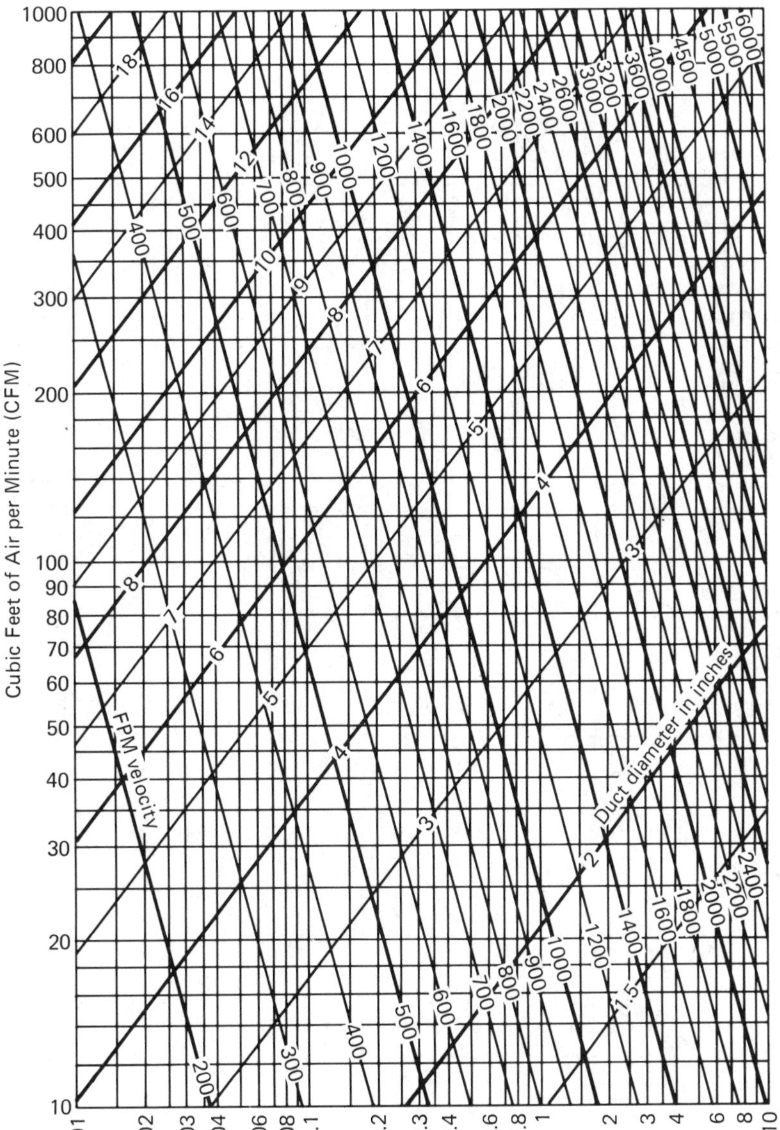

Figure 14–1

Friction loss in straight, aluminum sheet metal ducts for volumes 10 to 1000 CFM. (Based on standard air density of 0.0750 lb/ft^3 flowing through average, clean, round aluminum ducts having approximately 40 joints per 100 ft.) (Courtesy of Reynolds Metal Company.)

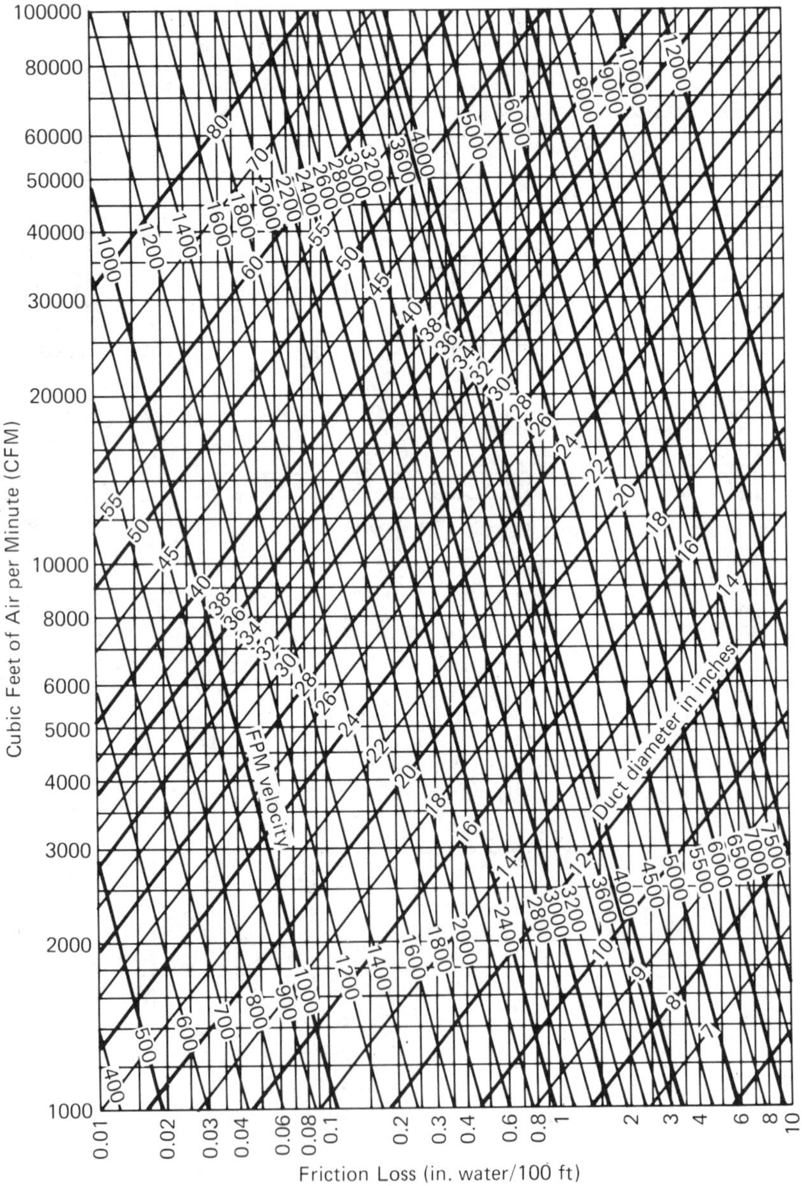

Figure 14–1a

Friction loss in straight, aluminum sheet metal ducts for volumes 1000 to 100,000 CFM. (Based on standard air density of 0.0750 lb/ft^3 flowing through average, clean, round aluminum ducts having approximately 40 joints per 100 ft.) (Courtesy of Reynolds Metal Company.)

either aluminum or galvanized sheet metal.

Referring to Figs. 14-1 and 14-1a, we may note that the friction loss shown on the horizontal scale is expressed in inches water gauge (in. WG) per 100 ft of straight duct. This will be referred to as the "friction rate" (f_{100}). The vertical scale is the air flow quantity in cubic feet per minute (cfm). Inclined lines from lower left to upper right are lines of constant duct diameter in inches. Inclined lines from upper left to lower right are lines of constant velocity in feet per minute (fpm). Although these charts are designed for standard air flowing, they may be used with little error for air temperatures between 50°F and 90°F. Also, a correction is not necessary for humidity or for small deviations in barometric pressure (not exceeding ±0.5 in. Hg).

Illustrative Problem 14-1

A sheet metal duct 150 ft long is 14-in. in diameter. The quantity of standard air to be delivered through the duct is 2000 cfm. Find (a) the friction loss in the duct and (b) the velocity of air flowing through the duct.

Solution:

(a) Referring to Fig. 14-1a, at the intersection of the horizontal line for 2000 cfm and the diagonal line for 14-in. diameter, drop vertically down to the friction rate scale and read 0.34 in. WG per 100 ft of length. For a total length of 150 ft,

$$\text{Friction Loss} = 0.34(150/100) = 0.51 \text{ in. WG}$$

(b) The air velocity in the round duct may be read from Fig. 14-1a at the same point of intersection as part a and is 1900 fpm.

Illustrative Problem 14-2

The static pressure available to overcome friction loss in a duct is 0.25 in. WG. The duct is 22 in. in diameter and is 315 ft long. Find (a) the maximum quantity of standard air that can flow through the duct and (b) the air flow velocity.

Solution:

(a) Friction rate = 100(0.25/315) = 0.079 (say 0.08 in. WG/100 ft). Referring to Fig. 14-1a, locate on the bottom (or top) friction rate scale the value of 0.08 and proceed vertically to the sloping line for 22 in. diameter. At this point of intersection read on the vertical scale 3000 cfm.
(b) The air flow velocity at this point of intersection is 1150 fpm.

Illustrative Problem 14-3

The maximum velocity of air in a duct is limited to 1600 fpm. The total length is 155 ft. If the total quantity of standard air to be delivered through the duct is 10,000 cfm, find (a) the required diameter of the duct and (b) the friction loss in the duct.

Solution:

(a) Referring to Fig. 14-1a, at the intersection of the horizontal line of 10,000 cfm and the diagonal velocity line of 1600 fpm, the size of the required round duct is 34 in. diameter.
(b) The friction loss per 100 ft of duct found in Fig. 14-1a is 0.086 in. WG per 100 ft of duct length.

$$\text{Total friction loss} = 0.086(155/100) = 0.133 \text{ in. WG}$$

14-4 CHANGES IN DUCT FRICTION LOSS WITH CHANGE IN AIR DENSITY

When the air flowing in ducts is at a density significantly different from that for which the design charts (Figs. 14-1 and 14-1a) were constructed, a correction for density must be made.

With a constant volume of air flowing (not a constant mass), the friction loss for a duct system varies directly as the density. Therefore, we may say

$$\frac{(h_L)_2}{(h_L)_1} = \frac{\rho_{a_2}}{\rho_{a_1}} \qquad (14\text{-}2)$$

where $(h_L)_1$ and ρ_{a_1} are the static head loss and density, respectively, of the air at condition 1, that is, standard air; and $(h_L)_2$ and ρ_{a_2} are the static head loss and density, respectively, at a condition 2.

From this, then, the mass flow changes with any variation in density, since Eq. (14-2) is based on volume flow through the duct. It is usually desirable to maintain a definite mass flow of air to absorb or supply sufficient heat in the conditioned space.

When this is the case, confusion can be avoided if the friction loss is first determined for a volume of standard air equal to the volume of actual nonstandard air. The friction loss is then modified by means of Eq. (14-2) to the loss for the nonstandard air.

The actual volume and actual pressure loss obtained by means of these computations are those for which a fan must be selected. This selection procedure will be discussed in a later section.

Illustrative Problem 14-4

Find the fan selection volume and actual friction loss to maintain a *constant-mass flow* corresponding to a standard air volume of 11,050 cfm in a 32-in. diameter round duct 500 ft long, when the air density is 0.0637 lbm/ft^3.

Solution:

The actual volume of air flowing at a density of 0.0637 lbm/ft^3, maintaining constant mass, is

$$11{,}050(0.0750/0.0637) = 13{,}000 \text{ cfm}$$

From Fig. 14-1a the friction rate with 13,000 cfm of standard air in a 32-in. round duct is 0.190 in. WG per 100 ft. The friction loss in 500 ft is

$$\begin{aligned} \text{Friction loss} &= 0.190(500/100) \\ &= 0.95 \text{ in. WG} \end{aligned}$$

By Eq. (14-2)

$$\begin{aligned} (h_L)_2 &= (h_L)_1\,(\rho_{a_2}/\rho_{a_1}) \\ &= (0.95)(0.0637/0.0750) \\ (h_L)_2 &= 0.81 \text{ in. WG} \end{aligned}$$

(which is the actual loss)

Illustrative Problem 14-5

A sheet metal duct 75 ft long is 18 in. in diameter. The quantity of standard air to be delivered through the duct is 3000 cfm. Find (a) the friction loss in the duct and (b) if the air temperature were increased to 120°F, what would be the friction loss?

Solution:

(a) Referring to Fig. 14-1a with 3000 cfm and an 18-in. round duct, find the friction rate of 0.22 in. WG per 100 ft. Therefore, the friction loss for a 75-ft length is

Friction loss = 0.22(75/100) = 0.165 in. WG

(b) Since there is no indication in the problem statement concerning constant-mass flow, constant-volume flow is assumed.

Only the temperature changes, and we know that the density varies inversely with absolute temperature. Therefore, Eq. (14-2) may be modified as follows:

$$\frac{(h_L)_2}{(h_L)_1} = \left(\frac{460 + 70}{460 + t}\right)$$

$$\text{Therefore, } (h_L)_2 = 0.165\left(\frac{460 + 70}{460 + 120}\right)$$
$$= 0.151 \text{ in. WG}$$

Density of air through a fan and duct differ. Occasionally installations are made in which a heating coil is installed between the fan and the duct work. As a result of this, the air flowing through the duct is at a different temperature and has a different density than the air flowing through the fan. Evidently, the mass of air flowing through the duct is exactly the same as the mass delivered by the fan. However, the volume of air flowing through the duct is greater than the volume of air flowing through the fan.

If the duct had been designed on the basis of the volume of standard air needed, the static pressure against which the fan must actually work will be different from the static pressure computed on the basis of standard air flowing in the duct. For a *constant mass of air flowing through the duct,* the static pressure against which the fan must work can be computed by means of the following relationship:

$$\frac{(h_L)_2}{(h_L)_1} = \frac{\rho_{a1}}{\rho_{a2}} \qquad \textit{(14-2a)}$$

where $(h_L)_1$ and ρ_{a1} are the static pressure and density, respectively, of standard air, and $(h_L)_2$ and ρ_{a2} are the actual static pressure and density of the air flowing through the duct.

Illustrative Problem 14-6

The static pressure loss in a duct with standard air flowing through the duct is 0.77 in. WG. Find the static pressure loss in the duct if the air flowing through the duct has a density of 0.0675 lb/ft^3, maintaining constant-mass flow. By Eq. (14-2a)

$$(h_L)_2 = (h_L)_1(\rho_{a_1}/\rho_{a_2})$$
$$= 0.77(0.075/0.0675)$$
$$(h_L)_2 = 0.856 \text{ in. WG}$$

For the condition of this problem a fan handling standard air should be selected for the volume of standard air needed, but against a static pressure of 0.856 in. WG, not 0.77 in. WG. Thus, if 15,000 cfm of standard air were needed for the conditions of the problem, the fan would be selected for 15,000 cfm, operating against a static pressure of 0.856 in. WG.

14-5 EQUIVALENT RECTANGULAR DUCTS

Rectangular ducts are used more frequently than round ducts. Although round ducts require the least metal in order to carry a given quantity of air, rectangular ducts are used in most installations because of space consideration. As a rule, rectangular ducts fit into available spaces in buildings and can be installed less conspicuously than round ducts.

A greater friction loss occurs in a rectangular duct than in a circular duct of equal cross-sectional flow area. If the friction loss per foot of length in a circular duct is to be made equal to that in a rectangular duct, fundamental expressions for the friction loss for the two types of ducts are equated; and, with suitable friction factors being used, the following expression is obtained:

$$D_e = 1.3\left[\frac{(HW)^5}{(H + W)^2}\right]^{1/8} = 1.3\frac{(HW)^{0.625}}{(H + W)^{0.25}} \tag{14-3}$$

where H and W represent height and width of a rectangular duct in feet or inches and D_e is the diameter (ft or in.) of a circular duct having the same friction loss per foot as a rectangular duct delivering the same quantity of air.

Table 14-2 gives the circular equivalents of rectangular ducts based on Eq. (14-3). It should be noted that the mean velocity in rectangular ducts is less than that in an equivalent round duct.

Rectangular duct dimensions are seldom quoted in fractions of inches because of the additional time and labor cost for fabrication, and also in using the design charts it is difficult to work with such accuracy. It has become common practice to specify one dimension of the duct (usually the vertical dimension) as 4, 8, 12, 14, 16, 18 or 20 in. and letting the duct width vary as required to obtain the flow area.

Illustrative Problem 14-7

If 4000 cfm of standard air flows through a 20-in. diameter duct, determine the friction loss for a 75-ft section of straight duct, the flow velocity in the round duct, the equivalent rectangular duct size, and the flow velocity in the rectangular duct.

Solution:

Referrring to Fig. 14-1a for 4000 cfm flowing through a 20-in. diameter duct, find V = 1880 fpm and the friction rate of 0.22 in. WG per 100 ft. For 75 ft of length the friction loss is 0.22(75/100) = 0.165 in. WG.

Referring to Table 14-2, an equivalent rectangular duct may be selected. We find that many different duct dimensions could be selected, three of which are 50 by 8 in. (D_e = 19.9 in.), 30 by 12 in. (D_e = 20.2 in.), or 22 by 16 in. (D_e = 20.4 in.). If we select the 30 by 12 in. duct the mean flow velocity in the rectangular duct would be

$$V = \frac{Q}{A} = \frac{4000(144)}{(30)(12)} = 1600 \text{ fpm}$$

Illustrative Problem 14-8

A duct 24 by 12 in. is to deliver 3000 cfm of standard air. Find the friction loss per 100 ft of duct length.

Solution:

Referring to Table 14-2, for a 24 by 12 in. duct, the equivalent round duct is found to be 18.3 in. Referring to Fig. 14-1a, for 3000 cfm and 18.3 in. diameter, find the friction loss is 0.20 in. WG per 100 ft of duct length.

Aspect Ratio (AR). When rectangular ducts are used, particular care should be exercised to maintain a low duct aspect ratio. Aspect ratio of a rectangular duct is the ratio of the long side to the short side of the duct. This ratio should be kept below 6 to 1 for the following reasons, all of which are economic:

1. Heat gain or loss to the air in the duct increase as the aspect ratio increases.
2. Material quantities increase with an increase in aspect ratio, as well as labor costs.
3. Operating costs increase because friction losses increase with increasing aspect ratios.

These facts can be illustrated in the following manner. Referring to Fig. 14-2, two duct

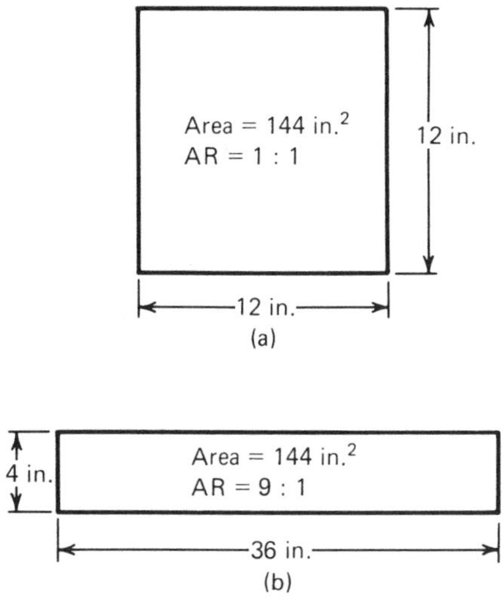

Figure 14–2

Rectangular duct cross-sections illustrating duct aspect ratio

cross sections are represented, each having the same cross-sectional area, 144 in.^{2} For section *a* the aspect ratio is 1:1 and the perimeter is 48 in. For section *b* the aspect ratio is 9:1 and the perimeter is 80 in. Therefore, the higher aspect ratio section requires more material per foot of length and has greater exposed surface area for heat transfer. Furthermore, the long side of the duct may need cross bracing to support the sheet metal.

Assume that a rectangular duct is to be sized to carry 5000 cfm of standard air at a velocity of 2000 fpm. This would require a flow area of 5000/2000 = 2.5 ft^2 = 360 in.^{2} The following table shows the effect on friction rate as the aspect ratio changes:

Rectangular Dimensions $H \times W$	AR	Equivalent Diam., D_e (Table 14-2)	Fraction Rate (Fig. 14-1a)
18 × 20	1 : 1	20.7	0.28
12 × 30	2.5 : 1	20.2	0.32
10 × 36	3.6 : 1	19.8	0.36
8 × 45	5.6 : 1	19.1	0.43
6 × 60	10 : 1	18.1	0.54

14-6 DYNAMIC (SHOCK) LOSSES IN DUCT WORK

Dynamic losses results from disturbances in the flow caused by fittings that cause a change in direction and/or area of the air flow path such as inlets, outlets, diffusers, nozzles, orifices, tees, elbows, and obstructions. Much data has been presented in past years in manufacturer's data books, handbooks on fluid flow, and scientific papers. In some cases the data are contradictory and limited to particular flow rates and shapes of fittings. Also, the data are reported in the format of the author's choice, resulting in presentation of loss as total equivalent length, additional equivalent length, equivalent length expressed as dimension multipliers, static regain, or velocity pressure (dynamic pressure) multipliers.

Here we will use the velocity pressure multipliers which may be found in Ref. 1. Selected local loss coefficients may be found in Tables 14-3 through 14-6. The dynamic loss caused by the fittings shown may be calculated by

$$\text{Dynamic Loss} \qquad (h_L)_d = C_o \times h_{v_o} \qquad \textit{(14-4)}$$

where $(h_L)_d$ is expressed in in. WG, h_{v_o} is the velocity head at the section *o*, and C_o is the local loss coefficient.

14-7 DYNAMIC LOSSES IN ELBOWS, TRANSITIONS, AND BRANCH TAKEOFFS

The elbows in duct systems must be correctly designed if the friction losses are to be

Table 14-3

Loss Coefficients for Duct Elbows (Round)*

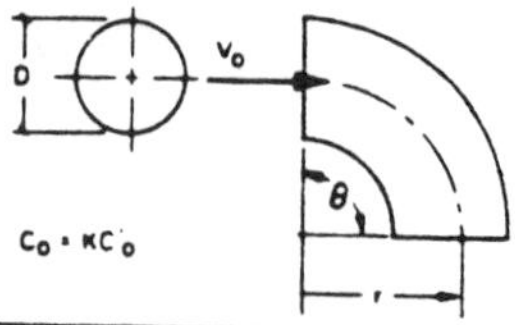

Coefficients for 90° Elbows:

r/D	0.5	0.75	1.0	1.5	2.0	2.5
C_o	0.71	0.33	0.22	0.15	0.13	0.12

For angles other than 90° multiply by the following factor:

θ	0	20	30	45	60	75	90	110	130	150	180
K	0	0.31	0.45	0.60	0.78	0.90	0.90	1.13	1.20	1.28	1.40

(a) Elbow, smooth radius, round.

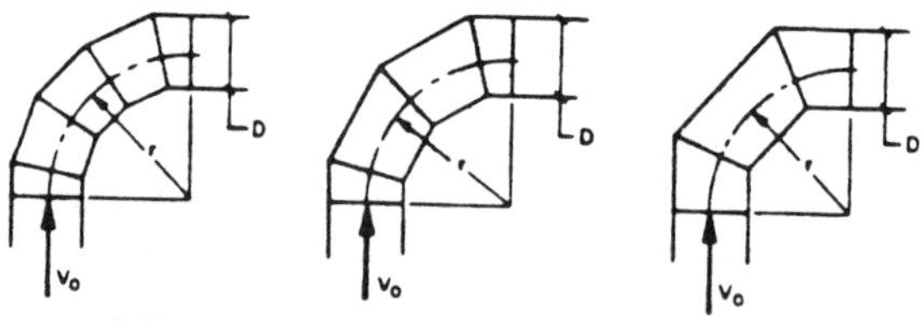

5-Piece **4-Piece** **3-Piece**

No. of Pieces	C_o				
	r/D				
	0.5	**0.75**	**1.0**	**1.5**	**2.0**
5	—	0.46	0.33	0.24	0.19
4	—	0.50	0.37	0.27	0.24
3	0.90	0.54	0.42	0.34	0.33

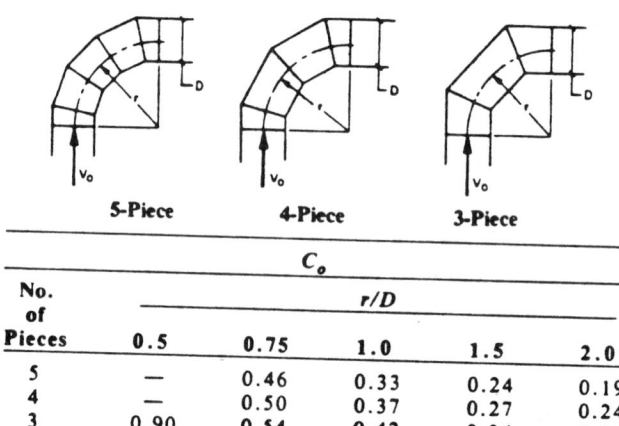

(b) Elbow, 90-deg. 3, 4 and 5-pieces, round

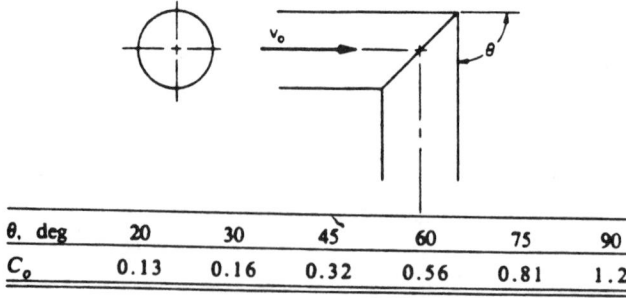

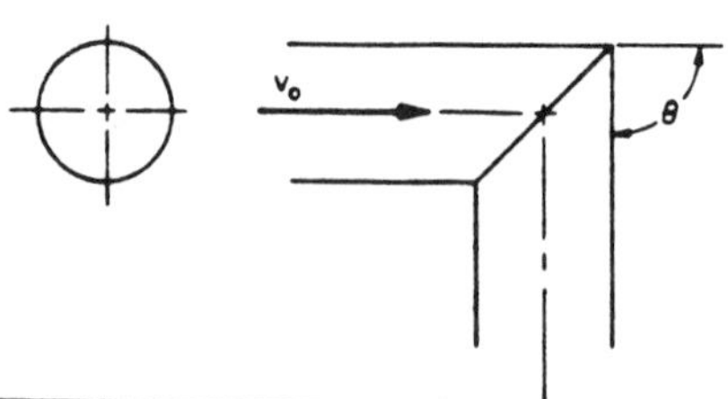

θ, deg	20	30	45	60	75	90
C_o	0.13	0.16	0.32	0.56	0.81	1.2

(c) Elbow, mitered, round

*From ASHRAE Handbook and Product Directory, 1977, Fundamentals, with permission of the American Society of Heating, Refrigerating and Air-Conditioning Engineers, Atlanta, GA.

Table 14-4

Local Loss Coefficients, Elbows (Rectangular)*

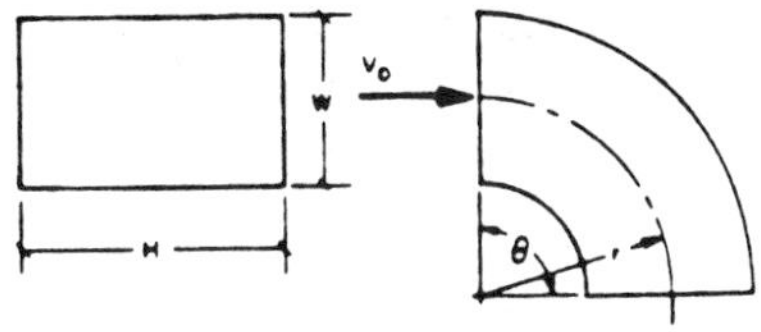

$C_o = K C_o$

Coefficients for 90° elbows:

	C_o										
	H/W										
r/W	0.25	0.5	1.0	1.5	2.0	3.0	4.0	5.0	6.0	7.0	8.0
0.5	1.3	1.0	0.93	0.95	0.99	1.1	1.2	1.3	1.4	1.5	1.6
0.75	0.61	0.46	0.39	0.38	0.39	0.41	0.43	0.46	0.49	0.51	0.54
1.0	0.36	0.26	0.21	0.20	0.20	0.20	0.21	0.22	0.23	0.23	0.24
1.5	0.18	0.12	0.09	0.08	0.08	0.07	0.07	0.08	0.08	0.08	0.08
2.0	0.11	0.07	0.05	0.04	0.04	0.04	0.04	0.04	0.04	0.04	0.04
2.5	0.07	0.04	0.03	0.03	0.02	0.02	0.02	0.02	0.02	0.02	0.02
3.0	0.05	0.03	0.02	0.02	0.02	0.01	0.01	0.01	0.01	0.01	0.01

For angles other than 90° multiply by the following factor:

θ	0°	20°	30°	45°	60°	75°	90°	110°	130°	150°	180°
K	0	0.31	0.45	0.60	0.78	0.90	1.0	1.13	1.20	1.28	1.40

(a) Elbow, smooth radius without vanes (Rectangular)

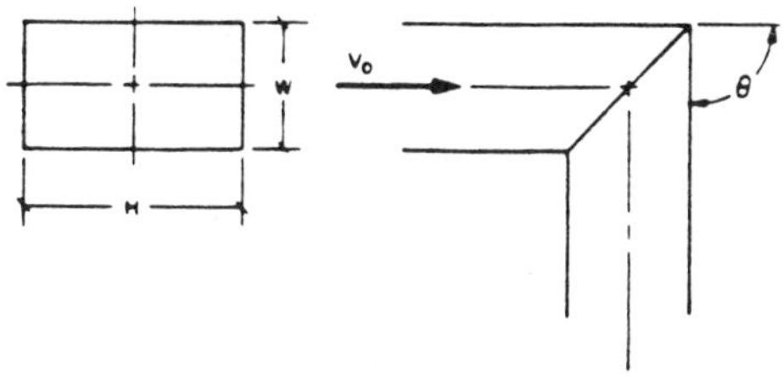

	C_o							
θ,	H/W							
deg	0.25	0.50	0.75	1.0	1.5	2.0	3.0	4.0
20	0.14	0.14	0.14	0.13	0.12	0.12	0.10	0.10
30	0.18	0.17	0.17	0.16	0.15	0.14	0.13	0.12
45	0.35	0.34	0.33	0.32	0.30	0.29	0.27	0.25
60	0.62	0.60	0.58	0.56	0.53	0.50	0.46	0.44
75	0.89	0.87	0.84	0.81	0.77	0.73	0.67	0.63
90	1.3	1,3	1.2	1.2	1.1	1.1	0.99	0.93

(b) Elbow, mitered (Rectangular)

**From ASHRAE* Handbook and Product Directory, 1977, Fundamentals, *with permission of the American Society of Heating, Refrigerating and Air-Conditioning Engineers, Atlanta, GA.*

Table 14-5

Local Loss Coefficients, Transitions*

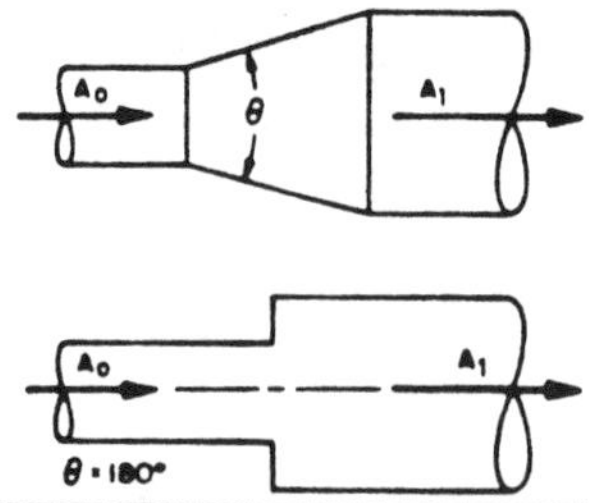

A_0/A_1	C_o, θ, degrees: 8	10	12	14	16	20	24	30	40	60	90	180
0	0.11	0.15	0.19	0.23	0.27	0.36	0.47	0.65	0.92	1.2	1.1	1.0
0.05	0.10	0.14	0.16	0.20	0.24	0.32	0.42	0.58	0.83	1.0	1.0	0.92
0.075	0.09	0.13	0.16	0.19	0.23	0.30	0.40	0.55	0.79	0.99	0.95	0.88
0.10	0.09	0.12	0.15	0.18	0.22	0.29	0.38	0.52	0.75	0.93	0.89	0.83
0.15	0.08	0.11	0.14	0.17	0.20	0.26	0.34	0.46	0.67	0.84	0.79	0.74
0.20	0.07	0.10	0.12	0.15	0.17	0.23	0.30	0.41	0.59	0.74	0.70	0.65
0.25	0.06	0.08	0.10	0.13	0.15	0.20	0.26	0.35	0.47	0.65	0.62	0.58
0.30	0.05	0.07	0.09	0.11	0.13	0.18	0.23	0.31	0.40	0.57	0.54	0.50
0.40	0.04	0.06	0.07	0.08	0.10	0.13	0.17	0.23	0.33	0.41	0.39	0.37
0.50	0.03	0.04	0.05	0.04	0.05	0.06	0.08	0.10	0.15	0.18	0.17	0.16
0.60	0.02	0.03	0.03	0.04	0.05	0.06	0.08	0.10	0.15	0.18	0.17	0.16

(a) Conical diffuser, Round.

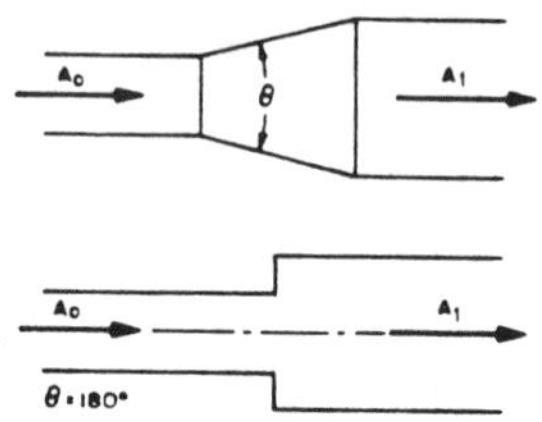

A_0/A_1	C_o, θ, degrees: 8	10	12	14	16	20	24	30	40	60	90	180
0	0.11	0.15	0.19	14	0.27	0.36	0.47	0.65	0.92	1.2	1.1	1.0
0.05	0.10	0.14	0.16	0.20	0.24	0.32	0.42	0.58	0.83	1.0	0.99	0.92
0.075	0.09	0.13	0.16	0.19	0.23	0.30	0.40	0.55	0.79	0.99	0.95	0.88
0.10	0.09	0.12	0.15	0.18	0.22	0.29	0.38	0.52	0.75	0.93	0.89	0.83
0.15	0.08	0.11	0.14	0.17	0.20	0.26	0.34	0.46	0.67	0.84	0.79	0.74
0.20	0.07	0.10	0.12	0.15	0.17	0.23	0.30	0.41	0.59	0.74	0.70	0.65
0.25	0.06	0.08	0.10	0.13	0.15	0.20	0.26	0.35	0.47	0.65	0.62	0.58
0.30	0.05	0.07	0.09	0.11	0.13	0.18	0.23	0.31	0.40	0.57	0.54	0.50
0.40	0.04	0.05	0.07	0.08	0.10	0.13	0.17	0.23	0.33	0.41	0.39	0.37
0.50	0.03	0.04	0.05	0.06	0.07	0.09	0.12	0.16	0.23	0.29	0.29	0.26
0.60	0.02	0.03	0.03	0.04	0.05	0.06	0.06	0.10	0.15	0.18	0.17	0.16

(b) Plane diffuser, Rectangular.

*From ASHRAE *Handbook and Product Directory*, 1977, *Fundamentals*, with permission of the American Society of Heating, Refrigerating and Air-Conditioning Engineers, Atlanta, GA.

Table 14-6

Local Loss Coefficients, Transitions*

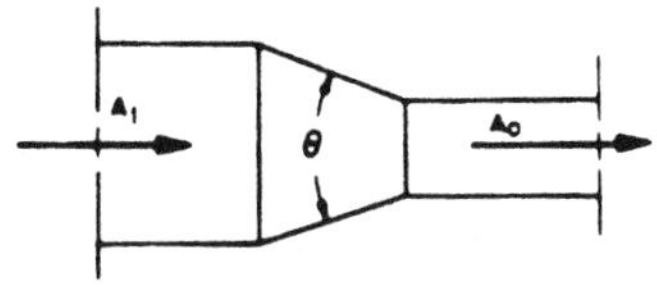

			C_o			
$\frac{A_o}{A_1}$			θ, degrees			
		1200	30	40	50	60
0.1	0	0.01	0.02	0.03	0.05	0.07
0.2	0	0.01	0.02	0.03	0.05	0.07
0.3	0	0.01	0.02	0.03	0.05	0.06
0.4	0	0.01	0.02	0.03	0.04	0.06
0.5	0	0.01	0.01	0.02	0.04	0.05
0.6	0	0.01	0.01	0.02	0.03	0.04
0.7	0	0	0.01	0.02	0.03	0.04
0.8	0	0	0.01	0.01	0.02	0.03
0.9	0	1	0	0.01	0.01	0.01

(a) Contractions, gradual, round and rectangular.

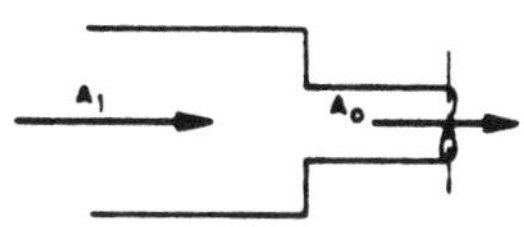

A_o/A_1	0.1	0.2	0.3	0.4	0.5	0.6	0.7	0.8	0.9
C_o	0.45	0.40	0.35	0.30	0.25	0.20	0.15	0.10	0.05

(b) Contractions, abrupt, rectangular/round to round.

**From ASHRAE* Handbook and Product Directory, 1977, Fundamentals, *with permission of the American Society of Heating, Refrigerating and Air-Conditioning Engineers, Atlanta, GA.*

held to a minimum. In addition to keeping the friction losses to a minimum, it is very important that the air flow be uniformly distributed over the cross section of the elbow if a register or diffuser supply grille is placed at the outlet of an elbow. Excessive friction loss in elbows is due to the fact that air is not uniformly distributed over the duct flow area. The portion of the airstream that travels along the outside edge of the elbow is deflected around the turn. However, the portion of the air traveling along the inside tends to continue in a straight line until it impinges against the airstream on the outer edge. As a result of the turbulence created, the dynamic loss in elbows is likely to be high unless this turbulence is prevented or reduced to a minimum.

Before proceeding with the methods of designing elbows, the meaning of a few terms that will be used must be understood. In straight lengths of horizontal duct work, the width *(W)* of the duct is the horizontal dimension of the duct work and the depth *(H)* is the vertical dimension. However, when discussing elbows, the terms *width* and *depth* of the elbow have a different meaning. The width *(W)* of an elbow is always the dimension of the elbow that lies in the same plane as the radius of the elbow. Thus, referring to Fig. (14-3) the width of elbow *A* is 20 in., and its depth is 10 in. For elbow *A* of this illustration, the conventional definition of width and depth of straight duct apply. However, for elbow *B* the width of the duct is 10 in. and its depth is 20 in. The width of an elbow is *always* the dimension in the same plane as the elbow radius.

A term that is widely used in elbow design is *radius ratio*. For a round duct the radius ratio is r/D. For a rectangular duct the radius ratio is found by dividing the centerline radius of an elbow by its width (r/W). Thus, for Fig. 14-3 the radius ratio of elbow *A* would be $(12 + 10)/20 = 1.1$ and, for elbow *B* it would be $(12 + 5)/10 = 1.7$.

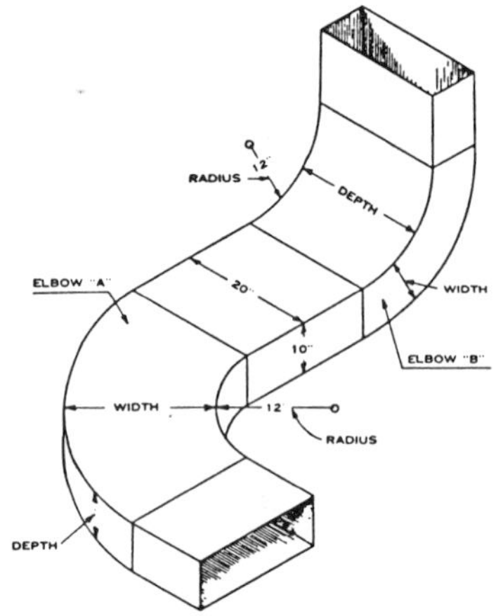

Figure 14–3

Duct elbow nomenclature

Another factor that is important in the design of rectangular elbows is the aspect ratio of the elbow. The aspect ratio for elbows and for straight ducts is defined differently. The aspect ratio of a straight duct is always equal to the long side divided by the short side of the rectangular duct, as previously noted. However, in the case of elbows the aspect ratio is always equal to the depth of the duct divided by its width (H/W). Thus, the aspect of elbow *A* of Fig. 14-3 is $10/20 = 0.5$. and, for elbow *B* the aspect ratio is $20/10 = 2.0$.

For elbow turns other than 90 degrees, the loss is not proportional to the angle. Correction factors, *K*, for loss at angles other than 90 degrees are shown in Tables 14-3(a) and 14-4(a).

The dynamic loss coefficient, C_o, is not affected by the roughness of the duct walls except in elbows. Here it affects the dynamic loss coefficient, thus introducing another friction factor besides dynamic loss and normal friction loss. To simplify the solution of pressure loss in elbows, the dynamic loss coefficient is applied to the velocity pressure head, Eqs. (13-4) or (13-6), and the friction in the elbow is taken care of by measuring the straight duct to the intersection of the elbow centerlines.

Static pressure losses in 90-degree and angular takeoffs for round ducts (Fig. 14-4) and for 90-degree takeoffs for rectangular ducts (Table 14-7) depend upon the relative flows in the main and branch. These takeoff losses are given as static pressure losses from the main to the branch. There is also a dynamic loss in the main (upstream to downstream) at the branch takeoff. This loss is normally very small in well-designed takeoffs, and can be neglected for small and medium-sized systems. However, for large, more complex systems, this loss should be investigated.

Illustrative Problem 14-9

Determine the dynamic pressure loss in a horizontal 90-degree elbow having a width of 12 in., a depth of 24 in., and a radius ratio of 0.75. Standard air flows through the duct at a flow rate of 4000 cfm.

Solution:

The aspect ratio of the elbow (H/W) = 24/12 = 2.0. The radius ratio is given as 0.75. From Table 14-4(a) we find C_o = 0.39.

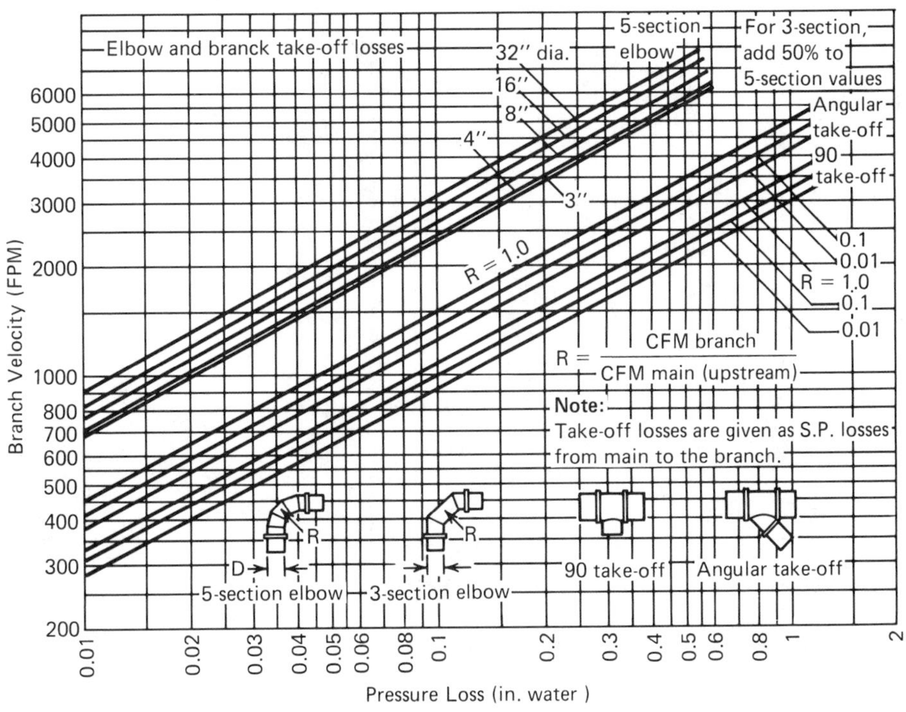

Figure 14–4

Elbow and branch take-off losses for round ducts (Reprinted from *Trane Air Conditioning Manual* with permission from The Trane Company, LaCrosse, Wisconsin.)

Table 14-7

Branch Take-Off Losses for Rectangular Ducts*

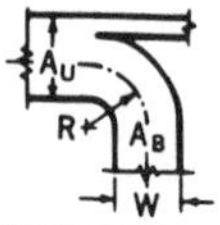

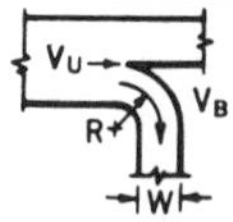

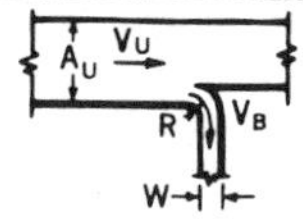

Branch Velocity, V_B, fpm	Aspect Ratio: H/W .25—.50		.75—3.0			4—6				
	Velocity Ratio: V_B/V_U 1.2	1.3	1.1	1.2	1.3	.9	1.0	1.1	1.2	1.3
	Area Ratio: A_B/A_U .7—1.2		.4—.6			.15—.3				
	Static Pressure Loss, Inches wg									
600	.011	.013	.007	.011	.018	.008	.011	.015	.018	.02
800	.02	.024	.013	.02	.032	.015	.02	.027	.032	.036
1000	.032	.038	.020	.032	.051	.023	.031	.042	.05	.057
1200	.046	.055	.028	.046	.073	.033	.045	.061	.073	.082
1400	.062	.074	.038	.062	.099	.045	.061	.082	.098	.111
1600	.081	.097	.05	.081	.129	.059	.08	.108	.129	.145
1800	.102	.123	.063	.102	.163	.074	.101	.136	.163	.183
2000	.126	.151	.078	.126	.20	.091	.125	.168	.20	.23
2100	.139	.167	.086	.139	.22	.101	.138	.185	.22	.25
2200	.153	.184	.095	.153	.24	.111	.151	.20	.24	.27
2300	.166	.20	.104	.166	.27	.121	.165	.22	.27	.30
2400	.182	.22	.113	.182	.29	.131	.18	.24	.29	.33
2500	.197	.24	.122	.197	.31	.142	.195	.26	.31	.35
2600	.21	.26	.133	.21	.34	.154	.21	.28	.34	.38
2700	.23	.28	.143	.23	.37	.166	.23	.31	.37	.41
2800	.25	.30	.153	.25	.39	.179	.24	.33	.39	.44
2900	.27	.32	.165	.27	.42	.192	.26	.35	.42	.48
3000	.28	.34	.176	.28	.45	.21	.28	.38	.45	.51
3100	.30	.36	.188	.30	.48	.22	.30	.40	.48	.54
3200	.32	.39	.20	.32	.52	.23	.32	.43	.52	.58
3300	.34	.41	.21	.34	.55	.25	.34	.46	.55	.62
3400	.36	.44	.23	.36	.58	.26	.36	.49	.58	.65
3500	.39	.46	.24	.39	.62	.28	.38	.51	.62	.69
3600	.41	.49	.25	.41	.65	.30	.40	.54	.65	.73
3700	.43	.52	.27	.43	.69	.31	.43	.58	.69	.78
3800	.46	.55	.28	.46	.73	.33	.45	.61	.73	.82
3900	.48	.58	.30	.48	.77	.35	.47	.64	.77	.86
4000	.51	.61	.31	.51	.81	.36	.50	.67	.81	.91
4100	.53	.64	.33	.53	.85	.38	.52	.71	.85	.95
4200	.56	.67	.35	.56	.89	.40	.55	.74	.89	1.00
4300	.58	.70	.36	.58	.93	.42	.58	.78	.93	1.05
4400	.61	.73	.38	.61	.97	.44	.60	.81	.97	1.10
4600	.67	.80	.41	.67	1.07	.48	.66	.89	1.06	1.20
4800	.73	.87	.45	.73	1.16	.53	.72	.97	1.16	1.30
5000	.79	.95	.49	.79	1.26	.57	.78	1.05	1.26	1.41
5200	.85	1.03	.53	.85	1.36	.62	.84	1.14	1.36	1.53
5400	.92	1.11	.57	.92	1.47	.67	.91	1.22	1.46	1.65
5600	.99	1.19	.61	.99	1.58	.72	.98	1.32	1.57	1.77
5800	1.06	1.27	.66	1.06	1.69	.77	1.05	1.41	1.69	1.90
6000	1.14	1.36	.70	1.14	1.81	.82	1.12	1.51	1.81	2.04

Example: Upstream duct, 24 by 12 inches, handling 5000 cfm; branch duct, 12 by 12, handling 3000 cfm; A_B/A_U = 0.5; H/W = 1.0; V_B/V_U = 3000/2500 = 1.2. Static pressure drop for 3000-fpm branch velocity = 0.28 inches wg.

Note: Radius ratio (R/W) should equal or exceed unity. For branch take-offs with lower velocity ratios (V_B/V_U) uses values for elbows of comparable aspect ratio (H/W) and radius ratio of 1.0. See Table 5.

Reprinted from Handbook of Air Conditioning, Heating and Ventilating, *2nd ed., with permission from The Industrial Press, New York, 1965.*

The velocity of flow in the elbow is

$$V = \frac{Q}{A} = \frac{4000(144)}{(24)(12)} = 2000 \text{ fpm}$$

By Eq. (13-6) $h_v = \left(\frac{V}{4005}\right)^2 = \left(\frac{2000}{4005}\right)^2$

$$h_v = 0.249 \text{ in. WG}$$

By Eq. (14-4)

$$(h_L)_d = C_o \times h_v = (0.39)(0.249)$$

$$(h_L)_d = 0.097 \quad \text{(say 0.10 in. WG)}$$

Illustrative Problem 14-10

A 16-in. diameter duct contains a five-piece, 90-degree elbow with a radius ratio of 2.0. Flow rate of standard air through the elbow is 3000 cfm. Determine the dynamic loss through the elbow by using (a) Table 14-3 and (b) Fig. 14-4.

Solution:

(a) The velocity of air flow is

$$V = \frac{Q}{A} = \frac{3000}{(\pi/4)(16/12)^2} = 2150 \text{ fpm}$$

or, from Fig. 14-1a, approximately 2150 fpm. By Eq. (13-6)

$$h_v = \left(\frac{V}{4005}\right)^2 = \left(\frac{2150}{4005}\right)^2 = 0.288 \text{ in. WG}$$

Using Table 14-3(b), for a five-piece elbow with $r/D = 2.0$, find $C_o = 0.19$. By Eq. (14-4),

$$(h_L)_d = C_o \times h_v = (0.19)(0.288)$$

$$= 0.0547 \quad \text{(say, 0.0055 in. WG)}$$

(b) For a flow velocity of 2150 fpm and a 16-in. round section, the static pressure loss for a five-piece, 90-degree elbow is read as 0.055 approximately.

The radius ratio used for the data of Fig. 14-4 is not indicated. If the data for Fig. 14-4 is to be compared with that of Table 14-3(b), one would have to assume that the radius ratio used for data in Fig. 14-4 was equal to 2.0. Such comparison is risky. As a suggestion, for future work regarding losses in round elbows, we shall refer to the data of Table 14-3.

Illustrative Problem 14-11

A 90-degree round branch takeoff occurs in a round duct main. The flow in the 20-in. main is 4000 cfm and in the 8-in. branch is 400 cfm. What is the static pressure loss from the main to the branch?

Solution:

Referring to Fig. 14-1, the branch velocity is 1150 fpm.

Referring to Fig. 14-4, we need the ratio of the flow in the branch to the flow in the main (upstream), or

$$R = \frac{\text{cfm Branch}}{\text{cfm Main (upstream)}} = \frac{400}{4000} = 0.10$$

In Fig. 14-4 with $R = 0.10$ and velocity in branch of 1150 fpm, find pressure loss equal to 0.125 in. WG.

Illustrative Problem 14-12

A rectangular duct has a 90-degree branch takeoff located where the upstream flow in the main is 5000 cfm. The branch flow is 2000 cfm. The duct dimensions upstream of the branch takeoff are 22 in. in width and 12 in. in depth. The branch duct is 8 in. wide by 12 in. deep. The radius ratio is 1.0. Determine the static pressure loss for the 2000 cfm branch takeoff.

<u>*Solution:*</u>

Use Table 14-7.

$$A_B/A_u = (8 \times 12)/(22 \times 12) = 0.36$$

$$\text{Branch } H/W = 12/8 = 1.5$$

$$V_B = (2000 \times 144)/(8 \times 12)$$

$$= 3000 \text{ fpm}$$

$$V_u = (5000 \times 144)/(22 \times 12)$$

$$= 2727 \text{ fpm}$$

$$V_B/V_u = 3000/2727 = 1.1$$

From Table 14-7, find loss in branch takeoff of 0.176 in. WG.

14-8 ELBOWS WITH SPLITTERS

When an elbow is followed by a duct whose length is at least four times the equivalent diameter of the elbow, no splitters are required if the radius ratio of the elbow has a value of 1.5. The friction loss in an ordinary elbow with a radius ratio of 1.5 is about the same as the friction loss in an elbow of smaller radius ratio with splitters. Hence, where space permits, an ordinary elbow of this radius ratio should be used.

Where space limitations make necessary the installation of an elbow having a radius ratio of less than 1.5, splitters should be installed in the elbow. For elbows with a small radius ratio, the installation of such splitters materially reduces the pressure loss.

When elbows discharge directly into the atmosphere, or are followed directly by a discharge grille or diffuser, splitters should always be used. The splitters help to distribute the air flow evenly across the discharge so that the air velocity leaving the elbow will be far more uniform.

The dimensions of an elbow cannot be changed because they are determined by the duct in which the elbow is installed. However, if the aspect ratio of an elbow is low, the aspect ratio may be increased by installing splitters as shown in Figs. 14-5 and 14-6. The effect of installing splitters in an elbow is to make a number of separate smaller elbows out of the original elbow. For example, for the left elbow of Fig. 14-5, the effect is exactly the same as though two separate elbows were provided and the air flow was divided into two unequal parts—each part flowing through its own elbow. Each of the smaller elbows will have a higher aspect ratio than the original elbow. This is also apparent in Fig. 14-6. In addition to improving the aspect ratio, splitters also increase the radius ratio. Thus, in Figs. 14-5 and 14-6 the radius ratio of each of the smaller elbows is higher than the radius ratio of the single original elbow.

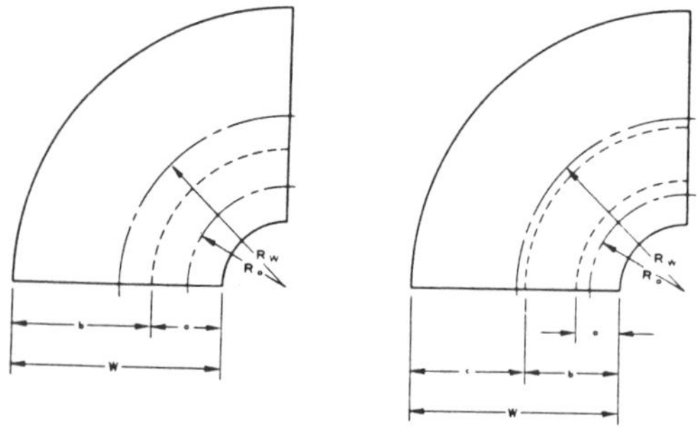

1 SPLITTER

R_w/W	a/W	$R_a/a = R_b/b$
0.75	0.309	1.31
1.00	0.364	1.86
1.25	0.394	2.40

2 SPLITTERS

R_w/W	a/W	b/W	$R_a/a = R_b/b = R_c/c$
0.75	0.178	0.303	1.91
1.00	0.219	0.319	2.76
1.25	0.244	0.325	3.56

Figure 14–5

Correct location of elbow splitters (Reprinted from *Trane Air Conditioning Manual* with permission from The Trane Company, LaCrosse, Wisconsin.)

Proper location of the splitters installed in an elbow is important. The data in Fig. 14-5 can be used to determine the location of the

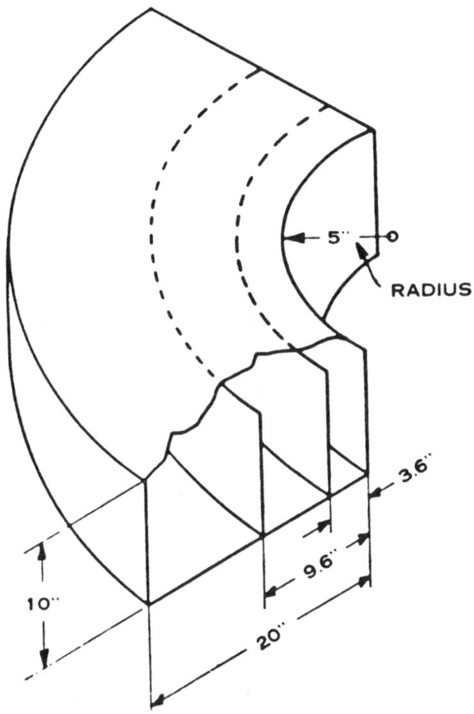

Figure 14–6

Improving aspect ratio with splitters

splitters. The use of one splitter in an elbow is quite common, and two splitters are used occasionally. The splitters are so located that the radius ratio of each of the separate elbows formed by the splitters is about the same. A word of caution should be noted here. When splitters are used in an elbow to improve the aspect ratio, there is a possibility that the radius ratio of each of the individual elbows formed may be considerably greater than 2. In such a case the friction loss of the elbow with splitters may be greater than the friction loss of a plain elbow due to the increased length of the elbow.

Where an elbow with a radius ratio of 1.5 can be installed, there is not a sufficient gain to warrant installing splitters in elbows. Splitters are used in elbows only where requirements of space or appearance makes it necessary to install an elbow with a small radius ratio.

Table 14-8A and 14-8B provide information for the dynamic loss coefficient for smooth, rectangular elbows with splitter vanes.

14-9 SQUARE (MITER) ELBOWS WITH VANES

Referring to Tables 14-3(c) and 14-4(b), we may note that loss coefficients for square, or miter, elbows are high compared with loss coefficients for elbows of a round heel radius. Frequently, however, because of requirements of appearance of space, a square elbow must be used. If this is the case, vanes should be installed as in Fig. 14-7 to keep turbulence losses as low as possible. In addition, when such elbows are used to discharge air directly to the atmosphere or through a grille, vanes are very valuable for uniformly distributing the air flow.

The space between each pair of vanes in Fig. 14-7 forms an individual elbow. Obviously, the aspect ratio of each of the individual elbows thus depends on the space between the vanes, and this space, in turn, depends upon the number of vanes used. As a rule, vanes are so spaced that the aspect ratio of each of the individual elbows formed by the vanes will be about 5. Thus, elbow *A* of Fig. 14-3 is 20 in. wide and 10 in. deep. If this elbow were made square, and the vanes were spaced 2 in. apart, the aspect ratio of each elbow formed would be $10/2 = 5$. Consequently, if the elbow were designed like the one in Fig. 14-7, the distance between each vane is 2 in., and since the elbow is 20 in. wide, the number of vanes required would be $(20/2) - 1 = 9$.

An extended lip is usually installed on each end of the vanes. This extension is usually made equal to about half the radius on the entering side and equal to the radius on the leaving side as shown in Fig. 14-7.

Table 14-8A

Coefficients for Elbows with One Splitter Vane*

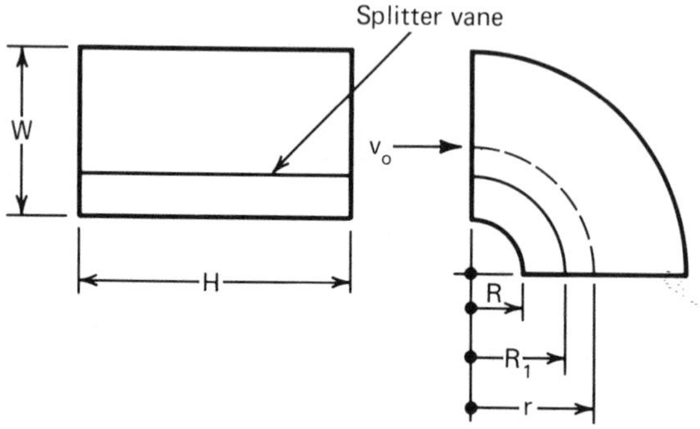

$C_o = KC'_o$
$R_1 = R/CR$

where

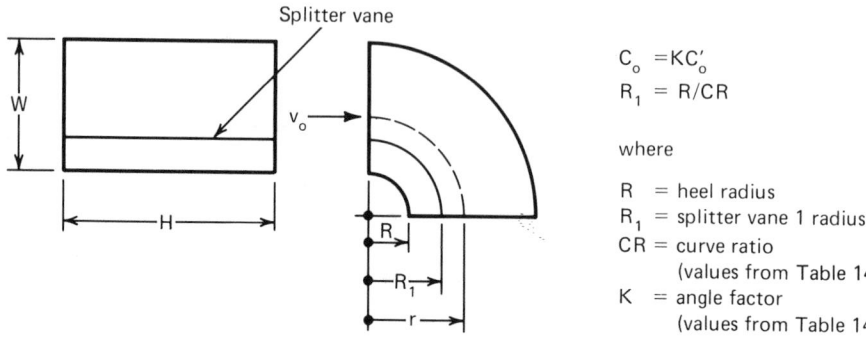

R = heel radius
R_1 = splitter vane 1 radius
CR = curve ratio
(values from Table 14-8A)
K = angle factor
(values from Table 14-4)

Table 14-8A
Coefficients for Elbows with One Splitter Vane*

C'_o

			H/W									
R/W	r/W	CR	0.25	0.5	1.0	1.5	2.0	3.0	4.0	5.0	6.0	7.0
0.05	0.55	0.218	0.52	0.40	0.43	0.49	0.55	0.66	0.75	0.84	0.93	1.0
0.10	0.60	0.302	0.36	0.27	0.25	0.28	0.30	0.35	0.39	0.42	0.46	0.49
0.15	0.65	0.361	0.28	0.21	0.18	0.19	0.20	0.22	0.25	0.26	0.28	0.30
0.20	0.70	0.408	0.22	0.16	0.14	0.14	0.15	0.16	0.17	0.18	0.19	0.20
0.25	0.75	0.447	0.18	0.13	0.11	0.11	0.11	0.12	0.13	0.14	0.14	0.15
0.30	0.80	0.480	0.15	0.11	0.09	0.09	0.09	0.09	0.10	0.10	0.11	0.11
0.35	0.85	0.509	0.13	0.09	0.08	0.07	0.07	0.08	0.08	0.08	0.08	0.09
0.40	0.90	0.535	0.11	0.08	0.07	0.06	0.06	0.06	0.06	0.07	0.07	0.07
0.45	0.95	0.557	0.10	0.07	0.06	0.05	0.05	0.05	0.05	0.05	0.06	0.06
0.50	1.00	0.577	0.09	0.06	0.05	0.05	0.04	0.04	0.04	0.05	0.05	0.05

*Reprinted from ASHRAE *Handbook and Product Directory,* 1977, *Fundamentals,* with permission of the American Society of Heating, Refrigerating and Air-Conditioning Engineers, Atlanta, GA.

Table 14-8B

Coefficients for Elbows with Two Splitter Vanes*

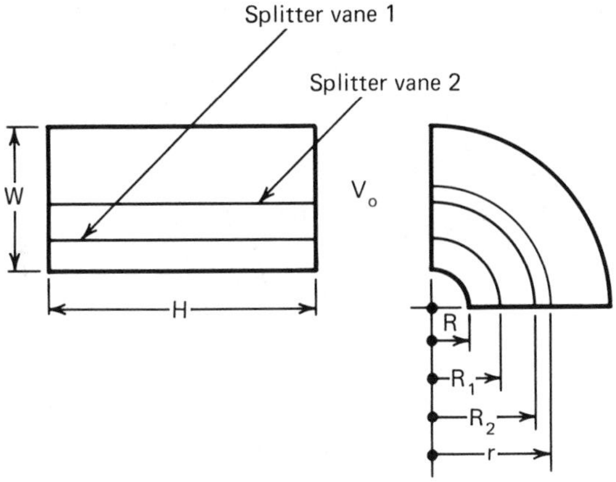

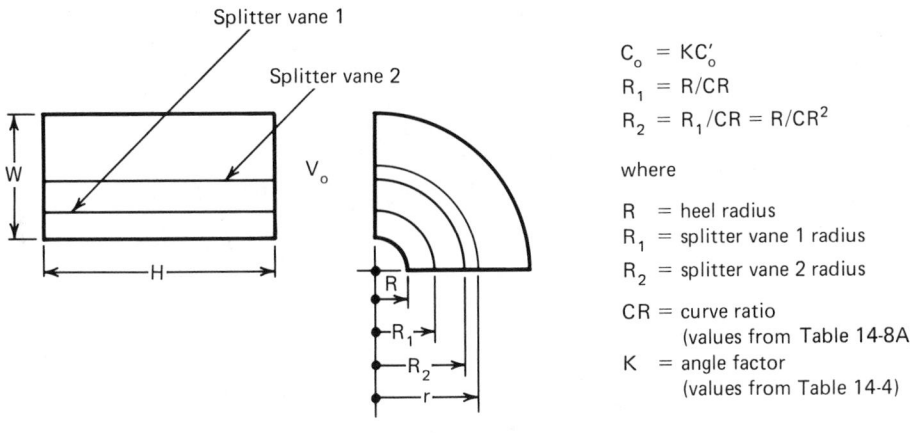

$C_o = KC'_o$
$R_1 = R/CR$
$R_2 = R_1/CR = R/CR^2$

where

R = heel radius
R_1 = splitter vane 1 radius
R_2 = splitter vane 2 radius
CR = curve ratio (values from Table 14-8A)
K = angle factor (values from Table 14-4)

Table 14-8B
Coefficients for Elbows with Two Splitter Vanes*

C'_o

			H/W									
R/W	r/W	CR	0.25	0.5	1.0	1.5	2.0	3.0	4.0	5.0	6.0	7.0
0.05	0.55	0.362	0.26	0.20	0.22	0.25	0.28	0.33	0.37	0.41	0.45	0.48
0.10	0.60	0.450	0.17	0.13	0.11	0.12	0.13	0.15	0.16	0.17	0.19	0.20
0.15	0.65	0.507	0.12	0.09	0.08	0.08	0.08	0.09	0.10	0.10	0.11	0.11
0.20	0.70	0.550	0.09	0.07	0.06	0.05	0.06	0.06	0.06	0.06	0.07	0.07
0.25	0.75	0.585	0.08	0.05	0.04	0.04	0.04	0.05	0.05	0.05	0.05	0.05
0.30	0.80	0.613	0.06	0.04	0.03	0.03	0.03	0.03	0.03	0.03	0.04	0.04
0.35	0.85	0.638	0.05	0.04	0.03	0.03	0.03	0.03	0.03	0.03	0.03	0.03
0.40	0.90	0.659	0.05	0.03	0.03	0.02	0.02	0.02	0.02	0.02	0.02	0.02
0.45	0.95	0.677	0.04	0.02	0.02	0.02	0.02	0.02	0.02	0.02	0.02	0.02
0.50	1.00	0.693	0.03	0.02	0.02	0.02	0.02	0.01	0.01	0.01	0.01	0.01

*Reprinted from ADHRAE *Handbook and Product Directory,* 1977, *Fundamentals,* with permission of the American Society of Heating, Refrigerating and Air-Conditioning Engineers, Atlanta, GA.

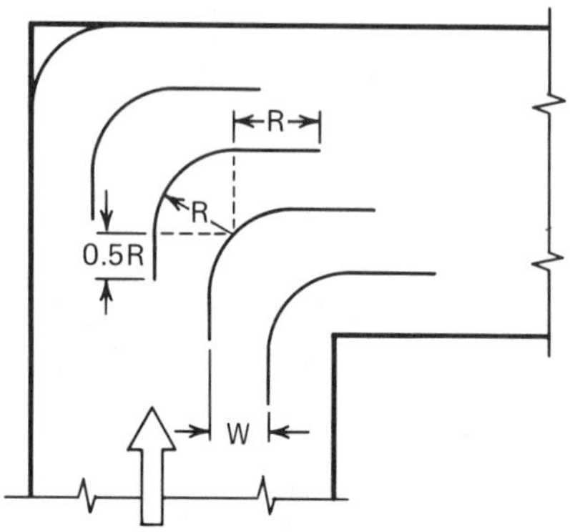

Figure 14–7

Turning vanes in rectangular, miter elbows (Reprinted from *Trane Air Conditioning Manual* with permission from The Trane Company, LaCrosse, Wisconsin.)

Because the blades illustrated in Fig. 14-7 are each drawn with a different center, the space between corresponding points of adjacent blades is not the same. Consequently, the velocity of the air will change as it flows through each of the individual elbows formed by adjacent blades. Because of this change in velocity, eddy losses occur which increase the loss of such elbows. It is possible, by using air foil-shaped blades, to maintain a uniform spacing and hence a uniform velocity. These blades are more expensive and frequently the gain by use of them is not sufficient to warrant the additional cost.

Ducturns are nonadjustable, 90-degree air turns designed to reduce the pressure loss in square duct elbows. Such devices, as manufactured by Tuttle & Bailey, have galvanized steel blades, are double walled and formed to assure that at any point on one blade is equidistant from the same point on an adjacent blade (see Fig. 14-8). This precise blade shape maintains constant duct area and assures constant air velocity around the elbow, eliminates loss due to velocity changes, and results in low friction loss, as seen in Table 14-9. The blade is rigid and eliminates the need for cross bracing or reinforcing rods. Roll-formed from a single sheet of metal, surfaces and edges are smooth and free from edge friction and blade turbulence. Blades assemble over precision-formed tenons on the sidepieces, which insures precise blade alignment, proper spacing, and predictable engineering performance. Blades and sidepieces are finished in 6-ft lengths. They are cut to size and assembled in the field. Assembly is simple and economical since no spe-

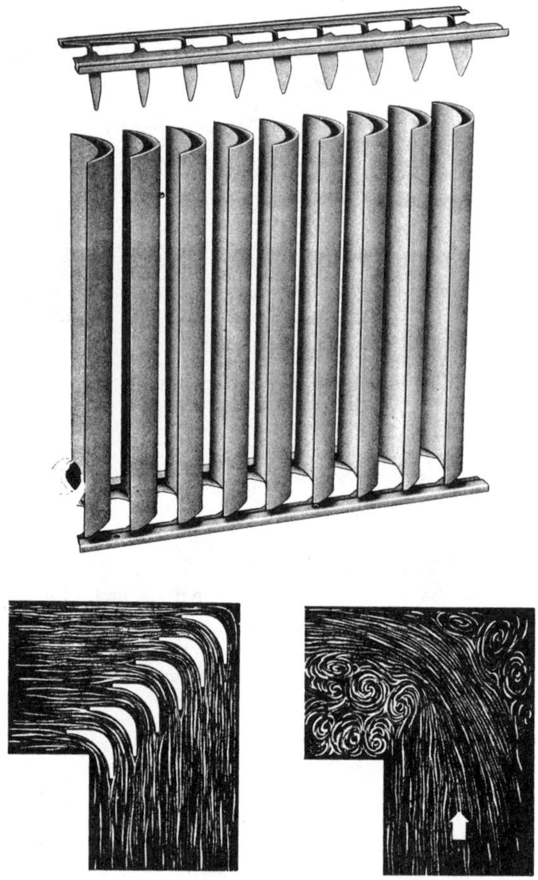

Figure 14–8

Ducturns (Courtesy of Tuttle & Bailey.)

Table 14-9

Pressure Loss in Ducturns, in. WG*

Type of Elbow	Duct Velocity (FPM)					
	500	1000	1500	2000	2500	3000
Ducturns	<0.01	0.02	0.04	0.07	0.11	0.16
No Vanes[a]	0.02	0.08	0.19	0.35	0.52	0.75

[a]See Table 14-4.

**Courtesy of Tuttle & Bailey, Division of Allied Thermal Corporation.*

cial tools and no fasteners are needed. The complete unit is screwed or riveted into the duct elbow.

14-10 DIVERGING AND CONVERGING DUCT SECTIONS

Special care must be exercised in making transition connections to and from equipment. Usually the cross-sectional area of the conditioning equipment (duct-mounted coils, etc.) will be greater than the cross-sectional area leading to or from the equipment, as shown in Fig. 14-9, necessitating a diverging section of duct work ahead of the equipment and converging section after it. Angle A should be made as small as possible in order to spread the air flow evenly across the face of the equipment and also to reduce the loss. When the angle is unduly large, the losses will be high because the air is unable to expand and fill the diverging section as rapidly as the metal sides diverge. Experiments have shown that the minimum loss is obtained when angle *A* is about 3 degrees. However, such small angles can rarely be used in actual installations because the length of the diverging section would be too great. An effort should be made to keep angle *A* less than 10 degrees. If it must be made larger because of space limitations, then splitters should be installed. These splitters will serve the double purpose of reducing the loss, and, at the same time, will distribute the air evenly over the face of the conditioning equipment. See Tables 14-5(a) and 14-5(b) for losses in diverging sections.

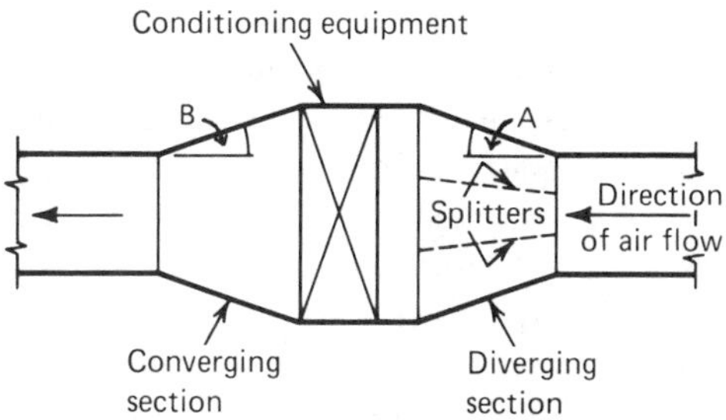

Figure 14–9

Transition pieces in ducts (Reprinted from *Trane Air Conditioning Manual* with permission from The Trane Company, LaCrosse, Wisconsin.)

The losses in converging sections are very much less than that of diverging sections. An airstream can be forced into a smaller area with almost no eddy loss if angle *B* of Fig. 14-9 can be made less than 15 degrees. If the angle is greater than 15 degrees, splitters could be used to reduce losses, but because of cost it is seldom done, unless angle *B* exceeds 30 degrees. See Tables 14-6(a) and 14-6(b) for losses in converging sections.

14-11 DUCT SIZING METHODS

The three most common methods used to determine velocities and pressure losses in ducts, and consequently duct sizes, are briefly summarized below:

a. velocity-reduction method

b. equal-friction-rate method

c. static-regain method

The choice of method depends upon the designer's choice, usually influenced by past experience and the size of the system. Small duct systems, where air flow velocities are 1000 fpm or less, are frequently designed on the basis of a modified velocity-reduction method. Design manuals published by the National Environmental Systems Contractors Association are available and discuss in detail the design of small-to-medium size duct systems that would be used in homes, shops, and small commercial establishments. Large systems using high-velocity air flow are most frequently designed by the static-regain method. Duct arrangements in between the small and large systems are nearly always designed by the equal friction-rate method. Sometimes a duct system will be designed by a combination of two methods. For instance, the trunk (main) will be laid out by the static-regain method and the branch ducts will be designed by the equal-friction-rate method.

There are no set rules concerning the air flow velocities to be used in duct systems. However, velocities recommended in Table 14-1 for small-capacity (5000 cfm) systems or velocity ranges shown in Fig. 14-10 for low-velocity and high-velocity systems having greater volume flow may be used as guides in selecting velocities for various systems.

14-12 THE VELOCITY-REDUCTION METHOD

The velocity-reduction method of duct design requires an arbitrary assignment of velocities to the various sections of the duct system. The highest velocity is chosen at the entrance to the duct system, immediately following the fan outlet, in accordance with velocities shown in Table 14-1 or Fig. 14-10. The velocities in succeeding sections are reduced as various branches are taken off the main duct. The velocity will be lowest at the end of the duct. Since the air quantity for each section of the duct is already known, assuming a velocity permits the flow area to be calculated, and the diameter of the duct may be easily calculated. A more direct method would be to use the flow volume (cfm) and assumed velocity (fpm) and enter Fig. 14-1 or 14-1a and determine the duct diameter and friction rate per 100 ft of duct.

The velocity-reduction method is empirical and requires good judgment by the designer to select appropriate velocities. In some cases, such as sizing exhaust duct systems, a modification of the velocity-reduction method is fastest and best suited for design purposes. This modified method, sometimes called *balanced-pressure-loss method,* which considers total pressure losses, makes it possible to design each branch run to have the same pressure loss from the fan in order that minimum dependence on dampering would be required.

Illustrative Problem 14-13

The supply duct work for a small office complex is shown in Fig. 14-11. Ceiling diffusers will be used with the duct system above the ceiling. Supply air volumes are represented at each outlet. Determine the following:

1. Size the duct work by the velocity-reduction method, using a uniform duct depth of 14 in. in the main and branches.
2. Calculate the friction loss for the system assuming that all branch takeoffs are 90-

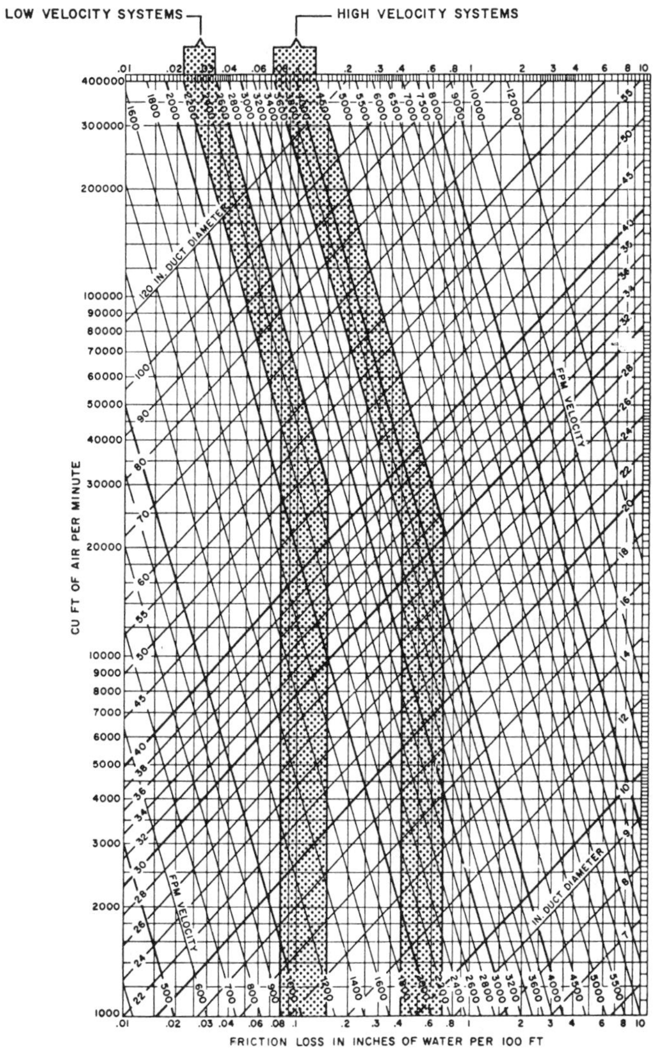

Figure 14–10

Recommended velocity ranges for air flow systems (Reproduced from ASHRAE *Handbook and Product Directory*, 1977 *Fundamentals*, with permission of the American Society of Heating, Refrigeration and Air-Conditioning Engineers, Atlanta, Georgia.)

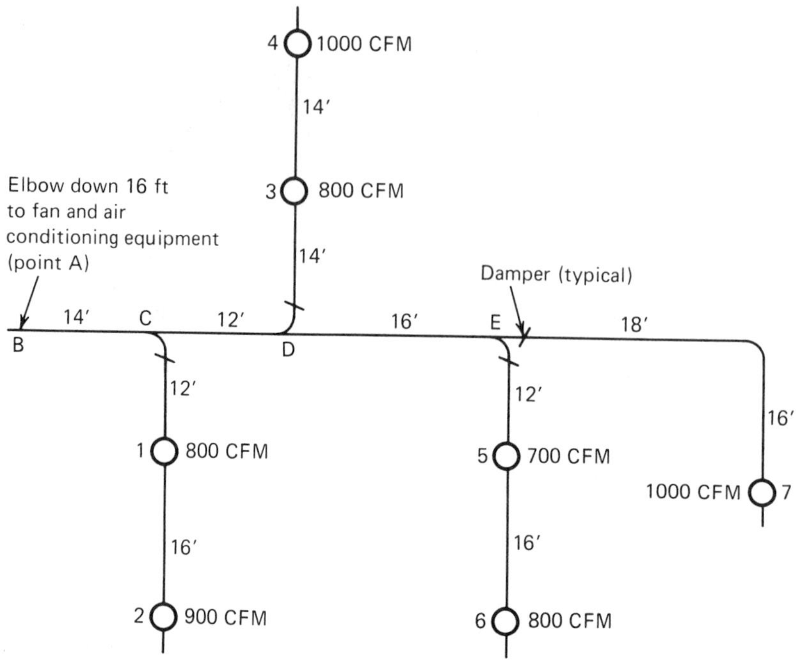

Figure 14–11

Diagram for Illustrative Problem 14-13 (Does not represent an actual system.)

degree rectangular elbows and that all elbows have a radius ratio of 1.0.

<u>*Solution:*</u>

Tabulation of various quantities is helpful in the solution of problems in duct design. Table 14-10 will be used for this problem.

Step 1: Tabulate the flow quantities in Table 14-10 for the various duct sections.

Step 2: Using Table 14-1 or Fig. 14-10, select velocities of various duct sections and enter in Table 14-10. Table 14-1 indicates a range of velocities for main ducts between 1000 and 1300 fpm (maximum 1100 to 1600 fpm). Figure 14-10, for a flow of 6000 cfm, indicates a range of 1400 to 1750 fpm for the main *ABC*. We shall select 1400 fpm for section *ABC* and reduce it to 1300 fpm for section *CD,* 1200 fpm for section *DE,* and 900 fpm for section *E*-7.

For the branch ducts, Table 14-1 recommends 600 to 900 fpm. We shall select 900 fpm for sections *E*-5, *D*-3, and *C*-1; and 800 fpm for sections 5-6, 3-4, and 1-2. These velocities are on the high side, but remember that they are for the round ducts. The air flow velocities in the rectangular ducts

Table 14-10

Tabulation of Results (Illustrative Problem 14-13)

Duct Section	Capacity (cfm)	Round Duct Velocity (fpm)	Round Duct Size (in.)	Rectangular Duct $W \times H$	Friction Rate (in. WG per 100 ft)
ABC	6000	1400	28.0	52 × 14	0.085
CD	4300	1300	25.0	40 × 14	0.085
DE	2500	1200	20.0	24 × 14	0.090
E–7	1000	900	14.3	12 × 14	0.085
E–5	1500	900	17.5	18 × 14	0.065
5–6	800	800	13.7	11 × 14	0.070
D–3	1800	900	19.3	24 × 14	0.059
3–4	1000	800	15.2	14 × 14	0.064
C–1	1700	900	19.0	22 × 14	0.060
1–2	900	800	14.3	12 × 14	0.066

selected for final design will be lower.

Step 3: Using Figs. 14-1 and 14-1a for the cfm and selected velocities, determine the round sizes for the various sections and the friction rate, and enter values in Table 14-10.

Step 4: Using Table 14-2, convert the round duct diameters into rectangular equivalents, maintaining a uniform duct depth of 14.0 in. Tabulate rectangular sizes in Table 14-10.

Step 5: Determine the maximum pressure loss for the system. The longest run (usually the one found to have the greatest loss) is *ABCDE* to outlet 7. In that section we have an elbow at *B* and one in section *E*-7. In addition we have the friction loss in the various duct sections.

For the elbow at B:

Radius ratio $(r/W) = 1.0$ (given)

Aspect ratio $(H/W) = 52/14 = 3.7$

From Table 14-4a, $C_o = 0.21$ (approximate)

The velocity in the rectangular duct is

$$V = Q/A = (6000)(144)/(52)(14) = 1187 \text{ fpm}$$

Assuming standard air, the velocity head is

$$h_v = (V/4005)^2 = (1187/4005)^2 = 0.0878 \text{ in. WG}$$

Therefore, the loss for the elbow is

$$(h_{L_d}) = C_o \times h_v = 0.21(0.0878) = 0.018 \text{ in. WG}$$

For the elbow between E and 7:

Radius ratio $(r/W) = 1.0$ (given)

Aspect ratio $(H/W) = 14/12 = 1.16$

From Table 14-4a, $C_o = 0.21$ (approximate)

The velocity in the rectangular duct is

$$V = Q/A = 1000(144)/(12)(14) = 857 \text{ fpm}$$

Assuming standard air, the velocity head is

$$h_v = (V/4005)^2 = (857/4005)^2 = 0.0458 \text{ in. WG}$$

Therefore the loss for the elbow is

$$(h_{Ld}) = C_o \times h_v = 0.21(0.0458) = 0.0096 \quad (\text{say } 0.010 \text{ in. WG})$$

The various duct sections (*ABCDE* to 7) have slightly different friction rates (in. WG per 100 ft), therefore we shall determine the friction loss in each section and sum the results:

$$ABC = 0.085(30/100) = 0.0255$$
$$CD = 0.085(12/100) = 0.0102$$
$$DE = 0.090(16/100) = 0.0144$$
$$E\text{-}7 = 0.085\ (34/100) = \underline{0.0289}$$
$$\text{Straight Duct Friction} = 0.0790 \text{ in. WG}$$

The total loss for the duct run *ABCDE* to 7 is

Straight duct friction	=	0.0790
Elbow loss at *B*	=	0.0180
Elbow loss *E*-7	=	0.0100
Total loss *ABCDE*-7	=	(say 0.110 0.107 in WG)

14-13 THE EQUAL-FRICTION-RATE METHOD

In the equal-friction-rate method of duct design, the design friction rate (in. WG per 100 ft) is maintained constant throughout the system. The design friction rate depends upon the allowable velocity in the duct system. In commercial installations this is determined wholly by considerations of noise. For ordinary commercial and industrial systems, the maximum velocities shown in Table 14-1 or Fig. 14-10 should not be exceeded.

Illustrative Problem 14-14

The supply duct work for a portion of an industrial system is shown in Fig. 14-12. Determine the rectangular size for each section of the system using the equal-friction-rate method. The duct depth will be maintained at 14 in. throughout the system. Standard air is flowing in the system at the flow rates indicated. Determine the maximum static pressure required at point *A* to flow the system. Assume that the static pressure loss at each of the supply outlets is 0.05 in. WG, *r*/*W* for elbows will be taken as 1.0.

<u>Solution:</u>

Step 1: Determine the uniform design friction rate. Table 14-1 allows a velocity range of 1300 to 2200 fpm for main ducts and 1000 to 1800 fpm for branch lines in industrial applications. Figure 14-10 allows comparable velocity ranges.

Referring to Fig. 14-1a with our total flow rate of 9000 cfm and a velocity 2200 fpm, the corresponding friction rate is 0.21 in. WG per 100 ft. Also, for 9000 cfm and a velocity of 1800 fpm, the corresponding friction rate is 0.125 in. WG per 100 ft. Since a vertical friction rate line of 0.150 in. WG per 100 ft appears on Fig. 14-1a, and because this

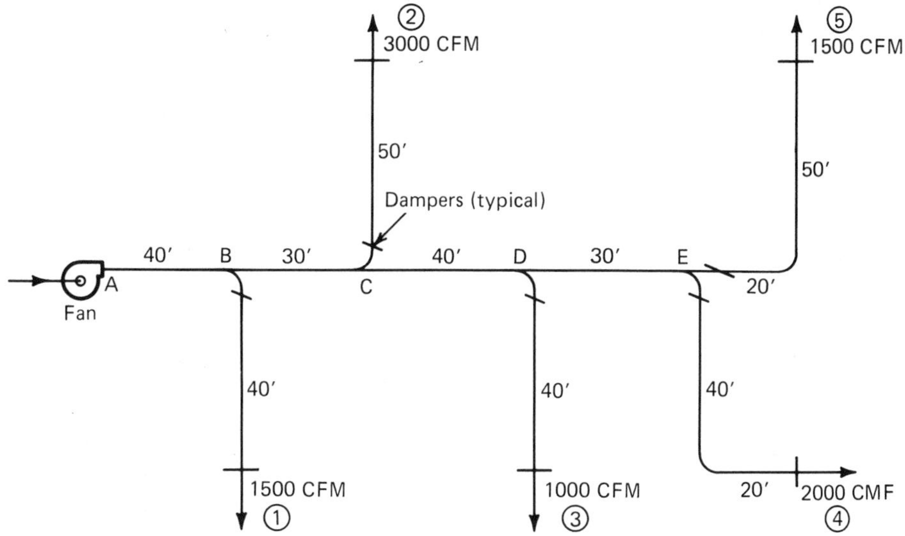

Figure 14–12

Diagram for Illustrative Problem 14-14 (Does not represent an actual system.)

Table 14-11

Tabulation of Results (Illustrative Problem 14-14)

Duct Section	Capacity (cfm)	Design Friction Rate (in. WG per 100 ft)	Round Duct Size (in.)	Rectangular Duct (*W* × *H*)	Rectangular Duct Velocity (fpm)	Velocity Head (in. WG)
AB	9000	0.150	29.2	58 × 14	1596	0.159
BC	7500		27.0	48 × 14	1607	0.161
CD	4500		22.5	32 × 14	1446	0.130
DE	3500		20.5	26 × 14	1385	0.120
E–5	1500		15.0	14 × 14	1102	0.076
E–4	2000		16.6	17 × 14	1210	0.091
D–3	1000		12.8	10 × 14	1028	0.066
C–2	3000		19.4	22 × 14	1402	0.123
B–1	1500		15.0	14 × 14	1102	0.076

rate is between the above two values, we shall select a design friction rate of 0.150 in. WG which will be held constant for the system.

Step 2: Construct a results table for the problem data.

Step 3: Tabulate flow quantities (cfm) and the design friction rate in Table 14-11.

Step 4: Using Figs. 14-1 and 14-1a with cfm and the constant design friction rate, find round duct diameters and insert in Table 14-11.

Step 5: Using Table 14-2, convert the round duct to rectangular equivalent sizes and insert results in Table 14-11.

Step 6: Calculate air flow velocity in each section of the duct system from $Q = AV$, and insert results in Table 14-11. Two reasons for this are to determine if the velocities are within limits and to determine the velocity heads in the sections for use as required in finding losses.

Step 7: Determine the greatest pressure loss for the system. It is difficult by inspection to establish whether the greatest loss occurs from point A to outlet 5 or to outlet 4. Therefore, we shall check each one.

(a) *Loss A to outlet 4:*

Branch Takeoff at E: Use of Table 14-7 requires that we know the branch velocity V_B, the aspect ratio H/W, velocity ratio V_B/V_u, and the area ratio A_B/A_u.

$V_B = 1210$ fpm: $V_u = 1385$ fpm (results table)

$H/W = 14/17 = 0.82$

$V_B/V_u = 1210/1385 = 0.87$

$A_B/A_u = (17)(14)/(26)(14) = 0.65$

From Table 14-7, find $(h_L)_d = 0.028$ in. WG (estimated).

Elbow between E and outlet 4: To find C_o from Table 14-4 we need r/W and H/W.

$r/W = 1.0$ (given)

$H/W = 14/17 = 0.82$

From Table 14-4(a), find $C_o = 0.22$ (approximate). Therefore, the elbow loss $(h_L)_d = C_o \times h_v = 0.22\,(0.091) = 0.02$ in. WG.

Duct loss $(h_L)_{A-4} = 0.150(200/100)$
$= 0.300$ in. WG

Total loss A to 4 $= 0.300 + 0.028 + 0.02$
$= 0.348$ in. WG

(b) *Loss A to outlet 5:*

Elbow between E and 5: $r/W = 1.0$ (given) and $H/W = 14/14 = 1.0$. From Table 14-4(a), find $C_o = 0.21$. Therefore, the elbow loss $(h_L)_d = 0.21(0.076) = 0.016$ in. WG

The duct loss A to 5, $(h_L)_{A-5} = 0.150\,(210/100) = 0.315$ in. WG

The total loss A to 5 $= 0.315 + 0.016 = 0.331$ in WG

The loss A to 4 is slightly higher than loss A to 5.

Therefore, the static pressure required at point A depends on the loss A to outlet 4.

Step 8: Determine the static pressure required at point A. The total pressure at A is equal to

$$h_{t_A} = \text{Duct loss A to 4} + \text{diffuser loss} + \text{velocity head in section } E\text{-4}$$

Therefore,

$$h_{tA} = 0.348 + 0.05 + 0.091$$
$$h_{tA} = 0.489 \text{ in. WG}$$

Since

$$h_{tA} = h_{s_A} + h_{v_A}, \text{ then } h_{s_A} = h_{t_A} - h_{v_A}, \text{or}$$
$$h_{s_A} = 0.498 - 0.159 = 0.330 \text{ in WG}$$

Use of the equal-friction-rate method requires that dampers be used in each of the branch lines because the available static pressure in the main at the branch takeoffs is higher than that required to flow in the branches. For example, the total pressure at point D will be equal to the total pressure at A minus the friction loss A to D, or

$$h_{t_D} = h_{t_A} - \text{losses}$$
$$= 0.489 - (0.150)(110/100)$$
$$h_{t_D} = 0.324 \text{ in. Wg}$$

The static pressure at D will be

$$h_{s_D} = h_{t_D} - (h_v)_{CD} = 0.324 - 0.130$$
$$= 0.194 \text{ in. WG}$$

If the pressure losses in branch D-3 total less than 0.194 in. WG, then a damper must be installed in the branch. An alternative to the use of a damper would be to resize branch D to 3 to use up the available pressure at D of 0.194 in. WG. This technique is called the *balanced-pressure method* and is sometimes employed in system design to reduce the necessity of severe dampering which otherwise might be required. The balanced-pressure method requires trial-and-error techniques that may be time consuming, and dampers should still be used to provide a means for final balancing of the system after it is installed. The following example illustrates this principle of balancing pressure losses in the system.

Illustrative Problem 14-15

Figure 14-13 shows a ventilation system with three supply outlets. Determine the required rectangular sizes of the duct sections such that the static pressure loss from the fan (point A) to each outlet will be approximately the same. The duct section AB is to have a depth of 18 in., and all other duct sections will have a depth of 14 in. Assume that all elbows and branch takeoffs have an $r/W = 0.5$. Start the problem by sizing the duct section $ABCD$ by the equal-friction-rate method using a design friction rate of 0.250 in. WG per 100 ft.

<u>*Solution:*</u>

Step 1: Determine duct sizes by equal-friction-rate method for duct run $ABCD$, as in table at top of following page.

Step 2: Determine pressure loss in section BC.

Because of the change in duct depth from section AB to section BC, an equal area transition (EAT) would normally be used. *For any equal area transition* the loss coefficient $C_o = 0.15$. Because of the 3000 cfm flow into branch BF, 7000 cfm flows into section BC. If the duct depth immediately following point B is 18 in., then, from Table 14-2 the width would be 27 in. Therefore, the EAT would be from 27 by 18 to 36 by 14, approximately.

$$V_{BC} = 7000(144)/(36)(14)$$
$$= 2000 \text{ fpm}$$
$$(h_v)_{BC} = (V/4005)^2$$

Duct Section	Capacity (cfm)	Design Friction Rate (in. WG/100 ft)	Round Duct Size (in.)	Rectangular Duct Size ($W \times H$)	Rectangular Duct Velocity (fpm)
AB	10,000	0.250	27.8	38 × 18	2100
BC	7,000	0.250	24.0	36 × 14	2000
CD	5,000	0.250	21.2	28 × 14	1837

$$= (2000/4005)^2$$

$$= 0.249 \text{ in. WG}$$

$$\text{Loss EAT} = C_o(H_v)_{BC} = (0.15)(0.249)$$

$$= 0.037 \text{ in. WG}$$

$$\text{Duct loss } BC = 0.250(100/100)$$

$$= 0.250 \text{ in. WG}$$

$$\text{Total loss } BC = 0.250 + 0.037$$

$$= 0.287 \text{ in. WG}$$

Step 3: Determine pressure loss for section CD. For the single elbow between C and D,

$$r/W = 0.5 \quad \text{(given)}$$

$$H/W = 14/28 = 0.5$$

From Table 14-4(a), $C_o = 1.0$. Velocity in section CD is, $V_{CD} = 5000(144)/(28)(14)$

$$V_{CD} = 1837 \text{ fpm}$$

$$(h_v)_{CD} = (1837/4005)^2 = 0.210 \text{ in. WG}$$

Elbow loss $= C_o(h_v)_{CD} = 1.0(0.210) =$ 0.210 in. WG

Duct loss $CD = 0.250(75/100) = 0.188$ in. WG

Total loss section $CD = 0.188 + 0.210 =$ 0.398 in. WG

Step 4: Make pressure loss in section CE equal to pressure loss in section CD (0.398 in. WG).

The procedure here is to assume a friction rate for section CE such that the loss will be equal to 0.398 in. WG

Noting that the length of section CE is about the same as section CD and there is one elbow, the major difference is the takeoff elbow at C.

As a *first trial,* in our trial-and-error solution, assume the friction rate at, say 0.20 in. WG. From Fig. 14-1a with 2000 cfm, find a trial diameter of 15.6 in. From Table 14-2, an equivalent rectangular size would be 15 by 14.

$$V_{CE} = 2000(144)/(15)(14) = 1371 \text{ fpm}$$

$$(h_v)_{CE} = (1371/4005)^2 = 0.117 \text{ in. WG}$$

For the *branch takeoff, using Table 14-7,*

Aspect ratio $H/W = 14/15 = 1.0$ (approximate)

Velocity ratio $V_B/V_u = 1371/2000 = 0.69$

Area ratio $A_B/A_u = (14)(15)/(36)(14) = 0.42$

The footnote of Table 14-7 recommends treating the takeoff as an elbow of $r/W = 1.0$. Re-

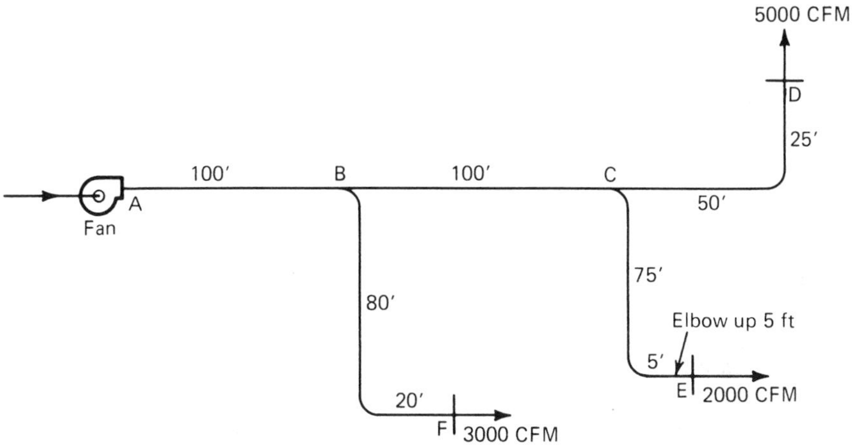

Figure 14–13

Diagram for Illustrative Problem 14-15 (Does not represent an actual system.)

ferring to Table 14-4(a) with $H/W = 1.0$, and $r/W = 1.0$, find $C_o = 0.21$.

$$\text{Branch takeoff loss} = C_o(h_v)_{CE} = 0.21(0.117) = 0.025 \text{ in. WG}$$

For each of the four elbows, using Table 14-4(a), results in the same loss as the takeoff, or 0.025 in. WG

Duct loss $CE = 0.20(85/100) = 0.170$ in. WG

Total loss section $CE = 0.170 + 4(0.025) = 0.270$ in. WG

This is significantly less than the desired value of 0.398 in. WG

For a *second trial,* assume a friction rate of 0.30 in. WG per 100 ft.

Using Fig. 14-1a, with a flow of 2000 cfm and friction rate of 0.30 in. WG per 100 ft, find $D = 14.5$ in. From Table 14-2 the equivalent rectangular size would be 12 by 14. This is still essentially a "square" cross section, and C_o will still be approximately 0.21.

$$V_{CE} = 2000(144)/(12)(14) = 1714 \text{ fpm}$$

$$(h_v)_{CE} = (1714/4005)^2 = 0.183 \text{ in. WG}$$

Loss in four elbows $= 4(0.21)(0.183) = 0.154$ in. WG

Friction loss $= 0.30\ (85/100) = 0.255$ in. WG

Total loss section $CE = 0.255 + 0.154 = 0.409$ in. WG

This loss is only $0.409 - 0.398 = 0.011$ in. WG above the loss of section *CD*. For the degree of accuracy with which we can work, this is satisfactory.

Step 5: Make the pressure loss in section *BF* equal to the loss in section

BCD or 0.287 + 0.398 = 0.685 in. WG.

NOTE: The procedure here would be essentially the same as for step 4. To use up the amount of loss (0.658 in. WG) may require rather high velocities. If this becomes a problem, it may be desirable to redesign the system using a lower friction rate than 0.25 in. WG per 100 ft for section *ABCD*.

14-14 THE STATIC-REGAIN METHOD

The intent of this method is to maintain the static pressure at practically a constant value throughout the system. The advantage of doing this is that static pressure determines the rate of discharge through outlets; hence, if the static pressure remains constant, the size of outlet for a given volume of discharge would also be constant. For installations such as hotels and hospitals, there would be an obvious advantage in having the same size outlet in a series of like rooms.

The static-regain method utilizes a velocity reduction at the end of each section of duct; the magnitude of the reduction being sufficient to provide a loss of velocity pressure equal to the loss of total pressure that occurred in the preceding section of duct. Thus, for a given length of duct the method of application would be as follows:

1. Determine the friction loss per 100 ft from friction charts (Figs. 14-1 or 14-1a).
2. Calculate the static pressure loss in the given length of duct by multiplying friction rate per 100 ft (part 1) by the duct length divided by 100.
3. For known velocity in the duct section calculate the velocity pressure $(h_v)_U$.
4. Calculate the required velocity pressure in the downstream section $(h_v)_D$ from

$$\text{SPR} = 0.5[(h_v)_U - (h_v)_D] \qquad (14\text{-}5)$$

or,

$$(h_v)_D = (h_v)_U - 2\ \text{SPR} \qquad (14\text{-}5a)$$

where SPR is the static pressure regain equal to the friction loss as determined in part 2; $(h_v)_U$ is the velocity pressure from part 3. The coefficient reflects the assumption that regain occurs at 50 percent efficiency. The velocity pressure $(h_v)_D$ is the necessary velocity pressure in the downstream section of the duct to provide complete regain of the static pressure lost in the upstream section.

5. Knowing $(h_v)_D$ one may easily calculate the necessary velocity in the downstream section and thus the required duct diameter.

The static-regain method is likely to be more time consuming than other duct design methods, but in many cases it will be found to justify the added effort through more effective air distribution. If this method is to be employed much of the time, it is recommended that one refer to Ref. 1.

A simple problem follows to illustrate the principle of static-pressure regain.

Illustrative Problem 14-16

Figure 14-14 represents a section of the main of a round duct system where three branch ducts are shown equally spaced along the main. The entering flow rate in this section of the main is 10,000 cfm at 2000 fpm, and each of the branches will carry 2000 cfm. The duct main *(ABCD)* will be maintained the same size throughout its length. Determine the values of the total, static, and velocity pressures at points *A*, *B*, *C*, and *D*. Assume the total pressure at station *A* is 1.00 in. WG.

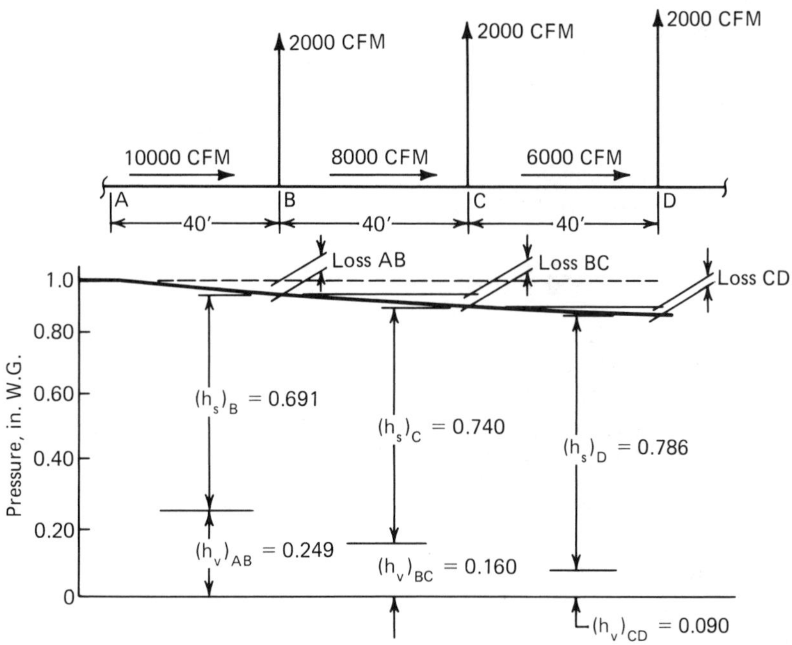

Figure 14–14

Diagram for Illustrative Problem 14-16 (Does not represent an actual system.)

Solution:

Step 1: Construct a summary table for data and results.

Line	Item	Section *AB*	Section *BC*	Section *CD*
1	Flow quantity (cfm)	10,000	8,000	6,000
2	Duct diameter (in.)	30.2	30.2	30.2
3	Flow velocity (fpm)	2,000	1,600	1,200
4	Friction rate (in. WG/100 ft)	0.150	0.10	.059
5	Friction loss (in. WG)	0.06	0.04	.024
		Station *B*	**Station *C***	**Section *D***
6	Velocity pressure (in. WG)	0.249	0.160	0.090
7	Total pressure (in. WG)	0.940	0.90	0.876
8	Static pressure (in. WG)	0.691	0.740	0.786

Step 2: Determine results to complete the table.

Referring to Fig. 14-1a, for an assumed velocity in section *AB* of 2000 fpm and a flow rate of 10,000 cfm, find the round duct diameter of 30.2 in. and a friction rate of 0.150. Insert values in results table.

The friction loss then is 0.150(40/100) = 0.06 in. WG

If the trunk main is maintained at a diameter of 30.2 in. throughout its length, follow this line of constant diameter on Fig. 14-1a; and at the intersection of 8000 and 6000 cfm, find the velocity and friction rate for the sections *BC* and *CD*. Enter these values in the results table.

Again, the friction loss (line 5) may be calculated as friction rate times length divided by 100. Enter values in results table.

Note that the volume flow rate decreases along the duct (line 1). Also, note that the velocity decreases along the duct (see line 3). This is because the duct size has been kept constant (line 2) even though the volume flow rate decreases. The friction rate for each section decreases (line 4), as well as the friction loss (line 5).

Since the velocity in each section (line 3) is known, the velocity pressures may be calculated.

$$(h_v)_{AB} = (V_{AB}/4005)^2 = (2000/4005)^2$$
$$= 0.249 \text{ in. WG}$$
$$(h_v)_{BC} = (V_{BC}/4005)^2 = (1600/4005)^2$$
$$= 0.160 \text{ in. WG}$$
$$(h_v)_{CD} = (V_{CD}/4005)^2 = (1200/4005)^2$$
$$= 0.090 \text{ in. WG}$$

Enter these values in summary table (line 6).

Now the total pressure (h_t) and friction should be calculated. The total pressure at station *A* is 1.00 in. WG (given). The friction loss in section *AB* was 0.06 in. WG; therefore, the total pressure at station *B* would be

$$(h_t)_B = (h_t)_A - (\text{loss})_{AB} = 1.00 - 0.06$$
$$= 0.94 \text{ in. WG}$$

Also,

$$(h_t)_C = (h_t)_B - (\text{loss})_{BC} = 0.94 - 0.04$$
$$= 0.90 \text{ in. WG}$$

and

$$(h_t)_D = (h_t)_C - (\text{loss})_{CC} = 0.90 - 0.024$$
$$= 0.876 \text{ in. WG}$$

Since by Eq. (13-8), $h_t = h_s + h_v$, the static pressures at *B, C,* and *D* may be calculated as follows:

$$(h_s)_B = (h_t)_B - (h_v)_{AB} = 0.940 - 0.249$$
$$= 0.691 \text{ in. WG}$$
$$(h_s)_C = (h_t)_C - (h_v)_{BC} = 0.90 - 0.160$$
$$= 0.740 \text{ in. WG}$$
$$(h_s)_D = (h_t)_D - (h_v)_{CD} = 0.876 - 0.090$$
$$= 0.786 \text{ in. WG}$$

Notice that along the main duct the velocity pressure decreased and the static pressure increased. In other words there was a *conversion* of velocity pressure to static pressure. This is called *static-pressure regain* (SPR). And here is the key to the static-regain method of duct design: Select the velocity in section *BC* such that the static pressure at station *C* is 0.691 in. WG, select the velocity in section *CD* such that the static pressure at station *D* is 0.691 in. WG.

Now it is evident what this means. The static pressure is 0.691 in. WG at each branch takeoff. Balancing the duct system should be simple; dampering will rarely be required. Some duct runs have outlets stubbed into the side or bottom of the duct (in place of the branch takeoffs). In such designs the static pressure behind each outlet is the same and the system is well balanced.

Theoretically, static pressure regain is independent of the manner in which the velocity is reduced, but the actual regain depends greatly on the physical configuration of the duct system. Problem 14-16 obtained static regain by using a constant-diameter duct with decreasing flow after each branch. Another technique would be to alter the duct diameter after each branch takeoff to obtain the required velocity reduction.

Illustrative Problem 14-17

Using Fig. 14-14 with the associated data and required calculated results from Problem 14-16, determine the required duct diameters in section *BC* and *CD* by use of Eq. (14-5a), if the diameter of section *AB* is 30.2 in.

Solution:

Diameter of Section *BC*: From Problem 14-16 the pressure loss in section *AB* was 0.06 in. WG, the velocity head $(h_v)_{AB} = 0.249$ in. WG.

The static-pressure regain in section *BC* must overcome losses in section *AB*. Therefore, by Eq. (14-5a)

$$(h_v)_{BC} = (h_v)_{AB} - 2(\text{SPR})$$

$$= 0.249 - 2(0.06)$$

$$(h_v)_{BC} = 0.129 \text{ in. WG}$$

or $V_{BC}/4005)^2 = 0.129$

$V_{BC} = 1398$ fpm (required velocity in section *BC* to regain static pressure)

Flow area section $BC = Q/V_{BC} = 8000/1498 = 5.6 \text{ ft}^2$

$A = \pi D^2/4 = 5.6$

$D_{BC} = 2.67 \text{ ft} = 32.0 \text{ in}$ (required diameter of section *BC*)

Diameter of Section CD: Referring to Fig. 14-1a with $D_{BC} = 32.4$ in. and $Q = 8000$ cfm, find friction rate of 0.07 in. WG per 100 ft. Therefore, the friction loss in section *BC* is (0.07)(40/100) = 0.028 in. WG. By Eq. (14-5a)

$$(h_v)_{CD} = (h_v)_{BC} - 2(\text{SPR})$$

$$= 0.129 - 2(0.028)$$

$$(h_v)_{CD} = 0.073 \text{ in. WG}$$

$$(V_{CD}/4005)^2 = 0.073$$

$$V_{CD} = 1082 \text{ fpm}$$

$$A_{CD} = 6000/1082 = 5.55 \text{ ft}^2$$

$$\pi D^2/4 = 5.55$$

$$D_{CD} = 2.66 \text{ ft}$$

$= 31.9$ in. (required diameter of section *CD*)

14-15 FANS

A fan, as it is considered here, is a gas flow-producing machine that operates on the same basic principles as a centrifugal pump or compressor. A fan is similar to these machines in that they all convert mechanical rotative energy, applied at their shafts, to gas (or fluid) energy. The conversion of mechanical rotative energy to gas energy is accomplished by means of a wheel (impeller), which imparts a spin to the gas. As a result of this spin, and, provided there is no entering spin, the tendency of the gas is always to leave the wheel with a forward spiral motion.

The difference between fans and pumps are readily apparent; however, the distinction between fans and compressors is not so obvious. Generally speaking, this distinction is one of output pressures with the lower-pressure machines called fans, while the higher-pressure machines are called compressors. Even this dividing line is not distinct.

Fans for heating, ventilating, and air-conditioning applications, even on high-velocity, high-pressure systems, rarely encounter more than 10 to 12 in. WG pressure. Industry standards on classes of fans have been established by the Air Moving and Conditioning Association (AMCA) as:

Class I: Maximum total pressure = 3-3/4 in. WG

Class II: Maximum total pressure = 6-3/4 in. WG

Class III: Maximum total pressure = 12-1/4 in. WG

14-16 TYPES OF FANS

Most commercial fans used in HVAC work may be placed in one of two general types based upon construction and air flow patterns. These two types are (1) *centrifugal* or *radial flow* and (2) *axial flow* and are illustrated in Fig. 14-15. Centrifugal fans have flow within the rotating wheel (rotor) that is substantially radial to the shaft, with the rotor operating in a scroll-type housing. Axial fans have flow within the wheel that is substantially parallel to the shaft and operates within a cylindrical or ring-type housing.

Centrifugal fans are designed to move air or gas over a wide volume range. Usual static pressures go up to 12 in. WG, but with special design may go as high as 60 in. WG. The rotor may have straight, forward curve, backward curve, radial-tip, or air foil type blades. The housing is sheet or case metal, and the rotor may have belt or direct drive.

Propeller fans are designed to move air from one enclosed space to another or from outdoors to indoors or vice versa in a wide range of volumes at low pressure (0 to 1 in. WG, static). The automatic shutter on discharge, shown in Fig. 14-15, is not part of the fan. It protects the fan from wind, rain, snow, and cold during shutdown. These fans have either belt or direct drive.

Tubeaxial fans have the axial-flow rotor in a cylinder and are designed to move a wide range of air volumes at medium pressures (¼ to 2½ in. WG, static). The fans may be mounted in the air duct and have either belt or direct drive. Rotor blades may be disc or airfoil type.

Vaneaxial fans have a set of air guide vanes mounted in a cylinder before or behind the airfoil-type rotor. It moves air over a wide range of volumes and pressures. Usual pressure range is ½ to 6 in. WG, static, but special designs can go as high as 60 in. WG, static. The rotor is often made of sheet metal but cast metal is also used. The fans have either belt or direct drive.

Fans are often designated as *boosters, blowers,* or *exhausters*. As they will be considered here, a booster is a fan with ducts connected to both inlet and outlet; a blower has a discharge duct only; an exhauster has an inlet duct only.

14-17 PRINCIPLE OF FAN OPERATION

All fans produce pressure by altering the velocity vector of the fluid flow. The fans produce pressure and/or flow due to the rotating blades of the impeller, which imparts kinetic energy to the fluid (air or gas) in the form of velocity changes. These velocity changes are resultants of tangential and radial velocity components in the case of centrifugal fans, and of axial and tangential velocity components in the case of axial-flow fans.

The various types of centrifugal fans that

CENTRIFUGAL

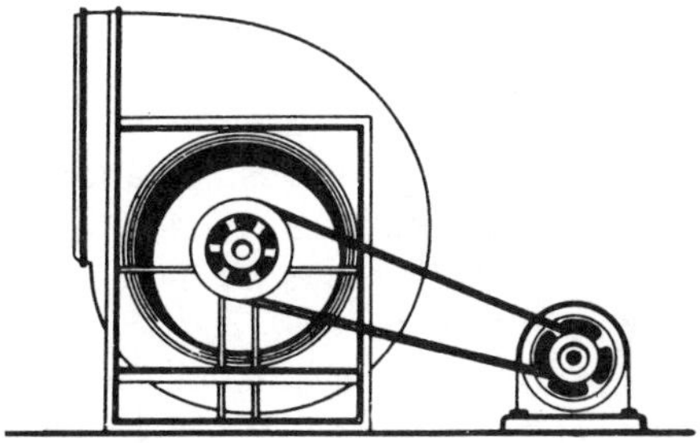

Designed to move air or gas over wide volume range. Usual static pressures go up to 25 in. water, special fans to 90 in. Wheel has straight, forward-curve, backward-curve, radial-tip or other type blades. Housing is sheet or cast metal with or without protective coat of rubber, lead, enamel, etc. Belt or direct drive

PROPELLER

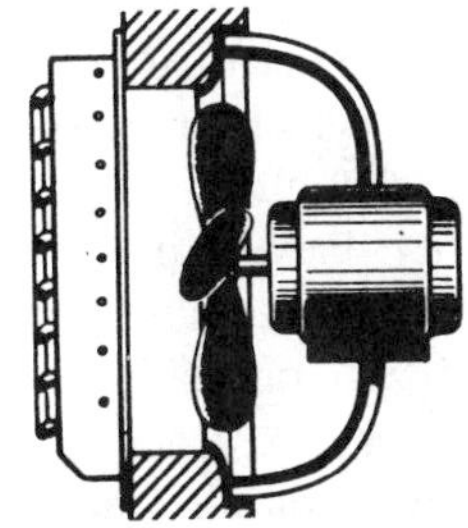

These are designed to move air from one enclosed space to another or from outdoors to indoors or vice versa in a wide range of volumes at low pressure (0 to 1 in. water, static). Automatic shutter on discharge is not part of fan. It protects fan from wind, rain, snow, cold during shutdown. Belt or direct drive

TUBEAXIAL

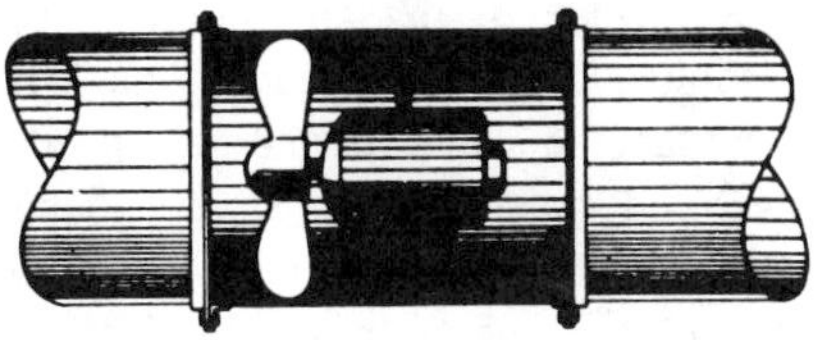

Axial-flow wheel in cylinder moves wide range of air or gas volumes at medium pressures (¼ to 2½ in. water, static). Belt or direct drive. Fan may be mounted in air or gas duct. Blades may be disk or airfoil type

VANEAXIAL

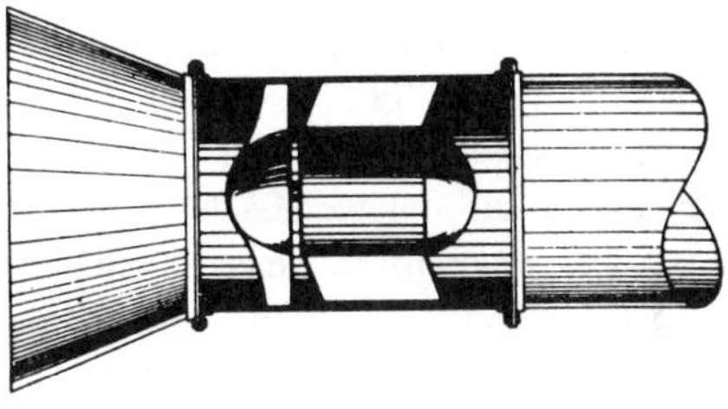

This fan has a set of air guide vanes mounted in cylinder before or behind airfoil-type wheel. It moves air over wide range of volumes and pressures. Usual pressure range is ½ to 6 in.; special designs can go up to 60 in. or higher. Wheel often made of sheet metal but cast metal is also used. Belt or direct drive

Figure 14–15

Types of fans: (a) Centrifugal (b) Propeller (c) Tubeaxial (d) Vaneaxial

are available differ from each other in structural details and in the details of their design and assembly. But the principle feature that distinguishes one type of centrifugal fan from another is the inclination of the impeller blades. Figure 14-16 shows the most common types of centrifugal fan elements. The fan impellers produce pressure from two related sources: (1) the centrifugal force created by the rotating air column enclosed between the blades and (2) the kinetic energy imparted to the air by virtue of its absolute velocity (V_t) leaving the impeller. This velocity in turn is a combination of rotative velocity (V_t) of the impeller and air velocity relative to the impeller (V_r). When the blades are inclined forward (Fig. 14-16c), these two velocities are cumulative; when the blades are inclined backward (Fig. 14-16a), oppositional. Backward-inclined blade fans are generally somewhat more efficient than forward-inclined blade fans, with radial blade fans between them.

Forward-curved blade fans operate at relatively low wheel peripheral velocities, V_t. The air absolute velocity, V_2, leaving the wheel is greater than the wheel peripheral speed. Forward-curved blade fans are very dependent upon the discharge casing design for conversion of velocity energy into static. This type of fan obtains about 20 percent of its static pressure from the wheel and about 80 percent from the casing. It needs a long duct run to slow down the air. The principle advantages of a forward-curved blade fan are that it occupies much less space and weighs less for a given duty, and is generally the least expensive.

Radial-blade fans operate with air speeds, V_2, about equal to the wheel peripheral velocities, V_t. The radial-blade fan operates at medium speeds, medium noise, and requires a fairly large space. In general, the peak efficiency is higher than for forward-curved blade fans. This type of fan is not as dependent upon the discharge casing for conversion of velocity pressure to static pressure. About 45 percent of its energy is converted to static pressure; about 55 percent of it into velocity pressure.

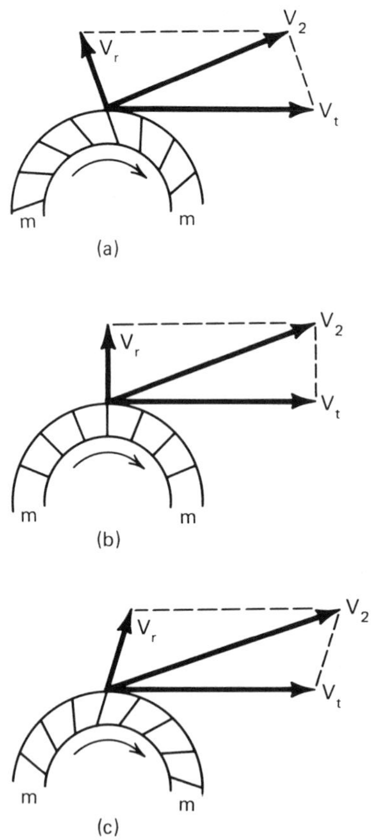

Figure 14–16

Centrifugal blade impeller types: (a) Backward-inclined (b) Radial (c) Forward-inclined

Backward-curved blade fans operate with air speed, V_2, lower than the peripheral wheel velocity and are driven at high speeds, which makes this type of fan most suitable for electric motor drive. This type of fan produces the highest efficiency (70 to 80 percent). It is much less dependent on the casing design for conversion of velocity pressure to static pressure because about 70 percent of the energy added develops static presure in the wheel. Its chief disadvantages are higher cost and larger size for a given duty.

Axial-flow fans produce pressure from the change in velocity of the air passing through the blades, with none being produced by centrifugal force.

14-18 FAN PERFORMANCE

Commercial fan performance data is calculated from laboratory tests conducted in accordance with the strict requirements of ASHRAE 51-75 plus AMCA 210-74. The ASHRAE standard specifies in detail the procedures and test setups to be used in testing the various types of fans and other air-moving devices. The codes establish uniform testing methods to determine a fan's flow rate, pressure, power, speed of rotation, and efficiency. The codes are designed for testing centrifugal and axial fans with various duct arrangements. The fans are tested in either the blower or exhauster configuration. Figure 14-17 shows a typical AMCA code test setup. When testing a fan, the Pitot tube makes a prescribed traverse of the cross-sectional area of the duct measuring velocity pressure, static pressure, and total pressure. From the results of these tests and from the computations outlined by the AMCA code, fan performance can be completely established.

The following terms are generally used when discussing fan performance:

Fan volume: The fan volume is defined as the volume, in cfm, passing through the fan outlet. In normal applications the volume leaving the fan is substantially equal to that entering since the change in specific volume of the air between fan inlet and outlet is negligible.

Fan outlet velocity: The outlet velocity of a fan is obtained by dividing the air volume by the fan outlet area. It is the average velocity that would occur at a point removed from the fan in a discharge duct having the same cross-sectional area as the fan outlet. It is important to note that the fan outlet velocity does not describe the velocity conditions that exist at the fan outlet since all fans have nonuniform outlet velocity. The fan outlet velocity is in reality a theoretical value; it is the velocity that would exist in the fan outlet if the velocity were uniform across the outlet area.

Fan velocity pressure: The fan velocity pressure is the pressure corresponding to the fan outlet velocity. The velocity pressure may be calculated by use of Eq. (13-4) or for standard air by Eq. (13-6).

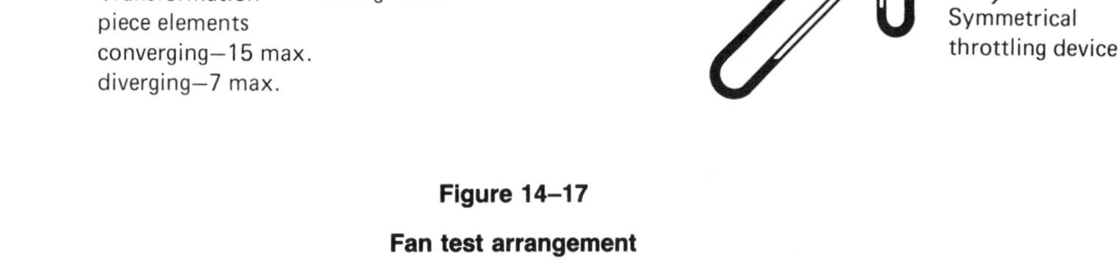

Figure 14–17

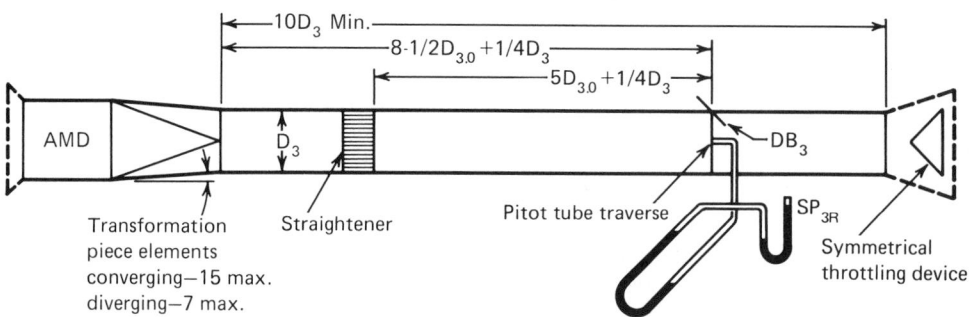

Fan test arrangement

Fan total pressure: The fan total pressure is the difference between the total pressure at the fan outlet and the total pressure at the fan inlet. The fan total pressure is a measure of the total mechanical energy added to the air or gas by the fan. Fan total pressure may be measured as illustrated in Fig. 14-18.

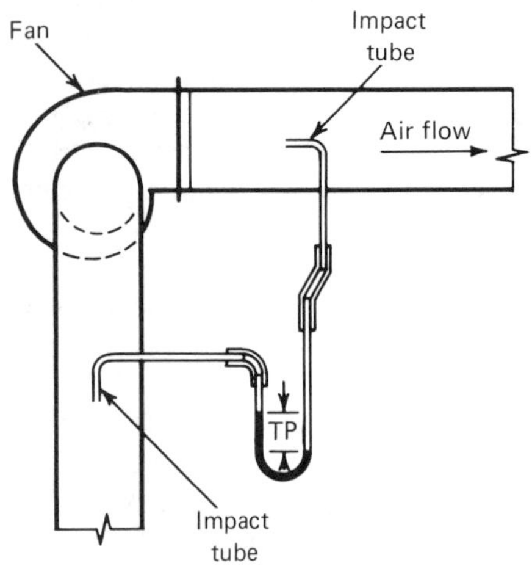

Figure 14–18

Measuring fan total pressure

Fan static pressure: The fan static pressure is the fan total pressure minus the fan velocity pressure. It can be calculated by subtracting the total pressure at the fan inlet from the static pressure at the fan outlet. It can be measured as illustrated in Fig. 14-19. By definition, the fan static pressure is

$$\text{Fan } h_s = (h_t)_o - (h_t)_i - (h_v)_o \qquad (14\text{-}6)$$

but from Eq. (13-8), rearranged, we have $(h_t)_o - (h_s)_o = (h_v)_o$; therefore, when substituted into Eq. (14-6) we have

$$\text{Fan } h_s = (h_s)_o - (h_t)_i \qquad (14\text{-}6a)$$

which is the method of measurement, where

h_s = static pressure

$(h_s)_o$ = static pressure at fan outlet

$(h_s)_i$ = static pressure at fan inlet

$(h_t)_o$ = total pressure at fan outlet

$(h_t)_i$ = total pressure at fan inlet

$(h_v)_o$ = velocity pressure at fan outlet

$(h_v)_i$ = velocity pressure at fan inlet

The definitions of fan total pressure and fan static pressure hold regardless of the manner of duct connection, that is, whether the fan is connected as a blower, booster, or exhauster.

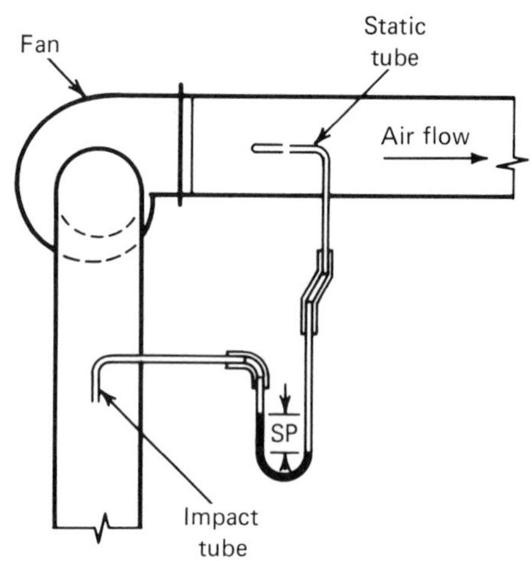

Figure 14–19

Measuring fan static pressure

Illustrative Problem 14-18

The following pressure measurements are reported on a fan connected as a booster that is, with inlet and outlet ducts.

At fan inlet: $(h_s)_i = -1.5$ in. WG (below atmospheric)

$$(h_v)_i = +0.25 \text{ in. WG}$$

$$\text{At fan outlet: } (h_s)_o = +0.75 \text{ in. WG}$$

$$(h_v)_o = +0.50 \text{ in. WG}$$

What is the fan static pressure?

Solution:

Fan static pressure = Fan total pressure minus fan velocity pressure.
By Eq. (14-6),

$$\text{Fan } h_s = (h_t)_o - (h_t)_i - (h_v)_o$$

$$\text{but, } (h_t)_i = -1.5 + 0.25 = -1.25 \text{ in. WG}$$
$$(h_t)_o = +0.75 + 0.50 = +1.25 \text{ in. WG}$$

therefore,

$$\text{Fan } h_s = 1.25 - (-1.25) - 0.50$$
$$= 2.0 \text{ in. WG}$$

or, by Eq. (14-6a),

$$\text{Fan } h_s = (h_s)_o - (h_t)_i = +0.75 - (-1.25)$$
$$= 2.0 \text{ in. WG}$$

Fan Horsepower: The theoretical horsepower required to drive a fan is the power that would be required if there were no losses in the fan, that is, if its efficiency were 100 percent. The theoretical horsepower to drive a fan can be calculated from

$$\text{Air horsepower (AHP)} = \frac{\text{cfm}[(h_t/12)(62{:}3)]}{33{,}000}$$

$$\text{AHP} = \frac{(\text{cfm})(h_t)}{6356} \qquad (14\text{-}7)$$

where cfm is the fan volume and h_t is the fan total pressure (in. WG).

The brake horsepower (BHP) is the power actually required to drive a fan. The brake horsepower is always larger than the theoretical horsepower because of energy losses in the fan. The brake horsepower required can be determined only by an actual test of the fan.

Fan efficiency. After testing a fan and determining the brake horsepower, both the *total efficiency* and *static efficiency* can be determined.

The total efficiency, or, as it is frequently called, the mechanical efficiency, is calculated by using the relationship

$$\text{Fan total efficiency } (\eta_f)_t = \frac{(\text{cfm})(h_t)}{(6356)(\text{BHP})} \qquad (14\text{-}8)$$

The fan static efficiency is calculated in a similar manner using fan static pressure instead of fan total pressure.

$$\text{Fan static efficiency } (\eta_f)_s = \frac{(\text{cfm})(h_s)}{(6356)(\text{BHP})} \qquad (14\text{-}9)$$

Fan total efficiency is the total output of useful energy divided by the power input. Fan static efficiency is the static energy output divided by the power input.

The total efficiency of a fan always provides a true indication of fan performance whenever the total pressure is known or can be accurately determined, while the static efficiency is not always satisfactory for measuring the actual work accomplished by the fan. For example, when the outlet velocity is higher than the duct velocity, a velocity pressure is available that may be converted to static pressure provided the duct system is properly designed. The static efficiency, as normally determined, would not take into account the increase in static pressure thus provided and therefore would not be an accurate protrayal of the fan performance. Again, when a fan operates at "free delivery," the static efficiency becomes zero and is therefore valueless.

In most circumstances, however, static efficiency is an entirely accurate indicator of fan performance. It has an added advantage in that the fan static pressure is known in practically every case whereas the fan total pressure frequently is not known. For that reason fan static efficiency is probably used more frequently than the fan total efficiency.

It is evident that with the fan efficiency known, the required brake horsepower may be obtained by rearranging Eqs. (14-8) and (14-9). It is important to note, however, that the proper efficiency should be used with the corresponding pressure. As indicated above, fan total efficiency must be used with fan total pressure and fan static efficiency with fan static pressure.

Illustrative Problem 14-19

A fan is to deliver 8000 cfm of standard air against a fan static pressure of 0.75 in. WG. The static efficiency of the fan is 73 percent. Find the brake horsepower required to drive the fan.

Solution:

By Eq. (14-9) $$(\eta_f)_s = \frac{(\text{cfm})(h_s)}{(6356)(\text{BHP})}$$

or

$$\text{BHP} = \frac{(\text{cfm})(h_s)}{6356(\eta_f)_s}$$

$$= \frac{8000(0.75)}{6356(0.73)}$$

$$\text{BHP} = 1.29 \text{ hp}$$

14-19 FAN PERFORMANCE CURVES

To aid in fan selection, fan performance curves (sometimes called *characteristic curves*) show the relationship between the quantity of air that a fan will deliver and the values of fan pressure, horsepower, efficiency and, occasionally, the noise level. Constant-speed performance curves are presented for a specific fan operating at a stated speed and handling air of a stated density.

Fan performance curves are obtained from a series of laboratory tests on a fan with various levels of restriction at the end of the test duct (see Fig. 14-17). The lines connecting the test points form the fan performance curve.

Typical performance curves for a backward-inclined air foil blade and forward-curved blade fan are shown in Figs. 14-20 and 14-21, respectively. Notice that in the figures the pressure-volume curves for both backward-inclined and forward-curved fans have a peak point at which the pressure is a maximum. To the right of this peak point, both types of fans have a steadily falling, steep pressure-volume performance curve. This steeply falling performance curve is desirable in fans. If a fan has a steep performance curve, a slight change in pressure from the one at which the fan is operating will not greatly affect the air delivery. For example, referring to the backward-inclined air foil fan represented in Fig. 14-20, if the static pressure is 4 in. WG, the fan will deliver 19,600 cfm. However, if the static pressure required of the fan should increase to 4.5 in. WG at 19,600 cfm, the air delivery will fall only to 18,200 cfm, at which point the fan will develop 4.5 in. WG static pressure. The procedure for determining this second point of operation is explained later in this chapter.

The forward-curved blade fan (Fig. 14-21) has a peak static pressure that corresponds to the region of maximum efficiency, whereas with the backward-inclined air foil fan, this maximum pressure occurs somewhat to the left of the region of maximum efficiency.

As shown in Figs. 14-20 and 14-21, the horsepower for both backward-inclined and forward-curved fans is a minimum at no delivery. The horsepower for the forward-curved

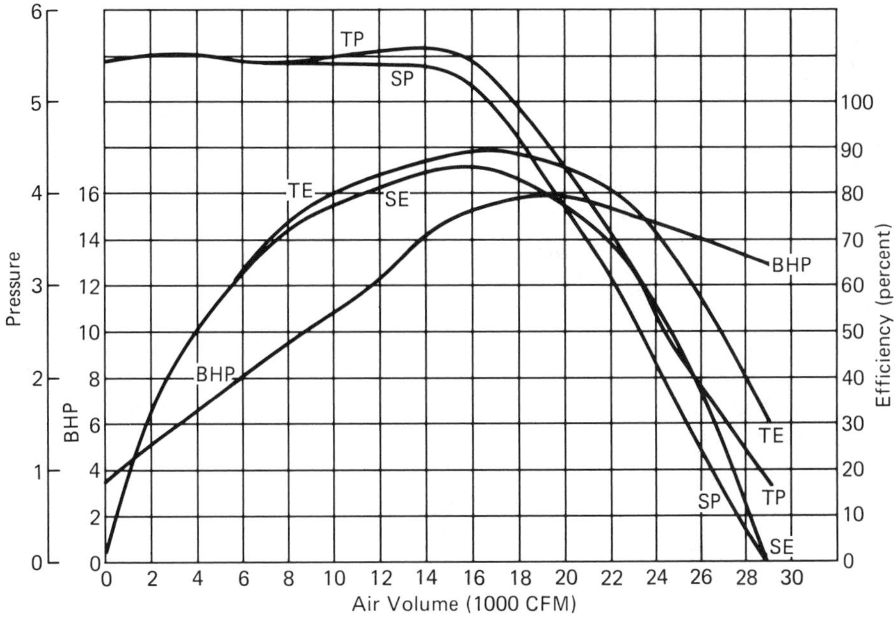

Figure 14–20

Sample performance curve. Backward inclined air foil blade fan. (36AF SW, 1042 RPM, 0.0750 lb/ft^3 air density.) (Courtesy of The Trane Company, LaCrosse, Wisconsin.) (Note: TP = total pressure; SP = static pressure; TE = total efficiency; SE = static efficiency; BHP = brake horsepower.)

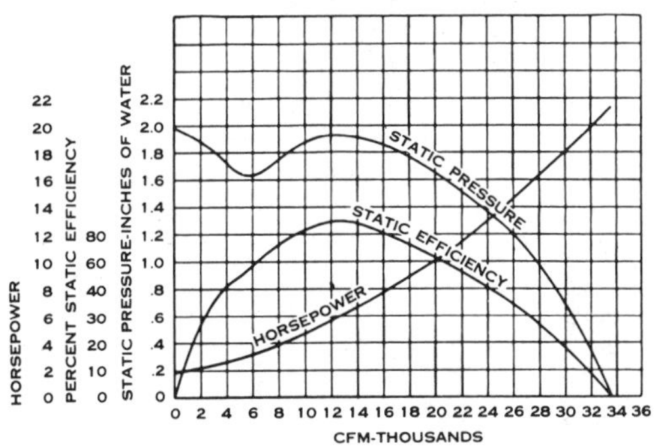

Figure 14–21

Sample performance curve. Forward curved blade fan (Courtesy of The Trane Company, LaCrosse, Wisconsin.)

fan increases continuously with increasing volume flow with maximum horsepower occuring at free delivery.

The horsepower for the backward-inclined air foil fan increases with increasing air flow only up to a point to the right of maximum efficiency and then gradually decreases. Fans with a horsepower curve of this shape are often referred to as having a nonoverloading horsepower characteristic because the maximum power that can be absorbed is usually not more than 10 percent above the power required at a normal selection point.

While the performance curve illustrated in Fig. 14-21 completely describes the fan performance, it is often useful to know the values of fan total pressure and fan total efficiency as shown in the curve of Fig. 14-20.

Performance curves of a typical propeller fan are illustrated in Fig. 14-22a and for a typical vaneaxial fan in Fig. 14-22b. As mentioned previously, a propeller fan has minimum horsepower requirements at free delivery. As can be seen from inspection of Fig. 9-20, this is distinctly different from a centrifugal fan. Tubeaxial fans and vaneaxial

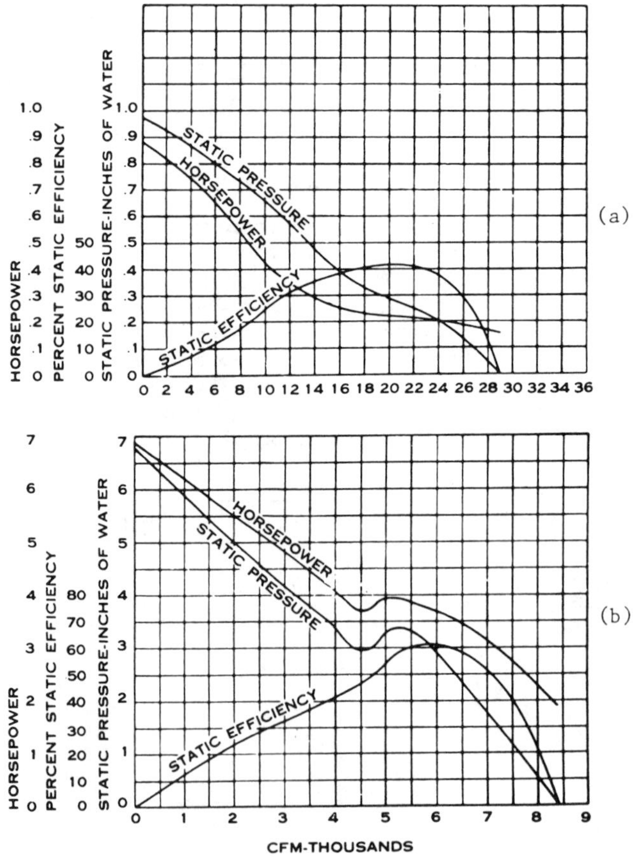

Figure 14–22

Axial-flow fan performance curves: (a) Propeller fan (b) Vaneaxial fan

fans have similar horsepower characteristics, as shown in Fig. 14-22b. This horsepower characteristic makes an axial fan a logical choice when large volumes at low pressures are necessary.

14-20 MULTIRATING FAN TABLES

The performance curves discussed in Section 14-19 are useful to show the general characteristics of fans operating at a single speed. In order to make the selection of fans as simple as possible, practically all manufacturers publish what is known as multirating tables. A separate table is provided for each diameter of wheel manufactured.

Such a table shows the speed (rpm) and horsepower required for various combinations of air quantity and static pressure for each size of wheel. Tables 14-12 and 14-13 are typical of such multirating tables. Table 14-12 is for a 24½-in. diameter, single width, belt-driven, backward-inclined utility fan, as manufactured by The Trane Company.

Table 14-13 is the multirating table for the geometrically similar fan, but having a wheel diameter of 36½-in. Knowing the air quantity and static pressure at which the fan is to operate, the speed and horsepower of the fan can be found directly in the table.

Illustrative Problem 14-20

Select a utility fan to deliver 12,000 cfm of standard air against a static pressure of 1.0 in. WG.

Solution:

Refer to Table 14-13. The capacity of 12,000 cfm is between 11,490 and 12,256 cfm. Interpolation is required to determine the operating rpm and BHP.

$$\text{For rpm} = 539 + \left(\frac{12{,}000 - 11{,}490}{12{,}256 - 11{,}490}\right)(557 - 539)$$

$$= 539 + \frac{510}{776}(18)$$

$$\text{rpm} = 551$$

$$\text{For BHP} = 2.83 + \left(\frac{12{,}000 - 11{,}490}{12{,}256 - 11{,}490}\right)(3.17 - 2.83)$$

$$\text{BHP} = 2.83 + \frac{510}{776}(0.34)$$

$$\text{BHP} = 3.05$$

14-21 FAN LAWS

Fortunately, for the application of fans, the performance of a fan at varying speeds, varying geometric size, and varying air density may be predicted by certain basic fan laws, when performance test data is available for a given fan. This test data is normally obtained at a certain fan speed and at standard air density and are similar to the fan characteristics shown in Figs. 14-20, 14-21, and 14-22.

The fan laws tell what happens to fan characteristics when (1) speed is varied while external system remains the same, (2) when fan size is changed in geometric proportion while point of rating is fixed, and (3) when air density changes. These laws apply to all fans. (Size laws apply *only* to geometrically similar fans.)

Summarized below are the fan laws important in determining the performance of all types of fans. The symbols used are as follows: Q = air volume, CFM; h = static, velocity or total pressure head, in. WG; hp = horsepower input.

Fan Law No. 1: Change in fan speed. (Air density and system constant).
Q varies in fan speed.

Table 14-12

Multirating Table (24BI Utility Fan)*

(Wheel Dia. 24½″ Tip Speed FPM = 6.41 x RPM Maximum RPM 1430 Outlet Area 3.46 Sq. Ft.)

VOL. CFM	OUT. VEL. FPM	1/8″ S.P.		1/4″ S.P.		3/8″ S.P.		1/2″ S.P.		3/4″ S.P.		1″ S.P.		1½″ S.P.		2″ S.P.		2½″ S.P.		3″ S.P.	
		RPM	BHP	RPM	BHP	RPM	BHP	RPM	BHP	RPM	BHP	RPM	BHP	RPM	BHP	RPM	BHP	RPM	BHP	RPM	BHP
2768	800	364	0.12	415	0.18	462	0.24	504	0.30	586	0.44										
3114	900	397	0.15	443	0.22	487	0.28	528	0.35	599	0.49	675	0.66								
3460	1000	431	0.19	473	0.26	514	0.34	552	0.41	622	0.56	684	0.73								
3806	1100	466	0.23	505	0.31	542	0.40	579	0.48	646	0.64	705	0.81	827	1.22						
4152	1200	502	0.29	537	0.37	571	0.47	606	0.56	670	0.73	728	0.91	834	1.31	946	1.77				
4498	1300	538	0.35	570	0.44	603	0.54	634	0.64	696	0.83	753	1.02	853	1.42	957	1.91				
4844	1400	575	0.42	605	0.52	635	0.62	664	0.73	723	0.95	778	1.14	876	1.57	965	2.03	1065	2.58		
5190	1500	612	0.51	640	0.60	667	0.71	696	0.83	751	1.07	803	1.28	899	1.72	985	2.19	1073	2.74	1163	3.32
5536	1600	649	0.60	675	0.70	701	0.82	728	0.94	779	1.19	831	1.44	924	1.89	1008	2.38	1086	2.91	1172	3.52
5882	1700	686	0.71	711	0.81	735	0.93	760	1.06	809	1.32	858	1.59	949	2.07	1031	2.58	1106	3.11	1180	3.70
6228	1800	723	0.82	747	0.93	770	1.05	793	1.19	840	1.47	886	1.75	974	2.27	1055	2.80	1129	3.35	1199	3.93
6574	1900	761	0.96	784	1.08	806	1.19	827	1.33	872	1.63	915	1.92	1001	2.49	1080	3.03	1152	3.60	1219	4.19
6920	2000	798	1.10	820	1.23	841	1.35	862	1.49	904	1.80	945	2.11	1028	2.73	1105	3.29	1177	3.88	1243	4.49
7266	2100	836	1.26	857	1.39	877	1.52	897	1.66	937	1.98	977	2.31	1056	2.97	1131	3.56	1201	4.16	1266	4.79
7612	2200	874	1.44	894	1.58	913	1.71	933	1.85	970	2.17	1009	2.52	1084	3.21	1157	3.85	1226	4.47	1291	5.13
7958	2300	912	1.63	931	1.77	950	1.92	968	2.06	1004	2.39	1041	2.74	1113	3.46	1185	4.18	1252	4.80	1316	5.47
8304	2400	950	1.84	968	1.99	987	2.14	1004	2.28	1038	2.62	1074	2.98	1143	3.73	1212	4.49	1278	5.16	1340	5.84
8650	2500	988	2.06	1006	2.22	1023	2.38	1040	2.52	1074	2.86	1107	3.24	1174	4.02	1240	4.81	1305	5.56	1366	6.24
8996	2600	1026	2.31	1043	2.47	1060	2.64	1076	2.79	1109	3.13	1140	3.51	1206	4.32	1269	5.14	1332	5.95	1392	6.65
9342	2700	1064	2.57	1081	2.74	1097	2.91	1113	3.08	1144	3.41	1175	3.81	1238	4.64	1298	5.48	1360	6.34	1419	7.10
9688	2800	1103	2.85	1118	3.03	1134	3.21	1150	3.38	1180	3.72	1209	4.12	1270	4.98	1328	5.85	1388	6.74		
10034	2900	1141	3.15	1156	3.34	1171	3.52	1186	3.71	1216	4.05	1244	4.45	1302	5.33	1359	6.23	1416	7.15		
10380	3000	1179	3.48	1194	3.67	1209	3.86	1223	4.05	1251	4.41	1280	4.81	1335	5.71	1392	6.65				

V-BELT DRIVE #1. RPM RANGE 364-466; #2. 461-542; #3. 543-720; #4. 681-778; #5. 744-898; #6. 876-1007; #7. 931-1103; #8. 1043-1151; #9. 1079-1281; #10. 1250-1450.

PERFORMANCE IS FOR UTILITY FAN WITH OUTLET DUCT. BHP DOES NOT INCLUDE LOSSES.

Reprinted with permission. Courtesy of The Trane Company, LaCrosse, Wisconsin. See manufacturer's catalog for complete data.

Table 14-13

Multirating Table (36BI Utility Fan)*
(Wheel diam. = 36½ in.; Top speed fpm = 9.55 × rpm; Max. rpm = 950; Outlet area = 7.66 sq. ft.)

(Wheel Dia. 36½″ Tip Speed FPM = 9.55 x RPM Maximum RPM 950 Outlet Area 7.66

VOL. CFM	OUT. VEL. FPM	1/8″ S.P.		1/4″ S.P.		3/8″ S.P.		1/2″ S.P.		3/4″ S.P.		1″ S.P.		1½″ S.P.		2″ S.P.		2½″ S.P.		3″ S.P.	
		RPM	BHP	RPM	BHP	RPM	BHP	RPM	BHP	RPM	BHP	RPM	BHP	RPM	BHP	RPM	BHP	RPM	BHP	RPM	BHP
6128	800	244	0.26	278	0.40	310	0.52	338	0.66	393	0.97										
6894	900	266	0.33	297	0.48	327	0.63	354	0.77	402	1.09	453	1.47								
7660	1000	289	0.41	317	0.58	345	0.75	370	0.91	417	1.24	459	1.61								
8426	1100	312	0.51	338	0.70	363	0.89	388	1.06	433	1.42	473	1.79	555	2.69						
9192	1200	336	0.63	360	0.82	383	1.03	406	1.24	449	1.61	488	2.01	560	2.89	635	3.92				
9958	1300	360	0.77	382	0.97	404	1.19	425	1.42	467	1.84	505	2.26	572	3.15	642	4.23				
10724	1400	385	0.93	405	1.14	425	1.37	445	1.62	485	2.10	521	2.53	587	3.47	647	4.50	715	5.72		
11490	1500	410	1.12	429	1.33	447	1.57	466	1.84	503	2.36	539	2.83	603	3.80	660	4.85	720	6.06	781	7.35
12256	1600	434	1.32	452	1.54	470	1.80	488	2.07	522	2.63	557	3.17	619	4.18	676	5.27	729	6.44	787	7.80
13022	1700	459	1.56	476	1.79	493	2.05	509	2.34	542	2.92	575	3.52	636	4.58	691	5.70	742	6.89	791	8.20
13788	1800	484	1.82	500	2.06	516	2.32	532	2.62	563	3.25	594	3.87	653	5.02	708	6.19	757	7.42	804	8.71
14554	1900	509	2.11	525	2.37	540	2.63	554	2.94	584	3.59	613	4.25	671	5.50	724	6.70	773	7.96	818	9.28
15320	2000	535	2.43	549	2.71	564	2.98	577	3.29	606	3.97	634	4.65	689	6.04	741	7.26	789	8.57	834	9.93
16086	2100	560	2.78	574	3.08	588	3.35	601	3.66	628	4.36	655	5.10	708	6.56	758	7.86	805	9.21	849	10.60
16852	2200	585	3.17	599	3.48	612	3.76	625	4.08	650	4.80	676	5.56	726	7.09	776	8.50	822	9.88	866	11.34
17618	2300	611	3.59	624	3.91	636	4.23	648	4.54	673	5.27	698	6.06	746	7.65	794	9.23	839	10.62	882	12.10
18384	2400	636	4.05	649	4.39	661	4.72	672	5.03	696	5.78	719	6.58	766	8.24	813	9.92	857	11.40	899	12.91
19150	2500	662	4.54	674	4.90	685	5.25	697	5.57	719	6.31	741	7.15	786	8.87	831	10.63	875	12.29	916	13.79
19916	2600	687	5.08	699	5.45	710	5.81	721	6.14	743	6.90	764	7.76	808	9.55	850	11.35	893	13.15	933	14.70
20682	2700	713	5.66	724	6.04	735	6.42	745	6.79	766	7.53	787	8.41	829	10.26	870	12.12	912	14.01		
21448	2800	738	6.28	749	6.68	760	7.07	770	7.46	790	8.21	810	9.10	851	11.00	890	12.92	930	14.90		
22214	2900	764	6.95	774	7.36	784	7.77	795	8.17	814	8.94	834	9.82	873	11.78	911	13.77				
22980	3000	790	7.66	800	8.09	809	8.51	819	8.93	838	9.72	857	10.61	894	12.60	932	14.69				

V-BELT DRIVE #1, RPM RANGE 233-304; #2, 303-362; #3, 350-422; #4, 370-418; #5, 419-502; #6, 485-580; #7, 552-648; #8, 649-694; #9, 643-746; #10, 746-850; #11, 826-955.

PERFORMANCE IS FOR UTILITY FAN WITH OUTLET DUCT. BHP DOES NOT INCLUDE LOSSES.

*Reprinted with permission. Courtesy of The Trane Company, LaCrosse, Wisconsin. See manufacturer's catalog for complete data.

h varies as fan speed squared.
hp varies as fan speed cubed.

Fan Law No. 2: Change in fan size. (Tip speed, air density, fan proportions constant; fixed rating).
Q varies as square of wheel diameter.
h remains constant.
rpm varies inversely as wheel diameter.
hp varies as square of wheel diameter.

Fan Law No. 3: Change in fan size. (rpm, air density, fan proportions constant; fixed rating).
Q varies as cube of wheel diameter.
h varies as square of wheel diameter.
Tip speed varies as wheel diameter.
hp varies as fifth power of wheel diameter.

Fan Law No. 4: Change in air density. (cfm, system, fan speed constant; fixed fan size).
Q is constant.
h varies as density.
hp varies as density.

Fan Law No. 5: Change in air density. (Constant pressure and system; fixed fan size, variable fan speed).
Q varies inversely as square root of the density.
h is constant.
rpm varies inversely as square root of the density.
hp varies inversely as square root of the density.

Fan Law No. 6: Change in air density. (Constant weight of air and constant system; fixed fan size, variable fan speed).
Q varies inversely as air density.
h varies inversely as air density.
rpm varies inversely as air density.
hp varies inversely as air density squared.

The best way to see how fan laws apply is to solve a few illustrative problems similar to those usually encountered.

Illustrative Problem 14-21

A fan delivers 10,000 cfm of air against a static pressure of 2.0 in. WG when the speed is 500 rpm and the horsepower input is 6.0 BHP. What speed, static pressure, and horsepower will be obtained for a delivery of 14,000 cfm?

Solution:

Fan law 1 applies.

$$\text{For speed: } \text{rpm}_2 = \text{rpm}_1\left(\frac{\text{cfm}_2}{\text{cfm}_1}\right)$$

$$\text{rpm}_2 = 500\left(\frac{14{,}000}{10{,}000}\right) = 700 \text{ rpm}$$

$$\text{For static pressure } (h_s)_2 = (h_s)_1\left(\frac{rpm_2}{rpm_1}\right)^2$$

$$(h_s)_2 = 2.0\left(\frac{700}{500}\right)^2$$

$$= 3.92 \text{ in. WG}$$

$$\text{For horsepower} \qquad \text{hp}_2 = \text{hp}_1\left(\frac{\text{rpm}_2}{\text{rpm}_1}\right)^3$$

$$\text{hp}_2 = 6.0 \left(\frac{700}{500}\right)^3$$

$$= 16.5 \text{ hp}$$

Illustrative Problem 14-22

A fan delivers 8000 cfm of air at 70°F and 29.92 in. Hg (density = 0.0750 lb/ft) against a static pressure of 2.0 in. WG when the speed is 600 rpm and the power input is 5.0. If the inlet air temperature is raised to 200°F (density – 0.602 lb/ft) but fan speed stays the same, what is the new static pressure and horsepower?

Solution:

Fan law 4 applies.

For static pressure $(h_s)_2 = (h_{s1})\left(\frac{\rho_{a_2}}{\rho_{a_1}}\right)$

$$(h_s)_2 = 2.0 \left(\frac{0.0602}{0.0750}\right)$$

$$= 1.6 \text{ in. WG}$$

For horsepower $\text{hp}_2 = \text{hp}_1 \left(\frac{\rho_{a_2}}{\rho_{a_1}}\right)$

$$\text{hp}_2 = 5.0 \left(\frac{0.0602}{0.0750}\right)$$

$$= 4.0 \text{ HP}$$

Illustrative Problems 14-23

For the fan of Problem 14-22, what speed would be required to give a static pressure of 2.0 in. WG at 200°F? What will be the cfm and power?

Solution:

Fan law 5 applies.

For rpm

$$\text{rpm}_2 = \text{rpm}_1 \sqrt{\frac{\rho_{a_1}}{\rho_{a_2}}}$$

$$\text{rpm}_2 = 600\sqrt{0.0750/0.0602}$$

$$= 670 \text{ rpm}$$

For cfm

$$\text{cfm}_2 = \text{cfm}_1 \sqrt{\frac{\rho_{a_1}}{\rho_{a_2}}}$$

$$\text{cfm}_2 = 8000 \sqrt{0.0750/0.0602}$$

$$= 8930 \text{ cfm}$$

For hp

$$\text{hp}_2 = \text{hp}_1 \sqrt{\frac{\rho_{a_1}}{\rho_{a_2}}}$$

$$\text{hp}_2 = 5.0 \sqrt{0.0750/0.0602}$$

$$= 5.58$$

14-22 SYSTEM CHARACTERISTIC

So far the flow of only a fixed quantity of air through a duct system has been discussed. However, the variation in the friction loss through a given duct system when various air quantities are forced through the duct is of interest. By computing the friction loss for the same duct with various air quantities and plotting all these points on a graph, a curve similar to that shown in Fig. 14-23a is obtained.

Notice as the air quantity increases, the friction loss at first rises slowly and then increases quite rapidly. A curve like the one in Fig. 14-23a is known as a system characteristic curve. It can be drawn not only for a duct of single diameter but for an entire system of duct work. Once the friction loss has been determined for a given quantity of air flowing through a complicated system of ducts consisting of different sizes, the friction loss for any other air quantity can be computed *approximately* by means of the following relationship:

$$\frac{(h_s)_2}{(h_s)_1} = \left(\frac{Q_2}{Q_1}\right)^2 \qquad \textit{(14-10)}$$

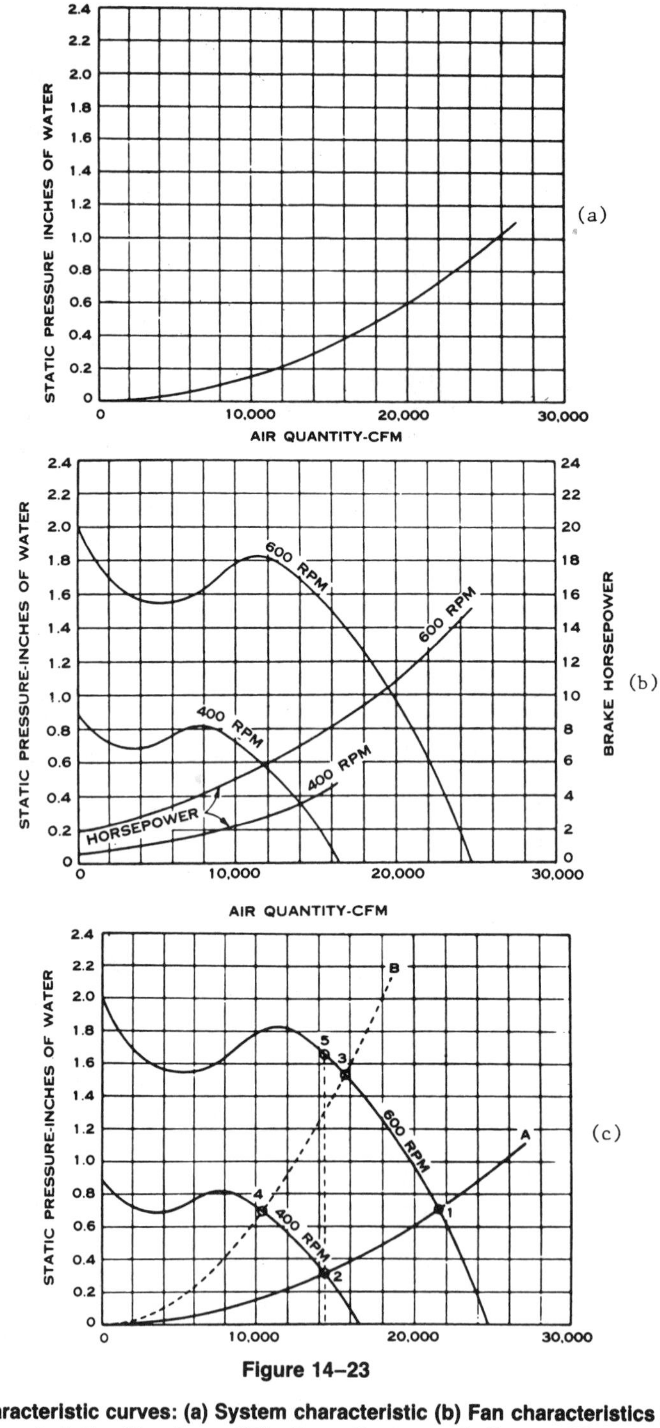

Figure 14–23

System and fan characteristic curves: (a) System characteristic (b) Fan characteristics (c) System and fan characteristics (Courtesy of The Trane Company, LaCrosse, Wisconsin.)

As is shown next, system characteristic curves are of value in helping to analyze the performance of a duct system and fan together.

14-23 COMBINATIONS OF FAN PERFORMANCE AND SYSTEM CHARACTERISTIC CURVES

The system characteristic curve of Fig. 14-23a shows the static pressure required to overcome the friction loss in a given duct system when delivering various air quantities. In order to deliver any given quantity of air, the fan must be so selected that it can deliver this air quantity against the friction of the duct system. Fans are ordinarily selected in this manner. However, air-conditioning systems do not always operate at constant fan speed nor is the resistance of the duct system constant. Very often, because of changes in the position of dampers, the system characteristic curve will change. Thus, the partial closing of a damper would mean that the friction loss of the duct system would increase. In other words the static pressure required to deliver a given quantity of air through the duct system would be increased because of the partial closing of the damper.

In order to visualize the results of changes in the operating conditions, a combination graph on which both fan and system characteristic curves are plotted is invaluable. As was the case with pipe and pump characteristic curves, the use of such combination graphs permit a great many otherwise complex problems to be analyzed in a readily understandable manner.

Fan performance curves of Fig. 14-23b have been plotted in Fig. 14-23c. These curves are for a fan running at 600 and 400 rpm. In addition, the system characteristic curve of Fig. 14-23a has been plotted on this same graph. The point at which the fan will operate on its characteristic when the fan is running at 600 rpm is found at the intersection of its performance curve for this speed and the system characteristic curve; in this case, point 1 of Fig. 14-23c. The fan will deliver 21,500 cfm and will operate against a static pressure of 0.70 in. WG as read at point 1. However, if the fan speed is reduced to 400 rpm, the quantity of air delivered by the fan will be reduced to 14,400 cfm and the static pressure will be 0.30 in. WG, as read at point 2.

If the resistance of a duct system is increased because of the closing of a damper or the clogging of air filters with dirt, the characteristic curve for the system will move up to some such position as indicated by the dotted line of Fig. 14-23c. In such a case the fan, when operating at 600 rpm, will deliver 15,600 cfm against a static pressure of 1.53 in. WG as read at point 3. If, however, the speed should be reduced to 400 rpm, the fan will deliver 10,500 cfm against 0.70 in. WG of static pressure as read at point 4.

Frequently, the volume of air delivered by a fan must be reduced. The fan volume can be reduced by (1) reducing the fan speed and (2) throttling the air flow by a damper.

The effect on the power consumption of a fan by reducing the air volume by either of these two methods can be studied advantageously by using the combination diagram. For example, assume that the volume of air is to be reduced from 21,500 cfm at point 1 of Fig. 14-23c to 14,400 cfm by reducing the fan speed from 600 to 400 rpm. The fan will, therefore, operate at point 2 of Fig. 14-23c. The brake horsepower required to operate the fan at point 1 of Fig. 14-23c is 12.3 hp as read in Fig. 14-23b. On the other hand the brake horsepower required to operate the fan at point 2 of Fig. 14-23c is 3.7 hp as read in Fig. 14-23b.

If, however, the fan is allowed to operate at constant speed and the air volume is reduced by partially closing a damper, it is evident that the system characteristic curve of Fig. 14-23c will move up on the fan curve. If the air volume is finally reduced to 14,400 cfm by closing a damper while the fan is still running at

600 rpm, the fan will finally operate at, say, point 5 of Fig. 14-23c. The power required to operate the fan at point 5 of Fig. 14-23c is 7.2 hp as read in Fig. 14-23b, compared to 3.7 hp obtained by speed reduction.

From the foregoing discussion it is evident that if the fan speed can be reduced without sacrificing the efficiency of the motor or without excessive losses in electric speed control equipment, more economical operation can be obtained at reduced air volumes by reducing the fan speed than by a throttling damper.

When a fan is connected to a given duct system, all points of intersection between the system characteristic curve and the fan performance curves at different speeds, such as points 1 and 2 of Fig. 14-23c, are called *corresponding points*. This statement would be strictly true if the friction in the system varied as the square of the air quantity, Eq. (14-10). In Fig. 14-23c, this is not strictly true. The friction varies as some exponent of the air quantity that is slightly less than the square of the air quantity. However, the "square" relationship is close enough for most practical purposes.

Let us use one more illustration to clarify the preceding discussion. Figure 14-24 shows the plot of system characteristic curves and fan performance curves. Again, the known system has friction characteristics that vary as the square of the volume flow producing a parabolic shaped curve, independent of fan type. Suppose that we wish to have a flow in the system of 11,000 cfm and the system needs ¾-in. WG static pressure to maintain the flow. If the system and fan plots are as shown in Fig. 14-24, we find that but one condition of static pressure and capacity satisfies both fan and system. This is point *A*, at the intersection of the two curves. It corresponds to 0.96 in. WG static pressure, 12,400 cfm, 3.08 hp, and 575 rpm. To obtain the exact flow needed (11,000 cfm, ¾-in. WG) the fan must be dampered or the fan speed changed.

At 11,000 cfm and 575 rpm, the fan pressure is 1.16 in. WG, point *C*. So if we use a damper, it must have a resistance equivalent to 1.16 − 0.75 = 0.41 in. WG at 11,000 cfm. The corresponding horsepower is 2.95. If we reduce the fan speed, instead of dampering, 510 rpm satisfies the condition; *A* moves back to *B*.

14-24 CENTRIFUGAL FAN DESIGNATIONS

In order to designate the direction in which a fan wheel will rotate, the designations clockwise and counterclockwise are used. To determine the direction in which the wheel of a double inlet fan wheel will rotate, the fan should be viewed from the pulley end of the shaft. To determine the direction in which the wheel of a single inlet fan will rotate, the fan should be viewed from the side opposite the inlet.

There are also a series of terms used in designating the air discharge of a fan. Fan discharge is designated as horizontal if the air leaves in a horizontal direction; as up blast if the air is discharged vertically upward from the fan; and as down blast if the air discharges vertically downward from the fan.

The location of the fan outlet is said to be either top or bottom, depending upon whether the air outlet is located above or below the shaft of the fan. For example, in a bottom horizontal discharge fan, the air outlet is located below the shaft of the fan and the air leaves in a horizontal direction. For example, the fan shown in Fig. 14-15a is said to have a top horizontal discharge because the air leaves in a horizontal direction from the top of the fan.

In addition to horizontal and vertical discharges, there are angular discharges. In a fan with an angular discharge, the air leaves at an angle of 45 degrees with the vertical and horizontal axes of the fan.

Centrifugal fans also have standardized no-

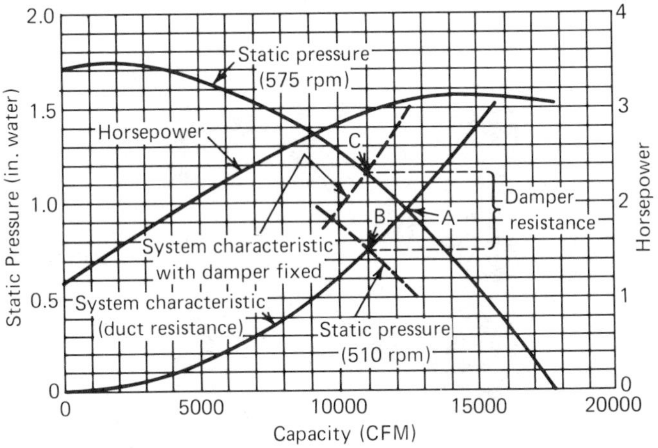

Figure 14–24

System characteristic is useful in predicting what will happen when a certain fan is connected to a known system. Effect of damper and speed reduction shown

menclature according to the arrangement of drive. These arrangements are illustrated in Fig. 14-26. Usually, for ventilation and general air-conditioning work, the belt-drive arrangements are preferred. This eliminates the necessity of selecting a direct-connected fan for a definite motor speed. The modern V-belt drive is inexpensive and reliable. Also, it permits fan speed adjustment on the installation if an adjustable-pitch motor sheave is used. Even with a fixed-speed drive, adjustments of speed can be made to meet changing conditions by substituting sheaves of different size.

Double-inlet fans, while usually less expensive for a given duty than single-inlet fans, should be located in a plenum. This is because duct connections to both inlets cannot be made unless the fan is furnished with inlet boxes. Thus, the majority of ventilating and air-conditioning centrifugal fans used are of the single-inlet type.

When a duct is connected to the inlet of a single-inlet fan, it is desirable to use drive arrangements other than 3 or 7. This is because with the other single-inlet drive arrangements there is no bearing in the fan inlet. A bearing located inside of the duct is more difficult to inspect and service than a bearing mounted on the drive side of the fan. Also, on smaller fans the bearing mounted in the inlet of single-inlet fans is detrimental to fan performance. Most manufacturers test and catalog single-inlet fans without an inlet bearing. The addition of an inlet bearing and bearing support presents additional air friction, which reduces fan output.

The effect of an inlet bearing on fan performance depends upon the size of the fan, the size of the bearing, and the design of the bearing support. This effect is not serious on large fans, and ratings on large single-inlet fans are usually made from tests with an inlet bearing.

Single-width (SW) or double-width (DW) fan. The available space, duct connections, air temperature, and degree of air contamination

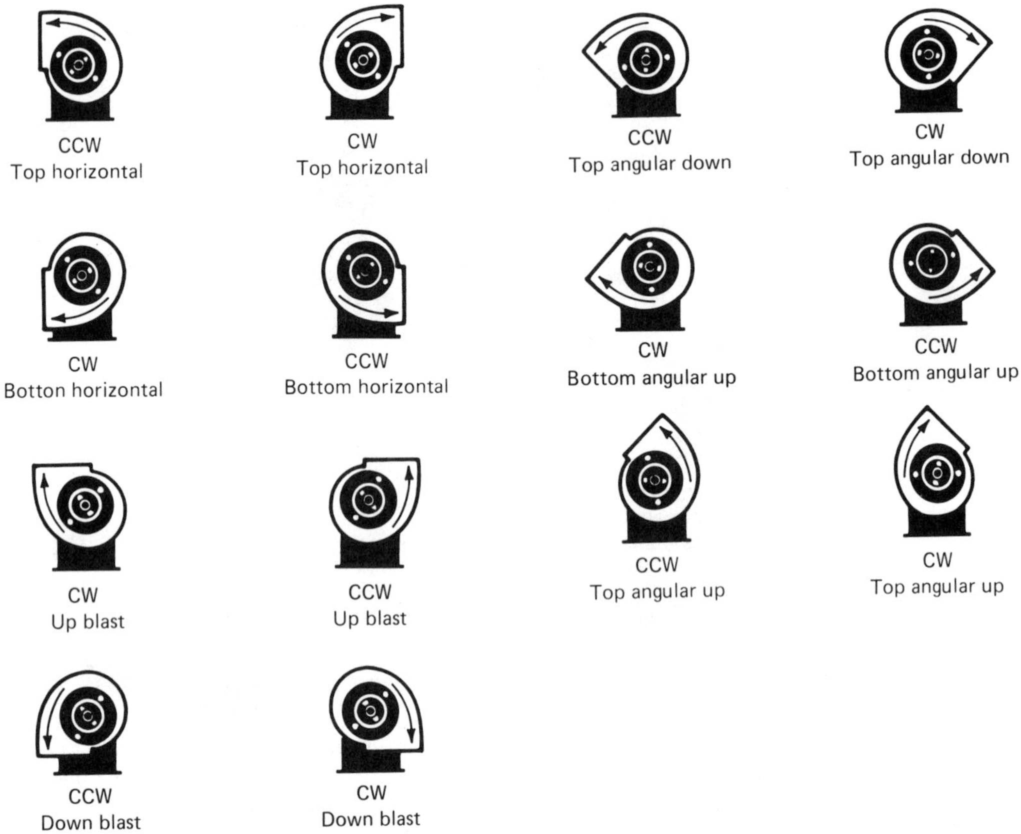

Figure 14–25

Fan rotation and discharge. Direction of rotation is found from drive side for either single- or double-inlet, or single- or double-width fans. Drive side of single-inlet fans is side opposite the inlet regardless of actual location of drive. For a fan inverted for ceiling suspension, direction of rotation and discharge is determined when fan is resting on floor. Designations above apply to all centrifugal fans

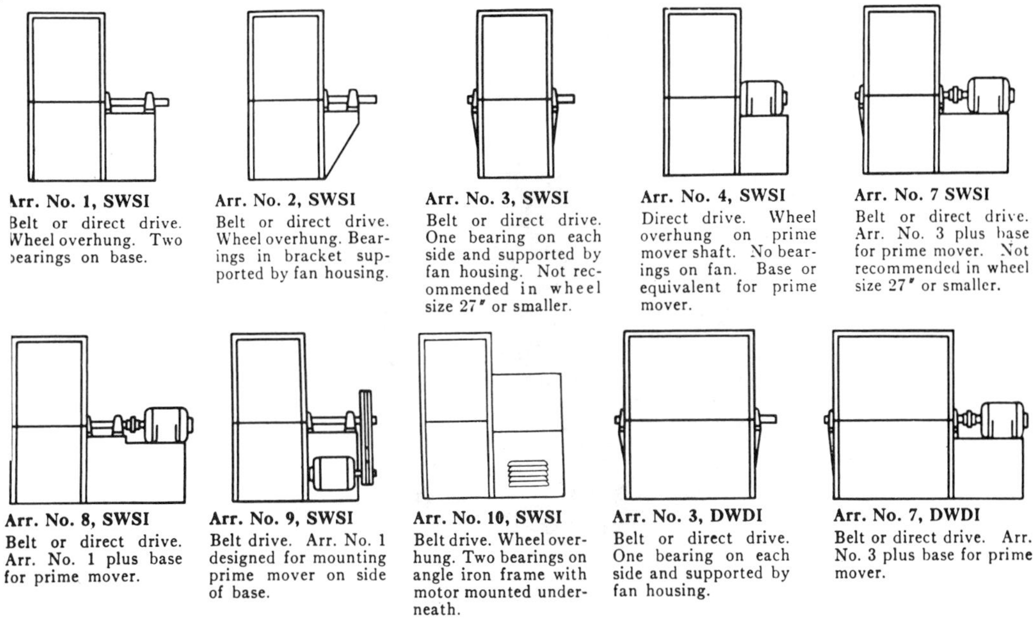

Arr. No. 1, SWSI
Belt or direct drive. Wheel overhung. Two bearings on base.
Arr. No. 2, SWSI
Belt or direct drive. Wheel overhung. Bearings in bracket supported by fan housing.
Arr. No. 3, SWSI
Belt or direct drive. One bearing on each side and supported by fan housing. Not recommended in wheel size 27" or smaller.
Arr. No. 4, SWSI
Direct drive. Wheel overhung on prime mover shaft. No bearings on fan. Base or equivalent for prime mover.
Arr. No. 7 SWSI
Belt or direct drive. Arr. No. 3 plus base for prime mover. Not recommended in wheel size 27" or smaller.
Arr. No. 8, SWSI
Belt or direct drive. Arr. No. 1 plus base for prime mover.
Arr. No. 9, SWSI
Belt drive. Arr. No. 1 designed for mounting prime mover on side of base.
Arr. No. 10, SWSI
Belt drive. Wheel overhung. Two bearings on angle iron frame with motor mounted underneath.
Arr. No. 3, DWDI
Belt or direct drive. One bearing on each side and supported by fan housing.
Arr. No. 7, DWDI
Belt or direct drive. Arr. No. 3 plus base for prime mover.

Figure 14–26

Drive arrangements. Standard arrangements for fan drive. When ordering or referring to a fan, you can easily identify by number its drive arrangement, width, and number of inlets. Thus, Arr. No. 1 SWSI means belt- or direct-driven unit with overhung wheel, two bearings on base, single width (SW), single inlet (SI). DW and DI mean double width and double inlet, respectively

must be considered in choosing a SW or DW fan. The cost of a DW fan will generally be less than an equivalent SW fan for the same duty. However, the DW fan is not used normally when inlet duct connections must be made or when bearings must be out of the airstream. Figure 14-27 illustrates a double-width backward inclined fan wheel.

Fan inlet and discharge conditions. The rated capacity of a fan can be achieved only if it is installed properly in the field. This includes unrestricted flow and uniform air flow to the fan inlet and proper discharge connections at fan outlet.

At the fan inlet, the following conditions will reduce the fan capacity: (1) spinning airstream, (2) nonuniform air distribution, and (3) insufficient space between fans or from the fan inlet to the wall.

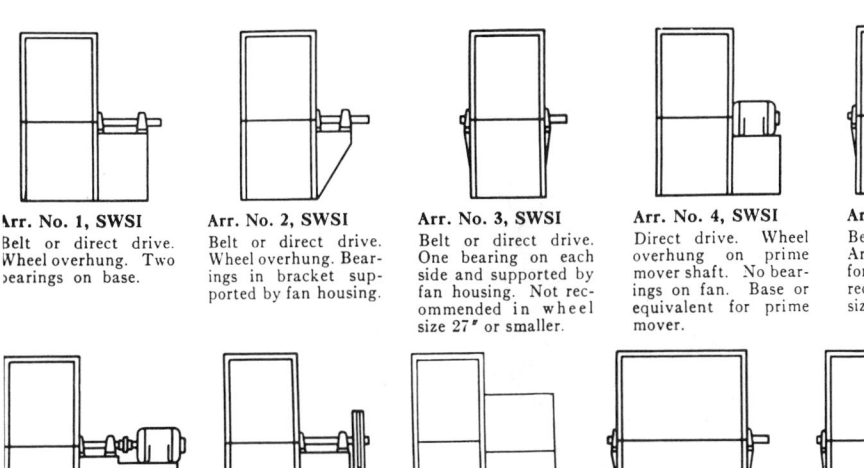

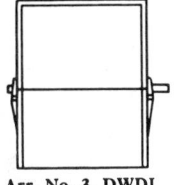

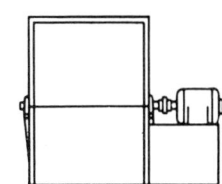

The spin of the air at the inlet, as shown in Fig. 14-28, may be due to the design of the duct connection at the inlet. This spin of the air into a fan is either forward or reverse spin with respect to the direction of rotation of the fan wheel, as shown in Fig. 14-29. If the spin is in the direction of wheel rotation, capacity will be reduced and the power consumption will drop. If the spin is counter to wheel rota-

Figure 14–27

Double width fan wheel (Courtesy of The Trane Company.)

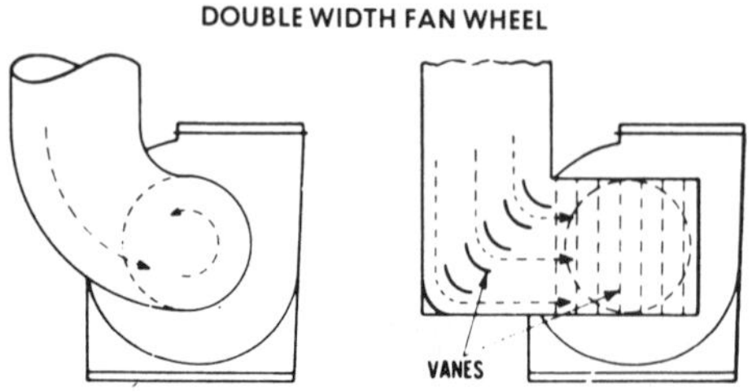

Figure 14–28

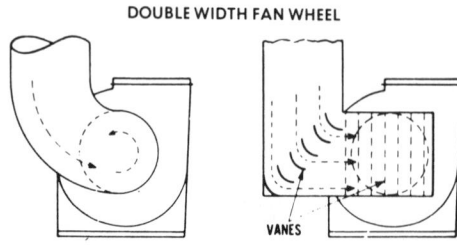

Compound curves on fan inlet duct (left) causes spin. Vanes in elbow and at fan inlet eliminate spin

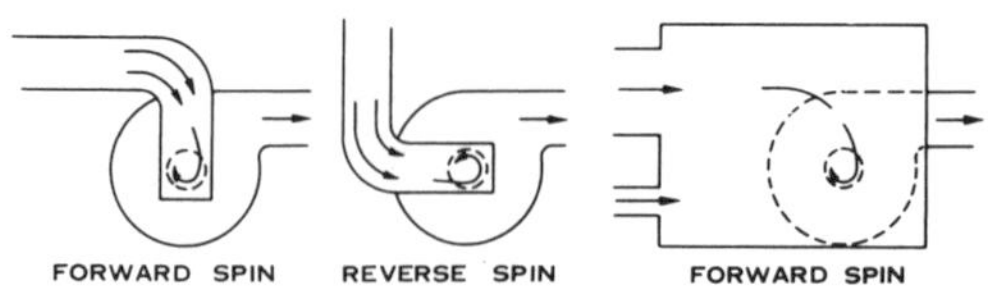

Figure 14–29

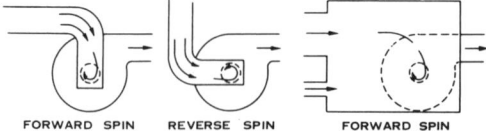

Examples of spin at fan inlet

tion, the power consumption will increase, the noise level will increase, but there will be practically no increase in pressure.

When there is nonuniform air flow into the fan inlet, as shown in Fig. 14-30, this may also be corrected by using guide vanes as shown. When there are no inlet ducts, fan capacity is affected when the space between fans in a multiple-fan installation or the space between fan inlet and the wall is too small. General practice is to locate multiple fans at least one fan diameter apart, and not to place a fan inlet closer than a distance of one-half fan diameter from the nearest obstruction. Test results, shown in Fig. 14-31, show that even these distances result in a slight decrease in capacity. Greater distances should be used if possible.

The discharge duct should be a straight section for at least several duct diameters to allow the conversion of fan energy from velocity pressure to static pressure.

When job conditions dictate that elbows be installed near the fan outlet, the loss of capacity and static pressure may be minimized by the use of vanes and proper direction of fan rotation relative to the direction of the bend in the elbow. The high-velocity side of the fan outlet should be directed at the outer radius of the elbow rather than the inside. Examples of correct elbow arrangements are shown in Fig. 14-32.

To summarize, Fig. 14-33 illustrates some hints on fan installation. Refer to them when installing a new fan or relocating an old one.

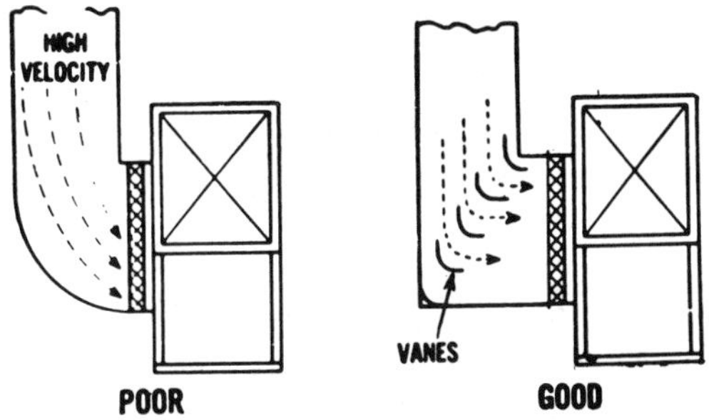

Figure 14–30

Uniform inlet distribution, through proper use of vanes, improves fan operation and reduces noise

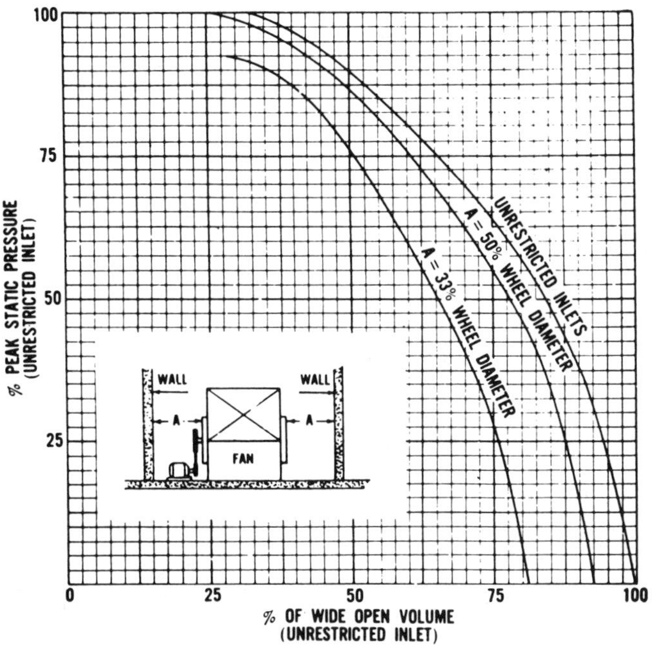

Figure 14–31

Space restriction reduces fan capacity as indicated by these curves from tests on double inlet fan (Courtesy of The Trane Company.)

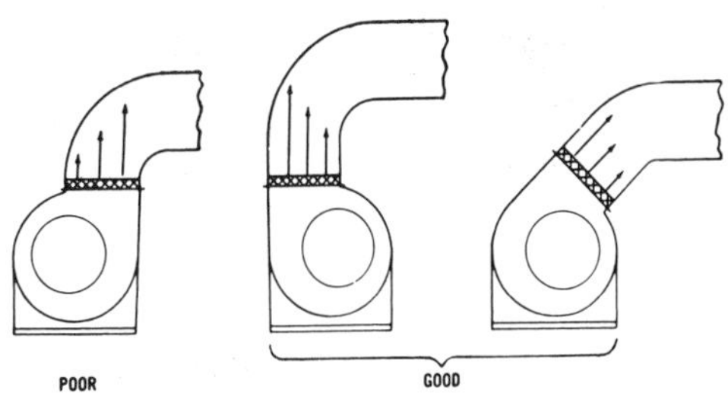

Figure 14–32

Proper discharge connections can minimize capacity reduction

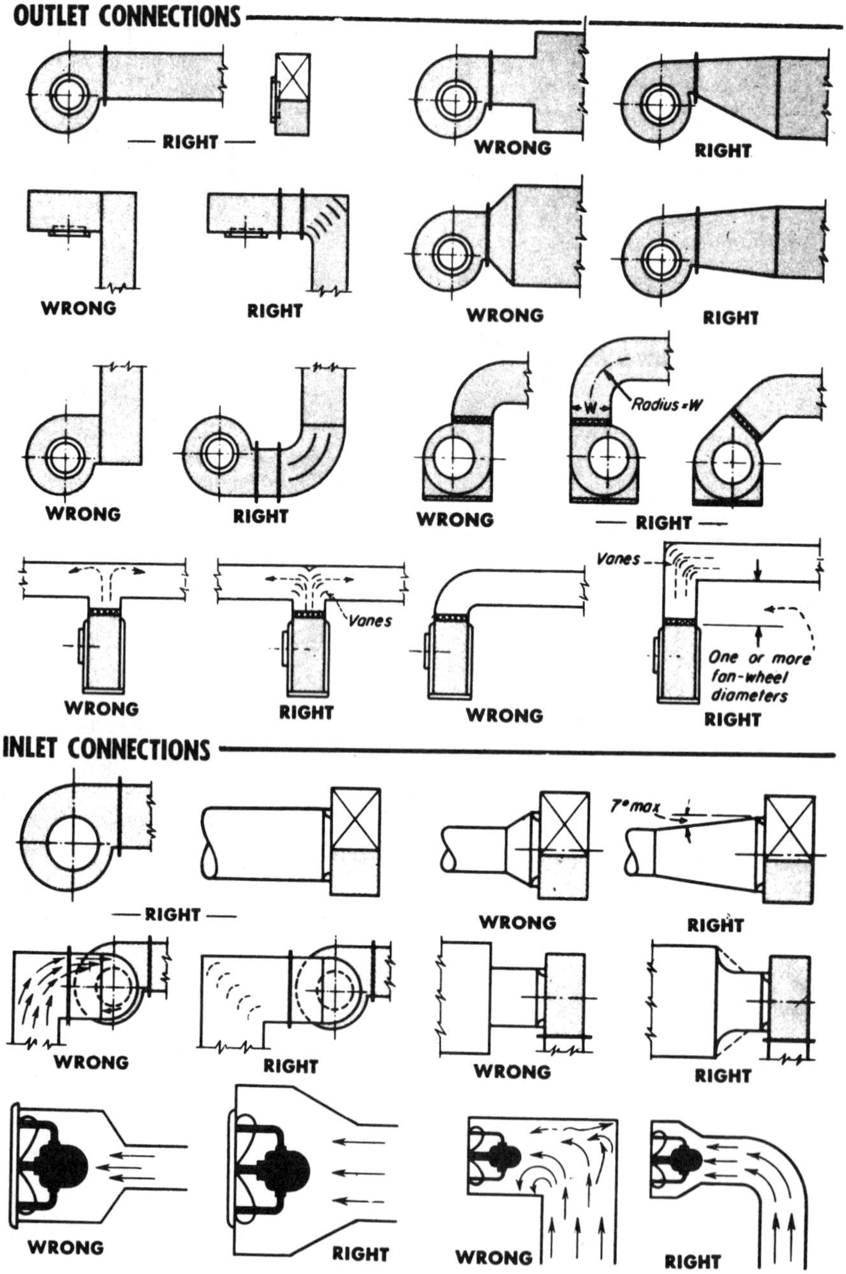

Figure 14–33

Right and wrong ways to install fans

CHAPTER 14

Review

14-1. A round, aluminum duct must be selected to handle 2000 cfm of standard air a maximum velocity of 1200 FPM. The duct is 40 ft long. Determine the duct diameter, friction rate, and friction loss for the duct section.

14-2. A section of duct is 65 ft long and 16 inches in diameter. If 2000 CFM of standard air is delivered by the duct, determine: (a) the velocity of air flowing; (b) the friction loss in the duct.

14-3. The static pressure available to overcome friction loss in a section of round duct 125 ft long is 0.20 inches W.G. Air quantity to be delivered is 3000 CFM (standard air). Determine: (a) the required duct diameter; and (b) the flow velocity.

14-4. A section of round aluminum duct 85 ft long is 16 in in diameter and carries 2500 CFM of air at 120 F. Determine friction loss in the section of duct assuming constant mass flow.

14-5. Determine the actual friction loss for a constant mass flow corresponding to a standard air volume of 4000 CFM in a 20-in. round duct 100 ft long, when the actual air density is 0.0650 lb/ft^3.

14-6. Figure 14-34 shows a sketch of a small ventilation system supplying standard air to four outlets in the quantities shown. Scale lengths are shown in feet. Using a friction rate of 0.10 in. W.G. per 100 ft, and the equal friction rate method, determine: (a) the round duct size for each section of duct shown; (b) the equivalent rectangular sizes for each section for equal friction and maintaining a duct depth of 8 in.; (c) which rectangular section has the greatest velocity, its magnitude, and is it a reasonable velocity?

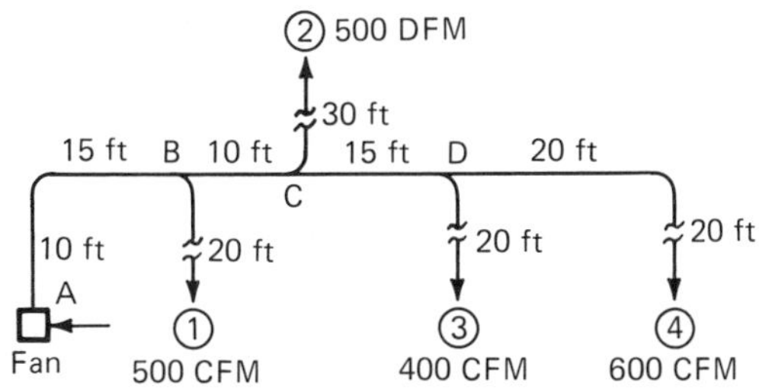

Figure 14–34

Schematic for Problem 14–6

14-7. Figure 14-35 shows a sketch of a small, warm air heating system for a house. The lengths of each section of duct are indicated in feet, diffusers have capacities and pressure drops indicated. Using a design friction rate of 0.06 in. W.G. per 100 ft and equal friction rate method, determine: (a) rectangular duct dimensions for each section for a uniform depth of 8 in.; (b) the static pressure required at point A. Assume all branch takeoffs are 90-degree elbows and all elbows have an r/w = 0.50.

14-8. (a) Figure 14-36 shows a portion of a ventilation system for a building. Duct section ABCD is to be sized by the equal friction rate method using a design friction rate of 0.20 in. W.G. per 100 ft, and duct depth of 16 in.
(b) Determine the rectangular duct size for section CE such that the pressure loss will be approximately equal to that in section CD, maintaining a 16-inch duct depth.
(c) What will be the required static pressure head at point A. (Assume each outlet grille requires 0.05 in. W.G. static pressure.)
(d) Refer to Table 14-12 and determine the required fan operating RPM and horsepower. Assume the pressure loss through the fan intake system and filter is 0.6 in. W.G., and the gradual expansion (fan outlet) has a 40-degree included angle.
NOTE: Assume all angles have an r/w = 0.5.

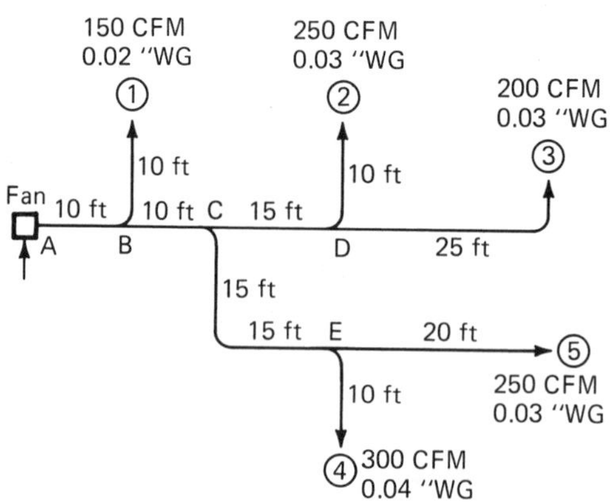

Figure 14–35

Schematic for Problem 14–7

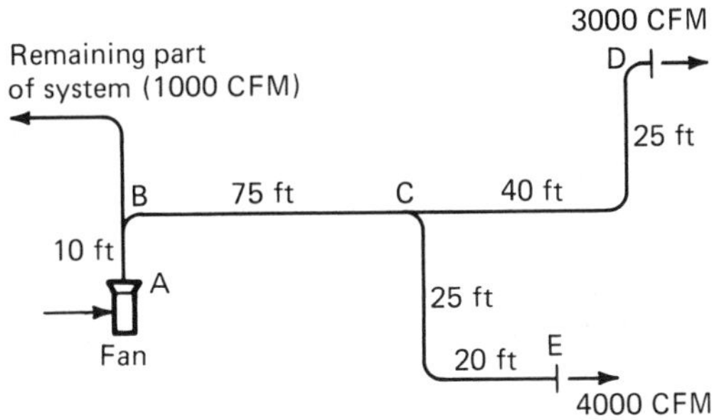

Figure 14–36

Schematic for Problem 14–8

14-9. A duct system has been design for a total air flow of 4875 CFM of 130°F air and a design friction rate of 0.06 in. W.G. per 100 ft. The velocity of air in the longest branch of the duct system was found to be 800 FPM, and at the start of the duct system 1225 FPM. The calculated equivalent length of the longest duct run was 200 ft. The static pressure loss in the return air system was 0.624 in. W.G. Determine: (a) the required static pressure rise across the fan; (b) select a fan for the system and determine the fan operating condition.

14-10. A fan delivers 12,000 CFM of standard air against a static pressure of 1.0 in. W.G. when operating at a speed of 400 RPM and requires an input of 4.0 H.P. If in the same installation, 15,000 CFM are desired, what would be the RPM, static pressure, and power required at this new flow rate?

14-11. The fan of Prob. 14-10 handles 12,000 CFM of standard air (ρ = 0.0750 lb/ft^3) against a static pressure of 1.0 in. W.G. rotating at 400 RPM. If the air temperature is increased to 200°F and the speed remains the same, what will be the static pressure and power?

14-12. If the fan of Problem 14-11 has its speed increased to produce a static pressure of 1.0 in. W.G. with 200°F air (same as with standard air), what will be its speed, capacity, and power?

14-13. If the speed of the fan of Pron. 14-12 is increased to deliver the same *weight* of air at 200°F as at 70°F, what will be the speed, capacity, static pressure, and power?

14-14. A fan develops 1.0 in W.G. static pressure and 0.4 in. W.G. velocity pressure when delivering 8000 CFM of 100°F air. The measured brakepower input to the fan was 2.0, and the speed was 1750 RPM. Calculate the fan static and mechanical efficiencies.

14-15. A fan is to be selected from a manufacturer's catalog to handle 70,000 pounds per hour of air at 90°F against a static pressure of 1.75 in. W.G. when the barometric pressure is 29.00 in. mercury. (a) What volume of air (CFM) and static pressure (in. W.G.) should be used to select a fan from a manufacturer's catalog; (b) If the manufacturer quotes a fan static efficiency of 72 percent, what brake horsepower will be required to drive the fan when handling the 90°F air?

14-16. A fan delivers 9192 CFM of standard air operating at 488 RPM, a brake horsepower of 2.01 against a static pressure of 1.0 in. W.G. Calculate the performance when the operating speed is changed to 450 RPM.

BIBLIOGRAPHY

1. ASHRAE *Handbook of Fundamentals,* American Society of Heating, Refrigerating and Air-Conditioning Engineers, New York, 1977.

2. ASHRAE *Handbook and Product Directory,* American Society of Heating, Refrigerating and Air-Conditioning Engineers, New York, 1979.

3. *Trane Air Conditioning Manual,* The Trane Company, LaCrosse, Wisconsin, 1965.

4. Burgess H. Jennings, *The Thermal Environment,* Harper & Row, New York, 1978.

5. *The Engineer's Reference Library, Power,* McGraw-Hill, New York, 1968.

6. *Aluminum Air Ducts,* Reynolds Metal Company, 1956.

7. McQuiston, Faye C., and Parker, Gerald D., *Heating, Ventilating, and Air Conditioning,* 2nd ed., John Wiley & Sons, Inc., New York, 1982.

8. Robert J. Aberbach, *Fans,* POWER Special Report, McGraw-Hill, Inc., New York, 1968.

Chapter 15

Fundamentals of Automatic Control and Control Components

15-1 INTRODUCTION

In previous chapters we have considered accepted methods for the design and layout of HVAC systems which will meet the requirements to maintain comfort conditions in buildings, if such systems are properly controlled. The HVAC system, its control system, and the building in which they are installed are inseparable parts of a whole. They must work together. The neglect of any one element may cause a partial or complete failure to provide for human comfort. Any well-designed HVAC system is useless without adequate control. Aslo, it should be stated that the best control system cannot overcome poor design of the HVAC system.

This, and the following chapter, will discuss automatic controls. In this limited space we will consider the fundamentals of automatic control and present some of the typical control systems in common use.

In our modern day living the term *automatic control* has come to be applied to almost everything we come in contact with, from simple household items such as an electric coffee pot to the more complicated automatic industrial machines capable of continuous operation without attention. As time has passed and technical developments have increased, so also has the need for automatic control of our complex machines and devices increased. The field of automatic control is a vast one and, like all of our other advancements, has become a series of highly specialized industries. One of these specialized industries is the field of automatic control of heating, ventilating, and air-conditioning equipment. It is a field that deals not only with human comfort and productivity but is concerned also with industry—by developing controls that increase safety, economy, and production.

Automatic controls can be used wherever a variable condition must be controlled. Almost any variable condition may be controlled automatically. That variable condition may be pressure, temperature, humidity, or rate or volume of flow, and it may exist in a gas, liquid, or solid. Of these four general conditions, some are controlled directly, others indirectly as a result of controlling another condition.

The purpose of an automatic control system is to provide stable operation of a process by maintaining the desired values of variable conditions. An automatic system is a collection of components, each with a definite function and designed to interact with the others, organized so that the system regulates itself. The control system senses disturbances and takes action to restore conditions to normal.

A control system is successful if the process consistently and economically produces an end product of good quality. Its success can be evaluated by how well it achieves three basic goals: (1) minimum deviation from normal conditions following a disturbance, (2) minimum time to restore conditions to normal, and (3) minimum offset due to changes in operating conditions.

Automatic control systems for heating, ventilating, and cooling fall into several basic categories according to application:

1. *The establishment of final conditions.* This is control of the end result in temperature, humidity, pressure, and so forth which the control system is designed to provide.
2. *Provision for safe operation.* Many applications require automatic controls to insure that temperatures, humidities, pressures, liquid levels, and so forth are kept within certain limits for safety reasons. Also, in many applications, it is required that safe conditions exist before mechanical equipment, such as a cold or heat generator, is permitted to operate. The control equipment is chosen to protect mechanical parts from damage due to freezing or overheating if the control system fails. This is referred to as ''fail-safe'' operation.
3. *The assurance of economical operation.* The correct application and proper correlation of a complete automatic control system provides proper operation of a control system at optimum operating cost.

15-2 GLOSSARY OF CONTROL SYSTEM TERMS AND DEFINITIONS

With the development of automatic controls there has also developed a series of ''terms'' that apply specifically to automatic systems and components. In order for us to follow more closely the description of control systems and components, we must become familiar with these terms and their definitions. Johnson Service Company suggests that it may be helpful to study these definitions within the following groupings:

1. SYSTEM

   Measured variable
   Controlled variable
   System feedback
   Closed-loop system
   Open-loop system
   Control agent

Controlled device
Controller
Load change

2. CONTROLLER

   Controller
   Controller feedback
   Ambient temperature
   Measuring element
   Input
   Summing point

Supply pressure
Supply voltage
Output signal
Direct acting
Reverse acting
Set point
Adjustment
Control point

3. CONTROL MODES

Controller
Two-position action
Differential
Proportional action
Offset
Proportional plus automatic reset action
Reset rate
Overshoot
Proportional plus rate action
Rate time

4. SENSITIVITY

Sensitivity
Gain
Bandwidth
Proportional band
Throttling range

5. CONTROLLED DEVICE

Controlled device
Normally closed
Normally open
Spring range

6. TRANSMITTER

Transmitter
Measuring element
Sensitivity
Working range
Range
Span

7. RATIO

Ratio
Auxiliary authority
Master
Submaster controller
Range of remote readjustment

The following glossary of terms and their definitions is extracted from Ref. 1 and 2, with permission from Honeywell, Inc.

Actuator: a device that converts a pneumatic or electrical signal to force, which produces movement. The movement may be rotary, linear, or a switching action. An actuator can effect a change in the controlled variable by operating the final control elements a number of times. Valves and dampers are examples of mechanisms that can be controlled by actuators.

Adjustment: the procedure required to produce the exact setting, response, or effect desired.

Ambient temperature: the temperature of the surrounding environment.

Amplifier: a device that receives an *input signal* from an independent source and delivers an *output signal* that is related and generally greater than the input signal.

Anticipating control: one which, by artificial means, is activated sooner than it would be without such means to produce a smaller differential of the controlled property. Heat and cool anticipaters are commonly used in thermostats.

Aquastat (Honeywell trademark): a thermostat used to control water temperature.

Auxiliary authority (Electronic Controllers): the amount of *measured variable* change at one *sensing element* compared to the change measured at the other sensing element, which causes the *output signal* to return to its original intermediate position, expressed in a percentage.

Auxiliary device: a component, added to a control system, which when actuated by the *output signal* from one or more controllers produces a desired function.

Averaging element: a thermostat sensing element which will respond to the average duct temperature.

Bandwidth (Electronic Controllers): the change in *controlled variable* that causes a full change in controller output.

Bimetallic element: one formed of two metals having different coefficients of thermal expansion such as used in temperature-indicating and controlling devices.

Bulb: a thermostat sensing element, usually remote, which will respond to the temperature in the immediate vicinity of the bulb.

Capacity index: the quantity of water, in gallons per minute at 60°F, that will flow through a given valve with a pressure drop of 1 psi. Also called *flow coefficient*.

Capillary tube: in refrigeration practice, a tube of small internal diameter used as a liquid refrigerant flow control or expansion device between high and low sides; also used to transmit pressure from the sensing bulb of some temperature controls to the operating element.

Changeover: the process of switching an air-conditioning system from heating to cooling, or vice versa.

Closeoff: the maximum allowable pressure drop to which a valve may be subjected while fully closed. Usually a function of the power available from the valve actuator.

Closed-loop system (HVAC): the arrangement of components to allow system *feedback*, i.e., a heating unit, valve, and thermostat arranged so that each component affects the other and can react to it.

Control agent: the medium manipulated by the control system to cause a change in the controlled medium. For example, suppose a heating coil through which steam is flowing is used to heat a room. The room thermostat is placed so that it measures temperature (controlled variable) in the room air (controlled medium) and operates a valve that regulates the flow (manipulated variable) of steam (control agent) through the heating coil. Heat from the coil is thus furnished to the room air. The most common control agents are: water (hot or cold), air (hot or cold), steam, electric current, and refrigerant.

Control point: the actual value of the *controlled variable* which the *controller* is maintaining at any given time, i.e., the actual temperature as measured.

Controlled device: the instrument that receives the controller's output signal and regulates the flow of the control agent. It is functionally divided into two parts: (1) **Operator:** receives the output signal and converts it into force; (2) **Regulator:** (valve body or damper) regulates the flow of the control agent.

Controlled medium: the substance (usually air, water, or steam) whose characteristics (such as temperature, pressure, flow rate, volume, concentration, etc.) are being controlled.

Controlled variable: that quantity or condition of the controlled medium that is measured and controlled. For example, temperature, pressure, flow rate, humidity, etc.

Controller: a device that senses and measures changes in the controlled variable and as a result meters energy in a usable form (output signal), which holds the controlled variable within predetermined limits.

Controller feedback: the change in the controller's output in response to a measured change in the controlled variable transmitted back to the controller input for evaluation.

Corrective action: control action that results in a change of the manipulated variable and is initiated by a deviation.

Damper: a device used to vary the volume of air passing through an air outlet, inlet, or duct.

Damper, opposed blade: alternate blades rotate in opposite directions. Provides an equal percentage flow characteristic—successive equal increments of rotation produce equal percentage increases in flow. Particularly useful for throttling applications where accurate control at low air flow is necessary.

Damper, parallel blade: all blades rotate in same direction. Provides a fairly linear air flow characteristic—the flow is nearly proportional to damper shaft rotation. Particularly useful in mixing applications where the sum of two air flows must be made constant.

Damper linkage: linkage used to connect motor to damper, usually consisting of a push rod, two crank arms, and two ball joints.

Desired value: the value of the controlled variable which it is desired to maintain.

Deviation: the difference between the set point and the value of the controlled variable at any instant.

Differential: the difference in values of the controlled variable that will activate a two-position controller to change an output of either maximum or zero to the opposite extreme, with no intermediate steps.

Differential, interstage: in a sequencing system the number of degrees of motor rotation from ON point of one stage to ON point of successive stage.

Differential, Stage: in a sequencing system the number of degrees of motor rotation from stage ON point to stage OFF point.

Differential gap: applies to two-position controllers. Also called "differential." It is the smallest range through which the controlled variable must pass in order to move the final control element from one to the other of its two possible positions. The difference between cut-in and cut-out temperatures, pressures, etc.

Direct acting: the output signal changing in the same direction the controlled or measured variable changes, i.e., an *increase* in the controlled or measured variable results in an *increased* output signal.

Economizer control: a system of ventilation control in which outdoor and return air dampers are controlled to maintain the proper mixed-air temperature for most economical operation. Also, outdoor air is used for "free cooling" when the outdoor air temperature is below the desired room air temperature.

Element: any device (such as a resistor, capacitor, transistor, transducer, heating coil, cooling coil, etc.) with terminals at which it may be connected directly to other devices. A component of a system.

Equal percentage characteristic: a valve characteristic which causes like movements of the valve stem at any point of the flow range to change existing flow an equal percentage, regardless of the existing flow quantity.

Face-and-bypass damper system: a heating system in which the mixed-air flow is divided into two duct sections, one through the coil (face) and the other around the coil (bypass). Dampers work in opposition (face damper closes while bypass damper opens, and vice versa) to regulate the amount of air which is heated. Used with either a steam or hot water coil. Demonstrates ON-OFF control of coil and modulating control of air flow across coil.

Feedback potentiometer: the potentiometer in the modulating motor which forms part of a bridge circuit along with the controller potentiometer and balancing relay in the motor. When the wiper on the controller potentiometer moves due to a change in the controlled variable, the motor runs and drives the wiper on the feedback potentiometer in the proper direction to rebalance the bridge circuit. When the circuit is balanced, the motor stops.

Final control element: that mechanism which directly acts to change the value of the manipulated variable. For example, a modulating valve.

Flow coefficient: *See* Capacity index.

High-limit control: a device which normally monitors the conditions of the controlled medium and interrupts system operation if the monitored condition becomes excessive.

Humidistat: a regulatory device, actuated by changes in humidity, used for the automatic control of relative humidity.

Hygrostat: same as Humidistat.

Lag: a delay in the effect of a changed condition at one point in a system, on some other condition to which it is related. Also, the delay in action of the sensing element of a control due to the time required for the sensing element to reach equilibrium with the property being controlled; i.e. temperature lag, flow lag, etc.

Limit: in a motor, a switch that cuts off power to the motor windings when the motor reaches its full-open position. Also, a device that monitors the condition of the controlled medium. *See* High-limit control, Low-limit control.

Limit shutdown: a condition in which the system has been stopped because the value of the temperature or pressure has exceeded a preestablished limit.

Linear characteristic: a control valve characteristic which results in equal volume changes for equal stem travel changes, regardless of the percentage of valve opening. Approximately a straight line when plotted on rectilinear coordinates.

Low-limit control: a device that normally monitors the condition of the controlled medium and interrupts system operation if the monitored condition drops below the desired minimum value.

Manipulated variable: that quantity or condition which is regulated by the automatic control system in such a manner that it causes the desired change in the controlled variable. The manipulated variable is a characteristic of the control agent.

Master: an instrument whose variable output is used to change the set point of a submaster controller. The master may be a humidistat, pressure controller, manual switch, transmitter, thermostat, etc.

Measured variable: the uncontrolled variable, such as temperature, relative humidity, or pressure sensed by the measuring element.

Measuring element: (1) when used in conjunction with a *pneumatic* controller, that part of the controller which measures a change in temperature, humidity, or pressure and converts this change into movement. (2) when used in conjunction with an *electronic* controller, that part of the controller which measures a change in the temperature, humidity, or pressure and converts this change into resistance to current flow.

Modulating: tending to adjust by increments and decrements; also one modified by variation of a second condition.

Modulating control: a mode of automatic control in which the action of the final control element is proportional to the deviation, from set point, of the controlled medium.

Modulating motor: an electric motor used to drive a damper, which can position the damper anywhere between fully open or fully closed in proportion to deviation of the controlled medium.

Morning warmup: an optional economizer application used in conjunction with automatic night setback to provide heating fuel economy. It is intended to raise space temperature to occupied conditions as soon as possible in the morning with the minimum use of fuel. The outdoor air dampers are kept closed until the building is warmed up. Also called "morning pickup."

Night setback: an optional economizer application used while a building is unoccupied, generally at night, to reduce heating expenses. The outdoor air dampers are closed to minimum position while the building space is maintained at a lower temperature.

Normally closed: applies to a controlled device that *closes* (stops the flow of control agent) when the signal applied to it is removed.

Normally open: applies to a controlled device that *opens* (allows the flow of the control agent) when the signal applied to it is removed.

Offset: the difference between the set point of the controller and the control point of the controlled variable caused by "load changes" affecting the system. Also, a sustained deviation between the actual control point and the set point under stable operating conditions. Offset is also referred to as "drift," "deviation," and "droop."

On-off control: a simple control system, consisting basically of a switch, in which the device being controlled is either fully on or fully off and no intermediate operating positions are available.

Open-loop system: the arrangement of components which will not allow system feedback.

Pneumatic: operated by air pressure.

Potentiometer: an electromechanical device having a terminal connected to each end of the resistive element and a third terminal connected to the wiper contact. The electrical input is divided as the contact moves over the element, thus making it possible to mechanically change the resistance.

Primary control: a device which directly or indirectly controls the control agent in response to needs indicated by the controller. Typically a motor, valve, relay, etc.

Primary element: the portion of the controller which first uses energy derived from the controlled medium to produce a condition representing the value of the controlled variable. For example, a thermostat bimetal.

Proportional action: an output signal changing in proportion to the amount of change in the controlled or measured variable.

Proportional band: the change in the controlled variable required to move the controlled device from one of its extreme limits of travel to the other. It is normally used in conjunction with recording and indicating controllers and is expressed in percent of the chart or scale range. Commonly used equivalents are "throttling range" and "modulating range."

Proportional plus automatic reset action: a combination of proportional action and a response which continually resets the control point back toward the set point to reduce the offset. *See* Reset rate.

Proportional plus rate action: a combination of proportional action and a response that precedes the normal proportional response. The combined response is proportional to the rate of change or speed with which the controlled variable deviates from the set point. *See* Rate time.

Rangeability: the ratio of the maximum controllable flow to the minimum controllable flow through a valve.

Rate time: the time in minutes that rate action response precedes normal proportional action response.

Reset: the process of automatically adjusting the control point of a given controller to compensate for changes in outdoor temperature. The hot deck control point is normally "reset" upward as the outdoor air temperature drops. The cold deck control point is normally "reset" downward as the outdoor air temperature increases.

Reset rate: the number of times per minute that the change made by the proportional action is duplicated by the reset action. It is usually expressed in "repeats per minute."

Reset ratio: the ratio of change in outdoor temperature to the change in control point temperature. For example, a 2 : 1 reset ratio means that the control point will increase 1 degree for every 2 degrees change in outdoor temperature.

Reverse acting: the output signal changing in the opposite direction the controlled or measured variable changes. For example, an *increase* in the controlled or measured variable results in a *decreased* output signal.

Sensing element: the first system element or group of elements that respond quantitatively to the controlled variable and perform the initial measurement operation.

Sensitivity: the change or number of psi the controller output changes per unit change in the controlled variable (Δpsi/degree temperature; Δpsi/percent relative humidity; Δpsi/psi pressure change of control agent).

Sequencer: a mechanical or electrical device that may be set to initiate a series of events and to make the events follow in sequence.

Sequencing control: a control which energizes successive stages of heating or cooling equipment as its sensor detects the need for increased heating or cooling capacity. May be electronic or electromechanical.

Set point: the point at which the controller is set. The desired controlled variable value which can be obtained by either an integral or remote adjustment of the controller.

Spring range: the range through which the signal applied must change to produce total movement of the controlled device from one extreme position to the other. **Normal spring range:** the change in applied signal that causes total movement when there is no external force opposing the operator. **Actual spring range:** the change in applied signal that operates the controlled device under actual conditions when it must overcome forces due to fluid flow, friction, etc., in addition to the normal spring range.

Submaster controller: a controller whose set point is automatically readjusted from a remote location. The set point is changed over a predetermined range by variations in an applied signal from a master. **Direct readjustment:** an increase in applied signal increases the set point of the submaster controller. **Reverse readjustment:** an increase in applied signal decreases the set point of the submaster controller.

Summing point: any point at which signals are added algebraically.

Supply pressure (Pneumatic): the energy source (compressed air) supplied to a controller and/or auxiliary device. It is usually a constant 15 to 20 psig, but, in special cases may be some other value.

Supply voltage: the energy source supplied to the controller and/or auxiliary device. Usually either 120 V AC or 24 V AC.

System feedback: the effect of the controller's response to a change in the controlled variable transmitted back to the controller.

Thermocouple: a device for measuring temperature utilizing the fact that an electromotive force is generated whenever two junctions of two dissimilar metals in an electric circuit are at different temperatures.

Thermostat: an instrument which responds to changes in temperature, and which directly or indirectly controls temperature.

Throttling range: the range of values of a proportional controller through which the controlled variable must pass to drive the final control element through its full operating range. Also called "proportional band." In a thermostat the temperature change required to drive the potentiometer wiper from one end to the other, typically 3°F to 5°F

Timed two-position action: a variation of two-position action in which the ON periods are prematurely shortened. Though this may be accomplished through a cam mechanism, it is usually accomplished in electric room thermostats by means of a heating element which is energized during the ON periods (heating) and OFF periods (cooling).

Transducer: an instrument that converts an input signal into an output signal, usually of another form; for example, electrical input to pneumatic output.

Transmitter: an instrument (not a controller) that transmits a directly functional signal in proportion to the variable measured. This proportion is factory set and is not to be adjusted.

Travel coefficient: the ratio between the flow at a given valve stem position and the flow through the valve in its wide-open position.

Two-position control: the type action in which the output signal is changed to either a maximum or minimum value with no intermediate steps. *See* On-off control.

Valve: any device in a pipe or tube that permits a flow in one direction only or regulates the flow of whatever is in the pipe by means of a flap, lid, plug, etc., acting to open or block passage.

Valve, automatic control: a valve that combines a valve body and a valve actuator or motor. A signal from some remote point can energize the actuator either to open or close the valve, or to proportion the rate of flow through the valve.

Valve, double-seated: has two seats and two discs arranged in such a way that in the closed position there is very little fluid pressure forcing stem toward the open or closed position. Requires less power to operate than a single-seated valve of the same size and port area.

Valve, modulating: a valve that can be positioned anywhere between fully on or fully off to proportion the rate of flow in response to a modulating controller. *See* Modulating control.

Valve, single-seated: has only one seat and one disc. Cheaper to construct than double-seated valve. Generally suitable for service requiring tight shutoff.

Valve, three-way diverting: has one inlet and two outlets and is double seated with a disc for each seat. Water entering the inlet port is diverted to either of the two outlet ports in any proportion desired by moving the valve stem. Can generally be used in mixing applications.

Valve, three-way mixing: has two inlets and one outlet and one disc and two seats. The proportion of the fluid leaving the outlet can be varied by moving the stem. Not suitable for diverting applications.

Valve, two-position: a valve that is either fully on or fully off with no positions between.

Valve flow characteristic: the relationship between the stem travel, expressed in percent, and the flow of fluid through the valve, expressed in percent of full flow. See Equal-percentage characteristic and linear characteristic.

Zoning: the practice of dividing a building into small sections for heating and cooling control. Each section is selected so that one thermostat can be used to determine its requirements.

15-3 COMPONENTS OF A CONTROL SYSTEM

To simplify the explanation of a control system, we can represent it in block-diagram form, Fig. 15-1. Each block represents an essential component of the *closed-loop system*. Figs. 15-1 through 15-6 and associated descriptive matter are from Ref. 1, courtesy of Honeywell, Inc.

The process. The major requirement of a control system is the *process*. If there is nothing to control, there is no control system and no need for controlling elements. An elaborate steam pressure control system is no longer a system if the steam is removed from the pipes. The process can take numerous forms, such as the flight of an airplane, the maintenance of a required temperature in a hospital operating room, or the operation of a power boiler.

DISTURBANCE-SENSING ELEMENT: The process is subject to a *disturbance* or external influence of some sort—a change in set point, supply, demand, or environment—or else there would be no need for a control system. The disturbance itself is an *independent variable* which cannot be changed by the system.

The disturbance causes a change in the *controlled variable,* which is the quantity or condition being controlled. The controlled variable is a characteristic of the *controlled medium,* which is the substance being controlled. For example, if water temperature is being controlled, the controlled variable is the temperature and the controlled medium is the water. In order to counteract the disturbance, it must first be detected and measured. The system must know how big a change in the controlled variable the disturbance caused. This is the job of the *disturbance-sensing element*. It is impossible for a human operator to tell the pressure within a pipe by looking at the outside of it, or to tell the temperature of a furnace accurately by observing the color of the flame. Fortunately, there are transducers of many kinds to get this information for the system. Pressure-, temperature-, and humidity-sensing

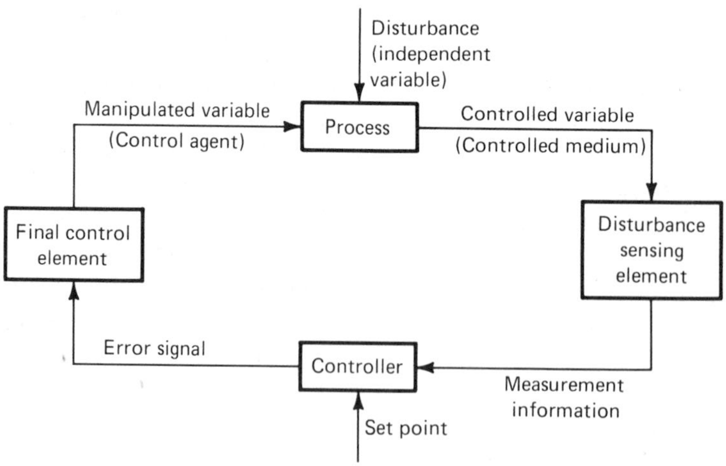

Figure 15–1

Simple block diagram of a closed loop control system. Each step responds to action of step before and effects step following

transducers are only a few of the many disturbance sensing elements available to provide *measurement information*.

CONTROLLER: The *controller* receives the *measurement information* from the disturbance-sensing element and interprets it to see how well the process is going. (In many cases the disturbance-sensing element may be a part of the controller, such as the bimetalic element in a thermostat.) A *set point* (or reference standard) is established in the controller. If conditions match this set point, the process is doing fine; but if a comparison of the measurement information with the set point shows a difference, the controller produces an *error signal* to initiate corrective action.

The set point may be manually set (for example, a temperature setting), in which case it will stay constant for a long period of time. Or it may be adjusted automatically (for example, a boiler system in which steam demand is the reference input and air flow is adjusted accordingly), in which case the reference standard is varying continuously.

FINAL CONTROL ELEMENT: The *final control element* is a mechanism that changes the value of the manipulated variable in response to the error signal from the controller. The *manipulated variable* is the quantity or condition (a characteristic of the control agent) that causes the desired change in the controlled variable. For example, if steam is flowing through a heating coil to heat a room, the steam is the control agent and its flow through the coil is the manipulated variable. The flow of steam varies in response to changes in the temperature (controlled variable) of the air (controlled medium) in the room.

The final control element causes the manipulated variable and control agent to exert a correcting influence on the process operation. Thus, the automatic control system is an error-sensitive, self-correcting system, often called a *closed-loop feedback control system*.

In summary, there are three basic parts that must be considered when putting together a closed-loop control system. They are:

1. *The control agent*—This source of energy supplied to the system can be either *hot* or *cold,* such as steam, hot water, heated air, chilled water, or chilled air.
2. *The controlled device*—A valve or damper can be either *normally open* (NO) or *normally closed* (NC) to regulate the flow of the control agent. It is chosen primarily for "fail-safe" operation as mentioned previously.
3. *The controller action*—A controller is furnished with either *direct action* or *reverse action*. This will allow the balance mentioned above.

The proper combination of these three parts must be applied or the closed-loop system will not operate.

The closed-loop control system is by far the most commonly used system. Nevertheless the *open-loop system* finds application in certain instances. For example, more heat is required to maintain a suitable room temperature as the outdoor air temperature drops. It is possible to arrange a thermostat to measure outdoor air temperature and cause the heat input of a building to increase as the outdoor air temperature decreases. The outdoor thermostat cannot measure the result of heat input to the room, hence there is no feedback. Figure 15-2 illustrates such a control system.

It should be clear that all control systems are comprised of one or more loops. Most of them will be closed loops, but some may be open loops or a combination of both. It is necessary to understand the difference in order to appreciate what results can be expected. When controllers are studied, it will be seen that they comprise one or more control loops within themselves.

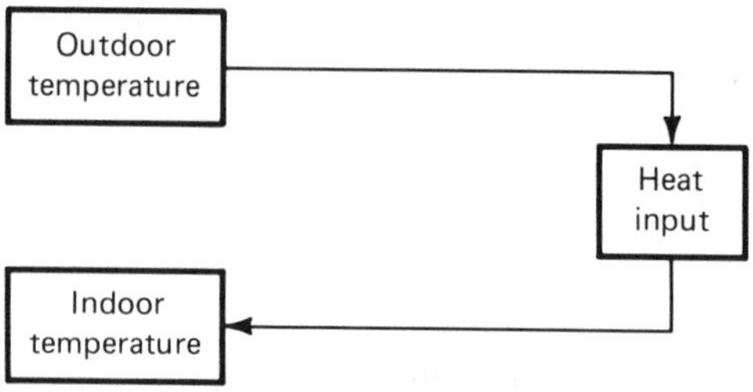

Figure 15–2

Open-loop system. No direct relation between two steps, hence no feedback

It should be pointed out that there are no perfect closed loops, that is, loops which must account only for the effect of the parts as described. In actual control conditions outside forces constantly work on the various parts to change the balance and set the loop cycle in operation to reestablish balance. If this were not the case, there would be no need for automatic control.

In conclusion, it can be said that a closed-loop system is one in which all parts have an effect on the next step in the loop and are affected by the action of the previous step. An open-loop system is one in which one or more of the steps has no direct effect or action imposed on the following step or is not affected by other steps in the loop.

15-4 MODES OF CONTROL (Control Action)

A mode of control is the manner by which a control system makes corrections in response to a disturbance. It relates the operation of the final control element to the measurement information provided by the disturbance-sensing element. Thus, it is generally a function of the controller. The proper matching of the mode to the process determines the overall performance of the control system. The basic modes are (1) on-off (two-position), (2) multiposition (multistage), (3) floating, and (4) proportioning (modulating). More complex variations of the proportioning mode include the addition of reset action, rate action, or both.

On-off (two-position). On-off control, as the name suggests, provides only 2 positions—either full ON or full OFF. There are no intermediate positions. When the controlled variable deviates a predetermined amount from the set point, the final control element moves to either of its extreme positions. The time travel of on or off is varied according to the load demand.

On-off control is the simplest mode of control, but it has definite disadvantages. It allows the controlled variable to vary over a range (sometimes a wide range) instead of letting it settle down to a near-steady condition. If this range becomes too narrow, the controller will wear itself out by continually switching on and off. Figure 15-3 illustrates schematically the system response and controller action.

Multiposition or multistage control. Multiposition control is an extension of the on-off control to two or more stages. Where the range between full on to full off is too wide to achieve the operation desired, multiple stages with much smaller ranges can be used. Each stage has only two positions (on or off), but there are as many positions as there are stages, resulting in a steplike operation. The greater the number of stages or steps, the smoother will be the operation. As the demand or load

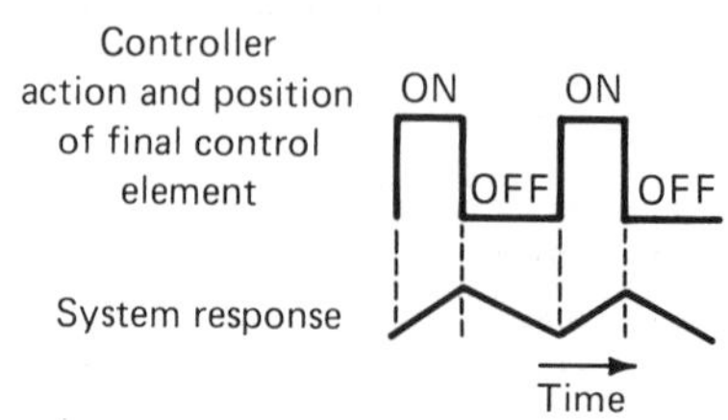

Figure 15–3

ON-OFF (2-position) mode of control

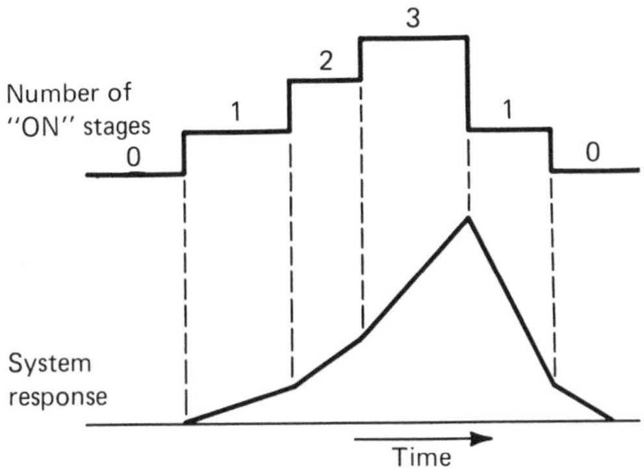

Figure 15–4

Multiposition (multistage) mode of control

increases, more stages are turned on. Normally, multiposition control provides for 2 to 10 operating stages. Two-stage firing of oil burners is quite common. Figure 15-4 illustrates schematically the system response and controller action.

Floating control. Floating control differs from on-off or multiposition control in that the final control element may assume any position between its extremes. The controller itself has two positions with a neutral zone in between. With the controller in the neutral zone, the final control element will not move. When a large enough disturbance occurs, the controller moves to one of its two positions. This causes the final control element to move in one direction or the other, depending on the position of the controller. The final element moves at a *constant speed.* Therefore, this mode is more properly called *single-speed floating control.* When the controller returns to the neutral zone, the final control element stops. It will stay at this position until the controller again moves to one of its two positions. Thus, this mode is called *floating* because the final control element comes to rest when the controller is floating between its two positions.

Floating control is used in applications with gradual load changes or where there is little lag between a disturbance and its detection. The speed of the final control element should be fast enough to keep pace with the most rapid load changes. If not, hunting (instability) and excessive cycling will occur. Figure 15-5 illustrates schematically the system response, controller action, and position of the final control element.

Proportioning or modulating control. In proportioning control, as in floating control, the final control element may assume any position between its extremes. Unlike floating control, the controller may also assume any position. It has no neutral zone, so any small disturbance will cause it to move. For each movement of the controller, there is a proportional amount of movement of the final control element. These movements can be made as frequently as disturbances occur. Proportional control produces a continuous linear relation between the amount of disturbance, controller action, and position of the final control element. Figure 15-6 illustrates schematically the system response, controller action, and position of the final control element.

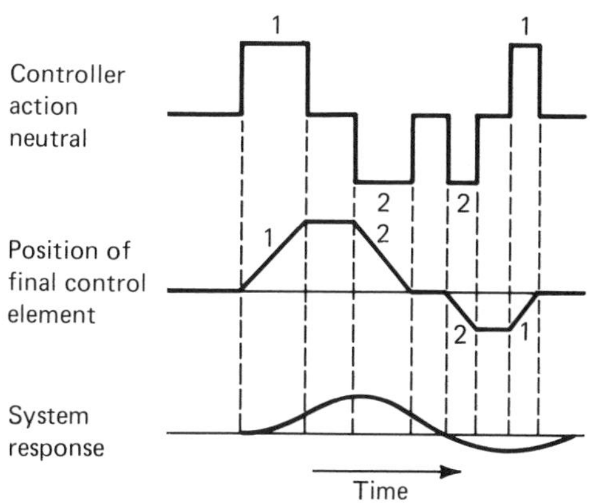

(1) and (2) indicate positions of controller

Figure 15–5

Floating mode of control

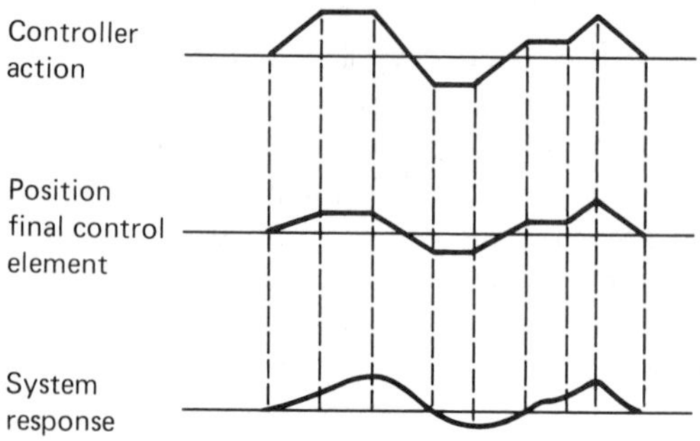

Figure 15–6

Proportioning (modulating) mode of control

More complex variations of proportioning control. On a sustained load change an *offset* can occur between the value of the controlled variable and the set point. Continual adjustment of the set point is necessary to keep the controlled variable at the same point throughout the load range of the equipment. In sophisticated controllers two types of automatic action are frequently added to assure the best possible control under the most severe process conditions.

With *reset action,* the error signal from the controller is varied in proportion to the offset and to the time that it persists. The final control element continues to move in a direction to correct the error. It will stop only when the error signal becomes zero, at which time the controlled variable is at the set point.

With *rate action,* the error signal from the controller is varied in proportion to the rate at which the disturbance takes place. It is used (1) to accelerate (speed up) the return of the controlled variable to the set point in a part of the process that is slow in responding and (2) to anticipate a disturbance that will occur and start corrective action before any change is detected.

15-5 TYPES OF CONTROL SYSTEMS

A control system can be classified by its source of power (energy). Control systems may be pneumatic, hydraulic, electromechanical, electronic, fluidic, or combinations of these.

A *pneumatic system* is operated by air pressure supplied by a compressor, and its components are connected by air lines. A *hydraulic system* is operated by movement and force of a liquid under pressure, and its components are connected by pipes and/or flexible tubes. An *electromechanical system* is operated by electricity, and its components are connected by electric circuits. An *electronic system* is also operated by electricity, but it uses electronic elements, such as transistors and vacuum tubes. In most cases it also uses an electronic amplifier to increase the minute voltage variations produced by the disturbance-sensing element to values required for operation of standard electromechanical controlled devices. *Fluidic systems* use air or gas and are all similar in operating principles to electronic as well as pneumatic systems. In reality, many control systems use combinations of these several power sources. An example is a system using an electronic-sensing element, an electromechanical controller, and a pneumatic valve.

In the following sections we will discuss the electromechanical, electronic, and pneumatic system components. These are the most commonly used for HVAC systems.

15-6 BASIC ELECTRICITY

Before studying electromechanical and electronic control systems and the required component devices, a brief review of basic electricity will be helpful to better understand the function of these systems. Reference 3 gives a very clear, brief discussion of basic electricity, and it is presented in the following sections with permission from Johnson Service Company.

Ohm's Law. There are three factors present in every electrical circuit: current, voltage and resistance.

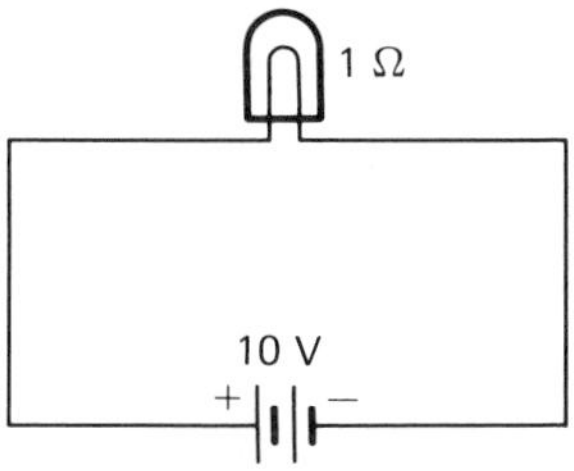

Figure 15–7

Light bulb in a battery circuit

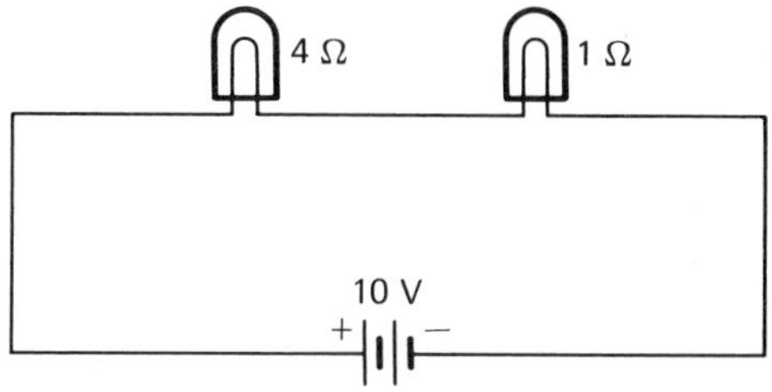

Figure 15–8

Series circuit

A common electrical power source is a battery. The battery in an electrical circuit is the same as the pump in a water circuit. Just as the pump produces a source of pressure measured in psi, the battery produces a source of electromotive force (EMF), measured in volts.

The amount of flow (gpm) in a water circuit depends on the pump pressure and the amount of restriction in the circuit. Similarly, the amount of current (amperes or ''amps'') in an electrical circuit depends on the battery voltage and the amount of resistance (ohms) in the circuit. The voltage is the force, the resistance is the opposition to the force, and the current is the flow or movement of electrons that results from the combination. Current is the flow rate of electrons per second past a point in the wire. One volt is required to force 1 ampere (A) of current through 1 ohm (Ω) of resistance. Expressed mathematically;

$$E(\text{voltage}) = I(\text{current in amps}) \times R(\text{resistance in ohms})$$

This is the equation on which all electric and electronic circuit theory is based and is called Ohm's law. It can be arranged as required to find one of the three factors, such as

$$E = IR \qquad I = E/R \qquad R = E/I$$

Figure 15-7 shows a light bulb connected across a 10-V battery. The light bulb has 1 ohm (Ω) of electrical resistance. To determine the current through the bulb, Ohm's law is stated and used as

$$I = E/R = 10\ \text{V}/\Omega = 10\ \text{A}$$

Thus, a current of 10 A is flowing through the bulb.

Series connections. Figure 15-8 shows two light bulbs connected in series across a battery. Two resistances are said to be connected in series when the current flowing through one resistance must also flow through the other. A series connection used to be used for Christmas tree lights; when one light bulb burns out, the circuit is broken and the whole string goes out. The total resistance *(R_T)* in a series circuit is the sum of the individual resistances:

$$R_T = R_1 + R_2 + R_3 \cdots$$

For Fig. 15-8, $R_T = 4\ \Omega + 1\ \Omega = 5\ \Omega$. To determine the current through the light bulbs, we have

$$I = E/R = 10\ \text{V}/5\ \Omega = 2\ \text{A}$$

Therefore, a current of 2 A is flowing through the bulbs.

The current is the same through all the resistances, but the voltage drop across each is dependent on the resistance value. When the current through a bulb and the resistance are known, Ohm's law can be used to determine the voltage drop across the bulb.

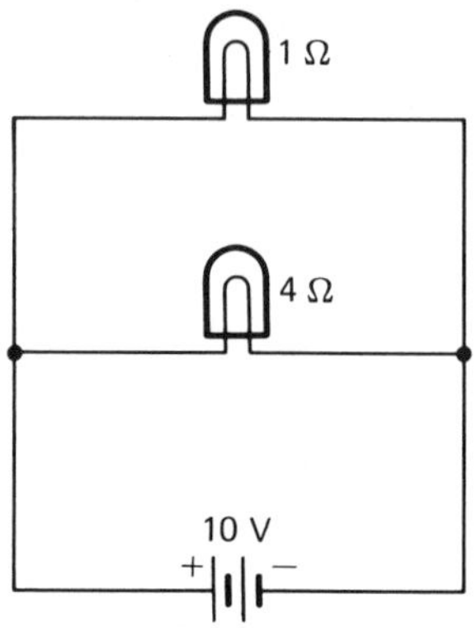

Figure 15–9

Parallel circuit

For the 1 ohm bulb, $E = IR$
$= 2 \times 1 = 2$ V

For the 4 ohm bulb, $E = 2 \times 4 = 8$ V

Parallel connections. Figure 15-9 shows two light bulbs connected in parallel across a battery. In parallel connection the voltage drops across the resistances will always be equal; the current flow through each resistance is dependent on the resistance values or

$$I_1 = E/R_1 = 10/4 = 2.5 \text{ A}$$

$$I_2 = E/R_2 = 10/1 = 10.0 \text{ A}$$

Both bulbs draw current from the battery, so the total current supplied is the current flow through both bulbs, or $I_T = I_1 + I_2 = 2.5 + 10.0 = 12.5$ A.

If Christmas tree lights were connected in parallel, one light burning out would not break the circuit because each bulb would have a direct connection to the power source.

The total resistance of the circuit can be determined by using the voltage across the circuit and the total current flow into the circuit, or $R_T = E/I_T = 10.0/12.5 = 0.8\ \Omega$. The total resistance of a parallel circuit is always less than the lowest resistance in the circuit and can be calculated as follows:

$$\frac{1}{R_T} = \frac{1}{R_1} + \frac{1}{R_2} + \frac{1}{R_3} + \cdots$$

The total current is the sum of the individual currents through each lamp:

$$I_T = I_1 + I_2 + I_3 + \cdots$$

Using Ohm's law, $I = E/R$

$$\frac{E_T}{R_T} = \frac{E_1}{R_1} + \frac{E_2}{R_2} + \frac{E_3}{R_3} + \cdots$$

But the voltage across each lamp is the same:

$$E_T = E_1 = E_2 = E_3 = \cdots$$

Substituting this into the current equation gives

$$\frac{E_T}{R_T} = \frac{E_T}{R_1} + \frac{E_T}{R_2} + \frac{E_T}{R_3}$$

In summary, for DC series circuits

a. the same current flows through all resistances

b. total voltage is the sum of all voltage drops across all resistances

c. total resistance is the sum of all resistances

In summary, for DC parallel circuits

a. the total current is the sum of all currents

b. voltage is the same across all resistances

c. reciprocal of the total resistance is the sum of the reciprocals of the individual resistances

d. total resistance is always less than the lowest resistance in the circuit.

<u>AC and DC power.</u> Electrical power can be supplied in two different forms, the difference being the characteristics of the current flow. A power source that causes current to flow in one direction only is referred to as a direct current (DC) source. A power source that causes current to flow alternately in one direction and then in the other is referred to as an alternating current (AC) source.

Batteries and automobile DC generators are common examples of DC electrical power sources. Normally, a DC current flow is thought of as a continuous unidirectional flow that is constant in magnitude; however, a pulsating current flow that changes in magnitude but not direction is also considered direct current.

The power supplied by power companies throughout the country is the most common example of an AC power source. If the magnitude of the current is recorded as it varies with time, the shape of the resultant curve is called the waveform. The waveform produced by the power companies' generators is a sine wave as shown in Fig. 15-10. When the waveform of an AC voltage or current passes through a complete set of positive and negative values, it completes a cycle. The frequency of an AC voltage or current is the number of cycles that occur in 1 sec; the frequency of the voltage supplied by U.S. power companies is 60 cycles per second (60 hertz).

All calculations using AC voltage are based on sine waves. There are four values of sine waves that are of particular importance.

1. *Instantaneous value:* The voltage or current in an AC circuit is continually changing. The value varies from zero to maximum and back to zero. If we measure the value at any given instant, we will obtain the instantaneous voltage or current value.

2. *Maximum value:* For two brief instants in each cycle the sine wave reaches a maximum value; one is a positive maximum, the other negative. This value is often referred to as the *peak* value. The two terms have identical meanings and are interchangeable.

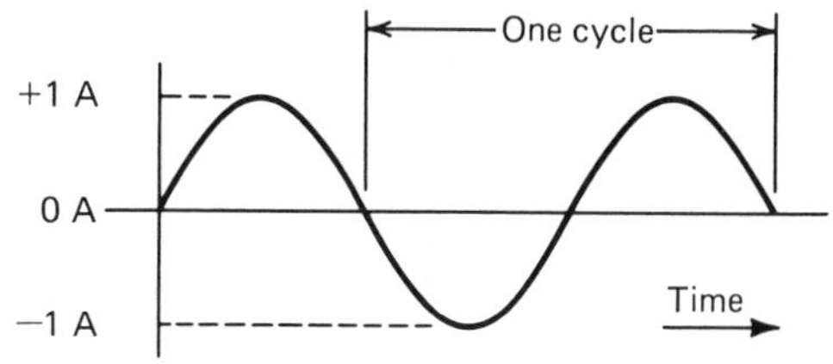

Figure 15–10

A.C. cycle wave form

3. *Average value:* Since the positive and negative halves of a sine wave are identical, the average value can be found by determining the area below the wave and calculating what DC value would enclose the same area over the same amount of time. For either the positive or negative half of the sine waves, the average value is 0.636 times the maximum value.

4. *Effective value* (rms): The effective value of an AC voltage or current required to provide the same average power or heating effect. Heating effect is independent of direction of electron flow, and therefore is the same for a full cycle as it is for a half cycle of a sine wave. The effective value of an AC sine wave is equal to 0.707 times the maximum value. Thus, an alternating current of a sine wave having a maximum value of 14 A produces the same amount of heat in a circuit that a direct current of 10 A produces.

The heating effect varies as the square of the voltage or current. If we square the instantaneous values of a voltage or current sine wave, we obtain a wave that is proportional to the instantaneous power or heating effect of the original sine wave. The average of this new waveform represents the average power that will be supplied. The square root of this average value is the voltage or current value that represents the heating effect of the original sine wave of voltage or current. This is the effective value of the wave, or the rms value (root-mean-square); the square root of the average of the squared waveform.

The effective values of voltage and current are more important than instantaneous, maximum, or average values. Most AC voltmeters and ammeters are calibrated in rms values.

The *phase* of an AC voltage refers to the relationship of its instantaneous polarity to that of another AC voltage. Figure 15-11 shows two AC sine waves in phase but unequal in amplitude. Figure 15-12 shows the two sine waves 45 degrees out of phase. Figure 15-13 shows the two sine waves 180 degrees out of phase. The length of a sine wave can be measured in angular degrees because each cycle is a repetition of the previous one, much like going in a circle. One complete cycle of a sine wave is 360 angular degrees.

Power in DC circuits. Whenever a force causes motion, work is performed. Electrical force is expressed as voltage, and when voltage causes a movement of electrons (current) from one point to another, energy is expended. The rate of work, or the rate of producing, transforming, or expending energy, is generally expressed in watts or kilowatts (1000 W). In a DC circuit 1 V forcing a current of 1 A through a 1-Ω resistance results in 1 W of power being expended. The equation for this is

$$P \text{ (watts)} = E \text{ (volts)} \times \text{I (amps)}$$

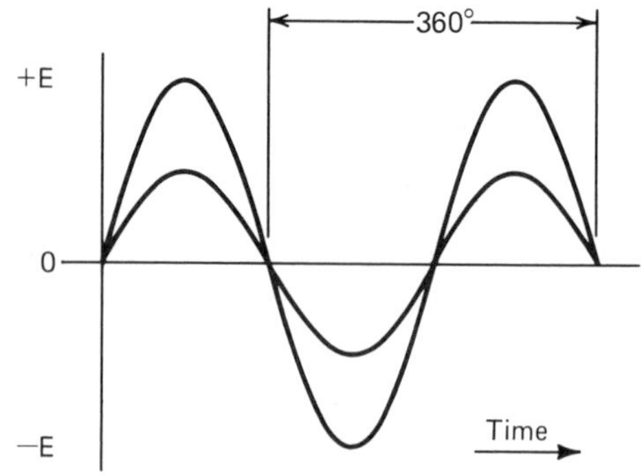

Figure 15–11

A.C. sine waves in phase but unequal in amplitude

To illustrate, what is the DC power used by the light bulbs in a series circuit such as Fig. 15-8?

The current is $I = E/R_T = 10/5 = 2$ A. Then from $P = E \times I$ we have, $P = 10 \times 2 = 20$ W of power.

The same calculations can be made for the light bulbs in a parallel circuit such as Fig. 15-9. The current is $I = E/R_T = 10/0.8 = 12.5$ A. Then the power is $P = E \times I = 10 \times 12.5 = 125$ W.

Power can be computed if any two of the three values of current, voltage, and resistance are known, as follows:

When resistance is unknown, $P = E \times I$

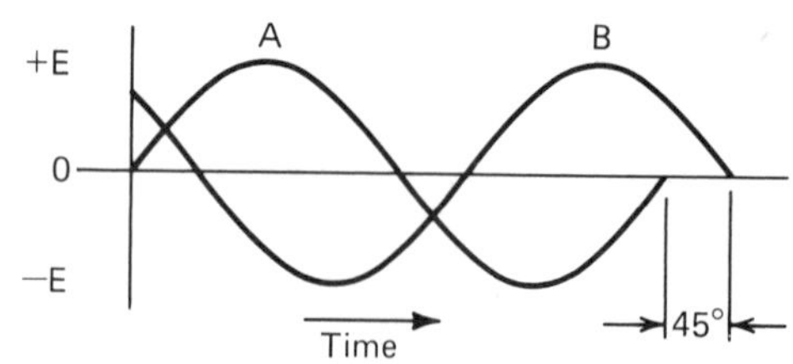

Figure 15–12

A.C. sine waves 45 degrees out of phase

When voltage is unknown, $P = (IR)I = I^2R$

When current is unknown, $P = E(E/R)$

$= E^2/R$

Power rating of equipment. Most electrical equipment is rated for both voltage and power (volts and watts). Electric lamps rated at 120 V are also rated in watts; they are usually identified by wattage rather than by voltage because the voltage is always assumed to be 120. The wattage rating of a light bulb or other electrical device indicates the rate at which electrical energy is changed into other forms of energy, such as light and heat. The greater amount of electrical power a lamp changes to light, the brighter the lamp will be; therefore, a 100-W bulb furnishes more light than a 75-W bulb.

Similarly, the power rating of motors, resistors, and other electrical devices indicate the rate at which the devices are designed to change electrical energy into some other form of energy. If the rated wattage is exceeded, the excess energy is usually converted to heat, and the equipment will overheat and perhaps be damaged. Some devices will have maximum DC voltage and current ratings instead of wattage; multiplied, these values give the effective wattage.

Resistors are rated in watts dissipation in addition to ohms of resistance; resistors of the same resistance value are available with different wattage ratings. Usually, carbon composition or ceramic resistors are rated from about 1⁄10 W to 2 W. Wire-wound resistors are used when a higher wattage rating is required. Generally, the larger the physical size of the resistor, the higher its wattage rating since a larger amount of material will absorb and give up heat more easily.

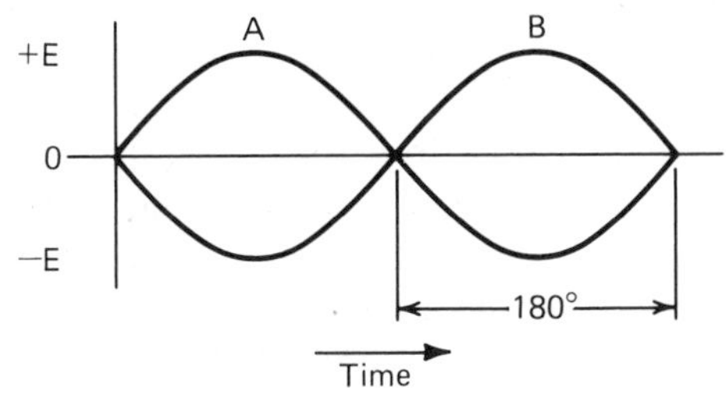

Figure 15–13

A.C. sine waves 180 degrees out of phase

Capacitance. Capacitance, like resistance, is a physical property of an electrical circuit. Resistance is opposition to current flow in an electrical circuit the same as friction is the opposition to motion in a mechanical system. Capacitance is the property of an electrical circuit (or component) that opposes any change in voltage across it. Capacitance, in an electrical circuit, allows electrons to be stored the same as a liquid or gas would be stored in a tank in a mechanical system. The capacity of the tank is rated in gallons or cubic feet; the capacity of a capacitor is rated in farads; a farad being the unit of capacitance. A capacitor has a capacitance of 1 farad when a voltage change of 1 V per second across its terminals produces a displacement current flow of 1 A. Figure 15-14 illustrates an electrical circuit that includes a capacitor. Because the usual size of a capacitor is only a small portion of 1 farad, the capacity is usually expressed in microfarads (millionths of a farad) or picofarads (millionths of a microfarad).

A capacitor has two conducting surfaces (material with low electrical resistance) separated by a dielectric (an insulating material with almost infinite electrical resistance).

Capacitors are used in electronic circuits for one of three basic purposes:

a. To couple an AC signal from one section of a circuit to another

b. To block out and/or stabilize any DC potential from some component

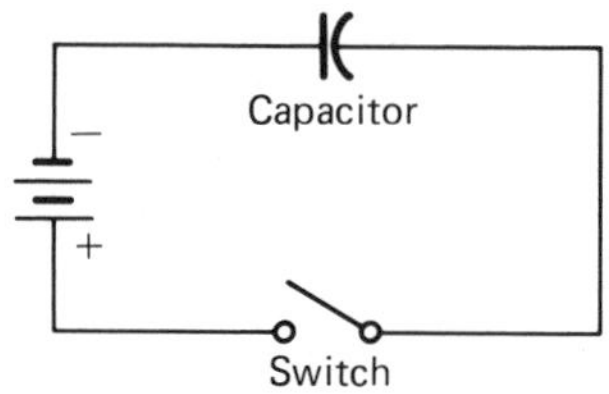

Figure 15–14

Circuit with capacitor

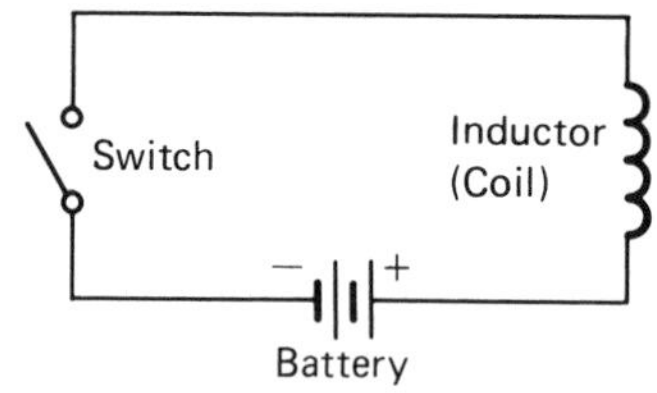

Figure 15–15

Circuit with inductor

c. To bypass or filter out the AC component of a complex wave

Inductance. Another factor to consider in electrical circuits is inductance. Inductance is the property of a circuit (or component) to oppose any change in current through it. When current attempts to change in an inductor, a voltage is self-induced in the coil. When discussing inductor, the word *coil* is often used to describe an inductor. A coil is a series of rings or a spiral of wire. Automobile and relay coils are common examples. The polarity of the induced voltage is such as to oppose the change in current. In this way inductance can be compared to kinetic inertia in a mechanical system, which tends to keep a body moving at a constant velocity. Figure 15-15 illustrates an electrical circuit that includes an inductor (coil).

Power in AC circuits. Ohm's law for an AC circuit states "the current flowing through an AC circuit is equal to the voltage impressed across that circuit divided by the impedance of the circuit." This is the same as Ohm's law for DC circuits, except that the word *impedance* is substituted for *resistance*. Impedance is the total opposition to the flow of current in an AC circuit offered by resistance, capacitance, and inductance. The ohm is the unit of impedance in an AC circuit. The three Ohm's law formulas for AC circuits are

$$I = \frac{E}{Z \text{ (impedance)}}$$

$$Z = \frac{E}{I} \qquad E = I \times Z$$

Power in a purely resistive AC circuit is calculated the same as in a DC circuit, $P = E \times I$.

When we add capacitance or inductance to the circuit, voltage × amperes does not indicate the actual wattage being taken from the power source. In a circuit, when capacitance predominates, the current will lead the voltage; where impedance predominates, the current will lag the voltage.. The measurement of the phase difference between current and voltage in a circuit is called *power factor* and is simply the cosine of the shared angle. When the current is greatly out of phase with the voltage, the power factor is said to be low. When the current is nearly in phase with the voltage, the power factor is said to be high. When the angle between current and voltage is 90 degrees, the power factor is zero. When the voltage is exactly in phase with the current, the power factor is unity. A high power factor is desirable in AC power systems because circuit losses are reduced. Even though the inductive or capacitive component in an AC circuit does not use any electrical power from the power source, it does draw current from the source. This extra current must flow through the generators, transformers, and wires that make up a power system and will cause extra losses in these devices. Also, the increased current may force the use of larger devices than would be necessary with a higher power factor, and this costs money. For these reasons power companies will penalize a customer with a lower power factor load by increasing the cost per kilowatt hour.

If two AC motors have the same horsepower rating, the electrical power used by both will be equal when doing the same amount of mechanical work. However, if the power factor of one motor is larger than the other, it will take less current to do that work. If the power factor of one motor is 0.9 and the other 0.7, but both are 1-hp (746 W), 120-V AC motors:

$$P = EI \times \text{power factor} = 746$$
$$746 = 120 \times I \times 0.9$$
$$I = 6.92 \text{ A}$$

and

$$P = EI \times \text{power factor} = 746$$
$$746 = 120 \times I \times 0.7$$
$$I = 8.9 \text{ A}$$

The above example shows why a higher power factor is desirable. A low power factor is a disadvantage in power circuits and should be corrected if possible. Since the properties of inductance and capacitance have the opposite affect in AC circuits, in highly inductive circuits (such as those containing many electric motors) placing large capacitors across the power line will raise the power factor.

Usually, electrical machinery is rated in VA (volt amperes) instead of watts to aid in determining the power factor. A wattmeter reading across the power lines will give the actual power taken from the source. The wattmeter reading divided by the full load VA for a motor is the power factor.

Semiconductors. An insulator has a very high resistance to current flow, and a conductor has a very low resistance. Therefore, as the name implies, a semiconductor has a "medium" resistance.

Semiconductors used in making diodes and transistors are basically germanium or silicon crystals with controlled amounts of impurities added. When arsenic or antimony is the added impurity, *N*-type semiconductor material is formed (an excess number of electrons). When galium or indium is the added impurity, *P*-type semiconductor material is made (a lack of electrons).

Rectifiers (Diodes). When *N*- and *P*-type materials are joined together, they form a rectifier, also referred to as a diode. The *PN* junction (diode) acts as a one-way valve to the flow of current. There is a forward, or low-resistance direction through the junction. Current flowing in the low-resistance direction is called forward bias (see Fig. 15-16); current flow in the opposite or high-resistance direction is called reverse bias (see Fig. 15-17).

Diodes are used as rectifiers, converting alternating current to direct current, and isolating one circuit from another circuit. A simple rectifier circuit is shown in Fig. 15-18. The output from the transformer is an AC voltage (see Fig. 15-19), but because the rectifier blocks current flow when the transformer output is negative, only pulsating DC voltage appears across the resistor.

Zener diode. Compared with pneumatic equipment, the zener diode is the equivalent of a

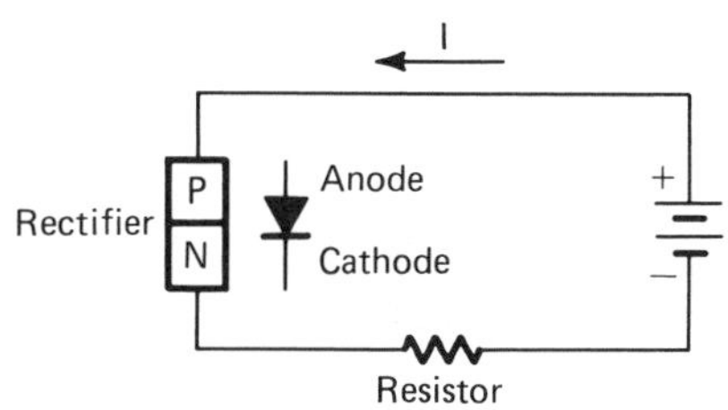

Figure 15–16

Forward bias

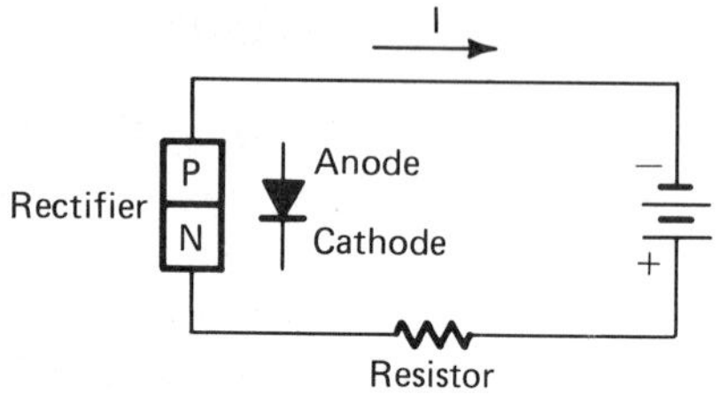

Figure 15–17

Reverse bias

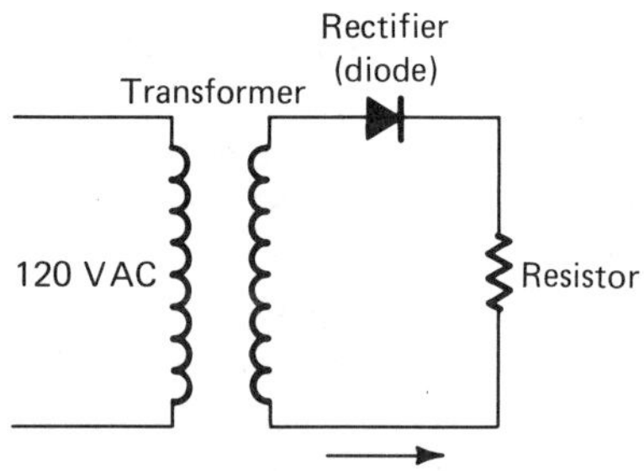

Figure 15–18

Simple rectifier circuit

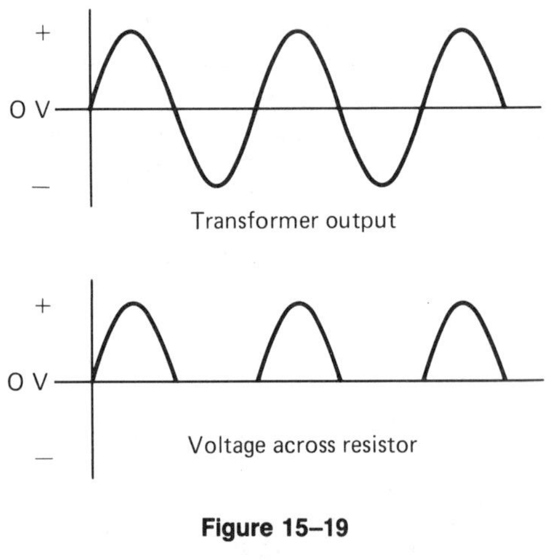

Figure 15–19

Rectifier action in circuit of Figure

nonadjustable pressure-reducing valve. It is used when a constant DC voltage is required.

When one polarity of voltage is applied to a rectifier, it blocks the flow of current. However, if the voltage is raised high enough, the rectifier breaks down, allowing current to flow. Normal rectifiers would be destroyed by this breakdown, but a zener diode is specially designed to operate in the breakdown region.

Figure 15-20 shows a typical zener diode circuit. The breakdown voltage on the diode is 10 V. As long as the battery voltage is 10 V or higher, the output across the diode will be 10 V. Any battery voltage above 10 V is dropped across the resistor so, if the voltage changes from 12 to 14 V, the voltage drop across the resistor changes from 2 to 4 V while the output voltage remains at 10 V.

Transistor. A transistor is a device used to amplify electronic signals. It consists of three layers of *P*- and *N*-type semiconductor material arranged in either of two ways, as shown in Fig. 15-21.

The theory of transistors has been known almost as long as vacuum tubes have been practical. However, it was not until 1948 that research laboratories developed a practical, high-quality transistor to meet production standards of cost and utility. Now the transistor is rapidly replacing the vacuum tube in many applications of electronics, plus opening up many more.

The transistor has many advantages over the vacuum tube:

1. Small size reduces the bulk and weight of electronic equipment.

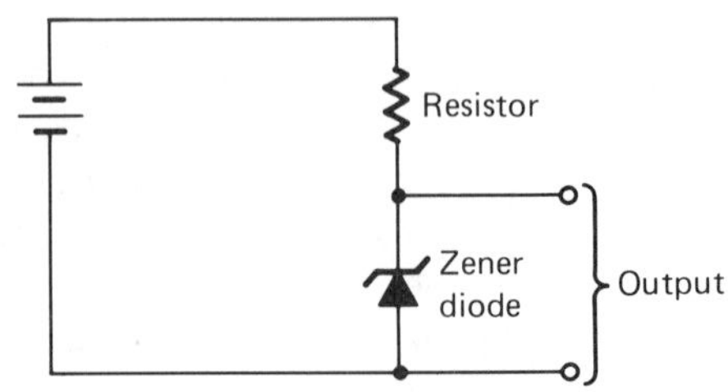

Figure 15–20

Lener diode circuit

2. Low power consumption means economical operation.
3. No warmup time results in faster equipment response.
4. Ruggedness enables it to better withstand mechanical shock and abuse.
5. Long life eliminates frequent maintenance and replacement.

Figure 15–21

Transistors

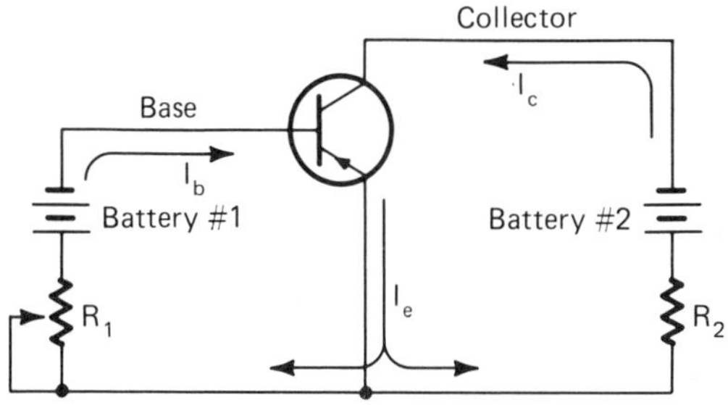

Figure 15–22

Simple transistor amplifier circuit

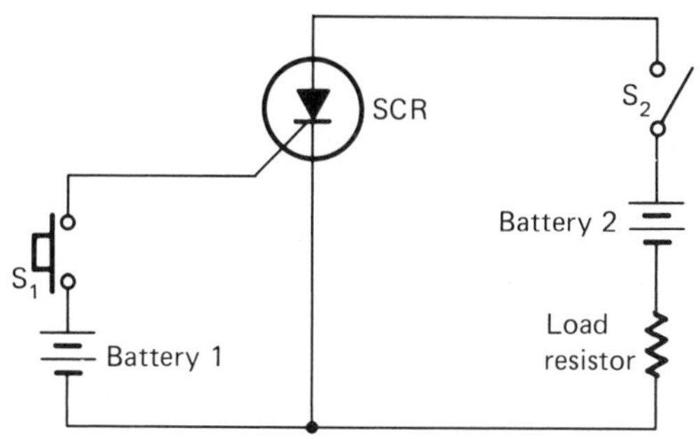

Figure 15–23

SCR circuit

Figure 15-22 shows a simple transistor amplifier. Battery 1 and adjustable resistor R_1 determine the input current to the transistor. When R_1 is high in resistance, the current flowing from the base to the emitter is very small. When the base-to-emitter current is small, the collector-to-emitter resistance appears as a very high resistance, limiting the current flow from battery 2 and limiting the voltage drop across R_2. As the resistance of R_1 is lowered, the current flowing through the base-to-emitter junction increases. As the base-to-emitter current is increased, the resistance of the transistor to collector to emitter is decreased. More current is flowing from battery 2 through R_2, and the voltage drop across R_2 is increased. A very small change in the current from battery 1 causes a large change in the current from battery 2. The ratio of the large change to the small change is defined as the *gain* of the transistor.

Silicon-controlled rectifier (SCR). The silicon controlled rectifier (SCR) is a four-layered *PNPN* device. The SCR can be defined as a high-speed semiconductor switch. It requires only a short voltage pulse to turn it "on," and it remains "on" as long as current is flowing through it.

Referring to Fig. 15-23, assume that the SCR is off (has a very high resistance); therefore, there is no current flowing through the resistor. When switch S_1 is closed for a time, just long enough to turn on the SCR (changes to a very low resistance), a current will flow through the resistor and SCR. The SCR will remain on until switch S_2 is opened. Opening S_2 stops the current flowing through the resistor and SCR, and the SCR will turn off. When S_2 is again closed, the resistance of the SCR remains high, and no current will flow through the resistor until S_1 is reclosed.

Figure 15-24 shows an SCR represented (a) schematically and (b) pictorially. The anode is the positive terminal, and the gate is the terminal used to turn the SCR on.

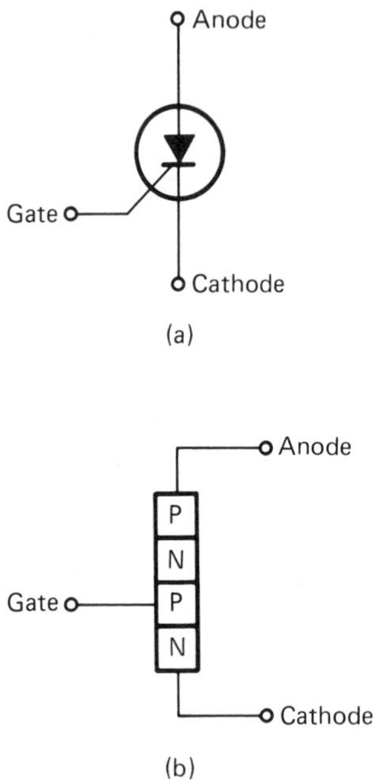

Figure 15–24

SCR symbol (a) Schematic (b) Pictorial

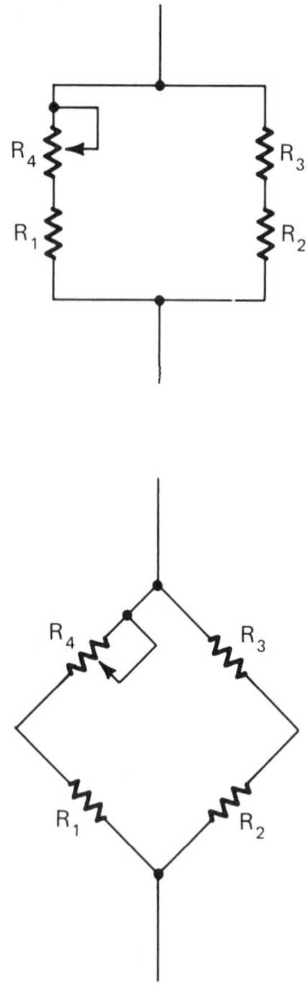

Figure 15–25

Series-parallel bridge circuit

BRIDGE THEORY: A bridge circuit is a network of resistances and capacitive or inductive impedances usually used to make precise measurements. The most common is the Wheatstone bridge, which consists of variable and fixed resistances. This is simply a series-parallel circuit, redrawn as shown in Fig. 15-25. The branches of the circuit forming the diamond shape are called legs.

If 10 V direct current were applied to the bridge in Fig. 15-26, one current would flow through R_1 and R_2 and another through R_3 and R_4. Since R_1 and R_2 are both fixed 1000-Ω resistors, the current through them is constant and each resistor will drop one-half of the battery voltage, 5 V.

However, as R_4 varies, the division of the battery voltage between R_3 and R_4 changes. In Fig. 15-26, 5 V is dropped across each resistor. The voltmeter senses the sum of the voltage drops across R_2 and R_3. Both are 5 V, but the R_2 drop is a $\pm$5-V drop and the R_3 drop is $\pm$5 V; they are opposite in polarity and cancel each other. This is called a "balanced" bridge. The relationship is usually expressed as a ratio of $R_1/R_2 = R_3/R_4$. The actual resistance values are not important; what is important is that this ratio is maintained and the bridge is balanced.

In Fig. 15-27 the variable resistor R_4 has changed to 950 Ω; the rest of the resistors are at the same value. Using Ohm's law, the voltage drop across R_4 is found to be 4.9 V; the

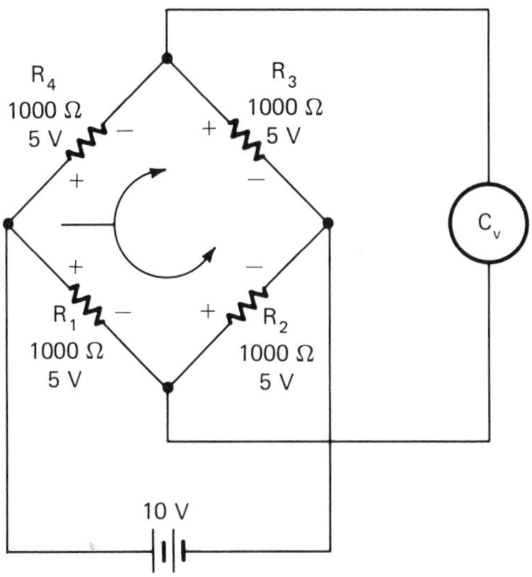

Figure 15–26

Balanced bridge

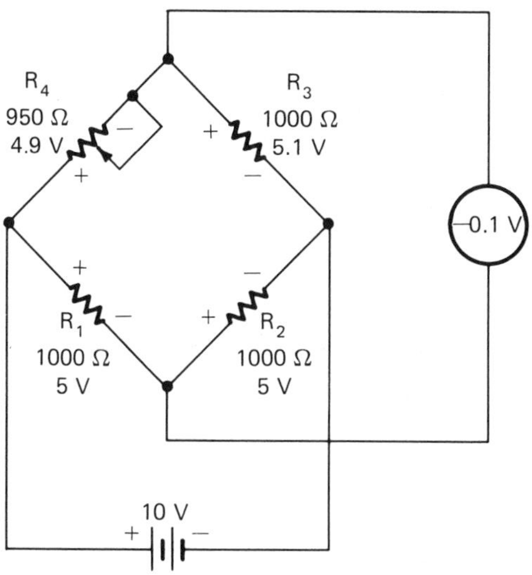

Figure 15–27

Bridge circuit, variable resistance reduced

rest of the voltage, 5.1 V, is dropped across R_3. In this illustration the voltmeter senses the algebraic sum of the voltage drops across R_2 and R_3, ± 5.0 V, and ± 5.1 V, registering a total of -0.1 V.

Conversely, in Fig. 15-28, the value of R_4 has changed to 1050 Ω; the voltage drop across R_5 is then 4.9 V. The voltmeter senses the sum of ± 5 V and ± 4.9 V, or $+0.1$ V.

When R_4 changes the same amount above or below the balanced bridge resistance, the magnitude of the DC output, measured by the voltmeter, is the same but the polarity is reversed.

15-7 ELECTRIC CONTROL SYSTEMS[1]

An electric control system consists of the basic control system components—disturbance-sensing element, controller, and final control element—interconnected by electrical wiring, relays, and amplifiers (where required). It uses electrical energy provided by an electric power supply.

Electric systems are used in all types of producing and processing industries and in HVAC

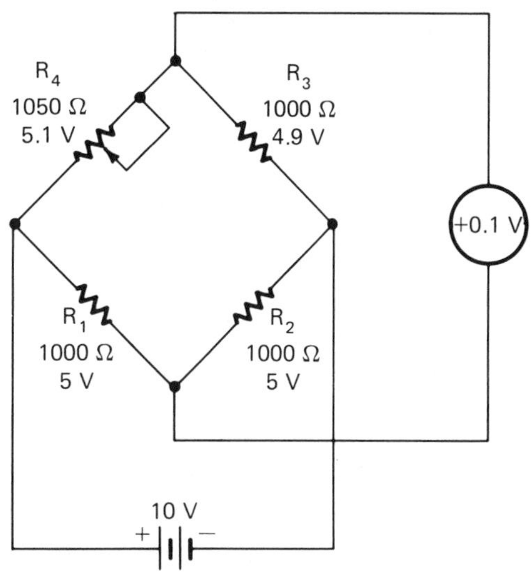

Figure 15–28

Bridge circuit, variable resistance increased

applications everywhere. They are particularly applicable when the final control element is electric or where relatively long distances separate the controller and the final control element.

Electromechanical versus electronic control systems. Electric control systems include both electromechanical and electronic systems. The main difference between them is that electronic systems use solid-state devices (vacuum tubes in older systems). Figure 15-29 shows comparable electromechanical and electronic devices. In an electronic system a thermister could replace a temperature-sensing bulb, a solid-state comparator circuit could replace a potentiometer, and a solid-state amplifier could replace a balancing relay. In actual practice many systems and components are *hybrid*—they use both electromechanical and electronic devices.

Electronic components have been developed for use in applications where the use of standard electromechanical components is difficult or inconvenient. They are intended to complement existing lines of control. Figures 15-29 through 15-56 with associated descriptive matter are from Ref. 1, courtesy of Honeywell, Inc.

15-8 DISTURBANCE SENSING ELEMENTS[1]

A disturbance-sensing element is a transducer that detects and measures changes in the controlled variable. Electromechanical elements, using moving parts, measure the motion produced by these changes. Electronic elements have no moving parts; they measure other characteristics. Table 15-1 lists the common disturbance-sensing elements used in electric control systems.

LIQUID-LEVEL SENSING ELEMENTS: A change in liquid level, which itself is produced by motion, is perhaps the simplest change to measure. An electromechanical float

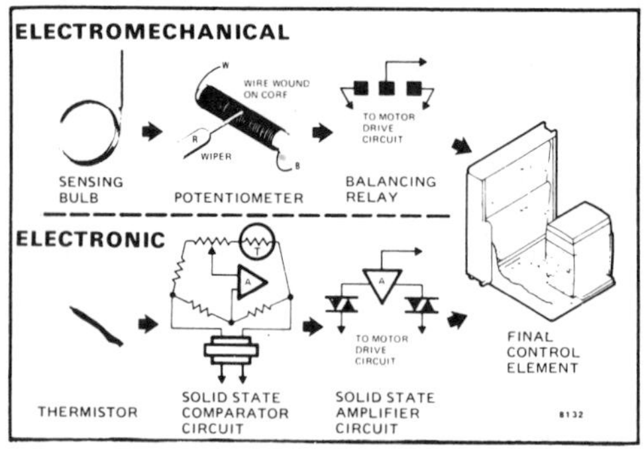

Figure 15–29

Comparison of electromechanical and electronic devices

mechanism is usually used. A common application is the low-water cutoff (LWCO) used on a steam heating boiler to prevent operation when the boiler is nearly dry. A sensitive pressure-sensing element can also be used.

PRESSURE-SENSING ELEMENTS: Pressure is defined as force per unit area. In electromechanical pressure-sensing elements, this force is easily applied to cause motion. Figure 15-30 illustrates some of the common pressure-sensing elements. Electronic pressure-sensing elements, such as piezoelectric crystals, are not commonly used in industrial or commercial applications. (Piezoelectric crystals, such as quartz or barium titanite, produce a voltage when subjected to mechanical stress such as compression, expansion, or twisting.) Flow rate (quantity per unit time), velocity, static pressure, and liquid level can also be measured using variations of pressure-sensing elements.

Flexible diaphram. A flexible diaphram is distorted by increased pressure (Fig. 15-30). A linkage connected to the diaphram translates this distortion into motion. One of the most common applications is in a gas pressure regulator used to maintain a constant gas pressure in a fuel-gas pipeline.

Table 15-1

Disturbance-Sensing Elements*

Controlled Variable	Electromechanical Elements	Electronic Elements
Liquid level	1. Float mechanism 2. Static pressure elements	—
Pressure	1. Flexible diaphram 2. Bellows 3. Pressure bell	1. Piezoelectric crystal (not commonly used)
Temperature	1. Bimetal element 2. Rod-and-tube element 3. Sealed bellows 4. Remote bulb 5. Fast-response element 6. Averaging element	1. Thermister 2. Resistance bulb 3. Thermocouple
Humidity	1. Nylon ribbon 2. Human hair 3. Wood 4. Leather, horn, silk, etc.	1. Hygroscopic (gold-foil grid)
Enthalpy	1. Remote bulb plus nylon ribbon	—
Flame Detection	1. Bimetal element	1. Flame rod (ionization) 2. Photoelectric cell 3. Phototube 4. Thermocouple
Smoke Opacity	—	1. Photoconductive cell

**From Ref. 1, courtesy of Honeywell, Inc.*

Bellows. A bellows is nothing more than a stack of washer-shaped diaphrams joined alternately at their outer and inner circumferences. The bellows allow more movement over a given range of pressure without increasing the circumference of the diaphrams.

The fluid can be admitted to the *inside* of the bellows; in which case, the other end of the bellows is closed by a solid diaphram and the linkage makes contact outside. Or, the fluid can exert pressure on the *outside* of the bellows in which case the bellows is enclosed in a sealed, bell-shaped housing; the bellows is filled with air, and the linkage makes contact inside.

One end of the bellows is firmly anchored. The other end (restrained by a spring) moves in or out as an increase in pressure causes the bellows to expand or contract. The mechanical linkage transmits this motion to a switch or potentiometer in the controller.

Pressure bell. A pressure bell is used for maximum sensitivity in the measurement of small changes in pressure (or vacuum) of air, or some other gas. An outer bell (inverted cup) is suspended from the linkage over an inner bell, which is fixed to the bottom of the case or tank. The buoyant force of the gas in the ring-shaped space between the bells causes the outer bell to ''float'' above the inner bell. The oil provides a nearly frictionless seal of the space between the bells. Changes in gas pres-

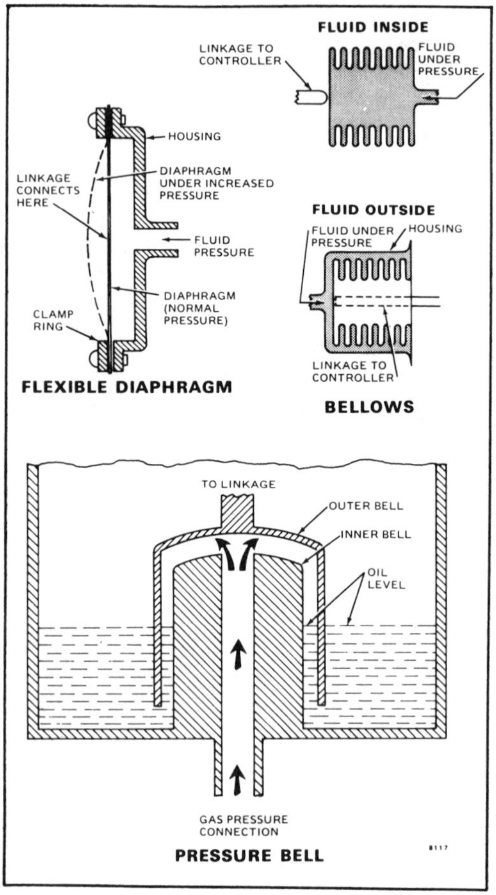

Figure 15–30

Electromechanical Pressure Sensing Elements

sure cause variations in the buoyant force, so the outer bell floats higher or lower. The linkage transmits these changes to the controller or measuring instrument.

In normal practice two identical bells are suspended from a beam like a pharmacist's scale. The weight of each bell is balanced by that of the other. One bell is subjected to a constant *reference* pressure, and the other bell to the pressure being measured. The difference of these pressures determines the position of the beam. This balanced system minimizes friction and the mechanical loads which must be moved, thus resulting in an extremely sensitive device.

ELECTROMECHANICAL TEMPERATURE-SENSING ELEMENTS: Electromechanical temperature-sensing elements operate on the familiar principle that all substances tend to expand or contract with increases or decreases in temperature. The rate of expansion and contraction (the coefficient of thermal expansion) is different for different

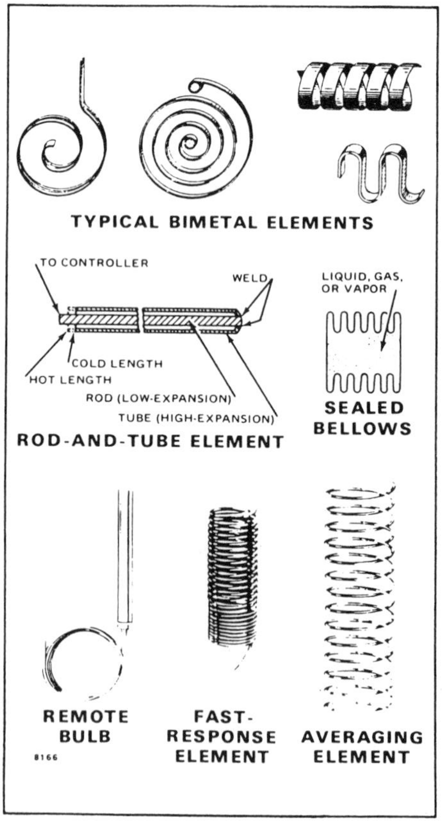

Figure 15–31

Electromechanical Temperature Sensing Elements

substances. This *differential expansion* causes motion which is measured. Electromechanical elements are large and bulky and respond relatively slowly to temperature changes. Figure 15-31 illustrates the more common electromechanical temperature-sensing elements.

Bimetal element. A bimetal element is composed of two thin layers or strips of dissimilar metals, welded or brazed together. Because the two metals have different coefficients of thermal expansion, the element bends as the temperature varies. The most common application of a bimetal element is in room thermostats.

A change in length of either strip, for a given temperature change, is always proportional to the original length of the strip (although the ratio varies for different metals). Therefore, the corresponding movement of the free end of the element is likewise proportional to the length of the element. Thus, a certain minimum length is required to provide a measurable movement for a small temperature change. To increase the sensitivity of bimetals, designers have resorted to curves and spirals, enabling much longer elements to be packaged in smaller spaces.

Rod-and-tube element. A rod-and-tube element consists of a high-expansion metal tube inside of which is a low-expansion rod with one end attached to the rear of the tube. As the temperature of the liquid or gas surrounding the tube changes, the length of the tube changes, causing the free end of the rod to move. This element is commonly used in certain types of insertion and immersion temperature controllers, particularly those mounted in boilers or storage tanks.

Sealed bellows. A bellows, similar to that used in pressure-sensing elements, is evacuated of air; filled with liquid, gas, or vapor; and sealed at both ends. The volume or pressure of the liquid, gas, or vapor changes as the temperature changes, causing the bellows to expand or contract lengthwise. A linkage connected to the free end of the bellows transmits this movement to the controller. This type of element is often used in room thermostats.

Remote bulb. A remote bulb is a capsule connected to a sealed bellows or diaphram by a capillary tube. The entire system—bulb, capillary, and bellows (or diaphram)—is filled with a liquid, vapor, or gas. Temperature changes at the bulb result in changes of volume and pressure in the fluid, which are transmitted to the bellows or diaphram through the capillary tubing. The remote bulb is useful where the temperature measuring point is at a distance from the desired location of the controller. It is usually provided with fittings to allow it to be installed in a pipe, duct, or tank.

Fast-response element. A fast-response element is a tightly coiled capillary that can be used in place of the bulb in a remote bulb element. The surface area to volume ratio is about seven times greater than that of a standard bulb, so its response time is seven times faster.

Averaging element. An averaging element is used in place of the bulb in a remote bulb element to obtain the *average* temperature in a duct. It is similar to the capillary except that it has a larger bore in order to hold the same amount of liquid as a standard bulb. The averaging element is distributed evenly over the cross section of the duct by winding it back and forth several times across the duct.

ELECTRONIC TEMPERATURE-SENSING ELEMENTS: Electronic temperature-sensing elements possess certain properties, such as resistance, which vary as the temperature changes. Generally, these elements are very small and can respond quickly to temperature changes. The speed of response to temperature changes is defined by the *time constant* of the sensing element—the shorter the

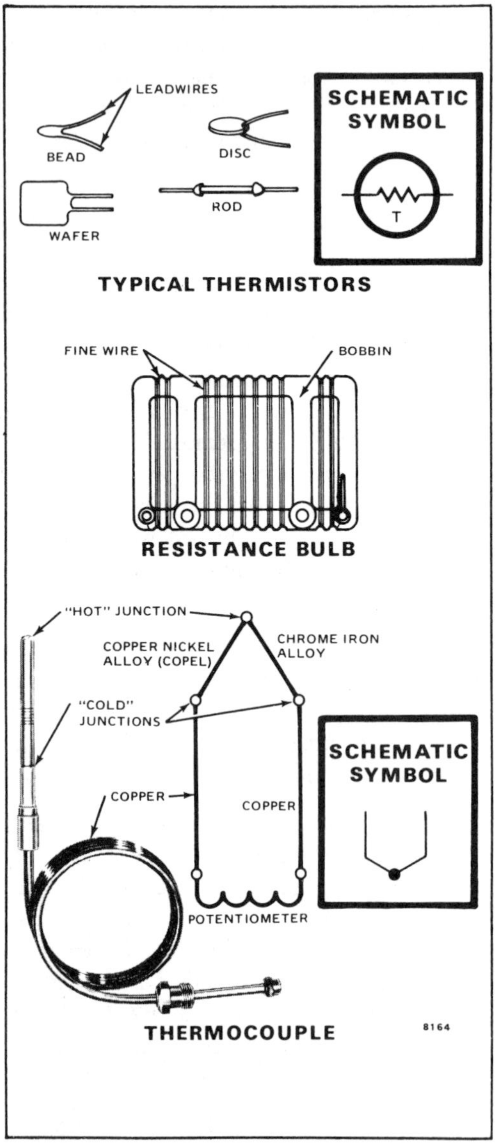

Figure 15–32

Electronic Temperature Sensing Elements

time constant, the faster the response. Figure 15-32 illustrates some of the commonly used electronic temperature-sensing elements.

Thermistor. A thermistor is a solid-state semiconductor, the electrical resistance of which varies with temperature. Its temperature coefficient of resistance is high, nonlinear, and negative—its resistance decreases as temperature increases, and vice versa.

The primary advantage of a thermistor is the large change in its resistance with a small change in temperature. It provides a large input signal, which makes the design of the controller easier and less expensive. Its small size is another advantage.

Disadvantages include its nonlinear temperature coefficient of resistance, which limits its operating temperature range to a linear portion of the curve—generally less than 800°F (427°C). Also, because of its negative temperature coefficient of resistance, it is not fail-safe; an open thermistor would cause the controller to keep calling for more heat. Another drawback is that the controller has to be calibrated to the individual sensing element, and sometimes it has to be recalibrated if aging changes the resistance of the thermistor.

Resistance bulb. A resistance bulb is a coil of fine wire wound around a bobbin. The resistance of the wire *increases* as the temperature increases. Thus, it would "fail-safe"—a break in the wire would indicate a maximum temperature and the heating system would shut down. Other advantages include a wide operating range [up to 1400°F (760°C)], interchangeability with other resistance bulbs, high output, excellent linearity, and excellent stability. Its primary disadvantage is its comparatively higher cost than the thermistor, although it is less expensive than the thermocouple.

Thermocouple. A thermocouple is a junction of two dissimilar metals joined at the point of heat application—the "hot" junction. A resulting millivoltage difference, directly proportional to the temperature at the "hot" junction, is developed across the free ends—the "cold" junctions. A potentiometer is usually used to measure the voltage. Different types of thermocouples are used, depending on the operating temperature range required and the operating environment.

The primary advantage of a thermocouple is the extreme temperatures it is capable of sensing [up to 3000°F (1649°C)]. One disadvantage is its low output, requiring a more expensive amplifier. Also, the controller must have a cold junction compensator to offset changes in ambient temperature.

ELECTROMECHANICAL HUMIDITY-SENSING ELEMENTS: Humidity is the moisture content in air. Electromechanical sensing elements used are made of *organic* materials that absorb moisture (stretch or swell) when the humidity increases, and release moisture (shrink) when the humidity decreases. Figure 15-33 illustrates some of the commonly used humidity-sensing elements.

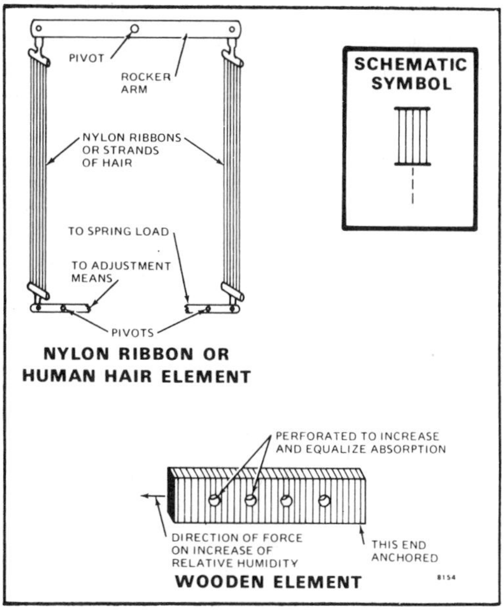

Figure 15–33

Electromechanical humidity sensing elements

<u>*Nylon ribbon or human hair element:*</u> Human hair has the property of changing length with changes in its moisture content. Because of its small diameter, it rapidly absorbs and dissipates moisture as humidity changes. The hairs used in humidity-sensing elements are carefully selected and matched to ensure the greatest possible uniformity. The strands are bunched, and several bunches are combined in a band or ribbonlike element.

Nylon has the same property. Because the manufacturing of nylon can be closely controlled, the selection and matching process required for human hair is eliminated, thus reducing the cost. Thus, most elements are now made of the less expensive nylon ribbon.

Although many mechanical arrangements are used, the one shown in Fig. 15-33 provides relatively large movement with a small change in humidity. The rocker arm at the top doubles the length of the nylon ribbon or hair for a given length of the element by, in effect, connecting the two ribbons in series. (Elements with only one ribbon and without the rocker are also used extensively.) As the humidity decreases, the air circulating over the nylon or hair absorbs some moisture, and the ribbons shorten. As the ribbon on the left shortens, it rotates the rocker arm counterclockwise. This shifts the top end of the right ribbon upward a distance equal to the amount that the left ribbon has shortened. Since the right ribbon has also shortened the same amount, the movement of its bottom end is *twice* the amount of shortening. This motion is transmitted through a lever mechanism to the controller.

<u>Wooden element.</u> A sticky desk drawer in humid weather demonstrates the principle of the wooden element. It consists of a strip of wood cut *across* the grain. One end of the strip is anchored, and the other is connected through a linkage to the controller. As the humidity increases, the wood absorbs moisture and swells; as the humidity decreases, the wood releases

moisture and shrinks. These changes in the length of the strip move the linkage to the controller. A plain wooden element exhibits considerable lag in moisture absorption and dissipation, and it is also subject to cracking in dry atmospheres. For these reasons wooden elements are seldom used in modern controls. However, they make a good teaching tool to demonstrate the principle of electromechanical humidity-sensing elements.

Other elements. Other materials that may be used in electromechanical humidity-sensing elements include leather, horn, and silk. All are more sensitive than wood.

ELECTRONIC HUMIDITY-SENSING ELEMENT: The electronic humidity-sensing element commonly used is a resistance element consisting of a thin slab of plastic or glass to which is fused strips of gold foil to form two interleaved, comblike grids (see Fig. 15-34). The grids are coated with a hygroscopic salt (one which readily absorbs moisture from the atmosphere). The salt forms a conductive path between the adjacent strips of foil in the two grids. The electrical resistance of the salt coating changes as it absorbs and releases moisture, depending on the humidity. Thus, the resistance between the two terminals changes. This change in electrical resistance is used to actuate the controller.

The large surface area and small volume of this element allow it to respond rapidly to humidity changes. The high resistance of the circuit (thousands of ohms) results in a relatively large change in resistance for a small change in humidity, thus providing high accuracy.

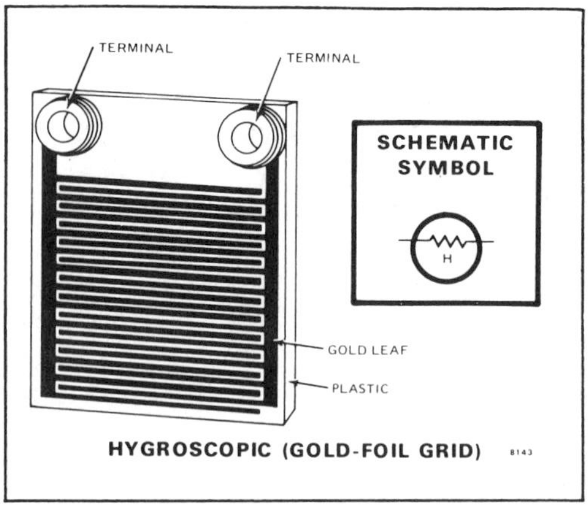

Figure 15–34

Electronic humidity sensing elements

ENTHALPY-SENSING ELEMENTS: Enthalpy is dependent on both temperature and humidity, so from a comfort standpoint, it is desirable to be able to control it. An enthalpy-sensing element simply combines temperature- and humidity-sensing elements in a single device. Usually, a remote bulb senses temperature and a nylon ribbon senses humidity. The controller combines the results and interprets them as enthalpy.

FLAME DETECTION: A flame detector may be looked upon as a form of temperature-sensing element since it is the heat of the flame which produces the ions and light detected. All flame safeguard systems require that a flame be detected to keep a burner in operation. If no flame were present, fuel could accumulate to form an explosive mixture.

Types of flame detectors commonly used are electronic. (A bimetal element, previously described, may be used on small, domestic burners.) Electronic flame detectors include flame rods, photoelectric cells, phototubes, and thermocouples (previously described).

Flame rod (ionization). A flame rod is simply a metal or ceramic rod inserted into the flame envelope to function as the positive electrode in a flame detection circuit. The ground electrode is usually the burner itself. When a flame is present, the heat causes molecules between the electrodes to collide with each other so forcibly as to knock some electrons out of the atoms, producing ions. This is called flame ionization. When the ions are present and a voltage is applied across the electrodes, a current will flow between them. Thus, a current indicates that a flame is present.

Photoelectric cell. A photoelectric cell is simply a device that is sensitive to light. When used in a flame detector, it is located so that it is sensitive to the light emitted by the flame. The cell can be photoconductive, photoemmisive, or photovoltaic. It can be sensitive to ultraviolet, infrared, or visible light rays.

A *photoconductive cell* is made of a material, such as cadmium sulfide or lead sulfide, the electrical resistance of which varies inversely with the intensity of the light that strikes it. Thus, if a flame were present, the resistance of the cell would be low and a measurable current would flow.

A *photoemissive cell* has a cathode coated with a material, such as caesium oxide, that emits electrons when light strikes it. When voltage is applied to the cell and a flame is present, electrons flow from the cathodes to the anode, and a current flows in the detection circuit. Rectifying photocells, which respond to visible light, are examples of photoemissive cells.

A *photovoltaic cell* generates a voltage when light strikes it. It is not used in flame detectors because the voltage generated is too small to be useful without more expensive amplification than that required by other types of flame detectors.

Phototube. A phototube is an electron tube containing a photocathode that releases electrons when exposed to light. When used in a flame detector, the tube is pointed at the flame so light from the flame falls upon the photocathode. When sufficient voltage is applied across the cathode and anode and a flame is present, current flows through the tube and through the external detection circuit.

SMOKE OPACITY: Smoke opacity is smoke density or blackness measured in Ringelmanns. Ringelmann numbers 0 and 5 represent 0 and 100 percent blackness, respectively. The sensing element used in a smoke detector does not detect smoke opacity directly. It detects the amount of light from a source which shines through the smoke. A photoconductive cell, made of cadmium sulfide, is used as a sensing element. The electrical resistance of the cell varies inversely with the intensity of the light that strikes it. Thus, the greater the smoke opacity, the less the light, the greater the resistance, and the less the current flowing in the detection circuit.

15-9 CONTROLLERS (1)

A controller receives the measurement information from the disturbance-sensing element and produces an error signal to initiate corrective action. The controller mechanism translates the measurement information into a form of energy that can be used by the control system. Electronic controllers are generally more complex than electromechanical controllers, but they provide *all* modes of control. Table 15-2 lists the common types of controllers used in electric control systems.

15-10 ELECTROMECHANICAL CONTROLLER[1]

An electromechanical controller directly utilizes the motion produced by a disturbance-sensing element to operate a switch or potentiometer. In many controllers the switch, in turn, energizes a relay to increase the current-handling capacity or to perform additional functions. An electromagnetic controller is relatively inexpensive and is generally used for simple on-off or proportioning control.

ELECTROMECHANICAL MECHANISMS: An electromechanical mechanism uses the motion produced by the sensing element to open or close an electric circuit, or to set up a varying resistance in an established circuit. Switches, relays, and potentiometers

Table 15-2

Controllers*

Modes of Control	Electromechanical Mechanisms	Electronic Mechanisms
On-off (two-position)	Mercury switches	Galvanometric (on-off only)
Multiposition	Snap-acting switches	Potentiometric
Floating	Mercury plunger relays	Resistance bridge
	Electromagnetic relays	
Proportioning (modulating)	Potentiometers	Potentiometric Resistance bridge
Reset		
Rate	—	

**From Ref. 1, courtesy of Honeywell, Inc.*

are the basic mechanisms used.

Mercury switches. A mercury switch is a glass tube in which fixed contacts and a pool of loose mercury are hermetically sealed (see Fig. 15-35). Insulated lead wires are attached to the contacts at the end(s) of the tube. The tube is usually mounted on a plate that is rotated by linkage from the sensing element. As the mounting plate rotates, the tube tilts and the mercury pool flows to the lower end to close or open the gap between the contacts. This action makes or breaks the contacts and any electrical circuits connected to the lead wires. The mechanical force needed to tilt the switch is low, so the make-or-break action is rapid and repeats accurately.

The contacts may be rated for currents from 0.1 A up to 10 A. They are arranged for spst (single-pole, single-throw) or spdt (single-pole, double-throw) switching action. Spdt switches make R to B when the value of the controlled variable falls, and make R to W when it rises.

Advantages of mercury switches include (1) sealed construction, so the contacts are protected from dirt, dust, and other contamination; (2) no moving parts (other than the mercury pool), so they provide maintenance-free, dependable operation; and (3) visible contacts, so a person can see if they are made or broken.

Limitations include the following, (1) they must be carefully leveled to obtain operation at the proper point(s); (2) in locations with a lot of vibration, the mercury slops around in the glass tube, causing contact chatter; and (3) mercury freezes at −35°F (− 37°C), so mercury switches cannot be used below this temperature.

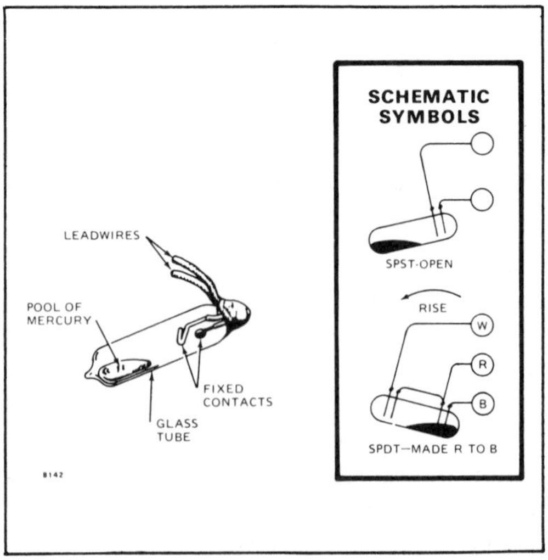

Figure 15–35

Mercury switches

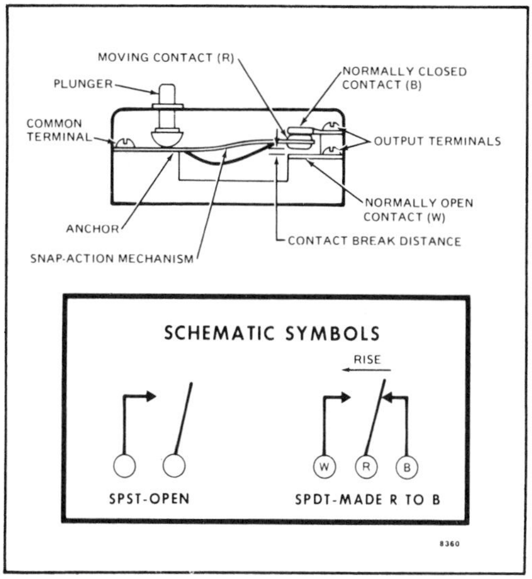

Figure 15–36

Snap-acting switches

Snap-acting switches. In a snap-acting switch, metal-to-metal contacts are made or broken by the movement of the sensing element against a reciprocating plunger (see Fig. 15-36). The plunger actuates a snap-action spring mechanism built into a plastic case. The moving contact snaps from one fixed contact to the other with repeatable accuracy. (An spst switch has only one fixed contact.) The contacts have about the same ratings as mercury switches (0.1 A up to 10 A) and are also arranged for spst or spdt switching action.

Advantages of snap-acting switches include: (1) they are fast and positive acting; (2) only a slight mechanical motion is needed for actuation, so they provide precise and accurate control where small tolerances are required; (3) no leveling is required (unlike mercury switches); and (4) they are unaffected by moderate vibration.

Limitations include: (1) they are not sealed (because of the plunger), so they are not as dust-tight as mercury switches; (2) they have lower DC ratings; and (3) the contacts are not visible, so an ohmmeter must be used to determine if they are open or closed.

Mercury plunger relays. In a mercury plunger relay (see Fig. 15-37) the contacts are pools of mercury into which electrodes have been inserted. A ferromagnetic plunger, actuated by a solenoid coil, accomplishes the switching. The use of mercury pools as contacts allows the switching of large currents—as high as 60 A. A mercury switch or snap-acting switch with a lower rating is usually used to energize the solenoid coil.

The plunger, mercury, and conducting electrodes are sealed in a glass tube surrounded by a solenoid coil. When the coil is deenergized, the plunger floats (partially submerged) on the main mercury pool. Inside the cylindrical plunger, just above the main pool, is a small mercury pool in a ceramic cup, supported by an insulated tungsten rod. When the solenoid coil is energized, the plunger moves downward, displacing the main mercury over the

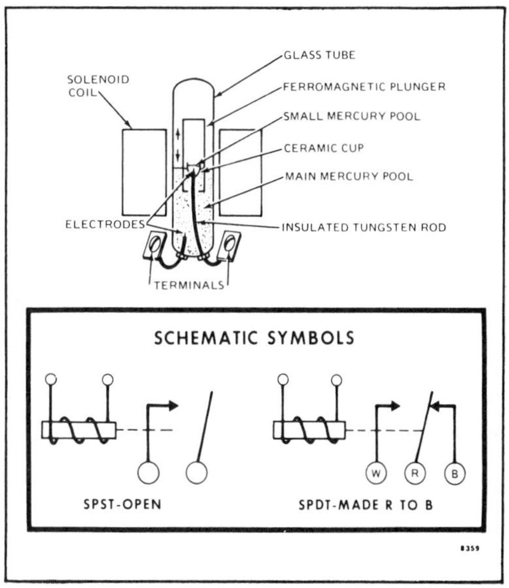

Figure 15–37

Mercury plunger relays

top of the ceramic cup. The two pools of mercury are now in contact and the switch is closed. When the coil is deenergized, the plunger moves upward and the surface of the main mercury pool falls below the top of the ceramic cup, breaking the contact and opening the switch.

Advantages of mercury plunger relays include (1) heavy current switching capacity; (2) sealed construction permitting operation in dirty and explosive environments; and (3) mercury-to-mercury contact, eliminating contact maintenance.

Electromagnetic relays. Electromagnetic relays (see Fig. 15-38) are merely electrically operated switches. Simple relays and relay combinations are extensively used to perform one or more of the following functions: (1) switching to line voltage or heavy current load when activated by a low-voltage controller of limited current-carrying capacity, (2) switching two or more separate load circuits from a single-pole controller switch, and (3) coordinating two or more control circuits to provide a desired sequence of load switching.

The coil-and-armature mechanism is similar to that used in the familiar doorbell or buzzer. It consists of (1) a magnetic circuit with fixed core, movable armature, and two air gaps; (2) one or more actuating coils that establish the magnetic flux; (3) one or more sets of contacts, with one contact in each set movable and coupled to the armature, and the others fixed; and (4) restoring springs and limit stops for armature positioning.

In the deenergized position shown for the spdt relay, the force of the restoring spring overcomes the magnetic force, and movable

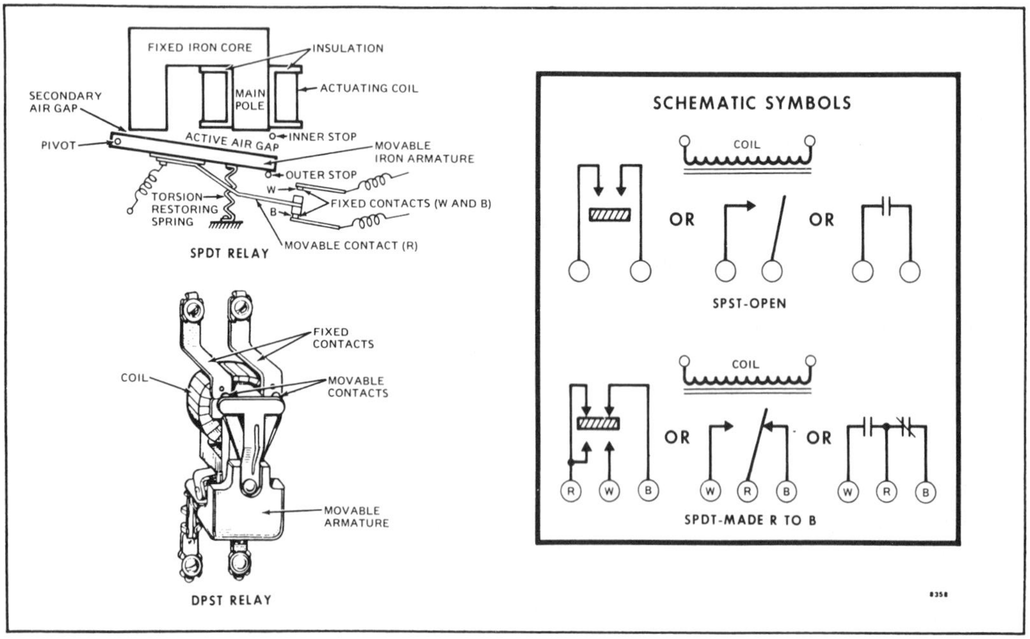

Figure 15–38

Electromagnetic relays

contact *R* is making contact with fixed contact *B*. When the actuating coil is energized, flux in the magnetic circuit increases and the magnetic force overcomes the spring force (plus friction). The armature moves to close the active air gap, and the movable contact *R* breaks with fixed contact *B* and makes with fixed contact *W*.

An spst relay can be either normally open (no *B* contact) or normally closed (no *W* contact). The dpst (double-pole, single-throw) relay shown can be represented schematically by two sets of normally open, spst relay contacts.

Potentiometers. A potentiometer (see Fig. 15-39) typically consists of fine wire wound around a core. A wiper rests on the windings and can move along them. The wire ends are connected to terminals labeled *W* and *B*. The wiper is connected to terminal *R*. As motion of the sensing element moves the wiper back and forth, the amount of resistance between *R* and *W* and between *R* and *B* changes. If the wiper moves toward *W*, resistance increases between *R* and *B* and decreases between *R* and *W*, and vice versa. The *R*, *W*, and *B* designations stand for Red, White, and Blue lead wire colors.

Potentiometers are used in proportioning (modulating) controllers. They determine how far the final control element must move to counteract a change in the controlled variable.

15-11 ELECTROMECHANICAL MODES OF CONTROL 1

Section 15-4 presented a brief discussion of the various modes of control. Here we will discuss some of the circuitry for electromechanical systems. Switches and relays are used for on-off (two-position), multiposition, and floating modes of control. Potentiometers are used for the proportioning (modulating) mode of control. The figures in the following discussions show line voltage circuits. Low-voltage devices would require a transformer if used in line voltage circuits.

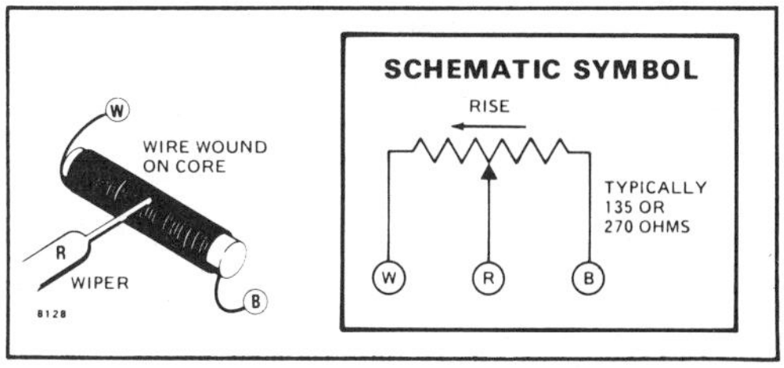

Figure 15–39

Potentiometer

On-Off (two-position) mode of control. In the on-off (two-position) mode of control, the controller is in one or the other of two possible positions, except for a brief period when it is passing in between. Each of these positions, in turn, determines which of two possible positions is occupied by the final control element. There are two values of the controlled variable that determine the position of the controller. Between these values is a zone, called *differential,* in which the controller cannot switch.

In the simplest mode of control, a spst controller is connected in a two-wire circuit (see Fig. 15-40). When the controlled variable rises to the higher of the two values (upper end of the differential), the controller contacts will either open or close, depending on the design. Let's assume they will close. This applies power to the final control element, which will be *electrically* driven to one of its two positions. The controller contacts will stay closed, and the final control element will remain in this position until the controlled variable falls to the lower of the two values (lower end of the differential). Then the controller contacts will open and the final control element will be *mechanically* driven (usually by a spring) to its other position. The controller contacts will stay open and the final control element will remain in this position until the controlled variable again rises to the high end of the differential. Had the controller been designed to *open* at the high end and *close* at the low end, the positions of the final control element would be reversed.

A spdt controller is used in a three-wire circuit (see Fig. 15-40). The final control element has three terminals corresponding to those of the controller. Usually, the movable contact of the controller is labeled *R*, and the fixed contacts are labeled *W* and *B*. The *R, W,* and *B* designations stand for red, white, and blue leadwire colors.

When the value of the controlled variable *rises* to the high end of the differential, the controller contacts will *break R* to *B* and *make R* to *W*. This applies power to the section of the final control element between terminals *R* and *W*. When this section is energized, the final control element will be *electrically* driven to a position which will *decrease* the value of the controlled variable. The controller contacts and final control element will remain in this position until the controlled variable *falls* to the low end of the differential. Then the controller contacts will *break R* to *W* and *make R* to *B*. This removes power from the *R-W* section and applies power to the *R-B* section of the final control element, which will now be *electrically* driven to a position that will *increase* the value of the controlled variable. Thus, the controlled variable is kept within the values determined by the differential of the controller.

If, as is often the case, a *heating* system of some type is being considered, the following little rhyme can be memorized to help remember the operation just described; "Red to Blue for BTU." This rhyme indicates that when the temperature falls and there is a call for heat (Btu's), the controller will make *R* to *B*, and vice versa. It also applies to other modes of control.

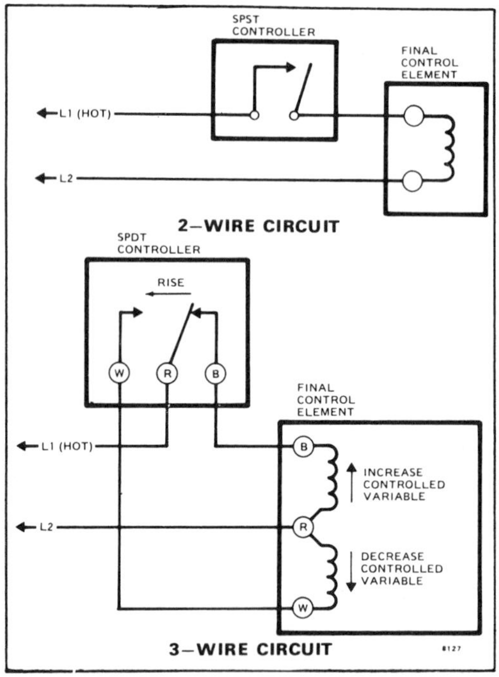

Figure 15–40

ON-OFF (2-Position) mode of control (Electromechanical)

Multiposition mode of control. The multiposition mode of control (Fig. 15-41) simply consists of two or more on-off stages that operate at different values of the controlled variable. In the example shown (Fig. 15-41), the mercury switches are all rotated together by the same sensing element. They are rotated *clockwise* as the value of the controlled variable falls, and *counterclockwise* as it rises. Because they are mounted at different angles, the switches *close* in sequence (stage 1 first and stage 5 last) as the value of the controlled variable falls. The final control elements are energized in the same order. Similarly, the switches open in the opposite sequence (stage 5 first and stage 1 last) and the final control elements are deenergized as the value of the controlled variable rises.

This mode of control allows as many stages to be energized as required to maintain the desired value of the controlled variable. Each stage may include an *entire* final control element that takes one of two positions as de-

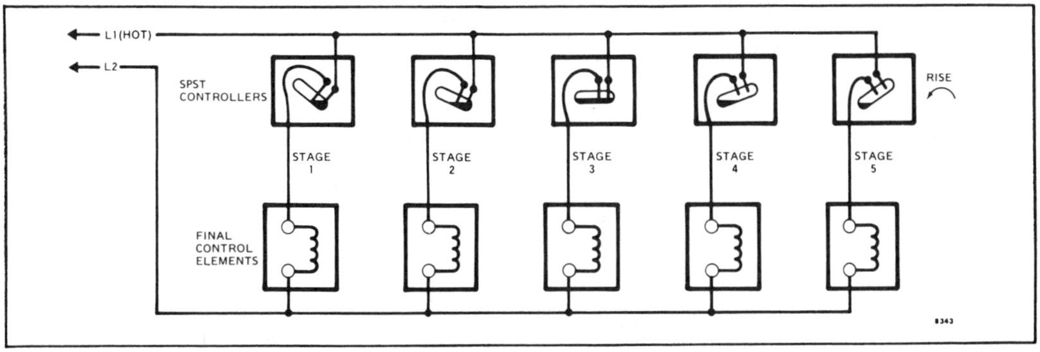

Figure 15–41

Multiposition mode of control (Electromechanical)

scribed for the on-off mode of control. Or, each stage may include a *section* of the mechanism of one final control element. In the latter case, as each stage is energized, the final control element will be driven to a different position.

Floating mode of control. The floating mode of control (Fig. 15-42) provides bidirectional action of the final control element anywhere over its full range. The controller uses an spdt switch that has a neutral zone in which no contact is made. As long as the controlled variable stays within this neutral zone, no contacts are closed, so no power is applied to the final control element and it stays where it is.

If the controlled variable *rises above* the neutral zone, the controller contacts will make *R* to *W*, applying power to the *R-W* section of the final control element. The final control element will be *electrically* driven in a direction that will *decrease* the controlled variable. When the controlled variable falls back into the neutral zone, the *R-W* contacts will break and the final control element will stop.

Similarly, if the controlled variable *falls below* the neutral zone, the controller contacts will make *R* to *B*, applying power to the *R-B* section of the final control element, and *electrically* driving it in a direction that will *increase* the controlled variable. When the controlled variable rises back into the neutral zone, the *R-B* contacts will break and the final control element will stop.

In summary, the final control element can occupy any position between its two extremes, and the controller "floats" to keep the controlled variable in the neutral zone. Floating

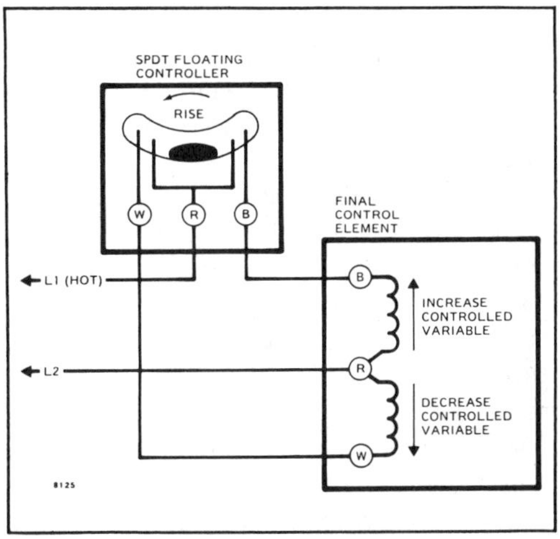

Figure 15–42

Floating mode of control (Electromechanical)

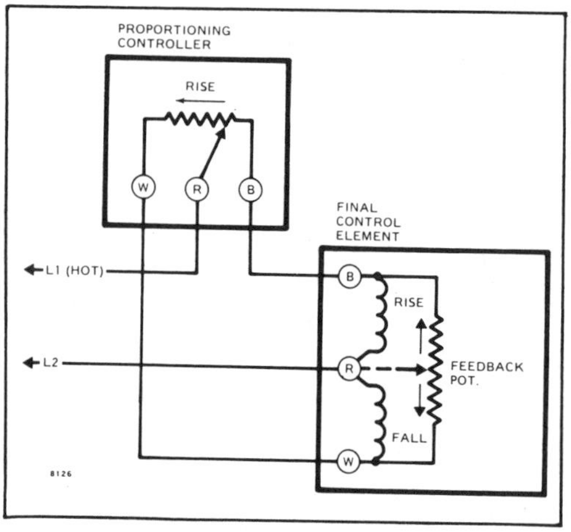

Figure 15–43

Proportioning (modulating) mode of control (Electromechanical)

control has largely been replaced by proportioning control, but it may be found in older systems.

Proportioning (modulating) mode of control. The proportioning mode of control (Fig. 15-43) is a refined version of floating control. The controller has a potentiometer instead of a switch, so it can *continuously vary* the position of the final control element, anywhere between its extremes, to keep the controlled variable at the desired value. This process is called *modulation*.

To simplify the explanation of proportioning control, much of the circuitry in the final control element has been omitted in Fig. 15-43. With the potentiometer wiper as shown, there is less resistance between terminals *R* and *B* of the controller than between *R* and *W*, so more current will flow in the *R-B* section of the final control element than in the *R-W* section. This will *electrically* drive the final control element in a direction to increase the controlled variable. Simultaneously, the wiper on a feedback potentiometer will drive toward the *W* end. When the *R-W* resistance of the feedback potentiometer equals the *R-B* resistance of the controller, the currents will be balanced and the final control element will stop.

If the controlled variable rises, the wiper will move toward the *W* end of the potentiometer, decreasing the resistance between terminals *R* and *W* of the controller. More current will flow in the *R-W* section of the final control element, *electrically* driving it in a direction to decrease the controlled variable. The wiper on the feedback potentiometer will drive to the *B* end until the *R-B* resistance of the feedback potentiometer equals the *R-W* resistance of the controller. The current will then be balanced and the final control element will stop.

Thus, for each position of the wiper on the controller, there is a corresponding position of the wiper on the feedback potentiometer and for the final control element. The final control element will be driven a distance proportional to the change in the controlled variable. This results in the most precise type of controller, used where the controlled variable must be kept within small tolerances.

15-12 ELECTRONIC CONTROLLERS[1]

An electronic controller compares the signal from a disturbance-sensing element to a standard reference to obtain an error signal. It then amplifies the error signal to provide enough energy to drive the final control element. Electronic controllers provide all modes of control, including reset and rate, plus the functions of alarm and limiting.

ELECTRONIC MECHANISMS: A basic electronic controller (see Fig. 15-44) consists of a null detector and amplifier, with input adjustments and an output switching element to drive the final control element. It utilizes solid-state components in place of electromechanical switches, although electromagnetic relays are frequently used as output switching elements.

The null *detector* compares the difference

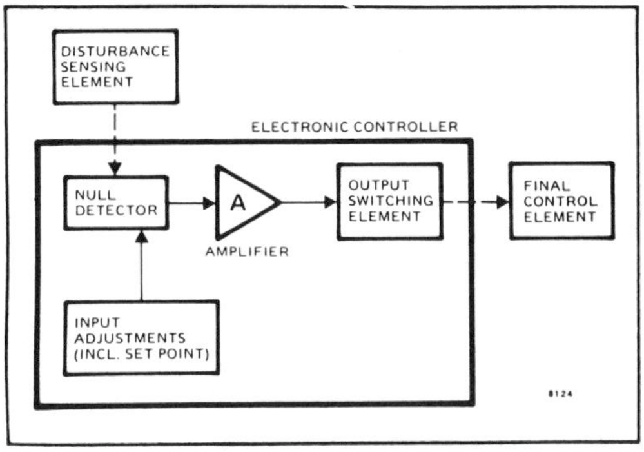

Figure 15–44

Basic electronic controller

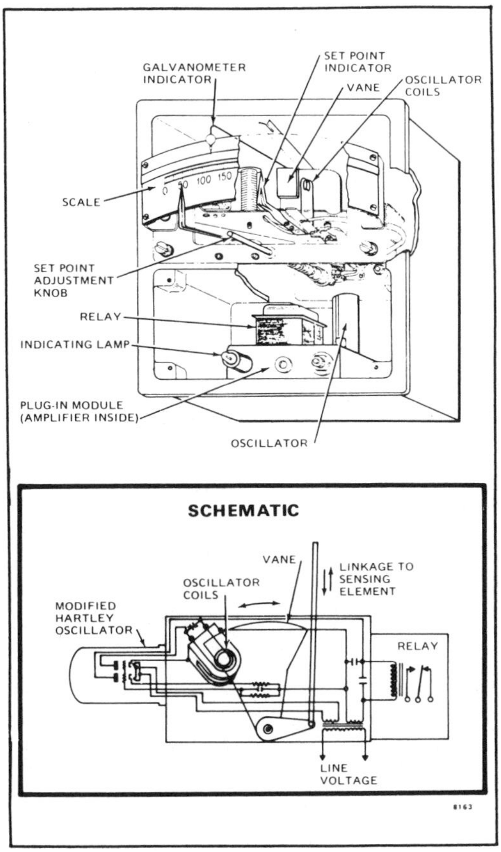

Figure 15–45

Galvanometric controller

between the signal from the disturbance-sensing element and the standard reference (set point). When they are equal, there is a balanced condition that usually results in zero output, called a null. The different mechanisms used to detect the null result in three types of electronic controllers—galvanometric, potentiometric, and resistance bridge. The resistance bridge controller will be emphasized because it is the most common.

Input adjustments include set point, differential, and proportioning range (throttling range) adjustments. Calibration adjustments may also be included. In most cases these adjustments take the form of a wire-wound potentiometer with a knob to adjust the position of the wiper.

The *solid-state amplifier* multiplies the output from the null detector to produce an amplified error signal. The amount of magnification is referred to as the *gain* of the amplifier. Most amplifiers also *discriminate* between signals of different polarity or phase. Discrimination determines if the signal to the amplifier is caused by an increase or a decrease in the controlled variable, so the error signal will drive the final control element in the proper direction.

The *output switching element* may be an electromagnetic relay, a saturable-core reactor, and SCR (silicon-controlled rectifier), or another device that drives the final control element. In the proportioning mode of control the amplifier may drive the final control element directly, eliminating the output switching element.

Galvonometric controller. A galvanometer (Fig. 15-45) is an instrument for measuring an electric current by measuring the mechanical motion produced by electromagnetic forces set up by the current. A galvanometric controller is basically a galvanometer that has been converted to a controller. It was introduced before it was technologically feasible to mass produce electronic amplifiers. It uses the *motion* produced by the sensing element to produce a signal that activates an spst or spdt relay. A galvanometric controller is limited to the on-off mode of control.

The basic mechanism is a vane-oscillator null detector. The lightweight metal vane is connected by a linkage to the sensing element. As it moves between (without touching) two oscillator coils, mutual inductance decreases and the modified Hartley oscillator is detuned. When the set point is reached (at the null), oscillation ceases. When oscillation stops, current is fed to the relay coil, causing precise snap action of the relay contacts. Vane motions as small as 0.002 in. (0.05 mm) initiate controller action. The set point adjustment knob moves the oscillator coil assembly upscale or downscale.

Potentiometric controller. A potentiometric controller (Fig. 15-46) operates on the null balance principle. The signal from the sensing element is balanced against a set point voltage in the input circuit to produce an error signal. The error signal is simplified and used to position the final control element to keep the controlled variable at the desired value. For precise control, this system is intrinsically more accurate than the galvanometer controller.

The potentiometric controller shown in Fig. 15-46 consists of a three-stage transisterized amplifier and an spdt electromagnetic relay mounted in a plug-in module. Linkage from the sensing element moves the wiper on the input potentiometer to a position representing the value of the controlled variable. In this position the voltage E_p is developed across the input potentiometer. The wiper on the reference potentiometer is manually adjusted to the desired set point, developing the voltage E_s. When the controlled variable differs from the set point, an error signal equal to the difference between E_p and E_s is produced and fed to the amplifier. The amplified error signal energizes the relay, which drives the final control element. When the controlled variable reaches the set point, E_p equals E_s and the relay is de-energized.

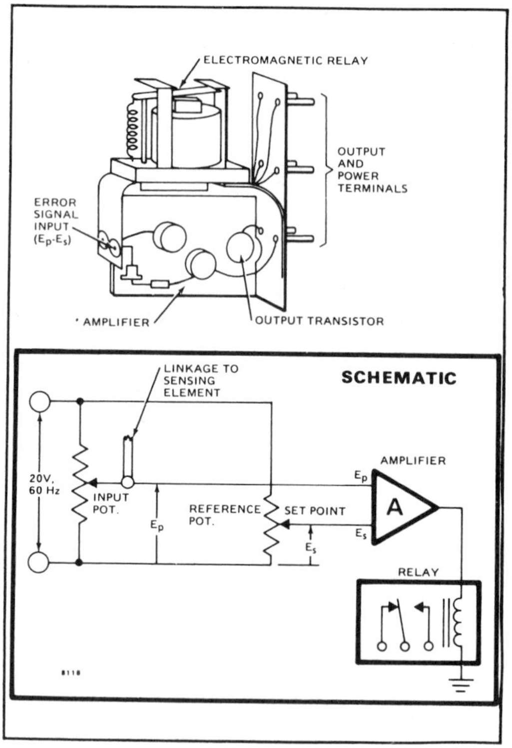

Figure 15–46

Potentiometric controller

Resistance bridge controller. The resistance bridge controller is by far the most used electronic controller. It also operates on the null balance principle. The input circuitry generates an error signal based on the difference between the resistance of an electronic sensing element and the set point resistance. The bridge circuit was discussed in Section 15-6 and most electronic controllers are adaptations of this circuit.

For control purposes the basic Wheatstone bridge is somewhat modified in order to provide signal to the controlled equipment. Figure 15-47 illustrates a typical modified Wheatstone

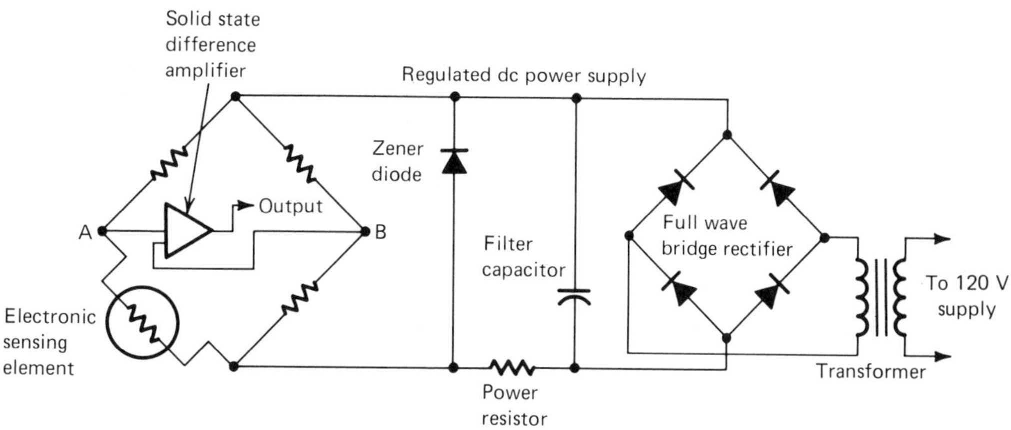

Figure 15–47

The modified Wheatstone Bridge

bridge circuit. The key changes are as follows:

1. The galvanometer (voltmeter of Figs. 15-26, 15-27, and 15-28) is replaced by a solid-state difference amplifier. The amplifier compares the voltage at point *A* with the reference voltage (set point) at point *B*.
2. The battery is replaced by a regulated DC power supply.
3. The R_4 resistor in the left half of the bridge is replaced by an electronic sensing element (varying resistance type, such as a thermistor).

The difference amplifier multiplies the voltage difference between points *A* and *B* of the bridge. The amount of magnification is referred to as the *gain* of the amplifier. For instance, a gain of 10 means that if the difference between *A* and *B* is 1.0 V, the output of the amplifier will be 10.0 V. The amplifier has two functions in the system. The first is to isolate the bridge from voltage changes in the rest of the control circuit. The second is to make the very small signal from the bridge strong enough to energize the output switching element (such as a relay coil or solid-state motor).

15-13 ELECTRONIC MODES OF CONTROL 1

The potentiometric and resistance bridge controllers can be used for *all* modes of control, including reset and rate. Galvanometric controllers are limited to the on-off mode of control. The figures in this section show line voltage circuits. Low-voltage devices would require a transformer if used in line voltage circuits.

On-Off (two-position) mode of control. In the on-off (two-position) mode of control (Fig. 15-48), the output switching element is a spst or spdt electromagnetic relay. The set point is adjusted so the relay switches at the null, or at some point on either side of the null. Operation is the same as for electromechanical on-off control discussed previously. Remember the rhyme—"Red to Blue for BTU."

Multiposition mode of control. The multiposition mode of control (Fig. 15-49) for electronic controllers is similar in concept to that

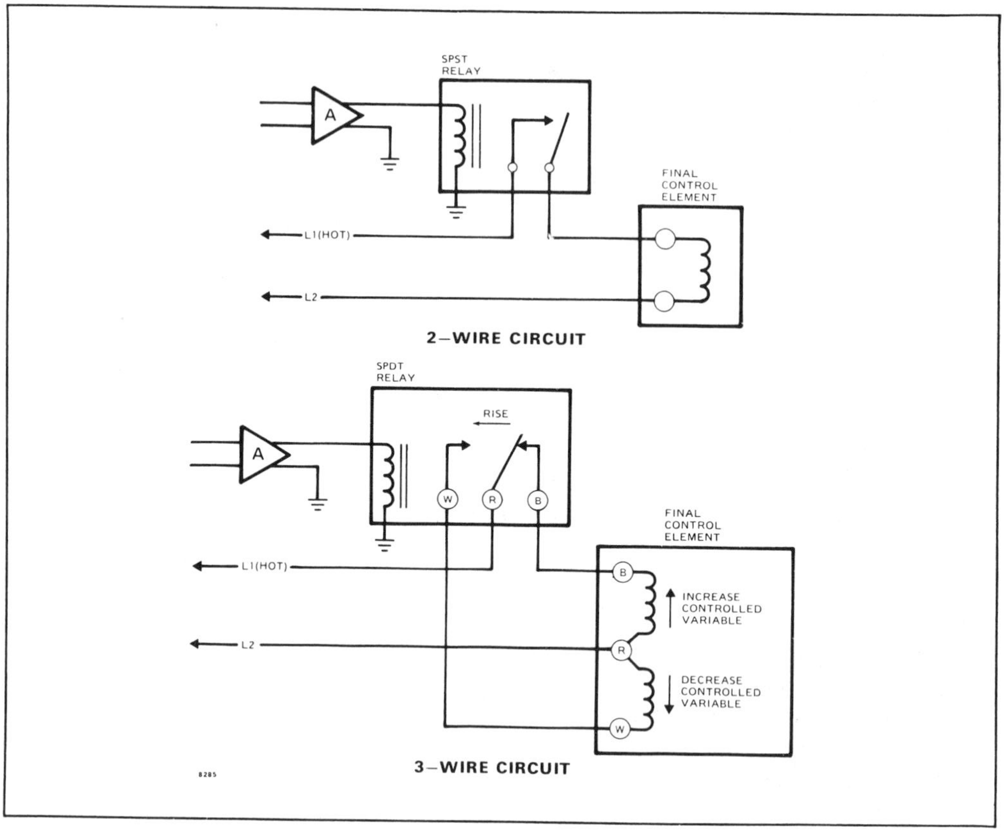

Figure 15–48

ON-OFF (2-Position) mode of control (Electronic)

for electromechanical controllers—two or more on-off stages operate at different values of the controlled variable. The error signal (amplifier output voltage) increases as the controlled variable departs from the set point. The spst relays, used as output switching elements, are set to pull in at different values of the amplifier output voltage. The relay switches close in sequence (stage 1 first and stage 5 last) as the controlled variable departs from the set point. The final control elements are energized in the same order. Thus, the farther the controlled variable departs from the set point, the higher the amplifier output voltage, and the more stages will be energized. For departure from the set point in the opposite direction (e.g., decreasing instead of increasing), a similar set of stages could be used. Polarity or phase discrimination in the amplifier would determine which set of stages to energize. In Fig. 15-49 departure from the set point and the resulting amplifier output are indicated by shading. For the departure shown, three stages are energized.

Floating mode of control. Although seldom used anymore, floating control (Fig. 15-50) can be accomplished in several ways. One

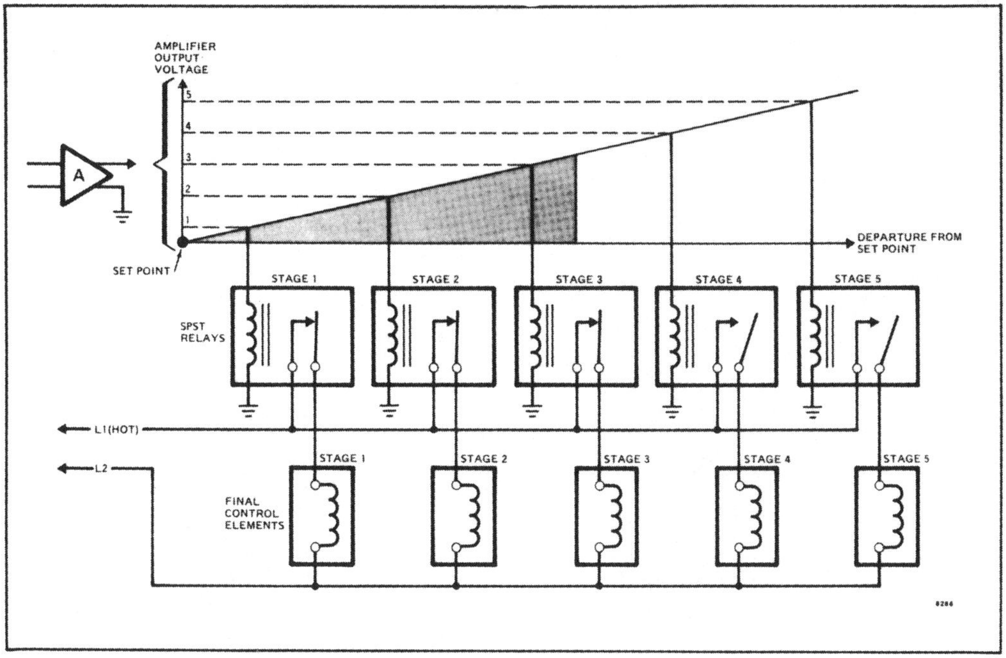

Figure 15–49

Multiposition mode of control (Electronic)

method is the use of two stages in a manner similar to multiposition control. The amplifier incorporates polarity or phase discrimination. As long as the controlled variable is "floating" in the neutral zone (shaded area) around the set point, neither relay is energized. If the controlled variable *falls below* the neutral zone, relay 1 will pull in, the switch will close, section *B-R* of the final control element will be energized, and it will drive in a direction to *increase* the controlled variable. When the controlled variable rises back into the neutral zone, relay 1 will drop out and the final control element will stop. Similarly, if the controlled variable *rises above* the neutral zone, relay 2 will pull in, section *W-R* will be energized, and the final control element will drive in a direction to *decrease* the controlled variable.

Proportioning (modulating) mode of control. When properly tuned, an electronic controller operating in the proportioning mode automatically adjusts its output to a value which will cause the controlled variable to stabilize at the set point. There are three types of proportioning control associated with electronic controllers—position proportioning, time proportioning, and current proportioning.

Position-proportioning control (Fig. 15-51), the most common, is the same type used with electromechanical controllers (see Fig. 15-43). The discriminator is actually part of the amplifier circuitry but is shown separately to explain the operation. The discriminator determines whether the controlled variable is rising or falling. If it is *rising above the set point,* relay 1 pulls in the *R-W* section of the final

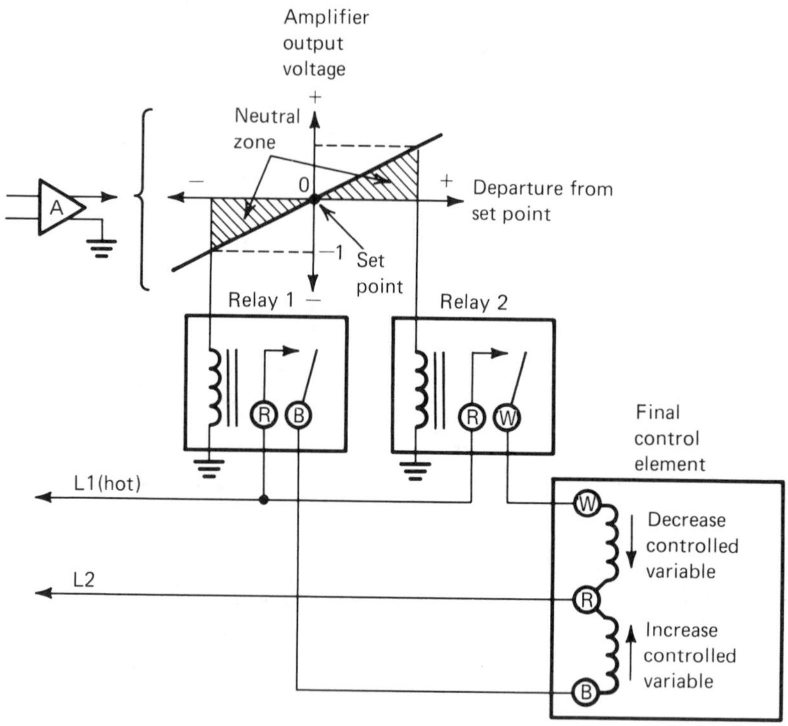

Figure 15–50

Floating mode of control (Electronic)

control element is energized. This electrically drives the final control element in a direction to *decrease* the controlled variable. Simultaneously, the wiper on the feedback potentiometer will drive the toward the *B* end. The feedback potentiometer is electrically connected into the null detector circuit of the controller. When the wiper has driven far enough to balance the null detector, the amplifier output is zero, relay 1 drops out, and the final control element stops.

Similarly, if the controlled variable is *falling below the set point,* relay 2 pulls in, and the *R-B* section of the final control element is energized. The final control element will be electrically driven in a direction to *increase* the controlled variable, and the feedback potentiometer wiper will drive toward the *W* end until the null detector is balanced. Then relay 2 will drop out, and the final control element will stop.

Thus, the final control element will be driven open or closed as necessary to keep the controlled variable at the set point. The graph in Fig. 15-51 for a typical controller shows the positions taken by the final control element as the controlled variable departs from the set point.

Time-proportioning control (Fig. 15-52) is available only on electronic controllers. Special circuitry, too complex to describe here, automatically switches the controller on and off, driving the final control element to either of two possible positions—fully open or fully closed. The ratio of on-time to off-time is proportional to the departure of the controlled var-

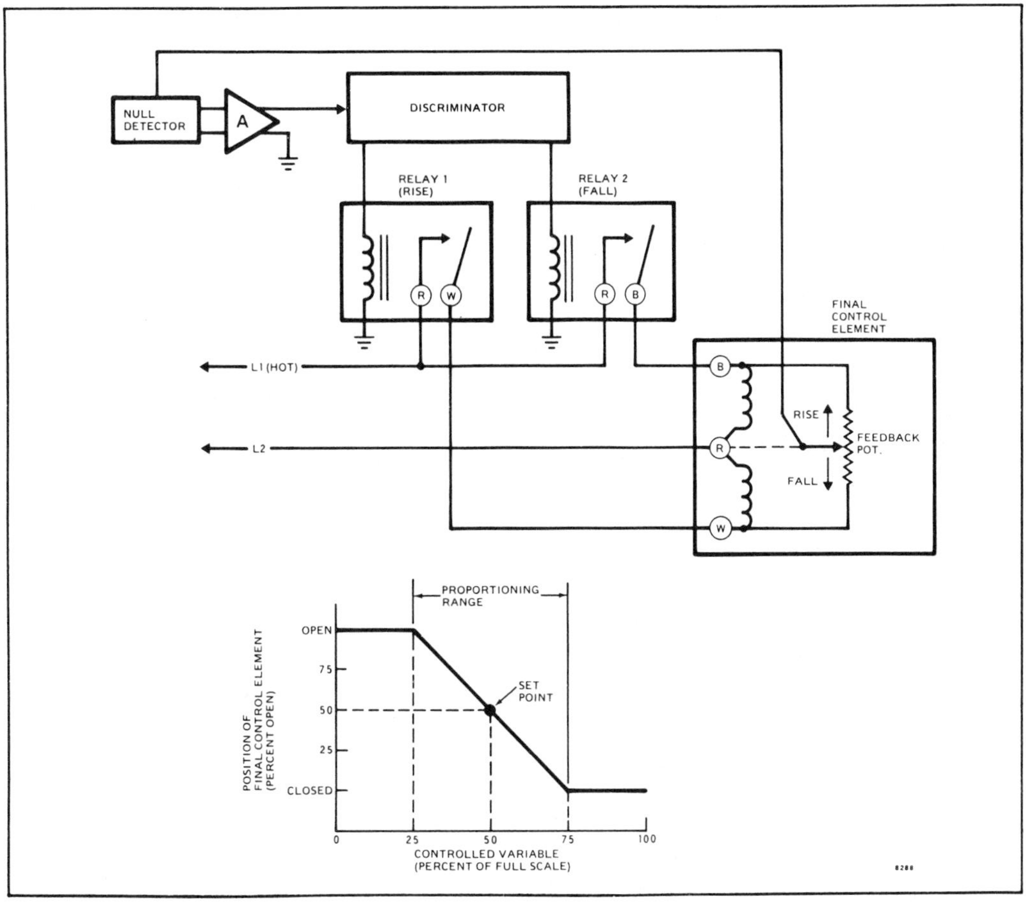

Figure 15–51

Position proportioning control (Electronic)

iable from the set point. The length of a complete cycle (on-time plus off-time) remains constant, but the percentage of on-time varies to keep the controlled variable at the set point.

In the example of Fig. 15-52 a complete cycle is 10 min. At *set point* the final control element is on as much as it is off—5 min each per cycle. *Below* the set point, it is on 7.5 min and off only 2.5 min. *Above* the set point, the periods are reversed—off 7.5 min and on 2.5 min.

The graph of Fig. 15-52 for a typical controller shows how the on-time varies as the controlled variable departs from the set point. Notice that outside of the proportioning range of the controller, the final control element will stay either on or off. As the controlled variable nears the set point, the on-time approaches 50 percent.

In *current-proportioning control,* the controller output is a DC milliamp signal that varies as the controlled variable changes. This

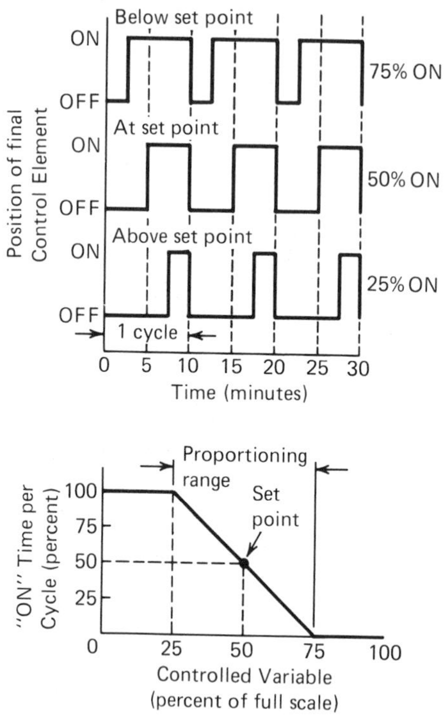

Figure 15–52

Time proportioning control (Electronic)

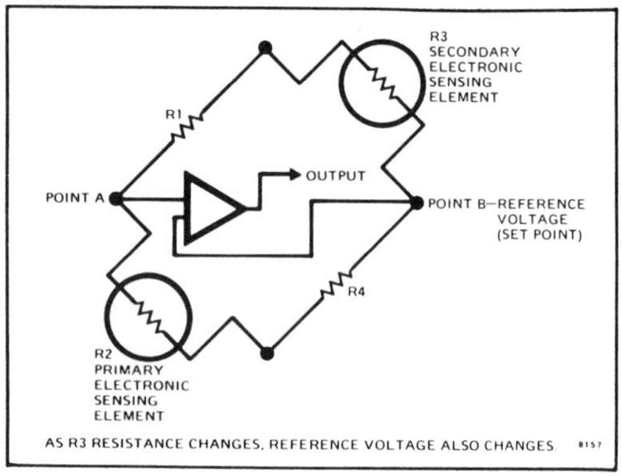

Figure 15–53

Automatic reset by adding a secondary electronic sensing element

signal drives a saturable core reactor, an SCR (silicon-controlled rectifier), a current relay, or a similar current-operated device.

RESET ACTION: A proportioning controller will not necessarily stabilize at the set point. The deviation from the set point—called *offset, droop,* or *drift*—may be small or large. If the offset were the same at all times, it could be remedied by a simple recalibration of the controller. However, it varies with the load conditions. Continual adjustment of the set point is necessary to keep the controlled variable at the same point throughout the load range of the equipment. There are two ways to eliminate offset—manual reset and automatic reset. Automatic reset is required for an automatic control system.

Manual reset is simply a potentiometer that can be adjusted to change the amplifier output in either a negative or positive direction. In effect, it recalibrates the controller to eliminate offset. Manual reset is most effective when there is no large changes in the process, ambient condition, line voltage, or load.

Automatic reset results if the potentiometer is replaced by a secondary electronic sensing element which automatically adjusts the set point as the load changes. This method is shown in Fig. 15-53 for a resistance bridge controller. Let us assume that the sensing elements are thermistors, so their resistance increases as the temperature decreases. Let us also suppose that *R*2 is sensing indoor temperature, and *R*3 is sensing outdoor temperature. If the outdoor temperature falls, the resistance of *R*3 increases. More voltage is dropped across *R*3, so the reference voltage at point *B* (which is actually the set point) decreases. To rebalance the bridge, the voltage at point *A* must also decrease. This requires a decrease in the resistance of *R*2, or an increase in the indoor temperature. Thus, as the outdoor temperature falls, the set point is automatically adjusted for the higher indoor temperature, and vice versa. This method is limited to systems in which the secondary sensing element can be located so that it detects changes in ambient conditions (as in the example), line voltage, the process, or the load—whatever is causing the offset.

Automatic reset can also be accomplished by adding a capacitor C (see Fig. 15-54), which continuously charges to a value proportional to the *offset*. This charge is added to the input error signal to automatically produce an amplifier output which has been corrected for offset. The manipulation of the signal in this manner is a form of advanced positive feedback called *integral action*. Integral action is proportional to the time constant RC and is initially calibrated by adjusting the resistor R.

RATE ACTION (Fig. 15-55): If a system has a long time delay between a change in the controlled variable and a corrective response by the controller, poor control may result. This time delay is called *lag*. A common cause of lag is having a sensing element located too far away from the controlled medium. If the lag is too great, the system becomes unstable.

The lag problem can be overcome by rate action (Fig. 15-55), also called *derivative action*. Rate action is delayed negative feedback, accomplished by adding a capacitor C which continuously changes to a value proportional to the *rate of change* of the input error signal. (Note that the *location* of the capacitor in an RC network determines how it charges, and hence the type of action—integral or derivative.) This charge is added to the input error signal to produce an amplifier output which has been corrected for lag. Rate action is proportional to the time constant RC and is initially calibrated by adjusting the resistor R.

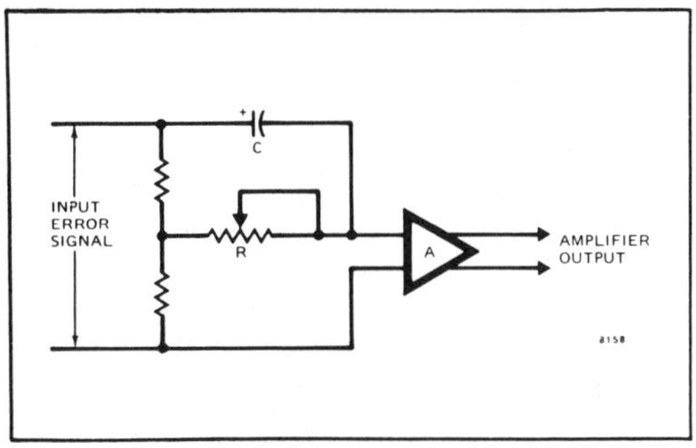

Figure 15–54

Automatic reset by integral action

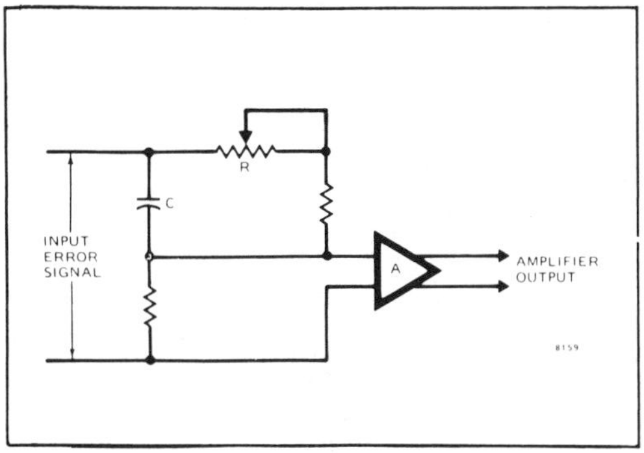

Figure 15–55

Rate action (derivative action)

THREE-MODE CONTROL: A proportioning controller with *both* reset and rate action is also called a three-mode (Fig. 15-56) controller (proportioning-plus-reset-plus-rate). This is the most sophisticated control mode, resulting in the best possible control under the most severe process conditions. In the examples (Fig. 15-56), the controlled variable, on startup, enters the proportioning range, overshoots the set point, and then undershoots below the set point. The curve is divided into four zones. The table shows the effect of adding reset and rate action, with the arrows indicating the direction in which each mode tends to change the controlled variable. Whenever the controlled variable approaches the set point, whether from below (zone 1) or from above (zone 3), rate action opposes reset in an effort

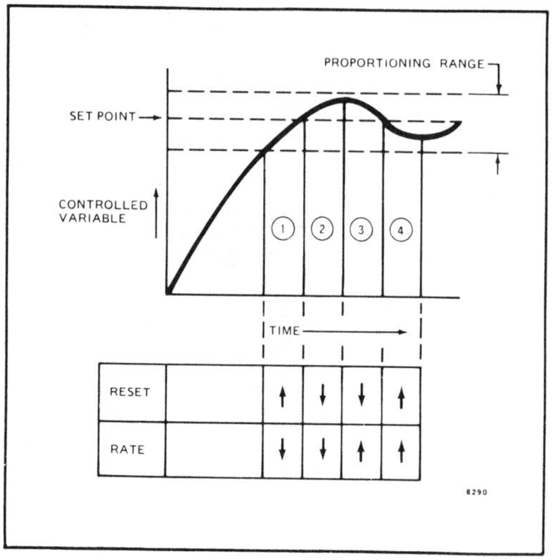

Figure 15–56

3-Mode control

to prevent overshoot and undershoot. After the set point has been passed, and with the controlled variable still moving further away from the set point (zones 2 and 4), all three control modes combine to minimize deviation from the set point.

15-14 PNEUMATIC CONTROLLERS (4)

In our discussion of electromechanical and electronic controllers, we discussed the disturbance-sensing element and the controller as two separate devices (which they are); however, we have seen that these two devices are frequently "packaged" in a single container which may be called simply a controller. Therefore, a controller has two main parts: the measuring or sensing element and the relay which produces the output signal. A pneumatic controller contains the same two parts.

The measuring or disturbance-sensing elements of a pneumatic controller are similar in most respects to electromechanical devices discussed previously, such as bimetal elements and remote bulbs for temperature, membrane or biwood element for humidity, and diaphram, bellows and Bourdon spring for pressure.

15-15 PNEUMATIC RELAY (4)

In all pneumatic controllers supply air is piped to the controller at a constant pressure of either 15 or 20 psig. This supply flow provides: (1) *volume,* to fill large areas within the controlled devices and connecting piping and (2) *pressure,* which provides the force to do the required work. The controller is designed to use the air internally through two separate circuits—pilot and volume amplifier.

Pilot circuit. The pilot circuit (Fig. 15-57) has a small volume and a reduced air flow that is restricted by a fixed orifice set to a value of 5 to 7 in. W.G. pressure and a flow of approximately 20 scim (standard cubic inches per minute) with the element or lid away from the control port. (Older model controllers are fitted with an adjusted orifice set to the same value.) Pilot pressure is then regulated by the position of the element relative to the control port, increasing to the same value as supply pressure with no flow when the port is fully closed. Figures 15-57 through 15-62 and associated descriptive matter are from Ref. 4, courtesy of Johnson Controls, Inc.

Volume amplifier circuit. The amplifier circuit is regulated by the action of the pilot circuit but has a much larger capacity since it admits supply air directly to the output line and the controlled device. It is therefore referred to as a volume amplifier.

There are four basic sections in a complete pneumatic relay (see Fig. 15-58): (1) the pilot chamber admits supply air through an orifice to the control port and pilot diaphram; (2) the exhaust chamber, between the pilot and output chambers, includes the exhaust seat assembly and exhaust port, which releases the excess output air to the atmosphere; (3) the output chamber incorporates a spring to oppose the pilot diaphram, the output diaphram, and an

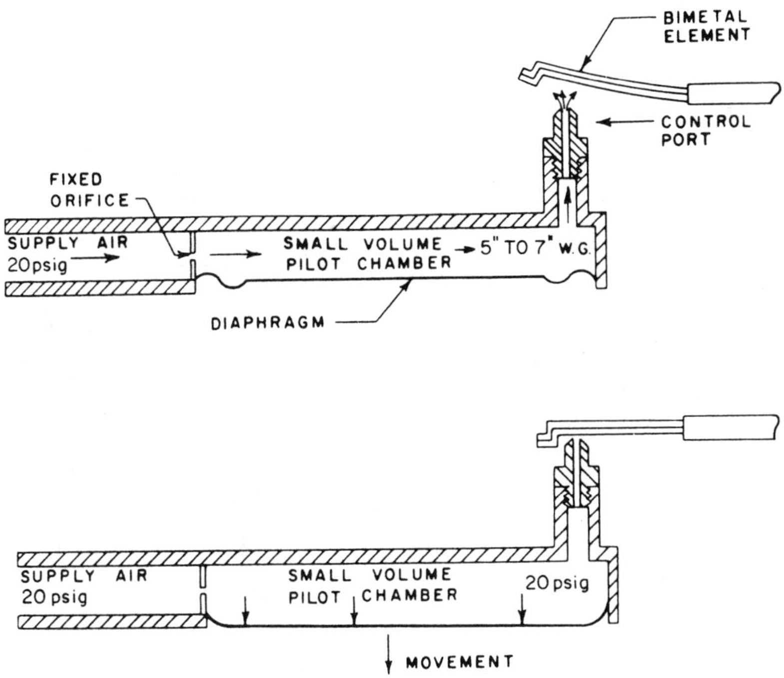

Figure 15–57

Pilot chamber action. The pilot chamber uses a restricted flow in a small area to respond quickly to changes dictated by the sensing element

outlet connection to provide output air to the controlled devices; and (4) the supply chamber includes the supply valve and supply valve seat assemblies which admit supply air to the output chamber.

The relay of a pneumatic controller translates the action of the sensing element into a useful output signal. The relay can be designed to produce proportional or two-position output, proportional being the more popular of the two. Any of the disturbance-sensing elements, involving motion, may be used for either mode.

Sequence of operation. In all controllers the sensing element senses and reacts to changes in the controlled variable. The sensing element is mounted or coupled to the relay so that it acts on a pilot circuit, modulating the pilot signal in proportion to changes in the controlled variable. The pilot signal applies its proportional signal to the amplifier which produces a useful signal. This signal is the output of the controller and is used to actuate the controlled device.

In a final step a feedback gain loop is included which allows the output signal to act on the pilot signal (see Fig. 15-59). Thus, the sensing element and relay together perform a total of five interrelated functions in sequence. The way the relay components are assembled determines whether the controller is proportional or two-position.

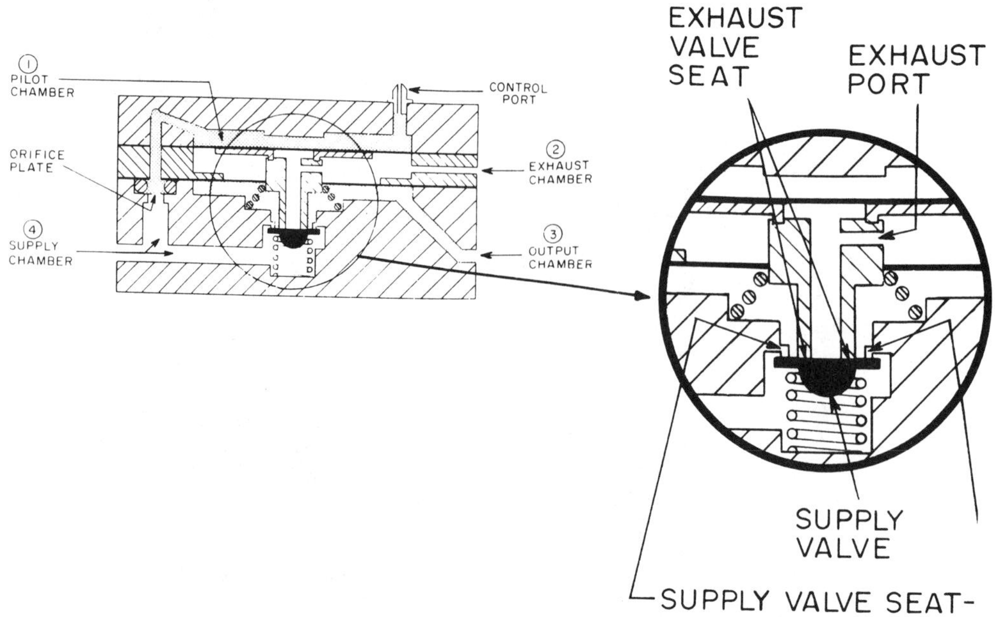

Figure 15–58

Four basic sections of a pneumatic relay. Series 4000 proportional relay (Courtesy of Johnson Controls, Inc.)

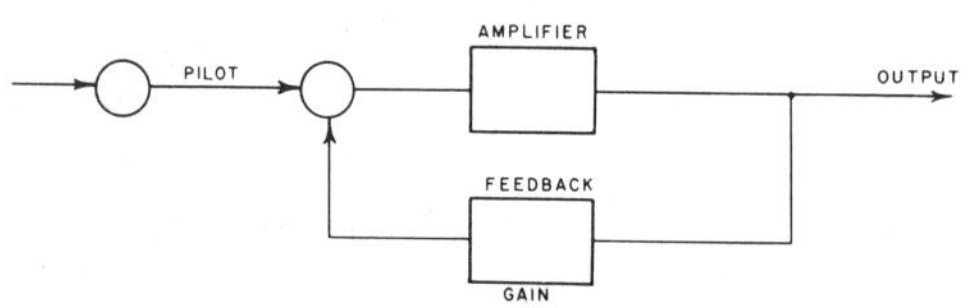

Figure 15–59

Relay components: Pilot, amplifier, output and feedback gain loop shown schematically

In Fig. 15-60 the various operating conditions are shown for a Johnson Controls 4000 series pneumatic controller. In Fig. 15-60a supply air is fed through the orifice into the pilot chamber. As the lid or sensing element closes against the control port, the pilot pressure increases gradually to maximum or supply pressure. The low-volume pilot pressure has several functions in the proportional relay. As pilot pressure increases, it moves a diaphram to close the exhaust valve seat, which prevents output air from exhausting to the atmosphere, Fig. 15-60b. A further increase in pilot pressure will then open the supply valve, to admit supply air pressure to the output air chamber and the output air line, Fig. 15-60c. The relay is designed with a high ratio to produce a large initial change in output pressure for a small change in pilot pressure caused by a minute movement of the lid against the control port. In the balanced condition, Fig. 15-60d, the exhaust seat is closed against the supply valve so that no output air can escape to the atmosphere and the supply valve is seated against the supply valve seat so that no supply air can enter

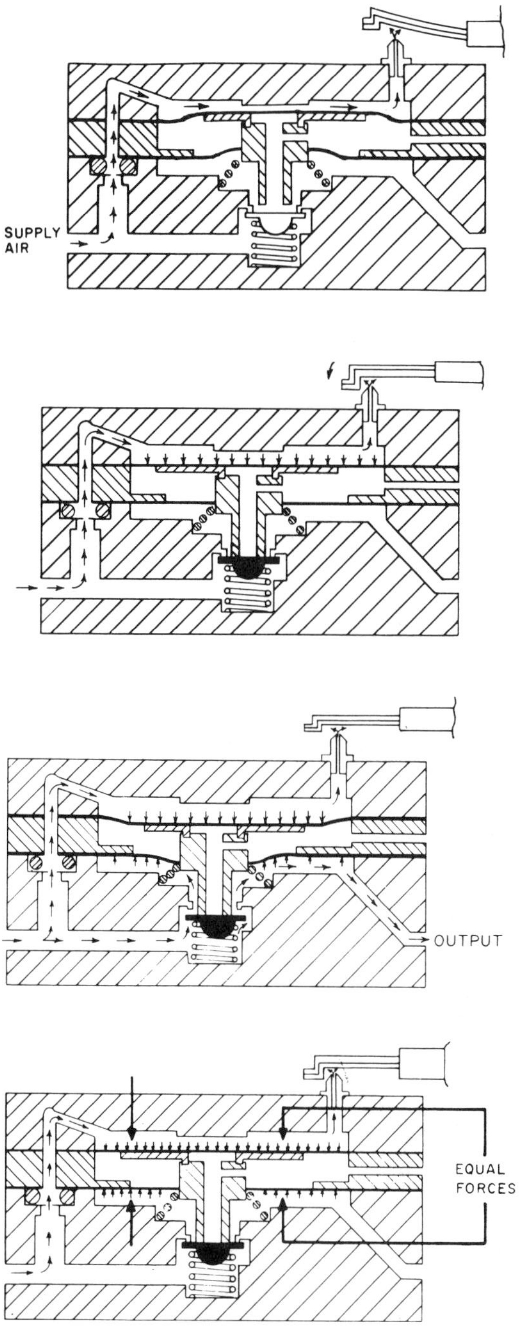

Figure 15–60

Operation of the 4000 Series proportional relay (Courtesy of Johnson Controls, Inc.)

the output chamber. The air pressure in the output chamber and air line will remain at the established pressure until the relay acts in response to a new change in lid position. The spring in the output chamber pushes against the pilot chamber diaphram so that pilot pressure must build up to about 3 psig before any action will occur.

Proportional pneumatic feedback action. *Negative* feedback produces the proportional action in a proportional controller. Negative feedback is arranged so that when a given force or pilot signal works in one direction, the resultant output force counteracts the original force and produces a balanced condition proportional to the value of the pilot signal. This can be referred to as a reverse-acting feedback loop. Figure 15-61 is a complete graph showing the output pressure changes that occur due to the 1.0 psig change in pilot signal. A formula is also provided to determine the input-to-output ratio. The relay shown in Fig. 15-61 is referred to as a 7:1 relay.

To observe how pneumatic feedback works in a proportional controller, assume that the sensitivity setting of a room thermostat is one point per degree Fahrenheit. When the bimetal element moves toward the control port in response to a 1 degree rise in temperature, the output pressure of the relay will increase 10 psig. This high volume is to provide a large flow to fill the system. However, as this flow fills the feedback chamber, pressure increases against the feedback diaphram in the controller, Fig. 15-62. The diaphram raises the feedback bar and sensitivity slider and moves the bimetal element away from the control port, cutting down the initial high output to a balance point of 1.0 psig output per 1.0 degree change in temperature.

This action is referred to as negative feedback in that when the sensing element moves toward the control port, due to a change in temperature the resulting increase in output pressure tends to move the sensing element

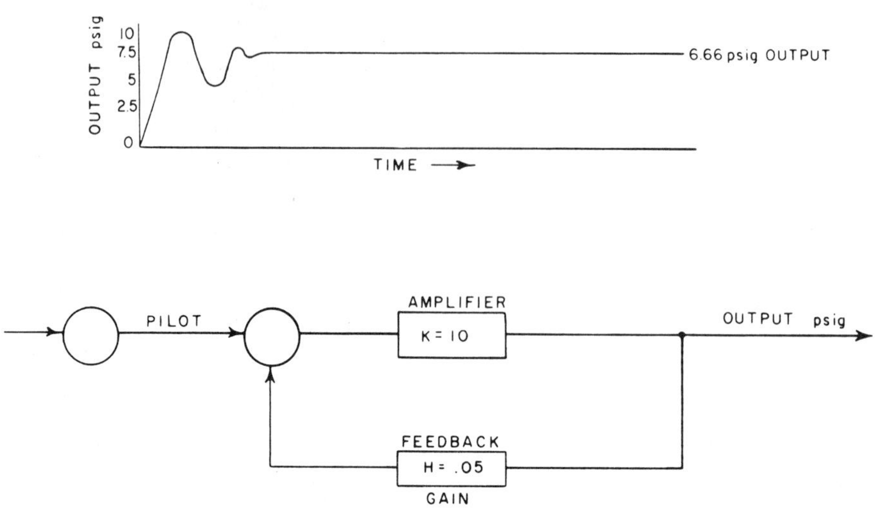

Figure 15–61

The cycle of response to any pilot signal change is similar to the graph. It has a high initial response to provide a high energy level with a relatively gradual settling or balancing at an output value with a precise relation to input signal. The resulting point of balance can be calculated by the formula:

$$p_0 = \frac{K}{1 + KH} = \frac{10}{1 + 10(0.05)} = 6.67 \text{ psig}$$

where:

p_0 = output pressure, psig
K = Forward loop gain (amplifier ratio
H = Feedback loop gain ratio

away from the control port. This counteraction provides a balanced condition for the output signal with respect to the temperature change measured at the sensing element.

Most proportional controllers used today incorporate negative penumatic feedback, with a *sensitivity adjustment* which transmits more or less feedback into the controller. If the sensitivity slider shown in Fig. 15-62 is moved toward the feedback diaphram, more of the diaphram movement would be transferred to the element, canceling out a greater portion of the change in output pressure initially created by the temperature change at the element. If the slider is moved away from the feedback diaphram, the diaphram movement and higher output pressure would have less canceling effect on the movement of the element. The result then is a greater change in output pressure in response to a given temperature change. It can be seen then that as more controller feedback is introduced, the sensitivity of the instru-

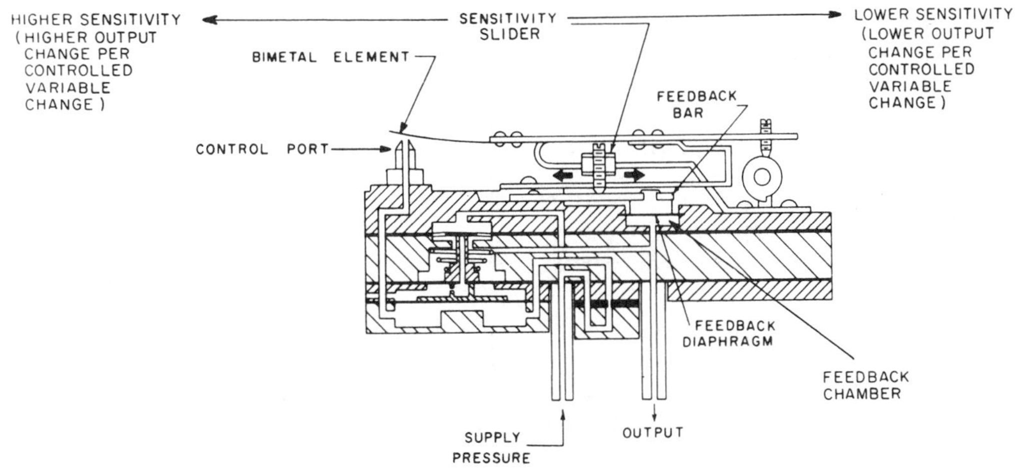

Figure 15–62

Cutaway drawing of a complete 4000 Series pneumatic proportional controller showing the controller feedback diaphram and sensitivity adjustment. (Courtesy of Johnson Controls, Inc.)

ment becomes lower, resulting in a smaller overall change in output pressure per unit change in the controlled variable. Conversely, as the amount of controller feedback is reduced, sensitivity increases, which allows a greater change in output pressure per unit change in the controlled variable.

The *action* of a proportional controller is determined by the way the sensing element is mounted. It can be mounted to bend or move toward the control port and thereby increase output pressure on an increase in measured variable (temperature, humidity, etc.). This combination is called *direct acting*. Or, the sensing element may be mounted to move away from the control port on an increase in the variable, allowing output pressure to drop. This combination is called *reverse acting*.

15-16 FINAL CONTROL ELEMENTS[1]

A final control element changes the value of the manipulated variable in response to the error signal from the controller. The manipulated variable and control agent then exert a correcting influence on the process operation to return the controlled variable to the set point.

The mechanism consists of two parts: (1) an actuator (sometimes called operator) and linkage, which translate the controller output signal into sufficient power and motion to operate the final control element and (2) the final control element, which is a device to adjust the value of the manipulated variable by controlling the flow of the agent. Mechanisms commonly used include motors, valves, and dampers.

Depending on the process, these mechanisms at times must work in extreme temperatures and pressures, be resistant to chemical action, be nearly maintenance free, and respond rapidly to error signals from the controller. The mechanisms vary, but in all cases they depend upon controller signals which are pneumatic, hydraulic, or electric in nature. Figures

15-63 through 15-72 and associated descriptive matter are from Ref. 1, courtesy of Honeywell, Inc.

15-17 ACTUATORS AND LINKAGES[1]

An *actuator* (sometimes called *operator*) is a device that starts and stops or varies the operation of the final control element in response to the controller. A *linkage* is a series of connecting rods with movable joints which transfer the motion of the actuator to the final control element. For instance, the linkage used to connect a motor to a damper usually consists of a pushrod, two crankarms, and two ball joints. The linkage can usually be adjusted to vary the response of the final control element to the actuator.

The actuator in an automatic control system may be electric, pneumatic, or hydraulic. Solenoids and motors are electric, diaphram types are pneumatic, and oil-operated cylinders are hydraulic. We will briefly discuss electric and pneumatic actuators in this section.

CHARACTERIZING METHODS: In most final control elements (valves and dampers), flow versus opening follows a square root relationship. This is disadvantageous for most control systems becuuse it makes it difficult to control the flow accurately, especially at low flow rates. For example, in one type of valve, a change from 0 to 10 percent valve opening results in a flow rate change from 0 to 31.6 percent of maximum. At the high end, a change from 90 to 100 percent valve opening results in a change of flow from 95 to 100 percent. It is also characteristic of many final control elements that they offer the greatest resistance to motion at the nearly closed position. Obviously, it is desirable to have a linear relationship between the error signal and the flow rate. This can be obtained by altering, or *characterizing,* the linkage or actuator.

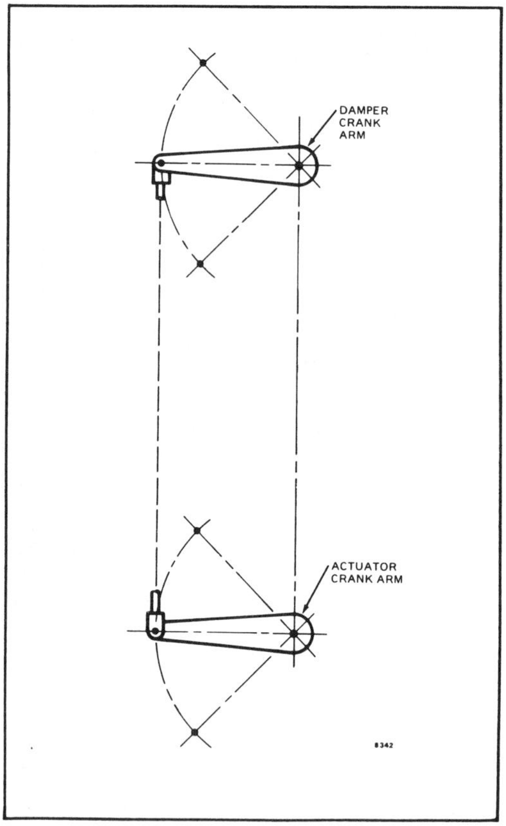

Figure 15–63

Parallel linkage

Linkage adjustment. One method of characterizing is by *adjusting the angularity of the linkage* between the actuator and the final control element. For example, let us look at a typical damper linkage as shown in Fig. 15-63. For a standard parallel linkage, the crank arms are parallel so the angle of travel of the damper crank arm is always the same as the angle of travel of the actuator crank arm. Without characterization, the square root relationship will exist between flow rate and error signal; that is, flow rate will change more when the error signal is small and the damper is just beginning to open.

If the linkage is adjusted as shown in Fig. 15-64, the angular rotation of the damper crank arm will approximate the square of the

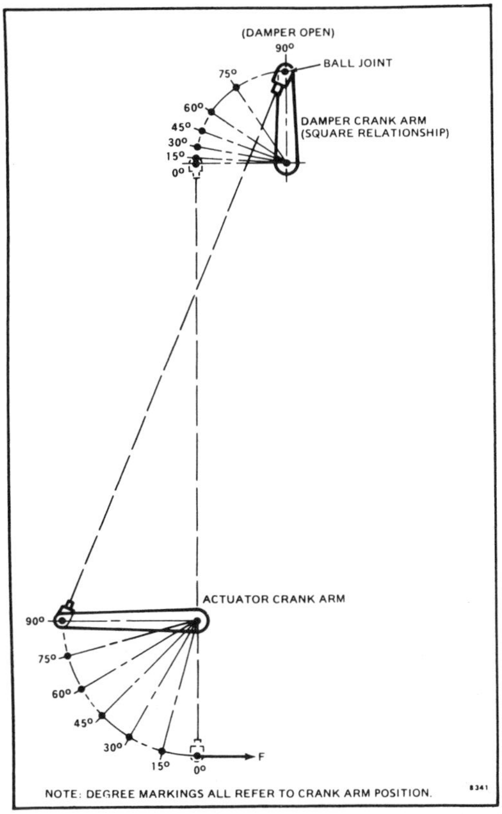

Figure 15–64

Characterized linkage

rotation of the actuator crank arm. This will result in a small amount of damper travel when it is just beginning to open, as shown by the dashed curve in Fig. 15-65. The *square* relationship of the characterized linkage will compensate for the *square root* relationship of the flow rate versus damper opening ($X^2 \cdot \sqrt{X} = X$), resulting in a *linear* relationship for flow rate versus error signal. If required, the linkage can be adjusted to obtain other curves, such as a square or even a fourth-power relationship between the flow rate and the error signal.

Cam shaping. For actuators with a cam-operated feedback mechanism, the actuator can be characterized by shaping the cam.

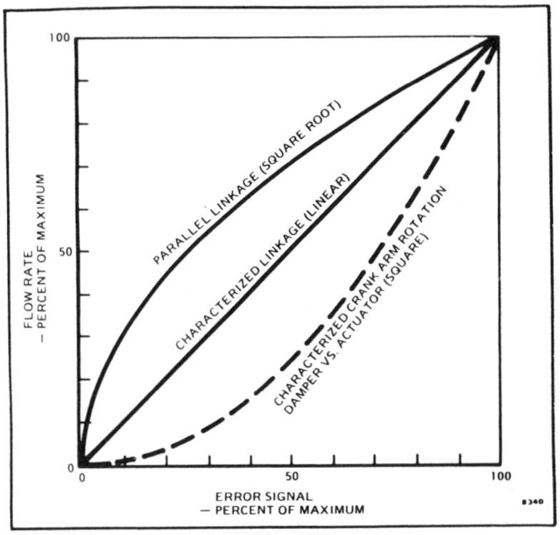

Figure 15–65

Result of characterizing the linkage

The *template method* of cam shaping is, as the name implies, the development of a template under actual operating conditions. At each of several values of error signal, the feedback mechanism is adjusted to obtain the desired position of the final control element. For each of these positions the position of the cam roller is marked on a template. A cam of the required shape can then be made from the template.

The *drafting method* allows the cam shape to be determined by a simple procedure that saves installing a template. The straight cam is left on the actuator and various error signals are introduced. The manipulated-variable condition is read from a meter and plotted against the error signal. Differences between this curve and the desired straight line can then be transferred to a drawing of the straight cam to plot the shape of the required cam face.

SOLENOIDS: A solenoid, Fig. 15-66, is simply a coil of wire which, when current flows through it, will act as a magnet to pull a movable iron core into the coil. The iron core is attached to the final control element. The

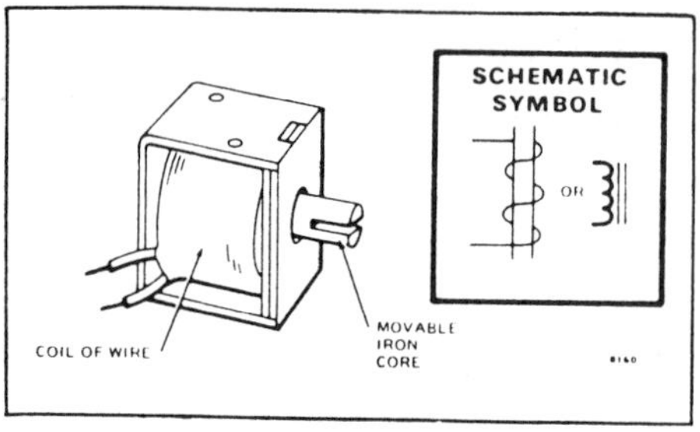

Figure 15–66

Solenoid

most common use of a solenoid as an actuator is with a valve. The iron core is attached to the stem of the valve. A disc on the stem is positioned in an orifice. When the solenoid is energized, the disc either opens or closes the orifice, depending on the valve model (normally closed or normally open).

The solenoid is limited to the on-off (two-position) mode of operation or to one stage in a multiposition mode. It provides fast response and is easily field replaceable.

CONTROL MOTORS: For heavier loads or for greater movement than a solenoid actuator can provide, specially designed electric motors are used. Single-phase induction motors are the most common. The magnetic action of the field coils *induces* currents in the rotor, causing it to turn. The rotor needs no direct electrical connections and so requires neither a commutator nor brushes.

Automatic control motors differ from ordinary small motors. Ordinary motors are designed to drive continuously rotating machinery, usually at high speed. Most of them consist only of field coils, an armature or rotor, and a shaft.

Control motors, on the other hand, must be capable of producing *slow* movement *from one definite position to another,* and they must hold this definite position under load. Therefore, besides the field coils and rotor, a control

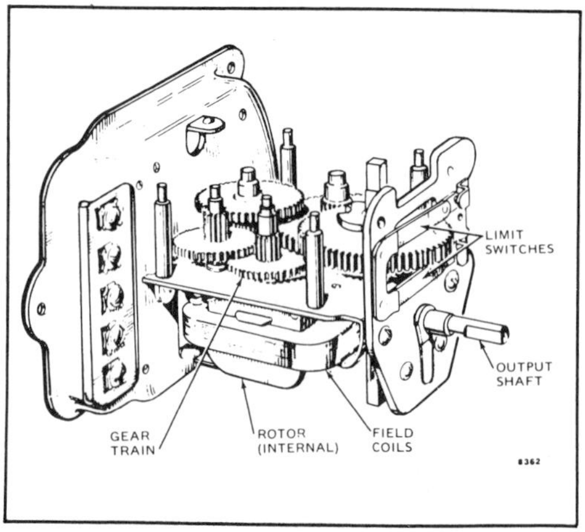

Figure 15–67

Typical control motor

motor, see Fig. 15-67, also normally includes: (1) a *gear train* between the rotor and the output shaft, for reducing speed and increasing torque and (2) mechanically operated switches, called *light switches,* for stopping at definite positions of the output shaft. A proportioning motor also includes a *balancing relay* and *feedback potentiometer* to proportion the travel of the motor to the change at the controller.

Unidirectional motors. A *unidirectional (non-reversible) motor,* see Fig. 15-68, is used for the on-off (two-position) mode of operation. Current through both field coils is always in phase, so the motor always runs in the same direction. A cam with only one node operates a maintaining switch. The cam is mounted on the motor shaft and turns with it. Both contacts, *A* and *B,* are closed when the motor is running. When the cam has rotated halfway, the node is at the bottom and switch *B* is open (shown by dashed line). Controller circuitry starts the motor again. When it has rotated another 180 degrees, the node is at the top of the cam and contact *A* is open. Thus, the motor is stopped after each half-turn of the motor shaft and crank arm.

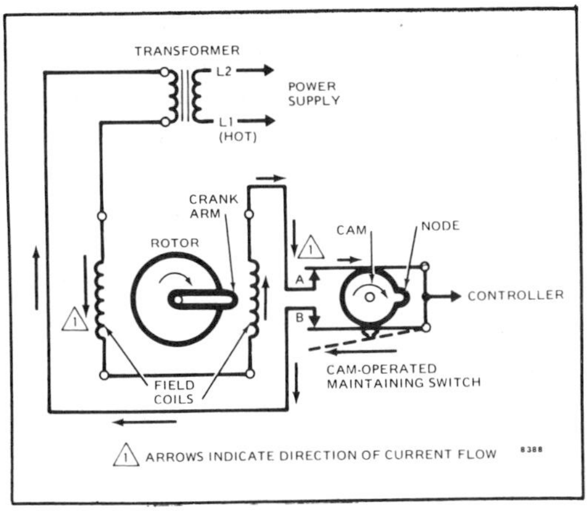

Figure 15–68

Unidirectional motor

<u>*Reversible motors.*</u> For floating and proportioning modes of operation, the motor must be able to reverse its direction. *Capacitor-type* and *shaded-pole-type* motors are the two types most often used in control applications.

In the *capacitor-type motor* (see Fig. 15-69) the two field coils are connected together at one end, and the other ends are connected through a capacitor. The capacitor causes the magnetic fields produced by the two coils to be out-of-phase, resulting in a rotating field that causes the rotor to turn.

The power source may be connected directly across either of the field coils. If the power source is connected between terminals *A* and *C*, power is supplied to coil 1 directly and to coil 2 through the capacitor. If the power source is connected between terminals *B* and *C*, power is supplied to coil 2 directly and to coil 1 through the capacitor. Consequently, the phase relationship between the coils is reversed, so the direction of rotation of the rotor reverses. This change in power source connections may be directly accomplished by the controller, or indirectly by a balancing relay in the motor.

In the *shaded-pole-type motor* (see Fig. 15-70) the stationary field coil is embedded in a laminated block of iron, called the stator. When the power is turned on, the field coil

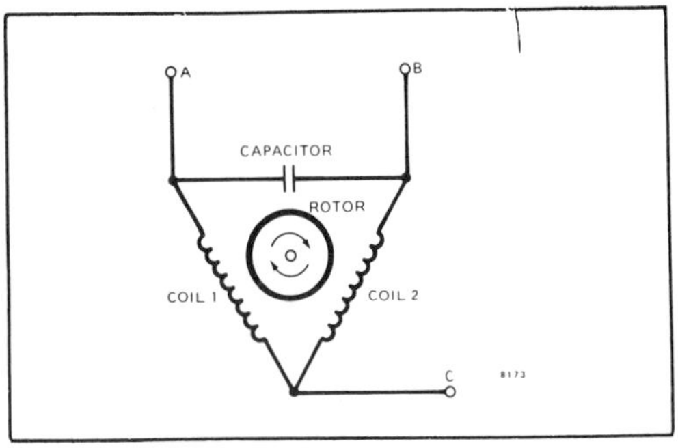

Figure 15–69

Capacitor type reversible motor

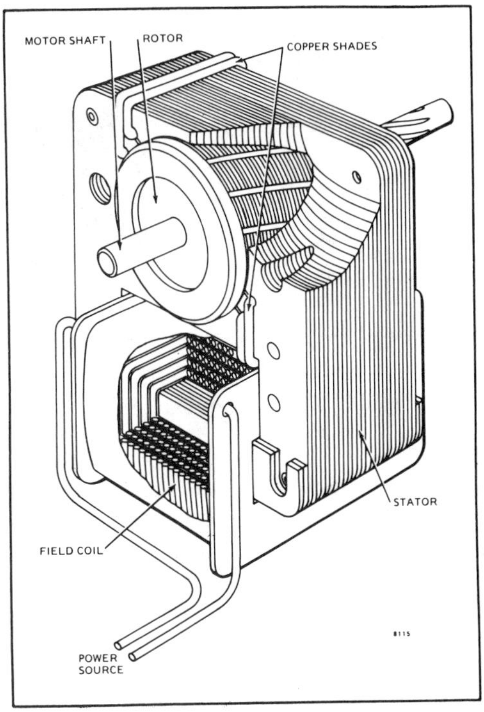

Figure 15–70

Shaded-pole type motor

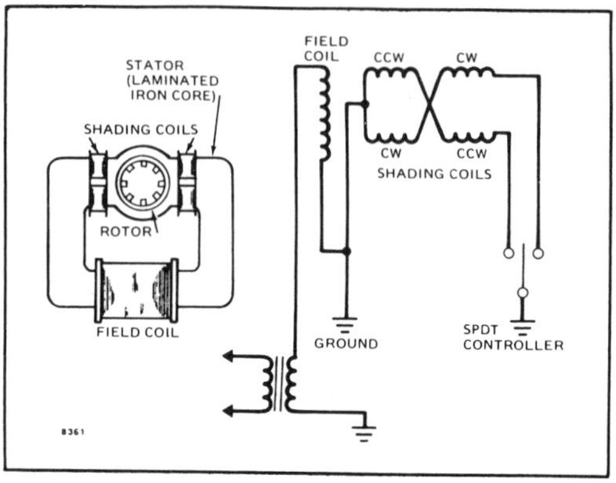

Figure 15–71

Reversing action in a shaded-pole type motor

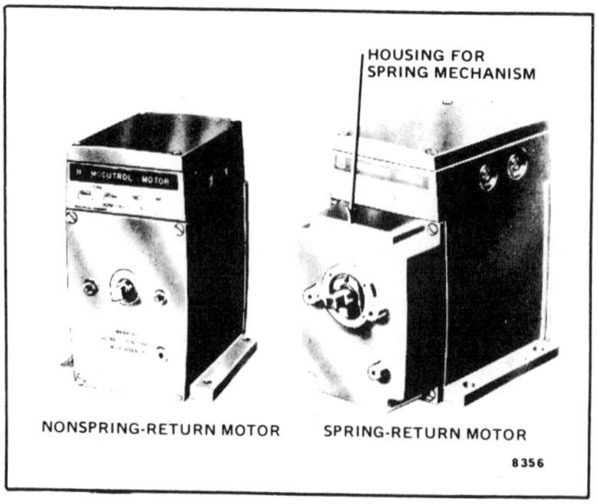

Figure 15–72

Identifying a spring-return motor

generates a magnetic field in the iron. The rotor moves freely in the hollowed-out stator. Heavy strips of copper, called shades, are attached to the poles of the stator on either side of the rotor. The copper magnetizes less rapidly than the iron in the stator. Therefore, the magnetic field of the shades always lags behind the field of the stator. This condition, combined with the swift reversal of the magnetic poles due to the alternating current of the power source, establishes a rotating magnetic field that causes the rotor to turn.

Reversing action is accomplished by using two pairs of coils for shades, see Fig. 15-71. Shortening one pair of coils will cause the magnetic field induced in the shades to *lag* behind the field of the stator, while shortening the other pair will cause it to *lead,* thus reversing the direction of the rotating magnetic field.

Spring-return motors. Many motors have a helical spring attached to the motor shaft to mechanically return the motor to the starting (closed) position when power is removed. The spring mechanism is contained in a housing, see Fig. 15-72, which projects from one end of the motor, identifying it as a spring-return type. When used in the on-off (two-position) mode, a spring-return motor only requires electrical power to drive it in one direction, so one less terminal or wire is needed. Spring-return motors are often used in the floating or proportioning mode, even when electrically driven in both directions, to provide fail-safe action in case of a power interruption.

Electronic motors. Electronic motors are similar to other reversible motors except that they have solid-state control circuits and solid-state switching circuits which control the direction of motor shaft rotation. Mechanical contacts are eliminated; therefore, these motors will not experience contact chatter, so vibration is not an important consideration during installation.

Electronic motors provide proportioning (modulating) or sequencing control of valves, dampers, step controllers, and economizers. They are used with central processors to control heating, cooling, and economizer operation in multizone control systems. They are also used with electronic sequencers to provide sequencing control of heating-cooling equipment in commercial or industrial applications.

Many electronic motors are designed for specific applications or with special functions, such as automatic reset with an adjustable reset ratio. For economy and convenience of installation, some of these motors combine the functions of an electronic controller and actuator. They include a null detector, input adjustments, an amplifier, and output switching elements. Most of these special motors are controlled by an electronic (thermistor type) sensing element which is one leg of a bridge circuit.

PNEUMATIC ACTUATORS: Pneumatic *damper actuators* (operators) of the piston type have a long, powerful straight stroke which requires no lever arrangements. Air from the controller is applied to the molded diaphram which has a positive seal to prevent air leakage, see Fig. 15-73. This air pressure expands the diaphram, forcing the piston and stem outward against the force of the spring. The movement of the piston varies proportionally with the air pressure applied to the diaphram. This air pressure, from the controller, varies over the full pressure range.

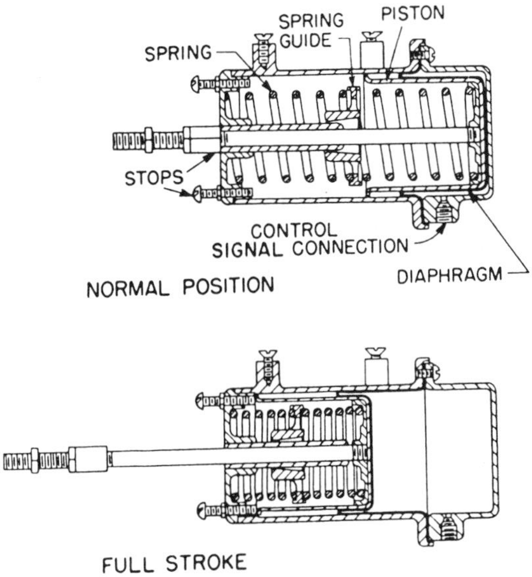

Figure 15–73

Components of a piston damper operator

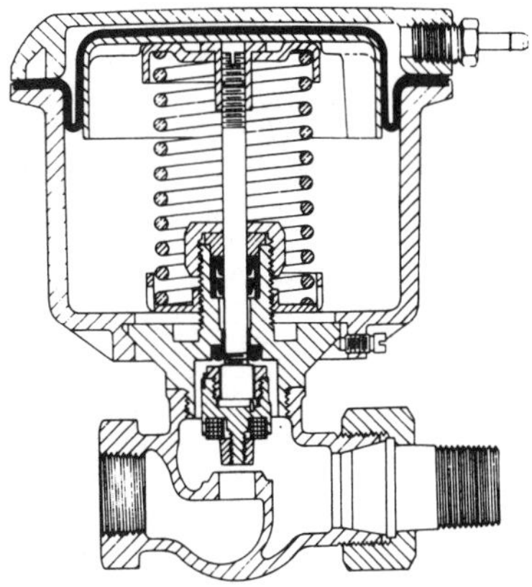

Figure 15–74

Piston top valve operator on a normally open valve

The spring returns the actuator to its normal position when the air pressure is removed from the diaphram. Full movement of the actuator can be restricted to set limits by using various spring ranges. Assume a spring range of 5 to 10 psig. With this spring the actuator is in its normal position when the air pressure applied to the diaphram is 5 psig or less. Between 5 and 10 psig the stroke will be proportional to the air pressure on the diaphram. Above 10 psig the actuator will be at its maximum stroke.

Damper actuators can be mounted on the damper frame and coupled directly to the damper blades. In some cases the actuator is mounted on the duct work and coupled to the damper blade axis through a crank arm and linkage arrangement (Section 15-16).

Pneumatic *valve actuators* generally furnished with valves are the *piston top,* Fig. 15-74, and the *rubber diaphram,* Fig. 15-75. The former is used mainly on small valves controlling terminal heating units and terminal air-conditioning units, while the latter is used to control large valves regulating flow through heating or cooling coils on central or zoned systems.

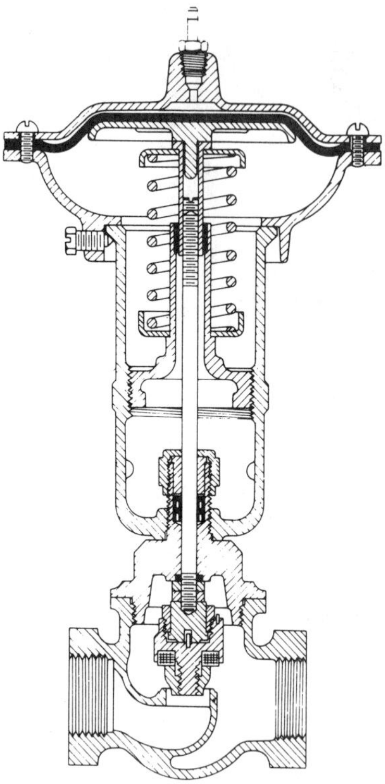

Figure 15–75

Rubber diaphram operator on a normally open valve

Valve actuators function in a manner similar to damper actuators. Air pressure, or hydraulic fluid, is applied to the diaphram, which expands and opposes the force of the spring. This causes the inner valve (stem and disc or plug) to move toward the seat of a normally open valve (Figs. 15-74 and 15-75), stopping the flow, or away from the seat of a normally closed valve, allowing flow. When air pressure is removed from the diaphram, the spring will return the inner valve to its normal position. Figures 15-73 through 15-75 and associated descriptive matter are from Ref. 4 courtesy of Johnson Controls, Inc.

15-18 CONTROL VALVES[1]

A control valve is any device by which the flow of liquid, air, other gas, loose bulk material, and so forth, may be started, stopped, or regulated by a movable part which opens or obstructs passage. The simplest definition is one used by valve designers—"an engineered obstruction in a pipe." A valve acts as a variable orifice in a fluid flow line based on some planned pattern of change in the rate of flow. In an automatic control system it is a final control element that directly changes the value of the manipulated variable by changing the rate of flow of the control agent.

Control valves are the most common and basic final control element in industry. They are widely used in regulating pressure, temperature (blending), and rate of flow of water, steam, chemicals, petroleum products, and just about anything that flows. They are used in processes, in power generation, in transmission pipelines—wherever flow of liquids and gases is controlled.

At this point in our work, we should review Chapter 11, Sections 11-12, 11-13, and 11-14 for some particular information on flow resistance in valves.

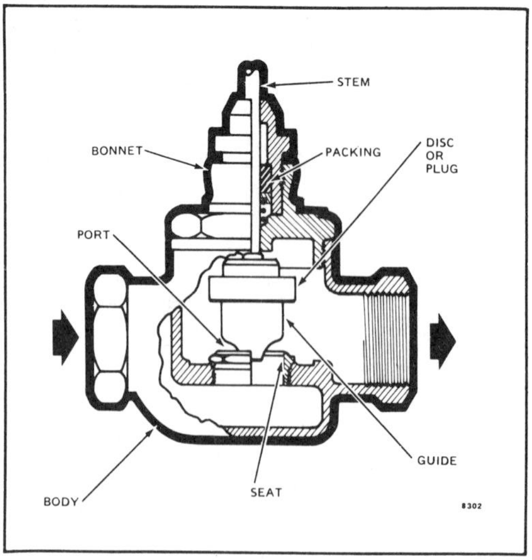

Figure 15–76

Components of a control valve

CONTROL VALVE COMPONENTS: Figure 15-76 is a sketch showing the various valve components defined in the following sections:

Body: The portion of the valve through which the medium flows.

Stem: The cylindrical shaft that is moved by the actuator and to which the disc is attached.

Packing: Material which seals the fluid flowing through the valve body, while permitting the valve stem to pass through as freely as possible. Standard packing is usually in the form of a nest of *V*-shaped rings of lubricated asbestos fiber or teflon.

Bonnet: The removable cover or top of the valve body (attached by screws or bolts). It provides a means for joining the actuator to the body, and it houses the packing. The bonnet assembly may include the valve trim.

Disc: The movable portion of a valve that comes into contact with the seat as the valve closes and that controls the flow of the medium through the valve. Three common types of control discs are the contoured, the *V* port, and the quick-opening.

Contoured disc: controls the flow by a shaped end, sometimes called a valve *plug* and is usually end-guided at the top or at the bottom (or both) of the valve body.

V-port disc: has a cylinder (called a skirt or guide) that rides up and down in the seat ring. The skirt guides the disc, and the shaped openings in the skirt vary the flow area.

Quick-opening disc: is so machined that maximum flow is quickly achieved when the disc lifts from the seat. It is either end-guided or guided by wings riding in the seat ring.

Some discs are built so that the part of the disc that contacts the seat is replaceable. This type is known as the renewable disc. A common example is a kitchen sink hot water faucet. Renewable discs are usually made of a composition material. Many valves having all metal or nonrenewable discs have to be "ground-in" to restore the seating surface, but valves with flat-type metal discs do not.

Guide: The part of the valve disc that keeps the disc aligned with the valve seat. Top or bottom guides on a valve disc usually do not influence flow; they merely perform a centering function. Valve guides often determine the valve flow characteristic. These guides are known as *characterized guides* or *skirted guides* and usually have notches or *V*'s cut into them to characterize the flow through the valve.

Plug: A term used to describe a disc that has a solid, shaped end (instead of a skirted guide) to characterize flow. In some valves the plug serves as a guide.

Seat: The stationary portion of a valve with which the movable portion (disc) comes in contact to stop the flow completely.

Port: The flow-controlling opening between the seat and disc of a valve. It does not refer to the body size nor to the end connection size of the valve. Regardless of the pipe size, the valve port size determines the flow. For example, a 1-in. valve with a ¾-in. reduced port size has the same flow-control as a ¾-in. valve with a standard port size.

Trim: All parts of a valve that are in contact with the flowing medium but are not part of the valve shell or casting. Seats, discs, packing rings, and so forth, are all trim components.

Figures 15-76 through 15-89 and associated descriptive matter are from Ref. 1, courtesy of Honeywell, Inc.

15-19 IMPORTANT VALVE CHARACTERISTICS[1]

To select the proper valve for a given application, several flow-related characteristics of the valve must be known.

VALVE FLOW CHARACTERISTIC: The factor that is most useful in selecting a valve type for a given application is the flow characteristic. This characteristic is the relationship that exists between the flow rate through the valve and the valve stem travel as the latter is varied from 0 to 100 percent. Different valves have different flow characteristics, depending primarily on internal construction. This flow relationship is usually shown in the form of a graph. The characteristic that is usually graphed is the "inherent flow characteristic," which is found under laboratory conditions and is the characteristic observed with constant pressure drop across the valve. However, the pressure drop across a valve in a system is not constant; it varies with flow and other changes in the system. We will discuss this in Chapter 16.

The three most common valve flow relationships are quick-opening, linear, and equal percentage. The ideal flow relationships are shown graphically in Fig. 15-77. The quick-opening valve plug gives the steepest initial flow characteristic possible. A small initial lifting of the plug produces a large increase in flow, and close to maximum flow is reached at a relatively low percentage of maximum stroke. This type of plug is generally used for on-off types of applications but may be used successfully in some linear valve applications. This is possible because of its initial linear characteristic at low percentage stem travel. The slope of this linear region is very steep, which produces a much higher gain than an ordinary linear plug and consequently is much more likely to be unstable.

The linear plug produces a characteristic with essentially a constant slope so that with a constant pressure drop across the valve the valve gain will be the same at any flow. This type of plug would be the easiest and most logical plug to use if everything else in the control loop were linear. However, this is not true in most systems, as will be shown in Chapter 16.

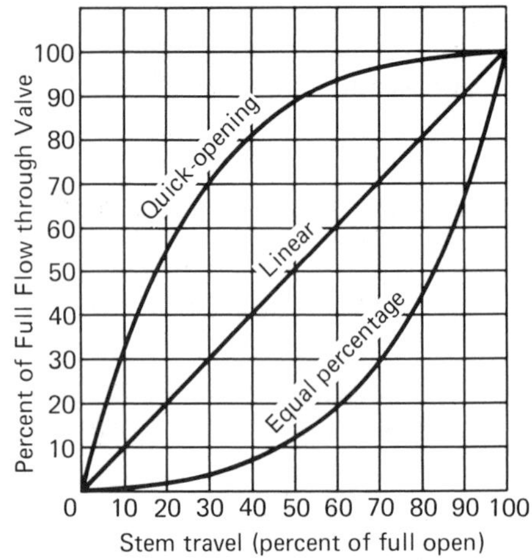

Figure 15–77

Valve Flow Characteristics

The third type of plug is the equal-percentage plug. The characteristic for this plug shows that for each unit step increase in valve stroke the flow increases an "equal percentage" over the previous flow. For example, if 30 percent stem lift produces 5 gpm flow and an increase of 10 percent lift, to 40 percent, produces 8 gpm, or a 60 percent increase over the previous 5 gpm, then a further stroke of 10 percent, to 50 percent stroke, now produces a 60 percent increase over the previous 8 gpm for a total flow of 12.8 gpm. Because of this constant relative increase in flow per unit stroke, the valve gain at a specified flow is independent of both the valve flow coefficient and the pressure drop. Because of this, system stability (at a specified flow) is independent of the valve size selected.

A valve may be *characterized* for the desired relationship between stem travel and flow by shaping the sides of the port and the openings in the skirted guide. A linkage between actuator and stem can also be characterized, as discussed previously in the section on actuators and linkages.

Valve gain. In the preceding discussion of

valve flow characteristics the term *gain* was used. Valve gain is the incremental change in flow rate produced by an incremental change in plug position. This gain is a function of valve size and type, plug configuration, and system operating conditions. The gain at any point in the stroke of a valve is equal to the slope of the characteristic curve at that point.

TRAVEL COEFFICIENT: Travel coefficient is the ratio between the flow at a given valve stem position and the flow through the valve at its wide-open-position. Travel coefficient is expressed as a decimal fraction. For example, if a valve with a lift of 1 in. passes 100 gpm fully open and passes 66 gpm at a valve lift of 0.5 in., this valve is said to have a 0.66 travel coefficient at 0.5 in. lift. Valve travel coefficients can be read directly from any plot showing valve flow characteristics.

RANGEABILITY: The ratio of the maximum *controllable* flow to the minimum *controllable* flow is called rangeability. For example, a valve with a rangeability of 50 to 1 having a total flow capacity of 100 gpm fully open can accurately control a flow as low as 2 gpm. Generally speaking, rangeabilities in the area of 50 to 1 or 40 to 1 are excellent for precision control. Valves with high rangeability are very expensive to manufacture since very close tolerances are required between the disc and the seat.

TURNDOWN: The ratio of the maximum *usable* flow to the minimum *controllable* flow is called turndown. Turndown is usually somewhat less than the rangeability, as the maximum usable or required flow is usually less than the total flow capacity of a valve. For instance, the rangeability example, after the 100-gpm valve has been applied to a job, it might turn out that the most flow you would ever need through the valve is 68 gpm. Since the minimum controllable flow is 2 gpm, the turndown of the valve is 34 to 1. In comparing rangeability and turndown, rangeability is the measure of the *predicted* stability of a control valve and turndown is the measure of the *actual* stability of the valve.

TIGHT SHUTOFF: A valve having tight shutoff has virtually no flow or leakage in its closed position. Generally speaking, only single-seated valves have tight shutoff. Double-seated valves can be expected to have 2 to 5 percent leakage in the closed position.

CAVITATION: Cavitation is a two-stage phenomenon which can greatly shorten the life of the valve trim in a control valve. Whenever a given quantity of liquid passes through a restricted area such as an orifice or a valve port, the velocity of the fluid increases. As the velocity increases, the static pressure decreases. If this velocity continues to increase, the pressure at the orifice will decrease below the vapor pressure of the liquid, and vapor bubbles will form in the liquid. This is the first stage of cavitation. Figure 15-78 shows schematically the variation of velocity and pressure as liquid flows through a restriction such as an orifice or valve port.

As the liquid moves downstream, the velocity decreases with a resultant increase in pressure. If the downstream pressure is maintained above the vapor pressure of the liquid, the voids or cavities of vapor will collapse or implode. This is the second stage of cavitation (see Fig. 15-78).

The second stage of cavitation is detrimental to valves. Because of the tremendous pressures created by these implosions (sometimes as high as 100,000 psi), tiny shock waves are generated in the liquid. If these shock waves strike the solid portions of the valve, they act as hammer blows on these surfaces. Repeated implosions on a minute surface will eventually cause fatigue of the metal surface and chip a portion of this surface off. Tests show that only those implosions close to the solid sur-

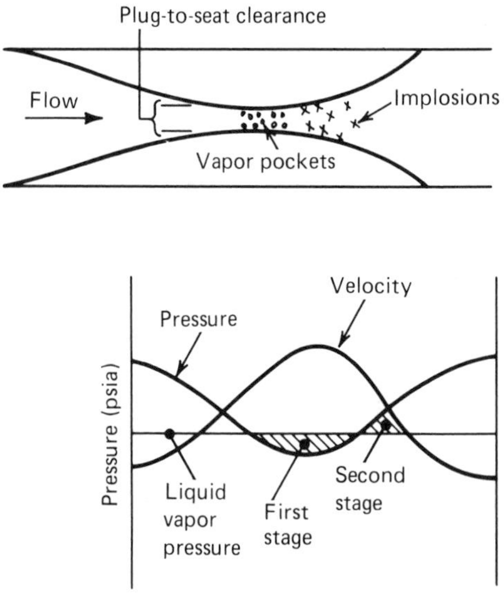

Figure 15–78

Variations of pressure and velocity for points along the flow stream through a restriction

faces of a valve act on the valve in this manner.

Low degrees of cavitation are tolerable in a control valve. Minimum damage to the valve trim and little variation in flow occur at these levels. However, there is a point where the increasing cavitation becomes detrimental to the valve trim and possibly to the valve body. Also at this point, the cavitation is beginning to "choke" the flow through the valve. At some point the flow rate will stay the same regardless of increases in pressure drop. The point at which cavitation becomes damaging can be expressed by

$$K_m = \frac{\Delta p_{max}}{p_1 - p_v} \qquad (15\text{-}1)$$

where K_m = valve recovery coefficient

p_1 = absolute inlet pressure (psia)

p_v = absolute vapor pressure of liquid (psia)

Δp_{max} = maximum allowable pressure drop (psi)

The valve recovery coefficient differs among the various types of valves. A conservative value from K_m is 0.5. Equation (15-1) can be used to estimate the allowable maximum pressure drop through a valve when warm liquids are flowing.

15-20 VALVE RATINGS[1]

To specify a valve for a given job, several ratings must be considered.

CAPACITY INDEX *(C_v):* Capacity index (also called flow coefficient in Chapter 11) is the quantity of water, in gallons per minute at 60°F (16°C), that flows through a given valve with a pressure drop of 1 psi. Once the C_v of the valve has been determined, the flow of any fluid through the same valve can be calculated, provided the fluid properties and the pressure drop through the valve are known. This will be illustrated in Chapter 16.

CLOSEOFF RATING: The close-off rating of a valve is the maximum allowable pressure drop to which the valve may be subjected while fully closed. This rating is usually a function of the power available from the valve actuator for holding the valve closed and is independent of the actual valve body rating. Structural parts, such as the stem, are sometimes the limiting factor. To illustrate, a valve having a close-off rating of 10 psi could operate with an upstream pressure of 40 psi and a downstream pressure of 30 psi. However, in applications where failure of the valve to close is hazardous, the maximum upstream pressure should not exceed the close-off rating regardless of the downstream pressure.

The close-off rating of a three-way valve is the maximum pressure difference between the outlet and either of the two inlets. For a three-way diverting valve it is the pressure difference between the inlet and either of the two outlets.

MAXIMUM FLUID PRESSURE AND TEMPERATURE: These ratings are limitations placed upon the valve by the maximum pressure and temperature to which the valve may be subjected. Determining factors may be packing, body material, disc material, or actuator limitations. For example, a valve having an actual body rating of 125 psig at 355°F (179°C) may have a packing with a 100 psi rating; this limits the valve to a fluid pressure of 100 psi maximum. The temperature of 115 psia steam is 338°F (170°C). This fact should limit the above valve to a maximum temperature rating of 338°F (170°C) rather than 355°F (179°C). However, if the valve has a composition disc with a maximum fluid temperature rating of 240°F (116°C), consider this temperature as the maximum fluid temperature.

PRESSURE DROP: Pressure drop of a valve is the difference between the upstream pressure and the downstream pressure of the fluid passing through a valve. Pressure drop across control valves may be assigned by the control system designer using values recommended by control valve manufacturers. We shall look at this in more detail in Chapter 16.

VALVE BODY RATING: This rating is the maximum pressure valve body can withstand.

The *nominal body rating* is the nominal pressure rating of the valve body (exclusive of packing, disc, etc.) expressed in psig. This is a specified rating (often cast on the valve body) based on the specified body material and universally accepted construction characteristics (such as wall thickness, flange dimensions, etc.) determined by specifications issued by approval bodies and other authorities having jurisdiction. It provides a convenient method of classifying the valve by pressure for identification purposes (see Chapter 11). Note that the *nominal* valve body rating is *not* the same as the *actual* valve body rating.

The actual body rating is the maximum safe permissible fluid pressure at a given temperature. Each nominal valve body rating has definite corresponding permissible pressures at various temperatures. Thus, the permissible pressure is usually dependent on the temperature. For instance, a 125 psi (nominal body rating) cast iron screwed body may have actual body ratings of 125 psi at 353°F (178°C) and 175 psi at 150°F (66°C). However, the maximum rating of the valve may be limited to a value below the body rating by the packing and disc material.

15-21 VALVE CLASSIFICATION BY TYPE OF CONSTRUCTION[1]

SINGLE-SEATED VALVE: A valve with only one seat and one disc is cheaper to construct than a double-seated valve. It is generally suitable for service requiring tight shutoff. Figure 15-79 shows schematically two types of single-seated valves. Since there is nothing in the single-seated valve to balance the force exerted by the fluid pressure against the disc, the single-seated valve requires more force to close than a double-seated valve of the same size.

DOUBLE-SEATED VALVE: A valve with two seats and two discs (Fig. 15-80) arranged so that in the closed position there is very little fluid pressure forcing the stem toward either the open or closed position. For a valve of given size and port area, a double-seated valve will always require less power to operate than a single-seated valve. An additional advantage of a double-seated valve is that it often has a

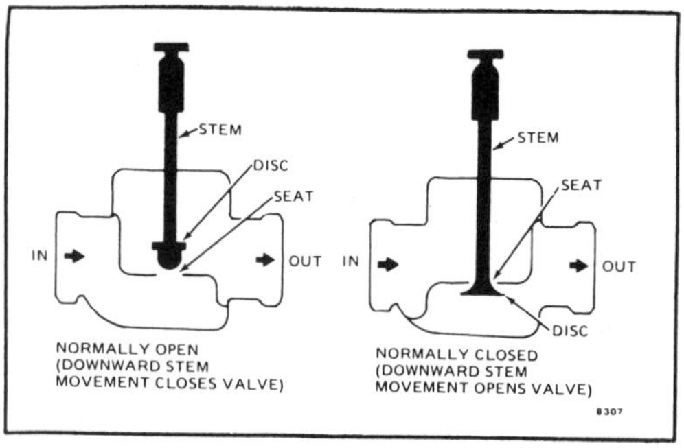

Figure 15–79

Single-seated valve

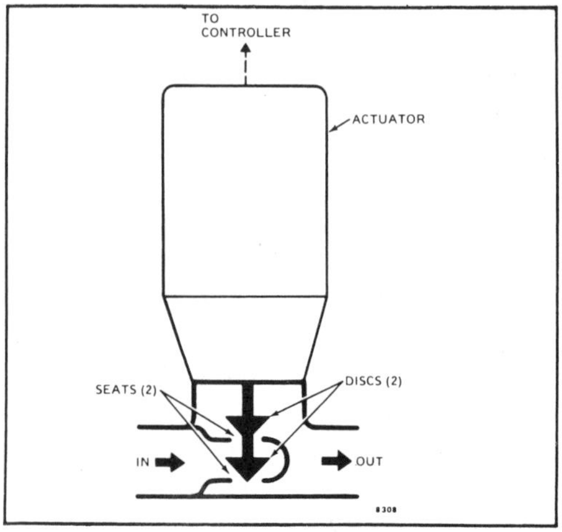

Figure 15–80

Double-seated valve

larger port area for a given pipe size. A disadvantage is that it does *not* have tight shutoff since both discs are rigidly connected. Changes in temperature of the fluid being controlled cause either the disc or the valve casting to expand, allowing one disc to seat before the other.

THREE-WAY MIXING VALVE: This type of valve always has two inlets and one outlet (Fig. 15-81). The proportion of fluid entering from each inlet may be controlled by

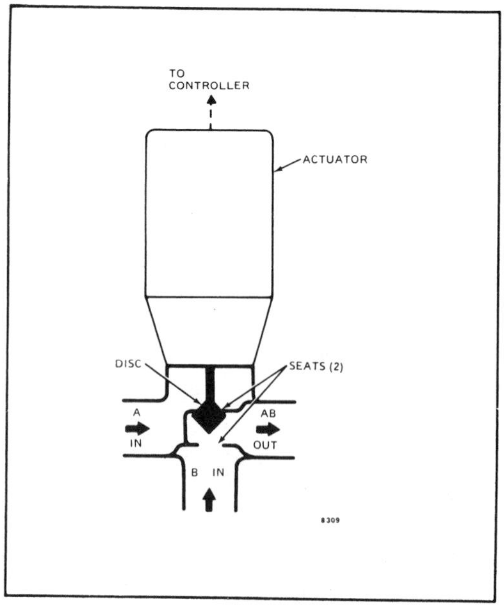

Figure 15–81

Three-way mixing valve

moving the valve stem. A valve designed for mixing service is not suitable for diverting applications because it has only one disc and two seats. If it is piped for diverting applications, the inlet pressure will slam the disc against the seat when it nears closing. This causes loss of control, oscillations, vibrations, excessive wear, and noise.

THREE-WAY DIVERTING VALVE: This type always has one inlet and two outlets (Fig. 15-82). A fluid entering the inlet port is diverted to either of the two outlet ports in any proportion desired by moving the stem. A valve designed for diverting service can generally be used in mixing applications also because it is double seated with a disc for each seat. When the diverting valve is piped for a mixing application, the two inlet pressures oppose and cancel each other. Under these conditions the pressures on the discs are balanced and the diverting valve responds much the same as a mixing valve.

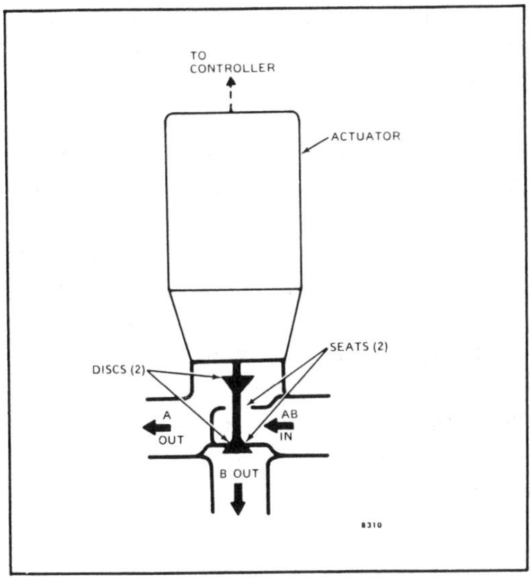

Figure 15–82

Three-way diverting valve

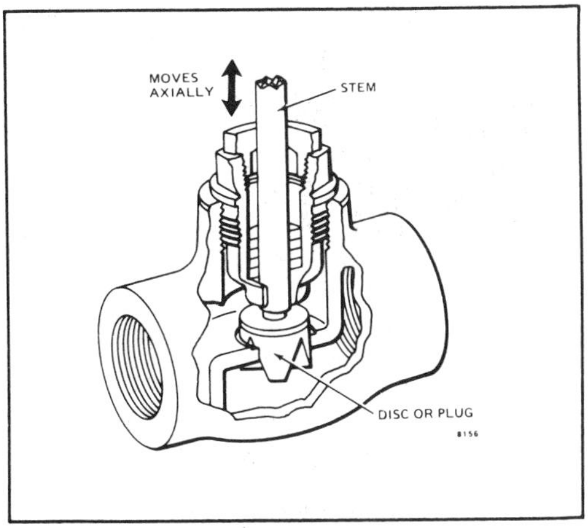

Figure 15–83

Typical sliding plug valve

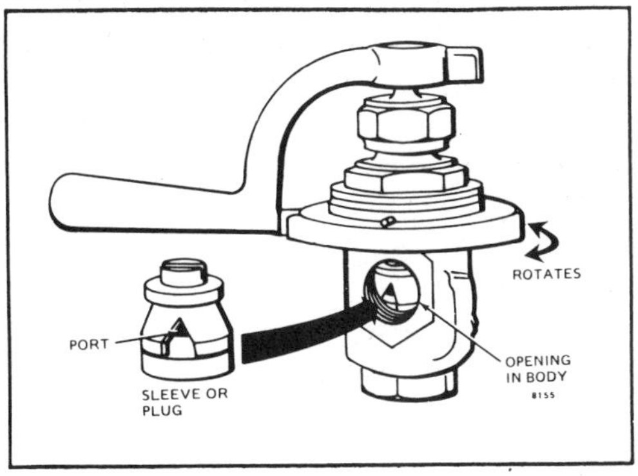

Figure 15–84

Typical rotary plug valve

15-22 VALVE CLASSIFICATION BY METHOD OF CONTROLLING FLOW[1]

SLIDING PLUG VALVE: In a sliding plug valve (Fig. 15-83), the disc or plug is attached to the end of the stem, and the orifice size is adjusted by moving the stem up or down (or in and out) along its axis. The valve flow characteristic is determined by the shape of the plug and port. A disc-shaped plug, as in an ordinary globe valve, gives a linear characteristic, but it has the disadvantage that its entire turndown range is expired within a very small range of axial movement. A *V*-port or a parabolic plug permits a more gradual change of valve opening and gives an equal-percentage characteristic.

ROTARY PLUG VALVE: A rotary plug valve (Fig. 15-84) consists of a ported sleeve or plug which is rotated past an opening in the body. The valve flow characteristic is determined by the shape of the port and the opening in the body. The figure shows a rotary plug valve with a round body opening and a conical plug with a *V*-shaped, tapered port, which gives an equal-percentage flow characteristic similar to that of a *V*-port sliding plug valve. A cylindrical plug with a straight-through rectangular port gives a characteristic approaching a straight line.

BUTTERFLY VALVE: A butterfly valve (Fig. 15-85) has a rotating hinged disc that somewhat resembles the opened wings of a

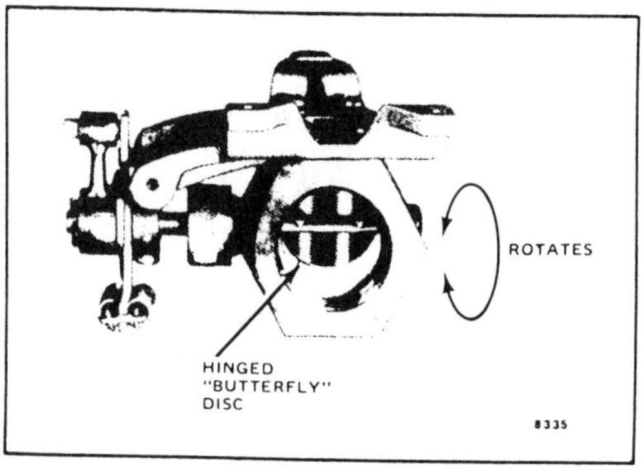

Figure 15–85

Typical butterfly valve

butterfly. The pressure drop across a wide-open butterfly valve is very small. The amount of leakage depends on the degree of machining of the disc and valve casing (which is the seat). The valve gives an equal-percentage flow characteristic similar to that of a *V*-port sliding plug valve. A butterfly valve has excellent high-volume flow characteristics, making it useful for close modulation of the supply of air or fuel gas to large furnaces. It is often used as a firing rate valve. This type of valve does not close tightly, so a separate valve must be used where tight shutoff is required.

15-23 DAMPERS[1]

Regulation of air flow is a vital factor in every HVAC control system as well as in many industrial processes—from small domestic systems to large industrial or commercial systems. These systems use dampers installed in air-carrying ducts to control air flow.

APPLICATIONS: Dampers are made in many shapes and sizes for installation in *round* or *rectangular* ducts. In individual ducts they are used for brute-force control of air volume. For *diverting* or *mixing action,* similar to three-way valves, two dampers are installed with synchronous operation in two-branch ducts. An example of this application is the simultaneous control of return air and outdoor air in a ventilation control system. The two dampers are connected so that one opens as the other closes. Another important application is the control of combustion air in a burner system.

Many types of electric, pneumatic, and hydraulic actuators are used to provide on-off (two-position), floating or proportioning (modulating) operation. Two or more actuators are often used to operate two or more dampers in unison or sequence, or one actuator may be used to operate several damper sections simultaneously.

BLADE DESIGNS: A *flap type* damper consists of one or more metal plates, each hinged at one edge, and all linked together for simultaneous operation (Fig. 15-86). The *splitter* damper is a single-blade, flap-type damper, usually located at a branch connection of a rectangular duct or outlet to divert flow into the branch. It is easy to operate but often causes irregular air flow (turbulence) in the duct. The *pinch* damper, or double-door type, is also a flap type used in horizontal ducts.

A *single-blade* damper is *center pivoted* for installation in rectangular ducts. The *butterfly* damper is also a single-blade, center-pivoted damper, but the blade is disc-shaped for use in round ducts. A *louver* damper is a series of center-pivoted blades with edges mating in the closed position. The louver damper is the most used for controlling inlet and exhaust flow because it provides more precise control.

Louver dampers. There are two types of louver dampers—parallel blade and opposed blade. Each type is designed to do a specific job.

Parallel-blade dampers (Fig. 15-87) are designed so that all blades rotate in the same direction. They are most commonly seen in ventilation applications because they are less expensive and easier to install. Also, they

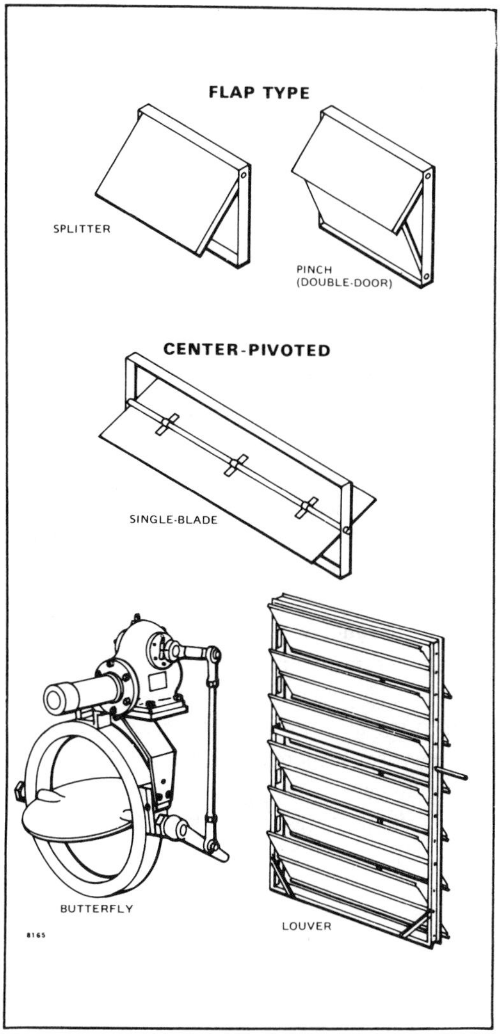

Figure 15–86

Damper blade designs

cause more turbulence and hence better mixing. Therefore, parallel-blade dampers perform better for most ventilation applications.

The air flow characteristic for parallel-blade dampers is nonlinear (see Fig. 15-88). Changes in damper blade rotation do *not* cause proportional changes in air flow. As the damper starts to open, a small rotational change

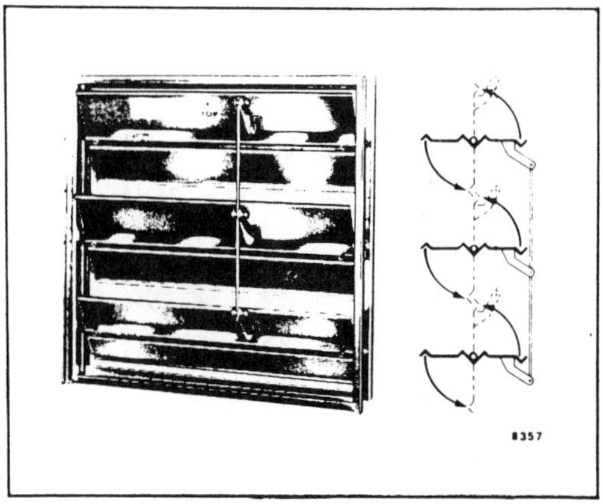

Parallel blade damper

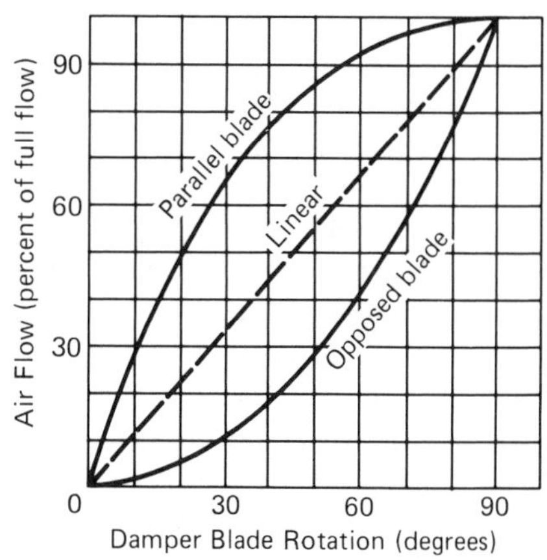

Figure 15–88

Air flow characteristics for parallel blade and opposed blade dampers

causes a large increase in air flow. Therefore, parallel-blade dampers should be adjusted so that the *actual* air flow meets system and code requirements. Improper adjustment of dampers can result in substantial waste of energy. Also, for better control from one end of the proportioning range of the controller to the other, the damper opening should be limited to a maximum of 60 degress. Further opening of the damper will provide only a small additional

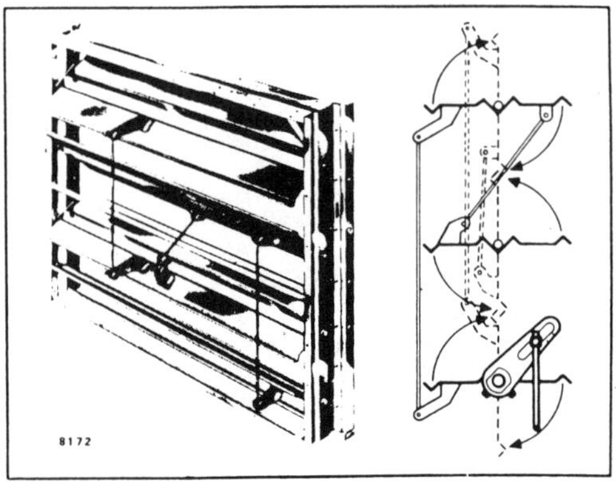

Figure 15–89

Opposed blade damper

amount of air.

Opposed-blade dampers (Fig. 15-89) are designed so that alternate blades rotate in *opposite* directions. They have a more linear flow characteristic and are intended for variable air volume (VAV) applications such as zone or individual room control. In many applications they tend to stratify the air, which results in poor mixing and less control.

Opposed-blade dampers have an equal-percentage air flow characteristic (see Fig. 15-88). Successive equal increments of rotation produce equal-percentage increases in flow. This allows close control at low air flow. Because these dampers are more expensive than parallel-blade dampers, they are usually used only where accurate control of low air flow is necessary.

DAMPER LEAKAGE: For the most economical operation, dampers must allow only the amount of air flow required by the system and by local codes. If dampers do not close tightly, the extra air which leaks into the system must be conditioned. Newer dampers have special seals and low-friction bearings to reduce leakage. Damper leakage is expressed in two ways: (1) percentage of the air flow through an open damper or (2) cubic feet of air per minute per square foot of damper area; both at a specified pressure differential.

DAMPER ACTUATION: Damper actuators are rated by the square feet of damper they can reliably operate. If the area of a damper (or combined dampers) is larger than the rating of the actuator, the installer has two options: (1) use an actuator with a larger rating or (2) use two or more smaller dampers with an actuator for each.

If the second option is chosen, the dampers must usually be driven in unison. A motor with auxiliary potentiometers can be used to drive others (called "slaving"). Also, some actuators can be driven in parallel from the appropriate controller.

Damper actuators may be mounted either outside or inside the air duct. When the actuator is mounted outside the duct, the axle driving the blades of the damper is extended out through the duct wall. When mounted inside, the actuator is mounted on the bottom of the duct. If the duct (or air passage) is stiff enough, the actuator may be bolted directly to it; otherwise a mounting bracket is used. A damper linkage is used to connect the actuator to the damper, as was discussed in Section 15-16.

CHAPTER 15

Review

15-1. Describe the function of each of the following components of an electric control system:

a. power supply
b. controller
c. primary control
d. high limit
e. low limit
f. manual controls

15-2. The primary element in a temperature controller is usually what?

15-3. Name a type of primary control that

a. controls system action directly
b. controls system action indirectly

15-4. A differential gap, or differential, occurs in (two-position/modulating) systems. How is it defined?

15-5. What is the purpose of reset in a proportional control system?

15-6. A delay in response from one part of a system to change occurring in another part of the system is called what? It is most obvious in (temperature/pressure) systems.

BIBLIOGRAPHY

1. ''Automatic Control Principles,'' Honeywell, Inc., Milwaukee, Wisconsin, 1972.

2. Commercial Air Conditioning Controls, ''Glossary,'' Honeywell, Inc., Milwaukee, Wisconsin, 1972.

3. Training Manual, ''Basic Electricity,'' Johnson Controls, Inc., Milwaukee, Wisconsin, 1980.

4. ''Fundamentals of Pneumatic Control,'' Johnson Field Training Handbook, Johnson Controls, Inc., Milwaukee, Wisconsin, 1979.

5. ''Fundamentals of: Heating and Cooling Systems, Pneumatic Control, Electronic Control,'' Training Manual, Johnson Controls, Inc., Milwaukee, Wisconsin, 1975.

Chapter 16

Automatic Control Systems

16-1 INTRODUCTION

The preceding chapter discussed the various control components, how they function with different types of energy, and the various modes of control. In this chapter we will consider the application of these components to form "control systems" for the control of HVAC. We shall consider only the elementary systems, with the understanding that the more complex systems are made up of combinations of these elementary systems.

We will deal primarily with control system function, with the assumption that this function can be accomplished by means of any of the various energy sources—electric (electromechanical or electronic), pneumatic, or fluidic.

Again let us repeat the definition of air conditioning as "the process of treating air so as to control simultaneously its temperature, humidity, cleanliness, and distribution." Air-conditioning systems are designed to meet the comfort and health needs of a building's occupants and also provide the correct environment for the equipment and processes in use. Occupant health and comfort are directly dependent on temperature, humidity, and cleanliness. In turn these quantities usually depend on air distribution.

The demands placed on an air-conditioning system vary with the design, construction, and use of the building in which they are installed. The equipment used to meet these demands may be classified into three general groups: (1) central-fan systems, (2) unitary systems, and (3) packaged systems.

Unitary and packaged sytems are, by and large, complete with controls when they leave the factory. At the most, some type of space thermostat may have to be added. Their application is mostly a matter of selecting the proper functions and capacity. Central-fan systems, on the other hand, are normally built up on the job site. Each system is specifically designed for the building in which it is to be used.

The majority of the following discussion and figures concerning basic control systems is from "Basic Control Systems," Commercial Air Conditioning Controls published by Honeywell, Inc. and used with permission.[1] It includes Figs. 16-1 through 16-26 with the associated descriptive matter.

16-2 CENTRAL-FAN SYSTEMS (Elementary)

Use of a central air-conditioning system requires some means of moving air throughout the building. "Ventilation," technically defined, is "the process of supplying or removing air by natural or mechanical means, to or from any space." Such air may or may not have been conditioned. This air may be either outdoor air, recirculated air, or a combination of the two, depending on the needs of the building and its occupants.

Many of the same principles that are used in central-fan systems are the same, or similar to, those used in unitary and packaged systems. The special features associated with unitary and packaged systems will be covered later in this chapter.

VENTILATION: The most elementary ventilation system is the one that recirculates building air. No special provision is made for the introduction of outdoor air except infiltration. This is the ventilation system typically used in a residential warm-air heating system. Sufficient fresh outdoor air normally leaks into a house to meet the needs of the occupants. The purpose of bringing in outdoor air is to replenish the oxygen supply and dilute airborne contamination. When these needs are fairly small, as in a house, they usually can be met by natural infiltration of outdoor air. Figure 16-1 illustrates schematically a duct system and fan handling recirculated air only.

In larger systems, however, definite arrangements normally must be made to mechanically bring outdoor air into the building. Building codes usually specify a definite amount of outdoor air to be used. One simple way of doing this would be to use a duct sized to admit the correct ratio of outdoor air (see Fig. 16-2). Theoretically, this system would do the job of getting outdoor air into the building. It does, however, have some practical limitations in actual practice.

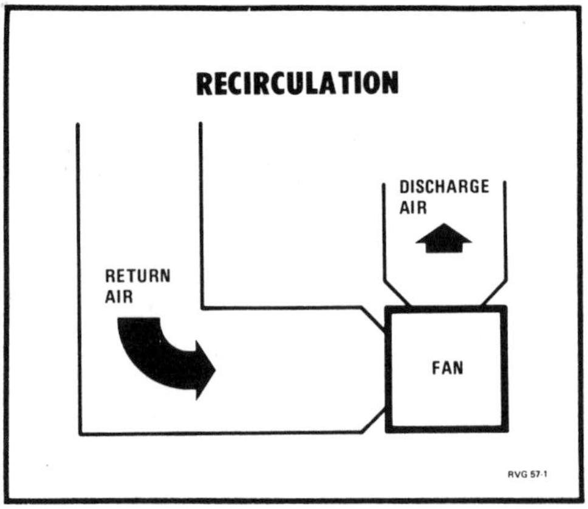

Figure 16–1

Recirculation

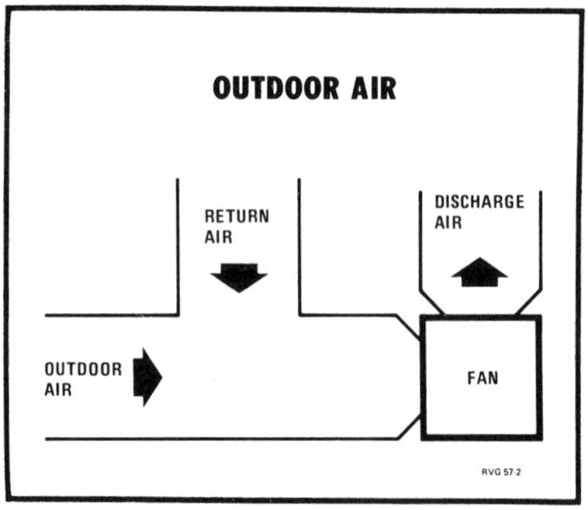

Figure 16–2

Outdoor Air

A practical system should have a means of closing the outdoor air entrance when the system is not in use. The system illustrated in Fig. 16-3 provides this feature. In this system a damper has been installed in the outdoor air inlet. The actuator that controls the damper is powered through the fan starter. When the fan is turned on, the damper opens; when the fan is turned off, power to the damper actuator is interrupted allowing the spring-loaded actuator to close the damper. Here, the amount of outdoor air brought in is still constant, but a means is provided to close the outdoor air entrance.

A further refinement to the ventilation system is the addition of a means to vary the amount of outdoor air brought in or a means to control the ratio of outdoor air to return air (see Fig. 16-4). In this system dampers are installed in both the return and outdoor air ducts. The dampers are connected together by a linkage so that as one opens the other closes. As in the previous system, the power to the damper actuator is controlled by the fan starter so

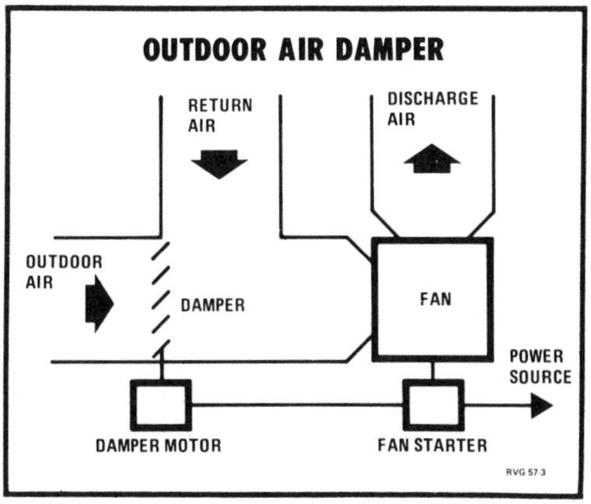

Figure 16–3

Outdoor Air Damper

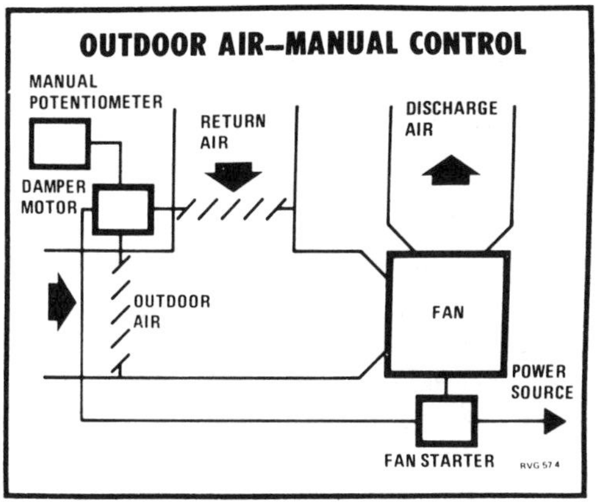

Figure 16–4

Outdoor Air-Manual Control

that the outdoor air damper will close when the system shuts down (also the return air damper opens). A manual potentiometer (electric) is used to control the position of the damper actuator, and thus the two dampers. The potentiometer can be located in a convenient location where the amount of outdoor air can be regulated to meet the needs of the occupants.

Exhaust. Some of the previous systems examined have brought outdoor air into the building, but so far we have not considered removing or exhausting the air. Typically, an exhaust fan and exhaust air damper are connected into the ventilation system, as indicated in Fig. 16-5. The ventilation system shown here is similar to Fig. 16-4 in that it provides a manually adjustable quantity of outdoor air. To this has been added an exhaust fan and exhaust damper. The provision of these added components allows the amount of air exhausted to be adjusted according to the amount of outdoor air used. Normally, it is desirable to maintain a slight positive pressure within the building. The relationship between the position of the outdoor air damper and the exhaust air damper controls the amount of positive pressure. The two damper actuators, for outdoor air and exhaust air, are connected so they operate together. They are also both spring loaded to close the dampers when the system is shut down.

The ventilation systems we have considered so far have had either fixed or manually adjustable arrangements for bringing in outdoor air. In many applications it is necessary or desirable to provide automatic control of the ventilation system. One common method is to put an automatic controller in the mixed airstream (see Fig. 16-6). It is used to adjust the outdoor and return air dampers to maintain a given mixed-air temperature. This system, a winter ventilation control, admits more outdoor air to lower the mixed-air temperature. The manual potentiometer permits a certain amount of outdoor air to enter regardless of the demands of the mixed-air controller.

The next step in the evolution of a ventilation system is to limit the amount of outdoor air used during the summertime. The system shown in Fig. 16-6 uses more and more outdoor air as the mixed-air temperature exceeds the controller set point. This may require additional cooling system operation to cool the hot outdoor air while already cool return air is being exhausted.

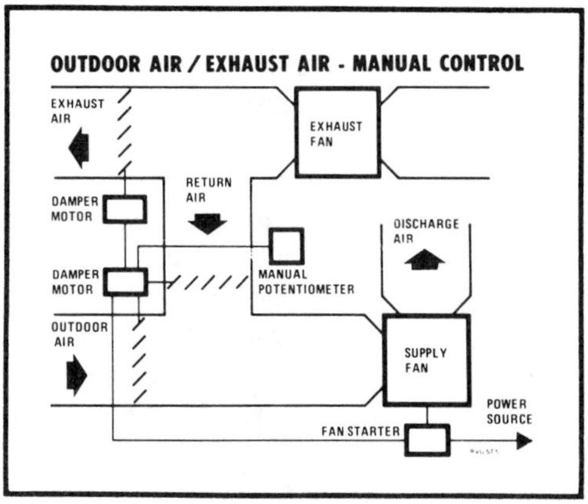

Figure 16–5

Outdoor Air/Exhaust Air—Manual Control

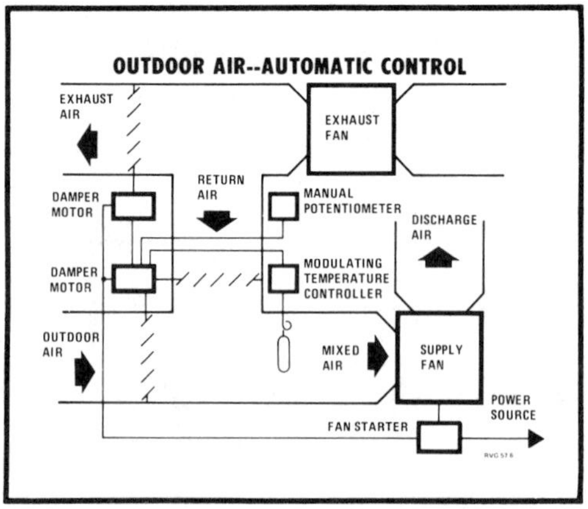

Figure 16–6

Outdoor Air—Automatic Control

The answer, then, is to limit the amount of outdoor air brought in when it is very warm outdoors. The system that does this is called an economizer ventilation control system and is shown schematically in Fig. 16-7. The only difference between Fig. 16-7 and Fig. 16-6 is the addition of a control sensor in the outdoor air. Now, when the outdoor air temperature rises too high, the outdoor air controller acts to close the outdoor air damper to the minimum position. The minimum position is set by the manual potentiometer.

Additional ventilation control functions and features may be provided in more complex systems. However, the systems covered here demonstrate the basic control functions that are necessary to ventilation systems. Additional control functions that tie the ventilation system into the heating and/or cooling system will be discussed later.

HEATING SYSTEMS: There are a number of heating systems used in central-fan-type air-conditioning systems. The ones to be considered in this discussion include:

a. steam coil

b. hot water coil

c. gas or oil-fired duct heater

d. gas direct-fired makeup air heater

e. electric duct heater

Methods of control. A control system may be designated by the controlled medium and the location of the sensors. Normally, the controlled medium will be hot water, steam, or air. The system sensors will be placed to control one of these media from one of the following locations:

1. *Space control.* The primary sensor (thermostat) is located somewhere in the controlled space. The controlled medium is adjusted to meet the requirements of the space.

2. *Control from discharge air* (medium control). The primary sensor is located in the discharge air duct. The controlled medium is adjusted to maintain constant discharge air temperature.

3. *Control from outdoor air.* The primary sensor is located either outdoors or in the duct brining in outdoor air. The control works as a changeover switch or else controls a preheater coil.

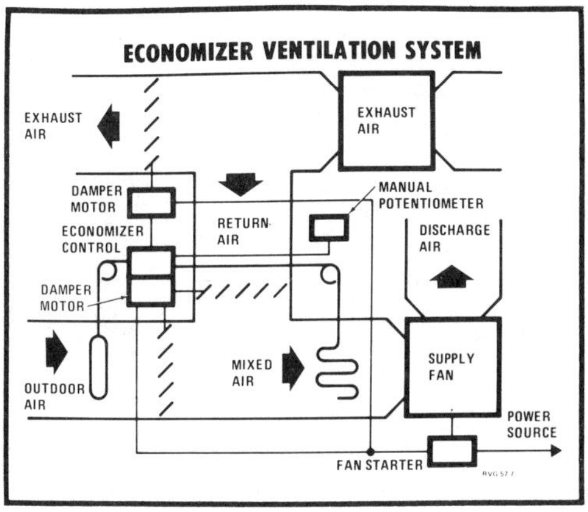

Figure 16–7

Economizer Ventilation System

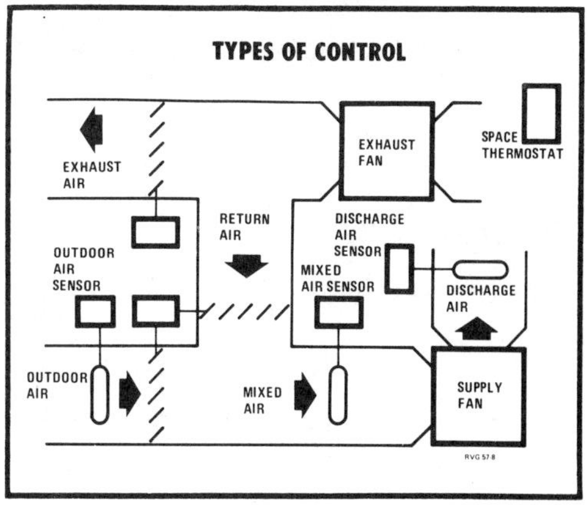

Figure 16–8

Types of Control

In typical systems two or more of these types of control may be combined; for example, space control of a steam valve with a discharge air low limit or control from discharge air with reset by outdoor air temperature. Figure 16-8 shows the relative locations of the control sensors.

<u>Steam coils.</u> Steam is generated in a boiler at a central location and distribution to one or several central systems within the building. Distribution is by means of supply and return mains at supply steam pressures of 2 to 15 psig, with 5 psig most common.

Normally, steam coils are designed for a maximum air temperature rise of 35°F through the coil. Two or more coils in series (in airstream) may be used where a greater temperature rise is required.

Steam coils may be used with either an on-off (two-position) control valve or a modulating control valve. A steam coil with a modulating control valve is used only in applications where the air entering the coil is always above freezing. This eliminates the possibility of coil freeze-up under light-load operating conditions.

One relatively simple heating system for a central fan system consists of a steam coil controlled by a sensor in the discharge air and a space thermostat, as illustrated in Fig. 16-9. In this illustration part of the ventilation system discussed previously is omitted for simplicity. This omitted part would have a means for bringing in outdoor air and exhausting air, using the associated controls already discussed.

In Fig. 16-9 the flow of steam to the coil is controlled by a modulating valve in the steam supply line. This valve is connected to a thermostat in the conditioned space and is run open or closed to maintain the desired space temperature. As a precaution against discharging uncomfortably cold air into the building, a low-limit sensor has been installed in the discharge airstream. The low-limit controller will partially open the steam valve, even though the space thermostat is not calling for heat, to prevent the discharge air temperature from dropping below 65°F to 70°F.

As stated before, since this system employs a modulating valve to control the steam coil, the temperature of the mixed air entering the coil must not be allowed to drop below freezing. In applications where an unusually large temperature rise through the coil is required, two steam coils may be used.

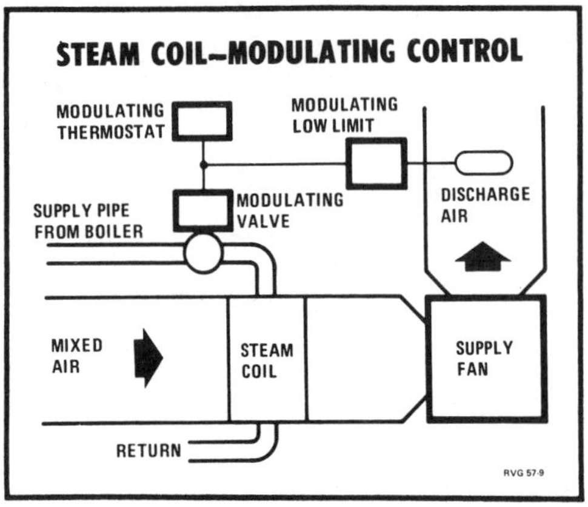

Figure 16–9

Steam Coil—Modulating Control

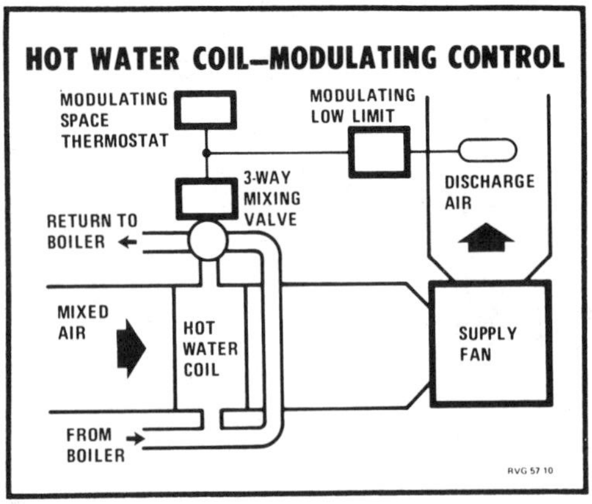

Figure 16–10

Hot Water Coil—Modulating Control

Water coils. As with the steam coils, the water coil is supplied by a central hot water boiler. Water temperatures commonly used range from 180°F to 220°F. Water coils may be controlled by either an on-off (two-position) or a modulating control valve. A system using a hot water coil controlled by space and discharge air temperature looks substantially the same as the steam coil system and is shown schematically in Fig. 16-10. In this case hot water is pumped from the boiler to the central system. The control valve is a modulating, three-way mixing valve. In other words it can adjust the proportion of the water allowed to go through the coil. As the space thermostat calls for more heat, the valve opens to let more supply water through the coil and less through the bypass. The low limit in the discharge air functions, as in the steam system, to prevent cold air from being blown into the building.

Face-and-bypass dampers. Another heating system using a hot water or steam coil is the face-and-bypass system, illustrated in Fig. 16-11. The new feature demonstrated by this sytem is on-off (two-position) control of the coil and modulating control of the air flow through the coil.

The mixed-air flow is divided into two duct sections, one of which contains the steam or hot water coil. Dampers at the entrance to each section are connected by linkage to work in opposition to each other. In other words, when one opens, the other closes. The actuator that drives the face-and-bypass dampers is controlled by a space thermostat. As the space temperature drops, the actuator drives the face damper open and the bypass damper closed. More air goes through the heating side of the duct, increasing the discharge air temperature.

The two-position control valve on the steam or hot water coil is controlled by an auxiliary switch on the modulating damper actuator. When the face damper (in front of the coil) starts to open, the valve opens. It remains fully open as long as the face damper is open any amount. As with the preceding heating systems controlled by space temperature, the modulating low limit prevents the temperature of the discharge air from getting too low.

Gas direct-fired makeup air heaters. A direct-fired makeup air heater is similar to a duct heater except that combustion takes place in the air being circulated. These heaters are fueled by gas and use 100 percent outdoor air.

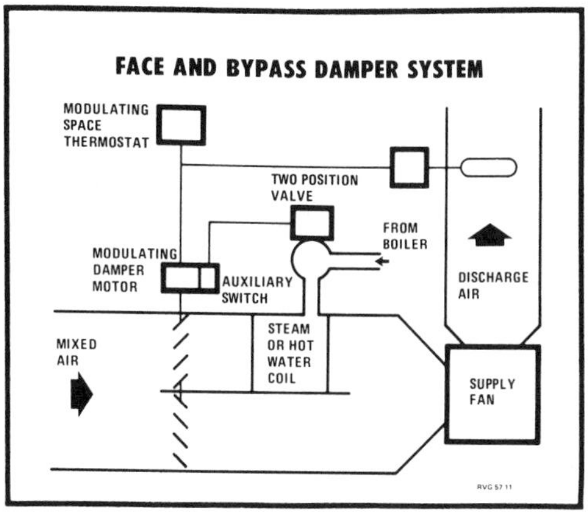

Figure 16–11

Face and Bypass Damper System

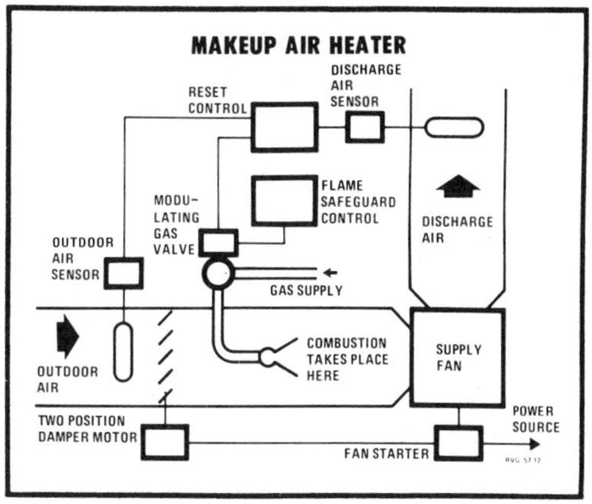

Figure 16–12

Makeup Air Heater

Products of combustion are mixed with the outdoor air and vented into the controlled space. Makeup air heaters are used to replace the air exhaust from a building to remove contamination such as that from some industrial process. Makeup air heaters are supplied by the manufacturer complete with controls, in a wide range of sizes. A typical schematic control setup is illustrated in Fig. 16-12.

Gas supply to the burner is controlled by a modulating valve that allows the output of the burner to be adjusted automatically. The gas valve, and in turn the burner output, are controlled to maintain a constant discharge air temperature. To compensate for increasing offset, as the system approaches full capacity, an outdoor reset control is sometimes used. The reset control raises the discharge air temperature control point as the outdoor air temperature drops. In this way the discharge air temperature remains the same whether the system is under light or heavy load.

Duct heaters. A duct heater, sometimes called duct furnace, is installed directly in the distribution duct of a central air-conditioning systems. They are supplied by the manufacturer complete with controls already installed. Duct heaters may be either gas or oil fired. Gas-fired units are easily vented through the roof of a single-story building and are most common in applications such as supermarkets, garages, and warehouses. Basically, a gas-fired duct heater is the same as a residential gas-fired furnace.

Electric heaters. Electric heaters consist of one or more stages of resistive heating elements in a duct or central furnace. One method of using electric heating in a central system is to put a heating element in the discharge duct leading to each room or zone. This heater is controlled by a space thermostat. Electric duct heaters of this type provide the advantage of individual room control and at the same time allow the use of the central air conditioning, air cleaning, outdoor ventilation air, and humidification.

Figure 16-13 illustrates a system using electric resistance heating. To prevent the discharge of very cold air to conditioned spaces, a heating element is sometimes installed at a central location. This heater brings the mixed-air temperature up to at least 65°F to 75°F when necessary. Even with the zone heaters turned off, the discharge air will not get uncomfortably cold.

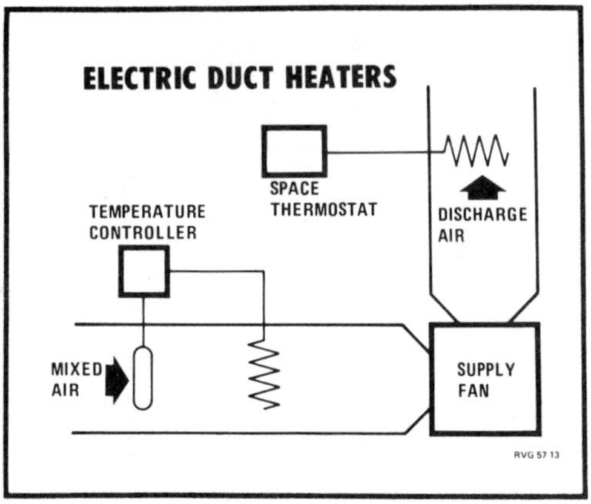

Figure 16–13

Electric Duct Heaters

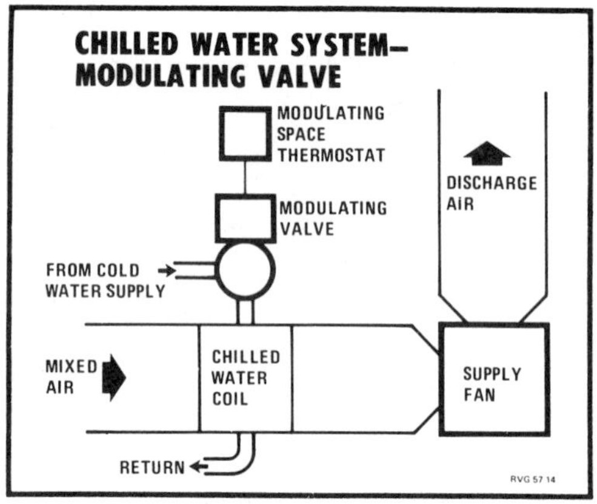

Figure 16–14

Chilled Water System—Modulating Valve

COOLING SYSTEMS: The means of cooling used in central-fan systems fall into two general categories: chilled water and direct expansion of a refrigerant.

A *chilled water* system customarily uses a mechanical cooling system (vapor compression) at a central location within the building. The vapor compression system (chiller) is used to cool the water which is circulated through piping to various parts of the building where it is used in central ventilation systems or some other type of cooling system. In a *direct expansion* system the mechanical cooling equipment is located at or near the ventilating system. The evaporator coil is mounted in the duct and directly cools the air that passes through it.

Modulating chilled water valve. One relatively simple chilled water system varies the air temperature by throttling the water flow through the coil, as illustrated in Fig. 16-14. A constant volume of air passes through the coil. A modulating-type valve in this system is controlled by a modulating space thermostat. This is the same type of thermostat used in a heating system; however, in this example, it is arranged to open the valve on a rise in space temperature, instead of the other way around.

As with hot water coils, as shown in Fig. 16-10, the same type of control system can be used with a three-way valve. Here the valve controls the relative amounts of chilled water going through the coil and bypass.

Face-and-bypass dampers. This cooling system (Fig. 16-15) is very similar to the face-and-bypass controlled heating system we have already discussed. The difference is the system's reaction to a change in space temperature. An increase in air temperature at the modulating space thermostat causes the actuator to drive the face damper toward the open position. As space temperature drops, the face damper closes and the bypass damper opens. When the face damper is fully closed, the auxiliary switch in the actuator opens and allows the two-position valve to close, stopping the flow of chilled water to the coil. When the face damper just starts to open again, in response to a need for cooling, the water valve opens all the way.

Direct expansion cooling systems. The least complicated type of central cooling system is the on-off control of a single compressor from

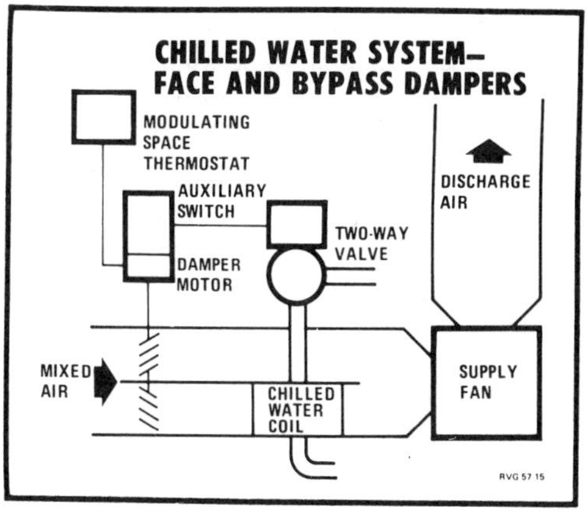

Figure 16–15

Chilled Water System—Face and Bypass Dampers

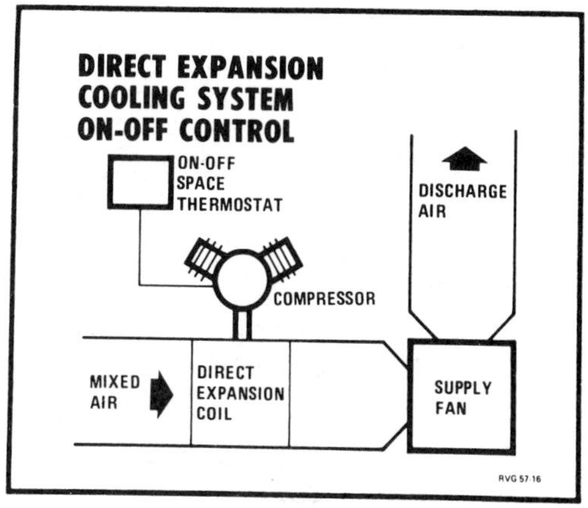

Figure 16–16

Direct Expansion Cooling System ON-OFF Control

space temperature, as illustrated in Fig. 16-16. Another direct expansion cooling system with additional capabilities is illustrated in Fig. 16-17. This system provides two stages of cooling by using two coils, two compressors, and a two-stage thermostat. The first-stage compressor is brought on when the thermostat first senses the requirement for cooling. The thermostat will cycle the first compressor on and off as long as it is able to meet the cooling needs of the space. When continuous first-stage operation can no longer keep the space cool, the second stage is brought on. This type of cooling system provides a means of adjusting the amount of cooling available to meet the needs of the conditioned space.

Other multiple-stage cooling systems made use of a means to control the capacity of the compressor. One common method is to bypass hot gas to the suction side of the compressor. In this way a single compressor, sized for the maximum load, can operate at reduced output for lighter loads.

When more than two stages of cooling are employed, it is common to use a sequencer to actuate additional cooling stages as the load increases. In the system illustrated in Fig. 16-18, four stages of cooling are used. The system

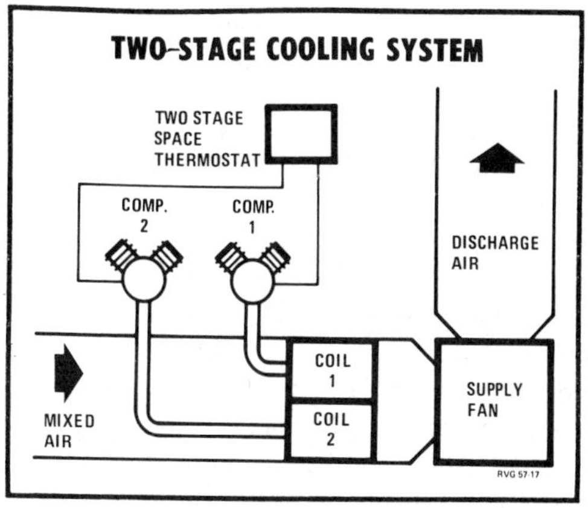

Figure 16–17

Two-Stage Cooling System

has two compressors, each with unloading, or capacity control. The four stages, in order of increasing capacity, are arrived at in the following manner:

a. stage one: compressor one at reduced capacity

b. stage two: compressor one at full capacity

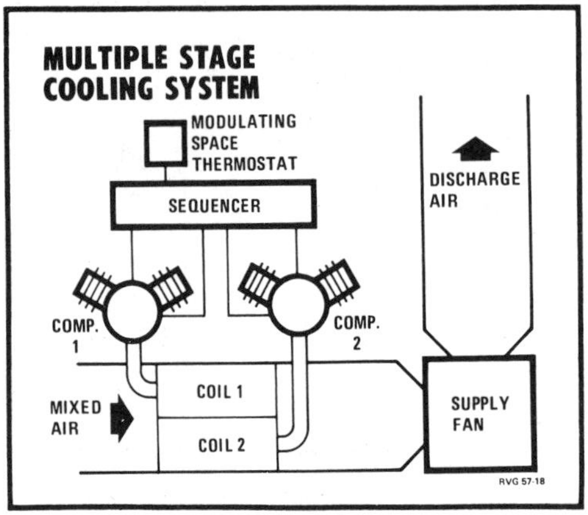

Figure 16–18

Multiple Stage Cooling System

c. stage three: compressor one at full capacity, and compressor two at reduced capacity

d. stage four: both compressors at full capacity

The control used here is a modulating space thermostat along with a sequencing device. As the cooling load in the conditioned space is sensed by the thermostat, the sequencer is advanced to bring on more stages of cooling. The exact interstage differential can be adjusted to meet the requirements of the system and controlled space.

HUMIDIFICATION: Humidification is the process of increasing the water vapor content of the conditioned air. As such it requires a source of heat (latent heat of evaporation). This heat may come from an external source, or from the air. In either case humidification has an effect on the total heat of the air, sometimes a significant effect on the dry-bulb temperature of the air (this was discussed in Chapters 4 and 5). It is important, therefore, that the heating system and the humidification system be designed to operate together.

Humidifiers are usually located downstream from a heating unit of some kind in a central system. This means that the air in contact with the humidifier will be warmer and able to absorb the amount of moisture necessary to meet the needs of the conditioned space. There are four basic types of humidifiers commonly used with central-fan systems:

a. open pan type

b. steam jet or injection type

c. water spray or atomizer type

d. water or air washer type

A *pan-type humidifier* consists of an open pan of water located in the duct of a central air-handling system where the water surface is open to the airstream. The pan also contains a means of heating the water, such as a steam coil, hot water coil, or an electric immersion coil, and a means for controlling the water temperature by varying the heat input. The pan must also have an automatic means for maintaining water level in the device. Rate of evaporation, humidifier capacity, is determined by the water temperature in the open pan. In the pan type of humidifier illustrated in Fig. 16-19, the steam coil in the water pan is controlled by a modulating valve. A modulating space humidistat is used to sense room humidity requirements and adjust the valve accordingly. The valve is interconnected to the system supply fan so that it will shut off the steam supply when the central system is off. Pan-type humidifiers have a limited capacity and are used only when humidification requirements are fairly small.

A *steam jet,* or injection type, humidifier discharges steam directly into the airstream. This system is capable of high capacity and accurate control and is illustrated in Fig. 16-20. A source of clean steam is very important to a steam jet humidifier. Boiler compounds in the

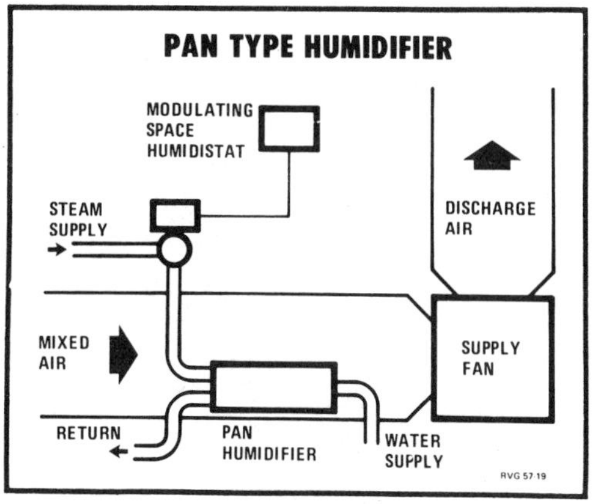

Figure 16–19

Pan Type Humidifier

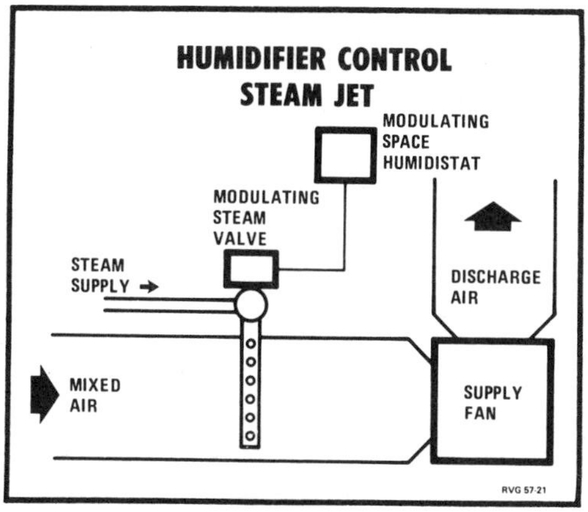

Figure 16–20

Humidifier Control Steam Jet

water will usually cause an objectionable odor. All the steam sprayed into the air is absorbed. Humidification by the steam jet method usually results in a slight rise in air dry-bulb temperature (1°F to 2°F). The steam flow is controlled by a modulating valve which, in turn, is controlled by a modulating space humidistat to maintain the desired relative humidity in the space. The steam valve is interconnected to the fan so that the valve will close when the fan is turned off.

A *water spray,* or atomizer, humidifier (Fig. 16-21) consists of a series of nozzles in the duct, through which water is sprayed into the air. The fine mist produced is completely absorbed by the air. Uniform water pressure is necessary to provide proper atomization of the water. For this reason this type of humidifier is controlled only on an on-off mode of control.

The evaporation of water removes a significant amount of sensible heat from the airstream. As the water spray is turned on and off, heating requirements vary widely. This makes close temperature control difficult.

The *air washer* type of humidifier, illustrated in Fig. 16-22, also sprays water into the airstream, but *is not* designed to evaporate all the water. Air leaving the washer is nearly saturated so that it is necessary to provide a means of *reheat* to get the correct temperature and humidity. The essential parts of an air washer include:

a. a mechanism (pump) for circulating and spraying water through the airstream
b. an eliminator to remove water droplets from the airstream before reaching the heating coil or fan
c. a heating coil which is used to adjust the discharge air temperature

Other arrangements of equipment in air washer types of humidifiers provide a preheat coil and/or heat exchanger to heat the spray water.

16-3 UNITARY SYSTEMS (Elementary)

A unitary system is a room unit used in conjunction with a central system. Unitary systems are usually classified according to the heating medium delivered to them by the central system. Three groups are:

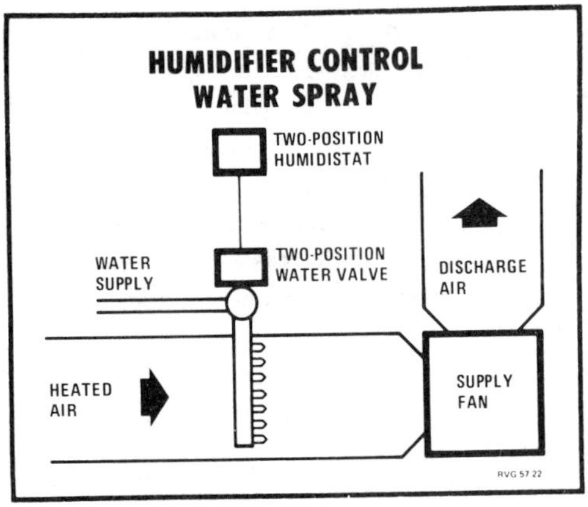

Figure 16–21

Humidifier Control Water Spray

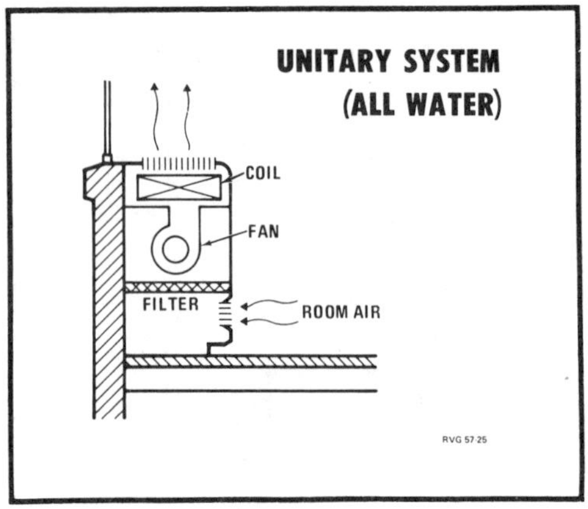

Figure 16–23

Unitary System (all Water)

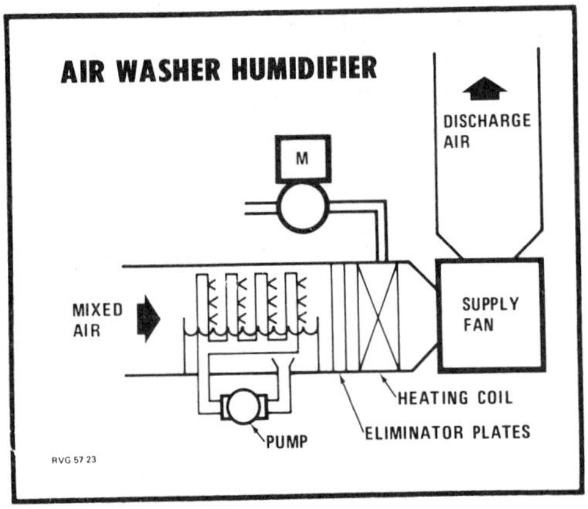

Figure 16–22

Air Washer Humidifier

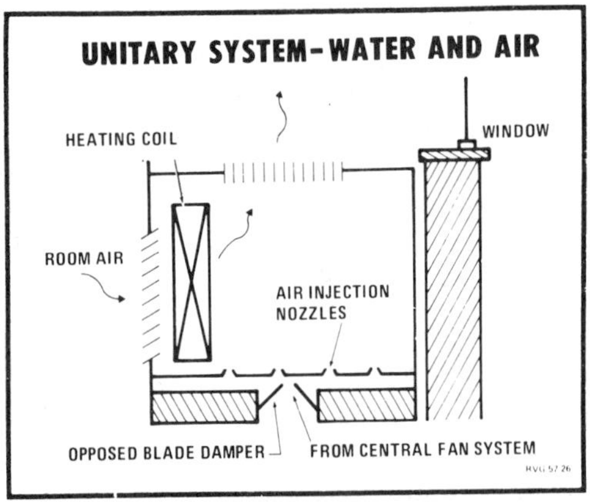

Figure 16–24

Unitary System-Water and Air

a. all water and no air

b. part water and part air

c. all air and no water

There may, or may not, be a central-fan system associated with the unitary equipment. A unitary system (Fig. 16-23), in effect, moves part of the equipment and some of the air-conditioning functions from the central location to the conditioned area. An average of this arrangement is that the air need not be returned to the central system for conditioning. The only air that must be distributed through the building is ventilation (outdoor) air. Sometimes the function is also accomplished by the unitary system, called a unit ventilator, eliminating the need for any central air distribution system.

ALL-WATER AND NO-AIR SYSTEMS: Typically called a *fan coil,* this type of

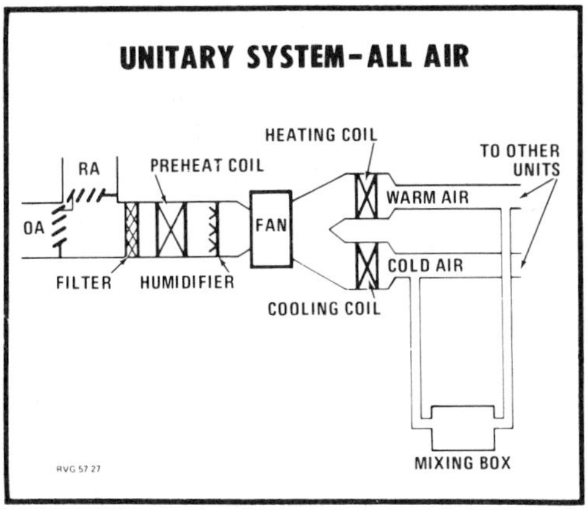

Figure 16–25

Unitary System-All Air

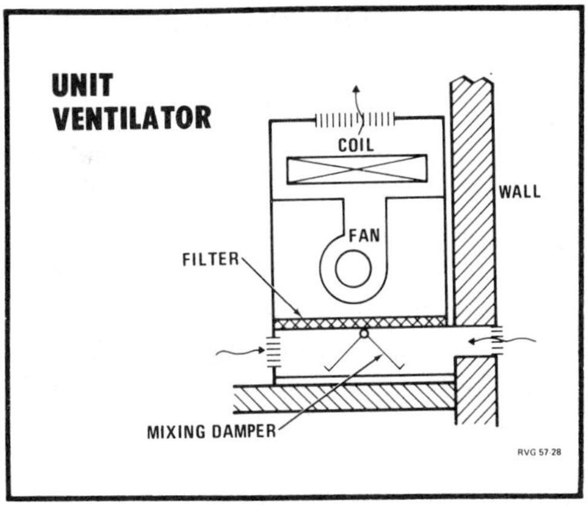

Figure 16–26

Unit Ventilator

unit is normally installed under a window, although many other arrangements are available. A common way of controlling a fan coil unit is to use a room thermostat to control a valve leading to the coil and operate the fan manually by a start and stop switch (frequently with speed selection). Quite a number of other control systems are also available. When the unit is used for both heating and cooling, a multiple-speed fan can be used to provide the required air circulation rate in each mode of operation.

PART-WATER AND PART-AIR SYSTEM: This type of unitary system (Fig 16-24) receives both air and hot water or steam from a central system. They are also called *induction systems,* either high pressure or low pressure. Air from the central fan system induces the flow of room air through the unit. No fan is used in the unit.

ALL-AIR AND NO-WATER SYSTEM: These systems condition the air at a central location and deliver it through two ducts, one cold and one warm, to a mixing unit in the room. The system (Fig. 16-25) is usually high pressure, high velocity, and frequently is called "high-velocity, dual duct." The mixing box is located in the room and mixes the hot and cold air as required to maintain the desired room temperature. A room thermostat controls a pair of coupled dampers in the two ducts to control the amount of hot and cold air admitted to the room.

Because distribution of condition air is at high pressure and high velocity, small size ducts can be used in the system, saving space and expense. However, silencers must be used to reduce the noise level of the high-velocity air.

UNIT VENTILATORS: Unit ventilators are similar to fan coil units but have the added feature of ventilation (outdoor air) capability, see Fig. 16-26. They generally contain a set of outdoor and return air dampers, coil, fan, and filters. The dampers are arranged to provide varying amounts of outdoor air for ventilation, depending on the load. The controls, usually supplied with the unit, are designed to provide a definite sequence of operation. The units are almost always located on outside walls under windows, but other arrangements are available. Some are designed to provide cooling, using chilled water.

16-4 PACKAGED SYSTEMS (Elementary)

The term *packaged systems* applies to a complete set of components for air conditioning, or one or more air-conditioning functions, manufactured and shipped as a unit for easy installation. Included are warm-air furnaces, boilers, and unit air conditioners. A *packaged multizone* unit is typically installed on a rooftop, and provides heating, cooling, and ventilation for a number of different zones within the building.

16-5 TYPICAL AUTOMATIC CONTROL SYSTEMS

In the preceding sections we have considered some of the possible elementary automatic control systems. In this section we will discuss some of the more commonly found complete systems using representative diagrams with descriptions of such things as special features and control sequences.

FORCED-CIRCULATION WARM-AIR FURNACE: The automatic controls for a forced-circulation warm-air furnace may fall in the category of packaged equipment because most of the controls are installed and wired at the factory, with the exception of the space thermostat. An almost endless number of control arrangements is possible, but the simplest and most practical is that shown in Fig. 16-27. The furnace may be gas fired or oil fired. The control components consist of:

1. *Room thermostat* which should be within the house in a central location where a representative temperature exists. The thermostat should be provided with a differential adjustment setting not to exceed about 1.0°F.
2. *Limit control* for shutting down burner whenever the bonnet air temperature becomes excessively high. The recommended setting is between 175°F and 200°F. In the normal control sequence the limit control is not called upon for action but is available in case of possible overheating of the furnace (safety control). Overheating may occur if air flow through the unit is restricted, perhaps by a partially clogged filter.
3. *Burner motor* is directly connected to the room thermostat circuit. Obviously, the

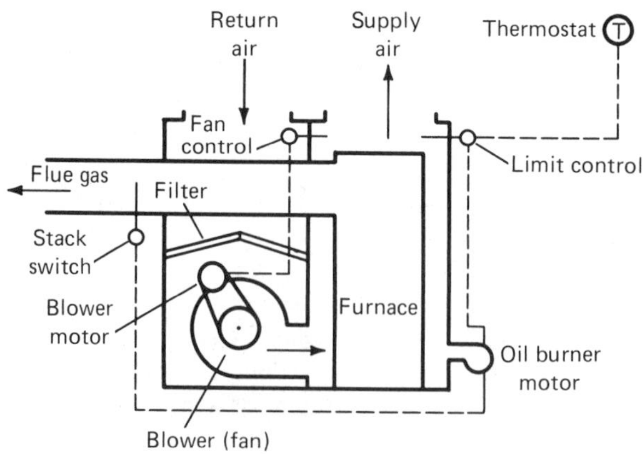

Figure 16–27

Control components for warm air heating system

frequency of operation of the burner is dependent upon the setting and sensitivity of the room thermostat. In the warm-air furnace system, unlike a hot water or steam system, there is little lag and at the same time no heat storage. In other words the heat must be generated only as needed and must be distributed as soon as it is generated. For this reason the sensitivity of the room thermostat must be acute, and the differential settings must be of the order of 0.5°F to 1.0°F. The purpose of the small differential setting is to cause frequent and short operations of the burner. The length of each burner cycle should not, however, be less than about 2 min. for a gas burner nor less than about 4 min for an oil burner because shorter operating periods might contribute to large combustion losses due to incomplete combustion at the beginning of each burner operation.

4. *Fan control* located in the supply air bonnet is a device that turns on the blower motor when the air temperature in the bonnet is high and turns off the blower motor when the bonnet air temperature is low. The recommended setting of the fan control is about 110°F for the cut-in point and about 85°F for the cut-out point. These settings are adjustable, and are of great importance in the successful operation of the entire system.

5. *Stack switch* is another safety device installed in the unit to delay the operation of the burner after a general power failure has occurred. If the power was immediately restored, and if the burner had been in operation, oil sprayed into the combustion chamber could explode.

The normal *control sequence* for the forced warm-air system would be as follows: when room thermostat calls for heat, the oil burner starts; when the bonnet air temperature reaches the cut-in point for the fan control, the fan starts; when room temperature reaches thermostat set point, the oil burner stops. The fan continues to operate until the bonnet air temperature drops to the cut-out point of the fan control.

A statement was made in the section concerning the fan control that settings were important in the successful operation of the system. With a forced-air system, the flow of air is not dependent upon the weather or the heating demands but is fixed only by the blower speed, the blower size, and the frictional restrictions in the duct system. It is possible with any given duct system to vary the air flow rate over a considerable range. Furthermore, it is possible to control over a wide range the number of hours that the blower will operate during the day. For example, on a day in which the outdoor air temperature is 30°F, the fan control can be adjusted so that the blower runs continuously during the 24-hr period, or for 16 hr, or for as little as 8 hours. This wide range of blower operating periods is dependent mainly upon the settings of the fan control, and to some extent upon the blower speed. Longer operating periods of the blower are desirable since, when the blower is operating, the control of distribution of heated air is possible; whereas when the blower is not operating, and the system is under the influence of gravity action, no control exists as far as air distribution to the various rooms is concerned.

The argument might be raised that, if continuous air circulation is desirable, the system should be operated to give continuous blower operation by replacing the fan control with a manual switch, which would be used to turn the blower on at the start of the heating season and have it run continuously during the season. Or, the fan control could be left in place and the cut-in point lowered to about 70°F. The two possible objections to this arrangement are:

a. that the blower does not need to be operated in extremely mild weather when little heat demand occurs, and the electrical cost of blower operation could be saved

b. that the temperature of the air issuing from the registers can be as low as room air temperature and might create objectionable drafts in front of the registers

With high sidewall registers, with floor registers, or with baseboard registers that discharge the air sharply upward toward the ceiling, this draft possibility does not exist. With such register installations the fan control cut-in settings can be lowered below 110°F, and such a lower setting would prove most advantageous from the standpoint of temperature control and uniform distribution of heated air.

In conclusion, the fan control should be set to give nearly continuous fan operation during the coldest weather, with intermittent oil burner operation controlled by the room thermostat.

FORCED-CIRCULATION HOT WATER SYSTEM: Figure 16-28 represents a schematic layout of the controls normally used for the simple hot water heating system for small residential applications. Again, as with the forced-circulation hot-air system, the primary controls come with the packaged boiler installed at the factory.

The normal *control sequence* for this system would be as follows: boiler water temperature is maintained at the set point established by the aquastat (burner control); when space thermostat calls for heat, the circulator starts, which forces water through the heating system components. The circulator is also under the control of the aquastat (circulator control), which prevents circulator operation if the boiler water is not up to the aquastat setting. When space temperature is at the space thermostat set point, the circulator stops.

The flow control valve is a mechanical device that prevents gravity circulation of the water in the system when the circulator is not operating. It is particularly necessary in a

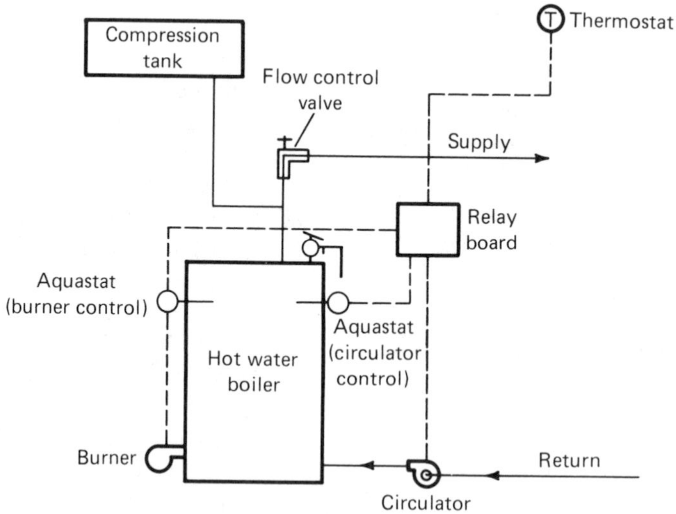

Figure 16–28

Control components for forced circulation hot water system

summer-winter system where the low-limit control aquastat (aquastat circulator control) maintains a relatively high boiler water temperature, whether heating is required or not.

ZONE CONTROL SYSTEMS: As the size of buildings increases, it becomes more difficult to provide good temperature regulation from a single thermostat. Therefore, it is advisable to split the building into two or more *zones*, each controlled by its own thermostat. The zones would be selected according to the conditioning (heating and/or cooling) load imposed by the type of occupancy or use and by the effects of different exposures to sun and prevailing winds. This might be and sometimes is done by providing a separate conditioning plant and control system for each zone. However, *zone control* implies the independent control of the conditioning medium (hot air, chilled air, hot water, chilled water, steam) to each zone from the same central conditioning plant.

The most accurate and satisfactory type of control system for any building, whether residence, school, apartment house, office or industrial building, is individual control of each conditioned space. A control system of this type includes, in each conditioned space, a thermostat which controls a heating and/or cooling device located in that space. Heating and cooling medium is supplied to these devices from a central plant. Thus, without regard to conditions in other spaces, the controller in each space regulates the amount of heating and/or cooling required by that space only. This form of control, primarily because of the number of devices required throughout the building, is usually most expensive. However, where maximum flexibility and the most accurate control are desired, individual space control provides the most satisfactory results.

Figure 16-29 illustrates the individual room control for a hot water heating unit. This type of control could be applied to each room throughout a building. It is designed for two-

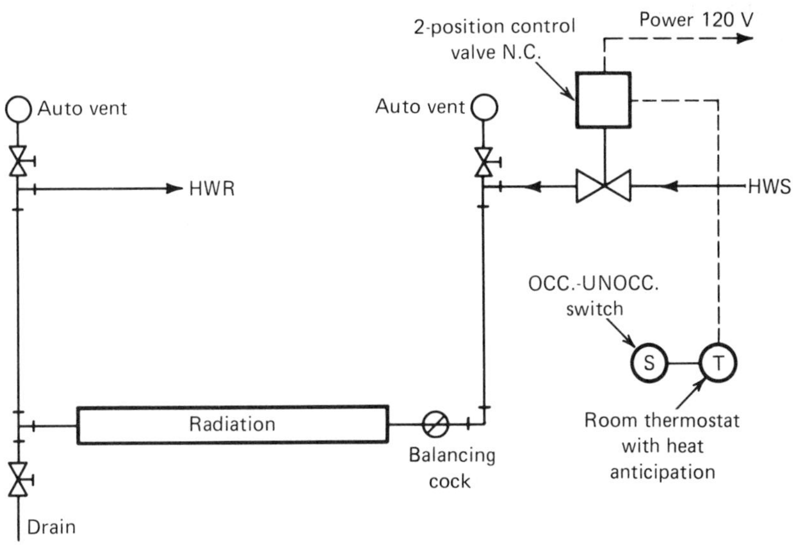

Figure 16–29

Control components for an individually controlled hot water radiation unit

position (on-off) control from a room thermostat. It has the special feature, provided by the *occ-unocc switch,* of permitting the radiation unit to be operated to maintain a lower room temperature when the space is unoccupied (nighttime, holidays, etc.). The sequence of operation would be as follows:

1. *Occupied:* A heat anticipating room thermostat will open or close the two-position motorized valve to control the room temperature at the set point.
2. *Unoccupied:* The heat anticipating room thermostat will control the two-position motorized valve to maintain a lower set point temperature (say 10°F below the occupied set point).

Zoned hot water heating system using multiple circulators or multiple-zone valves. Two popular methods for zone control of forced-circulation hot water systems for large residential or small commercial buildings are illustrated in Figs. 16-30 and 16-31. Both represent only two zones but could easily be expanded to include other zones.

Figure 16-30 represents a portion of a two-zone system using a separate thermostat and circulator for each zone. Basically, it represents two hot water systems connected to the boiler that operate independently of each other.

Figure 16-31 represents a portion of a two-zone system using a separate thermostat and two-position motorized valve for each zone and one circulator for the combined system. Each motorized control valve is provided with an auxiliary switch. The auxiliary switches are connected in parallel to the single circulating pump. A call for heat by either zone thermostat causes the respective zone valve to open and start the circulator. If neither zone is calling for heat, the two zone valves would be closed and the circulator stops.

TYPICAL RESET HOT WATER AND PUMP CONTROL (ELECTRONIC): Figure 16-32 illustrates a control system used to adjust the temperature of the supply water in a hot water heating system (upward as outdoor air temperature drops, downward as outdoor air temperature rises). This type of system is frequently used as part of the control system for hot water heating in commercial and indus-

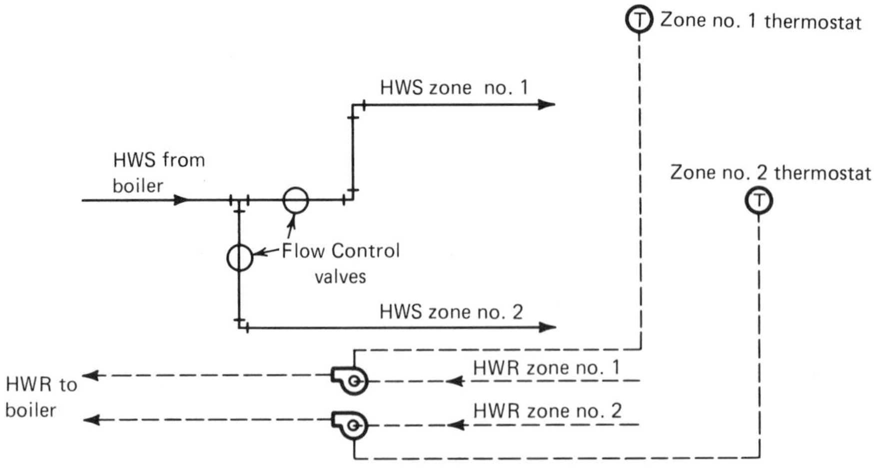

Figure 16–30

Zone control components for 2-zone hot water system (individual circulators)

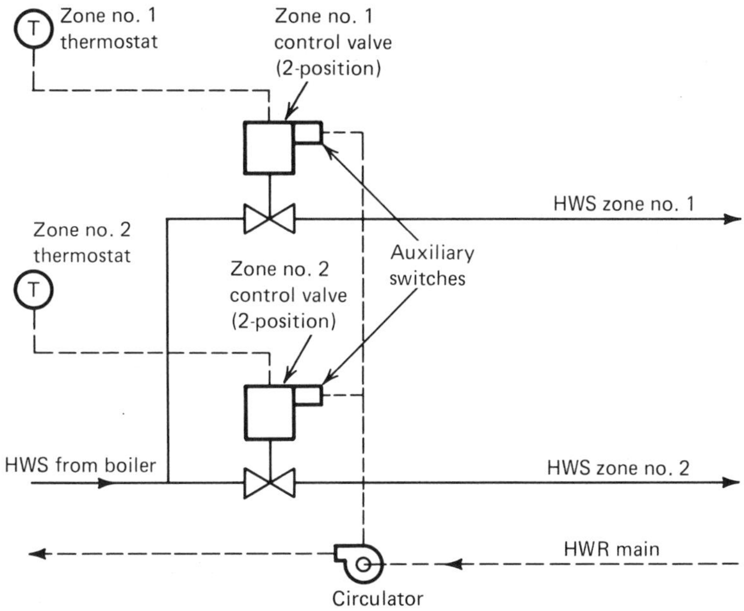

Figure 16–31

Zone control components for 2-zone hot water system (individual zone valves)

trial buildings. It is normally not used in residential applications because of the additional cost of the system. The sequence of operation of the system shown in Fig. 16-32 is as follows:

1. An outdoor sensor will reset an immersion sensor in the supply water to provide 80°F supply water at 70°F outdoor air temperature, increasing to 200°F supply water as outdoor temperature drops to −20°F. Bridge, amplifier, solid-state drive for proportioning three-way control valve and pump relay will be mounted in prewired panel.
2. Boiler water temperature is maintained at 200°F by aquastat recycling of oil burner.
3. Outdoor air temperature sensor will start hot water circulating pumps when outdoor air temperature falls below 70°F and stop pump when outdoor air temperature rises above 70°F.

CONTROL OF FAN COIL UNITS (TYPICAL): There are many types of fan coil units available, however, as the name implies, they all have a heat-transfer coil and a fan to circulate air through the coil and to distribute the conditioned air. Some units are designed for heating only, and some may be arranged for either heating or cooling.

Controls for fan coil units can either be (1) *on-off operation* of the fan or (2) *continuous fan operation* with modulation of the heating or cooling medium.

For on-off operation a room thermostat is used to start and stop the fan motor, illustrated in Fig. 16-33. A limit thermostat, often strapped to the supply or return pipe, prevents fan operation in the event that heat (heating unit) is not being supplied to the unit. An auxiliary switch that energizes the fan only when

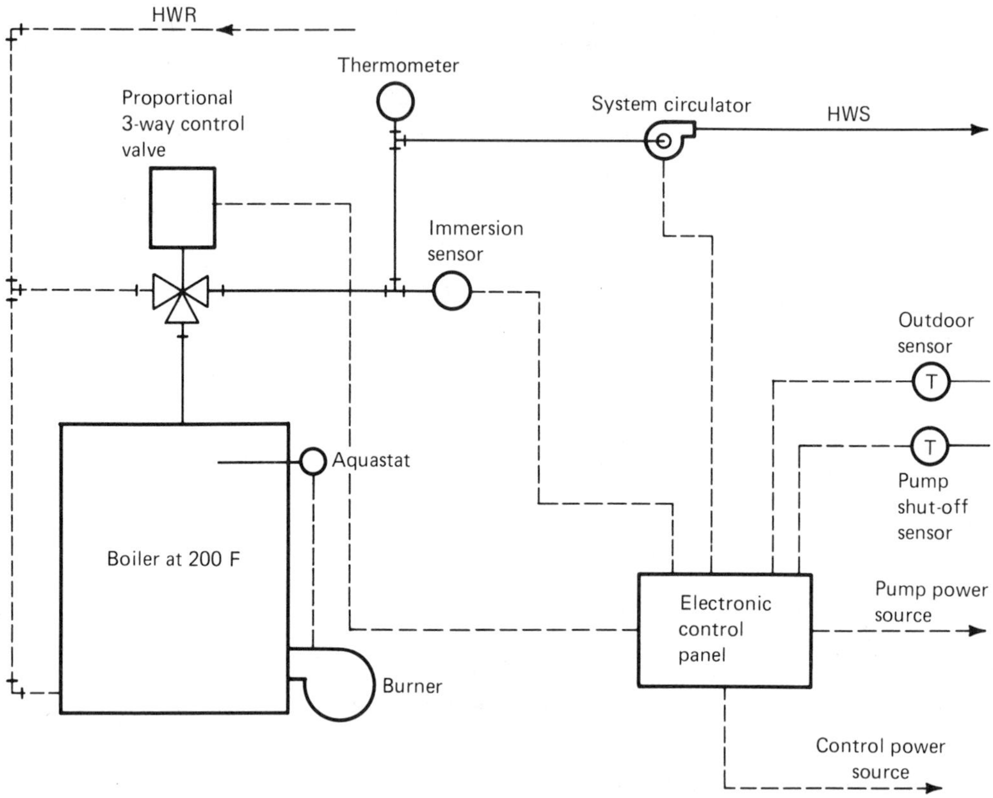

Figure 16–32

Reset hot water control system (electronic)

power is applied to open the motorized supply valve may also be used to prevent undesirable cool air being discharged by the unit.

Continuous fan operation eliminates the intermittent blasts of hot air resulting from on-off operation, as well as the stratification of temperature from floor to ceiling which often occurs during the off period. In this arrangement a proportioning room thermostat and valve modulates the heat supply to the coil, see Fig. 16-34, or a bypass around the heating element. A limit thermostat, or auxiliary switch on the control valve, stops the fan when heat is not available in the heating coil.

The unit controls shown in Figs. 16-33 and 16-34 would also apply if chilled water was supplied to the coils, assuming that the thermostat provided for heating-cooling changeover.

One type of control, illustrated in Fig. 16-35, used with blow-down unit heaters is designed to automatically return the warm air, which would normally stratify at the higher level, down to the occupied zone. Two thermostats, a two-position control valve, and an auxiliary switch are required. The lower thermostat is placed in the occupied zone and used to control the two-position supply valve to the heating element. The auxiliary switch is used to stop the fan when the supply valve is

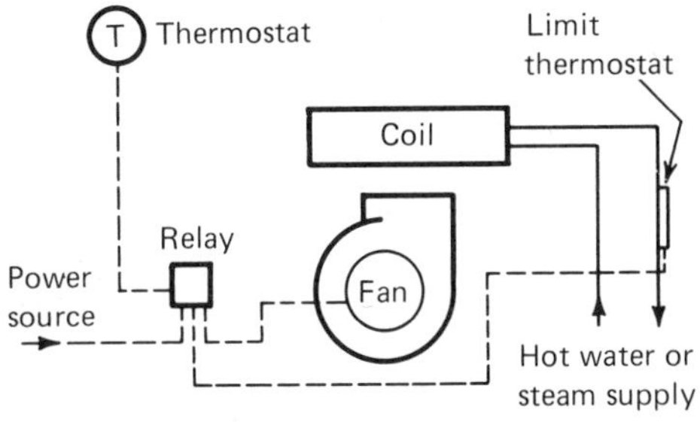

Figure 16–33

Control system for fan coil unit (cycle fan, constant heating medium flow)

closed. The higher thermostat is placed near the unit heater at the ceiling or roof level where the warm air tends to stratify. The lower thermostat will automatically close the steam (water) two-position valve when its set point is reached, but the higher thermostat will override the auxiliary switch to let the fan continue to run until the temperature at the higher level falls below a point sufficiently high to produce a heating effect.

CONTROL OF UNIT VENTILATORS (TYPICAL): For many years unit ventilators have been used for the winter heating-ventilating cycle for school classrooms. This has led to the general name *classroom unit ventilator*. They have also found application in many commercial and industrial buildings. With the addition of cooling coils and appropriate control, the unit is equipped for year-round control of the environment within building spaces.

The basic construction and essential operating elements of a typical unit ventilator are illustrated schematically in Fig. 16–36. The other types in common use are functionally the same, although they may differ slightly in the arrangement of dampers, coils, blower units, and so forth. For instance, a bypass damper may be used to bypass air around the coil when no heat is required.

Referring to Fig. 16–36, note that the unit ventilator has all the necessary elements to de-

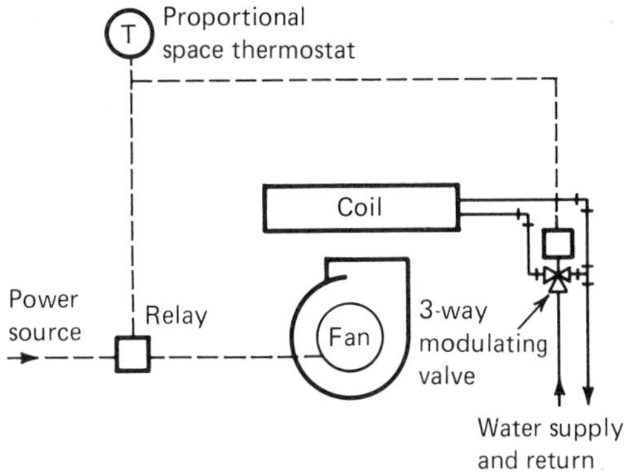

Figure 16–34

Control system for fan coil unit (constant fan operation, modulated heating medium temperature)

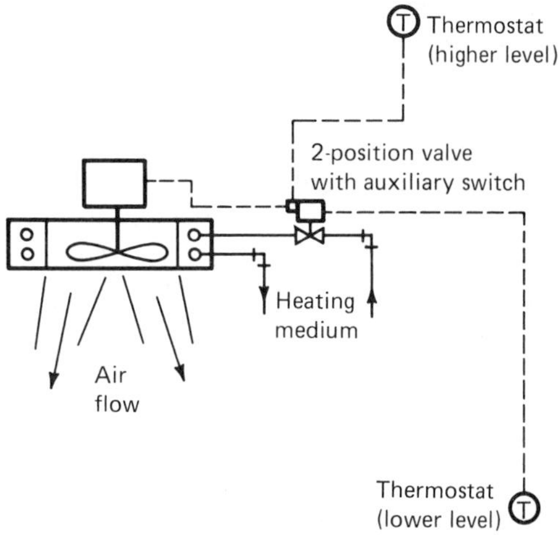

Figure 16–35

Control system for down-flow unit heater

liver air to the room at any temperature desired. At the bottom of the unit is found a dual damper, arranged so as to vary the relative volume of outdoor air and recirculated air admitted to the unit. As the outdoor air damper opens, the return air damper closes proportionally. By positioning the dampers, it is possible to admit any proportion from 0 to 100 percent

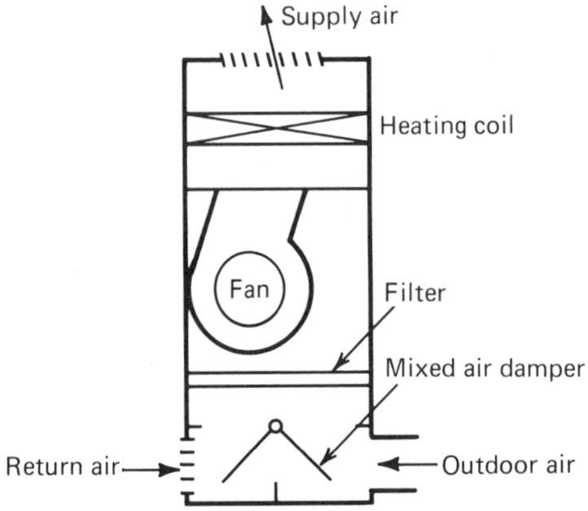

Figure 16–36

Unit ventilator components

of outdoor air to the unit. A blower unit, consisting of two or more centrifugal blowers, draws a constant volume of air through the unit. A heating coil, supplied with steam or hot water and equipped with a control valve, heats the air leaving the unit. Electric heating coils may also be used. It is obvious that changing the position of the heating coil and placing it below the fan (draw-through) does not alter its effect or change the control requirements for the unit.

Many different control cycles for unit ventilators are available. The principle difference in the various cycles pertains to the amount of outdoor air delivered to the room. Usually, a room thermostat simultaneously controls a coil valve, damper, or step controller to regulate the heat supply, and a damper to regulate the outdoor air supply. An airstream thermostat in the unit prevents the discharge of air below a desired minimum temperature. Unit ventilator control cycles provide the proper sequence for the following stages:

Warmup: All standard unit ventilator cycles automatically close the outdoor air damper whenever maximum heating capacity is required. Then, only room air is circulated, thereby providing full unit capacity for rapid warmup until the room temperature approaches the desired level.

Heating and Ventilating: As the room temperature rises into the operating range of the thermostat, ventilation is accomplished through the partial or complete opening of the outdoor air damper, depending on the cycle used. Auxiliary heating equipment (fin tube, if used) is shut off. As the room temperature continues to rise, the heat supply to the coil is throttled.

Cooling and Ventilating: When the room temperature rises above the set point level, cool air is discharged into the room. The room thermostat accomplishes this by throttling the heat supply, finally shutting it off, and then opening the outdoor air damper, as required, to prevent the room from overheating. The airstream thermostat frequently takes control during this stage to keep the discharge air temperature from falling below a set level. The unit cooling capacity is determined by the air delivery and difference between the outdoor air and room air temperatures. Cooling obtained in this manner is called "free cooling," "ventilation cooling," or "natural cooling."

ASHRAE Control Cycles: There are three basic cycles used, with variation, for unit ventilator control. They are referred to as ASHRAE Cycles I, II, and III.

ASHRAE Cycle I: During warmup, the outdoor air damper is closed, and unit handles 100 percent recirculated air. As room temperature approaches the thermostat set point, the outdoor air damper opens fully, and the unit handles 100 percent outdoor air. Unit capacity is then controlled by modulating the heating coil capacity.

ASHRAE Cycle II: This is the most widely used heating and ventilating cycle. During warmup, the outdoor air damper is closed and the unit handles 100 percent recirculated air. As the room temperature approaches the thermostat set point, the outdoor air damper opens to admit a predetermined minimum amount of outdoor air (usually between 25 to 50 percent). This minimum is determined by local code requirements and good engineering practice to provide adequate ventilation. Unit capacity is controlled by varying the heating coil output. If the room starts to overheat, the heating coil is first turned off, and then a gradually increasing amount of outdoor air is admitted until only outdoor air is being delivered.

This cycle usually incorporates a minimum discharge air temperature sensor that overrides the other controls to maintain an acceptable discharge temperature. When the outdoor air is very cold, the minimum air temperature control modulates the amount of outdoor air being delivered. This keeps the mixture temperature delivered to the room at 55°F to 60°F.

ASHRAE Cycle III: During warmup, the outdoor air damper is closed and the unit handles 100 percent recirculated air. As room temperature approaches thermostat set point, a thermostat on the entering side of the heating coil modulates the outdoor air damper to maintain a constant mixed-air temperature entering the heating coil. The unit heating coil heats as needed to maintain the room thermostat set point.

Figure 16-37 illustrates schematically the control system components for ASHRAE Cycle II.

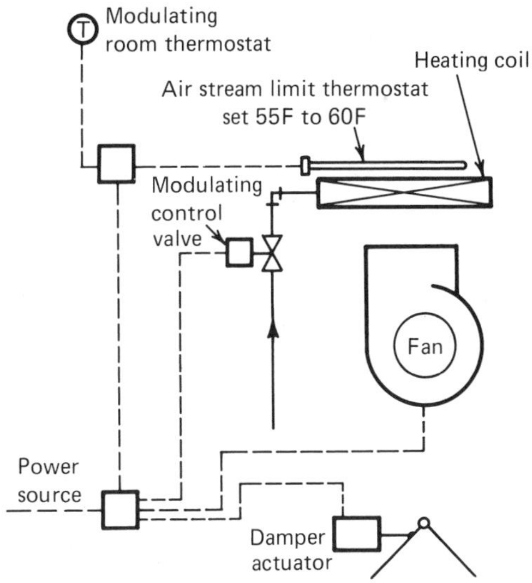

Figure 16–37

Control components for ASHRAE Cycle II

Cooling. Air-conditioning unit ventilators can utilize any of the ASHRAE control cycles; however, when cooling, they normally operate with a fixed minimum of outdoor air as determined by code requirements or good engineering practice. This fixed minimum can either be the same or different than that used when heating. When cooling, the amount should equal the minimum required to give good ventilation, as excessive amounts of outdoor air greatly increase the air-conditioning load and operating costs. Hydronic (chilled water) air-conditioning unit ventilators can have capacity control with either a modulating two- or three-way valve or face and bypass dampers.

Unit ventilators using direct expansion (DX) cooling coils are controlled by cycling the condensing unit in response to a room air thermostat. In addition, a frost protection device prevents frost buildup on the cooling coil during greatly decreased unit air delivery. This is a requirement because operation with a sizable portion of cool, wet outdoor air might have a tendency to form frost even though the unit ventilator and condensing unit are properly selected for the application design temperature.

Unit ventilator cooling control cycles should include provisions for cooling down a hot room on startup by closing the outdoor air damper and operating on 100 percent recirculated air.

Changeover. Changeover from heating-ventilating to cooling can be controlled a number of ways. Changeover to mechanical cooling should occur when ventilation cooling on 100 percent outdoor air does not overcome the design heat loads in the room. Usually this is in the neighborhood of 58°F outdoor air temperature.

Changeover can be done in individual units, as is common in four-pipe hydronic, hydronic cooling/electric heat, or direct expansion/electric heat applications. It can also be done by zones or the entire system on two-pipe hydronic systems.

Night cycle. For maximum economy it is general practice to maintain the building at reduced temperature (heating) at night, over weekends, and vacations. At such times the outdoor air dampers close and the unit handles only recirculated air. Changeover is usually done with a manual switch centrally located in the building. Some systems are arranged so that individual units may be manually returned to the occupied cycle by providing an override switch for such units. The remaining units would stay on the night cycle.

The unit capacity on the night cycle can be controlled by cycling the unit fan, or modulation of coil valve or face and bypass dampers. Occasionally, a system is designed to utilize the natural convective capacity of the unit ventilator along with auxiliary direct radiation, if available. The natural convective capacity of unit ventilators is approximately 10 percent of their normal capacity.

Reset water schedule. Unit ventilators with face and bypass control should be selected for minimum heat pickup in the bypass position. However, all unit ventilators using face and bypass control require water temperature reset systems if full ventilation cooling capacity is to be assured.

The reset schedule should provide water at approximately 100°F when the outside air temperature is 60°F. This will provide adequate heating capacity for mild temperature conditions yet prevent any significant loss of ventilation cooling capacity.

If a reset water schedule is not provided, automatic valves should be installed to shut off water to the coil when maximum ventilation cooling capacity is required.

Auxiliary radiation. Auxiliary radiation (usually fin pipe) is sometimes recommended to prevent window downdrafts when (1) single glazed windows are used, (2) windows extend over 40 percent or more of the window sill length, and (3) outdoor air temperatures are anticipated to be below 30°F for extended periods of room occupancy.

When all three conditions exist, heat loss through the windows at design conditions should be calculated and sufficient radiation installed below window sills to offset it. In heat loss calculations infiltration can be ignored as unit ventilator systems normally cause a room to be at a slightly positive pressure.

Careful control of auxiliary radiation is important if room overheating is to be prevented. During certain periods, sun load, combined with heat given off by lights and occupants, will more than offset the total heat loss even though the outdoor air temperature is low enough to require downdraft protection. Unless properly sized and controlled, auxiliary radiation heat, combined with heat from other sources, may be so great that it exceeds the ventilation cooling capacity of the unit ventilator, thus making it impossible to maintain comfortable room temperatures.

If full ventilation cooling capacity is required, auxiliary radiation should be separately controlled, either by temperature scheduling or water flow modulation. This can be done either on the basis of individual rooms or zones, although individual room control is preferred.

Electric heat. Electric heat, in either heating only or cooling-heating units, is normally controlled by a pneumatic or an electric step controller. This device individually energizes the electric heat elements as required to maintain comfort conditions. Thus, the steps of modulation equal the number of electric elements provided. Standard heating and ventilating cycles are available with all electric heating units.

16-6 CONTROL OF COMMERCIAL CENTRAL-FAN HEATING SYSTEMS

In Section 16-2 we discussed some of the basic features of the central-fan system. Central-fan systems for heating and air conditioning consist of many separate pieces of apparatus, designed to provide any or all of the following functions: (1) heating, (2) humidification, (3) ventilation, (4) distribution, (5) air cleaning, and (6) cooling by use of cool outdoor air whenever necessary.

There are many variations in the physical arrangement of equipment used to accomplish the functions outlined above. Selection of a method that will prove most satisfactory for a specific installation depends upon local practice, local climatic and economic factors, and good engineering judgment.

When more than one function is to be accomplished by a single system, particular care must be exercised in arranging and interlocking the equipment so as to provide a completely coordinated sequence of operation. The use of automatic temperature control provides a satisfactory and economical means of accomplishing this coordination.

Central-fan winter heating and ventilating units may be divided into two general types, namely: (1) *tempering systems* for ventilation only. In this type of system the mixture of outdoor air and return air is delivered at space temperature for ventilation only. The actual heat losses of the space are then provided for by means of direct radiation; and (2) *heating and ventilating systems* that utilize sufficient heat-transfer surface in the unit to provide heat not only to temper the mixture of outdoor and return air, but also to provide for the heat losses from the conditioned space.

Automatic control. Because of limited space here we shall consider only one of many possible control arrangements for a central-fan heating system. Pneumatic or electric-electronic systems may be used for central-fan systems.

Figure 16-38 illustrates a system for winter control of outdoor air quantity, with notable features of economy. Provisions have also been made for (1) closing the outdoor air damper when the fan stops, (2) a manually adjustable minimum quantity of outdoor air when the fan is in operation, (3) use of additional outdoor air for ventilation cooling when required, and (4) humidification. For purposes of illustration a heating coil with face and bypass dampers has been shown. This method of control may be used with any other method of providing heat. The sequence of operation is as follows:

1. When the fan is not operating, the outdoor air damper is automatically kept in the closed position.
2. During the pickup period, with the fan running, the outdoor air damper opens to a minimum position which may be manually adjusted by a minimum-position switch S_1. This switch may be located at any remote point desired for easy adjustment.
3. The steam valve V_1 and the face and bypass dampers are modulated to provide the necessary amounts of heat called for by the proportional return air controller T_2, which is adjustable by temperature adjuster A.
4. As the return air temperature rises, the return air controller T_2 gradually throttles the steam valve V_1 and opens the bypass damper to provide less heat to the space.

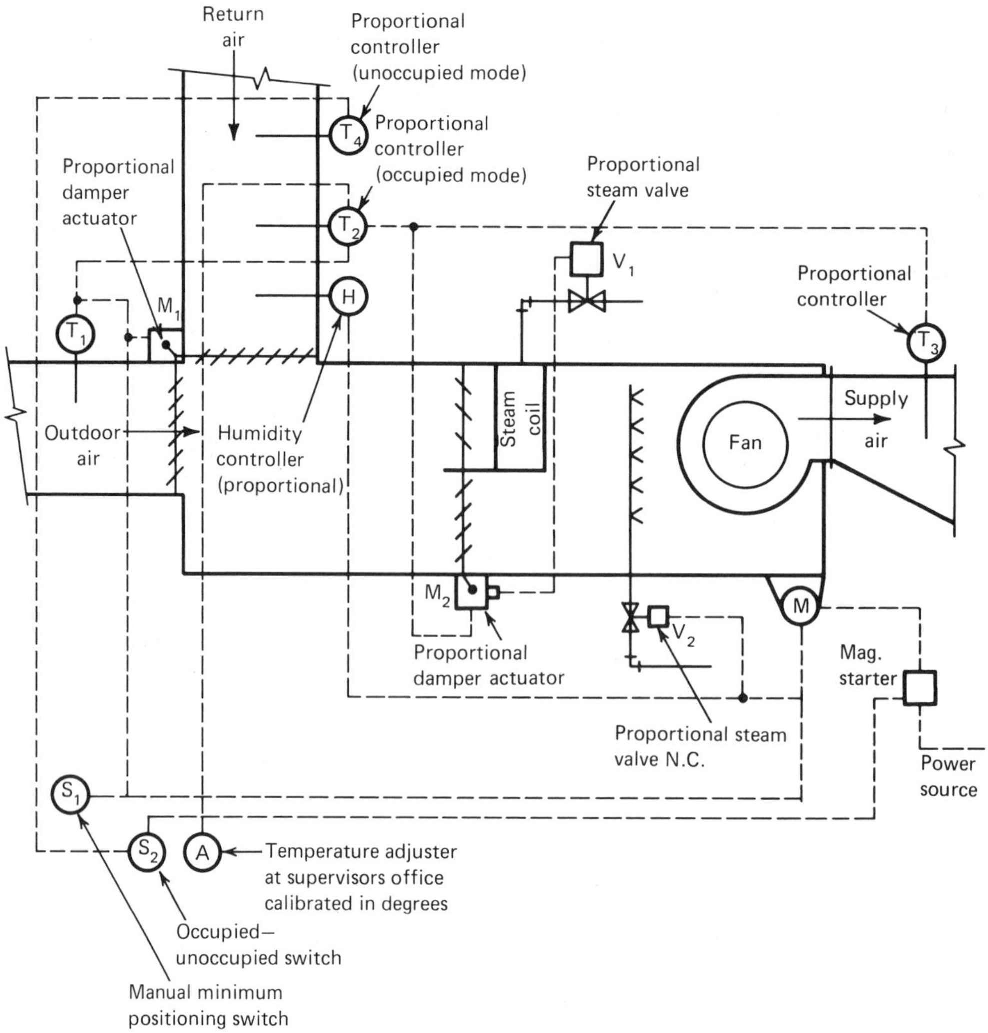

Figure 16–38

Control system for central fan heating and ventilating system

5. Should the return air thermostat T_2 become satisfied and thus call for a closed steam valve V_1, the discharge controller T_3 modulates the valve and the face and bypass dampers to provide just enough heat to maintain a minimum discharge temperature of, say, 65°F. This temperature should be sufficiently low to provide cooling when required. If, however, the temperature of the mixed outdoor and return air should be 65°F or higher, so that no heat is required to maintain a discharge temperature of 65°F or more, the proportional controller T_3 allows the steam valve V_1 to close tight.

6. If the return air temperature rises no further, the system then operates with the outdoor air damper at its minimum position and with all steam to valve V_1 shut off.

7. If the internal heat in the space increases sufficiently to cause overheating, the outdoor air damper actuator M_1 is operated at the command of the proportional return air controller T_2. The set point of T_1 is adjusted for gradual opening of the outdoor air damper on a rise in outdoor air temperature and to return the damper to minimum position as the outdoor air temperature falls to a point where the minimum percentage of outdoor air would be sufficient to maintain a discharge temperature of 65°F.

   For example, in a system designed for a minimum of 20 percent outdoor air, and with the return air at 70°F, the mixture of outdoor and return air would be at 65°F when the outdoor air temperature reached 45°F. However, as the temperature rises above 45°F, it is necessary to use more than the minimum amount of outdoor air if a 65°F discharge is to be maintained. For this condition, the outdoor air controller T_1 would be set to have the outdoor air damper wide open at 65°F and to close it gradually to the minimum position as the outdoor air temperature falls from 65°F to 45°F.

8. As long as the return air temperature does not indicate an overheated condition, the outdoor air damper is held at the minimum opening as set by the manual minimum-position switch S_1.

9. A proportional humidity controller H operates a proportional steam valve V_2 to control the steam flow for humidification. The valve is so interconnected with the fan motor circuit that it is tightly closed when the fan is stopped to prevent condensation of moisture in the ducts.

10. The manually operated occupied-unoccupied switch S_2 places the system under control of proportional controller T_4 for the unoccupied mode. Controller T_4 cycles the unit to maintain a reduced temperature in the space.

16-7 CONTROL OF YEAR-ROUND (SUMMER-WINTER) SYSTEMS

Where air-conditioning equipment is provided for year-round air conditioning, it is desirable to exercise particular care in the selection of automatic control equipment and to pay some additional attention to the automatic control aspect of the equipment design problem.

Changeover. The combination of winter and summer conditioning equipment into one installation introduces an additional control problem. If all elements of the air-conditioning system perform single functions, for example, if separate heating and cooling coils are used, the changeover from heating to cooling or from cooling to heating might conceivably be accomplished by allowing the separate heating and cooling controls to function independently. On a rise in temperature, say, the heating thermostat would shut off the flow of steam to the heating coil, and on a further rise the cooling thermostat would start the operation of the cooling equipment.

But it is not quite as simple as that in most cases. For one thing it is often necessary or desirable to utilize one piece of equipment in both the heating and the cooling cycle. A coil is sometimes used both for heating, with hot water as the medium, and for cooling, with chilled water. Means must then be provided for changing over from one source to the other. Again, the outdoor and return air dampers are normally used in both cycles, but the damper actuator must be positioned according to different schedules during the heating and cooling cycles. Provision must then be made for transferring command of the damper actuator from one controller schedule to the other.

In addition, even the single-purpose equipment elements may require special consideration of changeover means. Even if it is practical to have both heating and cooling capacity available at all times, suitable schedules of control sequence for heating and for cooling may not permit leaving the choice of heating or cooling operation to the independent operation of separate heating and cooling controllers. It is usually necessary, therefore, to provide for definite changeover, by means of suitable valves or switches, from one mode of operation to the other.

Manual versus automatic changeover. If it were always possible to determine that today is the last day of the heating season and tomorrow the cooling season begins, manual means of changeover would be entirely satisfactory. However, during late spring and early fall periods, there are times when the demands made upon the conditions system may change, from day to day or even during the day, from a heating to a cooling load and vice versa. During the mild seasons, it is often necessary that heat be supplied during the morning and evening hours, and at night if the building is occupied then; whereas temperature and humidity conditions during the day require cooling operation, or at least dehumidification. In some applications, in fact, where very large internal heat gains may occur intermittantly, this condition may occur during mild weather periods in the winter.

Even though a manual changeover can be accomplished easily, if it must be left to the judgment and alertness of the building engineer, some inefficiency and discomfort will often result. For these reasons it is generally desirable to provide for automatic changeover. Electric or pneumatic switches and valves may be operated automatically by thermostats or humidity controllers located within the conditioned space or outdoors according to the requirements of the specific application. These changeover controls may reverse the action of actuators and valves, transfer command of actuators or valves from one set of controllers to another, or divert the flow of a medium such as water, in order to convert the mode of operation from heating to cooling or vice versa.

If both the air-conditioning system and the control system are designed with automatic changeover in view, the inclusion of this feature involves only slight additional investment. In view of the gains in operating economy and in comfort of occupants, this investment may be considered self-liquidating.

YEAR-ROUND AIR-CONDITIONING SYSTEM (ELECTRONIC): Figure 16-39 illustrates only one of the many possible year-round air-conditioning systems which includes heating, ventilating, and cooling and the associated controls which could be used for a large space within a building (say a school auditorium). The heating is provided by a hot water coil controlled by a three-way valve. Cooling is provided by a direct expansion (DX) coil with two stages of cooling. The control system provides for certain special features for the specific application, such as

1. Manually selected occupied-unoccupied operation.
2. Fan operates continuously during occupied control sequence to provide good air circulation.
3. Fan cycles on-off during unoccupied control sequence to maintain lower space temperature.
4. Provides for cooling by using outdoor air (economy), which means a third cooling stage is available without the required use of refrigeration.

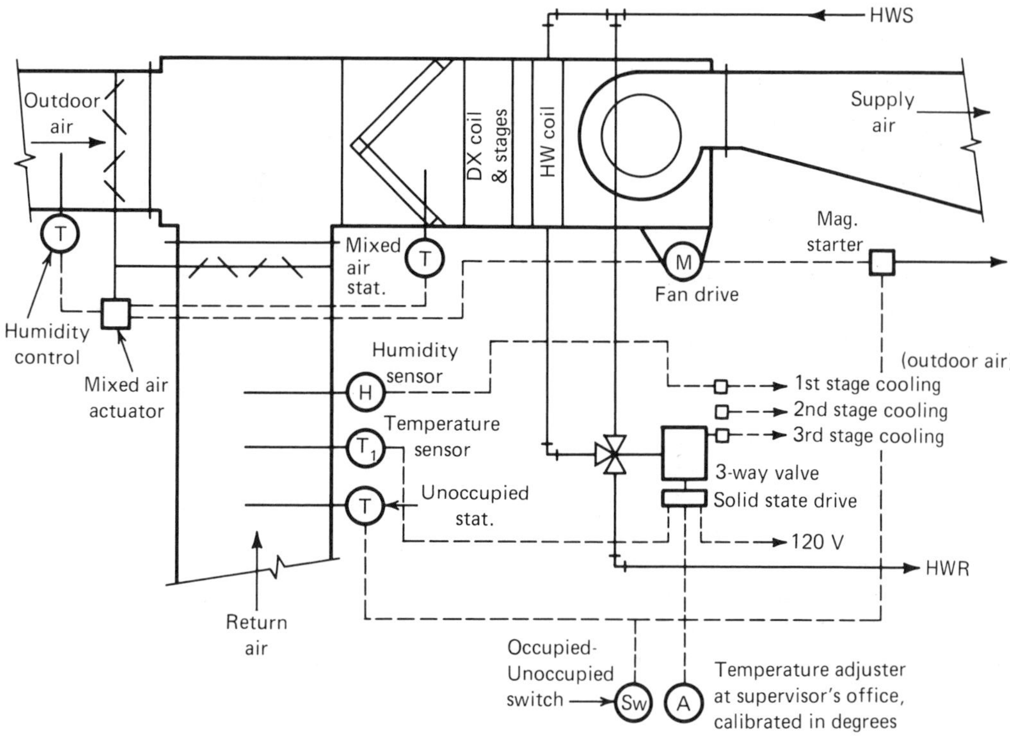

Figure 16–39

Control system for year-round air conditioning system (electronic)

5. Starting, stopping, occupied-unoccupied, and temperature control by single supervisor.

The control sequence for the system shown in Fig. 16-39 would be as follows:

Occupied. Occupied switch starts fan, which runs continuously. Mixed-air dampers move to maintain 65°F mixed-air temperature. Return air sensor T_1 positions three-way valve to maintain room temperature at setting of remote adjuster *A*. Three-way valve shall have a flat cam to provide dead-band between heating and cooling. When return air relative humidity rises above 50 percent, return air humidistat will energize one stage of DX cooling. When space requires cooling, first stage will open outdoor air damper subject to total heat of entering air (enthalpy control). A further call for cooling will energize the first and second DX cooling stages as required.

Unoccupied. Unoccupied switch energizes the unoccupied return air thermostat, which cycles the unit to maintain space temperature at 60°F.

16-8 SELECTION OF CONTROL VALVES[2]

In the preceding sections we have seen the use of control valves and dampers in air-conditioning systems. The selection of valves and dampers must be done with care and knowledge of their operating characteristics if we are to have a satisfactorily, controllable air-conditioning system. We shall first discuss the fundamentals of valve selection for heating-cooling applications. The benefits derived from proper selection of type of valve and its size are (1) greater *comfort* by providing even temperatures, (2) greater *economy* by using energy more efficiently, and (3) greater *life of equipment* by proper cycling.

A control valve regulates the flow of a liquid or gas in a system. This regulation is accomplished by varying the resistance which the valve introduced into the system as it is stroked. The increasing resistance causes the total system pressure drop to shift to the valve and in this way vary the flow in the system.

As the important function of the control valves in a system is becoming more fully realized, valves are receiving much more attention regarding their proper sizing and application. In the past many valves were installed line size, and no consideration was given to the controllability of the valve. This usually produced oversized valves that were only performing correctly when they were fully open or fully closed. This, in addition to valves installed backwards, is one of the reasons for the shock waves in older heating systems which produce the annoying hammering and knocking which is all too familiar.

The majority of the following discussion of control valves and their application is from Section Vb, ''Control Valves,'' Engineering Data, published by Johnson Controls, Inc. and used with permission.[3]

THEORY OF COILS: A control valve is used to vary the flow quantity of heating fluid to a heating coil. As the flow through the coil is varied, the heat output of the coil does not vary in a linear manner. The reason for this is different in steam coils than it is in water coils.

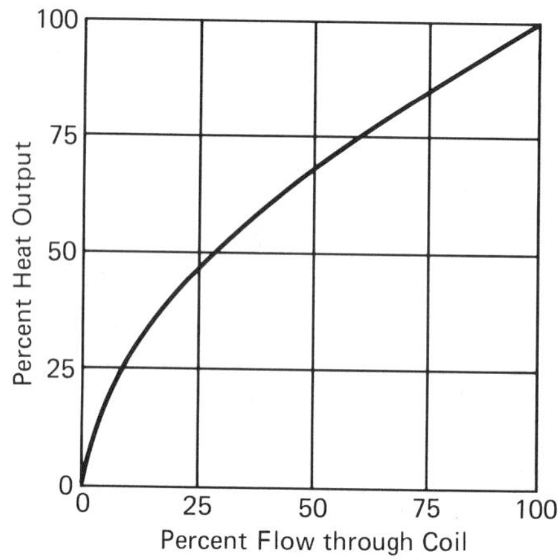

Figure 16–40

Non-linear characteristic of a heating coil

Water coils. The heat output of a water coil varies with flow in a manner as shown in Fig. 16-40. This is true because at full rated flow the water temperature drop (Δt) as it passes through the coil has a certain set value. But as less heat is required the control valve is partially closed and the flow through the coil is decreased. This decreases the water velocity through the coil tubes; the water than remains in the coil for a longer period of time and consequently has a larger temperature drop (Δt). For this reason each gallon of water flowing gives off more heat, and the total output of the coil is not linearly related to the flow of water through it. This nonlinearity can be compensated for by selecting a control valve characteristic that will offset the characteristic of the coil.

An *equal-percentage valve* has this type of characteristic. When the coil and equal-percentage valve are used together, the combined characteristic approaches linearity (see Fig. 16-41). This combination characteristic curve is arrived at by combining the individual characteristics of the heating coil and the equal-percentage plug. This characteristic approaches linear but is not quite so. When the effect of the increasing pressure drop across the valve is also added to this plot, the total characteristic

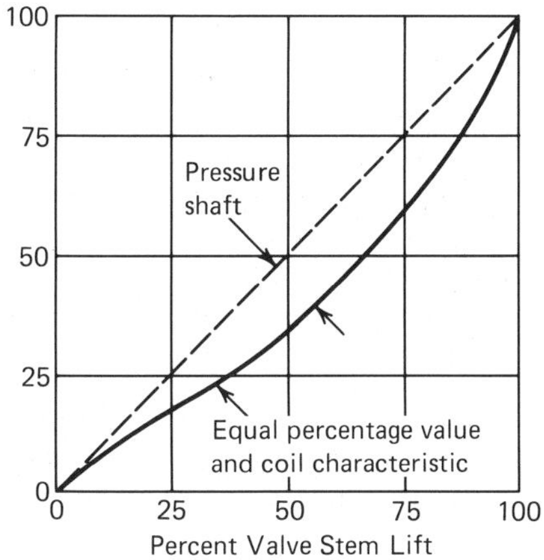

Figure 16–41

Total heat transfer characteristic for control valve and coil combination

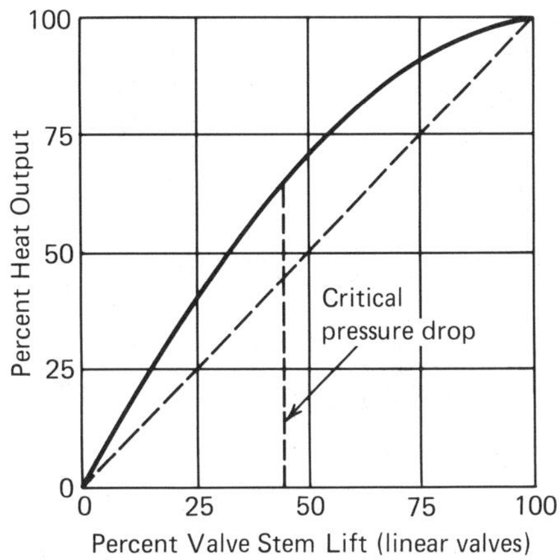

Figure 16–42

Heat output of steam coils when controlled by linear valves

goes slightly beyond linear. However, this characteristic is close enough to be considered linear from the controller's point of view.

What has been said about hot water coils can also be said for chilled-water coils, except that the heat transfer is in the other direction.

Steam coils. Steam coils can act the same way as water coils but for a different reason. A steam coil producing its maximum heat output is full of steam that condenses and gives off heat. As less heat is required, the control valve begins to throttle back and decrease the flow of steam. The steam that is in the coil condenses and the pressure in the coil drops; this causes a greater pressure drop across the valve and the steam velocity increases. Because of this, the quantity of steam supplied to the coil is almost as great as it was before, and the heat output has only decreased a small percentage for a large movement of the control valve. This action will continue until the critical pressure drop is attained across the valve. From this point on any further increase in pressure drop across the valve will not cause an increase in steam velocity; the maximum velocity was attained at the critical pressure drop. Because of the constant velocity under these conditions, the valve's performance will follow its inherent characteristic, and the heat output of the coil will resemble this characteristic. A *linear valve* on a steam coil will produce a total characteristic curve as shown in Fig. 16-42, and an equal-percentage valve will produce the curve of Fig. 16-43. The equal-percentage plug more closely resembles a linear characteristic. Because the desired total characteristic of the control valve and coil is linear, the equal-percentage plug would be more desirable.

The preferred way to match a steam control valve and coil would be to size the valve so that it has critical pressure drop across it under full-load conditions. In this way the flow of steam to the coil will depend entirely upon the free area between the valve plug and its seat. This area is determined by the characteristic of the valve and the position of the plug. Under these linear conditions the use of a linear valve characteristic would produce the desired system characteristic. Unfortunately, this can sometimes require a larger pressure drop than is allowable for the control valve.

The majority of heating systems must operate most of the time in a stage of low heating

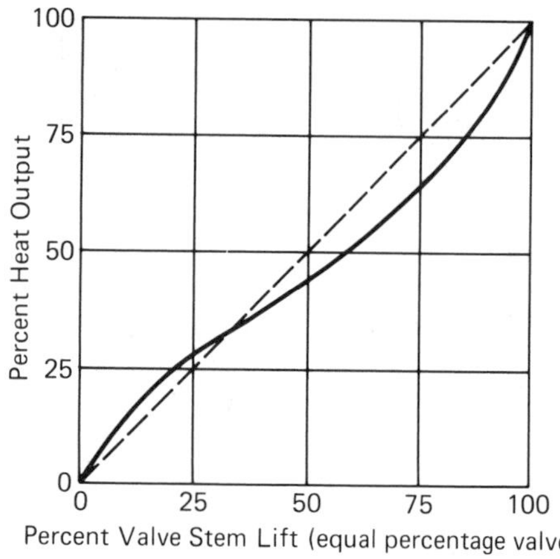

Figure 16–43

Heat output of steam coils when controlled by equal percentage valves

capacity. For this part of the valve's stroke the equal-percentage valve is quite nearly linear and possesses a relatively small valve gain. These facts make an equal-percentage valve very controllable and just as desirable as a linear valve for this type of application.

VALVE SIZING: Control valves must be sized correctly to perform the job for which they are intended. Undersized valves cannot deliver sufficient quantities for maximum-load conditions, and oversized valves attempt to perform correctly but must do so at the very end of their strokes where hunting or cycling is difficult to avoid. Oversizing is definitely the most prevalent because in every step of putting the system together safety factors are used and the final result is a system that is well oversized, control valve included. Correct sizing, however, is not very complicated. Once the correct characteristic is chosen, a few simple equations will give the desired flow coefficient for sizing the valve.

Valve flow coefficient. The first step in finding the size of a valve is to determine the flow coefficient (C_v) that is required for the system. Again, repeating the definition as "the number of gallons per minute of 60°F water that will flow through a *fully open valve* with a 1.0 psi pressure drop across it," the flow coefficient is determined by the construction of the valve and will not change. Identical valve sizes may have different flow coefficients if the body style or valve trim is different. The flow coefficient is probably the most useful piece of information necessary to size a valve.

Sizing water valves. The required flow coefficient (C_v) for a water control valve may be found from the following relationship:

$$C_v = \frac{\text{gpm}}{\sqrt{\Delta p}} \qquad (16\text{-}1)$$

where gpm = water flow required (gallons per minute)

C_v = valve flow coefficient

Δp = pressure drop between inlet and outlet of valve (psi)

The required water flow (gpm) for the coil should be available in the specifications or calculated from the heating or cooling requirements of the system.

Pressure drop across the valve (Δp). A valve and coil should be sized to produce a combined characteristic which is as close as possible to linear so that the controller can do an efficient job of controlling, as previously noted. A valve is matched with a coil based on the characteristic and flow coefficient of the valve. These valve parameters are determined at constant pressure drop and, consequently, the valve should be operated at as close to a constant pressure drop as is possible. However, as the valve closes, the total system pressure drop shifts to the valve. This hardly approaches a constant pressure drop. The best that can be done is to keep the *relative change* in pressure as low as possible.

For example, in a 20-psi system (Δp_2), if the open valve has a 7-psi drop (Δp_1) initially, then the change in Δp as the valve closes would be ($\Delta p_2 - \Delta p_1$) (100)/Δp_1 or (20 − 7) (100)/7 = 186 percent. A valve with an initial drop of 3 psi would have a relative change of (20 − 3)(100)/3 = 567 percent, which would produce a much larger shift in characteristic. Therefore, the valve pressure drop at maximum flow should be as large as is *practical* for the system. This means that other system components require certain values of pressure drop to operate efficiently. Often a maximum valve pressure drop is stated, and this value cannot be exceeded. When it is permissible to choose the pressure drop for the control valve, a value equal to 50 percent or greater of the pressure between the supply and return mains, should be selected.

For liquids other than water a correction for the difference in specific gravity of the liquid is necessary, and Eq. (16-1) becomes,

$$C_v = \frac{\text{gpm}\,\sqrt{\text{SG}}}{\sqrt{\Delta p}} \tag{16-2}$$

where SG is the specific gravity of the liquid.

Illustrative Problem 16-1

What is the flow capacity of a 1-in. water valve used in a water system where the design pressure drop across the valve is 15 psi and the valve manufacturer quotes a flow coefficient of 8.6 for the valve?

Solution:

By Eq. 16-1 $C_v = \dfrac{\text{gpm}}{\sqrt{\Delta p}}$

$$\text{or gpm} = C_v\sqrt{\Delta p} = 8.6\sqrt{15}$$

$$\text{gpm} = 33.3$$

Illustrative Problem 16-2

Water is to flow through a control valve at the rate of 98 gpm with a pressure drop of 4.0 psi. What flow coefficient would be required for the valve?

Solution:

By Eq. 16-1 $C_v = \dfrac{\text{gpm}}{\sqrt{\Delta p}} = \dfrac{98}{\sqrt{4.0}} = 49$

Select a valve size having a C_v = 49 from a valve manufacturer's catalog.

Sizing steam valves. The procedure for sizing steam valves is very similar to that for water valves. The formulas are almost the same except for a few slight variations. It is recommended that steam control valves be sized for full pressure drop when the inlet pressure is 10.0 psig or below. When the inlet pressure is above 10.0 psig, the critical pressure drop should be used in sizing the valve. The critical pressure drop is equal to 45 percent of the absolute inlet pressure. Therefore, the pressure drop would be

$$\Delta p = 0.45 p_1$$

where Δp = pressure drop used to determine flow coefficient

p_1 = absolute inlet pressure, psia

Although the specific guide above recommends specific pressure drops to use in determining the C_v for a steam valve, it would be better to look at a more general set of equations for other pressure drops and, if the steam is superheated (rarely in heating systems), we would have the following two relationships:
When p_2 is 0.45 of p_1, or less use

$$C_v = \frac{\dot{m}_s[1 + 0.0007\ (\text{deg. F superheat})]}{1.8 p_1} \tag{16-3}$$

When p_2 is more than 0.45 of p_1 use

$$C_v = \frac{\dot{m}_s[1 + 0.0007 \text{ (deg. F superheat)}]}{2.1\ (p_1 - p_2)(p_1 + p_2)} \tag{16-4}$$

where $\dot{m}_s$ = steam flow (lb/hr)

deg. F superheat = degrees of superheat

p_1 = absolute inlet pressure (psia)

p_2 = absolute outlet pressure (psia)

If the steam contains moisture (wet steam), it has a quality less than 100 percent, and the C_v value should be corrected for this condition. If the percent moisture is known, say 2 percent, then the quality is 1.0 − 0.2 = .98 or 98 percent. The C_v value obtained from either Eq. (16-3) or Eq. (16-4) should be corrected by multiplying by the square root of the quality.

Other saturated vapors to be controlled in HVAC systems include refrigerants. Equations (16-3) and (16-4) may be used if the appropriate constant appearing in the denominator is selected for the particular vapor. Such constants, K, appear in Table 16-1, and may be used in the following relationships:

When p_2 is 0.45 of p_1, or less

$$C_v = \frac{\dot{m}_v}{0.86K(p_1)} \tag{16-5}$$

When p_2 is more than 0.45 of p_1

$$C_v = \frac{\dot{m}_v}{K\sqrt{(p_1 - p_2)(p_1 + p_2)}}$$

where $\dot{m}_v$ is the mass of saturated vapor (lb/hr) and K is the constant from Table 16-1.

Table 16-1

Constants, *K*, for Saturated Vapors

Vapors	K
Freon 12	7.1
Freon 11	7.4
Freon 14	8.4
Freon 114	8.3
Dowtherm A	5.6
Ammonia	2.7

Illustrative Problem 16-3

A preheat steam coil requires 800 lbs/hr of saturated steam at a system pressure of 20.0 psig. What should be the flow coefficient for the control valve?

<u>*Solution:*</u>

Assume that p_2 will be 0.45 p_1 (critical pressure drop), and with no superheat, we have by Eq. (16-3)

$$C_v = \frac{\dot{m}_s}{1.8p_1} = \frac{800}{1.8(20 + 14.7)} = 12.8$$

Illustrative Problem 16-4

A steam control valve has a flow coefficient (C_v) of 20. It will be installed in a steam system where the initial pressure is 20 psig, and downstream from the valve the pressure is 15 psig. What maximum flow rate of saturated steam can be expected?

<u>*Solution:*</u>

Since p_2 is more than 0.45 of p_1 we use Eq. (16-4). By Eq. (16-4)

$$C_v = \frac{\dot{m}_s}{2.1\sqrt{(p_1 - p_2)(p_1 + p_2)}}$$

or

$$\dot{m}_s = C_v(2.1)\sqrt{p_1^2 - p_2^2}$$

$$= 20(2.1)\sqrt{(20 + 14.7)^2 - (15 + 14.7)^2}$$

$$\dot{m}_s = 754 \text{ lbs/hr}$$

Illustrative Problem 16-5

Determine the required flow coefficient (C_v) for a steam control valve having a maximum capacity of 1000 lb/hr of saturated steam with initial and final steam pressures of 20 psig and 15 psig, respectively.

Solution:

Since p_2 is more than 0.45 of p_1, we use Eq. (16-4).

By Eq. 16-4

$$C_v = \frac{\dot{m}_v}{2.1\sqrt{p_1^2 - p_2^2}}$$

$$= \frac{1000}{2.1\sqrt{(20 + 14.7)^2 - (15 + 14.7)^2}}$$

$$C_v = 26.5$$

Illustrative Problem 16-6

Determine the flow coefficient (C_v) for a steam control valve to pass 1221 lb/hr of saturated steam when the inlet pressure is 20 psig.

Solution:

Since the pressure drop is not specified, we will select the critical pressure drop, or $p_2 = 0.45p_1$, and use Eq. (16-3). By Eq. 16-3

$$C_v = \frac{\dot{m}_s}{1.8p_1} = \frac{1221}{1.8(20 + 14.7)}$$

$$C_v = 19.5$$

THREE-WAY VALVES: There are two basic types of three-way valves, as discussed previously: the mixing valve with two inlets and one outlet and the bypass valve with one inlet and two outlets.

There are also two types of applications to which these valves may be applied: a mixing application or a bypass application. The mixing valve can perform either application, but the bypass valve *can only* be used in a bypass application. These various combinations of three-way valve and piping arrangements for a coil are shown in Figs. 16-44, 16-45, and 16-46. The type of application that is chosen for a certain job will depend on the required fail-safe conditions and the way in which the rest of the system is piped. It is desirable to maintain as close as possible a constant system loss and pressure head in the system. This is possible if constant-flow and constant-friction losses are maintained. Three-way valves used in a bypass application are well suited for this because the flow through the valve is not stopped but just diverted through an alternate route back into the overall water system.

If this alternate route has an adjustable resistance (such as a balancing cock), it can be made to have the same resistance as the coil. In this way the resistance to system flow will be the same for full bypass flow as it is for full flow through the coil. However, for all conditions between these two extremes there exists two parallel paths for the system flow, and this greatly reduces the resistance. The maximum system flow will occur when the valve is in its

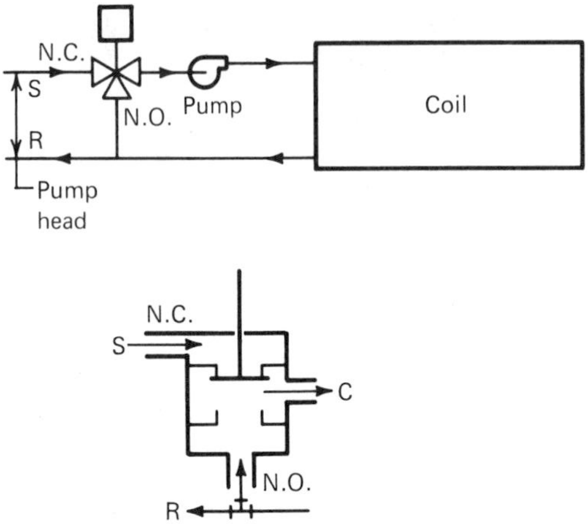

Figure 16–44

Mixing valve in mixing application piped N.C. (normally closed) to coil

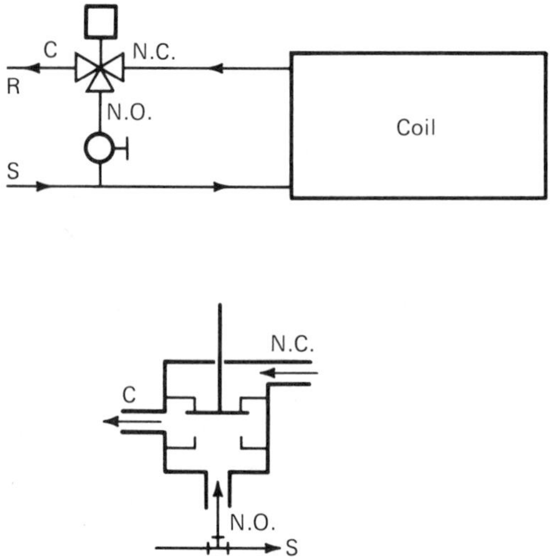

Figure 16–45

Mixing valve in bypass application piped N.C. (normally closed) to coil

midway position and one-half the flow is in each path. As the valve strokes in either direction from this middle position, the system resistance will increase and the flow will decrease and approach the minimum flow conditions which are present at either extreme of the valve stroke (see Fig. 16-47).

Again, the desired combined characteristic of the valve, coil, and bypass is linear as it was for the two-part valve and coil. But three-way valve characteristics will also shift from their inherent characteristics when they are subjected to this varying pressure drop across them. As before, the best way to combat this problem is to select a valve with as large a pressure drop as possible. This will keep the relative change in pressure across the valve small, resulting in less shift in the valve characteristic.

Mixing valves used in mixing application, Fig. 16-44, incorporate a pump that supplies water to the coil. This keeps a constant flow in the coil and thus eliminates the coil characteristic problem. Because of the constant flow through the coil, the heat output (heating coil) is directly related to the temperature of the water being supplied to the coil. The water temperature is regulated by mixing supply water with return water that has already been through the coil. This condition provides for very effective temperature control by the controller. But from the viewpoint of attempting to maintain constant overall system flow, this application is no better than two-way valve applications. At no-load conditions the control valve completely closes off to overall system flow and the pump is merely continuously circulating the same water through the coil. This causes great variations in system flow between full-load and no-load conditions.

Because the system pressure drop will shift to the valve as it is closed, the valve characteristic will also shift as it did for other valves. This cannot be eliminated, but the effects can be minimized. The same approach should be taken for mixing valves in mixing applications that we applied to all other valves; namely, the valve pressure drop must be as large as is practical so that the relative change in pressure

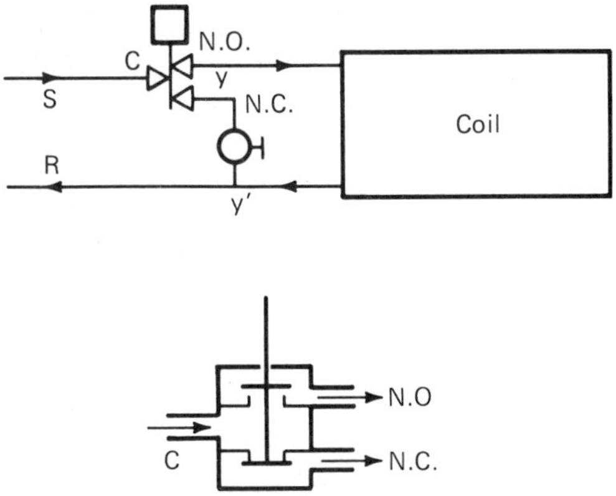

Figure 16–46

Parallel circuit: y through coil to y' bypass valve—N.O. (normally open) to coil

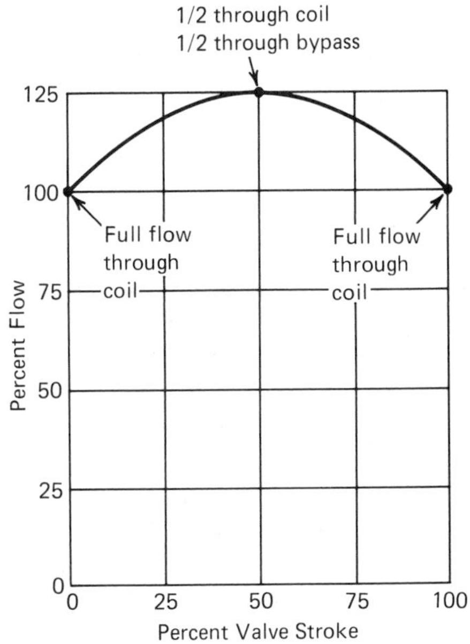

Figure 16–47

Three-way valve bypass application

across the valve will be kept as small as possible. This will make the shift of the valve characteristic relatively small.

VALVE SELECTION: This discussion has presented some of the basic theories behind control valves and their applications in systems. The topics covered should provide a better understanding of valves and also aid in their selection. Unfortunately, it is very seldom that the perfect valve for a certain application is available, and the next best valve must be selected.

As a general rule of thumb, because most system components are oversized, the next smaller flow coefficient (C_v) than that calculated will provide satisfactory results. Good judgment must be used, however, when selecting a smaller valve so that the maximum capacity of the coil is not greatly reduced. On the other hand choosing too large a valve can have very poor effects on controllability of the system as was discussed earlier.

16-9 SELECTION OF CONTROL DAMPERS[3]

A damper is a device that controls the flow of air in an air-conditioning system. The damper accomplishes air flow control by varying resistance to flow, just as a control valve does in a liquid circuit. This section deals with the multiple-blade damper in its basic design configurations, the parallel blade and the opposed blade. The same considerations that apply to valve selection, that is, correct size, flow characteristics, rangeability, and required pressure drop, also apply to damper selection.

Much has been published on the subject of valve selection and sizing. Instead of installing a valve the same size as the pipe, as was common practice not too many years ago, valves are now selected with regard to the function they are to perform. This was discussed in the preceding section. Manufacturers have given careful consideration to inner-valve construction. Most of them now provide a choice of flow characteristics ranging from quick-opening to linear to equal-percentage, or modifications of these, to meet various application requirements. They have published data regarding these characteristics, making it possible for the consulting engineer or control engineer in the field to make the best valve selection.

This has not been true with dampers. Damper design has changed very little. Published data regarding their performance characteristics have been extremely meager. Dampers are almost always the same size as the duct in which they are installed. As a result, performance of air flow control systems often leaves much to be desired. The following discussion of damper application is from Ref. 3, used with the permission of publisher, the Johnson Service Company. It includes Figs. 16-48 through 16-66 and associated descriptive matter.

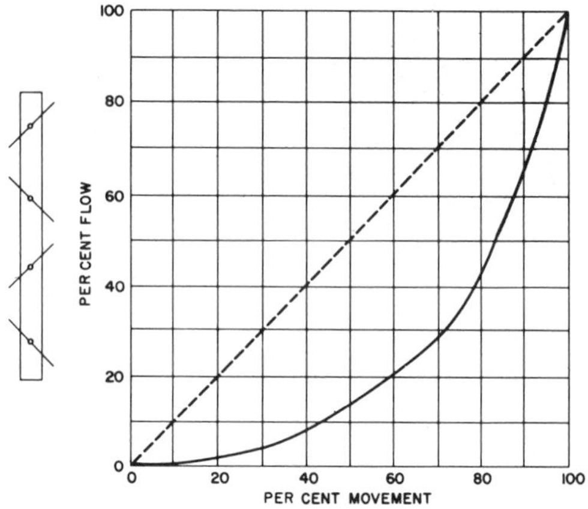

Figure 16–48

Parallel blade damper inherent characteristic curve. (Dashed line is hypothetical linear characteristic curve). Constant pressure drop

PRESENT DAMPER DESIGN: Dampers are manufactured in two basic styles: those with blades that rotate parallel to each other (Fig. 16-48) and those with blades that operate in an opposed manner (Fig. 16-49). Each has different flow characteristics. The application determines which should be used in each case for best control. Their design provides a very limited choice in characteristics.

The characteristic built into a damper is not necessarily the characteristic of the damper when installed in an actual system. The characteristic designed into the damper, here called the *inherent flow characteristic,* is represented by the curve "percent flow versus operator travel," which is produced with a constant pressure drop maintained across the damper (see Figs. 16-48 and 16-49). These are the basic characteristics designed into the dampers by virtue of the design of driving linkage, damper blade of leaf configuration, and other factors.

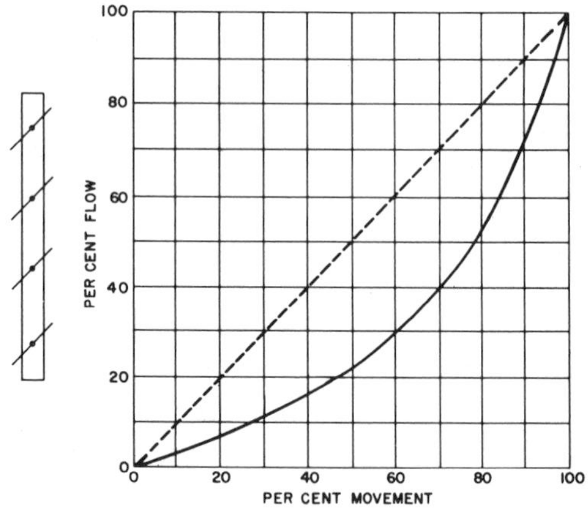

Figure 16–49

Opposed blade damper inherent characteristic curve. (Dashed line is hypothetical linear characteristic curve). Constant pressure drop

The *effective flow characteristic* is the actual flow characteristic of the damper obtained when the damper is applied to a given air-handling system; that is, it is the inherent flow characteristic, as altered by the varying pressure conditions in the system and other considerations. To select the inherent characteristic best suited to a particular flow control application, it is important to first decide what effective characteristic is most desirable for good control. Then, knowing how the system pressure variations will alter the inherent characteristic to produce an effective characteristic, the proper selection can be made.

THE IDEAL EFFECTIVE CHARACTERISTIC: Maximum control stability in a system occurs when there is a nearly linear relationship between the change in the controlled variable (error) as measured by the controller and the corrective action produced by the controlled device (damper). If, for example, a thermostat controls the temperature of a space by varying the volume of warm air admitted, each equal increment of space temperature change should result in an equal increment of change in air volume admitted to the space. Figure 16-50 shows how a linear flow characteristic produces air flow control through the entire travel of the damper (from closed to 100 percent open).

Since the relationship between the controller and damper operator, whether the control system is pneumatic, electric, or electronic, is essentially linear, it remains only to insure that the effective damper characteristic be relatively linear to have the entire control loop linear.

The way in which a damper is used in a system to a great extent determines how much the system will alter the inherent damper characteristic. It is this altered inherent characteristic, or effective characteristic, that is of importance to the end result. The following examples will illustrate why this is so.

Throttling control. For purposes of explanation only, assume a system shown in Fig. 16-50 wherein the space temperature is controlled by varying the volume of constant-temperature air admitted. Each equal increment in demand should result in equal increments of air volume change. The relationship between room temperature change and damper operator movement is essentially linear, as was mentioned previously. It remains to be seen how the relationship between operator movement and flow of air through the damper can be made as nearly linear as possible.

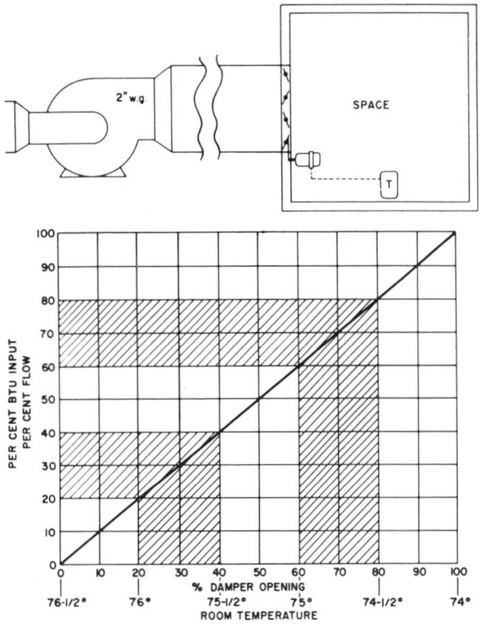

Figure 16–50

Variable volume temperature control system

At maximum flow, assume the fan pressure is 2.0 in. WG (inches of water gauge) and pressure drop across the wide open damper is 0.01 in. WG. Resistance of the duct system is then 1.99 in. WG. At the other extreme when the damper is completely closed, there is no flow and no duct loss, and the pressure drop across the damper is the full fan pressure of 2.0 in. WG, or 200 times what it was when wide open. (If the fan performance curve is taken into account, this ratio would be greater than 200 to 1.) Thus, as the damper gradually moves from wide open to closed, pressure drop across the damper gradually increases. Closing the damper to reduce flow is partially offset by this increase in pressure drop. With the increasing pressure drop partially offsetting each increment of damper closing, the inherent characteristic curve will shift upward. Figure 16-51 shows how a near linear inherent characteristic would shift upward in this application. The damper characteristic curves in Figs.

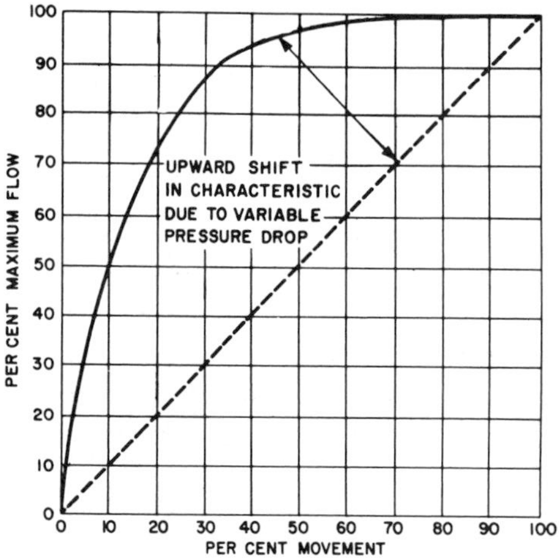

Figure 16–51

Shift in damper characteristic due to increasing pressure drop

16-48 and 16-49, however, are for *constant pressure drop*. If a damper were constructed with an inherent linear characteristic, as shown by the dashed lines of Figs. 16-48 and 16-49, variable pressure drop would bend the curves upward, and the effective characteristic would be far from linear in the system described (Fig. 16-51). Knowing this, it is best to select a damper with an inherent characteristic that falls well below the linear curve so that the increasing pressure drop, as the damper closes, will bend the curve upward toward linear rather than away from it.

A system with several volume control dampers, instead of the single damper used in the example, would behave exactly the same as shown in Fig. 16-50. Whenever the volume of air handled by a duct system or even a portion of a duct system varies, a variable pressure drop across the dampers results unless the system is specifically designed and controlled to prevent it. Variation of pressure drop will usually be quite large, so that, normally, a damper with an inherent characteristic curve that is the farthest below the linear curve should normally be selected. This is the opposed-blade damper of Fig. 16-49.

Face and bypass control. A face and bypass application must be treated much like the preceding example when dampers are applied. Although a constant pressure drop is maintained between points *A* and *B* of Fig. 16-52, causing it to resemble a mixing or constant pressure drop application, other factors must be analyzed. Looking at the face section alone, assume a pressure drop through the coil of 0.25 in. WG. Also assume a 0.10 in. WG pressure drop through the face damper in the wide-open position. When the face damper is wide open, the majority of the pressure drop in the face section is across the coil. But, as the face damper closes, the pressure drop shifts from the coil to the damper. At the near-closed position of the damper the majority of the pressure drop in the face section is across the face damper. In the preceding volume control example, the pressure drop shifted from the duct work to the damper; in this example it shifts from the coil to the face damper. This shift in pressure drop makes the problem of selection of the face damper much like the variable volume damper.

When analyzing the bypass section by itself, a study of Fig. 16-60 should be made. Figure 16-60 shows the resistance that is added by installing dampers smaller than the duct size. When the face and bypass dampers are each in a 50 percent open position, air is flowing through both the face and bypass section (air flow through 100 percent of duct area). When either damper is closed (100 percent face or 100 percent bypass), the pressure drop through the section is determined by the amount of total duct area being used for air flow. For example, with the face section closed, all the air must flow through the bypass section. The pressure drop added is that which is added by the blank-off section (closed face damper). Figure 16-60 shows that blanking off

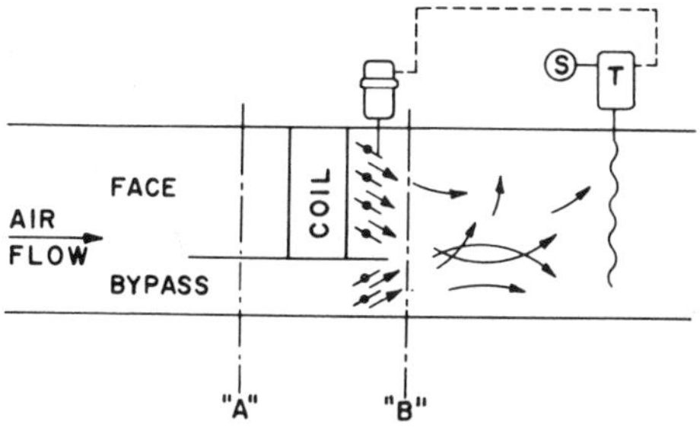

Figure 16–52

Face-and-bypass system

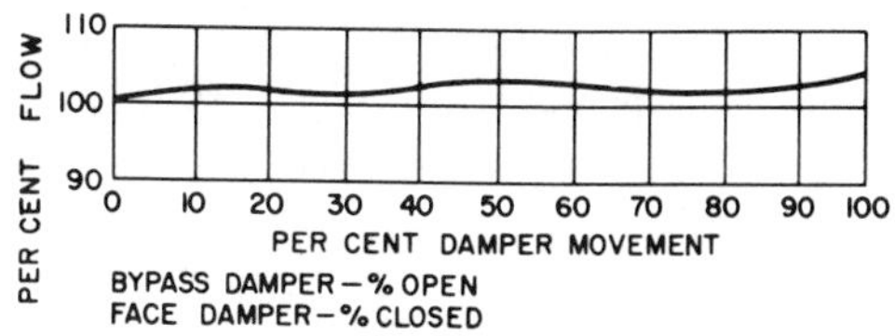

Figure 16–53

Percent flow versus damper position (face-and-bypass system)

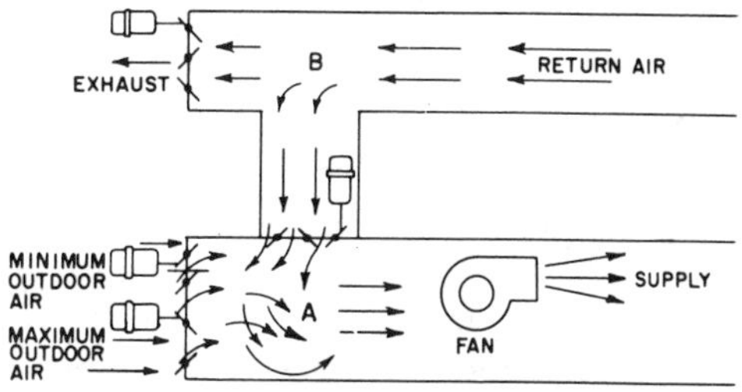

Figure 16–54

Typical outdoor, return and exhaust air mixing application. (Parallel blade damper application)

70 percent of the total duct area at an approach velocity of 500 fpm will add approximately 0.34 in. WG. On the other hand, with the bypass closed, 30 percent of the total duct area is blanked off and 0.02 in. WG pressure drop is added. Although the pressure drop across the entire face and bypass section is relatively constant, the preceding example shows how the pressure drop varies within the section. This makes it impossible to consider a face and bypass damper in the same manner as a constant pressure drop or mixing application.

Since the pressure drop across the dampers varies as the dampers are throttled, the inherent characteristic of the damper will be shifted. Truly linear dampers are therefore not the best choice. Figure 16-53 shows the total flow variations that can be expected if the proper characteristic is selected.

Mixing dampers. Application of mixing dampers often presents a considerably different problem from that described in the previous section. An outdoor air, return air, and exhaust damper combination could be this type of application. Figure 16-54 shows a combination of outdoor, return, and exhaust dampers in very close proximity to each other. No weather louvers, preheat coils, and so forth are considered. Though not typical of many systems, it demonstrates one of the few constant pressure drop applications. The minimum outdoor air quantity provides the required ventilation and keeps the building under a slightly positive pressure.

With the maximum outdoor air and exhaust dampers closed and the return air damper open, assume that the pressure at *A* (mixing chamber) is −0.10 in. WG and that the system capacity is at the desired value of 10,000 cfm, of which 2500 cfm is minimum outdoor air and 7500 cfm is recirculated air. If the pressure at *B* is +0.10 in. WG, the pressure drop from *B* to *A* is 0.20 in. WG. When the maximum outdoor air and exhaust dampers are open and the return air damper is closed, it is required that the system capacity remain at 10,000 cfm. It is then necessary to select the maximum outdoor air damper to handle 7500 cfm, with 0.10 in. WG pressure drop, and the exhaust damper and duct to handle 7500 cfm, with 0.10 in. WG pressure drop, thus achieving the desired capacity. Knowing the pressure losses in ducts and dampers for various air velocities, it is possible to design a system that will essentially meet these constant volume requirements.

Proper functioning of this type of air-conditioning system depends upon maintaining a constant rate of flow for all positions of the maximum outdoor air, return air, and exhaust dampers, not just the extreme positions mentioned. If the pressure drops through the dampers can be held at a constant value by correct design of the system and proper selection of damper sizes, as described above, and if the dampers have linear inherent characteristics, nearly constant air volumes should be expected for all damper positions. Hence a mixing application appears to require dampers of a relatively linear *inherent* characteristic if there is no appreciable changes in pressure drop to alter this inherent damper characteristic.

Unfortunately, very few systems are designed in the above manner. Abrupt expansion, abrupt conversion losses, minor duct losses, entrance protective device losses, and so forth, all provide losses that will tend to change the pressure drop across the damper as it modulates. Therefore, each of the dampers (outdoor air, return air, and exhaust) is studied separately and within its own system (as was done in the face and bypass example). It will be seen that truly linear characteristics for mixing applications are not necessary.

IMPORTANCE OF PROPER DAMPER SIZING: For many years it has been common practice to select the size of an automatic control valve to handle the required maximum flow at a pressure drop that will produce good controllability. An automatic damper is also a flow control device and, like a valve, it accomplishes its function by changing the flow area. It follows that it also should be sized for a pressure drop that will produce good controllability.

The damper, however, has not received this consideration. Much of this can be blamed on the fact that few manufacturers of dampers have had sufficient information about their damper characteristics and pressure drops to permit consulting engineers or control engineers to select dampers on this basis. For this reason it has been extremely unusual for dampers to be selected and sized with system controllability in mind. Instead, the damper is selected to fit the duct at a point where it is most convenient to install it. This has created serious problems because these dampers are usually oversized and overall system controllability suffers.

Why size dampers? It is important that the damper be sized to use a reasonable portion of the total system pressure drop in the open position so as to enable it to perform its control functions. If this is done, it is then possible to determine the change in pressure drop that will occur as the damper closes and thereby predict the shift in the inherent characteristics of the damper. Dampers with the proper characteristics can then be chosen so that the predictable change in pressure drop across the damper will shift the inherent characteristic curve to an effective characteristic curve which will prevent cycling and result in stable control.

To correctly size a damper, it is important to know the following information about the system in which it will be installed:

a. *Total pressure drop* in the portion of the system in which the damper is to be located or the total pressure drop which can be expected across the damper when in the closed position.

b. *The total volume* of air which the damper is expected to pass in the wide-open position.

c. *The inherent characteristics* of the damper which is to be selected for this system.

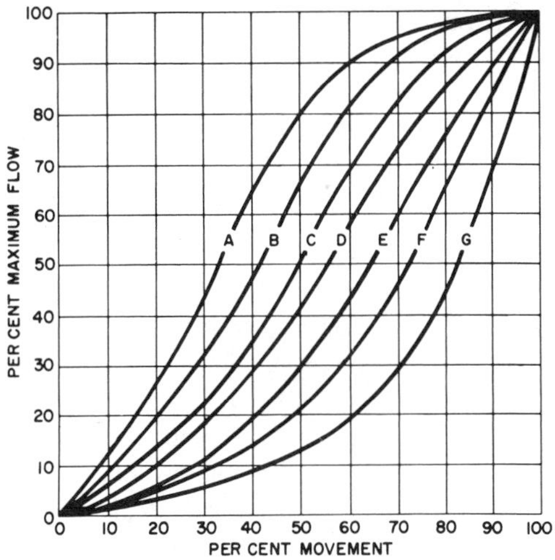

A-1
B-2
C-5
D-8
E-22
F-50

PER CENT OF TOTAL SYSTEM DROP THROUGH THE DAMPER IN THE WIDE OPEN POSITION

G- INHERENT CHARACTERISTIC OF THE DAMPER AT A CONSTANT PRESSURE DROP

Figure 16–55

Characteristics of an opposed blade damper (showing the effect of varying percent of system drop across the damper)

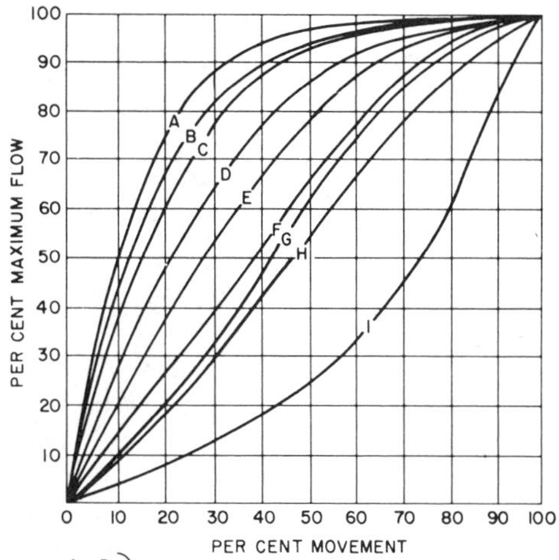

A-.5
B-.75
C-1.0
D-2.0
E-4.0
F-8.0
G-10.0
H-15.0

PER CENT OF TOTAL SYSTEM DROP THROUGH THE DAMPER IN THE WIDE OPEN POSITION

I - INHERENT CHARACTERISTIC OF THE DAMPER AT A CONSTANT PRESSURE DROP

Figure 16–56

Characteristics of a parallel blade damper (showing the effect of varying percent of system drop across the damper)

Figures 16-55 and 16-56 show the effect of variable pressure drops on the damper characteristics. These are the variations in pressure drop which occur, for example, when a damper is modulating to its closed position and the system resistance decreases, thus increasing the pressure drop across the damper. A study of the tables and the accompanying curves indicates the need for selecting the "wide-open" pressure drop by proper damper sizing. These curves are based on a particular range of damper sizes. As damper sizes change, these curves will very slightly.

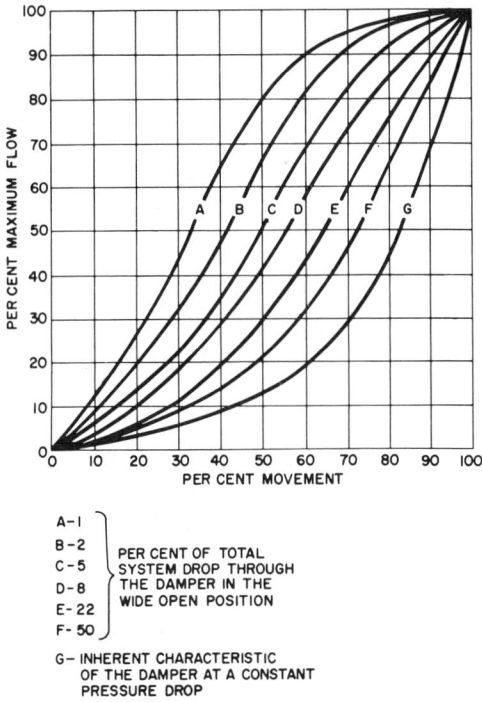

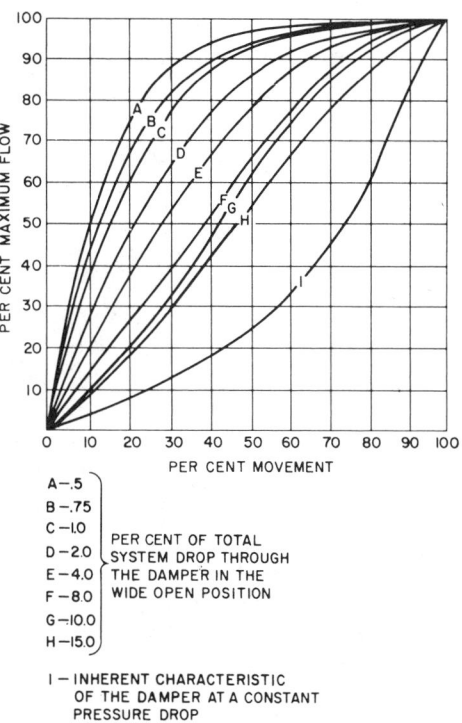

Curve *G* of Fig. 16-55 shows the inherent characteristic of an opposed-blade damper. It is the curve that results when the pressure drop across the damper remains constant regardless of damper position. For a throttling control application it was shown in Fig. 16-50 that the pressure drop across the damper increases from its wide-open value, at full flow, to full fan pressure when the damper is closed. Thus, if the wide-open pressure drop is 0.01 in. WG and the full fan pressure at no flow is 2.0 in. WG, the open damper resistance in percent of system resistance is 0.5 percent. Figure 16-55 shows the effect on the damper characteristic for various values of open damper resistance in percent of system resistance. Since a linear effective damper characteristic is desired for good control, curves *C* or *D* are best. Thus, a damper with an equal percentage inherent characteristic can be made essentially linear, in a throttling application, if it is sized so that its wide-open resistance falls in the range of 5 to 8 percent of the total system resistance.

For example, curve *A* of Fig. 16-55 is decidedly nonlinear. When 50 percent open, it has 80 percent of full open capacity. Most of its control is accomplished in less than one half of its travel. It would likely produce unstable control when near its closed position.

Figure 16-56 shows similar curves for a parallel-blade damper. For a throttling application its wide-open resistance would have to equal 10 to 15 percent of the total system resistance (curves *G* and *H*) to obtain a nearly linear effective characteristic. It is obvious that a damper with inherent equal percentage characteristics (opposed blade) requires less wide-open pressure drop to produce the desired effective characteristic under actual operation. As mentioned previously, Fig. 16-55 indicates that its wide-open resistance should be 5 to 8 percent of the total system resistance. Although this is a small percentage of the total system resistance, and should not be objectionable from the standpoint of added fan horsepower, it is a great deal more than can normally be expected if the dampers are sized the same as the duct.

To obtain the wide-open resistance required for proper control, the damper must usually be smaller than the duct. Correct damper sizing offers many advantages in addition to obtaining good control.

1. *Initial cost:* Correct damper sizing will generally mean dampers smaller than the duct. Dampers are usually priced by the square foot. Thus, the correctly sized damper will be less expensive. The smaller damper also requires less operating power and can reduce the cost of the damper operators.
2. *Leakage:* For a given damper construction and pressure drop, leakage is proportional to the damper area. A smaller damper will result in less leakage.
3. *Rangeability:* This is a term commonly applied to valves but is just as applicable to dampers. It is expressed as a ratio of maximum flow to minimum controllable flow. Reduced leakage decreases the uncontrollable flow and therefore increases the rangeability. This improves the controllability of the system.

Example of poor damper selection. A look at Fig. 16-56 shows why a parallel-blade damper is generally not suitable for throttling-type applications where the pressure drop across the damper increases appreciably as the damper (curve *I*) is more nearly linear than the inherent characteristics of an opposed-blade damper. To obtain the most nearly linear curve *H,* the damper would have to be sized to use up approximately 15 percent of the total system pressure drop in its wide-open position. If sized for 1 percent of the system pressure drop, a common occurrence, the damper would accomplish nearly 95 percent of its flow control in 50 percent of its travel (curve *C*). This would tend to produce unstable control and, since only about 50 percent of the total damper range is really effective, would multiply the effect of hysteresis and other losses.

Constant pressure drop applications. If the damper selected above is applied to a system where the pressure drop across the damper remains relatively constant, it can provide satisfactory control. Some mixing dampers are for this type of application. Since the pressure drop across the damper does not change appreciably, its inherent characteristics and effective actual characteristic are essentially the same. Sizing the wide-open damper to utilize a fair percentage of the total system pressure drop in this case is not as important as selecting the proper characteristic to produce constant total system flow.

DAMPER SIZING METHODS: Figures 16-55 and 16-56 indicate the percent of total system pressure that should be utilized by the damper in the wide-open position in order to avoid shifting the inherent characteristics to anything but the desired (near linear) characteristics. Figures 16-57 and 16-58 show the approach velocities that produce the pressure drops illustrated by the curves in Figs. 16-55 and 16-56. The approach velocity information is easier to use since it can be directly related to the duct area and, therefore, duct size if the cfm is known and no detailed pressure drop curves need be referred to.

Figure 16-59 shows the pressure drop that can be expected through a wide-open damper for various approach velocities and free area ratios. The free area ratio is the free area (total open area between blades and inside damper frame) divided by the total area of the damper. (Such data is usually available from the manufacturer.) The approach velocity is the velocity of the air in the duct approaching the complete damper area. Pressure drops in Fig. 16-59 are based on a duct and damper of the same size. From the information above it is obvious that selecting dampers to produce the desired characteristics, in most cases, adds very little pressure drop to the entire system. This drop

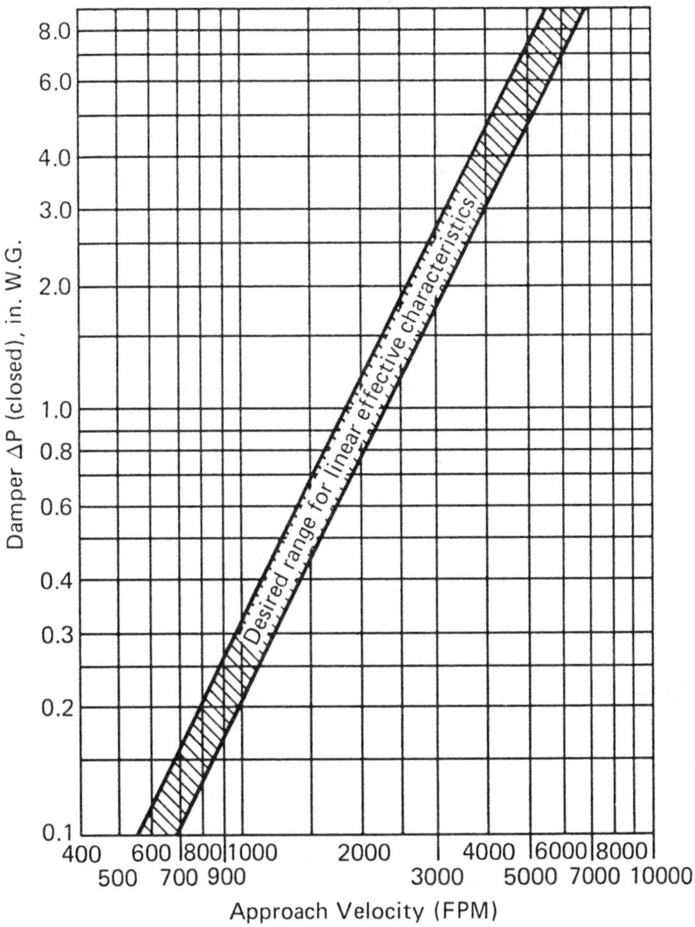

Figure 16–57

Desired pressure drops for opposed blade damper

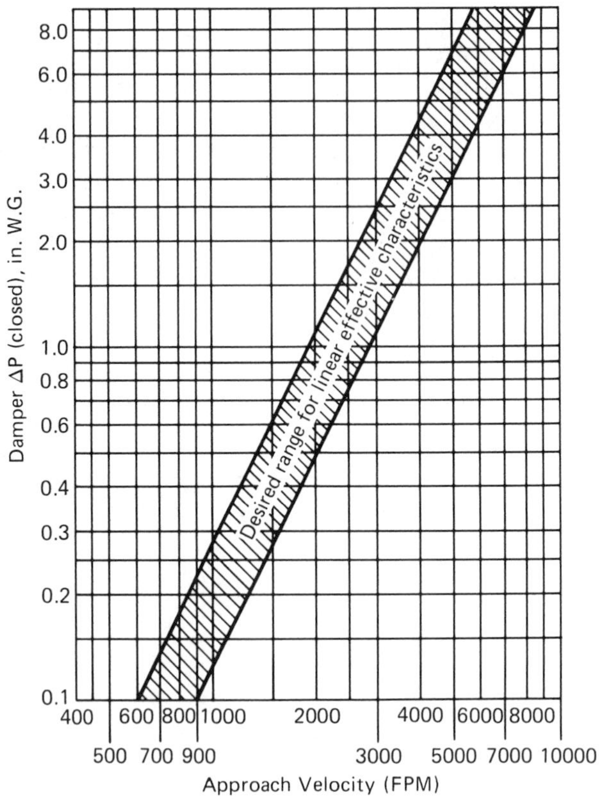

Figure 16–58

Desired pressure drops for parallel blade damper

should not be objectionable and will be offset by the superior system operation that can be expected.

If dampers are properly sized, a common question will be "where should they be located?" In most cases there is a conversion section within the duct system in which the properly sized damper could be located. In practically all cases it will mean moving the damper away from the coil (velocities of 400 to 600 fpm) to a location in the system where the velocities are higher. Moving the damper away from the coil will also produce better distribution of air flow across the coil. A more controversial answer to the question might be that it is often possible to size the duct, or install conversion duct sections, to fit the damper. There are also inexpensive and convenient means for making the properly sized damper fit the duct while adding little pressure drop.

APPLICATION OF PROPERLY SIZED DAMPERS: Selecting the damper with proper characteristics to do a good control job is relatively easy if characteristic curves are published by the damper manufacturers. Selecting a damper of proper size to match the selected characteristic to the system in which it is to be installed becomes feasible as damper manufacturers begin to take an engineering approach to the subject and publish the necessary test data. The only problem that remains is what to do if the properly sized damper is smaller than the duct in which it is to be installed. There are a number of approaches to this problem.

1. *Adapting the duct to fit the damper:* It is common practice for ducts to be designed to fit coils, filters, spray banks, and even "bug screens." In many cases these components are of no more importance to the proper functioning of the system than the dampers. Realizing this, the designing engineer will find cases where the duct can be made to conform to the dimensions of a properly sized damper.
2. *Damper location:* The usual practice of locating a face damper near the coil has some disadvantages. Air distribution through the coil is poor. When the damper is throttling, there is a tendency to produce a number of relatively high-velocity air jets through portions of the coil. This a major cause of freezing of preheat coils. In many cases there is a conversion section to reduce the duct size downstream from the coil. In this conversion section there is sometimes a location that provides the desired area for a correctly sized damper. If not, the damper may sometimes be located in the smaller distribution duct beyond the conversion section.
3. *Adapting the damper to fit the duct:* There will be a large number of cases where, regardless of the above suggestions, the properly sized damper will be smaller than the duct into which it is to be installed and some arrangements will be needed to accommodate the damper to the duct. In most installations it will be possible to make one dimension of the duct the same as that of the damper.

A simple blank-off plate, Fig. 16-60,

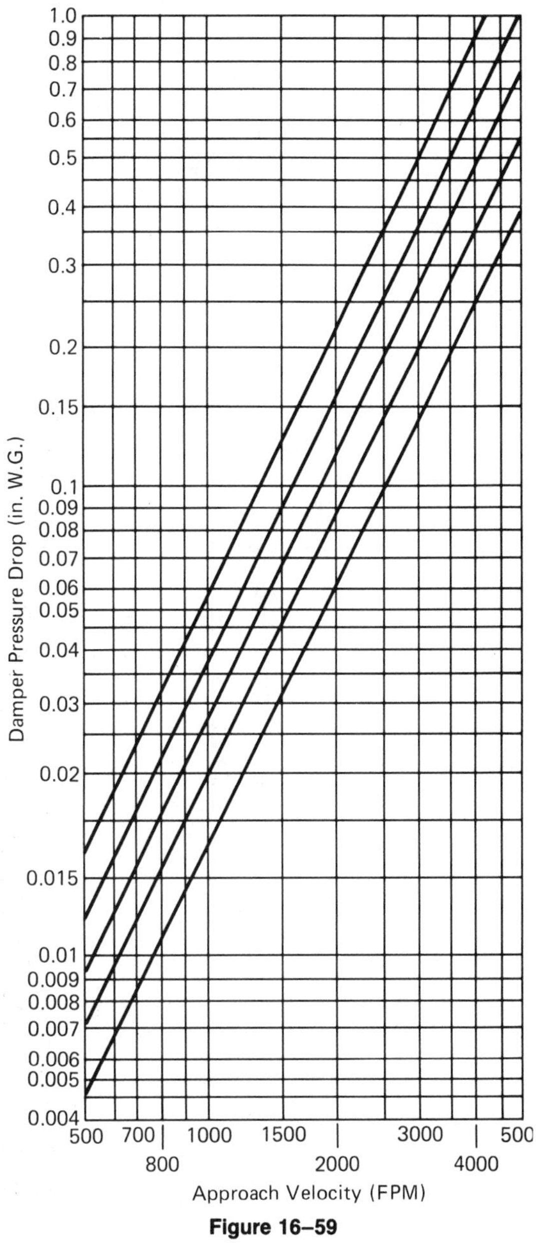

Figure 16–59

Pressure drop through wide open damper

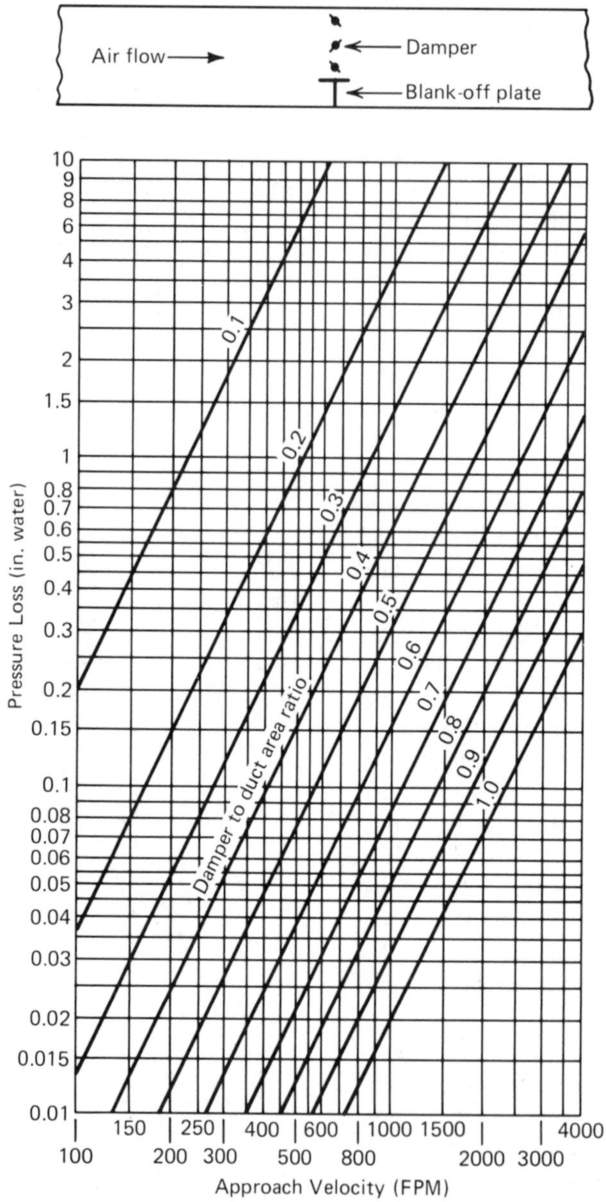

Figure 16–60

Effect of blank-off plate

is an easy way of filling the duct. While this arrangement creates more turbulence, and therefore a greater pressure loss than more elaborate arrangements, it has the advantage of being inexpensive. As Fig. 16-60 indicates, the pressure drops created by this arrangement are not excessive if the blank-off plate is less than 30 percent of the duct area. If greater reductions in duct area are required, more elaborate conversion sections must be used, unless damper sizes are being reduced purposely to introduce excessive pressure drops.

While adapting the damper to the duct system presents some problems, there are usually only a few dampers in even a fairly elaborate central air-conditioning system.

THROTTLING APPLICATION (OPPOSED-BLADE DAMPER WITH EQUAL-PERCENTAGE LINKAGE): Figure 16-55 shows how the inherent characteristics are shifted upward when a damper is installed in the system in which the pressure drop varies as the damper throttles the air flow. In order to shift the inherent characteristic to the desired nearly linear effective characteristic, the damper must be sized so that its wide-open pressure drop is a certain percent of the pressure drop across the damper when it is closed. It can be seen that curves *C* and *D* of Fig. 16-55 provide the most nearly linear effective characteristics. Figure 16-57 shows the damper velocities that produce the wide-open pressure drops necessary to create curves *C* and *D* in Fig. 16-55. This damper velocity information is to be used for sizing dampers for throttling applications. It can be directly related to the duct area and duct size if the cfm is known and no detailed pressure drop curves need be referred to.

The desired range band in Fig. 16-57 shows the approach velocities necessary to produce an effective characteristic curve which approaches curve *C* or *D* of Fig. 16-55. The vertical scale of Fig. 16-57 shows Δp damper (closed) in inches water gauge. (Δp is the differential pressure or pressure drop.) This is the maximum static pressure expected on the entering side of the damper when closed or the anticipated pressure drop across it when closed. In many systems this will be the maximum static pressure the fan can develop.

Illustrative Problem 16-7

Assume a system that has a total closeoff pressure of 1.0 in. WG and the system flow is 15,000 cfm. What size damper should be used?

Solution:

From Fig. 16-57 for a closeoff pressure of 1.0 in. WG, the desired damper velocity equals 1800 to 2300 fpm (use 2000 fpm for convenience).

Required damper area is

$$A = \frac{Q}{V} = \frac{15000}{2000} = 7.5 \text{ ft}^2$$

By referring to manufacturers' data, the rectangular dimensions of the damper may be determined. For example, a 30 by 36 in. damper has a nominal area of 7.5 ft^2. This damper could be selected for this application, and the duct should be sized to match the damper size. If the manufacturer of the above damper also quotes the free area as 5.73 ft^2, the free area ratio would be 5.73/7.50 = 0.76. Referring to Fig. 16-59, the pressure drop through the wide-open damper would be approximately 0.085 in. WG.

Smaller-than-duct-size dampers. When a damper is properly sized, it will often be smaller than duct size, as previously stated. A blank-off plate can be used in the majority of applications to adapt the smaller damper to the duct size. Figure 16-61 shows that when reducing the duct area by 30 percent or less, a blank-off plate can be used.

If the duct area is to be reduced by more than 30 percent, the pressure drop added by the blankoff is usually objectionable, see Fig. 16-60. Since the blank-off plate affects the characteristics of the damper, a larger percent of the total closeoff pressure must be used for the damper and blank-off combination. This percent is larger than is normally used for a duct size damper (see Fig. 16-65). Figure 16-61 shows the proper wide-open velocity for the dampers installed in ducts with various per-

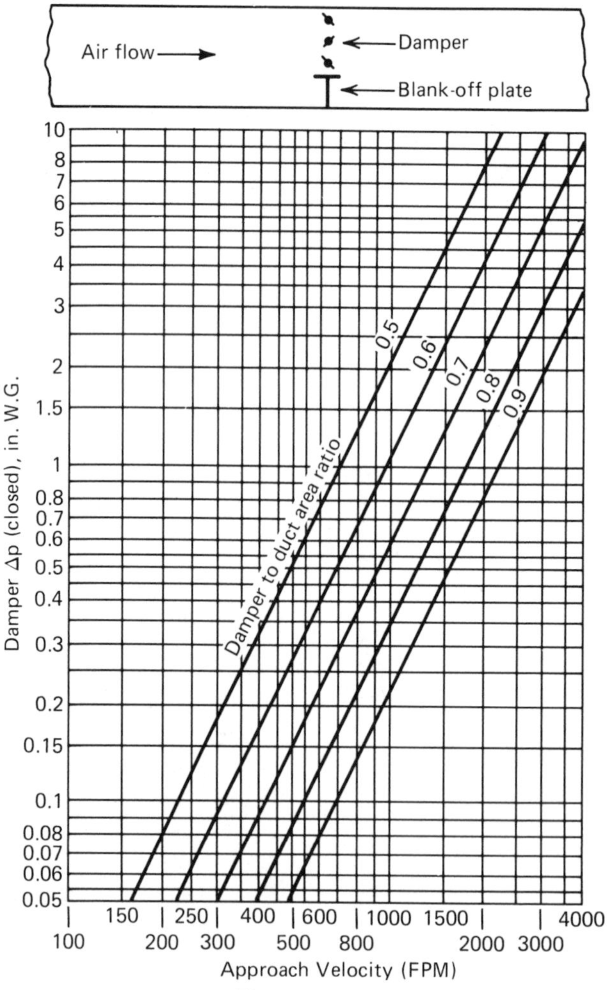

Figure 16–61

Desired pressure drop for opposed blade dampers with blank-off

cents of blank off. Using this chart will produce an actual effective characteristic close to curves *C* and *D* of Fig. 16-55 used for duct size dampers.

FACE-AND-BYPASS APPLICATION (PARALLEL-BLADE DAMPER WITH EQUAL-PERCENTAGE LINKAGE): The parallel-blade damper lends itself to the face-and-bypass application because of the better mixing effect downstream. Figure 16-62 shows the proper method of installing the parallel-blade damper to obtain the best mixing.

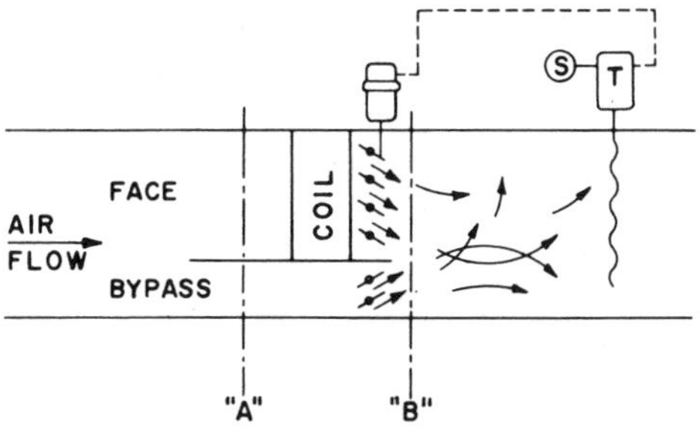

Figure 16–62

Face-and-bypass system (Repeat of Figure 16–52)

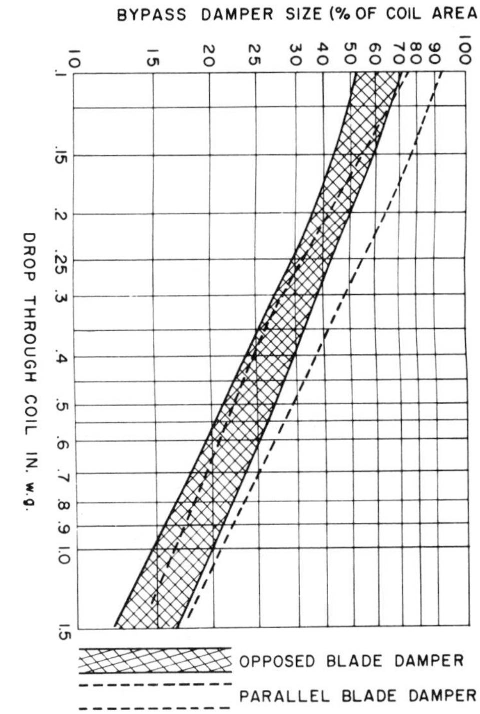

Figure 16–63

Bypass damper sizing curve

The face-and-bypass damper is treated like a throttling application because of the internal pressure changes within the face-and-bypass section. As stated previously, the pressure drop through the coil is transferred to the face damper as the face damper closes. The pressure drop across the bypass also changes due to the smaller-than-duct size effect. The face damper is sized to equal the area of the coil. By sizing the bypass damper from Fig. 16-63, 100 percent flow will be maintained through the face-and-bypass section for all positions of the face-and-bypass damper (see Fig. 16-53.)

Illustrative Problem 16-8

Assume a system of 5000 cfm total flow, a specified coil face velocity of 600 fpm and a 0.25 in. WG pressure drop through the coil. Determine the size of the face-and-bypass damper sections.

Solution:

The face damper will be coil size, or as close to coil size as possible. If the actual duct size is not known, size it for the specified velocity through the coil.

Face area $(A) = Q/V = 5000/600 = 8.33\ \text{ft}^2$

Checking manufacturers' data we could select a standard damper, say 30 by 42 in., giving a nominal area of 8.75 ft^2 for the face damper.

Using the parallel-blade damper, the bypass section and damper size should be about 40 percent of the face damper section (refer to Fig. 16-63). Therefore,

$$\text{Bypass damper area} = 0.40 \times 8.75$$
$$= 3.50 \text{ ft}^2$$

The bypass damper size would be 12 by 42 in., and its blade axis dimension (42 in.) is the same as the face damper for ease of installation.

MIXING APPLICATIONS: To properly size dampers for mixing applications such as outdoor air, return air, and exhaust dampers, each damper must be considered separately. Referring to Fig. 16-64 and assuming that the dampers will be selected to produce a constant system flow and, therefore, a constant-mixing plenum pressure, each damper must be considered separately in its own area of operation. The outdoor air damper must be considered as a throttling damper in the system from point *A* (atmospheric pressure) to point *B* (constant-mixing plenum pressure). The pressure drops in this system are the duct and entrance losses, the drop through the weather louver, and the drop through the damper. These pressure drops will shift to the damper as the flow is reduced and, therefore, shift the inherent characteristics of the outdoor air damper.

The return air damper in the system must be considered from point *C* (a constant pressure point in the return system) to point *B*. The losses in this section are the entrance and exit losses to that branch of the return system and the minor duct loss in that section, plus the loss of the return air damper. The exhaust damper is considered from point *C* to point *D* with similar pressure losses, which transfer to the damper as it closes.

Illustrative Problem 16-9

Assume a system as illustrated in Fig. 16-64. Total system volume delivered to the conditioned space is 10,000 cfm. The minimum outdoor air requirement for ventilation is 25 percent, or 2500 cfm. The return air fan has been selected to deliver 7500 cfm return air. The return fan is sized with sufficient discharge static pressure to overcome the duct friction between the fan discharge and the exhaust damper or the return damper, whichever is the greater. Determine size and characteristics of the outdoor air, return air, and exhaust dampers. (Typical pressure drops will be assumed as the problem progresses.)

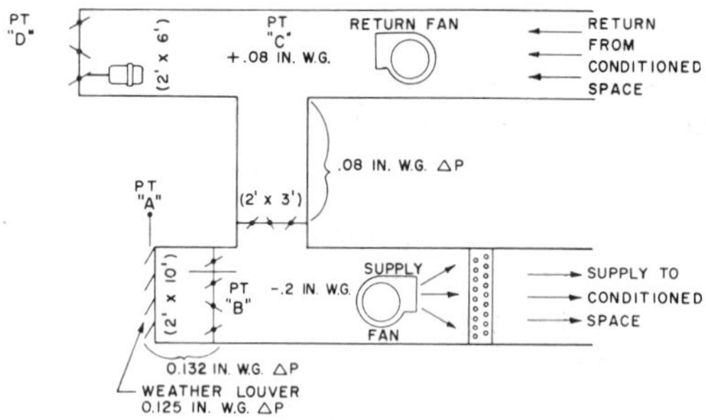

Figure 16–64

Typical constant pressure drop application

<u>*Solution:*</u>

<u>*Outdoor air damper.*</u> Reduced-size dampers are usually difficult to use on outdoor air intakes. Tapered conversion sections cannot be used since it is usually difficult to install them between the weather louver and the outdoor air damper due to space limitations in the fan room. Reducing the outdoor air damper with blank-off plates is not permitted in a close-coupled arrangement since the velocity will increase in certain areas of the weather louver and snow and rain will be drawn into the outdoor air intake. The outdoor air damper must, by necessity, be sized to match the weather louver size.

For Fig. 16-64 assume a normal velocity of 500 fpm through the weather louver at a pressure drop of 0.125 in. WG. To accommodate the 10,000 cfm at a velocity of 500 fpm, the area of the weather louver would be 10,000/500 = 20.0 ft^2. Assume the selected size of the weather louver to be 10.0 ft long by 2.0 ft high; the outdoor air damper will be selected to fit these dimensions. For this example the following modules are selected. One module will be a nominal 30 in. parallel to the blade axis by 24 in. perpendicular to the blade axis, and two modules will be 48 in. and 42 in. parallel to the blade axis by 24 in. perpendicular to the blade axis. The 30 by 24 in. module will be used as the 25 percent minimum outdoor air damper, and the remaining modules will be used as a maximum outdoor air damper.

Characteristics. Consider the system *A-B* as a small throttling system. The total pressure drop in this system is the sum of the loss in the outdoor weather louver, the losses due to the abrupt conversion at the entrance, and some small duct loss. The drop across the weather louver is 0.125 in. WG. The pressure drop across the outdoor air damper must be determined by referring to Fig. 16-59. Assume that a typical free area ratio of this damper to be 0.75, and with the approach velocity of 500 fpm, the pressure drop across the wide-open damper is about 0.007 in. WG. The total pressure drop between points *A* and *B* will then be (0.125 + 0.007) = 0.132 in. WG, plus the drop caused by conversion losses mentioned above, for a total of approximately 0.20 in. WG. This 0.20 in. WG will be the drop across the damper when it is closed or the total system resistance of system *A-B*. Figure 16-57 shows that with a system resistar.ce of 0.20 in. WG damper velocities between 775 and 1000 fpm will produce desired linear effective characteristics. The damper approach velocity in this problem is only 500 fpm.

A full-size damper with equal percentage characteristics will therefore not produce the desired linear characteristics but will approach an effective characteristic curve between curves *A* and *B* in Fig. 16-55. Since the effective characteristics of the outdoor air damper are not completely linearized, the return air damper must be selected and sized to produce a characteristic complimentary to that produced by the outdoor air damper.

Return air damper. Constant-mixing plenum pressure will assure a relatively constant pressure at the suction of the fan and therefore do a great deal to produce constant system volume. Proper sizing of the return air damper can produce this required constant-mixing plenum pressure.

The pressure drop in the system in which the return air damper throttles is from *B* (−0.20 in. WG) to *C* (+0.08 in. WG) or 0.28 in. WG. In an installation the exact operating pressure at points *B* and *C* are difficult to determine. The pressure drops of the components in the outdoor air section are available. In this example they are the damper and weather louver, which produce a drop of 0.132 in. WG. The return air damper should be sized to produce a similar drop.

The velocity in the return duct is 7500/2 × 3 = 1250 fpm. Figure 16-60 indicates that a duct velocity of 1250 fpm will produce a 0.132-in. WG drop if the damper is approximately 70 percent of the duct area. The damper area should therefore be 6 × 0.70 = 4.2 ft^2. A 2 by 2 ft damper with a nominal free area of 4.0 ft^2 would be selected.

Since the return fan was selected and the system balanced to overcome the losses in the return duct, the +0.08 in. WG is lost in moving the return air to the return damper. The pressure at the entrance to the damper is atmospheric or 0.0 in. WG. Therefore, the presssure in the mixing plenum will be equal to the pressure drop at the damper and the exit losses due to the abrupt expansion of the relatively high-velocity air. The damper pressure drop (wide open) as figured above is 0.132 in. WG. The abrupt expansion losses for the velocity ratio involved and the 1250-fpm return duct velocity (0.10 in. WG velocity pressure) is approximately 0.65 velocity heads or 0.065 in. WG. The total drop is 0.132 + 0.065 = 0.197 in. WG, which is very close to the 0.20 in. WG in the outdoor air duct.

The drop through the wide open damper is 0.132 in. WG. The drop across the closed damper is 0.28 in. WG. The percent of the total system drop through the damper in the wide-open position is 0.132/0.28 = 0.47 or 47 percent. The curves in Fig. 16-65 show that a wide-open pressure drop of 47 percent will produce a curve similar to curve *F*. Adding the effective characteristics of the outdoor and return air dampers, as is done in Fig. 16-66, produces a total system volume of near 100 percent at all positions of the dampers. The pressure at point *B* will also remain relatively constant at all damper positions, further assuring that total system volume will remain at a near constant value.

In the many systems tested, evaluated, and investigated, the return air damper sized in this manner produced from 45 to 75 percent of the total resistance of its return air section. This variation in resistance ratio has little effect on the actual effective characteristics of the return air damper (see Fig. 16-65) and hence the total volume of the system. In all cases where return air dampers were sized to produce the same wide-open pressure drop as is produced by the components (damper, weather louver, and at times, a preheat coil) in the outdoor air system at full flow, the total system flow varied less than 10 percent from the desired design flow.

If the return air damper is sized in the way described above, the velocity through it will be relatively high (about 1875 fpm). If the return duct at the entrance to the plenum covers the full width of the plenum, the high-velocity airstream from the return air will do an effective job of decreasing statification when it interrupts the low-velocity stream from the outdoor

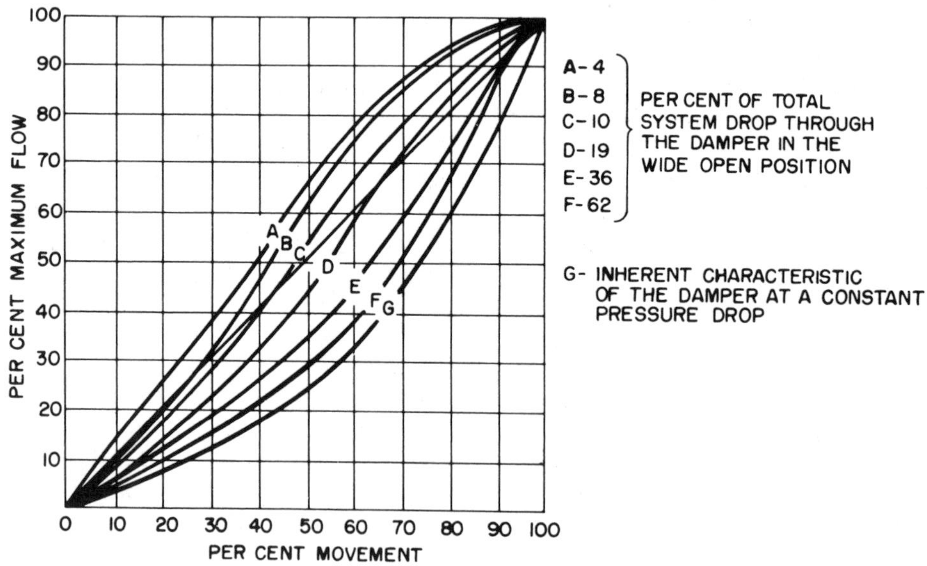

Figure 16–65

System characteristics for opposed blade damper with blank-off

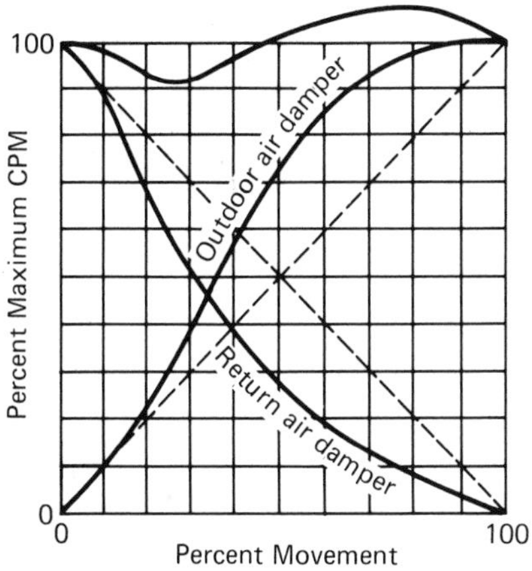

Figure 16–66

Desired effective characteristic curves for mixing damper application

air section. Two advantages are therefore gained by using this sizing method; constant total system volume and reduction in stratification problems.

Exhaust damper. When the return air damper is in the fully closed position, there is a 0.08 in. WG static pressure available to overcome duct friction, the loss in the exhaust damper, and exit loss due to protection screens, expansion of the air stream, and so forth. (The duct losses and exit losses are assumed to be approximately 0.070 in. WG) The exhaust damper must be sized to utilize the remaining available pressure, which is less than 0.01 in. WG. The exhaust damper will handle 7500 cfm maximum outdoor air while positive building pressure will be maintained with 2500 cfm minimum outdoor air.

For this example assume a duct-sized damper. Applying a duct-sized damper with nominal dimensions of 2.0 by 6.0 ft, and by referring to Fig. 16-59, it is found that the pressure drop through the wide-open damper of this size with an approach velocity of $7500/(2 \times 6) = 625$ fpm (duct velocity when damper is wide open) would be 0.007 in. WG for a free area ratio of 0.80. The duct-sized damper would be acceptable for this application. Again considering the damper within its own little throttling system (point *C* to point *D* of Fig. 16-64) and referring to Fig. 16-57 (desired approach velocities), it can be seen that this damper will produce near linear effective characteristics. The damper Δp (closed) will be 0.08 in. WG and the desired approach velocity for this duct-sized damper should be between 475 and 625 fpm. This means that the pressure drop should change the equal percentage inherent characteristic to a near linear effective characteristic. With the outdoor, return, and exhaust dampers sized and operated as described above, the damper and system flow characteristic would very closely approximate those of Fig. 16-66. By applying dampers as described, the total system volume is maintained very close to a constant 100 percent.

DAMPER LEAKAGE: Manufacturers of control dampers quote performance data in different ways, and it is necessary to carefully check such data to enable us to select properly. Data concerning damper leakage may be quoted as actual cfm leakage when damper closes against a given static pressure, or it may be quoted as percent leakage. The former method is preferred. The percent leakage may be defined differently by different manufacturers, and percent leakage means nothing unless the total volume on which this leakage is based is known. Some manufacturers refer to percent leakage as a percentage of the volume that could theoretically pass through the damper at a wide-open pressure drop equal to the closeoff pressure. Others may refer to it as the percent of volume that will pass through the damper for a given application.

CHAPTER 16

Review

16-1. Air conditioning involves the control of what four conditions of the air?

16-2. The basic difference between unitary or packaged systems and central-fan systems is what?

16-3. Define the following terms:
(a) ventilation
(b) discharge air
(c) face-and-bypass damper system?
(d) unitary system
(e) high velocity-double duct
(f) packaged multizone

16-4. What is the function of the outdoor air sensor in an economizer system?

16-5. What are three types of controlled mediums normally used in heating systems?

16-6. Name three typical locations for primary sensors.

16-7. What kind of space controller could you use to operate two compressors in a central-fan system?

16-8. When more than two stages of cooling are used, what device is used to operate the stages?

16-9. What are four basic types of humidifiers commonly used in central-fan systems?

16-10. Determine the flow coefficient (C_v) of a water control valve for a flow rate of 9000 gpm and a pressure loss of 4 psi.

16-11. Determine the pressure drop across a water control valve that has a flow coefficient (C_v) of 4500 and where the flow rate is 9000 gpm.

16-12. A water control valve has a flow coefficient (C_v) of 4500. What will be the flow rate in gpm through the valve for a pressure drop of 9.0 psi?

16-13. Benzine (SG = 0.69) is to flow through a control valve at the rate of 9000 gpm with an allowed pressure drop of 4.0 psi. What is the required flow coefficient of the valve?

16-14. A steam control valve has a flow coefficient (C_v) of 20. The inlet steam pressure is 20.0 psig and the outlet pressure is 18 psig. What will be the flow rate through the valve?

16-15. A steam control valve is to be selected for an application requiring a flow rate of 675 lb/hr of saturated steam. The initial steam pressure is 15 psig. What should be the flow coefficient (C_v) for the valve?

BIBLIOGRAPHY

1. Commercial Air Conditioning Controls, ''Basic Control Systems,'' Honeywell, Inc. Milwaukee, Wisconsin, 1972.

2. Engineering Data, Section Vb, ''Control Valves,'' Johnson Controls, Inc. Milwaukee, Wisconsin, 1980

3. Damper Manual, Johnson Controls, Inc. Milwaukee, Wisconsin, 1979.

Appendix A

Thermodynamic Properties of Steam and Dry Air

Table A-1*

Saturated Steam: Temperature

	Specific Volume			Internal Energy			Enthalpy			Entropy		
Temp. Fahr. t	.Press. Lbf. Sq. In. p	Sat. Liquid v_f	Sat. Vapor v_g	Sat. Liquid u_f	Evap. u_{fg}	Sat. Vapor u_g	Sat. Liquid h_f	Evap. h_{fg}	Sat. Vapor h_g	Sat. Liquid s_f	Evap. s_{fg}	Sat. Vapor s_g
32	*.08859*	*.016022*	*3305.*	*.01*	*1021.2*	*1021.2*	*.01*	*1075.4*	*1075.4*	*.00003*	*2.1870*	*2.1870*
32.018	.08866	.016022	3302.	.00	1021.2	1021.2	.01	1075.4	1075.4	.00000	2.1869	2.1869
35	.09992	.016021	2948.	2.99	1019.2	1022.2	3.00	1073.7	1076.7	.00607	2.1704	2.1764
40	.12166	.016020	2445.	8.02	1015.8	1023.9	8.02	1070.9	1078.9	.01617	2.1430	2.1592
45	.14748	.016021	203.7	13.04	1012.5	1025.5	13.04	1068.1	1081.1	.02618	2.1162	2.1423
50	.17803	.016024	1704.2	18.06	1009.1	1027.2	18.06	1065.2	1083.3	.03607	2.0899	2.1259
60	.2563	.016035	1206.9	28.08	1002.4	1030.4	28.08	1059.6	1087.7	.05555	2.0388	2.0943
70	.3632	.016051	867.7	38.09	995.6	1033.7	38.09	1054.0	1092.0	.07463	1.9896	2.0642
80	.5073	.016073	632.8	48.08	988.9	1037.0	48.09	1048.3	1096.4	.09332	1.9423	2.0356
90	.6988	.016099	467.7	58.07	982.2	1040.2	58.07	1042.7	1100.7	.11165	1.8966	2.0083
100	.9503	.016130	350.0	68.04	975.4	1043.5	68.05	1037.0	1105.0	.12963	1.8526	1.9822
110	1.2763	.016166	265.1	78.02	968.7	1046.7	78.02	1031.3	1109.3	.14730	1.8101	1.9574
120	1.6945	.016205	203.0	87.99	961.9	1049.9	88.00	1025.5	1113.5	.16465	1.7690	1.9336
130	2.225	.016247	157.17	97.97	955.1	1053.0	97.98	1019.8	1117.8	.18172	1.7292	1.9109
140	2.892	.016293	122.88	107.95	948.2	1056.2	107.96	1014.0	1121.9	.19851	1.6907	1.8892
150	3.722	.016343	96.99	117.95	941.3	1059.3	117.96	1008.1	1126.1	.21503	1.6533	1.8684
160	4.745	.016395	77.23	127.94	934.4	1062.3	127.96	1002.2	1130.1	.23130	1.6171	1.8484
170	5.996	.016450	62.02	137.95	927.4	1065.4	137.97	996.2	1134.2	.24732	1.5819	1.8293
180	7.515	.016509	50.20	147.97	920.4	1068.3	147.99	990.2	1138.2	.26311	1.5478	1.8109
190	9.343	.016570	40.95	158.0	913.3	1071.3	158.03	984.1	1142.1	.27866	1.5146	1.7932
200	11.529	.016634	33.63	168.04	906.2	1074.2	168.07	977.9	1145.9	.29400	1.4822	1.7762
210	14.125	.016702	27.82	178.10	898.9	1077.0	178.14	971.6	1149.7	.30913	1.4508	1.7599
220	17.188	.016772	23.15	188.17	891.7	1079.8	188.22	965.3	1153.5	.32406	1.4201	1.7441
230	20.78	.016845	19.386	198.26	884.3	1082.6	198.32	958.8	1157.1	.33880	1.3901	1.7289
240	24.97	.016922	16.327	208.36	876.9	1085.3	208.44	952.3	1160.7	.35335	1.3609	1.7143
250	29.82	.017001	13.826	218.49	869.4	1087.9	218.59	945.6	1164.2	.36772	1.3324	1.7001
260	35.42	.017084	11.768	228.64	861.8	1090.5	228.76	938.8	1167.6	.38193	1.3044	1.6864

**Tables A-1 through A-3 are abridged from* Steam Tables—Thermodynamic Properties of Water Including Vapor and Liquid Phases *by J. H. Keenan, P. G. Hill, and J. G. Moore, John Wiley and Sons, Inc., New York, 1969, with permission. From* Thermodynamics and Heat Power *by Irving Granet, Reston Publishing Company, Reston, Virginia, 1974.*

Table A-1 (Cont.)

	Specific Volume			Internal Energy			Enthalpy			Entropy		
Temp. Fahr.	Press. Lbf. Sq. In.	Sat. Liquid	Sat. Vapor	Sat. Liquid	Evap.	Sat. Vapor	Sat. Liquid	Evap.	Sat. Vapor	Sat. Liquid	Evap.	Sat. Vapor
t	p	v_f	v_g	u_f	u_{fg}	u_g	h_f	h_{fg}	h_g	s_f	s_{fg}	s_g
270	41.85	.017170	10.066	238.82	854.1	1093.0	238.95	932.0	1170.9	.39597	1.2771	1.6731
280	49.18	.017259	8.650	249.02	846.3	1095.4	249.18	924.9	1174.1	.40986	1.2504	1.6602
290	57.53	.017352	7.467	259.25	838.5	1097.7	259.44	917.8	1177.2	.42360	1.2241	1.6477
300	66.98	.017448	6.472	269.52	830.5	1100.0	269.73	910.4	1180.2	.43720	1.1984	1.6356
310	77.64	.017548	5.632	279.81	822.3	1102.1	280.06	903.0	1183.0	.45067	1.1731	1.6238
320	89.60	.017652	4.919	290.14	814.1	1104.2	290.43	895.3	1185.8	.46400	1.1483	1.6123
330	103.00	.017760	4.312	300.51	805.7	1106.2	300.84	887.5	1188.4	.47722	1.1238	1.6010
340	117.93	.017872	3.792	310.91	797.1	1108.0	311.30	879.5	1190.8	.49031	1.0997	1.5901
350	134.53	.017988	3.346	321.35	788.4	1109.8	321.80	871.3	1193.1	.50329	1.0760	1.5793
360	152.92	.018108	2.961	331.84	779.6	1111.4	332.35	862.9	1195.2	.51617	1.0526	1.5688
370	173.23	.018233	2.628	342.37	770.6	1112.9	342.96	854.2	1197.2	.52894	1.0295	1.5585
380	195.60	.018363	2.339	352.95	761.4	1114.3	353.62	845.4	1199.0	.54163	1.0067	1.5483
390	220.2	.018498	2.087	363.58	752.0	1115.6	364.34	836.2	1200.6	.55422	.9841	1.5383
400	247.1	.018638	1.8661	374.27	742.4	1116.6	375.12	826.8	1202.0	.56672	.9617	1.5284
425	325.6	.019014	1.4249	401.24	717.4	1118.6	402.38	802.1	1204.5	.59767	.9066	1.5043
450	422.1	.019433	1.1011	428.6	690.9	1119.5	430.2	775.4	1205.6	.6282	.8523	1.4806
475	539.3	.019901	.8594	456.6	662.6	1119.2	458.5	746.4	1204.9	.6586	.7985	1.4571
500	680.0	.02043	.6761	485.1	632.3	1117.4	487.7	714.8	1202.5	.6888	.7448	1.4335
525	847.1	.02104	.5350	514.5	599.5	1113.9	517.8	680.0	1197.8	.7191	.6906	1.4007
550	1044.0	.02175	.4249	544.9	563.7	1108.6	549.1	641.6	1190.6	.7497	.6354	1.3851
575	1274.0	.02259	.3378	576.5	524.3	1100.8	581.9	598.6	1180.4	.7808	.5785	1.3593
600	1541.0	.02363	.2677	609.9	480.1	1090.0	616.7	549.7	1166.4	.8130	.5187	1.3317
625	1849.7	.02494	.2103	645.7	429.4	1075.1	654.2	492.9	1147.0	.8467	.4544	1.3010
650	2205.	.02673	.16206	685.0	368.7	1053.7	695.9	423.9	1119.8	.8831	.3820	1.2651
675	2616.	.02951	.11952	731.0	289.3	1020.3	745.3	332.9	1078.2	.9252	.2934	1.2186
700	3090.	.03666	.07438	801.7	145.9	947.7	822.7	167.5	990.2	.9902	.1444	1.1346
705.44	3204.	.05053	.05053	872.6	0	872.6	902.5	0	902.5	1.0580	0	1.0580

Table A-2

Saturated Steam: Pressure

		Specific Volume		Internal Energy			Enthalpy			Entropy		
Press. Lbf./Sq.In. p	Temp. Fahr. t	Sat. Liquid v_f	Sat. Vapor v_g	Sat Liquid u_f	Evap. u_{fg}	Sat. Vapor u_g	Sat. Liquid h_f	Evap. h_{fg}	Sat. Vapor h_g	Sat. Liquid s_f	Evap. s_{fg}	Sat. Vapor s_g
.50	79.56	.016071	641.5	47.64	989.2	1036.9	47.65	1048.6	1096.2	.09250	1.9443	2.0368
1.0	101.70	.016136	333.6	69.74	974.3	1044.0	69.74	1036.0	1105.8	.13266	1.8453	1.9779
1.5	115.65	.016187	227.7	83.65	964.8	1048.5	83.65	1028.0	1111.7	.15714	1.7867	1.9438
2.0	126.04	.016230	173.75	94.02	957.8	1051.8	94.02	1022.1	1116.1	.17499	1.7448	1.9198
3.0	141.43	.016300	118.72	109.38	947.2	1056.6	109.39	1013.1	1122.5	.20089	1.6852	1.8861
4.0	152.93	.016358	90.64	120.88	939.3	1060.2	120.89	1006.4	1127.3	.21983	1.6426	1.8624
5.0	162.21	.016407	73.53	130.15	932.9	1063.0	130.17	1000.9	1131.0	.23486	1.6093	1.8441
7.5	179.91	.016508	50.30	147.88	920.4	1068.3	147.90	990.2	1138.1	.26297	1.5481	1.8110
10	193.19	.016590	38.42	161.20	911.0	1072.2	161.23	982.1	1143.3	.28358	1.5041	1.7877
14.696	211.99	.016715	26.80	180.10	897.5	1077.6	180.15	970.4	1150.5	.31212	1.4446	1.7567
15	213.03	.016723	26.29	181.14	896.8	1077.9	181.19	969.7	1150.9	.31367	1.4414	1.7551
20	227.96	.016830	20.09	196.19	885.8	1082.0	196.26	960.1	1156.4	.33580	1.3962	1.7320
25	240.08	.016922	16.306	208.44	867.9	1085.3	208.52	952.2	1160.7	.35345	1.3607	1.7142
30	250.34	.017004	13.748	218.84	869.2	1088.0	218.93	945.4	1164.3	.36821	1.3314	1.6996
35	259.30	.017078	11.900	227.93	862.4	1090.3	228.04	939.3	1167.4	.38093	1.3064	1.6873
40	267.26	.017146	10.501	236.03	856.2	1092.3	236.16	933.8	1170.0	.39214	1.2845	1.6767
45	274.46	.017209	9.403	243.37	850.7	1094.0	243.51	928.8	1172.3	.40218	1.2651	1.6673
50	281.03	.017269	8.518	250.08	845.5	1095.6	250.24	924.2	1174.4	.41129	1.2476	1.6589
55	287.10	.017325	7.789	256.28	840.8	1097.0	256.46	919.9	1176.3	.41963	1.2317	1.6513
60	292.73	.017378	7.177	262.06	836.3	1098.3	262.25	915.8	1178.0	.42733	1.2170	1.6444
65	298.00	.017429	6.657	267.46	832.1	1099.5	267.67	911.9	1179.6	.43450	1.2035	1.6380
70	302.96	.017478	6.209	272.56	828.1	1100.6	272.79	908.3	1181.0	.44120	1.1909	1.6321
75	307.63	.017524	5.818	277.37	824.3	1101.6	277.61	904.8	1182.4	.44749	1.1790	1.6265
80	312.07	.017570	5.474	281.95	820.6	1102.6	282.21	901.4	1183.6	.45344	1.1679	1.6214
85	316.29	.017613	5.170	286.30	817.1	1103.5	286.58	898.2	1184.8	.45907	1.1574	1.6165
90	320.31	.017655	4.898	290.46	813.8	1104.3	290.76	895.1	1185.9	.46442	1.1475	1.6119
95	324.16	.017696	4.654	294.45	810.6	1105.0	294.76	892.1	1186.9	.46952	1.1380	1.6076
100	327.86	.017736	4.434	298.28	807.5	1105.8	298.61	889.2	1187.8	.47439	1.1290	1.6034
105	331.41	.017775	4.234	301.97	804.5	1106.5	302.31	886.4	1188.7	.47906	1.1204	1.5995
110	334.82	.017813	4.051	305.52	801.6	1107.1	305.88	883.7	1189.6	.48355	1.1122	1.5957
115	338.12	.017850	3.884	308.95	798.8	1107.7	309.33	881.0	1190.4	.48786	1.1042	1.5921
120	341.30	.017886	3.730	312.27	796.0	1108.3	312.67	878.5	1191.1	.49201	1.0966	1.5886
125	344.39	.017922	3.588	315.49	793.3	1108.8	315.90	875.9	1191.8	.49602	1.0893	1.5853

Table A-2 (Cont.)

Press. Lbf./Sq.In. p	Temp. Fahr. t	Specific Volume Sat. Liquid v_f	Specific Volume Sat. Vapor v_g	Internal Energy Sat Liquid u_f	Internal Energy Evap. u_{fg}	Internal Energy Sat. Vapor u_g	Enthalpy Sat. Liquid h_f	Enthalpy Evap. h_{fg}	Enthalpy Sat. Vapor h_g	Entropy Sat. Liquid s_f	Entropy Evap. s_{fg}	Entropy Sat. Vapor s_g
130	347.37	.017957	3.457	318.61	790.7	1109.4	319.04	873.5	1192.5	.49989	1.0822	1.5821
135	350.27	.017991	3.335	321.64	788.2	1109.8	322.08	871.1	1193.2	.50364	1.0754	1.5790
140	353.08	.018024	3.221	324.58	785.7	1110.3	325.05	868.7	1193.8	.50727	1.0688	1.5761
145	355.82	.018057	3.115	327.45	783.3	1110.8	327.93	866.4	1194.4	.51079	1.0624	1.5732
150	358.48	.018089	3.016	330.24	781.0	1111.2	330.75	864.2	1194.9	.51422	1.0562	1.5704
160	363.60	.018152	2.836	335.63	776.4	1112.0	336.16	859.8	1196.0	.52078	1.0443	1.5651
170	368.47	.018214	2.676	340.76	772.0	1112.7	341.33	855.6	1196.9	.52700	1.0330	1.5600
180	373.13	.018273	2.533	345.68	767.7	1113.4	346.29	851.5	1197.8	.53292	1.0223	1.5553
190	377.59	.018331	2.405	350.39	763.6	1114.0	351.04	847.5	1198.6	.53857	1.0122	1.5507
200	381.86	.018387	2.289	354.9	759.6	1114.6	355.6	843.7	1199.3	.5440	1.0025	1.5464
225	391.87	.018523	2.043	365.6	750.2	1115.8	366.3	834.5	1200.8	.5566	.9799	1.5365
250	401.04	.018653	1.8448	375.4	741.4	1116.7	376.2	825.8	1202.1	.5680	.9594	1.5274
275	409.52	.018777	1.6813	384.5	733.0	1117.5	385.4	817.6	1203.1	.5786	.9406	1.5192
300	417.43	.018896	1.5442	393.0	725.1	1118.2	394.1	809.8	1203.9	.5883	.9232	1.5115
350	431.82	.019124	1.3267	408.7	710.3	1119.0	409.9	795.0	1204.9	.6060	.8917	1.4978
400	444.70	.019340	1.1620	422.8	696.7	1119.5	424.2	781.2	1205.5	.6218	.8638	1.4856
450	456.39	.019547	1.0326	435.7	683.9	1119.6	437.4	768.2	1205.6	.6360	.8385	1.4746
500	467.13	.019748	.9283	447.7	671.7	1119.4	449.5	755.8	1205.3	.6490	.8154	1.4645
550	477.07	.019943	.8423	458.9	660.2	1119.1	460.9	743.9	1204.8	.6611	.7941	1.4551
600	486.33	.02013	.7702	469.4	649.1	1118.6	471.7	732.4	1204.1	.6723	.7742	1.4464
700	503.23	.02051	.6558	488.9	628.2	1117.0	491.5	710.5	1202.0	.6927	.7378	1.4305
800	518.36	.02087	.5691	506.6	608.4	1115.0	509.7	689.6	1199.3	.7110	.7050	1.4160
900	532.12	.02123	.5009	523.0	589.6	1112.6	526.6	669.5	1196.0	.7277	.6750	1.4027
1000	544.75	.02159	.4459	538.4	571.5	1109.9	542.4	650.0	1192.4	.7432	.6471	1.3903
1250	572.56	.02250	.3454	573.4	528.3	1101.7	578.6	603.0	1181.6	.7778	.5841	1.3619
1500	596.39	.02346	.2769	605.0	486.9	1091.8	611.5	557.2	1168.7	.8082	.5276	1.3359
1750	617.31	.02450	.2268	634.4	445.9	1080.2	642.3	511.4	1153.7	.8361	.4748	1.3109
2000	688600	.02565	8813	662.4	404.2	1066.6	671.9	464.4	1136.3	.8623	.4238	1.2861
2250	652.90	.02698	.15692	689.9	360.7	1050.6	701.1	414.8	1115.9	.8876	.3728	1.2604
2500	668.31	.02860	.13059	717.7	313.4	1031.0	730.9	360.5	1091.4	.9131	.3196	1.2327
2750	682.46	.03077	.10717	747.3	258.6	1005.9	763.0	297.4	1060.4	.9401	.2604	1.2005
3000	695.52	.03431	.08404	783.4	185.4	968.8	802.5	213.0	1015.5	.9732	.1843	1.1575
3203.6	705.44	.05053	.05053	872.6	0	872.6	902.5	0	902.5	1.0580	0	1.0580

Table A-3

Properties of Superheated Steam

	Vapor											
p(t Sat.)	**14.696 (211.99)**				**20 (227.96)**				**30 (250.34)**			
t	**v**	**u**	**h**	**s**	**v**	**u**	**h**	**s**	**v**	**u**	**h**	**s**
Sat.	26.80	1077.6	1150.5	1.7567	20.09	1082.0	1156.4	1.7320	13.748	1088.0	1164.3	1.6996
150	*24.10*	*1054.5*	*1120.0*	*1.7090*	*17.532*	*1052.0*	*1116.9*	*1.6710*	*11.460*	*1047.3*	*1111.0*	*1.6185*
160	*24.54*	*1058.2*	*1125.0*	*1.7171*	*17.870*	*1056.0*	*1122.1*	*1.6795*	*11.701*	*1051.6*	*1116.5*	*1.6275*
170	*24.98*	*1062.0*	*1129.9*	*1.7251*	*18.204*	*1059.9*	*1127.2*	*1.6877*	*11.938*	*1055.8*	*1122.0*	*1.6364*
180	*25.42*	*1065.7*	*1134.9*	*1.7328*	*18.535*	*1063.8*	*1132.4*	*1.6957*	*12.172*	*1059.9*	*1127.5*	*1.6449*
190	*25.85*	*1069.5*	*1139.8*	*1.7405*	*18.864*	*1067.6*	*1137.4*	*1.7036*	*12.403*	*1064.0*	*1132.9*	*1.6533*
200	*26.29*	*1073.2*	*1144.7*	*1.7479*	*19.191*	*1071.4*	*1142.5*	*1.7113*	*12.631*	*1068.1*	*1138.2*	*1.6615*
210	*26.72*	*1076.9*	*1149.5*	*1.7553*	*19.515*	*1075.2*	*1147.5*	*1.7188*	*12.857*	*1072.1*	*1143.5*	*1.6694*
220	27.15	1080.6	1154.4	1.7624	*19.837*	*1079.0*	*1152.4*	*1.7762*	*13.081*	*1076.1*	*1148.7*	*1.6771*
230	27.57	1084.2	1159.2	1.7695	20.157	1082.8	1157.4	1.7335	*13.303*	*1080.0*	*1153.9*	*1.6847*
240	28.00	1087.9	1164.0	1.7764	20.475	1086.5	1162.3	1.7405	*13.523*	*1084.0*	*1159.0*	*1.6921*
250	28.42	1091.5	1168.8	1.7832	20.79	1090.3	1167.2	1.7475	*13.741*	*1087.9*	*1164.1*	*1.6994*
260	28.85	1095.2	1173.6	1.7899	21.11	1094.0	1172.1	1.7543	13.958	1091.7	1169.2	1.7064
270	29.27	1098.8	1178.4	1.7965	21.42	1097.7	1177.0	1.7610	14.173	1095.6	1174.2	1.7134
280	29.69	1102.4	1183.1	1.8030	21.73	1101.4	1181.8	1.7676	14.387	1099.4	1179.2	1.7202
290	30.11	1106.0	1187.9	1.8094	22.05	1105.0	1186.6	1.7741	14.600	1103.2	1184.2	1.7269
300	30.52	1109.6	1192.6	1.8157	22.36	1108.7	1191.5	1.7805	14.812	1106.9	1189.2	1.7334
310	30.94	1113.2	1197.4	1.8219	22.67	1112.4	1196.3	1.7868	15.023	1110.7	1194.1	1.7399
320	31.36	1116.8	1202.1	1.8280	22.98	1116.0	1201.0	1.7930	15.233	1114.4	1199.0	1.7462
330	31.77	1120.4	1206.8	1.8340	23.28	1119.7	1205.8	1.7991	15.442	1118.2	1203.9	1.7525
340	32.19	1124.0	1211.6	1.8400	23.59	1123.3	1210.6	1.8051	15.651	1121.9	1208.8	1.7586
350	32.60	1127.6	1216.3	1.8458	23.90	1126.9	1215.4	1.8110	15.859	1125.6	1213.6	1.7646
360	33.02	1131.2	1221.0	1.8516	24.21	1130.6	1220.1	1.8168	16.067	1129.3	1218.5	1.7706
370	33.43	1134.8	1225.7	1.8574	24.51	1134.2	1224.9	1.8226	16.273	1133.0	1223.3	1.7765
380	33.84	1138.4	1230.5	1.8630	24.82	1137.8	1229.7	1.8283	16.480	1136.7	1228.1	1.7822
390	34.26	1142.0	1235.2	1.8686	25.12	1141.4	1234.4	1.8340	16.686	1140.3	1233.0	1.7880
400	34.67	1145.6	1239.9	1.8741	25.43	1145.1	1239.2	1.8395	16.891	1144.0	1237.8	1.7936
420	35.49	1152.8	1249.3	1.8850	26.03	1152.3	1248.7	1.8504	17.301	1151.4	1247.4	1.8047
440	36.31	1160.1	1258.8	1.8956	26.64	1159.6	1258.2	1.8611	17.709	1158.7	1257.0	1.8155
460	37.13	1167.3	1268.3	1.9060	27.25	1166.9	1267.7	1.8716	18.116	1166.1	1266.6	1.8260
480	37.95	1174.6	1277.8	1.9162	27.85	1174.2	1277.2	1.8819	18.523	1173.4	1276.2	1.8364

Table A-3 (Cont.)

Vapor												
p(t Sat.)	14.696 (211.99)				20 (227.96)				30 (250.34)			
t	v	u	h	s	v	u	h	s	v	u	h	s
Sat.	26.80	1077.6	1150.5	1.7567	20.09	1082.0	1156.4	1.7320	13.748	1088.0	1164.3	1.6996
500	38.77	1181.8	1287.3	1.9263	28.46	1181.5	1286.8	1.8919	18.928	1180.8	1285.9	1.8465
520	39.59	1189.1	1296.8	1.9361	29.06	1188.8	1296.3	1.9018	19.333	1188.2	1295.5	1.8564
540	40.41	1196.5	1306.4	1.9457	29.66	1196.1	1305.9	1.9114	19.737	1195.5	1305.1	1.8661
560	41.22	1203.8	1315.9	1.9552	30.26	1203.5	1315.5	1.9210	20.140	1203.0	1314.8	1.8757
580	42.04	1211.2	1325.5	1.9645	30.87	1210.9	1325.2	1.9303	20.543	1210.4	1324.4	1.8851
600	42.86	1218.6	1335.2	1.9737	31.47	1218.4	1334.8	1.9395	20.95	1217.8	1334.1	1.8943
620	43.67	1226.1	1344.8	1.9827	32.07	1225.8	1344.5	1.9485	21.35	1225.3	1343.8	1.9034
640	44.49	1233.5	1354.5	1.9916	32.67	1233.3	1354.2	1.9575	21.75	1232.8	1353.6	1.9123
660	45.30	1241.0	1364.2	2.0004	33.27	1240.8	1363.9	1.9662	22.15	1240.4	1363.3	1.9211
680	46.12	1248.6	1374.0	2.0090	33.87	1248.3	1373.7	1.9749	22.55	1247.9	1373.1	1.9298
700	46.93	1256.1	1383.8	2.0175	34.47	1255.9	1383.5	1.9834	22.95	1255.5	1383.0	1.9384
720	47.75	1263.7	1393.6	2.0259	35.07	1263.5	1393.3	1.9918	23.35	1263.2	1392.8	1.9468
740	48.56	1271.4	1403.4	2.0342	35.66	1271.2	1403.2	2.0001	23.75	1270.8	1402.7	1.9551
760	49.37	1279.0	1413.3	2.0424	36.26	1278.8	1413.0	2.0082	24.15	1278.5	1412.6	1.9633
780	50.19	1286.7	1423.2	2.0504	36.86	1286.5	1423.0	2.0163	24.55	1286.2	1422.5	1.9714
800	51.00	1294.4	1433.1	2.0584	37.46	1294.3	1432.9	2.0243	24.95	1294.0	1432.5	1.9793
850	53.03	1313.9	1458.1	2.0778	38.96	1313.8	1457.9	2.0438	25.95	1313.5	1457.6	1.9988
900	55.07	1333.6	1483.4	2.0967	40.45	1333.5	1483.2	2.0627	26.95	1333.2	1482.8	1.0178
950	57.10	1353.5	1508.8	2.1151	41.94	1353.4	1508.6	2.0810	27.95	1353.2	1508.3	2.0362
1000	59.13	1373.7	1534.5	2.1330	43.44	1373.5	1534.3	2.0989	28.95	1373.3	1534.0	2.0541
1100	63.19	1414.6	1586.4	2.1674	46.42	1414.5	1586.3	2.1334	30.94	1414.3	1586.1	2.0886
1200	67.25	1456.5	1639.3	2.2003	49.41	1456.4	1639.2	2.1663	32.93	1456.2	1639.1	2.1215
1300	71.30	1499.3	1693.2	2.2318	52.39	1499.2	1693.1	2.1978	34.92	1499.1	1692.9	2.1530
1400	75.36	1543.0	1747.9	2.2621	55.37	1542.9	1747.9	2.2281	36.91	1542.8	1747.7	2.1833
1500	79.42	1587.6	1803.6	2.2912	58.35	1587.6	1803.5	2.2572	38.90	1587.5	1803.4	2.2125
1600	83.47	1633.2	1860.2	2.3194	61.33	1633.2	1860.1	2.2854	40.88	1633.1	1860.0	2.2407
1800	91.58	1727.0	1976.1	2.3731	67.29	1727.0	1976.1	2.3391	44.86	1726.9	1976.0	2.2944
2000	99.69	1824.4	2095.5	2.4237	73.25	1824.3	2095.4	2.3897	48.83	1824.2	2095.3	2.3450
2200	107.80	1924.8	2218.0	2.4716	79.21	1924.8	2218.0	2.4376	52.81	1924.7	2217.9	2.3929
2400	115.91	2028.1	2343.4	2.5170	85.17	2028.1	2343.3	2.4830	56.78	2028.0	2343.3	2.4383

Table A-3 (Cont.)

p (t Sat.)	40 (267.26)				50 (281.03)				60 (292.73)			
t	v	u	h	s	v	u	h	s	v	u	h	s
Sat.	10.501	1092.3	1170.0	1.6767	8.518	1095.6	1174.4	1.6589	7.177	1098.3	1178.0	1.6444
200	9.346	1064.6	1133.8	1.6243	7.370	1060.9	1129.1	1.5940	6.047	1057.1	1124.2	1.5680
210	9.523	1068.8	1139.3	1.6327	7.519	1065.4	1135.0	1.6029	6.178	1061.9	1130.5	1.5774
220	9.699	1073.0	1144.8	1.6409	7.665	1069.9	1140.8	1.6115	6.307	1066.6	1136.6	1.5864
230	9.872	1077.2	1150.3	1.6488	7.810	1074.2	1146.5	1.6198	6.432	1071.2	1142.6	1.5952
240	10.043	1081.3	1155.6	1.6565	7.952	1078.6	1152.1	1.6279	6.556	1075.7	1148.5	1.6036
250	10.212	1085.4	1161.0	1.6641	8.092	1082.8	1157.7	1.6358	6.677	1080.1	1154.3	1.6118
260	10.380	1089.4	1166.2	1.6714	8.231	1087.0	1163.1	1.6434	6.797	1084.5	1160.0	1.6198
270	10.546	1093.4	1171.4	1.6786	8.368	1091.1	1168.5	1.6509	6.915	1088.8	1165.6	1.6275
280	10.711	1097.3	1176.6	1.6857	8.504	1095.2	1173.9	1.6582	7.031	1093.0	1171.1	1.6351
290	10.875	1101.2	1181.7	1.6926	8.639	1099.2	1179.2	1.6653	7.146	1097.2	1176.5	1.6424
300	11.038	1105.1	1186.8	1.6993	8.772	1103.2	1184.4	1.6722	7.260	1101.3	1181.9	1.6496
310	11.200	1109.0	1191.9	1.7059	8.904	1107.2	1189.6	1.6790	7.373	1105.4	1187.3	1.6565
320	11.360	1112.8	1196.9	1.7124	9.036	1111.2	1194.8	1.6857	7.485	1109.5	1192.6	1.6634
330	11.520	1116.6	1201.9	1.7188	9.166	1115.1	1199.9	1.6922	7.596	1113.5	1197.8	1.6700
340	11.680	1120.4	-1206.9	1.7251	9.296	1119.0	1205.0	1.6986	7.706	1117.4	1203.0	1.6766
350	11.838	1124.2	1211.8	1.7312	9.425	1122.8	1210.0	1.7049	7.815	1121.4	1208.2	1.6830
360	11.996	1128.0	1216.8	1.7373	9.553	1126.7	1215.1	1.7110	7.924	1125.3	1213.3	1.6893
370	12.153	1131.7	1221.7	1.7432	9.681	1130.5	1220.1	1.7171	8.032	1129.2	1218.4	1.6955
380	12.310	1135.5	1226.6	1.7491	9.808	1134.3	1225.0	1.7231	8.139	1133.1	1223.5	1.7015
390	12.467	1139.2	1231.5	1.7549	9.935	1138.1	1230.0	1.7290	8.246	1136.9	1228.5	1.7075
400	12.623	1143.0	1236.4	1.7606	10.061	1141.9	1235.0	1.7348	8.353	1140.8	1233.5	1.7134
420	12.933	1150.4	1246.1	1.7718	10.312	1149.4	1244.8	1.7461	8.565	1148.4	1243.5	1.7249
440	13.243	1157.8	1255.8	1.7828	10.562	1156.9	1254.6	1.7572	8.775	1156.0	1253.4	1.7360
460	13.551	1165.2	1265.5	1.7934	10.811	1164.4	1264.4	1.7679	8.984	1163.6	1263.3	1.7469
480	13.858	1172.7	1275.2	1.8038	11.059	1171.9	1274.2	1.7784	9.192	1171.1	1273.2	1.7575
500	14.164	1180.1	1284.9	1.8140	11.305	1179.4	1284.0	1.7887	9.399	1178.6	1283.0	1.7678
520	14.469	1187.5	1294.6	1.8240	11.551	1186.8	1293.7	1.7988	9.606	1186.2	1292.8	1.7780
540	14.774	1194.9	1304.3	1.8338	11.796	1194.3	1303.5	1.8086	9.811	1193.7	1302.6	1.7879
560	15.078	1202.4	1314.0	1.8434	12.041	1201.8	1313.2	1.8183	10.016	1201.2	1312.4	1.7976
580	15.382	1209.9	1323.7	1.8529	12.285	1209.3	1323.0	1.8277	10.221	1208.8	1322.2	1.8071

Table A-3 (Cont.)

p (t Sat.)		40 (267.26)				50 (281.03)				60 (292.73)		
t	v	u	h	s	v	u	h	s	v	u	h	s
Sat.	10.501	1092.3	1170.0	1.6767	8.518	1095.6	1174.4	1.6589	7.177	1098.3	1178.0	1.6444
600	15.685	1217.3	1333.4	1.8621	12.529	1216.8	1332.8	1.8371	10.425	1216.3	1332.1	1.8165
620	15.988	1224.8	1343.2	1.8713	12.772	1224.4	1342.5	1.8462	10.628	1223.9	1341.9	1.8257
640	16.291	1232.4	1353.0	1.8802	13.015	1231.9	1352.4	1.8552	10.832	1231.5	1351.7	1.8347
660	16.593	1239.9	1362.8	1.8891	13.258	1239.5	1362.2	1.8641	11.035	1239.1	1361.6	1.8436
680	16.895	1247.5	1372.6	1.8977	13.500	1247.1	1372.0	1.8728	11.237	1246.7	1371.5	1.8523
700	17.196	1255.1	1382.4	1.9063	13.742	1254.8	1381.9	1.8814	11.440	1254.4	1381.4	1.8609
720	17.498	1262.8	1392.3	1.9147	13.984	1262.4	1391.8	1.8898	11.642	1262.0	1391.3	1.8694
740	17.799	1270.5	1402.2	1.9231	14.226	1270.1	1401.7	1.8982	11.844	1269.7	1401.2	1.8778
760	18.100	1278.2	1412.1	1.9313	14.467	1277.8	1411.7	1.9064	12.045	1277.5	1411.2	1.8860
780	18.401	1285.9	1422.1	1.9394	14.708	1285.6	1421.7	1.9145	12.247	1285.2	1421.2	1.8942
800	18.701	1293.7	1432.1	1.9474	14.949	1293.3	1431.7	1.9225	12.448	1293.0	1431.2	1.9022
850	19.452	1313.2	1457.2	1.9669	15.551	1312.9	1456.8	1.9421	12.951	1312.7	1456.4	1.9218
900	20.202	1333.0	1482.5	1.9859	16.152	1332.7	1482.2	1.9611	13.452	1332.5	1481.8	1.9408
950	20.951	1352.9	1508.0	2.0043	16.753	1352.7	1507.7	1.9796	13.954	1352.5	1507.4	1.9593
1000	21.700	1373.1	1533.8	2.0223	17.352	1372.9	1533.5	1.9975	14.454	1372.7	1533.2	1.9773
1100	23.20	1414.2	1585.9	2.0568	18.551	1414.0	1585.6	2.0321	15.454	1413.8	1585.4	2.0119
1200	24.69	1456.1	1638.9	2.0897	19.747	1456.0	1638.7	2.0650	16.452	1455.8	1638.5	2.0448
1300	26.18	1498.9	1692.8	2.1212	20.943	1498.8	1692.6	2.0966	17.449	1498.7	1692.4	2.0764
1400	27.68	1542.7	1747.6	2.1515	22.138	1542.6	1747.4	2.1269	18.445	1542.5	1747.3	2.1067
1500	29.17	1587.4	1803.3	2.1807	23.332	1587.3	1803.2	2.1561	19.441	1587.2	1803.0	2.1359
1600	30.66	1633.0	1859.9	2.2089	24.53	1632.9	1859.8	2.1843	20.44	1632.8	1859.7	2.1641
1800	33.64	1726.9	1975.9	2.2626	26.91	1726.8	1975.8	2.2380	22.43	1726.7	1975.7	2.2179
2000	36.62	1824.2	2095.3	2.3132	29.30	1824.1	2095.2	2.2886	24.41	1824.0	2095.1	2.2685
2200	39.61	1924.7	2217.8	2.3611	31.68	1924.6	2217.8	2.3365	26.40	1924.5	2217.7	2.3164
2400	42.59	2028.0	2343.2	2.4066	34.07	2027.9	2343.1	2.3820	28.39	2027.8	2343.1	2.3618

Table A-3 (Cont.)

p(t Sat.)		200 (381.86)				300 (417.43)				350 (431.82)				400 (444.70)		
t	v	u	h	s	v	u	h	s	v	u	h	s	v	u	h	s
Sat.	2.289	1114.6	1199.3	1.5464	1.5442	1118.2	1203.9	1.5115	1.3267	1119.0	1204.9	1.4978	1.1620	1119.5	1205.5	1 4856
400	2.361	1123.5	1210.8	1.5600	*1.4915*	*1108.2*	*1191.0*	*1.4967*								
410	2.399	1128.2	1217.0	1.5672	*1.5221*	*1114.0*	*1198.5*	*1.5054*								
420	2.437	1132.9	1223.1	1.5741	1.5517	1119.6	1205.7	1.5136	*1.2950*	*1112.0*	*1195.9*	*1.4875*				
430	2.475	1137.5	1229.1	1.5809	1.5805	1125.0	1212.7	1.5216	*1.3219*	*1117.9*	*1203.6*	*1.4962*	*1.1257*	*1110.3*	*1193.6*	*1.4723*
440	2.511	1142.0	1234.9	1.5874	1.6086	1130.3	1219.6	1.5292	1.3480	1123.7	1211.0	1.5045	*1.1506*	*1116.6*	*1201.7*	*1.4814*
450	2.548	1146.4	1240.7	1.5938	1.6361	1135.4	1226.2	1.5365	1.3733	1129.2	1218.2	1.5125	1.1745	1122.6	1209.6	1.4901
460	2.584	1150.8	1246.5	1.6001	1.6630	1140.4	1232.7	1.5436	1.3979	1134.6	1225.2	1.5201	1.1977	1128.5	1217.1	1.4984
470	2.619	1155.2	1252.1	1.6062	1.6894	1145.3	1239.1	1.5505	1.4220	1139.9	1232.0	1.5275	1.2202	1134.1	1224.4	1.5063
480	2.654	1159.5	1257.7	1.6122	1.7154	1150.1	1245.3	1.5572	1.4455	1145.0	1238.6	1.5346	1.2421	1139.6	1231.5	1.5139
490	2.689	1163.7	1263.3	1.6181	1.7410	1154.8	1251.5	1.5637	1.4686	1150.0	1245.1	1.5415	1.2634	1144.9	1238.4	1.5212
500	2.724	1168.0	1268.8	1.6239	1.7662	1159.5	1257.5	1.5701	1.4913	1154.9	1251.5	1.5482	1.2843	1150.1	1245.2	1.5282
510	2.758	1172.2	1274.2	1.6295	1.7910	1164.1	1263.5	1.5763	1.5136	1159.7	1257.8	1.5546	1.3048	1155.2	1251.8	1.5351
520	2.792	1176.3	1279.7	1.6351	1.8156	1168.6	1269.4	1.5823	1.5356	1164.5	1263.9	1.5610	1.3249	1160.2	1258.2	1.5417
530	2.826	1180.5	1285.0	1.6405	1.8399	1173 1	1275.2	1.5882	1.5572	1169.1	1270.0	1.5671	1.3447	1165.0	1264.6	1.5481
540	2.860	1184.6	1290.4	1.6459	1.8640	1177.5	1281.0	1.5940	1.5786	1173.7	1276.0	1.5731	1.3642	1169.8	1270.8	1.5544
550	2.893	1188.7	1295.7	1.6512	1.8878	1181.9	1286.7	1.5997	1 5998	1178.3	1281.9	1.5790	1.3833	1174.6	1277.0	1.5605
560	2.926	1192.7	1301.0	1.6565	1.9114	1186.2	1292.3	1.6052	1.6207	1182.8	1287.7	1.5848	1.4023	1179.2	1283.0	1.5665
570	2.960	1196.8	1306.3	1.6616	1.9348	1190.5	1297.9	1 6107	1.6414	1187 2	1293.5	1.5904	1.4210	1183.8	1289.0	1.5723
580	2.993	1200.8	1311.6	1.6667	1.9580	1194.8	1303.5	1.6161	1.6619	1191.6	1299.3	1.5960	1.4395	1188.4	1294.9	1.5781
590	3.025	1204.9	1316.8	1.6717	1.9811	1199.0	1309.0	1.6214	1.6823	1196.0	1304.9	1.6014	1.4579	1192.9	1300.8	1.5837
600	3.058	1208.9	1322.1	1.6767	2.004	1203.2	1314.5	1.6266	1.7025	1200.3	1310.6	1.6068	1.4760	1197.3	1306.6	1.5892
620	3.123	1216.9	1332.5	1.6864	2.049	1211 6	1325.4	1.6368	1.7424	1208.9	1321.8	1.6172	1.5118	1206.1	1318.0	1.5999
640	3.188	1224.9	1342.9	1.6959	2.094	1220.0	1336.2	1.6467	1.7818	1217.4	1332.8	1.6274	1.5471	1214.8	1329.3	1.6103
660	3.252	1232.8	1353.2	1.7053	2.139	1228.2	1347.0	1 6564	1.8207	1225.8	1343.8	1.6372	1.5819	1223.4	1340.5	1.6203
680	3.316	1240.8	1363.5	1.7144	2.183	1236.4	1357.6	1.6658	1 8593	1234.2	1354.6	1.6469	1.6163	1231.9	1351.6	1.6301
700	3.379	1248.8	1373.8	1.7234	2.227	1244.6	1368.3	1.6751	1.8975	1242.5	1365.4	1.6562	1.6503	1240.4	1362.5	1.6397
720	3.442	1256.7	1384.1	1.7322	2.270	1252.8	1378.9	1.6841	1.9354	1250.8	1376.2	1.6654	1.6840	1248.8	1373.4	1.6490
740	3.505	1264.7	1394.4	1.7408	2.314	1261.0	1389.4	1.6930	1.9731	1259.1	1386.9	1.6744	1.7175	1257.2	1384.3	1.6581
760	3.568	1272.7	1404.7	1.7493	2.357	1269.1	1400.0	1.7017	2 0104	1267.3	1397.5	1.6832	1.7506	1265.5	1395.1	1.6670
780	3.631	1280.6	1415.0	1.7577	2.400	1277.3	1410.5	1.7103	2.0476	1275.6	1408.2	1.6919	1.7836	1273.8	1405.9	1.6758

Table A-3 (Cont.)

p(t Sat.)	200 (381.86)				300 (417.43)				350 (431.82)				400 (444.70)			
t	v	u	h	s	v	u	h	s	v	u	h	s	v	u	h	s
Sat.	2.289	1114.6	1199.3	1.5464	1.5442	1118.2	1203.9	1.5115	1.3267	1119.0	1204.9	1.4978	1.1620	1119.5	1205.5	1 4856
800	3.693	1288.6	1425.3	1.7660	2.442	1285.4	1421.0	1.7187	2.085	1283 8	1418.8	1.7004	1.8163	1282.1	1416.6	1.6844
820	3.755	1296.6	1435.6	1.7741	2.485	1293.6	1431.5	1.7270	2 121	1292.0	1429.4	1.7088	1.8489	1290.5	1427.3	1.6928
840	3.818	1304.7	1446.0	1.7821	2.527	1301.7	1442.0	1.7351	2.158	1300.3	1440.0	1.7170	1.8813	1298.8	1438.0	1.7011
860	3.879	1312.7	1456.3	1.7900	2.569	1309.9	1452.5	1.7432	2.194	1308.5	1450.6	1.7251	1.9135	1307.1	1448.7	1.7093
880	3.941	1320.8	1466.7	1.7978	2.611	1318.1	1463.1	1.7511	2.231	1316.7	1461.2	1.7331	1.9456	1315.4	1459.4	1.7173
900	4.003	1328.9	1477.1	1.8055	2.653	1326.3	1473.6	1.7589	2.267	1325.0	1471 8	1.7409	1.9776	1323.7	1470.1	1.7252
920	4.064	1337.0	1487.5	1.8131	2.695	1334.5	1484.1	1.7666	2.303	1333.3	1482.5	1.7487	2.0094	1332.0	1480.8	1.7330
940	4.126	1345.2	1497.9	1.8206	2.736	1342.8	1494.7	1.7742	2.339	1341,6	1493.1	1.7563	2.0411	1340.4	1491.5	1.7407
960	4.187	1353.3	1508.3	1.8280	2.778	1351.1	1505.3	1.7817	2.375	1349.9	1503.7	1.7639	2.0727	1348.7	1502.2	1.7483
980	4.249	1361.6	1518.8	1.8353	2.819	1359.3	1515.8	1.7891	2.411	1358.2	1514.4	1.7713	2.1043	1357.1	1512.9	1.7558
1000	4.310	1369.8	1529.3	1.8425	2.860	1367.7	1526.5	1.7964	2.446	1366.6	1525.0	1.7787	2.136	1365.5	1523.6	1.7632
1020	4.371	1378.0	1539.8	1.8497	2.902	1376.0	1537.1	1.8036	2.482	1375.0	1535.7	1.7859	2.167	1373.9	1534.3	1.7705
1040	4.432	1386.3	1550.3	1.8568	2.943	1384.4	1547.7	1.8108	2.517	1383.4	1546.4	1.7931	2.198	1382.4	1545.1	1.7777
1060	4.493	1394.6	1560.9	1.8638	2.984	1392.7	1558.4	1.8178	2.553	1391.8	1557.1	1.8002	2.229	1390.8	1555.9	1.7849
1080	4.554	1403.0	1571.5	1.8707	3.025	1401.2	1569.1	1.8248	2.588	1400.2	1567.9	1.8072	2.261	1399.3	1566.6	1.7919
1100	4.615	1411.4	1582.2	1.8776	3.066	1409.6	1579.8	1.8317	2.624	1408.7	1578.6	1 8142	2.292	1407.8	1577.4	1.7989
1200	4.918	1453.7	1635.7	1.9109	3.270	1452.2	1633.8	1.8653	2.799	1451.5	1632.8	1 8478	2.446	1450.7	1631.8	1.8327
1300	5.220	1496.9	1690.1	1.9427	3.473	1495.6	1688.4	1.8973	2.974	1495.0	1687.6	1.8799	2.599	1494.3	1686 8	1.8648
1400	5.521	1540.9	1745.3	1.9732	3.675	1539.8	1743.8	1.9279	3.148	1539.3	1743.1	1.9106	2.752	1538.7	1742.4	1.8956
1500	5.822	1585.8	1801.3	2.0025	3.877	1584.8	1800.0	1.9573	3.321	1584.3	1799.4	1.9401	2.904	1583.8	1798.8	1.9251
1600	6.123	1631.6	1858.2	2.0308	4.078	1630.7	1857.0	1.9857	3.494	1630.2	1856.5	1.9685	3.055	1629.8	1855.9	1.9535
1800	6.722	1725.6	1974.4	2.0847	4.479	1724.9	1973.5	2.0396	3.838	1724.5	1973.1	2.0225	3.357	1724.1	1972.6	2.0076
2000	7.321	1823.0	2094.0	2.1354	4.879	1822.3	2093.2	2.0904	4.182	1822.0	2092.8	2.0733	3.658	1821.6	2092.4	2.0584
2200	7.920	1923.6	2216.7	2.1833	5.280	1922.9	2216.0	2.1384	4.525	1922.5	2215.6	2.1212	3.959	1922.2	2215.2	2.1064
2400	8.518	2026.8	2342.1	2.2288	5.679	2026.1	2341.4	2.1838	4.868	2025.8	2341.1	2.1667	4.260	2025.4	2340.8	2.1519

Table A-4

Properties of Dry Air*
(at 14.696 psia or 29.92 in. Hg)

Temp., F	Volume, Cu. Ft. per Lb.	Density, Lb. per Cu. Ft.	Specific Heat, Btu per Lb. per F	Cu. Ft. Warmed 1F By 1 Btu	Btu to Raise 1 Cu. Ft. 1F
−50	10.328	.09682	0.240	43.034	.02324
−40	10.580	.09452	0.240	44.084	.02268
−30	10.832	.09232	0.240	45.134	.02216
−20	11.084	.09022	0.240	46.184	.02165
−10	11.336	.08821	0.240	47.234	.02117
0	11.589	.08629	0.240	48.289	.02071
1	11.614	.08610	0.240	48.392	.02066
2	11.639	.08592	0.240	48.496	.02062
3	11.664	.08573	0.240	48.600	.02058
4	11.689	.08555	0.240	48.705	.02053
5	11.715	.08536	0.240	48.813	.02049
6	11.740	.08518	0.240	48.917	.02044
7	11.765	.08500	0.240	49.021	.02040
8	11.790	.08482	0.240	49.125	.02036
9	11.816	.08463	0.240	49.234	.02031
10	11.841	.08445	0.240	49.338	.02027
11	11.866	.08427	0.240	49.442	.02023
12	11.891	.08410	0.240	49.546	.02018
13	11.917	.08391	0.240	49.655	.02014
14	11.942	.08374	0.240	49.759	.02010
15	11.967	.08356	0.240	49.863	.02005
16	11.992	08339	0.240	49.967	.02001
17	12.017	08322	0.240	50.071	.01997
18	12.043	.08304	0.240	50.180	.01993
19	12.068	.08286	0.240	50.284	.01989
20	12.093	.08269	0.240	50.388	.01985
21	12.118	.08252	0.240	50.492	.01981
22	12.143	.08235	0.240	50.596	.01976
23	12.169	.08218	0.240	50.705	.01972
24	12.194	.08201	0.240	50.809	.01968
25	12.219	.08184	0.240	50.913	.01964
26	12.244	.08167	0.240	51.017	.01960
27	12.269	.08151	0.240	51.121	.01956
28	12.295	.08133	0.240	51.230	.01952
29	12.320	.08177	0.240	51.334	.01948
30	12.345	.08100	0.240	51.438	.01944
31	12.370	.08084	0.240	51.542	.01940
32	12.396	.08067	0.240	51.650	.01936
33	12.421	.08051	0.240	51.755	.01932
34	12.446	.08035	0.240	51.859	.01928
35	12.471	.08019	0.240	51.963	.01924
36	12.496	.08003	0.240	52.067	.01921
37	12.522	.07986	0.240	52.175	.01917

Table A-4 (Cont.)

Temp., F	Volume, Cu. Ft. per Lb.	Density, Lb. per Cu. Ft.	Specific Heat, Btu per Lb. per F	Cu. Ft. Warmed 1F By 1 Btu	Btu to Raise 1 Cu. Ft. 1F
38	12.547	.07970	0.240	52.280	.01913
39	12.572	.07954	0.240	52.384	.01909
40	12.597	.07938	0.240	52.487	.01905
41	12.622	.07923	0.240	52.592	.01901
42	12.648	.07906	0.240	52.700	.01898
43	12.673	.07891	0.240	52.805	.01894
44	12.698	.07875	0.240	52.909	.01890
45	12.723	.07860	0.240	53.013	.01886
46	12.748	.07844	0.240	53.117	.01882
47	12.774	.07828	0.240	53.225	.01879
48	12.799	.07813	0.240	53.330	.01875
49	12.824	.07798	0.240	53.434	.01871
50	12.849	.07783	0.240	53.538	.01868
51	12.874	.07768	0.240	53.642	.01864
52	12.900	.07752	0.240	53.750	.01860
53	12.925	.07737	0.240	53.855	.01857
54	12.950	.07722	0.240	53.959	.01853
55	12.975	.07707	0.240	54.063	.01850
56	13.001	.07692	0.240	54.171	.01846
57	13.026	.07677	0.240	54.275	.01842
58	13.051	.07662	0.240	54.380	.01839
59	13.076	.07648	0.240	54.484	.01835
60	13.101	.07633	0.240	54.588	.01832
61	13.127	.07618	0.240	54.696	.01828
62	13.152	.07603	0.240	54.800	.01825
63	13.177	.07589	0.240	54.905	.01821
64	13.202	.07575	0.240	55.009	.01818
65	13.227	.07560	0.240	55.113	.01814
66	13.253	.07545	0.240	55.221	.01811
67	13.278	.07531	0.240	55.325	.01808
68	13.303	.07517	0.240	55.430	.01804
69	13.328	.07503	0.240	55.534	.01801
70	13.353	.07489	0.240	55.638	.01797
71	13.379	.07474	0.240	55.746	.01794
72	13.404	.07460	0.240	55.850	.01791
73	13.429	.07447	0.240	55.955	.01787
74	13.454	.07433	0.240	56.059	.01784
75	13.480	.07418	0.240	56.167	.01780
76	13.505	.07405	0.240	56.271	.01777
77	13.530	.07391	0.240	56.375	.01774
78	13.555	.07377	0.240	56.480	.01771
79	13.580	.07364	0.240	56.584	.01767
80	13.606	.07350	0.240	56.692	.01764

Table A-4 (Cont.)

Temp., F	Volume, Cu. Ft. per Lb.	Density, Lb. per Cu. Ft.	Specific Heat, Btu per Lb. per F	Cu. Ft. Warmed 1F By 1 Btu	Btu to Raise 1 Cu. Ft. 1F
81	13.631	.07336	0.240	56.796	.01761
82	13.656	.07323	0.240	56.900	.01757
83	13.681	.07309	0.240	57.005	.01754
84	13.706	.07296	0.240	57.109	.01751
85	13.732	.07282	0.240	57.217	.01748
86	13.757	.07269	0.240	57.321	.01745
87	13.782	.07256	0.240	57.425	.01741
88	13.807	.07243	0.240	57.530	.01738
89	13.832	.07230	0.240	57.634	.01735
90	13.858	.07216	0.240	57.742	.01732
91	13.883	.07203	0.240	57.846	.01729
92	13.908	.07190	0.240	57.950	.01726
93	13.933	.07177	0.240	58.055	.01723
94	13.959	.07164	0.240	58.163	.01719
95	13.984	.07151	0.240	58.267	.01716
96	14.009	.07138	0.240	58.371	.01713
97	14.034	.07126	0.240	58.475	.01710
98	14.059	.07113	0.240	58.580	.01707
99	14.085	.07100	0.240	58.688	.01704
100	14.110	.07087	0.240	58.792	.01701
101	14.135	.07075	0.240	58.896	.01698
102	14.160	.07062	0.240	59.000	.01695
103	14.185	.07050	0.240	59.105	.01692
104	14.211	.07037	0.240	59.213	.01689
105	14.236	.07024	0.240	59.317	.01686
110	14.362	.06963	0.240	59.842	.01671
115	14.488	.06902	0.240	60.367	.01656
120	14.614	.06843	0.240	60.892	.01642
125	14.740	.06784	0.240	61.417	.01628
130	14.866	.06727	0.241	61.685	.01621
135	14.992	.06670	0.241	62.208	.01608
140	15.118	.06615	0.241	62.731	.01594
145	15.244	.06560	0.241	63.253	.01581
150	15.370	.06506	0.241	63.776	.01568
155	15.496	.06453	0.241	64.299	.01555
160	15.622	.06401	0.241	64.822	.01543
165	15.748	.06350	0.241	65.345	.01530
170	15.874	.06300	0.241	65.868	.01518
175	16.001	.06250	0.241	66.395	.01506
180	16.127	.06201	0.241	66.917	.01494
185	16.253	.06153	0.241	67.440	.01483

Table A-4 (Cont.)

Temp., F	Volume, Cu. Ft. per Lb.	Density, Lb. per Cu. Ft.	Specific Heat, Btu per Lb. per F	Cu. Ft. Warmed 1F By 1 Btu	Btu to Raise 1 Cu. Ft. 1F
190	16.379	.06105	0.241	67.963	.01471
195	16.505	.06057	0.241	68.485	.01460
200	16.631	.06013	0.241	69.008	.01449
205	16.757	.05967	0.241	69.531	.01438
210	16.883	.05924	0.241	70.054	.01428
215	17.009	.05879	0.241	70.577	.01417
220	17.135	.05834	0.242	70.806	.01422
225	17.262	.05793	0.242	71.331	.01420
230	17.388	.05750	0.242	71.851	.01392
235	17.514	.05711	0.242	72.372	.01382
240	17.640	.05669	0.242	72.893	.01372
245	17.766	.05627	0.242	73.413	.01362
250	17.892	.05590	0.242	73.934	.01353
255	18.018	.05549	0.242	74.455	.01343
260	18.144	.05513	0.242	74.975	.01334
265	18.270	.05473	0.242	75.496	.01325
270	18.396	.05435	0.242	76.017	.01315
275	18.522	.05400	0.242	76.537	.01307
280	18.648	.05362	0.242	77.058	.01298
285	18.774	.05328	0.243	77.259	.01294
290	18.900	.05291	0.243	77.778	.01286
295	19.026	.05255	0.243	78.296	.01278
300	19.153	.05222	0.243	78.819	.01269
310	19.405	.05152	0.243	79.856	.01252
320	19.657	.05086	0.243	80.893	.01236
330	19.909	.05023	0.243	81.930	.01221
340	20.161	.04960	0.244	82.627	.01210
350	20.413	.04900	0.244	83.660	.01195
360	20.665	.04838	0.244	84.693	.01181
370	20.918	.04870	0.244	85.730	.01665
380	21.170	.04724	0.245	86.408	.01157
390	21.422	.04669	0.245	87.437	.01144
400	21.674	.04615	0.245	88.465	.01130
410	21.926	.04560	0.246	89.130	.01122
420	22.178	.04509	0.246	90.154	.01109
430	22.430	.04458	0.246	91.179	.01097
440	22.683	.04409	0.246	92.207	.01084
450	22.935	.04359	0.247	92.854	.01077
460	23.187	.04312	0.247	93.874	.01065
470	23.439	.04266	0.248	94.512	.01058
480	23.691	.04221	0.249	95.145	.01051
490	23.943	.04177	0.249	96.157	.01040
500	24.195	.04134	0.249	97.169	.01029

Appendix B

Thermodynamic Properties of Refrigerants

Table B-1*

Ammonia

Temp.	Pressure		Volume	Density	Enthalpy from −40 F			Entropy from −40 F	
F	Abs. lb/in.²	Gage lb/in.²	Vapor ft³/lb	Liquid lb/ft³	Liquid Btu/lb	Vapor Btu/lb	Latent Btu/lb	Liquid Btu/lb F	Vapor Btu/lb F
t	p	p_d	v_g	$1/v_f$	h_f	h_g	h_{fg}	s_f	s_g
−100	1.24	*27.4	182.90	45.51	−61.5	571.4	632.9	−0.1579	1.6025
− 90	1.86	*26.1	124.28	45.12	−51.4	575.9	627.3	−0.1309	1.5667
− 80	2.74	*24.3	86.54	44.73	−41.3	580.1	621.4	−0.1036	1.5336
− 70	3.94	*21.9	61.65	44.32	−31.1	584.4	615.5	−0.0771	1.5026
− 65	4.69	*20.4	52.34	44.11	−26.0	586.6	612.6	−0.0642	1.4833
− 60	5.55	*18.6	44.73	43.91	−20.9	588.8	609.7	−0.0514	1.4747
− 55	6.54	*16.6	38.38	43.70	−15.7	591.0	606.7	−0.0382	1.4614
− 50	7.67	*14.3	33.08	43.49	−10.5	593.2	603.7	−0.0254	1.4487
− 45	8.95	*11.7	28.62	43.28	− 5.3	595.7	600.7	−0.0128	1.4363
− 40	10.41	*8.7	24.86	43.08	0.0	597.6	597.6	0.0000	1.4242
− 38	11.04	*7.4	23.53	42.99	2.1	598.3	596.2	.0051	.4193
− 36	11.71	*6.1	22.27	42.90	4.3	599.1	594.8	.0101	.4144
− 34	12.41	*4.7	21.10	42.82	6.4	599.9	593.5	.0151	.4096
− 32	13.14	*3.2	20.00	42.73	8.5	600.6	592.1	.0201	.4048
− 30	13.90	*1.6	18.97	42.65	10.7	601.4	590.7	0.0250	1.4001
− 28	14.71	0.0	18.00	42.57	12.8	602.1	589.3	.0300	.3955
− 26	15.55	0.8	17.09	42.48	14.9	602.8	587.9	.0350	.3909
− 24	16.42	1.7	16.24	42.40	17.1	603.6	586.5	.0399	.3863
− 22	17.34	2.6	15.43	42.31	19.2	604.3	585.1	.0448	.3818
− 20	18.30	3.6	14.68	42.22	21.4	605.0	583.6	0.0497	1.3774
− 18	19.30	4.6	13.97	42.13	23.5	605.7	582.2	.0545	.3729
− 16	20.34	5.6	13.29	42.04	25.6	606.4	580.8	.0594	.3686
− 14	21.43	6.7	12.66	41.96	27.8	607.1	579.3	.0642	.3643
− 12	22.56	7.9	12.06	41.87	30.0	607.8	577.8	.0690	.3600
− 10	23.74	9.0	11.50	41.78	32.1	608.5	576.4	0.0738	1.3558
− 8	24.97	10.3	10.97	41.69	34.3	609.2	574.9	.0786	.3516
− 6	26.26	11.6	10.47	41.60	36.4	609.8	573.4	.0833	.3474
− 4	27.59	12.9	9.991	41.52	38.6	610.5	571.9	.0880	.3433
− 2	28.98	14.3	9.541	41.43	40.7	611.1	570.4	.0928	.3393
0	30.42	15.7	9.116	41.34	42.9	611.8	568.9	0.0975	1.3352
2	31.92	17.2	8.714	41.25	45.1	612.4	567.3	.1022	.3312
4	33.47	18.8	8.333	41.16	47.2	613.0	565.8	.1069	.3273
6	35.09	20.4	7.971	41.07	49.4	613.6	564.2	.1115	.3234
8	36.77	22.1	7.629	40.98	51.6	614.3	562.7	.1162	.3195
10	38.51	23.8	7.304	40.89	53.8	614.9	561.1	0.1208	1.3157
12	40.31	25.6	6.996	40.80	56.0	615.5	559.5	.1254	.3118
14	42.18	27.5	6.703	40.71	58.2	616.1	557.9	.1300	.3081
16	44.12	29.4	6.425	40.61	60.3	616.6	556.3	.1346	.3043
18	46.13	31.4	6.161	40.52	62.5	617.2	554.7	.1392	.3006
20	48.21	33.5	5.910	40.43	64.7	617.8	553.1	0.1437	1.2969
22	50.36	35.7	5.671	40.34	66.9	618.3	551.4	.1483	.2933
24	52.59	37.9	5.443	40.25	69.1	618.9	549.8	.1528	.2897
26	54.90	40.2	5.227	40.15	71.3	619.4	548.1	.1573	.2861
28	57.28	42.6	5.021	40.06	73.5	619.9	546.4	.1618	.2825
30	59.74	45.0	4.825	39.96	75.7	620.5	544.8	0.1663	1.2790
32	62.29	47.6	4.637	39.86	77.9	621.0	543.1	.1708	.2755
34	64.91	50.2	4.459	39.77	80.1	621.5	541.4	.1753	.2721
36	67.63	52.9	4.289	39.67	82.3	622.0	539.7	.1797	.2686
38	70.43	55.7	4.126	39.50	84.6	622.5	537.9	.1841	.2652

Table B-1 (Cont.)

Temp.	Pressure		Volume	Density	Enthalpy from −40 F			Entropy from −40 F	
F	Abs. lb/in²	Gage lb/in²	Vapor ft³/lb	Liquid lb/ft³	Liquid Btu/lb	Vapor Btu/lb	Latent Btu/lb	Liquid Btu/lb F	Vapor Btu/lb F
t	p	p_d	v_g	$1/v_f$	h_f	h_g	h_{fg}	s_f	s_g
40	73.32	58.6	3.971	39.49	86.8	623.0	536.2	0.1885	1.2618
42	76.31	61.6	3.823	39.39	89.0	623.4	534.4	.1930	.2585
44	79.38	64.7	3.682	39.29	91.2	623.9	532.7	.1974	.2552
46	82.55	67.9	3.547	39.19	93.5	624.4	530.9	.2018	.2519
48	85.82	71.1	3.418	39.10	95.7	624.8	529.1	.2062	.2486
50	89.19	74.5	3.294	39.00	97.9	625.2	527.3	0.2105	1.2453
52	92.66	78.0	3.176	38.90	100.2	625.7	525.5	.2149	.2421
54	96.23	81.5	3.063	38.80	102.4	626.1	523.7	.2192	.2389
56	99.91	85.2	2.954	38.70	104.7	626.5	521.8	.2236	.2357
58	103.7	89.0	2.851	38.60	106.9	626.9	520.0	.2279	.2325
60	107.6	92.9	2.751	38.50	109.2	627.3	518.1	0.2322	1.2294
62	111.6	96.9	2.656	38.40	111.5	627.7	516.2	.2365	.2262
64	115.7	101.0	2.565	38.30	113.7	628.0	514.3	.2408	.2231
66	120.0	105.3	2.477	38.20	116.0	628.4	512.4	.2451	.2201
68	124.3	109.6	2.393	38.10	118.3	628.8	510.5	.2494	.2170
70	128.8	114.1	2.312	38.00	120.5	629.1	508.6	0.2537	1.2140
72	133.4	118.7	2.235	37.90	122.8	629.4	506.6	.2579	.2110
74	138.1	123.4	2.161	37.79	125.1	629.8	504.7	.2622	.2080
76	143.0	128.3	2.089	37.69	127.4	630.1	502.7	.2664	.2050
78	147.9	133.2	2.021	37.58	129.7	630.4	500.7	.2706	.2020
80	153.0	138.3	1.955	37.48	132.0	630.7	498.7	0.2749	1.1991
82	158.3	143.6	1.892	37.37	134.3	631.0	496.7	.2791	.1962
84	163.7	149.0	1.831	37.26	136.6	631.3	494.7	.2833	.1933
86	169.2	154.5	1.772	37.16	138.9	631.5	492.6	.2875	.1904
88	174.8	160.1	1.716	37.05	141.2	631.8	490.6	.2917	.1875
90	180.6	165.9	1.661	36.95	143.5	632.0	488.5	0.2958	1.1846
92	186.6	171.9	1.609	36.84	145.8	632.2	486.4	.3000	.1818
94	192.7	178.0	1.559	36.73	148.2	632.5	484.3	.3041	.1789
96	198.9	184.2	1.510	36.62	150.5	632.6	482.1	.3083	.1761
98	205.3	190.6	1.464	36.51	152.9	632.9	480.0	.3125	.1733
100	211.9	197.2	1.419	36.40	155.2	633.0	477.8	0.3166	1.1705
102	218.6	203.9	1.375	36.29	157.6	633.2	475.6	.3207	.1677
104	225.4	210.7	1.334	36.18	159.9	633.4	473.5	.3248	.1649
106	232.5	217.8	1.293	36.06	162.3	633.5	471.2	.3289	.1621
108	239.7	225.0	1.254	35.95	164.6	633.6	469.0	.3330	.1593
110	247.0	232.3	1.217	35.84	167.0	633.7	466.7	0.3372	1.1566
112	254.5	239.8	1.180	35.72	169.4	633.8	464.4	.3413	.1538
114	262.2	247.5	1.145	35.61	171.8	633.9	462.1	.3453	.1510
116	270.1	255.4	1.112	35.49	174.2	634.0	459.8	.3495	.1483
118	278.2	263.5	1.079	35.38	176.6	634.0	457.4	.3535	.1455
120	286.4	271.7	1.047	35.26	179.0	634.0	455.0	0.3576	1.1427
122	294.8	280.1	1.017	35.14	181.4	634.0	452.6	.3618	.1400
124	303.4	288.7	0.987	35.02	183.9	634.0	450.1	.3659	.1372

**From* Refrigeration and Air Conditioning *by R. C. Jordan and G. B. Priester, 2nd ed., Prentice-Hall, Inc. Englewood Cliffs, N.J., 1956.*

Table B-2[a]

Ammonia

Absolute Pressure in lb/in.² (Saturation Temperature in italics)

Temp. F	5			10			15			20		
	−63.11			*−41.34*			*−27.29*			*−16.64*		
t	*v*	*h*	*s*	*v*	*h*	*s*	*v*	*h*	*s*	*v*	*h*	*s*
Sat.	*49.31*	*588.3*	*1.4857*	*25.81*	*597.1*	*1.4276*	*17.67*	*602.4*	*1.3938*	*13.50*	*606.2*	*1.3700*
−50	51.05	595.2	1.5025									
−40	52.36	600.3	.5149									
−30	53.67	605.4	.5269	26.58	603.2	1.4420						
−20	54.97	610.4	.5385	27.26	608.5	.4542	18.01	606.4	1.4031			
−10	56.26	615.4	.5498	27.92	613.7	.4659	18.47	611.9	.4154	13.74	610.0	1.3784
0	57.55	620.4	1.5608	28.58	618.9	1.4773	18.92	617.2	1.4272	14.09	615.5	1.3907
10	58.84	625.4	.5716	29.24	624.0	.4884	19.37	622.5	.4386	14.44	621.0	.4025
20	60.12	630.4	.5821	29.90	629.1	.4992	19.82	627.8	.4497	14.78	626.4	.4138
30	61.41	635.4	.5925	30.55	634.2	.5097	20.26	633.0	.4604	15.11	631.7	.4248
40	62.69	640.4	.6026	31.20	639.3	.5200	20.70	638.2	.4709	15.45	637.0	.4356
50	63.96	645.5	1.6125	31.85	644.4	1.5301	21.14	643.4	1.4812	15.78	642.3	1.4460
60	65.24	650.5	.6223	32.49	649.5	.5400	21.58	648.5	.4912	16.12	647.5	.4562
70	66.51	655.5	.6319	33.14	654.6	.5497	22.01	653.7	.5011	16.45	652.8	.4662
80	67.79	660.6	.6413	33.78	659.7	.5593	22.44	658.9	.5108	16.78	658.0	.4760
90	69.06	665.6	.6506	34.42	664.8	.5687	22.88	664.0	.5203	17.10	663.2	.4856
100	70.33	670.7	1.6598	35.07	670.0	1.5779	23.31	669.2	1.5296	17.43	668.5	1.4950
110	71.60	675.8	.6689	35.71	675.1	.5870	23.74	674.4	.5388	17.76	673.7	.5042
120	72.87	680.9	.6778	36.35	680.3	.5960	24.17	679.6	.5478	18.08	678.9	.5133
130	74.14	686.1	.6865	36.99	685.4	.6049	24.60	684.8	.5567	18.41	684.2	.5223
140	75.41	691.2	.6952	37.62	690.6	.6136	25.03	690.0	.5655	18.73	689.4	.5312
150	76.68	696.4	1.7038	38.26	695.8	1.6222	25.46	695.3	1.5742	19.05	694.7	1.5399
160	77.95	701.6	.7122	38.90	701.1	.6307	25.88	700.5	.5827	19.37	700.0	.5485
170	79.21	706.8	.7206	39.54	706.3	.6391	26.31	705.8	.5911	19.70	705.3	.5569
180	80.48	712.1	.7289	40.17	711.6	.6474	26.74	711.1	.5995	20.02	710.6	.5653
190	......	......	......	40.81	716.9	.6556	27.16	716.4	.6077	20.34	715.9	.5736
200	......	......	......	41.45	722.2	1.6637	27.59	721.7	1.6158	20.66	721.2	1.5817
220	......	......	......	......	......	......	28.44	732.4	.6318	21.30	732.0	.5978
240	......	......	......	......	......	......	......	......	......	21.94	742.8	.6135

Temp. F	25			30			35			40		
	−7.96			*−0.57*			*5.89*			*11.66*		
Sat.	*10.96*	*609.1*	*1.3515*	*9.236*	*611.6*	*1.3364*	*7.991*	*613.6*	*1.3236*	*7.047*	*615.4*	*1.3125*
0	11.19	613.8	1.3616									
10	11.47	619.4	.3738	9.492	617.8	1.3497	8.078	616.1	1.3289			
20	11.75	625.0	.3855	9.731	623.5	.3618	8.287	622.0	.3413	7.203	620.4	1.3231
30	12.03	630.4	.3967	9.966	629.1	.3733	8.493	627.7	.3532	7.387	626.3	.3353
40	12.30	635.8	.4077	10.20	634.6	.3845	8.695	633.4	.3646	7.568	632.1	.3470
50	12.57	641.2	1.4183	10.43	640.1	1.3953	8.895	638.9	1.3756	7.746	637.8	1.3583
60	12.84	646.5	.4287	10.65	645.5	.4059	9.093	644.4	.3863	7.922	643.4	.3692
70	13.11	651.8	.4388	10.88	650.9	.4161	9.289	649.9	.3967	8.096	648.9	.3797
80	13.37	657.1	.4487	11.10	656.2	.4261	9.484	655.3	.4069	8.268	654.4	.3900
90	13.64	662.4	.4584	11.33	661.6	.4359	9.677	660.7	.4168	8.439	659.9	.4000
100	13.90	667.7	1.4679	11.55	666.9	1.4456	9.869	666.1	1.4265	8.609	665.3	1.4098
110	14.17	673.0	.4772	11.77	672.2	.4550	10.06	671.5	.4360	8.777	670.7	.4194
120	14.43	678.2	.4864	11.99	677.5	.4642	10.25	676.8	.4453	8.945	676.1	.4288
130	14.69	683.5	.4954	12.21	682.9	.4733	10.44	682.2	.4545	9.112	681.5	.4381
140	14.95	688.8	.5043	12.43	688.2	.4823	10.63	687.6	.4635	9.278	686.9	.4471
150	15.21	694.1	1.5131	12.65	693.5	1.4911	10.82	692.9	1.4724	9.444	692.3	1.4561
160	15.47	699.4	.5217	12.87	698.8	.4998	11.00	698.3	.4811	9.609	697.7	.4648
170	15.73	704.7	.5303	13.08	704.2	.5083	11.19	703.7	.4897	9.774	703.1	.4735
180	15.99	710.1	.5387	13.30	709.6	.5168	11.38	709.1	.4982	9.938	708.5	.4820
190	16.25	715.4	.5470	13.52	714.9	.5251	11.56	714.5	.5066	10.10	714.0	.4904
200	16.50	720.8	1.5552	13.73	720.3	1.5334	11.75	719.9	1.5148	10.27	719.4	1.4987
220	17.02	731.6	.5713	14.16	731.1	.5495	12.12	730.7	.5311	10.59	730.3	.5159
240	17.53	742.5	.5870	14.59	742.0	.5653	12.49	741.7	.5469	10.92	741.3	.5309
260	18.04	753.4	.6025	15.02	753.0	.5808	12.86	752.7	.5624	11.24	752.3	.5465
280	......	......	......	15.45	764.1	.5960	13.23	763.7	.5776	11.56	763.4	.5617
300	......	......	......	......	......	......	......	......	......	11.88	774.6	1.5766

[a] *From* Refrigeration and Air Conditioning *by R. C. Jordan and G. B. Priester, 2nd ed., Prentice-Hall, Inc., Englewood Cliffs, N.J., 1956.*

Table B-2 (Cont.)

Absolute Pressure in lb/in.² (Saturation Temperature in italics)

Temp. F	50 *21.67*			60 *30.21*			70 *37.70*			80 *44.40*		
t	*v*	*h*	*s*	*v*	*h*	*s*	*v*	*h*	*s*	*v*	*h*	*s*
Sat.	*5.710*	*618.2*	*1.2939*	*4.805*	*620.5*	*1.2787*	*4.151*	*622.4*	*1.2658*	*3.655*	*624.0*	*1.254*[illegible]
30	5.838	623.4	1.3046									
40	5.988	629.5	.3169	4.933	626.8	1.2913	4.177	623.9	1.2688			
50	6.135	635.4	1.3286	5.060	632.9	1.3035	4.290	630.4	1.2816	3.712	627.7	1.2619
60	6.280	641.2	.3399	5.184	639.0	.3152	4.401	636.6	.2937	3.812	634.3	.2745
70	6.423	646.9	.3508	5.307	644.9	.3265	4.509	642.7	.3054	3.909	640.6	.2866
80	6.564	652.6	.3613	5.428	650.7	.3373	4.615	648.7	.3166	4.005	646.7	.2981
90	6.704	658.2	.3716	5.547	656.4	.3479	4.719	654.6	.3274	4.098	652.8	.3092
100	6.843	663.7	1.3816	5.665	662.1	1.3581	4.822	660.4	1.3378	4.190	658.7	1.3199
110	6.980	669.2	.3914	5.781	667.7	.3681	4.924	666.1	.3480	4.281	664.6	.3303
120	7.117	674.7	.4009	5.897	673.3	.3778	5.025	671.8	.3579	4.371	670.4	.3404
130	7.252	680.2	.4103	6.012	678.9	.3873	5.125	677.5	.3676	4.460	676.1	.3502
140	7.387	685.7	.4195	6.126	684.4	.3966	5.224	683.1	.3770	4.548	681.8	.3598
150	7.521	691.1	1.4286	6.239	689.9	1.4058	5.323	688.7	1.3863	4.635	687.5	1.3692
160	7.655	696.6	.4374	6.352	695.5	.4148	5.420	694.3	.3954	4.722	693.2	.3784
170	7.788	702.1	.4462	6.464	701.0	.4236	5.518	699.9	.4043	4.808	698.8	.3874
180	7.921	707.5	.4548	6.576	706.5	.4323	5.615	705.5	.4131	4.893	704.4	.3963
190	8.053	713.0	.4633	6.687	712.0	.4409	5.711	711.0	.4217	4.978	710.0	.4050
200	8.185	718.5	1.4716	6.798	717.5	1.4493	5.807	716.6	1.4302	5.063	715.6	1.4136
210	8.317	724.0	.4799	6.909	723.1	.4576	5.902	722.2	.4386	5.147	721.3	.4220
220	8.448	729.4	.4880	7.019	728.6	.4658	5.998	727.7	.4469	5.231	726.9	.4304
240	8.710	740.5	.5040	7.238	739.7	.4819	6.187	738.9	.4631	5.398	738.1	.4467
260	8.970	751.6	.5197	7.457	750.9	.4976	6.376	750.1	.4789	5.565	749.4	.4626
280	9.230	762.7	1.5350	7.675	762.1	1.5130	6.563	761.4	1.4943	5.730	760.7	1.4781
300	9.489	774.0	.5500	7.892	773.3	.5281	6.750	772.7	.5095	5.894	772.1	.4933

Temp. F	90 *50.47*			100 *56.05*			120 *66.02*			140 *74.79*		
Sat.	*3.266*	*625.3*	*1.2445*	*2.952*	*626.5*	*1.2356*	*2.476*	*628.4*	*1.2201*	*2.132*	*629.9*	*1.2068*
50												
60	3.353	631.8	1.2571	2.985	629.3	1.2409						
70	3.442	638.3	.2695	3.068	636.0	.2539	2.505	631.3	1.2255			
80	3.529	644.7	.2814	3.149	642.6	.2661	2.576	638.3	.2386	2.166	633.8	1.2140
90	3.614	650.9	.2928	3.227	649.0	.2778	2.645	645.0	.2510	2.228	640.9	.2272
100	3.698	657.0	1.3038	3.304	655.2	1.2891	2.712	651.6	1.2628	2.288	647.8	1.2396
110	3.780	663.0	.3144	3.380	661.3	.2999	2.778	658.0	.2741	2.347	654.5	.2515
120	3.862	668.9	.3247	3.454	667.3	.3104	2.842	664.2	.2850	2.404	661.1	.2628
130	3.942	674.7	.3347	3.527	673.3	.3206	2.905	670.4	.2956	2.460	667.4	.2738
140	4.021	680.5	.3444	3.600	679.2	.3305	2.967	676.5	.3058	2.515	673.7	.2843
150	4.100	686.3	1.3539	3.672	685.0	1.3401	3.029	682.5	1.3157	2.569	679.9	1.2945
160	4.178	692.0	.3633	3.743	690.8	.3495	3.089	688.4	.3254	2.622	686.0	.3045
170	4.255	697.7	.3724	3.813	696.6	.3588	3.149	694.3	.3348	2.675	692.0	.3141
180	4.332	703.4	.3813	3.883	702.3	.3678	3.209	700.2	.3441	2.727	698.0	.3236
190	4.408	709.0	.3901	3.952	708.0	.3767	3.268	706.0	.3531	2.779	704.0	.3328
200	4.484	714.7	1.3988	4.021	713.7	1.3854	3.326	711.8	1.3620	2.830	709.9	1.3418
210	4.560	720.4	.4073	4.090	719.4	.3940	3.385	717.6	.3707	2.880	715.8	.3507
220	4.635	726.0	.4157	4.158	725.1	.4024	3.442	723.4	.3793	2.931	721.6	.3594
230	4.710	731.7	.4239	4.226	730.8	.4108	3.500	729.2	.3877	2.981	727.5	.3679
240	4.785	737.3	.4321	4.294	736.5	.4190	3.557	734.9	.3960	3.030	733.3	.3763
250	4.859	743.0	1.4401	4.361	742.2	1.4271	3.614	740.7	1.4042	3.080	739.2	1.3846
260	4.933	748.7	.4481	4.428	747.9	.4350	3.671	746.5	.4123	3.129	745.0	.3928
280	5.081	760.0	.4637	4.562	759.4	.4507	3.783	758.0	.4281	3.227	756.7	.4088
300	5.228	771.5	.4789	4.695	770.8	.4660	3.895	769.6	.4435	3.323	768.3	.4243

Table B-2 (Cont.)

Absolute Pressure in lb/in.² (Saturation Temperatures in italics)

Temp. F	160 *82.64*			180 *89.78*			200 *96.34*			220 *102.42*		
t	*v*	*h*	*s*	*v*	*h*	*s*	*v*	*h*	*s*	*v*	*h*	*s*
Sat.	*1.872*	*631.1*	*1.1952*	*1.667*	*632.0*	*1.1850*	*1.502*	*632.7*	*1.1756*	*1.367*	*633.2*	*1.1671*
90	1.914	636.6	1.2055	1.668	632.2	1.1853						
100	1.969	643.9	1.2186	1.720	639.9	1.1992						
110	2.023	651.0	.2311	1.770	647.3	.2123	1.567	643.4	1.1947	1.400	639.4	1.1781
120	2.075	657.8	.2429	1.818	654.4	.2247	1.612	650.9	.2077	1.443	647.3	.1917
130	2.125	664.4	.2542	1.865	661.3	.2364	1.656	658.1	.2200	1.485	654.8	.2045
140	2.175	670.9	.2652	1.910	668.0	.2477	1.698	665.0	.2317	1.525	662.0	.2167
150	2.224	677.2	1.2757	1.955	674.6	1.2586	1.740	671.8	1.2429	1.564	669.0	1.2281
160	2.272	683.5	.2859	1.999	681.0	.2691	1.780	678.4	.2537	1.601	675.8	.2394
170	2.319	689.7	.2958	2.042	687.3	.2792	1.820	684.9	.2641	1.638	682.5	.2501
180	2.365	695.8	.3054	2.084	693.6	.2891	1.859	691.3	.2742	1.675	689.1	.2604
190	2.411	701.9	.3148	2.126	699.8	.2987	1.897	697.7	.2840	1.710	695.5	.2704
200	2.457	707.9	1.3240	2.167	705.9	1.3081	1.935	703.9	1.2935	1.745	701.9	1.2801
210	2.502	713.9	.3331	2.208	712.0	.3172	1.972	710.1	.3029	1.780	708.2	.2896
220	2.547	719.9	.3419	2.248	718.1	.3262	2.009	716.3	.3120	1.814	714.4	.2989
230	2.591	725.8	.3506	2.288	724.1	.3350	2.046	722.4	.3209	1.848	720.6	.3079
240	2.635	731.7	.3591	2.328	730.1	.3436	2.082	728.4	.3296	1.881	726.8	.3168
250	2.679	737.6	1.3675	2.367	736.1	1.3521	2.118	734.5	1.3382	1.914	732.9	1.3255
260	2.723	743.5	.3757	2.407	742.0	.3605	2.154	740.5	.3467	1.947	739.0	.3340
270	2.766	749.4	.3838	2.446	748.0	.3687	2.189	746.5	.3550	1.980	745.1	.3424
280	2.809	755.3	.3919	2.484	753.9	.3768	2.225	752.5	.3631	2.012	751.1	.3507
290	2.852	761.2	.3998	2.523	759.9	.3847	2.260	758.5	.3712	2.044	757.2	.3588
300	2.895	767.1	1.4076	2.561	765.8	1.3926	2.295	764.5	1.3791	2.076	763.2	1.3668
320	2.980	778.9	.4229	2.637	777.7	.4081	2.364	776.5	.3947	2.140	775.3	.3825
340	3.064	790.7	.4379	2.713	789.6	.4231	2.432	788.5	.4099	2.203	787.4	.3978
360	.....	.....	.....	.....	.....	.....	2.500	800.5	.4247	2.265	799.5	.4127
380	.....	.....	.....	.....	.....	.....	2.568	812.5	.4392	2.327	811.6	.4273

Temp. F	240 *108.09*			260 *113.42*			280 *118.45*			300 *123.21*		
Sat.	*1.253*	*633.6*	*1.1592*	*1.155*	*633.9*	*1.1518*	*1.072*	*634.0*	*1.1449*	*0.999*	*634.0*	*1.1383*
110	1.261	635.3	1.1621									
120	1.302	643.5	.1764	1.182	639.5	1.1617	1.078	635.4	1.1473			
130	1.342	651.3	.1898	1.220	647.8	.1757	1.115	644.0	.1621	1.023	640.1	1.1487
140	1.380	658.8	.2025	1.257	655.6	.1889	1.151	652.2	.1759	1.058	648.7	.1632
150	1.416	666.1	1.2145	1.292	663.1	1.2014	1.184	660.1	1.1888	1.091	656.9	1.1767
160	1.452	673.1	.2259	1.326	670.4	.2132	1.217	667.6	.2011	1.123	664.7	.1894
170	1.487	680.0	.2369	1.359	677.5	.2245	1.249	674.9	.2127	1.153	672.2	.2014
180	1.521	686.7	.2475	1.391	684.4	.2354	1.279	681.9	.2239	1.183	679.5	.2129
190	1.554	693.3	.2577	1.422	691.1	.2458	1.309	688.9	.2346	1.211	686.5	.2239
200	1.587	699.8	1.2677	1.453	697.7	1.2560	1.339	695.6	1.2449	1.239	693.5	1.2344
210	1.619	706.2	.2773	1.484	704.3	.2658	1.367	702.3	.2550	1.267	700.3	.2447
220	1.651	712.6	.2867	1.514	710.7	.2754	1.396	708.8	.2647	1.294	706.9	.2546
230	1.683	718.9	.2959	1.543	717.1	.2847	1.424	715.3	.2742	1.320	713.5	.2642
240	1.714	725.1	.3049	1.572	723.4	.2938	1.451	721.8	.2834	1.346	720.0	.2736
250	1.745	731.3	1.3137	1.601	729.7	1.3027	1.478	728.1	1.2924	1.372	726.5	1.2827
260	1.775	737.5	.3224	1.630	736.0	.3115	1.505	734.4	.3013	1.397	732.9	.2917
270	1.805	743.6	.3308	1.658	742.2	.3200	1.532	740.7	.3099	1.422	739.2	.3004
280	1.835	749.8	.3392	1.686	748.4	.3285	1.558	747.0	.3184	1.447	745.5	.3090
290	1.865	755.9	.3474	1.714	754.5	.3367	1.584	753.2	.3268	1.472	751.8	.3175
300	1.895	762.0	1.3554	1.741	760.7	1.3449	1.610	759.4	1.3350	1.496	758.1	1.3257
320	1.954	774.1	.3712	1.796	772.9	.3608	1.661	771.7	.3511	1.544	770.5	.3419
340	2.012	786.3	.3866	1.850	785.2	.3763	1.712	784.0	.3667	1.592	782.9	.3576
360	2.069	798.4	.4016	1.904	797.4	.3914	1.762	796.3	.3819	1.639	795.3	.3729
380	2.126	810.6	.4163	1.957	809.6	.4062	1.811	808.7	.3967	1.686	807.7	.3878
400	.....	.....	.....	2.009	821.9	1.4206	1.861	821.0	1.4112	1.732	820.1	1.4024

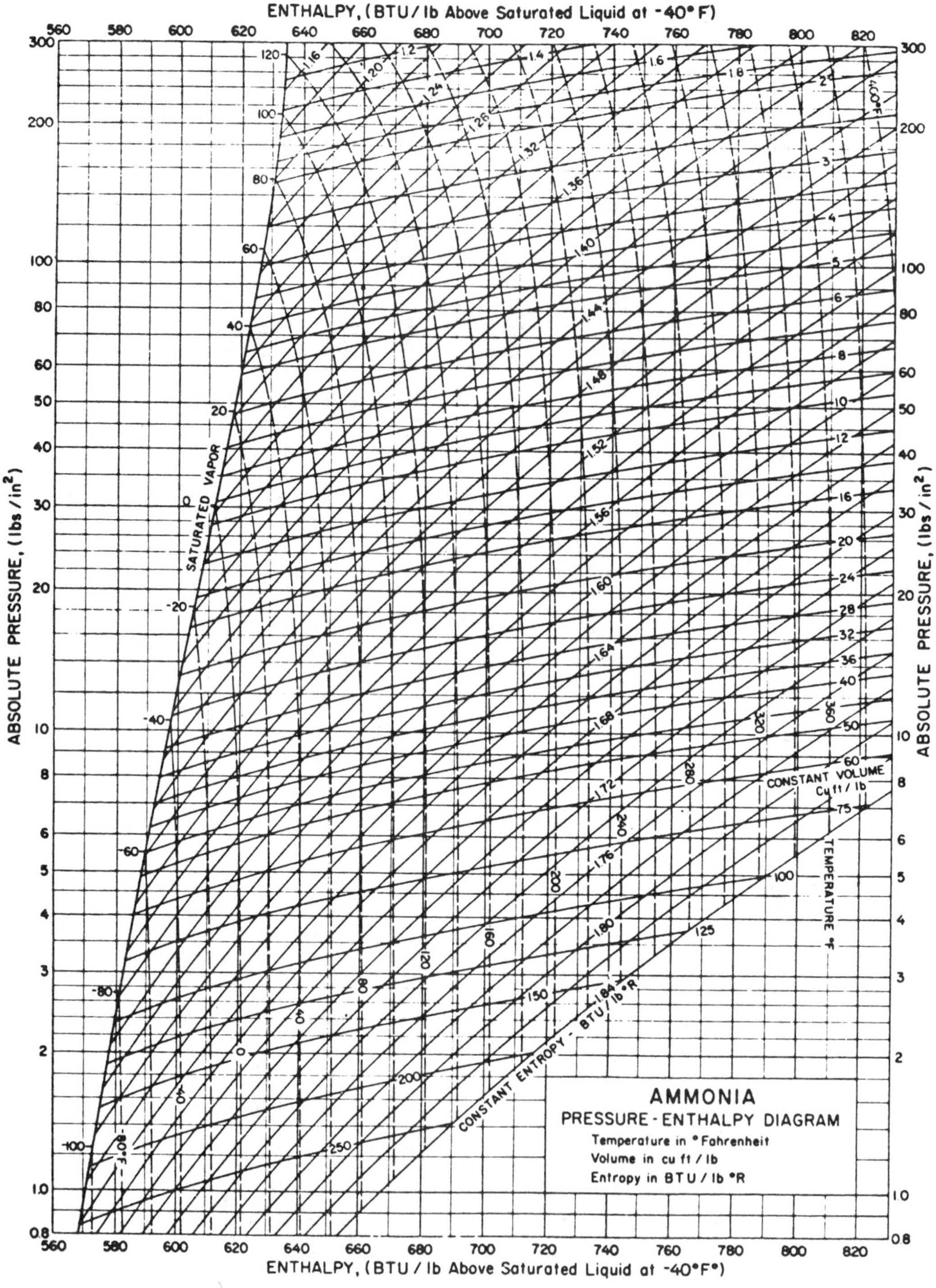
ENTHALPY, (BTU / lb Above Saturated Liquid at -40° F)
ABSOLUTE PRESSURE, (lbs / in²)
SATURATED VAPOR
CONSTANT VOLUME Cu ft / lb
TEMPERATURE °F
CONSTANT ENTROPY - BTU / lb °R
AMMONIA
PRESSURE - ENTHALPY DIAGRAM
Temperature in ° Fahrenheit
Volume in cu ft / lb
Entropy in BTU / lb °R
ENTHALPY, (BTU / lb Above Saturated Liquid at -40°F°)

Table B-3

Thermodynamic Properties of Freon-12[a]
(Properties of Saturated Liquid and Saturated Vapor)

TEMP.	PRESSURE		VOLUME cu ft/lb		DENSITY lb/cu ft		ENTHALPY Btu/lb			ENTROPY Btu/(lb)(° R)		TEMP.
°F	PSIA	PSIG	LIQUID v_f	VAPOR v_g	LIQUID $1/v_f$	VAPOR $1/v_g$	LIQUID h_f	LATENT h_{fg}	VAPOR h_g	LIQUID s_f	VAPOR s_g	°F
−40	9.3076	10.9709*	0.010564	3.8750	94.661	0.25806	0	72.913	72.913	0	0.17373	−40
−39	9.5530	10.4712*	0.010575	3.7823	94.565	0.26439	0.2107	72.812	73.023	0.000500	0.17357	−39
−38	9.8035	9.9611*	0.010586	3.6922	94.469	0.27084	0.4215	72.712	73.134	0.001000	0.17343	−38
−37	10.059	9.441*	0.010596	3.6047	94.372	0.27741	0.6324	72.611	73.243	0.001498	0.17328	−37
−36	10.320	8.909*	0.010607	3.5198	94.275	0.28411	0.8434	72.511	73.354	0.001995	0.17313	−36
−35	10.586	8.367*	0.010618	3.4373	94.178	0.29093	1.0546	72.409	73.464	0.002492	0.17299	−35
−34	10.858	7.814*	0.010629	3.3571	94.081	0.29788	1.2659	72.309	73.575	0.002988	0.17285	−34
−33	11.135	7.250*	0.010640	3.2792	93.983	0.30495	1.4772	72.208	73.685	0.003482	0.17271	−33
−32	11.417	6.675*	0.010651	3.2035	93.886	0.31216	1.6887	72.106	73.795	0.003976	0.17257	−32
−31	11.706	6.088*	0.010662	3.1300	93.788	0.31949	1.9003	72.004	73.904	0.004469	0.17243	−31
−30	11.999	5.490*	0.010674	3.0585	93.690	0.32696	2.1120	71.903	74.015	0.004961	0.17229	−30
−29	12.299	4.880*	0.010685	2.9890	93.592	0.33457	2.3239	71.801	74.125	0.005452	0.17216	−29
−28	12.604	4.259*	0.010696	2.9214	93.493	0.34231	2.5358	71.698	74.234	0.005942	0.17203	−28
−27	12.916	3.625*	0.010707	2.8556	93.395	0.35018	2.7479	71.596	74.344	0.006431	0.17189	−27
−26	13.233	2.979*	0.010719	2.7917	93.296	0.35820	2.9601	71.494	74.454	0.006919	0.17177	−26
−25	13.556	2.320*	0.010730	2.7295	93.197	0.36636	3.1724	71.391	74.563	0.007407	0.17164	−25
−24	13.886	1.649*	0.010741	2.6691	93.098	0.37466	3.3848	71.288	74.673	0.007894	0.17151	−24
−23	14.222	0.966*	0.010753	2.6102	92.999	0.38311	3.5973	71.185	74.782	0.008379	0.17139	−23
−22	14.564	0.270*	0.010764	2.5529	92.899	0.39171	3.8100	71.081	74.891	0.008864	0.17126	−22
−21	14.912	0.216	0.010776	2.4972	92.799	0.40045	4.0228	70.978	75.001	0.009348	0.17114	−21
−20	15.267	0.571	0.010788	2.4429	92.699	0.40934	4.2357	70.874	75.110	0.009831	0.17102	−20
−19	15.628	0.932	0.010799	2.3901	92.599	0.41839	4.4487	70.770	75.219	0.010314	0.17090	−19
−18	15.996	1.300	0.010811	2.3387	92.499	0.42758	4.6618	70.666	75.328	0.010795	0.17078	−18
−17	16.371	1.675	0.010823	2.2886	92.399	0.43694	4.8751	70.561	75.436	0.011276	0.17066	−17
−16	16.753	2.057	0.010834	2.2399	92.298	0.44645	5.0885	70.456	75.545	0.011755	0.17055	−16
−15	17.141	2.445	0.010846	2.1924	92.197	0.45612	5.3020	70.352	75.654	0.012234	0.17043	−15
−14	17.536	2.840	0.010858	2.1461	92.096	0.46595	5.5157	70.246	75.762	0.012712	0.17032	−14
−13	17.939	3.243	0.010870	2.1011	91.995	0.47595	5.7295	70.141	75.871	0.013190	0.17021	−13
−12	18.348	3.652	0.010882	2.0572	91.893	0.48611	5.9434	70.036	75.979	0.013666	0.17010	−12
−11	18.765	4.069	0.010894	2.0144	91.791	0.49643	6.1574	69.930	76.087	0.014142	0.16999	−11
−10	19.189	4.493	0.010906	1.9727	91.689	0.50693	6.3716	69.824	76.196	0.014617	0.16989	−10
− 9	19.621	4.925	0.010919	1.9320	91.587	0.51759	6.5859	69.718	76.304	0.015091	0.16978	− 9
− 8	20.059	5.363	0.010931	1.8924	91.485	0.52843	6.8003	69.611	76.411	0.015564	0.16967	− 8
− 7	20.506	5.810	0.010943	1.8538	91.382	0.53944	7.0149	69.505	76.520	0.016037	0.16957	− 7
− 6	20.960	6.264	0.010955	1.8161	91.280	0.55063	7.2296	69.397	76.627	0.016508	0.16947	− 6
− 5	21.422	6.726	0.010968	1.7794	91.177	0.56199	7.4444	69.291	76.735	0.016979	0.16937	− 5
− 4	21.891	7.195	0.010980	1.7436	91.074	0.57354	7.6594	69.183	76.842	0.017449	0.16927	− 4
− 3	22.369	7.673	0.010993	1.7086	90.970	0.58526	7.8745	69.075	76.950	0.017919	0.16917	− 3
− 2	22.854	8.158	0.011005	1.6745	90.867	0.59718	8.0898	68.967	77.057	0.018388	0.16907	− 2
− 1	23.348	8.652	0.011018	1.6413	90.763	0.60927	8.3052	68.859	77.164	0.018855	0.16897	− 1
0	23.849	9.153	0.011030	1.6089	90.659	0.62156	8.5207	68.750	77.271	0.019323	0.16888	0
1	24.359	9.663	0.011043	1.5772	90.554	0.63404	8.7364	68.642	77.378	0.019789	0.16878	1
2	24.878	10.182	0.011056	1.5463	90.450	0.64670	8.9522	68.533	77.485	0.020255	0.16869	2
3	25.404	10.708	0.011069	1.5161	90.345	0.65957	9.1682	68.424	77.592	0.020719	0.16860	3
4	25.939	11.243	0.011082	1.4867	90.240	0.67263	9.3843	68.314	77.698	0.021184	0.16851	4
5	26.483	11.787	0.011094	1.4580	90.135	0.68588	9.6005	68.204	77.805	0.021647	0.16842	5
6	27.036	12.340	0.011107	1.4299	90.030	0.69934	9.8169	68.094	77.911	0.022110	0.16833	6
7	27.597	12.901	0.011121	1.4025	89.924	0.71300	10.033	67.984	78.017	0.022572	0.16824	7
8	28.167	13.471	0.011134	1.3758	89.818	0.72687	10.250	67.873	78.123	0.023033	0.16815	8
9	28.747	14.051	0.011147	1.3496	89.712	0.74094	10.467	67.762	78.229	0.023494	0.16807	9
10	29.335	14.639	0.011160	1.3241	89.606	0.75523	10.684	67.651	78.335	0.023954	0.16798	10
11	29.932	15.236	0.011173	1.2992	89.499	0.76972	10.901	67.539	78.440	0.024413	0.16790	11
12	30.539	15.843	0.011187	1.2748	89.392	0.78443	11.118	67.428	78.546	0.024871	0.16782	12
13	31.155	16.459	0.011200	1.2510	89.285	0.79935	11.336	67.315	78.651	0.025329	0.16774	13
14	31.780	17.084	0.011214	1.2278	89.178	0.81449	11.554	67.203	78.757	0.025786	0.16765	14
15	32.415	17.719	0.011227	1.2050	89.070	0.82986	11.771	67.090	78.861	0.026243	0.16758	15

* Inches of mercury below one atmosphere

Table B-3 (Cont.)

TEMP. °F	PRESSURE PSIA	PRESSURE PSIG	VOLUME cu ft/lb LIQUID v_f	VOLUME cu ft/lb VAPOR v_g	DENSITY lb/cu ft LIQUID $1/v_f$	DENSITY lb/cu ft VAPOR $1/v_g$	ENTHALPY Btu/lb LIQUID h_f	ENTHALPY Btu/lb LATENT h_{fg}	ENTHALPY Btu/lb VAPOR h_g	ENTROPY Btu/(lb)(° R) LIQUID s_f	ENTROPY Btu/(lb)(° R) VAPOR s_g	TEMP. °F
15	32.415	17.719	0.011227	1.2050	89.070	0.82986	11.771	67.090	78.861	0.026243	0.16758	15
16	33.060	18.364	0.011241	1.1828	88.962	0.84544	11.989	66.977	78.966	0.026699	0.16750	16
17	33.714	19.018	0.011254	1.1611	88.854	0.86125	12.207	66.864	79.071	0.027154	0.16742	17
18	34.378	19.682	0.011268	1.1399	88.746	0.87729	12.426	66.750	79.176	0.027608	0.16734	18
19	35.052	20.356	0.011282	1.1191	88.637	0.89356	12.644	66.636	79.280	0.028062	0.16727	19
20	35.736	21.040	0.011296	1.0988	88.529	0.91006	12.863	66.522	79.385	0.028515	0.16719	20
21	36.430	21.734	0.011310	1.0790	88.419	0.92679	13.081	66.407	79.488	0.028968	0.16712	21
22	37.135	22.439	0.011324	1.0596	88.310	0.94377	13.300	66.293	79.593	0.029420	0.16704	22
23	37.849	23.153	0.011338	1.0406	88.201	0.96098	13.520	66.177	79.697	0.029871	0.16697	23
24	38.574	23.878	0.011352	1.0220	88.091	0.97843	13.739	66.061	79.800	0.030322	0.16690	24
25	39.310	24.614	0.011366	1.0039	87.981	0.99613	13.958	65.946	79.904	0.030772	0.16683	25
26	40.056	25.360	0.011380	0.98612	87.870	1.0141	14.178	65.829	80.007	0.031221	0.16676	26
27	40.813	26.117	0.011395	0.96874	87.760	1.0323	14.398	65.713	80.111	0.031670	0.16669	27
28	41.580	26.884	0.011409	0.95173	87.649	1.0507	14.618	65.596	80.214	0.032118	0.16662	28
29	42.359	27.663	0.011424	0.93509	87.537	1.0694	14.838	65.478	80.316	0.032566	0.16655	29
30	43.148	28.452	0.011438	0.91880	87.426	1.0884	15.058	65.361	80.419	0.033013	0.16648	30
31	43.948	29.252	0.011453	0.90286	87.314	1.1076	15.279	65.243	80.522	0.033460	0.16642	31
32	44.760	30.064	0.011468	0.88725	87.202	1.1271	15.500	65.124	80.624	0.033905	0.16635	32
33	45.583	30.887	0.011482	0.87197	87.090	1.1468	15.720	65.006	80.726	0.034351	0.16629	33
34	46.417	31.721	0.011497	0.85702	86.977	1.1668	15.942	64.886	80.828	0.034796	0.16622	34
35	47.263	32.567	0.011512	0.84237	86.865	1.1871	16.163	64.767	80.930	0.035240	0.16616	35
36	48.120	33.424	0.011527	0.82803	86.751	1.2077	16.384	64.647	81.031	0.035683	0.16610	36
37	48.989	34.293	0.011542	0.81399	86.638	1.2285	16.606	64.527	81.133	0.036126	0.16604	37
38	49.870	35.174	0.011557	0.80023	86.524	1.2496	16.828	64.406	81.234	0.036569	0.16598	38
39	50.763	36.067	0.011573	0.78676	86.410	1.2710	17.050	64.285	81.335	0.037011	0.16592	39
40	51.667	36.971	0.011588	0.77357	86.296	1.2927	17.273	64.163	81.436	0.037453	0.16586	40
41	52.584	37.888	0.011603	0.76064	86.181	1.3147	17.495	64.042	81.537	0.037893	0.16580	41
42	53.513	38.817	0.011619	0.74798	86.066	1.3369	17.718	63.919	81.637	0.038334	0.16574	42
43	54.454	39.758	0.011635	0.73557	85.951	1.3595	17.941	63.796	81.737	0.038774	0.16568	43
44	55.407	40.711	0.011650	0.72341	85.836	1.3823	18.164	63.673	81.837	0.039213	0.16562	44
45	56.373	41.677	0.011666	0.71149	85.720	1.4055	18.387	63.550	81.937	0.039652	0.16557	45
46	57.352	42.656	0.011682	0.69982	85.604	1.4289	18.611	63.426	82.037	0.040091	0.16551	46
47	58.343	43.647	0.011698	0.68837	85.487	1.4527	18.835	63.301	82.136	0.040529	0.16546	47
48	59.347	44.651	0.011714	0.67715	85.371	1.4768	19.059	63.177	82.236	0.040966	0.16540	48
49	60.364	45.668	0.011730	0.66616	85.254	1.5012	19.283	63.051	82.334	0.041403	0.16535	49
50	61.394	46.698	0.011746	0.65537	85.136	1.5258	19.507	62.926	82.433	0.041839	0.16530	50
51	62.437	47.741	0.011762	0.64480	85.018	1.5509	19.732	62.800	82.532	0.042276	0.16524	51
52	63.494	48.798	0.011779	0.63444	84.900	1.5762	19.957	62.673	82.630	0.042711	0.16519	52
53	64.563	49.867	0.011795	0.62428	84.782	1.6019	20.182	62.546	82.728	0.043146	0.16514	53
54	65.646	50.950	0.011811	0.61431	84.663	1.6278	20.408	62.418	82.826	0.043581	0.16509	54
55	66.743	52.047	0.011828	0.60453	84.544	1.6542	20.634	62.290	82.924	0.044015	0.16504	55
56	67.853	53.157	0.011845	0.59495	84.425	1.6808	20.859	62.162	83.021	0.044449	0.16499	56
57	68.977	54.281	0.011862	0.58554	84.305	1.7078	21.086	62.033	83.119	0.044883	0.16494	57
58	70.115	55.419	0.011879	0.57632	84.185	1.7352	21.312	61.903	83.215	0.045316	0.16489	58
59	71.267	56.571	0.011896	0.56727	84.065	1.7628	21.539	61.773	83.312	0.045748	0.16484	59
60	72.433	57.737	0.011913	0.55839	83.944	1.7909	21.766	61.643	83.409	0.046180	0.16479	60
61	73.613	58.917	0.011930	0.54967	83.823	1.8193	21.993	61.512	83.505	0.046612	0.16474	61
62	74.807	60.111	0.011947	0.54112	83.701	1.8480	22.221	61.380	83.601	0.047044	0.16470	62
63	76.016	61.320	0.011965	0.53273	83.580	1.8771	22.448	61.248	83.696	0.047475	0.16465	63
64	77.239	62.543	0.011982	0.52450	83.457	1.9066	22.676	61.116	83.792	0.047905	0.16460	64
65	78.477	63.781	0.012000	0.51642	83.335	1.9364	22.905	60.982	83.887	0.048336	0.16456	65
66	79.729	65.033	0.012017	0.50848	83.212	1.9666	23.133	60.849	83.982	0.048765	0.16451	66
67	80.996	66.300	0.012035	0.50070	83.089	1.9972	23.362	60.715	84.077	0.049195	0.16447	67
68	82.279	67.583	0.012053	0.49305	82.965	2.0282	23.591	60.580	84.171	0.049624	0.16442	68
69	83.576	68.880	0.012071	0.48555	82.841	2.0595	23.821	60.445	84.266	0.050053	0.16438	69
70	84.888	70.192	0.012089	0.47818	82.717	2.0913	24.050	60.309	84.359	0.050482	0.16434	70

Table B-3 (Cont.)

TEMP.	PRESSURE		VOLUME cu ft/lb		DENSITY lb/cu ft		ENTHALPY Btu/lb			ENTROPY Btu/(lb)(° R)		TEMP.
°F	PSIA	PSIG	LIQUID v_f	VAPOR v_g	LIQUID $1/v_f$	VAPOR $1/v_g$	LIQUID h_f	LATENT h_{fg}	VAPOR h_g	LIQUID s_f	VAPOR s_g	°F
70	84.888	70.192	0.012089	0.47818	82.717	2.0913	24.050	60.309	84.359	0.050482	0.16434	70
71	86.216	71.520	0.012108	0.47094	82.592	2.1234	24.281	60.172	84.453	0.050910	0.16429	71
72	87.559	72.863	0.012126	0.46383	82.467	2.1559	24.511	60.035	84.546	0.051338	0.16425	72
73	88.918	74.222	0.012145	0.45686	82.341	2.1889	24.741	59.898	84.639	0.051766	0.16421	73
74	90.292	75.596	0.012163	0.45000	82.215	2.2222	24.973	59.759	84.732	0.052193	0.16417	74
75	91.682	76.986	0.012182	0.44327	82.089	2.2560	25.204	59.621	84.825	0.052620	0.16412	75
76	93.087	78.391	0.012201	0.43666	81.962	2.2901	25.435	59.481	84.916	0.053047	0.16408	76
77	94.509	79.813	0.012220	0.43016	81.835	2.3247	25.667	59.341	85.008	0.053473	0.16404	77
78	95.946	81.250	0.012239	0.42378	81.707	2.3597	25.899	59.201	85.100	0.053900	0.16400	78
79	97.400	82.704	0.012258	0.41751	81.579	2.3951	26.132	59.059	85.191	0.054326	0.16396	79
80	98.870	84.174	0.012277	0.41135	81.450	2.4310	26.365	58.917	85.282	0.054751	0.16392	80
81	100.36	85.66	0.012297	0.40530	81.322	2.4673	26.598	58.775	85.373	0.055177	0.16388	81
82	101.86	87.16	0.012316	0.39935	81.192	2.5041	26.832	58.631	85.463	0.055602	0.16384	82
83	103.38	88.68	0.012336	0.39351	81.063	2.5413	27.065	58.488	85.553	0.056027	0.16380	83
84	104.92	90.22	0.012356	0.38776	80.932	2.5789	27.300	58.343	85.643	0.056452	0.16376	84
85	106.47	91.77	0.012376	0.38212	80.802	2.6170	27.534	58.198	85.732	0.056877	0.16372	85
86	108.04	93.34	0.012396	0.37657	80.671	2.6556	27.769	58.052	85.821	0.057301	0.16368	**86**
87	109.63	94.93	0.012416	0.37111	80.539	2.6946	28.005	57.905	85.910	0.057725	0.16364	87
88	111.23	96.53	0.012437	0.36575	80.407	2.7341	28.241	57.757	85.998	0.058149	0.16360	88
89	112.85	98.15	0.012457	0.36047	80.275	2.7741	28.477	57.609	86.086	0.058573	0.16357	89
90	114.49	99.79	0.012478	0.35529	80.142	2.8146	28.713	57.461	86.174	0.058997	0.16353	90
91	116.15	101.45	0.012499	0.35019	80.008	2.8556	28.950	57.311	86.261	0.059420	0.16349	91
92	117.82	103.12	0.012520	0.34518	79.874	2.8970	29.187	57.161	86.348	0.059844	0.16345	92
93	119.51	104.81	0.012541	0.34025	79.740	2.9390	29.425	57.009	86.434	0.060267	0.16341	93
94	121.22	106.52	0.012562	0.33540	79.605	2.9815	29.663	56.858	86.521	0.060690	0.16338	94
95	122.95	108.25	0.012583	0.33063	79.470	3.0245	29.901	56.705	86.606	0.061113	0.16334	95
96	124.70	110.00	0.012605	0.32594	79.334	3.0680	30.140	56.551	86.691	0.061536	0.16330	96
97	126.46	111.76	0.012627	0.32133	79.198	3.1120	30.380	56.397	86.777	0.061959	0.16326	97
98	128.24	113.54	0.012649	0.31679	79.061	3.1566	30.619	56.242	86.861	0.062381	0.16323	98
99	130.04	115.34	0.012671	0.31233	78.923	3.2017	30.859	56.086	86.945	0.062804	0.16319	99
100	131.86	117.16	0.012693	0.30794	78.785	3.2474	31.100	55.929	87.029	0.063227	0.16315	100
101	133.70	119.00	0.012715	0.30362	78.647	3.2936	31.341	55.772	87.113	0.063649	0.16312	101
102	135.56	120.86	0.012738	0.29937	78.508	3.3404	31.583	55.613	87.196	0.064072	0.16308	102
103	137.44	122.74	0.012760	0.29518	78.368	3.3877	31.824	55.454	87.278	0.064494	0.16304	103
104	139.33	124.63	0.012783	0.29106	78.228	3.4357	32.067	55.293	87.360	0.064916	0.16301	104
105	141.25	126.55	0.012806	0.28701	78.088	3.4842	32.310	55.132	87.442	0.065339	0.16297	105
106	143.18	128.48	0.012829	0.28303	77.946	3.5333	32.553	54.970	87.523	0.065761	0.16293	106
107	145.13	130.43	0.012853	0.27910	77.804	3.5829	32.797	54.807	87.604	0.066184	0.16290	107
108	147.11	132.41	0.012876	0.27524	77.662	3.6332	33.041	54.643	87.684	0.066606	0.16286	108
109	149.10	134.40	0.012900	0.27143	77.519	3.6841	33.286	54.478	87.764	0.067028	0.16282	109
110	151.11	136.41	0.012924	0.26769	77.376	3.7357	33.531	54.313	87.844	0.067451	0.16279	110
111	153.14	138.44	0.012948	0.26400	77.231	3.7878	33.777	54.146	87.923	0.067873	0.16275	111
112	155.19	140.49	0.012972	0.26037	77.087	3.8406	34.023	53.978	88.001	0.068296	0.16271	112
113	157.27	142.57	0.012997	0.25680	76.941	3.8941	34.270	53.809	88.079	0.068719	0.16268	113
114	159.36	144.66	0.013022	0.25328	76.795	3.9482	34.517	53.639	88.156	0.069141	0.16264	114
115	161.47	146.77	0.013047	0.24982	76.649	4.0029	34.765	53.468	88.233	0.069564	0.16260	115
116	163.61	148.91	0.013072	0.24641	76.501	4.0584	35.014	53.296	88.310	0.069987	0.16256	116
117	165.76	151.06	0.013097	0.24304	76.353	4.1145	35.263	53.123	88.386	0.070410	0.16253	117
118	167.94	153.24	0.013123	0.23974	76.205	4.1713	35.512	52.949	88.461	0.070833	0.16249	118
119	170.13	155.43	0.013148	0.23647	76.056	4.2288	35.762	52.774	88.536	0.071257	0.16245	119
120	172.35	157.65	0.013174	0.23326	75.906	4.2870	36.013	52.597	88.610	0.071680	0.16241	120
121	174.59	159.89	0.013200	0.23010	75.755	4.3459	36.264	52.420	88.684	0.072104	0.16237	121
122	176.85	162.15	0.013227	0.22698	75.604	4.4056	36.516	52.241	88.757	0.072528	0.16234	122
123	179.13	164.43	0.013254	0.22391	75.452	4.4660	36.768	52.062	88.830	0.072952	0.16230	123
124	181.43	166.73	0.013280	0.22089	75.299	4.5272	37.021	51.881	88.902	0.073376	0.16226	124
125	183.76	169.06	0.013308	0.21791	75.145	4.5891	37.275	51.698	88.973	0.073800	0.16222	125

Table B-3 (Cont.)

TEMP. °F	PRESSURE PSIA	PSIG	VOLUME cu ft/lb LIQUID v_f	VAPOR v_g	DENSITY lb/cu ft LIQUID $1/v_f$	VAPOR $1/v_g$	ENTHALPY Btu/lb LIQUID h_f	LATENT h_{fg}	VAPOR h_g	ENTROPY Btu/(lb)(°R) LIQUID s_f	VAPOR s_g	TEMP. °F
125	183.76	169.06	0.013308	0.21791	75.145	4.5891	37.275	51.698	88.973	0.073800	0.16222	125
126	186.10	171.40	0.013335	0.21497	74.991	4.6518	37.529	51.515	89.044	0.074225	0.16218	126
127	188.47	173.77	0.013363	0.21207	74.836	4.7153	37.785	51.330	89.115	0.074650	0.16214	127
128	190.86	176.16	0.013390	0.20922	74.680	4.7796	38.040	51.144	89.184	0.075075	0.16210	128
129	193.27	178.57	0.013419	0.20641	74.524	4.8448	38.296	50.957	89.253	0.075501	0.16206	129
130	195.71	181.01	0.013447	0.20364	74.367	4.9107	38.553	50.768	89.321	0.075927	0.16202	130
131	198.16	183.46	0.013476	0.20091	74.209	4.9775	38.811	50.578	89.389	0.076353	0.16198	131
132	200.64	185.94	0.013504	0.19821	74.050	5.0451	39.069	50.387	89.456	0.076779	0.16194	132
133	203.15	188.45	0.013534	0.19556	73.890	5.1136	39.328	50.194	89.522	0.077206	0.16189	133
134	205.67	190.97	0.013563	0.19294	73.729	5.1829	39.588	50.000	89.588	0.077633	0.16185	134
135	208.22	193.52	0.013593	0.19036	73.568	5.2532	39.848	49.805	89.653	0.078061	0.16181	135
136	210.79	196.09	0.013623	0.18782	73.406	5.3244	40.110	49.608	89.718	0.078489	0.16177	136
137	213.39	198.69	0.013653	0.18531	73.243	5.3965	40.372	49.409	89.781	0.078917	0.16172	137
138	216.01	201.31	0.013684	0.18283	73.079	5.4695	40.634	49.210	89.844	0.079346	0.16168	138
139	218.65	203.95	0.013715	0.18039	72.914	5.5435	40.898	49.008	89.906	0.079775	0.16163	139
140	221.32	206.62	0.013746	0.17799	72.748	5.6184	41.162	48.805	89.967	0.080205	0.16159	140
141	224.00	209.30	0.013778	0.17561	72.581	5.6944	41.427	48.601	90.028	0.080635	0.16154	141
142	226.72	212.02	0.013810	0.17327	72.413	5.7713	41.693	48.394	90.087	0.081065	0.16150	142
143	229.46	214.76	0.013842	0.17096	72.244	5.8493	41.959	48.187	90.146	0.081497	0.16145	143
144	232.22	217.52	0.013874	0.16868	72.075	5.9283	42.227	47.977	90.204	0.081928	0.16140	144
145	235.00	220.30	0.013907	0.16644	71.904	6.0083	42.495	47.766	90.261	0.082361	0.16135	145
146	237.82	223.12	0.013941	0.16422	71.732	6.0895	42.765	47.553	90.318	0.082794	0.16130	146
147	240.65	225.95	0.013974	0.16203	71.559	6.1717	43.035	47.338	90.373	0.083227	0.16125	147
148	243.51	228.81	0.014008	0.15987	71.386	6.2551	43.306	47.122	90.428	0.083661	0.16120	148
149	246.40	231.70	0.014043	0.15774	71.211	6.3395	43.578	46.904	90.482	0.084096	0.16115	149
150	249.31	234.61	0.014078	0.15564	71.035	6.4252	43.850	46.684	90.534	0.084531	0.16110	150
151	252.24	237.54	0.014113	0.15356	70.857	6.5120	44.124	46.462	90.586	0.084967	0.16105	151
152	255.20	240.50	0.014148	0.15151	70.679	6.6001	44.399	46.238	90.637	0.085404	0.16099	152
153	258.19	243.49	0.014184	0.14949	70.500	6.6893	44.675	46.012	90.687	0.085842	0.16094	153
154	261.20	246.50	0.014221	0.14750	70.319	6.7799	44.951	45.784	90.735	0.086280	0.16088	154
155	264.24	249.54	0.014258	0.14552	70.137	6.8717	45.229	45.554	90.783	0.086719	0.16083	155
156	267.30	252.60	0.014295	0.14358	69.954	6.9648	45.508	45.322	90.830	0.087159	0.16077	156
157	270.39	255.69	0.014333	0.14166	69.770	7.0592	45.787	45.088	90.875	0.087600	0.16071	157
158	273.51	258.81	0.014371	0.13976	69.584	7.1551	46.068	44.852	90.920	0.088041	0.16065	158
159	276.65	261.95	0.014410	0.13789	69.397	7.2523	46.350	44.614	90.964	0.088484	0.16059	159
160	279.82	265.12	0.014449	0.13604	69.209	7.3509	46.633	44.373	91.006	0.088927	0.16053	160
161	283.02	268.32	0.014489	0.13421	69.019	7.4510	46.917	44.130	91.047	0.089371	0.16047	161
162	286.24	271.54	0.014529	0.13241	68.828	7.5525	47.202	43.885	91.087	0.089817	0.16040	162
163	289.49	274.79	0.014570	0.13062	68.635	7.6556	47.489	43.637	91.126	0.090263	0.16034	163
164	292.77	278.07	0.014611	0.12886	68.441	7.7602	47.777	43.386	91.163	0.090710	0.16027	164
165	296.07	281.37	0.014653	0.12712	68.245	7.8665	48.065	43.134	91.199	0.091159	0.16021	165
166	299.40	284.70	0.014695	0.12540	68.048	7.9743	48.355	42.879	91.234	0.091608	0.16014	166
167	302.76	288.06	0.014738	0.12370	67.850	8.0838	48.647	42.620	91.267	0.092059	0.16007	167
168	306.15	291.45	0.014782	0.12202	67.649	8.1950	48.939	42.360	91.299	0.092511	0.16000	168
169	309.56	294.86	0.014826	0.12037	67.447	8.3080	49.233	42.097	91.330	0.092964	0.15992	169
170	313.00	298.30	0.014871	0.11873	67.244	8.4228	49.529	41.830	91.359	0.093418	0.15985	170
171	316.47	301.77	0.014917	0.11710	67.038	8.5394	49.825	41.562	91.387	0.093874	0.15977	171
172	319.97	305.27	0.014963	0.11550	66.831	8.6579	50.123	41.290	91.413	0.094330	0.15969	172
173	323.50	308.80	0.015010	0.11392	66.622	8.7783	50.423	41.015	91.438	0.094789	0.15961	173
174	327.06	312.36	0.015058	0.11235	66.411	8.9007	50.724	40.736	91.460	0.095248	0.15953	174
175	330.64	315.94	0.015106	0.11080	66.198	9.0252	51.026	40.455	91.481	0.095709	0.15945	175
176	334.25	319.55	0.015155	0.10927	65.983	9.1518	51.330	40.171	91.501	0.096172	0.15936	176
177	337.90	323.20	0.015205	0.10775	65.766	9.2805	51.636	39.883	91.519	0.096636	0.15928	177
178	341.57	326.87	0.015256	0.10625	65.547	9.4114	51.943	39.592	91.535	0.097102	0.15919	178
179	345.27	330.57	0.015308	0.10477	65.326	9.5446	52.252	39.297	91.549	0.097569	0.15910	179
180	349.00	334.30	0.015360	0.10330	65.102	9.6802	52.562	38.999	91.561	0.098039	0.15900	180

Table B-4

Thermodynamic Properties of Freon-12*
(Properties of Superheated Vapor)

Temp. F	Abs. Pressure 0.14 lb/in.^{2} Gage Pressure 29.64 in. vac. (Sat. Temp. −151.7 F)			Abs. Pressure 0.20 lb/in.^{2} Gage Pressure 29.51 in. vac. (Sat. Temp. −144.9 F)			Abs. Pressure 0.40 lb/in.^{2} Gage Pressure 29.11 in. vac. (Sat. Temp. −130.7 F)			Abs. Pressure 0.60 lb/in.^{2} Gage Pressure 28.70 in. vac. (Sat. Temp. −121.6 F)		
t	*v*	*h*	*s*	*v*	*h*	*s*	*v*	*h*	*s*	*v*	*h*	*s*
Sat.	*(194.91)*	*(60.656)*	*(0.20804)*	*(39.38)*	*(61.372)*	*(0.20449)*	*(72.756)*	*(62.897)*	*(0.19788)*	*(49.785)*	*(63.881)*	*(0.19419)*
−150	196.01	60.840	0.20864	......	......	....	....	......	....	......	......	....
−140	202.37	61.916	.21205	141.58	61.906	0.20617	....	......	....	......	......	....
−130	208.73	63.012	.21543	146.04	63.002	.20955	72.903	62.970	0.19810	......	......	....
−120	215.09	64.127	.21876	150.50	64.118	.21288	75.139	64.088	.20144	50.020	64.058	0.19472
−110	221.44	65.261	.22205	154.95	65.253	.21618	77.374	65.225	.20474	51.515	65.197	.19802
−100	227.80	66.414	0.22530	159.40	66.406	0.21943	79.607	66.381	0.20799	53.009	66.355	0.20128
− 90	234.15	67.585	.22851	163.85	67.578	.22264	81.839	67.554	.21121	54.502	67.530	.20451
− 80	240.50	68.774	.23169	168.30	68.768	.22582	84.070	68.746	.21439	55.993	68.723	.20769
− 70	246.85	69.981	.23482	172.75	69.975	.22896	86.299	69.954	.21753	57.483	69.934	.21084
− 60	253.20	71.206	.23793	177.19	71.200	.23206	88.527	71.181	.22064	58.972	71.161	.21395
− 50	259.54	72.447	0.24099	181.64	72.442	0.23513	90.755	72.424	0.22371	60.460	72.406	0.21702
− 40	265.89	73.706	.24403	186.09	73.701	.23816	92.982	73.684	.22675	61.947	73.667	.22006
− 30	272.23	74.980	.24703	190.53	74.976	.24117	95.207	74.960	.22976	63.433	74.944	.22307
− 20	278.58	76.271	.25000	194.97	76.267	.24414	97.433	76.252	.23273	64.919	76.238	.22605
− 10	284.92	77.578	.25294	199.42	77.574	.24708	99.657	77.560	.23567	66.404	77.546	.22899
0	291.27	78.901	0.25585	203.86	78.897	0.24998	101.88	78.884	0.23858	67.889	78.871	0.23190
10	297.61	80.239	.25873	208.30	80.235	.25286	104.11	80.223	.24146	69.373	80.210	.23479
20	303.95	81.591	.26158	212.74	81.588	.25571	106.33	81.576	.24431	70.857	81.565	.23764
30	310.30	82.959	.26440	217.18	82.955	.25854	108.55	82.945	.24714	72.340	82.934	.24046
40	316.64	84.341	.26719	221.62	84.338	.26133	110.77	84.327	.24993	73.823	84.317	.24326
50	322.98	85.737	0.26996	226.07	85.734	0.26410	113.00	85.724	0.25270	75.306	85.714	0.24603
60	329.32	87.147	.27270	230.51	87.144	.26684	115.22	87.135	.25544	76.789	87.126	.24877
70	335.66	88.570	.27541	234.95	88.567	.26955	117.44	88.559	.25816	78.271	88.550	.25149
80	342.01	90.007	.27810	239.39	90.004	.27224	119.66	89.996	.26084	79.753	89.988	.25417
90	348.35	91.457	.28076	243.82	91.454	.27490	121.88	91.447	.26351	81.235	91.439	.25684
100	354.69	92.920	0.28340	248.26	92.917	0.27754	124.10	92.910	0.26614	82.716	92.902	0.25948
110	361.03	94.395	.28601	252.70	94.393	.28015	126.32	94.386	.26876	84.197	94.378	.26209
120	367.37	95.882	.28860	257.14	95.880	.28274	128.54	95.874	.27135	85.679	95.867	.26468
130	373.71	97.382	.29116	261.58	97.380	.28530	130.77	97.374	.27391	87.160	97.367	.26725
140	380.05	98.893	.29370	266.02	98.891	.28784	132.99	98.885	.27645	88.641	98.879	.26979

**Courtesy of E. I. DuPont de Nemours and Company, Inc. Organic Chemicals Department "Freon" Products Division.*

Table B-4 (Cont.)

Temp. F	Abs. Pressure 0.80 lb/in.² Gage Pressure 28.29 in. vac. (Sat. Temp. −114.8 F)			Abs. Pressure 1.00 lb/in. Gage Pressure 27.88 in. vac. (Sat. Temp. −109.3 F)			Abs. Pressure 2.0 lb/in.² Gage Pressure 25.85 in. vac. (Sat. Temp. −90.7 F)			Abs. Pressure 3.0 lb/in.² Gage Pressure 23.81 in. vac. (Sat. Temp. −78.8 F)		
Sat.	*(38.051)*	*(64.624)*	*(0.19167)*	*(30.896)*	*(65.229)*	*(0.18977)*	*(16.195)*	*(67.276)*	*(0.18417)*	*(11.106)*	*(68.604)*	*(0.18114)*
−110	38.586	65.170	0.19324	......	....	....	......	....	....	......	....	....
−100	39.710	66.329	0.19651	31.730	66.303	0.19279	......	....	....	......	....	....
− 90	40.833	67.506	.19974	32.631	67.482	.19602	16.228	67.361	0.18440	......	....	....
− 80	41.954	68.701	.20292	33.531	68.679	.19922	16.634	68.567	.18762	......	....	....
− 70	43.074	69.913	.20608	34.429	69.892	.20237	17.139	69.788	.19079	11.375	69.683	0.18394
− 60	44.194	71.142	.20919	35.327	71.123	.20549	17.593	71.026	.19393	11.681	70.928	.18709
− 50	45.312	72.388	0.21227	36.223	72.370	0.20857	18.046	72.279	0.19703	11.986	72.188	0.19021
− 40	46.429	73.650	.21531	37.119	73.633	.21162	18.498	73.549	.20009	12.290	73.463	.19328
− 30	47.546	74.928	.21832	38.014	74.913	.21463	18.949	74.834	.20311	12.594	74.754	.19632
− 20	48.662	76.223	.22130	38.908	76.208	.21761	19.400	76.134	.20611	12.897	76.059	.19932
− 10	49.778	77.533	.22424	39.802	77.519	.22056	19.850	77.449	.20906	13.199	77.379	.20229
0	50.893	78.858	0.22716	40.695	78.845	0.22347	20.299	78.780	0.21199	13.500	78.714	0.20523
10	52.007	80.198	0.23004	41.587	80.186	.22636	20.748	80.125	.21488	13.802	80.063	.20813
20	53.121	81.553	.23290	42.480	81.542	.22922	21.197	81.484	.21775	14.102	81.426	.21100
30	54.235	82.923	.23572	43.372	82.912	.23204	21.645	82.858	.22058	14.403	82.803	.21384
40	55.348	84.307	.23852	44.263	84.297	.23484	22.093	84.245	.22339	14.703	84.194	.21665
50	56.461	85.705	0.24129	45.154	85.695	0.23761	22.540	85.646	0.22616	15.002	85.598	0.21944
60	57.574	87.116	.24403	46.045	87.107	.24036	22.988	87.061	.22891	15.302	87.015	.22219
70	58.686	88.541	.24675	46.936	88.533	.24307	23.435	88.489	.23163	15.601	88.445	.22492
80	59.799	89.9[illegible]0	.24944	47.826	89.971	.24576	23.881	89.930	.23433	15.900	89.889	.22762
90	60.911	91.431	.25210	48.716	91.423	.24843	24.328	91.384	.23700	16.198	91.345	.23029
100	62.023	92.895	0.25474	49.606	92.887	0.25107	24.774	92.850	0.23964	16.497	92.813	0.23293
110	63.134	94.371	.25736	50.496	94.364	.25368	25.220	94.329	.24226	16.795	94.293	.23556
120	64.246	95.860	.25995	51.386	95.853	.25628	25.666	95.819	.24485	17.093	95.785	.23815
130	65.357	97.361	.26251	52.275	97.354	.25884	26.112	97.322	.24742	17.391	97.289	.24073
140	66.468	98.873	.26506	53.165	98.867	.26139	26.558	98.836	.24997	17.689	98.805	.24327
150	67.579	100.397	0.26758	54.054	100.391	0.26391	27.003	100.361	0.25249	17.987	100.332	0.24580
160	68.690	101.932	.27007	54.943	101.926	.26640	27.449	101.898	.25499	18.284	101.869	.24830
170	69.801	103.478	.27255	55.832	103.472	.26888	27.894	103.445	.25747	18.582	103.418	.25078
180	70.912	105.035	.27500	56.721	105.030	.27133	28.340	105.003	.25992	18.879	104.977	.25324
190	......	....	....	......	....	....	28.785	106.572	.26236	19.176	106.547	.25567
200	......	....	....	......	....	....	29.230	108.151	0.26477	19.473	108.127	0.25808

Table B-4 (Cont.)

Temp. F	Abs. Pressure 5.0 lb/in.² Gage Pressure 19.74 in. vac. (Sat. Temp. −62.4 F)			Abs. Pressure 7.5 lb/in.² Gage Pressure 14.65 in. vac. (Sat. Temp. −48.1 F)			Abs. Pressure 10.0 lb/in.² Gage Pressure 9.56 in. vac. (Sat. Temp. −37.2 F)			Abs. Pressure 15 lb/in.² Gage Pressure 0.3 lb/in.² (Sat. Temp. −20.8 F)		
t	*v*	*h*	*s*	*v*	*h*	*s*	*v*	*h*	*s*	*v*	*h*	*s*
Sat.	*(6.9069)*	*(70.432)*	*(0.17759)*	*(4.7374)*	*(72.017)*	*(0.17501)*	*(3.6246)*	*(73.219)*	*(0.17331)*	*(2.4835)*	*(75.208)*	*(0.17111)*
−60	6.9509	70.729	0.17834	......	....	....	......	....	....	......	....	....
−50	7.1378	72.003	0.18149	......	....	....	......	....	....	......	....	....
−40	7.3239	73.291	.18459	4.8401	73.073	0.17755	......	....	....	......	....	....
−30	7.5092	74.593	.18766	4.9664	74.390	.18065	3.6945	74.183	0.17557	......	....	....
−20	7.6938	75.909	.19069	5.0919	75.719	.18371	3.7906	75.526	.17866	2.4885	75.131	0.17134
−10	7.8777	77.239	.19368	5.2169	77.061	.18673	3.8861	76.880	.18171	2.5546	76.512	.17445
0	8.0611	78.582	0.19663	5.3412	78.415	0.18971	3.9809	78.246	0.18471	2.6201	77.902	0.17751
10	8.2441	79.939	.19955	5.4650	79.782	.19265	4.0753	79.624	.18768	2.6850	79.302	.18052
20	8.4265	81.309	.20244	5.5884	81.162	.19556	4.1691	81.014	.19061	2.7494	80.712	.18349
30	8.6086	82.693	.20529	5.7114	82.555	.19843	4.2626	82.415	.19350	2.8134	82.131	.18642
40	8.7903	84.090	.20812	5.8340	83.959	.20127	4.3556	83.828	.19635	2.8770	83.561	.18931
50	8.9717	85.500	0.21091	5.9562	85.377	0.20408	4.4484	85.252	0.19918	2.9402	85.001	0.19216
60	9.1528	86.922	.21367	6.0782	86.806	.20685	4.5408	86.689	.20197	3.0031	86.451	.19498
70	9.3336	88.358	.21641	6.1999	88.247	.20960	4.6329	88.136	.20473	3.0657	87.912	.19776
80	9.5142	89.806	.21912	6.3213	89.701	.21232	4.7248	89.596	.20746	3.1281	89.383	.20051
90	9.6945	91.266	.22180	6.4425	91.166	.21501	4.8165	91.067	.21016	3.1902	90.865	.20324
100	9.8747	92.738	0.22445	6.5636	92.643	0.21767	4.9079	92.548	0.21283	3.2521	92.357	0.20593
110	10.055	94.222	.22708	6.6844	94.132	.22031	4.9992	94.042	.21547	3.3139	93.860	.20859
120	10.234	95.717	.22968	6.8051	95.632	.22292	5.0903	95.546	.21809	3.3754	95.373	.21122
130	10.414	97.224	.23226	6.9256	97.143	.22550	5.1812	97.061	.22068	3.4368	96.896	.21382
140	10.594	98.743	.23481	7.0459	98.665	.22806	5.2720	98.586	.22325	3.4981	98.429	.21640
150	10.773	100.272	0.23734	7.1662	100.198	0.23060	5.3627	100.123	0.22579	3.5592	99.972	0.21895
160	10.952	101.812	.23985	7.2863	101.741	.23311	5.4533	101.669	.22830	3.6202	101.525	.22148
170	11.131	103.363	.24233	7.4063	103.295	.23560	5.5437	103.226	.23080	3.6811	103.088	.22398
180	11.311	104.925	.24479	7.5262	104.859	.23806	5.6341	104.793	.23326	3.7419	104.661	.22646
190	11.489	106.497	.24723	7.6461	106.434	.24050	5.7243	106.370	.23571	3.8025	106.243	.22891
200	11.668	108.079	0.24964	7.7658	108.018	0.24292	5.8145	107.957	0.23813	3.8632	107.835	0.23135
210	11.847	109.670	.25204	7.8855	109.612	.24532	5.9046	109.553	.24054	3.9237	109.436	.23375
220	12.026	111.272	.25441	8.0051	111.215	.24770	5.9946	111.159	.24291	3.9841	111.046	.23614
230	12.205	112.883	.25677	8.1246	112.828	.25005	6.0846	112.774	.24527	4.0445	112.665	.23850
240	....	....	....	8.2441	114.451	.25239	6.1745	114.398	.24761	4.1049	114.292	.24085
250	....	....	....	......	....	....	6.2643	116.031	0.24993	4.1651	115.929	0.24317

Table B-4 (Cont.)

Temp. F	Abs. Pressure 20 lb/in.² Gage Pressure 5.3 lb/in.² (Sat. Temp. −8.1 F)			Abs. Pressure 26 lb/in.² Gage Pressure 11.3 lb/in.² (Sat. Temp. 4.1 F)			Abs. Pressure 32 lb/in.² Gage Pressure 17.3 lb/in.² (Sat. Temp. 14.4 F)			Abs. Pressure 40 lb/in.² Gage Pressure 25.3 lb/in.² (Sat. Temp. 25.9 F)		
Sat.	*(1.8977)*	*(76.397)*	*(0.16969)*	*(1.4835)*	*(77.710)*	*(0.16850)*	*(1.2198)*	*(78.793)*	*(0.16763)*	*(0.98743)*	*(80.000)*	*(0.16676)*
0	1.9390	77.550	0.17222	......	......	......	......	......	......	......	......	......
10	1.9893	78.973	.17528	1.5071	78.566	0.17033	......	......	......	......	......	......
20	2.0391	80.403	.17829	1.5468	80.024	.17340	1.2387	79.634	0.16939	......	......	......
30	2.0884	81.842	.18126	1.5861	81.487	.17642	1.2717	81.123	.17246	0.99865	80.622	0.16804
40	2.1373	83.289	.18419	1.6248	82.956	.17939	1.3042	82.616	.17548	1.0258	82.148	.17112
50	2.1858	84.745	0.18707	1.6632	84.432	0.18232	1.3363	84.113	0.17845	1.0526	83.676	0.17415
60	2.2340	86.210	.18992	1.7013	85.916	.18520	1.3681	85.616	.18137	1.0789	85.206	.17712
70	2.2819	87.684	.19273	1.7390	87.407	.18804	1.3995	87.124	.18424	1.1049	86.739	.18005
80	2.3295	89.168	.19550	1.7765	88.906	.19084	1.4306	88.639	.18707	1.1306	88.277	.18292
90	2.3769	90.661	.19824	1.8137	90.413	.19361	1.4615	90.161	.18987	1.1560	89.819	.18575
100	2.4241	92.164	0.20095	1.8507	91.929	0.19634	1.4921	91.690	0.19263	1.1812	91.367	0.18854
110	2.4711	93.676	.20363	1.8874	93.453	.19904	1.5225	93.227	.19535	1.2061	92.920	.19129
120	2.5179	95.198	.20628	1.9240	94.986	.20171	1.5528	94.771	.19803	1.2309	94.480	.19401
130	2.5645	96.729	.20890	1.9605	96.527	.20435	1.5828	96.323	.20069	1.2554	96.047	.19669
140	2.6110	98.270	.21149	1.9967	98.078	.20695	1.6127	97.883	.20331	1.2798	97.620	.19933
150	2.6573	99.820	0.21405	2.0329	99.637	0.20953	1.6425	99.451	0.20590	1.3041	99.200	0.20195
160	2.7036	101.380	.21659	2.0689	101.204	.21208	1.6721	101.027	.20847	1.3282	100.788	.20453
170	2.7497	102.949	.21910	2.1048	102.781	.21461	1.7017	102.611	.21100	1.3522	102.383	.20708
180	2.7957	104.528	.22159	2.1406	104.367	.21710	1.7311	104.204	.21351	1.3761	103.985	.20961
190	2.8416	106.115	.22405	2.1763	105.961	.21958	1.7604	105.805	.21600	1.3999	105.595	.21210
200	2.8874	107.712	0.22649	2.2119	107.563	0.22202	1.7896	107.414	0.21845	1.4236	107.212	0.21457
210	2.9332	109.317	.22891	2.2474	109.174	.22445	1.8187	109.031	.22089	1.4472	108.837	.21702
220	2.9789	110.932	.23130	2.2828	110.794	.22685	1.8478	110.656	.22330	1.4707	110.469	.21944
230	3.0245	112.555	.23367	2.3182	112.422	.22923	1.8768	112.289	.22568	1.4942	112.109	.22183
240	3.0700	114.186	.23602	2.3535	114.058	.23158	1.9057	113.930	.22804	1.5176	113.757	.22420
250	3.1155	115.826	0.23835	2.3888	115.703	0.23391	1.9346	115.579	0.23038	1.5409	115.412	0.22655
260	3.1609	117.475	.24065	2.4240	117.355	.23623	1.9634	117.235	.23270	1.5642	117.074	.22888
270	3.2063	119.131	.24294	2.4592	119.016	.23852	1.9922	118.900	.23500	1.5874	118.744	.23118
280	3.2517	120.796	.24520	2.4943	120.684	.24079	2.0209	120.572	.23727	1.6106	120.421	.23347
290	3.2970	122.469	.24745	2.5293	122.360	.24304	2.0495	122.251	.23953	1.6337	122.105	.23573

Table B-4 (Cont.)

Temp. F	Abs. Pressure 50 lb/in.² Gage Pressure 35.3 lb/in.² (Sat. Temp. 38.2 F)			Abs. Pressure 60 lb/in.² Gage Pressure 45.3 lb/in.² (Sat. Temp. 48.6 F)			Abs. Pressure 80 lb/in.² Gage Pressure 65.3 lb/in.² (Sat. Temp. 66.2 F)			Abs. Pressure 100 lb/in.² Gage Pressure 85.3 lb/in.² (Sat. Temp. 80.8 F)		
t	*v*	*h*	*s*	*v*	*h*	*s*	*v*	*h*	*s*	*v*	*h*	*s*
Sat.	*(0.79824)*	*(81.249)*	*(0.16597)*	*(0.67005)*	*(82.299)*	*(0.16537)*	*(0.50680)*	*(84.003)*	*(0.16450)*	*(0.40674)*	*(85.351)*	*(0.16389)*
40	0.80248	81.540	0.16655	....	....	....	....	....	....	....	....	....
50	0.82502	83.109	0.16966	0.67272	82.518	0.16580	....	....	....	....	....	....
60	.84713	84.676	.17271	.69210	84.126	.16892	....	....	....	....	....	....
70	.86886	86.243	.17569	.71105	85.729	.17198	0.51269	84.640	0.16571	....	....	....
80	.89025	87.811	.17862	.72964	87.330	.17497	.52795	86.316	.16885	....	....	....
90	.91134	89.380	.18151	.74790	88.929	.17791	.54281	87.981	.17190	0.41876	86.964	0.16685
100	0.93216	90.953	0.18434	0.76588	90.528	0.18079	0.55734	89.640	0.17489	0.43138	88.694	0.16996
110	.95275	92.529	.18713	.78360	92.128	.18362	.57158	91.294	.17782	.44365	90.410	.17300
120	.97313	94.110	.18988	.80110	93.731	.18641	.58556	92.945	.18070	.45562	92.116	.17597
130	.99332	95.695	.19259	.81840	95.336	.18916	.59931	94.594	.18352	.46733	93.814	.17888
140	1.0133	97.286	.19527	.83551	96.945	.19186	.61286	96.242	.18629	.47881	95.507	.18172
150	1.0332	98.882	0.19791	0.85247	98.558	0.19453	0.62623	97.891	0.18902	0.49009	97.197	0.18452
160	1.0529	100.485	.20051	.86928	100.176	.19716	.63943	99.542	.19170	.50118	98.884	.18726
170	1.0725	102.093	.20309	.88596	101.799	.19976	.65250	101.195	.19435	.51212	100.571	.18996
180	1.0920	103.708	.20563	.90252	103.427	.20233	.66543	102.851	.19696	.52291	102.257	.19262
190	1.1114	105.330	.20815	.91896	105.060	.20486	.67824	104.511	.19953	.53358	103.944	.19524
200	1.1307	106.958	0.21064	0.93531	106.700	0.20736	0.69095	106.174	0.20207	0.54413	105.633	0.19782
210	1.1499	108.593	.21310	.95157	108.345	.20984	.70356	107.841	.20458	.55457	107.324	.20036
220	1.1690	110.235	.21553	.96775	109.997	.21229	.71609	109.513	.20706	.56492	109.018	.20287
230	1.1880	111.883	.21794	.98385	111.655	.21471	.72853	111.190	.20951	.57519	110.714	.20535
240	1.2070	113.539	.22032	.99988	113.319	.21710	.74090	112.872	.21193	.58538	112.415	.20780
250	1.2259	115.202	0.22268	1.0159	114.989	0.21947	0.75320	114.559	0.21432	0.59549	114.119	0.21022
260	1.2447	116.871	.22502	1.0318	116.666	.22182	.76544	116.251	.21669	.60554	115.828	.21261
270	1.2636	118.547	.22733	1.0476	118.350	.22414	.77762	117.949	.21903	.61553	117.540	.21497
280	1.2823	120.231	.22962	1.0634	120.039	.22644	.78975	119.652	.22135	.62546	119.258	.21731
290	1.3010	121.921	.23189	1.0792	121.736	.22872	.80183	121.361	.22364	.63534	120.980	.21962
300	1.3197	123.618	0.23414	1.0949	123.438	0.23098	0.81386	123.075	0.22592	0.64518	122.707	0.22191
310	1.3383	125.321	.23637	1.1106	125.147	.23321	.82586	124.795	.22817	.65497	124.439	.22417
320	1.3569	127.032	.23857	1.1262	126.863	.23543	.83781	126.521	.23039	.66472	126.176	.22641
330	1.3754	128.749	.24076	1.1418	128.585	.23762	.84973	128.253	.23260	.67444	127.917	.22863
340	....	....	....	1.1574	130.313	.23980	.86161	129.990	.23479	.68411	129.665	.23083

Table B-4 (Cont.)

Temp. F	Abs. Press. 120 lb/in.² Gage Press. 105.3 lb/in.² (Sat. Temp. 93.3 F)			Abs. Press. 140 lb/in.² Gage Press. 125.3 lb/in.² (Sat. Temp. 104.4 F)			Abs. Press. 180 lb/in.² Gage Press. 165.3 lb/in.² (Sat. Temp. 123.4 F)			Abs. Press. 220 lb/in.² Gage Press. 205.3 lb/in.² (Sat. Temp. 139.5 F)		
Sat.	*(0.33886)*	*(86.459)*	*(0.16340)*	*(0.28964)*	*(87.389)*	*(0.16299)*	*(0.22276)*	*(88.857)*	*(0.16228)*	*(0.17917)*	*(89.937)*	*(0.16161)*
100	0.34655	87.675	0.16559	....	....	....	....	....	....	....	....	....
110	.35766	89.466	.16876	0.29548	88.448	0.16486	....	....	....	....	....	....
120	.36841	91.237	.17184	.30549	90.297	.16808	....	....	....	....	....	....
130	.37884	92.992	.17484	.31513	92.120	.17120	0.22863	90.179	0.16454	....	....	....
140	.38901	94.736	.17778	.32445	93.923	.17423	.23710	92.136	.16783	0.17957	90.043	0.16179
150	0.39896	96.471	0.18065	0.33350	95.709	0.17718	0.24519	94.053	0.17100	0.18746	92.156	0.16528
160	.40870	98.199	.18346	.34232	97.483	.18007	.25297	95.940	.17407	.19487	94.203	.16861
170	.41826	99.922	.18622	.35095	99.247	.18289	.26047	97.803	.17705	.20190	96.199	.17181
180	.42766	101.642	.18892	.35939	101.003	.18566	.26775	99.647	.17995	.20861	98.157	.17489
190	.43692	103.359	.19159	.36769	102.754	.18838	.27484	101.475	.18279	.21506	100.084	.17788
200	0.44606	105.076	0.19421	0.37584	104.501	0.19104	0.28176	103.291	0.18556	0.22130	101.986	0.18079
210	.45508	106.792	.19679	.38387	106.245	.19367	.28852	105.098	.18828	.22735	103.869	.18362
220	.46401	108.509	.19934	.39179	107.987	.19625	.29516	106.896	.19095	.23324	105.735	.18638
230	.47284	110.227	.20185	.39961	109.728	.19879	.30168	108.689	.19357	.23900	107.589	.18909
240	.48158	111.948	.20432	.40734	111.470	.20130	.30810	110.478	.19614	.24463	109.432	.19175
250	0.49025	113.670	0.20677	0.41499	113.212	0.20377	0.31442	112.263	0.19868	0.25015	111.267	0.19435
260	.49885	115.396	.20918	.42257	114.956	.20621	.32066	114.046	.20117	.25557	113.095	.19691
270	.50739	117.125	.21157	.43008	116.701	.20862	.32682	115.828	.20363	.26091	114.919	.19942
280	.51587	118.857	.21393	.43753	118.449	.21100	.33292	117.610	.20605	.26617	116.738	.20190
290	.52429	120.593	.21626	.44492	120.199	.21335	.33895	119.392	.20845	.27136	118.555	.20434
300	0.53267	122.333	0.21856	0.45226	121.953	0.21567	0.34492	121.174	0.21081	0.27648	120.369	0.20674
310	.54100	124.077	.22084	.45955	123.709	.21797	.35084	122.958	.21314	.28155	122.183	.20912
320	.54929	125.825	.22310	.46680	125.470	.22024	.35672	124.744	.21545	.28657	123.996	.21146
330	.55754	127.578	.22533	.47400	127.233	.22249	.36255	126.531	.21772	.29153	125.809	.21377
340	.56575	129.335	.22754	.48117	129.001	.22471	.36834	128.321	.21998	.29645	127.623	.21605
350	0.57393	131.097	0.22973	0.48831	130.773	0.22692	0.37409	130.113	0.22220	0.30134	129.438	0.21830
360	.58208	132.863	.23190	.49541	132.548	.22910	.37980	131.909	.22441	.30618	131.255	.22053
370	.59019	134.634	.23405	.50248	134.328	.23125	.38549	133.707	.22659	.31099	133.073	.22274
380	.59829	136.410	.23618	.50953	136.112	.23339	.39114	135.509	.22875	.31576	134.893	.22492
390	.60635	138.191	.23829	.51654	137.901	.23551	.39677	137.314	.23088	.32051	136.715	.22708
400	....	....	....	....	....	....	0.40237	139.122	0.23300	0.32523	138.540	0.22921
410	....	....	....	....	....	....	.40794	140.934	.23509	.32992	140.368	.23133

PRESSURE-ENTHALPY DIAGRAM

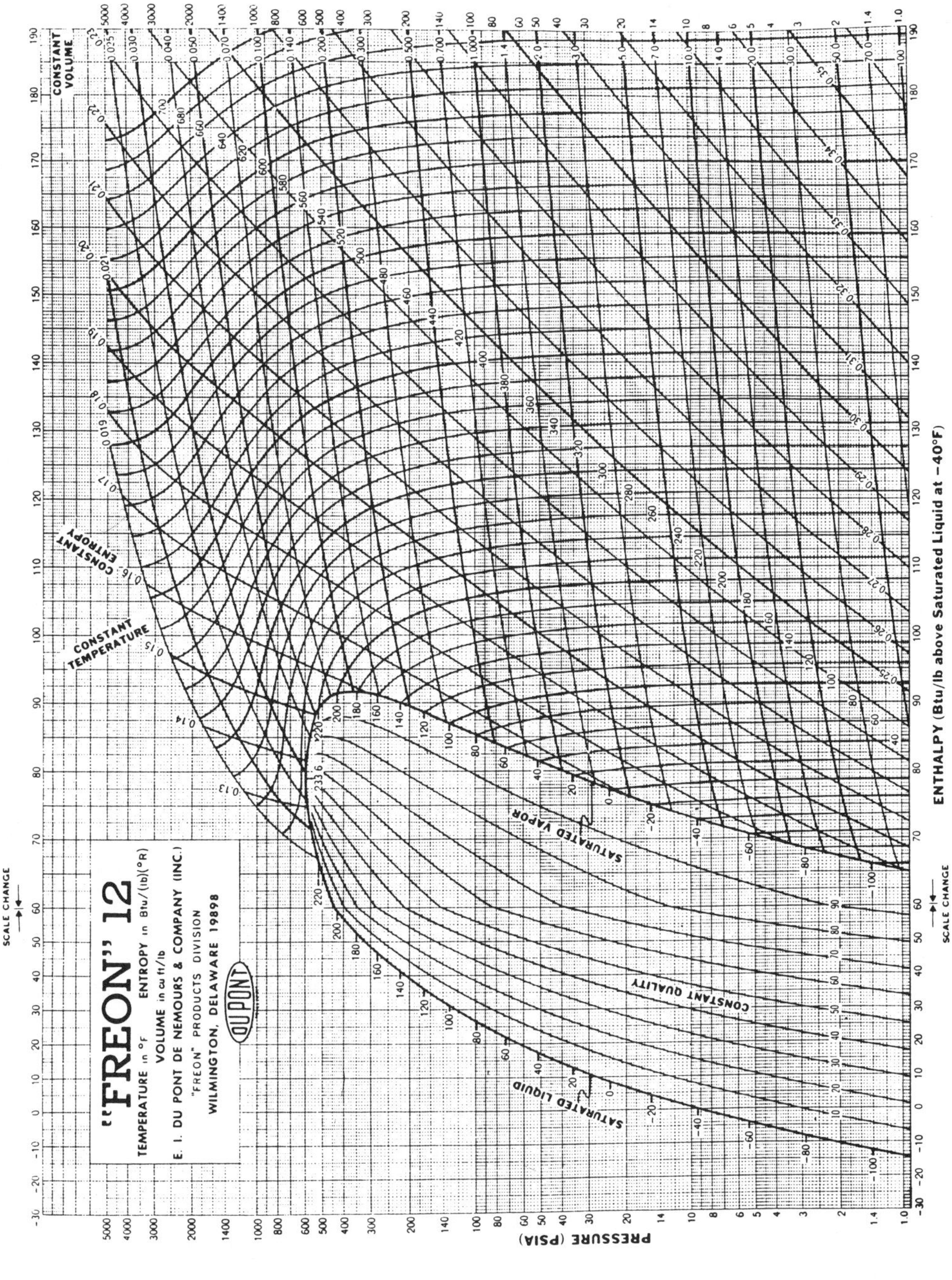

Table B-5

Thermodynamic Properties of Freon-22[a] (Properties of Saturated Liquid and Saturated Vapor)

TEMP.	PRESSURE		VOLUME cu ft/lb		DENSITY lb/cu ft		ENTHALPY Btu/lb			ENTROPY Btu/(lb)(°R)		TEMP.
°F	PSIA	PSIG	LIQUID v_f	VAPOR v_g	LIQUID $1/v_f$	VAPOR $1/v_g$	LIQUID h_f	LATENT h_{fg}	VAPOR h_g	LIQUID s_f	VAPOR s_g	°F
−45	13.354	2.732*	0.011298	3.7243	88.507	0.26851	−1.260	100.963	99.703	−0.00301	0.24046	−45
−44	13.712	2.002*	0.011311	3.6334	88.407	0.27523	−1.009	100.823	99.814	−0.00241	0.24014	−44
−43	14.078	1.258*	0.011324	3.5452	88.307	0.28207	−0.757	100.683	99.925	−0.00181	0.23982	−43
−42	14.451	0.498*	0.011337	3.4596	88.207	0.28905	−0.505	100.541	100.036	−0.00120	0.23951	−42
−41	14.833	0.137	0.011350	3.3764	88.107	0.29617	−0.253	100.399	100.147	−0.00060	0.23919	−41
−40	15.222	0.526	0.011363	3.2957	88.006	0.30342	0.000	100.257	100.257	0.00000	0.23888	−40
−39	15.619	0.923	0.011376	3.2173	87.905	0.31082	0.253	100.114	100.367	0.00060	0.23858	−39
−38	16.024	1.328	0.011389	3.1412	87.805	0.31835	0.506	99.971	100.477	0.00120	0.23827	−38
−37	16.437	1.741	0.011402	3.0673	87.703	0.32602	0.760	99.826	100.587	0.00180	0.23797	−37
−36	16.859	2.163	0.011415	2.9954	87.602	0.33384	1.014	99.682	100.696	0.00240	0.23767	−36
−35	17.290	2.594	0.011428	2.9256	87.501	0.34181	1.269	99.536	100.805	0.00300	0.23737	−35
−34	17.728	3.032	0.011442	2.8578	87.399	0.34992	1.524	99.391	100.914	0.00359	0.23707	−34
−33	18.176	3.480	0.011455	2.7919	87.297	0.35818	1.779	99.244	101.023	0.00419	0.23678	−33
−32	18.633	3.937	0.011469	2.7278	87.195	0.36660	2.035	99.097	101.132	0.00479	0.23649	−32
−31	19.098	4.402	0.011482	2.6655	87.093	0.37517	2.291	98.949	101.240	0.00538	0.23620	−31
−30	19.573	4.877	0.011495	2.6049	86.991	0.38389	2.547	98.801	101.348	0.00598	0.23591	−30
−29	20.056	5.360	0.011509	2.5460	86.888	0.39278	2.804	98.652	101.456	0.00657	0.23563	−29
−28	20.549	5.853	0.011523	2.4887	86.785	0.40182	3.061	98.503	101.564	0.00716	0.23534	−28
−27	21.052	6.536	0.011536	2.4329	86.682	0.41103	3.318	98.353	101.671	0.00776	0.23506	−27
−26	21.564	6.868	0.011550	2.3787	86.579	0.42040	3.576	98.202	101.778	0.00835	0.23478	−26
−25	22.086	7.390	0.011564	2.3260	86.476	0.42993	3.834	98.051	101.885	0.00894	0.23451	−25
−24	22.617	7.921	0.011578	2.2746	86.372	0.43964	4.093	97.899	101.992	0.00953	0.23423	−24
−23	23.159	8.463	0.011592	2.2246	86.269	0.44951	4.352	97.746	102.098	0.01013	0.23396	−23
−22	23.711	9.015	0.011606	2.1760	86.165	0.45956	4.611	97.593	102.204	0.01072	0.23369	−22
−21	24.272	9.576	0.011620	2.1287	86.061	0.46978	4.871	97.439	102.310	0.01131	0.23342	−21
−20	24.845	10.149	0.011634	2.0826	85.956	0.48018	5.131	97.285	102.415	0.01189	0.23315	−20
−19	25.427	10.731	0.011648	2.0377	85.852	0.49075	5.391	97.129	102.521	0.01248	0.23289	−19
−18	26.020	11.324	0.011662	1.9940	85.747	0.50151	5.652	96.974	102.626	0.01307	0.23262	−18
−17	26.624	11.928	0.011677	1.9514	85.642	0.51245	5.913	96.817	102.730	0.01366	0.23236	−17
−16	27.239	12.543	0.011691	1.9099	85.537	0.52358	6.175	96.660	102.835	0.01425	0.23210	−16
−15	27.865	13.169	0.011705	1.8695	85.431	0.53489	6.436	96.502	102.939	0.01483	0.23184	−15
−14	28.501	13.805	0.011720	1.8302	85.326	0.54640	6.699	96.344	103.043	0.01542	0.23159	−14
−13	29.149	14.453	0.011734	1.7918	85.220	0.55810	6.961	96.185	103.146	0.01600	0.23133	−13
−12	29.809	15.113	0.011749	1.7544	85.114	0.56999	7.224	96.025	103.250	0.01659	0.23108	−12
−11	30.480	15.784	0.011764	1.7180	85.008	0.58207	7.488	95.865	103.353	0.01717	0.23083	−11
−10	31.162	16.466	0.011778	1.6825	84.901	0.59436	7.751	95.704	103.455	0.01776	0.23058	−10
− 9	31.856	17.160	0.011793	1.6479	84.795	0.60685	8.015	95.542	103.558	0.01834	0.23033	− 9
− 8	32.563	17.867	0.011808	1.6141	84.688	0.61954	8.280	95.380	103.660	0.01892	0.23008	− 8
− 7	33.281	18.585	0.011823	1.5812	84.581	0.63244	8.545	95.217	103.762	0.01950	0.22984	− 7
− 6	34.011	19.315	0.011838	1.5491	84.473	0.64555	8.810	95.053	103.863	0.02009	0.22960	− 6
− 5	34.754	20.058	0.011853	1.5177	84.366	0.65887	9.075	94.889	103.964	0.02067	0.22936	− 5
− 4	35.509	20.813	0.011868	1.4872	84.258	0.67240	9.341	94.724	104.065	0.02125	0.22912	− 4
− 3	36.277	21.581	0.011884	1.4574	84.150	0.68615	9.608	94.558	104.166	0.02183	0.22888	− 3
− 2	37.057	22.361	0.011899	1.4283	84.042	0.70012	9.874	94.391	104.266	0.02241	0.22864	− 2
− 1	37.850	23.154	0.011914	1.4000	83.933	0.71431	10.142	94.224	104.366	0.02299	0.22841	− 1
0	38.657	23.961	0.011930	1.3723	83.825	0.72872	10.409	94.056	104.465	0.02357	0.22817	0
1	39.476	24.780	0.011945	1.3453	83.716	0.74336	10.677	93.888	104.565	0.02414	0.22794	1
2	40.309	25.613	0.011961	1.3189	83.606	0.75822	10.945	93.718	104.663	0.02472	0.22771	2
3	41.155	26.459	0.011976	1.2931	83.497	0.77332	11.214	93.548	104.762	0.02530	0.22748	3
4	42.014	27.318	0.011992	1.2680	83.387	0.78865	11.483	93.378	104.860	0.02587	0.22725	4
5	42.888	28.192	0.012008	1.2434	83.277	0.80422	11.752	93.206	104.958	0.02645	0.22703	5
6	43.775	29.079	0.012024	1.2195	83.167	0.82003	12.022	93.034	105.056	0.02703	0.22680	6
7	44.676	29.980	0.012040	1.1961	83.057	0.83608	12.292	92.861	105.153	0.02760	0.22658	7
8	45.591	30.895	0.012056	1.1732	82.946	0.85237	12.562	92.688	105.250	0.02818	0.22636	8
9	46.521	31.825	0.012072	1.1509	82.835	0.86892	12.833	92.513	105.346	0.02875	0.22614	9

*Inches of mercury below one atmosphere

Table B-5 (Cont.)

TEMP. °F	PRESSURE PSIA	PRESSURE PSIG	VOLUME cu ft/lb LIQUID v_f	VOLUME cu ft/lb VAPOR v_g	DENSITY lb/cu ft LIQUID $1/v_f$	DENSITY lb/cu ft VAPOR $1/v_g$	ENTHALPY Btu/lb LIQUID h_f	ENTHALPY Btu/lb LATENT h_{fg}	ENTHALPY Btu/lb VAPOR h_g	ENTROPY Btu/(lb)(°R) LIQUID s_f	ENTROPY Btu/(lb)(°R) VAPOR s_g	TEMP. °F
10	47.464	32.768	0.012088	1.1290	82.724	0.88571	13.104	92.338	105.442	0.02932	0.22592	10
11	48.423	33.727	0.012105	1.1077	82.612	0.90275	13.376	92.162	105.538	0.02990	0.22570	11
12	49.396	34.700	0.012121	1.0869	82.501	0.92005	13.648	91.986	105.633	0.03047	0.22548	12
13	50.384	35.688	0.012138	1.0665	82.389	0.93761	13.920	91.808	105.728	0.03104	0.22527	13
14	51.387	36.691	0.012154	1.0466	82.276	0.95544	14.193	91.630	105.823	0.03161	0.22505	14
15	52.405	37.709	0.012171	1.0272	82.164	0.97352	14.466	91.451	105.917	0.03218	0.22484	15
16	53.438	38.742	0.012188	1.0082	82.051	0.99188	14.739	91.272	106.011	0.03275	0.22463	16
17	54.487	39.791	0.012204	0.98961	81.938	1.0105	15.013	91.091	106.105	0.03332	0.22442	17
18	55.551	40.855	0.012221	0.97144	81.825	1.0294	15.288	90.910	106.198	0.03389	0.22421	18
19	56.631	41.935	0.012238	0.95368	81.711	1.0486	15.562	90.728	106.290	0.03446	0.22400	19
20	57.727	43.031	0.012255	0.93631	81.597	1.0680	15.837	90.545	106.383	0.03503	0.22379	20
21	58.839	44.143	0.012273	0.91932	81.483	1.0878	16.113	90.362	106.475	0.03560	0.22358	21
22	59.967	45.271	0.012290	0.90270	81.368	1.1078	16.389	90.178	106.566	0.03617	0.22338	22
23	61.111	46.415	0.012307	0.88645	81.253	1.1281	16.665	89.993	106.657	0.03674	0.22318	23
24	62.272	47.576	0.012325	0.87055	81.138	1.1487	16.942	89.807	106.748	0.03730	0.22297	24
25	63.450	48.754	0.012342	0.85500	81.023	1.1696	17.219	89.620	106.839	0.03787	0.22277	25
26	64.644	49.948	0.012360	0.83978	80.907	1.1908	17.496	89.433	106.928	0.03844	0.22257	26
27	65.855	51.159	0.012378	0.82488	80.791	1.2123	17.774	89.244	107.018	0.03900	0.22237	27
28	67.083	52.387	0.012395	0.81031	80.675	1.2341	18.052	89.055	107.107	0.03958	0.22217	28
29	68.328	53.632	0.012413	0.79604	80.558	1.2562	18.330	88.865	107.196	0.04013	0.22198	29
30	69.591	54.895	0.012431	0.78208	80.441	1.2786	18.609	88.674	107.284	0.04070	0.22178	30
31	70.871	56.175	0.012450	0.76842	80.324	1.3014	18.889	88.483	107.372	0.04126	0.22158	31
32	72.169	57.473	0.012468	0.75503	80.207	1.3244	19.169	88.290	107.459	0.04182	0.22139	32
33	73.485	58.789	0.012486	0.74194	80.089	1.3478	19.449	88.097	107.546	0.04239	0.22119	33
34	74.818	60.122	0.012505	0.72911	79.971	1.3715	19.729	87.903	107.632	0.04295	0.22100	34
35	76.170	61.474	0.012523	0.71655	79.852	1.3956	20.010	87.708	107.719	0.04351	0.22081	35
36	77.540	62.844	0.012542	0.70425	79.733	1.4199	20.292	87.512	107.804	0.04407	0.22062	36
37	78.929	64.233	0.012561	0.69221	79.614	1.4447	20.574	87.316	107.889	0.04464	0.22043	37
38	80.336	65.640	0.012579	0.68041	79.495	1.4697	20.856	87.118	107.974	0.04520	0.22024	38
39	81.761	67.065	0.012598	0.66885	79.375	1.4951	21.138	86.920	108.058	0.04576	0.22005	39
40	83.206	68.510	0.012618	0.65753	79.255	1.5208	21.422	86.720	108.142	0.04632	0.21986	40
41	84.670	69.974	0.012637	0.64643	79.134	1.5469	21.705	86.520	108.225	0.04688	0.21968	41
42	86.153	71.457	0.012656	0.63557	79.013	1.5734	21.989	86.319	108.308	0.04744	0.21949	42
43	87.655	72.959	0.012676	0.62492	78.892	1.6002	22.273	86.117	108.390	0.04800	0.21931	43
44	89.177	74.481	0.012695	0.61448	78.770	1.6274	22.558	85.914	108.472	0.04855	0.21912	44
45	90.719	76.023	0.012715	0.60425	78.648	1.6549	22.843	85.710	108.553	0.04911	0.21894	45
46	92.280	77.584	0.012735	0.59422	78.526	1.6829	23.129	85.506	108.634	0.04967	0.21876	46
47	93.861	79.165	0.012755	0.58440	78.403	1.7112	23.415	85.300	108.715	0.05023	0.21858	47
48	95.463	80.767	0.012775	0.57476	78.280	1.7398	23.701	85.094	108.795	0.05079	0.21839	48
49	97.085	82.389	0.012795	0.56532	78.157	1.7689	23.988	84.886	108.874	0.05134	0.21821	49
50	98.727	84.031	0.012815	0.55606	78.033	1.7984	24.275	84.678	108.953	0.05190	0.21803	50
51	100.39	85.69	0.012836	0.54698	77.909	1.8282	24.563	84.468	109.031	0.05245	0.21785	51
52	102.07	87.38	0.012856	0.53808	77.784	1.8585	24.851	84.258	109.109	0.05301	0.21768	52
53	103.78	89.08	0.012877	0.52934	77.659	1.8891	25.139	84.047	109.186	0.05357	0.21750	53
54	105.50	90.81	0.012898	0.52078	77.534	1.9202	25.429	83.834	109.263	0.05412	0.21732	54
55	107.25	92.56	0.012919	0.51238	77.408	1.9517	25.718	83.621	109.339	0.05468	0.21714	55
56	109.02	94.32	0.012940	0.50414	77.282	1.9836	26.008	83.407	109.415	0.05523	0.21697	56
57	110.81	96.11	0.012961	0.49606	77.155	2.0159	26.298	83.191	109.490	0.05579	0.21679	57
58	112.62	97.93	0.012982	0.48813	77.028	2.0486	26.589	82.975	109.564	0.05634	0.21662	58
59	114.46	99.76	0.013004	0.48035	76.900	2.0818	26.880	82.758	109.638	0.05689	0.21644	59
60	116.31	101.62	0.013025	0.47272	76.773	2.1154	27.172	82.540	109.712	0.05745	0.21627	60
61	118.19	103.49	0.013047	0.46523	76.644	2.1495	27.464	82.320	109.785	0.05800	0.21610	61
62	120.09	105.39	0.013069	0.45788	76.515	2.1840	27.757	82.100	109.857	0.05855	0.21592	62
63	122.01	107.32	0.013091	0.45066	76.386	2.2190	28.050	81.878	109.929	0.05910	0.21575	63
64	123.96	109.26	0.013114	0.44358	76.257	2.2544	28.344	81.656	110.000	0.05966	0.21558	64

Table B-5 (Cont.)

TEMP.	PRESSURE		VOLUME cu ft/lb		DENSITY lb/cu ft		ENTHALPY Btu/lb			ENTROPY Btu/(lb)(°R)		TEMP.
°F	PSIA	PSIG	LIQUID v_f	VAPOR v_g	LIQUID $1/v_f$	VAPOR $1/v_g$	LIQUID h_f	LATENT h_{fg}	VAPOR h_g	LIQUID s_f	VAPOR s_g	°F
65	125.93	111.23	0.013136	0.43663	76.126	2.2903	28.638	81.432	110.070	0.06021	0.21541	65
66	127.92	113.22	0.013159	0.42981	75.996	2.3266	28.932	81.208	110.140	0.06076	0.21524	66
67	129.94	115.24	0.013181	0.42311	75.865	2.3635	29.228	80.982	110.209	0.06131	0.21507	67
68	131.97	117.28	0.013204	0.41653	75.733	2.4008	29.523	80.755	110.278	0.06186	0.21490	68
69	134.04	119.34	0.013227	0.41007	75.601	2.4386	29.819	80.527	110.346	0.06241	0.21473	69
70	136.12	121.43	0.013251	0.40373	75.469	2.4769	30.116	80.298	110.414	0.06296	0.21456	70
71	138.23	123.54	0.013274	0.39751	75.336	2.5157	30.413	80.068	110.480	0.06351	0.21439	71
72	140.37	125.67	0.013297	0.39139	75.202	2.5550	30.710	79.836	110.547	0.06406	0.21422	72
73	142.52	127.83	0.013321	0.38539	75.068	2.5948	31.008	79.604	110.612	0.06461	0.21405	73
74	144.71	130.01	0.013345	0.37949	74.934	2.6351	31.307	79.370	110.677	0.06516	0.21388	74
75	146.91	132.22	0.013369	0.37369	74.799	2.6760	31.606	79.135	110.741	0.06571	0.21372	75
76	149.15	134.45	0.013393	0.36800	74.664	2.7174	31.906	78.899	110.805	0.06626	0.21355	76
77	151.40	136.71	0.013418	0.36241	74.528	2.7593	32.206	78.662	110.868	0.06681	0.21338	77
78	153.69	138.99	0.013442	0.35691	74.391	2.8018	32.506	78.423	110.930	0.06736	0.21321	78
79	155.99	141.30	0.013467	0.35151	74.254	2.8449	32.808	78.184	110.991	0.06791	0.21305	79
80	158.33	143.63	0.013492	0.34621	74.116	2.8885	33.109	77.943	111.052	0.06846	0.21288	80
81	160.68	145.99	0.013518	0.34099	73.978	2.9326	33.412	77.701	111.112	0.06901	0.21271	81
82	163.07	148.37	0.013543	0.33587	73.839	2.9774	33.714	77.457	111.171	0.06956	0.21255	82
83	165.48	150.78	0.013569	0.33083	73.700	3.0227	34.018	77.212	111.230	0.07011	0.21238	83
84	167.92	153.22	0.013594	0.32588	73.560	3.0686	34.322	76.966	111.288	0.07065	0.21222	84
85	170.38	155.68	0.013620	0.32101	73.420	3.1151	34.626	76.719	111.345	0.07120	0.21205	85
86	172.87	158.17	0.013647	0.31623	73.278	3.1622	34.931	76.470	111.401	0.07175	0.21188	86
87	175.38	160.69	0.013673	0.31153	73.137	3.2100	35.237	76.220	111.457	0.07230	0.21172	87
88	177.93	163.23	0.013700	0.30690	72.994	3.2583	35.543	75.968	111.512	0.07285	0.21155	88
89	180.50	165.80	0.013727	0.30236	72.851	3.3073	35.850	75.716	111.566	0.07339	0.21139	89
90	183.09	168.40	0.013754	0.29789	72.708	3.3570	36.158	75.461	111.619	0.07394	0.21122	90
91	185.72	171.02	0.013781	0.29349	72.564	3.4073	36.466	75.206	111.671	0.07449	0.21106	91
92	188.37	173.67	0.013809	0.28917	72.419	3.4582	36.774	74.949	111.723	0.07504	0.21089	92
93	191.05	176.35	0.013836	0.28491	72.273	3.5098	37.084	74.690	111.774	0.07559	0.21072	93
94	193.76	179.06	0.013864	0.28073	72.127	3.5621	37.394	74.430	111.824	0.07613	0.21056	94
95	196.50	181.80	0.013893	0.27662	71.980	3.6151	37.704	74.168	111.873	0.07668	0.21039	95
96	199.26	184.56	0.013921	0.27257	71.833	3.6688	38.016	73.905	111.921	0.07723	0.21023	96
97	202.05	187.36	0.013950	0.26859	71.685	3.7232	38.328	73.641	111.968	0.07778	0.21006	97
98	204.87	190.18	0.013979	0.26467	71.536	3.7783	38.640	73.375	112.015	0.07832	0.20989	98
99	207.72	193.03	0.014008	0.26081	71.386	3.8341	38.953	73.107	112.060	0.07887	0.20973	99
100	210.60	195.91	0.014038	0.25702	71.236	3.8907	39.267	72.838	112.105	0.07942	0.20956	100
101	213.51	198.82	0.014068	0.25329	71.084	3.9481	39.582	72.567	112.149	0.07997	0.20939	101
102	216.45	201.76	0.014098	0.24962	70.933	4.0062	39.897	72.294	112.192	0.08052	0.20923	102
103	219.42	204.72	0.014128	0.24600	70.780	4.0651	40.213	72.020	112.233	0.08107	0.20906	103
104	222.42	207.72	0.014159	0.24244	70.626	4.1247	40.530	71.744	112.274	0.08161	0.20889	104
105	225.45	210.75	0.014190	0.23894	70.472	4.1852	40.847	71.467	112.314	0.08216	0.20872	105
106	228.50	213.81	0.014221	0.23549	70.317	4.2465	41.166	71.187	112.353	0.08271	0.20855	106
107	231.59	216.90	0.014253	0.23209	70.161	4.3086	41.485	70.906	112.391	0.08326	0.20838	107
108	234.71	220.02	0.014285	0.22875	70.005	4.3715	41.804	70.623	112.427	0.08381	0.20821	108
109	237.86	223.17	0.014317	0.22546	69.847	4.4354	42.125	70.338	112.463	0.08436	0.20804	109
110	241.04	226.35	0.014350	0.22222	69.689	4.5000	42.446	70.052	112.498	0.08491	0.20787	110
111	244.25	229.56	0.014382	0.21903	69.529	4.5656	42.768	69.763	112.531	0.08546	0.20770	111
112	247.50	232,80	0.014416	0.21589	69.369	4.6321	43.091	69.473	112.564	0.08601	0.20753	112
113	250.77	236.08	0.014449	0.21279	69.208	4.6994	43.415	69.180	112.595	0.08656	0.20736	113
114	254.08	239.38	0.014483	0.20974	69.046	4.7677	43.739	68.886	112.626	0.08711	0.20718	114
115	257.42	242.72	0.014517	0.20674	68.883	4.8370	44.065	68.590	112.655	0.08766	0.20701	115
116	260.79	246.10	0.014552	0.20378	68.719	4.9072	44.391	68.291	112.682	0.08821	0.20684	116
117	264.20	249.50	0.014587	0.20087	68.554	4.9784	44.718	67.991	112.709	0.08876	0.20666	117
118	267.63	252.94	0.014622	0.19800	68.388	5.0506	45.046	67.688	112.735	0.08932	0.20649	118
119	271.10	256.41	0.014658	0.19517	68.221	5.1238	45.375	67.384	112.759	0.08987	0.20631	119

Table B-5 (Cont.)

TEMP. °F	PRESSURE PSIA	PSIG	VOLUME cu ft/lb LIQUID v_f	VAPOR v_g	DENSITY lb/cu ft LIQUID $1/v_f$	VAPOR $1/v_g$	ENTHALPY Btu/lb LIQUID h_f	LATENT h_{fg}	VAPOR h_g	ENTROPY Btu/(lb)(°R) LIQUID s_f	VAPOR s_g	TEMP. °F
120	274.60	259.91	0.014694	0.19238	68.054	5.1981	45.705	67.077	112.782	0.09042	0.20613	120
121	278.14	263.44	0.014731	0.18963	67.885	5.2734	46.036	66.767	112.803	0.09098	0.20595	121
122	281.71	267.01	0.014768	0.18692	67.714	5.3498	46.368	66.456	112.824	0.09153	0.20578	122
123	285.31	270.62	0.014805	0.18426	67.543	5.4272	46.701	66.142	112.843	0.09208	0.20560	123
124	288.95	274.25	0.014843	0.18163	67.371	5.5058	47.034	65.826	112.860	0.09264	0.20542	124
125	292.62	277.92	0.014882	0.17903	67.197	5.5856	47.369	65.507	112.877	0.09320	0.20523	125
126	296.33	281.63	0.014920	0.17648	67.023	5.6665	47.705	65.186	112.891	0.09375	0.20505	126
127	300.07	285.37	0.014960	0.17396	66.847	5.7486	48.042	64.863	112.905	0.09431	0.20487	127
128	303.84	289.14	0.014999	0.17147	66.670	5.8319	48.380	64.537	112.917	0.09487	0.20468	128
129	307.65	292.95	0.015039	0.16902	66.492	5.9164	48.719	64.208	112.927	0.09543	0.20449	129
130	311.50	296.80	0.015080	0.16661	66.312	6.0022	49.059	63.877	112.936	0.09598	0.20431	130
131	315.38	300.68	0.015121	0.16422	66.131	6.0893	49.400	63.543	112.943	0.09654	0.20412	131
132	319.29	304.60	0.015163	0.16187	65.949	6.1777	49.743	63.206	112.949	0.09711	0.20393	132
133	323.25	308.55	0.015206	0.15956	65.766	6.2674	50.087	62.866	112.953	0.09767	0.20374	133
134	327.23	312.54	0.015248	0.15727	65.581	6.3585	50.432	62.523	112.955	0.09823	0.20354	134
135	331.26	316.56	0.015292	0.15501	65.394	6.4510	50.778	62.178	112.956	0.09879	0.20335	135
136	335.32	320.63	0.015336	0.15279	65.207	6.5450	51.125	61.829	112.954	0.09936	0.20315	136
137	339.42	324.73	0.015381	0.15059	65.017	6.6405	51.474	61.477	112.951	0.09992	0.20295	137
138	343.56	328.86	0.015426	0.14843	64.826	6.7374	51.824	61.123	112.947	0.10049	0.20275	138
139	347.73	333.04	0.015472	0.14629	64.634	6.8359	52.175	60.764	112.940	0.10106	0.20255	139
140	351.94	337.25	0.015518	0.14418	64.440	6.9360	52.528	60.403	112.931	0.10163	0.20235	140
141	356.19	341.50	0.015566	0.14209	64.244	7.0377	52.883	60.038	112.921	0.10220	0.20214	141
142	360.48	345.79	0.015613	0.14004	64.047	7.1410	53.238	59.670	112.908	0.10277	0.20194	142
143	364.81	350.11	0.015662	0.13801	63.848	7.2461	53.596	59.298	112.893	0.10334	0.20173	143
144	369.17	354.48	0.015712	0.13600	63.647	7.3529	53.955	58.922	112.877	0.10391	0.20152	144
145	373.58	358.88	0.015762	0.13402	63.445	7.4615	54.315	58.543	112.858	0.10449	0.20130	145
146	378.02	363.32	0.015813	0.13207	63.240	7.5719	54.677	58.159	112.836	0.10507	0.20109	146
147	382.50	367.81	0.015865	0.13014	63.034	7.6842	55.041	57.772	112.813	0.10564	0.20087	147
148	387.03	372.33	0.015917	0.12823	62.825	7.7985	55.406	57.380	112.787	0.10622	0.20065	148
149	391.59	376.89	0.015971	0.12635	62.615	7.9148	55.774	56.985	112.758	0.10681	0.20042	149
150	396.19	381.50	0.016025	0.12448	62.402	8.0331	56.143	56.585	112.728	0.10739	0.20020	150
151	400.84	386.14	0.016080	0.12265	62.187	8.1536	56.514	56.180	112.694	0.10797	0.19997	151
152	405.52	390.83	0.016137	0.12083	61.970	8.2763	56.887	55.771	112.658	0.10856	0.19974	152
153	410.25	395.56	0.016194	0.11903	61.751	8.4011	57.261	55.358	112.619	0.10915	0.19950	153
154	415.02	400.32	0.016252	0.11726	61.529	8.5284	57.638	54.939	112.577	0.10974	0.19926	154
155	419.83	405.13	0.016312	0.11550	61.305	8.6580	58.017	54.515	112.533	0.11034	0.19902	155
156	424.68	409.99	0.016372	0.11376	61.079	8.7901	58.399	54.087	112.485	0.11093	0.19878	156
157	429.58	414.88	0.016434	0.11205	60.849	8.9247	58.782	53.652	112.435	0.11153	0.19853	157
158	434.52	419.82	0.016497	0.11035	60.617	9.0620	59.168	53.213	112.381	0.11213	0.19828	158
159	439.50	424.80	0.016561	0.10867	60.383	9.2020	59.557	52.767	112.324	0.11273	0.19802	159
160	444.53	429.83	0.016627	0.10701	60.145	9.3449	59.948	52.316	112.263	0.11334	0.19776	160
161	449.59	434.90	0.016693	0.10537	59.904	9.4907	60.341	51.858	112.199	0.11395	0.19750	161
162	454.71	440.01	0.016762	0.10374	59.660	9.6395	60.737	51.394	112.131	0.11456	0.19723	162
163	459.87	445.17	0.016831	0.10213	59.413	9.7915	61.136	50.923	112.060	0.11518	0.19696	163
164	465.07	450.37	0.016902	0.10054	59.163	9.9467	61.538	50.446	111.984	0.11580	0.19668	164
165	470.32	455.62	0.016975	0.098956	58.909	10.106	61.943	49.961	111.904	0.11642	0.19640	165
166	475.61	460.92	0.017050	0.097393	58.651	10.268	62.351	49.469	111.820	0.11705	0.19611	166
167	480.95	466.26	0.017126	0.095844	58.390	10.434	62.763	48.969	111.732	0.11768	0.19581	167
168	486.34	471.65	0.017204	0.094309	58.125	10.603	63.178	48.461	111.639	0.11831	0.19552	168
169	491.78	477.08	0.017285	0.092787	57.855	10.777	63.596	47.945	111.541	0.11895	0.19521	169
170	497.26	482.56	0.017367	0.091279	57.581	10.955	64.019	47.419	111.438	0.11959	0.19490	170
171	502.79	488.09	0.017451	0.089783	57.303	11.138	64.445	46.885	111.330	0.12024	0.19458	171
172	508.37	493.67	0.017538	0.088299	57.019	11.325	64.875	46.340	111.216	0.12089	0.19425	172
173	513.99	499.30	0.017627	0.086827	56.731	11.517	65.310	45.786	111.096	0.12155	0.19392	173
174	519.67	504.97	0.017719	0.085365	56.438	11.714	65.750	45.221	110.970	0.12222	0.19358	174

Table B-6

Thermodynamic Properties of Freon-22[a] (Properties of Superheated Vapor)

Temp. F.	Abs. Pressure .25 lb./in.² Gage Pressure 29.41 in. vac. (Sat. Temp. −150.9 F.)			Abs. Pressure .50 lb./in.² Gage Pressure 28.90 in. vac. (Sat. Temp. −137.2 F.)			Abs. Pressure .75 lb./in.² Gage Pressure 28.39 in. vac. (Sat. Temp. −128.4 F.)			Abs. Pressure 1.00 lb/in.² Gage Pressure 27.88 in. vac. (Sat. Temp. −121.9 F.)		
t	*v*	*h*	*s*	*v*	*h*	*s*	*v*	*h*	*s*	*v*	*h*	*s*
Sat.	*(151.8)*	*(87.26)*	*(0.2960)*	*(79.18)*	*(88.86)*	*(0.2853)*	*(54.19)*	*(89.89)*	*(0.2792)*	*(41.42)*	*(90.66)*	*(0.2750)*
−150	152.19	87.36	0.2962	......	......	......	......	......	......	......	......	......
−140	157.12	88.53	0.2999	......	......	......	......	......	......	......	......	......
−130	162.04	89.72	0.3036	80.96	89.71	0.2878	......	......	......	......	......	......
−120	166.96	90.93	0.3072	83.42	90.91	0.2914	55.58	90.90	0.2822	41.65	90.88	0.2756
−110	171.89	92.14	0.3107	85.89	92.13	0.2949	57.22	92.12	0.2857	42.89	92.10	0.2791
−100	176.81	93.38	0.3142	88.35	93.36	0.2984	58.86	93.35	0.2892	44.12	93.34	0.2826
− 90	181.74	94.62	0.3176	90.81	94.61	0.3018	60.51	94.60	0.2926	45.35	94.59	0.2860
− 80	186.66	95.89	0.3210	93.28	95.87	0.3052	62.15	95.86	0.2959	46.59	95.85	0.2894
− 70	191.59	97.16	0.3243	95.74	97 15	0.3085	63.79	97.14	0.2992	47.82	97.13	0.2927
− 60	196.51	98.46	0.3276	98.21	98.44	0.3118	65.44	98.43	0.3025	49.05	98.42	0.2960
− 50	201.43	99.76	0.3308	100.67	99.75	0.3150	67.08	99.74	0.3057	50.29	99.73	0.2992
− 40	206.35	101.08	.3340	103.13	101.07	.3182	68.73	101.06	.3089	51.52	101.05	.3024
− 30	211.28	102.42	.3371	105.60	102.41	.3213	70.37	102.40	.3121	52.75	102.39	.3055
− 20	216.20	103.77	.3403	108.06	103.76	.3245	72.01	103.75	.3152	53.99	103.74	.3087
− 10	221.12	105.14	.3433	110.52	105.13	.3275	73.65	105.12	.3183	55.22	105.11	.3117
0	226.04	106.52	0.3464	112.98	106.51	0.3306	75.30	106.50	0.3213	56.45	106.49	0.3148
10	230.97	107.91	.3494	115.45	107.90	.3336	76.94	107.89	.3244	57.68	107.88	.3178
20	235.89	109.32	.3524	117.91	109.31	.3366	78.58	109.30	.3273	58.92	109.29	.3208
30	240.81	110.75	.3553	120.37	110.74	.3395	30.22	110.73	.3303	60.15	110.72	.3237
40	245.73	112.19	.3582	122.83	112.18	.3424	81.86	112.17	.3332	61.38	112.16	.3266
50	250.66	113.64	0.3611	125.29	113.63	0.3453	83.51	113.62	0.3361	62.61	113.62	0.3295
60	255.58	115.11	.3639	127.76	115.10	.3482	85.15	115.09	.3389	63.84	115.08	.3324
70	260.50	116.59	.3668	130.22	116.59	.3510	86.79	116.58	.3417	65.08	116.57	.3352
80	265.42	118.09	.3696	132.68	118.09	.3538	88.43	118.08	.3446	66.31	118.07	.3380
90	270.34	119.61	.3724	135.14	119.60	.3566	90.07	119.59	.3473	67.54	119.58	.3408
100	275.27	121.13	0.8751	137.60	121.13	0.3593	91.72	121.12	0.3501	68.77	121.11	0.3435
110	280.19	122.68	.3779	140.07	122.67	.3621	93.36	122.66	.3528	70.00	122.66	.3463
120	285.11	124.24	.3806	142.53	124.23	.3648	95.00	124.22	.3555	71.24	124.21	.3490
130	290.03	125.81	.3833	144.99	125.80	.3675	96.64	125.80	.3582	72.47	125.79	.3517
140	294.95	127.40	.3859	147.45	127.39	.3701	98.28	127.38	.3609	73.70	127.38	.3543
150	299.87	129.00	0.3886	149.91	128.99	0.3728	99.92	128.99	0.3636	74.93	128.99	0.3570
160	......	......	......	152.37	130.61	.3754	101.56	130.60	.3662	76.16	130.60	.3596
170	......	......	......	154.83	132.24	.3780	103.21	132.24	.3688	77.39	132.23	.3622
180	......	......	......	......	......	......	104.85	133.38	.3714	78.62	133.88	.3684

[a]*Courtesy of E. I. du Pont deNemours and Company, Inc. Organic Chemicals Department "Freon" Products Division.*

Table B-6

Temp. F.	Abs. Pressure 2.0 lb/in.² Gage Pressure 25.85 in. vac. (Sat. Temp. −104.7 F)			Abs. Pressure 4.0 lb/in.² Gage Pressure 21.78 in. vac. (Sat. Temp. −85.4 F)			Abs. Pressure 6 lb/in.² Gage Pressure 17.71 in. vac. (Sat. Temp. −73.0 F)			Abs. Pressure 10 lb/in.² Gage Pressure 9.57 in. vac. (Sat. Temp. −55.8 F)		
t	*v*	*h*	*s*	*v*	*h*	*s*	*v*	*h*	*s*	*v*	*h*	*s*
Sat.	(21.71)	(92.72)	(0.2651)	(11.40)	(95.03)	(0.2558)	(7.823)	(96.52)	(0.2506)	(4.870)	(98.58)	(0.2444)
−100	22.00	93.29	0.2667	…	…	…	…	…	…	…	…	…
− 90	22.62	94.54	.2702	…	…	…	…	…	…	…	…	…
− 80	23.24	95.81	.2735	11.57	95.72	0.2576	…	…	…	…	…	…
− 70	23.86	97.08	.2768	11.88	97.00	.2609	7.886	96.91	0.2516	…	…	…
− 60	24.48	98.38	.2801	12.19	98.29	.2642	8.095	98.20	.2549	…	…	…
− 50	25.10	99.69	0.2833	12.50	99.60	0.2674	8.303	99.52	0.2581	4.943	99.35	0.2462
− 40	25.72	101.01	.2865	12.81	100.93	.2706	8.511	100.85	.2613	5.069	100.68	.2494
− 30	26.33	102.35	.2897	13.12	102.27	.2738	8.719	102.19	.2645	5.195	102.02	.2526
− 20	26.95	103.70	.2928	13.43	103.62	.2769	8.927	103.54	.2676	5.321	103.39	.2558
− 10	27.57	105.07	.2959	13.74	104.99	.2800	9.135	104.92	.2707	5.447	104.77	.2589
0	28.19	106.45	0.2990	14.05	106.38	0.2831	9.342	106.31	0.2738	5.573	106.16	0.2619
10	28.80	107.85	.3020	14.36	107.78	.2861	9.549	107.71	.2768	5.699	107.56	.2650
20	29.42	109.26	.3049	14.67	109.19	.2891	9.757	109.12	.2798	5.824	108.98	.2680
30	30.04	110.68	.3079	14.98	110.62	.2920	9.964	110.55	.2827	5.950	110.40	.2709
40	30.66	112.13	.3108	15.29	112.06	.2950	10.171	112.00	.2856	6.075	111.86	.2739
50	31.27	113.58	0.3137	15.60	113.52	0.2978	10.378	113.46	0.2885	6.200	113.32	0.2767
60	31.89	115.05	.3165	15.91	114.99	.3007	10.585	114.93	.2914	6.325	114.80	.2796
70	32.51	116.54	.3194	16.22	116.48	.3035	10.792	116.42	.2942	6.450	116.29	.2824
80	33.12	118.04	.3222	16.53	117.98	.3063	10.999	117.92	.2970	6.575	117.80	.2853
90	33.74	119.55	.3250	16.84	119.50	.3091	11.206	119.44	.2998	6.700	119.32	.2881
100	34.36	121.08	0.3277	17.15	121.03	0.3119	11.413	120.97	0.3026	6.824	120.85	0.2908
110	34.97	122.63	.3305	17.46	122.57	.3146	11.620	122.52	.3053	6.949	122.40	.2936
120	35.59	124.19	.3332	17.77	124.13	.3173	11.826	124.08	.3080	7.073	123.97	.2963
130	36.21	125.76	.3359	18.08	125.70	.3200	12.033	125.66	.3107	7.198	125.55	.2990
140	36.82	127.35	.3385	18.39	127.29	.3227	12.239	127.25	.3134	7.322	127.14	.3017
150	37.44	128.95	0.3412	18.69	128.90	0.3254	12.446	128.85	0.3161	7.447	128.74	0.3043
160	38.06	130.57	.3438	19.00	130.52	.3280	12.652	130.47	.3187	7.571	130.37	.3070
170	38.67	132.20	.3464	19.31	132.15	.3306	12.859	132.11	.3213	7.696	132.01	.3096
180	39.29	133.85	.3490	19.62	133.80	.3332	13.065	133.75	.3239	7.820	133.66	.3122
190	39.91	135.51	.3516	19.93	135.46	.3358	13.271	135.41	.3265	7.945	135.32	.3148
200	40.52	137.19	0.3542	20.24	137.14	0.3383	13.477	137.09	.3291	8.069	137.00	0.3173
210	…	…	…	20.55	138.84	.3409	13.683	138.79	.3316	8.193	138.70	.3199
220	…	…	…	20.86	140.54	.3434	13.889	140.50	.3341	8.441	140.41	.3224
230	…	…	…	…	…	…	14.096	142.22	.3366	8.317	142.13	.3250
240	…	…	…	…	…	…	…	…	…	8.565	143.87	.3275
250	…	…	…	…	…	…	…	…	…	8.689	145.63	0.3300

Table B-6 (Cont.)

Temp. F.	Abs. Pressure 20 lb/in.² Gage Pressure 5.3 lb/in.² (Sat. Temp. −29.4 F)			Abs. Pressure 40 lb/in.² Gage Pressure 25.3 lb/in.² (Sat. Temp. 1.5 F)			Abs. Pressure 60 lb/in.² Gage Pressure 45.3 lb/in.² (Sat. Temp. 21.8 F)			Abs. Pressure 80 lb/in.² Gage Pressure 65.3 lb/in.² (Sat. Temp. 37.4 F)		
t	*v*	*h*	*s*	*v*	*h*	*s*	*v*	*h*	*s*	*v*	*h*	*s*
Sat.	*(2.556)*	*(101.70)*	*(0.2365)*	*(1.334)*	*(105.17)*	*(0.2293)*	*(0.9061)*	*(107.32)*	*(0.2253)*	*(0.6856)*	*(108.86)*	*(0.2224)*
−20	2.616	103.00	0.2394	......	......	......	......	......	......	......	......	......
−10	2.681	104.38	.2426	......	......	......	......	......	......	......	......	......
0	2.745	105.78	0.2457	......	......	......	......	......	......	......	......	......
10	2.809	107.20	.2487	1.363	106.41	0.2319	......	......	......	......	......	......
20	2.873	108.63	.2517	1.396	107.87	.2350	......	......	......	......	......	......
30	2.937	110.07	.2547	1.430	109.34	.2380	0.9255	108.56	0.2278	......	......	......
40	3.001	111.52	.2577	1.463	110.82	.2410	.9491	110.07	.2308	0.6904	109.26	0.2232
50	3.065	113.00	0.2606	1.496	112.31	0.2439	0.9721	111.59	0.2338	0.7087	110.80	0.2263
60	3.128	114.48	.2634	1.529	113.81	.2469	.9950	113.10	.2368	.7269	112.32	.2293
70	3.192	115.98	.2663	1.562	115.33	.2498	1.0180	114.64	.2397	.7446	113.90	.2323
80	3.255	117.49	.2691	1.595	116.86	.2526	1.0406	116.20	.2426	.7625	115.48	.2352
90	3.319	119.02	.2719	1.628	118.40	.2555	1.0634	117.77	.2455	.7801	117.09	.2381
100	3.382	120.56	0.2747	1.660	119.96	0.2583	1.0858	119.35	0.2483	.7978	118.69	0.2410
110	3.445	122.12	.2775	1.693	121.53	.2610	1.1082	120.92	.2511	.8151	120.29	.2439
120	3.508	123.69	.2802	1.726	123.12	.2638	1.1305	122.53	.2539	.8327	121.91	.2467
130	3.571	125.27	.2829	1.758	124.72	.2665	1.1529	124.14	.2567	.8500	123.55	.2495
140	3.635	126.87	.2856	1.790	126.33	.2693	1.1751	125.77	.2594	.8673	125.19	.2522
150	3.698	128.49	0.2883	1.823	127.96	0.2719	1.1974	127.41	0.2621	.8844	126.84	0.2550
160	3.761	130.11	.2909	1.855	129.60	.2746	1.2193	129.06	.2648	.9014	128.51	.2577
170	3.824	131.76	.2936	1.887	131.25	.2773	1.2415	130.73	.2675	.9184	130.19	.2604
180	3.886	133.41	.2962	1.919	132.92	.2799	1.2634	132.41	.2702	.9356	131.89	.2630
190	3.949	135.08	.2988	1.951	134.60	.2825	1.2853	134.10	.2728	.9524	133.59	.2657
200	4.012	136.77	0.3013	1.983	136.30	0.2851	1.3071	135.81	0.2754	.9689	135.31	0.2683
210	4.074	138.47	.3039	2.016	138.01	.2877	1.3290	137.53	.2780	.9857	137.04	.2709
220	4.137	140.19	.3064	2.047	139.73	.2902	1.3509	139.26	.2805	1.0025	138.79	.2735
230	4.200	141.91	.3090	2.079	141.47	.2928	1.3726	141.01	.2831	1.0192	140.54	.2761
240	4.263	143.66	.3115	2.111	143.22	.2953	1.3943	142.77	.2856	1.0358	142.32	.2787
250	4.325	145.42	0.3140	2.143	144.99	0.2978	1.4159	144.55	0.2882	1.0524	144.10	0.2812
260	4.388	147.19	.3165	2.175	146.77	.3003	1.4375	146.33	.2907	1.0690	145.90	.2837
270	4.450	148.98	.3189	2.207	148.57	.3028	1.4591	148.14	.2932	1.0854	147.72	.2862
280	4.512	150.78	.3214	2.239	150.38	.3052	1.4806	149.97	.2956	1.1019	149.55	.2887
290	......	......	......	2.270	152.20	.3077	1.5021	151.81	.2981	1.1183	151.40	.2912
300	......	......	......	2.302	154.04	0.3101	1.5235	153.65	0.3005	1.1343	153.25	0.2936
310	......	......	......	2.334	155.89	.3125	1.5451	155.51	.3030	1.1507	155.12	.2961
320	......	......	......	......	......	......	1.5665	157.39	.3054	1.1671	157.00	.2985
330	......	......	......	......	......	......	1.5879	159.27	.3078	1.1834	158.90	.3009
340	......	......	......	......	......	......	......	......	......	1.1996	160.80	.3033

Table B-6 (Cont.)

Temp. F.	Abs. Pressure 100 lb/in.² Gage Pressure 85.3 lb/in.² (Sat. Temp. 50.4° F)			Abs. Pressure 120 lb/in.² Gage Pressure 105.3 lb/in.² (Sat. Temp. 61.5 F)			Abs. Pressure 140 lb/in.² Gage Pressure 125.3 lb/in.² (Sat. Temp. 71.3 F)			Abs. Pressure 160 lb/in.² Gage Pressure 145.3 lb/in.² (Sat. Temp. 80.1 F)		
t	*v*	*h*	*s*	*v*	*h*	*s*	*v*	*h*	*s*	*v*	*h*	*s*
Sat.	*(0.5505)*	*(110.03)*	*(0.2202)*	*(0.4586)*	*(110.91)*	*(0.2183)*	*(0.3919)*	*(111.60)*	*(0.2167)*	*(0.3411)*	*(112.06)*	*(0.2152)*
60	0.5651	111.51	0.2232	......	......	......	......	......	......	......	......	......
70	.5802	113.11	.2262	0.4698	112.28	0.2210	......	......	......	......	......	......
80	.5953	114.74	.2292	.4833	113.95	.2241	0.4022	113.08	0.2195	......	......	......
90	.6101	116.37	.2322	.4961	115.60	.2271	.4141	114.79	.2226	0.3518	113.90	0.2184
100	0.6249	118.00	0.2351	0.5089	117.27	0.2301	0.4257	116.48	0.2256	0.3627	115.64	0.2215
110	.6394	119.62	.2380	.5215	118.91	.2330	.4369	118.16	.2286	.3733	117.36	.2246
120	.6539	121.26	.2408	.5343	120.58	.2359	.4483	119.86	.2315	.3838	119.10	.2276
130	.6682	122.92	.2437	.5466	122.26	.2388	.4593	121.57	.2344	.3939	120.84	.2306
140	.6824	124.58	.2465	.5589	123.94	.2416	.4704	123.27	.2373	.4039	122.58	.2335
150	0.6965	126.25	0.2492	0.5711	125.63	0.2444	0.4812	124.99	0.2402	0.4138	124.33	0.2364
160	.7107	127.94	.2520	.5832	127.34	.2472	.4921	126.72	.2430	.4236	126.08	.2392
170	.7246	129.64	.2547	.5952	129.06	.2499	.5027	128.46	.2458	.4332	127.84	.2420
180	.7386	131.35	.2574	.6072	130.79	.2526	.5132	130.20	.2485	.4429	129.60	.2448
190	.7523	133.07	.2601	.6190	132.53	.2553	.5236	131.96	.2512	.4522	131.38	.2476
200	0.7660	134.80	0.2627	0.6307	134.27	0.2580	0.5340	133.73	0.2539	0.4615	133.17	0.2503
210	.7796	136.55	.2654	.6424	136.03	.2607	.5444	135.50	.2566	.4709	134.96	.2530
220	.7933	138.30	.2680	.6541	137.80	.2633	.5546	137.29	.2593	.4801	136.76	.2557
230	.8071	140.07	.2706	.6657	139.58	.2659	.5649	139.08	.2619	.4892	138.57	.2583
240	.8206	141.85	.2731	.6772	141.37	.2685	.5749	140.89	.2645	.4933	140.40	.2610
250	0.8343	143.64	0.2757	0.6888	143.17	0.2711	0.5850	142.70	0.2671	0.5073	142.23	0.2636
260	.8477	145.45	.2782	.7003	144.99	.2736	.5951	144.54	.2697	.5162	144.08	.2662
270	.8612	147.28	.2807	.7118	146.84	.2762	.6051	146.39	.2722	.5251	145.94	.2687
280	.8746	149.13	.2832	.7231	148.69	.2787	.6150	148.26	.2747	.5340	147.82	.2713
290	.8879	150.98	.2857	.7344	150.59	.2812	.6248	150.13	.2773	.5428	149.70	.2738
300	0.9011	152.84	0.2882	0.7456	152.43	0.2837	0.6347	152.02	0.2798	0.5516	151.60	0.2763
310	.9144	154.72	.2906	.7569	154.32	.2861	.6446	153.91	.2822	.5604	153.51	.2788
320	.9277	156.62	.2931	.7681	156.22	.2886	.6544	155.83	.2847	.5692	155.43	.2813
330	.9410	158.52	.2955	.7793	158.14	.2910	.6641	157.75	.2871	.5777	157.36	.2838
340	.9541	160.43	.2979	.7906	160.06	.2934	.6739	159.68	.2896	.5864	159.31	.2862
350	0.9674	162.36	0.3003	0.8018	162.00	0.2959	0.6835	161.63	0.2920	0.5950	161.26	0.2886
360	.9806	164.31	.3027	.8129	163.95	.2983	.6933	163.59	.2944	.6036	163.23	.2911
370	......	......	......	.8240	165.92	.3007	.7029	165.57	.2968	.6121	165.21	.2935
380	......	......	......	......	......	......	.7125	167.56	.2992	.6207	167.22	.2959
390	......	......	......	......	......	......	......	......	......	.6292	169.22	.2982

Table B-6

Temp. F	Abs. Pressure 180 lb/in.^{2} Gage Pressure 165.3 lb/in.^{2} (Sat. Temp. 88.2 F)			Abs. Pressure 200 lb/in.^{2} Gage Pressure 185.3 lb/in.^{2} (Sat. Temp. 95.6 F)			Abs. Pressure 220 lb/in.^{2} Gage Pressure 205.3 lb/in.^{2} (Sat. Temp. 102.5 F)			Abs. Pressure 260 lb/in.^{2} Gage Pressure 245.3 lb/in.^{2} (Sat. Temp. 115.0 F)		
t	*v*	*h*	*s*	*v*	*h*	*s*	*v*	*h*	*s*	*v*	*h*	*s*
Sat.	*(0.3012)*	*(112.60)*	*(0.2138)*	*(0.2690)*	*(112.93)*	*(0.2124)*	*(0.2425)*	*(113.17)*	*(0.2112)*	*(0.2013)*	*(113.42)*	*(0.2084)*
90	0.3031	112.93	0.2145	......	......	......	......	......	......	......	......	......
100	0.3135	114.73	0.2177	0.2734	113.75	0.2139	......	......	......	......	......	......
110	.3236	116.51	.2208	.2829	115.61	.2172	0.2498	114.62	0.2138	......	......	......
120	.3331	118.29	.2239	.2922	117.42	.2204	.2586	116.49	.2171	0.2066	114.45	0.2106
130	.3427	120.07	.2269	.3013	119.25	.2236	.2674	118.37	.2203	.2147	116.47	.2141
140	.3520	121.85	.2299	.3102	121.06	.2266	.2759	120.23	.2235	.2227	118.45	.2174
150	0.3611	123.63	0.2329	0.3189	122.89	0.2296	0.2843	122.11	0.2265	0.2303	120.44	0.2207
160	.3701	125.41	.2358	.3274	124.71	.2326	.2923	123.96	.2295	.2379	122.36	.2238
170	.3791	127.19	.2386	.3358	126.52	.2355	.3004	125.81	.2325	.2454	124.29	.2270
180	.3879	128.98	.2415	.3439	128.34	.2384	.3079	127.66	.2354	.2526	126.22	.2300
190	.3966	130.78	.2443	.3521	130.17	.2412	.3155	129.52	.2383	.2596	128.16	.2330
200	0.4051	132.59	0.2470	0.3599	132.00	0.2440	0.3229	131.39	0.2411	0.2663	130.10	0.2359
210	.4137	134.41	.2498	.3679	133.84	.2467	.3304	133.25	.2439	.2728	132.01	.2388
220	.4221	136.23	.2525	.3756	135.68	.2495	.3377	135.12	.2467	.2793	133.92	.2416
230	.4303	138.06	.2551	.3834	137.53	.2522	.3450	136.99	.2494	.2357	135.84	.2444
240	.4386	139.90	.2578	.3909	139.39	.2548	.3519	138.87	.2521	.2921	137.76	.2472
250	0.4468	141.75	0.2604	0.3985	141.26	0.2575	0.3590	140.76	0.2548	0.2984	139.70	0.2499
260	.4550	143.61	.2630	.4060	143.13	.2601	.3660	142.64	.2575	.3045	141.62	.2526
270	.4630	145.49	.2656	.4134	145.02	.2627	.3729	144.55	.2601	.3107	143.56	.2553
280	.4711	147.38	.2682	.4209	146.92	.2653	.3798	146.46	.2627	.3168	145.51	.2580
290	.4791	149.27	.2707	.4281	148.83	.2679	.3866	148.38	.2653	.3229	147.46	.2606
300	0.4870	151.17	0.2732	0.4351	150.75	0.2704	0.3933	150.31	0.2678	0.3287	149.43	0.2632
310	.4951	153.09	.2757	.4428	152.68	.2729	.4001	152.25	.2704	.3346	151.39	.2658
320	.5029	155.03	.2782	.4499	154.62	.2754	.4067	154.21	.2729	.3404	153.37	.2683
330	.5106	156.97	.2807	.4572	156.57	.2779	.4133	156.17	.2754	.3462	155.36	.2709
340	.5184	158.92	.2832	.4642	158.54	.2804	.4199	158.15	.2779	.3520	157.36	.2734
350	0.5263	160.89	0.2856	0.4713	160.51	0.2829	0.4264	160.13	0.2804	0.3577	159.36	0.2759
360	.5340	162.87	.2881	.4784	162.50	.2853	.4330	162.13	.2828	.3633	161.38	.2783
370	.5417	164.86	.2905	.4855	164.50	.2877	.4394	164.14	.2852	.3690	163.41	.2808
380	.5493	166.87	.2929	.4924	166.51	.2902	.4459	166.16	.2877	.3746	165.46	.2832
390	.5570	168.87	.2953	.4994	168.53	.2926	.4524	168.19	.2901	.3802	167.51	.2857
400	......	......	......	0.5064	170.57	0.2949	0.4588	170.24	0.2925	0.3858	169.57	0.2881
410	......	......	......	......	......	......	.4652	172.31	.2949	.3913	171.66	.2905
420	......	......	......	......	......	......	......	......	......	.3968	173.77	.2929

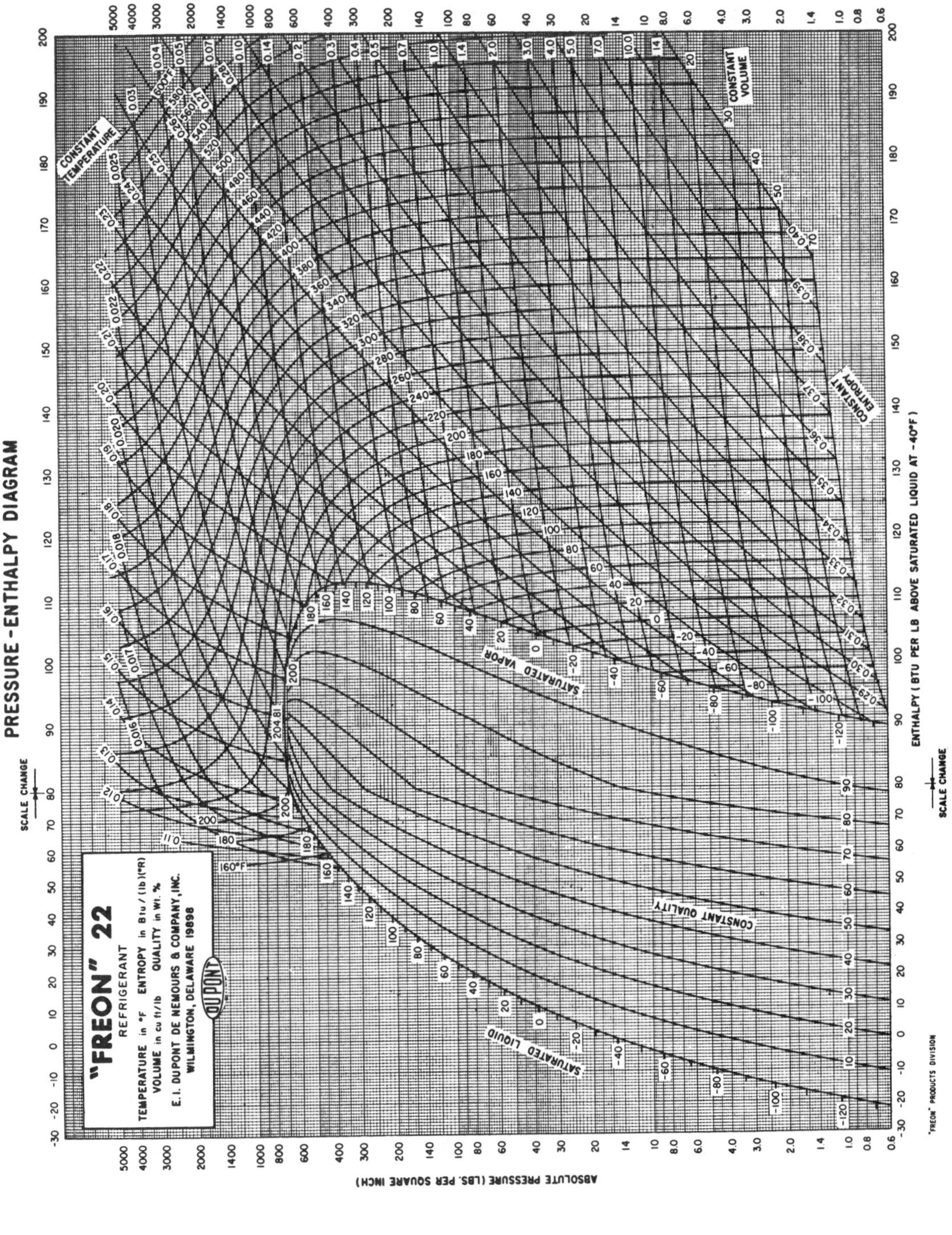
PRESSURE-ENTHALPY DIAGRAM
"FREON" 22
REFRIGERANT
TEMPERATURE in °F ENTROPY in Btu / (lb)(°R)
VOLUME in cu ft/lb QUALITY in Wt. %
E. I. DUPONT DE NEMOURS & COMPANY, INC.
WILMINGTON, DELAWARE 19898
DU PONT
CONSTANT TEMPERATURE
CONSTANT VOLUME
CONSTANT ENTROPY
CONSTANT QUALITY
SATURATED VAPOR
SATURATED LIQUID
SCALE CHANGE
ENTHALPY (BTU PER LB ABOVE SATURATED LIQUID AT -40°F)
ABSOLUTE PRESSURE (LBS. PER SQUARE INCH)
"FREON" PRODUCTS DIVISION

Appendix C

Weather Data

Table C-1

Outdoor Design Conditions (Winter and Summer)[a,b,*]

STATE AND STATION	WINTER		SUMMER		
	Latitude	DB 97½%	DB 2½%	WB 2½%	Outdoor Daily Range
ALABAMA					
Alexander City	33	20	94	78	21
Anniston AP	33	19	94	78	21
Auburn	32	25	96	79	21
Birmingham AP	33	22	94	78	21
Decatur	34	19	95	78	22
Dothan AP	31	27	95	80	20
Florence AP	34	17	95	78	22
Gadsden	34	20	94	77	22
Huntsville AP	34	17	95	77	23
Mobile AP	30	29	93	79	18
Mobile CO	30	32	94	79	16
Montgomery	32	26	95	79	21
Selma-Graig AFB	32	27	96	80	21
Talladega	33	19	95	78	21
Tuscaloosa AP	33	23	96	80	22
ALASKA					
Anchorage AP	61	−20	70	61	15
Barrow	71	−42	54	51	12
Fairbanks AP	64	−50	78	63	24
Juneau AP	58	− 4	71	64	15
Kodiak	57	12	66	60	10
Nome AP	64	−28	62	56	10
ARIZONA					
Douglas AP	31	22	98	69	31
Flagstaff AP	35	5	82	60	31
Fort Huachuca AP	31	28	93	68	27
Kingman AP	35	29	100	69	30
Nogales	31	24	98	71	31
Phoenix AP	33	34	106	76	27
Prescott AP	34	19	94	66	30
Tuscon AP	33	32	102	73	26
Winslow AP	35	13	95	65	32
Yuma AP	32	40	109	78	27
ARKANSAS					
Blytheville AFB	36	17	96	79	21
Camden	33	23	97	80	21
El Dorado AP	33	23	96	80	21
Fayetteville AP	36	13	95	76	23
Fort Smith AP	35	19	99	78	24
Hot Springs Nat. Pk.	34	22	97	78	22
Jonesboro	35	18	96	79	21
Little Rock AP	34	23	96	79	22
Pine Bluff AP	34	24	96	80	22
Texarkana AP	33	26	97	79	21
CALIFORNIA					
Bakersfield AP	35	33	101	71	32
Barstow AP	34	28	102	72	37
Blythe AP	33	35	109	77	28
Burbank AP	34	38	94	70	25
Chico	39	33	100	70	36
Concord	38	36	92	67	32
Covina	34	41	97	72	31
Crescent City AP	41	36	69	60	18
Downey	34	38	90	71	22
El Cajon	32	34	95	73	30
El Centro AP	32	35	109	80	34
Escondido	33	36	92	72	30
Eureka/Arcata AP	41	35	65	59	11
Fairfield-Travis AFB	38	34	94	69	34
Fresno AP	36	31	99	72	34
Hamilton AFB	38	35	85	68	28
Laguna Beach	33	39	80	68	18
Livermore	37	30	97	69	24
Lompoc, Vandenburg AFB	34	38	79	63	20
Long Beach AP	33	38	84	70	22
Los Angeles AP	34	43	83	68	15
Los Angeles CO	34	44	90	70	20
Merced-Castle AFB	37	32	99	72	36
Modesto	37	36	98	71	36
Monterey	36	37	79	63	20
Napa	38	34	92	68	30
Needles AP	34	37	110	75	27
Oakland AP	37	37	81	63	19
Oceanside	33	40	81	68	13
Ontario	34	34	97	71	36
Oxnard AFB	34	37	80	69	19
Palmdale AP	34	27	101	68	35
Palm Springs	33	36	108	78	35
Pasadena	34	39	93	70	29
Petaluma	38	32	90	68	31
Pomona CO	34	34	96	72	36
Redding AP	40	35	101	69	32
Redlands	34	37	96	71	33
Richmond	38	38	81	64	17
Riverside-March AFB	33	34	96	71	37
Sacramento AP	38	32	97	70	36
Salinas AP	36	35	85	65	24
San Bernadino, Norton AFB	34	33	98	73	38
San Diego AP	32	44	83	70	12
San Fernando	34	37	97	72	38
San Francisco AP	37	37	79	63	20
San Francisco CO	37	44	77	62	14
San Jose AP	37	36	88	67	26
San Luis Obispo	35	37	85	64	26
Santa Ana AP	33	36	89	71	28
Santa Barbara CO	34	36	84	66	24
Santa Cruz	37	34	84	65	28
Santa Maria AP	34	34	82	64	23
Santa Monica CO	34	45	77	68	16
Santa Paula	34	36	89	71	36
Santa Rosa	38	32	93	68	34
Stockton AP	37	34	98	70	37
Ukiah	39	30	96	69	40
Visalia	36	36	100	72	38
Yreka	41	17	94	66	38
Yuba City	39	34	100	70	36
COLORADO					
Alamosa AP	37	−13	82	61	35
Boulder	40	8	90	63	27
Colorado Springs AP	38	4	88	62	30
Denver AP	39	3	90	64	28
Durango	37	4	86	63	30
Fort Collins	40	− 5	89	62	28
Grand Junction AP	39	11	94	63	29
Greeley	40	− 5	92	64	29
La Junta AP	38	− 2	95	71	31
Leadville	39	− 4	73	55	30
Pueblo AP	38	− 1	94	67	31
Sterling	40	− 2	93	66	30
Trinidad AP	37	5	91	65	32
CONNECTICUT					
Bridgeport AP	41	8	88	76	18
Hartford, Brainard Field	41	5	88	76	22
New Haven AP	41	9	86	76	17
New London	41	8	86	75	16
Norwalk	41	4	89	76	19
Norwich	41	2	86	76	18
Waterbury	41	4	88	76	21
Windsor Locks, Bradley Field	42	2	88	75	22
DELAWARE					
Dover AFB	39	15	90	78	18
Wilmington AP	39	15	90	77	20
DISTRICT OF COLUMBIA					
Andrews AFB	38	16	91	77	18
Washington National AP	38	19	92	77	18
FLORIDA					
Belle Glade	26	39	91	79	16
Cape Kennedy AP	28	40	89	80	15
Daytona Beach AP	29	36	92	80	15

[a]See notes at end of table. [b]Temperature in degrees F.

Table C-1 (Cont.)

STATE AND STATION	WINTER Lati-tude	WINTER DB 97½%	SUMMER DB 2½%	SUMMER WB 2½%	SUMMER Outdoor Daily Range
Fort Lauderdale	26	45	90	80	15
Fort Myers AP	26	42	92	80	18
Fort Pierce	27	41	91	80	15
Gainesville AP	29	32	94	79	18
Jacksonville AP	30	32	94	79	19
Key West AP	24	58	89	79	9
Lakeland CO	28	39	93	79	17
Miami AP	25	47	90	79	15
Miami Beach CO	25	48	89	79	10
Ocala	29	33	94	79	18
Orlando AP	28	37	94	79	17
Panama City, Tyndall AFB	30	35	91	80	14
Pensacola CO	30	32	90	81	14
St. Augustine	29	35	92	80	16
St. Petersburg	28	42	91	80	16
Sanford	28	37	93	79	17
Sarasota	27	39	91	80	17
Tallahassee AP	30	29	94	79	19
Tampa AP	28	39	91	80	17
West Palm Beach AP	26	44	91	80	16
GEORGIA					
Albany, Turner AFB	31	30	96	79	20
Americus	32	25	96	79	20
Athens	34	21	94	77	21
Atlanta AP	33	23	92	77	19
Augusta AP	33	23	95	79	19
Brunswick	31	31	95	80	18
Columbus, Lawson AFB	32	26	96	79	21
Dalton	34	19	95	77	22
Dublin	32	25	96	79	20
Gainesville	34	20	92	77	21
Griffin	33	22	93	78	21
La Grange	33	20	94	78	21
Macon AP	32	27	96	79	22
Marietta, Dobbins AFB	34	21	93	77	21
Moultrie	31	30	95	79	20
Rome AP	34	20	95	77	23
Savannah-Travis AP	32	27	94	80	20
Valdosta-Moody AFB	31	31	94	79	20
Waycross	31	28	95	79	20
HAWAII					
Hilo AP	19	61	83	73	15
Honolulu AP	21	62	85	74	12
Kaneohe	21	61	83	73	12
Wahiawa	21	61	84	74	14
IDAHO					
Boise AP	43	10	93	66	31
Burley	42	8	93	66	35
Coeur d'Alene AP	47	7	91	65	31
Idaho Falls	43	−6	88	64	38
Lewiston AP	46	12	96	66	32
Moscow	46	1	89	63	32
Mountain Home AFB	43	9	96	66	36
Pocatello AP	43	−2	91	63	35
Twin Falls AP	42	8	94	64	34
ILLINOIS					
Aurora	41	−3	91	77	20
Belleville, Scott AFB	38	10	95	78	21
Bloomington	40	3	92	78	21
Carbondale	37	11	96	79	21
Champaign/Urbana	40	4	94	78	21
Chicago, Midway AP	41	1	92	76	20
Chicago, O'Hare AP	42	0	90	75	20
Chicago, CO	41	1	91	76	15
Danville	40	4	94	78	21
Decatur	39	4	93	78	21
Dixon	41	−3	91	77	23
Elgin	42	−4	90	76	21
Freeport	42	−6	90	77	24
Galesburg	41	0	92	78	22
Greenville	39	7	94	78	21

STATE AND STATION	WINTER Lati-tude	WINTER DB 97½%	SUMMER DB 2½%	SUMMER WB 2½%	SUMMER Outdoor Daily Range
Joliet AP	41	−1	92	77	20
Kankakee	41	1	92	77	21
LaSalle/Peru	41	1	93	77	22
Macomb	40	1	93	78	22
Moline AP	41	−3	91	77	23
Mt. Vernon	38	10	95	78	21
Peoria AP	40	2	92	77	22
Quincy AP	40	2	95	79	22
Rantoul, Chanute AFB	40	3	92	77	21
Rockford	42	−3	90	76	24
Springfield AP	39	4	92	78	21
Waukegan	42	−1	90	76	21
INDIANA					
Anderson	40	5	91	77	22
Bedford	38	7	93	78	22
Bloomington	39	7	92	78	22
Columbus, Bakalar AFB	39	7	92	78	22
Crawfordsville	40	2	93	77	22
Evansville AP	38	10	94	78	22
Fort Wayne AP	41	5	91	76	24
Goshen AP	41	0	90	76	23
Hobart	41	0	91	76	21
Huntington	40	2	92	76	23
Indianapolis AP	39	4	91	77	22
Jeffersonville	38	13	94	78	23
Kokomo	40	4	92	76	22
Layfayette	40	3	92	77	22
La Porte	41	0	91	76	22
Marion	40	2	91	76	23
Muncie	40	2	91	77	22
Peru, Bunker Hill AFB	40	1	89	76	22
Richmond AP	39	3	91	77	22
Shelbyville	39	6	92	77	22
South Bend AP	41	3	89	76	22
Terre Haute AP	39	7	93	78	22
Valparaiso	41	−2	90	76	22
Vincennes	38	9	94	78	22
IOWA					
Ames	42	−7	92	78	23
Burlington AP	40	0	92	78	22
Cedar Rapids AP	41	−4	90	76	23
Clinton	41	−3	90	77	23
Council Bluffs	41	−3	94	78	22
Des Moines AP	41	−3	92	77	23
Dubuque	42	−7	90	76	22
Fort Dodge	42	−8	92	77	23
Iowa City	41	−4	91	77	22
Keokuk	40	1	93	78	22
Marshalltown	42	−6	91	77	23
Mason City AP	43	−9	88	75	24
Newton	41	−5	93	77	23
Ottumwa AP	41	−2	93	78	22
Sioux City AP	42	−6	93	77	24
Waterloo	42	−8	89	76	23
KANSAS					
Atchison	39	2	95	78	23
Chanute AP	37	7	97	78	23
Dodge City AP	37	7	97	73	25
El Dorado	37	8	99	77	24
Emporia	38	7	97	77	25
Garden City AP	38	3	98	73	28
Goodland AP	39	4	96	70	31
Great Bend	38	6	99	76	28
Hutchinson AP	38	6	99	76	28
Liberal	37	8	100	73	28
Manhatten, Fort Riley	39	4	98	78	24
Parsons	37	9	97	78	23
Russell AP	38	4	100	76	29
Salina	38	7	99	76	26
Topeka	39	6	96	78	24
Wichita	37	9	99	76	23

Table C-1 (Cont.)

STATE AND STATION	WINTER Latitude	WINTER DB 97½%	SUMMER DB 2½%	SUMMER WB 2½%	SUMMER Outdoor Daily Range
KENTUCKY					
Ashland	38	10	92	76	22
Bowling Green AP	37	11	95	78	21
Corbin AP	37	9	91	77	23
Covington AP	39	8	90	76	22
Hopkinsville, Campbell AFB	36	14	95	78	21
Lexington AP	38	10	92	77	22
Louisville AP	38	12	93	78	23
Madisonville	37	11	94	78	22
Owensboro	37	10	94	78	23
Paducah AP	37	14	95	79	20
LOUISIANA					
Alexandria AP	31	29	95	80	20
Baton Rouge AP	30	30	94	80	19
Bogalusa	30	28	94	79	19
Houma	29	33	92	80	15
Layfayette AP	30	32	93	81	18
Lake Charles AP	30	33	93	79	17
Minden	32	26	96	80	20
Monroe AP	32	27	96	81	20
Natchitoches	31	26	97	80	20
New Orleans AP	30	35	91	80	16
Shreveport AP	32	26	96	80	20
MAINE					
Augusta AP	44	− 3	86	73	22
Bangor, Dow AFB	44	− 4	85	73	22
Caribou AP	46	−14	81	70	21
Lewiston	44	− 4	86	73	22
Millinocket AP	45	−12	85	72	22
Portland AP	43	0	85	73	22
Waterville	44	− 5	86	73	22
MARYLAND					
Baltimore AP	39	15	91	78	21
Baltimore CO	39	20	92	78	17
Cumberland	39	9	92	75	22
Frederick AP	39	11	92	77	22
Hagerstown	39	10	92	76	22
Salisbury	38	18	90	78	18
MASSACHUSETTS					
Boston AP	42	10	88	74	16
Clinton	42	2	85	74	17
Fall River	41	9	86	74	18
Framingham	42	3	89	74	17
Gloucester	42	6	84	73	15
Greenfield	42	− 2	87	74	23
Lawrence	42	1	88	74	22
Lowell	42	3	89	74	21
New Bedford	41	13	84	73	19
Pittsfield AP	42	− 1	84	72	23
Springfield, Westover AFB	42	2	88	74	19
Taunton	41	0	86	75	18
Worcester AP	42	1	87	73	18
MICHIGAN					
Adrian	41	4	91	75	23
Alpena AP	45	− 1	85	73	27
Battle Creek AP	42	5	89	74	23
Benton Harbor AP	42	3	88	74	20
Detroit Met. CAP	42	8	88	75	20
Escanaba	45	− 3	80	71	17
Flint AP	43	3	87	75	25
Grand Rapids AP	42	6	89	74	24
Holland	42	6	88	74	22
Jackson AP	42	4	89	75	23
Kalamazoo	42	5	89	75	23
Lansing AP	42	6	87	75	24
Marquette CO	46	− 4	86	71	18
Mt. Pleasant	43	1	87	74	24
Muskegon AP	43	8	85	74	21
Pontiac	42	4	88	75	21
Port Huron	43	3	88	74	21
Saginaw AP	43	3	86	75	23
Sault Ste. Marie AP	46	− 8	81	71	23
Traverse City AP	44	4	86	73	22
Ypsilanti	42	5	89	74	22
MINNESOTA					
Albert Lea	43	−10	89	76	24
Alexandria AP	45	−15	88	74	24
Bemidji AP	47	−28	84	72	24
Brainerd	46	−20	85	73	24
Duluth AP	46	−15	82	71	22
Fairbault	44	−12	88	75	24
Fergus Falls	46	−17	89	74	24
International Falls AP	48	−24	82	69	26
Mankato	44	−12	89	75	24
Minneapolis/St. Paul AP	44	−10	89	75	22
Rochester AP	44	−13	88	75	24
St. Cloud AP	45	−16	88	75	24
Virginia	47	−21	83	71	23
Willmar	45	−14	88	75	24
Winona	44	− 8	89	76	24
MISSISSIPPI					
Biloxi, Keesler AFB	30	32	92	81	16
Clarksdale	34	24	96	80	21
Columbus AFB	33	22	95	79	22
Greenville AFB	33	24	96	80	21
Greenwood	33	23	96	80	21
Hattiesburg	31	26	95	79	21
Jackson AP	32	24	96	78	21
Laurel	31	26	95	79	21
McComb AP	31	26	94	79	18
Meridian AP	32	24	95	79	22
Natchez	31	26	94	80	21
Tupelo	34	22	96	79	22
Vicksburg CO	32	26	95	80	21
MISSOURI					
Cape Girardeau	37	12	96	79	21
Columbia AP	39	6	95	78	22
Farmington AP	37	8	95	78	22
Hannibal	39	4	94	78	22
Jefferson City	38	6	95	78	23
Joplin AP	37	11	95	78	24
Kansas City AP	39	8	97	77	20
Kirksville AP	40	− 3	94	78	24
Mexico	39	3	94	78	22
Moberly	39	2	94	78	23
Poplar Bluff	36	13	96	79	22
Rolla	38	7	95	78	22
St. Joseph AP	39	3	95	78	23
St. Louis AP	38	8	95	78	21
St. Louis CO	38	11	94	78	18
Sedalia, Whiteman AFB	38	9	94	77	22
Sikeston	36	14	96	79	21
Springfield AP	37	10	94	77	23
MONTANA					
Billings AP	45	− 6	91	66	31
Bozeman	45	−11	85	60	32
Butte AP	46	−16	83	59	35
Cut Bank AP	48	−17	86	63	35
Glasgow AP	48	−20	93	67	29
Glendive	47	−16	93	69	29
Great Falls AP	47	−16	88	63	28
Havre	48	−15	87	64	33
Helena AP	46	−13	87	63	32
Kalispell AP	48	− 3	84	63	34
Lewiston AP	47	−14	86	63	30
Livingston AP	45	−13	88	62	32
Miles City AP	46	−15	94	69	30
Missoula AP	46	− 3	89	63	36

Table C-1 (Cont.)

STATE AND STATION	WINTER Latitude	WINTER DB 97½%	SUMMER DB 2½%	SUMMER WB 2½%	SUMMER Outdoor Daily Range
NEBRASKA					
Beatrice	40	1	97	77	24
Chadron AP	42	-9	95	70	30
Columbus	41	-3	96	76	25
Fremont	41	-3	97	77	22
Grand Island AP	41	-2	95	75	28
Hastings	40	1	96	75	27
Kearney	40	-2	95	75	28
Lincoln CO	40	0	96	77	24
McCook	40	0	97	72	28
Norfolk	42	-7	95	76	30
North Platte AP	41	-2	94	73	28
Omaha AP	41	-1	94	78	22
Scottsbluff AP	41	-4	94	69	31
Sidney AP	41	-2	92	69	31
NEVADA					
Carson City	39	7	91	61	42
Elko AP	40	-7	92	62	42
Ely AP	39	-2	88	59	39
Las Vegas AP	36	26	106	71	30
Lovelock AP	40	11	96	64	42
Reno AP	39	7	92	62	45
Reno CO	39	17	92	62	45
Tonopah AP	38	13	92	63	40
Winnemucca AP	40	5	95	62	42
NEW HAMPSHIRE					
Berlin	44	-15	85	71	22
Claremont	43	-9	87	73	24
Concord AP	43	-7	88	73	26
Keene	43	-8	88	73	24
Laconia	43	-12	87	73	25
Manchester, Grenier AFB	43	1	89	74	24
Portsmouth, Pease AFB	43	3	86	73	22
NEW JERSEY					
Atlantic City CO	39	18	88	77	18
Long Branch	40	13	91	76	18
Newark AP	40	15	91	76	20
New Brunswick	40	12	89	76	19
Paterson	40	12	91	76	21
Phillipsburg	40	10	91	76	21
Trenton CO	40	16	90	77	19
Vineland	39	16	90	77	19
NEW MEXICO					
Alamagordo, Holloman AFB	32	22	98	69	30
Albuquerque AP	35	17	94	65	27
Artesia	32	19	99	70	30
Carlsbad AP	32	21	99	71	28
Clovis AP	34	17	97	69	28
Farmington AP	36	9	93	65	30
Gallup	35	-1	90	63	32
Grants	35	-3	89	63	32
Hobbs AP	32	19	99	71	29
Las Cruces	32	23	100	69	30
Los Alamos	35	9	86	63	32
Raton AP	36	2	90	65	34
Roswell, Walker AFB	33	19	99	70	33
Santa Fe CO	35	11	88	63	28
Silver City AP	32	18	93	67	30
Socorro AP	34	17	97	66	30
Tucumcari AP	35	13	97	70	28
NEW YORK					
Albany AP	42	0	88	74	23
Albany CO	42	5	89	74	20
Auburn	43	2	87	73	22
Batavia	43	3	87	74	22
Binghamton CO	42	2	89	72	20
Buffalo AP	43	6	86	73	21
Cortland	42	-1	88	73	23
Dunkirk	42	8	86	74	18

STATE AND STATION	WINTER Latitude	WINTER DB 97½%	SUMMER DB 2½%	SUMMER WB 2½%	SUMMER Outdoor Daily Range
Elmira AP	42	5	90	73	24
Geneva	42	2	89	73	22
Glens Falls	43	-7	86	72	23
Gloversville	43	-2	87	73	23
Hornell	42	-5	85	72	24
Ithaca	42	0	88	73	24
Jamestown	42	5	86	73	20
Kingston	42	2	90	74	22
Lockport	43	6	85	74	21
Massena AP	45	-12	84	74	20
Newburgh-Stewart AFB	41	6	89	76	21
NYC - Central Park	40	15	91	76	17
NYC - Kennedy AP	40	21	87	76	16
NYC - LaGuardia AP	40	16	90	76	16
Niagara Falls AP	43	7	86	74	20
Olean	42	-3	85	72	23
Oneonta	42	-3	87	72	24
Oswego CO	43	6	84	74	20
Plattsburg AFB	44	-6	84	73	22
Poughskeepsie	41	3	90	75	21
Rochester AP	43	5	88	74	22
Rome-Griffiss AFB	43	-3	87	74	22
Schenectady	42	-1	88	73	22
Suffolk County AFB	40	13	84	75	16
Syracuse AP	43	2	87	74	20
Utica	43	-2	87	73	22
Watertown	44	-10	84	74	20
NORTH CAROLINA					
Asheville AP	35	17	88	74	21
Charlotte AP	35	22	94	77	20
Durham	36	19	92	77	20
Elizabeth City AP	36	22	91	79	18
Fayetteville, Pope AFB	35	20	94	79	20
Goldsboro, Seymour-Johnson AFB	35	21	92	79	18
Greensboro AP	36	17	91	76	21
Greenville	35	22	93	80	19
Henderson	36	16	92	78	20
Hickory	35	18	91	76	21
Jacksonville	34	25	92	80	18
Lumberton	34	22	93	80	20
New Bern AP	35	22	92	80	18
Raleigh/Durham AP	35	20	92	78	20
Rocky Mount	36	20	93	79	19
Wilmington AP	34	27	91	81	18
Winston-Salem AP	36	17	91	76	20
NORTH DAKOTA					
Bismarck AP	46	-19	91	72	27
Devil's Lake	48	-19	89	71	25
Dickinson AP	46	-19	93	70	25
Fargo AP	46	-17	88	74	25
Grand Forks AP	48	-23	87	72	25
Jamestown AP	47	-18	91	73	26
Minot AP	48	-20	88	70	25
Williston	48	-17	90	69	25
OHIO					
Akron/Canton AP	41	6	87	73	21
Ashtabula	42	7	87	75	18
Athens	39	7	91	76	22
Bowling Green	41	3	91	75	23
Cambridge	40	4	89	76	23
Chillicothe	39	9	91	76	22
Cincinnati CO	39	12	92	77	21
Cleveland AP	41	7	89	75	22
Columbus AP	40	7	88	76	24
Dayton AP	39	6	90	75	20
Defiance	41	1	91	76	24
Findlay AP	41	4	90	76	24
Fremont	41	3	90	75	24
Hamilton	39	8	92	77	22
Lancaster	39	5	91	76	23
Lima	40	4	91	76	24

Table C-1 (Cont.)

STATE AND STATION	WINTER		SUMMER		
	Latitude	DB 97½%	DB 2½%	WB 2½%	Outdoor Daily Range
Mansfield AP	40	3	89	75	22
Marion	40	6	91	76	23
Middletown	39	7	91	76	22
Newark	40	3	90	76	23
Norwalk	41	3	90	75	22
Portsmouth	38	9	92	76	22
Sandusky CO	41	8	90	75	21
Springfield	40	7	90	76	21
Steubenville	40	9	89	75	22
Toledo AP	41	5	90	75	25
Warren	41	4	88	74	23
Wooster	40	3	88	75	22
Youngstown AP	41	6	86	74	23
Zanesville AP	40	3	89	76	23
OKLAHOMA					
Ada	34	16	100	78	23
Altus AFB	34	18	101	76	25
Ardmore	34	19	101	78	23
Bartlesville	36	9	99	78	23
Chickasha	35	16	101	76	24
Enid-Vance AFB	36	14	100	77	24
Lawton AP	34	16	101	77	24
McAlester	34	17	100	78	23
Muskogee AP	35	16	99	78	23
Norman	35	15	99	77	24
Oklahoma City AP	35	15	97	77	23
Ponca City	36	12	100	77	24
Seminole	35	16	100	77	23
Stillwater	36	13	99	77	24
Tulsa AP	36	16	99	78	22
Woodward	36	8	101	74	26
OREGON					
Albany	44	27	88	67	31
Astoria AP	46	30	76	60	16
Baker AP	44	1	92	65	30
Bend	44	4	87	62	33
Corvallis	44	27	88	67	31
Eugene AP	44	26	88	67	31
Grants Pass	42	26	92	66	33
Klamath Falls AP	42	5	87	62	36
Medford AP	42	23	94	68	35
Pendleton AP	45	10	94	65	29
Portland AP	45	24	85	67	23
Portland CO	45	29	88	68	21
Roseburg AP	43	29	91	67	30
Salem AP	45	25	88	67	31
The Dalles	45	17	91	68	28
PENNSYLVANIA					
Allentown AP	40	5	90	75	22
Altoona CO	40	5	87	73	23
Butler	40	2	89	74	22
Chambersburg	40	9	92	75	23
Erie AP	42	11	85	74	18
Harrisburg AP	40	13	89	75	21
Johnstown	40	5	87	73	23
Lancaster	40	6	90	76	22
Meadville	41	4	86	73	21
New Castle	41	4	89	74	23
Philadelphia AP	39	15	90	77	21
Pittsburgh AP	40	9	87	74	22
Pittsburgh CO	40	11	88	74	19
Reading CO	40	9	90	76	19
Scranton/Wilkes-Barre	41	6	87	74	19
State College	40	6	87	73	23
Sunbury	40	7	89	75	22
Uniontown	39	8	88	74	22
Warren	41	1	87	73	24
West Chester	40	13	90	76	20
Williamsport AP	41	5	89	75	23
York	40	8	91	76	22
RHODE ISLAND					
Newport	41	11	84	74	16
Providence AP	41	10	86	75	19
SOUTH CAROLINA					
Anderson	34	22	94	76	21
Charleston AFB	32	27	92	80	18
Charleston CO	32	30	93	80	13
Columbia AP	34	23	96	79	22
Florence AP	34	25	94	79	21
Georgetown	33	26	91	80	18
Greenville AP	34	23	93	76	21
Greenwood	34	23	95	77	21
Orangeburg	33	25	95	79	20
Rock Hill	35	21	95	77	20
Spartanburg AP	35	22	93	76	20
Sumter-Shaw AFB	34	26	94	79	21
SOUTH DAKOTA					
Aberdeen AP	45	−18	92	75	27
Brookings	44	−15	90	75	25
Huron AP	44	−12	93	75	28
Mitchell	43	−11	94	76	28
Pierre AP	44	− 9	96	74	29
Rapid City AP	44	− 6	94	71	28
Sioux Falls AP	43	−10	92	75	24
Watertown AP	45	−16	90	74	26
Yankton	43	− 7	94	76	25
TENNESSEE					
Athens	33	18	94	76	22
Bristol-Tri City AP	36	16	90	75	22
Chattanooga AP	35	19	94	78	22
Clarksville	36	16	96	78	21
Columbia	35	17	95	78	21
Dyersburg	36	17	96	79	21
Greenville	35	14	91	75	22
Jackson AP	35	17	95	79	21
Knoxville AP	35	17	92	76	21
Memphis AP	35	21	96	79	21
Murfreesboro	35	17	94	78	22
Nashville AP	36	16	95	78	21
Tullahoma	35	17	94	78	22
TEXAS					
Abilene AP	32	21	99	75	22
Alice AP	27	34	99	80	20
Amarillo AP	35	12	96	71	26
Austin AP	30	29	98	78	22
Bay City	29	33	93	80	16
Beaumont	30	33	94	80	19
Beeville	28	32	97	80	18
Big Spring AP	32	22	98	73	26
Brownsville AP	25	40	92	80	18
Brownwood	31	25	100	75	22
Bryan AP	30	31	98	78	20
Corpus Christi AP	27	36	93	80	19
Corsicana	32	25	100	78	21
Dallas AP	32	24	99	78	20
Del Rio, Laughlin AFB	29	31	99	77	24
Denton	33	22	100	78	22
Eagle Pass	28	31	104	79	24
El Paso AP	31	25	98	69	27
Fort Worth AP	32	24	100	78	22
Galveston AP	29	36	89	81	10
Greenville	33	24	99	78	21
Harlingen	26	38	95	80	19
Houston AP	29	32	94	80	18
Houston CO	29	33	94	80	18
Huntsville	30	31	97	79	20
Killeen-Gray AFB	31	26	99	77	22
Lamesa	32	18	98	73	26
Laredo AFB	27	36	101	78	23
Longview	32	25	98	80	20

Table C-1 (Cont.)

STATE AND STATION	WINTER Lati-tude	WINTER DB 97½%	SUMMER DB 2½%	SUMMER WB 2½%	SUMMER Outdoor Daily Range
Lubbock AP	33	15	97	72	26
Lufkin AP	31	28	96	80	20
McAllen	26	38	100	79	21
Midland AP	32	23	98	73	26
Mineral Wells AP	32	22	100	77	22
Palestine CO	31	25	97	79	20
Pampa	35	11	98	72	26
Pecos	31	19	100	71	27
Plainview	34	14	98	72	26
Port Arthur AP	30	33	92	80	19
San Angelo, Goodfellow AFB	31	25	99	75	24
San Antonio AP	29	30	97	77	19
Sherman-Perrin AFB	33	23	99	78	22
Snyder	32	19	100	74	26
Temple	31	27	99	78	22
Tyler AP	32	24	97	79	21
Vernon	34	18	101	76	24
Victoria AP	28	32	96	79	18
Waco AP	31	26	99	78	22
Wichita Falls AP	34	19	100	76	24
UTAH					
Cedar City AP	37	6	91	64	32
Logan	41	7	91	65	33
Moab	38	16	98	65	30
Ogden CO	41	11	92	65	33
Price	39	7	91	64	33
Provo	40	6	93	66	32
Richfield	38	3	92	65	34
St. George CO	37	26	102	70	33
Salt Lake City AP	40	9	94	66	32
Vernal AP	40	− 6	88	63	32
VERMONT					
Barre	44	−13	84	72	23
Burlington AP	44	− 7	85	73	23
Rutland	43	− 8	85	73	23
VIRGINIA					
Charlottsville	38	15	90	77	23
Danville AP	36	17	92	77	21
Fredericksburg	38	14	92	78	21
Harrisonburg	38	9	90	77	23
Lynchburg AP	37	19	92	76	21
Norfolk AP	36	23	91	78	18
Petersburg	37	18	94	79	20
Richmond AP	37	18	93	78	21
Roanoke AP	37	18	91	75	23
Staunton	38	12	90	77	23
Winchester	39	10	92	76	21
WASHINGTON					
Aberdeen	47	27	80	61	16
Bellingham AP	48	18	74	65	19
Bremerton	47	29	81	66	20
Ellensburg AP	47	6	89	65	34

STATE AND STATION	WINTER Lati-tude	WINTER DB 97½%	SUMMER DB 2½%	SUMMER WB 2½%	SUMMER Outdoor Daily Range
Everett-Paine AFB	47	24	78	65	20
Kennewick	46	15	96	68	30
Longview	46	24	86	66	30
Moses Lake, Larson AFB	47	− 1	93	66	32
Olympia AP	47	25	83	65	32
Port Angeles	48	29	73	58	18
Seattle-Boeing Fld.	47	27	80	65	24
Seattle CO	47	32	79	65	19
Seattle-Tacoma AP	47	24	81	64	22
Spokane AP	47	4	90	64	28
Tacoma-McChord AFB	47	24	81	66	22
Walla Walla AP	46	16	96	68	27
Wenatchee	47	9	92	66	32
Yakima AP	46	10	92	67	36
WEST VIRGINIA					
Beckley	37	6	88	73	22
Bluefield AP	37	10	86	73	22
Charleston AP	38	14	90	75	20
Clarksburg	39	7	90	75	21
Elkins AP	38	5	84	73	22
Huntington CO	38	14	93	76	22
Martinsburg AP	39	10	94	77	21
Morgantown AP	39	7	88	74	22
Parkersburg CO	39	12	91	76	21
Wheeling	40	9	89	75	21
WISCONSIN					
Appleton	44	− 6	87	74	23
Ashland	46	−17	83	71	23
Beloit	42	− 3	90	76	24
Eau Claire AP	44	−11	88	74	23
Fond du Lac	43	− 7	87	74	23
Green Bay AP	44	− 7	85	73	23
La Crosse AP	43	− 8	88	76	22
Madison AP	43	− 5	88	75	22
Manitowoc	44	− 1	86	74	21
Marinette	45	− 4	86	72	20
Milwaukee AP	43	− 2	87	75	21
Racine	42	0	88	75	21
Sheboygan	43	0	87	74	20
Stevens Point	44	−12	87	73	23
Waukesha	43	− 2	89	75	22
Wausau AP	44	−14	86	72	23
WYOMING					
Casper AP	42	− 5	90	62	31
Cheyenne AP	41	− 2	86	62	30
Cody AP	44	− 9	87	60	32
Evanston	41	− 8	82	57	32
Lander AP	42	−12	90	62	32
Laramie AP	41	− 2	80	59	28
Newcastle	43	− 5	89	67	30
Rawlins	41	−11	84	61	40
Rock Springs AP	41	− 1	84	57	32
Sheridan AP	44	− 7	92	65	32
Torrington	42	− 7	92	67	30

EXPLANATION OF DESIGN CONDITIONS:

WINTER – 97½% indicates that the temperature will be at or above the design temperature shown 97½% of the time.

SUMMER – 2½% indicates that the temperature will exceed the design temperature shown only 2½% of the time.

OUTDOOR DAILY RANGE – The outdoor daily range of DB temperatures is the difference between the average maximum and average minimum temperatures during the warmest month at each location. Refer to page 39 when outdoor daily range is other than 20°.

Table C-2

Normal Degree-Days by Months for U.S. Cities*[a]

City	Jul	Aug	Sep	Oct	Nov	Dec	Jan	Feb	Mar	Apr	May	Jun	Total
ALABAMA													
Anniston-A	0	0	17	118	438	614	614	485	381	128	25	0	2820
Birmingham-A	0	0	13	123	396	598	623	491	378	128	30	0	2780
Mobile-A	0	0	0	28	219	376	416	304	222	47	0	0	1612
Mobile-C	0	0	0	23	198	357	412	290	209	40	0	0	1529
Montgomery-A	0	0	0	69	304	491	517	388	288	80	0	0	2137
Montgomery-C	0	0	0	55	267	458	483	360	265	66	0	0	1954
ARIZONA													
Flagstaff-A	49	78	243	586	876	1135	1231	1014	949	687	465	212	7525
Phoenix-A	0	0	0	22	223	400	474	309	196	74	0	0	1698
Phoenix-C	0	0	0	13	182	360	425	275	175	62	0	0	1492
Prescott-A	0	0	34	261	582	843	921	717	626	368	164	17	4533
Tuscon-A	0	0	0	24	222	403	474	330	239	84	0	0	1776
Winslow-A	0	0	20	274	663	946	1001	706	605	335	144	8	4702
Yuma-A	0	0	0	0	105	259	318	167	88	14	0	0	951
ARKANSAS													
Fort Smith-A	0	0	9	131	435	698	775	571	418	127	24	0	3188
Little Rock-A	0	0	10	110	405	654	719	543	401	122	18	0	2982
Texarkana-A	0	0	0	69	317	527	600	441	324	84	0	0	2362
CALIFORNIA													
Bakersfield-A	0	0	0	41	273	505	561	350	259	105	21	0	2115
Beaumont-C	0	0	18	103	298	487	574	473	437	286	146	18	2840
Bishop-A	0	0	55	253	564	803	840	664	546	319	140	38	4222
Blue Canyon-A	36	41	105	369	633	822	893	809	815	597	397	202	5719
Burbank-A	0	0	11	59	186	324	396	308	265	152	85	22	1808
Eureka-C	267	248	264	335	411	508	552	465	493	432	375	282	4632
Fresno-A	0	0	0	86	345	580	629	400	304	145	43	0	2532
Los Angeles-A	31	22	56	87	200	301	378	305	273	185	121	56	2015
Los Angeles-C	0	0	17	41	140	253	328	244	212	129	68	19	1451
Mount Shasta-C	37	46	165	434	705	939	998	787	722	549	357	174	5913
Oakland-A	84	77	76	157	336	508	552	400	360	282	212	119	3163
Red Bluff-A	0	0	0	59	319	564	617	423	336	177	51	0	2546
Sacramento-A	0	0	22	98	357	595	642	428	348	222	103	7	2822
Sacramento-C	0	0	17	75	321	567	614	403	317	196	85	5	2600
Sandberg-C	0	0	26	211	465	701	781	678	629	435	261	56	4243
San Diego-A	11	7	24	52	147	255	317	247	223	151	97	43	1574
San Francisco-A	144	136	101	174	318	487	530	398	378	327	264	164	3421
San Francisco-C	189	177	110	128	237	406	462	336	317	279	248	180	3069
San Jose-C	7	11	26	97	270	450	487	342	308	229	137	46	2410
Santa Catalina-A	21	11	24	77	168	311	375	328	344	264	205	121	2249
Santa Maria-A	98	94	111	157	262	391	453	370	341	276	229	152	2934

[a]The accompanying monthly normal number of degree-days for 335 weather stations are based on records for the 30-year period, 1921 to 1950, inclusive, and are those calculated by the U.S. Weather Bureau and released for publication late in 1953.

In the tables A indicates airport weather station, C city office station.

The degree-day normals are derived from the values for the monthly normal maximum and minimum temperature, the monthly mean temperature being the sum of the maximum and minimum divided by two. They are computed on the standard base of 65 degrees. The seasonal total is the total of the monthly figures.

The 30-year period covered is consistent with the term of years accepted by the World Meteorological Organization for climate normal.

Reproduced with permission from Handbook of Air Conditioning, Heating and Ventilating, *2nd Ed., The Industrial Press, New York, 1965.*

Table C-2 (Cont.)

City	Jul	Aug	Sep	Oct	Nov	Dec	Jan	Feb	Mar	Apr	May	Jun	Total
COLORADO													
Alamosa-A	64	121	309	648	1065	1414	1491	1176	1029	699	440	203	8659
Colorado Springs-A	8	21	124	422	777	1039	1122	930	874	555	307	75	6254
Denver-A	5	11	120	425	771	1032	1125	924	843	525	286	65	6132
Denver-C	0	5	103	385	711	958	1042	854	797	492	260	60	5673
Grand Junction-A	0	0	36	333	792	1132	1271	924	738	402	145	23	5796
Pueblo-A	0	0	74	383	771	1051	1104	865	775	456	203	27	5709
CONNECTICUT													
Bridgeport-A	0	0	66	334	645	1014	1110	1008	871	561	249	38	5896
Hartford-A	0	14	101	384	699	1082	1178	1050	871	528	201	31	6139
New Haven-A	0	18	93	363	663	1026	1113	1005	865	567	261	52	6026
DELAWARE													
Wilmington-A	0	0	47	282	585	927	983	876	698	396	110	6	4910
DISTRICT OF COLUMBIA													
Washington-A	0	0	37	237	519	837	893	781	619	323	87	0	4333
Washington-C	0	0	32	231	510	831	884	770	606	314	80	0	4258
Silver Hill Obs.	0	0	53	270	549	865	918	798	632	347	107	0	4539
FLORIDA													
Apalachicola-C	0	0	0	17	154	304	352	263	184	33	0	0	1307
Daytona Beach-A	0	0	0	0	83	205	245	187	137	11	0	0	868
Fort Myers-A	0	0	0	0	25	101	124	95	60	0	0	0	405
Jacksonville-A	0	0	0	16	148	309	331	247	169	23	0	0	1243
Jacksonville-C	0	0	0	11	129	276	303	226	154	14	0	0	1113
Key West-A	0	0	0	0	0	22	34	25	8	0	0	0	89
Key West-C	0	0	0	0	0	18	28	24	7	0	0	0	77
Lakeland-C	0	0	0	0	60	167	185	142	95	0	0	0	649
Melbourne-A	0	0	0	0	44	127	169	121	76	0	0	0	537
Miami-A	0	0	0	0	8	52	58	48	12	0	0	0	178
Miami-C	0	0	0	0	5	48	57	48	15	0	0	0	173
Miami Beach	0	0	0	0	0	37	43	34	9	0	0	0	123
Orlando-A	0	0	0	0	61	161	188	148	92	0	0	0	650
Pensacola-C	0	0	0	18	177	334	383	275	203	45	0	0	1435
Tallahassee-A	0	0	0	31	209	366	385	287	203	38	0	0	1519
Tampa-A	0	0	0	0	60	163	201	148	102	0	0	0	674
W. Palm Beach-A	0	0	0	0	7	62	85	61	33	0	0	0	248
GEORGIA													
Albany-A	0	0	0	39	242	427	446	333	236	40	0	0	1763
Athens-A	0	0	5	100	390	614	629	515	404	128	15	0	2800
Atlanta-A	0	0	8	110	393	614	632	512	404	133	20	0	2826
Atlanta-C	0	0	8	107	387	611	632	515	392	135	24	0	2811
Augusta-A	0	0	0	59	282	494	521	412	308	62	0	0	2138
Columbus-A	0	0	0	78	326	547	563	437	348	97	0	0	2396
Macon-A	0	0	0	63	280	481	497	391	275	62	0	0	2049
Rome-A	0	0	8	140	435	673	700	560	436	159	27	0	3138
Savannah-A	0	0	0	38	225	412	424	330	238	43	0	0	1710
Valdosta-A	0	0	0	32	203	366	386	290	210	38	0	0	1525
IDAHO													
Boise-A	0	0	135	389	762	1054	1169	868	719	453	249	92	5890
Lewiston-A	0	0	133	406	747	961	1060	815	663	408	222	68	5483
Pocatello-A	0	0	183	487	873	1184	1333	1022	880	561	317	136	6976
Salmon	22	55	292	592	996	1380	1513	1103	905	561	334	169	7922

Table C-2 (Cont.)

City	Jul	Aug	Sep	Oct	Nov	Dec	Jan	Feb	Mar	Apr	May	Jun	Total
ILLINOIS													
Cairo-C	0	0	28	161	492	784	856	683	523	182	47	0	3756
Chicago-A	0	0	90	350	765	1147	1243	1053	868	507	229	58	6310
Joliet-A	0	16	114	390	798	1190	1277	1084	893	519	233	64	6578
Moline-A	0	8	96	363	786	1181	1296	1075	862	453	199	45	6364
Peoria-A	0	11	86	339	759	1128	1240	1028	828	435	192	41	6087
Springfield-A	0	6	83	315	723	1066	1166	958	769	404	171	32	5693
Springfield-C	0	0	56	259	666	1017	1116	907	713	350	127	14	5225
INDIANA													
Evansville-A	0	0	59	215	570	871	939	770	589	251	90	6	4360
Fort Wayne-A	0	17	107	377	759	1122	1200	1036	874	516	226	53	6287
Indianapolis-A	0	0	79	306	705	1051	1122	938	772	432	176	30	5611
Indianapolis-C	0	0	59	247	642	986	1051	893	725	375	140	16	5134
South Bend-A	5	13	101	381	789	1153	1252	1081	908	531	248	62	6524
Terre Haute-A	0	5	77	295	681	1023	1107	913	725	371	145	24	5366
IOWA													
Burlington-A	0	0	83	336	765	1150	1271	1036	822	425	179	34	6101
Charles City-C	17	30	151	444	912	1352	1494	1240	1001	537	256	70	7504
Davenport-C	0	7	79	320	756	1147	1262	1044	834	432	175	35	6091
Des Moines-A	5	12	99	355	798	1203	1330	1092	868	438	201	45	6446
Des Moines-C	0	6	89	346	777	1178	1308	1072	849	425	183	41	6274
Dubuque-A	8	28	149	444	882	1290	1414	1187	983	543	267	76	7271
Sioux City-A	8	17	128	405	885	1290	1423	1170	930	474	228	54	7012
KANSAS													
Concordia-C	0	0	55	277	687	1029	1144	899	725	341	146	20	5323
Dodge City-A	0	0	40	262	669	980	1076	840	694	347	135	15	5058
Goodland-A	0	0	95	413	825	1128	1215	974	884	534	241	58	6367
Topeka-A	0	8	59	271	672	1017	1125	885	694	326	137	15	5209
Topeka-C	0	0	42	242	630	977	1088	851	669	295	112	13	4919
Wichita-A	0	0	32	219	597	915	1023	778	619	280	101	7	4571
KENTUCKY													
Bowling Green-A	0	0	47	215	558	840	890	739	601	286	98	5	4279
Lexington-A	0	0	56	259	636	933	1008	854	710	368	140	15	4979
Louisville-A	0	0	51	232	579	871	933	778	611	285	94	5	4439
Louisville-C	0	0	41	206	549	849	911	762	605	270	86	0	4279
LOUISIANA													
Baton Rouge-A	0	0	0	27	215	373	424	293	215	48	0	0	1595
Burrwood	0	0	0	0	81	225	303	226	169	29	0	0	1033
Lake Charles-A	0	0	0	22	218	353	416	284	210	40	0	0	1543
New Orleans-A	0	0	0	7	169	308	364	248	190	31	0	0	1317
New Orleans-C	0	0	0	5	141	283	341	223	163	19	0	0	1175
Shreveport-A	0	0	0	53	305	490	550	386	272	61	0	0	2117
MAINE													
Caribou-A	85	133	354	710	1074	1562	1745	1546	1342	909	512	201	10173
Eastport-C	141	136	261	521	798	1206	1333	1201	1063	774	524	288	8246
Portland-A	15	56	199	515	825	1237	1373	1218	1039	693	394	117	7681

Table C-2 (Cont.)

City	Jul	Aug	Sep	Oct	Nov	Dec	Jan	Feb	Mar	Apr	May	Jun	Total
MARYLAND													
Baltimore-A	0	0	43	256	558	884	942	820	651	360	97	0	4611
Baltimore-C	0	0	29	207	489	812	880	776	611	326	73	0	4203
Frederick-A	0	0	47	276	588	930	1001	865	673	368	106	0	4854
MASSACHUSETTS													
Boston-A	0	7	77	315	618	998	1113	1002	849	534	236	42	5791
Nantucket-A	22	34	111	372	615	924	1020	949	880	642	394	139	6102
Pittsfield-A	25	63	213	543	843	1246	1358	1212	1060	690	336	105	7694
MICHIGAN													
Alpena-C	50	85	215	530	864	1218	1358	1263	1156	762	437	135	8073
Detroit-Willow Run-A	0	10	96	393	759	1125	1231	1089	915	552	244	55	6469
Detroit-A	0	8	96	381	747	1101	1203	1072	927	558	251	60	6404
Escanaba-C	62	95	247	555	933	1321	1473	1327	1203	804	471	166	8657
Grand Rapids-A	14	29	144	462	822	1169	1287	1154	1008	606	301	79	7075
Grand Rapids-C	0	20	105	394	756	1107	1215	1086	939	546	248	58	6474
Lansing-A	13	33	140	455	813	1175	1277	1142	986	591	287	70	6982
Marquette-C	69	87	236	543	933	1299	1435	1291	1181	789	477	189	8529
Muskegon-A	26	48	152	462	795	1110	1243	1134	1011	642	350	116	7089
Sault Ste. Marie-A	109	126	298	639	1005	1398	1587	1442	1302	846	499	224	9475
MINNESOTA													
Duluth-A	56	91	298	651	1140	1606	1758	1512	1327	846	474	178	9937
Duluth-C	66	91	277	614	1092	1550	1696	1448	1252	801	487	200	9574
Internat'l Falls-A	70	118	356	716	1230	1733	1922	1618	1395	834	437	171	10600
Minneapolis-A	8	17	157	459	960	1414	1562	1310	1057	570	259	80	7853
Rochester-A	24	38	182	499	975	1426	1572	1316	1073	600	298	92	8095
Saint Cloud-A	32	53	225	570	1068	1535	1690	1439	1181	663	331	106	8893
Saint Paul-A	12	21	154	459	951	1401	1553	1305	1051	564	256	77	7804
MISSISSIPPI													
Jackson-A	0	0	0	69	310	503	535	405	299	81	0	0	2202
Meridian-A	0	0	0	90	338	528	561	413	309	85	9	0	2333
Vicksburg-C	0	0	0	51	268	456	507	374	273	71	0	0	2000
MISSOURI													
Columbia-A	0	6	62	262	654	989	1091	876	698	326	135	14	5113
Kansas City-A	0	0	44	240	621	970	1085	851	666	292	111	8	4888
Saint Joseph-A	0	5	49	265	681	1048	1175	930	716	326	127	14	5336
Saint Louis-A	0	0	45	233	600	927	1017	820	648	297	101	11	4699
Saint Louis-C	0	0	38	202	570	893	983	792	620	270	94	7	4469
Springfield-A	0	8	61	249	615	908	1001	790	632	295	118	16	4693
MONTANA													
Billings-A	8	20	194	497	876	1172	1305	1089	958	564	304	119	7106
Butte-A	115	174	450	744	1104	1442	1575	1294	1172	804	561	325	9760
Glasgow-C	14	30	244	574	1086	1510	1683	1408	1119	597	312	113	8690
Great Falls-A	24	50	273	524	894	1194	1311	1131	1008	621	359	166	7555
Havre-C	20	38	270	564	1023	1383	1513	1291	1076	597	313	125	8213
Helena-A	36	66	320	617	999	1311	1469	1165	1017	654	399	197	8250
Helena-C	51	78	359	598	969	1215	1438	1114	992	660	427	225	8126
Kalispell-A	47	83	326	639	990	1249	1386	1120	970	639	391	215	8055
Miles City-A	6	11	187	525	966	1373	1516	1229	1048	570	285	106	7822
Missoula-A	22	57	292	623	993	1283	1414	1100	939	609	365	176	7873

Table C-2 (Cont.)

City	Jul	Aug	Sep	Oct	Nov	Dec	Jan	Feb	Mar	Apr	May	Jun	Total
NEBRASKA													
Grand Island-A	0	6	84	369	822	1178	1302	1044	849	423	195	39	6311
Lincoln-A	0	12	82	340	774	1144	1271	1030	822	401	190	38	6104
Lincoln-C	0	7	79	310	741	1113	1240	1000	794	377	172	32	5865
Norfolk-A	0	17	122	422	903	1280	1417	1159	933	501	251	60	7065
North Platte-A	7	11	120	425	846	1172	1271	1016	887	489	243	59	6546
Omaha-A	0	5	88	331	783	1166	1302	1058	831	389	175	32	6160
Scottsbluff-A	0	0	137	456	867	1178	1287	1030	933	567	305	81	6841
Valentine-C	11	10	145	461	891	1212	1361	1100	970	543	288	83	7075
NEVADA													
Elko-A	6	28	229	546	915	1181	1336	1025	896	612	378	183	7335
Ely-A	22	44	228	561	894	1181	1302	1033	921	639	418	200	7443
Las Vegas-C	0	0	0	61	344	564	653	423	288	92	0	0	2425
Reno-A	27	61	165	443	744	986	1048	804	756	519	318	165	6036
Tonopah	0	5	96	422	723	995	1082	860	763	504	272	91	5813
Winnemucca-A	0	17	180	508	822	1085	1153	854	794	546	299	111	6369
NEW HAMPSHIRE													
Concord-A	11	57	192	527	849	1271	1392	1226	1029	660	316	82	7612
NEW JERSEY													
Atlantic City-C	0	0	29	230	507	831	905	829	729	468	189	24	4741
Newark-A	0	0	47	301	603	961	1039	932	760	450	148	11	5252
Trenton-C	0	0	55	285	582	930	1004	904	735	429	133	11	5068
NEW MEXICO													
Albuquerque-A	0	0	10	218	630	899	970	714	589	289	70	0	4389
Clayton-A	0	0	68	318	678	927	995	795	729	420	184	24	5138
Raton-A	17	36	148	431	798	1104	1203	924	834	543	292	87	6417
Roswell-A	0	0	8	156	501	750	787	566	443	185	28	0	3424
NEW YORK													
Albany-A	0	24	139	443	780	1197	1318	1179	989	597	246	50	6962
Albany-C	0	6	98	388	708	1113	1234	1103	905	531	202	31	6319
Bear Mountain-C	0	25	119	409	753	1110	1212	1098	921	561	244	59	6511
Binghamton-A	16	63	192	518	834	1228	1342	1215	1051	672	318	88	7537
Binghamton-C	0	36	141	428	735	1113	1218	1100	927	570	240	48	6556
Buffalo-A	16	30	122	433	753	1116	1225	1128	992	636	315	72	6838
New York-La Guard.-A	0	0	28	250	546	908	992	907	760	447	141	10	4989
New York-C	0	0	39	263	561	908	995	904	753	456	153	18	5050
New York-Central Pk.	0	0	31	250	552	902	1001	910	747	435	130	7	4965
Oswego-C	20	39	139	430	738	1132	1249	1134	995	654	355	90	6975
Rochester-A	9	34	133	440	759	1141	1249	1148	992	615	289	54	6863
Schenectady-C	0	19	137	456	792	1212	1349	1207	1008	597	233	40	7050
Syracuse-A	0	29	117	396	714	1113	1225	1117	955	570	247	37	6520
NORTH CAROLINA													
Asheville-C	0	0	50	262	552	769	794	678	572	285	105	5	4072
Charlotte-A	0	0	7	147	438	682	704	577	449	172	29	0	3205
Greensboro-A	0	0	29	202	510	772	806	672	528	241	50	0	3810
Hatteras-C	0	0	0	63	244	481	527	487	394	171	25	0	2392
Raleigh-A	0	0	16	149	438	701	732	613	477	202	41	0	3369
Raleigh-C	0	0	10	118	387	651	691	577	440	172	29	0	3075
Wilmington-A	0	0	0	73	288	508	533	463	347	104	7	0	2323
Winston-Salem-A	0	0	28	182	492	756	797	666	519	232	49	0	3721

Table C-2 (Cont.)

City	Jul	Aug	Sep	Oct	Nov	Dec	Jan	Feb	Mar	Apr	May	Jun	Total
NORTH DAKOTA													
Bismarck-A	29	37	227	598	1098	1535	1730	1464	1187	657	355	116	9033
Devils Lake-C	47	61	276	654	1197	1668	1866	1576	1314	750	394	137	9940
Fargo-A	25	41	215	586	1122	1615	1795	1518	1231	687	338	101	9274
Williston-C	29	42	261	605	1101	1528	1705	1442	1194	663	360	138	9068
OHIO													
Akron-Canton-A	0	17	83	378	738	1082	1166	1033	884	537	235	50	6203
Cincinnati-A	0	6	77	295	648	973	1029	871	732	392	149	23	5195
Cincinnati-C	0	0	42	222	567	880	942	812	645	314	108	0	4532
Cincinnati-Abbe Obs.	0	0	56	263	612	930	989	846	682	347	132	13	4870
Cleveland-A	0	10	75	340	699	1057	1132	1019	874	531	223	46	6006
Cleveland-C	0	9	60	311	636	995	1101	977	846	510	223	49	5717
Columbus-A	0	8	69	337	693	1032	1094	946	781	444	180	31	5615
Columbus-C	0	0	59	299	654	983	1051	907	741	408	153	22	5277
Dayton-A	0	5	74	324	693	1032	1094	941	781	435	179	39	5597
Sandusky-C	0	0	66	327	684	1039	1122	997	853	513	217	41	5859
Toledo-A	0	12	102	387	756	1119	1197	1056	905	555	245	60	6394
Youngstown-A	0	19	83	355	732	1085	1163	1030	877	534	241	53	6172
OKLAHOMA													
Oklahoma City-A	0	0	14	154	480	769	865	650	490	182	40	0	3644
Oklahoma City-C	0	0	12	149	459	747	843	630	472	169	38	0	3519
Tulsa-A	0	0	18	152	462	750	856	644	485	173	44	0	3584
OREGON													
Astoria-A	138	111	146	338	537	691	772	613	611	459	357	222	4995
Baker-C	25	47	255	518	852	1138	1268	972	837	591	384	200	7087
Burns-C	10	37	219	552	855	1156	1274	946	809	552	349	159	6918
Eugene-A	33	34	144	381	591	756	831	624	567	423	270	125	4779
Meacham-A	88	102	294	605	903	1113	1243	1008	961	717	527	327	7888
Medford-A	0	0	77	326	624	822	862	627	552	381	207	69	4547
Pendleton-A	0	0	104	353	717	921	1066	795	614	386	197	51	5204
Portland-A	25	22	116	319	585	750	856	658	570	396	242	93	4632
Portland-C	13	14	85	280	534	701	791	594	515	347	199	70	4143
Roseburg-C	14	10	98	288	531	694	744	563	508	366	223	83	4122
Salem-A	21	23	113	326	588	744	825	622	564	408	249	91	4574
Sexton Summit	88	69	169	456	714	877	905	801	797	621	450	270	6217
Troutdale-A	33	31	131	335	591	766	874	664	574	405	256	115	4775
PENNSYLVANIA													
Allentown-A	0	9	89	366	690	1051	1132	1019	840	495	164	25	5880
Erie-C	0	17	76	352	672	1020	1128	1039	911	573	273	55	6116
Harrisburg-A	0	0	69	308	630	964	1051	921	750	423	128	14	5258
Park Place-C	14	57	173	484	807	1200	1277	1142	998	648	290	85	7175
Philadelphia-A	0	0	47	269	573	902	986	879	704	402	104	0	4866
Philadelphia-C	0	0	33	219	516	856	933	837	667	369	93	0	4523
Pittsburgh-Allegheny-A	0	6	78	336	678	1004	1073	955	784	447	167	27	5555
Pittsburgh-Gr. Pitt.-A	0	20	94	377	720	1057	1116	986	818	486	195	36	5905
Pittsburgh-C	0	0	56	298	612	924	992	879	735	402	137	13	5048
Reading-C	0	5	57	285	588	936	1017	902	725	411	123	11	5060
Scranton-C	0	18	115	389	693	1057	1141	1028	859	516	196	35	6047
Williamsport-A	0	16	101	377	699	1057	1132	1005	828	477	181	25	5898

Table C-2 (Cont.)

City	Jul	Aug	Sep	Oct	Nov	Dec	Jan	Feb	Mar	Apr	May	Jun	Total
RHODE ISLAND													
Block Island-A	6	21	88	330	591	927	1026	955	865	603	335	96	5843
Providence-A	0	26	107	381	672	1035	1125	1019	874	570	258	58	6125
Providence-C	0	7	68	330	624	986	1076	972	809	507	197	31	5607
SOUTH CAROLINA													
Charleston-A	0	0	0	52	270	456	472	379	281	63	0	0	1973
Charleston-C	0	0	0	34	214	410	445	363	260	43	0	0	1769
Columbia-A	0	0	0	82	338	558	566	468	340	83	0	0	2435
Columbia-C	0	0	0	76	308	524	538	443	318	77	0	0	2284
Florence-A	0	0	0	94	347	574	588	487	334	83	0	0	2507
Greenville-A	0	0	10	131	411	648	673	552	442	161	32	0	3060
Spartanburg-A	0	0	7	136	414	654	670	549	436	152	26	0	3044
SOUTH DAKOTA													
Huron-A	10	16	149	472	975	1407	1597	1327	1032	558	279	80	7902
Rapid City-A	32	24	193	500	891	1218	1361	1151	1045	615	357	148	7535
Sioux Falls-A	16	21	155	472	984	1414	1575	1274	1023	558	276	80	7848
TENNESSEE													
Bristol-A	0	0	58	239	576	815	818	697	576	274	95	0	4148
Chattanooga-A	0	0	24	169	477	710	725	588	467	179	45	0	3384
Knoxville-A	0	0	33	179	498	744	760	630	500	196	50	0	3590
Memphis-A	0	0	17	126	432	673	725	574	427	139	24	0	3137
Memphis-C	0	0	13	98	392	639	716	574	423	131	20	0	3006
Nashville-A	0	0	22	154	471	725	778	636	498	186	43	0	3513
TEXAS													
Abilene-A	0	0	5	98	350	595	673	479	344	113	0	0	2657
Amarillo-A	0	0	37	240	594	859	921	711	586	298	99	0	4345
Austin-A	0	0	0	30	214	402	484	322	211	50	0	0	1713
Big Springs-A	0	0	0	75	316	577	639	454	314	105	0	0	2480
Brownsville-A	0	0	0	0	59	159	219	106	74	0	0	0	617
Corpus Christi-A	0	0	0	0	113	252	330	192	118	6	0	0	1011
Dallas-A	0	0	0	53	299	518	607	432	288	75	0	0	2272
Del Rio-A	0	0	0	26	188	371	419	235	147	21	0	0	1407
El Paso-A	0	0	0	70	390	626	670	445	330	110	0	0	2641
Fort Worth-A	0	0	0	58	299	533	622	446	308	90	5	0	2361
Ft. Worth-A. Carter Fld.	0	0	0	57	299	524	619	432	326	81	0	0	2338
Galveston-A	0	0	0	0	132	286	362	249	176	28	0	0	1233
Galveston-C	0	0	0	0	131	271	356	247	176	30	0	0	1211
Houston-A	0	0	0	7	181	321	394	265	184	36	0	0	1388
Houston-C	0	0	0	0	162	303	378	240	166	27	0	0	1276
Laredo-A	0	0	0	0	91	215	270	134	71	0	0	0	781
Lubbock-A	0	0	23	173	492	756	812	613	481	201	36	0	3587
Palestine-C	0	0	0	45	260	440	531	368	265	71	0	0	1980
Port Arthur-A	0	0	0	20	218	349	406	274	211	39	0	0	1517
Port Arthur-C	0	0	0	8	170	315	381	258	181	27	0	0	1340
San Angelo-A	0	0	0	72	280	502	556	378	257	62	0	0	2107
San Antonio-A	0	0	0	25	201	374	462	293	190	34	0	0	1579
Victoria-A	0	0	0	0	131	277	352	209	143	14	0	0	1126
Waco-A	0	0	0	44	251	459	557	385	263	66	0	0	2025
Wichita Falls-A	0	0	5	115	404	657	756	538	394	140	16	0	3025

City	Jul	Aug	Sep	Oct	Nov	Dec	Jan	Feb	Mar	Apr	May	Jun	Total
UTAH													
Blanding	0	0	100	409	792	1079	1190	913	800	510	272	73	6138
Milford-A	0	0	114	462	828	1147	1277	955	800	516	269	77	6445
Salt Lake City-A	0	0	88	381	771	1039	1194	885	741	453	233	81	5866
Salt Lake City-C	0	0	61	330	714	995	1119	857	701	414	208	64	5463
VERMONT													
Burlington-A	19	47	172	521	858	1308	1460	1313	1107	681	307	72	7865
VIRGINIA													
Cape Henry-C	0	0	0	120	366	648	698	636	512	267	60	0	3307
Lynchburg-A	0	0	49	236	531	808	846	722	584	289	82	5	4153
Norfolk-A	0	0	9	152	408	688	729	644	500	265	59	0	3454
Norfolk-C	0	0	5	118	354	636	679	602	464	220	41	0	3119
Richmond-A	0	0	33	210	498	791	828	708	550	271	66	0	3955
Richmond-C	0	0	31	181	456	750	787	675	529	254	57	0	3720
Roanoke-A	0	0	50	233	543	806	840	722	588	289	81	0	4152
WASHINGTON													
Ellensburg-A	13	17	176	496	849	1116	1268	949	753	504	296	105	6542
Kelso-A	85	84	186	409	636	784	856	652	605	453	316	173	5239
Northhead	239	205	234	341	486	636	704	585	598	492	406	285	5211
Olympia-A	91	83	207	434	645	794	868	700	660	498	338	183	5501
Omak	0	46	194	533	921	1212	1352	1061	781	453	222	59	6834
Port Angeles-A	233	226	303	459	603	719	772	652	645	519	422	297	5850
Seattle-C	49	45	134	329	540	679	753	602	558	396	246	107	4438
Seattle-Tacoma-A	75	70	192	412	633	781	862	675	636	477	307	155	5275
Spokane-A	17	28	205	508	879	1113	1243	988	834	561	330	146	6852
Stampede Pass	251	260	414	701	1002	1203	1280	1064	1063	837	636	438	9149
Tacoma-C	66	62	177	375	579	719	797	636	595	435	282	143	4866
Tattosh Island-C	295	288	315	406	528	648	713	610	629	525	437	330	5724
Walla Walla-C	0	0	93	308	675	890	1023	748	564	338	171	38	4848
Yakima-A	0	7	150	446	807	1066	1181	862	660	408	205	53	5845
WEST VIRGINIA													
Charleston-A	0	0	60	250	576	834	887	750	632	310	110	8	4417
Elkins-A	9	31	122	412	726	995	1017	910	797	477	224	53	5773
Huntington-C	0	0	35	210	549	803	837	728	570	251	85	5	4073
Parkersburg-C	0	0	56	272	600	896	949	826	672	347	119	13	4750
Petersburg-C	0	5	72	308	654	942	967	820	667	384	133	14	4966
WISCONSIN													
Green Bay-A	32	58	183	515	945	1392	1516	1336	1132	696	347	107	8259
LaCrosse-A	11	20	152	447	921	1380	1528	1280	1035	552	250	74	7650
Madison-A	13	31	150	459	891	1302	1423	1207	1008	579	272	82	7417
Madison-C	10	30	137	419	864	1287	1417	1207	1011	573	266	79	7300
Milwaukee-A	20	32	134	428	831	1218	1336	1142	983	621	351	109	7205
Milwaukee-C	11	24	112	397	795	1184	1302	1117	961	606	335	100	6944
WYOMING													
Casper-A	13	24	231	577	951	1225	1324	1095	1011	660	381	146	7638
Cheyenne-A	33	39	241	577	897	1125	1225	1044	1029	717	462	173	7562
Lander-A	7	23	244	632	1050	1383	1494	1179	1045	687	396	163	8303
Rock Springs-A	20	32	266	648	1038	1349	1457	1182	1110	735	443	193	8473
Sheridan-A	27	41	239	578	957	1271	1392	1170	1035	645	387	161	7903

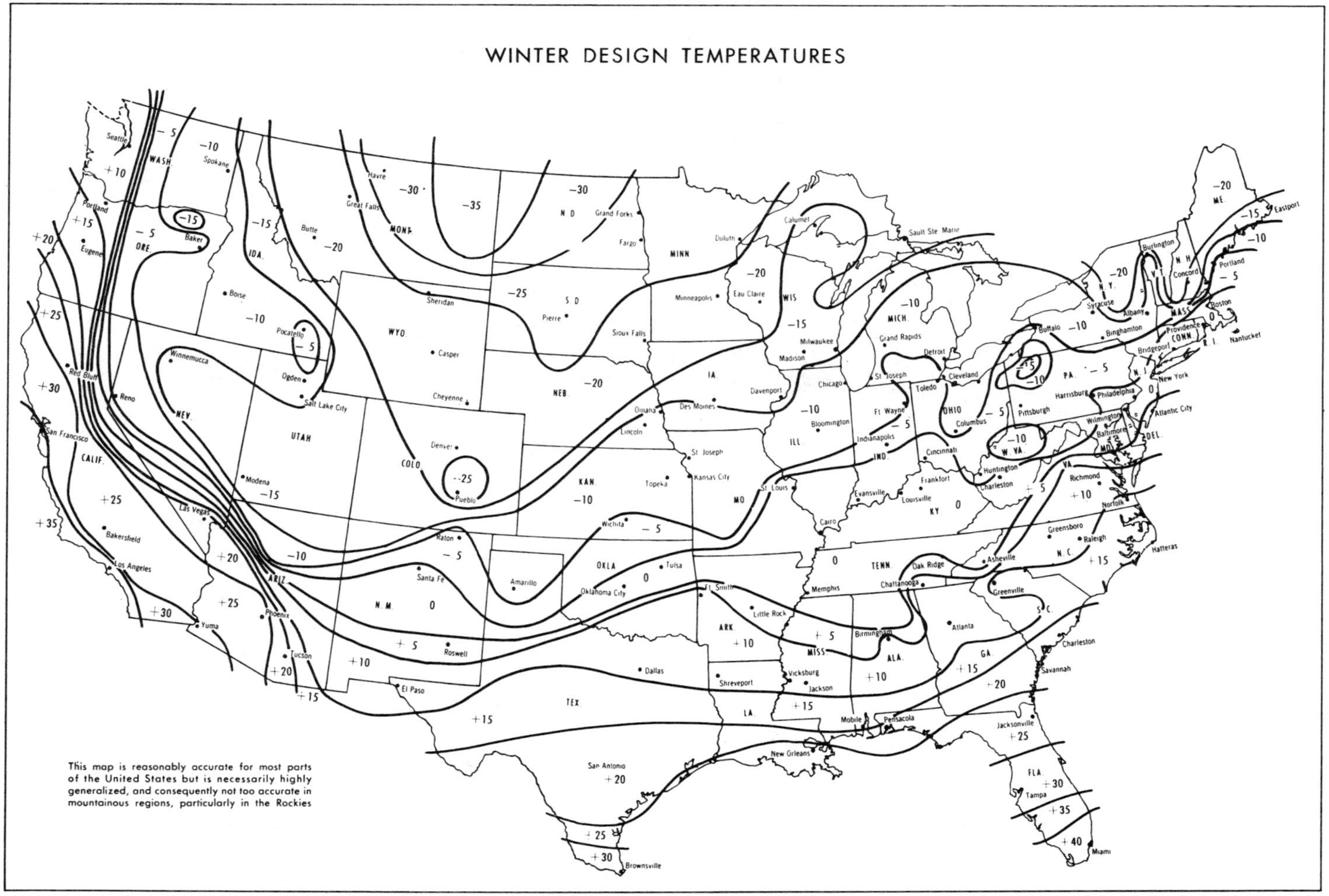

Fig. C-1. Winter Design Temperatures. Reproduced from *Handbook of Air Conditioning, Heating and Ventilating*, 2nd ed., The Industrial Press, New York, 1965.

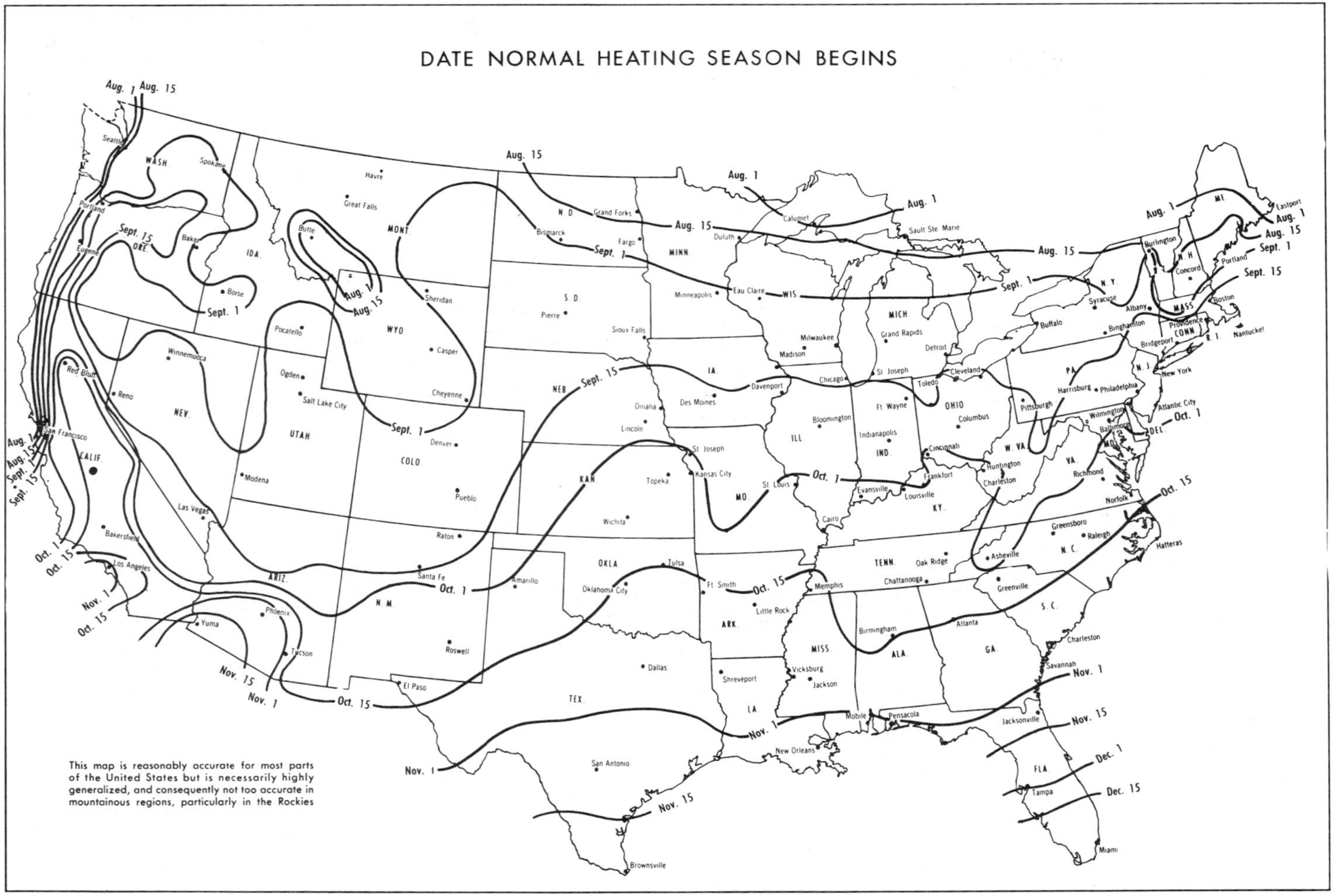

Fig. C-2. Date Normal Heating Season Begins. Reproduced with permission from *Handbook of Air Conditioning, Heating, and Ventilating*, 2nd ed., The Industrial Press, New York, 1965.

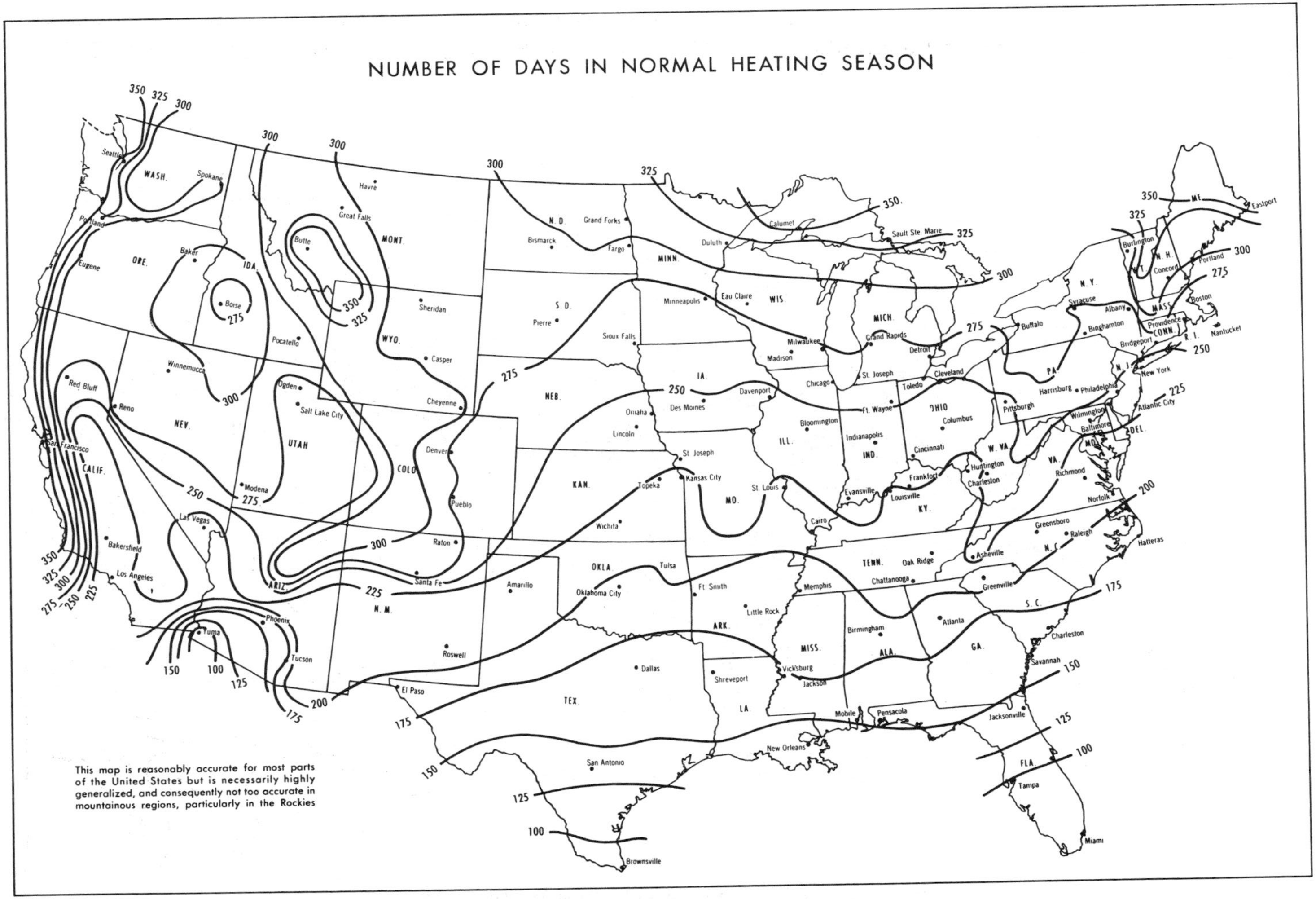

Fig. C-3. Number of Days in Normal Heating Season. Reproduced with permission from *Handbook of Air Conditioning, Heating and Ventilating*, 2nd ed., The Industrial Press, New York, 1965.

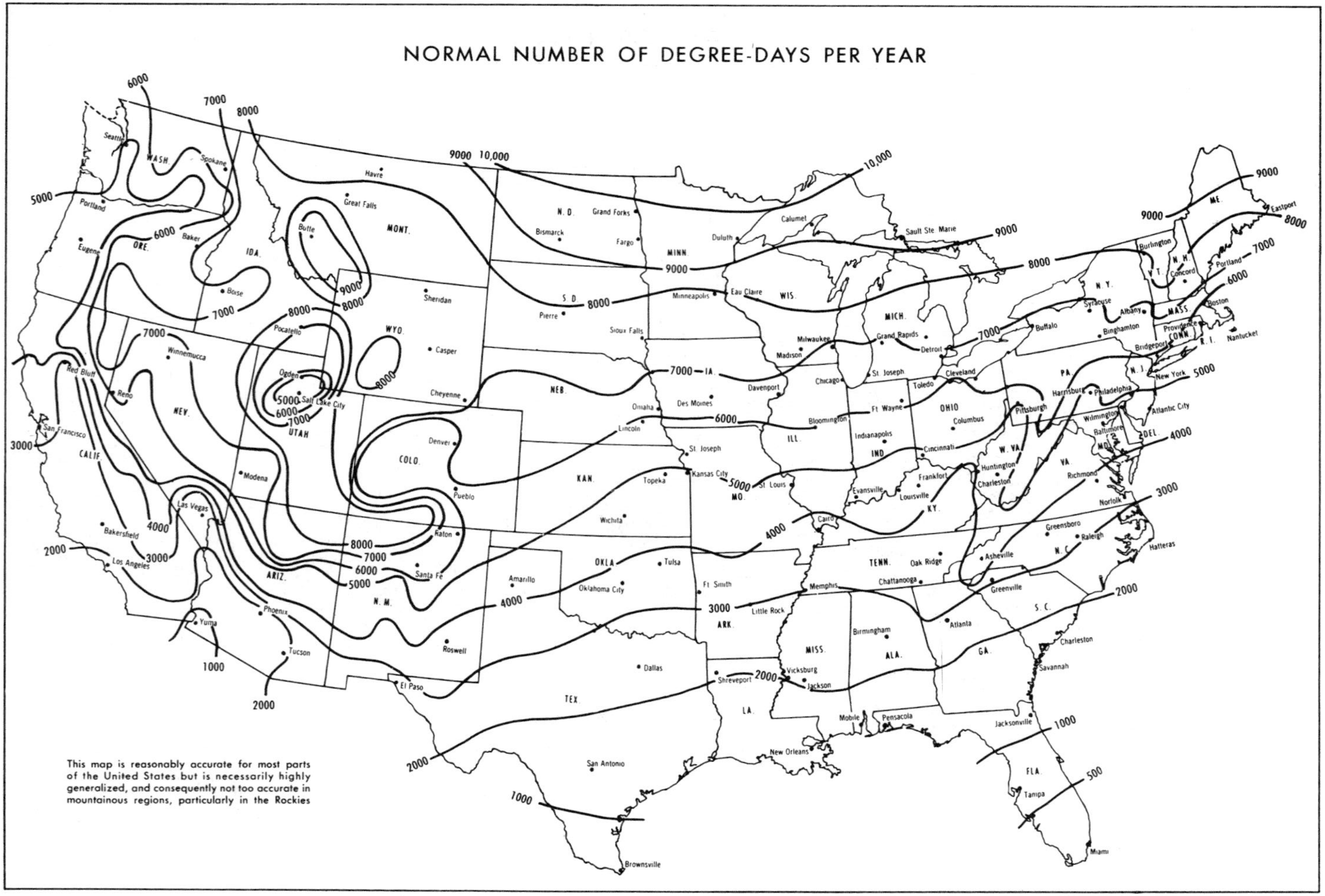

Fig. C-4. Normal Number of Degree-Days per Year. Reproduced with permission from *Handbook of Air Conditioning, Heating and Ventilating*, 2nd ed., The Industrial Press, New York, 1965.

Appendix D

Heat Losses Through Pipe

Table D-1

Heat Losses from Bare Steel Pipe*

HORIZONTAL PIPES											
Diameter of Pipe, Inches	Temperature of Pipe, Deg. F.										
	100	120	150	180	210	240	270	300	330	360	390
	Temperature Difference, Pipe to Air, Deg. F.										
	30	50	80	110	140	170	200	230	260	290	320
	Heat Loss per Lineal Foot of Pipe, Btu per Hour										
½	13	22	40	60	82	106	133	162	193	227	265
¾	15	27	50	74	100	131	163	199	238	280	325
1	19	34	61	90	123	160	199	243	292	343	399
1¼	23	42	75	111	152	198	248	302	362	427	496
1½	27	48	85	126	173	224	280	343	410	483	563
2	33	59	104	154	212	275	344	420	503	594	692
2½	39	70	123	184	252	327	410	502	600	709	827
3	46	84	148	221	303	393	493	601	721	852	994
3½	52	95	168	250	342	444	556	680	816	964	1125
4	59	106	187	278	381	496	621	759	911	1076	1257
5	71	129	227	339	464	603	755	924	1109	1311	1532
6	84	151	267	398	546	709	890	1088	1306	1544	1806
8	107	194	341	509	697	906	1137	1391	1671	1977	2312
10	132	238	420	626	857	1114	1399	1714	2060	2437	2852
12	154	279	491	732	1003	1305	1640	2009	2415	2860	3346
14	181	326	575	856	1173	1527	1918	2350	2826	3347	3918
16	203	366	644	960	1314	1711	2149	2634	3168	3753	4395
18	214	385	678	1011	1385	1802	2266	2777	3339	3958	4635
20	236	426	748	1115	1529	1990	2501	3066	3690	4373	5123

VERTICAL PIPES											
Diameter of Pipe, Inches	Temperature of Pipe, Deg. F.										
	100	120	150	180	210	240	270	300	330	360	390
	Temperature Difference, Pipe to Air, Deg. F.										
	30	50	80	110	140	170	200	230	260	290	320
	Heat Loss per Lineal Foot of Pipe, Btu per Hour										
½	11	20	35	52	71	93	116	142	170	201	235
¾	14	25	44	65	89	116	145	177	213	252	294
1	17	31	55	81	111	145	181	222	266	315	368
1¼	22	39	69	103	141	183	230	281	337	398	465
1½	25	45	79	118	161	210	263	321	386	456	532
2	31	56	99	147	201	262	328	401	481	569	665
2½	37	68	120	178	244	317	397	486	583	687	805
3	46	83	146	217	297	386	484	592	710	839	980
3½	52	94	166	248	339	440	552	676	810	958	1119
4	59	106	187	279	382	496	622	760	912	1078	1259
5	72	131	231	344	472	612	768	939	1126	1331	1555
6	86	156	275	410	562	729	915	1119	1342	1587	1853
8	112	203	358	534	731	950	1191	1456	1747	2065	2412
10	140	254	447	667	913	1186	1487	1818	2181	2578	3012
12	166	301	530	790	1081	1404	1761	2154	2584	3054	3567
14	195	354	624	930	1273	1653	2073	2536	3042	3596	4200
16	221	400	705	1051	1438	1868	2343	2865	3437	4063	4745
18	234	425	748	1115	1526	1982	2486	3040	3648	4311	5036
20	260	472	831	1239	1696	2203	2763	3378	4053	4791	5596

Table D-2

Heat Losses from Bare Tarnished Copper Tube*

HORIZONTAL TUBES

Nominal Diameter of Tube, Inches	Temperature of Tube, Deg. F.										
	100	120	150	180	210	240	270	300	330	360	390
	Temperature Difference, Tube to Air, Deg. F.										
	30	50	80	110	140	170	200	230	260	290	320
	Heat Loss per Lineal Foot of Tube, Btu per Hr.										
¼	4	8	14	21	29	37	46	56	66	77	88
⅜	6	10	18	28	37	48	60	72	85	99	114
½	7	13	22	33	45	59	72	88	104	121	139
⅝	8	15	26	39	53	68	85	102	121	141	163
¾	9	17	30	45	61	79	97	117	139	162	187
1	11	21	37	55	75	97	120	146	173	201	232
1¼	14	25	45	66	90	117	145	175	207	242	279
1½	16	29	52	77	105	135	167	203	241	281	324
2	20	37	66	97	132	171	212	257	305	356	411
2½	24	44	78	117	160	206	255	310	367	429	496
3	28	51	92	136	186	240	297	360	428	501	578
3½	32	59	104	156	212	274	340	412	490	573	662
4	36	66	118	174	238	307	381	462	550	644	744
5	43	80	142	212	288	373	464	561	669	783	905
6	51	93	166	246	336	432	541	656	776	915	1059
8	66	120	215	317	435	562	699	848	1010	1184	1372
10	80	146	260	387	527	681	848	1031	1227	1442	1670
12	94	172	304	447	621	802	999	1214	1446	1699	1969

VERTICAL TUBES

Nominal Diameter of Tube, Inches	Temperature of Tube, Deg. F.										
	100	120	150	180	210	240	270	300	330	360	390
	Temperature Difference, Tube to Air, Deg. F.										
	30	50	80	110	140	170	200	230	260	290	320
	Heat Loss per Lineal Foot of Tube, Btu per Hr.										
¼	3	6	10	15	21	27	34	41	49	57	66
⅜	4	8	14	21	28	36	45	55	65	77	88
½	5	10	17	26	35	46	57	69	82	96	111
⅝	6	12	21	31	42	54	68	82	98	114	132
¾	7	14	24	36	49	64	79	96	114	134	155
1	10	18	31	46	63	82	102	123	147	172	198
1¼	12	21	38	57	77	100	125	151	180	210	243
1½	14	25	45	67	91	118	147	178	212	248	287
2	18	33	59	88	120	155	192	233	277	325	375
2½	22	41	73	109	148	191	238	288	343	402	464
3	27	49	87	129	176	227	283	343	408	478	552
3½	31	57	101	150	204	264	328	398	474	554	641
4	35	64	114	171	232	300	374	453	539	631	729
5	43	80	142	212	288	373	464	561	669	783	905
6	52	96	170	253	344	445	554	670	798	934	1080
8	69	127	226	337	458	592	737	892	1063	1244	1438
10	86	158	281	419	570	737	917	1110	1322	1548	1789
12	103	189	336	501	682	881	1097	1328	1582	1851	2140

**Reprinted with permission from* Handbook of Air Conditioning, Heating and Ventilating, *2nd ed., The Industrial Press, New York, 1965.*

Table D-3*

HEAT LOSS THROUGH PIPE INSULATION — ¾ INCH STEEL PIPE

Insulation Conductivity k	Temperature Difference, Pipe to Air, Deg. F										
	30	50	80	110	140	170	200	230	260	290	320
	Heat Loss per Lineal Foot of Bare Pipe, Btu per Hour										
	15	27	50	74	100	131	163	199	238	280	325
	Heat Loss per Lineal Foot of Insulated Pipe, Btu per Hour										
1 Inch Thick Insulation — ¾ Inch Pipe											
0.20	2	6	7	10	13	16	18	21	24	27	29
0.25	3	6	9	12	16	19	23	26	29	33	36
0.30	4	7	11	15	19	23	27	31	35	39	43
0.35	5	8	12	17	21	26	30	35	40	44	49
0.40	5	9	14	19	24	29	34	39	44	50	55
0.45	6	10	15	21	27	32	38	44	49	55	61
0.50	6	10	17	23	29	35	42	48	54	60	67
0.55	7	11	18	25	32	38	45	52	59	65	72
0.60	7	12	19	27	34	41	48	56	63	70	77
1½ Inch Thick Insulation — ¾ Inch Pipe											
0.20	2	4	6	8	10	12	14	16	19	21	23
0.25	3	5	7	10	13	16	18	21	24	27	29
0.30	3	5	9	12	15	19	22	25	28	32	35
0.35	4	6	10	14	18	21	25	29	33	37	40
0.40	4	7	11	16	20	24	29	33	37	41	46
0.45	5	8	13	17	22	27	32	37	41	46	51
0.50	5	9	14	19	25	30	35	40	46	51	56
0.55	6	10	15	21	27	32	38	44	49	55	61
0.60	6	10	16	23	29	35	41	47	53	59	66
2 Inch Thick Insulation — ¾ Inch Pipe											
0.20	2	3	5	7	9	11	13	15	17	19	21
0.25	2	4	6	9	11	14	16	18	21	23	26
0.30	3	5	8	11	13	16	19	22	25	28	31
0.35	3	6	9	12	16	19	22	26	29	32	36
0.40	4	6	10	14	18	21	25	29	33	37	40
0.45	4	7	11	16	20	24	28	32	37	41	45
0.50	5	8	12	17	22	26	31	36	40	45	50
0.55	5	8	14	19	24	29	34	39	44	49	54
0.60	5	9	15	20	26	31	37	42	48	53	59
2½ Inch Thick Insulation — ¾ Inch Pipe											
0.20	2	3	5	6	8	10	12	13	15	17	19
0.25	2	3	5	7	9	11	13	15	17	19	21
0.30	3	4	7	10	12	15	17	20	23	25	28
0.35	3	5	8	11	14	17	20	23	26	29	32
0.40	3	6	9	13	16	20	23	26	30	33	37
0.45	3	6	10	14	18	22	26	29	33	37	41
0.50	4	7	11	16	20	24	28	33	37	41	45
0.55	5	8	12	17	22	26	31	36	40	45	50
0.60	5	8	13	18	24	29	34	39	44	49	54

**Reprinted with permission from* Handbook of Air Conditioning, Heating and Ventilating, *2nd ed., The Industrial Press, New York, 1965.*

Table D-3 (Cont.)

HEAT LOSS THROUGH PIPE INSULATION—1 INCH STEEL PIPE

Insulation Conductivity k	Temperature Difference, Pipe to Air, Deg. F										
	30	50	80	110	140	170	200	230	260	290	320
	Heat Loss per Lineal Foot of Bare Pipe, Btu per Hour										
	19	34	61	90	123	160	199	243	292	343	399
	Heat Loss per Lineal Foot of Insulated Pipe, Btu per Hour										
1 Inch Thick Insulation — 1 Inch Pipe											
0.20	3	5	8	12	15	18	21	24	27	30	34
0.25	4	6	10	14	18	23	26	30	34	37	41
0.30	5	8	12	17	21	26	30	35	40	44	49
0.35	5	9	14	19	24	30	35	40	45	50	56
0.40	6	10	16	22	27	33	39	45	51	57	63
0.45	7	11	17	24	30	37	43	50	56	63	69
0.50	7	12	19	26	33	40	47	55	62	69	76
0.55	8	13	20	28	36	44	51	59	67	74	82
0.60	8	14	22	30	39	47	55	63	72	80	88
1½ Inch Thick Insulation — 1 Inch Pipe											
0.20	2	4	7	9	12	14	17	19	22	24	27
0.25	3	5	8	11	15	18	21	24	27	30	33
0.30	4	6	10	14	17	21	25	29	32	36	40
0.35	4	7	11	16	20	24	29	33	37	41	46
0.40	5	8	13	18	23	27	32	37	42	47	52
0.45	5	9	14	20	25	31	36	41	47	52	58
0.50	6	10	16	22	28	33	39	45	51	57	63
0.55	6	11	17	24	30	37	43	49	56	62	69
0.60	7	12	19	26	32	39	46	53	60	67	74
2 Inch Thick Insulation — 1 Inch Pipe											
0.20	2	4	6	8	10	13	15	17	19	21	24
0.25	3	5	7	10	13	15	18	21	23	26	29
0.30	3	5	9	12	15	18	21	25	28	31	34
0.35	4	6	10	14	17	21	25	29	32	36	40
0.40	4	7	11	15	20	24	28	32	37	41	45
0.45	5	8	13	17	22	27	31	36	41	46	50
0.50	5	9	14	19	24	29	35	40	45	50	55
0.55	6	9	15	21	26	32	38	43	49	55	60
0.60	6	10	16	23	29	35	41	47	53	59	66
2½ Inch Thick Insulation — 1 Inch Pipe											
0.20	2	3	5	7	9	11	13	15	17	19	21
0.25	3	4	7	10	12	15	17	20	23	25	28
0.30	3	5	8	11	14	16	19	22	25	28	31
0.35	3	6	9	12	16	19	22	26	29	32	36
0.40	4	6	10	14	18	22	26	29	33	37	41
0.45	4	7	11	16	20	24	29	33	37	41	46
0.50	5	8	13	17	22	27	32	36	41	46	51
0.55	5	9	14	19	24	29	34	40	45	50	55
0.60	6	9	15	21	26	32	37	43	49	54	60

Table D-3 (Cont.)

HEAT LOSS THROUGH PIPE INSULATION — 1¼ INCH STEEL PIPE

Insulation Conductivity k	Temperature Difference, Pipe to Air, Deg. F										
	30	50	80	110	140	170	200	230	260	290	320
	Heat Loss per Lineal Foot of Bare Pipe, Btu per Hour										
	23	42	75	111	152	198	248	302	362	427	496
	Heat Loss per Lineal Foot of Insulated Pipe, Btu per Hour										
1 Inch Thick Insulation — 1¼ Inch Pipe											
0.20	4	6	10	14	17	21	25	28	32	36	39
0.25	5	8	12	17	21	26	30	35	39	44	48
0.30	5	9	14	19	25	30	35	41	46	51	57
0.35	6	10	16	22	28	35	41	47	53	59	65
0.40	7	11	18	25	32	39	46	52	59	66	73
0.45	8	13	20	28	35	43	50	58	66	73	81
0.5c	8	14	22	30	39	47	55	63	72	80	88
0.55	9	15	24	33	42	50	59	68	77	86	95
0.60	10	16	26	35	45	54	64	73	83	92	102
1½ Inch Thick Insulation — 1¼ Inch Pipe											
0.20	3	5	8	11	14	16	19	22	25	28	31
0.25	4	6	10	13	17	20	24	27	31	35	38
0.30	4	7	11	16	20	24	28	32	37	41	45
0.35	5	8	13	18	23	28	33	37	42	47	52
0.40	6	9	15	20	26	31	37	42	48	53	59
0.45	6	10	16	23	29	35	41	47	53	59	66
0.50	7	11	18	25	32	38	45	52	59	65	72
0.55	7	12	20	27	34	42	49	56	64	71	78
0.60	8	13	21	29	37	45	53	61	69	77	84
2 Inch Thick Insulation — 1¼ Inch Pipe											
0.20	2	4	7	9	12	14	17	19	21	24	26
0.25	3	5	8	11	14	17	20	23	27	30	33
0.30	4	6	10	13	17	21	24	28	32	35	39
0.35	4	7	11	16	19	24	28	32	37	41	45
0.40	5	8	13	18	22	27	32	37	42	47	51
0.45	5	9	14	20	25	30	36	41	46	52	57
0.50	6	10	16	22	27	33	39	45	51	57	63
0.55	6	11	17	24	30	36	43	49	56	62	68
0.60	7	12	19	26	32	39	46	53	60	67	74
2½ Inch Thick Insulation — 1¼ Inch Pipe											
0.20	2	4	6	8	10	12	15	17	19	21	23
0.25	3	5	7	10	13	16	18	21	24	26	29
0.30	3	5	9	12	15	19	22	25	28	32	35
0.35	4	6	10	14	18	21	25	29	33	37	40
0.40	4	7	11	16	20	24	29	33	37	41	46
0.45	5	8	13	18	22	27	32	37	42	46	51
0.50	5	9	14	19	25	30	35	41	46	51	57
0.55	6	10	15	21	27	33	39	44	50	56	62
0.60	6	11	17	23	29	36	42	48	55	61	67

Table D-3 (Cont.)

HEAT LOSS THROUGH PIPE INSULATION—1½ INCH STEEL PIPE

Insulation Conductivity *k*	Temperature Difference, Pipe to Air, Deg. F										
	30	50	80	110	140	170	200	230	260	290	320
	Heat Loss per Lineal Foot of Bare Pipe, Btu per Hour										
	27	48	85	126	173	224	280	343	410	483	563
	Heat Loss per Lineal Foot of Insulated Pipe, Btu per Hour										
1 Inch Thick Insulation — 1½ Inch Pipe											
0.20	4	7	11	15	19	23	27	31	35	39	43
0.25	5	8	13	18	23	28	32	37	42	47	52
0.30	6	10	16	21	27	33	39	45	50	56	62
0.35	7	11	18	24	31	38	44	51	58	64	71
0.40	7	12	20	27	35	42	50	57	65	72	80
0.45	8	14	22	30	39	47	55	63	72	80	88
0.50	9	15	24	33	42	51	60	69	78	87	96
0.55	10	16	26	36	46	55	65	75	85	94	104
0.60	10	17	28	38	49	59	70	80	90	100	111
1½ Inch Thick Insulation — 1½ Inch Pipe											
0.20	3	5	8	12	15	18	21	24	27	30	34
0.25	4	7	10	14	18	22	26	30	34	38	42
0.30	5	8	12	17	22	26	31	35	40	45	49
0.35	5	9	14	19	25	30	35	41	46	51	57
0.40	6	10	16	22	28	34	40	46	52	58	64
0.45	7	11	18	25	31	38	45	51	58	65	71
0.50	7	12	20	27	34	41	49	56	63	71	78
0.55	8	13	21	29	37	45	53	61	69	77	85
0.60	9	14	23	32	40	49	57	66	75	83	92
2 Inch Thick Insulation — 1½ Inch Pipe											
0.20	3	4	7	10	12	15	18	20	23	25	28
0.25	3	6	9	12	15	19	22	25	29	32	35
0.30	4	7	10	14	18	22	26	30	34	38	42
0.35	5	8	12	17	21	26	30	35	44	49	53
0.40	5	9	14	19	24	29	34	40	45	50	55
0.45	6	10	15	21	27	33	38	44	50	56	61
0.50	6	11	17	23	30	36	43	49	55	62	68
0.55	7	12	18	25	32	39	46	53	60	67	74
0.60	8	13	20	28	35	43	50	58	65	73	80
2½ Inch Thick Insulation — 1½ Inch Pipe											
0.20	2	4	6	9	11	13	16	18	21	23	25
0.25	3	5	8	11	14	17	20	23	25	28	31
0.30	4	6	9	13	16	20	23	27	30	34	37
0.35	4	7	11	15	19	23	27	31	35	39	44
0.40	5	8	12	17	22	26	31	35	40	45	49
0.45	5	9	14	19	24	29	34	40	45	50	55
0.50	6	10	15	21	27	32	38	44	49	55	61
0.55	6	10	17	23	29	35	42	48	54	60	67
0.60	7	11	18	25	32	38	45	52	59	65	72

Table D-3 (Cont.)

HEAT LOSS THROUGH PIPE INSULATION — 2½ INCH STEEL PIPE

Insulation Conductivity *k*	Temperature Difference, Pipe to Air, Deg. F										
	30	50	80	110	140	170	200	230	260	290	320
	Heat Loss per Lineal Foot of Bare Pipe, Btu per Hour										
	39	70	123	184	252	327	410	502	600	709	827
	Heat Loss per Lineal Foot of Insulated Pipe, Btu per Hour										
1 Inch Thick Insulation — 2½ Inch Pipe											
0.20	5	9	14	20	25	31	36	42	47	52	58
0.25	7	11	18	24	31	38	44	51	57	64	71
0.30	8	13	21	29	36	44	52	60	68	75	83
0.35	9	15	24	33	42	50	59	68	77	86	95
0.40	10	17	27	37	47	57	67	77	87	97	107
0.45	11	18	29	40	52	63	74	85	96	107	118
0.50	12	21	33	45	57	70	82	94	107	119	131
0.55	13	22	35	48	61	74	87	100	113	126	139
0.60	14	23	37	51	65	79	93	106	120	134	148
1½ Inch Thick Insulation — 2½ Inch Pipe											
0.20	4	7	11	15	19	24	28	32	36	40	44
0.25	5	9	14	19	24	29	34	39	44	50	55
0.30	6	10	16	22	28	35	41	47	53	59	65
0.35	7	12	19	26	33	40	47	54	61	68	75
0.40	8	13	21	29	37	45	53	61	69	77	84
0.45	9	15	23	32	41	50	59	67	76	85	94
0.50	10	16	26	35	45	55	64	74	83	93	103
0.55	10	17	28	38	49	59	70	80	91	101	112
0.60	11	19	30	41	53	64	75	86	98	109	120
2 Inch Thick Insulation — 2½ Inch Pipe											
0.20	3	6	9	13	16	20	23	26	30	33	37
0.25	4	7	11	16	20	24	29	33	37	41	46
0.30	5	9	14	19	24	29	34	39	44	49	54
0.35	6	10	16	22	27	33	39	45	51	57	63
0.40	7	11	18	24	31	38	44	51	58	64	71
0.45	7	12	20	27	35	42	50	57	64	72	79
0.50	8	14	22	30	38	46	55	63	71	79	87
0.55	9	15	24	33	42	50	59	68	77	86	95
0.60	10	16	26	35	45	55	64	74	84	93	103
2½ Inch Thick Insulation — 2½ Inch Pipe											
0.20	3	5	8	11	14	17	20	23	26	29	32
0.25	4	6	10	14	18	21	25	29	33	36	40
0.30	4	7	12	16	21	25	30	34	39	43	48
0.35	5	9	14	19	24	29	35	40	45	50	55
0.40	6	10	16	22	27	33	39	45	51	57	63
0.45	7	11	18	24	31	37	44	50	57	64	70
0.50	7	12	19	27	34	41	48	55	63	70	77
0.55	8	13	21	29	37	45	53	61	69	77	84
0.60	9	14	23	31	40	49	57	66	74	83	92

Table D-3 (Cont.)

HEAT LOSS THROUGH PIPE INSULATION — 2 INCH STEEL PIPE

Insulation Conductivity *k*	Temperature Difference, Pipe to Air, Deg. F										
	30	50	80	110	140	170	200	230	260	290	320
	Heat Loss per Lineal Foot of Bare Pipe, Btu per Hour										
	33	59	104	154	212	275	344	420	503	594	692
	Heat Loss per Lineal Foot of Insulated Pipe, Btu per Hour										
1 Inch Thick Insulation — 2 Inch Pipe											
0.20	5	8	13	17	22	27	32	36	41	46	51
0.25	6	10	15	21	27	32	39	44	50	56	62
0.30	7	11	18	25	32	39	45	52	59	66	73
0.35	8	13	21	29	36	44	52	60	68	75	83
0.40	9	15	23	32	41	49	58	67	76	84	93
0.45	10	16	26	35	45	55	64	74	83	93	103
0.50	10	18	28	39	49	60	70	81	91	102	112
0.55	11	19	30	42	53	64	76	87	99	110	121
0.60	12	20	32	45	57	69	81	93	106	118	130
1½ Inch Thick Insulation — 2 Inch Pipe											
0.20	4	6	10	13	17	21	24	28	32	35	39
0.25	5	8	12	17	21	26	30	35	39	44	48
0.30	5	9	14	20	25	30	36	41	46	52	57
0.35	6	10	16	23	29	35	41	47	53	59	66
0.40	7	12	18	25	32	39	46	53	60	67	74
0.45	8	13	21	28	36	44	51	59	67	75	82
0.50	9	15	23	32	41	49	58	67	75	84	93
0.55	9	15	25	34	43	52	61	71	80	89	98
0.60	10	17	26	36	46	55	66	76	83	96	106
2 Inch Thick Insulation — 2 Inch Pipe											
0.20	3	5	8	11	14	17	20	23	27	30	33
0.25	4	6	10	14	18	22	25	29	33	37	41
0.30	4	8	12	17	21	26	30	35	39	44	48
0.35	5	9	14	19	24	30	35	40	45	50	56
0.40	6	10	16	22	28	33	39	45	51	57	63
0.45	7	11	18	24	31	37	44	51	57	64	70
0.50	7	12	19	27	34	41	48	56	63	70	77
0.55	8	13	21	29	37	45	53	61	69	77	84
0.60	9	14	23	31	40	49	57	66	74	83	92
2½ Inch Thick Insulation — 2 Inch Pipe											
0.20	3	5	7	10	13	15	18	21	23	26	29
0.25	3	6	9	12	16	19	22	26	29	32	36
0.30	4	7	11	15	19	23	27	31	35	39	43
0.35	5	8	12	17	22	26	31	35	40	45	49
0.40	5	9	14	19	25	30	35	40	46	51	56
0.45	6	10	16	22	27	33	39	45	51	57	63
0.50	6	11	17	24	30	37	43	50	56	63	69
0.55	7	12	19	26	33	40	48	54	61	68	76
0.60	8	13	21	29	37	45	53	61	69	77	85

Table D-3 (Cont.)

HEAT LOSS THROUGH PIPE INSULATION — 3 INCH STEEL PIPE

Insulation Conductivity *k*	Temperature Difference, Pipe to Air, Deg. F										
	30	50	80	110	140	170	200	230	260	290	320
	Heat Loss per Lineal Foot of Bare Pipe, Btu per Hour										
	46	84	148	221	303	393	493	601	721	852	994
	Heat Loss per Lineal Foot of Insulated Pipe, Btu per Hour										
1 Inch Thick Insulation — 3 Inch Pipe											
0.20	6	11	17	23	30	36	42	49	55	61	68
0.25	8	13	21	28	36	44	52	60	67	75	83
0.30	9	15	24	33	43	52	61	70	79	88	97
0.35	10	17	28	38	49	59	69	80	90	101	111
0.40	12	19	31	43	54	66	78	89	101	113	124
0.45	13	21	34	47	60	73	86	99	111	124	137
0.50	14	23	37	51	65	79	93	107	121	135	149
0.55	15	25	40	55	71	86	101	116	131	146	161
0.60	16	27	43	59	76	92	108	124	140	156	173
1½ Inch Thick Insulation — 3 Inch Pipe											
0.20	5	8	13	18	22	27	32	37	41	46	51
0.25	6	10	16	22	27	33	39	45	51	57	63
0.30	7	12	19	26	33	39	46	53	60	67	74
0.35	8	13	22	30	39	46	54	62	70	78	86
0.40	9	15	24	33	42	51	60	69	78	87	96
0.45	10	17	27	37	47	57	67	77	87	97	107
0.50	11	18	29	40	51	62	73	84	95	106	117
0.55	12	20	32	44	56	68	80	92	104	116	128
0.60	13	21	34	47	60	73	86	99	112	124	137
2 Inch Thick Insulation — 3 Inch Pipe											
0.20	4	7	11	15	18	22	26	30	34	38	42
0.25	5	8	13	18	23	28	33	38	43	47	52
0.30	6	10	15	21	27	33	39	45	51	56	62
0.35	7	11	18	25	31	38	45	52	58	65	72
0.40	8	13	20	28	36	43	51	58	66	74	81
0.45	8	14	23	31	40	48	57	65	74	82	91
0.50	9	16	25	34	44	53	62	72	81	90	100
0.55	10	17	27	37	48	58	68	78	88	98	109
0.60	11	18	29	40	51	62	73	84	95	106	117
2½ Inch Thick Insulation — 3 Inch Pipe											
0.20	3	6	9	12	16	19	23	26	29	33	36
0.25	4	7	11	16	20	24	28	33	37	41	45
0.30	5	8	13	19	24	29	34	39	44	49	54
0.35	6	10	16	22	27	33	39	45	51	57	63
0.40	7	11	18	24	31	38	44	51	58	64	71
0.45	7	12	20	27	35	42	50	57	64	72	79
0.50	8	14	22	30	38	46	55	63	71	79	87
0.55	9	15	24	33	42	51	60	69	78	87	96
0.60	10	16	26	36	45	55	65	74	84	94	103

Table D-3 (Cont.)

HEAT LOSS THROUGH PIPE INSULATION — 4 INCH STEEL PIPE

Insulation Conductivity *k*	Temperature Difference, Pipe to Air, Deg. F										
	30	50	80	110	140	170	200	230	270	300	320
	Heat Loss per Lineal Foot of Bare Pipe, Btu per Hour										
	59	106	187	278	381	496	621	759	911	1076	1257
	Heat Loss per Lineal Foot of Insulated Pipe, Btu per Hour										
1 Inch Thick Insulation — 4 Inch Pipe											
0.20	8	13	20	28	33	43	51	59	66	74	82
0.25	10	16	25	35	44	54	63	73	82	92	101
0.30	11	19	30	41	52	63	74	86	97	108	119
0.35	13	21	34	47	59	72	85	98	110	123	136
0.40	14	24	38	52	66	81	95	109	123	137	152
0.45	16	26	42	58	73	89	105	121	136	152	168
0.50	17	28	45	62	80	97	114	131	148	165	182
0.55	18	31	49	67	86	104	123	141	159	178	196
0.60	20	33	53	72	92	112	132	151	171	191	211
1½ Inch Thick Insulation — 4 Inch Pipe											
0.20	6	10	15	21	27	33	38	44	50	56	61
0.25	7	12	19	26	33	40	47	55	62	69	76
0.30	8	14	22	28	39	48	56	64	73	81	90
0.35	10	16	26	36	45	55	65	74	84	94	103
0.40	11	18	29	40	51	62	73	84	95	106	116
0.45	12	20	32	44	56	69	81	93	105	117	129
0.50	13	22	35	49	62	75	88	101	115	128	141
0.55	14	24	38	53	67	81	96	110	124	139	153
0.60	15	26	41	57	72	88	103	118	134	149	165
2 Inch Thick Insulation — 4 Inch Pipe											
0.20	5	8	13	17	22	27	31	36	41	46	50
0.25	6	10	16	21	27	33	39	45	51	57	62
0.30	7	12	19	26	32	39	46	53	60	67	74
0.35	8	13	21	29	37	45	53	61	69	77	85
0.40	9	15	24	33	42	51	60	69	79	88	97
0.45	10	17	27	37	47	57	67	78	88	98	108
0.50	11	19	30	41	52	63	74	85	96	107	118
0.55	12	20	32	44	56	69	81	93	105	117	129
0.60	13	22	35	48	61	74	87	101	114	127	140
2½ Inch Thick Insulation — 4 Inch Pipe											
0.20	4	7	11	15	19	23	27	31	35	39	44
0.25	5	8	13	18	24	29	34	39	44	49	54
0.30	6	10	16	22	28	34	40	46	52	58	64
0.35	7	12	19	26	32	39	46	53	60	67	74
0.40	8	13	21	29	37	45	52	60	68	76	84
0.45	9	15	24	32	41	50	59	68	76	85	94
0.50	10	16	26	36	45	55	65	75	84	94	104
0.55	11	18	28	39	49	60	71	81	92	102	113
0.60	11	19	31	42	53	65	76	88	99	111	122

Index

T

U

V

W